PRINCIPLES OF INORGANIC CHEMIS

PRINCIPLES OF INORGANIC CHEMISTRY

THIRD EDITION

G.S. SODHI, *Ph.D.*
Associate Professor, Department of Chemistry
Coordinator, Forensic Science Unit
S.G.T.B. Khalsa College, University of Delhi
Delhi

Viva Books
New Delhi | Mumbai | Chennai | Kolkata | Bengaluru | Hyderabad | Kochi | Guwahati

Publisher's note

First Published 2011
Reprinted 2013, 2014
Second Revised & Enlarged Edition 2015
Third Edition 2018

Viva Books Private Limited

- 4737/23, Ansari Road, Daryaganj, New Delhi 110 002
  Tel. 011-42242200, 23258325, 23283121, Email: vivadelhi@vivagroupindia.net
- 76, Service Industries, Shirvane, Sector 1, Nerul, Navi Mumbai 400 706
  Tel. 022-27721273, 27721274, Email: vivamumbai@vivagroupindia.net
- Megh Tower, Old No. 307, New No. 165, Poonamallee High Road, Maduravoyal, Chennai 600 095
  Tel. 044-23780991, 23780992, 23780994, Email: vivachennai@vivagroupindia.net
- B-103, Jindal Towers, 21/1A/3 Darga Road, Kolkata 700 017
  Tel. 033-22816713, Email: vivakolkata@vivagroupindia.net
- 7, Sovereign Park Aptts., 56-58, K. R. Road, Basavanagudi, Bengaluru 560 004
  Tel. 080-26607409, Email: vivabangalore@vivagroupindia.net
- 101-102, Moghal Marc Apartments, 3-4-637 to 641, Narayanguda, Hyderabad 500 029
  Tel. 040-27564481, Email: vivahyderabad@vivagroupindia.net
- First Floor, Beevi Towers, SRM Road, Kaloor, Kochi 682 018
  Tel. 0484-2403055, 2403056, Email: vivakochi@vivagroupindia.net
- 232, GNB Road, Beside UCO Bank, Silpukhuri, Guwahati 781 003
  Tel. 0361-2666386, Email: vivaguwahati@vivagroupindia.net

www.vivagroupindia.com

ISBN: 978-93-87692-98-5

Published by Vinod Vasishtha for Viva Books Private Limited, 4737/23, Ansari Road, Daryaganj, New Delhi 110 002.

Printed and bound by Raj Press, R-3, Inderpuri, New Delhi 110 012.

SUB-0005

PREFACE TO THE THIRD EDITION

Principles of Inorganic Chemistry affirms to a comprehensive study of the elements and their characteristics. It encompasses a detailed description of a broad spectrum of inorganic compounds, including complexes and organometallic derivatives, with emphasis on their structures, reactivities and biochemical properties. It disseminates information in a systematic and scientific manner on almost every section of inorganic chemistry.

Three new chapters, *viz.* 'Inorganic Polymers', 'Group 12 Elements' and 'Qualitative Analysis' have been incorporated in this edition. Important concepts like black body radiation, stationary states, formal charge, corrosion and theories of hydrogen bonding find a place in the present text. Applications of radioactivity have been expanded to include airport security, preservation of artifacts, safeguarding of food products and detection of forgeries. The industrial importance of organometallic catalysis has been contextualized in an easy-to-learn format.

We endeavor to offer a rigorous description of the fundamental principles on which inorganic chemistry is based, and that too in a style and organization typical of the updated curriculum prescribed for professional courses.

G.S. Sodhi

PREFACE TO THE SECOND EDITION

Like any other discipline, Inorganic Chemistry too grows and evolves. With the passage of time, numerous novel inorganic compounds have been synthesized. Their structures have been firmly established and correlated with their characteristics, reaction kinetics and biochemical activities. As a consequence, inorganic chemistry no longer remains a compartmentalized subject, but permeates almost every sphere of human activity.

The second edition of the text has been designed keeping in view the ubiquitous nature and wide expanse of inorganic chemistry. Two new chapters, 'Redox Reactions' and 'Bioinorganic Chemistry', have been added to the contents. The schedule of inorganic compounds has been augmented by including sodium thiosulfate, basic beryllium acetate, phosphine, ozone, besides others.

The classifications, nitrides, phosphazenes, oxides and polysulfates have added to the list of inorganic assortments. The concepts underlying Ketelaar triangles, fluxional behaviour of organometallics and chemical methods of distinguishing geometrical isomers have been expounded. The molecular orbital theory has been extended to include polyatomic molecules, while additional factors governing the stability and lability of complexes have been explicated.

At the time of this writing, many new inorganic compounds must be on way to characterization, many new facts and theories must be in the process of being unearthed. I hope to include these in a future edition of this book!

G.S. Sodhi

PREFACE TO THE FIRST EDITION

The majority of textbooks on Inorganic Chemistry place considerable emphasis on theory and concepts, but do not give due importance to properties and reactions of elements and their compounds. The present text endeavors to seek a balance between these two segments with the aim to impart a coherent foundation to the study of inorganic chemistry. It thus strives to fill a real need for the critical and comprehensive account of the chemistry of elements. The overall goal has been to present a systematic discussion of the important principles and facts of modern inorganic chemistry.

The content focuses on trends in synthesis and characterization, as well as on structure and bonding of the chemical species in relation to their congeners. Several advanced topics, such as molecular orbital theory, relativistic effect, identification of isotopes, fullerenes, organometallics and inorganic reaction mechanisms, find a place in this book.

Each chapter of the text is largely self-contained, beginning with the basic aspects of the relevant topic and proceeding to a more rigorous treatment. This makes it easy for students and teachers to select the material they require for a particular course. In addition, every chapter includes a number of worked examples. Christened case study, each such example deals with a challenging and often tricky concept relevant to that chapter. In fact, an effort has been made to integrate the principles and applications of inorganic chemistry throughout the textbook.

The book forms a versatile and modular basis for different areas that come within the ambit of inorganic chemistry. It endeavors to convey the core material within the framework of chemical concepts, and in an understandable and enjoyable manner. A considerable effort has been expended to make it a student-oriented text and it is therefore anticipated that they will find the book useful, not only for reference, but also for learning.

G.S. Sodhi

CONTENTS

ABBREVIATIONS

ADP	adenosin diphosphate
ATP	adenosin triphosphate
BCC	body-centered cubic
CCP	cubic close-packed
CFSE	crystal field stabilization energy
CRT	cathode ray tube
DNA	deoxyribonucleis acid
DOTA	1,4,7,10-tetraazacyclododecane -N^{I}, N^{II}, N^{III}, N^{IIII} -tetraacetic acid
EA	electron affinity
EAN	effective atomic number [rule]
EDTA	ethylenediamine tetraacetic acid
EMF	Electromotive force
HCP	hexagonal close-packed
H_4Pdta	1,2-propylenediaminetetraaceticacid
HSAB	hard and soft acid base [theory]
IE	ionization energy
IUPAC	International Union of Pure and Applied Chemistry
Laser	light amplification by stimulated emission of radiation
LGO	ligand group orbital
NADP	Nicotinamide adenine dinucleotide phosphate
NSF	Nephrogenic systemic fibrosis
STP	standard temperature and pressure
VSEPR	valence shell electron pair repulsion [theory]

1. ATOMIC STRUCTURE

Introduction

Chemistry is the science of structural determination of matter. To elucidate the structure of matter, it is first necessary to determine precisely the structure of atoms that compose it. The quest to know the composition of atom has taken us on a long road, from classical mechanics to quantum mechanics. These two schools of thought are not altogether contradictory, but reconcile over several aspects; the most important being that an atom is composed of three fundamental particles: electrons, protons and neutrons. We begin this chapter by describing the characteristics of these particles, then proceed to outline the outcome of classical mechanics and finally highlight the importance of quantum mechanics.

The Electron

The fact that electrolysis of ionic compounds caused their dissociation into constituent elements suggested that atoms are held together in a molecule by attractions that are electrical in nature. This implied that atoms contain units of electrical charge in the form of subatomic particles. These units, as we now know, are called electrons.

The most convincing evidence for the existence of electrons within the atom came from the discovery of *cathode rays*. These rays may be produced in a sealed gas tube called cathode ray tube (CRT), endowed with two electrodes. The tube contains a gas at an extremely low pressure. When high voltage is applied, cathode rays emerge from the cathode and travel towards the anode. These rays are actually a beam of electrons, which are made to move parallel to each other with the aid of a collimator (Fig. 1.1). This beam is detected by observing the glow on a fluorescent screen placed behind the anode. A small object placed in the path of cathode ray beam casts a shadow (Fig. 1.2). This shows that cathode rays travel in a straight line.

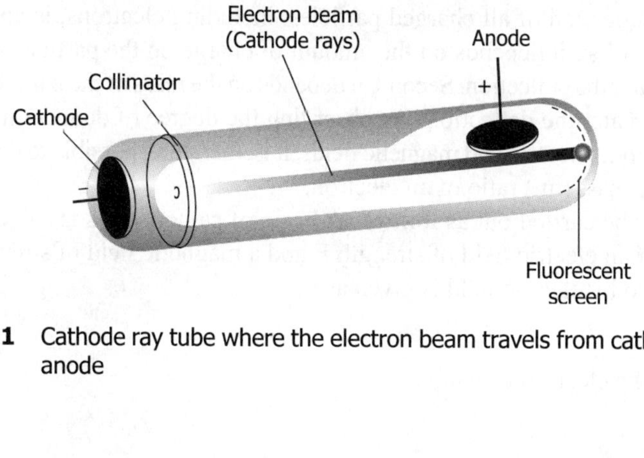

Fig. 1.1 Cathode ray tube where the electron beam travels from cathode to anode

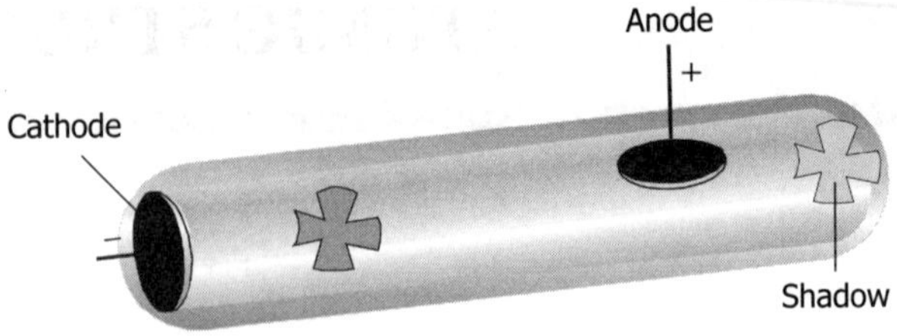

Fig. 1.2 Cathode ray tube showing the shadow of an object placed in the path of cathode rays

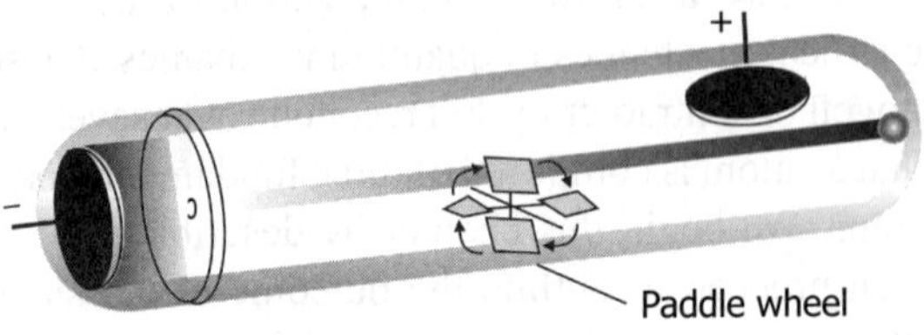

Fig. 1.3 Cathode rays rotating a paddle wheel

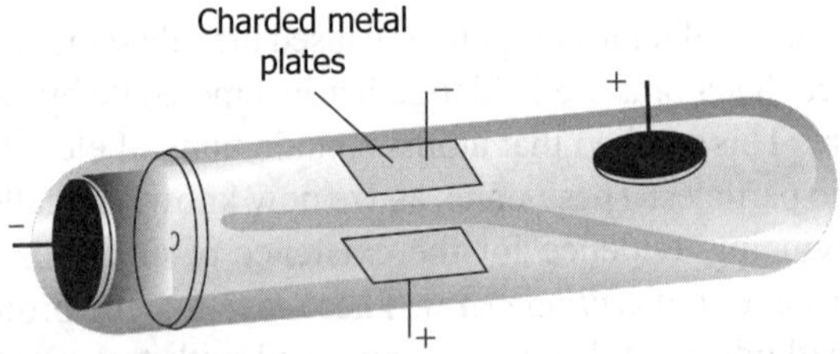

Fig. 1.4 Cathode ray tube showing deflection of cathode rays in an electric field

A small paddle wheel placed in the beam's path rotates, revealing that cathode rays are made up of material particles (Fig. 1.3). These material particles are negatively charged. This may be demonstrated by their deflection towards the positive plate in an electric field (Fig. 1.4). Similarly, the cathode rays are also deflected in a magnetic field (Fig. 1.5). In this case, the deflection is at right angle to the applied field. The degree of deflection of all charged particles, including electrons, is controlled by two factors. First, it depends on the amount of charge on the particle—higher the charge, greater the deflection. Second, it depends on the mass of the particle—smaller the mass, greater the deflection. By observing the degree of deflection of cathode rays in different electric and magnetic fields, it is, therefore, possible to determine the charge (e) to mass (m_e) ratio of an electron.

This may be carried out as follows. A beam of cathode rays is subjected to the influence of an electric field of strength E and a magnetic field of strength H. The force exerted by electric field is given as,

$$F_{elec} = eE \tag{1.1}$$

where e is the electronic charge.

The force exerted by magnetic field is given as,

$$F_{mag} = Hev \tag{1.2}$$

where v is the velocity of cathode ray particles.

When either field is applied alone, the cathode rays deflect from their straight path, the deflection caused by electric field being opposite to that caused by magnetic field.

Suppose, first an electric field is applied to cause a deflection δ from the straight course of the cathode rays. The electric force, given by equation (1.1) a related to the Newton's second law of motion,

$$F_{elec} = eE = m_e$$

or $$a = \frac{eE}{m_e} \tag{1.3}$$

where a is the acceleration due to gravity of a cathode ray particle.

The amount of deflection is related to the acceleration due to gravity by the expression,

$$\delta = \frac{1}{2}at^2 \tag{1.4}$$

where t is the time taken by a cathode ray particle to traverse length l of the cathode ray tube.

From equation (1.3) and (1.4),

$$\delta = \frac{1}{2}\left(\frac{eE}{m_e}\right)t^2 \tag{1.5}$$

The velocity of a cathode ray particle is given as.

$$v = \frac{l}{t}$$

or $$t = \frac{l}{v} \tag{1.6}$$

From equations (1.5) and (1.6),

$$\delta = \frac{1}{2}\left(\frac{eE}{m_e}\right)\left(\frac{l}{v}\right)^2 \tag{1.7}$$

Next, the magnetic field is applied to the cathode ray tube so as to bring back the deflected beam to its original, horizontal pathway. Under these conditions, the force exerted by electric field becomes equal to that exerted by magnetic field.

$$F_{elec} = F_{mag} \tag{1.8}$$

From equations (1.1), (1.2) and (1.8), we get,

$$eE = HeV$$

or $$v = \frac{E}{H} \tag{1.9}$$

Combining equations (1.7) and (1.9),

$$\frac{e}{m_e} = \frac{2\delta}{l^2}\left(\frac{E}{H^2}\right) \tag{1.10}$$

Knowing the values of δ, *l*, E and H, e/m_e may be calculated.
Numerically,

$$\frac{e}{m_e} = 1.7588 \times 10^{11}\ \text{C kg}^{-1} \tag{1.11}$$

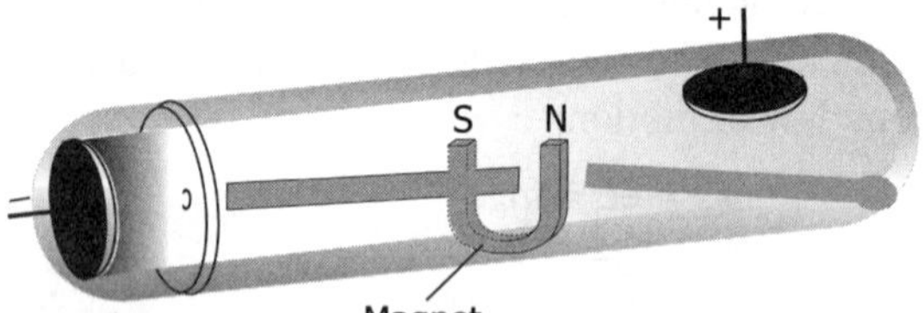

Fig. 1.5 Cathode ray tube showing deflection of cathode rays in a magnetic field

This ratio is the same regardless of the type of gas in the CRT or the composition of electrodes—a proof that electrons are fundamental particles present in all atoms.

To find out the mass of an electron, the value of its charge must be known. This may be elucidated from the *Millikan oil drop experiment.* In this experiment, tiny droplets of oil produced by an atomizer are allowed to pick up electrons produced by the action of X-rays on the molecules of gases present in the immediate vicinity of the atomizer. The charged oil droplets are then allowed to fall through an upper plate to a lower plate.

The force with which an oil droplet falls under the influence of gravity is given by,

$$F = m'g$$

where m′ is the mass of oil droplet and g is the acceleration due to gravity.

However, such a free fall can occur only in a vacuum. In the oil drop chamber, the air offers resistance to the falling drop by a force, F_{res}

$$F_{res} = 6\pi\eta r$$

where r is the radius of the oil droplet and η is the viscosity of air.

The net force with which the oil droplet falls down is given as,

$$F_{net} = F - F_{res}$$

$$F_{net} = m'g - 6\pi\eta r \tag{1.12}$$

Next, the upper plate is imparted a positive charge and the lower plate a negative charge (Fig. 1.6). Now the rate of fall of oil droplets decreases as these are repelled by the lower plate. As the voltage is increased, a stage comes when electrical and gravitational forces on the drops exactly balance each other, and the droplets remain stationary.

At this stage, the force due to electric field (F_{elec} = eE, equation 1.1) is equal to the force with which the oil droplet falls.

$$F_{elec} = F_{net}$$

or
$$eE = m'g - 6\pi\eta r$$

$$e = \frac{m'g - 6\pi\eta r}{E} \qquad (1.13)$$

The numerical value of the charge on an electron has been determined to be,

$$e = 4.8 \times 10^{-10} \text{ esu} = 1.60219 \times 10^{-19} \text{ C} \qquad (1.14)$$

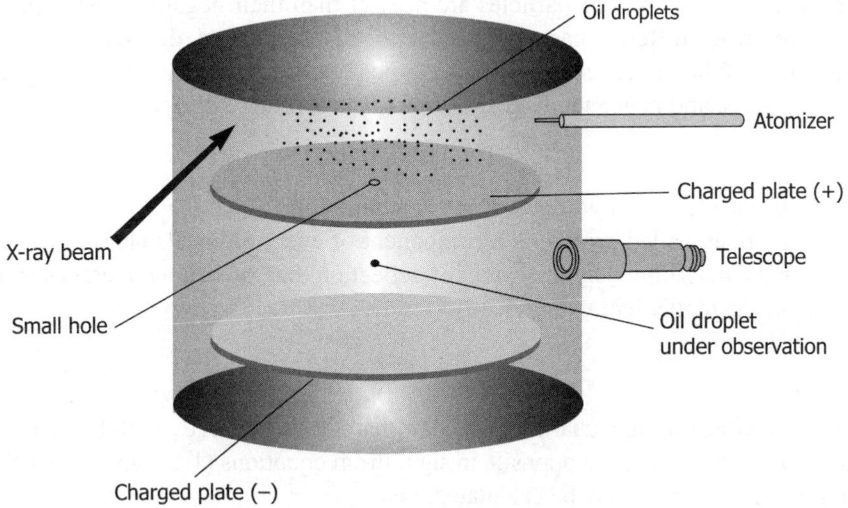

Fig. 1.6 Millikan oil drop experiment

From equations (1.11) and (1.14), the mass of an electron may be calculated, i.e.,

$$m_e = 9.10952 \times 10^{-31} \text{ kg} \qquad (1.15)$$

This is 1/1836 the mass of a hydrogen atom, the lightest of all atoms.

The Proton

The CRT can also be manipulated to generate a beam of positively charged particles that travel from anode to cathode. When a perforated cathode is used, a beam emerges out in the form of a channel or canal (Fig. 1.7). For this reason, the positively charged particles are called *canal rays*.

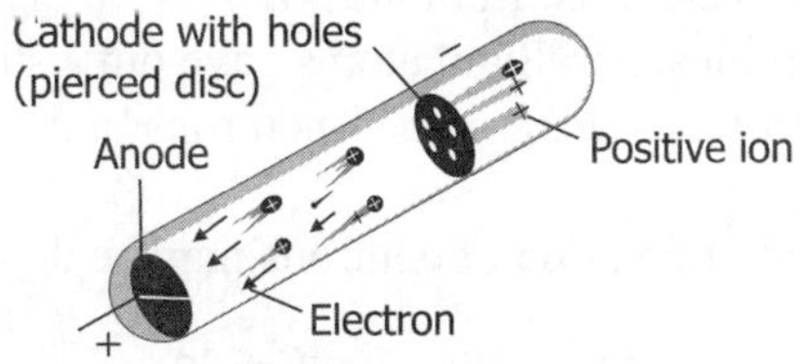

Fig. 1.7 Cathode ray tube showing canal rays (Electrons move towards the anode and positive ions towards the cathode.)

Initially, cathode rays are generated in the tube. These rip apart electrons of the gaseous atoms, present at extremely low pressure inside the tube.

$$X(g) \rightarrow X^+(g) + e^-$$

The positively charged ions move towards the cathode in the form of canal rays, while the electrons move towards the anode. The canal rays are deflected by electric and magnetic fields, but in the opposite direction from cathode rays. The charge to mass ratio of positively charged ions is much smaller than that of electrons, implying that the positively charged particles are heavier than their negative counterparts. Moreover, when different gases are used in the discharge tube, the charge to mass ratio varies. When hydrogen gas is used, a positively charged particle having the smallest mass, and hence the largest charge to mass ratio, is formed.

$$H \rightarrow H^+ + e^-$$

(In a discharge tube, hydrogen exists in atomic form.)

This particle, called proton, is a component of every atom. Its charge to mass ratio may be determined by the extent of deflection that the beam undergoes in an electric or magnetic field. Numerically,

$$\frac{e}{m_p} = 9.5791 \times 10^7 \text{ C kg}^{-1} \qquad (1.16)$$

The proton has a charge equal in magnitude to that of the electron i.e., 1.60219×10^{-19} C [from equation (1.14)], but opposite in sign. From equations (1.14) and (1.16), the mass of a proton (m_p) may be calculated, i.e.,

$$m_p = 1.672 \times 10^{-27} \text{ kg} \qquad (1.17)$$

This value, as has been stated earlier, is 1836 times larger than the mass of an electron [equation (1.15)].

The Neutron

Since atoms are electrically neutral, it is obvious that the number of electrons should be equal to the number of protons. However, when the mass of an atom was calculated on the basis of the number of electrons and protons present in it, the value was found to be smaller than the observed mass. This suggested that some other fundamental particles must be present in the atom. Moreover, to maintain electrical neutrality, these particles must be uncharged. The proof of existence of these entities emerged from the observation that on being bombarded with high-energy α-particles, beryllium atoms gave out a stream of novel, neutral particles. These new particles were named neutrons and are represented by the symbol n.

The reaction leading to liberation of neutrons may be depicted as follows:

$$^{9}_{4}Be + \underset{\alpha\text{-particle}}{^{4}_{2}He} \rightarrow {}^{12}_{6}C + {}^{1}_{0}n$$

We shall learn more about such reactions in Chapter 4, which deals with nuclear chemistry. At present, suffice it to say that the mass of a neutron, m_n, is slightly larger than that of a proton.

$$m_n = 1.6749 \times 10^{-27}\ kg \tag{1.18}$$

The Nucleus

With the discovery of neutrons, the picture of an atom became appreciably clear. Atoms consist of a small, dense nucleus surrounded by electrons at relatively great distances. The nucleus contains protons and neutrons. The nucleus of hydrogen atom is however an exception, as it contains only one proton and no neutron. The number of protons or the number of electrons is called *atomic number.* The total number of protons and neutrons inside the nucleus is called *mass number.*

Rutherford Model of Atom

The atomic model propounded by Rutherford was based on the *gold leaf* or *gold foil experiment.* In this experiment, a very thin leaf of gold was bombarded with α-particles originating from a radioactive source. A fluorescent zinc sulfide screen was placed behind the gold foil to record the scattering of α-particles as scintillations (Fig. 1.8).

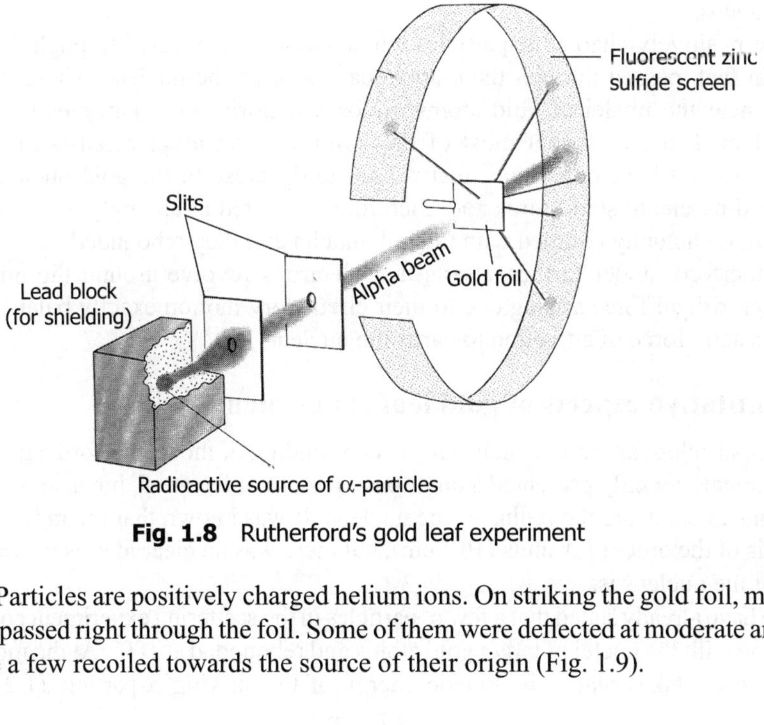

Fig. 1.8 Rutherford's gold leaf experiment

α-Particles are positively charged helium ions. On striking the gold foil, most of them passed right through the foil. Some of them were deflected at moderate angles, while a few recoiled towards the source of their origin (Fig. 1.9).

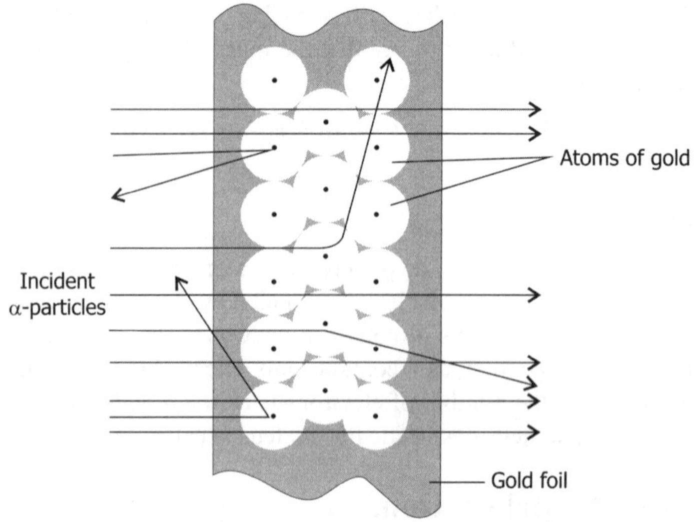

Fig. 1.9 Deflection pattern of α-particles

Rutherford concluded from this experiment that an atom consists of a small, dense nucleus with large empty space around it. All the positive charge of an atom and most of its mass is concentrated in the nucleus. The units of positive charge (protons) inside the nucleus is equal to the units of negative charge (electrons) outside the nucleus. The electrons are thinly distributed in the empty space surrounding the nucleus.

The positively charged α-particles which passed undeflected through the gold foil, in fact, passed through the empty space around the nucleus. These did not come near the nuclei of gold atoms. Since a majority of α-particles remained undeflected, it shows that most of the volume of an atom consists of empty space. Some of the α-particles that passed fairly close to the gold nuclei, were repelled by electrostatic force and, therefore, deflected at an angle. A few of the α-particles actually collided with the gold nuclei and they rebounded.

Rutherford model further asserts that electrons revolve around the nucleus. The centrifugal force arising due to their circulatory motion exactly balances the electrostatic force of attraction towards the nucleus.

Quantitative aspects of gold leaf experiment

The α-particle scattering, which was the core finding of the Rutherford's gold leaf experiment, not only presented a qualitative picture of the atom, but also provided a means to calculate the radius of the nucleus. It was known that the radius of an atom is of the order of Å units (10^{-8} cm), but there was no clear idea as to what the size of the nucleus is.

We have already stated that a few α-particles in the scattering experiment collided head-on with the nuclei of target gold atoms and rebounded at 180°. At the moment when recoil takes place, the kinetic energy of the striking α-particle (1/2 mv^2) becomes equal to the potential energy $\left(\frac{Z_1Z_2e^2}{r_0}\right)$.

That is, $$\frac{1}{2}mv^2 = \frac{Z_1Z_2e^2}{r_0}$$

Here, m is the mass of α-particle = 6.68×10^{-24} g
v is the velocity of α-particle = 1.6×10^{9} cm s^{-1}
Z_1 is the atomic number of α-particle = 2
Z_2 is the atomic number of gold = 79
e is the electronic charge = 4.8×10^{-10} esu

The factor r_0 is the distance of closest approach between the nucleus of the α-particle and that of the gold atom with which it strikes. It may be taken as the internuclear distance.

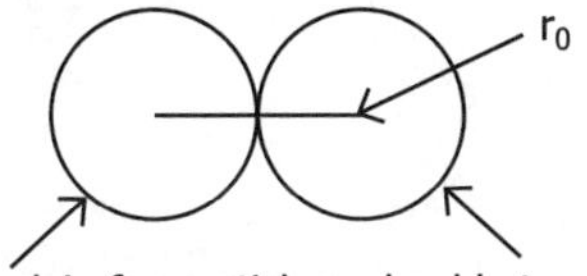

Nuclei of α-particle and gold atom

$$\frac{1}{2}(6.68 \times 10^{-24})\,(1.6 \times 10^{9})^2 = \frac{2 \times 79 \times (4.8 \times 10^{10})^2}{r_0}$$

$$r_0 = 4.25 \times 10^{-12} \text{ cm}$$

Radius of the gold nucleus = $\frac{1}{2} \times 4.25 \times 10^{-12}$ or 2×10^{-12} cm (approximately)

The Rutherford gold leaf experiment proved that the radius of nuclei varies in the range 10^{-12} to 10^{-13} cm. It also brought to light how small the nucleus is as compared to the atom. The ratio between the volume of atom (V_{atom}) and volume of nucleus ($V_{nucleus}$) may be calculated as follows.

$$\frac{V_{atom}}{V_{nucleus}} = \frac{\frac{4}{3}\pi r^3_{atom}}{\frac{4}{3}\pi r^3_{nucleus}}$$

$$\frac{V_{atom}}{V_{nucleus}} = \frac{(10^{-8})^3}{(10^{-12})^3} = 10^{12}$$

$$V_{atom} = 10^{12}\,(V_{nucleus})$$

Thus the nucleus, in which virtually all the mass of the atom is concentrated, occupies 10^{-12} the volume of the atom.

Electromagnetic Radiation

A probe into the characteristics of electromagnetic radiations is relevant to atomic structure for two reasons. First, as we shall see later, electrons behave like electromagnetic radiations. Second, and more importantly, when an atom in an excited state is allowed to come to the ground state, electromagnetic radiation is emitted. Analysis of this radiation throws light on the atomic architecture.

Electromagnetic radiation travels through space in wave motion (Fig. 1.10). A wave is characterized by the following features:

1. Wavelength

The distance between two successive crests or two successive troughs of a wave is called its wavelength. It is designated as λ.

2. Amplitude

The amplitude of a wave is its height. It is generally designated as a. The intensity (I) or brightness of a radiation is directly proportional to the square of its amplitude, i.e.,

$$I \propto a^2 \tag{1.19}$$

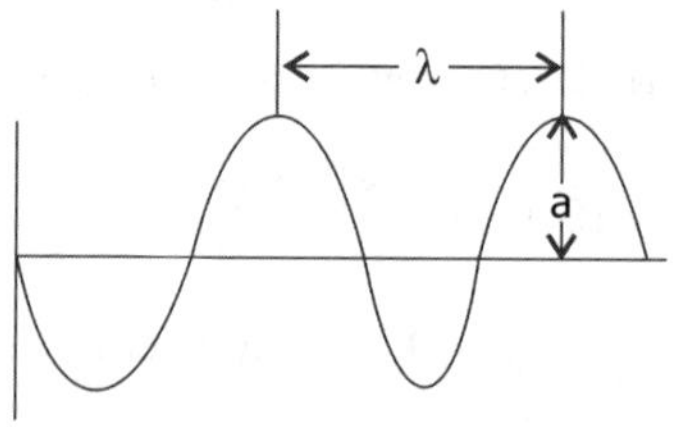

Fig. 1.10 Electromagnetic radiation travels through space in wave motion

3. Velocity

The distance travelled by a wave in unit time is called its velocity or speed. It is designated as v. In case of light waves, the velocity is represented as c. Numerically,

$$c = 3 \times 10^8 \text{ m s}^{-1} \tag{1.20}$$

In vacuum, all waves, irrespective of their wavelength, travel with the speed of light.

4. Frequency

The number of waves that pass through a given point in one second is called the frequency of the radiation. It is usually denoted as υ and expressed as cycles per second or simply as s^{-1}. One cycle per second is called one Hertz (Hz). The frequency and wavelength of a wave are related to each other by the expression,

$$\lambda\upsilon = c$$

or

$$\upsilon = \frac{c}{\lambda} \tag{1.21}$$

The wavelength and frequency are, therefore, inversely proportional to each other.

$$\lambda \propto \frac{1}{\upsilon} \tag{1.22}$$

5. Wave number

The number of waves in one centimeter is called wave number. It is designated as $\bar{\upsilon}$ and is related to wavelength by the expression,

$$\bar{\upsilon} = \frac{1}{\lambda} \tag{1.23}$$

Its relationship to frequency may be deduced by substituting the value of λ from equation (1.21) in equation (1.23).

$$\bar{\upsilon} = \frac{\upsilon}{c} \tag{1.24}$$

Wave Nature of Light

Light is usually considered to have wave character. A beam of light propagates through space in the form of a wave. If the wavelength of a light beam is known, its frequency and wave number may be determined. Knowing the amplitude, intensity may be elucidated. Above all, light shows the properties of reflection, refraction and diffraction—all indicative of its wave behavior. A further proof of the wave nature of light comes from the dispersion phenomenon that manifests itself in the form of a continuous spectrum. We will study more about continuous spectrum in a later section of this chapter (*see* p. 14).

Particle Nature of Light

Although light is usually described in terms of wave behavior, it may also be demonstrated as being composed of particles. The idea that light is made up of particles formed the basis of *Planck's quantum theory.* According to this theory, each particle of light is called a *photon.* Each photon possesses a specific amount of energy called *quantum.* The amount of energy (E) possessed by a photon depends on the color of light. It is directly proportional to the frequency of light.

$$E \propto \upsilon$$

or

$$E = h\upsilon$$

Here, h is a constant called *Planck's constant.* Its value is 6.63×10^{-34} J s.

A high-energy radiation has high frequency. This means that for a high-energy beam, a large number of waves pass through a specific point in one second. Therefore, the wavelength of such a radiation must be short [equation (1.22)].

Planck's quantum theory marked the beginning of a new school of thought—the concept of *quantum mechanics*. It represented a significant break from the concept of classical mechanics. The latter considered the electron as a particle and light as a wave. Quantum mechanics considers that both electron and light have dual character. These can behave as particles as well as waves. Rutherford atomic theory was based on classical mechanics and for this reason it is now considered trivial.

Proofs of particle nature of light

Photoelectric effect

The most convincing evidence for the particle nature of light comes from *photoelectric effect.* It is a phenomenon in which a beam of light falls on the surface of a metal and causes emission of electrons. This is analogous to one billiard ball striking another and removing it from the surface of the table. The most reactive metals, for example, alkali metals, show the most pronounced photoelectric effect. The electrons released as a result of photoelectric effect are called *photoelectrons.*

Two fallouts of photoelectric effect need to be emphasized:

1. The velocity with which the photoelectrons are emitted is directly proportional to the frequency of light impinged on the metal surface. It is independent of the intensity of light.

$$v \propto \upsilon \tag{1.25}$$

2. The number of photoelectrons emitted is directly proportional to the intensity of light striking the metal surface. It is independent of the frequency of light.

Classical mechanics fails to explain photoelectric effect. It suggested that the number of photoelectrons emitted depends on the energy, i.e., on the frequency of light. High-energy light would emit more electrons than low-energy light. Further, the number of photoelectrons emitted depends on the duration for which the light is allowed to fall on the surface. Longer the duration of illumination, greater the number of electrons emitted.

The explanation for photoelectric effect is provided by *Einstein's concept of light quanta*. The term "light quanta" itself suggests that the concept is based on the principles of quantum mechanics. A photon of light collides with an electron near the surface of metal. As a result, the electron gets knocked out of the surface. When the intensity of light increases, the number of photons striking a given area of the metal surface per second also increases. Therefore, a larger number of photoelectrons are emitted. Further, during collision, the photon transfers its energy to the electron. The energy of photon must be greater than that required to liberate an electron from the attractive influence of the nucleus, only then will the electron be released. This means,

$$h\upsilon > \omega$$

where $h\upsilon$ is the energy of photon and ω is the work required to remove an electron from the metal surface. The residual energy, i.e., the difference between energy of a photon and the binding energy of an electron is picked up by the emitted electron. It enables the electron to escape from the metal surface and hence equals the kinetic energy of an electron, i.e.,

$$\frac{1}{2}mv^2 = h\upsilon - \omega \tag{1.26}$$

where $\frac{1}{2}mv^2$ (m is the mass and v is the velocity of an electron) is the kinetic energy of an electron. Since m, h and ω are constants, it implies that the velocity of a photoelectron is directly proportional to the frequency (or energy) of impinging light [equation (1.25)].

The expression (1.26) is called *Einstein's photoelectric equation.*

Case Study 1.1 When an alkali metal is bombarded with ultraviolet radiation of wavelength 300 nm, the ejected electrons have a kinetic energy of 98.17 kJ mol^{-1}. Calculate the work function of the alkali metal.

Given: Kinetic energy of one mole of ejected electrons = 98.17 kJ or 98.17×10^3 J

$$\text{Kinetic energy of one electron} = \frac{98.17 \times 10^3}{N} = \frac{98.17 \times 10^3 \text{ J}}{6.023 \times 10^{23}} = 1.63 \times 10^{-19} \text{ J}$$

[N is the Avogadro number]

This value corresponds to the term $\frac{1}{2}mv^2$ in equation (1.26)

Given: Wavelength of striking radiation, $\lambda = 300$ nm or 300×10^{-9} m

With reference to equation (1.21), the frequency of radiation, $\upsilon = \frac{c}{\lambda}$

$$\upsilon = \frac{3 \times 10^8 \text{ m s}^{-1}}{300 \times 10^{-9} \text{ m}} = 1 \times 10^{15} \text{ s}^{-1}$$

$$h\upsilon = (6.63 \times 10^{-34} \text{ J s})(1 \times 10^{-15} \text{ s}^{-1}) = 6.63 \times 10^{-19} \text{ J}$$

Rewriting equation (1.26)

$$\omega = h\upsilon - \frac{1}{2}mv^2 = (6.63 \times 10^{-19} \text{ J}) - (1.63 \times 10^{-19} \text{ J}) = 5.0 \times 10^{-19} \text{ J}$$

Work required to remove the electron form the alkali metal = 5.0×10^{-19} J

Compton effect

Another proof of the particle nature of electromagnetic radiation comes from the change in wavelength of X-rays when these are scattered upon collision with elements of low mass numbers. This phenomenon is called Compton effect. X-rays are a form of electromagnetic radiation and are, therefore, made up of photons.

X-rays of a specific wavelength, on striking a surface, get scattered. The scattered radiation has a longer wavelength than the original beam. Actually, the X-rays collide with the electrons of the element of which the surface is made of. The collision is elastic, i.e., the momentum and energy are conserved during the impact. The momentum of the electron increases, while that of the photon decreases by an equal amount. Similarly, the energy of the electron increases, while that of the photon decreases by an equal amount. The energy of photon is given by the equation,

$$E = h\upsilon = h\frac{c}{\lambda}$$

Since h and c are constants, a decrease in energy will occur only if the wavelength increases. Due to this reason, the scattered X-rays have a longer wavelength than incident X-rays. After the collision, the photon moves in one direction, the electron in the opposite direction.

Case Study 1.2 The frequency of violet light is 7.31×10^{14} s^{-1}. Calculate the wavelength and energy of an individual photon comprising this radiation.

The wavelength of a radiation is related to the frequency by the expression,

$$\upsilon = \frac{c}{\lambda}$$

$$\upsilon = 7.31 \times 10^{14}\ s^{-1},\ c = 3 \times 10^{8}\ m\ s^{-1}$$

$$\lambda = \frac{3 \times 10^{8}\ m\ s^{-1}}{7.31 \times 10^{14}\ s^{-1}} = 4.10 \times 10^{-7}\ m$$

$$\text{since } 1\ nm = 10^{-9} m$$

$$\upsilon = 410\ nm$$

The energy of a photon is related to the frequency of light by the expression,

$$E = h\upsilon$$
$$= (6.63 \times 10^{-34}\ Js)\ (7.31 \times 10^{14}\ s^{-1})\ (\text{Since } h = 6.63 \times 10^{-34}\ J\ s)$$
$$= 4.84 \times 10^{-19}\ J$$

The wavelength of violet light is 410 nm and the energy of its photon is 4.84×10^{-19} J.

Electromagnetic Spectrum

Suppose a beam of white light is allowed to pass through a prism. On emerging out of the prism, the beam bends or refracts. The amount of refraction depends upon the wavelength. A radiation with shorter wavelength refracts more than one with a longer wavelength. White light is composed of seven electromagnetic components. Each has a different color (VIBGYOR) and a different wavelength. Each, therefore, refracts to a varying degree. As a result, when a beam of white light passes through a prism, it disperses into a wide band of colors. This range of colors is called electromagnetic spectrum. The spectrum is an array of rainbow colors, one following the other, with no blank spaces in between the colors. Violet merges into indigo, indigo into blue, and so on (Fig 1.11). Due to this reason, electromagnetic spectrum is also called *continuous spectrum.*

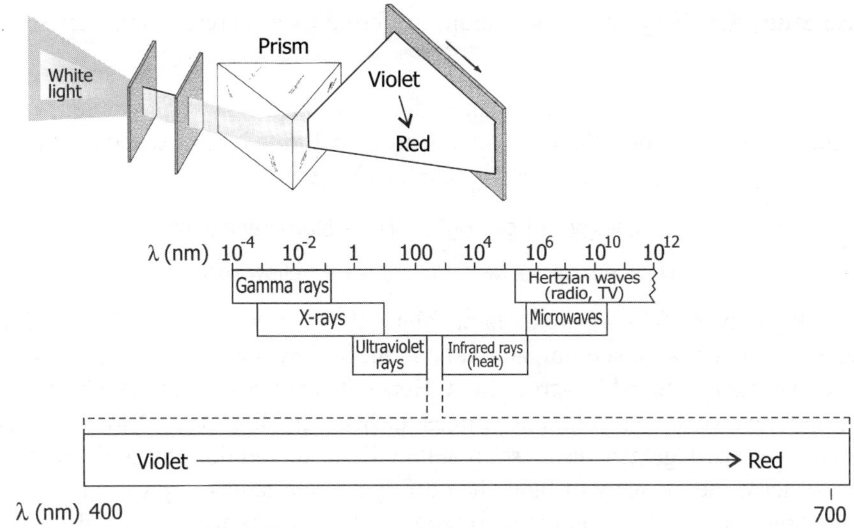

Fig. 1.11 Electromagnetic spectrum

Case Study 1.3 Calculate the frequencies of light at the extremities of the visible component of electromagnetic spectrum.

As shown in Fig. 1.11, the visible region of electromagnetic spectrum traverses from the violet end (having wavelength 400 nm) to the red end (having wavelength 700 nm).

We refer to equation (1.21) and substitute the value 3×10^8 ms^{-1} for velocity of light, c.

At violet end (by convention also called blue end): $\lambda = 400$ nm or 400×10^{-9} m

$$\upsilon = \frac{3 \times 10^8 \text{ m s}^{-1}}{400 \times 10^{-9} \text{ m}} = 7.5 \times 10^{14} \text{ s}^{-1}$$

At red end: $\lambda = 700$ nm or 700×10^{-9} m

$$\upsilon = \frac{3 \times 10^8 \text{ m s}^{-1}}{700 \times 10^{-9} \text{ m}} = 4.2 \times 10^{14} \text{ s}^{-1}$$

The frequency of light at violet end is 7.5×10^{14} s^{-1}, while at red end is 4.2×10^{14} s^{-1}.

White light is also called visible light as it can be detected by human eye. However, there are many other forms of electromagnetic radiations apart from visible light. Some have wavelengths shorter than that of the violet component of visible light. Others have wavelengths longer than that of the red component of visible light. In fact, visible light is only a very small portion of the electromagnetic spectrum (Fig 1.11). The constituents of electromagnetic radiation having wavelengths shorter or longer than visible light cannot be detected by human eye. Nevertheless, these may be recorded photographically by detectors. These also show continuous spectra.

Case Study 1.4 Why are sodium lamps preferred over mercury lamps for street lighting?

Both sodium lamps and mercury lamps function on the same principle: When bomborded by electrons, the metal atoms are transformed into excited states and on reverting to their ground states, these emit visible light.

$$Hg + energy \rightarrow Hg^*; \; Hg^* \rightarrow Hg + \text{blue-green light}$$

$$Na + energy \rightarrow Na^*; \; Na^* \rightarrow Na + \text{yellow light}$$

The light given off by a mercury lamp falls in the blue-green region of the visible spectrum, while that of sodium lamp falls in the yellow region. Yellow light has a longer wavelength than blue-green light. Hence it is not easily scattered by fog or atmospheric dust. As a result, it penetrates further than bluish light. Moreover, the human eye shows greater response to yellow light. In addition, for a given input of electricity, the intensity of light shed off by sodium lamps in greater than that emitted by mercury lamps. For these reasons, sodium lamps are considered a better choice for street lighting than mercury lamps.

Atomic Spectrum

When the atoms of an element are heated in an electric arc, they are transformed from ground energy state to a thermally excited state. This is because they absorb heat energy from the electric arc. The atoms are not stable in the high energy states and, therefore, tend to return to the ground state. On reverting to their original state, they emit energy in the form of radiation. When this radiation is passed through a prism, a spectrum is obtained. It is called atomic spectrum.

Unlike electromagnetic spectrum, atomic spectrum is not displayed in the form of wide bands. Rather, it is in the form of sharp lines. Hence, these are also called *line spectra.* Each element has a unique line spectrum. Typical line spectra of some elements are shown in Fig. 1.12.

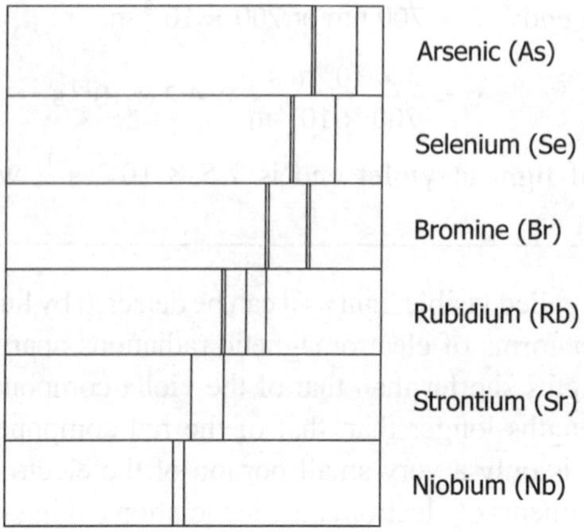

Fig. 1.12 Typical line spectra of some elements

Each line in the atomic spectrum corresponds to a *specific wavelength.* In contrast, each band of electromagnetic spectrum corresponds to a *range of wavelengths.* Therefore, electromagnetic spectrum is continuous in nature, with one band merging into another, but atomic spectrum is discontinuous, with one line separated from the other.

This means that an atom falls from the excited state to the ground state in several stages. At each stage, a radiation is emitted, whose wavelength is different from that of the radiation emitted in the previous or next stage. An atomic spectrum is a photographic recording of separated wavelengths.

Case Study 1.5 What is the energy of one mole of a radiation having a frequency of 9×10^5 Hz?

The energy of one photon of the radiation may be calculated by the equation,

$$E = h\upsilon$$

Here h is the Planck's constant $= 6.63 \times 10^{-34}$ J s

The frequency, $\upsilon = 9 \times 10^5$ Hz (or s^{-1})

$$E = (6.63 \times 10^{-34}\ \text{J s})\ (9 \times 10^5\ \text{s}^{-1})$$
$$= 5.96 \times 10^{-28}\ \text{J}$$

One mole of the radiation would contain. Avogadro number (6.023×10^{23}) of photons. That is,

$$NE = (6.023 \times 10^{23})\ (5.96 \times 10^{-28}\ \text{J})$$
$$= 3.58 \times 10^{-4}\ \text{J}$$

The energy of one mole of the said radiation is 3.58×10^{-4} J

Failure of Rutherford Model

Rutherford pictured an atom as containing a small, dense, positively charged nucleus, with a number of light, negatively charged electrons revolving around it. The rapid motion of electrons around the nucleus ensures that the electrostatic force of attraction between the positive and negative charges is exactly counter-balanced by the centrifugal force due to planetary rotation. However, an atom based on this model would be extremely unstable. According to the classical electromagnetic theory, a particle which is extremely small, is charged and is in constant motion—and an electron satisfies all the three criteria—would lose kinetic energy. This means that an electron would follow a spiral path of decreasing curvature until it eventually falls into the nucleus (Fig. 1.13).

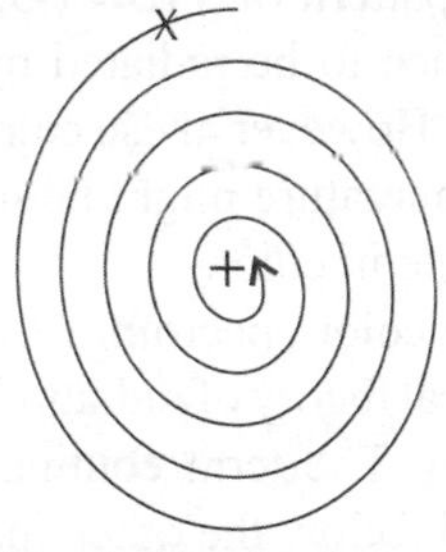

Fig. 1.13 Rutherford model of a revolving electron that follows a spiral path and falls into the nucleus

There is yet another problem with Rutherford theory. The loss of kinetic energy of an electron should manifest itself in the form of electromagnetic radiations. Since there is continuous loss of kinetic energy, there should be an unabated emission of electromagnetic waves. When these waves are passed through a prism, a continuous spectrum, similar to the band spectrum of visible light, should be obtained. At a specific time, the electron emits radiation of one wavelength and a fraction of second later, it emits another radiation whose wavelength is infinitesimally longer than the previous one. This way, a large number of wavelengths would merge with each other to give continuous bands in the spectrum (Fig. 1.11).

However, when radiations emitted by an atom are passed through a prism, a discontinuous spectrum is obtained. As has already been stated, this type of spectrum is displayed in the form of lines—each line representing a specific wavelength, separated from the other. Accordingly, it is called a line spectrum.

The line spectrum clearly indicates that an electron does not lose kinetic energy continuously, i.e., electromagnetic radiations are not emitted uninterrupted. Such radiations are emitted only when an electron jumps from a higher energy level to a lower energy level and not when it follows a path of decreasing curvature towards the nucleus. The concept of electron jumping from one level to another is explained on the basis of Bohr theory, which is certainly an improvement over the Rutherford model.

Box 1.1

The Concept of Black Body Radiation

We have already stated the Planck's hypothesis that light is made up of discrete particles called photons (p. 11). Further, each photon possesses a fixed amount of energy. Planck arrived at this conclusion by studying the color of the radiations emanating from a heated black body.

A true black body is one which absorbs radiations of all colors, and reflects none. Nevertheless, a black body is a very efficient emitter of radiation. For this reason it is often referred to as a complete radiator or an ideal radiator.

When a black body is heated, it first radiates red color. As the temperature is raised, the color of the body turns orange and then yellow. If we follow the color pattern of VIBGYOR in a reverse manner, after yellow the next color sequence to be radiated by a black body should be green, blue, indigo and violet. However, these colors do not emanate from a black body however high its temperature might be increased. Rather, after yellow, the body appears to be white in color.

The color spectrum of the radiating black body seems puzzling since the classical theory of radiation predicts that emission should occur in a continuous manner. The term continuous here means that one wavelength of radiation should follow the other, and then the next. If this would have been true, then on increasing the temperature of the black body, yellow should have been followed by green, blue, indigo and violet.

Contd...

However, from the black body experiment it became clear that the energy can be radiated only discontinuously and that too in discrete quantities. The smallest packet of energy that can be radiated in form of an electromagnetic ray is called a *quantum*. A single quantum possesses energy equivalent to hγ, where h is called Planck's constant and υ is the frequency of the emanating radiation. Energy is always emitted in multiples of hυ, that is, 2 hυ, 3 hυ, etc., but never as 2.37 hυ, 3.28 hυ, etc.

If an atom that makes up the black body can radiate only a specific quantity of energy, it follows from the same consideration, that it can absorb only a fixed amount of energy. This means that the energy of atoms is quantized. The Bohr model of atom, discussed in the next section, is based on this premise.

Bohr Model of Atom

To overcome objections to the Rutherford model, as well as to explain the discontinuous nature of atomic spectrum, Bohr propounded that the electrons revolve around the nucleus in special types of orbits called *stationary states.* As long as an electron revolves in a specific stationary state, it does not emit energy. Energy change occurs only when the electron moves from one stationary state to another. If it is promoted to a higher stationary state, energy is absorbed. In this case, the energy must be imparted to the electron from some source, such as an electric arc. When the electron falls to a lower stationary state, energy is emitted spontaneously.

The electron can move from one stationary state to another but cannot be placed in between two stationary states. Suppose the energy difference between two successive stationary states is 1×10^{-18} J. The electron will move from the lower to the higher state only if it acquires energy equal to, or greater than 1×10^{-18} J. If the electron absorbs energy equivalent to 0.9×10^{-18} J, it will not be excited. It will continue to remain in the lower level state. This means that an electron will not acquire energy that would place it between the permissible stationary states.

Thus, an electron can occupy only discrete stationary states. When it moves from one stationary state to another, there is discrete absorption or emission of energy. If an electron falls from the third to the second stationary state, it will emit a specific amount of radiant energy, the wavelength of which would appear as a line in the atomic spectrum. Now, if the same electron falls from the second to the first energy level, it will again emit discrete radiant energy, the wavelength of which would be different from that of the earlier transition. The second line in the atomic spectrum would, therefore, be placed at a different position than the first line. As a result, a discontinuous spectrum will be obtained.

The Bohr model of atom is based on the following postulates:

1. An atom has a number of stable orbits, called stationary states, in which an electron can reside without emitting energy.
2. An electron may absorb energy from a source and get excited to a higher stationary state. It would not be stable in the excited state. When it reverts to its original state, it would emit radiation. The energy of this radiation

corresponds to the energy difference between the higher and lower stationary states. It is given by Planck's quantum theory,

$$\Delta E = h\upsilon$$

If E_2 is the energy of higher stationary state and E_1 of lower stationary state, then,

$$\Delta E = E_2 - E_1$$

$$\text{or} \quad h\upsilon = E_2 - E_1 \qquad (1.27)$$

3. The angular momentum of an electron (mvr) in any stationary state is an integral multiple of $h/2\pi$, i.e.,

$$mvr = \frac{nh}{2\pi} \qquad (1.28)$$

where h is Planck's constant, m is the mass of an electron, v is its velocity, r is the radius of stationary state in which it is revolving and n is an integer having values 1, 2, 3...

Bohr's picture of hydrogen atom

The structure of a hydrogen atom is the simplest of all elements. It comprises a nucleus with charge Ze, around which an electron, with charge e, revolves in a circular orbit of radius r. The electrostatic force of attraction, F_e, between the nucleus and the electron is given as,

$$F_e = \frac{Ze.e}{r^2} = \frac{Ze^2}{r^2} \qquad (1.29)$$

The centrifugal force, F_c, of an electron having mass m and revolving around the nucleus with velocity v is given as,

$$F_c = \frac{mv^2}{r}$$

For an electron to remain stable in its orbit, the electrostatic force must be balanced by the centrifugal force, i.e.,

$$\frac{Ze^2}{r^2} = \frac{mv^2}{r} \qquad (1.30)$$

$$\text{or} \quad v = \sqrt{\frac{Ze^2}{mr}}$$

The velocity of an electron may also be obtained from equation (1.28):

$$v = \frac{nh}{2\pi mr} \qquad (1.31)$$

Equations (1.30) and (1.31) may be equated.

$$\sqrt{\frac{Ze^2}{mr}} = \frac{nh}{2\pi mr}$$

$$\text{or} \quad r = \frac{n^2h^2}{4\pi^2 mZe^2} \qquad (1.32)$$

Since hydrogen atom has only one proton in its nucleus, therefore Z = 1. The radius of the smallest orbit of hydrogen (n = 1) is given as,

$$r_0 = \frac{h^2}{4\pi^2 me^2} \tag{1.33}$$

The radii of the second, third, etc. orbits may be similarly calculated.

Bohr's model enables not only to calculate the radii of various orbits of a hydrogen atom, but also to assess the energy of an electron in each orbit. The kinetic energy of an electron is given as,

$$\text{Kinetic energy} = \frac{1}{2}mv^2 \tag{1.34}$$

The potential energy of an electron depends on the electrostatic force of attraction that the nuclear charge exerts on it. As the distance between the nucleus and the electron increases, the potential energy decreases and ultimately becomes zero at infinite distance of separation. The potential energy of an electron may be elucidated by integrating the electrostatic force of attraction [equation (1.29)],

$$\text{Potential energy} = \int \frac{Ze^2}{r^2} = \frac{-Ze^2}{r} \tag{1.35}$$

The total energy of an electron, E, is the sum of kinetic energy [equation (1.34)] and potential energy [equation (1.35)].

$$E = \frac{1}{2}mv^2 + \left(\frac{-Ze^2}{r}\right) \tag{1.36}$$

Dividing equation (1.30) by 2 and simplifying, we get:

$$\frac{Ze^2}{2r} = \frac{mv^2}{2} \tag{1.37}$$

From equations (1.36) and (1.37),

$$E = \frac{-Ze^2}{2r} \tag{1.38}$$

Substituting the value of r from equation (1.32) into equation (1.38),

$$E_n = \frac{-2\pi^2 Z^2 e^4 m}{n^2 h^2} \tag{1.39}$$

where E_n is the energy of an electron in the nth orbit.

Bohr's explanation for hydrogen spectrum

According to Bohr theory, energy is radiated when an electron moves from an outer stationary state to an inner stationary state. The term stationary state may now be used interchangeably with orbit. Thus, when an electron falls from the second orbit to the first, the energy change is given by equation (1.27). If we designate the energy of outer orbit as E_o and that of inner orbit as E_i, equation (1.27) may be re-written in a more generalized form as,

$$h\upsilon = E_o - E_i \tag{1.40}$$

where h is Planck's constant and υ is the frequency of the radiation emitted when an electron falls to a lower stationary state. An electron falls to a lower level only after excitation to a higher level. The excitation may be carried out by allowing the electron to absorb energy from an external source, such as an electric arc. Once it absorbs energy, it is promoted to the second, third, fourth or even higher stationary states.

A higher energy system is never stable. Therefore, the electron in excited state tends to revert to the lowest energy state. The lowest energy state of an electron is called *ground state.* In the case of a hydrogen atom, the first orbit, i.e., the stationary state with n = 1 represents the ground state. However, as discussed earlier, it will not revert to the ground state in one step, but in a number of steps. For example, an electron excited to the fourth orbit will first fall to the third orbit then to the second and finally to the first orbit. At each stage, a radiation of specific wavelength is emitted and is identified as a distinct line in the spectrum of the atom.

Transitions of an electron to the innermost orbit represented by n = 1, from n_2, n_3, n_4, etc. orbits give rise to the first, second, third, etc. lines of the *Lyman series*, respectively. The wavelengths of the lines of Lyman series lie in the ultraviolet region. Transitions of electrons to the second orbit (n = 2) from outer orbits give a collection of lines called *Balmer series.* The wavelengths of these lines lie in the visible region. Similarly, transitions from higher orbits to the third orbit (n = 3) give the *Paschen series* of lines, the wavelengths of which lie in the infrared region (Fig. 1.14).

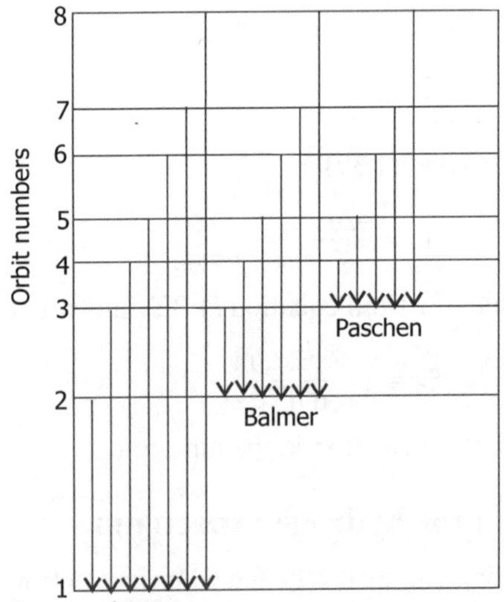

Fig. 1.14 Line spectra of a hydrogen atom

The frequency of each line of each series may be calculated as follows:

Equation (1.39) represents the energy, E_n, of an electron in the nth orbit. The difference between the energies of an electron in two orbit may be elucidated by combining equations (1.39) and (1.40). This gives,

$$h\upsilon = E_{no} - E_{ni} = -\frac{2\pi^2 Z^2 e^4 m}{h^2}\left(\frac{1}{n_o^2} - \frac{1}{n_i^2}\right)$$

or $$h\upsilon = E_{no} - E_{ni} = \frac{2\pi^2 Z^2 e^4 m}{h^2}\left(\frac{1}{n_i^2} - \frac{1}{n_o^2}\right)$$

$$\upsilon = \frac{2\pi^2 Z^2 e^4 m}{h^3}\left(\frac{1}{n_i^2} - \frac{1}{n_o^2}\right) \quad (1.41)$$

On substituting the value of various constants, equation (1.41) may be simplified to,

$$\upsilon = 3.29 \times 10^{15}\left(\frac{1}{n_i^2} - \frac{1}{n_o^2}\right) s^{-1} \quad (1.42)$$

The wavelength of various lines may be calculated by substituting the value of frequency from equation (1.41) in equation (1.21).

Similarly, the wave number of each line may be determined by substituting the value of frequency from equation (1.41) in equation (1.24).

$$\bar{\upsilon} = \frac{2\pi^2 Z^2 e^4 m}{h^3 c}\left(\frac{1}{n_i^2} - \frac{1}{n_o^2}\right) \quad (1.43)$$

The term $2\pi^2 Z^2 e^4 m/h^3 c$ in equation (1.43) is a constant. It is called *Rydberg constant* and is designated as R. Thus equation (1.43) may be re-written as,

$$\bar{\upsilon} = R\left(\frac{1}{n_i^2} - \frac{1}{n_o^2}\right) \quad (1.44)$$

For hydrogen atom (Z = 1), the value of R = 109700 cm^{-1}. As is clear from equation (1.44), the wave number of any spectral line is a combination of two terms. The first term, R is a constant, while the second, $\left(\frac{1}{n_i^2} - \frac{1}{n_o^2}\right)$ varies throughout each spectral series. This concept is called *Ritz Combination principle.*

Case Study 1.6 Calculate the wave number of the first three lines of the Balmer series in the spectrum of hydrogen atom.

The wave numbers of the concerned lines may be calculated from equation (1.44).

For Balmer series, $n_i = 2$

For the next three lines, n_o = 3, 4, 5

Rydberg constant, R = 109700 cm^{-1}

For the first line : $\bar{\upsilon} = 109700\left[\frac{1}{2^2} - \frac{1}{3^2}\right] = 15236.11\ cm^{-1}$

For the second line : $\bar{\upsilon} = 109700\left[\frac{1}{2^2} - \frac{1}{4^2}\right] = 20568.75\ cm^{-1}$

For the third line : $\bar{\upsilon} = 109700\left[\frac{1}{2^2} - \frac{1}{5^2}\right] = 23037\ cm^{-1}$

The wave numbers of the first three lines of Balmer series in the spectrum of hydrogen atom are 15236.11 cm^{-1}, 20568.75 cm^{-1} and 23037 cm^{-1}

Failure of Bohr model

Bohr theory was successful in explaining the main characteristics of hydrogen spectrum as well as the spectrum of species having only one electron, such as singly ionized helium and doubly ionized lithium. However, it failed to predict the energy states of many-electron atoms. The wavelengths theoretically predicted for various electronic transitions of complicated atoms did not agree with those obtained from actual spectra.

Bohr theory represents an interface between classical mechanics and quantum mechanics. At the time Bohr propounded his atomic model, light was known to possess dual nature. The fundamental equation of Planck's quantum theory, $E = h\upsilon$, reflects the dual characteristic of radiation, i.e., frequency is associated with wave motion, while energy is associated with particle concept involving mass and velocity. Bohr used this equation, but applied it to study the radiations emitted when an electron transitioned from higher to lower levels. However, as far as an electron is concerned, Bohr, like Rutherford, treated it as a particle. The idea of an electron jumping from one stationary state to another can be grasped only in terms of particle behavior. The fact that an electron also possesses dual characteristic came to be known more than a decade after Bohr forwarded his atomic theory.

Case Study 1.7 Calculate the frequency, wavelength and wave number of a line in the hydrogen spectrum that corresponds to an electronic transtition from n = 3 level to n = 2 level.

We know that the frequency of a line corresponding to a transition in hydrogen spectrum is given as,

$$\upsilon = 3.29 \times 10^{15} \left(\frac{1}{n_i^2} - \frac{1}{n_o^2}\right) s^{-1}$$

Here, $n_i = 2$, $n_o = 3$

$$\upsilon = 3.29 \times 10^{15} \left(\frac{1}{2^2} - \frac{1}{3^2}\right)$$

$$= 4.569 \times 10^{14}\ s^{-1}$$

The wavelength is related to the frequency by the equation,

$$\lambda = \frac{c}{\upsilon}$$

$$\lambda = \frac{3 \times 10^8}{4.569 \times 10^{14}}\ m \quad [\text{Since } c = 3 \times 10^8\ m\ s^{-1}]$$

$$= 6.562 \times 10^{-7}\ m$$

$$= 656.2\ nm$$

The wave number is related to the frequency by the equation,

$$\bar{\upsilon} = \frac{\upsilon}{c}$$

$$= \frac{4.569 \times 10^{14}}{3 \times 10^8}\ \frac{s^{-1}}{m\ s^{-1}}$$

$$= 1.523 \times 10^6 \text{ m}^{-1}$$
$$= 15{,}230 \text{ cm}^{-1}$$

Thus, for a line corresponding to electronic transition from n = 3 to n = 2 in a hydrogen spectrum,

$$\text{Frequency} = 4.569 \times 10^{14} \text{ s}^{-1}$$
$$\text{Wavelength} = 656.2 \text{ nm}$$
$$\text{Wave number} = 15{,}230 \text{ cm}^{-1}$$

Case Study 1.8 Using Bohr model of atom, calculate the ionization energy of a hydrogen atom.

The change in energy involved in the transition of an electron from an inner orbit (n_i) to an outer orbit (n_o) is given as,

$$E_{no} - E_{ni} = \frac{2\pi^2 Z^2 e^4 m}{h^2}\left(\frac{1}{n_i^2} - \frac{1}{n_o^2}\right)$$

An increase in principal quantum number from n = 1 to n = ∞ corresponds to the ionization of an atom (*see* Chapter 2, p. 62). The energy change is equivalent to the ionization energy (IE).

$$IE = \frac{2\pi^2 Z^2 e^4 m}{h^2}\left(\frac{1}{n_1^2} - \frac{1}{n_\infty^2}\right)$$

Mass of an electron, $m = 9.1 \times 10^{-28}$ g

Charge on an electron, $e = 4.8 \times 10^{-10}$ esu

Planck's constant, $h = 6.63 \times 10^{-27}$ erg s

For hydrogen atom, Z = 1

Substituting these values in the above equation,

$$IE = 2.18 \times 10^{-11} \text{ erg atom}^{-1} = 2.18 \times 10^{-18} \text{ J atom}^{-1}$$
$$= 2.18 \times 10^{-18} \text{ N J mol}^{-1}$$

where N is the Avogadro number, 6.023×10^{23}

$$IE = 1.312 \times 10^6 \text{ J mol}^{-1}$$
$$= 1312 \text{ kJ mol}^{-1}$$

The calculated value of ionization energy of a hydrogen atom is 1312 kJ mol^{-1}.

Particle Nature of Electron

The fact that electron is a particle is quite apparent. It has mass and moves with a specific velocity. Therefore, it has momentum associated with it. Above all, as already stated, a beam of electrons in a CRT rotates a paddle wheel. The atomic models based on classical mechanics treated electron as a particle. As these models failed to accomplish an accurate picture of atom, it was evident that the laws of classical physics do not apply to microparticles. A particle is always constrained within finite boundaries. An electron, however, cannot be confined to the boundaries of fixed energy levels. It dissipates well outside these levels. It has dual character. Like light, it can behave both as a particle and as a wave.

Wave Nature of Electron

The wave nature of electron was demonstrated by *Davisson and Germer experiment.* A beam of electrons, impinged on different planes of a nickel crystal, showed diffraction. This type of behavior is possible only for waves. The diffraction patterns that emerged from nickel grating thus conclusively proved that electrons behave as waves.

The wave nature of electron marked the advent of a new vista, the fascinating field of *quantum mechanics,* also called *wave mechanics.* It represented quite a break from the earlier concept of classical mechanics. It asserted that though large objects (such as golf balls) obey the laws of classical mechanics, such as Newton's laws of motion, very small objects (such as electrons) do not.

Quantum mechanics finds its expression in the following facts:

- Electrons show dual behavior (de Broglie principle).
- Being a wave, the path of an electron around the nucleus cannot be accurately determined (Heisenberg uncertainty principle).
- The motion of an electron is described in terms of a wave equation (Schrödinger wave equation).
- Various electronic arrangements in an atom can be described (quantum numbers).

We will discuss these concepts in the succeeding sections of this chapter.

de Broglie Principle

On the basis of the fact that light has dual nature, de Broglie suggested that a small, moving particle, such as an electron, would also behave both as a particle and as a wave. This hypothesis came to be known as the *wave theory of matter.*

The energy of a photon can be expressed in terms of Planck's quantum theory,

$$E = h\upsilon$$

Further, the energy of a particle with mass m is given in terms of Einstein's theory of relativity,

$$E = mc^2 \tag{1.45}$$

Combining the above two equations, we get:

$$h\upsilon = mc^2$$

where h is Planck's constant, υ is the frequency of a wave and c is the velocity of light.

$$h\frac{\upsilon}{c} = mc \tag{1.46}$$

From equation (1.21), $\frac{\upsilon}{c} = \frac{1}{\lambda}$

Therefore, $\frac{h}{\lambda} = mc$

or $$\lambda = \frac{h}{mc} \quad (1.47)$$

If a particle of mass m is an electron and if it is moving with velocity v, equation (1.47) may be re-written as,

$$\lambda = \frac{h}{mv} \quad (1.48)$$

Equation (1.48) is the mathematical form of de Broglie principle. The wavelength of the matter wave, λ is often called *de Broglie wavelength.* This type of wavelength is applicable only to microparticles, such as electrons, and not to macroparticles, such as golf balls.

Thus, according to de Broglie's concept, an electron is no longer a point mass moving in a stationary state. Rather, it is a standing wave train, surrounding the nucleus in a circular form (Fig. 1.15). If the circle occupied by the wave has a radius r, then the circumference (2πr) would contain an integral number (n) of wavelengths, i.e.,

$$2\pi r = n\lambda \quad (1.49)$$

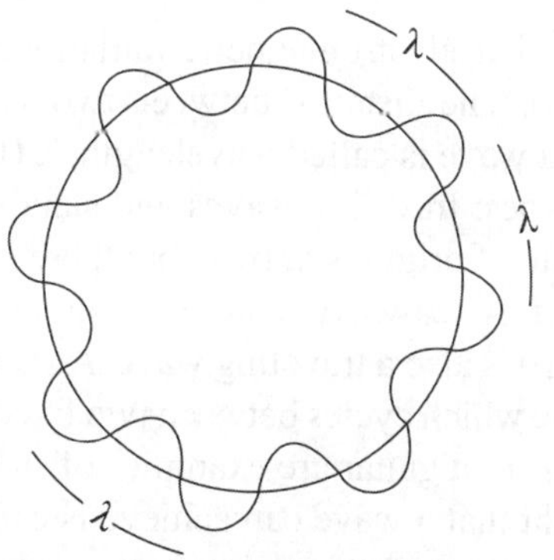

Fig. 1.15 Electron waves engulf the nucleus in a circular form

Substituting the value of λ from equation (1.48) into equation (1.49),

$$2\pi r = \frac{nh}{mv}$$

or $$mvr = \frac{nh}{2\pi}$$

where mvr represents the *angular momentum* of an electron. The wave mechanical picture of de Broglie thus reaffirms Bohr's postulate that angular momentum of an electron is an integral multiple of h/2π.

Case Study 1.9 Calculate the de Broglie wavelength of an electron traveling at 10% the speed of light.

Mass of an electron = 9.1×10^{-31} kg

Velocity of the electron is 10% the speed of light $= \left(\frac{10}{100}\right)(3 \times 10^8)$

$= 3 \times 10^7 \text{ m s}^{-1}$

Momentum of an electron, mv $= (9.1 \times 10^{-31})(3 \times 10^7)$

$= 2.73 \times 10^{-23} \text{ kg m s}^{-1}$

Planck's constant, h = 6.63×10^{-34} J s or kg m^2 s^{-1}

$$\lambda = \frac{h}{mv} = \frac{6.63 \times 10^{-34}}{2.73 \times 10^{-23}} \frac{\text{kg m}^2 \text{ s}^{-1}}{\text{kg m s}^{-1}}$$

$$= 2.424 \times 10^{-11} \text{ m}$$

$$= 0.02424 \text{ nm}$$

The de Broglie wavelength of the electron is 0.02424 nm.

Box 1.2

The Concept of Stationary States

We had earlier stated that electromagnetic radiations travel through space in wave motion (p. 10). The distance between two successive crests or two successive troughs of a wave is called wavelength, λ (Fig. 1.10).

Waves are of two types: traveling waves and standing waves. A *traveling wave* is one whose point of origin is known, but there is no clarity as to where the vibrations wane off. A sea wave is an example of a traveling wave. The wave shown in Fig. 1.10 is also a traveling wave. A *standing wave*, also called a *stationary wave* is one which cycles between two fixed points. The vibrations generated by the strings of a guitar are examples of stationary waves.

The maximum height that a wave can achieve is called amplitude, a. As is clear from Fig. 1.10, there are certain points in the pathway of the wave where the amplitude is zero. These points are called *nodes*. The two fixed points between which a standing wave travels are also nodes. So are the positions at which the string of a guitar is attached. The distance between two successive nodes is half the wavelength, that is, λ/2. Suppose a standing wave traverses a distance *l* between two fixed points, p and q (Fig. 1.16). As shown in Fig. 1.16a, it completes one half-wavelength between p and q. There are only two nodes in its pathway, and these are the fixed points, p and q.

Therefore, $l = \lambda/2$.

In Fig. 1.16b, the wave completes two half-wavelengths between p and q. There are now three nodes in its pathway, p, x and q. The distance between p and x is λ/2.; the distance between x and q is also λ/2.

Contd...

Therefore, $l = 2\left(\frac{\lambda}{2}\right)$.

In Fig. 1.16c, where the wave completes three half-wavelengths, there are four nodes, p, x, y and q.

Here $\overline{px} = \overline{xy} = \overline{yq} = \frac{\lambda}{2}$

Therefore, $l = 3\left(\frac{\lambda}{2}\right)$.

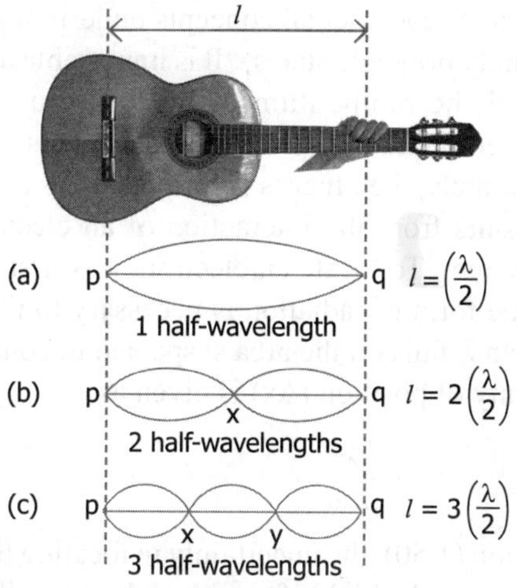

Fig. 1.16 Motion of a standing wave

Thus for a standing wave, the only allowed vibrations are those for which $l = n\,(\lambda/2)$, where n is a whole number. When n is a fraction, say, for example, 2.5, the wave dwindles away rapidly due to overlapping of crests and troughs.

Let us once again return to the example of guitar, in which the strings vibrate between fixed points and produce a standing wave. When plucked, a string creates only a specific tone and not a mixture of tones. Likewise, when an electron moves in an orbit of fixed radius, it can have only a speafic wavelength or a speafic energy, that is, the energy of electron is quantized. It was such an orbit that Bohr designated as a stationary state.

In an earlier section we had stated that the wave nature of electron was not established at the time Bohr proposed his atomic model (p. 24). Yet, while discussing the concept of stationary states, we are suggesting that an electron possesses wave motion. The reason is that Bohr had *arbitrarily assumed* the existence of such pathways in which the electron could move around the nucleus without losing or gaining energy. He could not provide an explanation as to why such pathways exist. Nor could he explain why the energy of an electron in a stationary state is quantized. It was only after de Broglie proved that electrons possess wave-like motion that the arbitrary assumption of Bohr received credence. It has therefore become customary to associate stationary waves with stationary states.

Heisenberg Uncertainty Principle

If electron is visualized as a particle, as was considered earlier by classical mechanics, it is possible to constraint it within finite boundaries. However, according to quantum mechanics, electron behaves as a wave, and a wave tends to dissipate throughout space. Moreover, the properties of a particle, such as its position and momentum, are easy to determine. For a wave, these properties defy determination. Since electron has dual nature, a deterministic model of an atom is impossible to construct. A justification for this assertion is provided by Heisenberg uncertainty principle, one of the most fundamental concepts underlining the quantum theory.

Heisenberg uncertainty principle states, "It is impossible to determine accurately both the position and the momentum of an electron simultaneously." The uncertainties in the determination of these two parameters vary inversely. If one is determined fairly accurately, the other is determined less accurately.

The uncertainty results from the interaction of an electron under observation with the measuring system. To locate an electron, it is first necessary to visualize it. To visualize it, some form of radiation is necessary to illuminate it. Suppose a radiation of wavelength λ falls on the area suspected to contain the electron, at an angle α. The uncertainty of position (Δx) is given as,

$$\Delta x \approx \frac{\lambda}{\sin\alpha} \tag{1.50}$$

According to equation (1.50), the uncertainty in locating the position will be low (or certainty of location will be high) if wavelength of the illuminating radiation is short. This may be qualitatively explained as follows: an electron can be located by an electromagnetic wave only if it occupies the space spanned by two successive crests or troughs. If the wavelength of the radiation is long, it would be difficult to locate the electron. Thus, uncertainty of the position will increase. If, on the other hand, the wavelength of illuminating radiation is short, a very small space will be spanned by two successive crests or troughs and it would be relatively easy to locate the electron (Fig. 1.17). The uncertainty of position in this case will decrease or certainty of position will increase.

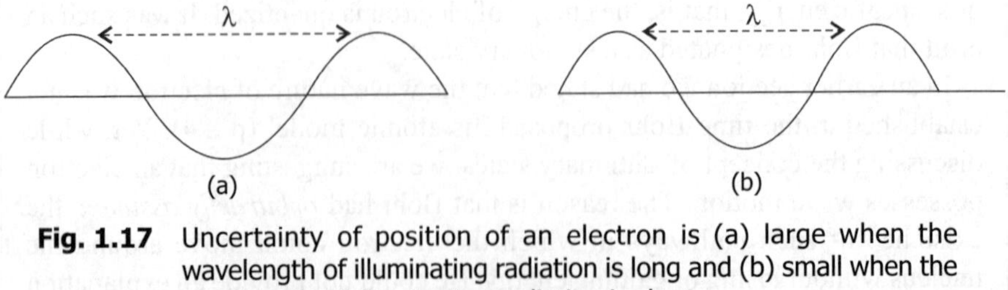

Fig. 1.17 Uncertainty of position of an electron is (a) large when the wavelength of illuminating radiation is long and (b) small when the wavelength of illuminating radiation is short

Therefore, if an electron has to be precisely located, it is best to use gamma- rays as illuminating light. Gamma-rays have an extremely short wavelength. However, this raises another problem. Shorter the wavelength of a radiation, higher is its frequency or energy. This means that gamma-rays possess tremendously high energy.

According to Compton effect, when a photon strikes an electron, the collision is elastic. The electron gains momentum, the photon loses it. The photons are scattered thereafter. The increase in momentum of an electron can be assessed accurately if the degree of scattering is small. High-energy radiations, such as gamma-rays, are scattered more than low-energy ones. Therefore, when the electron is illuminated by short-wavelength, high-energy gamma-rays, the uncertainty of its momentum increases.

Mathematically, the uncertainty of momentum (Δp) is given as,

$$\Delta p \approx \frac{h}{\lambda}\sin\alpha \tag{1.51}$$

Combining equations (1.50) and (1.51),

$$\Delta x.\ \Delta p \approx \left(\frac{\lambda}{\sin\alpha}\right)\left(\frac{h}{\lambda}\sin\alpha\right) \approx h \tag{1.52}$$

Equation (1.52) is the mathematical form of Heisenberg uncertainty principle. It can be clearly seen from this expression that to decrease the uncertainty of position (or to increase the certainty of position), the value of wavelength (λ) should be as small as possible. However, shorter the wavelength, higher is the uncertainty of momentum.

Case Study 1.10 An electron is revolving around the nucleus of an atom with a velocity of $7.29 \times 10^6 \pm 1\%$ m s^{-1}. Calculate the uncertainty of its position.

The uncertainty of velocity is 1%, that is, 0.01 of 7.29×10^6 m s^{-1}.

$$\Delta v = (7.29 \times 10^6 \text{ m s}^{-1})(0.01) = 7.29 \times 10^4 \text{ m s}^{-1}$$

Mass of electron $= 9.1 \times 10^{-31}$ kg

Uncertainty of momentum, $\Delta p = m\Delta v = (9.1 \times 10^{-31}\text{ kg})(7.29 \times 10^4 \text{ m s}^{-1})$

$$h = 6.63 \times 10^{-34} \text{ kg m}^2 \text{ s}^{-1}$$

Uncertainty in position, Δx may be calculated by substituting the value, of Δp and h in equation (1.52)

$$\Delta x \approx \frac{h}{\Delta p}$$

$$\approx \frac{(6.63 \times 10^{-34} \text{ kg m}^2 \text{ s}^{-1})}{(9.1 \times 10^{-31}\text{ kg})(7.29 \times 10^4 \text{ m s}^{-1})}$$

$$\approx 1 \times 10^{-8} \text{ m}$$

The diameter of an atom is of the order of 10^{-10} m. The uncertainty of electron position is 100 times greater than the diameter of the atom.

Schrödinger Wave Equation

Since an electron behaves like a wave, its motion ought to be described by a wave equation. The motion of optic and sound waves is described by a second-order differential equation of the type,

$$\frac{\delta^2\phi}{\delta x^2}+\frac{\delta^2\phi}{\delta y^2}+\frac{\delta^2\phi}{\delta z^2}=\frac{1}{c^2}\cdot\frac{\delta^2\phi}{\delta t^2} \tag{1.53}$$

where c is the velocity of light, t is the time and ϕ is the amplitude of the wave along x, y and z coordinates.

It is pertinent to assume that equation (1.53) is applicable to electron waves as well. The amplitude of a wave is a function of both time (t) and space (x, y, z coordinates).

$$\phi = f\,(x, y, z, t) \tag{1.54}$$

The amplitude function ϕ may also be represented as product of two functions, one dependent on space, $\psi_{(x,\,y,\,z)}$ and the other dependent on time, $\psi_{(t)}$, i.e.,

$$\phi = \psi_{(x,\,y,\,z)}\cdot\psi_{(t)} \tag{1.55}$$

The time-dependent function can have several values. However, the most common one is an exponential value,

$$\psi_{(t)} = e^{2\pi\upsilon it} \tag{1.56}$$

where υ is the frequency of the wave and i is the imaginary number, $\sqrt{-1}$. Substituting the value of $\psi_{(t)}$ from equation (1.56) into (1.55),

$$\phi = \psi_{(x,\,y,\,z)}\cdot e^{2\pi\upsilon it} \tag{1.57}$$

For convenience, the space-dependent wave function, $\psi_{(x,\,y,\,z)}$ may be written without mentioning the x, y and z coordinates. Equation (1.57) now becomes,

$$\phi = \psi\,.\,e^{2\pi\upsilon it} \tag{1.58}$$

Second-order differentiation of equation (1.58) with respect to x, y and z coordinates gives the following equations:

$$\frac{\delta^2\phi}{\delta x^2}=\frac{\delta^2\psi}{\delta x^2}\,.\,e^{2\pi\upsilon it} \tag{1.59}$$

$$\frac{\delta^2\phi}{\delta y^2}=\frac{\delta^2\psi}{\delta y^2}\,.\,e^{2\pi\upsilon it} \tag{1.60}$$

$$\frac{\delta^2\phi}{\delta z^2}=\frac{\delta^2\psi}{\delta z^2}\,.\,e^{2\pi\upsilon it} \tag{1.61}$$

First-order differentiation of equation (1.58) with respect to time gives:

$$\frac{\delta\phi}{\delta t}=(2\pi\upsilon i)\cdot e^{2\pi\upsilon it}.\psi$$

while the second-order differential gives:

$$\frac{\delta^2\phi}{\delta t^2}=(2\pi\upsilon i)^2\cdot e^{2\pi\upsilon it}.\psi$$

$$\frac{\delta^2\phi}{\delta t^2}=-\,4\,\pi^2\upsilon^2\cdot e^{2\pi\upsilon it}.\psi \tag{1.62}$$

$$(i^2=-1)$$

Substituting the identities of equations (1.59)–(1.62) into equation (1.53).

$$\left(\frac{\delta^2\psi}{\delta x^2}+\frac{\delta^2\psi}{\delta y^2}+\frac{\delta^2\psi}{\delta z^2}\right).\, e^{2\pi\, \upsilon i t} = \frac{-4\,\pi^2\upsilon^2}{c^2}\, e^{2\pi\, \upsilon i t}.\, \psi$$

$$\frac{\delta^2\psi}{\delta x^2}+\frac{\delta^2\psi}{\delta y^2}+\frac{\delta^2\psi}{\delta z^2} = \frac{-4\,\pi^2}{\lambda^2}.\psi \tag{1.63}$$

$$(\lambda = \frac{c}{\upsilon})$$

Since $\lambda = \frac{h}{mv}$ [de Broglie relation, equation (1.48)],

$$\frac{\delta^2\psi}{\delta x^2}+\frac{\delta^2\psi}{\delta y^2}+\frac{\delta^2\psi}{\delta z^2} = -4\pi^2\,\frac{m^2\upsilon^2}{h^2}\,.\,\psi \tag{1.64}$$

Equation (1.64) may be re-written as,

$$\frac{\delta^2\psi}{\delta x^2}+\frac{\delta^2\psi}{\delta y^2}+\frac{\delta^2\psi}{\delta z^2} = \frac{-8\pi^2 m}{h^2}.\frac{1}{2}mv^2\,\psi \tag{1.65}$$

where $\frac{1}{2}mv^2$ denotes the kinetic energy of an electron. The total energy of an electron, E, is the sum of kinetic energy, T, and potential energy, V.

$$E = T + V$$

or

$$T = \frac{1}{2}mv^2 = E - V \tag{1.66}$$

From equations (1.65) and (1.66),

$$\frac{\delta^2\psi}{\delta x^2}+\frac{\delta^2\psi}{\delta y^2}+\frac{\delta^2\psi}{\delta z^2} = \frac{-8\,\pi^2 m}{h^2}\,(E - V).\psi$$

or

$$\frac{\delta^2\psi}{\delta x^2}+\frac{\delta^2\psi}{\delta y^2}+\frac{\delta^2\psi}{\delta z^2}+\frac{8\pi^2 m}{h^2}\,(E - V).\psi = 0 \tag{1.67}$$

To simplify, the differential terms in equation (1.67) are replaced by Laplacian operator, $\nabla^2\psi$.

$$\nabla^2\psi + \frac{8\,\pi^2 m}{h^2}\,(E - V).\psi = 0 \tag{1.68}$$

Equation (1.67) or (1.68) is called Schrödinger wave equation. The dependence of wave function on all three coordinates of space suggests that quantum mechanical model considers the charge of an electron as distributed about the nucleus in three dimensions. It is not confined to plane orbits as in Bohr model.

Quantum Numbers

According to the uncertainty principle, it is not possible to locate an electron precisely. Therefore, the only option is to say where the probability of finding the electron is high and where it is low. A region where there is a high probability of finding an electron is called an *orbital.* Each orbital can hold a maximum of two electrons.

An orbital is a component of the energy level of an atom. The characteristics of the energy levels (their size and shape) and the state of an electron occupying them (its orientation in space as well as with respect to other electrons) are described in terms of a set of numbers called quantum numbers. There are a total of four quantum numbers. These are discussed below.

1. Principal quantum number

Principal quantum number describes the size of an orbital and is designated as n. It also identifies the shell or level to which an electron belongs. It is a positive integer.

$$n = 1, 2, 3...$$

For example, in the orbitals 1s and 2s, the principal quantum numbers are 1 and 2, respectively. The size of a 2s orbital is larger than that of a 1s orbital.

2. Azimuthal quantum number

Azimuthal quantum number, also called *subsidiary quantum number,* is denoted by l. It gives the shape of an orbital that an electron occupies. Within each energy level, defined by the principal quantum number (n), l may take values from zero to (n – 1).

$$\begin{array}{llllll} l = 0 & 1 & 2 & 3 & ...(n-1) \\ \quad s & p & d & f & \end{array}$$

The notations s, p, d and f arise out of spectral terms—sharp, principal, diffuse and fundamental. The number of values that l can have reflects the number of subshells or sublevels. For example, when $l = 0$, there is only one value for azimuthal quantum number (i.e., zero). Therefore, there is only one sublevel. When $l = 1$, there are two values for azimuthal quantum number (i.e., zero and 1). There are now two sublevels—s and p. Similarly, for $l = 2$, there are a total of three values (i.e., zero, 1 and 2) and, therefore, three sublevels—s, p and d.

3. Magnetic quantum number

The magnetic quantum number, m_l represents the orientation of sublevels in space under the influence of an applied magnetic field. Its values range from $-l$, through zero, to $+l$. A maximum of $(2l + 1)$ values may be assigned to magnetic quantum number.

$$m_l = -l...0...+l; \text{ Maximum} = 2l + 1$$

For $l = 0$, $2l + 1 = 1$, i.e., only one value is possible for magnetic quantum number: $m_l = 0$. This means that the sublevel can have only one orientation in the magnetic field. In fact, the s sublevel is aligned along the magnetic field. For $l = 1$, $2l + 1 = 3$, i.e., three values are possible for magnetic quantum number: $m_l = -1, 0, +1$. Each value corresponds to an orbital of p sublevel (Fig. 1.18). These are designated as p_x, p_y and p_z. These have three orientations mutually perpendicular to each other.

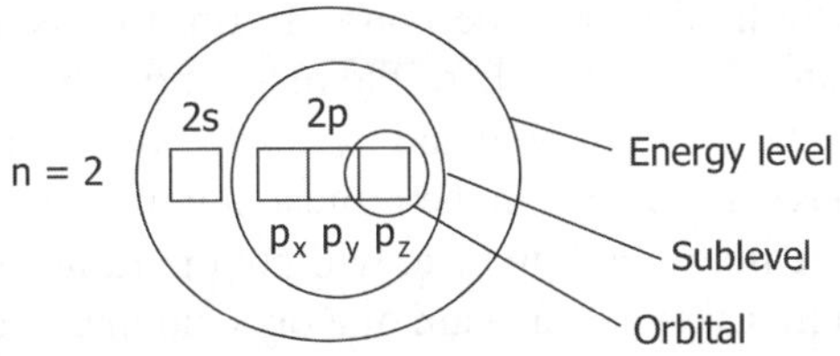

Fig. 1.18 Orbitals and sublevels are components of an energy level

Similarly, for $l = 2$, the d sublevel can have five orientations with respect to magnetic field. Accordingly, there are five orbitals in d sublevel. There are seven orbitals in f sublevel.

4. Spin quantum number

The spin quantum number, m_s, describes the direction of spin of an electron in an orbital (Fig. 1.19). The spin may be clockwise, for which the assigned value of m_s, = +½, or anti-clockwise for which m_s=−½. Since electrons are negatively charged, their spin motion produces magnetic fields. The field generated by one electron interacts with that of the other. When two electrons are in the same orbital, their spins are always opposite, i.e., one electron has a spin quantum number of +½ and the other −½.

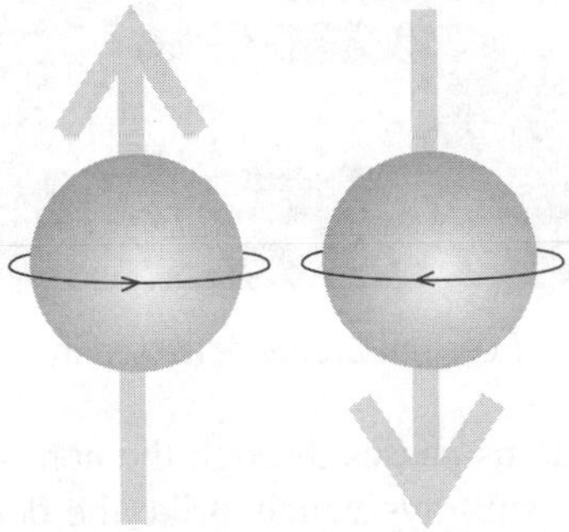

Fig. 1.19 Electron spins

Experimental evidence for electron spin

That an electron has two types of spins–clockwise and anti-clockwise–may be proved by *Stern-Gerlach experiment*, wherein the deflection of a beam of atoms of a particular element in a magnetic field is studied. The deflection, in turn, is observed only if the following two conditions are met.

1. The atom of the concerned element should have an unpaired electron in its valence shell. A spinning electron behaves as a bar magnet. (if two electrons are paired-up in the valence shell, their spin moments cancel out, and so do their magnetic characteristics). Thus placing such an atom in a magnetic field is analogous to placing a bar magnet in a magnetic field.

2. The magnetic field should be non-uniform, that is, the strength of the north pole and the south pole of the magnet that creates the field should be different. If the north pole and the south pole have the same strength, and a bar magnet is placed in their midst, it will experience an equal force from both the poles, and, in turn, will not orient itself in the direction of either of them. Likewise, a beam of atoms moving in a uniform magnetic field will not be deflected, but travel in a straight line. However, when the magnetic field is non-uniform, the beam of atoms having an unpaired electron in their valence shell will be deflected either towards north pole or towards south pole.

In Stern-Gerlach experiment, a beam of either alkali metal atoms or of silver atoms is passed through a magnetic field. These metal atoms are preferred because their atoms have only one unpaired valence electron and therefore the results are easy to interpret. As shown in Fig. 1.20, cesium is heated to its boiling point in a furnace and the beam of emerging atoms is passed through a series of collimators. The latter are the slits which narrow down the width of the beam.

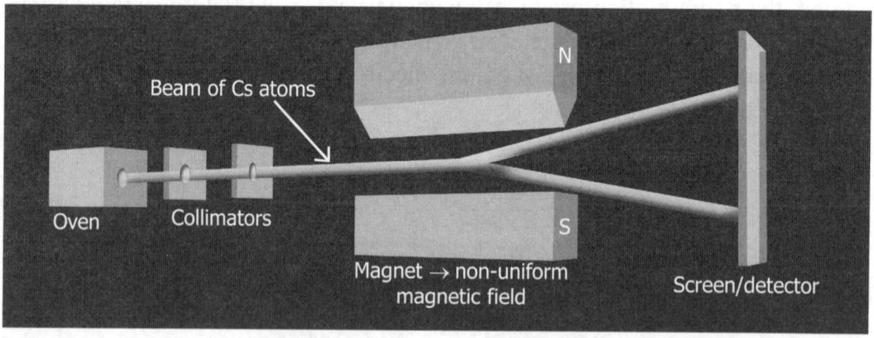

Fig. 1.20 A beam of cesium atoms deflected in a magnetic field

As the ray of cesium atoms passes through the non-uniform magnetic field, it splits up into two distinct sub-rays which strike the detector screen at different points. One sub-ray is oriented towards the north pole, the other towards the south pole. This shows that the magnetic moment due to the single valence electron can take up two orientations—clockwise ($s = +1/2$) and anti-clockwise ($s = -1/2$).

Case Study 1.11 How many atomic orbitals are possible for the energy level $n = 4$?

For $n = 4$, $l = 0, 1, 2, 3$

For $l = 0$, $m_l = 0$; there is one 4s orbital.

For $l = 1$, $m_l = +1, 0, -1$; there are three 4p orbitals.

For $l = 2$, $m_l = +2, +1, 0, -1, -2$; there are five 4d orbitals.

For $l = 3$, $m_l = +3, +2, +1, 0, -1, -2, -3$; there are seven 4f orbitals.

Total number of atomic orbitals $= 1 + 3 + 5 + 7 = 16$

Transformation of Coordinates

In Schrödinger equation, the wave function, $\psi_{(x, y, z)}$ is written in terms of cartesian coordinates. These coordinates represent the x-, y- and z-axis in a three-dimensional system and indicate the orientation of nucleus of an atom in space. However, now that we are aware of the particle–wave nature of electrons, we are more interested in knowing their distribution around the nucleus, rather than knowing the orientation of nucleus in space. Since the electrons are distributed spherically around the nucleus, it is pertinent to transform the cartesian coordinates into spherical polar coordinates. These are represented as r, θ and ϕ in Fig. 1.21. Here, r is the distance between the nucleus and electron, indicated by line segment OP. Further, θ is the angle which the line OP makes with respect to z-axis, while ϕ is the angle which the projection of line OP in xy plane (i.e., OM) makes with x-axis. The (x, y, z) coordinates are changed into (r, θ, ϕ) coordinates as follows:

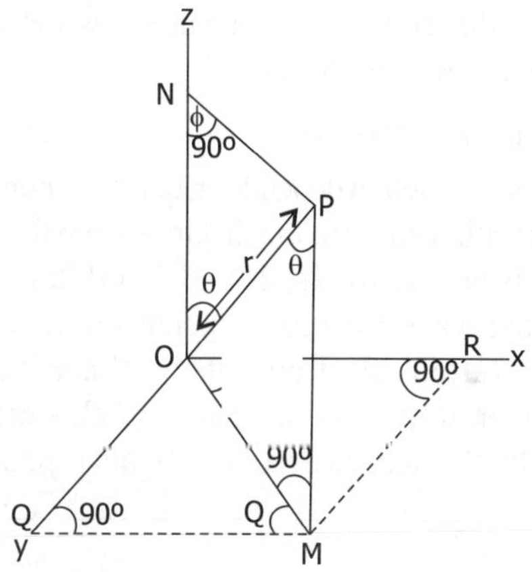

Fig. 1.21 Transformation diagram for spherical coordinates

In ΔMOP,

$$\sin\theta = \frac{OM}{OP}$$

$$OM = OP \sin\theta = r \sin\theta \quad (1.69)$$

In ΔMOR,

$$\cos\phi = \frac{OR}{OM}$$

$$OR = OM \cos\phi = r \sin\theta \cos\phi \quad (1.70)$$

In ΔMOQ,

$$\sin\phi = \frac{OQ}{OP}$$

$$OQ = OM \sin\phi = r \sin\theta \sin\phi \quad (1.71)$$

In ΔNOP,

$$\cos\theta = \frac{ON}{OP}$$

$$ON = OP \cos\theta = r \cos\theta \quad (1.72)$$

In equations (1.70), (1.71) and (1.72), OR, OQ and ON represent the x-, y- and z-axis, respectively. Therefore, these equations may be re-written as,

$$x = r \sin\theta \cos\phi \tag{1.73}$$

$$y = r \sin\theta \sin\phi \tag{1.74}$$

$$z = r \cos\theta \tag{1.75}$$

By using these transformation equations, a wave function $\psi_{(r,\ \theta,\ \Phi)}$, which represents the orientation of electrons around the nucleus, can be obtained.

Significance of Wave Function

The wave function involving polar coordinates, $\psi_{(r,\ \theta,\ \Phi)}$ may be split into a product of two wave functions, called *radial function* and *angular function.*

$$\psi_{(r,\ \theta,\ \Phi)} = \psi_{Radial}\ \psi_{Angular}$$

The radial function is determined by distance r, while the angular function is determined by angles θ and ϕ. Accordingly,

$$\psi_{(r,\ \theta,\ \phi)} = \psi_{(r)}\ \psi_{(\theta,\phi)}$$

The radial part of the wave function depends on quantum numbers n and *l*. It gives the distribution of electrons with respect to the distance from the nucleus. The variations in radial function with distance r for 1s, 2s and 2p orbitals are shown in Fig. 1.22.

The angular part of the wave function depends on the quantum numbers *l* and m_l. It gives the shape of various electronic states. Since it does not depend on the principal quantum number, it means that shapes of all s orbitals (1s, 2s,...) would be the same. So would be the shapes of all p orbitals (2p, 3p,...).

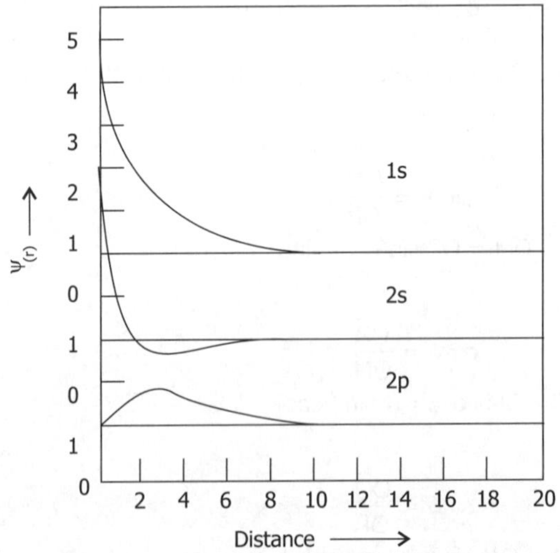

Fig. 1.22 Variations in radial function with distance for 1s, 2s and 2p orbitals

The angular dependence of s, p and d orbitals is shown in Fig. 1.23. These are the plots of wave function $\psi_{(\theta,\ \phi)}$ with respect to angles θ and ϕ. The s orbital is independent of angles θ and ϕ. Irrespective of the value of angles, the angular part of wave function is constant and, therefore, spherically symmetrical. The p orbitals, on the other hand, show definite geometry. The angular plots of three p orbitals are in the form of identical dumbbell patterns, mutually perpendicular to each other. Similarly, the d orbitals show double dumbbell patterns. The area enclosed by these geometric patterns is called *boundary diagram.* The area within the boundary possesses almost 90% of the electron density.

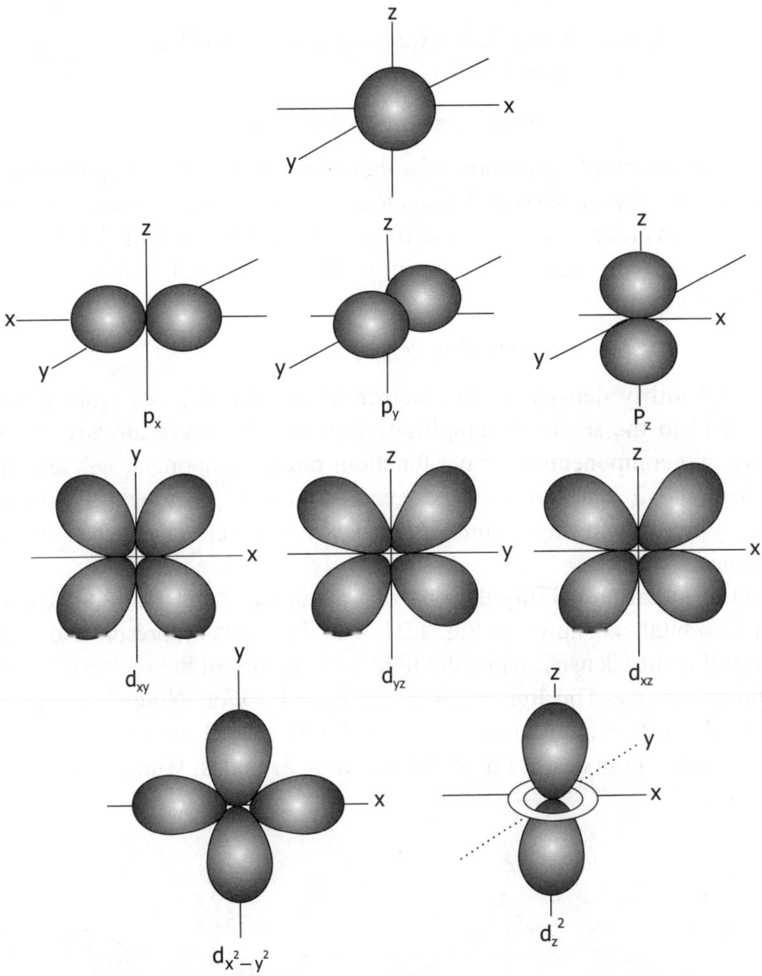

Fig. 1.23 Angular dependence of s, p and d orbitals

Significance of Probability Concept

We have already drawn an analogy between light wave and electron wave. The intensity of a light wave (I) is directly proportional to the number of photons of light in a given volume element. The number of photons in a given volume element is defined as *photon probability density*. This means,

$$I \propto \text{Photon probability density} \tag{1.76}$$

Further, the intensity of light is mathematically given by the expression,

$$I = \frac{c}{4\pi}a^2 \tag{1.77}$$

where c is the velocity of light and a is the amplitude of the wave. Since velocity of light is a constant parameter,

$$I \propto a^2$$

This relationship has already been expressed as equation (1.19).

From equations (1.19) and (1.76),

$$\text{Photon probability density} \propto a^2$$

Thus, the square of amplitude of a light wave is directly proportional to the probability density of photon. By analogy it may be stated that the square of amplitude of an electron wave is directly proportional to the probability density of an electron. In Schrödinger wave equation, the amplitude function is designated as ψ. Therefore,

$$\text{Electron probability density} \propto \psi^2 \tag{1.78}$$

The probability density of an electron at any point in an atom is directly proportional to the square of amplitude function. We have already stated that there are two components of wave function: radial component, $\psi_{(r)}$ and angular component, $\psi_{(\theta, \phi)}$. The square of these components, $\psi^2_{(r)}$ and $\psi^2_{(\theta, \phi)}$ gives *radial probability density* and *angular probability density*, respectively. Each has a different significance.

A plot of radial probability density, $\psi^2_{(r)}$ against distance from nucleus for 1s, 2s and 3s orbitals is shown in Fig. 1.24. The dark areas represent the region of greater probability density, while the light areas represent the region of negligible probability density. The light regions are called *nodes*. Note that a 1s orbital has no node, a 2s orbital has one node and a 3s orbital, two nodes. Thus, a ns orbital has (n – 1) nodes, a np orbital has (n – 2) nodes, while a nd orbital has (n – 3) nodes.

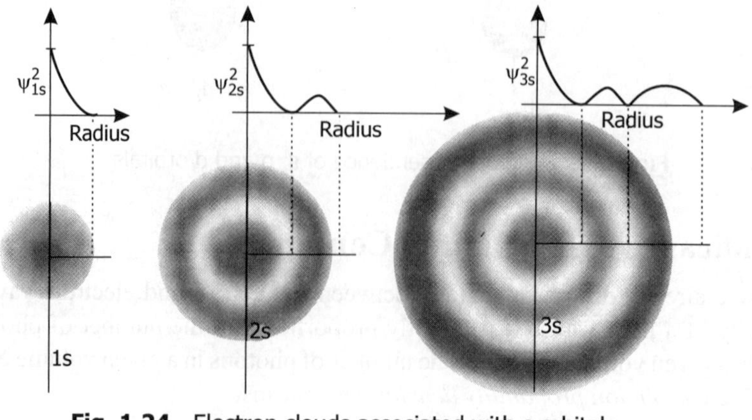

Fig. 1.24 Electron clouds associated with s orbitals

Electron distribution may be expressed in a better way in terms of radial probability function, rather than radial probability density. The term *radial probability function* signifies the probability of finding an electron in a spherical shell between the distance r and r + dr from the nucleus (Fig. 1.25).

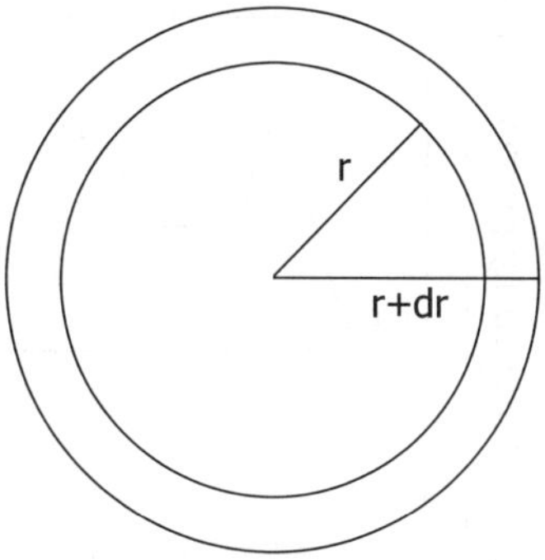

Fig. 1.25 Radial probability function is the probability of finding an electron in a spherical shell between distance r and r + dr

Mathematically,

$$\text{Radial probability function} = \text{Radial probability density} \times \text{Volume of radial shell} \quad (1.79)$$

$$\text{Volume of radial shell} = \begin{bmatrix}\text{Volume of shell} \\ \text{of radius (r + dr)}\end{bmatrix} - \begin{bmatrix}\text{Volume of shell} \\ \text{of radius r}\end{bmatrix}$$

$$= \frac{4}{3}\pi (r + dr)^3 - \frac{4}{3}\pi r^3$$

$$= \frac{4}{3}\pi (r^3 + 3r^2dr + 3rdr^2 + dr^3 - r^3)$$

Neglecting the terms $3rdr^2$ and dr^3 (which are very small) and simplifying,

$$\text{Volume of radial shell} = 4\pi r^2 dr \quad (1.80)$$

$$\text{Radial probability density} = \psi^2_{(r)} \quad (1.81)$$

From equations (1.79), (1.80) and (1.81),

$$\text{Radial probability function} = 4\pi r^2 d^2 \psi^2 dr$$

A plot of radial probability function with respect to distance for 1s, 2s and 2p orbitals is shown in Fig. 1.26. The curves show that the radial probability function is zero at the nucleus. Its value increases with distance, reaches a maximum, and then drops. This maximum corresponds to the most probable value. The distance corresponding to the most probable value is called *distance of maximum probability.*

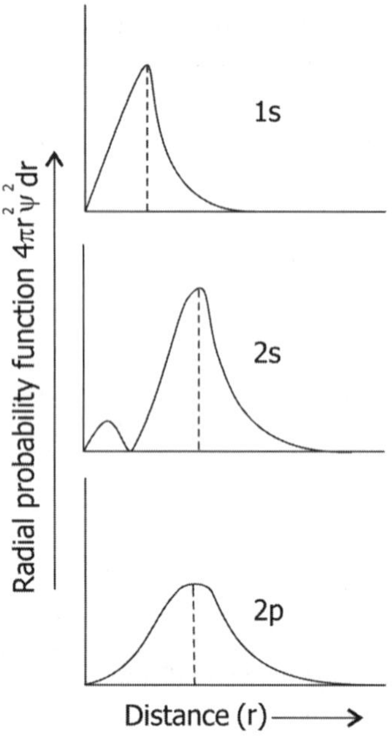

Fig. 1.26 A plot of radial probability function with respect to distance for 1s, 2s and 2p orbitals

The radial probability function curve for 2s orbital shows two maxima, the smaller one near the nucleus, the larger one at a greater distance from the nucleus. In between the two maxima, there is a region of zero probability and it represents a node or a nodal point. The distance of maximum probability for 2s orbital is larger than that of 1s orbital. This indicates that the 2s orbital is larger and farther away from the nucleus as compared to the 1s orbital. As a result, it has higher energy than 1s orbital. Further, the distance of maximum probability for a 2p electron is slightly less than that of a 2s electron. However, the smaller maximum in 2s curve is near the nucleus. This shows that the electron in 2s orbital spends some of its time in the vicinity of the nucleus. It is an indication that a 2s electron penetrates closer to the nucleus and is, therefore, held more tightly than a 2p electron. It also implies that a 2s electron is more stable and has lower energy than a 2p electron.

The radial probability density or radial probability function indicates how the electron distribution varies with distance from the nucleus, whereas the angular probability density, $\psi^2_{(\theta,\ \phi)}$, indicates how electron distribution varies within the area covered by boundary diagram of an orbital. Qualitatively, the angular probability distribution is proportional to the length of a line drawn from the center of the orbital and extended towards the circumference of the boundary figure. For s orbital, the length of the line is the same irrespective of the direction. This means that the angular probability density is same at every point on the circumference. Therefore, s orbitals are referred to as spherically symmetrical and not simply spherical. For p orbitals, the situation is different (Fig. 1.27).

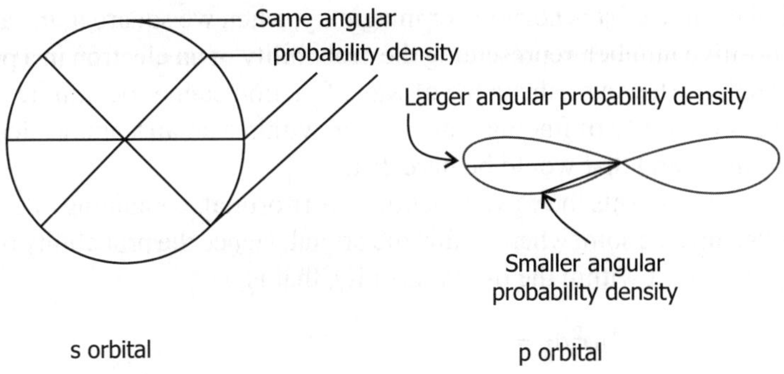

Fig. 1.27 Variation in angular probability density within an orbital

The angular probability is maximum at the tip of the lobes and decreases as one moves towards the center. An end-to-end overlap involves that region of the orbitals where the angular probability distribution is maximum. A lateral overlap involves that region where the angular probability distribution is not maximum (Fig. 1.28). For this reason, a σ-bond is stronger than a π-bond.

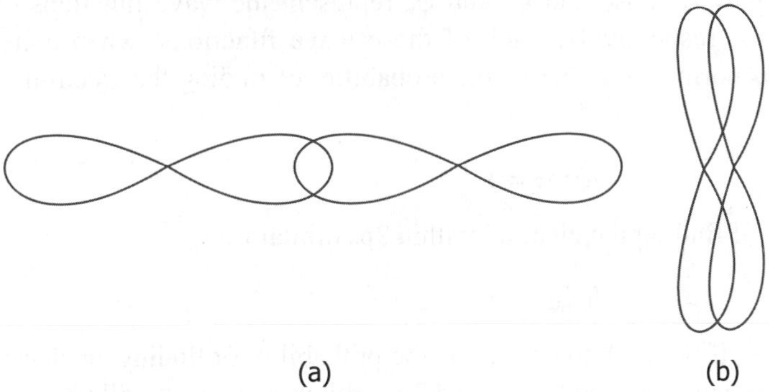

Fig. 1.28 Types of overlapping (a) end-to-end and (b) lateral

Normalized and Orthogonal Wave Functions

If we wish to solve the Schrödinger wave equation (1.68) we must know the value of wave function, ψ. The value of ψ, in turn, will depend on the values of x, y and z coordinates. This means that ψ may have more than one value. However, once we fix the coordinates x, y and z, then at that particular point, ψ will have only one value. This value may be positive (+p), negative (–p) or imaginary ($\sqrt{-p}$). What is important is not the value of wave function, but the probability of finding an electron in an orbital represented by that wave function. As equation (1.78) shows, the electron probability density is the square of the wave function, that is, ψ^2. More precisely,

$$\text{Probability} = \psi_{(x,\,y,\,z)}\,\psi_{(x,\,y,\,z)}\,dxdydz = \int \psi^2 d\tau$$

where dτ represents the volume element spanned by coordinates x, y and z.

Even if the value of ψ is negative or imaginary, when we square it, the answer is always a positive number, representing the probability of an electron in a particular volume element. However, the value of wave function cannot be infinite. If that is so, then the probability of finding that electron within a small volume element too will be infinity – and that would be unrealistic.

Suppose ψ_a represents the wave function of a 1s orbital, containing one electron. This electron must be somewhere within the orbital. Hence the probability of finding it at any given point within the orbital is unity, that is,

$$\int \psi_a^2 d\tau = 1 \tag{1.82}$$

Now suppose the 1s orbital with wave function ψ_a overlaps with another 1s orbital with wave function ψ_b. The probability of finding the electron is the region of overlap is given by,

$$\int \psi_a \psi_b d\tau = 1 \tag{1.83}$$

The wave functions which satisfy equations (1.82) and (1.83) are called normalized wave functions.

Let us now suppose that ψ_a and ψ_b represent the wave functions of $2p_x$ and $2p_y$ orbitals, respectively. Each of these wave functions, when considered in isolation is normalized, that is, the probability of finding the electron within $2p_x$ orbital is,

$$\int \psi_a^2 d\tau = 1$$

white that of finding the electron within $2p_y$ orbital is,

$$\int \psi_b^2 d\tau = 1$$

However, if we wish to determine the probability of finding an electron within a common space spanned by $2p_x$ and $2p_y$ orbitals, the result will be,

$$\int \psi_a \psi_b d\tau = 0 \tag{1.84}$$

The reason is that there is no common space between $2p_x$ and $2p_y$ orbitals. As shown in Fig. 1.29, the two orbitals do not overlap. They are said to be orthogonal to each other.

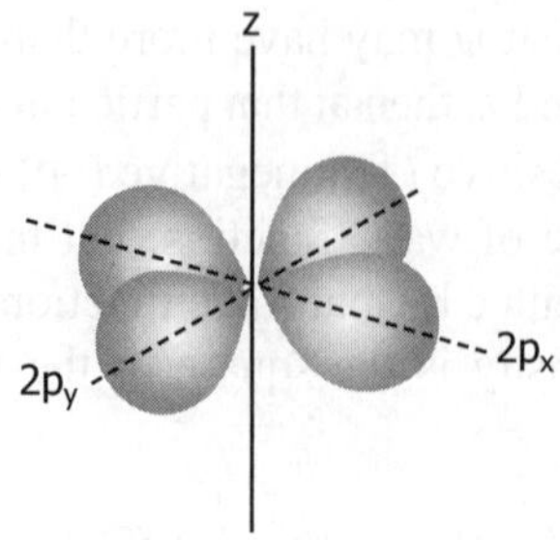

Fig. 1.29 The $2p_x$ and $2p_y$ orbitals are orthogonal to each other, with no common area of overlap

Likewise, if we integrate the wave function of a 2s orbital (ψ_{2s}) with the wave function of 4s orbital (ψ_{4s}), in the hope of finding an electron in the integrated volume element, we will get,

$$\int \psi_{2s}\psi_{4s}d\tau = 0 \tag{1.85}$$

The difference in energy of 2s and 4s orbitals is so large that the two will not overlap and there would be no common area in which the electron of any one of these orbitals could reside.

The wave functions which satisfy equations (1.84) and (1.85) are called orthogonal wave functions.

Case Study 1.12 What does the *Unsöld's theorem* state?

Unsöld's theorem states that just as the single s orbital is spherically symmetrical, the total electron density of a set of three p orbitals, five d orbitals or seven f orbitals is also spherical. Thus, while it might appear that argon, with a filled set of s and p orbitals would have a non-symmetrical and lumpy electron cloud, the total probability distribution is perfectly spherical.

Pauli Exclusion Principle

Pauli exclusion principle states that no two electrons in the same atom can have the same value for all four quantum numbers. Even if two electrons occupy an orbital of the same size, shape and orientation in space, these will differ in their spin quantum number values. For example, in the case of two electrons in the 2s orbital, designated as A and B, the values of quantum numbers are:

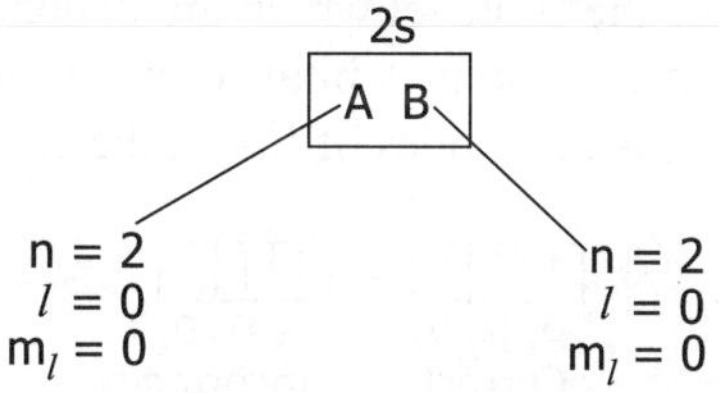

Both the electrons have the same value for principal, azimuthal and magnetic quantum numbers. If the two are to be accommodated in the same orbital, these must have different values for spin quantum number.

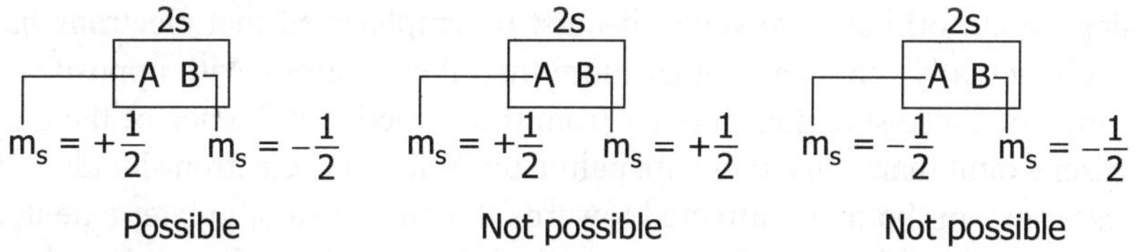

Accordingly, the two electrons would pair-up in the 2s orbital, one spinning clockwise, and the other anti-clockwise.

If we compare an electron in 1s orbital with that in 2s orbital, even then Pauli exclusion principle is obeyed, as the two have different values of principal quantum number.

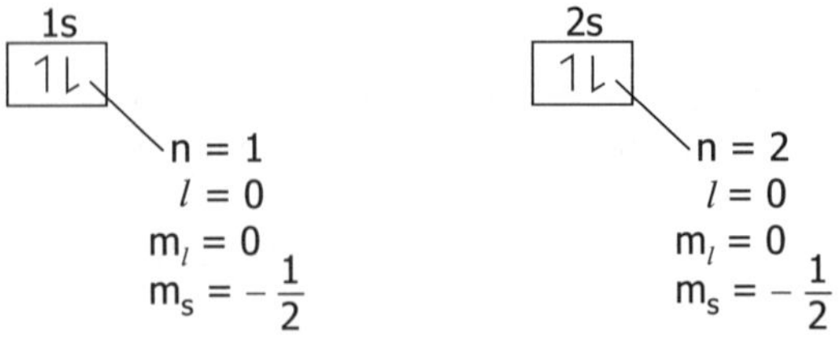

Similarly, an electron in $2p_x$ orbital differs in magnetic quantum number from that in $2p_y$ orbital.

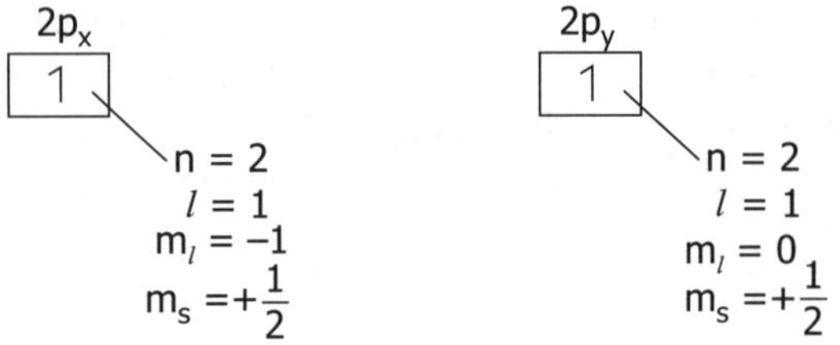

Hund's Rule of Maximum Multiplicity

According to Hund's rule, electrons must occupy all the orbitals of a given sublevel singly before pairing begins. The unpaired electrons have parallel spins. This implies that as far as possible electrons tend to remain unpaired in a sublevel. Hund's rule applies to those orbitals of the sublevel which are degenerate, i.e., have the same energy. For example, the p_x, p_y and p_z orbitals of p sublevel are degenerate. For a p^3 system, there are two possible modes of filling the electrons.

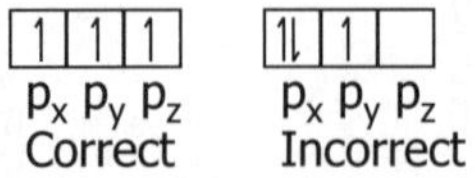

The correct mode is the one in which the three electrons are spread out, one in each orbital. The conventional explanation of Hund's rule is that there is more repulsion between two electrons in the same orbital than when these are in two degenerate orbitals. However, it must be emphasized that electrons having same spin actually repel each other more than the electrons with opposite spins. Nevertheless, these shield each other from the attractive influence of the nucleus to a lesser extent than when these are paired up. When one electron shields another to a lesser extent, both are attracted towards the nucleus to a greater extent. As a result, the energy of the atom decreases and stability increases. Thus, when electrons with parallel spins occupy degenerate orbitals, the nuclear attraction experienced by these increases, the overall energy decreases, and, in turn, the stability of the atom increases.

Case Study 1.13 What is *exchange energy?*

Exchange energy is the energy expended when an electron with a particular spin quantum number, say s = +1/2, jumps from one degenerate orbital to another and yet retains the same spin quantum number.

If the number of electrons with parallel spins is small, there is a greater possibility of such an exchange. Under these circumstances, exchange energy will be low. Yet, since exchange energy has to expended, however low its value might be, it contributes to the instability of the configuration and hence of the atom.

If, on the other hand, the number of electrons with parallel spins in a degenerate set of orbitals is large, the possibility of exchange is small. Under these circumstances, the exchange energy will have a higher value. If the value is so high that it prevents the exchange, the atom will gain stability. Mathematically, exchange energy E_{ex} is given by the expression,

$$E_{ex} = \frac{n(n-1)}{2}k$$

where n is the number of electrons having parallel spins and k is a constant.

For n = 1, E_{ex} = Zero; for n = 2, E_{ex} = k; for n = 3, E_{ex} = 3k, and so on.

The number of parallel spins is maximum for half-filled and completely filled subshells. The exchange energy is therefore so high for such configurations that the exchange actually does not take place. This results in the special stability associated with p^3 and p^6; d^5 and d^{10}; and f^7 and f^{14} configurations. There is no possibility of exchange in an unfilled subshell too. Therefore, special stability is also associated with p^o, d^o and f^o systems.

Aufbau Principle

The building-up of electronic configuration of different elements is dictated by Aufbau principle. The build-up is a stepwise process. The start is made with the hydrogen atom, which has only one electron in 1s orbital. Its configuration is $1s^1$. By adding one electron, we get the configuration of the next element, helium. Since each orbital can accommodate only two electrons, the configuration of helium is $1s^2$. The next higher energy level is 2s. The electron of the next element, lithium, is added to this orbital. Hence, the configuration of lithium is $1s^2 2s^1$. The electron added while moving from one element to the next is called *differentiating electron.* It is this electron which makes the configuration of an element different from that of the element preceding it.

Aufbau principle states that the electron which distinguishes an element from its predecessor, enters the lowest energy atomic orbital. There are two rules governing Aufbau principle:

1. The differentiating electron will occupy that orbital which has a lower value of principal quantum number, n. Thus, 2s will be occupied before 3s; 4d will be occupied before 5d. This is because energy of the orbital increases with increasing principal quantum number.

2. For the same value of n, the differentiating electron will occupy that orbital which has a lower value of (n + l). For example, both 2s and 2p orbitals have the same value of n, but (n + l) value of former is smaller than that of the latter.

	n	l	(n + l)
2s	2	0	2
2p	2	1	3

Therefore, the electrons will be first filled up in 2s orbital and then in 2p orbital. Figure 1.30 is a helpful aid for writing the electronic configuration of various elements in accordance with Aufbau principle.

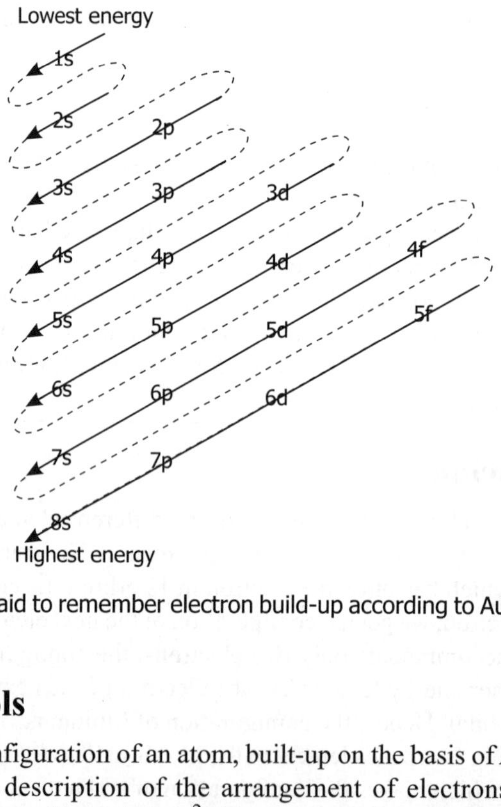

Fig. 1.30 An aid to remember electron build-up according to Aufbau principle

Term Symbols

The electronic configuration of an atom, built-up on the basis of Aufbau principle, is an incomplete description of the arrangement of electrons in the orbitals. For example, the configuration $2p^2$ does suggest that the electrons are placed in an energy level having principal quantum number 2 and azimuthal quantum number 1. However, it does not clarify which values of the magnetic quantum number, among the possibilities, +1, 0, –1 may be assigned to the two electrons. Nor does it specify the value of spin quantum number, out of the possibilities, $+\frac{1}{2}, -\frac{1}{2}$.

In fact, an atom may have several different states of total orbital angular momenta, as represented by the magnetic quantum number and spin angular momenta, as represented by the spin quantum number. Each state corresponds to the occupation of orbitals with different values of magnetic quantum number (m_l) by electrons with different values of spin quantum number (m_s). Such distinct arrangement of electrons is called a *microstate* of the electronic configuration.

The total number of microstates, N that arise when x electrons are placed in y orbitals of equivalent energy is mathematically expressed as,

$$N = \frac{(2y)!}{x!(2y - x)!}$$

For a single electron (x = 1), occupying a p sublevel (y = 3, corresponding to p_x, p_y, p_z orbitals), the total number of microstates is six. These are represented in Fig. 1.31.

	p_x	p_y	p_z
(a)	↑		
(b)		↑	
(c)			↑
(d)	↓		
(e)		↓	
(f)			↓

Fig. 1.31 Microstates for p^1 configuration

All the six microstates of p^1 configuration are equal in energy. The reason being that the energy of a lone electron in not perturbed, as the two remaining orbitals of the sublevel are vacant. However, this in not the case with p^2 configuration. Even though the p orbitals are degenerate, the two electrons interact with each other. The interactions are of two types: spin interactions and orbital angular momentum interactions. Fifteen microstates are possible for this system (x = 2, y = 3, therefore N = 15). These are shown in Fig. 1.32.

	p_x	p_y	p_z
(a)	↑	↑	
(b)		↑	↑
(c)	↑		↑
(d)	↓	↓	
(e)		↓	↓
(f)	↓		↓

	p_x	p_y	p_z
(g)	↑	↓	
(h)		↑	↓
(i)	↑		↓
(j)	↓	↑	
(k)		↓	↑
(l)	↓		↓

	p_x	p_y	p_z
(m)	↑↓		
(n)		↑↓	
(o)			↑↓

Fig. 1.32 Microstates for p^2 configuration

These microstates may be divided into three sets. The first set comprises microstates (a) – (f), in which the two electrons have same spins. The second set comprises microstates (g) – (l), in which the electrons have opposed spins. The third set of microstates (m) – (o) has electrons with opposed spins placed in the same orbital.

Under these circumstances, it is pertinent to elucidate which set represents the ground state of the atom and which set or sets correspond to the excited state(s). To achieve this, two parameters need to be determined.

1. Using the values of spin quantum numbers, $m_s = +\frac{1}{2}, -\frac{1}{2}$, the total spin quantum number, M_s is computed. Mathematically,

$$M_S = \Sigma m_s$$

M_s has values ranging from +S … 0 …–S. The maximum value of S is selected.

$$S = \text{Maximum value of } M_S$$

The spin multiplicity, (2S + 1) is then calculated.

2. Using the values of magnetic quantum number for p subshell, $m_l = +1, 0, -1$, the total orbital angular momentum, M_L is computed. Mathematically,

$$M_L = \Sigma m_l$$

M_L has values ranging from +L … 0 … –L. The maximum value of L is selected.

$$L = \text{Maximum value of } M_L$$

The value of L defines the electronic state of the atom.

L =	0	1	2	3	4	5
State =	S	P	D	F	G	H

The electronic state of the atom, in turn, is characterized by particular values of energy, orbital angular momentum and spin angular momentum. Each electronic state is designated by a term symbol, which is expressed as ^{2S+1}L

When S = 0, 2S + 1 = 1 the electronic state is referred to as ‘singlet’.

When $S = \frac{1}{2}$, 2S + 1 = 2, the electronic state is referred to as ‘doublet’.

When S = 1, 2S + 1 = 3, the electronic state is referred to as ‘triplet’.

When $S = \frac{3}{2}$, 2S + 1 = 4, the electronic state is referred to as ‘quartet’.

And so on.

We refer back to Fig. 1.32 and note that in the first set,

- For microstates (a) and (d) $\Sigma m_l = 1$
- For microstates (b) and (e) $\Sigma m_l = -1$
- For microstates (c) and (f) $\Sigma m_l = 0$

$$M_L = 1, 0, -1$$

L (the largest value of M_L) = 1

Electronic state = P

Even without involving the microstates, we may arrive at the same conclusion.

$$M_L = (m_{l1} + m_{l2}), (m_{l1} + m_{l2} - 1), (m_l + m_{l2} - 2) \ldots\ldots$$

For two electrons in p subshell,

$$M_L = (1 + 0),\quad (1 + 0 - 1),\quad (1 + 0 - 1)$$

$$M_L = 1, 0, -1$$

$$L = 1 \quad \text{or} \quad \text{P electronic state}$$

For each of the microstate (a) – (f),

$$S = m_{s1} + m_{s2} = +\frac{1}{2} + \left(+\frac{1}{2}\right) = 1$$

$$2S + 1 = 3$$

The terms symbol is ^3P and is referred to as "triplet P".
In the second set,

- For microstates (g) and (j), $\Sigma m_l = 1$
- For microstates (h) and (k), $\Sigma m_l = -1$
- For microstates (i) and (l), $\Sigma m_l = 0$

$$M_L = 1, 0, -1$$

$$L = 1$$

Electronic state = P
For each of the microstate (g) – (l),

$$S = +\frac{1}{2} + \left(-\frac{1}{2}\right) = 0$$

$$2S + 1 = 1$$

The term symbol is ^1P, referred to an "singlet P".
In the third set,

- For microstate (m), $\Sigma m_l = 2$
- For microstate (n), $\Sigma m_l = 0$
- For microstate (o), $\Sigma m_l = -2$

$$M_L = 2, 0, -2$$

$$L = 2$$

Electronic state = D
For each of the microstate (m) – (o),

$$S = +\frac{1}{2} + \left(-\frac{1}{2}\right) = 0$$

$$2S + 1 = 1$$

The term symbol is ^1D, referred to as "singlet D".

For deciding which term symbol represents the ground state, we follow the following two rules, which may be considered as a more sophisticated expression of Hund's concept of maximum multiplicity.

Rule 1. The ground state is one which has the highest value of spin multiplicity.

Rule 2. If several states have highest spin multiplicity, the most stable (and ground state) is one that has the highest value of total orbital angular momentum, L.

For the p^2 system, the microstates of the first set, (a) – (f), designated by the term symbol ^3p represent the ground state; the microstates of the second set, (g) – (*l*), designated by the term symbol ^1P represent the excited state; the microstates of the third set, (m) – (o) are irrelevant as these violate the Hund's rule of maximum multiplicity.

We now formulate term symbols for a few representative cases.

(i) s^1 system. For example, hydrogen atom.

$$m_l = 0; L = 0; S = \frac{1}{2}; 2S + 1 = 2$$

The term symbol is ^2S, "doublet S".

(ii) s^2 system. For example, helium atom.

$$m_l = 0; L = 0; S = 0; 2s + 1 = 1$$

The term symbol is ^1S, "singlet S".

(iii) p^1 system. For example, boron atom.

$$m_l = +1, 0, -1; L = 1; S = \frac{1}{2}; 2S + 1 = 2$$

The term symbol is ^2P, "doublet P".

(iv) p^2 system. For example, carbon atom.
We have elucidated that the term symbol for this system is ^3P, "triplet P".

(v) d^1 system. For example, scandium atom.

$$m_l = +2, +1, 0, -1, -2; L = 2; S = \frac{1}{2}; 2S + 1 = 2$$

The term symbol is ^2D, "doublet D".

Case Study 1.14 Deduce the term symbol(s) for the configuration s^1p^1.

In the section of term symbols, we stated two rules, according to which, electronic state having the highest value of spin multiplicity constitutes the ground state. In case the spin multiplicities of two states are same, then the one having a higher value of total orbital angular momentum, L constitutes the ground state. However, it must be emphasized that these rules, being a part of Hund's concept, are applicable to degenerate orbitals.

In the configuration, s^1p^1 the two orbitals, s and p are not degenerate. Hence more than one ground state may be possible for this system.

We first compute the values of Σm_l

	s	p	Σm_l
m_l	0	+1	1
m_l	0	0	0
m_l	0	−1	−1

L = 1; the electronic state is P. Next we compute the values of Σm_s

	s	p	
	[↑]	[↑][ ][ ]	
m_s	$+\frac{1}{2}$	$+\frac{1}{2}$	$\Sigma m_s = 1;\ S = 1;\ 2S + 1 = 3$
	[↑]	[↓][ ][ ]	
m_s	$+\frac{1}{2}$	$-\frac{1}{2}$	$\Sigma m_s = 0;\ S = 0;\ 2S + 1 = 1$

Two term symbols are possible for s^1p^1 configuration: ^3P or "triplet P" and ^1P or "singlet P".

2. PERIODIC TABLE

Introduction

All the existing elements are arranged in an extended form of periodic table. This table has been designed in accordance with the modern periodic law, which states that properties of elements are periodic functions of their atomic numbers. This means that if we arrange the elements in the order of increasing atomic number, we periodically encounter elements that have similar physical and chemical properties. The extended form (also called long form) of the periodic table is such an arrangement.

The Classification

Figure 2.1 provides a pictorial representation of the modern periodic table. The vertical columns of the periodic table are called *groups*, while the horizontal rows are called *periods*. There are eighteen groups and seven periods in the periodic table. Two separate rows of rare earth metals are arranged below the main table. The periodic table may be divided into four blocks depending on the valence shell electronic configuration of the elements.

1. s-Block elements

The first two groups of the periodic table constitute the s-block. Group 1 contains alkali metals, with valence shell electronic configuration ns^1 and Group 2 contains alkaline earth metals, with valence shell electronic configuration ns^2.

2. Representative elements

Members of Groups 13 to 18 are called representative elements or *p-block elements*. Their valence electrons are in s and p orbitals. Group 18 consists of noble gases which have completely filled s and p valence shells, ns^2np^6 (except helium with $1s^2$ configuration) and are therefore non-reactive.

3. Transition elements

Members of Groups 3 to 12 are called transition elements. The properties of these elements are transitional between those of s-block elements on their left and p-block elements on their right. They have partially filled d orbitals in the valence shell and are therefore also called *d-block elements*.

Metals

Metalloids

Non-metals

Group 1	2	3	4	5	6	7	8	9	10	11	12	13	14	15	16	17	18
Hydrogen 1 **H** 1.00797																	Helium 2 **He** 4.0026
Lithium 3 **Li** 6.942	Beryllium 4 **Be** 9.0122											Boron 5 **B** 10.811	Carbon 6 **C** 12.01115	Nitrogen 7 **N** 14.0067	Oxygen 8 **O** 15.9994	Fluorine 9 **F** 18.9984	Neon 10 **Ne** 20.183
Sodium 11 **Na** 22.9898	Magnesium 12 **Mg** 24.312											Aluminium 13 **Al** 26.9815	Silicon 14 **Si** 28.086	Phosphorus 15 **P** 30.9738	Sulfur 16 **S** 32.064	Chlorine 17 **Cl** 35.453	Argon 18 **Ar** 39.948
Potassium 19 **K** 39.102	Calcium 20 **Ca** 40.08	Scandium 21 **Sc** 44.956	Titanium 22 **Ti** 47.90	Vanadium 23 **V** 50.942	Chromium 24 **Cr** 51.996	Manganese 25 **Mn** 54.9380	Iron 26 **Fe** 55.847	Cobalt 27 **Co** 58.9332	Nickel 28 **Ni** 58.70	Copper 29 **Cu** 63.54	Zinc 30 **Zn** 65.37	Gallium 31 **Ga** 69.72	Germanium 32 **Ge** 72.59	Arsenic 33 **As** 74.9216	Selenium 34 **Se** 78.96	Bromine 35 **Br** 79.909	Krypton 36 **Kr** 83.80
Rubidium 37 **Rb** 85.4678	Strontium 38 **Sr** 87.62	Yttrium 39 **Y** 88.9059	Zirconium 40 **Zr** 91.224	Niobium 41 **Nb** 92.9064	Molybdenum 42 **Mo** 95.94	Technetium 43 **Tc** (98)	Ruthenium 44 **Ru** 101.07	Rhodium 45 **Rh** 102.9055	Palladium 46 **Pd** 106.4	Silver 47 **Ag** 107.870	Cadmium 48 **Cd** 112.40	Indium 49 **In** 114.82	Tin 50 **Sn** 118.69	Antimony 51 **Sb** 121.75	Tellurium 52 **Te** 127.60	Iodine 53 **I** 126.9044	Xenon 54 **Xe** 131.30
Cesium 55 **Cs** 132.905	Barium 56 **Ba** 137.34	Lanthanum 57 ***La** 138.91	Hafnium 72 **Hf** 178.49	Tantalum 73 **Ta** 180.948	Tungsten 74 **W** 183.85	Rhenium 75 **Re** 186.2	Osmium 76 **Os** 190.2	Iridium 77 **Ir** 192.2	Platinum 78 **Pt** 195.09	Gold 79 **Au** 196.967	Mercury 80 **Hg** 200.59	Thallium 81 **Tl** 204.37	Lead 82 **Pb** 207.19	Bismuth 83 **Bi** 208.980	Polonium 84 **Po** (209)	Astatine 85 **At** (210)	Radon 86 **Rn** (222)
Francium 87 **Fr** (223)	Radium 88 **Ra** (226)	Actinium 89 ****Ac** (227)	Unnilquadium 104 **Unq** (261)	Unnilpentium 105 **Unp** (262)	Unnilhexium 106 **Unh** (263)	Unnilseptium 107 **Uns** (262)	Unniloctium 108 **Uno** (265)	Unnilennium 109 **Une** (266)									

* Lanthanoid Series

Cerium 58 **Ce** 140.12	Praseodymium 59 **Pr** 140.907	Neodymium 60 **Nd** 144.24	Promethium 61 **Pm** (145)	Samarium 62 **Sm** 150.35	Europium 63 **Eu** 151.96	Gadolinium 64 **Gd** 157.25	Terbium 65 **Tb** 158.924	Dysprosium 66 **Dy** 162.50	Holmium 67 **Ho** 164.930	Erbium 68 **Er** 167.26	Thulium 69 **Tm** 168.934	Ytterbium 70 **Yb** 173.04	Lutetium 71 **Lu** 174.97

** Actinoid Series

Thorium 90 **Th** 232.038	Protactinium 91 **Pa** 231.0359	Uranium 92 **U** 238.03	Neptunium 93 **Np** (237)	Plutonium 94 **Pu** 244)	Americium 95 **Am** (243)	Curium 96 **Cm** (247)	Berkelium 97 **Bk** (247)	Californium 98 **Cf** (251)	Einsteinium 99 **Es** (252)	Fermium 100 **Fm** (257)	Mendelevium 101 **Md** (258)	Nobelium 102 **No** (259)	Lawrencium 103 **Lr** (260)

Fig. 2.1 Modern periodic table

There are three series of transition elements:

(i) 3d or 1st series, with valence shell electronic configuration $3d^n4s^2$.
(ii) 4d or 2nd series, with valence shell electronic configuration $4d^n5s^2$.
(iii) 5d or 3rd series, with valence shell electronic configuration $5d^n6s^2$.

4. Inner transition elements

Also called f-block elements, the inner transition elements have their valence electrons in f orbitals. These are arranged in two rows below the main periodic table. The first row is of lanthanoids, which have partially filled 4f orbitals. The second row is of actinoids, which have partially filled 5f orbitals.

Size of Elements

The term size, when applied to elements, refers to the radius of their atoms or ions. Size depends on the following factors:

1. Bond order.
2. Oxidation number of the atom in the concerned molecule.
3. Degree of ionic or covalent character of the bond type.
4. Coordination number of the atom.
5. Repulsive forces due to other atoms not directly bonded.

Atomic radii

The term atomic radii encompasses metallic radii for metals, alloys or intermetallic compounds and covalent radii for non-metals or for elements forming covalent bonds. Literally, radius means the distance from the center of the nucleus to the outermost shell of the atom. In Fig. 2.2, O is the center of the nucleus and OA is the atomic radius.

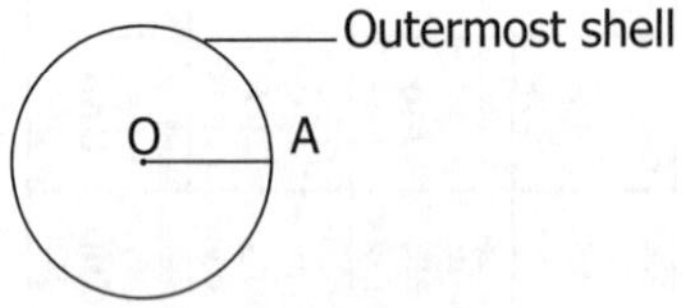

Fig. 2.2 Atomic radius

Technically, however, it is not possible to measure this distance in an isolated atom. According to Heisenberg uncertainty principle, it is impossible to precisely locate the outermost electron. This implies that the distance from the outermost electron to the center of the nucleus cannot be precisely elucidated. Nevertheless, the internuclear distance between two adjacent atoms in a crystal or a gaseous molecule can be determined. If it is assumed that atoms in the crystal or the gaseous molecule behave as effective spheres which are barely in contact with each other, the distance between the centers of two nuclei gives the sum of atomic radii of the two atoms.

For example, to calculate the atomic radius of fluorine, the intermolecular distance in a gaseous fluorine molecule is measured. Assuming that the two fluorine atoms in the molecule touch each other, the atomic radius is half the internuclear distance. As the internuclear distance in fluorine is 128 pm, the atomic radius is taken as 64 pm (Fig. 2.3a).

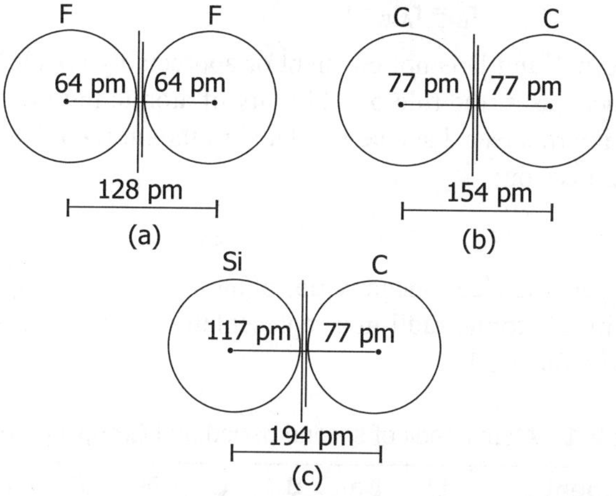

Fig. 2.3 Calculation of the atomic radii of from internuclear distance (a) fluorine molecule (b) diamond crystal and (c) silicon carbide molecule

Similarly, the carbon–carbon distance in a diamond crystal is 154 pm (Fig. 2.3b). The atomic radius of carbon is half this value, i.e., 77 pm. In the case of silicon carbide, the silicon–carbon bond length is 194 pm. The atomic radius of silicon may be obtained by subtracting the atomic radius of carbon (77 pm) from the silicon–carbon internuclear distance. This gives a value of 117 pm (Fig. 2.3c).

While comparing the periodicity of size among various elements, it is important to ensure that the environment around the atoms is as similar as possible. For example, if the carbon–oxygen single bond distance and the atomic radius of carbon are known, the atomic radius of oxygen may be obtained by subtraction.

Internuclear C–O distance	= 143 pm
Atomic radius of carbon	= 77 pm
Hence, atomic radius of oxygen	= 66 pm

However, the same criterion cannot be applied if the carbon–oxygen internuclear distance is measured in a compound having carbon–oxygen double bond, C=O. The fluorine molecule, F_2, is suitable for calculating atomic radius of fluorine atom, but similar calculations cannot be extended to a nitrogen molecule, N_2, for calculating the atomic radius of nitrogen. A triple bond holds the two nitrogen atoms together in a nitrogen molecule. Atoms connected by multiple bonds are distorted from sphericity and, therefore, the concept of radius loses significance. Hence, to calculate the atomic radius of nitrogen, the internuclear distance is measured in a compound in which two nitrogen atoms are linked by a single bond.

The use of the above method for calculation of atomic radii is applicable as long as the two atoms have same or nearly same electronegativities. Suppose two atoms, A and B, vary widely in electronegativity values. Then, if the internuclear distance (r_{A-B}) and atomic radius of A (r_A) are known, the atomic radius of B (r_B) *cannot* be calculated from the relation:

$$r_B = r_{A-B} - r_A$$

The bond between A and B is not covalent or appreciably covalent. It is polar in nature and in such cases, the rule of additivity of atomic radii does not apply. In such cases, the internuclear distance is related to the individual atomic radius by the following expression:

$$r_{A-B} = r_A + r_B - 9\,|\chi_A - \chi_B|$$

where χ_A and χ_B are the electronegativities of atoms A and B, respectively.

The periodicity of atomic radii in a representative period and a representative group is shown in Table 2.1.

Table 2.1 Atomic radii of second period and Group 17 elements

Element	**Li**	**Be**	**B**	**C**	**N**	**O**	**F**	**Ne**
Atomic radius (pm)	152	111	88	77	70	66	64	160
							Cl	
							99	
							Br	
							114	
							I	
							100	

As seen from Table 2.1, there is a progressive decrease in atomic radii along a period from left to right. Within a period, the electrons are added in the same principal quantum number valence shell. However, as the atomic number increases, the magnitude of nuclear charge also increases. Consequently, the outermost electrons are held progressively more tightly with increase in atomic number. On moving from top to bottom in a group, the atomic radii increase. The atomic number and hence the nuclear charge also increase down a group. However, within a group the valence electrons are not added in the same principal quantum number shell. Rather more and more electron layers are incorporated on moving down the group, and the valence electrons occupy progressively higher principal quantum number shells. This, in turn, causes a progressive expansion in size.

Some exceptions are notable in these general trends. The atomic radius of helium, the last element of the first period, is larger than that of fluorine. In fact, each noble gas is bigger in atomic size as compared to the corresponding halogen. Due to their inert nature, the noble gases do not form ionic or covalent bonds with other elements or among themselves. It is therefore not possible to have the values of internuclear distances from which the atomic radii of noble gases may be calculated. The best that can be done is to measure the distance when two atoms of a noble gas are in closest proximity to one another (Fig. 2.4). However, this distance is larger than the internuclear distance between two bonded atoms.

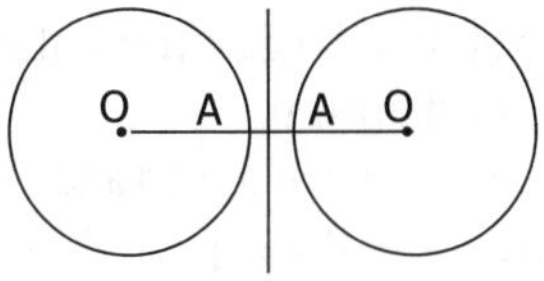

Fig. 2.4 Calculation of the atomic radii of noble gases

The atomic radius is taken to be half the value of the distance between the nuclei of two atoms that are very close to each other, but not linked by a chemical bond. Obviously, this value will be higher than expected. Actually, this is not the covalent radius, as no covalent bond exists between the two atoms of the noble gas. It is called *van der Waals radius*, named after van der Waal forces of attraction that bring the noble gas atoms close to each other. Unlike covalent or metallic radii, van der Waals radii are non-bonded radii. Their values are on the higher side.

The atomic radii of Group 13 elements are given in Table 2.2.

Table 2.2 Atomic radii of Group 13 elements

Element	Atomic radius (pm)
B	88
Al	125
Ga	125
In	150
Tl	155

Although there is an expected increase in size on moving from boron to aluminium, the atomic radii of aluminium and gallium are the same. The 3d transition series intervenes between these two elements. The valence shell configuration of 3d transition metals is $3d^n4s^2$. The penultimate d shell cannot shield the ultimate s shell effectively from the attractive influence of the nucleus. The screening power varies in the order, $s > p > d > f$, and d orbitals are relatively poor shielders. For this reason, the outermost 4s shell of the transition metals is drawn close to the nucleus and there is decrease in size on traversing the 3d series. This decrease negates the expected increase in atomic radius on moving from aluminium to gallium. Across the period 4, this decrease is observed from gallium to krypton. It is called *scandide contraction.* It is so named since it is caused by the incorporation of a new orbital, 3d, which first appears in scandium.

The atomic radius of thallium is slightly larger than that of indium. This increase is much less than that expected by incorporation of an additional shell of electrons.

On moving from indium to thallium, one has to pass through 14 lanthanoid elements. The valence shell configuration of lanthanoids is $4f^n6s^2$. Since the screening ability of f shell is the poorest of all, the outermost 6s shell is held very tightly by the nucleus. As the atomic number of the lanthanoid metals increases, the outermost shell is drawn closer and closer to the nucleus. Thus with increasing atomic number, there is a consistent decrease in the size of lanthanoid atoms. This decrease, called *lanthanoid contraction* nullifies, to a large extent, the expected increase in atomic radii on moving from indium to thallium.

The lanthanoids also intervene between the 4d and 5d series of transition elements. As a result of lanthanoid contraction, the expected increase in size on moving from a 4d element to a 5d element of the same group is not observed.

4d Series	Zr	Nb
Atomic radius (pm)	145	134
5d Series	Hf	Ta
Atomic radius (pm)	144	134

The size of 4d and 5d transtition metals, therefore, remains nearly constant within each group.

Ionic radii

The radii of atoms are changed by the addition or loss of electrons. When electrons are removed from an atom, a cation is formed. The radius of a cation is smaller than that of the neutral atom. In a neutral atom, the number of protons in the nucleus is exactly same as the number of orbital electrons. However, in a cation, the number of electrons is less than the number of protons. The effective nuclear charge of the cation is, therefore, more than that of the neutral atom. This results in the remaining electrons being held more tightly by the nucleus, causing, in turn, a shrinkage in size. As more electrons are removed, the effective nuclear charge increases progressively. Therefore, greater the charge on the cation, smaller is its ionic radius.

Element	Na	Fe
Atomic radius (pm)	186	117
Ion	Na^+	Fe^{2+}
Ionic radius (pm)	102	78
Ion	—	Fe^{3+}
Ionic radius (pm)	—	64

When an anion is formed, one or more electrons are added to the neutral atom. The number of valence shell electrons now exceeds the number of protons in the nucleus. The effective nuclear charge decreases and, therefore, electrons in the outermost shell are held less tightly. This causes the electron cloud to expand

and the ionic radius to increase. The radius of an anion is larger than that of the corresponding neutral atom. As the negative charge on the anion increases, the ionic radius progressively increases.

Element	Cl	O	N
Covalent radius (pm)	99	66	70
Ion	Cl^-	O^{2-}	N^{3-}
Ionic radius (pm)	184	140	138

The periodic trends of ionic radii are same as those of atomic radii. The ionic radii increase down the group because extra shells of electrons are added.

Ion	Li^+	Na^+	K^+
Ionic radius (pm)	76	102	138

The ionic radii decreases on moving from left to right across a period partly due to the increase in nuclear charge and partly due to the increase in the positive charge on the ion.

Ion	Na^+	Mg^{2+}	Al^{3+}
Ionic radius (pm)	102	72	53

The calculation of ionic radius, however, is not as straight forward as that of atomic radius. The internuclear distance (r_{C-A}) between the cation (C) and anion (A) in an ionic compound can be measured (Fig. 2.5). It is equal to the sum of the radius of the cation (r_C) and the anion (r_A), i.e.,

$$r_{C-A} = r_C + r_A \quad (2.1)$$

However, the cation is invariably smaller than the anion. The problem, therefore, is how to split the internuclear distance into the radius of cation and the radius of anion.

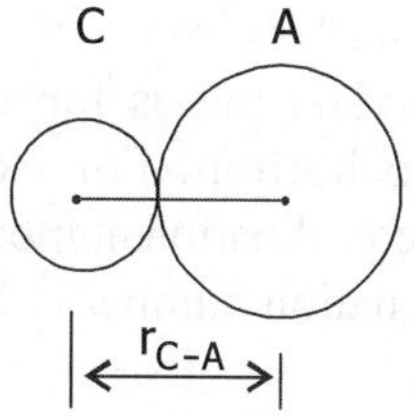

Fig. 2.5 Calculation of ionic radii

The radius of an ion is inversely proportional to the effective nuclear charge, i.e.,

$$r_C \propto \frac{1}{Z_C - \sigma_C} \quad (2.2)$$

$$r_A \propto \frac{1}{Z_A - \sigma_A} \quad (2.3)$$

Z_C and Z_A are the actual nuclear charges of the cation and anion, respectively. The corresponding screening constants are σ_C and σ_A. From equations (2.2) and (2.3),

$$\frac{r_C}{r_A} = \frac{Z_A - \sigma_A}{Z_C - \sigma_C} \tag{2.4}$$

The two unknowns, r_C and r_A may be calculated using equations (2.1) and (2.4).

Crystal radii

Crystal radius refers to the size of a cation or an anion within a crystal lattice. Although it is a measure of the size of an ion, yet crystal radius is different form ionic radius. We have seen that ionic radius of a cation C is calculated on the basis of its nuclear charge (Z_C) and its distance of separation from the anion A (r_{C-A}). The crystal radius, on the other hand, depends on the coordination number of a particular ion in a lattice. Thus a metal ion may have a coordination number of 6 in the crystal of its chloride, and a coordination number of 4 in the crystal of its iodide. The crystal radius of the metal ion will have a different value in each case.

In context of crystal radii, the following trends are noteworthy.

1. There is an increase in crystal radii with increasing coordination number. For example, the crystal radius of lithium ion in 73 pm when the coordination number is 4 and 90 pm when the coordination number is 6. Likewise, the crystal radius of sodium ion is 113, 116 and 132 pm for coordination numbers 4, 6 and 8, respectively.
2. Crystal radii of anions is greater than that of isoelectronic cations. For example, the crystal radius of fluoride ion is 119 pm, while that of isoelectronic sodium ion is 116 pm.
3. With increasing cationic charge, the crystal radius decreases. Thus the crystal radius of titanium(III) ion is 81 pm, while that of titanium(IV) ion is 75 pm.
4. Within a group, the crystal radius for a specific coordination number increases form top to bottom. For example, the crystal radius of magnesium(II) ion for coordination number 6 in 86 pm and for calcium(II) ion for the same coordination number is 114.

Ionization Energy

Ionization energy, also called *ionization potential*, is the minimum amount of energy required to remove the most loosely bound electron from an isolated gaseous atom. It is possible to remove more than one electron from most atoms. Accordingly, the first ionization energy is the energy required to remove the first electron and convert the atom into a univalent cation, the second ionization energy is the energy required to remove the second electron and convert the atom into a divalent cation, and so on. Ionization is an endothermic phenomenon. Energy must be supplied to the electron so that it can overcome the attractive influence of the nucleus and exit from the atom.

Ionization energy depends on the following factors:

1. Atomic size.
2. Nuclear charge.
3. Screening effect of inner shells.
4. The type of orbital in which the valence shell electron resides.

In a small atom, the outermost electron is held more tightly than in a large atom. Therefore, the ionization energy decreases as the size of the atom increases. In two nearly equal sized atoms, the one with a larger nuclear charge will possess a higher ionization energy. If the inner shells screen the outermost electron effectively from nuclear attraction, the ionization energy will be small. If screening is poor, the outermost electron is strongly attracted by the nuclear charge and the ionizaion energy will be concomitantly high. The screening effect depends on the symmetry of the orbital and varies in the order, s > p > d > f. Therefore, in atoms with s orbitals constituting the penultimate shell, the ionization energy is low; where f orbitals are the penultimate shell, the ionization energy is high. Ionization energy also depends on the penetration power of the orbital in which the outermost electron resides. Each orbital has a tendency to penetrate nearer the nucleus. The penetration power follows the order s > p > d > f. Therefore, an s-electron is held more tightly than a p-electron. Similarly, a p-electron is attracted by the nucleus to a larger extent as compared to a d-electron which, in turn, is held more tightly than an f-electron.

Along a period, the ionization energies tend to increase from left to right. This tendency may be attributed to rising nuclear charge as the atomic number increases. Further, in a period, the atomic radii decrease from left to right. These two factors—increase in nuclear charge and decrease in size—serve to hold the outermost electron more tightly.

The first ionization energies of elements in the second period are given in Table 2.3.

Table 2.3 First ionization energies of elements in the second period

Element	Ionization energy (kJ mol^{-1})
Li	520
Be	899
B	801
C	1086
N	1403
O	1314
F	1681
Ne	2080

Two irregularities are noted. Boron has a higher nuclear charge and smaller atomic radius than beryllium. However, its ionization energy is less than that of beryllium. The latter has the configuration $1s^22s^2$. Ionization of a completely filled 2s shell requires higher energy since extra work has to be done to unpair the electrons before removing one of these.

The second irregularity occurs at nitrogen, which has a higher ionization energy than oxygen. Nitrogen has a stable half-filled p shell in its electronic configuration. It requires comparatively more energy to remove an electron from a stable shell. Conversely, oxygen with the configuration $1s^22s^22p^4$, readily parts away with one electron to achieve a stable half-filled 2p shell. This causes its ionization energy to be less than expected.

Down a group, the ionization energies decrease. Two factors account for this trend. First, on moving from top to bottom, the atomic size increases and, therefore, the distance between the nucleus and the outermost electron increases. Second, the number of electronic shells increases on descending the group. The inner shells act as a screen between the nucleus and the outermost electron, thereby decreasing the force of attraction between the two. The combined outcome of the two factors is that the force holding the valence electrons decreases, resulting in a decrease in ionization energies.

There is a regular decrease in ionization energies on descending Groups 1 and 2. However, in Group 13, the decrease is irregular. The first ionization energy data are given in Table 2.4.

Table 2.4 Ionization energies of Group 13 elements

Element	Ionization energy (kJ mol^{-1})
B	801
Al	577
Ga	579
In	558
Tl	589

Between boron and aluminium the expected decrease in the first ionization energies is observed. However, gallium has higher ionization energy than aluminium. Between aluminium and gallium, there are ten elements of the first transition series, having the configuration $3d^n4s^2$. The penultimate d shell is unable to screen the outer 4s electrons effectively from nuclear attraction. Therefore, a progressive decrease in size is observed while traversing the 3d series elements. As a result, the size of gallium is almost the same as that of aluminium. Further, gallium has a higher nuclear charge than aluminium. The net effect is that its ionization energy is higher than expected. Similarly, the lanthanoid elements intervene between indium and thallium. The lanthanoids have the configuration $4f^n6s^2$. The penultimate f shell is a poor shielder, so that the 6s electrons are held tightly by the nuclei. This causes the size of atoms to decrease along the lanthanoid series (lanthanoid contraction). As a result of lanthanoid contraction, the expected increase in size on moving from indium to thallium is not observed. Therefore, the ionization energy of thallium is more than expected.

Group 1 metals have the lowest and noble gases have the highest ionization energies in their respective periods. Removal of the second electron from Group 1 elements require, on an average, ten times more energy than that required for removal of the first electron. These metals have just one valence electron which is removed quite easily. However, the second electron has to be removed from a completed octet and, therefore, requires exceptionally high energy.

Case Study 2.1 Although the second ionization energy of magnesium is double than the first ionization energy, yet on heating with chlorine, magnesium forms magnesium(II) chloride and not magnesium(I) chloride. Why?

The second ionization energy of magnesium is almost double than the first ionization energy. After one electron has been removed, the effective nuclear charge of the cation increases and its size gets reduced. Thus, magnesium(I) ion is smaller than neutral magnesium atom. Therefore, the remaining electrons are held more tightly and consequently second ionization energy is higher than the first. Despite this, the reaction between magnesium and chlorine yields magnesium(II) chloride and not even a trace of magnesium(I) chloride. Due to smaller size and higher charge, magnesium(II) ion forms a more stable lattice than magnesium(I) ion. Therefore, greater amount of lattice energy is released when magnesium(II) chloride is formed than if magnesium(I) chloride were to form. The lattice energy compensates for the extra ionization energy required to form magnesium(II) ion.

Electron Affinity

The energy change involved when an electron is added to an isolated gaseous atom to form an anion is called electron affinity (EA) or *affinity energy*. In some cases, energy is absorbed when an electron is added to an atom, while in others, energy is released. As in conventional chemical reactions, a positive value is assigned to electron affinity when energy is absorbed (endothermic) and a negative value when energy is released (exothermic).

The electron affinities of beryllium and chlorine may be represented as follows:

$$Be(g) + e^- \rightarrow Be^-(g) \qquad EA = +241 \text{ kJ mol}^{-1}$$

$$Cl(g) + e^- \rightarrow Cl^-(g) \qquad EA = -348 \text{ kJ mol}^{-1}$$

The first reaction is endothermic and the second, exothermic.

When one electron is added to an isolated oxygen atom, energy is released.

$$O(g) + e^- \rightarrow O^-(g) \qquad EA = -142 \text{ kJ mol}^{-1}$$

However, the addition of a second electron is accompanied by absorption of energy.

$$O^- + e^- \rightarrow O^{2-} \qquad EA = +702 \text{ kJ mol}^{-1}$$

This is because the second electron has to be forced into the valence shell of oxygen in which seven other electrons are already present. Oxygen atom is small in size and, therefore, strong electron–electron repulsions must be overcome before this second electron is accommodated in the valence shell. Despite this, the oxide ion exists as a divalent anion (O^{2-}) and not a monovalent one (O^{-}). The extra energy expended in the formation of the divalent anion is later compensated by the energy released during lattice formation. The interaction of a divalent oxide ion with a cation yields a higher lattice energy than a similar interaction involving a monovalent oxide ion.

Across a period, EA increases from left to right. The atomic size decreases in the same direction and, therefore, it becomes easier for the nucleus to attract the incoming electron. Electron affinities of second period elements and Group 17 elements are given in Table 2.5.

Table 2.5 Electron affinities of second period and Group 17 elements

Element	Li	Be	B	C	N	O	F	Ne
Electron affinity (kJ mol^{-1})	– 57	241	– 23	– 121	31	– 142	– 333	99
							Cl	
							–348	
							Br	
							–343	
							I	
							–314	

The increase in EA value from left to right across a period is not regular. Noteworthy exceptions are beryllium and nitrogen. Beryllium, with electronic configuration $1s^22s^2$, has a positive value of EA because its valence shell is already complete. Thus, the incoming electron can be accommodated only in the high energy 2p shell. Nitrogen, having the configuration $1s^22s^22p^3$, possesses a stable half-filled set of 2p orbitals. Addition of another electron would lead to a relatively less stable configuration. Fluorine has the most negative value of EA in the period since it achieves a stable noble gas configuration after acquiring one electron. Neon with a completed octet has no tendency to capture an electron and, therefore, possesses a positive value of EA.

Down a group, EA decreases because the atomic size increases. As new layers of electrons are built-up, the nucleus finds it difficult to attract the electron that is being added to the atom. However, the EA of fluorine (–333 kJ mol^{-1}) is lower than that of chlorine (–348 kJ mol^{-1}). This discrepancy may be attributed to the extremely small size of fluorine. With seven electrons already spread over a small volume, the incoming eighth electron faces stiff repulsion. Chlorine, with the same number of valence electrons, has a relatively large size. The valence electrons are spread over a comparatively larger volume and, therefore, the incoming electron is easily accommodated.

Electronegativity

Electronegativity is the tendency of an atom, *within a molecule,* to attract electrons towards itself. In a covalent molecule, such as hydrogen fluoride, the shared pair of electrons is not midway between the hydrogen and fluorine. Since fluorine has a greater nuclear charge than hydrogen, the bonding pair drifts more close to fluorine and away from hydrogen. As a result, fluorine acquires a small negative charge (δ^-) and hydrogen a small positive charge (δ^+).

$$H^{\delta+}—F^{\delta-}$$

Electronegativity is quite similar to EA in the sense that both the properties reflect the ability of an atom to capture electrons. Therefore, the periodic trends of electronegativity are same as those of EA. As shown in Table 2.6, electro-negativity increases from left to right in a period and decreases from top to bottom in a group.

Table 2.6 Electronegativities of second period and Group 17 elements

Element	Li	Be	B	C	N	O	F
Electronegativity	1.0	1.5	2.0	2.5	3.0	3.5	4.0
							Cl 3.0
							Br 2.8
							I 2.5

The nucleus of a smaller atom can attract electrons more strongly. Hence, small atoms have high electronegativities. Fluorine has the highest electronegativity, followed by oxygen.

Electronegativity and EA differ from each other in two aspects. First, electronegativity indicates the ability of an atom within a molecule to attract electrons, while electron affinity indicates a similar characteristic of an isolated atom. Second, EA demands that the acquired electron actually enters the valence shell of the concerned atom. Electronegativity, however, stipulates only a shift of electrons towards the concerned atom.

Variation of electronegativity

The electronegativity value of any element is not constant. It is defined by the environment of the atom of the concerned element. Two factors determine the extent of attraction of electrons by an atom: (i) charge and (ii) hybridization of the atom.

An atom which has acquired a positive charge or a partial positive charge will tend to attract electrons to it more readily as compared to a neutral atom. A negatively charged atom possesses reduced ability to attract electrons as compared to a neutral atom.

Hybridization affects electronegativity to the extent of degree of s character in the hybrid orbital. The s orbitals have low energy and great electron-attracting power. Therefore, those hybrid orbitals which have greater s character, have a greater ability to withdraw electrons. For example, in the case of carbon, the variation in electronegativity values with hybridization is as follows:

Hybridization	sp^3	sp^2	sp
% s character	25	33	50
Electronegativity	2.48	2.75	3.29

The commonly accepted value of 2.5 is based on tetrahedral or sp^3 hybridization.

Case Study 2.2 Predict the products formed in the reaction of aqueous potassium hydroxide with (a) methyl iodide and (b) trifluoromethyl iodide.

In methyl iodide, the iodine atom induces a positive charge on carbon. For this reason, the hydroxide ion attacks the methyl group, forming alcohol.

$$CH_3I + OH^- \longrightarrow CH_3OH + I^-$$

In case of trifluoromethyl iodide, initially the three fluorine atoms induce a positive charge on carbon. The electronegativity of carbon now becomes more than that of iodine. As a result, the carbon withdraws the pair of electrons that it shares with iodine towards itself. The positive charge, therefore, is induced on iodine which becomes the target for hydroxide ion attack.

$$CF_3I + OH^- \longrightarrow CF_3H + OI^-$$

Trifluoromethane is formed in this case.

Electronegativity scales

The electronegativity values of elements may be expressed on a number of arbitrary scales. The important ones are listed below:

1. Pauling scale

Suppose two elements, A and B, having different electronegativities, combine to form a molecule A–B, by the reaction:

$$A\text{–}A + B\text{–}B \longrightarrow 2A\text{–}B$$

The compound A–B is more stable than A–A and B–B because it exists in two canonical forms.

$$A\text{–}B \longleftrightarrow A^{\delta+}\text{–}B^{\delta-} \text{ (or } B^{\delta+}\text{–}A^{\delta-})$$

This means that the molecule A–B has greater bond energy than the geometric mean of the bond energies of A–A and B–B molecules, i.e.,

$$E_{A\text{–}B} > (E_{A\text{–}A} \cdot E_{B\text{–}B})^{1/2}$$

The extra bond energy of the molecule A–B may be expressed as:

$$\Delta = E_{A-B} - (E_{A-A} \cdot E_{B-B})^{1/2}$$

Suppose χ_A is the electronegativity of A and χ_B the electronegativity of B. The difference in electronegativity values is related to the extra bond energy (expressed as kJ mol^{-1}) by the equation:

$$\chi_A - \chi_B = 0.1017\ (\Delta)^{1/2}$$

If one of the atoms is hydrogen, the equation may be written as:

$$\chi_A - \chi_H = 0.1017\ (\Delta)^{1/2}$$

Pauling assigned an arbitrary value of 2.05 to the electronegativity of hydrogen, i.e., $\chi_H = 2.05$. The electronegativities of other elements may thus be calculated as shown in Table 2.7.

Table 2.7 Pauling electronegativity values

Bond A–H	**Δ (kJ mol^{-1})**	**0.1017 (Δ)$^{1/2}$ (=$\chi_A - \chi_H$)**	χ_A
C–H	24.3	0.50	2.50
H–Cl	102.3	1.02	3.07
N–H	105.9	1.04	3.09

If atoms A and B have similar electronegativities, then the bond A–B will be predominantly covalent. As the difference between the electronegativities of A and B increases, the degree of polar character also increases.

2. Mulliken scale

According to Mulliken, electronegativity represents the average of the IE and EA of an atom, e.g., for atom A,

$$\chi_A = \frac{IE + EA}{2}$$

Mulliken electronegativities, χ_M are related to Pauling values, χ_P by the equation:

$$\chi_P = 0.336\ (\chi_M - 0.615)$$

It is evident that the values calculated on Mulliken scale are directly proportional to the values calculated on Pauling scale.

3. Allred–Rochow scale

On Allred–Rochow scale, electronegativity is considered as the force of attraction between a nucleus and an electron at a distance equivalent to the covalent radius of the atom. If $Z_{eff}e$ is the nuclear charge, e is the electronic charge and r_{cov} is the covalent radius of the atom, the force of attraction, F is given by Coulomb's law:

$$F = \frac{Z_{eff} e.e}{r_{cov}^2} = \frac{Z_{eff} \cdot e^2}{r_{cov}^2}$$

Here, Z_{eff} is the actual protonic charge, Z minus the screening constant, σ.

$$Z_{eff} = Z - \sigma$$

The F values may be converted into electronegativity values on Pauling scale using the relationship:

$$\chi = 0.794 + \frac{0.759\ Z_{eff}}{r_{cov}^2}$$

4. Jaffé scale

Jaffé calculated electronegativity as a function of partial charge on an atom. The energy of an atom in a combined state is given by the quadratic equation:

$$E = aq + bq^2$$

where q is the partial charge on the atom.

Electronegativity, according to Jaffé, is defined as the change in energy with respect to charge, i.e.,

$$\frac{dE}{dq} = a + 2bq$$

Electronegativity of a neutral atom is given by a. The parameter b is called *charge coefficient* and represents the rate of change of electronegativity with charge. In general, large atoms have smaller values of b, while small atoms have larger values.

Case Study 2.3 What is the oxidation state of phosphorus in phosphine?

On Pauling scale, the electronegativities of phosphorus and hydrogen, 2.1 and 2.05 respectively, are quite close to each other. The electronegativity value of Pauling scale are empirical in nature. The scale has been devised relative to an arbitrary value assigned to the atom of a particular element.

On the other hand Mulliken or Jaffé scales have been devised on the basis of the energy of valence electrons of the concerned atom in a molecule. The energy scales are considered better as these take into account the parameters which fluctuate on changing the structure of the molecule. On Mulliken/Jaffé scale, the electronegativities of phosphorus and hydrogen are 2.41 and 2.25 respectively.

Since phosphorus is more electronegative than hydrogen, its oxidation state in phosphine, PH_3 is (–III)

---

5. Sanderson scale

The ability of an atom to attract electrons that it shares with the adjacent atom in a molecule depends upon its nuclear charge. Greater the nuclear charge, greater is the tendency to pull the electron cloud towards itself, and away from the neighboring atom. The nuclear charge, in turn, depends on the compactness of the concerned atom. A small, compact atom can withdraw the shared electrons more towards itself than a comparatively large atom.

Sanderson electronegativity (S) is expressed in terms of the compactness of an atom. It is defined as the ratio of the average electron density (ED) of an atom to the electron density of the nearest inert gas atom (ED_i). That is,

$$S = \frac{ED}{ED_i}$$

The electron density reflects the average number of electrons per unit volume of the atom. For an atom having an atomic number Z and a non-polar covalent radius r, the electron density is given as,

$$ED = \frac{Z}{\frac{4}{3}\pi r^3} \text{ or } \frac{Z}{4.19\, r^3}$$

Suppose we wish to calculate the Sanderson electronegativity of fluorine, for which atomic number, Z is 9 and covalent radius, r is 71 pm.

$$ED = \frac{9}{4.19(71)^3} = 6.0 \times 10^{-6}\ \text{pm}^{-3} \qquad (2.5)$$

The nearest noble gas atom is neon, for which Z is 10 and r is 131 pm. (the Van der Waals radii of nitrogen, oxygen and fluorine are 150, 140 and 135 pm, respectively; by extrapolation, the corresponding value of neon is taken to be 131 pm).

$$ED_i = \frac{10}{4.19(131)^3} = 1.06 \times 10^{-6}\ \text{pm}^{-3} \qquad (2.6)$$

Sanderson electronegativity of fluorine may be obtained by dividing equation (2.5) by (2.6).

$$S_F = \frac{6.0\times10^{-6}\ (\text{pm}^{-3})}{1.06\times10^{-6}\ (\text{pm}^{-3})} = 5.66$$

Likewise, the electron density of sodium, for which Z is 11 and r is 154 pm, may be calculated.

$$ED = \frac{11}{4.19\ (154)^3} = 7.18 \times 10^{-7}\ \text{pm}^{-3} \qquad (2.7)$$

The noble gas nearest to sodium, in terms of atomic number, is neon. Hence Sanderson electronegativity of sodium may be elucidated by dividing equation (2.7) by (2.6).

$$S_{Na} = \frac{7.18\times10^{-7}\ (\text{pm}^{-3})}{1.06\times10^{-6}\ (\text{pm}^{-3})} = 0.67$$

The concept of Sanderson electronegativity lends support to the *principle of electronegativity equalization*, which may be explained as follows. When two atoms, let us say A and B, having different electronegativity values combine, the electron cloud flows from the electropositive element (A) to the electronegative atom (B).

$$A^{\delta+} \longrightarrow B^{\delta-}$$

The atom A acquires a partial positive charge, the atom B, a partial negative charge. As a consequence, A now begins to attract the electron cloud, while B tends to shed it off. The electron cloud, therefore, flows in the reverse direction.

$$A^{\delta+} \longleftarrow B^{\delta-}$$

Ultimately, an equilibrium state is achieved, under which both the atoms acquire an equal electronegativity. This equilibrium value, S_{eqlm} is intermediate between the electronegativities of A and B. It corresponds to the geometric mean of the electronegativity values of A and B.

Thus for a molecule AB, the equilibrium electronegativity is given as,

$$S_{eqlm} = (S_A S_B)^{1/2}$$

For a molecule AB_2,

$$S_{eqlm} = (S_A S_B S_B)^{1/3}$$

For AB_3,

$$S_{eqlm} = (S_A S_B S_B S_B)^{1/4}$$

and so on.

We have already calculated the Sanderson electronegativity values of sodium and fluorine. Let us now elucidate the equilibrium electronegativity in a molecule of sodium fluoride, NaF.

$$S_{eqlm} = (S_{Na} S_F)^{1/2}$$

$$= (0.67 \times 5.66)^{1/2} = 1.95$$

The equilibrium electronegativity is intermediate between the corresponding parameter for sodium and fluorine.

Case Study 2.4 Calculate the Sanderson electronegativity of magnesium (Z = 12; r = 136 pm) and the equilibrium electronegativity for magnesium fluoride.

We first calculate the electron density of magnesium atom.

$$ED = \frac{12}{4.19\,(136)^3} = 1.13 \times 10^{-6}\ pm^{-3} \qquad (2.8)$$

The electron density for neon, which is the nearest inert gas to magnesium, is given by equation (2.6). Dividing equation (2.8) by (2.6), we get,

$$S_{Mg} = \frac{1.13 \times 10^{-6}\ (pm^{-3})}{1.06 \times 10^{-6}\ (pm^{-3})} = 1.06$$

The Sanderson electronegativity of fluorine, as calculated above, is 5.66. The equilibrium electronegativity of magnesium fluoride, MgF_2 may be given as,

$$S_{eqlm} = (S_{Mg}\, S_F S_F)^{1/3}$$
$$= (1.06 \times 5.66 \times 5.66)^{1/3} = 3.23$$

The Sanderson electronegativity of magnesium is 1.06. The equilibrium electronegativity of magnesium fluoride is 3.23.

Box 2.1

Bent Rule

Bent rule states that more electronegative substituents prefer those hybrid orbitals which have less s character, while less electronegative (or more electropositive) substituents prefer hybrid orbitals with more s character.

For example, in phosphorus pentahalides, the central atom is sp^3d hybridized. An sp^3d hybrid orbital may be considered as a combination of sp^2 and pd hybrids. The former forms trigonal, equatorial bonds; the latter, perpendicular, axial bonds. Thus in case of mixed halides of phosphorus, such as PCl_3F_2, the highly electronegative fluorine atoms will be directed towards the pd hybrids, while the relatively less electronegative chlorine atoms will be directed towards the sp^2 hybrids. In accordance with Bent rule, the chlorine atoms will occupy equatorial positions and fluorine atoms will occupy axial positions in the trigonal bipyramidal structure (Fig. 2.6).

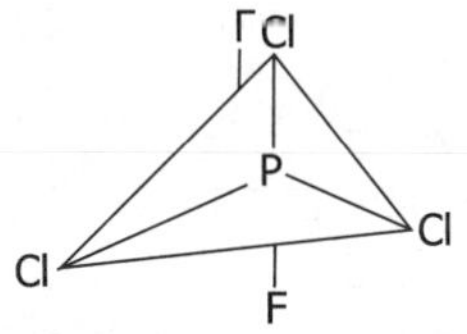

Fig. 2.6 Structure of PCl_3F_2

The s orbitals have a high tendency to attract electrons towards themselves. Therefore, these tend to attract less electronegative substituents. Conversely, the more electronegative substituents tend to get closer towards those orbitals which have lesser ability to attract electrons. The tendency of more electronegative substituents to seek out orbitals with less s character is called *apicophilicity.*

Slater Rules

The penultimate shell of an atom shields the outermost electron(s) from nuclear attraction. The extent of shielding is defined by *shielding constant*, also called *screening constant*. Slater rules are a means for quantitative calculations of screening constants. The rules are as follows:

1. The valence electrons are grouped in the order:
   (1s) (2s, 2p) (3s, 3p) (3d) (3f) (4s, 4p) (4d) (4f).......

Electron being shielded is in (ns, np) group

2. Electrons in any group to the right of (ns, np) group do not contribute to shielding.
3. All other electrons within the (ns, np) group shield the outermost electron to the extent of 0.35 each.
4. The electrons in the (n – 1) shell shield to an extent of 0.85 each.
5. Complete shielding is provided by the electrons in (n – 2) or lower groups. This means that electrons in (n – 2) or lower groups contribute to the extent of 1.00 each to the shielding constant.

Electron being shielded is in (nd) or (nf) group

Rules 1 and 2 remain the same, but rules 3–5 are replaced by:

6. All the electrons within the (nd) or (nf) group shield to an extent of 0.35 each.
7. All electrons in the groups lying to the left of (nd) or (nf) group shield completely, i.e., to the extent of 1.00 each.

Case Study 2.5 Calculate the shielding constants for 4s and 3d electrons of zinc.

The electrons of zinc may be grouped as follows:

$(1s)^2\ (2s, 2p)^8\ (3s, 3p)^8\ (3d)^{10}\ (4s)^2$

For 4s electrons, shielding constant is given by:

$\sigma = (2 \times 1.00) + (8 \times 1.00) + (8 \times 0.85) + (10 \times 0.85) + (1 \times 0.35)$

$\sigma = 25.65$

Atomic number of zinc = 30

Effective nuclear charge, $z^* = z - \sigma$

$= 30 - 25.65$

$= 4.35$

For 3d electron, shielding constant is given by:

$\sigma = (2 \times 1.00) + (8 \times 1.00) + (8 \times 1.00) + (9 \times 0.35)$

$\sigma = 21.15$

Effective nuclear charge, $z^* = 30 - 21.15$

$= 8.85$

The shielding constants for 4s and 3d electrons of zinc are 25.65 and 21.15, respectively.

Relativistic Effect

The term relativistic effect is the outcome of a combination of Einstein theory of relativity and Bohr model of atom. It offers an explanation for the following three important anomalies in the periodic table:

1. Chemical reactivity increases down a group, yet mercury is less reactive than zinc and cadmium.

2. The ionization energy decreases down a group, yet in Group 13, thallium(I) is more stable than thallium(III); and in Group 14, lead(II) is more stable than lead(IV)—an anomaly referred to as *inert pair effect*.
3. The strength of metallic bond decreases down a group among s- and p-block elements, yet it increases down the group among d-block elements.

According to the theory of relativity, the mass m of a particle (read electron) increases from its rest mass, m_0, when its velocity, v, approaches the speed of light, c. Mathematically,

$$m = \frac{m_0}{\sqrt{1-\left(\frac{v}{c}\right)^2}} \tag{2.9}$$

From Bohr model of atom, discussed in Chapter 1, we pick up equation (1.28),

$$mvr = \frac{nh}{2\pi}$$

and equation (1.30),

$$\frac{Ze^2}{r^2} = \frac{mv^2}{r}$$

where r is the radius of the orbit n in which the electron revolves, h is Planck's constant, Z is the atomic number of the element and e is the electronic charge. By combining and rearranging the above two equations, we get two new equations, one for the value of velocity:

$$v = Ze^2\left(\frac{2\pi}{nh}\right) \tag{2.10}$$

and other for the value of radius:

$$r = \frac{Ze^2}{mv^2} \tag{2.11}$$

According to equation (2.10), the value of v increases with Z. The value of (v/c) in equation (2.9) also increases. For example,

when n = 1 and Z = 1, then v/c = 0.0073
when n = 1 and Z = 80, then v/c = 0.58

With increasing value of v/c, the denominator in expression (2.9) decreases and the value of m increases. Equation (2.11) shows that a higher value of m leads to reduction in r, the radius of the orbit. This reduction of orbit's radius is referred to as *relativistic contraction*. For a particular principle quantum number shell, s orbitals experience the most severe relativistic contraction. The p orbitals, although relatively distant from the nucleus, also experience relativistic contraction, albeit to a lesser extent. Whenever relativistic contraction occurs (and it occurs in heavier elements, having higher Z values), the s and p orbitals shield the d orbitals to a greater degree than if there would have been no reduction in the value of r. As a result, the d orbitals, instead of contracting, tend to expand—a phenomenon called *relativistic expansion*. The relativistic contraction of s orbitals for elements with higher atomic number leads to a stronger attraction between s electrons and the nucleus. Accordingly, the ionization energies are higher than that expected for elements whose valence electrons reside in 6s orbital. For this reason, mercury is relatively inert, while thallium and lead show inert pair effect.

As we have already stated and as shall be discussed further in Chapter 21, the transition elements are catalogued in three groups of the periodic table, the 3d, 4d and 5d groups, depending on the orbital in which the valence electrons reside. Due to the relativistic effects, all the d orbitals expand, but the 5d orbitals, being farthest from the nucleus, expand to the greatest extent, followed by the 4d and 3d orbitals. As a result, the overlap between two 5d orbitals of neighbouring metal atoms in a crystal lattice is better than the analogous overlap between two 4d orbitals which, in turn, is better than the overlap between two 3d orbitals. This means that the metal–metal bond is extremely strong in 5d elements, moderate in 4d elements and relatively weak in 3d elements. This is reflected in the exceptionally high melting points and boiling points of the 5d transition metals.

The relativistic effect also explains the variation in color of Group 11 members. Thus while copper and gold have yellowish hue, silver is white. For explaining this variation, we have to invoke the band theory of metallic bonding which we have discussed in Chapter 3, p. 152. At this stage, suffice it to state that the energy difference between the 3d band and s-p band in copper is quite small. Hence electrons may be excited easily by absorbing blue component of visible light. The complementary color, yellow is reflected. In silver, the energy difference between the 4d band and s-p band lies in the ultraviolet region, and hence the white color. For gold, the relativistic effect comes into play. The combined effect of relativistic expansion of 5d shell and the relativistic contraction of 4s/4p shells is that the energy gap between the 5d band and s-p band gets reduced and again falls in the visible range. Therefore, the color of gold is similar to that of copper.

Case Study 2.6 In Group 10, the electronic configuration of the elements are as follows. Ni: [Ar]$3d^8 4s^2$; Pd: [Kr]$4d^{10} 5s^0$; and Pt: [Xe]$5d^9 6s^1$ Explain the trend.

The electronic configuration of nickel is on expected lines. It can be worked out on the basic of Aufbau principle (*see* Chapter 1, p. 47). In palladium, the energy difference between 4d and 5s orbitals is less than what it is between the 3d and 4s orbitals of nickel. Hence the 5s electrons migrate to 4d orbital, giving a stable, fully-filled d shell. In platinum, the 5d shell undergoes relativistic expansion, the 6s shell, relativistic contraction, and the two come to have comparable energy. One electron from 6s migrates to 5d shell.

Why in this case the electronic configuration is not $5d^{10} 6s^0$? Had it been so, there would have been no unpaired electron in the valence shell, the metal-metal bonding would have been extremely weak and platinum, in all probability would have been a liquid metal like mercury. Between unfavorable binding energy and unfavorable electronic configuration, platinum chooses the latter (No doubt, palladium with $4d^{10}5s^0$ configuration is not a liquid, we should not lose sight of the fact that relativistic effect appears in platinum and not in palladium.)

Box 2.2

The Third Liquid Element

Of all the known elements, eleven are gases under ordinary conditions of temperature and pressure. These include the six noble gases—helium, neon, argon, krypton, xenon and radon—and five others—hydrogen, nitrogen, oxygen, fluorine and chlorine. Only two elements are liquids at 298 K: Mercury and bromine. The third liquid element is *possibly* francium.

Francium is a highly radioactive element; its most stable isotope, francium-223, has a half-life of only 22 minutes. When this isotope disintegrates, so much heat is evolved that the decaying element itself vaporizes. It is, therefore, not possible to isolate weighable quantities of francium and study its properties. Only periodic trends shed light on the characteristics of this element.

Table 2.8 correlates the atomic number of alkali metals with their melting points.

Table 2.8 Atomic numbers and melting points of alkali metals

Element	Li	Na	K	Rb	Cs	Fr
Atomic number	3	11	19	37	55	87
Melting point (K)	453.6	370.9	336.7	312.4	301.7	–

A plot of atomic numbers against melting points of alkali metals is shown in Fig. 2.7.

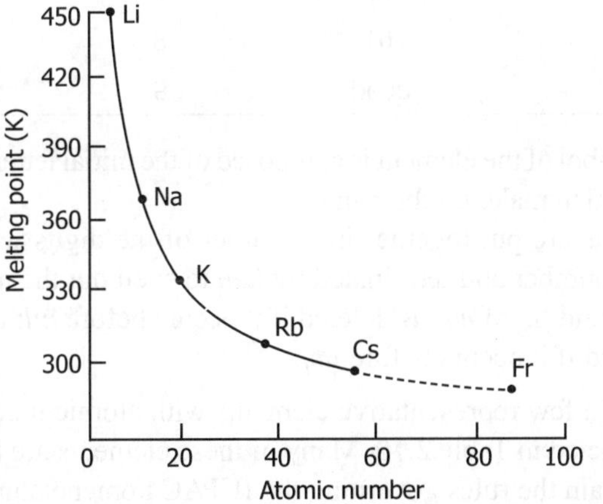

Fig. 2.7 Variation of melting points of alkali metals with atomic numbers

Extrapolation of the curve predicts that the melting point of francium would be nearly 296 K, which would make it a liquid under natural conditions.

Nomenclature of Elements with Atomic Numbers Greater than 100

The elements having atomic numbers greater than 100 are not well characterized. These are radioactive and hence unstable in nature. Although it is feasible to prepare some of these, it is generally not possible to isolate them in weighable quantities.

The nomenclature of these elements, as of the others, are recommended and approved by International Union of Pure and Applied Chemistry (IUPAC). Elements with atomic numbers of 101 to 103 have trivial names, and corresponding two-letter symbols approved by IUPAC. These are, Mendelevium (Md), Nobelium (No) and Lawrencium (Lr), respectively. For naming the subsequent elements, the IUPAC proposed the following rules.

1. The names should be short and simple.
2. The names should end with *ium*.
3. The symbols for the systematically named elements should consist of three letters. This is necessary because any systematically derived set of two-letter symbols will tend to duplicate some of the two-letter symbols of elements of atomic numbers less than 104.
4. The symbols should be derived directly from the atomic numbers using the numerical roots shown in Table 2.9

Table 2.9 Numerical roots used for nomenclature of elements

Numerical	Root	Numerical	Root
0	nil	5	pent
1	un	6	hex
2	bi	7	sept
3	tri	8	oct
4	quad	9	enn

5. The symbol of the element is composed of the initial letters of the numerical roots which make up the name.
6. The roots are put together in the order of the digits which make up the atomic number and terminated by *ium* to spell out the name.
7. The second 'n' of *enn* is deleted if it occurs before *nil*; the 'i' of *bi* and *tri* is deleted if it occurs before *ium*.

The names of a few representative elements with atomic numbers greater than hundred are depicted in Table 2.10. Many of these elements are fictitious, yet they are listed to explain the rules governing the IUPAC nomenclature.

Table 2.10 Nomenclature of some representative elements of atomic numbers more than 100

Atomic number	Name	Symbol
101	Mendelevium	Md
102	Nobelium	No
103	Lawrencium	Lr
104	Unnilquadium	Unq
105	Unnilpentium	Unp
106	Unnilhexium	Unh
107	Unnilseptium	Uns
108	Unniloctium	Uno
109	Unnilennium	Une
110	Ununnilium	Uun
111	Unununium	Uuu
112	Ununbium	Uub
113	Ununtrium	Uut
114	Ununquadium	Uuq
115	Ununpentium	Uup
116	Ununhexium	Uuh
117	Ununseptium	Uus
118	Ununoctium	Uuo
119	Ununennium	Uue
120	Unbinilium	Ubn
121	Unbiunium	Ubu
130	Untrinilium	Utn

3. CHEMICAL BONDING

Introduction

Atoms of most elements, except those of noble gases, have a tendency to combine with each other to form molecules. This combination takes place through redistribution of the valence electrons of the concerned atoms, and is called *chemical bonding*. The forces of attraction, which hold the atoms together in a molecule, are called *chemical bonds.* The formation of chemical bonds is governed by the fact that, in general, a molecule has lower energy than its constituent atoms. This is because each atom in a molecule normally has a stable, noble gas configuration.

There are two major types of chemical bonds: *ionic bonds*, formed by transfer of electrons from one chemical entity to another, and *covalent bonds*, formed by sharing of one or more pairs of electrons between two atoms. These bond types, however, represent two extremes—all bonds have partial ionic and partial covalent characteristics.

In addition, there is *metallic bonding*, which accounts for the strong forces of attraction prevalent in a metal crystal. There are also weak bonds, collectively called *van der Waals forces*, which generally exist between two or more molecules. *Hydrogen bonds* too are weak attractive forces that may exist within a molecule or between the molecules.

Ionic Bond

Ionic bonding results from the transfer of one or more electrons from one atom or group of atoms to another. Such type of bonding occurs most feasibly between metals with low ionization energies and non-metals with high electron affinities.

The formation of sodium chloride (NaCl) may be explained by this transfer mechanism.

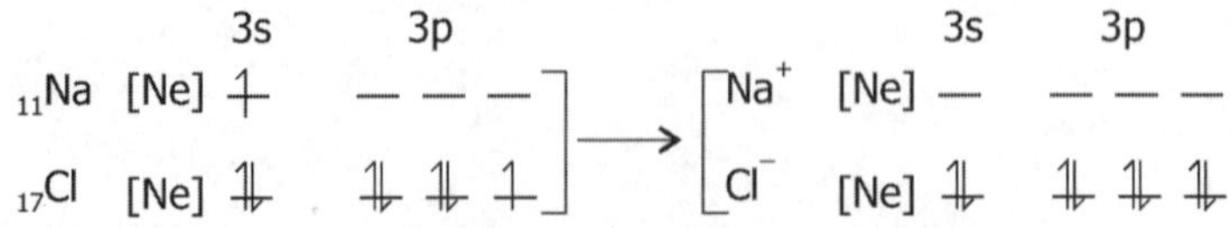

Na atom loses one electron to form sodium ion (Na^+), while Cl atom captures this electron to form chloride ion (Cl^-). The electrostatic attraction between the two oppositely charged ions holds them together in the molecule.

The electrostatic attractions in ionic compounds are governed by Coulomb's law. It states that the force of attraction between two oppositely charged particles (in this case, ions) is directly proportional to the magnitude of charges and inversely proportional to the square of distance between their centers. This implies that greater the charge on the ions and smaller the size of the ions, stronger the ionic bonding. The strong electrostatic forces of attraction among ions in ionic compounds account for their relatively high melting and boiling points. Melting points and boiling points are also related to the coordination number of the crystal structure (*see* the section on radius ratio concept, p. 97). Ionic bonds are non-directional in nature, with each cation being surrounded by a specific number of anions and vice versa. This means that the electrostatic field of ions extends in all directions. In case of NaCl, the ions are arranged in a regular pattern to give a cubic lattice (Fig. 3.1) in which each Na^+ is surrounded by six Cl^-, and each Cl^- is surrounded by six Na^+.

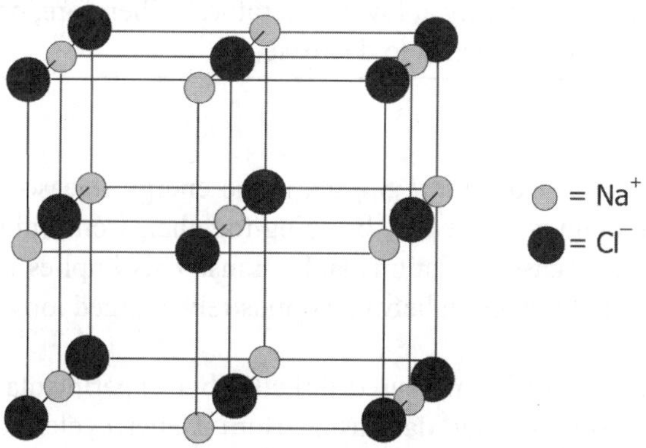

Fig. 3.1 Crystal structure of sodium chloride

In solid state, ionic compounds are poor conductors of electricity. However, in fused state, or in aqueous solution, they become good conductors. Under these conditions, the constituent ions are not restricted to stationary positions, as in a solid lattice. The freely moving ions permit electrical conductance.

Ionic compounds are soluble in polar solvents. A salt dissolves in aqueous medium, since the positive and negative partial charges in a water molecule interact with the anion and cation, respectively in the crystal lattice. As a result, the force of attraction between the cation and the anion is reduced and the compound solubilizes.

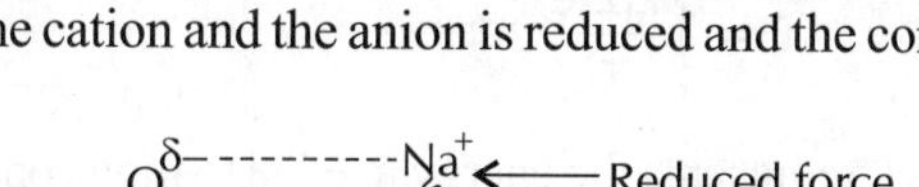

This implies that the hydration energy exceeds the lattice energy (*see* p. 82), causing the compound to dissolve in water.

In case of barium sulfate ($BaSO_4$), the charges on its cation and anion are double than that on sodium and chloride ions. The lattice energy is, therefore, so high that it cannot be exceeded by hydration energy. Hence, $BaSO_4$ is insoluble in water, even though it is ionic in nature.

As a result of strong electrostatic forces of attraction, ionic crystals are quite hard. Hardness increases with decreasing interionic distances and increasing charges on the counter ions. Moreover, ionic crystals are neither malleable nor plastic, since the sliding of one layer of ions over another would bring like charges together. This, in turn, would increase the forces of repulsion within the crystal, resulting in its rupture.

Case Study 3.1 The solubility of sodium sulfate in water increases up to 305 K, and thereafter decreases. Why?

Generally, the solubility of an ionic compound in water increases with an increase in temperature. Therefore, as the temperature increases, the solubility of sodium sulfate also increases. However, this trend continues till 305 K and then reverses.

Dissolution of sodium sulfate in water is an exothermic phenomenon. Exothermic reactions occur more efficiently at low temperatures. Therefore, beyond 305 K, the solubility of sodium sulfate tends to decrease.

Lattice energy

Lattice energy, also called *crystal energy*, is the energy released when one mole of an ionic compound is formed by bringing together, from infinite distance, the necessary number of gaseous cations and anions. This implies that setting up of electrostatic forces of attraction between oppositely charged ions is an exothermic process.

Lattice energy cannot be measured directly, but experimental values may be obtained from thermodynamic data using Born–Haber cycle (*see* next section, p. 84). Theoretically, lattice energy may be calculated by assuming that a cation with charge Z_1e and an anion with charge Z_2 e, separated by a distance r, are attracted in accordance with Coulomb's law. The force of attraction is given by:

$$F = \frac{Z_1 e \, Z_2 e}{r^2} \qquad (3.1)$$

The potential energy of this system may be calculated by integrating equation (3.1) between the limits of r and infinity.

$$E = \int_r^\infty \frac{Z_1 Z_2 e^2}{r^2} \, dr \qquad (3.2)$$

Equation (3.2), in fact, represents the energy needed to move one ion from an infinite distance to a distance r from the second ion. On solving, the value of energy may be expressed as:

$$E = -\frac{Z_1 Z_2 e^2}{r} \qquad (3.3)$$

Equation (3.3) represents the electrostatic energy between one cation and one anion. However, in an ionic crystal, each cation is surrounded by a definite number of anions, and similarly, each anion is surrounded by a fixed number of cations. As a result of so many interactions, the electrostatic energy has a value different than that given by equation (3.3).

This difference may arise both due to attractions and repulsions from the surrounding ions. A particular ion is attracted not only by a single ion of opposite charge, but by a large number of counter ions surrounding it. The sum total of attractive forces is given by a constant called *Madelung constant,* A, which depends upon the number and arrangement of the neighboring ions around a particular counter ion. The repulsive forces in a crystal, on the other hand, originate due to cation–cation and anion–anion interactions. The summation of repulsive forces is equivalent to B/r^n, where B and n are constants, called *repulsion coefficient* and *Born exponent,* respectively.

After incorporating the total attractive and repulsive forces, the lattice energy of an ion pair, as given by equation (3.3), may now be expressed as:

$$U = -\frac{AZ_1Z_2e^2}{r} + \frac{B}{r^n} \tag{3.4}$$

For one mole of the ionic crystal, the total lattice energy, U_o is given by:

$$U_0 = -\frac{NAZ_1Z_2e^2}{r} + \frac{NB}{r^n} \tag{3.5}$$

where N is the Avogadro number.

The equilibrium distance between ions is determined by the balance between the attractive and repulsive forces. For attaining the maximum value of lattice energy, equation (3.5) is differentiated with respect to distance r and then equated to zero

$$\frac{dU_0}{dr} = \frac{NAZ_1Z_2e^2}{r^2} - \frac{nNB}{r^{n+1}} = 0$$

$$B = \frac{AZ_1Z_2e^2r^{n-1}}{n} \tag{3.7}$$

Substituting the value of B from equation (3.7) in equation (3.5).

$$U_0 = -\frac{NAZ_1Z_2e^2}{r}\left(1 - \frac{1}{n}\right) \tag{3.8}$$

Equation (3.8) is called *Born–Landé equation.* According to this equation, when the distance between the ion pairs is small, the lattice energy is correspondingly high. Thus, lithium fluoride (LiF) has a higher lattice energy than lithium iodide (LiI), since the interionic distance in LiI is larger than that in LiF.

Born–Landé equation also predicts that higher the product of charges, greater is the lattice energy. Although LiF and magnesium oxide (MgO) have almost the same interionic distances, the lattice energy of the latter is substantially higher. The charges on magnesium and oxide ions are double than those on lithium and fluoride ions.

Born–Haber cycle

As discussed in the previous section, lattice energy is essentially the energy which holds an ionic crystal together. It is a measure of the stability of an ionic compound. Greater the lattice energy, more is the stability of the electrovalent compound.

Although lattice energy is an important parameter for studying the characteristics of ionic compounds, it is not possible to measure it directly. It can, however, be evaluated by dissecting the heat of formation of a crystal into a number of energy functions and applying the Hess' law of constant heat summation. In this manner, it is possible to devise a cyclic procedure which enables algebraic calculation of lattice energy of a crystal. This cyclic procedure is called Born–Haber cycle.

With reference to NaCl, the Born–Haber cycle may be set out as in Fig. 3.2.

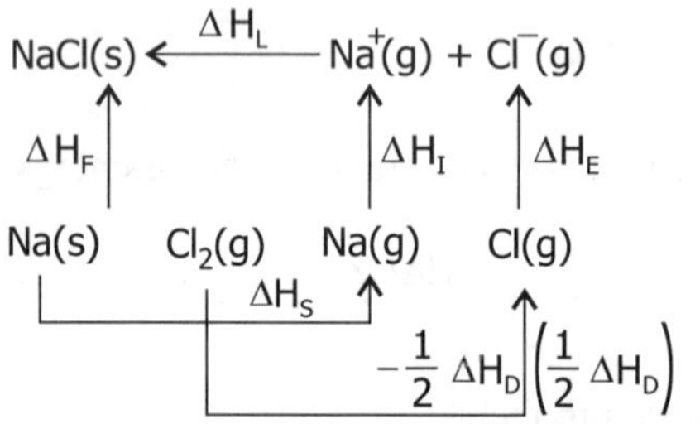

Fig. 3.2 Born–Haber cycle for sodium chloride

The steps constituting the cyclic procedure may be outlined as follows:

Step	Reaction	Enthalpy change
1. Conversion of solid sodium into gaseous sodium atoms (Sublimation energy of sodium)	$Na(s) \rightarrow Na(g)$	$\Delta H_s = 109$ kJ mol^{-1}
2. Conversion of gaseous sodium atoms into gaseous sodium ions (Ionization energy of sodium)	$Na(g) \rightarrow Na^+(g) + e^-$	$\Delta H_I = 496$ kJ mol^{-1}
3. Conversion of molecular gaseous chlorine into gaseous chlorine atoms (Dissociation energy of chlorine)	$½Cl_2(g) \rightarrow Cl(g)$	$½\Delta H_D = 122$ kJ mol^{-1}
4. Conversion of gaseous chlorine atoms into gaseous chloride ions (Electron affinity of chlorine)	$Cl(g) + e^- \rightarrow Cl^-(g)$	$\Delta H_E = -348$ kJ mol^{-1}
5. Conversion of gaseous cations and gaseous anions into solid sodium chloride (Lattice energy of sodium chloride)	$Na^+(g) + Cl^-(g) \rightarrow NaCl(s)$	$\Delta H_L = ?$
6. Formation of solid sodium chloride from constituent atoms in natural states (Heat of formation of sodium chloride)	$Na(s) + ½Cl_2(g) \rightarrow NaCl(s)$	$\Delta H_F = -411$ kJ mol^{-1}

Applying Hess' law,

$$\Delta H_F = \Delta H_S + ½\Delta H_D + \Delta H_I + \Delta H_E + \Delta H_L$$

$$-411 \text{ kJ mol}^{-1} = 109 \text{ kJ mol}^{-1} + 122 \text{ kJ mol}^{-1} + 496 \text{ kJ mol}^{-1} + (-348 \text{ kJ mol}^{-1}) + \Delta H_L$$

$$\Delta H_L = -790 \text{ kJ mol}^{-1} \text{ of sodium chloride}$$

The Kapustinskii equation

The Born–Landé equation (3.8) may be used for computing the lattice energy of an ionic compound provided the value of the Madelung constant, A, is known. As has been stated, Madelung constant depends on the arrangement of cations and anions in the crystal lattice of a compound. This means that Born–Landé equation can be used only if the lattice structure is known beforehand.

To overcome this problem, Kapustinskii modified the Born–Landé equation by multiplying and dividing it by υ, the total number of ions per unit formula of the compound. For example, the value of υ for sodium chloride is 2, for calcium chloride is 3. Thus,

$$U_0 = -\frac{NZ_1Z_2e^2\upsilon}{r}\left(1-\frac{1}{n}\right)\frac{A}{\upsilon}$$

The ratio between Madelung constant and the total number of ions per unit formula, $\frac{A}{\upsilon}$ is a near constant for all ionic lattices. Its commonly accepted value is 0.874. Similarly, the commonly accepted value of the Born exponent, n, is 9. On substituting these values as well as of Avogadro number, N (6.023×10^{23}) and electronic charge (4.8×10^{-10}), the above equation may be simplified to,

$$U_0 = -1.078 \times 10^5 \left(\frac{Z_1Z_2\upsilon}{r}\right) \text{ erg mol}^{-1}$$

$$= -1.078 \times 10^{-5} \left(\frac{Z_1Z_2\upsilon}{r}\right) \text{ kJ mol}^{-1}$$

The above equation is called Kapustinskii equation. It gives a fairly accurate value of lattice energy of an ionic compound even if its crystal structure is not known. Another important application of this equation is to calculate radii of polyatomic anions such as sulphate, carbonate and borate. For example, the lattice energy of potassium perchlorate ($KClO_4$) is known from thermochemical data, and using this value, the radius of the perchlorate anion may be calculated as follows:

$$U_0 = -591 \text{ kJ mol}^{-1}$$

$$\upsilon = 2$$

$$Z_1 = Z_2 = 1 \text{ (unipositive cation and uninegative anion)}$$

Substituting these values in Kapustinskii equation, we get,

$$r = 3.64 \times 10^{-8} \text{ cm or } 364 \text{ pm}$$

Further, $r = r_c + r_a$

The ionic radius of potassium, $r_c = 133$ pm

Therefore, the ionic radius of perchlorate ion, $r_a = 364 - 133 = 231$ pm

The radii calculated in this manner are termed *thermochemical radii.*

Case Study 3.2 The thermochemical radii of potassium and nitrate ions are 137 and 189 pm respectively. Calculate the lattice energy of potassium nitrate using Kapustinskii equation.

Numerical value of charge on potassium ion (Z_1) = 1
Numerical value of charge on nitrate ion (Z_2) = 1
Total number of ions, $\upsilon = 1 + 1 = 2$

Radius of cation, $r_c = 137$ pm
Radius of anion, $r_a = 189$ pm

Inter ionic distance,
$$r = r_c + r_a$$
$$= 137 + 189 \text{ pm}$$
$$= 326 \text{ pm}$$
or 3.26×10^{-8} cm

Substituting these values in Kapustinskii equation, we get,

$$U_o = -1.078 \times 10^{-5}\left(\frac{2}{3.26 \times 10^{-8}}\right)$$
$$= -661.3 \text{ kJ mol}^{-1}$$

The lattice energy of potassium nitrate is -661.3 kJ mol^{-1}

Degree of ionic character

The formation of an ionic compound may be depicted in three steps:

1. The metal atom releases one or more electrons and is converted into a cation. For example,

$$Na(g) \rightarrow Na^+(g) + e^-$$

This is an endothermic process, as energy, in the form of ionization potential, is required to completely remove the electron against the attractive forces of the nucleus.

2. The non-metal atom accepts the released electron(s), and is converted into an anion. For example,

$$Cl(g) + e^- \rightarrow Cl^-(g)$$

This is an exothermic process, as energy, in the form of electron affinity, is released when the non-metal atom acquires a more stable state by attaining the noble gas configuration.

3. The gaseous cations and anions condense to form a solid, ionic crystal. For example,

$$Na^+(g) + Cl^-(g) \rightarrow Na^+Cl^-(s)$$

This is an exothermic process, as energy, in the form of lattice energy, is released during crystal formation.

The degree of ionic character depends upon the amount of energy necessary to remove the electron from the metal atom and the amount of energy released when this electron is added to the non-metal atom. Thus, the degree of ionic character depends upon the difference between the ionization energy of the metal and the electron affinity of the non-metal. When this difference is small, it may be easily compensated by energy released in the form of lattice energy. In such a case, the compound is essentially ionic. However, when this difference is so large that it cannot be compensated by lattice energy, the compound is essentially covalent.

For example, let us compile and compare the ionization energy–electron affinity difference for sodium chloride, beryllium(II) chloride ($BeCl_2$) and boron(III) chloride (BCl_3).

	NaCl	$BeCl_2$	BCl_3
Ionization energy (kJ mol^{-1})	496	1757	3659
Electron affinity (kJ mol^{-1})	348	2 × 348	3 × 348
Difference (kJ mol^{-1})	148	1061	2615
	Ionic	Partially ionic	Covalent

We have seen in the section dealing with Born–Haber cycle that during the formation of one mole of NaCl, energy equivalent to 790 kJ is released as lattice energy. Out of this amount, 148 kJ is utilized to compensate the difference between the ionization energy of sodium and electron affinity of chlorine. The remaining is released as heat. Since the ionization energy–electron affinity difference is duly compensated, therefore, NaCl is an ionic compound.

At the other extreme, BCl_3 is essentially a covalent compound, as the difference between the third ionization energy of boron and electron affinity of three chlorine atoms is too high to be covered by lattice energy. $BeCl_2$ has intermediate character. The difference between the second ionization energy of beryllium and the electron affinity of two chlorine atoms is partially compensated due to lattice build-up. The evidence for the partial ionic character is provided by the fact that $BeCl_2$ is soluble in both water and benzene.

We shall now take the examples of hydrogen chloride (HCl) and aluminium(III) chloride ($AlCl_3$) to study the concept further.

	HCl	$AlCl_3$
Ionization energy (kJ mol^{-1})	1311	2744
Electron affinity (kJ mol^{-1})	348	3 × 348
Difference (kJ mol^{-1})	963	1700
	Covalent	Covalent

The difference in both the cases is too large to be compensated by lattice energy. Hence, HCl in gaseous state and $AlCl_3$ in solid state are both covalent. Nevertheless, when these two compounds are dissolved in water they dissociate into ions showing ionic behavior. In an aqueous solution, solvation energy overcomes the difference between ionization energy and electron affinity. *Solvation energy* is the energy released when one mole of a substance interacts with the solvent (water in this case). Smaller the size of the cation, greater its interaction with the solvent and higher its solvation energy. The hydrogen ion and aluminium(III) ion are very small in size. The high degree of solvation allows them to exist as discrete ions in an aqueous medium.

This concept may be extended for elucidating the degree of ionic character in two or more compounds even in the absence of quantitative data. We know from periodic properties that ionization energy of potassium is less than that of sodium (*see* Chapter 2, p. 62). While comparing the degree of ionic character in NaCl and KCl, we can, therefore, predict that the ionization energy–electron affinity difference shall be less for KCl. This means that KCl has greater ionic character than NaCl.

Similarly, fluorine has higher electron affinity than iodine. Sodium fluoride will, therefore, have smaller ionization energy–electron affinity difference, and greater ionic character, as compared to sodium iodide.

The degree of ionic character may also be predicted by a set of rules called *Fajans' rules*, which are detailed in the next section. However, whether one follows the concept of ionization energy–electron affinity difference or Fajans' rules, the end results are the same.

Fajans' rules

Ionic and covalent bondings represent two extreme bond forms. Most compounds are neither completely ionic nor completely covalent, and possess intermediate bond types. This transitional behavior may be explained in terms of polarizing effects of the ions.

Suppose two oppositely charged ions, A^+ and B^- are approaching each other from a large distance of separation. As these reach an equilibrium distance, the cation tends to distort the electron cloud of the anion. Two factors contribute to this distortion. First, if cation is small, the nuclear charge of A^+ can draw the electron cloud of B^- towards itself. If A^+ is large, it finds it difficult to reach out to the valence electrons of B^- (Fig. 3.3).

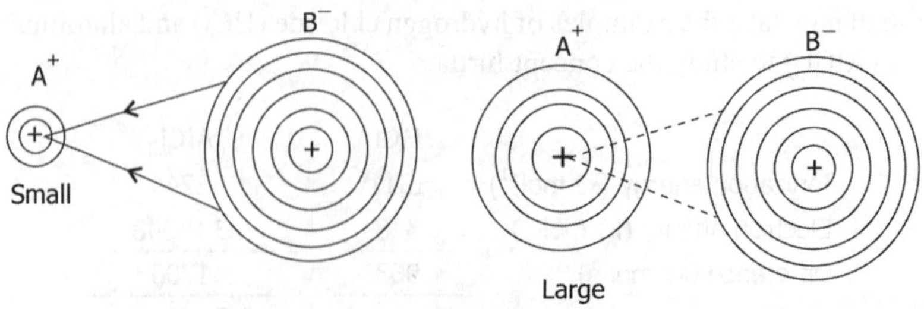

Fig. 3.3 A small cation has higher polarizing power than a large cation

Second, if an anion is large, the outermost electrons of the anion are far away from the nucleus and, therefore, susceptible to be pulled towards the nucleus of the cation. However, if the anion is small, its nucleus will hold the electron cloud tightly and will not permit the nucleus of the cation to deform it (Fig. 3.4).

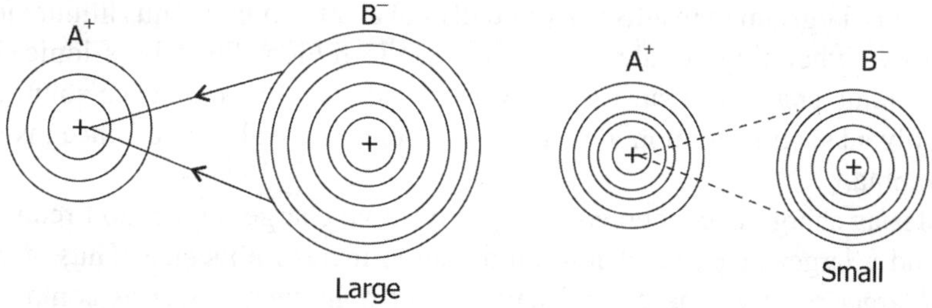

Fig. 3.4 A large anion has higher polarizability than a small anion

Thus, a small cation can perturb the valence electron cloud of a large anion. With respect to the cation, this deformation is called *polarizing power*, while with respect to the anion, it is called *polarizability*. When the polarizing power of a cation is large and so is the polarizability of the anion, the electrons of the anion are drawn towards the cation. The electrons would not completely leave B^- and get transferred to A^+. In such a case, there would be no bond formation between A and B. Nevertheless, the electron cloud of the anion shall be pulled to the middle of A and B. The presence of electron cloud in between A and B, even though to a meager extent, implies sharing of electron density, i.e., covalent character (Fig. 3.5).

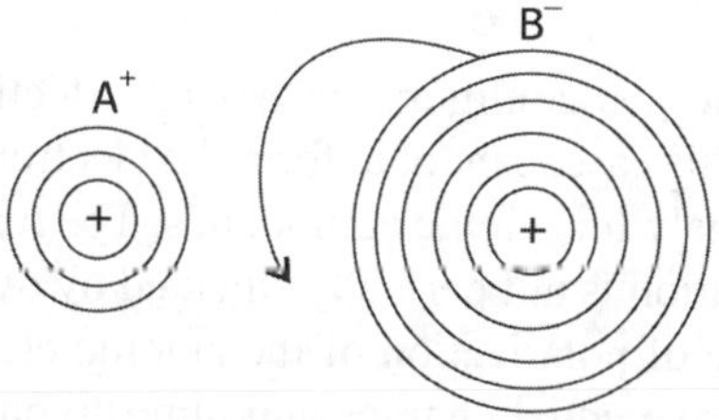

Fig. 3.5 When a cation has high polarizing power and an anion is highly polarizable, the electron density of the anion is drawn in between the two ions, resulting in covalent character

Fajans' rules qualitatively assess the extent of polarization occuring in an ion pair, thus analyzing the extent of ionic or covalent character of the resulting bond type. The conditions for predicting the bond types are given in Table 3.1.

Table 3.1 Conditions for predicting bond types

Ionic	Covalent
1. Large cation	1. Small cation
2. Small anion	2. Large anion
3. Low charge on the cation and/or anion	3. High charge on the cation and/or anion
4. Cation has noble gas configuration	4. Cation does not have noble gas configuration

We have already explained that a small cation possesses high polarizing power and a large anion possesses high polarizability. Hence, a combination of a small cation and a large anion results in essentially covalent bonding. Thus lithium iodide has an overwhelming covalent character, while lithium fluoride is ionic. NaCl has less ionic character than KCl. Silver chloride is ionic and hence soluble in a polar solvent such as ammonia, while silver iodide is covalent and hence insoluble in ammonia.

As far as charge is concerned, a larger positive charge on a cation reduces its size, and a larger negative charge on an anion increases its size. Thus, a cation with a larger positive charge will have higher polarizing power as compared to a cation with a smaller positive charge. Similarly, an anion with a large negative charge shall possess high polarizability as compared to its congener with a small negative charge. Moreover, if the charges on counter ions are large, then according to Coulomb's law, there will be a greater degree of attraction between the nucleus of the cation and the electron cloud of the anion. This would result in a greater degree of polarization and hence covalency.

Tin(II) ion has a smaller charge and larger size as compared to tin(IV) ion. Thus, tin(II) chloride is expected to show a greater degree of ionic character as compared to tin(IV) chloride. This is borne out of the observation that tin(II) chloride is a solid, with relatively high melting point and boiling point. On the other hand, tin(IV) chloride is a liquid in its natural state and soluble in organic solvents. Likewise, manganese(II) oxide is an ionic solid, while manganese(VII) oxide is a covalent liquid.

A cation with a noble gas configuration has an effectively shielded nucleus. The nuclear charge, in this case, cannot deform the electron cloud very efficiently. Hence, such cations largely form ionic compounds. The nucleus of a cation with a non-noble gas configuration is not perfectly shielded by overlying orbitals. It thus causes a greater degree of polarization of the anionic electron cloud. Potassium ion and copper(I) ion have equal charges and almost equal sizes. Yet potassium chloride is soluble in water, while copper(I) chloride is not. This is because potassium ion has a noble gas configuration and hence its compounds show typical ionic behavior. The copper(I) ion, on the other hand, has a non-noble gas configuration and, therefore, a high polarizing power. Its compounds are more on the covalent side.

Case Study 3.3 How does lattice energy reflect the degree of ionic character of a compound?

The agreement between the experimental lattice energy and the value calculated using Born-Haber cycle reflects the extent to which a compound is ionic in nature. Closer the values of experimental and calculated values of lattice energy, greater is the ionic character of the compound.

For example, consider the values of lattice energy for lithium fluoride and lithium iodide.

Compound	**Lithium fluoride**	**Lithium iodide**
Calculated lattice energy (kJ mol^{-1})	1029	730
Experimental lattice energy (kJ mol^{-1})	1030	757

According to Fajans' rules, lithium fluoride is a truly ionic compound, while lithium iodide has appreciable covalent character. The calculated and experimental values of lattice energy of the former are in close agreement while for the latter these values are widely divergent.

Stoichiometric defects in ionic crystals

Stoichiometric compounds are those ionic compounds in which the number of cations and anions are present in a fixed ratio. For example, in a crystal of the type A^+B^-, the number of cations is exactly equal to the number of anions. Moreover, theoretically each ion is surrounded by a three-dimensional pattern of a specific number of counter ions. No site which should be occupied by a cation or an anion is vacant. However, such an ideal arrangement exists only at absolute zero. As the temperature increases, thermal agitations push a few ions out of the lattice. Some sites become unoccupied, so that there are either cation vacancies, or anion vacancies, or both. On increasing the temperature, the number of such vacancies increases. At room temperature, the crystal array is no longer regular and the structure becomes defective. Two types of defects exist in stoichiometric ionic crystals: Schottky defects and Frenkel defects.

1. **Schottky defects**

   A Schottky defect consists of an equal number of vacant cation sites and anion sites (Fig. 3.6). This means that in a crystal which contains Schottky defects, both the cation and the anion lattices are incomplete.

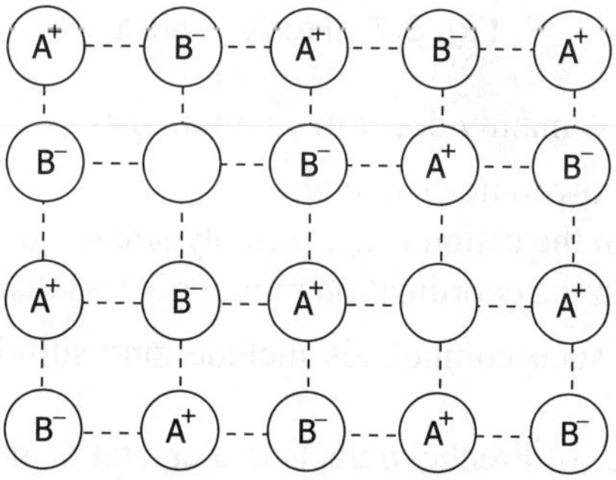

Fig. 3.6 Schottky defect

Schottky defects mainly occur in compounds—

- having appreciably high ionic character.
- in which the size of cation is comparable to that of anion.
- in which the coordination number is greater than six.

Examples of such compounds include sodium chloride, potassium bromide and cesium chloride.

The number of Schottky defects in a crystal is given by the expression,

$$n_s = Ne^{-Ws/2RT} \tag{3.9}$$

where n_s is the number of Schottky defects in 1 cm^3 of the crystal at temperature T (Kelvin), N is the number of ion pairs per cm^3 of an ideal crystal containing no defects, W_s is the work required to create a Schottky defect and R is the gas constant.

2. Frenkel defects

A Frenkel defect consists of a vacant lattice site, with the ion, which ideally should have occupied that position, getting accommodated in an interstitial location (Fig. 3.7). Both cations and anions can migrate to the interstices. However, since cations are smaller in size, they can be more easily squeezed into an alternate site. Thus, most of the time, crystals showing Frenkel defects have incomplete cation lattice. A cation at an interstitial position comes very close to other cations. The closeness of similar charges increases the dielectric constant of the crystal.

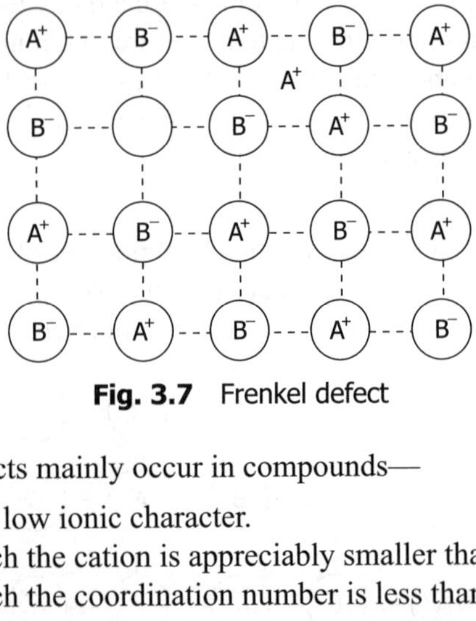

Fig. 3.7 Frenkel defect

Frenkel defects mainly occur in compounds—

- having low ionic character.
- in which the cation is appreciably smaller than the anion.
- in which the coordination number is less than six.

Examples of such compounds include zinc sulfide, silver bromide and silver iodide.

The number of Frenkel defects in a crystal is given by the expression:

$$n_f = \sqrt{NN^1} e^{-W_f/2RT} \quad (3.10)$$

where n_f is the number of Frenkel defects in 1 cm^3 of the crystal, N^1 is the number of interstitial sites that are available in 1 cm^3 of the crystal, W_f is the work required to create a Frenkel defect. Rest of the notations represent the same parameters as in equation (3.9).

Case Study 3.4 If the work required to create Schottky defects in a crystal of sodium chloride is 167 kJ mol^{-1}, find out the number of such defects at 773 K (density of sodium chloride = 2.17 g cm^{-3}).

$$W_s = 167 \text{ kJ mol}^{-1} = 1{,}67{,}000 \text{ J mol}^{-1}$$

$$R = 8.3143 \text{ J deg}^{-1}\text{mol}^{-1}$$
$$T = 773 \text{ K}$$

Gram molecular weight of sodium chloride = 58.5 g

Density of sodium chloride = 2.17 g cm^{-3}

Molecular volume of sodium chloride = $\frac{58.5}{2.17} = 26.95 \text{ cm}^3$

A perfect crystal of sodium chloride having a molecular volume of 26.95 cm^3 contains Avogadro number (6.023×10^{23}) of sites for each ion pair.

Number of sites for ion pairs per cm^3 of an ideal crystal,

$$N = \frac{6.023 \times 10^{23}}{26.95} = 2.23 \times 10^{22}$$

$$n_s = (2.23 \times 10^{23})\, e^{\frac{-167000}{2 \times 8.3143 \times 773}}$$

$$n_s = 5 \times 10^{16}$$

There are 5×10^{16} Schottky defects in 1 cm^3 crystal of sodium chloride.

Consequences of stoichiometric defects

Theoretically, a crystal can have both Schottky and Frenkel defects, but generally only one of them predominates. The concentration of the two defects will be the same only if the energies required to create them are comparable, and this is rarely so. Normally, the energy required to create Schottky defects is less than that required to create Frenkel defects.

The presence of defects in a crystal alters some of its physical properties. When the density of a crystal containing Schottky defects is measured, it is lower than expected. The reason is that due to missing cations and anions, the mass in a unit volume of the crystal decreases. In crystals with Frenkel defects, however, the ions do not go missing. They just occupy alternative positions. Hence, the density does not change. The vacant sites lower the lattice energy, and hence the stability of the crystal. In extreme cases, where there are large number of vacant sites, the lattice may partially collapse.

Ionic crystals having stoichiometric defects conduct small amount of electricity by an ionic mechanism. As one ion moves into an adjacent vacant site, it vacates its own site, to which another ion migrates. If this process continues, the vacant site migrates from one end of the crystal to the other, which is equivalent to movement of an ion in the opposite direction. This type of conduction is called *intrinsic semiconduction.* A crystal containing Frenkel defects has vacancies only at the cation sites. Intrinsic semiconduction, therefore, involves movement of cations alone. A crystal containing Schottky defects, on the other hand, has vacancies at both the cation and the anion sites. Intrinsic semiconduction, in this case, involves migration of both types of ions throughout the crystal lattice.

Non-stoichiometric defects in ionic crystals

Non-stoichiometric compounds are those ionic compounds in which the ratio of the number of atoms of one kind to the number of atoms of another kind does not correspond to the ideal whole number suggested by the molecular formula. Such compounds do not obey the law of constant composition. Rather, they exist over a range of chemical compositions. For example, iron(II) sulfide which is normally written as FeS is actually a non-stoichiometric compound—its composition being close to $Fe_{0.9}S$.

Such an imbalance of composition originates if the crystal structure possesses defects. These defects are called *non-stoichiometric defects* and occur in addition to the stoichiometric defects. These are similar to stoichiometric defects in the sense that here too there are vacant sites as well as ions in interstitial positions.

Non-stoichiometric defects fall into two broad types: (i) those in which there is a metal excess, and (ii) those in which there is a metal deficiency. Of course, electrical neutrality is maintained by either having extra electrons in the lattice, or by the presence of a few metal ions in variable valency.

1. **Metal excess due to anion vacancies**

   Metal excess in an ionic crystal may arise when an anion leaves the lattice and its position is occupied by an extra electron to maintain the electrical balance (Fig. 3.8). Non-stoichiometric defect of this type originates in crystals which would have Schottky, rather than Frenkel defects. For example, when NaCl is heated with sodium vapors, its lattice loses some of the chloride ions, as a result of which its composition may be represented as $NaCl_{1-\delta}$, where δ is a small fraction. Non-stoichiometric NaCl is yellow in color, while non-stoichiometric KCl is lilac. Anionic sites occupied by electrons are referred to as *F-centers*. The higher the number of F-centers in a crystal, the greater the intensity of its color.

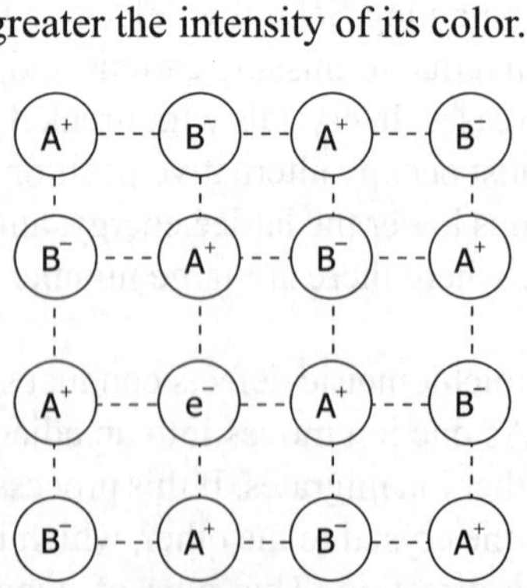

Fig. 3.8 Metal excess defect due to anion vacancies

2. **Metal excess due to interstitial cations**

   Metal excess defects may also occur when an extra cation occupies an interstitial position and, to maintain electrical balance, an electron that was previously associated with the anion, occupies another interstitial position (Fig. 3.9). Non-stoichiometric defect of this type originates in crystals which have Frenkel, rather than Schottky defect. Examples include zinc oxide and cadmium oxide, the composition of which may be represented as $Zn_{1+\delta}O$ and $Cd_{1+\delta}O$, respectively, where δ is a small fraction.

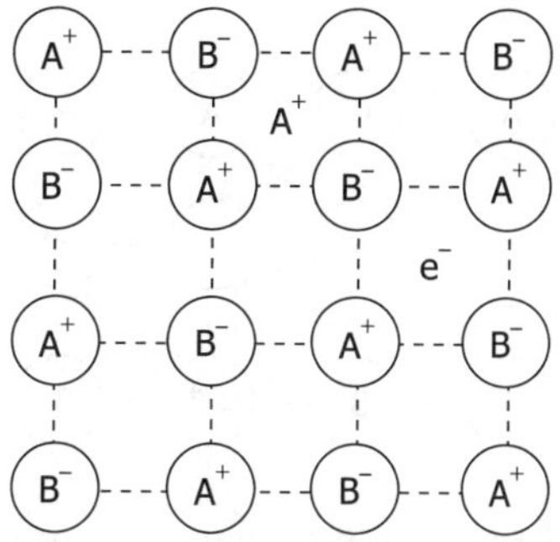

Fig. 3.9 Metal excess defect due to interstitial cations

3. **Metal deficiency due to cation vacancies**

In this case, the lattice acquires additional atoms of the non-metal, which gain electrons to become anions. These electrons are obtained from the metal ions in the lattice which, in turn, are oxidized to a higher valence state. Figure 3.10 shows a pictorial representation of such a lattice. The number of anions in the lattice is increased without a concomitant increase in the number of cations. The crystal, therefore, possesses a complete anion lattice but an incomplete cation lattice. This type of defects are possible in transition metal salts, since these metals show variable valency. Iron(II) oxide and iron(II) sulfide are examples of compounds which show metal deficiency defects involving cationic vacancies. Their compositions may be represented as $Fe_{1-\delta}O$ and $Fe_{1-\delta}S$, respectively, δ being a small fraction.

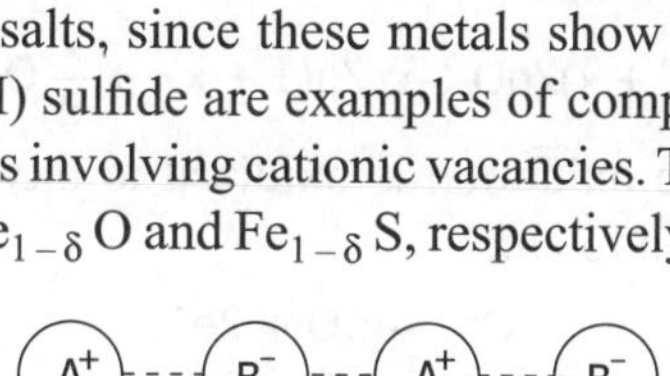

Fig. 3.10 Metal deficiency defect due to cation vacancies

4. **Metal deficiency due to interstital anions**

In this case, the lattice acquires additional atoms of non-metals which are transformed into anions by gaining electrons as a result of oxidation of some of the metal ions. The additional anions occupy interstitial positions (Fig. 3.11). The crystal thus has complete cation and anion lattices, with no vacant sites to compensate for interstitial anions. However, as anions are larger in size than the cations, it is difficult to squeeze them at interstitial positions. Due to this reason, metal deficiency defects due to interstitial anions serve only academic interest.

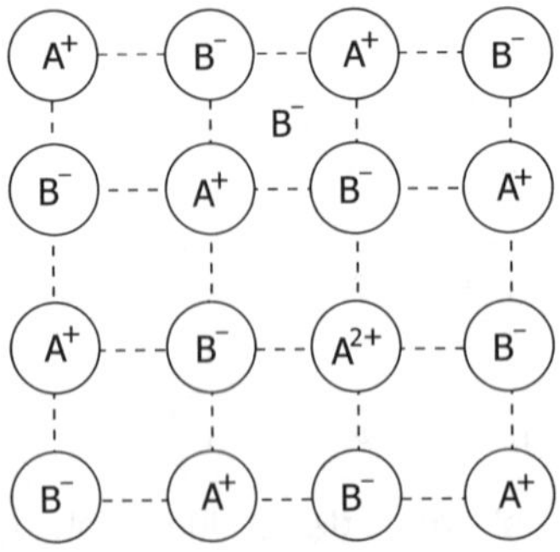

Fig. 3.11 Metal deficiency defect due to interstitial anions

Case Study 3.5 Zinc oxide is yellow when hot and white when cold. Explain.

The stoichiometry of zinc oxide is dependent on the vapor pressure of oxygen over the solid phase. Under conditions of normal temperature and pressure, it has a 1:1 stoichiometric ratio between the cation and the anion. However, on heating, the vapor pressure of oxygen over the solid phase decreases and zinc oxide begins to lose oxygen.

$$(1 + x)ZnO \rightarrow Zn(1 + x) + \frac{x}{2} O_2$$

In this transformation, the oxide ion is oxidized to oxygen and the electrons so released reduce zinc(II) ion to zinc.

$$O^{2-} \rightarrow O + 2e^-$$
$$O + O \rightarrow O_2$$
$$Zn^{2+} + 2e^- \rightarrow Zn$$

As the oxygen "boils off", zinc oxide acquires a non-stoichiometric composition. The excess zinc atoms now occupy the interstitial sites. At 1100 K, the excess of zinc over oxygen is only 7×10^{-3}%. The non-stoichiometric zinc oxide is yellow in color.

On cooling, the vapor pressure of oxygen over the solid phase increases. The molecular oxygen is reduced to oxide ions,

$$O_2 + 4e^- \rightarrow 2O^{2-}$$

The electrons for reduction are provided by zinc atoms housed in the interstitial sites.

$$Zn \rightarrow Zn^{2+} + 2e^-$$

The zinc(II) ions come to their normal positions in the lattice, the stoichiometry is restored, and the color changes to white.

The reversible change of color which occurs with respect to temperature is called *thermochromism*.

Consequences of non-stoichiometric defects

The metal excess types of non-stoichiometric compounds have free electrons trapped in the crystal lattice (Figs. 3.8 and 3.9). These electrons are not static, but can traverse throughout the crystal. When the electrons move from one position to another, they conduct electricity. However, the amount of current is very small, since the number of defects, and hence the number of free electrons, in a lattice is quite small. For this reason, such defect materials are called *semiconductors*. Further, the current is being carried by electrons, and not by ions—the normal mode of electrical conduction. The metal excess non-stoichiometric compounds are thus designated as *n-type semiconductors,* i.e., those conductors which carry a small amount of current by normal electron conduction mechanism.

The metal deficient non-stoichiometric compounds are also semiconductors, but their mechanism of conduction is different from that of metal excess compounds. These crystals contain metal ions in different oxidation states (Figs. 3.10 and 3.11). Under a potential difference, an electron can shift from a metal ion in a lower valence state to that in a higher valence state. As a result, the electrons move throughout the lattice and conduct a small amount of current. As the electron moves in one direction, the higher valent positive ion, called a *positive hole*, appears to be moving in the opposite direction, although, in reality, the cation does not move at all. The metal deficient non-stoichiometric compounds are thus designated as *p-type semiconductors, i.e.,* those conductors which carry a small amount of current by positive hole conduction mechanism.

Case Study 3.6 What thermodynamic considerations are involved in creation of stoichiometric defects?

Stoichiometric defects involve an interplay of two thermodynamic parameters—enthalpy and entropy. When a crystal is formed from gaseous ions, a large amount of heat is released as enthalpy of formation. Simultaneously, however, a large unfavorable entropy change occurs, because the randomly moving cations and anions are now held in a rigid, well-ordered crystal lattice. The introduction of defects in the lattice reduces the enthalpy of formation, but increases the entropy favorably, since the crystal becomes less ordered. The balance between enthalpy and entropy determines the concentration of defects.

Thermodynamic considerations are, therefore, important in creating stoichiometric defects. For this reason, these defects are often called *thermodynamic defects.*

Radius ratio concept

In their solid states, ionic compounds have a close-packed arrangement of cations and anions. The positive and negative ions are held together by strong electrostatic forces of attraction. However, within the same lattice, repulsive forces among like-charged ions also exist. As a result, a structure finally prevails in which the forces of attraction dominate over the forces of repulsion. This, in turn, becomes possible if in the crystal lattice, a particular ion is surrounded by a large number of oppositely charged ions. The number of counter ions surrounding a specific ion is called *coordination number.* In general, higher the coordination number, greater the stability of the ionic compound.

The structure adopted by an ionic crystal depends on the coordination number, which, in turn, depends on the ratio of the radius of the cation (r_c) to the radius of the anion (r_a). Thus the radius ratio concept, also called the *radius ratio rule*, is a guide to determine the most stable arrangement of cations and anions in the crystal lattice of an ionic solid.

For a particular coordination number, the ratio r_c/r_a is not fixed. Rather, there is a range within which that coordination number is adopted by the ions. Below that range and above that range, a different coordination number, and hence a different geometry, exists. The range of radius ratio, representing a particular coordination number is called *limiting radius ratio.* Table 3.2 lists the limiting radius ratios for different coordination numbers.

Table 3.2 Limiting radius ratios for various coordination numbers

Coordination number	Limiting radius ratio	Geometry	Example
3	0.155–0.225	Plane triangle	Boron(III) oxide
4	0.225–0.414	Tetrahedral	Zinc sulfide
6	0.414–0.732	Octahedral	Sodium chloride
8	0.732–0.999	Cubic	Cesium chloride

The reason why the limiting radius ratios have a range of values will be evident if we consider, as a representative case, coordination number three geometry. Since the cations are invariably smaller than the anions, coordination number three implies that a small positive ion, A^+ is surrounded by three negative ions, X^-. At one extreme, this type of geometry may be depicted in Fig. 3.12a, where the positive ion is touching all three negative ions, but the negative ions are not touching each other. This is a very favorable arrangement. There are only forces of attraction between the cation and the anions, but there are no repulsive forces among the anions.

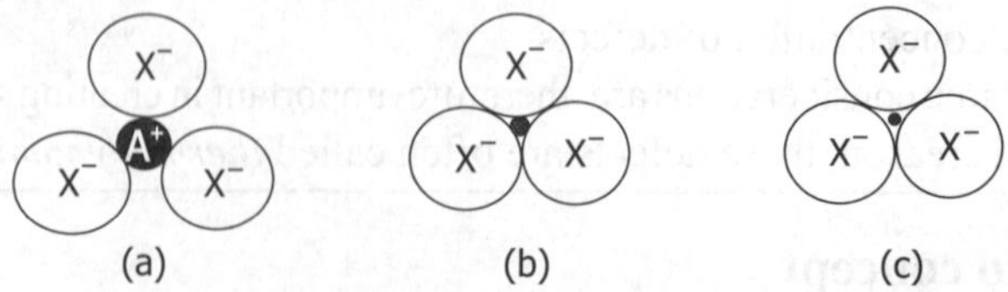

Fig. 3.12 Possible arrangements for plane triangle geometry. (a) Cation in contact with three anions, (b) cation in contact with three anions, and three anions in contact with each other, and (c) cation not in contact with anions, but three anions in contact with each other.

At the other extreme, the cation A^+ becomes smaller, so that it still touches the three anions, X^-. However, now these three anions also touch each other (Fig. 3.12b). This situation is unfavorable, as in addition to the forces of attraction between the cation and the anions, there are also forces of repulsion among the anions. The radius ratio, r_c/r_a for the arrangement in Fig. 3.12b is less than that for the arrangement in Fig. 3.12a, since the size of the cation is reduced. If we lower the radius ratio still further, either by decreasing the size of the cation or by increasing the size of the anions, the forces of repulsion will dominate over the forces of attraction. This pattern is shown in Fig. 3.12c, where the cation is no longer in contact with the anions, but the anions are still touching each other.

Thus, the model shown in Fig. 3.12b represents a limiting case of radius ratio. If the ratio is lowered below the value represented by this model, the coordination number as well as the geometry will change. A linear geometry with coordination number two would become more favorable. On the other hand, if we try to increase the radius ratio beyond the value represented by the arrangement 3.12a either by increasing the size of cation or by decreasing the size of anions, the coordination number and geometry would change once again. Obviously, a larger cation prefers to have a higher coordination number since that will impart more stability to the crystal lattice. Therefore, the cation A^+ would prefer to have a tetrahedral arrangement of four X^- anions around it, rather than a plane triangular arrangement of three anions.

The arrangement of ions shown in Fig. 3.12a thus represents the higher extreme in the range of radius ratio values, while the arrangement shown in Fig. 3.12b represents the lower extreme. The limiting radius ratio values are obtained from geometrical calculations.

At times, the structure of a compound does not comply with radius ratio calculations. For example, potassium fluoride possesses rock salt (i.e., sodium chloride) structure, yet its radius ratio, 0.73, is same as that of cesium chloride structure. Such discrepancies arise because while calculating the radius ratio, it is assumed that the ions are rigid spheres, which actually they are not. We now take up the geometrical calculations for various coordination numbers listed in Table 3.2.

1. **Coordination number 3 (plane triangle)**

   The plane triangle arrangement is shown in Fig. 3.13, where a smaller cation with radius r_c is in contact with three larger anions, each with radius r_a. In equilateral Δ ABC,

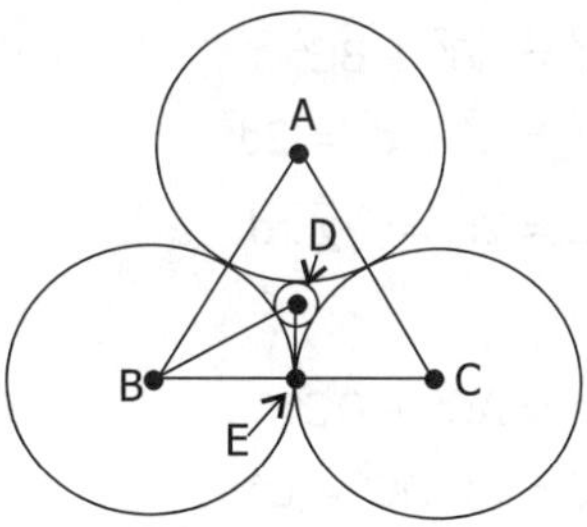

Fig. 3.13 Calculation of limiting radius ratio for coordination number 3

$$AB = BC = AC = 2r_a$$

$$BE = r_a \quad (3.11)$$

$$BD = r_c + r_a \quad (3.12)$$

$$\angle ABC = 60^o$$

$$\angle DBE = 30^o$$

$$\cos 30^o = \frac{BE}{BD} \quad (3.13)$$

Substituting the value of BE and BD from equations (3.11) and (3.12), respectively in equation (3.13),

$$0.866 = \frac{r_a}{r_c + r_a} \qquad (\cos 30^o = 0.866)$$

or
$$\frac{r_c + r_a}{r_a} = \frac{1}{0.866}$$

$$\frac{r_c}{r_a} = \frac{1}{0.866} - 1 = 0.155$$

The limiting radius ratio for plane triangle geometry is 0.155.

2. **Coordination number 4 (tetrahedral)**

The tetrahedral arrangement of four anions around a cation, inscribed in a cube of length d, is shown in Fig. 3.14.

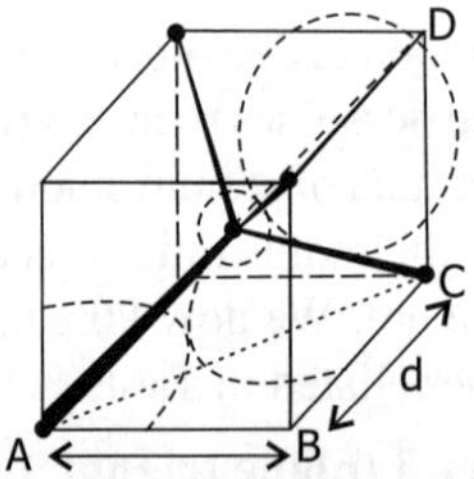

Fig. 3.14 Calculation of limiting radius ratio for coordination number 4

The diagonal AC, on the bottom face = $2r_a$

In ΔABC,

$$AC^2 = AB^2 + BC^2$$

$$AC^2 = d^2 + d^2 = 2d^2 \quad (3.14)$$

$$AC = 2r_a = \sqrt{2} \cdot d \quad (3.15)$$

In ΔADC,

$$AD^2 = AC^2 + DC^2 \quad (3.16)$$

$$DC = d; \; DC^2 = d^2$$

Substituting the value of AC^2 from equation (3.14) into equation (3.16),

$$AD^2 = 2d^2 + d^2 = 3d^2$$

$$AD = \sqrt{3} \cdot d$$

Further, $AD = 2r_c + 2r_a$

Therefore, $2r_c + 2r_a = \sqrt{3} \cdot d$

$$2r_c = \sqrt{3} \cdot d - 2r_a \quad (3.17)$$

Substituting the value of $2r_a$ from equation (3.15) into equation (3.17),

$$2r_c = \sqrt{3} \cdot d - \sqrt{2} \cdot d \quad (3.18)$$

Dividing equation (3.18) by equation (3.15),

$$\frac{r_c}{r_a} = \frac{\sqrt{3}d - \sqrt{2}d}{\sqrt{2}d} = 0.225$$

The limiting radius ratio for tetrahedral geometry is 0.225.

3. **Coordination number 6 (octahedral)**
In octahedral geometry, a cation is in contact with six anions. Four of these are in the plane of the cation, while the remaining two are above and below this plane. A cross-section of octahedral coordination is shown in Fig. 3.15. Only four planar anions are shown in this figure, as the out-of-plane anions have no role in the calculation of limiting radius ratio.

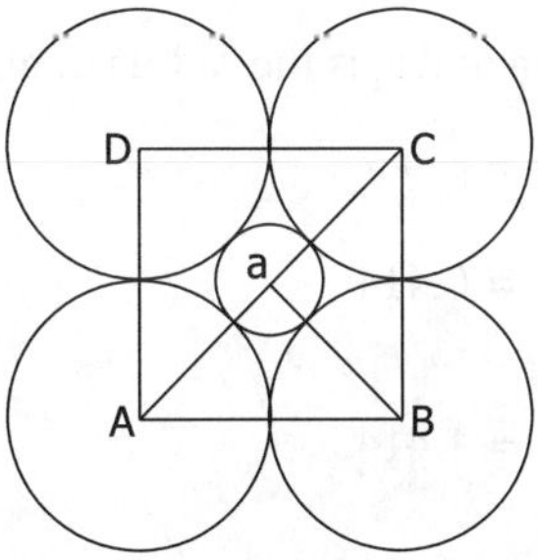

Fig. 3.15 Calulation of limiting radius ratio for coordination number 6

If the two out-of-plane anions are removed to infinite distance from the cation, the resulting geometry will be square planar. Strictly, Fig. 3.15 represents square planar arrangement because, as we have already stated, we have ignored the anions which are above and below the plane of the cation. Therefore, the limiting radius ratio value for square planar geometry is same as that for octahedral geometry.

$$AB = BC = 2r_a \quad (3.19)$$

$$AC = 2r_c + 2r_a \quad (3.20)$$

In ΔABC,

$$AC^2 = AB^2 + BC^2 \quad (3.21)$$

From equations (3.19), (3.20) and (3.21),

$$(2r_c + 2r_a)^2 = (2r_a)^2 + (2r_a)^2 = 8r_a^2$$

$$2r_c + 2r_a = 2\sqrt{2}\, r_a$$

$$r_c + r_a = \sqrt{2}\, r_a$$

$$\frac{r_c}{r_a} + 1 = \sqrt{2}$$

$$\frac{r_c}{r_a} = 0.414$$

The limiting radius ratio for octahedral as well as for square planar geometries is 0.414.

Case Study 3.7 The interionic distance in lithium iodide and potassium iodide is 300 pm and 353 pm respectively. Calculate the ionic radii of iodide and potassium ions.

In lithium iodide, the small lithium ion is coordinated to four large iodide ions in a square planar arrangement. Since the four iodide ions, being large, are in contact with each other, we may apply the radius ratio concept valid for coordination number 6 (ignoring the two out-of-plane anions).

For such an arrangement,

$$\frac{r_c}{r_a} = 0.414$$

where r_c is the radius of cation and r_a is the radius of anion. The interionic distance,

$$d = r_c + r_a$$

Hence, $$\frac{d - r_a}{r_a} = 0.414$$

$$\frac{d}{r_a} = 1.414$$

For lithium iodide, d = 300 pm

$$r_a = \frac{300}{1.414} = 212 \text{ pm}$$

In potassium iodide, the size of cation is quite large as compared to that in lithium iodide. As a result, there is no anion-anion interaction. We may, therefore, consider the interionic distance as arising from the combination of a single cation and a single anion.

$$r_{kI} = 353 \text{ pm}$$

$$r_{I^-} = 212 \text{ pm}$$

$$r_{k^+} = r_{kI} - r_{I^-}$$

$$= 353 - 212 = 141 \text{ pm}$$

The ionic radius of potassium ion is 141 pm, while that of iodide ion is 212 pm.

---

4. Coordination number 8 (cubic)

In the cubic geometry, each of the eight corners of the cube represents the center of an anion. However, only three anions, at positions A, B and D, are shown in Fig. 3.16 to make the diagram simple and clear.

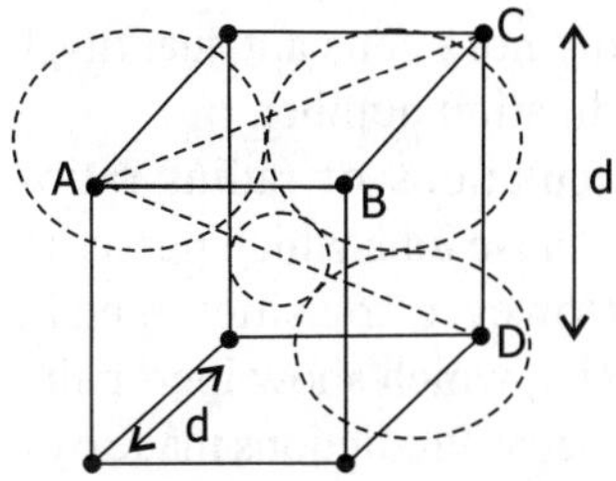

Fig. 3.16 Calculation of limiting radius ratio for coordination number 8

$$AB = DC = 2r_a = d \tag{3.22}$$

$$r_a = \frac{d}{2}$$

$$AD = 2r_c + 2r_a \tag{3.23}$$

In ΔABC,

$$AC^2 = AB^2 + BC^2$$

$$AC^2 = d^2 + d^2 = 2d^2 \tag{3.24}$$

In ΔADC,

$$AD^2 = AC^2 + DC^2 \tag{3.25}$$

Putting the value of AC^2 from equation (3.24) into equation (3.25),

$$AD^2 = 2d^2 + d^2 = 3d^2$$

$$AD = \sqrt{3}.d \tag{3.26}$$

Substituting the value of AD from equation (3.23) into equation (3.26),

$$2r_c + 2r_a = \sqrt{3}\,.\,d$$

$$r_c + r_a = \frac{\sqrt{3}}{2}.d \tag{3.27}$$

Dividing equation (3.27) by equation (3.22),

$$\frac{r_c + r_a}{r_a} = \frac{\sqrt{3}}{2}.d.\frac{2}{d}$$

$$\frac{r_c}{r_a} + 1 = \sqrt{3}$$

$$\frac{r_c}{r_a} = 0.732$$

The limiting radius ratio for cubic geometry is 0.732.

Limitations of radius ratio concept

The limiting radius ratio method is a good guide for predicting the plausible geometries of ionic crystals, but it is not an altogether reliable concept. The following parameters dilute the significance of radius ratio approach:

1. The radius of an ion is not a constant quantity. It varies with several factors, including the coordination number.
2. The geometric calculations of radius ratios assume that the ions are spherical in shape. Those ions which possess noble gas configurations are indeed spherical. However, transition metal ions or metal ions, such as thallium(I) and tin(II), which show inert pair effect, occassionally have a non-spherical symmetry. Predictions made by radius ratio rules often fail in such cases.
3. In radius ratio calculations, the ions are supposedly non-elastic in nature. Ideally, they should merely touch similar or counter ions in the lattice. However, in real terms, a close-packed structure involves a certain degree of compression of ions against each other, as a consequence of which the lattice gets distorted.
4. Radius ratios are valid if the forces of attraction between the ions are exclusively electrostatic in nature. This may be true in case of compounds such as sodium chloride or cesium chloride. However, as we have seen in the section on Fajans' rule (*see* p. 88), compounds such as zinc sulfide, which are considered ionic in nature, do possess some degree of covalent character as well. As the covalent character of the compound increases, the predictions made by radius ratio calculations become less accurate.

Covalent Bond

A covalent bond is formed when two atoms share one or more pairs of electrons. For example, in the formation of chlorine molecule, Cl_2, each constituent atom parts away with one of the seven valence electrons for sharing.

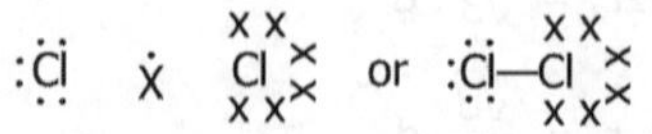

As a result, each Cl atom acquires a stable, eight-electron configuration. The shared pair of electrons is attracted by the nuclei of two Cl atoms. As a result, the two Cl atoms are held together in the molecule.

The covalent bond is directional in nature, i.e., the shared pair of electrons always lie in between the two atoms that enter into covalent bonding. Therefore, although the force of attraction between the two atoms is comparable to that between the two counter ions held together by ionic bonding, the mutual attractions between two or more molecules of a covalent compound are very weak. Due to this reason, covalent compounds are generally liquids or gases, and normally possess low melting or boiling points. Due to the absence of charges, they are non-conducting in nature.

Due to lack of charge separation in their molecules, most of the covalent compounds are insoluble in polar solvents such as water. These are, however, soluble in non-polar solvents—a characteristic controlled by thermodynamic considerations. Suppose 1 ml of ether is added to 1 ml of benzene. If the two compounds are miscible, then a larger volume of space becomes available to the molecules of benzene originally present in 1 ml of the compound. As a result, entropy of the system increases, leading to more thermodynamic stability. Thus, benzene is soluble in ether or, in general, covalent compounds are soluble in non-polar solvents.

Case Study 3.8 Why does a soft drink normally spill out on opening the bottle?

The solubility of a gas in a liquid depends on two factors:

a. It is directly proportional to the partial pressure of the gas above the liquid – *Henry's law.*
b. It is inversely proportional to temperature.

The soft drink in carborated beverages is sealed under high pressure of carbon dioxide. When the seal is opened, the pressure of the gas is reduced. As a result, its solubility decreases and, therefore, it tends to escape out. In this process, it throws out a part of the liquid too. If the bottle is kept under cold conditions, a larger amount of carbon dioxide remains in soluble state than if it is kept at room temperature. Therefore on opening the container, the evolution of the gas is relatively slow. Hence, lesser amount of drink will spill out if the bottle is refrigerated than if it is not.

Covalent bonding may be explained on the basis of two concepts: *Valence bond theory* and *molecular orbital theory*. The former is based on the pairing and neutralization of opposed electron spins, while the latter assumes that the molecule as a whole has electronic energy levels which are different from the atomic energy levels.

Valence bond theory

The simplest case of covalent bonding is reflected in the formation of diatomic hydrogen molecule. Each of the constituent atom has the electronic configuration $1s^1$. The probability density of this one electron is spherically distributed about the hydrogen nucleus.

According to the valence bond theory, as the two H atoms approach each other, the electron of one is attracted by the nucleus of the other. The degree of attraction increases as the two atoms are brought closer and closer. Concomitantly, the energy of the interacting H atoms decreases with respect to that of two isolated H atoms. At some distance, a minimum energy level is reached. This minimum energy, $-435\,\text{kJ mol}^{-1}$, corresponds to the bond energy of the hydrogen molecule. The internuclear distance, 74 pm, at which the energy minimum occurs corresponds to the bond length (*see* p. 139) in the hydrogen molecule.

Attempts to bring the two H atoms still closer result in repulsion between the two nuclei. The two electrons also repel each other. As the repulsive forces dominate over attractive forces, the energy of the interacting H atoms increases.

In Fig. 3.17, curve (a) shows a plot of energy as a function of internuclear distance. The electrons in the two H atoms possess opposite spins. The minimum in the curve represents the optimum distance between the atoms, and the maximum energy released due to covalent bonding. In Fig. 3.17, curve (b) shows the plot when the electrons in the two H atoms possess parallel spins. There is no energy minimum in this case. Moreover, the energy of this hypothetical molecule is higher than that of two isolated H atoms. This implies that when two H atoms with electrons of parallel spins are brought together, covalent bonding does not take place.

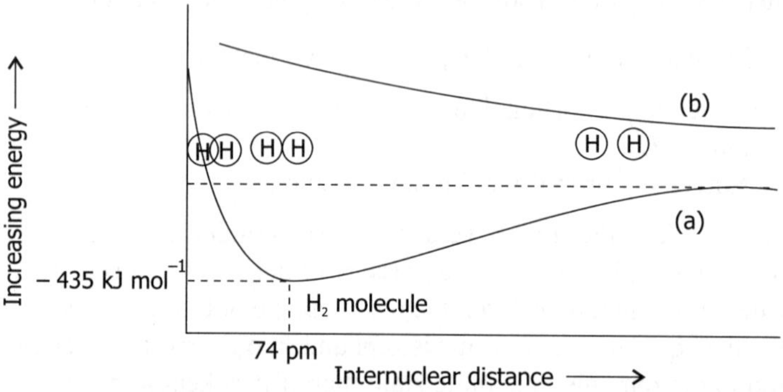

Fig. 3.17 Potential energy curve for the formation of a hydrogen molecule from two hydrogen atoms with (a) opposite spin electrons and (b) parallel spin electrons

As a result of its spin, each electron possesses a magnetic dipole, which causes it to behave like a small magnet. When two electrons of opposite spins are brought together within atomic distances of each other, the mutual neutralization of their magnetic dipole moments results in attraction between the corresponding atoms. On the other hand, the magnetic moment of two electrons with parallel spins oppose each other, and their approach leads to repulsion between the atoms. It is, therefore, pertinent to assert that a stable covalent bond between two H atoms can be formed only when the spins of two electrons are opposite to each other.

In quantum mechanical terms, when two atoms approach each other, their valence electrons exhibit the property of wave motion. When the spins of the electrons are opposite, an overlapping of waves occurs. This, in turn, causes an increase in electron density in the region of overlap. Each electron now has an equal probability of being found on either atom, and each nucleus may be associated with both the electrons. The binding force of a covalent bond may thus be considered as the electrostatic attraction of the nuclei for the electron cloud dispersed between them.

Hybridization

Hybridization refers to the mixing of two or more orbitals that have comparable energies, followed by their splitting into an equal number of degenerate orbitals. The new orbitals are called *hybrid orbitals*. The energy of hybrid orbitals is always lower than that of the parent orbitals. Hence hybrid orbitals are more stable as compared to the atomic orbitals from which they are created. The geometry of a molecule is determined by the number and type of the hybrid orbitals around its central atom. For example, in a beryllium(II) chloride ($BeCl_2$) molecule, the central atom has the following ground state valence shell configuration:

	1s	2s	2p		
Be (ground state)	⇅	⇅			

Be does not possess any unpaired electron in the ground state, and theoretically this suggests that the metal is inert. However, practically, Be is quite reactive and shows an oxidation state of II in its compounds. This is possible only if we formulate an excited valence shell configuration of Be, in which one of the electrons from 2s orbital is promoted to the 2p orbital.

	1s	2s	2p		
Be (excited state)	⇅	1	1		

Let us consider the valence shell configuration of the substituent Cl atom.

	3s	3p		
Cl (ground state)	⇅	⇅	⇅	1
		$3p_x$	$3p_y$	$3p_z$

$BeCl_2$ molecule may be formed by the overlap of 2s orbital of Be with $3p_z$ orbital of one Cl atom and by the overlap of 2p orbital of the same Be with $3p_z$ orbital of another Cl atom. The picture that emerges from this scenario is as follows:

$$Cl \overset{2s—3p_z}{\text{———}} Be \overset{2p—3p_z}{\text{———}} Cl$$

The two Be–Cl bonds, according to this consideration, are built-up from the wave functions of different atomic orbitals. These are, therefore, expected to have different bond lengths and different bond energies. This theoretical treatment is contrary to the observed facts. It has been determined experimentally that both the Be–Cl bonds are absolutely identical, having same bond length and same bond energy.

We then delve upon another model which discards the involvement of individual atomic orbitals of the central atom in bonding. Rather it theorizes the mixing of valence shell atomic orbitals of the central atom before envisaging an overlap with the orbitals of the substituents. Here, the concept of hybridization comes into picture.

According to the new model, the individual 2s and 2p orbitals of Be merge with each other and then split-up into two new orbitals, called sp hybrid orbitals (Fig. 3.18).

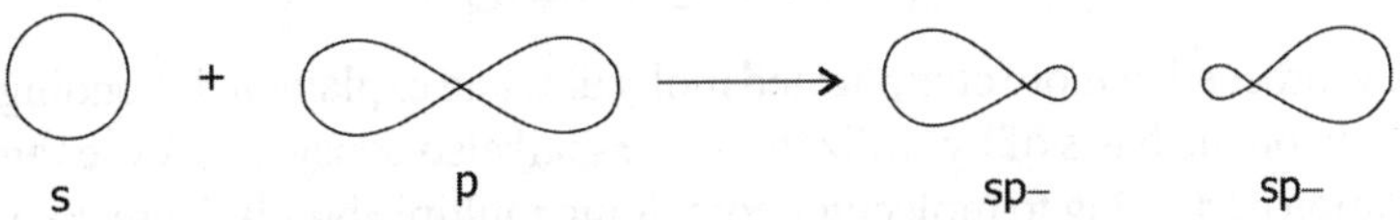

Fig. 3.18 Formation of sp hybrid orbitals

The two newly formed hybrid orbitals have identical energy, which is less than the energies of individual 2s and 2p orbitals (Fig. 3.19). This indicates that beryllium(II) chloride formed by using hybrid orbitals of Be is more stable than the hypothetical molecule that would have formed by using the individual 2s and 2p orbitals of the central atom. Moreover, the energy that was initially required to promote the 2s electron of Be to the 2p orbital has been duly compensated by the energy released during the formation of hybrid orbitals.

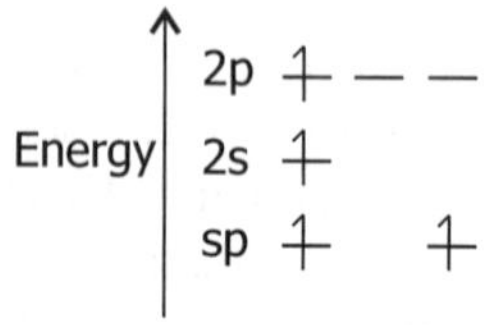

Fig. 3.19 Energy pattern of atomic orbitals and hybrid orbitals

Initially the 2s and 2p orbitals each had one unpaired electron. These are transferred to the hybrid orbitals. Therefore, each of the sp hybrid orbital now has a single, unpaired electron, in accordance with Hund's rule. These hybrid orbitals now overlap with the $3p_z$ orbital of each chlorine atom to form beryllium(II) chloride molecule.

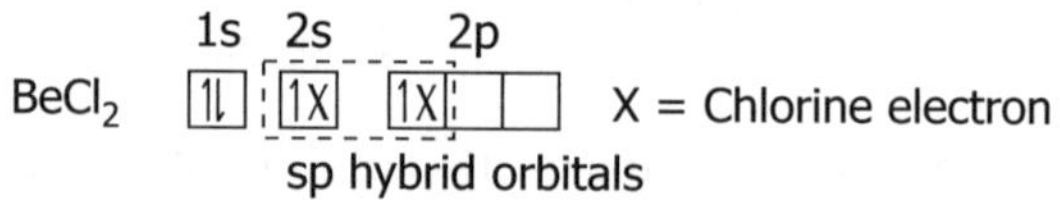

Since both the Be–Cl bonds have sp–$3p_z$ character, these are identical in every respect. The results of the hybridization model match with experimental findings. Theoretically, the two Cl atoms may assume infinite orientations around Be.

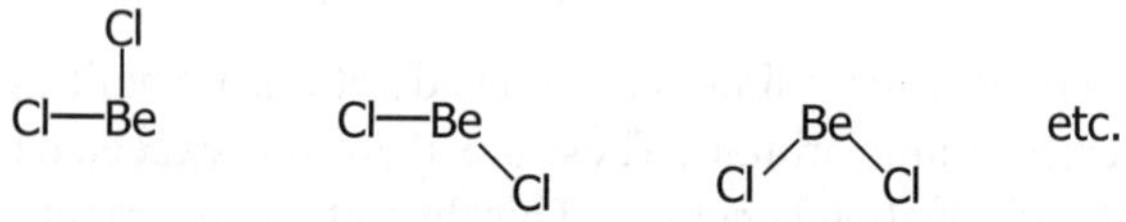

However, each Be–Cl bond is made up of two electrons, also called the bond pairs. The bond pair of one Be–Cl bond repels that of the other Be–Cl bond. The two Cl atoms, therefore, adopt that orientation around Be which has minimum repulsion. A linear configuration, with a bond angle of 180°, is the best bargain.

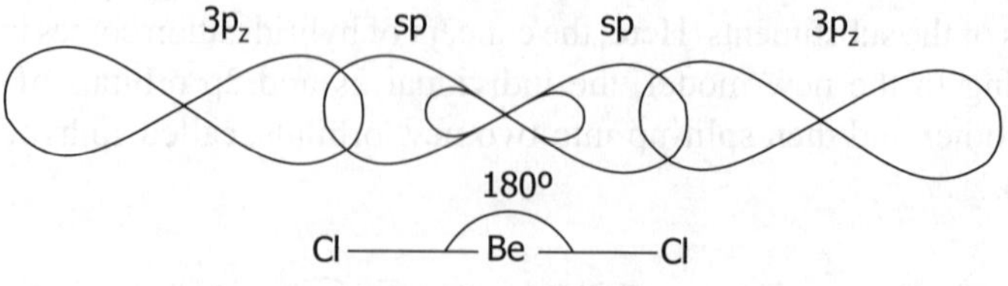

We now proceed to more complicated molecules and explain their bonding as well as geometry on the basis of hybridization. We shall also consider a few examples of hybridization pertaining to molecules containing multiple bonds. Here, two aspects must be taken into consideration. First, a multiple bond does not alter the overall geometry of the molecule. Second, the orbitals involved in π-bond formation (such bonds are formed by lateral overlap of orbitals) are not included in hybridization. Finally, we will consider cases of a few simple anions. Here, electrons equivalent to the anionic charge are added to the valence shell of that atom on which the charge resides.

Boron(III) chloride (BCl_3) Since B forms bonds with three Cl atoms, a minimum of three unpaired electrons must be present on the central atom. The formation of sp^2 hybrid orbitals is shown in Fig. 3.20. The repulsions between

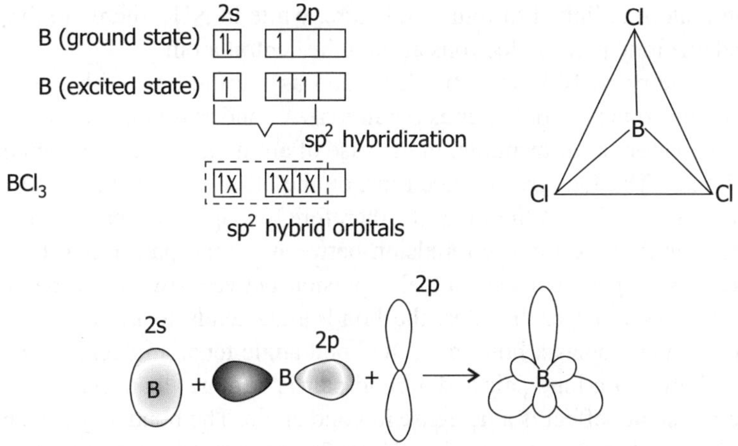

Fig. 3.20 Formation of sp^2 hybrid orbitals

the bond pairs are minimized if the three Cl atoms coordinate to the B atom at an angle of 120°. The geometry is plane triangular.

Methane (CH_4) The valence shell electronic configuration of the central atom, C, in the excited state is depicted as follows:

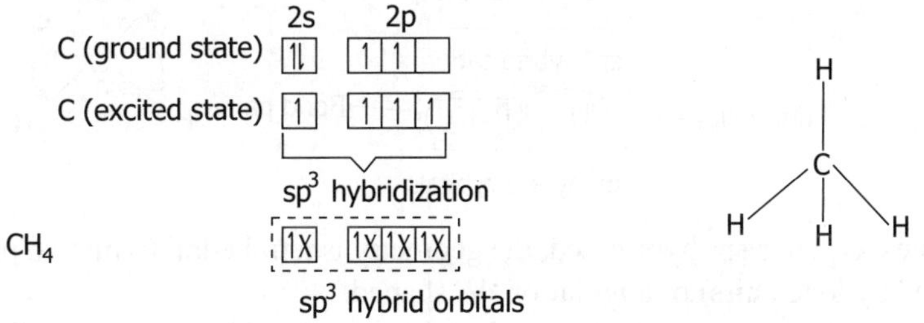

There is a tetrahedral coordination of four H atoms around C, with a bond angle of 109° 28′.

Ammonia (NH_3) The central atom, N, in ammonia has three unpaired electrons in its ground state valence shell electronic configuration. Since the N atom has to combine with three H atoms, there is no need for formulating an excited state configuration of the central atom in this case.

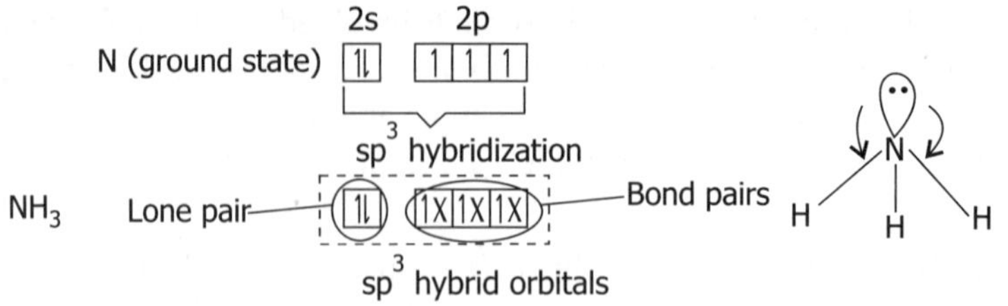

Similar to C atom in methane (CH_4), the N atom of ammonia is sp^3 hybridized. However, there is one important difference between methane and ammonia. In CH_4, the central atom is linked to four bond pairs, while in NH_3, there are three bond pairs and one lone pair of electrons around the central atom.

The geometry of NH_3 may be described as tetrahedral, with one position occupied by a lone pair, or better as pyramidal. A bond pair of electrons is trapped between two atoms; for example, in the case of ammonia it lies between nitrogen and hydrogen. The lone pair of electrons, on the other hand, is bound to an atom at one end, but is free at the other. It, therefore, occupies a greater volume than the bond pair. As a result, the repulsion between a bond pair and a lone pair of electrons is stronger than the mutual repulsion between two bond pairs. When two bond pairs repel each other, the bond angle tends to increase. However, when a lone pair repels a bond pair, the bond angle tends to decrease. Since the repulsion between a lone pair and a bond pair is stronger than that between two bond pairs, the net effect is a decrease in bond angle. The bond angle in ammonia is 107° 3′, less than the ideal value of 109° 28′ expected for a regular tetrahedral geometry.

Water (H_2O) The O atom of a water molecule has two unpaired electrons in its ground state, sufficient to affect coordination number 2 required for water. This obviates the necessity for excited state configuration.

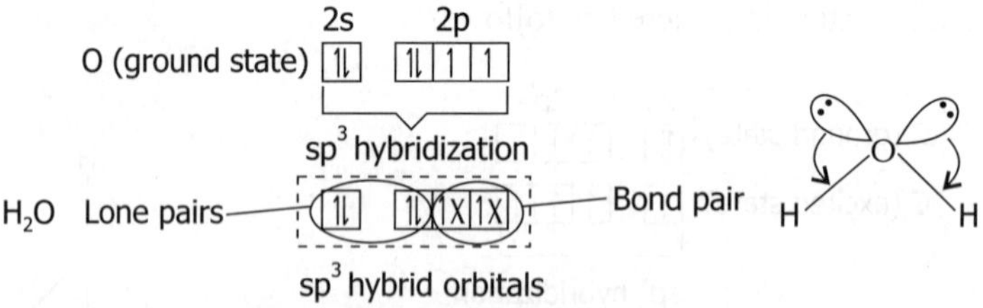

Since the oxygen is sp^3 hybridized, the geometry is tetrahedral (with two positions occupied by lone pairs) or angular or 'V' shaped.

In this case, three repulsions have to be taken into consideration for predicting the bond angle: lone pair–lone pair repulsions (lp–lp), lone pair–bond pair repulsions (lp–bp) and bond pair–bond pair repulsions (bp–bp). According to *Sidwick–Powell theory,* the lone pair–lone pair repulsions are the strongest, followed by lone pair–bond pair and bond pair–bond pair repulsions, i.e.,

lp–lp > lp–bp > bp–bp

As the two lone pairs on oxygen repel each other strongly, these come closer to each of the bond pairs. Next these repel the bond pairs, closing down the bond angle to 104°5′.

Case Study 3.9 What is the evidence that oxygen in water is sp^3 hybridized?

In its ground state, the oxygen has two unpaired electrons in the p shell and each of these could have overlapped with the 1s orbital of a hydrogen atom to form a water molecule. However, the p orbitals of an atom are mutually oriented at 90° with respect to each other. If the individual p orbitals of oxygen had been involved in bonding, the bond angle of water would have been nearly 90°. The fact that the actual bond angle is close to the value expected for tetrahedral geometry, supports the existence of sp^3 hybridization.

Invoking sp^3 hybrid orbitals in bonding is advantageous on two counts. First, the hybrid orbitals have lower energy—and more stability—than the individual atomic orbitals. Second, if p orbitals are involved in bonding, and the bond angle is 90°, the repulsions between the two bonds would be relatively large. At the observed bond angle of 104° 5′, the repulsion is markedly less.

Phosphorus(V) chloride (PCl_5) The coordination number of P in PCl_5 is 5 and, therefore, in its excited state it ought to have five unpaired electrons.

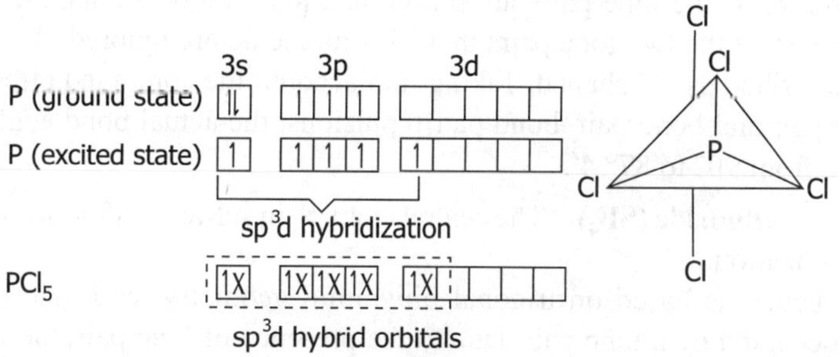

The five sp^3d hybrid orbitals around P are directed along the corners of a trigonal bipyramid skeleton. The three edges of the triangle are designated as equatorial positions. The three equatorial Cl atoms, along with the central P are in the same plane. The remaining two Cl atoms, above and below the plane of the triangle, are designated as axial Cl atoms. The bond angles subtended by equatorial and axial Cl atoms are 120° and 90°, respectively.

As with the other molecules, the bond pairs in phosphorus(V) chloride too repel each other. However repulsions at 90° are more pronounced than repulsions at 120°. An equatorial chlorine has two nearest neighbors—the two axial Cl atoms— at 90°. An axial chlorine has three nearest neighbors—the three equatorial Cl atoms—at 90°. The axial Cl atoms are thus repelled more than the equatorial Cl atoms. As a result, P–Cl (axial) bond length is greater than the P–Cl (equatorial) bond length.

Chlorine trifluoride (ClF_3) The central atom, Cl, in chlorine trifluoride undergoes sp^3d hybridization. Of the five positions, three, occupied by F atoms, represent the bond pairs, while the remaining two are occupied by lone pairs of electrons.

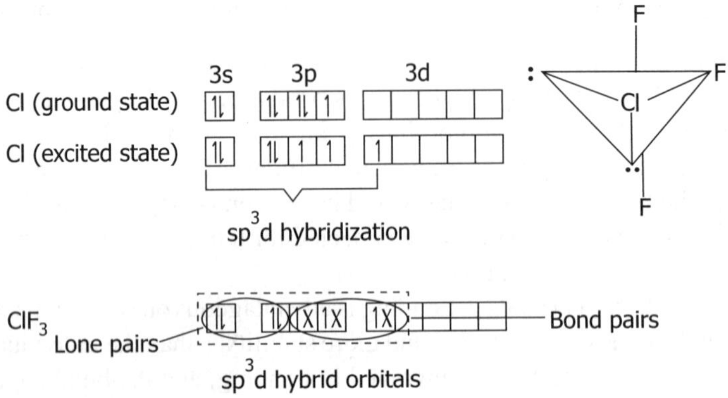

The geometry is trigonal bipyramid, with two positions occupied by lone pairs. The non-bonding electrons in a trigonal bipyramid geometry always occupy equatorial positions. This is becuase at this position, a lone pair is repelled by two neighbors (the axial groups) at 90°. However, if the lone pair were to occupy an axial position, it would be repelled by three neighbors (the equatorial groups) at 90°. Therefore, if the lone pairs are placed at equatorial positions, the molecule is more stable. If the two lone pairs in a ClF_3 molecule are ignored, the geometry may be described as 'T' shaped. Taking into account the combined effect of lone pair–lone pair and bond pair–bond pair repulsions, the actual bond angle in ClF_3 is reduced from 90° to 87° 4′.

Sulfur tetrafluoride (SF_4) The central atom, S, in sulfur tetrafluoride undergoes sp^3d hybridization.

The structure is based on trigonal bipyramid geometry, with one equatorial position occupied by a lone pair. Due to the presence of lone pair, the equatorial bond angle is reduced from 120° to 101° 6′, while the axial bond angle is reduced from 180° to 173° 1′.

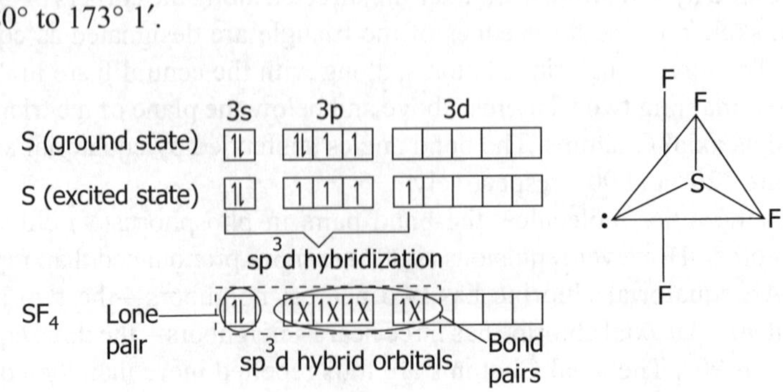

Sulfur hexafluoride (SF_6) For the formation of sulfur hexafluoride, two electrons in the valence shell of the central atom, S, must be promoted to higher energy levels, so that a total of six unpaired electrons are available for bonding.

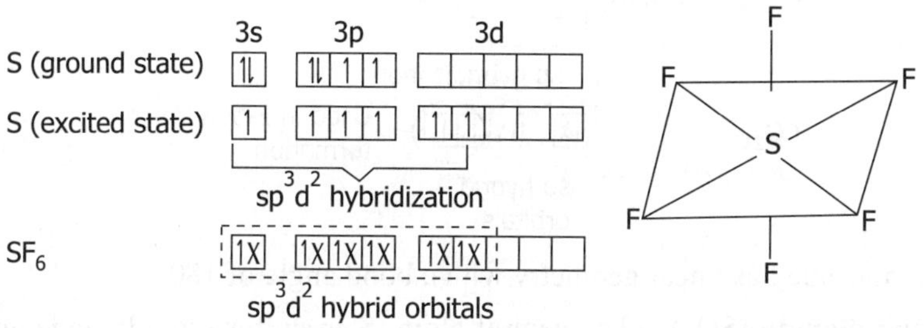

The six sp^3d^2 hybrid orbitals are directed along the apex of a regular octahedron, where they overlap with the F orbitals. All the bond angles are 90°.

Bromine pentafluoride (BrF_5) The central atom, Br, in bromine pentafluoride is sp^3d^2 hybridized.

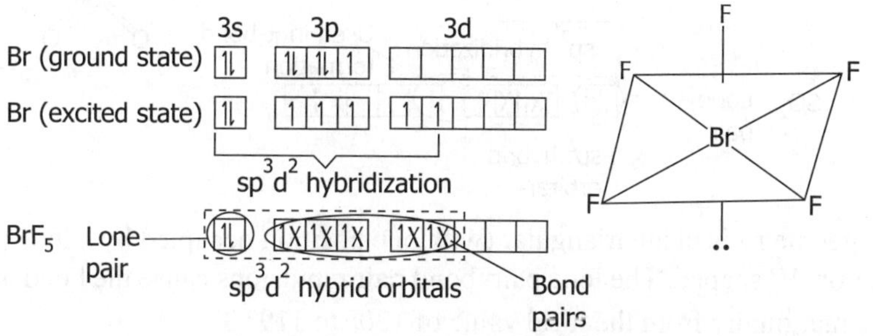

The geometry is octahedral, with one position occupied by a lone pair. Alternately, it may be described as square pyramid. Br is slightly below the basal plane of four F atoms. The presence of lone pair causes the bond angle to be less than 90°.

Iodine heptafluoride (IF_7) The only example of sp^3d^3 hybridization is manifested by IF_7. The central atom, I, has the following ground state and excited state configurations:

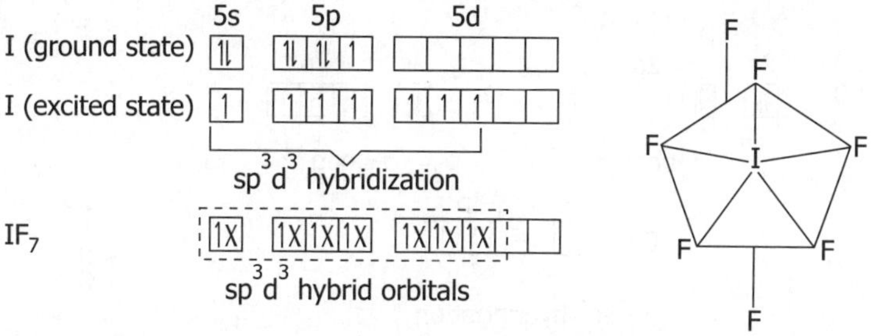

The structure is pentagonal bipyramid. The equatorial bond angles are 72°, the axial, 90°.

Carbon dioxide (CO_2) The C atom undergoes sp hybridization in carbon dioxide, forming two σ- and two π-bonds.

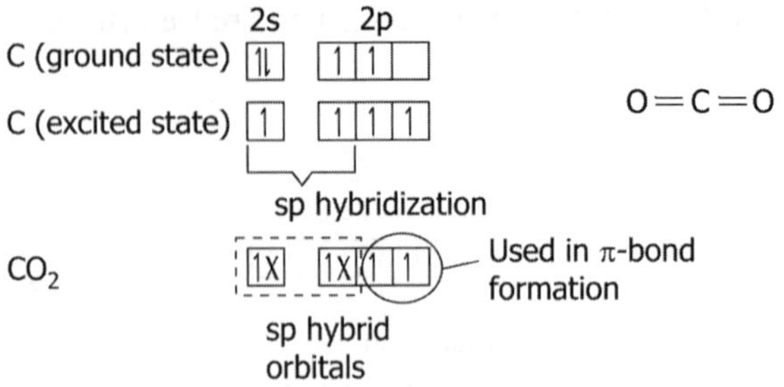

The molecule has linear geometry with a bond angle of 180°.

Sulfur dioxide (SO_2) The central atom, S in sulfur dioxide undergoes sp^2 hybridization.

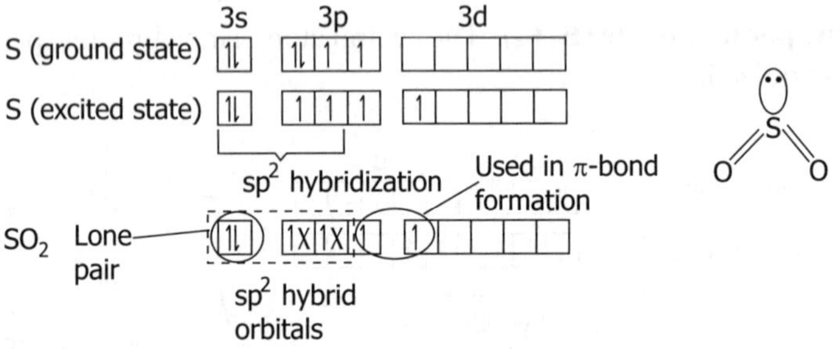

The geometry is plane triangular (with one position occupied by a lone pair) or angular or 'V' shaped. The lone pair–bond pair repulsions cause the bond angle to deviate marginally from the ideal value of 120° to 119° 3′.

Carbonate anion (CO_3^{2-}) In CO_3^{2-}, the C atom is surrounded by three O atoms. Two of these, connected to the C atom by single bonds, carry a unit negative charge each, while the third one is neutral, and is held by a double bond to the central atom.

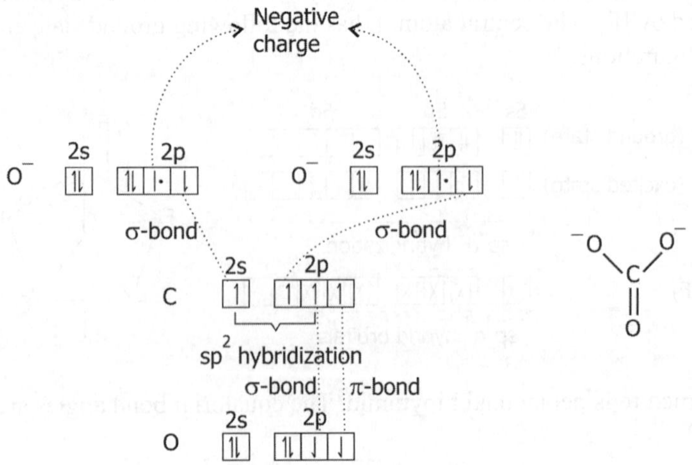

The carbon is sp^2 hybridized and the geometry of carbonate anion is plane triangular.

Sulfate anion (SO_4^{2-}) The S in sulfate anion is four-coordinated. Two neutral oxygen atoms are bound to it by double bonds and the remaining two, carrying a unit negative charge each, are bound by single bonds.

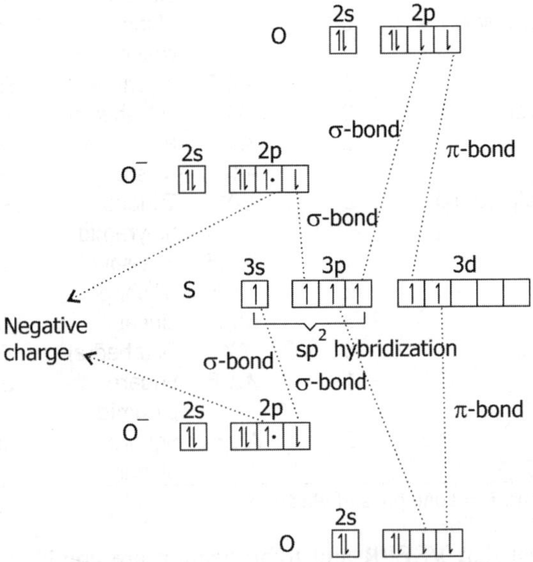

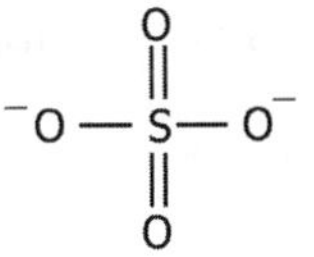

The central S atom is sp^3 hybridized and the geometry is tetrahedral.

Valence shell electron pair repulsion theory

Like hybridization, the valence shell electron pair repulsion (VSEPR) theory is a component of valence bond approach to chemical bonding. It is based on the premise that each valence shell pair of electrons, whether bonding or non-bonding, of the central atom, is significant in predicting the geometry of the relevant molecule or ion.

The valence electron pairs are arranged about the central atom in such a way that repulsions among these are minimum. This means that the regions of high electron density around the central atom should be as far apart as possible. For example, two regions of high electron density are most stable on the opposite side of the central atom, giving a linear arrangement. Three regions are most stable when these are arranged at the corners of an equilateral triangle, giving plane trigonal arrangement. Thus, the arrangement of regions of high electron density around the central atom determines the geometry of the molecule.

In valence shell electron pair repulsion theory, a molecule may be represented by the composition, AX_mE_n, where A is the central atom, X are the substituent atoms or groups and E are the number of lone pairs of electrons. The total number of positions around the central atom, occupied by substituents and the lone pairs of electrons, is called *steric number*. For the molecule, AX_mE_n, the steric number is (m + n). The steric number for water molecule, AX_2E_2 is four, while that for xenon tetrafluoride, AX_4E_2 is six.

Some common examples of molecular geometries predicted on the basis of VSEPR model are set out in Table 3.3.

Table 3.3 Molecular geometries according to VSEPR concept

Number of electron pairs	Arrangement	Number of lone pairs	Type	Shape	Example
2	Linear	0	AX_2	Linear	$BeCl_2$
3	Plane triangular	0	AX_3	Plane triangular	BCl_3
		1	AX_2E	'V' shape	$SnCl_2$
4	Tetrahedral	0	AX_4	Tetrahedral	CH_4
		1	AX_3E	Pyramidal	NH_3
		2	AX_2E_2	'V' shape	H_2O
5	Trigonal bipyramid	0	AX_5	Trigonal bipyramid	PCl_5
		1	AX_4E	See-saw	SF_4
		2	AX_3E_2	'T' shape	ClF_3
		3	AX_2E_3	Linear	XeF_2
6	Octahedral	0	AX_6	Octahedral	SF_6
		1	AX_5E	Square pyramid	BrF_5
		2	AX_4E_2	Square planar	XeF_4

A = central atom; X = substituent; E = Lone pairs of electrons

These examples manifest that VSEPR and hybridization are complementary concepts. The VSEPR theory predicts the direction of the regions of high electron densities, while hybridization predicts the number, type and orientation of hybrid orbitals around the central atom. The end result is the explanation of the observed shapes of the chemical species.

Molecular orbital theory

Valence bond theory explains the formation of covalent bonds as a result of sharing of electrons in overlapping orbitals of different atoms. These orbitals may be pure atomic orbitals or hybridized ones. Molecular orbital theory predicts that atomic orbitals on different atoms combine to form a set of new orbitals. These new orbitals are called *molecular orbitals*. The electrons contained in the molecular orbitals belong to the molecule as a whole.

The major success of molecular orbital theory is the prediction of paramagnetic behavior of oxygen molecule. The valence bond theory, on the other hand, had predicted oxygen to be diamagnetic. Magnetic studies reveal that oxygen indeed is paramagnetic, possessing two unpaired electrons. Thus, the results of molecular orbital theory, and not the valence bond theory, match with the experimental observation.

Linear combination of atomic orbitals

The formation of molecular orbitals from atomic orbitals is explained on the basis of a concept called linear combination of atomic orbitals (LCAO). The results of the LCAO model are qualitative in nature and are derived from the principles of quantum mechanics.

According to quantum mechanical description, each atomic orbital is represented by a wave function. When atomic orbitals combine to yield molecular orbitals, it is in fact their wave functions that combine. Electron waves have positive (crest) or negative (trough) phases or amplitudes. The two waves may combine constructively or destructively.

Figure 3.21 shows constructive combination of two waves that are in phase. The new wave has a greater amplitude as compared to that of the combining waves.

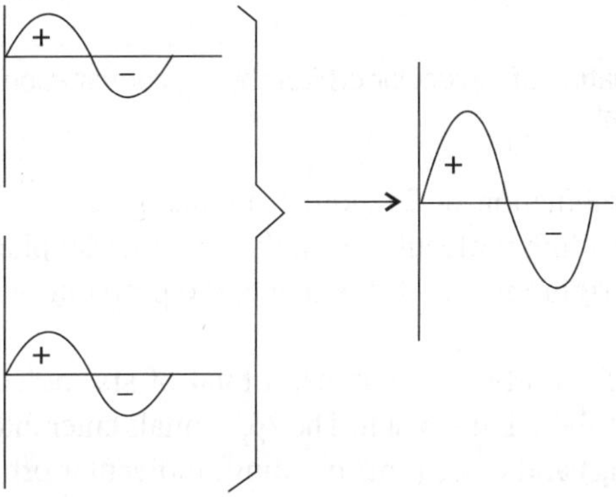

Fig. 3.21 In-phase combination of waves

Figure 3.22 shows destructive combination of waves that are out of phase. The new wave has zero amplitude.

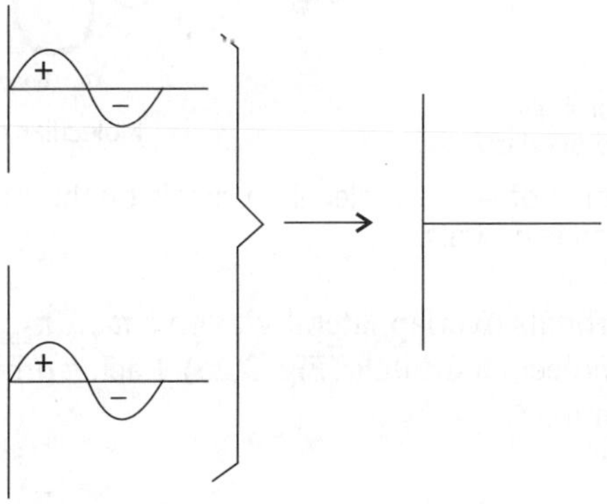

Fig. 3.22 Out-of-phase combination of waves

In the same manner, when two atomic orbitals undergo in-phase addition, their electron density lies in between the two nuclei and a low energy bonding molecular orbital is produced. When these undergo out-of-phase addition, the electron density in the region between the nuclei gets substantially reduced and a high energy anti-bonding molecular orbital is formed. The combination of two atomic orbitals thus produces two molecular orbitals—one bonding and the other anti-bonding (Fig. 3.23). For example, the combination of two 1s orbitals of different atoms gives a bonding molecular orbital, designated as σ_{1s} and an anti-bonding molecular orbital, designated as σ^*_{1s}.

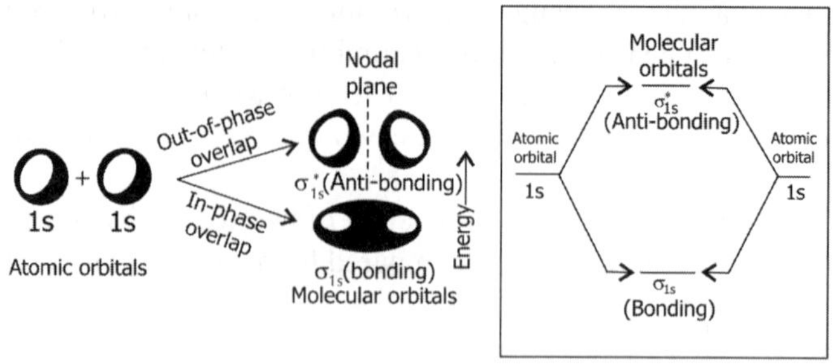

Fig. 3.23 Formation of molecular orbitals by the combination of two 1s atomic orbitals

Similarly, the combination of 2s atomic orbitals gives σ_{2s} and σ^*_{2s} molecular orbitals. The anti-bonding molecular orbitals have a nodal plane which bisects the internuclear axis at right angle. At this plane, the probability of finding electrons is zero.

When two sets of 2p orbitals combine, a total of six molecular orbitals—three bonding and three anti-bonding—result. The $2p_x$ orbitals enter into end-to-end overlap to give σ_{2px} (bonding) and σ^*_{2px} (anti-bonding) molecular orbitals (Fig. 3.24).

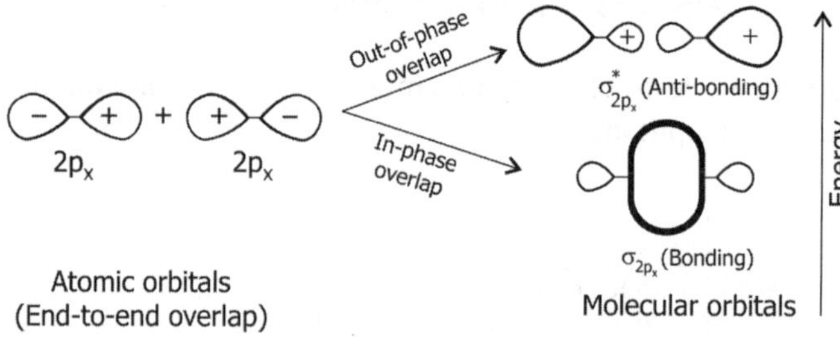

Fig. 3.24 Formation of sigma molecular orbitals by the combination of two $2p_x$ atomic orbitals

The $2p_y$ and $2p_z$ orbitals overlap laterally to give π_{2py}, π_{2pz} (bonding) and π^*_{2py}, π^*_{2pz} (anti-bonding) molecular orbitals (Fig. 3.25). Each π orbital has a nodal plane along the internuclear axis.

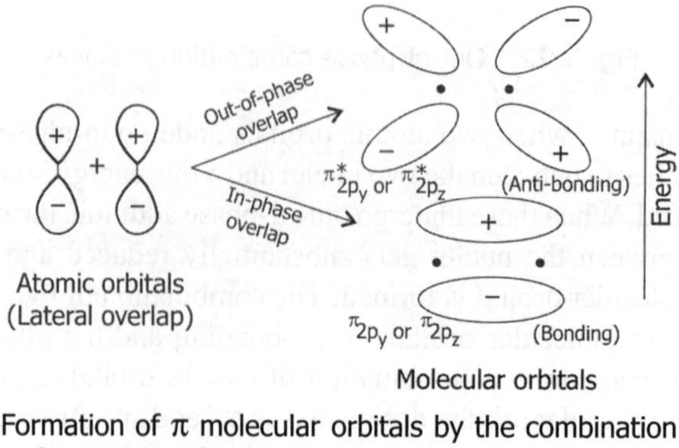

Fig. 3.25 Formation of π molecular orbitals by the combination of two $2p_y$ or $2p_z$ atomic orbitals

The LCAO model is based on the premise that the energies of the atomic orbitals involved in the construction of molecular orbitals are equal or nearly equal. It is also essential that the distance between the combining atoms should be such that effective mixing of their orbitals can take place. If the distance is too large, there will be no interaction between the atomic orbitals; and if too short, the repulsion between the two nuclei and/or the two electrons will preclude the formation of molecular orbitals.

Homonuclear diatomic molecules Now that we know how the atomic orbitals combine to give molecular orbitals, we may construct the molecular orbital energy level diagram for simple homonuclear molecules. This is shown in Fig. 3.26. Note that π_{2py} and π_{2pz} orbitals have lower energy than σ_{2px} orbital. However, as the mass number of the constituent atoms increases, the σ_{2px} orbital is stabilized as compared to the π_{2py} and π_{2pz} orbitals. Therefore, for molecules such as oxygen and fluorine, σ_{2px} molecular orbital has lower energy than π_{2py} and π_{2pz}. Hence, while writing the configuration of these molecules, we shall follow Fig. 3.27.

While using the molecular orbital diagram, the same rules are observed as for writing the electronic configuration of atoms. The molecular orbital with lower energy is filled before that with higher energy (Aufbau principle, *see* Chapter 1, p. 47). When molecular orbitals are degenerate (as π_{2py} and π_{2pz} or π^*_{2py} and π^*_{2pz}), the electrons remain unpaired as far as possible (Hund's rule, *see* Chapter 1, p. 46). Two electrons can be accommodated in a single molecular orbital only if their spins are opposite (Pauli exclusion principle, *see* Chapter 1, p. 45).

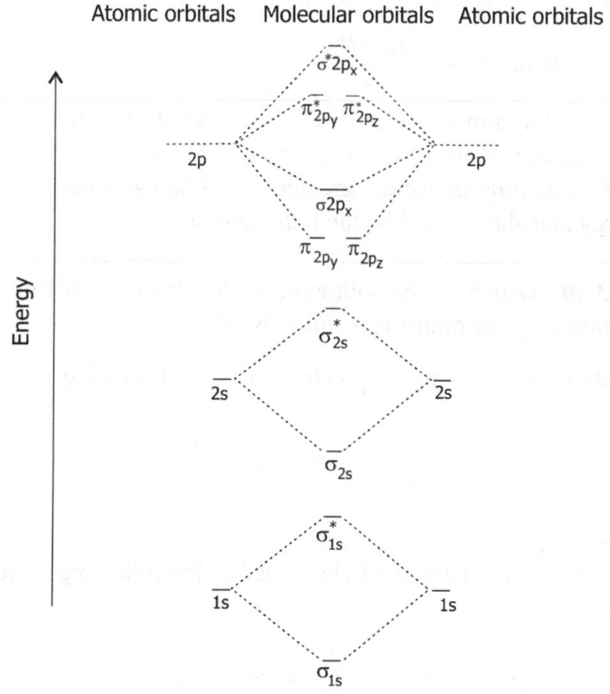

Fig. 3.26 Molecular orbital diagram for hydrogen through nitrogen

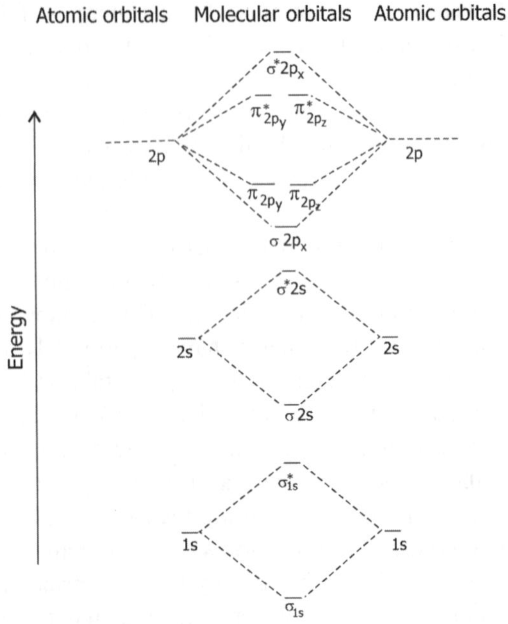

Fig. 3.27 Molecular orbital diagram for oxygen through neon

Molecular orbital theory enables us to elucidate the *bond order* (the number of covalent bonds holding the two atoms together) of a molecule.

$$\text{Bond order} = \frac{N_b - N_a}{2}$$

where N_b is the total number of electrons in the bonding molecular orbitals and N_a is the total number of electrons in anti-bonding molecular orbitals. Higher the bond order of a diatomic molecule, greater would be its stability, greater will be the bond energy and shorter will be the bond length.

Case Study 3.10 Which of the following molecular ions will be more stable: dinitrogen cation (N_2^+) or dinitrogen anion (N_2^-)?

Dinitrogen cation, N_2^+, having 13 electrons has the following molecular orbital configuration:

$$\sigma_{1s}^2\sigma_{1s}^{*2}\sigma_{2s}^2\sigma_{2s}^{*2}\pi_{2p_y}^2\pi_{2p_z}^2\sigma_{2p_x}^1$$

Its bond order is 2.5.

Dinitrogen anion, N_2^-, having 15 electrons has the following molecular orbital configuration:

$$\sigma_{1s}^2\sigma_{1s}^{*2}\sigma_{2s}^2\sigma_{2s}^{*2}\pi_{2p_y}^2\pi_{2p_z}^2\sigma_{2p_y}^2\pi_{2p_y}^{*1}$$

Its bond order is also 2.5.

Although the stability of a molecular species is directly proportional to its bond order, in this case, the dinitrogen anion is less stable. This is because the last electron occupies the anti-bonding molecular orbital of this entity. The dinitrogen cation has no electron in a high-energy anti-bonding orbital and, therefore, is more stable.

Heteronuclear diatomic molecules The electronic configuration of heteronuclear diatomic molecules may be written by counting the total number of electrons present in each constituent atom. For example, in carbon monoxide (CO) molecule, C contributes six and O contributes eight electrons and, therefore, there are a total of 14 electrons. The configuration is similar to that of nitrogen molecule, which also has 14 electrons.

However, in the case of heteronuclear molecules, the molecular orbital diagram has to be modified. The reason is that the corresponding orbitals of two different elements do not have same energy. For example, the 2s orbital of O has lower energy than the analogous orbital of C. The atomic orbitals of more electronegative element are lower in energy than the corresponding orbitals of less electronegative element. In case the electronegativity values of the two elements are nearly the same, the atomic orbitals of the one that possesses higher number of protons will have lower energy. The molecular orbital diagram for CO thus becomes skewed, as shown in Fig. 3.28.

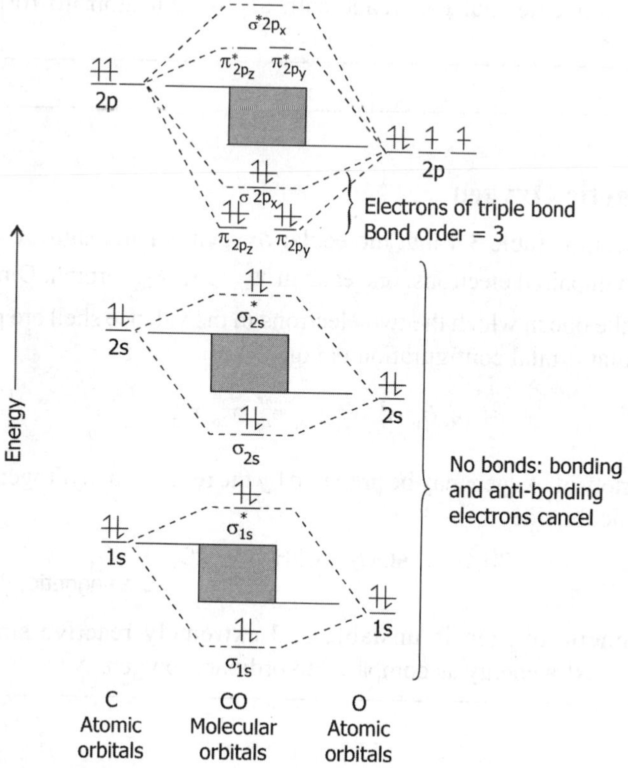

Fig. 3.28 Molecular orbital diagram for carbon monoxide

A molecular orbital imbibes the characteristics of that atomic orbital to which it is closer in energy. Thus, in case of CO, the bonding molecular orbitals resemble the atomic orbitals of O, while the anti-bonding molecular orbitals resemble the atomic orbitals of C.

The electronic configurations of a few molecules/molecular ions, along with their bond orders are given in Table 3.4.

Case Study 3.11 Although the oxygen molecule (O_2) has two unpaired electrons, yet two such molecules do not combine to form a tetra-atomic species (O_4). Why?

If electron pairing were the only criterion dictating the formation of covalent compounds, it might be expected that two oxygen molecules would combine to form an O_4 molecular entity. However, this does not happen because the two unpaired electrons in the oxygen molecule are in anti-bonding orbitals. No combination can take place between two anti-bonding electrons of one oxygen molecule with two anti-bonding electrons of another oxygen molecule. Nevertheless, combination of an oxygen molecule is possible with another oxygen atom, since the latter would furnish two bonding electrons. This combination leads to the formation of ozone. Similarly, nitric oxide molecule, having an unpaired electron in anti-bonding orbital, cannot combine with another nitric oxide molecule, but can react with an oxygen atom to form nitrogen dioxide molecule.

Box 3.1

Diamagnetic Oxygen

We have seen in Table 3.4 that molecular oxygen is paramagnetic in nature, having two unpaired electrons, one each in $\pi^*_{2p_y}$ and $\pi^*_{2p_z}$ orbital. Diamagnetic oxygen is the one in which the two electrons in the valence shell are paired up, the molecular orbital configuration being,

$$\sigma^2_{1s}\,\sigma^{*2}_{1s}\,\sigma^2_{2s}\,\sigma^{*2}_{2s}\,\sigma^2_{2p_x}\,\pi^2_{2p_y}\,\pi^2_{2p_z}\,\pi^{*2}_{2p_y}$$

This variety of oxygen may be prepared by the reaction of hydrogen peroxide and chlorine dioxide.

$$2H_2O_2 + 4ClO_2 \rightarrow 2H_2O + 2Cl_2 + \underset{\text{Diamagnetic}}{5O_2}$$

Diamagnetic oxygen is unstable and extremely reactive since it has 95 kJ mol^{-1} extra energy as compared to ordinary oxygen.

Table 3.4 Molecular orbital configurations of some chemical species

Chemical species	No. of electrons	Configuration	Bond order
H_2^+	1	σ_{1s}^1	0.5
H_2	2	σ_{1s}^2	1
HeH	3	$\sigma_{1s}^2\ \sigma_{1s}^{*1}$	0.5
He_2 (hypothetical)	4	$\sigma_{1s}^2\ \sigma_{1s}^{*2}$	Zero
Li_2	6	$\sigma_{1s}^2\ \sigma_{1s}^{*2}\ \sigma_{2s}^2$	1
B_2	10	$\sigma_{1s}^2\ \sigma_{1s}^{*2}\ \sigma_{2s}^2\ \sigma_{2s}^{*2}\ \pi_{2p_y}^1\ \pi_{2p_z}^1$	1
C_2	12	$\sigma_{1s}^2\ \sigma_{1s}^{*2}\ \sigma_{2s}^2\ \sigma_{2s}^{*2}\ \pi_{2p_y}^2\ \pi_{2p_z}^2$	2
N_2/CO/NO^+	14	$\sigma_{1s}^2\ \sigma_{1s}^{*2}\ \sigma_{2s}^2\ \sigma_{2s}^{*2}\ \pi_{2p_y}^2\ \pi_{2p_z}^2\ \sigma_{2p_x}^2$	3
NO	15	$\sigma_{1s}^2\ \sigma_{1s}^{*2}\ \sigma_{2s}^2\ \sigma_{2s}^{*2}\ \pi_{2p_y}^2\ \pi_{2p_z}^2\ \sigma_{2p_x}^2\ \pi_{2p_y}^{*1}$	2.5
O_2	16	$\sigma_{1s}^2\ \sigma_{1s}^{*2}\ \sigma_{2s}^2\ \sigma_{2s}^{*2}\ \sigma_{2p_x}^2\ \pi_{2p_y}^2\ \pi_{2p_z}^2\ \pi_{2p_y}^{*1}\ \pi_{2p_z}^{*1}$	2
O_2^-	17	$\sigma_{1s}^2\ \sigma_{1s}^{*2}\ \sigma_{2s}^2\ \sigma_{2s}^{*2}\ \sigma_{2p_x}^2\ \pi_{2p_y}^2\ \pi_{2p_z}^2\ \pi_{2p_y}^{*2}\ \pi_{2p_z}^{*1}$	1.5
O_2^{2-}/F	18	$\sigma_{1s}^2\ \sigma_{1s}^{*2}\ \sigma_{2s}^2\ \sigma_{2s}^{*2}\ \sigma_{2p_x}^2\ \pi_{2p_y}^2\ \pi_{2p_z}^2\ \pi_{2p_y}^{*2}\ \pi_{2p_z}^{*2}$	1
Ne_2 (hypothetical)	20	$\sigma_{1s}^2\ \sigma_{1s}^{*2}\ \sigma_{2s}^2\ \sigma_{2s}^{*2}\ \sigma_{2p_x}^2\ \pi_{2p_y}^2\ \pi_{2p_z}^2\ \pi_{2p_y}^{*2}\ \pi_{2p_z}^{*2}\ \sigma_{2p_x}^{*2}$	Zero

Polyatomic molecules In polyatomic molecules, the molecular orbitals are constructed from linear combinations of those atomic orbitals that have same symmetry, much the same way as for diatomic molecules. Moreover, like diatomic molecules, a combination of N atomic orbitals yields N molecular orbitals. Yet the molecular orbital picture of polyatomic molecules differs from that of diatomic molecules in an important aspect. Before the molecular orbitals are constructed, the atomic orbitals of the substituent atoms combine with each other to form group orbitals. The latter then overlap with those atomic orbitals of central atom with which their symmetry matches, giving rise to the molecular orbitals. A single group orbital or a single molecular orbital is designated as a or b; a doubly degenerate as e; and a triply degenerate as t.

Beryllium dihydride (BeH_2). The molecular orbitals of beryllium dihydride, a linear triatomic molecule, are constructed from the 1s orbitals on the two substituent hydrogens and the 2s and one of the 2p orbitals on the central beryllium atom. Prior to the formation of molecular orbitals, the 1s orbitals of the two hydrogens combine with each other to give two group orbitals. These are designated as a and a′ and are shown on the right of Fig. 3.29. These enter into combination with the 2s and 2p orbitals on beryllium, shown on the left of Fig. 3.29.

Since four orbitals are involved in bonding, four molecular orbitals are formed. These are depicted in the center of Fig. 3.29. The symmetry of group orbital a matches with 2s atomic orbital of beryllium. This combination gives rise to two molecular orbital—2^{σ}_{g} bonding molecular orbital and 3^{σ}_{g} anti-bonding molecular orbital. The group orbital a′ combines with the $2p_z$ orbital of beryllium, as the two have the same symmetry. This combination leads to the formation of bonding molecular orbital 1^{σ}_{g} and anti-bonding molecular orbital 2^{σ}_{u}. The symmetry of $2p_x$ and $2p_y$ orbitals of beryllium does not match with any of the group orbital and hence these remain non-bonding.

The molecular orbital configuration of beryllium dihydride is, $2\sigma_g^2\ 1\sigma_g{}^2$.

Ammonia (NH_3). The molecular orbital diagram of ammonia is shown in Fig. 3.30. On the left are the four valence orbitals of nitrogen—one 2s and three 2p orbitals. On the right are the three group orbitals formed by the combination of 1s orbital on each of the three hydrogen atoms. The singly degenerate group orbital, a is lower in energy as compared to the doubly degenerate group orbitals, e. The three group orbitals of hydrogen overlap with the four atomic orbitals of nitrogen to form seven molecular orbitals of ammonia.

The combination of group orbital a of hydrogens with 2s and $2p_z$ orbitals of nitrogen gives rise to three molecular orbitals—the $1a_1$, $2a_1$ and $3a_1$. Of these, $2a_1$ molecular orbital has an extremely small component of group orbital a. For this reason we have refrained from connecting $2a_1$ and a orbitals in Fig. 3.30.

The two degenerate group orbitals e on hydrogens have appropriate symmetry to overlap with $2p_x$ and $2p_y$ orbitals on nitrogen. This combination gives rise to four molecular orbitals—two degenerate bonding molecular orbitals 1e and two degenerate anti-bonding molecular orbitals 2e.

The energies of the seven molecular orbitals varies in the order,

$$1a_1 < 1e = 1e < 2a_1 < 2e = 2e < 3a_1$$

A molecule of ammonia has three pairs of bonding electrons and one non-bonding electron pair. Its molecular orbital configuration is, $1a_1^2\ 1e^4\ 2a^2{}_1$.

Sulfur hexafluoride (SF_6). The molecular orbital diagram of sulfur hexafluoride is shown in Fig. 3.31. The combination of $2p_z$ orbitals on each fluorine results in six group orbitals. In order of increasing energy, these are: one a orbital, two degenerate e orbitals and three degenerate t orbitals. These combine with the four valence orbital of sulfur—one 3s and three 3p.

The ligand orbital a has appropriate symmetry to overlap with 3s orbital of sulfur. This combination produces a bonding molecular orbital $1a_1$ and an anti-bonding molecular orbital $2a_1$. The three degenerate group orbitals t of fluorines combine with the three 3p orbitals of sulfur to yield six molecular orbitals—three degenerate bonding molecular orbitals, designated as 1t and three degenerate anti-bonding molecular orbitals, designated as 2t. The doubly degenerate e group orbitals do not have the appropriate symmetry to combine with any of the nitrogen's valence orbitals. These, therefore, remain non-bonding and are designated as 1e.

A molecule of sulfur hexafluoride has a total of twelve electrons (six contributed by sulfur and one by each fluorine). Its molecular orbital configuration is, $1a_1^2\ 1t^6\ 1e^4$.

Case Study 3.12 Which molecule is expected to have a higher electron affinity: C_2 or O_2?

The hypothetical molecule C_2 will have a higher electron affinity than the oxygen molecule. The reason being that C_2 will add on the additional electron in a low-energy σ_{2px} molecular orbital. The oxygen molecule, on the other hand will gain an electron in a high-energy π_{2py} orbital.

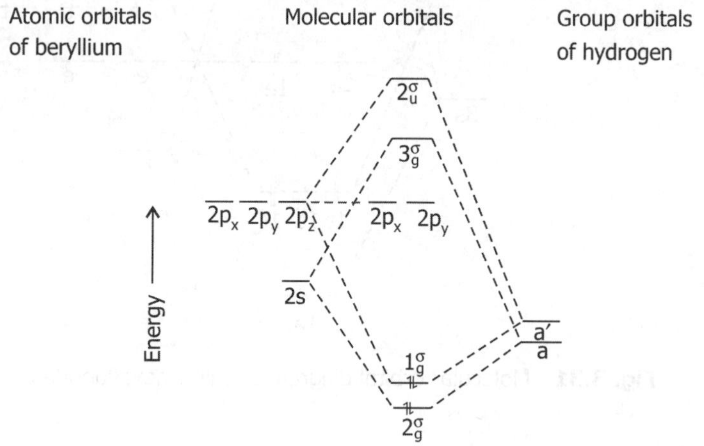

Fig. 3.29 Molecular orbital diagram of beryllium dihydride

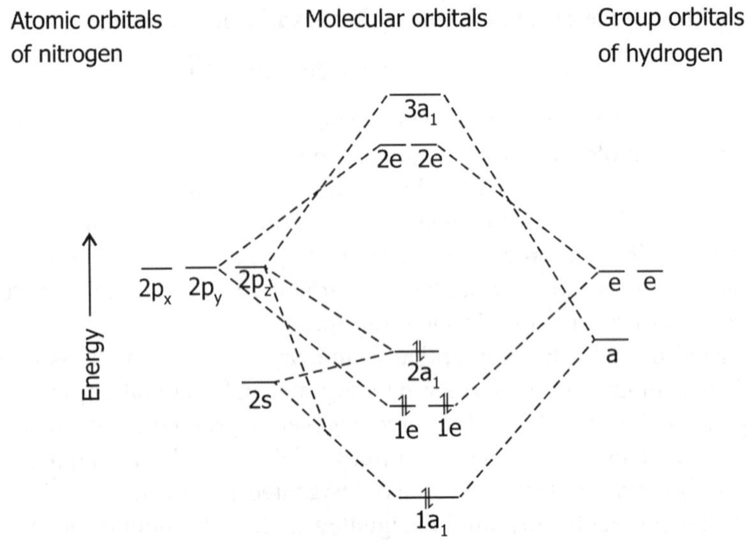

Big. 3.30 Molecular orbital diagram of ammonia

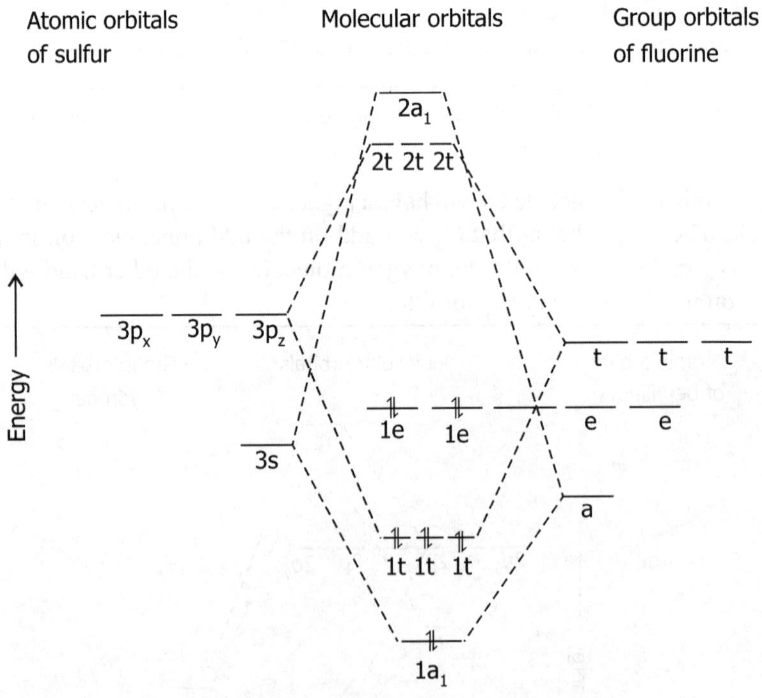

Fig. 3.31 Molecular orbital diagram of sulfur hexafluoride

Box 3.2

Coordinate Bond

A normal covalent bond is formed between two atoms, A and B by mutual sharing of an electron pair. One of the electron in the shared pair is contributed by atom A, while the other is contributed by atom B.

$$A^{\times} + {}^{\cdot}B \rightarrow A \overset{\cdot}{\times} B \text{ (or A—B)}$$

The same combination may be achieved in yet another way. The atom A may not contribute any electron for sharing, while atom B may contribute both the electrons.

$$A + :B \rightarrow A:B \text{ (or } A \leftarrow B\text{)}$$

A bond formed by sharing of two electrons, both of which are contributed by one constituent atom is called a *coordinate bond* or a *dative bond.*

In case of a hypothetical molecule, AB cited above, B is an electron-rich species, while A is an electron-deficient species. The electron-rich species possess one or more lone pairs of electrons and are designated as *donors*. The electron-deficient species are referred to as *acceptors*. Such bonds are represented by arrows, pointing from donor towards the acceptor atom.

A coordinate bond is different from a covalent bond only in the mode of formation. Once formed, the two types of bonds are indistinguishable.

The combination of an ammonia molecule and a proton to yield ammonium ion takes place through coordinate bond formation. The nitrogen of ammonia has a lone pair of electrons and acts as a donor. The electron-deficient proton acts as an acceptor.

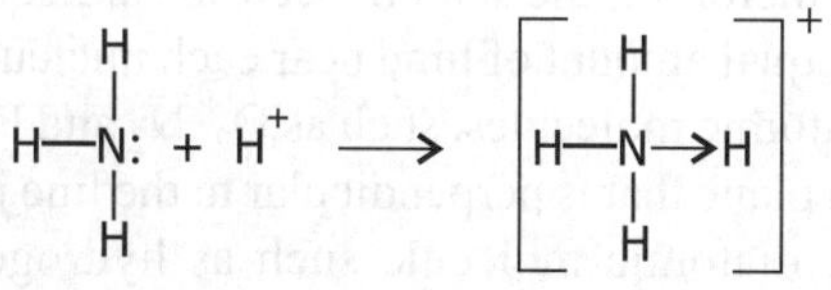

Similarly, the electron-deficient boron in boron(III) fluoride, accepts an electron pair, through a coordinate bond, from an electron-rich fluoride ion.

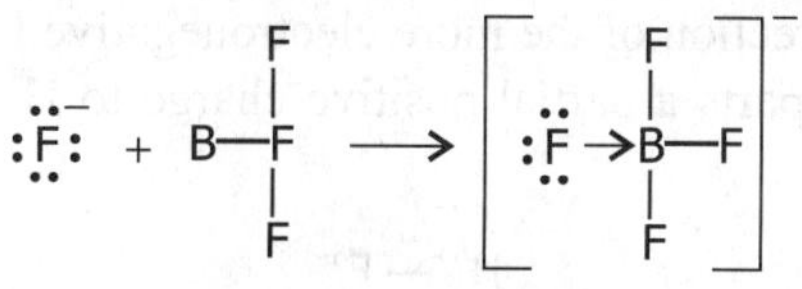

Once a coordinate bond is formed, the donor species continues to have a stable octet, while the acceptor species acquires a more stable electronic arrangement.

Case Study 3.13 What is *Isosterism*?

It is observed that molecules or ions, having same number of atoms, same number of electrons and same charges show similarities in physical properties. Such chemical species are called *isosteres* and this phenomena is called isosterism. Nitrogen (N_2) and carbon monoxide (CO) constitute an isosteric group. Several of their physical parameters are comparable and are listed in Table 3.5.

Table 3.5 Physical constants of isosteric compounds

Property	Nitrogen	Carbon monoxide
Critical temperature (K)	146	151
Critical pressure (atm)	33	35
Density of liquid (g cm^{-3})	0.796	0.793
Viscosity at 293 K (poise × 10^6)	166	163
Thermal conductivity (at 373 K)	0.0566	0.0543

The cyanide ion (CN^-) has same number of atoms and electrons as nitrogen and carbon monoxide. However, since it carries a unit negative charge, its physical constants are not too close to that of nitrogen and carbon monoxide.

Carbon dioxide (CO_2) and nitrous oxide (N_2O) constitute another isosteric group and azide ion (N_3^-) and cyanate ion (CNO^-) yet another.

Polar covalent bonds

Covalent bonds may be either polar or non-polar. In a *non-polar covalent bond*, such as that in a hydrogen molecule (H_2), the shared pair of electrons is equidistant from the two constituent H atoms. This is because both the H atoms have same electronegativity and, therefore, the shared electrons are equally attracted to both H atoms. They spend equal amount of time near each nucleus. In this, as well as in other homonuclear diatomic molecules, such as O_2, N_2 and F_2, the electron density is symmetrical about a plane that is perpendicular to the line joining the two nuclei.

In a heteronuclear, diatomic molecule such as hydrogen fluoride (HF), the constituent atoms—H and F—do not have equal electronegativities. F is more electronegative, therefore, attracts the shared pair of electrons more strongly than H, which is less electronegative. In this case, the electron density is not symmetrical; it is distorted in the direction of the more electronegative F atom. The small drift of electron density imparts a partial positive charge to H and a partial negative charge to F.

$$H^{\delta+}—F^{\delta-}$$

Covalent bonds in which the electron pairs are unequally shared are called *polar covalent bonds*. Such bonds may be considered as intermediate between non-polar covalent bonds and ionic bonds. In fact, the bond polarity is often referred to as *partial ionic character*. Greater the difference between the electronegativities of the constituent atoms, higher will be the partial ionic character. Mathematically, the variation in per cent ionic character (P) with difference in electronegativity is given by *Hannay–Smith equation,*

$$P = 16\,(\Delta\chi) + 3.5\,(\Delta\chi)^2 \tag{3.28}$$

where $\Delta\chi$ is the difference in electronegativities of two atoms, A and B, i.e., $(\chi_A - \chi_B)$. For example, the electronegativities of chlorine (Cl) and hydrogen (H) are 3.0 and 2.1, respectively. Therefore, for hydrogen chloride (HCl) molecule,

$$P = 16\,(0.9) + 3.5\,(0.9)^2 = 17.2\% \qquad (\Delta\chi = 0.9)$$

Thus, an HCl molecule has approximately 17% ionic character.

Dipole moment

The separation of charges in a polar covalent bond creates an electric dipole moment. Dipole means two poles—referring to the positive and negative centers in the molecule. Dipole moment expresses the polarity of bonds on a numerical scale. The extent of dipole moment may be measured by placing the molecule between two plates and applying a voltage. If the molecule is polar, it will orient itself so that the positive center is aligned towards the negative plate and the negative center towards the positive plate (Fig. 3.32). This will cause the voltage between the plates to fall drastically. On the other hand, a non-polar molecule will not cause a marked decrease in the voltage between the plates.

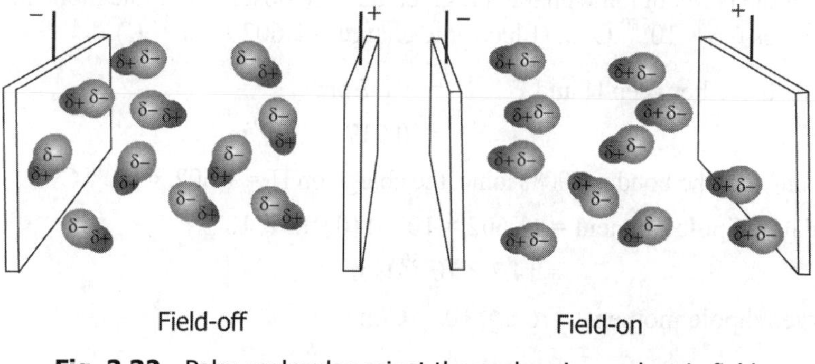

Fig. 3.32 Polar molecules orient themselves in an electric field

Dipole moment (μ) is defined as the product of bond distance and the charge on any one atom. Its earlier unit was Debye (D), expressed in terms of esu-cm. Since bond distance and charge are of the order of 10^{-8} cm and 10^{-10} esu, respectively, $1\text{D} = 10^{-18}$ esu-cm. The modern unit of dipole moment is Coulomb meter or C m. In terms of Coulomb, charge is of the order of 10^{-19}, while in terms of meters, bond distances measure out in the order of 10^{-10}. Therefore, in terms of Coulomb meters, the dipole moment is expressed in the order of 10^{-30} C m. Further,

$$1\text{D} = 3.33 \times 10^{-30}\ \text{C m}$$

There are two broad applications of dipole moment. First, it helps in elucidating the per cent ionic character of a polar bond and second, it sheds light on the structure of a compound having partial covalent bonds.

1. **Determination of ionic character** The measured dipole moment of a molecule may be used to determine the degree of its ionic character. For example, consider the following data for an HCl molecule:

   Charge on H, if the bond is 100% ionic $= 1.602 \times 10^{-19}$ C

   Bond length $= 1.29 \times 10^{-10}$ m

   Calculated dipole moment $= (1.602 \times 10^{-19})\,(1.29 \times 10^{-10})$

   $= 20.67 \times 10^{-30}$ C m

   Experimental dipole moment $= 3.43 \times 10^{-30}$ C m

   If dipole moment is 20.67×10^{-30} C m, % ionic character = 100

   If dipole moment is 3.43×10^{-30} C m,

   $$\% \text{ ionic character} = \frac{3.43 \times 10^{-30}}{20.67 \times 10^{-30}} \times 100 = 16.6\%$$

   The ionic character of HCl molecule is approximately 17%. The same result is obtained by the application of Hannay–Smith equation (*see* p. 129).

Case Study 3.14 The bond distance in hydrogen fluoride (HF) molecule is 91.7pm. Find out the per cent ionic character, given that the observed dipole moment of the molecule is 6.6×10^{-30} C m. (Electronic charge $= 1.602 \times 10^{-19}$ C).

Bond distance between H and F $=$ 91.7 pm

$= 0.917 \times 10^{-10}$ m

Assuming that the bond is 100% ionic, the charge on H $= 1.602 \times 10^{-19}$ C

Calculated dipole moment $= (1.602 \times 10^{-19})\,(0.917 \times 10^{-10})$

$= 14.7 \times 10^{-30}$ C m

Observed dipole moment $= 6.6 \times 10^{-30}$ C m

$$\text{Per cent ionic character} = \frac{6.6 \times 10^{-30}}{14.7 \times 10^{-30}} \times 100 = 44.8\%$$

The HF molecule has approximately 45% ionic character.

---

2. **Determination of structure** Diagramatically, dipole moment of a molecule, or of a specific bond in a molecule, is indicated by drawing an arrow, so that the head points towards the negative end and the crossed-tail towards the positive end.

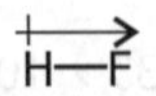

The measured value of dipole moment of a polyatomic compound reflects the resultant of bond moments present in the molecule. If there is more than one polar bond in a molecule, the overall dipole moment will depend on the orientation of these moments with respect to each other. The orientations may supplement each other, resulting in a high dipole moment, or may oppose each other, resulting in cancellation of dipoles.

For example, in CO_2 molecule, each C–O bond is polar due to the electronegativity difference between C and O. However, the molecule, as a whole, has zero dipole moment. This indicates that the molecule is linear and the polarities cancel each other.

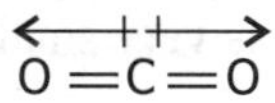

Water, on the other hand, has net positive dipole moment, indicating that the polarities of the two oxygen–hydrogen bonds supplement each other. This, in turn, is indicative of the angular shape of water molecule.

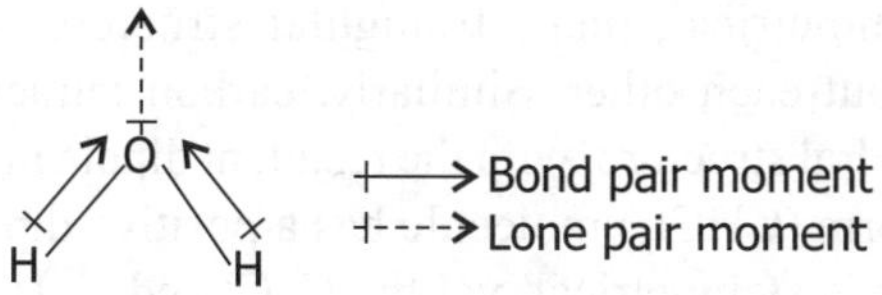

In fact, the dipole moment of water is 6.12×10^{-30} C m. This value includes not only the dipole moments of two O–H bonds, but also the dipole moment of the two lone pairs of electrons. Thus, the resultant dipole of a molecule is made up of two moments: the bond pair moment and the lone pair moment. The lone pair moment is directed away from the molecular skeleton.

In case of water, both the bond pair moment and the lone pair moment are aligned in the same direction. As a result, the net dipole moment of water is large. In other cases, the bond pair moment and lone pair moment may be aligned opposite to each other. The net dipole moment of such molecules is less than expected.

Consider, for example, ammonia (NH_3) and nitrogen trifluoride (NF_3). The former has an appreciably high dipole moment as compared to the latter. In NH_3, the electronegativity of N atom is higher than that of H atoms. The bond pair and the lone pair moments are aligned in the same direction and, therefore, reinforce each other.

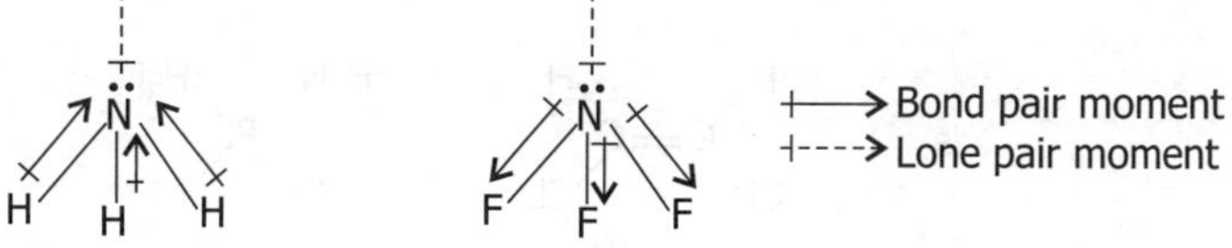

NH_3 NF_3

In NF_3, F atom is more electronegative than the central N atom. Therefore, the bond pair moments are aligned in one direction, while the lone pair moment is aligned in the opposite direction. The net effect is the near cancellation of the two moments.

Case Study 3.15 The dipole moment of carbon monoxide (CO) molecule is too small. Why?

Although there is an appreciable difference between the electronegativities of C and O, the dipole moment of CO is only 0.3×10^{-30} C m. To satisfy the octet rule, O donates a pair of electrons to C through a coordinate bond. The direction of coordinate bond is reverse of that expected from the high electronegativity of O. As a result, the net dipole moment is very small.

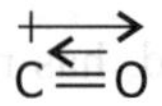

The dipole moment of boron(III) chloride (BCl_3) is zero, showing that the molecule has a symmetrical, plane triangular structure. The dipoles of each B–Cl bond cancel out each other. Similarly, carbon tetrachloride (CCl_4) has a symmetrical, tetrahedral structure, with the resultant dipole moment of zero. On the other hand, chloroform ($CHCl_3$) molecule has a positive dipole moment, because though the molecule is symmetrical, yet the C–Cl and C–H bonds have different degrees of polarities.

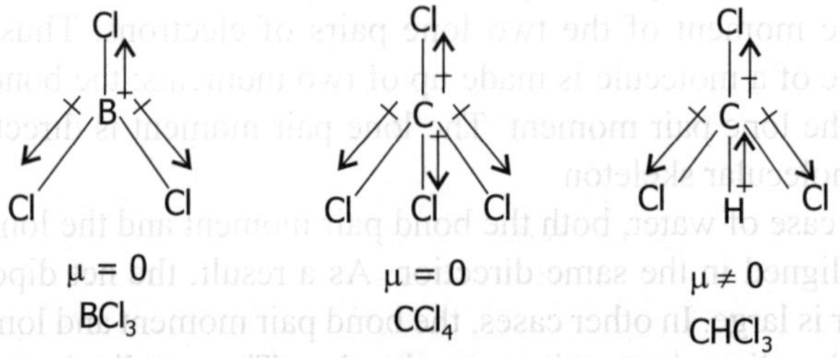

Dipole moment studies endeavor to distinguish between *cis* and *trans* isomers. The *trans* isomer invariably has zero or nearly zero dipole moment, while the *cis* isomer has a positive value of dipole moment.

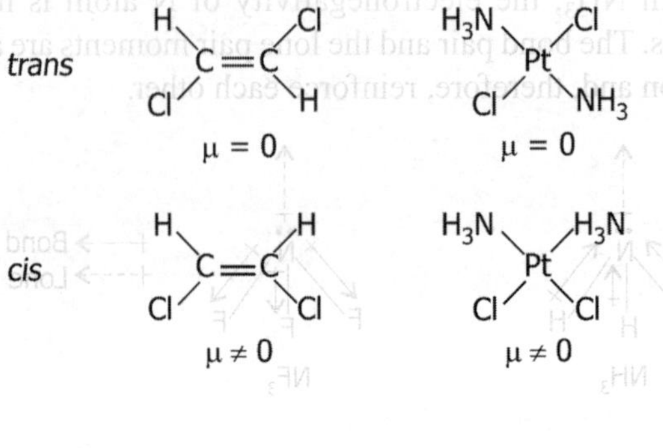

However, while making a distinction between *cis* and *trans* isomers, especially in metal complexes, on the basis of dipole moment studies, one should exercise caution. Consider the following example:

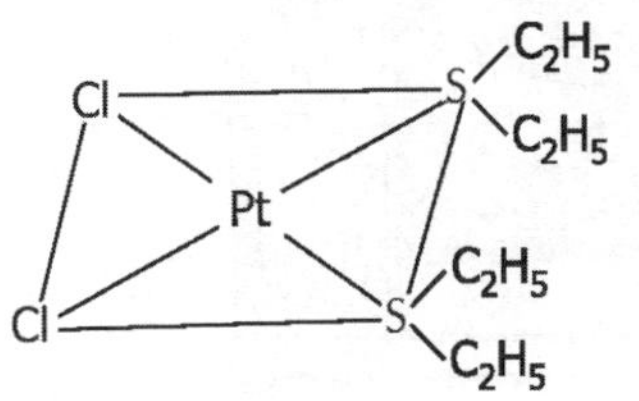

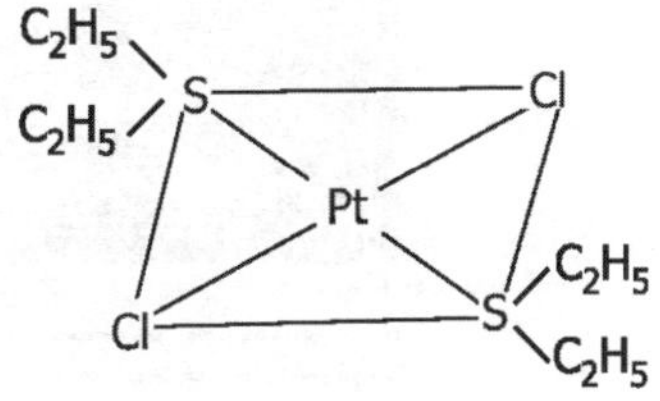

$\mu = 31.6 \times 10^{-30}$ C m
cis

$\mu = 8 \times 10^{-30}$ C m
trans

The dipole moment of the *trans* isomer, no doubt, is less than that of *cis* analog, but it is nowhere near zero. The lone pair moment arising out of the non-bonding electrons on sulfur of diethylthioether group contributes to the relatively large dipole moment of *trans* isomer.

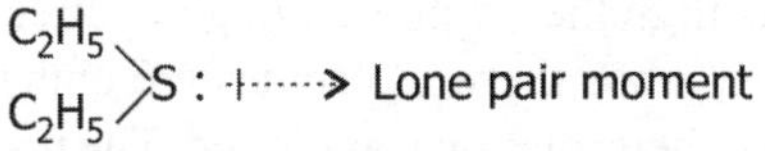

Case Study 3.16 The bond distance in lithium hydride (LiH) molecule is 159.5 pm and its dipole moment is 19.5×10^{-30} C m. Calculate the fractional charges on Li and H. (Electronic charge, $e = 1.602 \times 10^{-19}$ C)

Bond distance (r) = 159.5 pm = 1.595×10^{-10} m

Dipole moment (μ) = 19.5×10^{-30} C m

Charge = q.e = q. 1.602×10^{-19}

$\mu = (qe)r$

$$q = \frac{\mu}{er} = \frac{19.5 \times 10^{-30}}{1.602 \times 10^{-19} \times 1.595 \times 10^{-10}} = 0.76$$

The partial charge on Li is + 0.76 and that on H is –0.76.

Box 3.3

Dipole Moment and Microwave Cooking

In Chapter 1 we had learnt about electromagnetic radiations. Microwaves are a component of electromagnetic spectrum (Fig. 1.11), having a frequency between 3×10^8 Hz and 3×10^{11} Hz. During the past few years, microwave ovens have emerged as an ubiquitous household appliance for heating and cooking food (Fig. 3.33). The energy normally required for processing food items is about 3×10^9 Hz and it lies within the frequency range of microwaves.

Contd...

Fig. 3.33 Microwave ovens process food faster than conventional ovens

The microwave radiations, generated within the chamber of the oven are absorbed by those molecules in the foodstuff which possess dipole moments. The most common chemical entity in food items that absorbs microwaves are the water molecules which, being polar, possess appreciable dipole moment. On acquiring extra energy from the microwaves, the water and other polar molecules tend to rotate, so as to align their dipoles along the direction of propagation of microwaves. However, since these molecules are embedded within the solid matrix of the foodstuff, their rotation is inhibited. The resistance to free rotation results in friction between polar molecules on one hand, and between polar molecules and the non-polar food matrix on the other. The friction, in turn, appears as heat which warms up or cooks the food. The microwave food holder is made of a non-polar material and hence does not get heated-up.

In conventional ovens, the heat traverses through the food product by conduction. This means that the heat travels stepwise from one layer of the food to the next. This is a slow process. In microwave ovens, the polar molecules in all the layers of food stuff get heated up simultaneously. Hence the food processing is faster in microwave ovens than in conventional ovens.

Resonance

The concept of resonance states that if all the properties of a chemical species cannot be explained by a single molecular structure, it is pertinent to formulate alternative structures which, in totality, cover all the characteristics of the substance in question. The various possible structures, called *resonating structures* or *canonical forms,* are segregated from each other by double head arrows. The actual structure is a hybrid of the canonical forms. It is called *resonance hybrid* and cannot be pictorially represented, i.e., cannot be drawn on paper.

For example, the carbon dioxide molecule may be truly represented by the following three canonical forms:

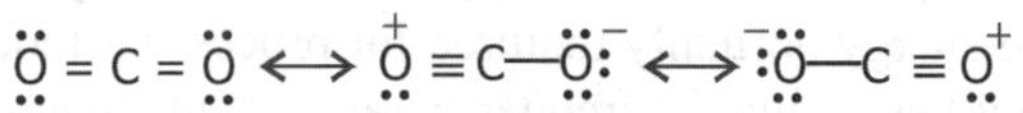

(I) (II) (III)

Structure (I), formulated on the basis of valence bond theory, explains virtually all the properties of carbon dioxide. However, the C–O bond length in carbon dioxide is less than expected—a characteristic that structure (I) fails to explain. Structures (II) and (III) involve a C≡O triple bond and their contribution to the resonance hybrid serves to emphasize why the C=O double bond in carbon dioxide has a shorter bond length.

The actual structure of carbon dioxide is neither one of the above possibilities, nor an average of all three possibilities. If it had been one of the possibilities say, for example, structure (I), then there would have been no need to formulate structures (II) and (III) in the first place. If it had been an average of all three canonical forms, each structure would have explained one-third characteristics of carbon dioxide. All the properties of carbon dioxide are deservedly explained by structure (I), except, of course, the abnormality in C–O bond length, for which structures (II) and (III) have been invoked. This means that the resonance hybrid of carbon dioxide is not made up of equal parts of structures (I), (II) and (III). If it had been so, an attempt could have been made to draw its framework. Instead, it is composed of a very high content of structure (I) and an infinitesimally small content of structures (II) and (III). Therefore, it defies pictorial depiction.

The following canonical forms (IV–VI) represent resonance in carbonate ion (CO_3^{2-}).

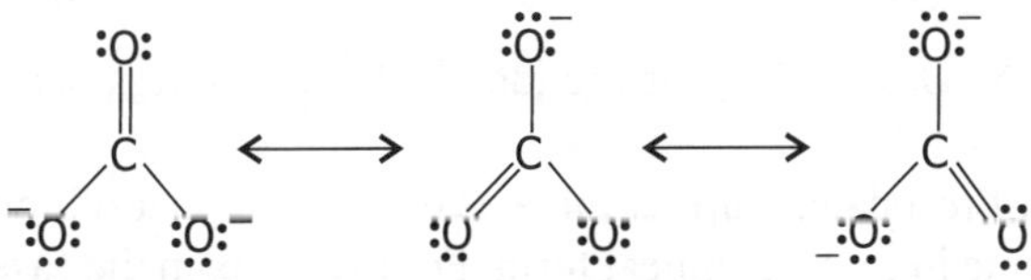

(IV) (V) (VI)

Structure (IV), considered in isolation, suggests that there are two long and one short C–O bond lengths in the carbonate ion. However, all three bond lengths in carbonate ion have been found to be identical. Their value is in between that of C–O single bond and C=O double bond, indicating partial double bond character. These properties can be explained only if canonical forms (V) and (VI) are incorporated along with canonical form (IV) in the resonance hybrid of carbonate ion.

While writing the resonance structures of any molecule or ion, the relative positions of the atoms should not be changed. Thus in each canonical form of carbon dioxide, the constituent atoms are arranged in the order O—C—O. The resonating structures arise by transfer of non-bonding electron pairs from one atom of the molecule to another. This transfer is possible only if the molecule is planar. Non-planar structures either contribute less or do not contribute at all in building-up the resonance hybrid. The more the non-planarity of the resonating structure, the lesser is its contribution to the hybrid.

The canonical forms that do not carry any charge are more stable than those in which there are positive and negative centers. For example, extra energy must be imparted to structures (II) and (III) of carbon dioxide so as to overcome the attraction between two oppositely charged O atoms. Structure (I), on the other hand, is neutral, has lower energy and hence greater stability.

The difference between the energy of the most stable canonical form, structure (I) in case of carbon dioxide, and that of the resonance hybrid is called *resonance energy.* Greater the resonance energy of a compound, more is its stability.

Among the structures which possess polar charges, those contributing forms are more stable in which the negative charge is concentrated on the electronegative element and positive charge resides on the electropositive element. This is yet another reason for low stability, and hence smaller contribution, of structures (II) and (III) of carbon dioxide, as in these canonical forms one of the O atoms is positively charged.The carbonate ion is, however, appreciably stable because the negative charge resides on the highly electronegative O atoms, although some amount of stability is lost since extra energy has to be imparted to all the resonating forms, so as to prevent repulsions between two negative O atoms. Similarly, in the case of nitrate ion (NO_3^-), negative charge is distributed on more electronegative O atoms, while positive charge is concentrated on the relatively less electronegative N.

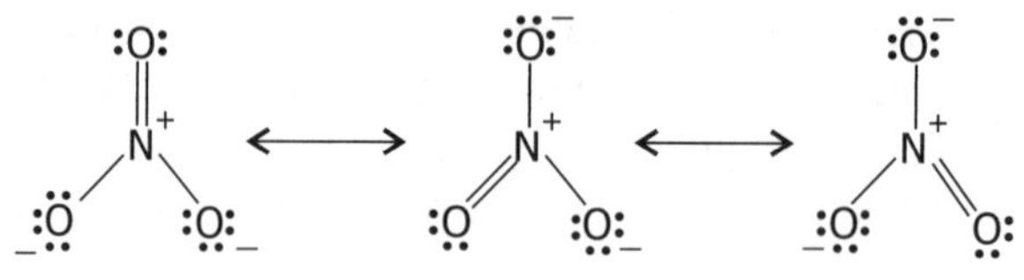

(VII) (VIII) (IX)

All the three N–O bond lengths are identical, and possess partial double bond character.

The concept of resonance further dictates that the number of unpaired electrons should be the same in each canonical form. For example, in the case of nitric oxide (NO_2),

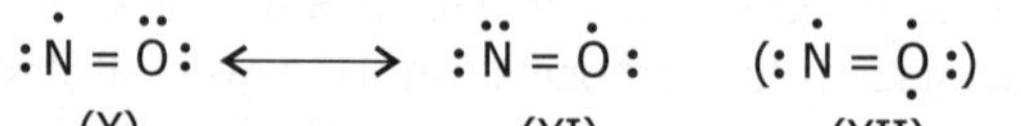

(X) (XI) (XII)

canonical forms (X) and (XI), each with a single unpaired electron contribute to the resonance hybrid, while structure (XII), with three unpaired electrons, does not.

Case Study 3.17 The cyanate ion is stable but the isomeric fulminate ion is not. Why?

The cyanate ion, OCN^- is a resonance hybrid of three canonical forms.

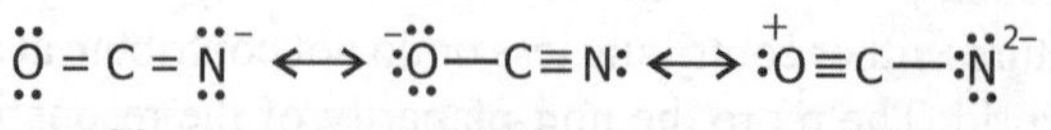

(I) (II) (III)

Structure (II) contributes to the maximum extent to the resonance hybrid as the most electronegative atom, oxygen has a negative charge in this canonical form. This is followed by structure (I) in which the nitrogen atom had the negative charge. The contribution of structure (III) is too little as the electronegative oxygen bears a positive charge. The combined contribution of the three resonating structures is responsible for the stability of cyanate ion.

Theoretically, three cannonical forms may be drawn for the isomeric fulminate ion, CNO^-.

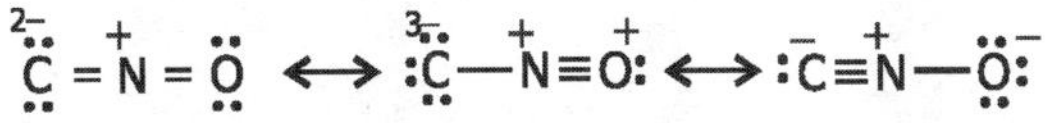

(IV) (V) (VI)

None of these structures contribute to the stability of fulminate ion as each has a negative charge on carbon, the least electronegative of the constituent atoms. For this reason fulminate ion is unstable. The only common ionic fulminates are of mercury and silver—and both are explosive in nature.

Box 3.4

A Blend of Valence Bond Theory and Molecular Orbital Theory

The valence bond theory and the molecular orbital theory use different pathways to explain the same concepts. It is therefore but natural to ask as to which theory is better. Valence bond theory is better because of its simplicity; molecular orbital theory is better because of its accuracy. Best of all, however, is a combination of the two theories.

Valence bond theory suffers from two shortcomings: firstly, it fails to make correct prediction about the paramagnetic nature of molecules such as oxygen; and secondly, for molecules like ozone, no one structure based on valence bond theory provides an accurate picture. Therefore, the concept of resonance has to be invoked. The first problem occurs rarely, but the second is more common. While writing the resonating structures of a molecule, the best bargain is to use a combination of valence bond theory and molecular orbital theory—the former to describe the formation of σ bonds and the latter to describe the formation of π bonds.

Consider, for example, ozone. According to valence bond theory, ozone is a resonance hybrid of two structures, both of which have two oxygen-oxygen σ bonds and one oxygen-oxygen π bond.

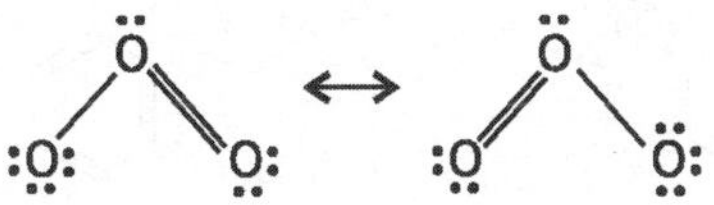

I II

Structure (I) has an extra lone pair of electrons in the p orbital on the left-hand oxygen atom and a π bond to the right-hand oxygen. Structure (II) has an extra lone pair of electrons in the p orbital on the right-hand oxygen and a π bond to the left-hand oxygen (Fig. 3.34). The actual structure is a hybrid of structures (I) and (II), in which the four electrons (one lone pair and one π bond forming pair) are delocalized over the entire skeleton of the molecule.

Contd...

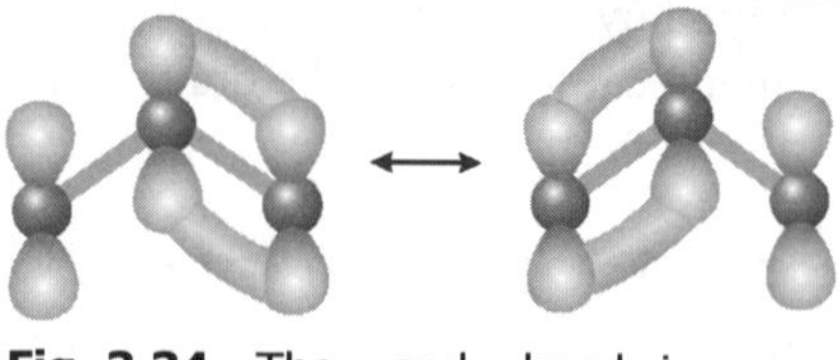

Fig. 3.34 The σ and π bonds in ozone

Valence bond theory gives a good description of the oxygen-oxygen σ bonds but a poor description of bonding involving π orbitals, whereby electrons are spread out over the entire molecule. The molecular orbital theory, on the other hand, does the reverse.

Thus the canonical forms of any molecule are best described by a blend of valence bond theory and molecular orbital theory. The formation of σ bonds is explained in valence bond terminology while the delocalization of the π cloud is explained in molecular orbital terminology.

Case Study 3.18 What does the term *steric inhibition of resonance* signify?

Steric inhibition of resonance implies that the contribution of various canonical forms to the resonance hybrid decreases in proportion to the steric over-crowdedness in the molecule. Consider, for example, aniline molecule, a resonance hybrid of the following canonical forms:

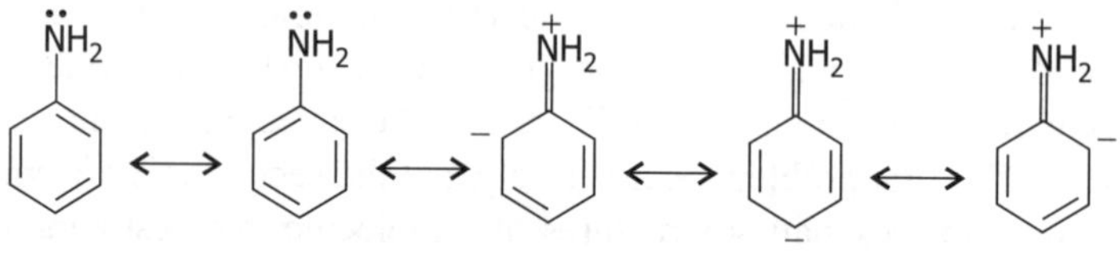

(I) (II) (III) (IV) (V)

Structures (III–V) make aniline a weak base. When a mineral acid is added to aniline, the proton (H^+) is repelled by positively charged nitrogen in these structures.

Now consider N-methylaniline (VI).

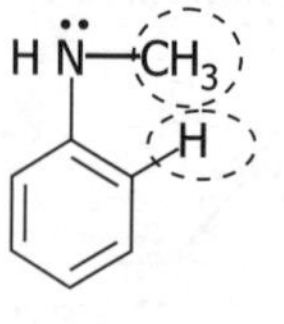

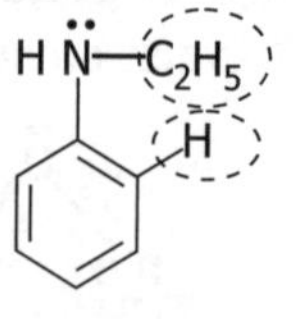

(VI) (VII)

The steric interaction between the N-methyl group and the *ortho*-hydrogen atom makes the aromatic ring puckered. As a result, the molecule is now not completely planar and the contribution of resonating structures, corresponding to canonical forms (III–V) of aniline, decreases. N-methylaniline is more basic than aniline. In N-ethylaniline (VII), the steric interaction is more pronounced since the N-ethyl group is bulkier than the N-methyl function. The contribution of canonical forms decreases further. The basic character increases concomitantly.

Bond length

Bond length refers to the average distance between the nuclei of two covalently bonded atoms in a molecule. The need for considering *average distance* arises because a covalent bond is not static in nature. The atoms united by the covalent bond vibrate with respect to each other. At one instant, the bond is stretched, while at another, it is compressed. The mean distance between the two atoms reflects a state of maximum bond energy and minimum repulsion.

Ideally, the bond length should be the sum of the covalent radii of the combining atoms. For example, the covalent radius of carbon is 77 pm, while the carbon–carbon bond distance is 154 pm. The same analogy holds good for silicon–silicon and silicon–carbon bond lengths, as shown by the following data:

	Covalent radius (pm)		Bond length (pm)
C	77	C—C	154
Si	117	Si—Si	234
		Si—C	194

However, for most of the bonds, the picture is not so simple. The actual bond length deviates appreciably from that calculated on the basis of additive rule. Some important factors which cause this deviation are outlined below:

1. **Bond order** As the number of bonds holding the two atoms together increases, the bond strength increases and the bond length decreases. This is because the combining atoms are held more tightly when these are united by multiple bonds, than when connected by a single bond. Thus, the bond distance between two atoms of an element X would vary in the order:

$$-X\equiv X- \; < \; -X=X- \; < \; -X-X-$$

This is borne out by the following data:

Bond	Bond length (pm)	Bond	Bond length (pm)
C–C	154	N–N	145
C=C	134	N=N	124
C≡C	120	N≡N	110

2. **Lone electron pairs** The F–F, O–O and N–N bond distances are longer than expected. This has been attributed to strong repulsive interactions between the lone pairs of electrons on adjacent atoms. Since F, O and N are relatively small in size, the lone pairs on adjacent atoms are quite close and, therefore, repel each other more strongly than in, say, Cl–Cl or S–S bonds.

For example, the larger than expected N–N bond length in hydrazine is restored to its normal size by protonation of lone pairs and formation of hydrazinium ion, $[H_3N—NH_3]^{2+}$. Similarly, the N–O bond distance in hydroxylamine is larger than that in trimethylamine oxide. In the latter case, there is no lone pair of electrons on nitrogen.

Compound	Bond	Bond length (pm)
$H_2\ddot{N}{-}NH_2$	$\ddot{N}{-}N$	147
$[H_3N{-}NH_3]^{2+}$	N–N	140
$H_2\ddot{N}{-}OH$	N–O	146
$(CH_3)_3N^+ \rightarrow O^-$	N–O	136

3. **Electronegativity difference** As the electronegativity difference between the two constituent atoms of the bond increases, the bond length gets reduced. This is because with greater difference between the electronegativities of the uniting atoms, the bond acquires polar character. The polar structure contributes to the resonance stability of the bond. For example,

$$C{-}F \longleftrightarrow \overset{\delta+}{C}{-}\overset{\delta-}{F}$$

The electrostatic attraction between the two atoms occurs in addition to the attraction prevailing in a normal covalent bond. The polar contribution thus strengthens the bond and, in turn, makes the bond length small.

Bond	Bond length (Calculated) (pm)	Bond length (Experimental) (pm)
C–Br	191	188
C–F	149	136
B–Br	202	195
B–F	160	142

Mathematically, the bond length r_{A-B} between two atoms A and B is related to the electronegativity difference given by *Schomaker and Stevenson equation,*

$$r_{A-B} = r_A + r_B - 9\,|\chi_A - \chi_B| \tag{3.29}$$

Here, r_A and χ_A are the covalent radius and electronegativity of atom A, respectively while r_B and χ_B are the corresponding parameters for atom B. This equation clearly signifies that as the electronegativity difference between the atoms A and B increases, the bond length gets reduced below the value expected from simple addition of the covalent radii.

4. **pπ-dπ interactions** Let us calculate the C–F and Si–F bond lengths by applying Schomaker and Stevenson equation (3.29) to the following data:

Bond	$r_{(A)}$ (pm)	$r_{(B)}$ (pm)	χ_A	χ_B	r_{A-B} (Calculated) $r_A + r_B - 9\lvert\chi_A - C_B\rvert$ (pm)	r_{A-B} (Experimental) (pm)
C–F	77	72	2.5	4.0	135	136
Si–F	117	72	1.8	4.0	169	156

It is seen that the calculated value for C–F bond length matches with the experimental value, while that of Si–F bond length does not. The actual Si–F bond distance is less than that calculated on the basis of Schomaker and Stevenson equation. This has been attributed to the overlapping of a vacant d orbital of Si with a filled p orbital of F. Such an overlap, called pπ-dπ interaction, occurs in addition to the normal σ overlap between Si and F atoms. Thus, pπ-dπ interactions impart multiple bond character to the existing bond. Multiple bonding strengthens—and shortens—the bond length. Carbon does not have d orbitals and, therefore, pπ-dπ interactions are not possible for C–F bond. The latter does not possess multiple bond character and hence its bond length has normal dimensions.

5. **Degree of s character** The hybrid orbitals that have a greater s character are better endowed to overlap with a substituent than those which have a lesser s character. More effective overlap means a stronger—and shorter—bond. Thus the C–H bond length decreases if the hybrid orbitals of C have more s character and vice versa.

Compound	Hybridization	s character	r_{C-H} (pm)
H—C≡C—H	sp	50%	105.7
$H_2C{=}CH_2$	sp^2	33%	107.9
CH_4	sp^3	25%	109.4

In a trigonal bipyramid molecule, the central atom is sp^3d hybridized. This type of hybridization may be considered as a combination of two components: (i) sp^2 hybrid orbitals, which are directed towards equatorial positions and (ii) pd hybrid orbitals, which are directed towards axial positions ($sp^3d = sp^2 + pd$). The component directed towards equatorial substituents has 33% s character, while that directed towards axial substituents has no s character. As a result, the hybrid orbitals of the central atom have a more effective overlap with equatorial groups than that with the axial groups. Accordingly, the axial bond lengths are larger than equatorial bond lengths. The following data supports this assertion:

Compound	Axial bond length (pm)	Equatorial bond length (pm)
PCl_5	210	204
PF_5	157	153
ClF_3	171	162
BrF_3	181	172

Case Study 3.19 Why does a dioxygen difluoride (O_2F_2) molecule possess a very short O–O bond length?

The O–O bond distance in O_2F_2 is extremely short. In fact, it is comparable to the bond length in oxygen molecule (122 pm), where the two constituent atoms are held together by a double bond. The short bond length in O_2F_2 may be explained on the basis of *hyperconjugation*. The latter shifts the non-bonding electrons of O to F and, in turn, imparts double bond character to O–O bond.

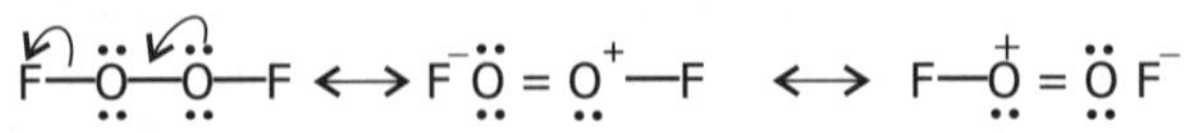

Since a double bond holds the two O atoms in O_2F_2 molecule together, the bond length is less than expected (*see also* Chapter 18, under oxygen fluorides, p. 552).

Bond angle

The angle subtended by two adjacent bonds in a molecule is called bond angle. It is an important physical parameter, as it sheds light on the three-dimensional arrangement of orbitals about the atom common to the two bonds. This, in turn, determines the symmetry and shape of the molecule. A slight deviation from the ideal value of bond angle distorts the symmetry and alters the shape of the molecule. As a result, the physical and chemical properties of the substance are profoundly affected.

Here are enunciated some important factors which influence the values of bond angles.

1. **Lone pairs** We have already seen in the section on hybridization that presence of one or more lone pairs of electrons on the central atom reduces the bond angle from the ideal value. Hence, although ammonia and water molecules exhibit tetrahedral geometry, their bond angles are less than that of methane. The repulsion between two lone pairs is stronger than that between a lone pair and a bond pair which, in turn, is stronger than that between two bond pairs. As a result, the lone pairs induce the bond angles to narrow down.

   Two lone pairs in chlorine trifluoride and one lone pair in bromine pentafluoride cause the bond angle in both the compounds to be less than 90°. Further, the following example shows that two non-bonding electrons cause a greater deviation in bond angle than a single non-bonding electron:

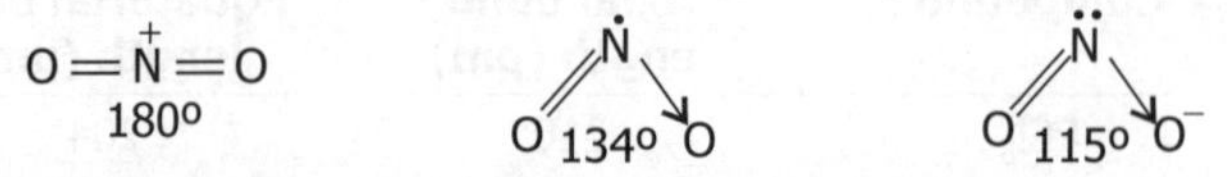

2. **Electronegativity of substituents** The electronegativity of the substituents, relative to that of the central atom, has a marked influence on the bond angles.

   As the electronegativity of the substituents increases, the bonded pairs of electrons are pulled away from the central atom. When the bonding electrons are near the substituents, the mutual distance between them is large; when the bonding electrons are near the central atom, the mutual distance between them is small. Accordingly, the repulsion between the bond pairs is lessened—and so is the bond angle—as the electronegativity of the substituent increases.

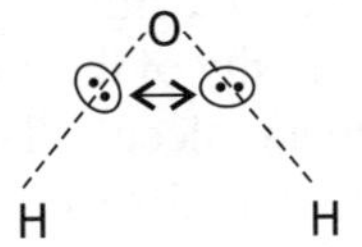

Smaller distance between bonded pairs results in more repulsion, causing the bond angle to increase.

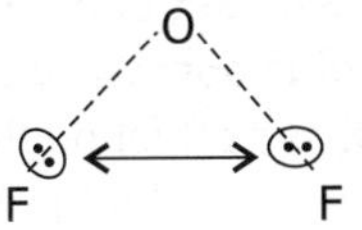

Larger distance between bonded pairs results in less repulsion, causing the bond angle to decrease.

H_2O 104° 5′	F_2O 103° 2′
NH_3 107° 3′	NF_3 102°
PBr_3 101° 5′	PCl_3 100°

Conversely, as the electronegativity of the central atom decreases, the bond angle also decreases.

H_2O 104° 5′	H_2S 92°
NH_3 107° 3′	PH_3 93° 3′

The variation in bond angles with electronegativities of the substituents may also be explained by invoking another concept. Consider, once again, water and oxygen difluoride molecules. In both the compounds, the central atom, O, carries two lone pairs of electrons and is sp^3 hybridized. However, O tends to direct the s character of the hybrid orbitals towards the lone pairs and p character towards the substituents (H or F). This is because s orbital has a greater electronegativity and is consequently oriented towards the non-bonding electron pairs. The p orbitals have lesser electronegativity and are, therefore, aligned towards the substituents. Greater the electronegativity of the substituents, greater will be the amount of p character that they tend to extract from the hybrid orbitals. This, in turn, will cause a greater segregation of s and p character of the hybrid orbitals.

The F atoms in OF_2 are able to draw more p character towards themselves than H atoms in water. Since the three p orbitals in an atom are oriented at 90° with respect to each other, the bond angle in OF_2 is closer to this figure. In case of water, the s and p character of the hybrid orbitals are not altogether differentiated. Therefore, the bond angle is more close to the tetrahedral value of 109° 28′ than to 90°.

3. **Multiple bonds** The presence of multiple bonds in a molecule does not affect its stereochemistry. However, these do change the bond angles. The reason for this is that a double bond containing four electrons, occupies more space than a single bond containing two electrons. Likewise, a triple bond, with six bonding electrons, occupies a still larger space than even the double bond. The greater electron density of the multiple bonds exerts a repulsive effect on the other bonds in a molecule. This causes the angle involving the double bond to increase. Simultaneously, the other bonds in the molecule are pushed closer and the bond angles subtended by them are reduced.

For example, although the bond angle in ethene and its derivatives ($X_2C{=}CX_2$) is expected to be 120°, the ∠XCC is above this value, while the ∠XCX is below this value. The same effect is observed in the derivatives of formaldehyde (CXOX) which, like ethene, are expected to have a plane triangular geometry.

$H_2C{=}CH_2$	$F_2C{=}CH_2$
∠XCX = 117° 7′	109° 3′
∠XCC = 122°	125° 3′
$H_2C{=}O$	$F_2C{=}O$
∠XCX = 115° 8′	108°
∠XCO = 122° 1′	126°

In case of tetrahedral geometry, the bond angle is reduced below the ideal value of 109° 28′, in case there is a double bond in the molecule.

$Cl_3P{=}O$	$F_3P{=}O$
∠XPO = 103° 6′	102° 5′
$Br_3P{=}S$	$Cl_3P{=}S$
∠XPS = 106°	100° 5′

4. **pπ-dπ interactions** As has been stated in the section on bond length, pπ-dπ interactions arise when the vacant d orbitals of an atom overlap with the filled p orbitals of another atom in the molecule. The charge density gets transferred from the p to d orbitals. The bond order now becomes greater than one, although it is certainly less than two. Nevertheless, the increase in bond order is accompanied by an increase in electron density, and the bond angle is affected in much the same way as in molecules possessing true multiple bonds.

We have seen that as the electronegativity of substituents increases, the bond angle decreases. Thus, the variation in bond angle follows the order:

$$NH_3 > NF_3$$
$$POCl_3 > POF_3$$

However, when a comparison is made between the bond angles of dichlorine oxide and water, this order is reversed. The same applies to the variation in bond angles of phosphine and trifluorophosphine. The observed bond angles are as follows:

Cl_2O	H_2O
110° 8′	104° 5′
PF_3	PH_3
104°	93° 3′

In both the examples, the compound with more electronegative substituent has a larger, rather than smaller, bond angle. Due to the pπ-dπ interactions, the following canonical forms contribute to the structure of dichlorine oxide, indicating that the charge density is transferred from O to Cl:

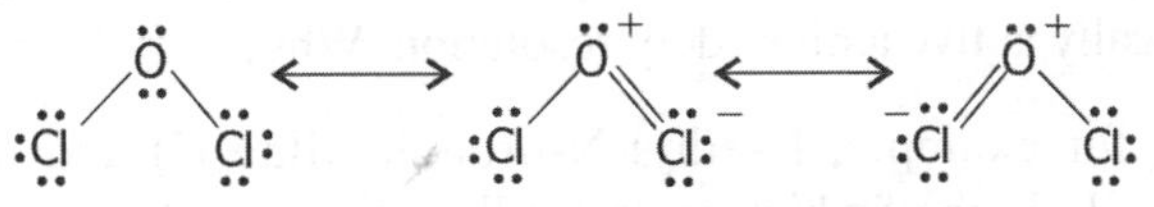

(I) (II) (III)

The increase in electron density about the chlorine–oxygen bond (Cl^-=O^+), in structures II and III, induces a strong repulsive effect on the other chlorine–oxygen bond (Cl–O). This causes the bond angle to open up. Due to absence of d orbitals in hydrogen, analogous structures are not possible in water.

The pπ-dπ interactions are not possible in phosphine as well. However, trifluorophosphine can have such interactions, giving rise to structures such as:

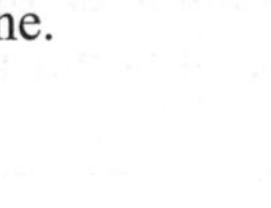

The larger size and high charge density of the double bond increases the bond angle relative to phosphine.

A similar effect is observed in arsenic trifluoride, the bond angle of which is greater than that in arsine.

AsF_3	AsH_3
102°	91° 8′

5. **Dipole inducement** In its gaseous state, the barium fluoride (BaF_2) molecule has essentially covalent bonding. Ideally, its structure should be linear. However, in this geometry, the dipole moment would be zero. If the molecule acquires an angular shape, the dipole moment would have a positive value.

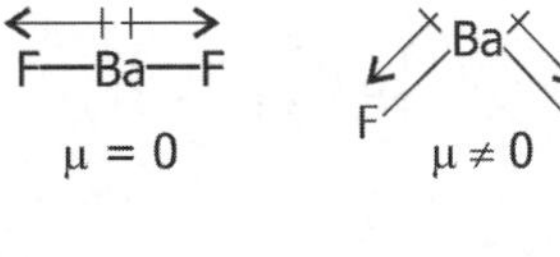

We have seen that ionic character in a polar molecule increases with dipole moment. Thus, if BaF_2 molecule attains a 'V' shape, its bond energy, and hence stability, increases due to contribution from polar structure, in which Ba has a partial positive charge and F atoms have partial negative charges.

For this reason, the observed angle of BaF_2 is less than 180°. As the bond angle gets smaller, the dipole moment, and hence the stability of the molecule, increases. However, there is a limit to which the bond angle can be reduced. At very low values, the two F atoms start repelling each other. The actual bond angle represents a balance between dipole stabilization and electrostatic repulsion.

Case Study 3.20 Optically active phosphines can be resolved into enantiomers, while optically active amines defy resolution. Why?

Consider, for example, N-ethyl-N-propylaniline (I) and the corresponding phosphine (II), both of which are optically active.

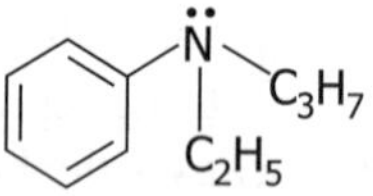

(I)

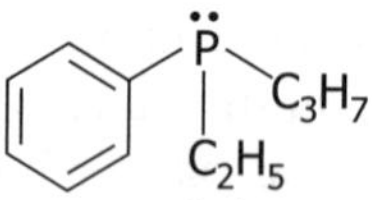

(II)

The D-form of the amine undergoes inversion to L-form, resulting in an optically inactive racemic mixture. The phosphine too undergoes inversion, but with difficulty. This is due to the bond angles in these compounds. During inversion, the molecule is transformed from pyramidal geometry to plane trigonal geometry and then reverts to pyramidal shape. Its bond angle has to momentarily increase to about 120°—the requirement for the transitional, plane triangular geometry. In the phosphine derivative, the bond angles are small. Therefore, it has to expend more energy to increase the bond angle and to undergo inversion. This means that the barrier to inversion is high. As a result, the resolution into D- and L-forms is possible. For the amine derivative, the bond angles are large and consequently the barrier to inversion is low—so low that racemization takes place faster than resolution. Hence, it is not possible to isolate optically active isomers of amines.

Formal charge

The formal charge on an atom in a molecule is the total number of valence electrons of that atom minus the number of electrons it 'possesses' in the molecule. The term 'possesses' here means the sum of all the unshared valence electrons on that atom and half of its shared valence electrons.

$$\text{Formal charge of an atom} = V - \left(u + \frac{S}{2}\right)$$

where V is the number of valence electrons; u is the number of unshared electrons; and S is the number of shared electrons on that atom. Let us consider the molecule of ozone, O_3.

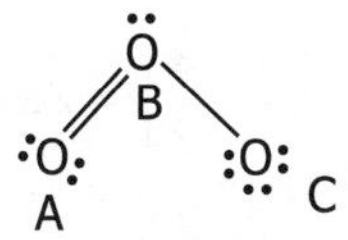

The formal charge on each of oxygen atoms A, B and C may be calculated as follows.

$$O_A : 6 - \left[4 + \frac{4}{2}\right] = \text{Zero}$$

$$O_B : 6 - \left[2 + \frac{6}{2}\right] = +1$$

$$O_C : 6 - \left[6 + \frac{2}{2}\right] = -1$$

Sum of formal charges of all atoms = Zero + (+1) + (–1) = Zero

For assigning formal charges, the following two rules may be taken into consideration.

1. In a neutral molecule, the sum of formal charges is zero.

   We have already seen that the sum of formal charges of all the constituent atoms of ozone is zero. Let us consider the example of nitrous oxide, N_2O.

The formal charges on its atoms may be calculated as follows.

$$\text{Left hand N} : 5 - \left[2 + \frac{6}{2}\right] = \text{Zero}$$

$$\text{Middle N} : 5 - \left[0 + \frac{8}{2}\right] = +1$$

$$O : 6 - \left[6 + \frac{2}{2}\right] = -1$$

Sum of formal charges of all atoms = Zero + (+1) + (–1) = Zero

2. In a polyatomic ion, the sum of formal charges of all atoms in equal to the overall charge on the ion.

   Let us consider the carbonate anion, CO_3^{2-}.

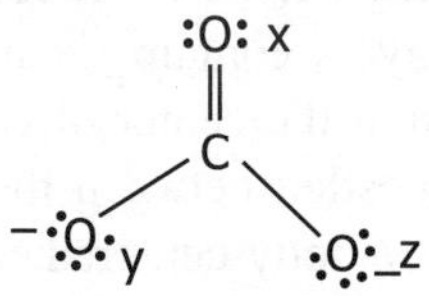

The formal charges on the constituent atoms are as follows:

$$C : 4 - \left[0 + \frac{8}{2}\right] = \text{Zero}$$

$$O_x : 6 - \left[4 + \frac{4}{2}\right] = \text{Zero}$$

$$O_y : 6 - \left[6 + \frac{2}{2}\right] = -1$$

$$O_z : 6 - \left[6 + \frac{2}{2}\right] = -1$$

Sum of formal charges = Zero +Zero + (–1) + (–1) = –2

The sum of formal charges is thus equivalent to the overall charge on the carbonate anion.

Case Study 3.21 Calculate the formal charges of all the atoms of cyanate ion.

The cyanate ion may be displayed as,

$$\left[\underset{\cdot\cdot}{\ddot{N}} = C = \underset{\cdot\cdot}{\ddot{O}}\right]^-$$

The formal charges may be calculated as follows.

$$N : 5 - \left[4 + \frac{4}{2}\right] = -1$$

$$C : 4 - \left[0 + \frac{8}{2}\right] = \text{Zero}$$

$$O : 6 - \left[4 + \frac{4}{2}\right] = \text{Zero}$$

Sum of formal charges (–1) is equivalent to the overall charge on the anion.

Metallic Bond

The metals possess the following physical properties:

1. **Close-packed structures** Metals have two types of close-packed arrangements: cubic close-packed (CCP) and hexagonal close-packed (HCP). Each has a coordination number of 12 in the metallic lattice. In CCP arrangement, the unit cell consists of eight metal atoms at the corners of a cube and one metal atom at the center of each face of the cube. In an HCP system, the unit cell consists of three layers of atoms. The top and bottom layers contain six metal atoms at the corners of a hexagon and one atom at the center of each hexagon. The middle layer contains three atoms nestled between the atoms of the top and bottom layers. In a few metals, a body-centered cubic (BCC) arrangement with a coordination number of 8 prevails. In such cases, a metal atom is placed

at each corner of a unit cubic cell, while one metal atom occupies the center of the cube.

2. **Conductivity** Metals are good conductors of electricity and heat. The flow of current is due to the movement of electrons through the conductor. This implies that the conductor must possess some mobile electrons which pass easily from one atom to another. Further, there is a close relationship between conduction of electricity and conduction of heat; the mobile electrons which carry electric current also play an important role in thermal conductivity. It has been observed that metals which are good conductors of electricity are also good conductors of heat. According to *Wiedemann–Franz law* (*see* Case Study 3.22, p. 150), the ratio of thermal conductivity to electric conductivity at ordinary temperatures is constant for all metals.

   The conductivity of metals decreases with increasing temperature. As has already been stated, conductivity arises due to the flow of electrons. This movement of electrons, however, is restricted due to their collision with the atomic core. With a rise in temperature, the core begins to vibrate with a greater amplitude. This means that the core now covers a wider expanse. As a result, the flow of electrons is impeded to a greater extent. The outcome is a fall in conductivity.

3. **Luster** Metals are bright, shiny and highly reflective. The metallic luster arises because the mobile electrons in the lattice absorb visible light. The energy acquired as a result of this absorption, momentarily promotes the free electrons to excited states. The energy is almost immediately re-emitted in the form of visible light when the electrons drop back to their original, ground state.

4. **Malleability and ductility** Metals are malleable, i.e., they may be deformed into various shapes, and ductile, i.e., they may be drawn into wires. As we shall see in the sea of electrons theory described below, these characteristics too are dependent on the mobile electrons. Fewer the number of mobile electrons, greater the malleability and ductility of the concerned metal.

5. **Cohesive forces in the lattice** Metals possess large cohesive forces in their lattice. These hold the metal atoms together in a metallic lattice. The degree of cohesion is reflected in the melting and boiling points of the metals. On descending a group in the periodic table, the melting and boiling points decrease, indicating that these parameters vary inversely with size. On moving from left to right in a period, these parameters increase (among metals only), showing their dependence on the number of valence electrons.

6. **Formation of alloys** Metals form alloys. A metal can be mixed with another metal, often over a wide range of compositions, to give a variety of alloys. These alloys find a wide spectrum of industrial applications. The properties of alloys are discussed later in this chapter.

Case Study 3.22 What is Wiedemann–Franz law?

Wiedemann–Franz law is of fundamental importance in the theory of metallic state, as it demonstrates a connection between two characteristic properties of metals—electrical conductivity (λ) and thermal conductivity (χ). It states that electrical and thermal conductivities are proportional to each another. This means that a metal having high electrical conductivity would also have high thermal conductivity and vice versa. Further, at a particular temperature T, the factor $\lambda/\chi T$ is constant for all metals.

Theories of Bonding in Metals

The physical properties of metals, cited above, present a paradox. These reflect the presence of a large number of valence electrons in metal atoms. This is specifically true when we consider the high coordination number, conductivity and luster of metals. However, the metals have only a few valence electrons. Group 1, 2 and 13 metals have just one, two and three valence electrons, respectively, but show very high conductivities. In fact, the non-metals have much larger number of valence electrons than the metals. The theories on metallic bonding attempt to explain the physical characteristics of metals in terms of a meager number of valence electrons. There are two such theories.

1. Sea of electrons theory

According to this theory, a metallic lattice consists of rigid positive ions, arranged in a close-packed array, with their valence electrons spread throughout the crystal. In the close-packed arrangement, most of the rigid spheres are so arranged that each is in contact with six other analogs in the same layer (Fig. 3.35). A second layer is formed by placing the center of spheres on the interstices (A) of the first layer. The second layer is shown as dotted spheres in Fig. 3.35 and its interstices are marked B. In this layer too, each sphere is in contact with six neighbors. A third layer may be similarly constituted by placing the center of its sphere either on depressions A or on depressions B. A three-dimensional network results. A sphere in any one layer has a large number of neighbors, not only in its own layer, but in the layers above and below it. For this reason, each atom in a metallic lattice has a high coordination number.

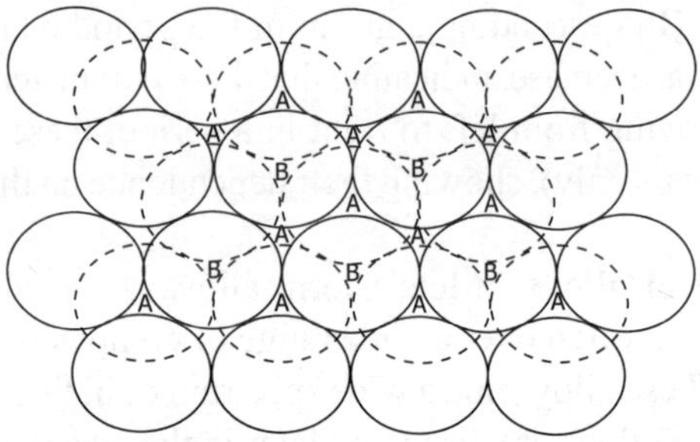

Fig. 3.35 Close-packed arrangement of atoms in a metal crystal

The valence electrons of each positive sphere are distributed among the interstices (A or B) of a particular layer. This serves to hold together the spheres within each layer (Fig. 3.36).

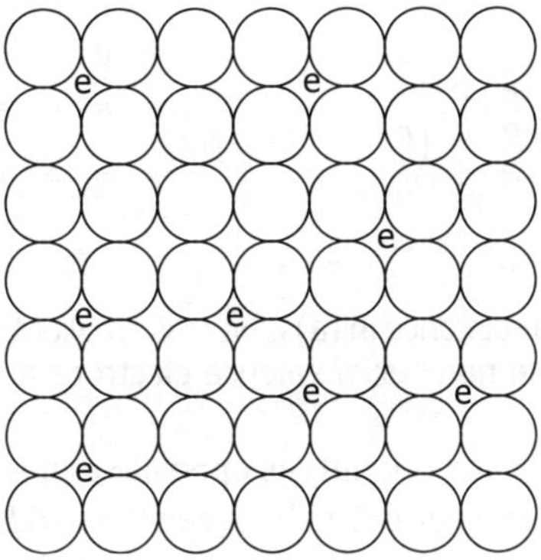

Fig. 3.36 Distribution of electrons in the interstices of a layer of atoms in a metallic lattice

The electrons also occupy the space between the two layers and, therefore, serve to overcome repulsions between two positive planes (Fig. 3.37).

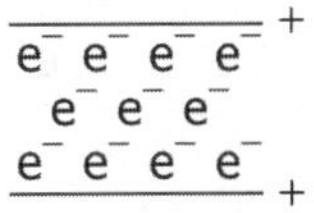

Fig. 3.37 Distribution of electrons between two layers of atoms in a metallic lattice

The distribution of electrons, whether in the interstices or in between the two layers, is random. Hence, the electrons in the crystal cannot be identified with the parent metal atom. For this reason, these electrons are referred to as *sea of electrons* or *electron gas.* Since the electrons are free from the influence of parent metal atoms, these can easily conduct electricity. Moreover, these can be easily excited by the visible component of atmospheric light and hence the shining surface of the metals.

The force of cohesion depends on how strongly two adjacent layers of the metallic lattice are held together. In a crystal of Na, each atom contributes only one electron to the sea of electrons. The two layers are loosely held. In case of Mg, two valence electrons per atom are contributed to the sea of electrons. Therefore, the two planes are held relatively tightly. With Al, where there are three valence electrons, the force of cohesion is still larger. The melting points and boiling points vary in the order,

Aluminium > Magnesium > Sodium

Down the group, the number of electrons in the sea of electrons remains the same. Lithium (Li) contributes one valence electron per atom, and so does cesium (Cs). However, the internuclear distance in a Cs lattice (r′) is larger than that in a Li lattice (r) (Fig. 3.38). Since the force of attraction is inversely proportional to the square of distance of separation, the layers of Li are held more tightly than those of Cs. The force of cohesion, therefore, decreases down the group and so do the melting and boiling points.

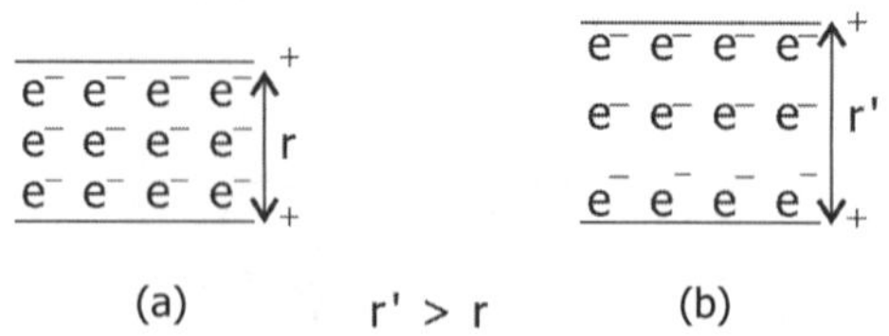

Fig. 3.38 Internuclear distance in (a) lithium is smaller than that in (b) cesium, although the number of valence electrons is same

The trend of malleability and ductility is opposite to that of cohesive force. In the case of Na, there are a fewer electrons in between the two layers as compared to that in Mg. It is, therefore, more easy to glide one layer with respect to the other in Na than in Mg. Hence it requires lesser force to deform Na than it requires to deform Mg. In general, greater the number of valence electrons, lesser the malleability and ductility of the metal. Therefore, Na is more malleable and ductile than Mg which, in turn, is more malleable and ductile than Al (Fig. 3.39).

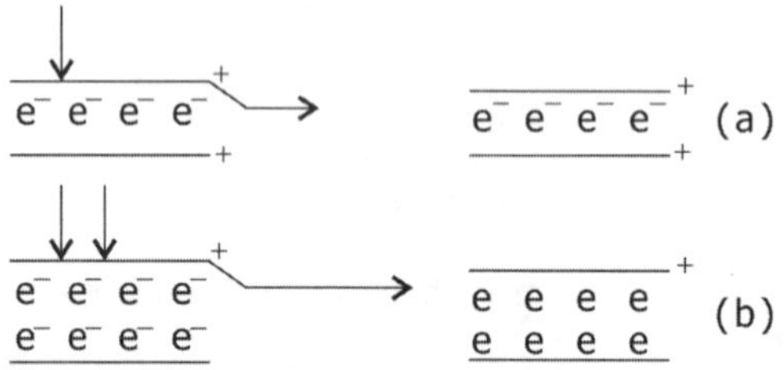

Fig. 3.39 Lesser force is required to deform (a) Na than to deform (b) Mg

2. Band theory

This theory of metallic bonding is based on the molecular orbital concept. According to this concept, when two atomic orbitals combine, two molecular orbitals are formed, one having lower energy than the parent atomic orbitals—called bonding molecular orbital, and the other having higher energy than the parent atomic orbitals—called anti-bonding molecular orbital.

The molecular orbital concept may be extended to Na, the valence shell configuration of which, in elemental state, may be expressed as,

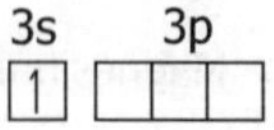

The combination of two 3s atomic orbitals of two Na atoms produces two molecular orbitals, one bonding and one anti-bonding (Fig. 3.40). The two valence electrons occupy the bonding molecular orbital.

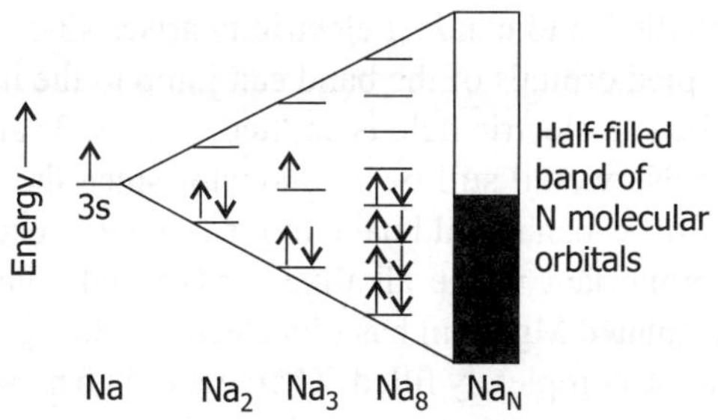

Fig. 3.40 Band of orbitals resulting from the interaction of 3s orbitals in a crystal of sodium

Suppose three Na atoms combine to form Na_3. Three 2s orbitals will combine to form three molecular orbitals—one bonding, one anti-bonding and one non-bonding. The energy of the non-bonding molecular orbital is in between that of bonding and anti-bonding orbitals (Fig. 3.40) Two of the three valence electrons from three atoms would occupy the bonding molecular orbital and one would occupy the non-bonding molecular orbital. Likewise, if N atomic orbitals interact, N molecular orbitals are formed.

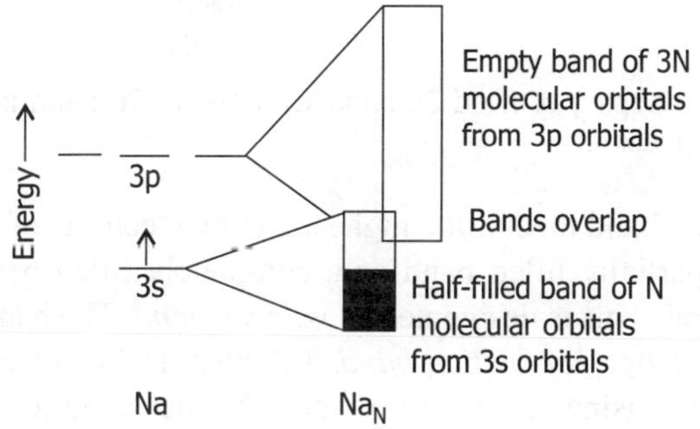

Fig. 3.41 Overlapping of half-filled 3s band with empty 3p band in a crystal of sodium

In a crystal containing one mole of Na atoms, the interaction of 6.023×10^{23} (Avogadro number) atomic orbitals produces 6.023×10^{23} molecular orbitals. This results in a series of closely spaced molecular orbitals—both bonding, σ_{3s} and anti-bonding, σ^*_{3s}—comprising a nearly continuous band which encompasses the crystal as a whole. Since one mole of Na atoms contributes 6.023×10^{23} valence electrons, 6.023×10^{23} orbitals in the band are half-filled.

The vacant 3p atomic orbitals of the Na atoms also interact to form a broad band of $3 \times 6.023 \times 10^{23}$ orbitals (there are three subshells in a p shell). The 3s and 3p orbitals are quite close in energy, so that the two bands of molecular orbitals overlap with each other (Fig. 3.41). The two overlapping bands contain $4 \times 6.023 \times 10^{23}$ molecular orbitals and only 6.023×10^{23} electrons. Since each orbital can hold two electrons, the overlapping bands are only one-eighth full.

The ability of metallic Na to conduct electricity arises since the highest energy electrons in the occupied orbitals of the band can jump to the unoccupied orbitals of the same band, when an electric field is applied. Even if 3s and 3p bands would not have overlapped, Na would still be a conductor since the electrons from the occupied, lower-half of 3s band could have moved to the unoccupied, upper-half. Such an overlap is important in the alkaline earth metals. For example, the 3s atomic orbital of an isolated Mg atom has two electrons. In a crystal of Mg, the 3s band would therefore be completely filled. There would be no scope of promoting the electrons to higher level molecular orbitals and Mg would have behaved as an insulator. Nevertheless, Mg is a good conductor of electricity at room temperature since the 3s and 3p bands overlap and electrons from the highest occupied molecular orbitals jump to lower unoccupied molecular orbitals (Fig. 3.42).

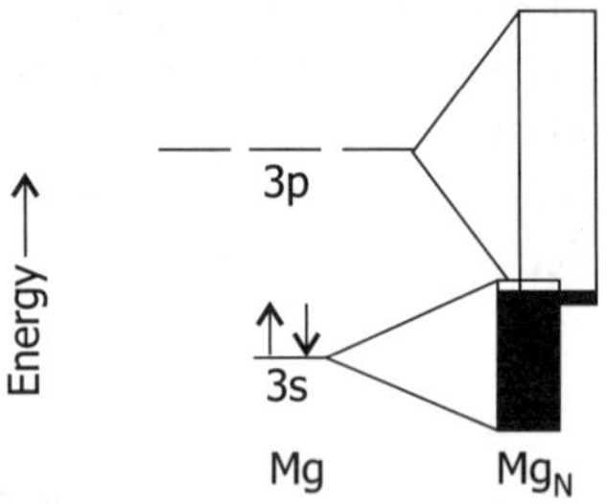

Fig. 3.42 Overlapping of filled 3s band with empty 3p band in a crystal of magnesium

According to the band theory, the highest energy electrons of metallic crystals occupy either a partially filled band or a completely filled band that overlaps with an empty band. This is designated as *valence band*. The band into which the electrons jump is called *conduction band*. The electrical conductivity of a metal decreases with increasing temperature since thermal agitation of metal atoms impede the flow of electrons into conduction band when the electric field is applied. The metals also conduct heat. On absorbing heat, the electrons become thermally excited and are promoted to conduction band. The reverse process is accompanied by release of heat. Metals have a lustrous appearance because mobile electrons can absorb a wide range of radiant energy when they jump to the conduction band. On reverting to their original position, they emit photons of visible light, making the surface shiny.

Case Study 3.23 The melting points of chromium and zinc are 2180 K and 693 K, respectively. Use band theory to explain the difference.

The electronic configuration of chromium is (Ar) $3d^5 4s^1$, while that of zinc is (Ar) $3d^{10} 4s^2$. In case of chromium the five d orbitals of 3d shell and one s orbital of 4s shell of one metal atom overlap with analogous orbitals of adjacent metal atom to give six bonding and six anti-bonding molecular orbitals. The two chromium atoms will have a total of twelve electrons and all of these will be accommodated in bonding molecular orbitals. The two zinc atoms will have a total of twenty four electrons and will occupy both the bonding and anti-bonding molecular orbitals (Fig. 3.43).

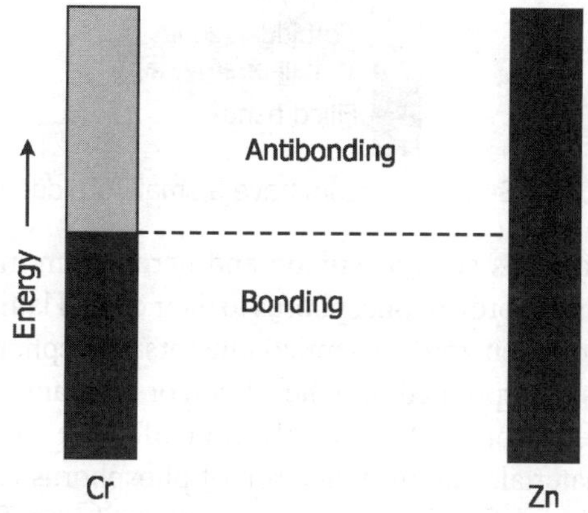

Fig. 3.43 Bands of bonding and anti-bonding molecular orbitals in chromium and zinc

The melting points depend on the occupancy of the bonding and anti-bonding molecular orbitals. Strong bonding and consequently a high melting point are expected for chromium because only the bonding molecular orbitals are occupied, while all the anti-bonding molecular orbitals are vacant. Weak bonding and a low melting point are observed for zinc since both the bonding and the anti-bonding molecular orbitals are occupied.

Insulators and semiconductors Insulators, such as diamond and phosphorus, do not conduct electricity since their highest energy electrons occupy completely filled bands of molecular orbitals that are separated from the conduction band by a large energy gap. This energy gap, called *forbidden zone,* is too large for electrons to jump to reach the conduction band (Fig. 3.44). Elements such as silicon, which are semiconductors, have completely filled bands that are only slightly below the empty bands. The forbidden zone in such cases is small (Fig. 3.45). They do not conduct electricity at low temperatures. However, a small increase in temperature is sufficient to raise the electrons to the conduction band. Therefore, they act as conductors at higher temperatures. This behavior is opposite to that of metals.

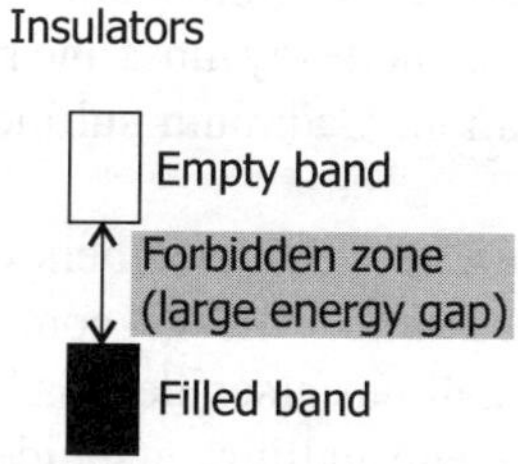

Fig. 3.44 Insulators have a large forbidden zone

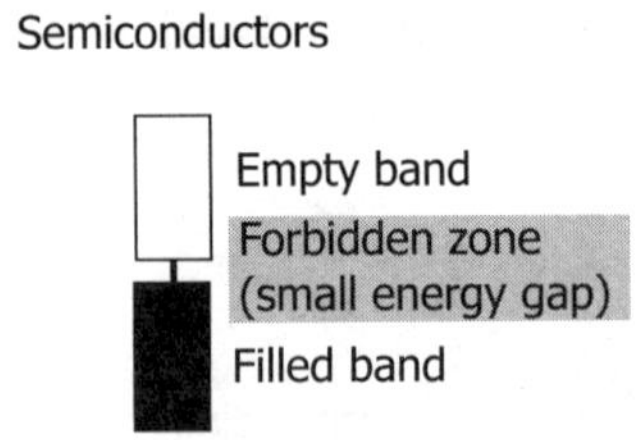

Fig. 3.45 Semiconductors have a small forbidden zone

Conduction in materials such as silicon and germanium may be enhanced by adding small amounts of boron or phosphorus to their crystal lattices. These additives narrow down the forbidden zone in semiconductors. Phosphorus has five valence electrons. When it is incorporated into the silicon or germanium structure—which have four valence electrons—it uses only four of these to become covalently bound to the host material. The fifth electron of phosphorus can be swayed by an electric field. Phosphorus thus behaves as an electron donor. The net effect is that the conductivity of silicon or germanium increases.

When boron, with three valence electrons, is doped in a crystal of silicon or germanium, it too becomes covalently bound to the parent lattice. However, in this case, only three of the four valence electrons of silicon or germanium are involved in bonding with boron. The fourth remains free. As the temperature is increased, the free electron enters the vacant orbitals of boron. The latter thus acts as an electron acceptor. In consequence, an electron that is two atoms away fills the newly created vacancy in silicon atom. The result is a cascade effect, whereby the electron moves throughout the lattice. Therefore, less energy is required to make a material like silicon or germanium conduct electricity if the crystal contains an electron donor like phosphorus or an electron acceptor like boron.

Box 3.5

The "II-VI" and "III-V" Compounds

The 1:1 compounds formed between members of Group 12 and 16 are called "II-VI" compounds. These are so named since formerly, Group 12 of the periodic table was designated as Group IIB, while Group 16 was designated as Group VI. Examples include, zinc sulfide (ZnS), cadmium selenide (CdSe) and cadmium telluride (CdTe). These being isostructural with the semiconductors, silicon and germanium, behave as semiconductors in their own right. Zinc sulfide is also widely used in cathode ray tubes and radar screens as it fluoresces upon being struck by radiation. Cadmium sulfide is a yellow pigment that shows phosphoroscence.

The 1:1 compounds formed between members of Group 3 and Group 15 (earlier called Group V) are called "III-V" compounds. Examples include, gallium phosphide (GaP), gallium arsenide (GaAs) and indium antimonide (InSb). Gallium phosphide and gallium arsenide are widely used as light

Contd...

emitting diodes (LEDs) in pocket calculators, wrist watches and scientific instrument displays. In these devices, the "III-V" compounds convert electrical energy into light. The color of light depends on the energy gap between the filled and the empty bands. For example, gallium arsenide emits infrared light; gallium phosphide emits green light. These compounds are also used in the reverse process of converting light energy into electrical energy. For example, indium antimonide finds application as a photoconductive detector of infrared radiation.

Case Study 3.24 In household wiring aluminium cables are used, although copper has greater conductivity. Why?

The conductivity of aluminium is only 60% that of copper. Yet, aluminium cables are more often used than copper cables for household wiring. This is because the relatively low density of aluminium enables thicker wires to be used for electrical connections. For the same conductivity, the weight of an aluminium wire is only half that of a copper wire.

Alloys

When two or more metals are melted together and the mixture is allowed to solidify, the resulting product is called an alloy. The properties of alloys are generally different from those of the parent metals of which they are composed. Alloys may be classified into three types, depending on the characteristics of constituent elements.

1. **Alloys in which the constituents form a solid solution** When two metals that are miscible at all concentrations, both in liquid and solid states, are allowed to crystallize from the melt, a solid solution of one component in the other results. In fact, there is a complete range of solid solutions, from 100% of one constituent to 100% of the other. In such solid solutions, atoms of one metal replace those of the other at random sites in the lattice (Fig. 3.46). Two metals, whose atomic radii do not differ by more than 15%, and which possess the same number of valence electrons are miscible in all proportions. If one atom is significantly larger than the other, its induction in the metallic lattice shall produce a strain and the solid solution will contain only a meager amount of bigger atoms. Moreover, the two metals should possess similar electropositive character and same crystal structure.

   For example, sodium and potassium are chemically similar and possess identical crystal structures (BCC). However, the radius of sodium (186 pm) and potassium (235 pm) differ by almost 20% and for this reason the two metals do not form a solid solution. On the other hand, nickel and copper possess similar electropositive character, identical crystal structures (CCP) and their atomic radii differ by only 2.3% (nickel, 125 pm; copper, 128 pm). For this reason, nickel may replace a few atoms of copper to yield a solid solution type of alloy.

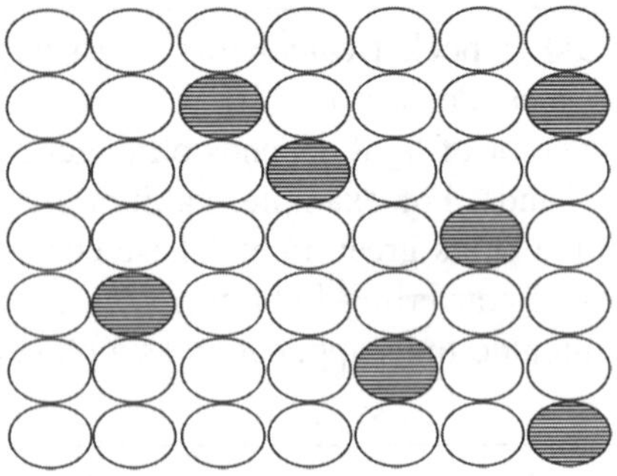

Fig. 3.46 In solid solutions, atoms of one metal are randomly replaced by those of another

---

Case Study 3.25 What is 18-8?

It is a form of steel. The latter is an alloy of iron containing 1–1.5% carbon. Depending on the end use, different kinds of steel contain varying amounts of manganese, chromium, nickel and/or tungsten. For example, stainless steel contains chromium in addition to iron and carbon. The 18-8 steel is so named because it contains 18% chromium and 8% nickel, the rest being iron and carbon. It is stain resistant, has aesthetic luster and is not affected by alkaline solutions, implying that it can be cleaned with soaps and detergents. Due to these reasons, it is used for making household utensils.

---

2. **Alloys in which the constituents form intermetallic compounds** The two metals in an alloy may combine to form one or more intermediate phases, also called intermetallic compounds. In these compounds, the two metals maintain a specific stoichiometric ratio. The crystal structures of the intermediates, however, are different from those of the parent metals. A metal rarely forms an intermetallic compound with another metal of the same group of periodic table. Moreover, it combines with either all members of another group or with none.

   Three different phases of intermetallic compounds have been identified: β, γ and ε phases. These possess a specific valence electron to number of atoms ratio. This ratio is called *Hume–Rothery ratio* and has the values, 21:14 (or 3:2), 21:13 and 21:12 (or 7:4) for the β, γ and ε phases, respectively. For brass, the compositions CuZn, Cu_5Zn_8 and $CuZn_3$ represent the β, γ and ε phases, respectively. In general, the β alloys are strong, ductile and conducting in nature. The γ and ε analogs are brittle, hard and non-conducting.
3. **Alloys in which constituents crystallize separately** There are certain alloys in which the individual components neither mix to form a solid solution nor combine to form intermetallic compounds. The alloy in such cases is composed of an intimate mixture of small crystals of the constituent metals in a relatively pure state. The freezing point of the

alloy is lower than the individual freezing points of the constituents. Zinc and cadmium, when melted together and allowed to solidify form this type of alloy.

Case Study 3.26 What are *zintl phases*?

When two metals, differing greatly in their electronegativities, combine by transfer of electrons from more electropositive to less electropositive species, several intermediate compositions, called zintl phases, result. The more electropositive metal is generally a Group 1 or 2 element, while the less electropositive metal is generally a transition element. Examples of zintl phases include $CaZn_2$ and KCd. Due to a larger difference in the electronegativities of component metals, the reactions leading to the zintl phase alloys are more exothermic than those which result in the formation of Hume-Rothery alloys.

Box 3.6

Ketelaar Triangles

A Ketelaar triangle—also called *Arkel-Ketelaar triangle*—is a plot of difference in electronegativities ($\Delta\chi$) of elements in a binary compound against the average electronegativity (χ_{mean}) of the two elements that constitute the molecule of the compound. The vertices of the triangle represent ionic, covalent and metallic bonding.

Ionic bonding is characterized by a large difference in electronegativity. A large difference reflects that the electronegativity of one element is high and that of the other is low. As a result, the average of the electronegativities has an intermediate value. For example, the electronegativity values of cesium and fluorine are 0.80 and 4.00, respectively. Therefore, for cesium fluoride, $\Delta\chi$ is 3.20 and χ_{mean} is 2.40. This compound occupies the ionic apex of the ketelaar triangle, shown in Fig. 3.47.

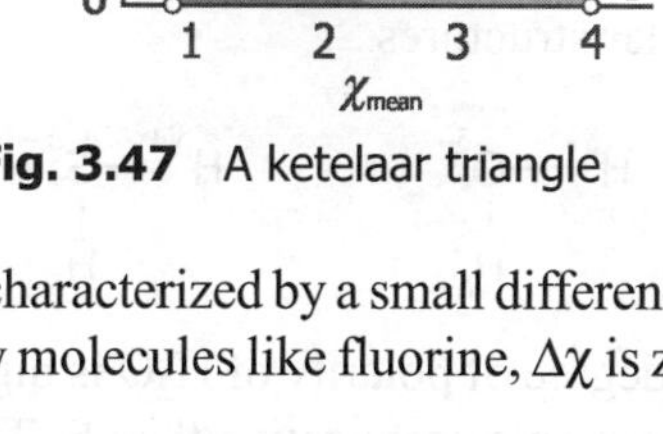

Fig. 3.47 A ketelaar triangle

Covalent bonding is characterized by a small difference in electronegativities. For homonuclear binary molecules like fluorine, $\Delta\chi$ is zero. However, the χ_{mean}

Contd...

for such molecule is very high. For fluorine it is 4.0. It follows that the covalent region of the triangle is the lower, right-hand vertex.

Metallic bonding is also characterized by a small or zero electronegativity difference, and therefore, also lies towards the base of the triangle. In metallic bonding, however, the electronegativities are low and so are the values of χ_{mean}. As a result, metallic bonding occupies the lower, left-hand corner of the triangle. In the present example, this vertex is occupied by the metallic lattice of cesium, for which $\Delta\chi$ is zero and χ_{mean} is 0.80.

Hydrogen Bond

Hydrogen bond is defined as a bridge in which a hydrogen atom links two highly electronegative elements, such as fluorine, oxygen or nitrogen.

.....X—H.....X—H.....X—H..... X—

(X = F, O or N)

It is not a bond in true sense of the term. Rather, it is a form of fairly strong dipole–dipole interaction.

A molecule in which an H atom is covalently bound to a highly electronegative element is polar in nature. For example, in hydrogen fluoride (HF), the electron pair which links the two atoms is unequally shared. It is displaced more towards F atom and away from H atom. This is because F has much higher electronegativity than H. As a result, F acquires a small negative charge, while H acquires a small positive charge.

$$H^{\delta+}—F^{\delta-}$$

The negative center of one molecule attracts the positive center of the other.

$$.....H^{\delta+}—F^{\delta-}.....H^{\delta+}—F^{\delta-}.....H^{\delta+}—F^{\delta-}$$

These dipole–dipole attractions lead to the property of association.

Factors affecting hydrogen bonding

Two factors determine the extent of hydrogen bonding.

1. **Electronegativity of the atom other than hydrogen** The more electronegative the element to which H is bound, the stronger is the hydrogen bonding. Both water and hydrogen sulfide (H_2S) are polar in nature, with similar structures.

However, the degree of polarity of H_2O is higher than that of H_2S. This is because O is more electronegative than S. This means that O in water is able to draw the shared pairs of electrons more towards itself than S in hydrogen sulfide can do. This also means that the quantum of negative charge on O and positive charges on H atoms in water is higher than the corresponding charges on S and H atoms in H_2S.

$$\delta+ > \delta'+$$

$$\delta- > \delta'-$$

The higher partial charges on water molecule enable it to get associated with other water molecules through hydrogen bonding.

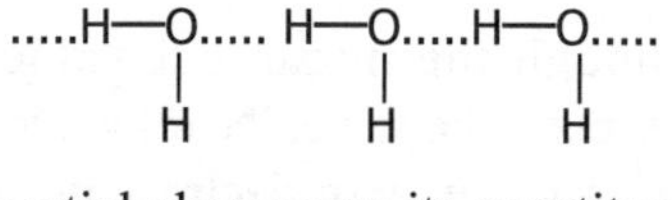

Due to lower partial charges on its constituent atoms, H_2S molecules are not linked to each other through hydrogen bonding.

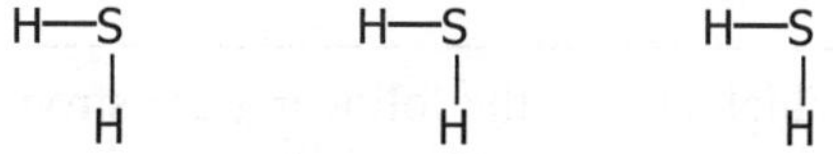

Hydrogen bonding thus, does not exist in all polar molecules. It prevails only when H is bound to a highly electronegative element such as F, O or N.

Yet, the relatively high dielectric constant (160 at 273 K), melting point (260 K) and boiling point (298 K) of liquid hydrogen cyanide point to the existence of hydrogen bonding in this compound. If the electronegativity difference between carbon and hydrogen is considered, the C–H bond does not possess high enough polar character to enter into hydrogen bonding. Nevertheless, hydrogen cyanide may be represented by the following canonical forms.

(I) (II)

In canonical form (II), carbon has a formal positive charge and, therefore, its electronegativity increases to a level that it induces a positive charge on hydrogen through inductive effect. This, in turn, results in a hydrogen-bonded structure, N...H–C

2. **Size of the atom other than hydrogen** Smaller the size of the electronegative atom to which H is attached, greater is the extent of hydrogen bonding. Consider the example of ammonia (NH_3) and hydrogen chloride (HCl). Both molecules are polar in nature. The polarity of H–Cl and H–N bonds is nearly equal, because Cl and N atoms have almost the same values of electronegativities.

$$H^{\delta+}-Cl^{\delta-} \qquad H^{\delta'+}-N^{\delta-}$$

Therefore, the quantum of charges on the two atoms is nearly equal.

$$\delta+ \approx \delta'+$$
$$\delta- \approx \delta'-$$

Yet NH_3 possesses hydrogen bonding, while HCl does not. This may be explained on the basis of the small size of N atom as compared to Cl. The partial charge on the N of NH_3 is spread out over a small volume element, while the same amount of partial charge is dispersed over a relatively large Cl atom in the case of HCl.

Thus, even though the numerical value of charge on both the electronegative atoms is the same, the effective charge on N is more. This implies that there is a less intense electrostatic field around Cl atom, due to which it cannot attract the partial positive charge on the H of the adjacent HCl molecule.

Case Study 3.27 Which among the following has stronger hydrogen bonding: ethyl alcohol or phenol; water or formic acid?

Increasing the electronegativity of the atom bound to hydrogen, increases the strength of hydrogen bond. In phenol, the canonical forms such as,

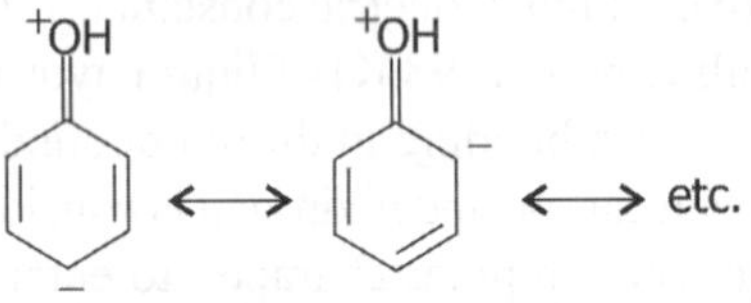

contribute to the structure. In these resonating structures, O bears a positive charge, and hence is more electronegative than neutral O of ethyl alcohol (C_2H_5OH). Therefore, phenol forms stronger hydrogen bonds than ethyl alcohol.

In formic acid, resonating structures, such as,

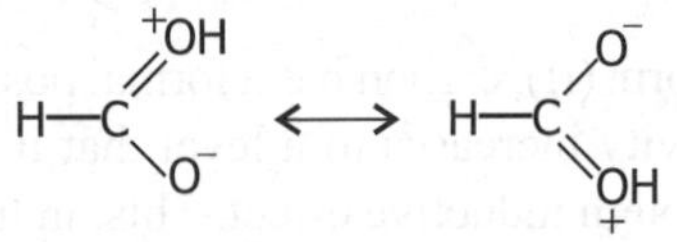

have positively charged oxygen attached to the proton. Therefore, formic acid has stronger hydrogen bonding than water.

Properties affected by hydrogen bonding

The presence of hydrogen bonding in a compound alters a number of physical parameters.

1. **State** A molecule capable of entering into hydrogen bonding is associated with several other molecules of its type. Therefore, apart from the covalent forces of attraction which link the atoms in a molecule, there are additional

intermolecular forces of attraction arising out of dipole–dipole interactions (Fig. 3.48).

H—Y H—Y H—Y (a)

H—X...H—X...H—X (b)

Fig. 3.48 (a) Molecules having covalent forces of attraction and (b) molecules having hydrogen bonding in addition to covalent bonds

These additional attractive forces tend to transform the substance from a gaseous state to the liquid state, or from a liquid state to the solid state. For this reason, water, which shows the property of association due to hydrogen bonding, is a liquid, while hydrogen sulfide (H_2S), in which there is no hydrogen bonding, and hence no association, is a gas. As may be anticipated, ice has a greater degree of hydrogen bonding than liquid water. Similarly, hydrogen fluoride (HF), which shows hydrogen bonding, is a liquid in its natural state, while HCl, in which there is no hydrogen bonding, is a gas.

2. **Melting point and boiling point** When a substance showing hydrogen bonding is to be melted or boiled, a greater amount of energy is required to overcome the additional dipole–dipole attractions. This means that compounds in which hydrogen bonding exists have higher melting and boiling points than their counterparts with no hydrogen bonding. Thus, HF and HCl boil at 292 K and 188 K, respectively.

   We have stated that higher the electronegativity of the atom covalently linked to hydrogen, stronger will be the hydrogen bonding. This implies that HF should have stronger hydrogen bonding than water because the electronegativity of F is more than that of O. Yet the boiling point of water (373 K) is much higher than that of HF (292 K). This may be explained on the basis of the fact that statistically, there are far larger number of hydrogen bonds in water than in hydrogen fluoride. Water has two hydrogen atoms per molecule, while hydrogen fuoride has only one.

3. **Solubility** The maxim, *like disolve like*, is generally meant to imply that ionic compounds are soluble is polar solvents and covalent compounds are soluble in non-polar solvents. However, there are exceptions to this rule. Alcohol, for example, is completely miscible with water. This is because alcohol can enter into mixed hydrogen bonding with water (Fig. 3.49).

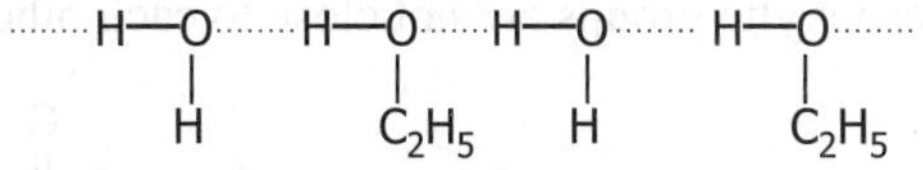

Fig. 3.49 Mixed hydrogen bonding between water and alcohol

Figure 3.49 may be an over-simplification. The arrangement of alternate water and alcohol molecules is too symmetrical to be realistic. The idea is to show that mixed hydrogen bonding does exist between water and alcohol—and this is responsible for their mutual miscibility. Sugar, with five hydroxyl groups, shows similar interactions in aqueous solution.

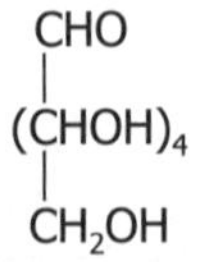

It is thus readily soluble in water, even though it is a typical covalent compound.

4. **Strength of acids and bases** Hydrogen bonding profoundly affects the acid or base strength of many organic derivatives. For example, trimethylammonium hydroxide (a) is a weaker base as compared to tetramethylammonium hydroxide (b). This is becuase in the former, the OH group is hydrogen-bonded to the amine group. As a result, it ionizes with difficulty.

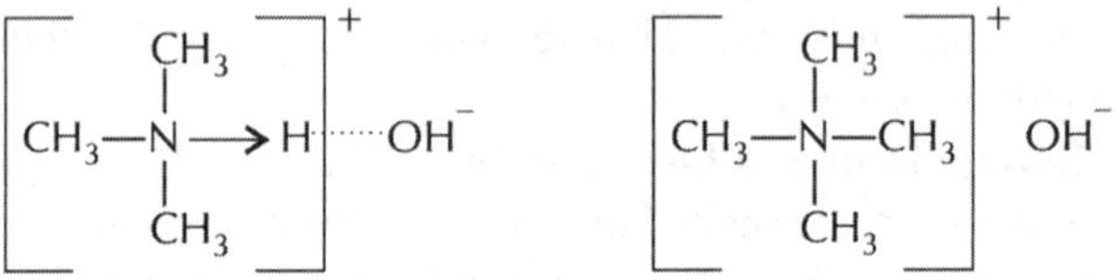

(a) (b)

In tetramethyl derivative, hydrogen bonding is not possible and, therefore, the OH group ionizes easily, making it a stronger base.

Maleic acid and fumaric acid, being geometrical isomers, are expected to have equal acid strengths. However, this has not been observed. Maleic acid is a *cis* compound and loses its first proton easily. The resulting maleate ion is stabilized by hydrogen bonding.

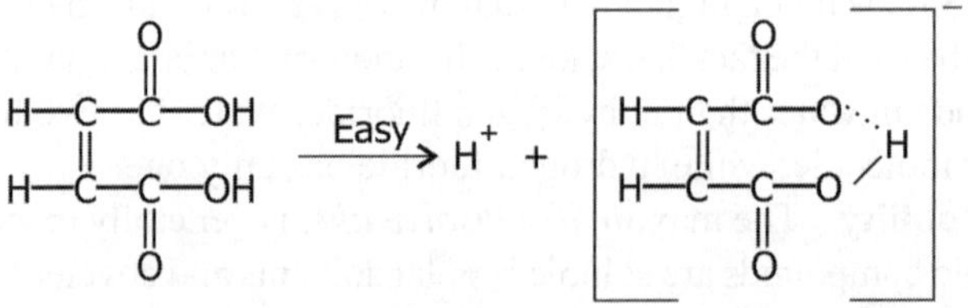

Maleic acid Maleate ion

This stability induces the above reaction to proceed rapidly. Hydrogen bonding is not possible in fumaric acid, which is a *trans* compound, since the two carboxylic groups are not close to each other.

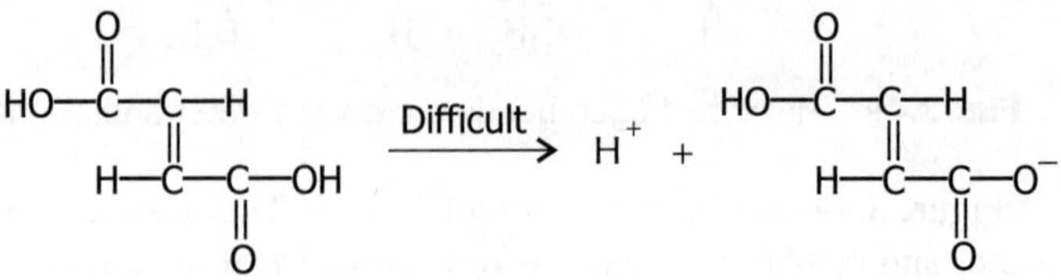

Fumaric acid Fumarate ion

However, the second proton dissociates more easily from fumarate ion. In maleate ion, the proton is involved in hydrogen bonding and, therefore, ionizes with difficulty.

5. **Dimerization of carboxylic acids** Hydrogen bonding is responsible not only for association among molecules of simple hydrides, but also for dimerization of several carboxylic acids. Thus, acetic acid or benzoic acid exist as dimers, the carboxylic group of one molecule is linked to the other by hydrogen bonds.

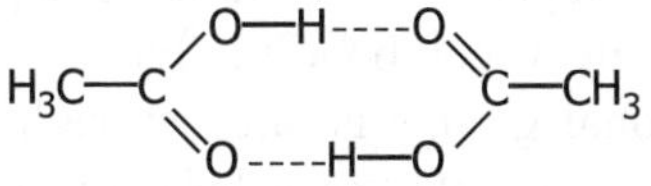

The hydrogen bonds in the dimers are so strong that they do not cleave even in the gas phase.

Types of hydrogen bonding

There are two types of hydrogen bondings: intermolecular and intramolecular.

Intermolecular hydrogen bonding is one that exists between two or more molecules. H_2O or HF show the property of association since a large number of their respective molecules are linked together through hydrogen bonding.

Intramolecular hydrogen bonding is one which exists within the same molecule. Examples include *ortho* derivatives of disubstituted aromatic compounds, such as *o*-nitrophenol.

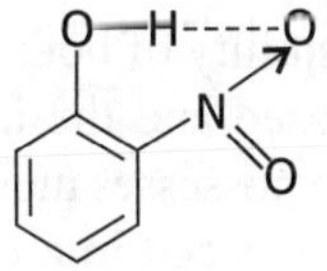

When compared to the corresponding *meta*- or *para*-derivatives, *ortho*-nitrophenol is less acidic. The reason is that due to hydrogen bonding, the proton is trapped in between two O atoms and, therefore, it cannot dissociate so easily.

Case Study 3.28 Which technique is used to separate *o*-nitrophenol from *p*-nitrophenol?

When phenol is nitrated, it gives a mixture of *o*- and *p*-nitrophenols (phenolic group is *ortho*/*para* directing). It is the concept of hydrogen bonding which dictates the type of technique to be used for separating the two products. The *p*-nitrophenol derivative has intermolecular hydrogen bonding.

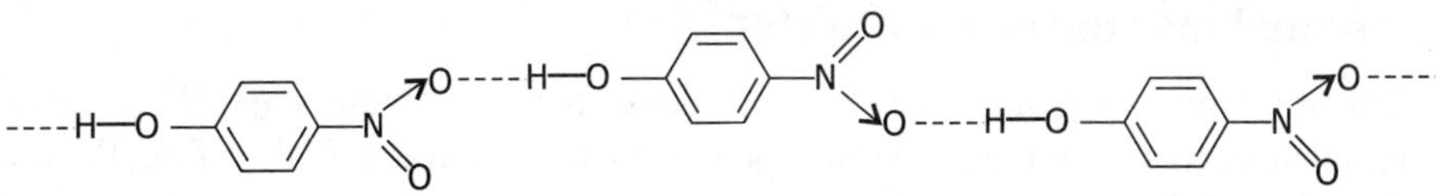

Since the molecules of *p*-nitrophenol are associated by hydrogen bonding, it is non-volatile. In case of *o*-nitrophenol, lack of association is responsible for a low boiling point (373 K at 70 mm pressure). Thus, while the *ortho* derivative is steam-distillable, the *para* one is not. A current of steam is passed through the mixture of *o*- and *p*-nitrophenols. The *ortho* derivative volatilizes along with steam and distills out. The non-volatile *para* derivative remains behind.

It is quite obvious that only intermolecular hydrogen bonding can take place in *p*-nitrophenol. The nitro and phenolic functional groups are widely separated to be linked by intramolecular hydrogen bonding. Even in *m*-nitrophenol, the distance between the two functional groups is large enough to preclude intramolecular hydrogen bonding. The presence of nitro and phenolic functional groups on adjacent C atoms in *o*-nitrophenol permits intramolecular hydrogen bonding.

However, in case of *o*-nitrophenol, there is a distinct possibility for the occurrence of intermolecular hydrogen bonding as well. Yet, the molecules of *o*-nitrophenol are never linked by intermolecular hydrogen bonding.

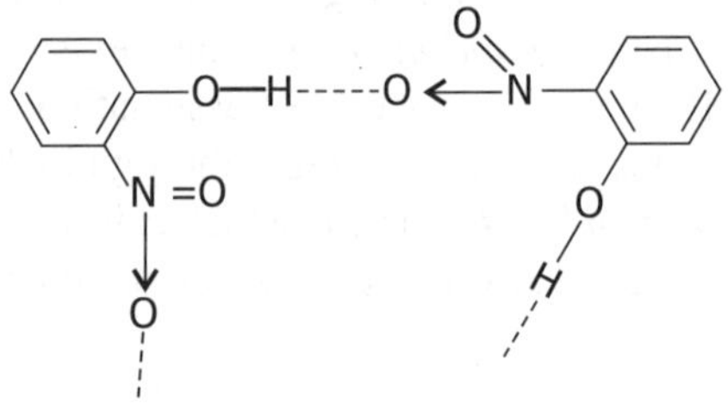

In cases where there is a possibility of both inter- and intramolecular hydrogen bondings, the former is suppressed and the latter prevails. A compound having intramolecular hydrogen bonding possesses more thermodynamic stability. Suppose one mole of such a substance is placed in a closed system. A total of Avogadro number of particles of the substance (6.023×10^{23}) may be identified in this system. This is possible because each molecule retains its individual identity. On the other hand, the number of particles in one mole of a substance having intermolecular hydrogen bonding will be less than the Avogadro number. This is because each molecule does not retain its individuality, as it will be associated with several other entities of its type.

As a result, there will be a greater degree of disorderness in the intramolecular hydrogen-bonded system. Such a system will have a greater positive value of change in entropy, ΔS, as compared to the one which invokes intermolecular hydrogen-bonded molecules. A greater degree of randomness or a greater positive value of change in entropy implies greater thermodynamic stability.

Anomalous character of water

The solid state of a substance is more dense than the liquid state. Water shows anomalous character. Ice has a lower density than liquid water. It, therefore, floats on the water surface and freezing of aquatic systems occurs from surface downwards.

If water had behaved like other substances, the aquatic systems would have frozen from bottom up. As a result, the aquatic life in oceans, rivers or lakes would have perished in the winters. Hydrogen bonding is responsible for this unusual property of water.

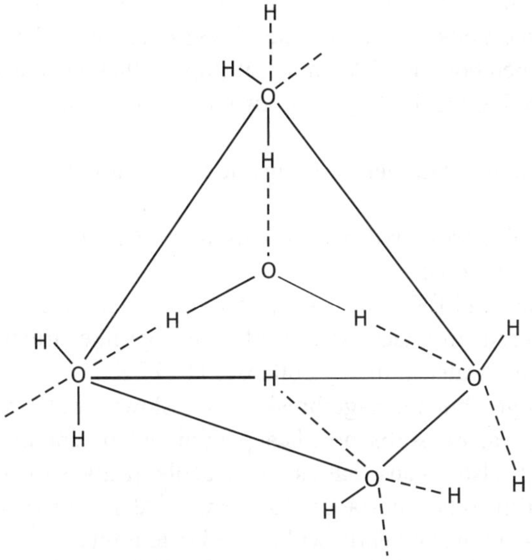

Fig. 3.50 Structure of ice

In ice, each water molecule is hydrogen-bonded to four other water molecules. Two of the hydrogen bonds are formed through the two H atoms covalently bound to O, while the other two are formed through the two lone pairs on O atom.

The tetrahedral coordination around each water molecule results in a cage-like structure. Due to cage formation, the volume is unusually large and, therefore, density (= mass/volume) is abnormally low. As the temperature is increased, some of the hydrogen bonds break and the cage-like structure tends to become dismantled. The volume progressively decreases and density consistently increases. At 277 K, the cage-like structure completely breaks down. The volume is now at a minimum and density at a maximum. Beyond 277 K, thermal expansion of water increases the volume. The density once again registers a decline. Thus, *water has the maximum density at 277 K.*

Many gaseous molecules, such as sulfur dioxide, chlorine or methane, under high pressure, may be trapped in the cage-like structure of solidified water. The resulting entities are called *clathrate hydrates*. No chemical bond exists between the host cage and the guest molecules. Only weak van der Waals forces hold the trapped gases inside the cage.

The methane hydrate is stable at atmospheric pressure and low temperature. When a match is held to it, the methane content of the hydrate catchers fire. This demonstration is called *burning ice*.

Case Study 3.29 Why doesn't the cage-like structure of ice completely collapse at 273 K?

Let us assume that the only intermolecular forces between water molecules are the hydrogen bonds. The heat of fusion of ice is 5.88 kJ mol^{-1}. This means that when one mole of solid ice melts into liquid water, 5.88 kJ of heat is released. The average energy of a hydrogen bond is 21 kJ mol^{-1}. It implies that when energy equivalent to 21 kJ is imparted to ice, hydrogen bonds uniting one mole of water molecules are broken down.

With 21 kJ of energy, hydrogen bonds in one mole of water molecules are broken down.

With 5.88 kJ of energy, hydrogen bonds in 5.88/21 or 0.28 mole of water molecules are broken down.

This proves that at 273 K, only 28% of the hydrogen bonds in ice are broken down, while 72% remain intact. As a result, the cage-like structure continues to persist even in liquid water. It finally collapses at 277 K.

Since the hydrogen-bonded cage breaks down slowly, water has high specific heat, meaning that water absorbs more heat per unit of temperature rise than almost any other liquid. It also means that as water cools, it gives off more heat to its surroundings than any other substance. As a result, water acts as a large thermostat, confining the temperatures on earth within moderate limits.

Theories of hydrogen bonding

We have seen that hydrogen bonding is an attractive interaction between the hydrogen atom of a molecule, X—H (where X is an atom possessing high electronegativity) and another electronegative atom, Y of the same (intramolecular) or different (intermolecular) molecule (Fig. 3.51a)

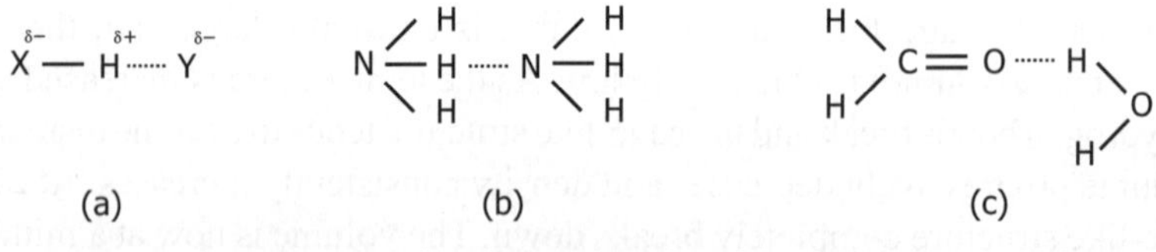

Fig. 3.51 Representation of hydrogen bonding (a) which may occur between molecules of same type (b) or of different types (c)

X and Y may be the fragments of the same chemical species, for example, hydrogen bonding in ammonia (Fig. 3.51b) or of different chemical species, for example hydrogen bonding between formaldehyde and water (Fig. 3.51c). As compared to an ionic or a covalent bond, hydrogen bond is much more complicated since three types of forces are involved in its formation. We now examine the characteristics of these forces.

1. **Electrostatic interactions** The forces of attraction which contribute to the largest extent (almost 90%) in the formation of hydrogen bonds are electrostatic in nature. These arise due to ionic attractions between

small positive charge on hydrogen and small negative charge on X and Y neighbors (Fig. 3.51a). Such interactions may also be classed as dipole-dipole attractions and are one of the forms of van der Waals forces (*see* next section).

2. **Charge-transfer interactions** As shown in Fig. 3.51a, the proton in a hydrogen-bonded system acquires a partial positive charge (δ+), while the electronegative fragment, Y acquires a partial negative charge (δ–). These charges may be theoretically computed. However, the experimentally determined charges are less than the calculated ones. This may be explained on the premise that a part of the negative charge on Y flows to H. Since charge flow is akin to atoms that are linked by sharing of electrons, it may be advocated that the hydrogen bond has a covalent component as well. The evidence for this is provided by the observation that the water molecules which form the cage-like structure of ice (Fig. 3.50) are oriented in specific directions—and directionality is a characteristic of covalent bonds.
3. **Dispersion interactions** As we shall see in the next section, dispersion interactions are short-range forces which arise when a charged end of an atom or a molecule comes in vanity of oppositely charged end of another atom or molecule. The same may happen in a hydrogen-bonded system (Fig. 3.52a).

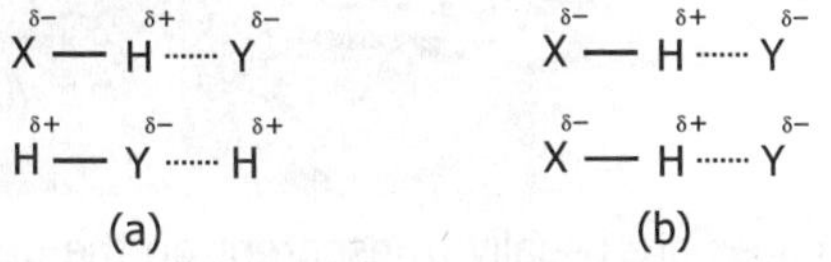

Fig. 3.52 Dispersion interactions can lead to attraction (a) or repulsion in hydrogen-bonded molecules

However, if the chemical entities are squeezed too close to each other, the similar charges may come in vicinity, leading to repulsion (Fig. 3.52b). The repulsive van der Waals forces, which are discussed in the next section, also arise due to orientation of like charges in proximity to each other. Thus while electrostatic and charge transfer interactions are a source of attraction between molecules, the dispersion interactions may be either attractive or repulsive in nature. Accordingly, these may supplement or reduce the strength of hydrogen bond.

Box 3.7

The Concept Behind Ice Skating

We have already described that due to its hydrogen-bonded, cage-like structure, solid ice is less dense than liquid water. According to Le Chatelier principle, when pressure is applied to a system involving two states of the same substance, the equilibrium shifts towards that phase which has a higher density. This means that the application of pressure on ice-water system would push the equilibrium towards the liquid state. *Contd...*

$$H_2O_{(s)} \xrightleftharpoons{\text{High pressure}} H_2O_{(l)}$$

While skating, the pressure exerted by the roller wheels tend to melt the surface ice. Thus, immediately below the wheel of the skating blades, there is a layer of liquid water, and not solid ice. Although this layer is supercooled—its temperature may be as low as 243 K—yet the person glides on liquid water, which offers almost negligible friction as compared to the solid ice. This makes ice skating smoother—and enjoyable.

Case Study 3.30 Why ice cubes are opaque on the inside?

Even though water is transparent, yet on freezing it yields ice cubes which are hazy on the inside (Fig. 3.53).

Fig. 3.53 Ice cubes are usually transparent on the outside and hazy on the inside

Natural water contains a host of salts (mainly of calcium and magnesium) and gases (mainly carbon dioxide) in dissolved state. These soluble constituents lower the freezing point of water below the normal value of 273 K. When water begins to solidify, the dissolved impurities tend to move deeper inside. As a result, the peripheral layers of ice are made of pure water and hence are transparent. The inner layers solidly at a lower temperature and when they do, a matrix of salts and gases get dispersed in that section. For this reason, the inside of ice cubes appear opaque.

van der Waals Forces

van der Waals forces encompass several types of attractive interactions among atoms and molecules. All types are quite weak as compared to typical ionic or covalent forces of attraction. Nevertheless, these are important in explaining several physical properties of chemical species. The major types of van der Waals attractions are now described.

Dipole–dipole attractions

The dipole moment of two or more molecules may interact with each other, so as to cause an alignment of oppositely charged centers of the molecules. This results in attraction among the molecules. Rather than existing in isolation, the molecules are now associated with each other. A molecular association of this type leads to a higher melting point or boiling point than would be normally expected from segregated molecules.

Hydrogen bonding is a form of dipole–dipole attraction. It causes the molecules of HF or H_2O to associate. But for this type of dipole–dipole interactions, water would have existed in gaseous form and life would not have originated on the earth. We have already discussed hydrogen bonding earlier in this chapter. Other molecules which are associated through dipole–dipole interactions include bromine fluoride (BrF) and nitrogen trifluoride (NF_3) (Fig. 3.54). In the case of NF_3, the boiling point would have been lower by 60°, had there been no dipole–dipole attractions.

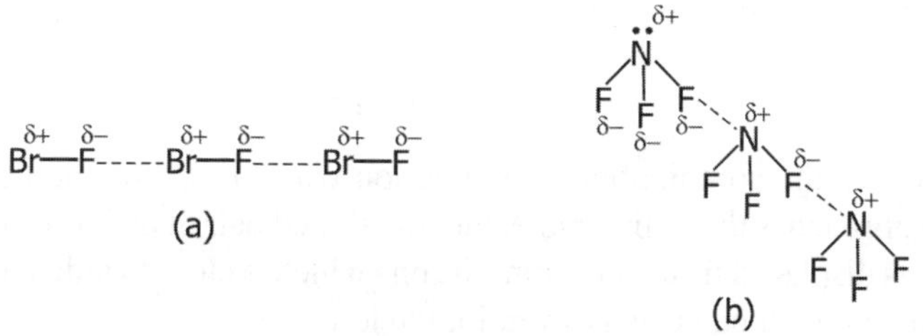

Fig. 3.54 Dipole–dipole attractions in (a) bromine fluoride and (b) nitrogen trifluoride

Dipole–dipole interactions vary inversely to the fourth power of distance of separation, $1/r^4$. These are effective only over very short distances. All such types of attractions, including hydrogen bonding, are to a large extent directional in nature. At higher temperatures, the kinetic energy of the molecules increases, leading to random orientations. Therefore, the strength of dipole–dipole interactions decreases with increasing temperature.

Ion-dipole attractions

An ion, either a cation or an anion, may interact with a polar molecule through weak electrostatic forces of attraction. This type of interaction is called ion-dipole attraction. The most common example of ion-dipole attraction is the combination of a proton (a cation) with a water molecule to produce hydronium ion.

$$H^+ + H_2O \rightarrow H_3O^+$$

Yet another example is the combination of fluoride ion (an anion) with a hydrogen fluoride molecule to yield hydrogendifluoride ion.

$$F^- + HF \rightarrow F^- \ldots\ldots H^{\delta+} - F^{\delta-} \text{ or}$$

$$[F \ldots\ldots H - F]^-$$

The ion-dipole bond obeys Coulomb's law which means that greater the quantum of charges on the interacting atoms, stronger is the bond.

Ion-induced dipole attractions

Large symmetrical molecules, in which the nuclei of constituent atoms are at a relatively large distance from electrons, tend to become less symmetrical when brought in contact with an ion. As the symmetry is lowered, the center of positive charge does not coincide with the center of negative charge. The molecule acquires dipole moment. This is not a natural dipole. It has arisen since an ion has been brought in close vicinity of the molecule. Hence it is called *induced dipole*. The bonding between the ion and the molecule is called ion-induced dipole attractions.

Take the case of iodine molecule. The shared pair of electrons is equidistant from the two iodine atoms. Hence the molecule is symmetrical. It has no natural dipole moment. Being covalent in nature, it is sparingly soluble in water. However, when iodide ions are added to water containing iodine crystals, the latter are solubilized. This is due to the formation of triodide ion.

$$I_2 + I^- \rightarrow I_3^-$$

What induces the combination of iodide ion with the iodine molecule? As the iodide ion approaches the iodine molecule, the shared pair of electrons in the iodine molecule gets displaced from one atom (the one which is close to iodide ion) towards the other (the one which is away from iodide ion).

I^- $\qquad\qquad$ $I : I$
No dipole moment

I^- $\qquad$ $I^{\delta+} : I^{\delta-}$
Induced dipole moment

I^---------- $I^{\delta+}—I^{\delta-}$
Ion-induced dipole attraction

Thus, iodide ion induces a dipole moment in the large iodine molecule. The source of bonding is the attraction between the negatively charged iodide ion and the induced electric dipole of the iodine molecule.

London forces

When a noble gas is cooled under pressure, it first liquifies and then solidifies. As it is transformed from gaseous to liquid state, and then from liquid to solid state, the forces of attraction between its atoms increase. The atoms of noble gas cannot—and do not—form ionic or covalent bonds with each other; there is no possibility of coordinate or metallic bonding either. The forces which hold the atoms of noble gases together in liquid or solid state are called London forces.

London forces are weak attractive forces that are important only over extremely short distances, since these vary inversely to the seventh power of internuclear separation. These are the only kind of intermolecular forces present among symmetrical, non-polar molecules such as carbon dioxide, bromine, hydrogen and monoatomic species, such as noble gases.

Even the symmetrical molecules or neutral atoms possess a dipole moment which fluctuates in magnitude and direction. The fluctuating dipole, in turn, is attributed to the electronic motion within the molecule or the atom. As a result of the electronic motion, there is a continuous change in the charge intensities within the particle. During these fluctuations, the electronic charge may became concentrated at one end of the particle, while the nucleus may be displaced to the other (Fig. 3.55).

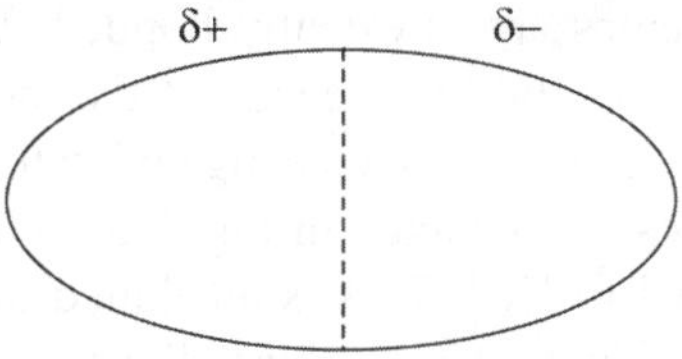

Fig. 3.55 Instantaneous dipole moment

This type of dipole moment is called *instantaneous dipole moment*, as it exists only for a fraction of a second. Next moment it may become oriented in the opposite direction, so that the net dipole is zero (Fig. 3.56).

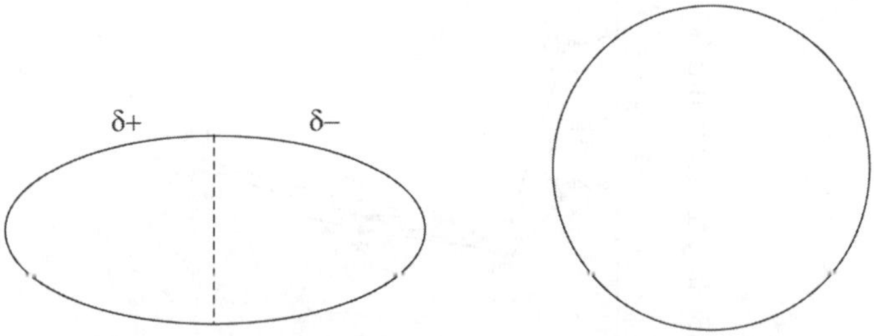

(a) Instantaneous dipole (b) No dipole

Fig. 3.56 Change in orientation of dipole resulting in net zero dipole

An atom or a molecule possessing instantaneous dipole can induce a similar fluctuation in another atom or molecule, when the latter is brought close to it. The positive end of one particle can increase the negative charge intensity of the adjacent particle. The positive and negative ends on adjacent atoms attract each other and these attractions are termed London forces (Fig. 3.57). Alternately, these are referred to as *instantaneous dipole–instantaneous induced dipoles* or *dispersion forces*.

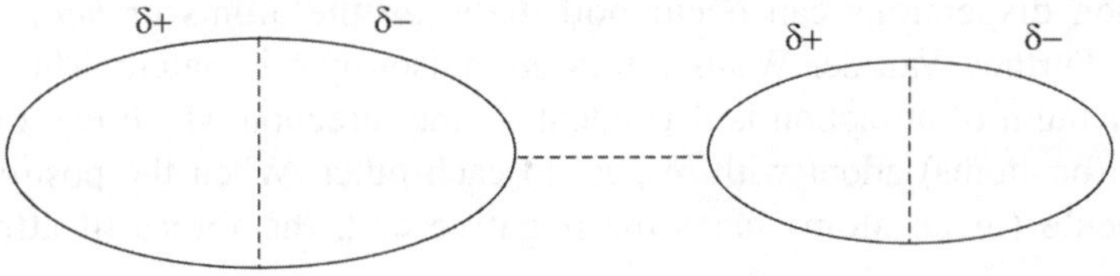

Instantaneous dipole Instantaneous induced dipole

Fig. 3.57 London forces

The probability of instantaneous dipole increases with increasing number of electrons and size of atoms or molecules. The strength of London forces, therefore, also increases in the same direction. Thus, the melting/boiling points of the noble gases increase from top to bottom. The ease of liquefaction or solidification also increases in the same order. The first member, helium, is so small that its chances of developing instantaneous dipole are quite meager. Hence, helium can be liquified but not solidified.

All the halogen molecules are covalently bonded. Yet F_2 and Cl_2 are gases, Br_2 is a liquid, while I_2 is a solid. As the size of the molecule increases, London forces of attraction increase, causing the change of state. The variations in boiling points of hydrogen halides are depicted in Fig. 3.58. HF has an abnormally high boiling point as compared to HCl. This is attributed to the additional attractive forces in HF as a result of hydrogen bonding. For the remaining compounds the boiling points are in the order, HCl < HBr < HI. Thus, as the number of electrons and molecular size increases, London forces of attraction also increase, causing, in turn, the boiling points to increase. The same trend is observed for the hydrides of Group 15 and 16 as well.

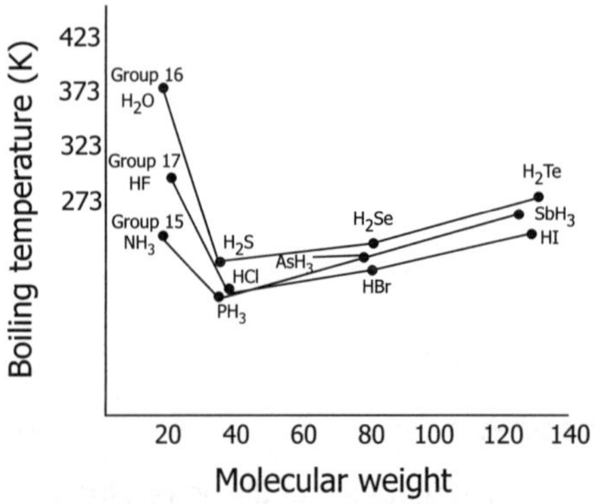

Fig. 3.58 Variation in the boiling points of hydrides of Groups 15, 16 and 17 with molecular weight

Repulsive interactions

We have seen that van der Waals forces are intermolecular in nature, that is, these are perceived *between* the molecules, and not *within* a molecule. The London dispersions can occur both *between* the atoms or *between* the molecules. Further, Van der Waals forces are anisotropic in nature. This means that the quantum of attraction is dependent on the direction which two or more molecules (or atoms) adopt with respect to each other. When the positive end of a molecule (or an atom) faces the negative end, the forces of attraction are at a peak.

However, these very forces can be repulsive in nature too. At extremely high pressures, the molecules (or atoms) are forced to come so close to each other that the ends with like charges may become oriented in vicinity of each other. In case of atoms, the nuclei may be forced into close proximity. In either case, the forces of repulsion begin to emerge under these circumstances. If the sum of attractive forces exceed the sum of repulsive force, the molecules (or atoms) will attract each other. If it is the other way round, the forces of repulsion will dominate.

Case Study 3.31 What is *Trouton constant*?

Trouton constant is the heat of vaporization of a liquid in cal mol^{-1} at its boiling point, divided by the boiling point in Kelvin. It is thus the entropy of vaporization at the boiling point and represents the degree of association in the liquid state in comparison to the gaseous state. Greater the Trouton constant, greater is the degree of association. For example, the Trouton constants for methane, ammonia and water are 17.5, 23.3 and 28.3 cal mol^{-1} K^{-1}, respectively.

4. NUCLEAR CHEMISTRY

Introduction

Nuclear chemistry deals with the study of atomic nucleus—its nature, stability and characteristics. It also encompasses the study of nuclear reactions, i.e., the reactions which involve changes in the composition of nuclei. In some nuclei, these changes may occur spontaneously. Such nuclei are unstable and are said to manifest the phenomenon of *natural radioactivity*. In other nuclei, these changes may be affected artificially—a course of action which falls under the purview of *artificial radioactivity*. The first part of this chapter deals with the characteristics of atomic nucleus, while the later dwells on the concept of radioactivity.

The Atomic Nucleus

An atom consists of a nucleus of small volume bearing positive charge, Ze, surrounded by a cloud of Z electrons, moving in an immensely large volume. The size of the nucleus is very small as compared to the overall size of the atom, which may be gauged from the following analogy: if the radius of the nucleus is 1 cm, the radius of the atom would be 1 km. The actual radius of the nucleus is measured in terms of femtometers, fm, where 1 fm = 10^{-15} m. Yet it is in the nucleus where most of the mass of an atom is concentrated. Due to its large mass and small volume, the density of the nucleus is very high, of the order of 2.5×10^{14} g cm^{-3}.

The nucleus contains positively charged protons and neutral neutrons. Therefore, it has a net positive charge. A neutron has a mass of 1.00893 atomic mass units (amu). This mass is relative to the mass of an oxygen atom taken as 16.000 amu. A neutron is designated as ${}_{0}^{1}n$, where the subscript indicates the charge and the superscript indicates the mass to the nearest whole number. A proton carries a positive charge of one unit and has a mass of 1.00758 amu. Hence, it is represented as ${}_{1}^{1}p$. Each neutron and proton has a spin of $\pm\frac{1}{2}$. The neutrons and protons are collectively referred to as *nucleons*.

The extranuclear electron is very light as compared to the neutron or proton. Its mass is just 0.0005486 amu and for all practical purposes is taken as zero. This explains why the mass of an atom is considered same as the mass of its nucleus. An electron carries a unit negative charge. It is symbolized as ${}_{-1}^{0}e$. Some general characteristics of the subatomic particles—neutron, proton and electron—are listed in Table 4.1.

The number of protons in the nucleus is equal to the number of extranuclear electrons, thus making the atom overall neutral. This number is called *atomic number* and is designated as Z. The total number of neutrons and protons in the nucleus of an atom is called *mass number* and is designated as A. An atom X having atomic number Z and mass number A is represented as,

$$^{A}_{Z}X$$

Table 4.1 General characteristics of subatomic particles

Particle	Symbol	Charge	Mass (amu)	Spin
Neutron	n	0	1.00893	± 1/2
Proton	p	+	1.00758	± 1/2
Electron	e	–	0.0005486	± 1/2

This atom has Z protons and A nucleons in its nucleus. Accordingly, the number of neutrons in X are (A – Z). For example, the oxygen atom has eight protons and eight neutrons in its nucleus. Therefore, its atomic number and mass number are 8 and 16, respectively. It is represented as,

$$^{16}_{8}O$$

In fact, the mass of this atom is fixed as a standard at 16.000 amu to elucidate the masses of other atoms in terms of atomic mass units.

Atoms of the same element, whose nuclei contain the same number of protons but different numbers of neutrons are called *isotopes.* Thus, isotopes have same atomic number but different mass numbers. Their physical properties differ but their chemical characteristics are similar. Atoms having same mass number but different atomic numbers are termed *isobars.* These are the atoms of different elements and, therefore, show divergent physical as well as chemical properties. Atoms having same number of neutrons but different number of protons are called *isotones.* These too are atoms of different elements, having dissimilar physical and chemical properties. Some examples of isotopes, isobars and isotones are given in Table 4.2.

Table 4.2 Some common isotopes, isobars and isotones

Atomic species	Examples
Isotopes	$^{1}_{1}H$, $^{2}_{1}H$ (or $^{2}_{1}D$), $^{3}_{1}H$ (or $^{3}_{1}T$)
	$^{16}_{8}O$, $^{17}_{8}O$, $^{18}_{8}O$
	$^{235}_{92}U$, $^{238}_{92}U$, $^{239}_{92}U$
Isobars	$^{76}_{32}Ge$, $^{76}_{34}Se$
	$^{130}_{52}Te$, $^{130}_{54}Xe$, $^{130}_{56}Ba$
	$^{210}_{81}Tl$, $^{210}_{82}Pb$, $^{210}_{83}Bi$, $^{210}_{84}Po$
Isotones	$^{30}_{14}Si$, $^{31}_{15}P$

The nucleus of elements may or may not be stable. Even among the isotopes, the nuclear stabilities are widely divergent. The stability of a nucleus depends not only on the number of nucleons, but also on their arrangement. The arrangement of nucleons, in turn, determines how effectively the neutrons and protons are bound together in the nucleus.

Binding energy of the nucleus

Binding energy refers to the amount of energy that holds the nucleons together in the nucleus. Higher the binding energy, more strongly are the nucleons held together and more stable is the nucleus. Binding energy is not provided to the nucleus by an outside source. Rather, some of the mass of the nucleus is converted into energy to bind the protons and the neutrons.

This means that the actual (or experimental) mass of the nucleus is less than its calculated (or theoretical) mass. The difference between calculated mass and actual mass of the nucleus is called *mass defect*. The fraction of the nuclear mass, which is equivalent to the mass defect, is the one that is converted into binding energy.

The calculated mass of the nucleus is given by the following expression:

$$\text{Calculated mass} = M_pZ + M_n(A - Z) \quad (4.1)$$

where M_p is the mass of a proton, Z is the atomic number or the total number of protons, M_n is the mass of a neutron and A is the mass number. Thus, the first term on the right hand side of expression (4.1), (M_pZ) gives the total mass of all the protons, while the second term [$M_n(A-Z)$] gives the total mass of all the neutrons.

If M is the actual mass of the nucleus, then mass defect, Δm, is mathematically expressed as,

$$\Delta m = [M_pZ + M_n(A - Z)] - M \quad (4.2)$$

The relation between mass defect (Δm) and binding energy (E) is given by Einstein equation,

$$E = \Delta mc^2 \quad (4.3)$$

where c is the velocity of light (3×10^8 ms^{-1}). The value of mass defect, calculated from equation (4.1) is expressed in terms of amu. It may be converted into kilograms by the following relation:

$$1 \text{ amu} = 1.661 \times 10^{-27} \text{ kg}$$

Substituting the value of mass defect in kg and velocity of light in m s^{-1} in equation (4.3),

$$E = \Delta m.\ 1.661 \times 10^{-27} (2.998 \times 10^8)^2$$

or

$$E = \Delta m.\ 1.4929 \times 10^{-10} \text{ J} \quad (4.4)$$

It is customary to express binding energy in electron volts (eV) or million electron volts (MeV), rather than in Joules.

$$1 \text{ MeV} = 1.602 \times 10^{-13} \text{ J} \quad (4.5)$$

From equations (4.4) and (4.5),

$$E = \Delta m. \frac{1.4929 \times 10^{-10}}{1.602 \times 10^{-13}} \text{ MeV}$$

or $$E = \Delta m. 931 \text{ MeV} \tag{4.6}$$

Equation (4.6) gives the total binding energy of the nucleus. A more useful parameter is binding energy per nucleon.

$$\text{Binding energy per nucleon} = \frac{\Delta \text{ m. } 931}{\text{Mass number}} \text{ MeV}$$

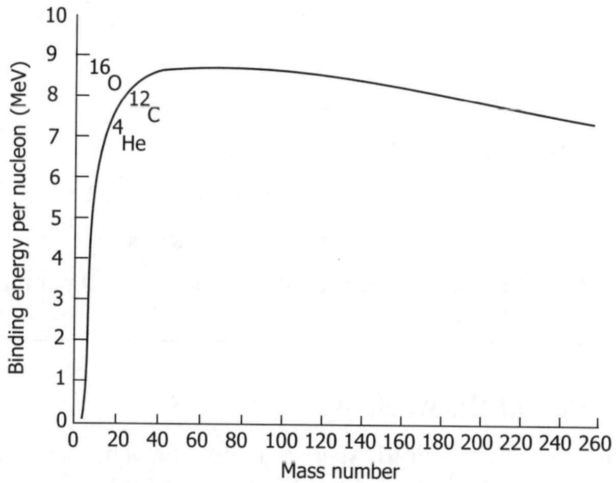

Fig. 4.1 Nuclear binding energy curve

A plot of binding energy per nucleon against mass number is shown in Fig. 4.1. In this plot, there is a large central region with mass numbers ranging from 25 to 40, where the binding energy per nucleon is nearly constant at 8.5 MeV. The lower binding energy of the nuclei with lower masses is attributed to the lesser interaction between the nucleons. Nuclear forces are short-range interactions. When the number of nucleons is small, they are spread out within the nucleus as far apart from each other as possible. Their mutual attraction is then reduced.

Helium-4, carbon-12 and oxygen-16 nuclei, even though relatively light, possess higher than expected binding energy. An explanation for their additional stability is provided in Box 4.1 (p. 188). The low binding energies of heavier nuclei are attributed to coulombic repulsions between a large number of protons.

Case Study 4.1 Calculate the energy released (in kJ mol^{-1}) in the nuclear reaction,
$^{3}_{1}H + ^{2}_{1}H \rightarrow ^{4}_{2}He + ^{1}_{0}n$

(Atomic masses: $^{3}_{1}H$ = 3.016 amu, $^{2}_{1}H$ = 2.014 amu, $^{4}_{2}He$ = 4.003 amu and $^{1}_{0}n$ = 1.009 amu).

For the nuclear reaction,

$$^{3}_{1}H + ^{2}_{1}H \rightarrow ^{4}_{2}He + ^{1}_{0}n$$

Total mass of reactants = 3.016 + 2.014 = 5.03 amu

Total mass of products = 4.003 + 1.009 = 5.012 amu

Mass defect, Δm = 0.018 amu

Energy released per helium atom = Δm. 931

= 0.018 × 931 = 16.76 MeV

= $16.76 \times 1.602 \times 10^{-16}$ kJ atom^{-1}

(Since 1 MeV = 1.602×10^{-16} kJ)

= 2.68×10^{-15} kJ atom^{-1}

Energy released per mole of helium = 2.68×10^{-15} N

(N = Avogadro's number = 6.023×10^{23})

= $2.68 \times 10^{-15} \times 6.023 \times 10^{23}$

= 1.61×10^{9} kJ mol^{-1}

The energy released in the nuclear reaction is 1.61×10^{9} kJ per mole of helium formed.

Packing fraction of the nucleus

When atomic masses are determined experimentally, their values are found to deviate slightly from whole numbers. This deviation from integral mass number may be expressed in terms of packing fraction, which is mathematically defined as follows:

$$\text{Packing fraction} = \frac{\text{Atomic mass} - \text{Mass number}}{\text{Mass number}} \times 10^{4} \qquad (4.7)$$

Here, atomic mass is the actual mass of the nucleus and mass number is the total number of protons and neutrons in the nucleus. The atomic mass is close to a whole number, while mass number is the nearest whole number. The first term in equation (4.7) has an extremely small value and to obtain a convenient figure, it is multiplied by 10,000.

The majority of elements possess a negative value of packing fraction, although for a few the value is positive. For example, for cobalt, the packing fraction is negative.

Atomic mass = 58.93

Mass number = 59

$$\text{Packing fraction} = \frac{58.93 - 59}{59} \times 10^{4} = -11.86$$

Hydrogen nucleus has a positive value of packing fraction.

Atomic mass = 1.008

Mass number = 1

$$\text{Packing fraction} = \frac{1.008 - 1}{1} \times 10^4 = +80$$

Nuclei with negative packing fractions are stable, while those with positive packing fractions are relatively less stable. When plotted against mass numbers, the packing fractions for stable isotopes of various elements, with the exception of helium-4, carbon-12 and oxygen-16, fall on a smooth curve (Fig. 4.2). The curve drops from hydrogen with a packing fraction of +80 to about zero for elements with mass numbers between 16 and 20 and then passes through a flat minima in the vicinity of mass number 50. Thereafter, it rises linearly and finally, at mass number 200, it again becomes positive. The instability of the elements lying on the last part of the curve is manifested in their radioactivity.

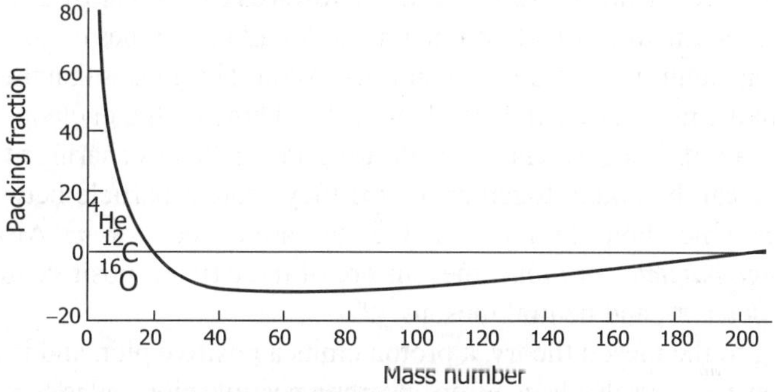

Fig. 4.2 Packing fraction curve

The packing fraction curve (Fig 4.2) is inverse of the binding energy curve (Fig. 4.1). Iron and nickel occupy the maxima in the binding energy curve and minima in the packing fraction curve. Accordingly, these atoms possess maximum nuclear stability. These elements are largely concentrated in the core of the earth. Helium-4, carbon-12 and oxygen-16 nuclei are more stable than their immediate neighbors. For an explanation, *see* Box 4.1.

Case Study 4.2 The masses of the three isotopes of oxygen are as follows:

Oxygen-16 = 16.0000 amu ; Oxygen-17 = 17.0045 amu; and Oxygen-18 = 18.0049 amu. The abundance of these three isotopes in naturally occurring oxygen is 99.76%, 0.04% and 0.20%, respectively. Find out the atomic mass of oxygen.

The relative abundance of the isotopes are —

Oxygen-16: 99.76% = 0.9976

Oxygen-17: 0.04% = 0.0004

Oxygen-18: 0.20% = 0.0020

Isotope	Atomic mass		Relative abundance		Weighted mean
Oxygen-16	16.0000	×	0.9976	=	15.9616
Oxygen-17	17.0045	×	0.0004	=	0.0068
Oxygen-18	18.0049	×	0.0020	=	0.0360
			Total	=	16.0044

The atomic mass of oxygen is 16.0044 amu.

Meson theory of nuclear forces

The forces that bind the nucleons together in a nucleus are many times greater than those which hold the atoms together in a molecule. For example, two protons, separated by a nuclear distance from each other, undergo electrostatic repulsion to the order of 6 tonnes of force, while these are attracted towards each other by 240 tonnes of force. Moreover, the force of attraction is same for all three types of nucleon pairs: proton–proton, neutron–neutron and neutron–proton. This means that the attractive forces within the nucleus are independent of the charge on the nucleons.

Just as two atoms in a covalent molecule are held together by sharing of electrons, two nucleons can be bound together only if they share a particle between them. The particles which hold the nucleons together are called *mesons.* Also referred to as *exchange particles* or *pions,* mesons are of three types: positive mesons, π^+, negative mesons, π^-, and neutral mesons, π^0.

According to the meson theory, a proton emits a positive pion and is converted into a neutron. Conversely, the neutron absorbs a positive pion and is converted into a proton. The net result is that the neutron and proton are held together.

$$P \rightleftharpoons \pi^+ + n$$

Similarly, a neutron emits a negative pion and gets converted into a proton which, in turn, takes up a negative pion and gets reconverted into a neutron. This phenomenon also binds the neutron with the proton.

$$n \rightleftharpoons \pi^- + p$$

It is thus evident that positive and negative mesons act as exchange particles for holding neutrons and protons together in the nucleus. Two other types of interactions within the nucleus, neutron–neutron attractions and proton–proton attractions, have also to be accounted for. These involve sharing of neutral pions.

$$P \rightleftharpoons \pi^0 + P$$

$$n \rightleftharpoons \pi^0 + n$$

Sharing of neutral mesons forges the attraction between two protons and two neutrons.

The conversion of protons into neutrons or vice versa by exchange of positive and negative mesons is a first-order reaction. Greater the mass number, higher is the rate of conversion.

Characteristics of nuclear forces

Nuclear forces are also called *exchange forces* because these result from an exchange of energy or an exchange of matter between the nucleons. These forces act along the line joining any two nucleons. These are short-range forces, operating over very short distances, and show great variations with small changes in distance. As shown in Fig. 4.3, nuclear forces show maximum attraction at a distance of 80 fm between the two nucleons. At smaller distances, the forces of attraction decrease and at about 50 fm, turn into forces of repulsion. At larger distances, these tend to approach zero (at about 400 fm).

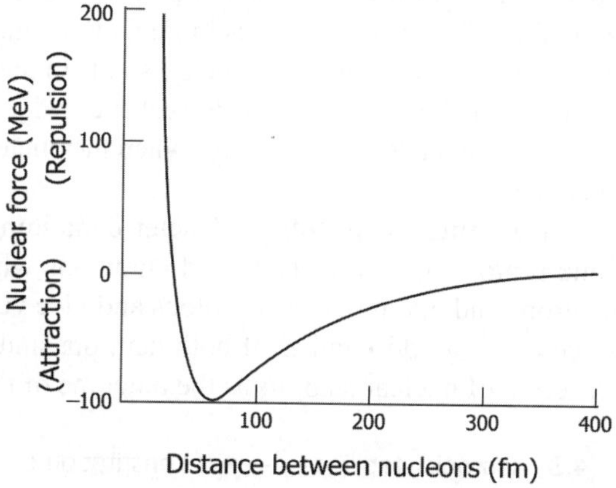

Fig. 4.3 Variation in nuclear forces with distance between nucleons

The magnitude of nuclear forces also depends on the direction of spins of the nucleons. For example, take the case of deuterium. This atom has one proton and one neutron in its nucleus. Each nucleon has a spin of ½. Since the net spin of deuterium nucleus is unity, it is evident that the proton occupies one energy level and the neutron occupies another energy level. If these nucleons had coupled together in a single energy level, the resultant spin of the deuterium nucleus would have been zero (Fig. 4.4).

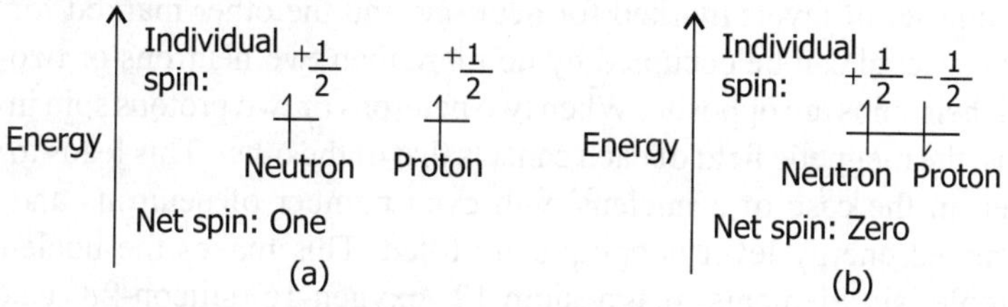

Fig. 4.4 Magnitude of nuclear forces when (a) the neutron and proton in deuterium nucleus occupy different energy levels and (b) the neutron and proton are coupled in the same energy level

Despite having an odd number of each nucleon the deuterium nucleus is quite stable. This is attributed first to the fact that the neutron to proton ratio is unity and second, to the observation that the net spin of deuterium is unity and not zero.

Shell model of the nucleus

Just as electrons are arranged in different energy levels outside the nucleus, nucleons are arranged in different energy levels within the nucleus. There are two sets of energy shells in the nucleus, one occupied by neutrons and the other by protons. As the mass number increases, the spacing between the proton levels becomes more than the spacing between neutron levels. This is essential since with increasing mass number, the number of neutrons becomes more than the number of protons. Therefore, there must be a greater number of energy shells in the neutron set than in the proton set. When the nucleons occupy the lowest energy levels, these are said to be in the ground state. The nucleus emits γ-rays when the nucleons fall from the excited to the ground state.

Even–odd criteria of nuclear stability Nuclei containing even numbers of both neutrons and protons are more stable and numerous whereas those with even number of neutrons and odd number of protons and vice versa are relatively less stable. Nuclei containing odd number of both neutrons and protons are rare (Table 4.3). This criterion of nuclear stability is the outcome of the shell model.

Table 4.3 Isotopic stabilities vis-à-vis constitution of nuclei

Neutrons	Protons	Number of stable isotopes
Even	Even	162
Odd	Even	56
Even	Odd	52
Odd	Odd	4

The preferential stability of even–even nuclei is associated with the requirements of the Pauli exclusion principle. The nucleons occupy specific energy levels in the nucleus, one set of levels marked for neutrons and the other marked for protons. Each energy level can be occupied by no more than two neutrons or two protons, provided their spins are opposite. When two neutrons or two protons spin in opposite directions, the magnetic field of each cancels that of the other. This leads to stability. Moreover, in the case of a nucleus with even number of neutrons and protons, each occupied energy level is completely filled. This makes the nucleus stable. For example, six elements, magnesium-12, oxygen-16, silicon-28, calcium-40, titanium-48 and iron-56, each containing even number of neutrons and even number of protons, constitute 80% of the earth's crust.

The nuclei in which the neutron–proton number is even–odd or odd–even, the highest occupied level in either neutron or proton set is incomplete. Therefore, such nuclei are less stable than even–even cases. The nuclei in which the neutron–proton number is odd–odd, the highest occupied level in both neutron and proton set is incomplete and, therefore, such isotopes are rare. In fact, there are only four stable nuclear species of odd–odd type. These are deuterium-2, lithium-6, boron-10 and nitrogen-14. They owe their stability to the presence of an equal number of neutrons and protons in their nucleus.

Case Study 4.3 Why is beryllium-8 species non-existent?

The hypothetical beryllium-8 nucleus would have contained four neutrons and four protons. The neutron to proton ratio in this species would have been unity. Moreover, there are even number of neutrons and even number of protons. It is, therefore, expected that this isotope should be very stable. However, it does not exist in nature. This is because helium-4 nucleus is much more stable than beryllium-8 nucleus. As a result, beryllium-8 nucleus immediately splits up into two helium-4 nuclei.

$$^{8}_{4}\text{Be} \rightarrow 2\,^{4}_{2}\text{He}$$

It has been observed that elements with even number of protons are more abundant, more stable and have a wider range of isotopes as compared to those with odd number of protons. Similarly, elements with even number of neutrons are more abundant, more stable and have a wider range of isotopes as compared to those with odd number of neutrons.

Magic numbers Another fallout of the shell model is the concept of magic numbers. Just as certain atoms containing a specific number of electrons exhibit exceptional stability, certain nuclei containing a specific number of nucleons possess additional stability. This number is called *magic number* or *shell number*. The magic number nucleons—2, 8, 20, 50, 82 and 126—represent a closed shell in the nucleus.

The elements having the largest number of stable isotopes are those containing 20 and 50 protons or 20, 50 and 82 neutrons. For example, tin, with atomic number 50, forms ten stable isotopes. Three of them have odd mass numbers (115, 117 and 119). Apart from tin, no other element has more than two stable isotopes with odd mass numbers. Tin is an exception since its number of protons corresponds to magic number 50.

The isotopes which are more abundant in the universe contain magic number of either protons or neutrons or both. Examples include oxygen-16, calcium-40 and lead-208. Such nuclei have very small cross-section for neutron capture, i.e., these do not add additional neutrons, since then the magic number rule is violated. If an isotope has a neutron in excess of magic number, it readily emits the neutron and disintegrates. For example,

$$\underset{\text{51 neutrons}}{^{87}_{36}\text{Kr}} \longrightarrow \underset{\text{50 neutrons}}{^{86}_{36}\text{Kr}} + {}^{1}_{0}\text{n}$$

If a radioactive element disintegrates by emission of α-particles, the energy change during the nuclear reaction is exceptionally large if the decay product belongs to the magic number category. For example,

$$^{212}_{84}Po \rightarrow ^{208}_{82}Pb + ^{4}_{2}He$$

128 neutrons 84 protons	126 neutrons 82 protons	α-particle

Isotopes such as lead-208, which possess magic number of both neutrons and protons, are more stable as compared to isotopes having magic number of only neutrons (e.g., xenon-136; n = 82) or only protons (e.g., tin-115; p = 50).

Liquid drop model of the nucleus

The liquid drop model compares the atomic nucleus with a drop of liquid. There are a large number of similarities between the two.

1. The radius of a nucleus, r, is directly proportional to the cube root of mass number, A.

$$r \propto (A)^{1/3} \tag{4.8}$$

The volume of a sphere, V—read nucleus—is given as,

$$V = \frac{4}{3}\pi r^3$$

or

$$V \propto (r)^3 \tag{4.9}$$

From equations (4.8) and (4.9),

$$V \propto (A^{1/3})^3$$

or

$$V \propto A \tag{4.10}$$

This implies that the volume of the nucleus is directly proportional to the total number of nucleons present in it. The volume of a drop of liquid is also proportional to the number of atoms or molecules contained in it.

2. The density of a nucleus is nearly constant, and so is that of a liquid drop.
3. The binding energy of a nucleus increases with increasing number of nucleons. The net chemical binding energy within a liquid drop also varies with the number of atoms or molecules within it.
4. Nuclear forces are short-range forces. As a result, each nucleon is strongly held by those in its immediate vicinity, but is relatively unaffected by those which are not its immediate neighbors. This is analogous to the cohesive force of surface tension within a liquid drop. Under the influence of surface tension, the liquid drop attains spherical shape. The atomic nucleus is also spherical. Further, there are dissipative forces acting between the constituent atoms or molecules within the liquid drop. These correspond to the electrostatic proton–proton repulsions within the nucleus.
5. Very large drops of a liquid are not stable for long, since the force of surface tension progressively becomes weak. The large drop dissociates into smaller droplets. In case of atomic nuclei, the size of the nucleus increases with

mass number. For very heavy elements, the size of the nuclei reaches a critical value and gets deformed. A slight provocation is enough to cause its disintegration (Fig. 4.5).

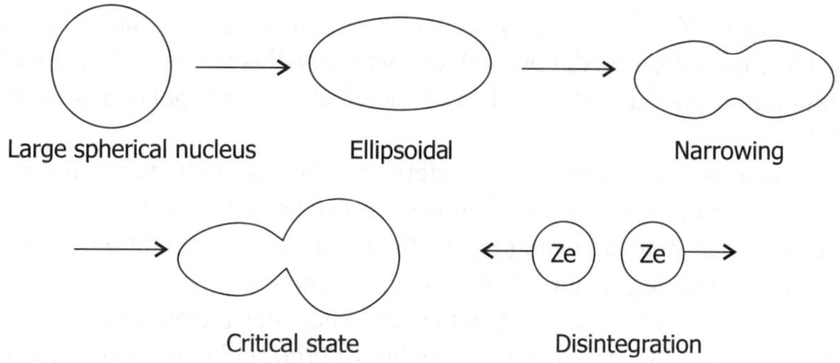

Fig. 4.5 Mechanism of disintegration of a large nucleus according to the liquid drop model

6. The provocation for disintegration is provided when a projectile particle strikes the target nucleus and is captured by it. Since nuclear forces are short-range interactions, the energy of the projectile particle is very rapidly distributed among all the nucleons. For a particular nucleon, the extra energy may be enough to cause its escape. This process of disintegration is analogous to the slow evaporation of atoms or molecules from the surface of a liquid drop.

Neutron–proton ratio criteria of nuclear stability According to the liquid drop model, the stability of a nucleus is determined by the number of protons and neutrons present in it. For elements with atomic numbers up to 20, the nucleus contains an equal number of neutrons and protons, i.e., the neutron to proton ratio is unity. If the number of neutrons is plotted against the number of protons, and a line (OA) passing through the origin is drawn at an angle of 45°, the points representing the first 20 elements lie on this line (Fig. 4.6).

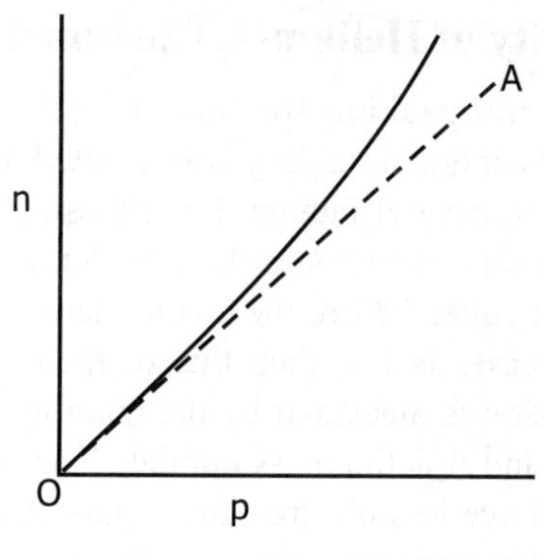

Fig. 4.6 A plot of number of neutrons (n) against number of protons (p)

As the atomic number increases, the ratio of neutrons to protons tends to become greater than unity. To maintain stability, the number of neutrons must be more than the number of protons. Each proton repels, and is repelled, by all other protons in the nucleus. The electrostatic repulsion increases with atomic number and is given by the expression, Z^2/r, where Z is the atomic number and r is the radius of the nucleus. The liquid drop model propounds that the radius varies as $(A)^{1/3}$, where A is the mass number (equation 4.8). Hence the electrostatic repulsion is expressed as $Z^2/(A)^{1/3}$.

To overcome the increasing electrostatic repulsions with increasing atomic number, the nuclei must contain neutrons in excess of protons. For the heaviest stable nuclei, such as lead-208 and bismuth-209, the neutron to proton ratio is nearly 1.6. In elements with still higher atomic numbers, the excessive neutrons are unable to overcome the electrostatic repulsions which exist between the protons. The binding energy becomes low and, as a result, the nuclei undergo spontaneous disintegration. Therefore, the elements in which the neutron to proton ratio is greater than 1.6 are radioactive in nature.

Case Study 4.4 Beryllium-9 is stable, but boron-9 is not. Why?

The composition of the two nuclei is as follows:

$^{9}_{4}Be$: 5 neutrons, 4 protons

$^{9}_{5}B$: 4 neutrons, 5 protons

The neutron to proton ratio is not unity in either case. However, beryllium-9 has an additional neutron, which tends to overcome the electrostatic repulsions between the protons. As a result, the isotope is stable. On the other hand, boron-9 has an extra proton. The electrostatic repulsions in the nucleus are large enough to make it unstable.

Box 4.1

Exceptional Stability of Helium-4, Carbon-12 and Oxygen-16

The binding energies and packing fractions of helium-4, carbon-12 and oxygen-16 reveal that exceptional stability is associated with these three nuclei. This may be attributed to several factors. First, these atoms have even–even combination of protons and neutrons in their nuclei. Second, the number of protons and neutrons is equal. Third, the electrostatic repulsion between the protons in these three cases is less than that of their immediate neighbors. The electrostatic repulsion is measured by the quantity, $Z^2/(A)^{1/3}$, where Z is the number of protons and A is the mass number. The following calculations show that these nuclei have less electrostatic repulsion within their atoms as compared to their next heavier analogs:

Contd...

$$^{4}_{2}\text{He}: \frac{(2)^2}{(4)^{1/3}} = 2.53 \qquad ^{6}_{3}\text{Li}: \frac{(3)^2}{(6)^{1/3}} = 4.97$$

$$^{12}_{6}\text{C}: \frac{(6)^2}{(12)^{1/3}} = 15.78 \qquad ^{14}_{7}\text{N}: \frac{(7)^2}{(14)^{1/3}} = 20.33$$

$$^{16}_{8}\text{O}: \frac{(8)^2}{(16)^{1/3}} = 25.39 \qquad ^{19}_{9}\text{F}: \frac{(9)^2}{(19)^{1/3}} = 30.45$$

For helium-4 and oxygen-16, there is an additional reason for exceptional stability. The number of neutrons, as well as the number of protons, in each case corresponds to the magic number (2 and 8, respectively).

Natural Radioactivity

Two parameters characterize the heavier elements, i.e., elements possessing atomic numbers more than 83. First, the ratio of neutrons to protons in their nuclei is greater than 1.6. Second, their binding energy is less than the elements immediately preceding them. Both the parameters imply that the nuclei of heavy elements are unstable. They undergo decay and are transformed into relatively more stable products, which may be an isotope of the original heavy atom or an altogether different element. A few of the lighter elements, for example, carbon-14 and potassium-40, also undergo natural decay. In any case, the disintegration is accompanied by emission of radiations. The spontaneous emission of radiation by an element is called *radioactivity.* The concerned element is called *radioactive element.* It is sometimes also referred to as *radioisotope* or *radionucleide.* The emissions accompanying the decay of radioactive elements are called *radioactive emanations.* There are three types of radioactive emanations—*alpha* (α) *rays, beta* (β) *rays* and *gamma* (γ) *rays.*

Alpha rays are charged helium particles. These carry two units of positive charge and four units of atomic mass (represented as $^{4}_{2}$He). Being positively charged, these are deflected towards the negative pole in an electrical field. However, the deflection is small since α-particles are quite heavy. Their initial velocity is approximately one-tenth the velocity of light. Their penetration power is not very high. In fact, these are absorbed by an aluminium foil of 0.1 mm thickness. Nevertheless, the absorption power of α-particles depends not only on the thickness of the sheet but also on the nature of the metal it is made of. Higher the mass number of the metal, higher is the absorption power. Thus, lead-208 will absorb α-particles more easily than tin-119. This means that a beam of α-ray which can just penetrate through a tin foil would be stopped by a lead foil of equal thickness. The range of α-particles depends on the nature of the radioactive elements which emit them. Those elements which decay rapidly emit α-particles with longer ranges. Thus thorium-228, with a half-life of approximately 2 years emits α-particles with a range of 4 cm, while polonium-214, with a half-life of less than 1 second produces α-particles with a range of 7 cm.

Beta rays are high-speed electrons. Being negatively charged, these are deflected towards the positive pole in an electric field and are designated as $_{-1}^{0}e$. Their velocities are spread over a very large range—the upper limit approaching the velocity of light. The fast moving particles are designated as *hard beta rays,* while those with lesser speeds are called *soft beta rays.* As a result of their phenomenal velocities, β-rays have far greater penetrating power than α-rays. Here too the penetration power depends on the nature of metal—greater the mass number of the metal, lesser is the penetration power of beta rays. Thus, a lead foil of 1 mm thickness has the same effectiveness of absorbing β-particles as an aluminium foil of 5 mm thickness.

Gamma rays are radiations with short wavelength and very high energy. They are not deflected by an electrical field, and this indicates the absence of charge on these rays. Moreover, γ-rays do not have any mass and, therefore, are not considered as particles. These possess shorter wavelength, and hence higher energy, than X-rays. Accordingly, their penetration power is greater than that of X-rays—a reason for which these are often called *hard rays.* Like α- and β-rays, γ-rays are more effectively absorbed by heavier metals. Thus a beam of γ-ray may penetrate 25 cm of iron, but only 8 cm of lead of equal thickness. Gamma rays travel with the speed of light. According to the shell model of atomic nucleus, γ-rays arise when a nucleon falls from a higher energy state to a lower energy state within the nucleus. The γ-rays emitted by a specific isotope have a discrete set of energies. This shows that the nuclear energy levels are quantized, much like the atomic energy levels.

Disintegration theory

The atoms of very heavy elements, in whose nuclei the ratio of neutrons to protons exceeds 1.6, are unstable. They disintegrate into a new element, the nucleus of which has a relatively low neutron to proton ratio and, therefore, is comparatively more stable. Disintegration may occur in either of the two ways described below:

1. When an element emits an α-particle, the mass number of the decay product is reduced by four units and the atomic number by two units. As a result, the new product is displaced two groups to the left of the parent element in the periodic table. For example,

$$^{226}_{88}Ra \rightarrow {}^{222}_{86}Rn + \underset{\alpha\text{-particle}}{{}^{4}_{2}He}$$

2. When an element emits β-particle, the mass number of the decay product remains the same, but its atomic number increases by one unit. As a result, the decay product is displaced one group to the right of the parent element in the periodic table. For example,

$$^{210}_{82}Pb \rightarrow {}^{210}_{83}Bi + \underset{\beta\text{-particle}}{{}^{0}_{-1}e}$$

Since β-particle is synonymous with electron, it is pertinent to explain how this type of emission takes place from the nucleus. The electrons are located outside the nucleus. However, β-emission involves the conversion of a neutron into proton and electron.

$$\text{Neutron} \rightarrow \text{proton} + \text{electron} + \text{nutrino}$$

The electron escapes the atom as β-particle, leaving one extra proton in the nucleus, due to which the atomic number increases. The nutrinos are energy particles. They ensure that the energy of the reaction, involving transformation of neutron into proton and electron, is conserved.

These two rules of the disintegration theory are also called *group displacement laws*. These are so named because the position of the decay product in the periodic table is displaced with respect to the position of the parent element.

Case Study 4.5 How many α- and β-particles would be emitted during the disintegration of thorium-232 to lead-208?

Let the number of α- and β-particles emitted during the disintegration of thorium-232 to lead-208 be x and y, respectively. The overall nuclear reaction may then be represented as follows:

$$^{232}_{90}\text{Th} \rightarrow {}^{208}_{82}\text{Pb} + x\,^{4}_{2}\text{He} + y\,^{0}_{-1}\text{e}$$

Balancing mass numbers on both sides,

$$232 = 208 + 4(x) + 0(y)$$

$$x = 6$$

Balancing atomic numbers on both sides,

$$90 = 82 + 2(x) + \{-1(y)\}$$

Substituting x = 6, we get,

$$y = 4$$

Thorium-232 disintegrates into lead-208 by emission of six α- and four β-particles.

Rate of disintegration

The group displacement laws represent the tendency of radioactive elements to disintegrate. Different elements disintegrate at different rates. The rate of a nuclear reaction depends on the nature of the element undergoing disintegration as well as on the number of atoms of that element present at a given time. In fact, the quantity of the radioactive element that disappears in unit time is directly proportional to the total amount of that element present, i.e.,

$$\frac{-dN}{dt} \propto N \qquad (4.11)$$

where N is the total number of atoms of the radioactive element present at time t. The term dN/dt is the rate of disappearance of the element. The negative sign signifies that the number of atoms decreases with time. From equation (4.11),

$$\frac{-dN}{dt} = \lambda N \tag{4.12}$$

where λ is a constant called *radioactive constant* or *decay constant.* Equation (4.12) may be rearranged as,

$$\frac{dN}{N} = -\lambda\, dt \tag{4.13}$$

Integrating equation (4.13) between the limits of N_0, the number of atoms present at time t = 0 (initial time) and N, the number of atoms present at time t = t (final time),

$$\int_{N_0}^{N} \frac{dN}{N} = -\lambda \int_{t=0}^{t=t} dt$$

$$\ln N \Big]_{N_0}^{N} = -\lambda t \Big]_{0}^{t}$$

$$\ln N - \ln N_0 = -\lambda t$$

$$\ln \frac{N}{N_0} = -\lambda t \tag{4.14}$$

$$\frac{N}{N_0} = e^{-\lambda t}$$

$$N = N_0 e^{-\lambda t} \tag{4.15}$$

Equation (4.14) shows that radioactive decay follows first-order kinetics. This means that the radioactivity decreases in an exponential manner. If radioactivity of an element is plotted against time, a logarithmic curve is obtained, which cuts the x-axis at infinity (Fig. 4.7), i.e., as disintegration proceeds, the number of atoms of the radioactive element decreases, and so does the rate of decay.

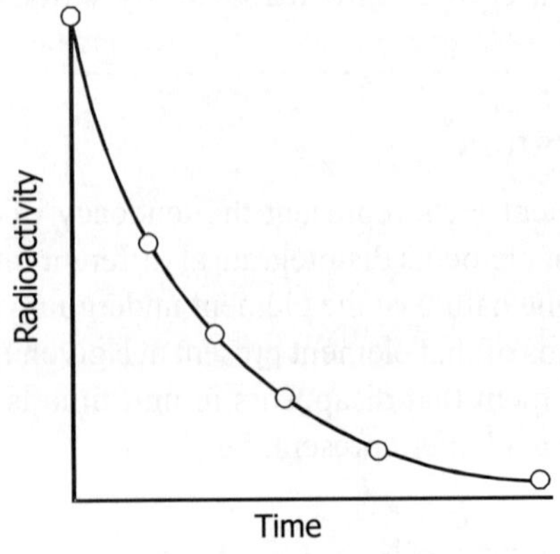

Fig. 4.7 Exponential nature of radioactive decay

The *half-life period* of an element—the time required for a given amount of sample to disintegrate to one-half of its initial amount—may be elucidated from equation (4.14). At half-life period, $t_{1/2}$, the ratio between N and N_0 is 1:2. Substituting these values in equation (4.14),

$$\ln \frac{1}{2} = -\lambda t_{1/2}$$

$$2.303 \log 2 = \lambda t_{1/2}$$

$$t_{1/2} = \frac{0.693}{\lambda} \qquad (4.16)$$

By knowing the radioactive constant of a given sample, its half-life period may be calculated from equation (4.16).

The unit of radioactivity is *Becquerel.* A Becquerel, abbreviated as Bq, is equal to one disintegration per second.

Case Study 4.6 The half-life of a radioactive element is 100 s. Calculate the time required for 1 g of the element to disintegrate to 0.01 g.

The half-life period of a radioactive substance ($t_{1/2}$) is related to the decay constant (λ) by the expression,

$$t_{1/2} = \frac{0.693}{\lambda}$$

$$t_{1/2} = 100 \text{ s}$$

$$\lambda = \frac{0.693}{100} = 6.93 \times 10^{-3} \text{ s}^{-1}$$

Further, the amount of radioactive substance (N) at a given time (t) is related to the amount initially present (N_0) by the expression,

$$N = N_0 e^{-\lambda t}$$

or

$$\frac{N_0}{N} = e^{\lambda t}$$

$$2.303 \log \frac{N_0}{N} = \lambda t$$

$$t = \frac{2.303}{\lambda} \log \frac{N_0}{N}$$

Here, N_0 = 1 g, N = 0.01 g and $\lambda = 6.93 \times 10^{-3}$ s^{-1}

$$t = \frac{2.303}{6.93 \times 10^{-3}} \log \frac{1}{0.01}$$

$$t = 664.6 \text{ s}$$

It will take 664.6 s for the substance to reduce from 1 g to 0.01 g.

Radioactive equilibrium

If an isotope with a long half-life disintegrates, a number of elements are successively formed. After a period of time, the rate of formation of each element becomes equal to the rate of its decay. A condition known as *radioactive equilibrium* sets in.

For example, consider radium, a moderately long-lived element, which slowly disintegrates into radon by the emission of an α-particle.

$$^{226}_{88}Ra \rightarrow ^{222}_{86}Rn + ^{4}_{2}He$$

If N_1 atoms of radium and N_2 atoms of radon are present in the system, at a particular time t, then the rate of disintegration of radium is given as:

$$\frac{-dN_1}{dt} = \lambda_1 N_1 \quad (4.17)$$

where λ_1 is the decay constant for radium.

The rate of disintegration of radon is given as:

$$\frac{-dN_2}{dt} = \lambda_2 N_2$$

where λ_2 is the decay constant for radon.

Equation (4.17) represents not only the rate of disintegration of radium but also the rate of formation of radon. Under the conditions of radioactive equilibrium,

Rate of formation of radon = Rate of disappearance of radon

$$\frac{-dN_1}{dt} = \frac{-dN_2}{dt}$$

or

$$\lambda_1 N_1 = \lambda_2 N_2$$

$$\frac{N_1}{N_2} = \frac{\lambda_2}{\lambda_1} \quad (4.18)$$

With the passage of time, radon starts decaying into polonium which, in turn, starts disintegrating into lead, at a rate which depends on the number of atoms of the concerned element as well as on its nature. Ultimately, the rate of formation of each element becomes equal to the rate of its decay.

The mathematical expression (4.18) is called *law of successive disintegration.* It is applicable to any two members of a disintegration series (*see* next section), whether these are adjacent to each other or not.

Case Study 4.7 A given sample of a radioisotope shows an activity of 9600 counts min^{-1} at time t_0 and 8000 counts min^{-1} after one hour. Determine its half-life.

The radioactivity measured in terms of counts min^{-1} is directly proportional to the number of nuclei of the concerned isotope. We first elucidate the radioactive constant, λ by re-writing equation (4.15) as,

$$\log \frac{N_0}{N} = \frac{\lambda t}{2.303}$$

Here, $N_0 = 9600$, $N = 8000$, $t = 1$ h

$$\lambda = \frac{2.303 \log (9600/8000)}{1}$$

$$\lambda = \frac{2.303 \log 1.2}{1} = 0.182$$

Substituting the value of λ in equation (4.16), we get,

$$t_{1/2} = \frac{0.693}{0.182} = 3.80 \text{ h}$$

The half-life of the given sample is 3.80 hours.

Radioactive disintegration series

A heavy radioactive element decays, by the emission of α- or β-particles, into an isotope which is more stable than itself. However, the decay product need not be completely stable. It may also possess radioactive properties and decay into yet another isotope. This process continues till an element is formed which shows no further disintegration. The successive steps of one radioactive element disintegrating into another, till a non-radioactive, stable element is formed constitute a *disintegration series*, also called a *decay series*.

There are four radioactive disintegration series—three natural and one artificial. The natural disintegration series are called thorium (Table 4.4), uranium (Table 4.5) and actinium series (Table 4.6), while the artificial one is referred to as neptunium series (Table 4.7). These are designated as follows:

Thorium 4n series

Neptunium (4n + 1) series

Uranium (4n + 2) series

Actinium (4n + 3) series

Table 4.4 Thorium disintegration series

Radioactive element	Isotope	Radiation	Half-life
Thorium	^{232}Th	α	1.39×10^{16} y
Radium	^{228}Ra	β	6.7 y
Actinium	^{228}Ac	β	6.13 h
Thorium	^{228}Th	α	1.90 y
Radium	^{224}Ra	α	3.64 d
Radon	^{220}Rn	α	54.5 s
Polonium	^{218}Po	α and β	0.16 s
Lead	^{212}Pb	β	10.6 h
Astatine	^{216}At	α	54 s
Bismuth	^{212}Bi	α and β	60.5 s
Polonium	^{212}Po	α	3×10^{-7} s
Thallium	^{203}Ti	β	3.1 min
Lead	^{208}Pb (Stable)		

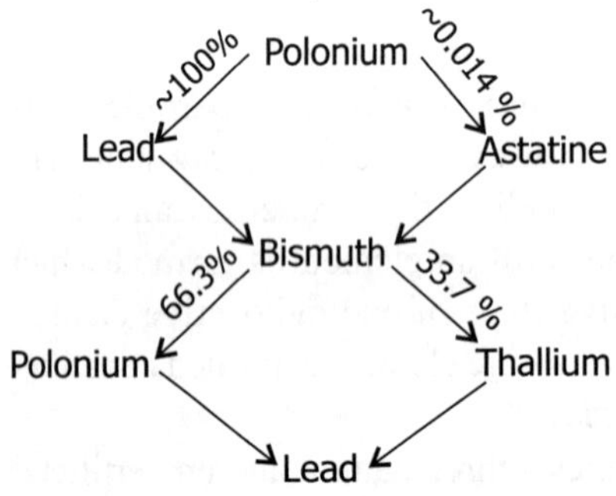

Table 4.5 Uranium disintegration series

Radioactive element	Isotope	Radiation	Half-life
Uranium	^{238}U	α	4.5×10^{9} y
↓ Thorium	^{234}Th	β	24.5 d
↓ Protactinium	^{234}Pa	β	1.14 min
↓ Uranium	^{234}U	α	2.67×10^{5} y
↓ Thorium	^{230}Th	α	8.3×10^{4} y
↓ Radium	^{226}Ra	α	1.62×10^{3} y
↓ Radon	^{222}Rn	α	3.82 d
↓ Polonium	^{218}Po	α and β	3.05 min
Lead (99.96%)	^{214}Pb	β	26.8 min
Astatine (0.04%)	^{218}At	α	2 s
Bismuth	^{214}Bi	α and β	19.7 min
Polonium (99.96%)	^{214}Po	α	1.5×10^{-4} s
Thallium (0.04%)	^{210}Ti	β	1.32 min
Lead	^{210}Pb	β	22 y
↓ Bismuth	^{210}Bi	α and β	4.85 d
Polonium (~100%)	^{210}Po	α	138.3 d
Thallium (~10^{-5}%)	^{206}Ti	β	4.23 min
Lead	^{206}Pb (Stable)		

Table 4.6 Actinium disintegration series

Radioactive element	Isotope	Radiation	Half-life
Uranium	^{235}U	α	7.07×10^{3} y
↓ Thorium	^{231}Th	β	24.6 h
↓ Protactinium	^{231}Pa	α	3.2×10^{4} y
↓ Actinium	^{227}Ac	α and β	21.7 d
Thorium (98.8%)	^{227}Th	α	18.9 d
Francium (1.2%)	^{223}Fr	β	21 min
Radium	^{223}Ra	α	11.2 d
↓ Radon	^{219}Rn	α	3.92 s
↓ Polonium	^{215}Po	α	1.82×10^{-3} s
↓ Lead	^{211}Pb	β	3.61 min
↓ Bismuth	^{211}Bi	α and β	2.61 min
Polonium (99.68%)	^{211}Po	α	5×10^{-3} s
Thallium (0.32%)	^{207}Ti	β	4.76 min
Lead	^{207}Pb (Stable)		

Table 4.7 Neptunium disintegration series

Radioactive element	Isotope	Radiation	Half-life
Plutonium ↓	^{241}Pu	β	13 y
Americium ↓	^{241}Am	α	470 y
Neptunium ↓	^{237}Np	α	2.20×10^{6} y
Protactinium ↓	^{233}Pa	β	27.4 d
Uranium ↓	^{233}U	α	1.62×10^{5} y
Thorium ↓	^{229}Th	α	7.34×10^{3} y
Radium ↓	^{225}Ra	β	14.8 d
Actinium ↓	^{225}Ac	α	10 d
Francium ↓	^{221}Fr	α	4.8 min
Astatine ↓	^{217}At	α	0.02 s
Bismuth (98% → Polonium; 2% → Thallium)	^{213}Bi	α and β	47 min
Polonium (98%) → Lead	^{213}Po	α	4.2×10^{-6} s
Thallium (2%) → Lead	^{209}Ti	β	2.2 min
Lead ↓	^{209}Pb	β	3.3 h
Bismuth	^{209}Bi (Stable)		

The thorium, neptunium and actinium series are so named because an isotope of these elements has the longest half-life in the respective series. The actinium series was once thought to originate with actinium-227—the isotope having the longest half-life among the series members. However, it has since been discovered that the series actually initiates with a very long-lived uranium-235 isotope.

The numbers 4n, 4n+1, 4n+2 and 4n+3 indicate that members of the concerned series have mass numbers divisible either by 4 or by 4 with remainder 1, 2 and 3, respectively. The thorium, uranium and actinium series end with an isotope of lead—lead-208, lead-206 and lead-207, respectively—while the nepturium series terminates at bismuth-209.

These is no cross-linking between the four series. This means that an isotope of one series will not decay to an isotope belonging to another series. This is expected since the mass change takes place only on emission of α-particles (no mass change occurs when β-particles are emitted). The resultant change is of four mass units and the mass number of each family member is either 4n, 4n+1, 4n+2 or 4n+3.

Nevertheless, the series do not follow straight chain disintegrations. Branched disintegrations occur in each series. In such cases, a part of the isotope decays into one product by α-particle emission and the remaining part decays into another product by β-particle emission. The two isotopes of branched disintegrations then further decay into a common product.

Case Study 4.8 Is it true that neptunium disintegration series differs significantly from thorium, uranium and actinium series?

When compared to thorium, uranium and actinium disintegration series, the neptunium series is indeed significantly different, as the following aspects reveal:

1. The members of thorium, uranium and actinium series are naturally occurring isotopes. The members of neptunium series, except the last element, bismuth, do not occur in nature. These have been synthesized artificially by transmutation of elements.
2. The end product of neptunium series is bismuth, whereas the end products of the natural radioactive series are isotopes of lead.
3. The neptunium series does not contain any gaseous element. Other series contain isotopes of radon.
4. Neptunium series contains isotopes of francium and astatine as successive members of the main chain. In other series, these elements appear only as minor, branched decay products.
5. Branched disintegration occurs more frequently in the natural disintegration series than in the neptunium series.

Detection of radiation

Different types of radiations emanating from radioactive materials may be detected by two means: cloud chamber and Geiger-Müller counter. We now discuss the working of these radiation detectors.

1. Cloud chamber

Also called *Wilson chamber,* the cloud chamber serves to trace the pathway of emitted radiation. The chamber is made up of glass and contains supersaturated vapors of water or alcohol. One variety of a cloud chamber is shown in Fig. 4.8.

Fig. 4.8 A cloud chamber

The radioactive material is placed inside the chamber. The α-, β- or γ-rays, given out by the radioisotope, ionize the molecules of the components of air trapped inside the vessel. The resulting ions act as condensation nuclei for vapors of water or alcohol. That is, the vapors of water or alcohol condense around the ions generated by the interaction between air and radiation. As a result, trails of mist, called *fog tracks* are produced. The fog tracks indicate the direction in which the emitted radiation travels. These resemble the condensation trails of high-flying jets, and are shown in Fig. 4.9.

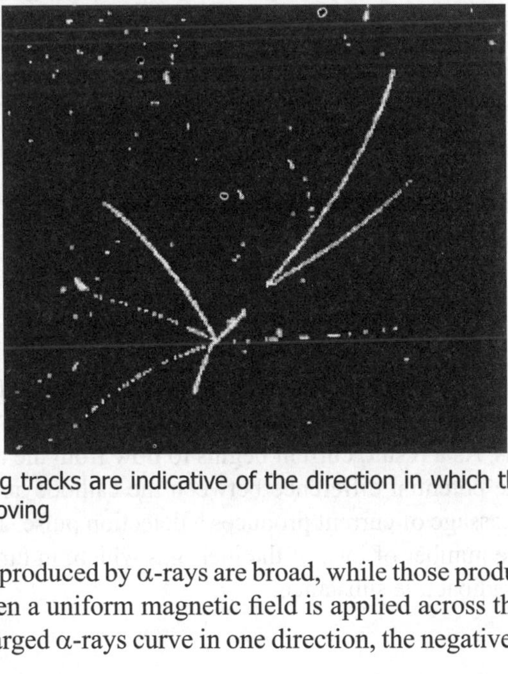

Fig. 4.9 Fog tracks are indicative of the direction in which the radiation is moving

The fog tracks produced by α-rays are broad, while those produced by β-rays are thin. Further, when a uniform magnetic field is applied across the cloud chamber, the positively charged α-rays curve in one direction, the negatively charged β-rays in opposite direction. The pathway of α-rays under the influence of a magnetic field is shown in Fig. 4.10.

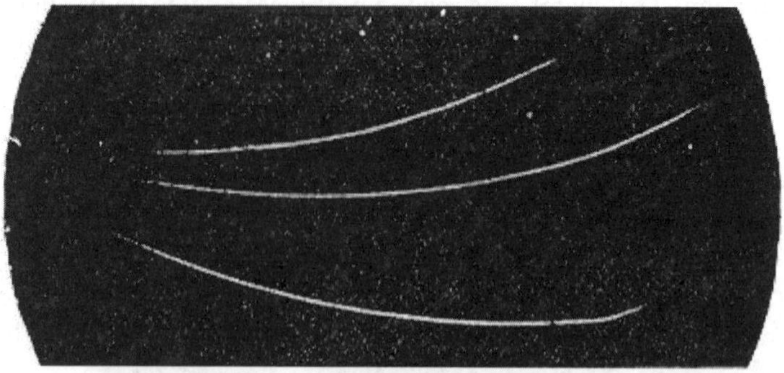

Fig. 4.10 Curved pathway of α-rays in a cloud chamber under the influence of a magnetic field

2. Geiger-Müller counter

A Geiger-Müller counter not only detects radiation, but also measures the amount of radioactively in a sample. It consists of a tube with a wire running along its axis. The tube acts as a cathode, the wire as an anode. The tube is filled with an inert gas like helium, neon or argon at low pressure, to which a high voltage is applied. A schematic diagram of a Geiger-Müller counter is displayed in Fig. 4.11.

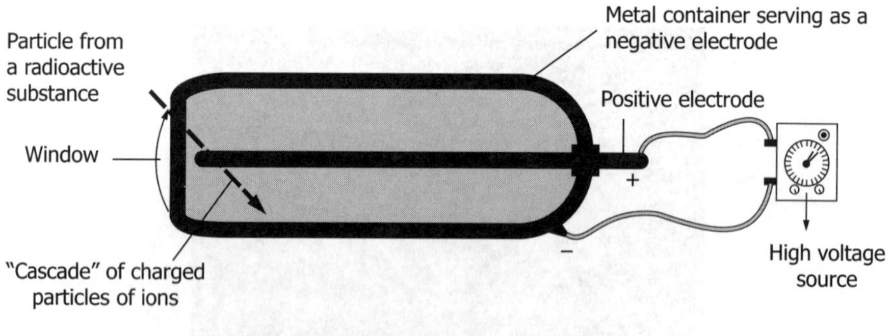

Fig. 4.11 A Geiger-Müller counter

A window in the counter allows the radiation (α-, β- or γ-rays) from a radioactive source to percolate into the tube. A particle or a photon of the radiation ionizes the inert gas atoms. As a result, current begins to flow from the anodic wire to the cathodic tube. The potential difference between the cathode and the anode is so adjusted that the passage of current produces a detection pulse. The pulse count is proportional to the number of ions of the inert gas which, in turn, is proportional to the amount of radioactive substance.

Box 4.2

Representation of a Nuclear Reaction

Just as the number of atoms on both sides of a molecular chemical equation and the number of atoms and charges on both sides of an ionic chemical equation must be balanced, the mass numbers and atomic numbers on both sides of an equation representing a nuclear reaction must also be balanced. Consider, for example, the following nuclear reaction:

$$^{19}_{9}F + ^{1}_{0}n \rightarrow ^{16}_{7}N + ^{4}_{2}He$$

The total mass number on the left side of the equation is 20 and so is on the right side. Likewise, the atomic number is 9 on both sides of the equation. The same reaction may be abbreviated as:

$$^{19}_{9}F\ (n, \alpha)\ ^{16}_{7}N$$

This implies that fluorine-19 is the target nucleus, neutrons are the projectiles, nitrogen-16 is the product nucleus and α-particles are released as emanations. Another way to highlight the essence of the process is to say that nitrogen-16 is prepared from fluorine-19 by (n, α) reaction.

Induced nuclear reactions

A nuclear reaction can be accomplished only if the projectile nucleus has sufficient energy, not only to strike the nucleus of the target atom, but also to cause exchange of nucleons. This energy threshold is achieved by a device called *nuclear accelerator* or *particle accelerator*. As the name implies, this equipment increases the momentum of the projectile to a level that when it strikes the target, a set of new products is formed.

Nuclear accelerators are of several types, the oldest and the most common one being the *cyclotron*. It consists of two flat and hollow, semi-circular vessels having the shape of alphabet 'D'. In fact, these two compartments are called 'dees'. The two dees are separated by a passage to which a voltage of several thousand volts is applied. The entire assembly is encased in a magnetic field (Fig. 4.12).

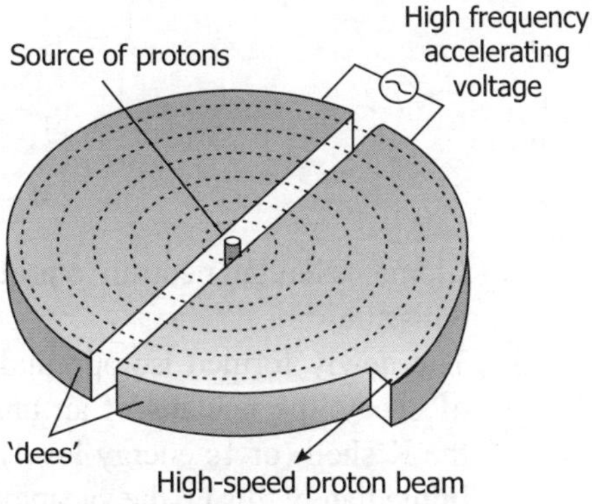

Fig. 4.12 Representation of a cyclotron

The projectile—let us assume, a proton—is injected into the central passage. The magnetic field induces it to follow a circular path and enter into one of the dees. As it completes its course in that dee, the polarity of the applied voltage reverses and the proton is inducted into the second dee. This time its radius of curvature increases, and so does its speed. The cycle is repeated many times, and each time the projectile shuttles between the two dees, its speed progressively rises and it follows a spiral path away from the center. Finally, it oozes out of the rim of the chamber and strikes the target. The resulting products are directed to instruments to confirm their identity.

Artificial Radioactivity

When atomic nuclei are bombarded with different types of projectiles, novel isotopes are produced. This process is called *artificial transmutation of elements.* Most of the isotopes synthesized in this way are unstable and radioactive. They tend to disintegrate as soon as they are formed. The radioactivity so produced is called artificial radioactivity. It is also referred to as *induced radioactivity.* The decay resulting from artificial radioactivity is usually a one-step process, i.e., the radioisotope disintegrates into a stable isotope which does not show any tendency to decay further. Artificial radioactivity manifests itself generally in one of the following forms:

1. The newly formed isotope undergoes positron emission.

$$^{24}_{12}\mathrm{Mg} + {}^{4}_{2}\mathrm{He} \rightarrow {}^{27}_{14}\mathrm{Si} + {}^{1}_{0}\mathrm{n}$$

$$^{27}_{14}\mathrm{Si} \rightarrow {}^{27}_{13}\mathrm{Al} + {}^{0}_{+1}\mathrm{e}$$

$$^{10}_{5}\mathrm{B} + {}^{4}_{2}\mathrm{He} \rightarrow {}^{13}_{7}\mathrm{N} + {}^{1}_{0}\mathrm{n}$$

$$^{13}_{7}\mathrm{N} \rightarrow {}^{13}_{6}\mathrm{N} + {}^{0}_{+1}\mathrm{e}$$

Elements having neutron to proton ratio less than unity undergo positron emission.

2. The newly formed isotope undergoes electron emission (β-decay).

$$^{9}_{4}\mathrm{Be} + {}^{2}_{1}\mathrm{H} \rightarrow {}^{10}_{4}\mathrm{Be} + {}^{1}_{1}\mathrm{H}$$

$$^{10}_{4}\mathrm{Be} \rightarrow {}^{10}_{5}\mathrm{B} + {}^{0}_{-1}\mathrm{e}$$

$$^{31}_{15}\mathrm{B} + {}^{1}_{0}\mathrm{n} \rightarrow {}^{28}_{13}\mathrm{Al} + {}^{4}_{2}\mathrm{He}$$

$$^{28}_{13}\mathrm{Al} \rightarrow {}^{28}_{14}\mathrm{Si} + {}^{0}_{-1}\mathrm{e}$$

Elements having neutron to proton ratio more than unity undergo electron emission.

3. The newly formed isotope undergoes K-electron capture. In this mode of decay, the nucleus of an unstable isotope captures an electron from the K shell (or 1s energy level). An electron from a higher energy level immediately fills up the vacancy. Since this electron falls from a higher to a lower shell, energy is emitted in the form of X-rays.

$$^{48}_{22}\mathrm{Ti} + {}^{2}_{1}\mathrm{H} \rightarrow {}^{49}_{23}\mathrm{V} + {}^{1}_{0}\mathrm{n}$$

$$^{49}_{23}\mathrm{V} + {}^{0}_{-1}\mathrm{e} \rightarrow {}^{49}_{22}\mathrm{Ti} + \text{X-rays}$$

When the electron is absorbed by the nucleus, a proton is converted into a neutron. This increases the neutron to proton ratio reducing the electrostatic repulsion among the protons and thereby increasing the stability.

4. The newly formed isotope may possess a nuclear isomer. The two isomers then decay with different half-life periods. *Nuclear isomers* are those having same atomic number and mass number but different arrangements of neutrons and protons in the shell structure of the nucleus. The nucleons are in the ground state in one isomer and in the excited state in the other. As a result, they possess dissimilar half-lives.

For example, when bromine-79 absorbs a neutron, it is transformed into bromine-80. The latter disintegrates by emission of an electron.

$$^{79}_{35}\text{Br} + {}^{1}_{0}\text{n} \rightarrow {}^{80}_{35}\text{Br}$$

$$^{80}_{35}\text{Br} \rightarrow {}^{80}_{36}\text{Kr} + {}^{0}_{-1}\text{e}$$

However, measurement of rates of disintegration indicates two half-life periods of 18 min and 4.2 h. This shows that bromine-80 exists as two nuclear isomers. The one with shorter half-life has the nucleons arranged in ground state of the nuclear shell, while the other with longer half-life has nucleons in the excited state (Fig. 4.8).

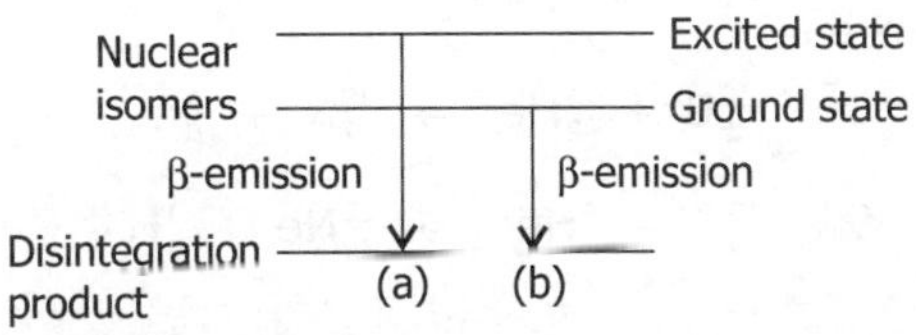

Fig. 4.8 Decay from (a) excited state takes a longer time than that from (b) ground state

Nuclear reactions are of various types. The two most important ones are fission and fusion reactions (pp. 206–208). The other nuclear reactions fall in the following categories:

a. Projectile-capture reactions

These types of reactions may involve neutrons or protons as projectiles.

(i) Neutrons as the projectiles. The neutron capture is followed by emission of γ-rays. For example,

$$^{85}_{37}\text{Rb} + {}^{1}_{0}\text{n} \rightarrow {}^{86}_{37}\text{Rb} + \gamma$$

$$^{40}_{18}\text{Ar} + {}^{1}_{0}\text{n} \rightarrow {}^{41}_{18}\text{Ar} + \gamma$$

(ii) Protons as the projectiles. Proton capture is possible only if the target nucleus has low mass number. For example,

$$^{12}_{6}\text{C} + {}^{1}_{1}\text{H} \rightarrow {}^{13}_{7}\text{N} + \gamma$$

b. Projectile-capture particle-emission reactions

In these reactions, a projectile penetrates the nucleus which, in turn, ejects a particle. The majority of nuclear reactions fall in this category. Such nuclear reactions may be induced by the following:

(i) Neutrons, e.g.,

$$^{14}_{7}N + ^{1}_{0}n \rightarrow ^{12}_{6}C + ^{3}_{1}H$$

$$^{27}_{13}Al + ^{1}_{0}n \rightarrow ^{24}_{11}Na + ^{4}_{2}He$$

(ii) Protons, e.g.,

$$^{11}_{5}B + ^{1}_{1}H \rightarrow ^{11}_{6}C + ^{1}_{0}n$$

$$^{14}_{7}N + ^{1}_{1}H \rightarrow ^{11}_{6}C + ^{4}_{2}He$$

(iii) Deuterons, e.g.,

$$^{31}_{15}P + ^{2}_{1}H \rightarrow ^{32}_{15}P + ^{1}_{1}H$$

$$^{12}_{6}C + ^{2}_{1}H \rightarrow ^{10}_{5}B + ^{4}_{2}He$$

(iv) α-particles, e.g.,

$$^{10}_{5}B + ^{4}_{2}He \rightarrow ^{13}_{7}N + ^{1}_{0}n$$

$$^{19}_{9}F + ^{4}_{2}He \rightarrow ^{22}_{10}Ne + ^{1}_{1}H$$

c. Spallation reactions

When high-speed projectiles, possessing energies of the order of 400 MeV, strike a heavy target nucleus, a relatively lighter nucleus gets cleaved from it. This type of reactions are called spallation reactions. The lighter nucleus generally has an atomic number of 10 to 20 units lower than the target nucleus. This differentiates spallation reactions from other induced nuclear reactions where generally the atomic number of the product differs from that of the target by 2 to 3 units. As we shall see in the next section, the atomic numbers of reactant and product nuclei vary greatly in fission reactions too. However, nuclear fission is a self-sustaining chain reaction. Spallation reactions are not self-sustaining. The fragmentation of copper-63 nucleus by an α-particle is an example of spallation reaction.

$$^{63}_{29}Cu + ^{4}_{2}He \rightarrow ^{37}_{17}Cl + 14^{1}_{1}H + 16^{1}_{0}n$$

Nuclear fission

Nuclear fission involves fragmentation of a heavy nucleus into two medium-sized nuclei. The process is accompanied by the release of a large amount of energy. Neutrons are used as projectiles for fission. The nuclear reaction also releases 2 or

3 neutrons, which strike a fresh target of the heavy nucleus. In this manner a chain reaction ensues. A typical nuclear fission reaction may be represented as follows:

$$^{235}_{92}U + ^{1}_{0}n \rightarrow ^{141}_{56}Ba + ^{92}_{36}Kr + 2 \text{ or } 3\ ^{1}_{0}n + 200 \text{ MeV}$$

The mechanism of fission reaction may be explained on the basis of the liquid drop model of atomic nucleus. According to this model, the nucleus acquires a spherical shape, just like a drop of liquid, because the attractive forces due to binding energy are balanced by repulsive forces between the protons. For very heavy nuclei, with atomic numbers greater than 90, the binding energy tends to decrease and the forces of repulsion tend to dominate. In such cases, the nucleus becomes deformed and elongated—just like a very large drop of a liquid. Immediately before the fission, the elongated nucleus attains a dumbbell shape (Fig. 4.5), with each lobe being positively charged. The electrostatic repulsion between the two lobes causes cleavage of the nucleus. Immediately after fission, the two fragments recede from each other and collapse to their equilibrium, spherical shapes.

In the fission reaction cited above, the neutron to proton ratio of the target uranium atom is higher than those of the products, barium and krypton atoms. Nevertheless, this ratio in the product nuclei is still higher than those of stable elements. Therefore, barium and krypton atoms continue to decay further by β-emissions to attain stability.

Whereas uranium-235 can be fissioned by slow-moving neutrons, uranium-238 can be fissioned only by fast neutrons. The ease of fission is profoundly affected by the even–odd criteria of nuclear stability. In uranium-235 nucleus, there are an odd number of neutrons and an even number of protons. The odd–even combination is relatively less stable. In uranium-238, the number of neutrons and protons are both even. The stability in this case is higher. For this reason, uranium-235 can be fissioned by slow neutrons, but uranium-238 requires fast neutrons.

An isotope like uranium-235, which can be easily fissioned, is called *fissile isotope.* The one which cannot be fissioned, or which can be fissioned with difficulty is called *fertile isotope.* Uranium-238 is a fertile isotope. Naturally occuring uranium contains only 0.7% of uranium-235. Before carrying out its fission, the concentration of uranium-235 in natural uranium is increased to about 3% by a process called *enrichment.* A mixture of uranium-235 and uranium-238 isotopes is converted to their respective hexafluorides, UF_6. Both the hexafluorides are volatile in nature. These are then separated and concentrated by gaseous diffusion method, described in a subsequent section. The uranium-235 isotope gets enriched at the cost of uranium-238 analog.

The fission of uranium releases a large amount of energy which may be used for producing electricity. This topic is discussed in detail in the section on applications of radioactivity (p. 215).

Nuclear fusion

Nuclear fusion involves unification of two lighter nuclei to form a relatively heavy nucleus. The fusing entities have to overcome the electrostatic repulsions between the protons of their nuclei. These potential barriers are immensely large and can be crossed-over only at extremely high temperatures (of the order of 10^7 K). At such high temperatures, the kinetic energies of the reactant nuclei are so large that these override their mutual repulsion barriers and fuse with each other. However, once these barriers are crossed, fusion of nuclei takes place, and a large amount of energy is liberated. A typical nuclear fusion reaction may be represented as follows:

$^{2}_{1}$üü§üü$^{3}_{1}$ü§üü§ü $^{4}_{2}$ + $^{1}_{0}$ +

In this reaction, the nuclei of deuterium and tritium fuse to yield more stable helium.

Another example is the fusion of tritium and hydrogen nuclei to yield helium.

$$^{3}_{1}H + ^{1}_{1}H \rightarrow ^{4}_{2}He + \gamma + 20 \text{ MeV}$$

Fusion is possible only among very light nuclei. Heavier elements have larger number of protons and the repulsion barriers are concomitantly larger.

The stars, including the sun, derive their energy from nuclear fusion. In a simplified form, the reaction may be represented as follows:

ü§üüüü§ü^{1}ü§ü + $^{0}_{+1}$ +

Positron

This reaction requires a temperature of about 2×10^7 K. Such high temperatures can be easily reached in the cores of heavenly bodies. However, if a nuclear fusion reaction is to be conducted artificially, then a fission reaction must precede it. The high temperatures generated by fission shall carry the fusion through.

Case Study 4.9 Which nuclear reaction is more energy efficient: fission or fusion?

For the fission reaction,

$$^{235}_{92}U + ^{1}_{0}n \rightarrow ^{141}_{56}Ba + ^{92}_{36}Kr + 2 \text{ or } 3\ ^{1}_{0}n + 200 \text{ MeV}$$

the amount of energy released corresponds to 236 mass units.

236 mass units release 200 MeV of energy.

One mass unit releases $\frac{200}{236} = 0.847$ MeV of energy.

For the fusion reaction,

$$^{2}_{1}H + ^{3}_{1}H \rightarrow ^{4}_{2}He + ^{1}_{0}n + 17.5 \text{ MeV}$$

The amount of energy released corresponds to 5 mass units.

5 mass units release 17.5 MeV of energy.

One mass unit releases $\frac{17.5}{5} = 3.5$ MeV of energy.

$$\frac{\text{Energy released per unit mass in fusion reaction}}{\text{Energy released per unit mass in fission reaction}} = \frac{3.5}{0.847} = 4.1$$

Thus, on an average, fusion reactions release four times more energy than fission reactions. This is true not only for the two reactions cited in this case study, but also for other examples of fission and fusion reactions.

Identification of Isotopes

A small variation occurs in the mass numbers of isotopes of the same element. Yet, if the isotopes are to be identified, the small mass differences have to be detected. This is done with the aid of an instrument called *mass spectrograph*. It is based on the principle that when charged ions pass first through an electric field and then through a magnetic field, they undergo deflection. The extent of deflection is a function of charge to mass ratio (e/m) of the isotope. A mass spectrograph is shown in Fig. 4.13.

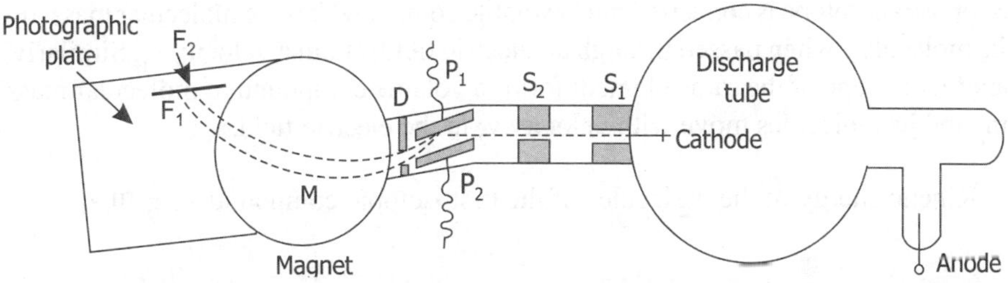

Fig. 4.13 A mass spectrograph

The isotopes in cationic forms emanate from a narrow opening in the cathode of a discharge tube. These are sorted into a thin beam by two parallel slits, S_1 and S_2. On emerging from these slits, the beam enters an electric field maintained by parallel charged plates, P_1 and P_2. The electric field accelerates the cations to an almost equal extent. Thus, all the particles constituting the beam have approximately the same kinetic energy. This implies that the lighter particles have more speed than the heavier ones. On coming out of the plates P_1 and P_2, the beam breaks into two rays—the component made of heavier and slow-moving isotopes is deflected downwards, while the component made of lighter and fast-moving isotopes is deflected upwards.

The rays pass through the diaphragm D and emerge into a magnetic field, M, maintained at right angle to the electric field. Here too the heavier particles are deflected downwards and lighter particles upwards. The rays then impinge on a photographic plate. The particles having a particular value of e/m strike at one point (F_1) whereas those with another value of e/m focus at a different point (F_2). The impingements on the photographic plate are seen as straight lines, whose lengths are proportional to the concentration of the isotope in the original mixture.

Separation of Isotopes

The degree to which two isotopes may be separated from each other in a one-step process is called *separation factor*. It is the ratio of the relative concentration of one isotope after its separation to its relative concentration before separation. Mathematically, the separation factor s is given by the equation:

$$s = \frac{n_2}{n_1}\,\frac{N_1}{N_2}$$

where n_1 and n_2 are the number of moles of isotope 1 and 2, respectively before separation, while N_1 and N_2 are the corresponding values after separation. Often, the separation process has to be repeated several times to achieve a meaningful separation. For example, if the separation is achieved by repeating the cycle x times, the overall separation factor is expressed as s^x.

We now describe two important methods for segregating isotopes.

1. Gaseous diffusion method

Suppose an isotope is converted into a volatile compound having molecular mass m_1. Its molecules, when passed through an electric field, attain a velocity v_1. Similarly, another isotope of the same element forms a volatile compound of molecular mass m_2 and its molecules move with velocity v_2 in the electric field.

Kinetic energy of the molecules of the first isotopic compound $= \frac{1}{2} m_1 v_1^2$

Kinetic energy of the molecules of the second isotopic compound $= \frac{1}{2} m_2 v_2^2$

Now, if a mixture of the two compounds is passed through an electric field, the kinetic energy of the molecules of both the compounds are raised to an equal value. Thus,

$$\frac{1}{2} m_1 v_1^2 = \frac{1}{2} m_2 v_2^2$$

or

$$\frac{v_1}{v_2} = \sqrt{\frac{m_2}{m_1}}$$

Since molecular weight is directly proportional to vapor density,

$$\frac{v_1}{v_2} = \sqrt{\frac{d_2}{d_1}}$$

where d_1 and d_2 are the vapor densities of the first and second constituent of the mixture, respectively. This implies that the rates at which different gaseous components diffuse through a porous barrier (a screen permeable to isotopes), at a particular temperature, are inversely proportional to the square roots of their vapor densities. That is, the component having lower molecular mass (or vapor density) will diffuse through a porous barrier more rapidly than the one with a higher molecular mass (or vapor density).

The gas which crosses the barrier is called *diffusate*. It should be removed immediately, otherwise an equilibrium will soon be established and the concentration of each component would be the same on both sides of the barrier.

In practical terms, a large number of porous barriers are arranged in series. Only half of the diffusate that crosses barrier number 1 is allowed to proceed to barrier number 2. The other half returns to mix with the entering gas. This procedure is repeated at each barrier. At every stage, the diffusate becomes enriched with the lighter isotope. The separation factor progressively increases, and so does the efficiency of separation.

2. Centrifugal method

Under the influence of gravity, the atoms or molecules tend to move downwards in proportion to their masses. If a mixture of two gaseous isotopic compounds of the same element is subjected to a gravitational field, the heavier analog will concentrate in the direction of the field by moving down at a more rapid pace than its lighter counterpart. This leads to separation of the constituents of the mixture.

In the gaseous diffusion method, the degree of separation depends on the square root of the molecular masses of the isotopic compounds, whereas in the centrifugal method, it is dictated by the mass difference of the isotopes. Greater the difference, more efficient is the separation. The uranium-235 and uranium-238 isotopes, having a mass number difference of 3 units can be separated more efficiently than hydrogen and deuterium, with a mass number difference of only one unit. Further, the difference between the mass numbers of the two isotopes is constant and hence the degree of separation is not dependent on the nature of the gaseous compound whose vapors are subjected to gravitational field.

In the gravitational method, a mixture of gaseous compounds of the isotopes of the same element is injected in a tall cylindrical centrifuge. Pseudogravitational force about one lakh times the natural gravity is applied. As the cylinder is rotated, the compound of the heavier isotope moves downwards along the periphery, while that of the lighter isotope rises above along the axis.

Applications of Radioactivity

Radioactivity offers a wide range of applications to the modern civilization. It owes its significance to a broad array of isotopes which, in turn, are benevolent because of the following reasons:

- Their rates of decay are known. This parameter assists in determining the age of artifacts.
- They can be converted into other isotopes by artificial transmutation. Power generation by fission reaction is the most important application of radioactivity which falls under this category.

- They emit radiations continuously. By far, the largest number of applications of radioactivity, including those in the field of agriculture and medicine, come under this purview. Two isotopes of the same element—one radioactive and the other non-radioactive—show similar chemical properties. Therefore, if a radioactive isotope is added to its non-radioactive counterpart, the mixture shows uniform chemical properties. However, the radioactive isotope is more easy to detect because it emits radiations on a continuous basis. The radiations can be detected and amplified by a *Geiger-Müller counter* (*see* p. 202). The number of disintegrations and their rates can then be elucidated. Thus, a radioisotope may be introduced at one point of a physical, chemical or biological system and its activity may be measured in different parts of the system using a Geiger-Müller counter. Such a radioisotope is called a *tracer* and this method is called *tracer technique.*

We now describe some important applications of radioactivity which permeate virtually every sphere of human life.

1. Carbon dating

The age of artifacts of organic origin may be estimated by a technique called *radiocarbon dating* or simply carbon dating. The technique relies on the fact that carbon-14 isotope is produced continuously in the upper atmosphere by the bombardment of nitrogen atoms with cosmic ray neutrons.

$$^{14}_{7}N + ^{1}_{0}n \rightarrow ^{14}_{6}C + ^{1}_{1}H \tag{4.19}$$

The carbon-14 isotope is then oxidized to radioactive carbon dioxide by atmospheric oxygen.

$$^{14}C + O_2 \rightarrow ^{14}CO_2$$

This carbon dioxide, along with non-radioactive carbon dioxide, is consumed by plants in photosynthesis. It then enters the animal tissue via the food chain. As a result, a fraction of carbon content in living organisms is in the form of carbon-14 isotope. In the organism's body, it disintegrates by β-emission.

$$^{14}_{6}C \rightarrow ^{14}_{7}N + ^{0}_{-1}e \tag{4.20}$$

Thus, carbon-14 is continuously added to the bodies of living organisms (reaction 4.19) and is continuously lost (reaction 4.20). There is an equilibrium in the concentration of radiocarbon isotope in living beings throughout their life span. Once the organism dies, further assimilation of carbon-14 in the body stops, but its decay continues with a half-life of approximately 5700 years. The rate of decay, however, decreases with time. The activity per gram of carbon in an archeological sample is compared with the activity of carbon in a fresh sample. This comparison reveals the length of time that has elapsed since the death of the organism, which, in turn, gives the estimated age of the artifact.

Case Study 4.10 A piece of wood taken from a cave temple is found to have carbon-14 activity only 0.636 times that of a fresh wood sample. Estimate the age of the wood. The half-life of carbon-14 is 5700 years.

The value of decay constant (λ) may be calculated from the expression,

$$t_{1/2} = \frac{0.693}{\lambda}$$

where $t_{1/2}$, the half-life of carbon-14, is 5700 y.

$$\lambda = \frac{0.693}{5700} = 1.21 \times 10^{-4}\ y^{-1} \quad (4.21)$$

The carbon-14 activity of the ancient sample, N is 0.636 times that of the fresh sample, N_0, i.e.,

$$N = 0.636\ N_0 \quad (4.22)$$

According to the law of radioactive disintegration,

$$N = N_0\ e^{-\lambda t}$$

(t is the time over which disintegration occurs)

$$\text{or} \quad \frac{N_0}{N} = e^{\lambda t}$$

$$2.303 \log \frac{N_0}{N} = \lambda t$$

$$t = \frac{2.303}{\lambda} \log \frac{N_0}{N} \quad (4.23)$$

Substituting the values of λ and N from equations (4.21) and (4.22), respectively in equation (4.23),

$$t = \frac{2.303}{1.21 \times 10^{-4}} \log \left(\frac{N_0}{0.636\ N_0} \right)$$

$$t = 3740\ y$$

The wood sample found in the cave temple is 3740 years old.

2. Age of earth and minerals

The age of the most ancient minerals or rocks provides some idea about the age of the earth. The age of minerals, in turn, is reflected in the amount of radioactive elements such as uranium or thorium present in them, relative to the non-radioactive lead. Naturally occurring uranium and thorium decay into stable isotopes of lead through a series of α- and β-emissions. Knowing the ratio of uranium to lead or thorium to lead in a particular mineral, and knowing the half-lives of the radioisotopes, the age of the natural substance may be estimated.

Thorium is present in minerals in larger quantities than uranium. Thorium-232 has the longest half-life among all the isotopes of this element—about 1.3×10^{10} years. Consequently, the age of earth is less than this figure.

The longest-lived isotope of neptunium has a half-life of 2×10^6 years. This isotope is not found in nature, implying that the age of earth is more than 4×10^6 years. The longest-lived isotope of uranium, the one with mass number 235, on the other hand, has a half-life of 4.5×10^9 years. This element is present in minerals in moderate amounts. From these values, the age of earth is estimated to be close to $2–3 \times 10^9$ years.

Case Study 4.11 A sample of uranium ore contains 4.64 mg of uranium-238 and 1.22 mg of lead-206. Estimate the age of the ore. The half-life of uranium-238 isotope is 4.51×10^9 years.

In this example, an assumption is to be made that the entire lead-206 content in the ore has resulted from the disintegration of uranium-238.

The decay constant, λ of uranium-238 may be calculated from the expression,

$$t_{1/2} = \frac{0.693}{\lambda}$$

where $t_{1/2}$, the half-life of uranium-238, is 4.51×10^9 y.

$$\lambda = \frac{0.693}{4.51 \times 10^9} = 1.54 \times 10^{-10}\ y^{-1}$$

Amount of lead-206 in the sample = 1.22 mg

The amount of uranium-238 that has disintegrated into lead-206 may be calculated as follows:

206 mg of lead-206 is obtained from disintegration of 238 mg of uranium-238.

1.22 mg of lead-206 is obtained from disintegration of $\frac{238}{206} \times 1.22 = 1.41$ mg of uranium-238.

Amount of uranium-238 that has disintegrated = 1.41 mg

Amount of uranium-238 that is present in the ore = 4.64 mg

Total amount of uranium originally present in the ore = (4.64 + 1.41) = 6.05 mg

i.e., $N_0 = 6.05$ mg, $N = 4.64$ mg.

From the radioactive disintegration law,

$$N = N_0\, e^{-\lambda t}$$

(t is the time over which disintegration has occurred)

$$\text{or} \quad t = \frac{2.303}{\lambda} \log \frac{N_0}{N}$$

$$t = \frac{2.303}{1.54 \times 10^{-10}} \log \left(\frac{6.05}{4.64}\right)$$

$$= 1.7 \times 10^9\ y$$

The age of the ore is 1.7×10^9 y.

---

3. Power generation

The fission of uranium-235 nucleus by slow neutrons,

$$^{235}_{92}U + ^{1}_{0}n \rightarrow ^{141}_{56}Ba + ^{92}_{36}Kr + 2 \text{ or } 3\ ^{1}_{0}n + 200 \text{ MeV}$$

(a reaction we have already cited in the section on nuclear fission), releases 200 MeV energy, which may be converted into electrical energy. Power produced by nuclear energy is unmatched to any other energy resource. Theoretically, one ton of uranium can give as much electrical energy as 3 million tones of coal or 12 million barrels of oil. Further, not only the reserves of coal and oil are fast dwindling, but also burning of these fossil fuels pollutes the environment. Nuclear energy is a clean and abundant resource for power generation.

The fission reaction is carried out in a special type of device called *nuclear reactor* (Fig. 4.14). In the reactor, uranium-235 is placed in the form of rods, each measuring about 3.5 m. These are called *fuel rods.*

The fuel rods are, in fact, pellets of uranium dioxide, $^{235}UO_2$ (or uranium(IV) oxide) stacked one above the other. We have already stated in the section of nuclear fission, that enriched uranium is obtained by gaseous diffusion method as uranium hexafluoride, $^{235}UF_6$ (or uranium(VI) fluoride). In order to convet it into uranium dioxide the following procedure is followed. Enriched uranium hexafluoride is treated with ammonia and water.

$$2UF_6 + 14NH_3 + 7H_2O \rightarrow (NH_4)_2U_2O_7 + 12NH_4F$$

The resulting product, ammonium diuranate is heated with hydrogen, whereby it is reduced to uranium dioxide.

$$(NH_4)_2U_2O_7 + 2H_2 \rightarrow 2UO_2 + 3H_2O + 2NH_3$$

Uranium dioxide is obtained in powder form and is subsequently fabricated into pellets.

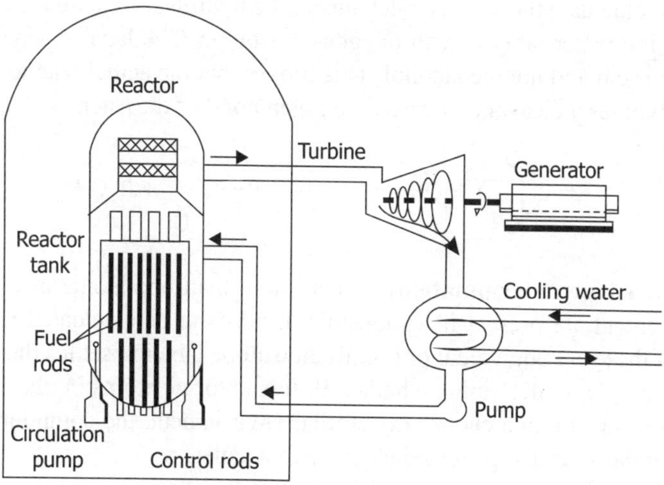

Fig. 4.14 A nuclear reactor

The fission of uranium-235 is a self-sustaining reaction. The neutrons produced as end products can act as projectiles and keep the reaction going. However, it is very important to keep the number of neutrons under control. To achieve this, each fuel rod is flanked by claddings of cadmium on both sides. Cadmium absorbs the extra neutrons and keeps the reaction within the limits of safety.

$$^{113}_{48}Cd + ^{1}_{0}n \rightarrow ^{114}_{48}Cd + \gamma$$

The cadmium rods are called *control rods.*

Not only the number of neutrons need to be controlled, but their speed too must be regulated. Since slow neutrons are required for the fission reaction, the projectiles are first passed through a *moderator*, which reduces their velocity. The most commonly used moderators are graphite and heavy water. The reactor also uses water or liquid sodium as a *coolant.* A coolant absorbs the heat generated during the fission reaction. The heat energy is then used to perform useful work of turning a turbine and producing electricity.

4. Chemical investigations

As the following examples show, radioactivity assists in chemical investigations.

a. **Reaction mechanisms** There are two possible mechanisms for the hydrolysis of esters.

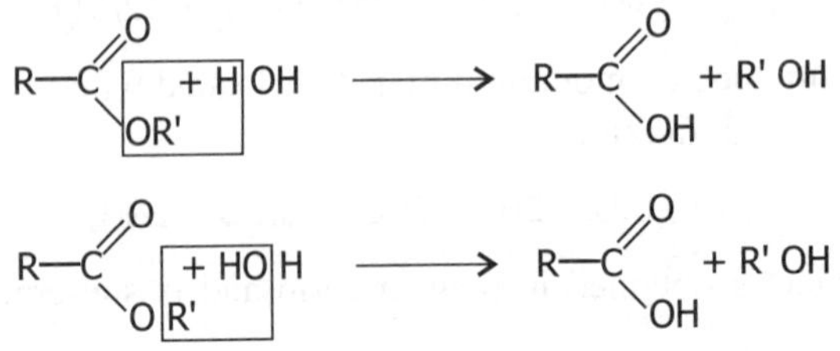

To elucidate the correct mechanism, the hydrolysis reaction is carried out using water labeled with oxygen-18 isotope. The labeled oxygen enters the acid and not the alcohol. This proves that the actual reaction pathway involves a cleavage of carbon–oxygen bond of the ester.

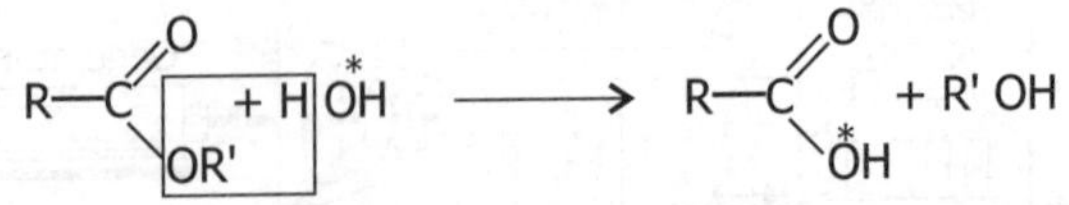

b. **Structure determination** When more than one atoms of a particular element are present in a molecule, it needs to be elucidated whether all of these are equivalent or not. Radioisotope labeling settles the issue. For example, to determine whether all five chlorine atoms in phosphorus(V) chloride are in a chemically similar environment, the compound may be synthesized using labeled chlorine molecules.

$$PCl_3 + {}^{*}Cl_2 \rightarrow P^{*}Cl_5$$

If all five chlorine atoms are in a similar environment, the two labeled chlorine atoms would rapidly exchange with three non-labeled ones. Finally, all five chlorine atoms would exhibit radioactive character. However, when phosphorus(V) chloride is hydrolyzed,

$$P^{*}Cl_5 + H_2O \rightarrow POCl_3 + 2HCl^{*}$$

the radioactivity remains with hydrogen chloride and not with phosphorus oxotrichloride. This shows that two chlorine atoms are in a different environment as compared to the other three. This is also confirmed by other methods of structure determination, which reveal that two chlorine atoms are axial and three are equatorial in the trigonal bipyramid skeleton of phosphorus(V) chloride molecule.

Case study 4.12 What evidence can be offered to prove that the two sulfur atoms in thiosulfate ion are not in an identical environment?

The proof that the two sulfur atoms in the thiosulfate ion are not identical can be deduced by heating labeled sulfur, ^{35}S (or *S) with an aqueous solution of sodium sulfite.

$$SO_3^{2-} + {}^{*}S \rightarrow {}^{*}SSO_3^{2-}$$

If the labeled thiosulfate ion, so formed, is treated with dilute hydrochloric acid, it decomposes into sulfur dioxide and sulfur.

$${}^{*}SSO_3^{2-} + 2H^{+} \rightarrow {}^{*}S + SO_2 + H_2O$$

All the radioactivity returns to the sulfur and none is passed on to sulfur dioxide. This shows that the two sulfur atoms in thiosulfate ion are not identical.

---

5. Agricultural applications

The following applications of radioactivity in the field of agriculture are noteworthy:

a. The distribution of minerals absorbed by roots and transferred to different parts of the plant can be traced by the technique of *radioautography*. In this technique, the conventional plant minerals are mixed with small amounts of a radioactive substance. For example, the sulfate ions containing radioactive sulfur-35 isotope are added to the soil along with normal sulfates. Small photographic plates are held in front of different parts of the plants. The part of the plant which receives the radioactive sulfate emits radiations and darkens the photographic plate. Thus, the movement of minerals and their re-distribution in various parts of the plant are ascertained. This method also figures out the period in the life cycle of the plant when it needs the minerals most.

b. Phosphate fertilizers are added to the soil for healthy growth of plants. Phosphates are also a natural content of the soil. Addition of phosphates, containing radioactive phosphorus-32 isotope, makes it possible to determine the extent to which a plant may select phosphates from native soil and from the chemical fertilizers.

c. Certain elements such as boron, zinc, manganese and molybdenum are required for the proper growth of plants in extremely minute quantities. These are called *micronutrients*. The amount of these substances in a specific plant sample is so small that conventional methods of chemical analyses fail to detect them. However, if radioactive analogs of these micronutrients are taken up by the plant, it is not only possible to determine the amount absorbed, but also to identify the tissue in which these become concentrated. This, in turn, elucidates their functions in plant growth.

d. Fungicides containing isotopic elements reveal the mechanism by which the fungal pest is killed. The technique also indicates whether any damage is being caused to the crop itself. For example, sulfur was considered to be an effective fungicide to protect lemon plants. However, experiments performed with sulfur-35 isotope reveal that the fungus is, no doubt, destroyed, but the lemon fruit also becomes contaminated with the fungicide.

e. Chemical insecticides, such as DDT, persist in the environment for a very long time. These are toxic to humans and animals. Over long durations, the insects tend to become immune to them. A better way to deal with insect pests is to rely on radiologic technique rather than on conventional insecticides. In this technique, the male insects are exposed to γ-rays, which damage their reproductive cells and make them sterile. The insect population soon declines. Radiological method has proved to be very effective in controlling screw-worm fly pest.

6. Clinical applications

The applications of radioactivity in the area of medicine fall in two broad categories: (a) for disease diagnosis and (b) for disease treatment.

a. **As diagnostic agents** The following are some of the important examples where radioisotopes are used as diagnostic agents:

(i) Radioactive sodium chloride, containing sodium-24 isotope, is used to study obstructive vascular diseases in the extremities of the body. Normally, when sodium chloride is injected into the blood stream, it gets absorbed and then moves in about half an hour from forearms to the feet. If the patient has impaired circulation, the movemnt of sodium chloride will take a longer time. Moreover, sodium-24 isotope tends to get concentrated at the site of obstruction, and thus assists in locating the problem.

(ii) Thallium-201 isotope tends to concentrate in healthy heart tissue, while technetium-99 concentrates in damaged heart tissue. The two are injected together to assess the extent of problem in heart patients.

(iii) Iodine-131 concentrates in thyroid gland, liver and brain. It can gauge the level of thyroid problem as well as the growth of liver and brain tumors.

b. As therapeutic agents Radioisotopes also assist in disease treatment, as the following examples highlight:

(i) Radioactive phosphorus in the form of phosphate is used to control polycythemia vera—a condition in which the bone marrow produces excess red blood cells. The radioactive phosphorus-32 gets preferentially absorbed in the bone marrow and emits β-rays, which destroy excessive red blood corpuscles.

(ii) Radioactive iodine gets absorbed in thyroid gland and retards hyperthyrodism—a disorder in which the gland becomes overactive. Radioactive iodine destroys the abnormal tissue of the gland and thus cures the disease.

(iii) Cobalt-60 is used to irradiate malignant tumors with γ-rays. Cobalt-60 is prepared by bombarding nickel-60 isotope with neutrons.

$$^{60}_{28}\text{Ni} + ^{1}_{0}\text{n} \rightarrow ^{60}_{27}\text{Co} + ^{1}_{1}\text{H}$$

It is embedded in the cancerous tissue, where it decays with the emission of both β- and γ-rays.

$$^{60}_{27}\text{Co} \rightarrow ^{60}_{28}\text{Ni} + ^{0}_{-1}\text{e} + \gamma$$

The γ-rays destroy cancer cells very rapidly.

Case Study 4.13 A cancer patient undergoing radiotherapy is given a dose of 3.42 μg cobalt-60. How much isotope will remain in his body after 30 years. The half-life of cobalt-60 is 5.27 years.

The decay constant, λ, of cobalt-60 may be calculated using the relation,

$$t_{1/2} = \frac{0.693}{\lambda}$$

$$t_{1/2} = 5.27 \text{ y}$$

$$\lambda = \frac{0.693}{5.27} = 0.131 \text{ y}^{-1}$$

According to the law of radioactive disintegration,

$$t = \frac{2.303}{\lambda} \log\left(\frac{N_0}{N}\right)$$

$$N_0 = 3.42 \ \mu\text{g}; \ t = 30 \text{ y}$$

$$30 = \frac{2.303}{0.131} \log\left(\frac{3.42}{N}\right)$$

$$\log\left(\frac{3.42}{N}\right) = 1.706$$

$$\frac{3.42}{N} = 50.87$$

$$N = 0.067\ \mu g$$

The amount of cobalt-60 that would remain in the patient's body after 30 years is 0.067 μg.

Box 4.3

Brachytherapy

Brachytherapy is a curative procedure to combat cancer. It is a substitute for chemotherapy. Titanium capsules, each having the size of a rice grain, containing iodine-125, palladium-103 or cesium-131 are embedded in the cancerous tissue. These isotopes have a short half-life of 60, 17 and 10 days, respectively. The radiations emanating from the capsules destroy the malignant cells.

Brachytherapy has several advantages. Firstly, since the capsules are seated inside the diseased tissue (*brachy* in Greek means closeness), a high dose of radiation is delivered to the cancer cells in a short time. Secondly, the normal, healthy cells are not affected by the background radiation. Thirdly, very little radiation passes out of the patient's body, as a result of which the family members are not exposed to harmful rays. Fourthly, the therapeutic isotopes have short half-lives and hence decay completely within a few months. Fifthly, the decay products are non-radioactive and for this reason there is no need to surgically remove the titanium capsules after the treatment session.

7. Biological applications

Radioisotopes play a profound role in the field of biology, as shown by the following examples:

a. **Cell dynamics** Studies with isotopic tracers have shown that the constituents of cells, such as proteins, carbohydrates and fats are not static entities. These constituents are in a state of constant flux, being consistently exchanged with extracellular components.
b. **Transport of ions across cell membranes** The mechanism of movement of ions through the cell membranes as well as their rate of transportation may be elucidated by employing isotopic tracers. The ions are tagged with radioactive elements and their movements are monitored.
c. **Metabolism** Carbon labeling in sugars, fatty acids and amino acids helps reveal the mechanism by which these biomolecules metabolize. Calcium and hemoglobin metabolisms can be similarly studied.
d. **Volume of blood** The volume of blood in an individual can be determined by a technique called *isotopic dilution.* In this method, 1 ml of the person's

blood is withdrawn and treated with a salt of either sodium-24 or iron-59 isotope. The activity of 0.1 ml of this blood sample is measured and the remaining 0.9 ml is reinjected into the person's body.

Suppose the initial activity in the 0.1 ml of blood = a_i

Then the activity in the 0.9 ml of blood $= \frac{a_i}{0.1} \times 0.9$

$= x$ (say)

This activity gets homogeneously distributed in the entire blood stream within a few hours. Another 1 ml of the blood is again withdrawn and activity is measured.

Suppose the final activity in 1 ml of blood $= a_f$

Suppose the total volume of blood $= y$ ml

Final activity in the entire blood $= y\, a_f$

$$y\, a_f = x$$

$$y = \frac{x}{a_f}$$

The total volume of blood in a person can thus be determined. An average person has 5–6 L of blood.

Case Study 4.14 One milliliter blood of a person was labeled with iron-59. A 0.1 ml sample of the treated blood showed an activity of 14,000 counts per minute. The remaining 0.9 ml of blood was reinjected and after adequate time for homogenization, 1 ml of blood was withdrawn again. It showed an activity of 240 counts in 10 min. Find the volume of blood in the person.

A 0.1 ml sample of labeled blood showed an activity of 14,000 counts per min.

Activity in 0.9 ml of labeled blood $= \frac{14000}{0.1} \times 0.9 = 126{,}000$ counts per min

Suppose the volume of blood in the person = y ml

Activity in 1ml of blood after the second withdrawal = 240 counts in 10 min

= 24 counts per min

Total activity in the entire blood = 24y

$$24y = 12{,}6000$$

$$y = 5250 \text{ ml} = 5.250 \text{ L}$$

The amount of blood in the person is 5.250 L.

---

8. Industrial applications

Some of the important applications of radioactivity in the industrial sector are discussed below:

a. **Level monitoring** The level of a liquid, such as water or oil, in a storage tank can be measured from the exterior by making a γ-ray emitting radioisotope float on the surface of the fluid. Cesium-137 and cobalt-60 isotopes are generally used for this purpose. A Geiger-Müller counter fitted outside the tank registers the γ-rays when it is at the same height as the level of the liquid inside the tank.

b. **Thickness monitoring** The thickness of a sheet of rubber, plastic or metal may be monitored by a β-ray emitter isotope, such as strontium-90. The sheet is made to pass between the β-ray source and the detector. A change in thickness produces change in radioactivity. If the thickness is more than desired, the β-ray signal gets feeble; if less than desired, it gets intense.

c. **Friction monitoring** Two surfaces in a machine, on rubbing against each other, tend to get worn out due to friction. Lubricating the surfaces retards friction. Nevertheless, friction still occurs, although it is delayed. The particles which chip-off fall in the lubricating oil. The extent of friction is proportional to the amount of these particles. However, the amount of particles is so meager that it is virtually impossible to determine it quantitatively. To gauge the rate of friction, the surfaces are irradiated and the level of radioactivity is determined in the lubricating oil. Higher the radioactive count of the oil, greater is the extent of friction.

d. **Gamma radiography** The technique of gamma radiography, also called *gammagraphy* is used to detect microcracks in machine parts. A photographic film is placed behind the part desired to be checked, while a γ-ray emitter, such as iridium-192, is placed in front of it. Wherever there is a crack, the photographic film gets darkened.

9. Analytical applications

The applications of radioactivity also extend to analytical methods, as the following examples demonstrate:

a. **Solubility of sparingly soluble salts** Some inorganic salts, such as lead iodide, possess extremely meager solubility in water. Conventional methods of analysis cannot determine the trace amount of salt that is solubilized. To elucidate the solubility, the concerned salt is prepared by incorporating a radioisotope. For example, potassium iodide, containing radioactive iodine, is added to a solution of lead acetate.

$$(CH_3COO)_2Pb + 2KI^* \rightarrow 2CH_3COOK + PbI_2^*$$

The precipitated lead iodide now contains the radioactive iodine. The precipitate is collected, washed repeatedly with water to remove unreacted potassium iodide, and then dried. Its activity is determined. Next, a weighed quantity of lead iodide is suspended in a fixed volume of water. The suspension is filtered and the activity of the filtrate is determined. From this activity, the amount of lead iodide that has dissolved in water may be calculated.

b. **Neutron activation analysis** The technique of neutron activation analysis is used to determine the quantity of a trace element in a bulk material, such as minerals, ores and soils. The concerned element is transformed into a radioisotope by bombarding it with neutrons. Suppose the amount of the element in the material is W g, and it gives an activity of A after a specific time. A standard material containing known amount of the same element is then irradiated with neutrons for the same time and the activity is determined. Suppose the standard contains W_0 g of the trace element and shows an activity of A_0. The weight of the element in the unknown sample is determined from the relation,

$$\frac{W}{W_0} = \frac{A}{A_0}$$

10. Airport security

All airports are endowed with baggage scanning facilities at the security check counters (Fig. 4.15). A bomb detector is an integral component of the scanning equipment. It uses californium-252 isotope as a source of neutrons.

$$^{252}_{98}Cf \rightarrow {}^{251}_{98}Cf + {}^{1}_{0}n$$

The released neutrons are absorbed by the nitrogen-14 content of explosive substances. In the course of this reaction, γ-rays are emitted.

$$^{14}_{7}N + {}^{1}_{0}n \rightarrow {}^{15}_{7}N + \gamma$$

By examining the level and patterns of γ-ray emissions, an investigator can ascertain whether the baggage contains an unscrupulous item.

Fig. 4.15 Baggage scanning facility at an airport

However, certain legal commodities like wood and leather too have appreciable amount of nitrogen-14. Nevertheless, in such cases, the possibility of manual checking is always open.

11. Preservation of food products

Gamma irradiation of cash crops and food items increases their shelf-life. For example, it retards the sprouting of potatoes and onions.

Food products tend to become stale, and ultimately rot, due to invasion by microorganisms. Heating food items kills the microorganisms, while refrigeration slows their growth. However, the best bargain is to expose the eatables to γ-rays. These damage the biomolecules within the cells of microorganisms, thereby killing them. The process is more cost-effective than either canning or freezing the food products.

12. Preservation of artifacts

Pieces of art, such as paintings, may be preserved by exposing these to radiation. The works of art tend to decay with time since microorganisms and insects damage the canvas, framework or wooden structures of the artifacts. Gamma radiation kills the microorganisms and insects, retarding the degradation of such adornments.

13. Detection of forged paintings

Over a millennium, artists have used lead-based pigments for creating paintings. Such pigments contain a small amount of radioactive lead-210 isotope, which decays with a relatively short half-life. In paintings which are a few centuries old, an infinitesimally small amount of radioactive lead would be present. However, in relatively new paintings, the amount of this radioisotope would be appreciably high. If a person tries to sell a painting which is a few decades old and claims that it is a few hundred years old, his trickery may be exposed by measuring the amount of radioactivity in the piece of art.

14. Smoke detectors

A smoke detector consists of an ionization chamber having two oppositely charged plates. A small amount (about 2×10^{-4} g) of americium-241 isotope is placed inside the chamber. This isotope has a half-life of 470 years and disintegrates with the emission of α-particles.

$$^{241}_{95}\mathrm{Am} \rightarrow \, ^{237}_{93}\mathrm{Np} + \, ^{4}_{2}\mathrm{He}$$

The α-particles, in turn, ionize the oxygen and nitrogen molecules that are present inside the chamber as components of air.

$$O_2 \rightarrow O_2^+ + e^-$$
$$N_2 \rightarrow N_2^+ + e^-$$

The resulting oxygen and nitrogen molecular cations are attracted towards the negative plate and the electrons towards the positive plate of the ionization chamber. A small amount of current is generated, which is detected by the electronic circuitry.

When smoke from an arson episode enters the ionization chamber, the soot particles become attached to the molecular oxygen and nitrogen cations and neutralize their charges. The circuitry senses the drop in current across the plates and sets off the warning alarm.

Case Study 4.15 A museum wishes to analyze a piece of ruby for chromium content. What should be the preferred method of analysis?

Ruby crystals contain 0.1–0.4% chromium. When present at this level, the element can be determined by any conventional method. However, conventional methods destroy a part of the sample. The museum would not desire to lose a precious artifact. Therefore, it should allow only neutron activation analysis on ruby. This is a non-destructive method, that is, the sample would remain unchanged at the end of analysis. Moreover, chromium has a large neutron capture cross-section. This means that chromium readily absorbs neutrons and is converted into a radioisotope whose activity can be easily measured.

$$^{50}_{24}\mathrm{Cr} + ^{1}_{0}\mathrm{n} \rightarrow ^{51}_{24}\mathrm{Cr} + \gamma$$

Neutron activation analysis is well adapted to study the contents of precious stones, jewelry, ancient coins and archeological samples.

5. ACIDS AND BASES

Introduction

There are several theories of acids and bases, each based on a different concept. Each defines acids and bases in a different manner. No theory is completely ideal and none is less than perfect. The applicability of a particular theory depends on the type of chemical system under consideration. We dwell on three of the most important theories of acids and bases. These are: Bronsted–Lowry theory, Lewis theory, and hard and soft acid base theory.

Bronsted–Lowry Theory

According to Bronsted–Lowry theory, an acid is a proton donor, while a base is a proton acceptor. An acid–base reaction involves the transfer of proton from an acid to a base. The process of neutralization is the formation of salt and solvent. When applied to aqueous system, the theory takes consideration of the self-ionization reaction of water.

$$H_2O + H_2O \rightleftharpoons H_3O^+ + OH^-$$

Accordingly, a substance which increases the concentration of hydronium ion (H_3O^+) is an acid and the one which decreases it is a base.

The interaction between hydrogen chloride and water is an acid–base reaction, since the former donates a proton to the latter. Therefore, hydrogen chloride acts as an acid and water as a base.

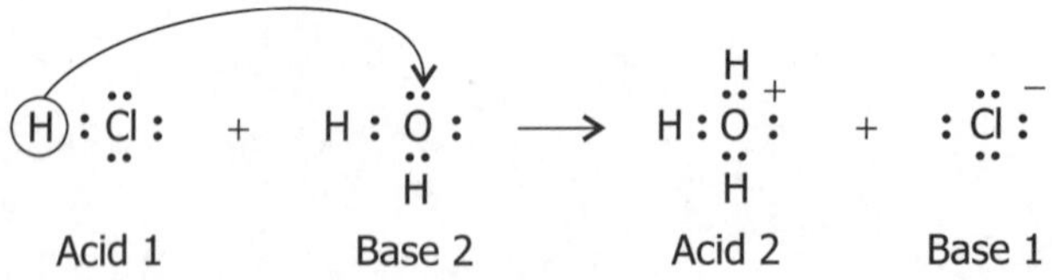

Hydrogen chloride is completely ionized in water. Therefore, the above reaction is spontaneous and proceeds to completion, implying that hydrogen chloride is a strong acid.

Every proton donor, however, is not a strong acid. For example, ionization of hydrogen fluoride in water occurs only to a slight extent. The reaction,

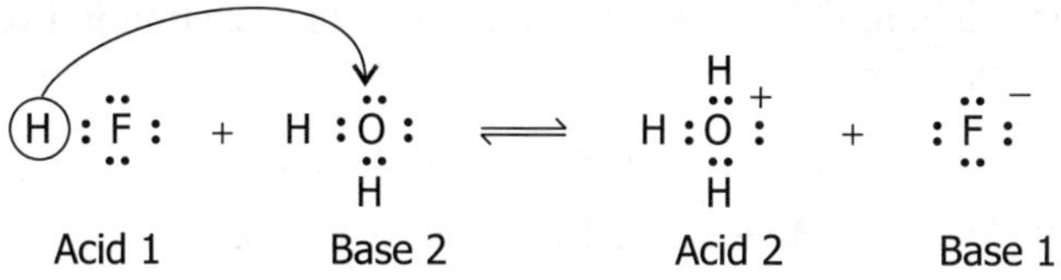

is reversible, implying that hydrogen fluoride is a weaker acid as compared to hydrogen chloride. In the reverse reaction, hydronium ion acts as an acid or proton donor, while the fluoride ion acts as a base or proton acceptor.

When a Bronsted–Lowry acid donates a proton, the remaining moiety is termed as a *conjugate base*. For example, chloride ion is the conjugate base of hydrogen chloride.

$$HCl \rightarrow H^+ + Cl^-$$

Acid   Conjugate base

When a Bronsted–Lowry base accepts a proton, the resulting moiety is termed as a *conjugate acid*. For example, hydronium ion is the conjugate acid of water.

$$H_2O + H^+ \rightarrow H_3O^+$$

Base   Conjugate acid

Any acid–base reaction involves two acids and two bases. The combinations involving an acid and its conjugate base or a base and its conjugate acid are collectively called *conjugate pairs*. As mentioned earlier, a typical acid–base reaction is written as:

$$\text{Acid 1} + \text{Base 2} \rightleftharpoons \text{Acid 2} + \text{Base 1}$$

The conjugate acid and base are designated by the same number. This means that acid 1 and base 1 form one conjugate pair, while acid 2 and base 2 form another conjugate pair.

When hydrogen fluoride releases its proton in aqueous solution, there are two bases which compete to accept it: water and fluoride ion. The fact that the hydrogen fluoride–water reaction is reversible indicates that the proton is predominantly taken up by fluoride ion, producing unionized hydrogen fluoride. Therefore, fluoride ion is a stronger base than water.

When hydrogen chloride releases its proton in aqueous solution, there are two bases which compete to accept it: water and chloride ion. The fact that the hydrogen chloride–water reaction is irreversible, and proceeds to completion, indicates that the proton is predominantly taken up by water. Therefore, water is a stronger base than chloride ion.

We, therefore, conclude—

1. The conjugate base of a strong acid is a weak base.
2. The conjugate base of a weak acid is a strong base.

When ethoxide ion, $C_2H_5O^-$ reacts with water, the reaction is irreversible and proceeds to completion.

$$\underset{\text{Base 1}}{C_2H_5O^-} + \underset{\text{Acid 2}}{H_2O} \rightarrow \underset{\text{Acid 1}}{C_2H_5OH} + \underset{\text{Base 2}}{OH^-}$$

Ethoxide ion can accept a proton from two acids: water and ethanol. The proton is taken up exclusively from water, indicating that the conjugate acid, ethanol is an extremely weak acid.

When ammonia reacts with water, the reaction is reversible; it does not proceed to completion.

$$\underset{\text{Base 1}}{NH_3} + \underset{\text{Acid 2}}{H_2O} \rightleftharpoons \underset{\text{Acid 1}}{NH_4^+} + \underset{\text{Base 2}}{OH^-}$$

The hydroxide ion can accept a proton from two acids: water and ammonium ion. It accepts it from ammonium ion, since it is a stronger acid than water. This causes the reaction to proceed, at least in part, from right to left.

We, therefore, further conclude—

3. The conjugate acid of a strong base is a weak acid.
4. The conjugate acid of a weak base is a strong acid.

While reacting with hydrogen chloride, water acts as a base; while reacting with ammonia, it acts as an acid. Therefore, water is both a weak acid and a weak base. It can donate a proton as well as accept it. It is *amphoteric*, or more precisely *amphiprotic*, in nature.

Case Study 5.1 Is it possible to prepare sodium ethoxide by the reaction of sodium hydroxide and ethanol?

In case it is possible to prepare sodium ethoxide by the action of sodium hydroxide on ethanol, the reaction may be indicated as follows:

$$\underset{\text{Acid 1}}{C_2H_5OH} + \underset{\text{Base 2}}{NaOH} \rightarrow \underset{\text{Base 1}}{C_2H_5ONa} + \underset{\text{Acid 2}}{H_2O}$$

In this reaction, ethanol acts as an acid and donates a proton.

$$\underset{\text{Acid}}{C_2H_5OH} \rightarrow \underset{\text{Conjugate base}}{C_2H_5O^-} + H^+$$

The hydroxide ion acts as a base and accepts a proton.

$$\underset{\text{Base}}{OH^-} + H^+ \rightarrow \underset{\text{Conjugate acid}}{H_2O}$$

Water is the conjugate acid of a strong base, hydroxide ion. It is, therefore, a weak acid. Yet, it is a stronger acid than ethanol. Since ethanol is an extremely weak acid, its conjugate base, the ethoxide ion, is an extremely strong base— much more stronger than hydroxide ion.

Such a strong base will seek out a proton from either of the two acids in the reaction system: water or ethanol. Since water is the stronger of the two acids, it will react with sodium ethoxide, causing the reaction to proceed from right to left.

Therefore, sodium ethoxide cannot be prepared by the reaction of sodium hydroxide and ethanol.

Binary acids

The binary acids, also called *hydro acids*, are those whose molecules are made up of two types of atoms, one being hydrogen. Water and hydrogen chloride are examples of hydro acids. For the hydro acids derived from the elements of any given group of a periodic table, the strength increases from top to bottom. For example,

Group 16: $H_2O < H_2S < H_2Se < H_2Te$

Group 17: $HF < HCl < HBr < HI$

This order is reverse of the ionic character. Although hydrogen fluoride is more ionic than hydrogen iodide, it is a weaker acid. This is because acid character, no doubt, depends on ionic character, but it is more dependent on the ease with which a proton is released. This means that the acid strength is predominantly determined by the strength of the bond that links the proton with the remaining atom in the molecule. According to Coulomb's law, the force of attraction between two charged atoms is inversely proportional to the square of the distance of separation. When the interatomic distance is small, the force of attraction is large and, therefore, it is difficult to release the proton. Conversely, if the interatomic distance is large, the force of attraction is small and, therefore, it is easy to release the proton. Hence for determining the acid strength, interatomic distance plays a more important role than ionic character. For this reason, hydrogen fluoride, with a smaller bond length, is a weak acid, while hydrogen iodide, with a larger bond length is a strong acid. Similarly, hydrogen telluride is a stronger acid than water.

Across a period of the periodic table, the acid strength increases from left to right. For example,

$$NH_3 < H_2O < HF$$

The electronegativity of the elements other than hydrogen increases from left to right. This causes the ionic character to increase in the same direction. Hydrogen fluoride has more ionic character than water which, in turn, has more ionic character than ammonia. Greater the ionic character, more easily will the molecule dissociate into constituent ions. Therefore, hydrogen fluoride releases the proton more easily than water, and ammonia does not release it at all.

Oxyacids

Oxyacids are those compounds in which the proton is attached to an oxygen atom, and the latter is bound to a central atom. A typical oxyacid may be represented as:

E : O : H

The strength of oxyacids primarily depends on the size and electronegativity of the central atom, E. If this atom is small and highly electronegative, it will tend to attract the electron pair, which it shares with oxygen, towards itself. The oxygen, in turn, will pull the electron pair that it shares with hydrogen so close to itself that the proton is released. Symbolically,

$$E \leftarrow : \quad O \leftarrow : \,|\, H \dashrightarrow H^+$$

The central atoms, for example, nitrogen, phosphorus, sulfur or chlorine, are less electronegative than oxygen. Yet these are able to displace electronic charge away from oxygen and towards themselves. This is because the central atoms are linked to other oxygen atoms as well through multiple bonds. The electromeric effect induces a positive charge on the central atoms, and makes them more electronegative than neutral oxygen.

Consider, for example, sulfuric acid, whose structure is represented as follows:

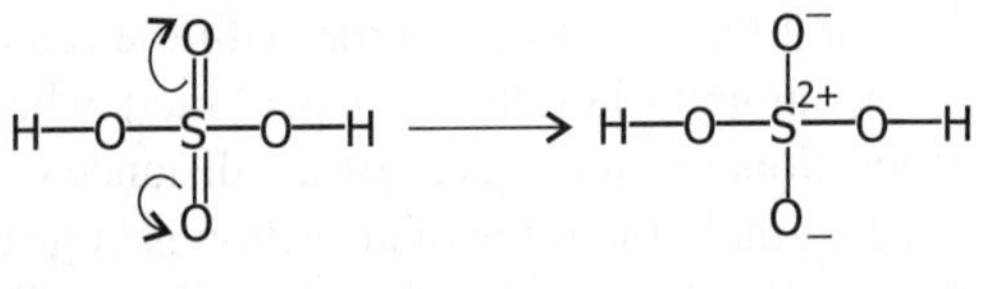

The electromeric effect induces a positive charge on sulfur making it more electronegative than oxygen.

Selenic acid, H_2SeO_4, is a weaker acid than sulfuric acid since selenium has larger size and smaller electronegativity than sulfur. The oxygen–hydrogen bond in selenic acid is more difficult to break than the corresponding bond in sulfuric acid. Therefore, sulfuric acid releases the proton more easily than selenic acid. For similar reason, perbromic acid, $HBrO_4$, is a weaker acid than perchloric acid, $HClO_4$.

There may exist a number of oxyacids for a particular central atom. In such cases, the one having a higher oxidation state of the central atom and a larger number of oxygen atoms is the stronger acid. Thus for oxyacids of chlorine, the strength varies in the following order:

VII		V		III		I
$HClO_4$	>	$HClO_3$	>	$HClO_2$	>	$HClO$

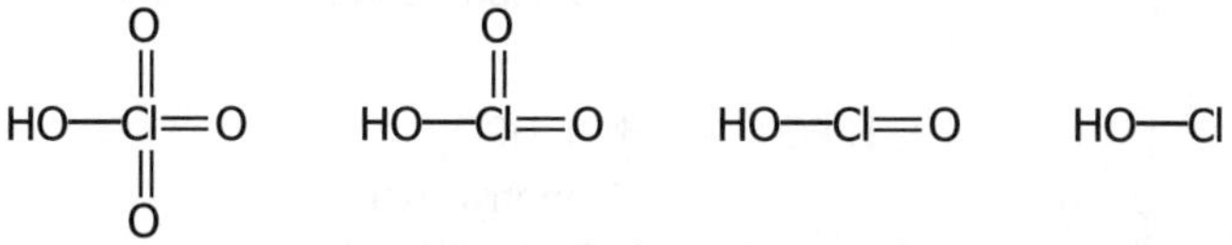

Perchloric acid	Chloric acid	Chlorous acid	Hypochlorous acid

When the oxidation state of the central atom is high, it is able to attract the pair of electrons that it shares with oxygen more towards itself. This, in turn, causes the oxygen–hydrogen bond to weaken leading to the release of proton.

For oxyacids of sulfur and nitrogen, the order of acid strength is:

VI		IV
H_2SO_4	>	H_2SO_3
Sulfuric acid		Sulfurous acid
V		III
HNO_3	>	HNO_2
Nitric acid		Nitrous acid

Those acids which have more than one ionizable hydrogen per molecule are called *polyprotic acids*. Sulfuric acid is a polyprotic acid—and to be more specific, it is a diprotic acid as it dissociates into two protons per molecule in aqueous solution.

Step 1: $H_2SO_4 \rightarrow H^+ + HSO_4^-$
Hydrogen sulfate ion

Step 2: $HSO_4^- \rightarrow H^+ + SO_4^{2-}$
Sulfate ion

The ionization in step 1 is more easy than that in step 2. In the first step, a proton is released from a neutral molecule, while in the second step, it is released from a negative ion.

Phosphoric acid, (H_3PO_4) is a triprotic acid.

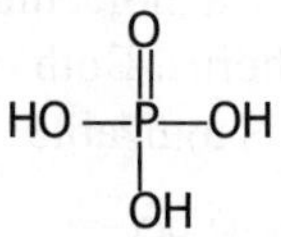

It can successively release three ionizable protons.

$$H_3PO_4 \rightarrow H^+ + H_2PO_4^-$$
Dihydrogen phosphate ion

$$H_3PO_4^- \rightarrow H^+ + HPO_4^{2-}$$
Monohydrogen phosphate ion

$$HPO_4^{2-} \rightarrow H^+ + PO_4^{3-}$$
Phosphate ion

Although phosphorus acid has the composition H_3PO_3, it is a diprotic acid.

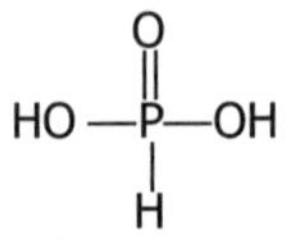

It has only two ionizable protons—the ones which are linked to phosphorus *through* the oxygen atoms. The proton which is directly linked to phosphorus is not ionizable. The strength of the phosphorus–hydrogen bond is very high as compared to that of the oxygen–hydrogen bond. For this reason, the hydrogen directly attached to phosphorus does not dissociate.

$$H_3PO_3 \rightarrow H^+ + H_2PO_3^-$$
Dihydrogen phosphite ion

$$H_3PO_3^- \rightarrow H^+ + HPO_3^{2-}$$
Monohydrogen phosphite ion

Hypophosphorus acid, H_3PO_2, is a monobasic acid.

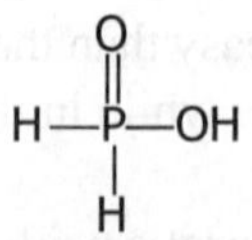

There are three hydrogen atoms in a molecule of hypophosphorus acid, of which two are directly attached to phosphorus. Both of them are non-ionizable. The one which is attached through oxygen is ionizable.

$$H_3PO_2 \rightarrow H^+ + H_2PO_2^-$$
Hypophosphite ion

The variation of acid strength among oxyacids of phosphorus follows the order:

I		III		V
H_3PO_2	>	H_3PO_3	>	H_3PO_4

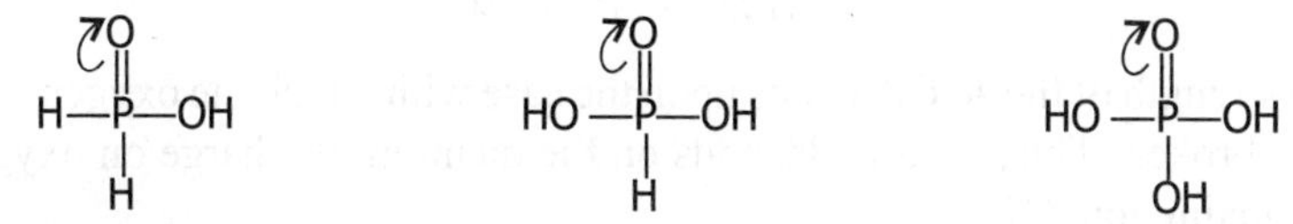

Hypophosphorus acid   Phosphorus acid   Phosphoric acid

This order is reverse of what is expected. In phosphoric acid, the number of oxygen atoms is the largest and the central atom has the highest oxidation state, yet it is the weakest acid in the series. The observed order may be explained on the basis of electromeric effect, which induces a positive charge on the central atom of each member of the series.

The positive charge induced on the central atom is same in each case, because all the three oxyacids have only one phosphorus–oxygen double bond. By virtue of the positive charge, phosphorus becomes more electronegative than oxygen(s) bound to it by single bond(s). As a result, the bonding electron pair shared between phosphorus and hydroxyl oxygen atoms is displaced towards the central atom. This displacement is maximum for hypophosphorus acid, since the central atom is attached to only one hydroxyl group and, therefore, has to attract only one pair of electrons. In case of phosphorus acid, the central atom has to attract two electron pairs from two hydroxyl groups. The displacement is relatively less. In phosphoric acid, the central atom has to attract three shared pairs of electrons from three hydroxyl groups and, accordingly, the displacement is least in this case. The more close the shared pair is drawn towards the central atom, the more easy it is for the proton to dissociate and more strong is the acid. As a result, hypophosphorus acid is the strongest and phosphoric acid is the weakest in this series.

While making a comparison between phosphoric acid and sulfuric acid, two factors need be emphasized. Firstly, sulfur in sulfuric acid has an oxidation state of VI, while phosphorus in phosphoric acid has an oxidation state of V. Secondly sulfuric acid has two oxo groups, phosphoric acid has only one. Therefore, variation of acid strength follows the order:

$$H_2SO_4 > H_3PO_4$$

Likewise, in chromic acid, H_2CrO_4 and permanganic acid, $HMnO_4$, the oxidation states of the central metal ions are VI and VII respectively. Moreover, the latter has one more oxo group than the former. The order of acid strength is:

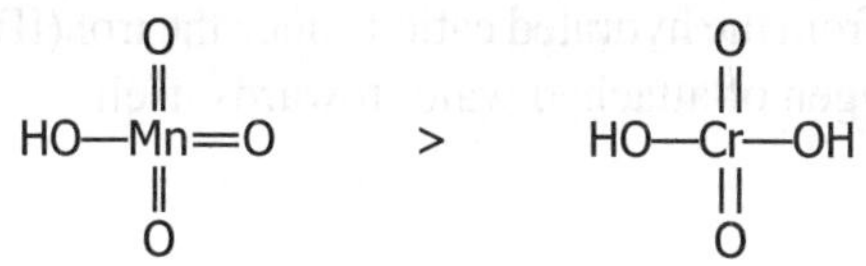

Case Study 5.2 Explain the variation in strength of hypohalous acids, HOX (X = Cl, Br or I).

Hypohalous acids dissociate in aqueous solution in accordance to the following reaction:

$$HOX \rightarrow H^+ + OX^-$$

The strength of the acid depends upon the ease with which the oxygen–hydrogen bond is broken. This, in turn, depends on the quantum of charge on oxygen atom of hypohalite ion, OX^-.

$$[H]^+—[O—X]^-$$

If the charge on oxygen is large, it is strongly attracted to hydrogen and the proton dissociates with difficulty. The acid is weak. Conversely, if the charge on oxygen is small, the attraction for hydrogen is relatively weak and the proton readily dissociates. The acid is strong.

The unit negative charge on hypohalite ion is shared by two atoms: oxygen and halogen. However, the sharing is not equal—it depends on the electronegativity of the halogen. Iodine is the least electronegative of all the halogens. Therefore, the unit negative charge on hypoiodite ion is almost entirely concentrated on oxygen. The force of attraction between oxygen and hydrogen is large and, therefore, hypoiodous acid is a relatively weak acid.

Chlorine has quite high electronegativity. An appreciable fraction of the uninegative charge on hypochlorite ion is concentrated on chlorine. This means, the quantum of charge on oxygen of hypochlorite ion is less than what it was on hypoiodite ion. The force of attraction between oxygen and hydrogen is concomitantly less and, therefore, hypochlorous acid is relatively strong.

The electronegativity of bromine is intermediate between that of chlorine and iodine. Accordingly, the strength of hypobromous acid is intermediate between that of hypochlorous and hypoiodous acids.

The variation of strength follows the order:

$$HOCl > HOBr > HOI$$

Hydrated cation acids

The species which, in aqueous solution, furnish a proton by dissociation of water molecule bound to a metal ion are called hydrated cation acids. Hydrated iron(II) ion, for example, will release a proton in accordance with the following reaction:

$$[Fe(H_2O)_6]^{2+} + H_2O \rightleftharpoons [Fe(H_2O)_5(OH)]^+ + H_3O^+$$

The proton is lost from the hydrated cation, since the iron(II) ion tends to displace the lone pairs on oxygen of attached water towards itself.

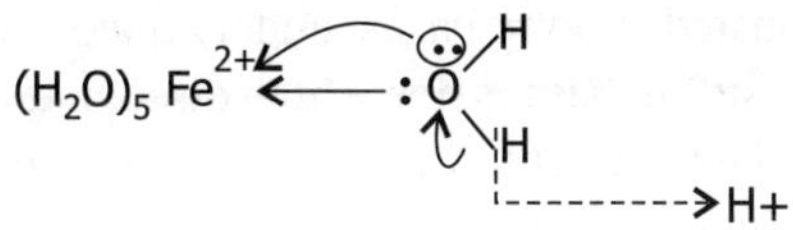

This induces the oxygen to attract the electron pair that it shares with hydrogen towards itself. As a result, the oxygen–hydrogen bond gets weakened and the proton dissociates.

Smaller the size of and/or higher the charge on the metal ion in the hydrated cation, stronger is the acid. A smaller cation with a higher charge can displace the lone pairs on oxygen of water more towards itself and, consequently, facilitates the release of proton more efficiently than a larger cation with a lower charge. Therefore, the hydrated iron(III) ion is a stronger acid than hydrated iron(II) ion, while hydrated cobalt(III) ion is a stronger acid than hydrated cobalt(II) ion. Similarly, hydrated aluminium(III) cation is a stronger acid than hydrated gallium(III) cation.

$$[Fe(H_2O)_6]^{3+} > [Fe(H_2O)_6]^{2+}$$

$$[Co(H_2O)_6]^{3+} > [Co(H_2O)_6]^{2+}$$

$$[Al(H_2O)_6]^{3+} > [Ga(H_2O)_6]^{3+}$$

$$\xrightarrow{\hspace{3cm}}$$

Decreasing acid strength

Case Study 5.3 Arrange the following hydrated cation acids in decreasing order of their strength: $[Sc(H_2O)_6]^{3+}$, $[Mn(H_2O)_6]^{2+}$ and $[Ni(H_2O)_6]^{2+}$.

The scandium(III) ion has a higher charge and a smaller size as compared to manganese(II) and nickel(II) ions. Its hydrated cation acid is, therefore, the strongest in the given series. The nickel(II) ion is smaller than the manganese(II) ion and hence the hydrated cation of the former is more acidic than that of the latter. The order of decreasing acid strength is:

$$[Sc(H_2O)_6]^{3+} > [Ni(H_2O)_6]^{2+} > [Mn(H_2O)_6]^{2+}$$

Bases

The strongest Bronsted–Lowry base that can exist in aqueous solution is the hydroxide ion, OH^-. Species such as ethoxide ion ($C_2H_5O^-$) and amide ion (NH_2^-) are, no doubt, stronger bases than hydroxide ion, but only if their reactions are studied in non-aqueous solvents. However, the acid–base reactions are usually carried out in aqueous medium. In the presence of water, even these strong bases are hydrolyzed to hydroxide ions.

$$C_2H_5O^- + H_2O \rightarrow C_2H_5OH + OH^-$$

$$NH_2^- + H_2O \rightarrow NH_3 + OH^-$$

For this reason, Bronsted–Lowry bases, under aqueous conditions, encompass only metal hydroxides. A strong base is one which dissociates into a higher concentration of hydroxide ions. The concentration of hydroxide ions, in turn, depends on the size and charge of the metal ion.

A metal ion having a large size and/or a small charge will form an ionic hydroxide. In water, it will be completely ionized, giving a high concentration of hydroxide ions and, therefore, would be a strong base. On the other hand, the hydroxide of a metal with a small size and a high charge, would possess appreciable covalent character. It will be incompletely dissociated in water, giving a low concentration of hydroxide ions and, therefore, would be a weak base. On moving from left to right in a period of the periodic table, the base strength of hydroxides decreases.

$$NaOH > Mg(OH)_2 > Al(OH)_3$$

In fact, aluminium hydroxide is amphoteric in nature.

Down a group, the base strength of hydroxides increases.

$$KOH > NaOH > LiOH$$

If an element exists in more than one oxidation state then the lower valent hydroxide is more basic.

$$\overset{I}{Tl(OH)} > \overset{III}{Tl(OH)_3}$$

Case Study 5.4 On adding a solution of potassium cyanide to aqueous aluminium(III) sulfate, a precipitate of aluminium(III) hydroxide is formed. Explain.

When an aqueous solution of potassium cyanide is added to aluminium(III) sulfate solution, the product obtained is aluminium(III) hydroxide and not aluminium(III) cyanide. This may be explained on the basis of Bronsted–Lowry theory. Potassium cyanide is a salt of a strong base (KOH) and a weak acid (HCN). Therefore, its aqueous solution is alkaline in nature, with a reasonably high concentration of hydroxide ions. When this solution is added to aqueous aluminium(III) sulfate, two ions compete to combine with aluminium(III) ions: cyanide ions and hydroxide ions. The cyanide ions are a conjugate base of hydrocyanic acid, the hydroxide ions of water. Hydrocyanic acid is a stronger acid than water; its conjugate base, cyanide ions, is weaker as compared to hydroxide ions, the conjugate base of water. The reaction with the stronger base prevails, yielding aluminium(III) hydroxide.

Leveling and differentiating effects

The strength of a Bronsted–Lowry acid or base depends on the nature of solvent in which it is dissolved. In aqueous solution, for example, species such as perchloric acid and hydrogen chloride are completely ionized. The reaction between the acids and water proceeds to completion.

$$HClO_4 + H_2O \rightarrow H_3O^+ + ClO_4^-$$

$$\underset{\text{Acid 1}}{HCl} + \underset{\text{Base 2}}{H_2O} \rightarrow \underset{\text{Acid 2}}{H_3O^+} + \underset{\text{Base 1}}{Cl^-}$$

The reaction is irreversible because water is a strong base as compared to perchlorate or chloride ions. Since both acids dissociate into protons to a complete extent, no variation is observed in their strength. Water is said to have leveled the strength of these acids. This phenomenon is called *leveling effect.*

Suppose these two acids are then dissolved in acetic acid solvent. The perchloric acid will still be completly ionized.

$$\underset{\text{Acid 1}}{HClO_4} + \underset{\text{Base 2}}{CH_3COOH} \rightarrow \underset{\text{Acid 2}}{CH_3COOH_2^+} + \underset{\text{Base 1}}{ClO_4^-}$$

In this reaction, acetic acid accepts a proton from perchloric acid and, therefore, acts as a base. It is a stronger base than perchlorate ion and for this reason, the reaction is irreversible. Perchloric acid is a strong acid.

The reaction between hydrogen chloride and acetic acid, however, does not proceed to completion.

$$\underset{\text{Acid 1}}{HCl} + \underset{\text{Base 2}}{CH_3COOH} \rightleftharpoons \underset{\text{Acid 2}}{CH_3COOH_2^+} + \underset{\text{Base 1}}{Cl^-}$$

The base strengths of acetic acid and chloride ion are comparable. Therefore, the reaction is reversible. Hydrogen chloride is only partially ionized. It is a weak acid.

Acetic acid solvent thus differentiates between the strength of perchloric acid and hydrogen chloride.This phenomenon is called *differentiating effect.*

Nitric acid and oxalic acid show a variation in strength when dissolved in water. Nitric acid is completely ionized and acts as a strong acid.

$$\underset{\text{Acid 1}}{HNO_3} + \underset{\text{Base 2}}{H_2O} \rightarrow \underset{\text{Acid 2}}{H_3O^+} + \underset{\text{Base 1}}{NO_3^-}$$

Oxalic acid is only partially ionized and acts as a weak acid.

$$\underset{\text{Acid 1}}{H_2C_2O_4} + \underset{\text{Base 2}}{2H_2O} \rightleftharpoons \underset{\text{Acid 2}}{2H_3O^+} + \underset{\text{Base 1}}{C_2O_4^{2-}}$$

Water exerts the differentiating effect.

However, if these two acids are dissolved in liquid ammonia, their strengths get even. Both are completely ionized in liquid ammonia and both act as strong acids.

$$HNO_3 + NH_3 \rightarrow NH_4^+ + NO_3^-$$

$$\underset{\text{Acid 1}}{H_2C_2O_4} + \underset{\text{Base 2}}{2NH_3} \rightarrow \underset{\text{Acid 2}}{2NH_4^+} + \underset{\text{Base 1}}{C_2O_4^{2-}}$$

Liquid ammonia exerts the leveling effect.

Case Study 5.5 Under which conditions would sulfuric acid behave as a base?

In aqueous solution, sulfuric acid is completely ionized.

$$\underset{\text{Acid 1}}{H_2SO_4} + \underset{\text{Base 2}}{H_2O} \rightarrow \underset{\text{Acid 2}}{H_3O^+} + \underset{\text{Base 1}}{HSO_4^-}$$

Two factors drive the above reaction to completion. First, sulfuric acid is a far better proton donor than water and second, hydrogen sulfate ion is a very weak base as compared to water. As a result, sulfuric acid is a strong acid.

However, sulfuric acid would behave as a base if it is dissolved in a still stronger acid. The only species that acts as an acid in the presence of sulfuric acid is hydrogen tetrakis(hydrogensulfato)borate(III), $H[B(HSO_4)_4]$.

$$\underset{\text{Acid 1}}{H[B(HSO_4)_4]} + \underset{\text{Base 2}}{H_2SO_4} \rightarrow \underset{\text{Acid 2}}{H_3SO_4^+} + \underset{\text{Base 1}}{[B(HSO_4)_4]^-}$$

In this reaction, sulfuric acid is a proton acceptor and, therefore, a base.

Lewis Theory

According to Lewis theory, an acid is a species which can accept electron pairs, while a base is a species which can donate electron pairs. Thus the acid is an electron-deficient atom, ion or molecule, while the base is an electron-rich chemical entity. An example of Lewis acid is boron(III) fluoride, BF_3. This compound has three bond pairs of electrons around boron and to achieve the stable octet needs one more. It is an electron-deficient moiety. On the other hand, ammonia is an electron-rich molecule. It has a non-bonding pair of electrons on nitrogen. It is an example of Lewis base.

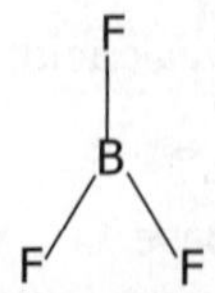

Boron(III) fluoride
Electron-deficient
Lewis acid

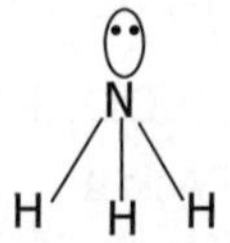

Ammonia
Electron-rich
Lewis base

Neutralization involves donation of electron pair from the base to the acid, through a coordinate bond.

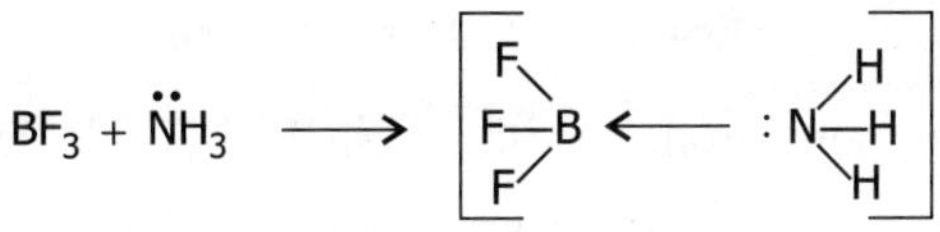

Acid Base Adduct of boron(III) fluoride–ammonia

Thus under Lewis concept, the process of neutralization is the formation of a coordinate bond between the electron donor (base) and the electron acceptor (acid).

In the formation of hydrated cobalt(III) ion, the metal ion acts as a Lewis acid, while the water molecules act as Lewis bases.

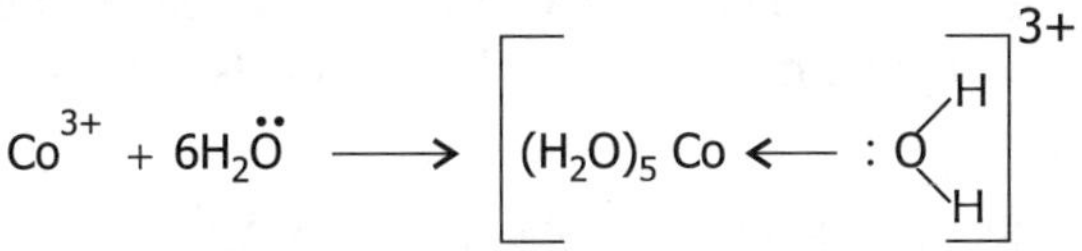

Acid Base Hydrated cobalt(III) ion

The oxygen of each water molecule donates one non-bonding pair of electrons to cobalt(III) ion.

The proton (H^+) is a Bronsted–Lowry acid, but can also act as a Lewis acid since it is electron-deficient.

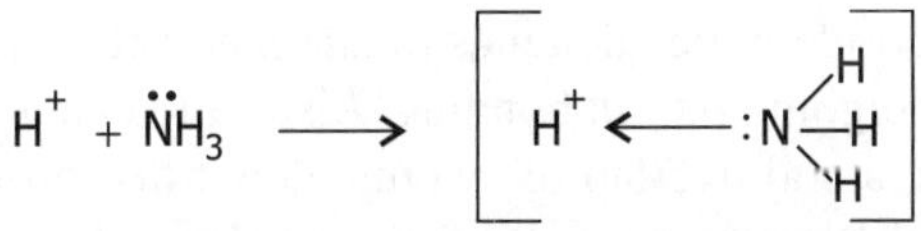

Acid Base Ammonium ion

Water is a Bronsted–Lowry base, but can also act as a Lewis base since it is electron-rich.

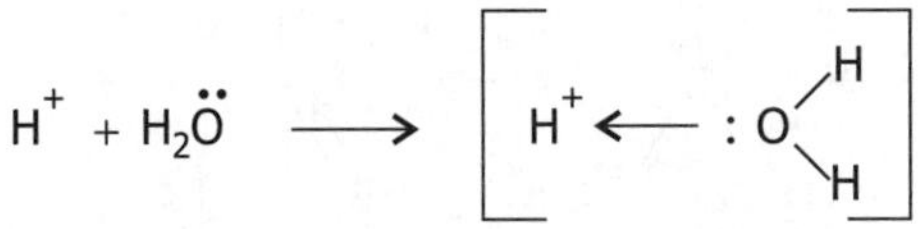

Acid Base Hydronium ion

Case Study 5.6 What are consolvating agents?

A substance which increases the acidity of a solution of acidic substance is called a consolvating agent. Boron(III) fluoride serves as a consolvating agent for hydrogen fluoride. Solutions of hydrogen fluoride and boron(III) fluoride are more acidic than solutions containing same concentrations of either compound. As hydrogen fluoride dissociates in aqueous solution,

$$HF(aq) \rightleftharpoons H^+(aq) + F^-(aq)$$

fluoride ions interact more strongly with boron(III) fluoride than with water.

$$BF_3 + F^- \rightarrow BF_4^-$$

Since fluoride ions are removed, hydrogen fluoride further dissociates into H^+ ions, thus increasing the acid character of the solution.

Acids

Boron(III) halides are electron-deficient molecules and, therefore, act as Lewis acids. These may accept an electron pair from Lewis bases such as ammonia, diethyl ether or trimethylamine. In the resulting adducts, boron has a stable octet configuration around it.

$$BX_3 + (C_2H_5)_2O \rightarrow [X_3B \leftarrow O(C_2H_5)_2]$$

$$\underset{\text{Acid}}{BX_3} + \underset{\text{Base}}{(CH_3)_3N} \rightarrow \underset{\text{Adduct}}{[X_3B \leftarrow N(CH_3)_3]}$$

The Lewis acid strength of boron(III) halides varies in the following order:

$$BI_3 > BBr_3 > BCl_3 > BF_3$$

The boron(III) fluoride molecule tends to attain an octet even in the absence of a Lewis base by undergoing pπ–pπ bonding. Also called *back bonding*, the pπ–pπ interaction involves lateral overlap of a completely filled orbital on fluorine with an empty orbital on boron. As a result of this overlap, a lone pair of electrons on fluorine is donated to boron; the octet of the latter gets completed. The π-bond is delocalized throughout the molecule.

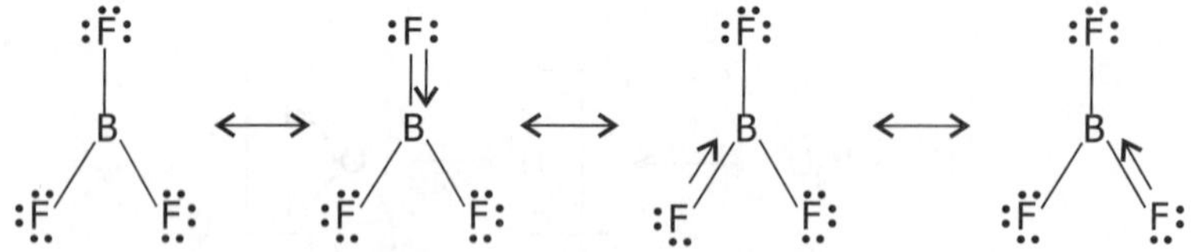

(I) (II) (III) (IV)

As a result of pπ–pπ bonding, the boron–fluorine bond length in boron(III) fluoride is less than expected and the bond energy is more than expected.

Resonance structures (II–IV) cannot accept an electron pair from a Lewis base. These must revert to structure (I) before an acid–base reaction can take place. For this reason, boron(III) fluoride reacts slowly with a Lewis base and, therefore, is a weak acid. In boron(III) chloride, the boron–chlorine bond distance is longer than the boron–fluorine bond distance in boron(III) fluoride. With increasing bond length, the chances of lateral overlap, leading to pπ–pπ bonding, decrease. The contribution of structures (II–IV) in case of boron(III) chloride is, therefore, less relative to boron(III) fluoride. As a result, boron(III) chloride reacts more rapidly with a Lewis base. It is a stronger Lewis acid than boron(III) fluoride. The boron–bromine bond distance in boron(III) bromide is still longer and its acid strength is concomitantly greater. For boron(III) iodide, the boron–iodine bond distance is so long that canonical forms (II–IV) do not exist. Its reaction with a Lewis base is the fastest. It is the strongest Lewis acid in the series.

Case Study 5.7 Comment on the strength of boric acid.

The composition of boric acid, H_3BO_3, suggests it is a Bronsted–Lowry oxyacid. However, it is a Lewis acid. A better way to indicate its composition is to depict it as $B(OH)_3$.

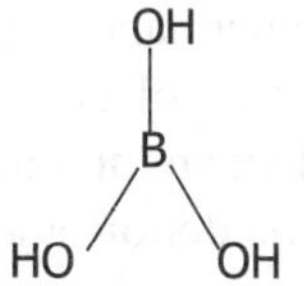

There are six bonding electrons around boron and to complete its octet, it needs two more. It abstracts a hydroxide ion, OH^- from water and completes its octet by forming tetrahydroxoborate(III) ion, $[B(OH)_4]^-$.

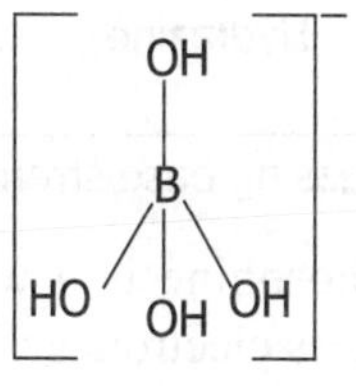

In this process a proton is released.

$$B(OH)_3 + 2H_2O \rightleftharpoons [B(OH)_4]^- + H_3O^+$$

It is, therefore, evident that boric acid does not furnish a proton from its own molecule. Rather, it accepts hydroxide ion from a water molecule, thereby adding the leftover proton to the aqueous solution. The above reaction is reversible. It does not proceed to completion, showing that boric acid is a weak acid. The titration of boric acid with sodium hydroxide does not give a sharp end-point.

However, if a polyhydroxy organic compound, having hydroxyl groups on adjacent carbon atoms, is added to the titration mixture, the strength of boric acid is remarkably enhanced. It gives a sharp end-point with the alkali. The polyhydroxy compound, for example, glycol, reacts with tetrahydroxoborate(III) ion and removes it from the above equilibrium reaction.

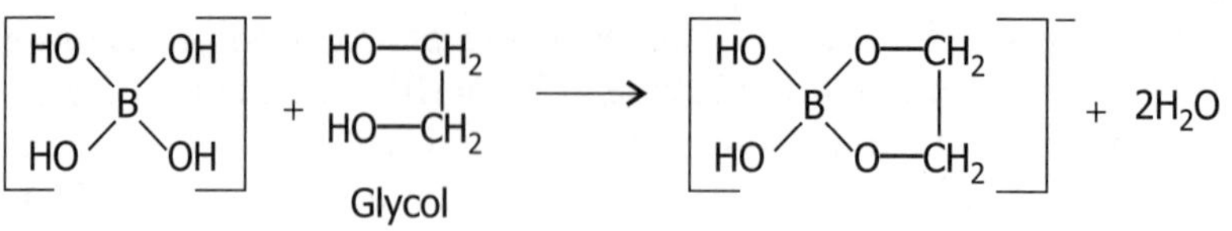

In accordance with the Le Chatelier's principle, as one of the products of an equilibrium system is removed, the reaction proceeds more towards the right. As a result, the concentration of protons increases, and so does the strength of boric acid.

Bases

As compared to ammonia, methylamine is a stronger Lewis base. The methyl group is electron-releasing and, therefore, induces negative charge on the nitrogen. The nitrogen can then donate the lone pair of electrons more easily. The electron-withdrawing substituents, on the other hand, induce a positive charge on nitrogen. The nitrogen then holds the lone pair of electrons more tightly. The compound behaves as a weaker Lewis base than ammonia. For example, hydroxylamine and nitrogen trifluoride are poor electron donors and hence weaker bases as compared to ammonia.

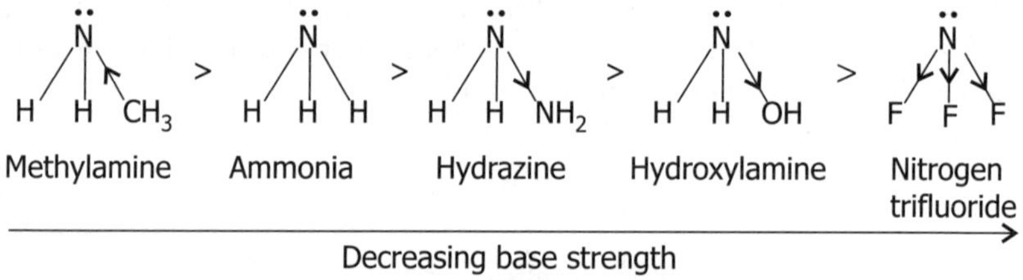

As compared to ammonia, phosphine is a weaker Lewis base. This seems paradoxical since phosphorus is less electronegative than nitrogen and, therefore, should be a better electron donor. However, the bond angle in phosphine is smaller than that in ammonia. In the case of ammonia, all the four orbitals around nitrogen— three directed towards the hydrogen atoms and the fourth housing the lone pair—are sp^3 type. In phosphine, because of smaller bond angle, the orbitals directed towards the substituents—the three hydrogen atoms in this case—tend to have more p character, while the orbital occupied by lone pair of electrons tends to have more s character. A lone pair in s orbital is held more tightly than the one in sp^3 hybrid orbital. It cannot be donated easily.

$$sp^3 \xrightarrow{\text{Small bond angle}} \underset{\text{Lone pair}}{s} + \underset{\text{Substituents}}{p^3}$$

Moreover, the s orbitals are larger in size as compared to sp^3 orbitals. Moreover, the 3s orbital on phosphorus is larger in size than the 2s orbital on nitrogen. Therefore, the lone pair of electrons in phosphine is more diffused than the non-bonding pair of electrons in ammonia. This means that before phosphine reacts with a Lewis base, it has to re-orient itself to such a position that the lone pair of electrons point towards the vacant orbitals of the acceptor atom. The electron donor–electron acceptor relationship is hindered, making phosphine a weaker Lewis base.

$$\ddot{N}H_3 > \ddot{P}H_3$$

Case Study 5.8 Carbon monoxide behaves as a Lewis base towards diborane, but not towards boron(III) fluoride. Why?

Boron(III) fluoride is a stronger Lewis acid than diborane, yet carbon monoxide reacts with the latter and not with the former.

$$B_2H_6 + 2CO \rightarrow 2[BH_3 \leftarrow CO]$$

This is because the product formed as a result of acid–base reaction is stabilized by hyperconjugation.

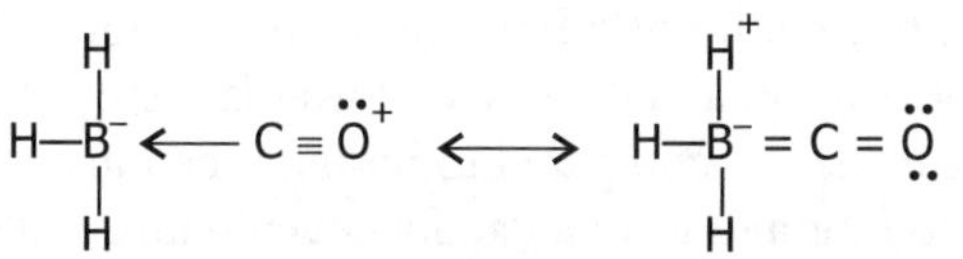

Since F^+ ion does not exist, this phenomenon is not possible with boron(III) fluoride.

The base strength is also affected by the type of hybridization which the orbitals of donor atom adopt. As the s character of the orbitals increases, the electronegativity of the donor atom also increases and its ability to act as a Lewis base decreases. For example, the sp^3 hybridized nitrogen on ammonia has less s character as compared to sp hybridized nitrogen in acetonitrile. This means that nitrogen is more electronegative in acetonitrile than in ammonia. As a result, ammonia is a stronger Lewis base, while acetonitrile is an extremely weak Lewis base.

	$\ddot{N}H_3$	$\gg CH_3C\ddot{N}$
Hybridization on nitrogen	sp^3	sp
s character of nitrogen	25%	50%

Case Study 5.9 What do the terms *B-strain*, *F-strain* and *I-strain* signify?

The terms B-strain, F-strain and I-strain refer to the phenomena which affect the strength of Lewis acids or bases or both. B-strain means *back strain*. It arises when the substituents on the Lewis base are bulky, and due to steric over-crowdedness, the electron donation to the Lewis acid is retarded. For example, triethylamine shows B-strain relative to ammonia.

Similarly, the presence of bulky substituents on Lewis acid reduces its strength due to steric hinderance. This phenomenon is called F-strain or *front strain*. Trimethylaluminium shows F-strain relative to aluminium(III) chloride.

I-strain means *internal strain*. The term applies to Lewis bases having exocyclic structures. For example, in case of lactams, as the size of the ring is reduced, the internal bond angles decrease and the hybrid orbitals of cyclic atoms have less s character. In turn, a greater s character is placed on exocyclic oxygen atom. The electronegativity of the oxygen atom increases and electron donor property decreases. As a result of I-strain, the base strength follows the order:

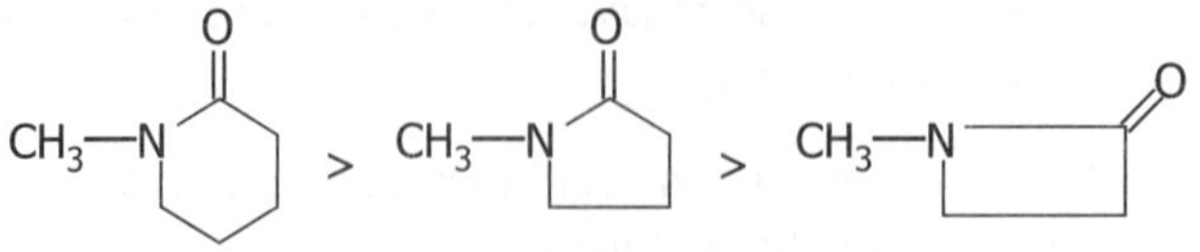

2,6-Dimethylpyridine is a stronger base towards protons than pyridine. Yet pyridine acts as a Lewis base towards trimethylboron(III), forming an adduct. However, 2,6-dimethylpyridine does not form an adduct with trimethylboron(III). Steric overcrowdedness around the small boron atom precludes the acid–base reaction between 2,6-dimethylpyridine and trimethylboron(III). Nevertheless, 2,6-dimethylpyridine behaves as a Lewis base towards triethylaluminium(III). Since aluminium is larger in size as compared to boron, it can increase its coordination number from three to four and form an adduct even with a bulky Lewis base.

Box 5.1

Superacid Medium

Superacid medium refers to a system whereby a strong acid is transformed to a more stronger proton donor. Such systems use a Lewis base to enhance the strength of a Bronsted–Lowry acid. Hence, these may be placed at the interface of Lewis theory and Bronsted–Lowry theory of acids and bases.

For example, the acid strength of hydrogen fluoride (a Bronsted–Lowry acid) may be increased to a substantial extent by adding antimony(V) fluoride (a Lewis base).

$$2HF + SbF_5 \rightarrow [H_2F]^+ [SbF_6]^-$$

Fluorosulfuric acid is a strong proton donor, yet its strength increases manifold in the presence of antimony(V) fluoride.

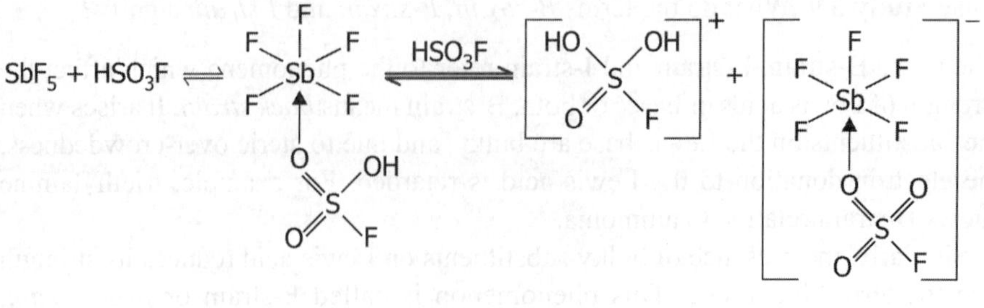

Contd...

The above acid is called *magic acid*, so named because of its ability to dissolve candle wax.

A still stronger acid results if a mixture of sulfur trioxide and antimony(V) fluoride in the ratio of 3:1 is added to fluorosulfuric acid.

$$3SO_3 + SbF_5 + 2HSO_3F \rightarrow [H_2SO_3F]^+ + [SbF_2(SO_3F)_4]^-$$

This solution reacts with alkanes to yield carbonium ions.

$$R_3CH + [H_2SO_3F]^+ \rightarrow R_3CH_2^+ + HSO_3F$$

Hard and Soft Acid Base Theory

The hard and soft acid base theory, or HSAB theory, is based on the following classification:

1. Hard acids are those metal ions which have a small size and/or high charge. They are polarizing in nature. Examples include lithium(I), beryllium(II), aluminium(III) and titanium(IV). They are also referred to as Type (a) metal ions.
2. Soft acids are those metal ions which have a large size and/or low charge. They have negligible polarizing ability. Examples include silver(I), mercury(II) and palladium(II). They are also referred to as Type (b) metal ions.
3. Hard bases are those species whose donor atom tends to be small and slightly polarizable. Examples include ammonia and fluoride ion.
4. Soft bases are those species whose donor atom tends to be large and highly polarizable. Examples include phosphine and iodide ion.
5. Hydrogen ion, H^+ is a hard acid, while the hydride ion, H^- is a soft base.

The HSAB theory is based on the concept that hard acids prefer to combine with hard bases, while soft acids prefer to combine with soft bases. This concept is called *Pearson's principle.* According to this principle, the tendency of the donor atoms to combine with hard acids, or Type (a) metal ions, varies in the order,

$$N > P > As$$
$$O > S > Se$$
$$F > Cl > Br$$

The tendency of donor atoms to combine with soft acids, or Type (b) metal ions, varies in the reverse order,

$$N < P < As$$
$$O < S < Se$$
$$F < Cl < Br$$

It is thus evident that phosphine, thioether and iodide ion have a greater tendency to coordinate with soft metal ions, while ammonia, ether and fluoride ion have a preference for hard metal ions. For example,

$$\underset{\text{Soft–hard}}{CH_3HgF} + \underset{\text{Hard–soft}}{LiI} \rightarrow \underset{\text{Soft–soft}}{CH_3HgI} + \underset{\text{Hard–hard}}{LiF}$$

Silver chloride and silver iodide are both soluble in saturated aqueous potassium iodide, but insoluble in saturated potassium chloride. The silver halides react with a saturated solution of potassium iodide to yield a soluble complex, potassium diiodoargentate(I).

$$AgCl + 2KI \rightarrow K[AgI_2] + KCl$$

$$AgI + KI \rightarrow K[AgI_2]$$

As has been stated, silver(I) ion is a soft acid. It prefers to combine with iodide ions which, being larger in size, act as a soft base. The reaction between soft silver(I) ion and relatively hard chloride ions fails to yield the corresponding dichloro complex of silver.

Case Study 5.10 Comment on the stability of silver(II) oxide, AgO.

Although oxidation state II is unusual for silver, yet silver(II) oxide exists—and is quite stable. This is attributed to the hard nature of oxide ion. Since hardness of an element usually increases with oxidation state, an unconventional, high-valence state can be stabilized by combining that element with an exceptionally strong hard base. The hard–hard combination is responsible for the existence and stability of silver(II) oxide. For the same reason, potassium tetraoxoferrate(VI), $K_2[FeO_4]$, in which iron has an unusual oxidation state of VI, is also stable.

Hardness and softness are relative terms. In alkali metal group, cesium(I) ion is soft relative to lithium(I) ion. Thallium(I) ion is soft as compared to thallium(III) ion. The addition of soft substituents imparts softness to a chemical entity, while addition of hard substituents increases its hardness. Thus boron is hard in boron(III) fluoride, but soft in boron(III) hydride.

$$\underset{\text{Soft–hard}}{R_2S.BF_3} + \underset{\text{Ether}}{R_2O} \rightarrow \underset{\text{Hard–hard}}{R_2O.BF_3} + R_2S$$

$$\underset{\text{Hard–soft}}{R_2O.BH_3} + \underset{\text{Thioether}}{R_2S} \rightarrow \underset{\text{Soft–soft}}{R_2S.BH_3} + R_2O$$

Similarly, nitrogen of ammonia is hard, while that of pyridine is soft.

$$\underset{\text{Hard–soft}}{H_3N.BH_3} + \underset{\text{Pyridine}}{C_5H_5N} \rightarrow \underset{\text{Soft–soft}}{C_5H_5N.BH_3} + NH_3$$

Case Study 5.11 Should lead(II) ion be considered a hard acid or a soft acid ?

Ions with electronic configuration $4d^{10}5s^2$ or $5d^{10}6s^2$ are borderline cases as far as hard or soft acid behavior is concerned. Lead(II) ion, with electronic configuration $5d^{10}6s^2$ too has intermediate behavior. When one substituent is bound, it acts as a hard acid, but when four or six groups are attached, it acts as a soft acid. When lead(II) ion has to combine with an anion, it prefers a hard one; when it has to combine with several anions, it prefers soft ones. Thus, PbF^+ is more stable than $PbCl^+$. On the other hand, $PbCl_4^{2-}$ exists while PbF_4^{2-} does not.

The driving force for hard–hard interactions is the formation of a strong ionic bond, supplemented by a gain in lattice energy. The interaction between a hard acid and a hard base is typically electrostatic. Since the size of the donor atom in hard base and acceptor atom in hard acid is small, the ionic bond formation from hard–hard combination is accompanied by release of a significant amount of lattice energy.

The driving force for soft–soft interactions is the formation of covalent bonds, supplemented by π-bonding. The soft donor atoms are polarizable in nature. According to Fajans' rules, these tend to form covalent compounds. Moreover, soft metal ions, such as copper(I), silver(I) or palladium(II), form compounds with appreciable covalent character.

The benzene ring, with its rich π-electron cloud acts as a soft base. It can combine with a soft acid, such as silver(I) ion by donating its π-electrons.

Thus silver perchlorate is soluble in benzene, but not in saturated hydrocarbon solvents

Case Study 5.12 What is symbiosis?

If a central atom, linked to hard substituents, were to expand its coordination number, it would prefer to combine further with hard species. Conversely, a central atom linked to soft substituents would prefer to coordinate further with soft species. This phenomenon is called symbiosis.

Suppose a mixture of fluoride and hydride ions are added to boron(III) fluoride. The boron atom would prefer to tie-up with the hard fluoride ion rather than the soft hydride ion.

$$BF_3 + F^- \rightarrow BF_4^-$$

If, on the other hand, the mixture of fluoride and hydride ions is added to diborane, the soft hydride ion would preferentially get coordinated to boron.

$$B_2H_6 + 2H^- \rightarrow 2BH_4^-$$

In a competitive situation, the following reaction will prevail:

$$BF_3H + BH_3F^- \rightarrow BF_4^- + BH_4^-$$

Similarly,

$$CF_3H + CH_3F \rightarrow CF_4 + CH_4$$

Yet another example of symbiosis is the irreversible reaction,

$$(C_2H_5)_3SiI + AgBr \rightarrow (C_2H_5)_3SiBr + AgI$$

where the soft silver ion prefers to bond with soft iodide ion, while the hard silicon prefers the relatively hard bromine.

The concept of symbiosis thus emphasizes that hard substituents impart hardness to the donor or acceptor atom, while soft substituents induce softness in the concerned atom.

6. NON-AQUEOUS SOLVENTS

Introduction

Water has long been designated as the universal solvent. This epitaph emanates from the fact that water is easily available, non-toxic, easy to handle and has a wide liquid range. Above all, it has unusual solvating characteristics. Most ionic compounds are soluble in water. Even some of the covalent compounds dissolve in water.

Water owes its unique solvating ability to two properties: large dipole moment and high dielectric constant. A compound with a large dipole moment possesses polar character. The partial charges in a polar molecule reduce the force of attraction between the cation and the anion of the ionic compound that is intended to be dissolved. As a result, the compound enters solution. For example, when sodium chloride is added to water, the sodium ions are attracted towards the negative center, while the chloride ions are attracted towards the positive center of the solvent molecule.

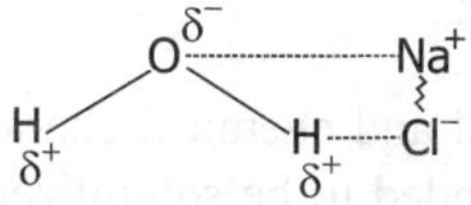

The effective force between sodium and chloride ions is weakened and, therefore, sodium chloride is solubilized.

The force of attraction (F) between oppositely charged ions in solution phase is given by the expression:

$$F = \frac{1}{\varepsilon} \cdot \frac{q_1 q_2}{r^2}$$

where q_1 and q_2 are the charges on the cation and anion, respectively, r is the distance of separation between the two counter ions and ε is the dielectric constant of the solvent. The above equation shows that the force of attraction between the cation and anion diminishes with increasing value of dielectric constant. This, in turn, increases the ease of dissolution. Therefore, higher the dielectric constant of the solvent, greater is its power to dissolve ionic compounds.

The values of dipole moment and dielectric constant of water are listed in Table 6.1. A comparison is also made with two other solvents, liquid ammonia and liquid sulfur dioxide. For these two solvents, commonly called non-aqueous solvents, the values of dipole moments and dielectric constants are less than that of water, yet these are moderately high. For this reason, fewer ionic compounds are soluble in liquid ammonia and liquid sulfur dioxide than in water. It is also possible that a given salt is completely soluble in water, but partially soluble in these non-aqueous solvents.

Table 6.1 Dipole moment and dielectric constant of different solvents

Solvent	**Dipole moment $\times 10^{-30}$ (C m)**	**Dielectric constant**
Water	6.12	78.5 (at 298 K)
Liquid ammonia	4.86	22 (at 240 K)
Liquid sulfur dioxide	5.36	17 (at 256 K)

This means that the difference in water, on the one hand, and liquid ammonia or liquid sulfur dioxide on the other, is a matter of degree only. Even if fewer salts are soluble in non-aqueous solvents or a particular salt is less soluble, the types of reactions that occur in aqueous and non-aqueous solvents are essentially the same. We shall now list some typical reactions that occur in aqueous solutions and compare them with corresponding reactions in non-aqueous solutions.

It must be emphasized that besides liquid ammonia and liquid sulfur dioxide, there are other non-aqueous solvents, such as liquid hydrogen fluoride, liquid hydrogen cyanide and liquid hydrogen sulfide. However, their utility is too meager to merit a comparison with aqueous medium.

Box 6.1

Types of Solvents

Depending on their physical and chemical characteristics, as well as on the nature of solute that is intended to be solubilized, various types of solvents are identified.

Protic solvents are those which furnish protons, for example, water.

Aprotic solvents are those which do not furnish protons, for example, carbon tetrachloride.

Acid solvents are those which have acidic properties, for example, liquid hydrogen fluoride.

Basic solvents are those which have alkaline properties, for example, liquid ammonia.

Amphoteric solvents are those which have both acidic and basic nature, for example, water.

Neutral solvents are those which are neither acidic nor basic in nature, for example, ethyl alcohol.

Ionizing solvents are those which are polar in nature and undergo self-ionization, for example, liquid sulfur dioxide.

Non-ionizing solvents are those which are non-polar and do not undergo self-ionization, for example, benzene.

Acid–Base Reactions

Acid–base reactions are significant for those solvents which undergo self-ionization, producing an acidic and an alkaline species. The neutralization of the acidic and alkaline species yields salt and solvent.

Water The self-ionization of water may be indicated as follows:

$$\underset{\text{Acid 1}}{H_2O} + \underset{\text{Base 2}}{H_2O} \rightleftharpoons \underset{\text{Acid 2}}{H_3O^+} + \underset{\text{Base 1}}{OH^-}$$

Accordingly, a compound which furnishes hydronium ions, H_3O^+ (or a proton, H^+) in aqueous medium behaves as an acid, while the one which furnishes hydroxide ions, OH^- behaves as a base. Hydrochloric acid is an acid and sodium hydroxide is a base. The neutralization results is the formation of sodium chloride (salt) and water (solvent).

$$\underset{\text{Acid}}{HCl} + \underset{\text{Base}}{NaOH} \rightarrow \underset{\text{Salt}}{NaCl} + \underset{\text{Solvent}}{H_2O}$$

The hydroxide ion may undergo further ionization.

$$\underset{\text{Acid 1}}{H_2O} + \underset{\text{Base 2}}{OH^-} \rightleftharpoons \underset{\text{Acid 2}}{H_3O^+} + \underset{\text{Base 1}}{O^{2-}}$$

Therefore, even the species which furnish oxide ions, O^{2-} are alkaline in nature. Sodium oxide is a base, while sulfuric acid is an acid in aqueous medium. The neutralization reaction is indicated as below:

$$\underset{\text{Acid}}{H_2SO_4} + \underset{\text{Base}}{Na_2O} \rightarrow \underset{\text{Salt}}{Na_2SO_4} + \underset{\text{Solvent}}{H_2O}$$

Liquid ammonia For liquid ammonia, the self-ionization reaction,

$$\underset{\text{Acid 1}}{NH_3} + \underset{\text{Base 2}}{NH_3} \rightleftharpoons \underset{\text{Acid 2}}{NH_4^+} + \underset{\text{Base 1}}{NH_2^-}$$

indicates that those species which furnish ammonium ions in liquid ammonia behave as acids, while those which furnish amide ions behave as alkalis. Accordingly, ammonium chloride, in liquid ammonia, is an acid and potassium amide, in liquid ammonia, is an alkali. The neutralization reaction produces potassium chloride (salt) and ammonia (solvent).

$$\underset{\text{Acid}}{NH_4Cl} + \underset{\text{Base}}{KNH_2} \rightarrow \underset{\text{Salt}}{KCl} + \underset{\text{Solvent}}{2NH_3}$$

The amide ion may undergo further ionization in liquid ammonia.

$$\underset{\text{Acid 1}}{NH_3} + \underset{\text{Base 2}}{NH_2^-} \rightleftharpoons \underset{\text{Acid 2}}{NH_4^+} + \underset{\text{Base 1}}{NH^{2-}}$$

$$\underset{\text{Acid 1}}{NH_3} + \underset{\text{Base 2}}{NH^{2-}} \rightleftharpoons \underset{\text{Acid 2}}{NH_4^+} + \underset{\text{Base 1}}{N^{3-}}$$

Therefore, even the imide ions, NH^{2-} and nitride ions, N^{3-} behave as alkalis in liquid ammonia. The following reactions support this assertion:

$$\underset{\text{Acid}}{2NH_4NO_3} + \underset{\text{Base}}{PbNH} \rightarrow \underset{\text{Salt}}{Pb(NO_3)_2} + \underset{\text{Solvent}}{3NH_3}$$

$$\underset{\text{Acid}}{3NH_4NO_3} + \underset{\text{Base}}{BiN} \rightarrow \underset{\text{Salt}}{Bi(NO_3)_3} + \underset{\text{Solvent}}{4NH_3}$$

Liquid sulfur dioxide The following is the self-ionization reaction of liquid sulfur dioxide:

$$\underset{\text{Acid 1}}{SO_2} + \underset{\text{Base 2}}{SO_2} \rightleftharpoons \underset{\text{Acid 2}}{SO^{2+}} + \underset{\text{Base 1}}{SO_3^{2-}}$$

Those compounds which yield thionyl ions (SO^{2+}) in liquid sulfur dioxide are acids, while those which yield sulfite ions (SO_3^{2-}) are bases. Thus, thionyl chloride, in liquid sulfur dioxide, behaves as an acid and sodium sulfite, in liquid sulfur dioxide, behaves as a base. The neutralization yields sodium chloride (salt) and sulfur dioxide (solvent).

$$\underset{\text{Acid}}{SOCl_2} + \underset{\text{Base}}{Na_2SO_3} \rightarrow \underset{\text{Salt}}{2NaCl} + \underset{\text{Solvent}}{2SO_2}$$

Case Study 6.1 What is the evidence that self-ionization of liquid ammonia yields ammonium and amide ions.

If labeled ammonium chloride, $^{15}NH_4Cl$ (or $^*NH_4Cl$) is dissolved in liquid ammonia, and the solvent evaporated, the activity of radioactive nitrogen is evenly distributed between ammonium chloride and ammonia. The labeled ammonium chloride will ionize in liquid ammonia as follows:

$$^*NH_4Cl \rightleftharpoons {}^*NH_4 + Cl^-$$

The radioactive nitrogen is passed on from ammonium ion to liquid ammonia. This provides evidence for the self-ionization reaction of liquid ammonia.

Precipitation Reactions

A reaction in which all the reactants are soluble in the solvating medium, and so are all the products except one is called a precipitation reaction. The insoluble product precipitates out of the solution.

Water Consider the following reaction in aqueous medium:

$$\underset{\text{Soluble}}{AgNO_3} + \underset{\text{Soluble}}{NaCl} \rightarrow \underset{\text{Insoluble}}{AgCl\downarrow} + \underset{\text{Soluble}}{NaNO_3}$$

Both the reactants, silver nitrate and sodium chloride are soluble in water. One of the products, sodium nitrate is also soluble in water, while the other, silver chloride is insoluble. The insoluble silver chloride is obtained as a white precipitate.

Liquid ammonia Most of the iodides, nitrates and thiocyanates are soluble in liquid ammonia. All ammonium salts, irrespective of the anions, are also soluble. Based on this information, the following examples represent precipitation reactions in liquid ammonia:

$$\underset{\text{Soluble}}{NH_4Cl} + \underset{\text{Soluble}}{NaI} \rightarrow \underset{\text{Insoluble}}{NaCl\downarrow} + \underset{\text{Soluble}}{NH_4I}$$

$$\underset{\text{Soluble}}{(NH_4)_2S} + \underset{\text{Soluble}}{2AgNO_3} \rightarrow \underset{\text{Insoluble}}{Ag_2S\downarrow} + \underset{\text{Soluble}}{2NH_4NO_3}$$

Liquid sulfur dioxide The nitrates, nitrites and acetates are all soluble in liquid sulfur dioxide. Most of the thallium(I) and mercury(II) salts are also soluble. In addition, all the sulfites, irrespective of the cations, are soluble. Thionyl chloride is completely miscible with liquid sulfur dioxide. Following are examples of precipitation reactions in liquid sulfur dioxide:

$$\underset{\text{Soluble}}{2TlCl} + \underset{\text{Soluble}}{K_2SO_3} \rightarrow \underset{\text{Insoluble}}{2KCl\downarrow} + \underset{\text{Soluble}}{Tl_2SO_3}$$

$$\underset{\text{Soluble}}{2CH_3COOAg} + \underset{\text{Soluble}}{SOCl_2} \rightarrow \underset{\text{Insoluble}}{2AgCl\downarrow} + \underset{\text{Soluble}}{(CH_3COO)_2SO}$$

Box 6.2

Solution of Metals in Liquid Ammonia

The alkali metals, and to a lesser extent the alkaline earth metals, as well as europium and ytterbium of lanthanoid series, dissolve to a very high concentration in liquid ammonia. The solution contains solvated metal ions and solvated electrons. The solvated electrons are distributed over a cavity in the solvent of radius 300–340 pm formed by the displacement of 2 or 3 ammonia molecules.

$$M + (x + y)NH_3 \rightleftharpoons [M(NH_3)_x]^+ + [e(NH_3)_y]^-$$

On evaporating ammonia, the equilibrium of the above reaction shifts to the left. The electrons return to their parent cations and the entire amount of metal that was originally dissolved may be reclaimed.

Contd...

There are three unique characteristics of such solutions.

1. Color

Dilute solutions of these metals in liquid ammonia are blue in color. The color is attributed to the absorption of visible light by solvated electrons. The concentrated solutions are copper-bronze colored, due to the formation of metal ion clusters. On standing, the color of the solution fades out, since the impurities, such as iron oxide, catalyze the conversion of metal into amide.

$$2M + 2NH_3 \rightarrow 2MNH_2 + H_2$$

2. Conductivity

Dilute solution of a metal in liquid ammonia has conductivity comparable to that of an aqueous solution of a typical ionic compound. The concentrated solution has conductivity approaching that of a pure metal. In either case, conductivity is attributed to the presence of solvated electrons.

3. Reducing nature

The presence of free electrons make the solutions powerful reducing agents. In the following example, nickel(II) ion is reduced to nickel(0) when added to a solution of sodium in liquid ammonia:

$$Na_2[Ni(CN)_4] + 2Na \xrightarrow{NH_3} Na_4[Ni(CN)_4]$$

A solution of potassium in liquid ammonia reduces tetrammineplatinum(II) bromide to tetrammineplatinum(0).

$$[Pt(NH_3)_4]Br_2 + 2K \xrightarrow{NH_3} [Pt(NH_3)_4] + 2KBr$$

Similarly, elemental sulfur, when added to a solution of sodium in liquid ammonia, is reduced to sulfide ion.

$$S + 2Na \xrightarrow{NH_3} Na_2S$$

Complexation Reactions

Complexation reactions involve the combination of a metal ion with a fixed number of ligand moieties. The metal ion and the ligands lose their typical physical and chemical characteristics.

Water Quite a few complexation reactions take place in water. For example, addition of potassium cyanide to silver nitrate initially results in the formation of insoluble silver cyanide. However, subsequent addition of excess potassium cyanide, results in the formation of a soluble complex, potassium dicyanoargentate(I).

$$AgNO_3 + KCN \rightarrow \underset{\text{Insoluble}}{AgCN} + KNO_3$$

$$AgCN + KCN \rightarrow \underset{\text{Soluble}}{K[Ag(CN)_2]}$$

In some reactions, water acts as a solvent as well as a ligand. For example, dissolution of ferrous sulfate in water yields the complex, hexaquairon(II) sulfate.

$$FeSO_4 + 6H_2O \rightarrow [Fe(H_2O)_6]SO_4$$

Liquid ammonia The reaction between zinc nitrate and potassium amide, in liquid ammonia, first precipitates the insoluble zinc amide.

$$Zn(NO_3)_2 + 2KNH_2 \rightarrow \underset{\text{Insoluble}}{Zn(NH_2)_2} + 2KNO_3$$

If excess of potassium amide is added, zinc amide is transformed into soluble potassium tetramidozincate(II) complex.

$$Zn(NH_2)_2 + 2KNH_2 \rightarrow \underset{\text{Soluble}}{K_2[Zn(NH_2)_4]}$$

At times, liquid ammonia may act both as a solvent and a ligand. For example, the reaction between sodium chloride and calcium nitrate in liquid ammonia gives octamminecalcium(II) chloride.

$$2NaCl + Ca(NO_3)_2 + 8NH_3 \rightarrow [Ca(NH_3)_8]Cl_2 + 2NaNO_3$$

Liquid sulfur dioxide When sodium sulfite is added to aluminium chloride, in liquid sulfur dioxide medium, insoluble aluminium sulfite is initially formed.

$$2AlCl_3 + 3Na_2SO_3 \rightarrow \underset{\text{Insoluble}}{Al_2(SO_3)_3} + 6NaCl$$

Further addition of sodium sulfite yields the soluble sodium trisulfitoaluminate(III) complex.

$$Al_2(SO_3)_3 + 3Na_2SO_3 \rightarrow \underset{\text{Soluble}}{2Na_3[Al(SO_3)_3]}$$

Case Study 6.2 Is it possible to synthesize complexes having metal ions with unusual oxidation states in liquid ammonia solvent?

Coordination compounds having central metal ions with unusual and low oxidation states may be synthesized in liquid ammonia, in presence of sodium. For example, the reaction between palladium(II) and cyanide ions first yields tetracyanopalladate(II) ion.

$$Pd^{2+} + 4CN^- \xrightarrow{NH_3} [Pd(CN)_4]^{2-}$$

Subsequently, it is reduced to tetracyanopalladate(0) ion.

$$Na(NH_3) \rightarrow Na^+ + e^-(NH_3)$$

$$[Pd(CN)_4]^{2-} + 2e^-(NH_3) \rightarrow [Pd(CN)_4]^{4-} + 2NH_3$$

Similarly, hexacyanocobaltate(III) ion is reduced to tetracyanocobaltate(I) ion.

$$[Co(CN)_6]^{3-} + 2Na \xrightarrow{NH_3} [Co(CN)_4]^{3-} + 2NaCN$$

Redox Reactions

Redox reactions, also called *oxidation–reduction reactions*, are those in which one reactant is oxidized, while the other is reduced.

Water A neutral solution of potassium permanganate acts as a strong oxidizing agent. For example, it oxidizes hydrogen sulfide to sulfur.

$$2KMnO_4 + 4H_2S \rightarrow 2MnS + K_2SO_4 + 4H_2O + S$$

Liquid ammonia The oxidizing power of potassium permanganate is weaker in liquid ammonia than in water. Yet, it oxidizes amide ions to nitrogen, and itself gets reduced to potassium manganate, K_2MnO_4.

$$6KMnO_4 + 6KNH_2 \rightarrow 6K_2MnO_4 + 4NH_3 + N_2$$

In liquid ammonia, iodine oxidizes potassium hexamidostannate(II) to potassium hexamidostannate(IV), and itself gets reduced to iodide ions.

$$K_4[Sn(NH_2)_6] + I_2 \rightarrow K_2[Sn(NH_2)_6] + 2KI$$

Liquid sulfur dioxide Antimony(V) chloride is reduced to antimony(III) chloride and potassium hexachloroantimonate(III) by potassium iodide in liquid sulfur dioxide solvent.

$$6KI + 3SbCl_5 \rightarrow 2K_3[SbCl_6] + SbCl_3 + 3I_2$$

Case Study 6.3 When zinc is added to a solution of sodium amide in liquid ammonia, hydrogen gas is liberated. On adding ammonium iodide, a white precipitate appears which dissolves in excess of ammonium iodide. Explain the sequence of reactions.

Zinc reacts with sodium amide in liquid ammonia solvent to yield sodium tetramidozincate(II) complex, along with hydrogen.

$$Zn + 2NaNH_2 + 2NH_3 \rightarrow Na_2[Zn(NH_2)_4] + H_2(g)\uparrow$$

On adding ammonium iodide, the complex decomposes to a white precipitate of zinc amide.

$$[Zn(NH_2)_4]^{2-} + 2NH_4^+ \rightarrow Zn(NH_2)_2 + 4NH_3$$

When excess of ammonium iodide is added, zinc amide solubilizes to produce tetramminezinc(II) iodide.

$$Zn(NH_2)_2 + 2NH_4I \rightarrow [Zn(NH_3)_4]I_2$$

Solvolysis Reactions

A reaction in which the solvent also acts as a reactant is called a solvolysis reaction.

Water When water is the solvent, the solvolysis reactions are more precisely referred to as *hydrolysis* reactions. For example:

$$CuSO_4 + 5H_2O \rightarrow CuSO_4.5H_2O$$

$$P_4O_{10} + 6H_2O \rightarrow 4H_3PO_4$$

Liquid ammonia Analogous reactions take place in liquid ammonia and are sometimes referred to as *ammonolysis* reactions.

$$AuCl + 12NH_3 \rightarrow AuCl.12NH_3$$

$$AlI_3 + 6NH_3 \rightarrow AlI_3.6NH_3$$

The reaction of ammonia with nitrosyl chloride in liquid ammonia solvent gives nitrosyl amide,

$$NH_3 + NOCl \rightarrow NO(NH_2) + HCl$$

while that with mercurous chloride gives a mixture of elemental mercury, amidochloromercury(II) and ammonium chloride.

$$2NH_3 + Hg_2Cl_2 \rightarrow Hg + [Hg(NH_2)Cl] + NH_4Cl$$

Silicon(IV) chloride undergoes solvolysis reaction in presence of liquid ammonia to yield tetramidosilicon(IV) at temperatures below the freezing point.

$$SiCl_4 + 4NH_3 \xrightarrow{< 270\ K} [Si(NH_2)_4] + 4HCl$$

The product successively decomposes into diamidoimidosilicon(IV) and silicon(IV) nitride as the temperature is raised.

$$[Si(NH_2)_4] \xrightarrow{273\ K} \underset{\text{Diamidoimidosilicon(IV)}}{[Si(NH)(NH_2)_2]} + NH_3$$

$$3[Si(NH)(NH_2)_2] \xrightarrow{1500\ K} \underset{\text{Silicon(IV) nitride}}{Si_3N_4} + 5NH_3$$

Case Study 6.4 What is meant by the term *ammono acid*?

Phosphorus oxotrichloride on hydrolysis yields phosphoric acid and on ammonolysis yields phosphoramide.

$$OPCl_3 + 6H_2O \rightarrow \underset{\text{Phosphoric acid}}{OP(HO)_3} + 3H_2O + 3HCl$$

$$OPCl_3 + 6NH_3 \rightarrow \underset{\text{Phosphoramide}}{OP(NH_2)_3} + 3NH_4Cl$$

Due to the structural resemblance between phosphoric acid and phosphoramide, the latter is often referred to as ammono acid.

Liquid sulfur dioxide Solvates of metal salts are formed in liquid sulfur dioxide. For example:

$$SrI_2 + 4SO_2 \rightarrow SrI_2.4SO_2$$

$$LiI + 2SO_2 \rightarrow LiI.2SO_2$$

The reaction between phosphorus(V) chloride and sulfur dioxide, in liquid sulfur dioxide medium, leads to the formation of phosphorus oxotrichloride and thionyl chloride.

$$PCl_5 + SO_2 \rightarrow POCl_3 + SOCl_2$$

Likewise, niobium(V) chloride is transformed to niobium oxotrichloride.

$$NbCl_5 + SO_2 \rightarrow NbOCl_3 + SOCl_2$$

The reaction between liquid sulfur dioxide and diethylzinc(II), to yield zinc sulfite and diethylsulfoxide, too comes under the category of solvolysis.

$$(C_2H_5)_2Zn + 2SO_2 \rightarrow ZnSO_3 + (C_2H_5)_2SO$$

Liquid sulfur dioxide undergoes yet another solvolysis reaction with hydroxylamine.

$$NH_2OH + SO_2 \rightarrow NH_2(SO_2)OH$$

This reaction is carried out in a sealed, thick-walled borosilicate glass tube, called *Carius tube* (Fig 6.1). The hydroxylamine is placed in the tube, which is then cooled to 228 K. Thereafter, liquid sulfur dioxide is introduced into the tube. Next the tube is sealed and allowed to warm to room temperature. The reaction takes several days to complete.

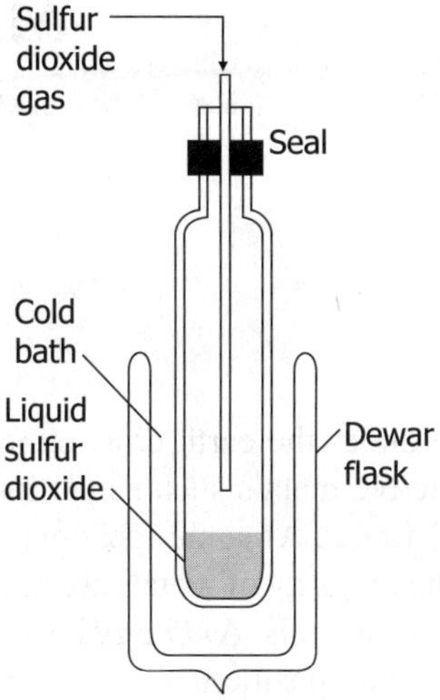

Fig. 6.1 A Carius tube

Case Study 6.5 Outline the reaction of tungsten(VI) chloride with liquid sulfur dioxide.

The solvolysis reaction between liquid sulfur dioxide and tungsten(VI) chloride yields thionyl chloride ($SOCl_2$) and tungsten oxotetrachloride ($WOCl_4$).

$$WCl_6 + SO_2 \rightarrow WOCl_4 + SOCl_2$$

7. METALLURGY

Introduction

Metals are found in nature under the earth crust either in the elemental state or in a combined state. Unreactive metals such as platinum and gold are dispersed through rocks in elemental forms. At some locations these are found in clumps called *nuggets*. However, the majority of metals are buried under the earth crust in chemically combined states as minerals. A *mineral* is a naturally occurring substance with a characteristic range of compositions.

A mineral from which a metal can be isolated economically is called an *ore*. Many minerals contain metals either in too small a concentration, or in too difficult a form to allow extraction. Such minerals do not serve as economic sources of metals. These are, therefore, not referred to as ores. For example, ordinary clay contains substantial amount of aluminium. However, extraction of aluminium from clay is not cost-effective. On the other hand, aluminium can be more easily, and more economically, extracted from bauxite. Hence, bauxite is an ore of aluminium, but clay is not.

In bauxite ore, aluminium exists as oxide. Other metals may be found as sulfides, halides or phosphates in their respective ores. Table 7.1 lists different types of ores along with their examples.

Table 7.1 Different types of ores

Ore type	Examples (composition)
Native ores	Gold, Iridium, Osmium, Palladium, Platinum and Silver
Sulfide ores	Argentite (Ag_2S), Chalcopyrite ($CuFeS_2$), Cinnabar (HgS), Copper glance (Cu_2S), Galena (PbS), Iron pyrites (FeS_2), Nickel blende (NiS) and Zincblende (ZnS)
Oxide ores	Bauxite ($Al_2O_3.2H_2O$), Cassiterite (SnO_2), Cuprite (Cu_2O), Hematite (Fe_2O_3), Magnetite (Fe_3O_4), Pyrolusite (MnO_2), Wolframite [$(Fe, Mn)WO_4$] and Zincite (ZnO)
Carbonate ores	Calamine ($ZnCO_3$), Cerussite ($PbCO_3$), Dolomite ($MgCO_3.CaCO_3$), Limestone ($CaCO_3$), Magnesite ($MgCO_3$), Malachite ($CuCO_3.Cu(OH)_2$) and Siderite ($FeCO_3$)
Sulfate ores	Anglesite ($PbSO_4$), Barite ($BaSO_4$), Epsom salt ($MgSO_4.7H_2O$) and Gypsum ($CaSO_4.2H_2O$)
Halide ores	Carnallite ($KCl.MgCl_2.6H_2O$), Cryolite (Na_3AlF_6), Fluorspar (CaF_2), Horn silver ($AgCl$), Rock salt ($NaCl$) and Sylvite (KCl)

Phosphate ores	Pyromorphite $[3Pb_3(PO_4)_2.PbCl_2]$ and Rock phosphate $[Ca_5(PO_4)_3(OH, F, Cl)]$
Silicate ores	Asbestos $[Mg_3Si_2O_5(OH)_4]$, Feldspar $[(K, Na)AlSi_3O_8]$, Hemimorphite $[Zn_4Si_2O_7(OH)_2.H_2O]$ and Mica $[KAl_2(Si_3Al)O_{10}(OH, F)_2]$, Zircon $[ZrSiO_4]$

Case Study 7.1 The common ores of nickel and copper are sulfides. However, aluminium is extracted from its oxide and calcium from its carbonate. Why?

As discussed in Chapter 5, the sulfide ion is a soft base, while carbonate and oxide ions are hard bases. The nickel(II) and copper(II) ions possess considerable softness and hence under natural conditions these prefer to combine with sulfide ions. On the other hand, aluminium and calcium ions possess considerable hardness and hence these occur as oxide and carbonate under the earth crust.

Correlation between Type of Ore and Standard Electrode Potential of Metal

Whether a particular metal would be dispersed in the natural environment in an uncombined, elemental state, or in a combined state, depends primarily on its standard electrode potential, or, to be more specific, on its *reduction potential*. The reduction potential also dictates that when a metal occurs in a combined state, with which anion would it possibly be associated in the lithosphere or in the hydrosphere.

We have discussed the concept of reduction potential in Chapter 8 and listed the values of this physical parameter for a few chosen elements in Table 8.1 (p. 285). In the present context, suffice it to state that greater the negative value of reduction potential, more reactive is the metal. Conversely, a smaller negative value or a positive value of reduction potential means that the metal is relatively unreactive.

To gauge the elemental composition of an ore on the basis of reduction potential, it is pertinent to classify the metals into the following four categories.

1. Highly reactive metals

The reduction potential of these metals is –2.0 V or less. The alkali and alkaline earth metals (except beryllium) come under this category. For example, the reduction potential of sodium and magnesium are –2.71 V and –2.38 V, respectively. Such metals largely occur as halide and sulfate ores. The bonding between the metal ion and the counter ion is so strong that extraction is carried out by electrolysis of fused salts.

2. Moderately reactive metals

Under this classification are listed the metals for which the reduction potential values are in the range –2.0 V to +0.5 V. Examples include zinc, iron, tin and lead. These occur mainly as sulfide and oxide ores.

3. Weakly reactive metals

The metals having reduction potential between +0.5 V and 1.0 V are considered as weakly reactive. These may occur both in combined and uncombined states, depending upon the characteristics of the geologic site. For example, mercury is embedded in igneous rocks as native metal at some locations and as cinnabar (HgS) at others. Silver too comes under this category.

4. Unreactive metals

The metals for which the reduction potential values are + 1.0 V or more are so inert that these do not interact with the chemical entities present in minerals. Hence these occur in native states. Gold and platinum are examples of such metals.

Overview of Isolation Operation

Metallurgy is the science and technology of isolating metals from their ores. Strictly speaking, the extraction of each metal is unique. Operational steps involved in the isolation procedures are different for different metals. However, the following four broad steps may be generalized as being applicable to most metallurgies:

1. Concentration of ore.
2. Conversion of ore into oxide.
3. Reduction of treated ore to crude metal.
4. Purification of metal.

We now dwell upon the details of each step.

Concentration of Ore

An ore is invariably accompanied by earthy impurities such as clay, stones, sand, mica and feldspar. These impurities are called *gangue*. Concentration of ore, also called *ore dressing* or *ore pre-treatment* or *beneficiation*, involves removal of gangue. Some of the important methods for concentrating ores are discussed below.

1. Hydraulic washing

Hydraulic washing is based on the principle that in a finely divided state, the density of ore is greater than that of gangue. This type of ore pre-treatment is carried out in a conical reservoir called *hydraulic classifier* (Fig. 7.1). The pulverized ore is added from the top of the reservoir, while a stream of water, under high pressure, is introduced from the bottom. The gangue, being lighter, is carried by the hydraulic stream to the upper end of the classifier and the ore, being heavier, collects at the lower end.

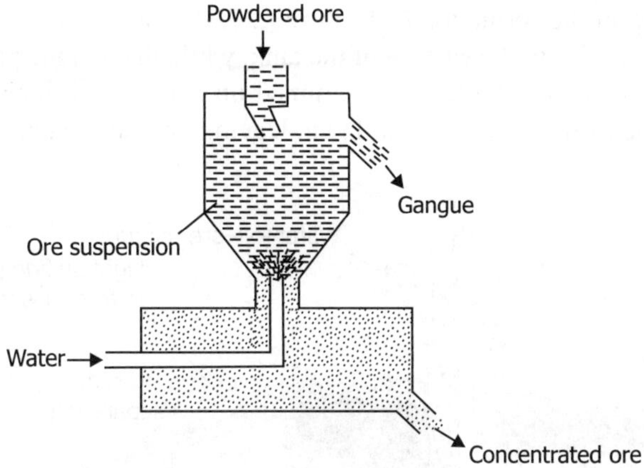

Fig. 7.1 A hydraulic classifier

2. Magnetic separation

Magnetic ores, such as magnetite and wolframite, may be separated from non-magnetic gangue by a magnetic separator. The separator consists of a moving belt, made of leather or brass, which passes over two wheels. One of the wheels is magnetized, while the other in not (Fig. 7.2). The powdered ore is added on the belt. As it passes over the magnetized wheel, the gangue falls down, but the magnetic ore sticks to the belt. When the belt loses contact with the magnetized wheel, the ore too falls down. Thus, the ore collects in one heap and gangue in another.

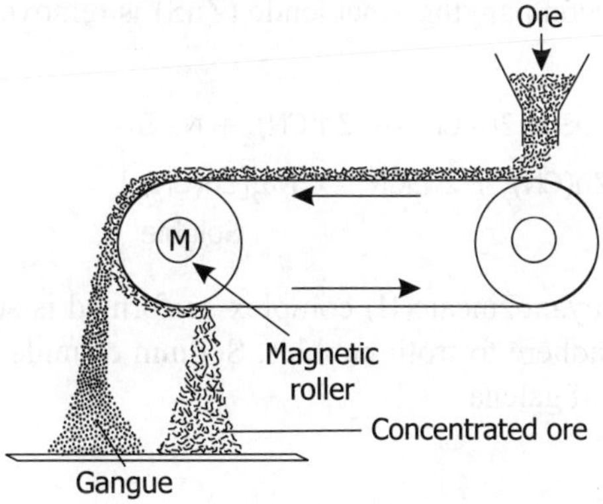

Fig. 7.2 A magnetic separator

3. Froth flotation

The technique of froth flotation is used largely for the concentration of sulfide ores. It is based on the principle that the ore is preferentially wetted by oil, while the gangue is preferentially wetted by water. Finely divided ore is introduced in a tank of water containing about 1% pine or eucalyptus oil. Air is blown to agitate water, resulting in the formation of froth (Fig. 7.3). The ore particles stick to the froth bubbles and rise to the surface of the tank, while the gangue particles remain at the floor of the tank. The foam is skimmed-up from the surface and subjected to pressure. The froth bubbles break down, leaving behind the concentrated ore.

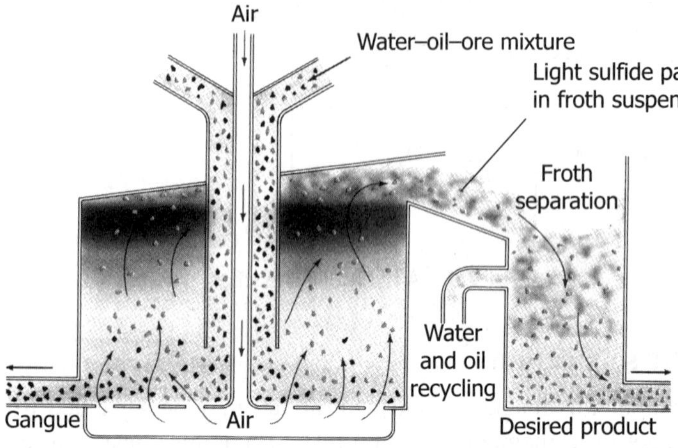

Fig. 7.3 Froth flotation process

Some additives enhance the efficiency of froth flotation process. For example, *froth stabilizers*, such as cresols, increase the lifetime of foam. *Depressants* are added when two sulfide ores occur together in nature and one is to be concentrated at the cost of other. For example, during froth flotation treatment of galena (PbS), the accompanying zincblende (ZnS) is removed by adding sodium cyanide.

$$ZnS + 2NaCN \rightarrow Zn(CN)_2 + Na_2S$$

$$Zn(CN)_2 + 2NaCN \rightarrow \underset{\text{Soluble}}{Na_2[Zn(CN)_4]}$$

The sodium tetracyanozincate(II) complex so formed is soluble in water and, therefore, does not adhere to froth bubbles. Sodium cyanide acts as a depressant during ore dressing of galena.

4. Leaching

In leaching, the ore is converted into a soluble compound, which is then separated from insoluble impurities by filtration. In this case the impurities include not only gangue but also chemical entities which are not solubilized during the leaching operation. For example, bauxite ($Al_2O_3.2H_2O$) is always accompanied by iron(III) oxide (Fe_2O_3) impurity. The pulverized ore is treated with a concentrated solution of sodium hydroxide. While bauxite is converted into soluble sodium tetrahydroxoaluminate(III) complex, $Na[Al(OH)_4]$, iron(III) oxide remains unaffected.

$$Al_2O_3.2H_2O + 2NaOH + H_2O \rightarrow \underset{\text{Soluble}}{2Na[Al(OH)_4]}$$

On filtration, the iron(III) oxide impurity as well as gangue remain in the residue, but sodium tetrahydroxoaluminate(III) complex passes into the filtrate. The complex is decomposed into aluminium(III) hydroxide by passing carbon dioxide gas through it.

$$2Na[Al(OH)_4] + CO_2 \rightarrow 2Al(OH)_3 + Na_2CO_3 + H_2O$$

On ignition, aluminium(III) hydroxide yields alumina.

$$2Al(OH)_3 \xrightarrow{\text{Heat}} Al_2O_3 + 3H_2O$$

The alumina so obtained is devoid of iron(III) oxide impurity, and hence can be processed for extraction of aluminium. The removal of unwanted iron(III) oxide from bauxite ore by leaching technique is called *Bayer's process.*

Conversion of Ore into Oxide

After concentration, the ore is converted into oxide, so that it may be easily reduced to metal in the subsequent steps. There are two methods by which the ore is converted into oxide: calcination and roasting. The ores which are already in their anhydrous oxide form, for example, rutile, are not subjected to calcination/roasting. After pre-treatment, these are directly reduced.

1. Calcination

The process in which an ore is heated in the absence of air is called calcination. For example, limestone, on calcination, yields calcium oxide.

$$CaCO_3 \xrightarrow{\text{Heat}} CaO + CO_2$$

Similarly, on calcination, dolomite gives a mixture of magnesium and calcium oxides.

$$MgCO_3.CaCO_3 \xrightarrow{\text{Heat}} MgO + CaO + 2CO_2$$

Calcination also removes any organic matter that adheres to the ore after concentration. It also expels moisture or water of crystallization. For example,

$$Fe_2O_3.3H_2O \rightarrow Fe_2O_3 + 3H_2O$$

2. Roasting

The process in which an ore is heated in the presence of air is called roasting. During this operation, the temperature is kept below the melting point of the ore. The sulfide ores are the ones that are generally subjected to roasting.

$$2ZnS + 3O_2 \rightarrow 2ZnO + 2SO_2$$
$$2PbS + 3O_2 \rightarrow 2PbO + 2SO_2$$

The impurities which form volatile oxides are also driven off during roasting. For example,

$$C + O_2 \rightarrow CO_2\uparrow$$
$$S + O_2 \rightarrow SO_2\uparrow$$
$$P_4 + 5O_2 \rightarrow P_4O_{10}\uparrow$$
$$As_4 + 5O_2 \rightarrow As_4O_{10}\uparrow$$

Reduction of Treated Ore to Crude Metal

The roasted or calcined ore is reduced to crude metal by a number of methods. Important ones are highlighted below.

1. Smelting

The reduction of oxide ore using carbon (in the form of coke) or carbon monoxide is called smelting. This is the most common form of reduction employed in metallurgical operations.

$$Fe_2O_3 + 3C \rightarrow 2Fe + 3CO$$
$$ZnO + CO \rightarrow Zn + CO_2$$

Case Study 7.2 Is it possible to reduce a sulfide ore directly with carbon?

Normally, a sulfide ore is first roasted in air to yield the corresponding oxide, which is then reduced with carbon. For example,

$$2ZnS + 3O_2 \rightarrow 2ZnO + 2SO_2$$
$$2ZnO(s) + C(s) \rightarrow 2Zn(s) + CO_2(g) \quad (7.1)$$

The sulfide may also be directly reduced with carbon.

$$2ZnS(s) + C(s) \rightarrow 2Zn(s) + CS_2(l) \quad (7.2)$$

Since carbon disulfide is a liquid and carbon dioxide is a gas, reaction (7.1) has a greater positive value of entropy as compared to reaction (7.2). Therefore, reaction (7.1) is thermodynamically more efficient than reaction (7.2). Hence, it is always preferred to first convert the sulfide ore into oxide and then reduce it, rather than carrying out its direct reduction.

Magnesium oxide, obtained from the calcination of magnesite, can also be reduced by carbon in an electric arc furnace. The reduction reaction, however, is reversible in nature.

$$MgO + C \rightleftharpoons Mg + CO$$

It proceeds from left to right above 1500 K, and from right to left below this temperature. If the reaction mixture is allowed to cool slowly, the magnesium vapors and carbon monoxide react to re-form magnesium oxide. For this reason, the contents are cooled extremely rapidly by injecting a stream of hydrogen. This technique is called *shock cooling.*

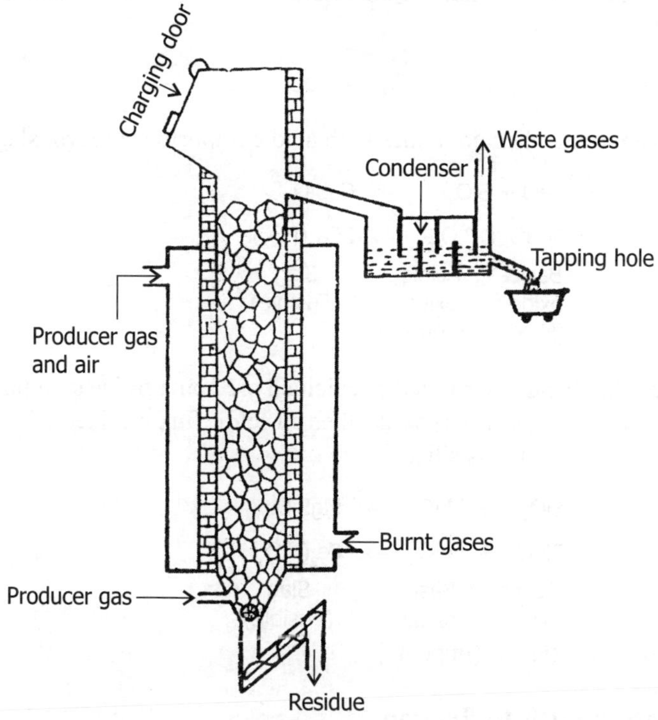

Fig. 7.4 A vertical retort furnance for smelting

Reduction of magnesium oxide is more easily carried out by using calcium carbide, rather than carbon. The reaction,

$$MgO + CaC_2 \rightarrow Mg + CaO + 2C$$

takes place at a much lower temperature and is not reversible.

Almost a similar problem confronts the smelting of zincblende, ZnS. The ore is roasted in a blast furnace to yield zinc oxide which may then be reduced by coke. However, the reduction takes place near the boiling point of zinc, as a result of which a part of the metal is lost due to vaporization. The problem is overcome by spraying metallic lead from the top of the furnace. Zinc vapors cool down to liquid state which, in turn, gets dissolved in lead. Zinc is recovered from the solid solution by vacuum distillation.

Smelting is carried out in a vertical retort, lined with fire-clay bricks (Fig. 7.4). The charge consists of the roasted or calcined ore, carbon and a flux. A *flux* is a compound that combines with impurities to form a fusible component, which is immiscible with molten metal. The fusible component is called *slag*. It is lighter than the molten metal and, therefore, floats on its surface, from where it may be withdrawn.

If the roasted or calcined ore contains acidic impurities, such as silica or phosphorus(V) oxide, then a basic flux is mixed with the charge. An example of basic flux is calcium carbonate. In the retort, it decomposes into calcium oxide, a basic oxide.

$$CaCO_3 \rightarrow CaO + CO_2$$

The calcium oxide then combines with acidic impurities to form slag.

$$CaO + SiO_2 \rightarrow CaSiO_3$$

$$6CaO + P_4O_{10} \rightarrow 2Ca_3(PO_4)_2$$

Basic oxide (Flux)	Acidic oxide (Impurity)	Slag (Fusible)

If, on the other hand, the roasted or calcined ore contains basic impurities, such as magnesium oxide or iron(III) oxide, then an acidic flux is mixed with the charge. An example of acidic flux is silica.

$$SiO_2 + MgO \rightarrow MgSiO_3$$

$$3SiO_2 + Fe_2O_3 \rightarrow Fe_2(SiO_3)_3$$

Acidic oxide (Flux)	Basic oxide (Impurity)	Slag (Fusible)

2. Reduction with hydrogen

Hydrogen is not a very effective reducing agent for obtaining metals from their oxides. Nevertheless, zinc and copper oxides may be reduced to their respective metals by hydrogen.

$$ZnO + H_2 \rightarrow Zn + H_2O$$

$$CuO + H_2 \rightarrow Cu + H_2O$$

The oxides of aluminium, magnesium and calcium, however, cannot be reduced by hydrogen. Magnetite may be reduced by this method but an equilibrium composition is readily established.

$$Fe_3O_4 + 4H_2 \rightleftharpoons 3Fe + 4H_2O$$

3. Air reduction process

Roasting of cinnabar yields metallic mercury, rather than mercury(II) oxide.

$$HgS + O_2 \rightarrow Hg + SO_2$$

This method is called air reduction process and is carried out in a special type of heating system called *Idrian furnace* (Fig. 7.5). The furnace is equipped with condensing chambers, where the vapors of mercury are transformed into liquid state. Air reduction process actually occurs in two steps. First, the formation of oxide,

$$2HgS + 3O_2 \rightarrow 2HgO + 2SO_2$$

and second, its decomposition into the metal.

$$2HgO \rightarrow 2Hg + O_2$$

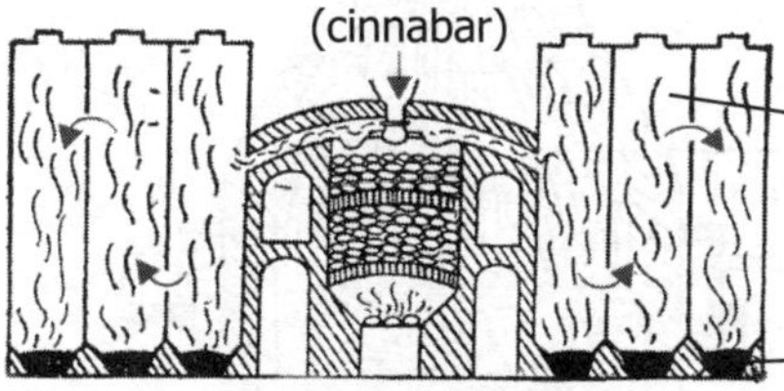

Fig. 7.5 Idrian furnace for extraction of mercury

4. Alumino–thermite process

Also called the *Goldschmidt process*, the alumino–thermite method uses aluminium to reduce those metal oxides which have substantial covalent character. These oxides are not easily reduced by carbon or carbon monoxide. Examples include chromium(III) and manganese(IV) oxides. These are mixed with aluminium powder and ignited in a closed crucible (Fig. 7.6). The ignition is triggered off by burning a combination of magnesium ribbon and barium peroxide. The heat, so liberated, initiates the reduction reaction.

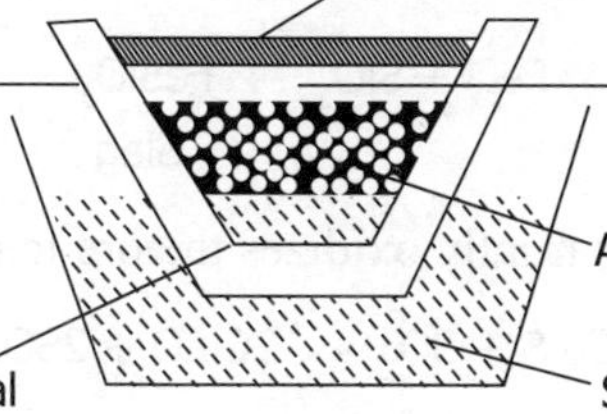

Fig. 7.6 Alumino–thermite reduction of chromium(III) oxide

$$Cr_2O_3 + 2Al \rightarrow Al_2O_3 + 2Cr$$

$$3Mn_3O_4 + 8Al \rightarrow 4Al_2O_3 + 9Mn$$

The reduction reaction itself is exothermic in nature. The heat generated during the conversion of metal oxide to metal is enough to sustain the reaction, even if the initial heat source is later cut off.

5. Bessemerization

The term bessemerization refers to self-reduction of partially roasted ore. It is carried out in a pear-shaped furnace, called *Bessemer converter* (Fig. 7.7). The converter is lined inside with a coating of magnesium oxide and is provided with tuyeres to introduce hot air and sand (flux).

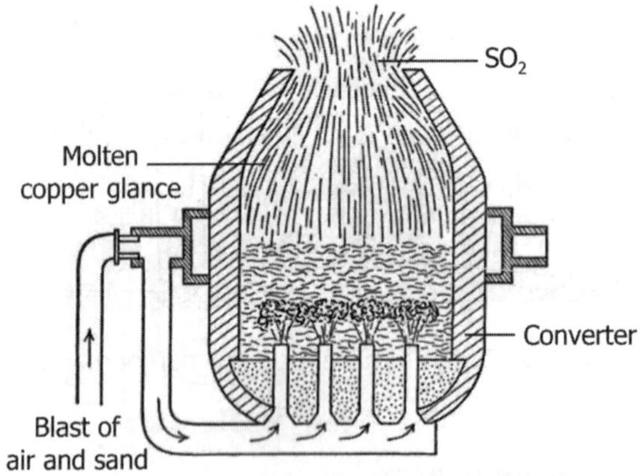

Fig. 7.7 A Bessemer converter

Copper is obtained by bessemerization. The common ore of copper, copper glance (Cu_2S), is introduced into the Bessemer converter in molten form. It invariably contains iron(II) sulfide as impurity. As hot air is blown in the furnace, the iron(II) sulfide is converted into iron(II) oxide.

$$2FeS + 3O_2 \rightarrow 2FeO + 2SO_2$$

The iron(II) oxide then combines with sand to form fusible slag, which is skimmed off from the surface.

$$FeO + SiO_2 \rightarrow \underset{\text{Slag}}{FeSiO_3}$$

The hot blast of air also partially oxidizes the ore to copper(I) oxide.

$$2Cu_2S + 3O_2 \rightarrow 2Cu_2O + 2SO_2$$

The copper(I) oxide then reacts with the unconverted copper(I) sulfide. This self-reduction reaction produces metallic copper.

$$Cu_2S + 2Cu_2O \rightarrow 6Cu + SO_2$$

Since the reaction is exothermic, copper remains in molten state. As sulfur dioxide gas escapes out of melted copper, blisters are formed on the surface of metal. Thus, copper obtained after bessemerization is often referred to as *blister copper.*

Lead may also be obtained by self-reduction method. The ore, galena (PbS), is incompletely roasted to lead(II) oxide.

$$2PbS + 3O_2 \rightarrow 2PbO + 2SO_2$$

The lead(II) oxide, so formed, and the unreacted lead(II) sulfide are then bessemerized.

$$PbS + 2PbO \rightarrow 3Pb + SO_2$$

6. Reduction by electrolysis

Oxides of alkali and alkaline earth metals may be reduced by carbon at extremely high temperatures. At such temperatures, the metals invariably combine with carbon to yield carbides. Hence, carbon is not a good reducing agent for the oxides of Group 1 and 2 metals. Moreover, the metals react with water and hence the aqueous solution of their salts cannot be reduced by displacement reactions. Further, the metals are more electropositive than aluminium and hence their oxides cannot be reduced by alumino–thermite process.

This leaves out electrolysis as the only suitable reduction method. The ore is converted into a metal halide, which is then fused and subjected to electrolysis. Sometimes, the ore itself is in halide form. For example, sodium exists in nature as rock salt—chemically, sodium chloride. In such cases, the ore is concentrated, fused and then electrolyzed. The metal gets deposited at the cathode.

The electrolysis of fused rock salt is carried out in an electrolytic equipment called *Down's cell* (Fig. 7.8). The electrolyte is a fused mixture of sodium chloride and calcium chloride. The latter is added to lower the melting point of the former from about 1100 K to 900 K. Lowering the melting temperature serves two purposes. First, it is more economical to run the plant at a lower temperature. Second, and more importantly, at a higher temperature, some of the sodium will be lost due to volatilization. If electrolysis is conducted at a lower temperature, almost the entire amount of sodium is recovered.

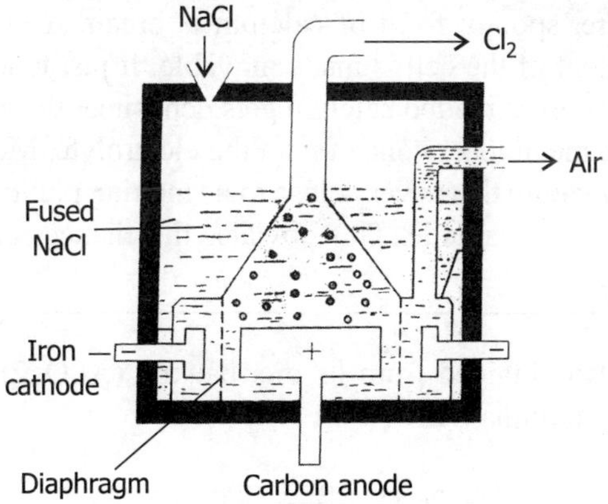

Fig. 7.8 Down's cell

During electrolysis, chlorine is released at the anode and sodium is deposited at the cathode.

At anode: $Cl^- \rightarrow Cl + e^-$

$$Cl + Cl \rightarrow Cl_2$$

At cathode: $Na^+ + e^- \rightarrow Na$

A diaphragm is used to partition the cathode from the anode. Otherwise, the sodium and chlorine are likely to re-combine to form sodium chloride. The discharge potential of calcium and sodium are quite close to each other and, therefore, some amount of calcium is also deposited at the cathode. However, the two metals are immiscible with each other. Sodium, being lighter, floats over calcium and is siphoned off.

Potassium cannot be manufactured by electrolytic reduction of fused potassium chloride. Potassium has a lower boiling point than sodium, and potassium chloride has a higher melting point than sodium chloride. Therefore, the temperature at which the electrolytic reduction of potassium chloride takes place is high enough to vaporize the potassium liberated at the cathode. As a result, potassium is lost due to volatilization. For this reason, potassium is obtained by electrolysis of fused potassium hydroxide. The melting temperature of potassium hydroxide is much lower than that of potassium chloride. Thus, electrolysis is carried out at a relatively low temperature and the entire amount of potassium, deposited at the cathode, is recovered.

Case Study 7.3 In the metallurgy of calcium, why is the electrolytic reduction cell endowed with an adjustable cathode?

Calcium is produced by the electrolysis of fused calcium chloride, mixed with calcium fluoride. Best results are obtained if the temperature of the electrolyte is nearly the same as the melting point of calcium and if the melted metal is not brought in contact with the electrolyte. When the fused mixture is heated above the melting point of calcium, considerable amount of metal diffuses into the electrolyte. If it is heated below its melting point and the metal mixes with the melt, an impure, spongy form of calcium is obtained. For this reason, the cathodic compartment of the cell is made movable. It just touches the surface of the electrolyte. As soon as melted calcium gets deposited, the level of the cathode is raised so that the metal loses contact with the electrolyte. Moreover, calcium is deposited at a temperature that is very close to its melting point; it solidifies almost as soon as it reaches the cathode. This obviates the chances of its dripping back into the electrolyte.

Aluminium is also obtained from its ore, bauxite ($Al_2O_3.2H_2O$). The latter is initially calcined to alumina (Al_2O_3).

$$Al_2O_3.2H_2O \xrightarrow{\text{Heat}} Al_2O_3 + 2H_2O$$

However, alumina has an extremely high lattice energy and hence cannot be easily melted. Therefore, it is dissolved in molten cryolite (Na_3AlF_6) and then electrolyzed (Fig. 7.9).

At anode: $O^{2-} \rightarrow O + 2e^-$

$O + O \rightarrow O_2$

At cathode: $Al^{3+} + 3e^- \rightarrow Al$

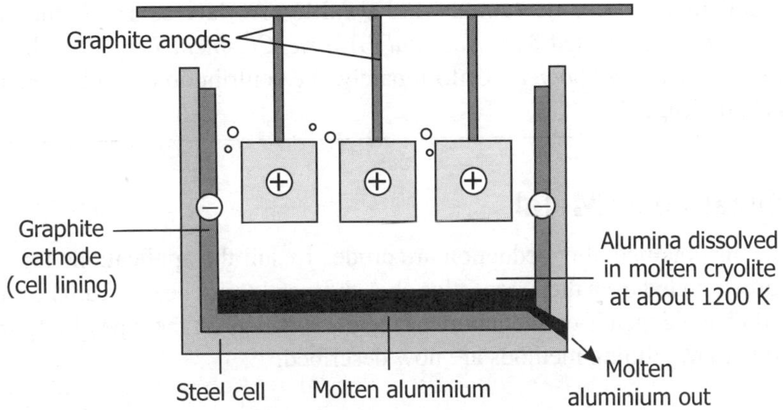

Fig. 7.9 Electrolytic reduction of alumina

Aluminium, being heavy, sinks to the bottom of the electrolytic cell. This method of recovering aluminium is called *Hall process.*

Box 7.1

The Resplendent Aluminium

Before the advent of Hall process, aluminium used to be extracted from clay. This procedure was so costly that the price of aluminium was comparable to that of gold. The metal, no doubt, was admired because of its lightness, strength and durability, yet its utility was restricted to the elite. In fact, Napoleon III of France was so fascinated by aluminium that he had a set of tableware made of it to be used in place of gold utensils for his most important banquets. Thanks to the Hall process, named after Charles Martin Hall, aluminium transitioned from royal palaces to household of commoners.

One day in 1885, Charles Martin Hall was attending his chemistry class at Oberlin College in USA, wherein his teacher, Professor Frank Fanning Jewett, displayed a sample of aluminium and remarked, 'If anyone should invent a process by which aluminium could be made on a commercial scale, not only would he be a benefactor to the world, but would also be able to lay up for himself a great fortune'. At this, Hall turned to one of his classmates and said, 'I am going after that metal'.

Contd...

Hall started his work on extraction of aluminium from bauxite in a backyard laboratory. In February 1886, he standardized the methodology which we have cited above, and which since then is named Hall process. In this endeavor, he was greatly assisted by his elder sister, Julia Brainerd Hall. She also helped him establish the first large-scale aluminium production plant in Pittsburgh. It was earlier called *Pittsburgh Reduction Company*, but later renamed as *Aluminium Company of America (ALCOA)*.

Hall indeed became famous and wealthy. As late as 1997, the Hall process was designated as a *National Historic Chemical Landmark* by the *American Chemical Society*. Unfortunately, the contribution of Julia remains unacknowledged!

Purification of Metal

The metals obtained after reduction are crude. To suit the applications for which these are extracted from their respective ores, the metals must be refined. The method of purification depends on the nature of metal as well as on the type of impurities. The important refining methods are now described.

1. Liquation

The method of liquation is employed in those cases where the metal has a relatively low melting point and the impurities are infusible. The crude metal is placed on the sloping hearth of a furnace and heated (Fig. 7.10). The molten metal floats down the inclined hearth, while the infusible impurities are left behind. Tin is refined by liquation.

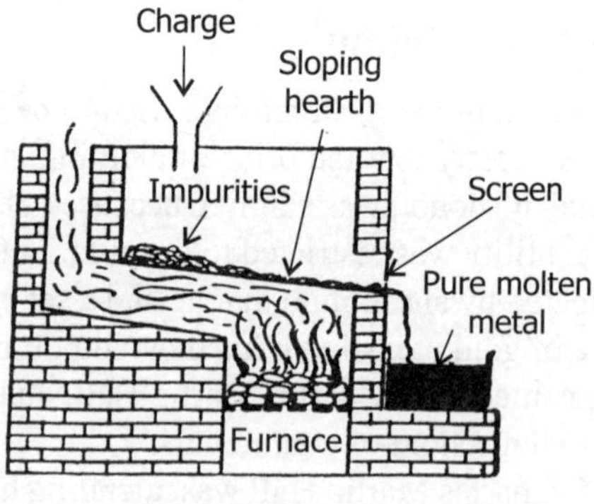

Fig. 7.10 Liquation

2. Distillation

The method of distillation is employed in those cases where the metal has a lower boiling point than the impurities. The crude metal is heated to boiling in a closed retort. The vapors of the metal are allowed to pass into a receiver, where they are condensed by cooling. The non-volatile impurities are left behind in the retort. Zinc, cadmium and mercury are refined by distillation.

3. Selective oxidation

If the impurities are more prone to oxidation as compared to the metal that is desired to be extracted, the technique of selective oxidation is applied. The impurity of lead in silver is removed by this technique. Impure silver is placed in a boat-shaped crucible called *cupel.* When a blast of hot air is passed, the lead content is easily oxidized to lead monoxide (PbO). The latter, being very light, is carried away by the air current, leaving pure silver behind. This method is also called *cupellation.*

4. Selective dissolution

In selective dissolution method, the crude metal is treated with a medium which selectively dissolves the impurities. For example, lead invariably contains a small amount of silver as impurity. While silver is soluble in molten zinc, lead is not. Therefore, a small amount of zinc (1–2%) is added to crude, molten lead. Silver dissolves in zinc and the mixture floats on fused lead. On cooling, the zinc–silver layer solidifies first, and is scrapped from the top. It is re-melted and distilled. Zinc is recovered from the distillate and silver from the residue. This technique is called *Parkes process.*

5. Hydrometallurgy

The term hydrometallurgy implies a combination of leaching and displacement reactions. It is applicable to metals such as gold and silver, which exist in native states. The crude metal is transformed into a solution, while the impurities, which remain behind in solid state, are removed by filtration. For example, when silver is treated with aqueous sodium cyanide, in the presence of air, for a prolonged period of time, it is leached into the solution as soluble sodium dicyanoargentate(I) complex.

$$4Ag + 8NaCN + 2H_2O + \underset{\text{Air}}{O_2} \rightarrow \underset{\text{Soluble}}{4Na[Ag(CN)_2]} + 4NaOH$$

Pure silver is obtained from the complex by displacement reaction with zinc.

$$2Na[Ag(CN)_2] + Zn \rightarrow Na_2[Zn(CN)_4] + 2Ag$$

6. Vapor phase refining

In vapor phase refining, the crude metal is converted into a compound that can be easily vaporized. The vapors are segregated from the non-volatile, solid impurities and then decomposed into metal. The metal so obtained is devoid of all impurities. There are two important purification methods based on vapor phase refining technique: Mond process and van Arkel process.

a. Mond process

The vapor phase refining of nickel is carried out by Mond process. On treating with carbon monoxide, nickel forms tetracarbonylnickel(0)—a compound which can be easily volatilized at 316 K. The impurities either do not react with carbon monoxide or, if they do, form non-volatile carbonyls. The vapors of tetracarbonylnickel(0) are separated from the impurities. The temperature is then raised to about 473 K. At this temperature, the vapors decompose into metallic nickel and carbon monoxide.

$$Ni(CO)_4 \xrightarrow{473\text{ K}} Ni + 4CO$$

The released carbon monoxide may be re-used to purify more nickel.

b. van Arkel process

Titanium and zirconium are refined by van Arkel process. The crude metals are heated with iodine at about 500 K and converted into volatile iodides.

$$M + 2I_2 \xrightarrow{500\text{ K}} MI_4 \quad (M = Ti \text{ or } Zr)$$

The vapors of the iodides are separated from non-volatile impurities and passed over a tungsten wire, heated to 1000 K. The iodides decompose into pure metals and iodine. The latter may be recycled.

$$MI_4 \rightarrow M + 2I_2$$

Case Study 7.4 Comment on the possibility of using Ti as a structural material.

Ti possesses high strength to weight ratio. It is resistant to corrosion and is a relatively abundant metal. All these characteristics suggest the possibility of using Ti in place of iron as structural material. Practically, however, this is not possible for two reasons. First, at high temperatures used in metallurgical processes, Ti forms brittle interstitial compounds with carbon and nitrogen. Therefore, its extraction from its ore is not as simple as that of iron. Second, and as a consequence of the first reason, it is expensive too.

---

7. Zone refining

The technique of zone refining is used when metals with purity of the order of 99.999% are desired for such applications as solar cells and semiconductor devices. It is based on the principle that impurities are more soluble in molten metal than in solid metal. Mainly silicon and germanium are refined by this method.

The metal to be purified is cast in the form of a rod. A circular heater is fitted at one end of the rod and slowly moved to the other end (Fig. 7.11). The heater melts a zone of the metal and the impurities become concentrated in this zone. As the heater traverses the rod, the melted band too moves forward. At each region, the pure metal crystallizes out, while the impurities pass on into the melt. The process is repeated several times. The end of the rod where the impurities become concentrated is scrapped.

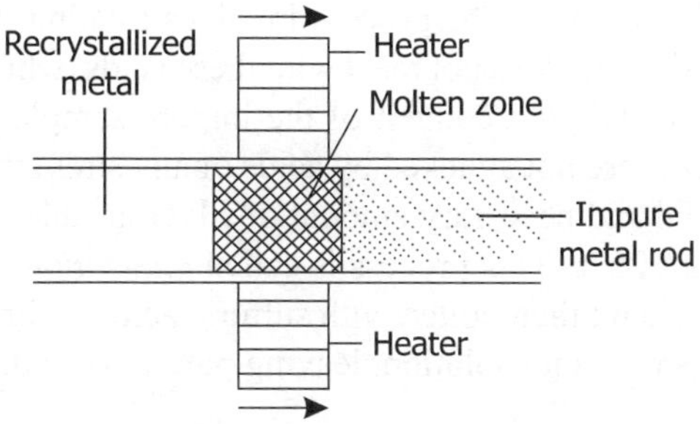

Fig. 7.11 Zone refining

8. Electrolytic refining

By far, the most important method of purifying metals is electrolytic refining. A large number of metals, including chromium, copper, nickel, tin and zinc are refined by this method. The impure metal is made the anode and a thin strip of pure metal is made the cathode. A soluble salt of the metal serves as the electrolyte. For example, in the electrolytic refining of copper, an acidified solution of copper sulfate is the electrolyte (Fig. 7.12). When current is passed, the metal content of anode is oxidized and becomes dissolved in the electrolyte. The impurities are not affected.

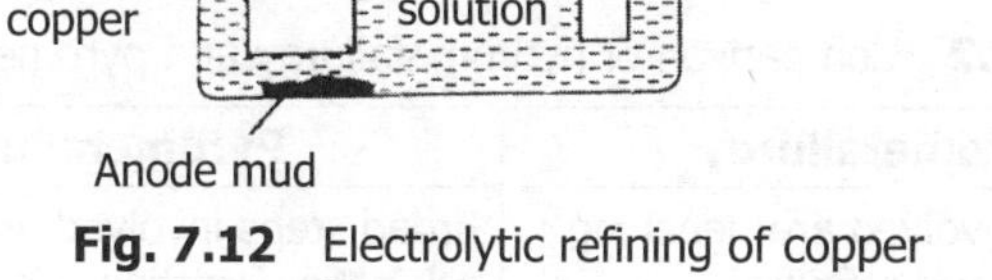

Fig. 7.12 Electrolytic refining of copper

$$\text{At anode: } Cu \rightarrow Cu^{2+} + 2e^-$$

From electrolyte, the metal ions travel to the cathode, where these are reduced to elemental state.

$$\text{At cathode: } Cu^{2+} + 2e^- \rightarrow Cu$$

The pure metal gets deposited on the cathode from where it is subsequently scrapped. When the entire amount of metal at anode is deposited at the cathode, the noble impurities (such as silver and platinum) fall down as *anode mud*, also called *anode sludge*. These costly metals are then extracted from anode mud.

9. Parting

In the extraction of gold, the copper and silver impurities, if present, are removed by a technique called parting. The process involves treating the crude gold with sulfuric or nitric acid. Gold does not react with these acids, while copper and silver dissolve. However, if the gold content of the impure sample is more than 30%, even copper and silver are not attacked by acids of any strength. In such cases, the crude metal in melted and necessary amount of silver is added so as to reduce the gold content to about 25%. This process is called *quartation*. The resulting alloy is granulated in water and then boiled with sulfuric acid or nitric acid. The copper and silver impurities pass into solution, leaving pure gold undissolved.

Case Study 7.5 Normally it is considered undesirable to refine vanadium to a high degree of purity. Why?

Vanadium, in an extremely pure state, is rare since, like titanium, it is quite reactive. It forms interstitial compounds with carbon, nitrogen and oxygen at temperatures which are normally used in the refining step of metallurgical operations. In any case, extremely pure vanadium has hardly any practical applications. Its chief commercial use is in a steel alloy called *ferrovanadium*. This type of steel is highly ductile and shock resistant. The quality of ferrovanadium is not compromised even if vanadium used in its production is not very pure.

Box 7.2

Hydrometallurgy and Pyrometallurgy: A Comparison

Hydrometallurgy is the process of extracting metals at ordinary temperatures by leaching the ores with aqueous solutions. Pyrometallurgy is the process of extracting metals by processes requiring high temperatures. Each technique has its own merits and demerits, which are compared in Table 7.2.

Table 7.2 Comparison of hydrometallurgy and pyrometallurgy

Hydrometallurgy	Pyrometallurgy
Broad steps involved are: leaching, displacement and refining	Broad steps involved are: calcination/roasting, smelting, refining
Requires low energy inputs	Requires high energy inputs
Plants may be set up in compact space	Plants require extensive space
Low-grade ores may be processed	Ore should be duly concentrated before being processed
Low level of air and water pollution	Extensive air and water pollution
Only a limited number of metals, for example, gold and silver, may be extracted	A very large number of metals, including copper and zinc may be extracted

Thermodynamics of Metallurgical Processes

The choice of a reducing agent for converting an oxide into a metal during metallurgical operations is determined by thermodynamic factors. The driving force for any chemical reaction, including the reduction processes that we will discuss now, is controlled by decrease in free energy (ΔG) associated with the transformation. Greater the negative value of change in free energy, more spontaneous is the reaction. The free energy change, in turn, depends on the change in entropy (ΔS) during the reaction.

$$\Delta G = \Delta H - T\Delta S$$

Here, ΔH is the change in enthalpy of the reaction occurring at temperature T. At a specific temperature, higher the positive value of ΔS, greater is the decrease in ΔG and more easy it is to carry out the reaction.

The suitability of a reducing agent in a particular metallurgical process is dictated by *Ellingham diagram*. The latter is a plot of the free energy change accompanying the conversion of an element into its oxide against temperature. Ellingham curves for some important elements are depicted in Fig. 7.13.

In general, the curves tend to slope upwards with increasing temperature. This means that at higher temperatures, ΔG for oxide formation becomes less negative. This is due to the formation of a metal oxide:

$$2M(s) + O_2(g) \rightarrow 2MO(s)$$

No gaseous product is formed. Rather gaseous oxygen is consumed. As a result, ΔS has a low positive or a negative value. Accordingly, ΔG tends to increase.

Exceptions occur in a few cases. For carbon, the reaction,

$$2C(s) + O_2(g) \rightarrow 2CO(g)$$

indicates that one mole of a gaseous reactant (oxygen) is consumed and two moles of a gaseous product (carbon monoxide) are released. Overall, the change in entropy is positive and, therefore, change in free energy decreases. The curve slopes downwards.

Consider the oxidation of lead to lead monoxide.

$$2Pb + O_2 \rightarrow 2PbO$$

Lead monoxide is very light and is easily carried away with the oxygen stream. In fact, it is this property which affects the removal of lead impurities from silver during cupellation. Due to the tendency of lead monoxide to become a part of the gaseous stream, the entropy of the reaction has a high positive value and the slope of the lead curve follows a downward direction.

In the Ellingham diagram (Fig. 7.13), the oxides of the top two elements, silver and mercury, acquire a positive value of change in free energy at high temperatures. A compound with a positive free energy is unstable and spontaneously decomposes into its constituent elements. For this reason, the two oxides decompose into their respective metals and oxygen above a certain temperature. Therefore, no chemical reducing agents are required for converting these oxides into metals. We have already seen this phenomenon in the air reduction process of cinnabar.

Suppose we wish to know which of the following reactions would be possible.

a. Reduction of aluminium(III) oxide with chromium.
b. Reduction of chromium(III) oxide with aluminium.

At a fixed temperature, let us say 1773 K, the change in free energy (as obtained from the Ellingham diagram) for the formation of aluminium(III) oxide and chromium(III) oxide is indicated as follows:

$$2Al + \frac{3}{2}O_2 \rightarrow Al_2O_3; \quad \Delta G = -900 \text{ kJ mol}^{-1} \tag{7.3}$$

$$2Cr + \frac{3}{2}O_2 \rightarrow Cr_2O_3; \quad \Delta G = -500 \text{ kJ mol}^{-1} \tag{7.4}$$

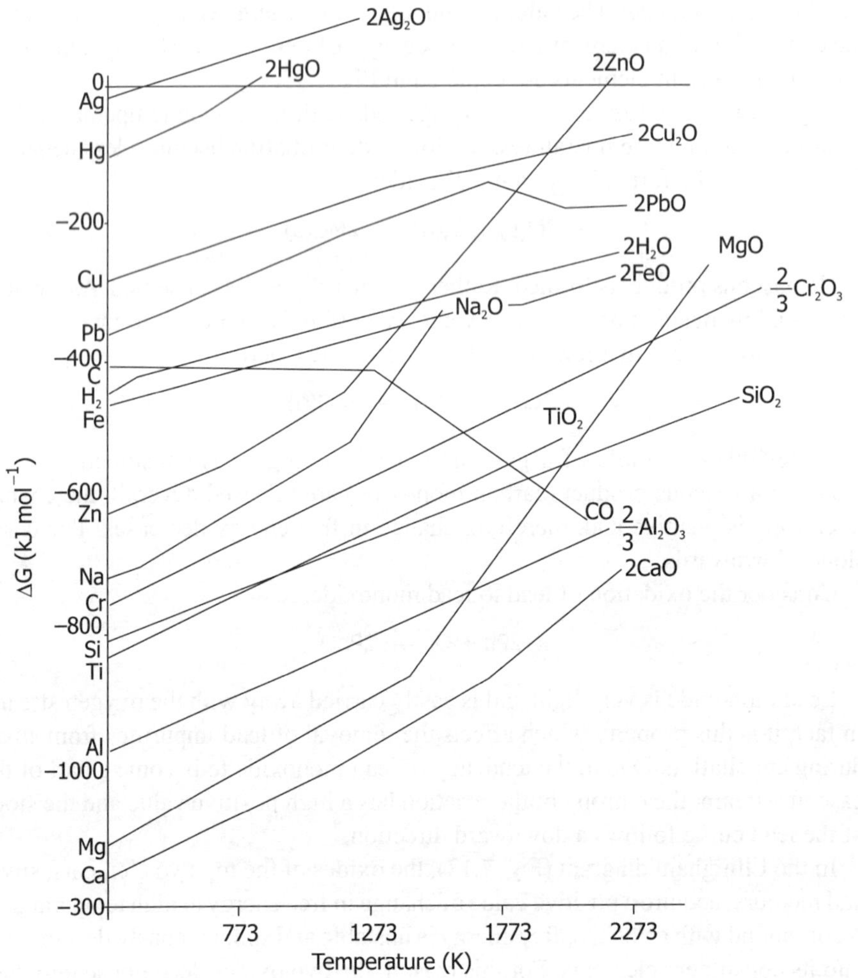

Fig. 7.13 Ellingham diagram

The reaction for the reduction of aluminium(III) oxide with chromium may be formulated by subtracting equation (7.3) from equation (7.4).

$$Al_2O_3 + 2Cr \rightarrow Cr_2O_3 + 2Al; \quad \Delta G = 400 \text{ kJ mol}^{-1}$$

This reaction is not feasible since the change in free energy has a positive value.

The reaction for the reduction of chromium(III) oxide with aluminium may be formulated by subtracting equation (7.4) from equation (7.3).

$$Cr_2O_3 + 2Al \rightarrow Al_2O_3 + 2Cr; \quad \Delta G = -400 \text{ kJ mol}^{-1}$$

This reaction is feasible since the change in free energy has a negative value.

We conclude that while it is possible to reduce chromium(III) oxide with aluminium, it is not possible to reduce aluminium(III) oxide with chromium. Further, in the Ellingham diagram, the aluminium curve is below the chromium curve. It may, therefore, be generalized that an oxide can be reduced by an element below it in the Ellingham diagram, but not by an element above it.

Case Study 7.6 Which of the following reactions is plausible?

a. Reduction of magnesium oxide with silicon.
b. Reduction of silicon dioxide with magnesium.

Below 2000 K, the magnesium curve is below the silicon curve. However, above 2000 K, the silicon curve lies below the magnesium curve. Thus, magnesium can reduce silicon dioxide to silicon below 2000 K.

$$SiO_2 + 2Mg \rightarrow 2MgO + Si$$

Silicon can reduce magnesium oxide above 2000 K.

$$2MgO + Si \rightarrow SiO_2 + 2Mg$$

Hence, both the reduction reactions are feasible.

It is also evident from the Ellingham diagram that hydrogen can reduce zinc(II) oxide only above 1400 K, but carbon can reduce it at about 1250 K. On the other hand, copper(I) oxide may be reduced by hydrogen at 700 K. Further, carbon can reduce iron(II) oxide only above 700 K.

Case Study 7.7 Is it possible to reduce rutile, TiO_2, with carbon?

A reference to Ellingham diagram shows that the carbon curve comes below the titanium curve at about 1900 K. This means that rutile can be reduced with carbon only if the temperature is above 1900 K. At such a high temperature, titanium dioxide is, no doubt, reduced, but instead of forming titanium metal, a non-stoichiometric titanium carbide results. Titanium carbide is an interstitial compound. Carbon atoms are accommodated in the interstices of titanium lattice from where the chances of their expulsion are remote. To avoid these problems, the rutile ore is treated with carbon and chlorine at red heat.

$$TiO_2 + C + 2Cl_2 \rightarrow TiCl_4 + CO_2$$

No carbide is formed in this case, even though the temperature is quite high. Interstitial compounds are a characteristic of solid lattices. However, titanium(IV) chloride, formed in the above reaction, is a liquid. It is purified by distillation and then heated to about 1000 K with magnesium.

$$TiCl_4 + 2Mg \rightarrow Ti + 2MgCl_2$$

This method of preparing titanium is called *Kroll process.* Titanium obtained by this process may be purified by van Arkel vapor phase refining method.

8. REDOX REACTIONS

Introduction

A substantial number of inorganic reactions involve oxidation of one species and simultaneous reduction of another. Such reactions, in which oxidation and reduction go hand-in-hand, are called redox reactions. These reactions occur by transfer of electrons from one chemical entity to another. Electron gain leads to reduction, while electron loss leads to oxidation. The species which supplies electrons is called the *reducing agent* or the *reductant*, the one which accepts electrons is called *oxidizing* agent or *oxidant*. Further, oxidation corresponds to increase in oxidation state of the concerned element; the reduction involves a decrease in the oxidation state of the particular element.

Half Reactions

The reaction between copper(II) ions and zinc is one of the most cited examples of redox classification.

$$Cu^{2+} + Zn \rightarrow Zn^{2+} + Cu \qquad (8.1)$$

In this reaction, copper(II) ions are reduced to elemental copper and elemental zinc is oxidized to zinc(II) ions. Two electrons required for the reduction of copper(II) ion are provided by zinc. In order to reveal the transference of electrons from zinc to copper(II) ions, it is pertinent to split reaction (8.1) into the following two half reactions.

$$Zn \rightarrow Zn^{2+} + 2e^- \qquad (8.2)$$

$$Cu^{2+} + 2e^- \rightarrow Cu \qquad (8.3)$$

The oxidized and reduced species in a particular half reaction are referred to as a *redox couple*. A redox couple is written in the format, oxidized species/reduced species. Thus in half reaction (8.2), the redox couple is Zn^{2+}/Zn, while in the half reaction (8.3), the redox couple is Cu^{2+}/Cu.

Standard Electrode Potential

Since the half reactions (8.2) and (8.3), or for that matter all other half reactions, accomplish by transference of electrons, their tendency to proceed in a particular direction can be elucidated electrically. Moreover, like all other chemical changes, the half reactions too follow thermodynamic considerations, that is, these will occur spontaneously only if the change in free energy, ΔG has a negative value. The change in free energy is expressed as,

$$\Delta G = -nFE^0 \quad (8.4)$$

where n is the number of electrons involved in a half reaction, F is the Faraday's constant (96500 Coulomb mol^{-1}) and E^0 is the standard electrode potential. The latter is the potential difference across a redox couple, M^{n+}/M, when the concentration of the ionic species, M^{n+} is 1M and the temperature is 298 K. Its units is volts (V).

Let us consider the half reaction (8.2) in which zinc is oxidized to zinc(II) ions. The standard electrode potential for this reaction is +0.76 V. To be more specific, the standard electrode potential in this case is the oxidation potential. The half reaction (8.2) may now be elaborated to include the oxidation potential.

$$Zn \rightarrow Zn^{2+} + 2e^- \qquad E^0 = +0.76\ V \quad (8.5)$$

For the reverse reaction, that is reduction of zinc(II) ions to zinc, the standard electrode potential is referred to as reduction potential. Its numerical value is the same as that of oxidation potential, but the sign is reversed.

$$Zn^{2+} + 2e^- \rightarrow Zn \qquad E^0 = -0.76\ V \quad (8.6)$$

As has been stated, the change in free energy for a spontaneous reaction ought to be negative. When we refer to equation (8.4), it becomes evident that the value of free energy change can be negative only if the standard electrode potential has a positive value. Thus the half reaction (8.5), in which zinc is oxidized to zinc(II) ions, is spontaneous in nature; the half reaction (8.6), in which zinc(II) ions are reduced to elemental zinc, on the other hand, is non-spontaneous.

As already discussed, a half reaction may be written with respect to oxidation potential,

$$M \rightarrow M^{n+} + ne^- \qquad E^0 = \text{oxidation potential}$$

or with respect to reduction potential

$$M^{n+} + ne^- \rightarrow M \qquad E^0 = \text{reduction potential}$$

If the oxidation potential has a positive sign, the reduction potential will have a negative sign and vice versa.

By convention, the standard electrode potentials are expressed for reduction half reactions. The reduction potentials of a few elements are listed in Table 8.1. The first element, lithium has a very high negative value of reduction potential. This indicates that there is very little tendency for the reaction,

$$Li^+ + e^- \rightarrow Li \qquad E^0 = -3.04\ V$$

to proceed from left to right. Conversely, there is a very high tendency for the reaction to proceed in reverse direction,

$$Li \rightarrow Li^+ + e^- \qquad E^0 = +3.04\ V$$

This implies that lithium is a very strong reducing agent.

Table 8.1 Reduction potentials of a few elements

Half reaction	Reduction potential E^0 (V)	
$Li^+ + e^- \rightarrow Li$	–3.04	very strong reducing agents/very weak oxidizing agents
$K^+ + e^- \rightarrow K$	–2.92	
$Ba^{2+} + 2e^- \rightarrow Ba$	–2.90	
$Na^+ + e^- \rightarrow Na$	–2.71	
$Mg^{2+} + 2e^- \rightarrow Mg$	–2.38	
$Zn^{2+} + 2e^- \rightarrow Zn$	–0.76	
$Fe^{2+} + 2e^- \rightarrow Fe$	–0.44	
$Ni^{2+} + 2e^- \rightarrow Ni$	–0.25	
$Sn^{2+} + 2e^- \rightarrow Sn$	–0.14	
$Pb^{2+} + 2e^- \rightarrow Pb$	–0.13	
$2H^+ + 2e^- \rightarrow H_2$	0.00	
$Cu^{2+} + 2e^- \rightarrow Cu$	+0.34	
$Ag^+ + e^- \rightarrow Ag$	+0.80	
$Hg^{2+} + 2e^- \rightarrow Hg$	+0.85	
$Os^{2+} + 2e^- \rightarrow Os$	+0.85	
$Pd^{2+} + 2e^- \rightarrow Pd$	+0.95	
$Pt^{2+} + 2e^- \rightarrow Pt$	+1.18	
$Au^{3+} + 3e^- \rightarrow Au$	+1.50	very strong oxidizing agents/very weak reducing agents

The reduction potential for copper(II) ion has a positive value.

$$Cu^{2+} + 2e^- \rightarrow Cu \qquad E^0 = +0.34 \text{ V}$$

This means that copper(II) ion has a very high tendency to get reduced, that is, it acts as a good oxidizing agent.

We may thus generalize that—

Elements having a high negative value of reduction potentials are good reducing agents and those having a high positive value of reduction potentials are good oxidizing agents.

Case Study 8.1 On what principle do photochromatic lenses work?

Photochromatic lenses are tailor-made to respond to changing light intensity. At night or when indoors, these are colorless, when exposed to sunlight, these turn tinted (Fig. 8.1). This change in hue is brought about by a light sensitive redox reaction between silver chloride and copper(I) chloride. Both these chemicals are incorporated into the matrix of glass from which the lenses are produced.

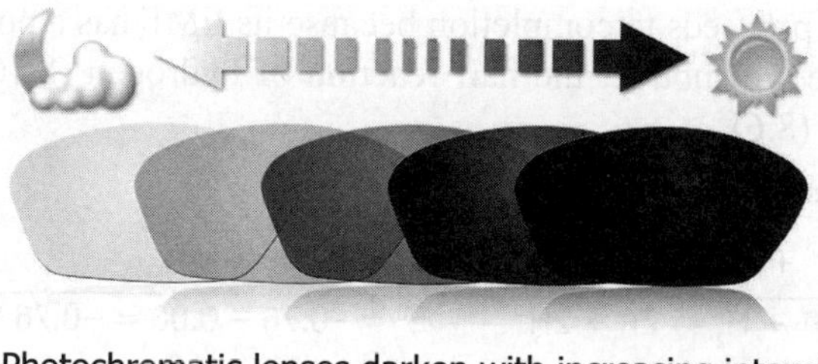

Fig. 8.1 Photochromatic lenses darken with increasing intensity of light

When sunlight falls on lenses, the chloride ions (from either or both compounds) are oxidized to chlorine atoms.

$$Cl^- \xrightarrow{h\nu} Cl + e^- \quad (8.7)$$

The electrons, so released, reduce silver(I) ions to silver.

$$Ag^+ + e^- \rightarrow Ag \quad (8.8)$$

The finely divided, black silver particles produced in reaction (8.8) serve two purposes. Firstly, these darken the lenses and secondly, they reflect most of the light and prevent it from causing irritation to the eyes.

The chlorine atoms formed in reaction (8.7) oxidize copper(I) ions and are themselves reduced to chloride ions.

$$Cl + Cu^+ \rightarrow Cu^{2+} + Cl^- \quad (8.9)$$

When the intensity of light declines, the copper(II) ions, formed in reaction (8.9), oxidize the silver atoms formed in reaction (8.8).

$$Cu^{2+} + Ag \rightarrow Ag^+ + Cu^+$$

As a result, the transparency of the lenses is restored.

Electromotive Force

The standard electrode potential—whether reduction or oxidation—for a half reaction cannot be determined directly. It has to be elucidated relative to a reference half reaction. The reference half reaction is chosen to be that of hydrogen for which, as has been indicated in Table 8.1, the reduction potential is taken to be zero.

$$2H^+ + 2e^- \rightarrow H_2 \qquad E^0 = 0.00 \text{ V} \quad (8.10)$$

All other electrode potentials are scaled relative to hydrogen electrode.

If we subtract the half reaction of zinc (8.6) from the half reaction of hydrogen (8.10), we get,

$$\begin{array}{ll} 2H^+ + 2e^- \rightarrow H_2 & E^0 = 0.00 \text{ V} \\ Zn^{2+} + 2e^- \rightarrow Zn & E^0 = -0.76 \text{ V} \\ \hline Zn + 2H^+ \rightarrow Zn^{2+} + H_2 & E^0 = 0.00 - (-0.76) = +0.76 \text{ V} \end{array} \quad (8.11)$$

The net potential of reaction (8.11) is +0.76 V. The resultant electrode potential of the overall reaction is called electromotive force and is abbreviated as EMF. It is defined as the difference in electric potential between two different electrodes. Reaction (8.11) proceeds to completion because its EMF has a positive value.

Now suppose we subtract the half reaction of hydrogen (8.10) from the half reaction of zinc (8.6).

$$\begin{array}{ll} Zn^{2+} + 2e^- \rightarrow Zn & E^0 = -0.76 \text{ V} \\ 2H^+ + 2e^- \rightarrow H_2 & E^0 = 0.00 \text{ V} \\ \hline Zn^{2+} + H_2 \rightarrow Zn + 2H^+ & E^0 = -0.76 - 0.00 = -0.76 \text{ V} \end{array} \quad (8.12)$$

The EMF of reaction (8.12) has a negative value. It is, therefore, a non-spontaneous reaction and implies that zinc(II) ions cannot be reduced by hydrogen.

Case Study 8.2 Given the following data on reduction potentials, predict whether zinc will reduce copper(II) ions or vice versa.

$$Zn^{2+} + 2e^- \rightarrow Zn \qquad E^0 = -0.76\ V$$

$$Cu^{2+} + 2e^- \rightarrow Cu \qquad E^0 = +0.34\ V$$

If we subtract Zn^{2+}/Zn couple from Cu^{2+}/Cu couple, we get,

$$Cu^{2+} + Zn \rightarrow Zn^{2+} + Cu \qquad E^0 = +0.34 - (-0.76) = +1.10\ V$$

This reaction is feasible as its EMF has a positive value.
The reverse reaction,

$$Zn^{2+} + Cu \rightarrow Cu^{2+} + Zn \qquad E^0 = -0.76 - (+0.34) = -1.10\ V$$

will not proceed as its EMF has a negative value.

Redox Reactions and Color

The blue hue of sapphires is attributed to a redox reaction between two natural contaminants in the gemstone. The base material of sapphire is *corundum,* Al_2O_3 which contains titanium(IV) and iron(II) ions as impurities. Within the lattice of *corundum*, the following redox reaction takes place.

$$Ti(IV) + Fe(II) \rightarrow Ti(III) + Fe(III)$$

The energy required to carry out this reaction is provided by visible radiation. We shall learn in Chapter 21, that when absorption of visible light leads to transition of electrons (from one energy level to another, or from one species to another), the chemical entity in question appears to be colored. The redox reaction between titanium(IV) and iron(II) ions, cited above, occurs when orange and/or yellow components of visible light are absorbed by the crystals of sapphire. As a fallout, blue light is reflected (*see* Table 26.2, p. 728). Hence the blue shade of sapphire.

Box 8.1

Deterioration of the Statue of Liberty

The Statue of Liberty is a colossal sculpture which was dedicated on October 28, 1886 at the Liberty Island off the New York harbor. Before it celebrated its hundredth anniversary it had to be closed down for renovation (Fig. 8.2). Its deterioration was attributed mainly to a redox reaction.

Contd...

Fig. 8.2 The statue of liberty was closed to the public from 1984 to 1986 for renovation

The copper sheathings of the statue were mounted on iron support bars. As time progressed, copper slowly, and in parts, got oxidized to copper(II) ions. Since iron is more active than copper, the following oxidation-reduction reaction ensued.

$$2Fe + 3Cu^{2+} \rightarrow 2Fe^{3+} + 3Cu$$

As a result, the iron armatures got corroded and the copper skin turned patchy. It was not that the designers of the statue were unaware that such a reaction could disfigure the sculpture. To prevent this possibility, they segregated the copper and iron components by an asbestos lining. Unfortunately, the asbestos insulation could not withstand the corrosive sea winds and, over a period of time, broke down at several places. This brought the two metals in contact with each other, resulting in the defacement of the statue.

Case Study 8.3 What is *galvanic shock*?

Galvanic shock refers to dental pain experienced by a person whose one of the teeth has been treated with a filling and a covering has been placed either on the adjacent tooth or on its counterpart in the opposite jaw. It is also called *galvanic pain*.

Galvanic shock is attributed to a redox reaction that takes place between the chemical entities of dental cap and dental filling. The cap is usually made of gold, while the filling is an amalgam of mercury, copper and tin. Gold is the material of choice for the tooth covering because it is inert in nature (*see* Box 21.3, p. 629). However, if a person's diet is rich in acidic foods like fruit juices and pickles, gold slowly converts into gold(III) ion. Since tin in the dental filling is more active than gold, the following oxidation-reduction reaction takes place when the two teeth come in contact with each other.

$$3Sn + 2Au^{3+} \rightarrow 2Au + 3Sn^{2+}$$

The dental filling slowly dissolves, leaving a metallic taste in the patient's mouth. Simultaneously, the electrolytes present in the saliva create a galvanic cell between the two teeth. A small amount of current is generated which stimulates the nerves in the dental pulp, causing pain. However, the pain is short-lived and normally does not require a medical treatment.

Redox Reactions and Removal of Tarnish

Silver items often get tarnished due to the formation of a black, silver sulfide layer on their surface. One method of bringing back the sheen is to treat the article with a mild acid. However, in this process some of the silver is lost. A better option is to reduce silver sulfide back to silver, thereby avoiding the loss. The reduction is carried out by aluminium, using household items.

An aluminium foil is spread out on the inner surface of a pan. The silver article is placed inside the pan and enough water is added to cover it. A few grams of baking soda are added. Due to its ionic nature, baking soda assists in transference of electrons from one surface to another. The pot is placed on a hot plate and water is boiled. The following redox reaction occurs.

$$3Ag_2S + 2Al \rightarrow 6Ag + Al_2S_3$$

The silver sulfide coating gets removed without the loss of silver.

Latimer Diagrams

A Latimer diagram is an arrangement of different oxidation states of a particular element. The highest oxidation state is at the extreme left, followed by successively decreasing oxidation states, the lowest one occupying the extreme right end. Two adjacent species are separated by an arrow, over which the numerical value of electromotive force (EMF, in volts), involved in that particular transformation is indicated.

Latimer diagrams are more important for transition elements since each member of this category has more than one oxidation state (*see* Chapter 21, p. 605). Further, for each element, the type and number of species encountered in a Latimer diagram depends on whether the reaction sequence is being investigated in acidic medium or in alkaline medium. The EMF values too are different for acidic and alkaline media.

The Latimer diagrams for chromium, manganese and iron in acidic solution are shown in Figs. 8.3, 8.4 and 8.5, respectively. The corresponding diagrams for the same elements in alkaline solution are depicted in Figs. 8.6, 8.7 and 8.8, respectively.

$$Cr_2O_7^{2-} \xrightarrow{+1.38\ V} Cr^{3+} \xrightarrow{-0.42\ V} Cr^{2+} \xrightarrow{-0.90\ V} Cr^0$$

Fig. 8.3 Latimer diagram of chromium in acidic medium

$$MnO_4^- \xrightarrow{+0.56\ V} MnO_4^{2-} \xrightarrow{+2.26\ V} MnO_2 \xrightarrow{+0.95\ V} Mn^{3+} \xrightarrow{+1.51\ V} Mn^{2+} \xrightarrow{-1.18\ V} Mn^0$$

Fig. 8.4 Latimer diagram of manganese in acidic medium

$$FeO_4^{2-} \xrightarrow{+2.20\ V} Fe^{3+} \xrightarrow{+0.77\ V} Fe^{2+} \xrightarrow{-0.44\ V} Fe^0$$

Fig. 8.5 Latimer diagram of iron in acidic medium

$$CrO_4^{2-} \xrightarrow{-0.11\ V} Cr(OH)_3 \xrightarrow{-1.10\ V} Cr(OH)_2 \xrightarrow{-1.40\ V} Cr^0$$

Fig. 8.6 Latimer diagram of chromium in alkaline medium

$$MnO_4^- \xrightarrow{+0.56\ V} MnO_4^{2-} \xrightarrow{+0.27\ V} MnO_4^{3-} \xrightarrow{+0.96\ V} MnO_2 \xrightarrow{+0.15\ V}$$

$$Mn_2O_3 \xrightarrow{-0.25\ V} Mn(OH)_2 \xrightarrow{-1.56\ V} Mn^0$$

Fig. 8.7 Latimer diagram of manganese in alkaline medium

$$FeO_4^{2-} \xrightarrow{+0.72\ V} Fe(OH)_3 \xrightarrow{-0.56\ V} Fe(OH)_2 \xrightarrow{-0.88\ V} Fe^0$$

Fig. 8.8 Latimer diagram of iron in alkaline medium

A pair of species on the Latimer diagram is called a *couple*. For example, in Fig. 8.3, the dichromate ion-chromium(III) ion, $Cr_2O_7^{2-} - Cr^{3+}$ form a couple. Likewise, in Fig. 8.8, the tetraoxoferrate(VI) ion – iron(III) hydroxide, $FeO_4^{2-} - Fe(OH)_3$ constitute a couple. A couple need not be composed of adjacent species. Thus the permanganate ion-manganese(IV) oxide, $MnO_4^- - MnO_2$ in Fig. 8.4 may be designated as a couple.

Latimer diagrams incorporate a wealth of information in a compact and concise form. The following are their important applications.

1. Composing the half reactions for conversion of one species into another

The arrows between any two species in the Latimer diagram indicate the direction of a half reaction. For example, the extract, $MnO_4^- \xrightarrow{+0.56\ V} MnO_4^{2-}$, in Fig. 8.4 indicates, in short form, that for the reduction of permanganate ion to manganate ion, the EMF is +0.56 V. The actual half reaction may be elaborated as,

$$MnO_4^- + e^- \rightarrow MnO_4^{2-} \qquad E^0 = +0.56\ V$$

Half reactions may be written for non-adjacent couples as well, taking care that in such formulations, both the number of atoms and the charges ought to be balanced. For example, the half reaction for the conversion of permanganate ions to manganese(II) ions in acidic solution (Fig. 8.4) may be formulated by including H^+/H_2O species.

$$MnO_4^- + 8H^+ + 5e^- \rightarrow Mn^{2+} + 4H_2O$$

Likewise, the conversion of dichromate ions to chromium(III) ions in acidic medium (Fig. 8.3) may be written as,

$$Cr_2O_7^{2-} + 14H^+ + 6e^- \rightarrow 2Cr^{3+} + 7H_2O$$

With reference to Fig. 8.5, the half reaction for the transformation of tetraoxoferrate(VI) ions to iron(III) ions, in acidic medium, may be compiled as,

$$FeO_4^{2-} + 8H^+ + 3e^- \rightarrow Fe^{3+} + 4H_2O \qquad (8.13)$$

In alkaline medium, the half reaction is compiled by including the H_2O/OH^- redox couple. For example, in the Latimer diagram depicted in Fig. 8.8, the half reaction for the conversion of tetraoxoferrate(VI) ion to iron(III) hydroxide may be shown as,

$$FeO_4^{2-} + 4H_2O + 3e^- \rightarrow Fe(OH)_3 + 5OH^- \qquad (8.14)$$

It is also pertinent that the number of electrons involved in a half reaction must match with the change in oxidation number of the metal ion. For example, in reactions (8.13) and (8.14), iron is being reduced from (VI) to (III) oxidation state. Hence 3 electrons are required per mole of tetraoxoferrate(VI) ion to accomplish these half reactions.

Case Study 8.4 Refer to Fig. 8.7 and compose the half reaction for converting permanganate ions to manganese(IV) oxide.

Fig. 8.7 is the Latimer diagram of manganese in alkaline medium. Therefore, for composing a half reaction, one has to involve the H_2O/OH^- redox couple in the concerned transformation. As is the practice, the number of atoms and the charges must also be balanced.

The half reaction in question may be written as,

$$MnO_4^- + 2H_2O + 3e^- \rightarrow MnO_2 + 4OH^-$$

---

2. Stability of oxidation states

A Latimer diagram helps in determining the relative stabilities of different oxidation states of a particular element. There are three sources of instability of a species in aqueous, acidic or basic solution.

- The species may reduce water and itself get oxidized.
- The species may oxidize water and itself get reduced.
- The species may disproportionate. This means that it acts as a good oxidizing agent as well as a good reducing agent. Two ions combine with each other, one gets oxidized, the other gets reduced.

a. Reducing species in acidic medium The metal or the metal ion may reduce hydronium ion and get oxidized. The half reaction for reduction of hydronium ion may be written as,

$$2H_3O^+ + 2e^- \rightarrow 2H_2O + H_2 \qquad E^0 = 0.00\ V \qquad (8.15)$$

(This is a modified form of reaction (8.10)).

Suppose a neutral metal, M in acidic medium is oxidized to M(I) ion.

$$M \xrightarrow{Acid} M^+ + e^- \qquad (8.16)$$

In a Latimer diagram, the ions are listed from higher to lower oxidation states (for example, M(II), M(I), M). If the couple $M^+ \rightarrow M$ has an EMF of +x V, then reaction (8.16) may be re-written as,

$$M \xrightarrow{Acid} M^+ + e^- \qquad E^0 = -x\ V \qquad (8.17)$$

Multiplying reaction (8.17) by 2 and combining it with reaction (8.15), we get,

$$2H_3O^+ + 2e^- \rightarrow 2H_2O + H_2 \qquad E^0 = 0.00\ V$$

$$M \rightarrow M^+ + e^-\ ] \times 2 \qquad E^0 = -x\ V$$

$$2M + 2H_3O^+ \rightarrow 2M^+ + 2H_2O + H_2 \qquad E^0 = 0.00 + (-x) = -x\ V$$

(E^0 is the standard electrode potential; its value is constant and not dependent on stoichiometry).

This reaction is non-spontaneous and hence will not occur.

Now suppose that the value of EMF for the $M^+ \rightarrow M$ couple is –x V. Reaction (8.17) may be written as,

$$M \xrightarrow{Acid} M^+ + e^- \qquad E^0 = +x\ V \qquad (8.18)$$

Multiplying reaction (8.18) by 2 and combining it with reaction (8.15), we get,

$$2H_3O^+ + 2e^- \rightarrow 2H_2O + H_2 \qquad E^0 = 0.00\ V$$

$$M \rightarrow M^+ + e^-\ ] \times 2 \qquad E^0 = +x\ V$$

$$2M + 2H_3O^+ \rightarrow 2M^+ + 2H_2O + H_2 \qquad E^0 = 0.00 + x = +x\ V$$

This reaction will occur spontaneously as its EMF has a positive value.

Hence it may be generalized that—

All negative EMFs result in reduction of hydronium ions by the species right of that EMF value.

In the Latimer diagram of manganese (Fig. 8.4), it is evident that Mn^0 is the reducing species as it is placed to the right of a negative EMF. This may be confirmed by the following calculations:

$$Mn^{2+} + 2e^- \rightarrow Mn \qquad E^0 = -1.18\ V$$

or

$$Mn \rightarrow Mn^{2+} + 2e^- \qquad E^0 = +1.18\ V \qquad (8.19)$$

Combining reactions (8.15) and (8.19), we get,

$$\begin{array}{ll} 2H_3O^+ + 2e^- \rightarrow 2H_2O + H_2 & E^0 = 0.00 \text{ V} \\ Mn \rightarrow Mn^{2+} + 2e^- & E^0 = +1.18 \text{ V} \\ \hline Mn + 2H_3O^+ \rightarrow Mn^{2+} + 2H_2O + H_2 & E^0 = 0.00 + 1.18 = +1.18 \text{ V} \end{array}$$

Elemental manganese is readily oxidized to manganese(II) ions, which means that it is a reducing species.

b. Oxidizing species in acidic medium A metal ion, M(I) may oxidize water to oxygen and itself get reduced to M. The half reaction for oxidation of water under acidic conditions is written as,

$$2H_2O \rightarrow 4H^+ + O_2 + 4e^- \quad E^0 = -1.23 \text{ V} \tag{8.20}$$

The reduction of metal ion may be represented as,

$$M^+ + e^- \xrightarrow{\text{Acid}} M \quad E^0 = +x \text{ V} \tag{8.21}$$

Multiplying reaction (8.21) by 4 and combining it with reaction (8.20), we get,

$$\begin{array}{ll} 2H_2O \rightarrow 4H^+ + O_2 + 4e^- & E^0 = -1.23 \text{ V} \\ M^+ + e^- \rightarrow M \,] \times 4 & E^0 = +x \text{ V} \\ \hline 4M^+ + 2H_2O \rightarrow 4M + 4H^+ + O_2 & E^0 = (x - 1.23) \text{ V} \end{array}$$

This reaction will occur spontaneously only if the value of x is greater than 1.23 V, so that the EMF may have a positive value.

Hence it may be generalized that—

All EMF values more positive than +1.23 V result in the oxidation of water by the species to the left of that EMF value.

In the Latimer diagram of manganese, depicted in Fig. 8.4, we note that for the couple, $Mn^{3+} \rightarrow Mn^{2+}$, the EMF has a value greater than +1.23 V. Hence manganese(III) ion ought to act as an oxidizing species. This may be confirmed by the following calculations:

$$Mn^{3+} + e^- \rightarrow Mn^{2+} \quad E^0 = 1.51 \text{ V} \tag{8.22}$$

Multiplying equation (8.22) by 4 and combining it with equation (8.20), we get

$$\begin{array}{ll} 2H_2O \rightarrow 4H^+ + O_2 + 4e^- & E^0 = -1.23 \text{ V} \\ Mn^{3+} + e^- \rightarrow Mn^{2+} \,] \times 4 & E^0 = +1.51 \text{ V} \\ \hline 4Mn^{3+} + 2H_2O \rightarrow 4Mn^{2+} + 4H^+ + O_2 & E^0 = (-1.23 + 1.51) \text{ V} = +0.28 \text{ V} \end{array}$$

It is evident that manganese(III) ion has the potential to oxidize acidic water to oxygen and in this process it itself gets reduced to manganese(II) ion.

Likewise, the EMF for the conversion of manganate ion, MnO_4^{2-} to manganese (IV) oxide, MnO_2 has a higher value than +1.23 V. Hence manganate ions too act as oxidizing species.

Case Study 8.5 In acidic solutions, is there a tendency of iron(II) ions to get reduced to elemental iron?

With reference to Fig. 8.5, the half reaction for reduction of iron(II) to elemental iron may be shown as,

$$Fe^{2+} + 2e^- \rightarrow Fe \qquad E^0 = -0.44\ V$$

Reduction of iron(II) ion must be accompanied by simultaneous oxidation of H^+/H_2O, couple for which the half reaction is presented in equation (8.20). When we combine the two reactions, we get,

$$2H_2O \rightarrow 4H^+ + O_2 + 4e^- \qquad E^0 = -1.23\ V$$

$$Fe^{2+} + 2e^- \rightarrow Fe\ ] \times 2 \qquad E^0 = -0.44\ V$$

$$2Fe^{2+} + 2H_2O \rightarrow 2Fe + 4H^+ + O_2 \qquad E^0 = -1.67\ V$$

Since the EMF for this reaction has a negative value, it will not occur.

Reduction of a metal ion to elemental metal in acidic solution will take place only if, firstly, the EMF of the half reaction has a positive value of electrode potential and secondly, if the numerical value of electrode potential is greater than 1.23 V. None of these two conditions are met in the conversion of iron(II) ion to iron(0). There is, therefore, no possibility for this transformation to take place.

c. Reducing species in alkaline medium The half reaction for the reduction of water, in presence of an alkali, is written as,

$$2H_2O + 2e^- \rightarrow 2OH^- + H_2 \qquad E^0 = -0.83\ V \qquad (8.23)$$

Suppose in a Latimer diagram, the couple $M^+ \rightarrow M$ has a EMF of $-x$ V. The half reaction would be written as,

$$M^+ + e^- \xrightarrow{\text{Alkali}} M \qquad E^0 = -x\ V \qquad (8.24)$$

Although, as the metal (or one of its ions) is acting as a reducing species, reaction (8.24) may be written in reverse manner.

$$M \xrightarrow{\text{Alkali}} M^+ + e^- \qquad E^0 = +x\ V \qquad (8.25)$$

Multiplying reaction (8.24) by 2 and subtracting it from equation (8.23), we get,

$$2H_2O + 2e^- \rightarrow 2OH^- + H_2 \qquad E^0 = -0.83\ V$$

$$M^+ + e^- \rightarrow M] \times 2 \qquad E^0 = -x\ V$$

$$2M + 2H_2O \rightarrow 2M^+ + 2OH^- + H_2 \qquad E^0 = (-0.83 - (-x))\ V$$

Such a reaction will be spontaneous only if the numerical value of x is greater than 0.83.

Hence it may be generalized that—

All EMF values greater than 0.83 will result in reduction of water in alkaline medium by the species to the right of that value.

For example, in the Latimer diagram of iron in alkaline medium (Fig. 8.8), elemental iron (Fe^0) is the reducing species. This may be confirmed by the following calculations.

$$Fe + 2OH^- \rightarrow Fe(OH)_2 + 2e^- \qquad E^0 = +0.88\ V \tag{8.26}$$

Combining reaction (8.23) with reaction (8.26), we get,

$$\begin{array}{ll} 2H_2O + 2e^- \rightarrow 2OH^- + H_2 & E^0 = -0.83\ V \\ Fe + 2OH^- \rightarrow Fe(OH)_2 + 2e^- & E^0 = +0.88\ V \\ \hline Fe + 2H_2O \rightarrow Fe(OH)_2 + H_2 & E^0 = 0.05\ V \end{array}$$

Although, the EMF value is very small, yet it is positive and proves that elemental iron may act as a mild reducing agent.

Likewise, in the Latimer diagram of manganese in alkaline medium (Fig. 8.7), elemental manganese (Mn^0) is a reducing species.

d. Oxidising species in alkaline medium A metal ion may oxidize hydroxide ions to oxygen and may itself get reduced. The relevant equations for this reaction are,

$$\begin{array}{ll} 4OH^- \rightarrow 2H_2O + O_2 + 4e^- & E^0 = -0.40\ V \\ M^+ + e^- \rightarrow M\ ] \times 4 & E^0 = +x\ V \\ \hline 4M^+ + 4OH^- \rightarrow 4M + 2H_2O + O_2 & E^0 = (x - 0.40)\ V \end{array} \tag{8.27}$$

This reaction will occur spontaneously only if the value of x is greater than + 0.40 V.

Hence it may be generalized that—

All EMFs more positive than 0.40 V result in the oxidation of hydroxide ions by the species to the left of that value.

If we refer to the Latimer diagram of manganese in alkaline solution (Fig. 8.7), we note that permanganate ion acts as an oxidizing species. This may be substantiated by the following calculations:

$$\begin{array}{ll} 4OH^- \rightarrow 2H_2O + O_2 + 4e^- & E^0 = -0.40\ V \\ MnO_4^- + e^- \rightarrow MnO_4^{2-}\ ] \times 4 & E^0 = +0.56\ V \\ \hline 4MnO_4^- + 4OH^- \rightarrow 4MnO_4^{2-} + 2H_2O + O_2 & E^0 = +0.16\ V \end{array}$$

Since the EMF of the above reaction has a positive value, the reaction proceeds spontaneously. Likewise, the tetraoxomanganese(V) ion, MnO_4^{3-} too is oxidizing in nature.

Case Study 8.6 Is tetraoxoferrate(VI) ion, FeO_4^{2-} more stable in acidic medium or in alkaline medium?

The Latimer diagrams of iron in acidic and alkaline media are shown in Figs. 8.5 and 8.8, respectively. In both the cases, the tetraoxoferrate(VI) ion occupies the extreme left position, which means that it cannot be further oxidized (in any case, iron does not possess an oxidation state higher than (VI)). The stability of tetraoxoferrate(VI) ion therefore depends on its tendency to get reduced. Higher this tendency, greater is its instability.

The half reaction for conversion of tetraoxoferrate(VI) ions to iron(III) ions in acidic medium has been formulated in equation (8.13), while the half reaction for the oxidation of water in acidic medium is written as equation (8.20). On combining these two half reactions, we get,

$$2H_2O \rightarrow 4H^+ + O_2 + 4e^- \,] \times 3 \qquad E^0 = -1.23 \text{ V}$$

$$FeO_4^{2-} + 8H^+ + 3e^- \rightarrow Fe^{3+} + 4H_2O \,] \times 4 \qquad E^0 = +2.20 \text{ V}$$

$$4FeO_4^{2-} + 20H^+ \rightarrow 4Fe^{3+} + 10H_2O + 3O_2 \qquad E^0 = +0.97 \text{ V} \qquad (8.28)$$

The positive value of the resultant EMF implies that the free energy change for this reaction has a negative value which, in turn, means that the reaction is spontaneous. Thus tetraoxoferrate(VI) ion will not be stable in acidic medium.

The half reaction for the transformation of tetraoxoferrate(VI) ion to iron(III) hydroxide in alkaline medium has been compiled in equation (8.14), while the half reaction for the oxidation of water in basic medium is written in equation (8.27). Multiplying equation (8.14) by 4 and half equation (8.27) by 3 and combining the two reactions, we get,

$$4OH^- \rightarrow 2H_2O + O_2 + 4e^- \,] \times 3 \qquad E^0 = -0.40 \text{ V}$$

$$FeO_4^{2-} + 4H_2O + 3e^- \rightarrow Fe(OH)_3 + 5OH^- \,] \times 4 \qquad E^0 = +0.72 \text{ V}$$

$$4FeO_4^{2-} + 10H_2O \rightarrow 4Fe(OH)_3 + 8OH^- + 3O_2 \qquad E^0 = +0.32 \text{ V} \qquad (8.29)$$

No doubt, reaction (8.29) too has a positive EMF, yet the value is much smaller than that for the corresponding reaction in acidic medium. Reaction (8.29) is therefore less spontaneous than reaction (8.28), meaning that tetraoxoferrate(VI) ion is relatively stable in alkaline medium. This has also been corroborated in the discussions on chemistry of iron in oxidation state (VI) (*see* Chapter 27, p. 773).

e. Disproportionation reactions It is possible that a particular metal ion may disproportionate in aqueous solution. Such an ion is both a good reducing agent and oxidizing agent. For such reactions it may be generalized that—

If the gradual reduction of EMF values from left to right is disrupted, the species to the left of that EMF value is prone to disproportionation.

For example, in Fig. 8.4, we note that the EMF value for the transformation of permanganate ion (MnO_4^-) to manganate ion (MnO_4^{2-}) is +0.56 V. Ideally the next EMF, for the conversion of manganate ion to manganese(IV) oxide (MnO_2), should have been less than +0.56 V, which it is not. This shows that the manganate ion would undergo disproportionation and this may be substantiated as follows:

The half reaction for reduction of manganate ion is written as,

$$MnO_4^{2-} + 4H^+ + 2e^- \rightarrow MnO_2 + 2H_2O \qquad E^0 = +2.26\ V \tag{8.30}$$

while its oxidation is depicted as,

$$MnO_4^{2-} \rightarrow MnO_4^- + e^- \qquad E^0 = -0.56\ V \tag{8.31}$$

Multiplying equation (8.31) by 2 and combining it with equation (8.30), we get,

$$\begin{array}{ll} MnO_4^{2-} + 4H^+ + 2e^- \rightarrow MnO_2 + 2H_2O & E^0 = +2.26\ V \\ MnO_4^{2-} \rightarrow MnO_4^- + e^-\ ] \times 2 & E^0 = -0.56\ V \\ \hline 3MnO_4^{2-} + 4H^+ \rightarrow 2MnO_4^- + MnO_2 + 2H_2O & E^0 = +1.70\ V \end{array} \tag{8.32}$$

The positive valve of EMF for reaction (8.32) shows that manganate ion undergoes both oxidation (to permanganate ion) and reduction (to manganese(IV) oxide).

Referring to Fig. 8.4 again, we note that a break in the decreasing values of EMF occurs at manganese(III) ion. Hence the latter too is prone to disproportionation.

Case Study 8.7 Comment on the stability of manganese(II) ion in aqueous acidic solution.

When we refer to the Latimer diagram of manganese in acidic medium (Fig. 8.4), we note that the reduction of manganese(II) ion to elemental state has a negative value of EMF and hence is non-spontaneous.

$$Mn^{2+} + 2e^- \rightarrow Mn \qquad E^0 = -1.18\ V \tag{8.33}$$

The oxidation of manganese(II) to manganese(III) too is not feasible.

$$Mn^{2+} \rightarrow Mn^{3+} + e^- \qquad E^0 = -1.51\ V \tag{8.34}$$

On multiplying equation (8.34) by 2 and adding it to equation (8.33), we get,

$$\begin{array}{ll} Mn^{2+} + 2e^- \rightarrow Mn & E^0 = -1.18\ V \\ Mn^{2+} \rightarrow Mn^{3+} + e^-\ ] \times 2 & E^0 = -1.51\ V \\ \hline 3Mn^{2+} \rightarrow 2Mn^{3+} + Mn & E^0 = -2.69\ V \end{array} \tag{8.35}$$

Since reaction (8.35) too has a negative EMF value, it means that manganese (II) ion does not undergo disproportionation

Thus manganese(II) ion undergoes neither oxidation, nor reduction, nor disproportionation spontaneously, implying that it remains stable in aqueous acidic solution.

---

f. Skip EMF The electrode potential required for the conversion of one species into a non-adjacent species on the Latimer diagram is referred to as the skip EMF. Consider a hypothetical Latimer diagram shown in Fig. 8.9.

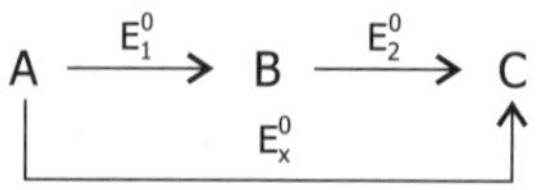

Fig. 8.9 A hypothetical Latimer diagram depicting skip EMF

The skip EMF for the non-adjacent A–C couple, marked as E_x^0, cannot be calculated from the standard potentials E_1^0 and E_2^0 as the latter are non-additive parameters. Nevertheless, the free energies are additive quantities and thus allow the elucidation of skip EMF.

Suppose the conversion of A into B involves a transference of n_1 electrons. With reference to equation (8.4), the free energy change, ΔG_1 may be written as,

$$\Delta G_1 = -n_1 F\, E_1^0 \tag{8.36}$$

Likewise, suppose the conversion of B into C involves an exchange of n_2 electrons. The free energy change, ΔG_2 would be,

$$\Delta G_2 = -n_2 F\, E_2^0 \tag{8.37}$$

Adding equations (8.36) and (8.37), we get,

$$\Delta G_1 + \Delta G_2 = \Delta G_x = -F[n_1 E_1^0 + n_2 E_2{}^0] \tag{8.38}$$

Where ΔG_x is the free energy change for converting A to C. It is related to the skip EMF, E_x by the expression,

$$\Delta G_x = -(n_1 + n_2)FE_x^0 \tag{8.39}$$

From equations (8.38) and (8.39), we get,

$$-F[n_1 E_1^0 + n_2 E_2^0] = -(n_1 + n_2)FE_x^0$$

or

$$E_x^0 = \frac{n_1 E_1^0 + n_2 E_2^0}{(n_1 + n_2)} \tag{8.40}$$

The product of number of electrons exchanged and EMF, nE^0 is called *volt-equivalents*. Equation (8.40) may be written in the following form,

$$E_x = \frac{\text{Summation of volt-equivalents}}{\text{Summation of electrons exchanged}} \tag{8.41}$$

We now modify the Latimer diagram of manganese in acidic medium, shown in Fig. 8.4. The revised diagram, as depicted in Fig. 8.10, includes the skip EMF for the non-adjacent permanganate ion-manganese(II) ion ($MnO_4^- - Mn^{2+}$) couple.

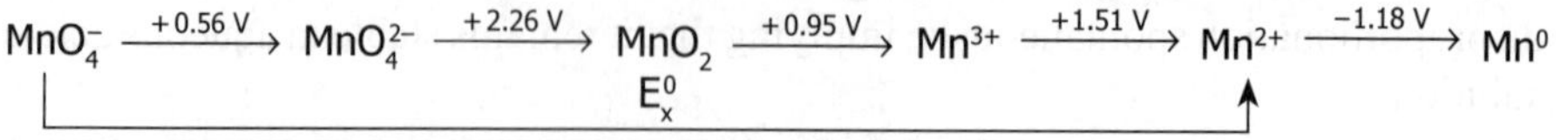

Fig. 8.10 Modified Latimer diagram of manganese in acidic medium

The conversion of permanganate ions into manganese(II) ions may be split into several steps that are listed in Table 8.2.

Table 8.2 Calculation of skip EMF for $MnO_4^- - Mn^{2+}$ couple

Step	n	E^0 (V)	nE^0 (V)
$MnO_4^- \rightarrow MnO_4^{2-}$	1	+0.56	0.56
$MnO_4^{2-} \rightarrow MnO_2$	2	+2.26	4.52
$MnO_2 \rightarrow Mn^{3+}$	1	+0.95	0.95
$Mn^{3+} \rightarrow Mn^{2+}$	1	+1.51	1.51
Summation	5	–	7.54

$$\text{skip EMF} = \frac{7.54}{5} = 1.508 \text{ V}$$

Case Study 8.8 Calculate the skip EMF for the iron(III)–iron(0) ($Fe^{3+} - Fe^0$) couple in acidic medium.

Refer to the Latimer diagram of iron in acidic medium, shown in Fig. 8.5, and segregate the conversion of iron(III) to iron(0) into two steps, listed in Table 8.3.

Table 8.3 Calculation of skip EMF for $Fe^{3+} - Fe^0$ couple

Step	n	E^0 (V)	nE^0 (V)
$Fe^{3+} \rightarrow Fe^{2+}$	1	+0.77	0.77
$Fe^{2+} \rightarrow Fe^0$	2	−0.44	−0.88
Summation	3	–	−0.11

$$\text{Skip EMF} = -\frac{0.11}{3} = -0.036 \text{ V}$$

Ebsworth-Frost Diagrams

The Ebsworth-Frost diagrams, often (and imprecisely) referred to as simply *Ebsworth diagrams* or *Frost diagrams* are representations in which volt-equivalents of M^{n+}/M^0 couples are plotted against the oxidation states of the concerned element. The volt-equivalents are taken along the vertical axis, while the oxidation states, in increasing order from left to right, are taken along the horizontal axis.

Construction of Ebsworth-Frost diagrams

By re-arranging equation (8.4), we note that the volt-equivalents are related to change in free energy by the expression,

$$nE^0 = -\frac{\Delta G}{F} \tag{8.42}$$

The vertical axis of an Ebsworth-Frost diagram, which is usually shown in terms of volt-equivalents, may also be depicted in terms of $(-\Delta G/F)$.

Fig. 8.11 shows the Ebsworth-Frost diagram of manganese. The vertical axis may be constructed by referring to Fig. 8.4, the Latimer diagram of manganese in acidic medium. We make a start with the lowest oxidation state of the element, manganese(0), for which, by convention, the free energy is taken to be zero. Hence, as shown in Fig. 8.11, the volt-equivalent, nE^0 for manganese(0) is zero. We now proceed step by step and calculate the volt-equivalents for each of the Mn^{n+}/Mn^0 couple.

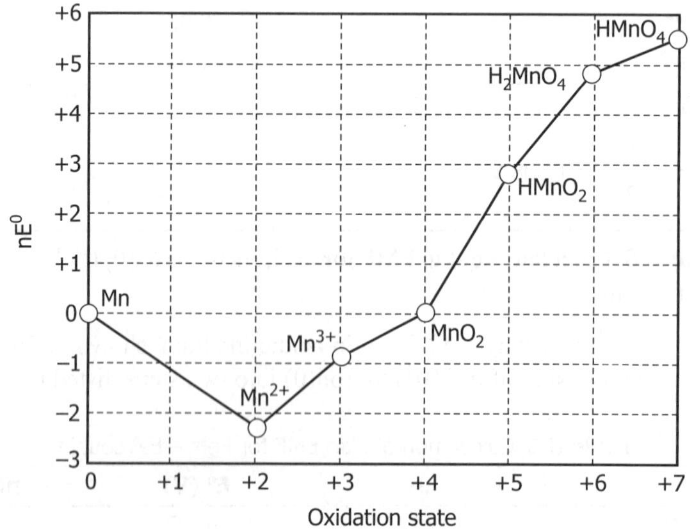

Fig. 8.11 The Ebsworth-Frost diagram of manganese

Step 1. $Mn^{2+} \rightarrow Mn^0$

$Mn^{2+} + 2e^- \rightarrow Mn^0$ $\quad E^0 = -1.18$ V; n = 2; $\quad nE^0 = -2.36$ V

Step 2. $Mn^{3+} \rightarrow Mn^0$

$Mn^{2+} + 2e^- \rightarrow Mn^0$ $\quad nE^0 = -2.36$ V

$Mn^{3+} + e^- \rightarrow Mn^{2+}$ $\quad E^0 = +1.51$ V; n = 1; $\quad nE^0 = +1.51$ V

$Mn^{3+} + 3e^- \rightarrow Mn^0$ $\quad nE^0 = -0.85$ V

Step 3. $MnO_2 \rightarrow Mn^0$

$Mn^{3+} + 3e^- \rightarrow Mn^0$ $\quad nE^0 = -0.85$ V

$MnO_2 + 4H^+ + e^- \rightarrow Mn^{3+} + 2H_2O$ $\quad E^0 = +0.95$ V; n = 1; $\quad nE^0 = +0.95$ V

$MnO_2 + 4H^+ + 4e^- \rightarrow Mn^0 + 2H_2O$ $\quad nE^0 = +0.10$ V

Step 4. $MnO_4^{2-} \rightarrow Mn^0$

$MnO_2 + 4H^+ + 4e^- \rightarrow Mn^0 + 2H_2O$ $\quad nE^0 = +0.10$ V

$MnO_4^{2-} + 4H^+ + 2e^- \rightarrow MnO_2 + 2H_2O$ $\quad E^0 = +2.26$ V; n = 2; $\quad nE^0 = +4.52$ V

$MnO_4^{2-} + 8H^+ + 6e^- \rightarrow Mn^0 + 4H_2O$ $\quad nE^0 = +4.62$ V

Step 5. $MnO_4^- \rightarrow Mn$

$MnO_4^{2-} + 8H^+ + 6e^- \rightarrow Mn^0 + 4H_2O$		$nE^0 = +4.62$ V
$MnO_4^- + e^- \rightarrow MnO_4^{2-}$	$E^0 = +0.56$ V; n = 1	$nE^0 = +0.56$ V
$MnO_4^- + 8H^+ + 7e^- \rightarrow Mn^0 + 4H_2O$		$nE^0 = +5.18$ V

When the volt-equivalents of various Mn^{n+}/M^0 couples are plotted against the oxidation state of manganese in the relevant species, the curve which emerges sheds light on the relative stability of different compounds/ions of this element.

Applications of Ebsworth-Frost diagrams

The Ebsworth-Frost diagrams offer the following applications:

1. Gauging the most stable oxidation state It is evident from equation (8.4) that the species having a greater negative value of volt-equivalents (nE^0) will possess a greater negative value of change in free energy, ΔG and hence will be more stable than its congeners on the Ebsworth-Frost diagram. Fig. 8.11 reveals that the manganese(II) ion has the greatest negative value of volt-equivalents as compared to the other manganese species. We also note that manganese(II) ion occupies the lowest position on the Ebsworth-Frost diagram.

Thus it may be generalized that—

The oxidation state occupying the lowest position on the Ebsworth-Frost diagram is the most stable one.

As discussed in Case Study 8.7, the same conclusion was reached by analyzing the Latimer diagram data of manganese in acidic medium. In Chapter 21 we will learn that the stability associated with manganese(II) ion is attributed to the half-filled, $3d^5$ configuration of the ion.

Case Study 8.9 Given the Latimer diagram of titanium in acidic medium,

$$TiO^{2+} \xrightarrow{+0.10\ V} Ti^{3+} \xrightarrow{-0.37\ V} Ti^{2+} \xrightarrow{-1.63\ V} Ti$$

construct the Ebsworth–Frost diagram.

To construct the Ebsworth–Frost diagram from the Latimer diagram, we may proceed as follows.

Step 1. $Ti^{2+} \rightarrow Ti^0$

$Ti^{2+} + 2e^- \rightarrow Ti^0$	$E^0 = -1.63$ V; n = 2;	$nE^0 = -3.26$ V

Step 2. $Ti^{3+} \rightarrow Ti^0$

$Ti^{2+} + 2e^- \rightarrow Ti^0$		$nE^0 = -3.26$ V
$Ti^{3+} + e^- \rightarrow Ti^{2+}$	$E^0 = -0.37$ V; n = 1;	$nE^0 = -0.37$ V
$Ti^{3+} + 3e^- \rightarrow Ti^0$		$nE^0 = -3.63$ V

Step 3. $TiO^{2+} \rightarrow Ti^0$

$Ti^{3+} + 3e^- \rightarrow Ti^0$ $\qquad nE^0 = -3.63\ V$

$TiO^{2+} + 2H^+ + e^- \rightarrow Ti^{3+} + H_2O \qquad E^0 = +0.10\ V;\ n = 1; \qquad nE^0 = +0.10\ V$

$TiO^{2+} + 2H^+ + 4e^- \rightarrow Ti^0 + H_2O \qquad nE^0 = -3.53\ V$

The Ebsworth–Frost diagram, shown in Fig. 8.12 has been constructed by plotting volt-equivalents against the oxidation states of titanium.

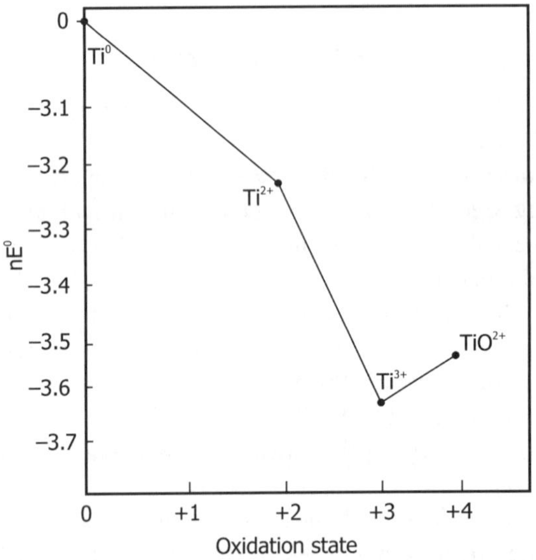

Fig. 8.12 Ebsworth – Frost diagram of titanium

As is evident from this diagram, Ti^{3+} is the most stable oxidation state of titanium.

---

2. Gauging the species prone to disproportionation In our discussion on the Latimer diagram of manganese in acidic medium (Fig. 8.4), we had concluded that manganese(III) ions (Mn^{3+}) and manganate ions (MnO_4^{2-}) are prone to disproportionation. To reach the same conclusion for manganese(III) ion through Ebsworth-Frost diagram, we write half reactions for its conversion to the species to its right (Mn^{2+}) and the species to its left (MnO_2) on the Latimer diagram (Fig. 8.4), elucidating the volt-equivalents in each case.

$$Mn^{3+} + e^- \rightarrow Mn^{2+} \qquad E^0 = +1.51\ V;\ n = 1 \qquad nE^0 = +1.51\ V \qquad (8.43)$$

$$MnO_2 + 4H^+ + e^- \rightarrow Mn^{3+} + 2H_2O \qquad E^0 = +0.95\ V \qquad (8.44)$$

Reversing equation (8.44), we get,

$$Mn^{3+} + 2H_2O \rightarrow MnO_2 + 4H^+ + e^- \qquad E^0 = -0.95\ V;\ n = 1; \qquad nE^0 = -0.95\ V \qquad (8.45)$$

Combining equation (8.43) and (8.45), we get,

$$Mn^{3+} + e^- \rightarrow Mn^{2+} \qquad nE^0 = +1.51\ V$$

$$Mn^{3+} + 2H_2O \rightarrow MnO_2 + 4H^+ + e^- \qquad nE^0 = -0.95\ V$$

$$2Mn^{3+} + 2H_2O \rightarrow MnO_2 + Mn^{2+} + 4H^+ \qquad nE^0 = +0.55\ V \qquad (8.46)$$

Substituting the value of volt-equivalents (nE^0) of reaction (8.46) in equation (8.4) and taking the value of Faraday's constant F as 96500 coulombs, we get,

$$\Delta G = -530.75\ kJ\ mol^{-1}$$

The negative value of change in free energy implies that reaction (8.46) is spontaneous in nature, indicating thereby that manganese(III) ion readily disproportionates.

We also note in Fig. 8.11, that the manganese(III) ion, as well as the manganate ion lie above the line joining the preceding and succeeding species. Thus we may generalize that—

A species on an Ebsworth-Frost diagram is unstable with respect to disproportionation if it lies above the line connecting the species before it and after it.

Case Study 8.10 What is *conproportionation*?

Conproportionation is opposite of disproportionation, that is, the combination of a low and a high oxidation state to yield an intermediate oxidation state. A good example of conproportionation is found in the chemistry of silver. The latter forms compounds in oxidation states (I), (II) and (III).

A mixed silver(I)–silver(III) oxide, having the composition $Ag^{I}[Ag^{III}]O_2$ may be prepared by the reaction of silver(I) oxide with peroxodisulfuric acid.

$$Ag_2O + H_2S_2O_8 + H_2O \rightarrow Ag^{I}[Ag^{III}]O_2 + 2H_2SO_4$$

The mixed metal oxide, when treated with perchloric acid, is converted to tetraquasilver(II) perchlorate.

$$Ag^{I}[Ag^{III}]O_2 + 4HClO_4 + 6H_2O \rightarrow 2[Ag(H_2O)_4](ClO_4)_2$$

The above reaction may be outlined as,

$$Ag(I) + Ag(III) \rightarrow 2Ag(II)$$

so as to express conproportionation in a simplified manner.

Corrosion

Corrosion refers to the deterioration of metals by a set of redox reactions that occur naturally on their surface. It is a broad term since there is a long schedule of metals that corrode under natural conditions. Yet, its most familiar example is the rusting of iron. This may be attributed to two reasons. Firstly, deterioration of iron entails heavy economic loses as this metal is used in a large chunk of societal amenities, household utilities and machinery components. Secondly, the mechanism of rusting is well established and may be extended to corrosion of other metals. Therefore, it is pertinent to discuss the concept of corrosion with reference to iron.

Mechanism of rusting

The electrochemical process that ultimately culminates in rusting initiates on a section of iron surface which serves as anode. Here iron is oxidized to iron(II) ions.

$$Fe \rightarrow Fe^{2+} + 2e^- \quad (8.47)$$

As elemental iron leaves the surface, a pit is created at the anodic region. The electrons released as a result of reaction (8.47), traverse through the metal to another region of its surface that serves as cathode. Here they reduce aerial oxygen.

$$O_2 + 4H^+ + 4e^- \rightarrow 2H_2O \quad (8.48)$$

The protons (H^+ ions) required for reaction (8.48) are produced by dissolution of atmospheric carbon dioxide in water droplets adsorbed on the items surface.

$$CO_2 + H_2O \rightarrow H_2CO_3$$

The carbonic acid, so produced, dissociates into protons.

$$H_2CO_3 \rightleftharpoons H^+ + HCO_3^-$$

Hence that section of the metallic surface acts as cathode where not only the concentration of oxygen is high, but an appreciable amount of moisture, containing dissolved carbon dioxide, is also present.

By combining equations (8.47) and (8.48), we get,

$$2Fe + O_2 + 4H^+ \rightarrow 2Fe^{2+} + 2H_2O \quad (8.49)$$

The chemical changes occurring on the surface of iron do not stop here. The iron(II) ions formed in reaction (8.49), though originally produced at anode, get solubilized in water and are dispersed throughout the surface. These are then oxidized to hydrated iron(III) oxide.

$$4Fe^{2+} + O_2 + (4 + 2x)H_2O \rightarrow 2Fe_2O_3{\cdot}xH_2O + 8H^+ \quad (8.50)$$

The overall redox reaction may be formulated by combining equations (8.49) and (8.50).

$$4Fe + 3O_2 + 2xH_2O \rightarrow 2Fe_2O_3{\cdot}xH_2O \quad (8.51)$$

It is the hydrated iron oxide—iron(III) oxide with variable number of water of crystallization—that in layman's language is referred to as *rust*. The mechanism of rusting is set out in Fig. 8.13.

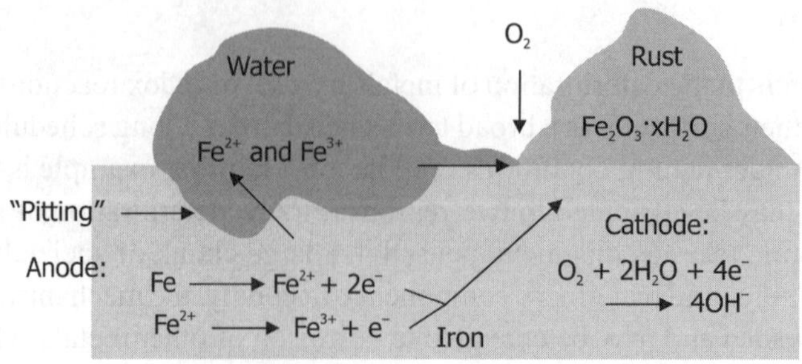

Fig. 8.13 Mechanism of rusting

Reaction (8.51) clearly reveals that rusting is induced by oxygen and moisture. Conversely, in absence of oxygen and moisture, rusting either does not occur or is delayed till the time these two chemical entities somehow seep to the iron article. As reaction (8.49) shows, corrosion proceeds rapidly under acidic conditions. Further, presence of electrolytes increase the chances of rusting as these assist in migration of ions across the surface. For this reason, automobiles rust at a fast pace in frigid areas because sodium chloride is frequently sprayed on highways to clear off snow.

Prevention of rusting

Rusting may be prevented or prolonged by the following techniques.

1. Painting The most simple and trivial method to prevent corrosion is to paint the surface of the article. Paint coating is impermeable to oxygen and moisture and therefore mitigates the reaction of metal with atmospheric constituents. However, if the paint is scratched or dented to expose even a small area of bare metal, rust begins to form under the coating.

2. Alloying Iron rusts, but its alloy, stainless steel does not. Stainless steel contains chromium, besides a host of other elements like carbon, nickel and manganese. When exposed to natural conditions, a thin layer of chromium oxide is formed on the surface of the stainless steel item. This layer protects the item from further damage.

3. Passivation To shield a metal against corrosion by an overlying, thin layer of an inert material, usually an oxide, is called passivation. It is of two types.

a. A thin layer of oxide may form spontaneously—and naturally—over the metal. For example, when aluminium is exposed to air, a coat of insoluble alumina is formed on its surface. This layer is smooth and non-porous and thus protects the metal underneath. For this reason aluminium kitchenware do not corrode; nor do aeroplanes. As has been stated earlier, even rust is a metal oxide. Yet, it fails to protect iron from further damage because, being a hydrated oxide with variable composition, it is rough and porous.

b. A microcoating of an oxide may be applied on the metallic surface. For example, iron may be passivated by treatment with nitric acid. A smooth and impermeable layer of iron oxide is formed on the surface that prevents the metal from reacting with air and moisture. Citric acid has proven to be a better passivating agent than nitric acid.

   The iron and steel parts of firearms, such as rifles and pistols, are passivated by treatment with a solution of potassium nitrate and sodium hydroxide at about 400 K. A thin, protective layer of magnetite (Fe_3O_4) is formed over these parts. This process is called *bluing*, so named since the passivated components acquire a blue-black shade.

4. Cathodic protection We have seen that rusting of an item initiates when a section of iron surface acts as an anode. It is reaction (8.47), the oxidation of elemental iron to iron(II) ions, which triggers off rusting. If the item in question is brought in contact with a more active metal, that is, a metal having a greater negative value of reduction potential than iron, then rusting can be prevented. Under these circumstances, iron acts as a cathode.

The process in which the metal that has to be protected from corrosion is made the cathode and is placed in conjunction with a more reactive metal that acts as the anode is called cathodic protection. Usually, magnesium and zinc are the metals of choice for cathodic protection. As shown in Table 8.1, both these metals have a greater negative reduction potential than iron. This means that instead of iron, these metals undergo anodic oxidation.

$$Mg \rightarrow Mg^{2+} + 2e^-$$

$$Zn \rightarrow Zn^{2+} + 2e^-$$

As a result, these are consumed in preference to iron. For this reason they are referred to as *sacrificial anode*. Thus the hulls of ships and underground water pipes are coupled with magnesium to combat corrosion of the base metal (Fig. 8.14).

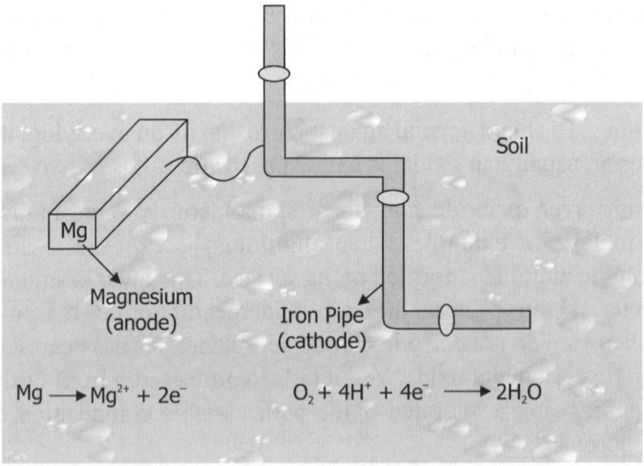

Fig. 8.14 Sacrificial magnesium anode to protect underground water pipes

Sacrificial anodes are also used to protect oil rigs, underground tanks and water heaters.

Case Study 8.11 What is *patina*?

Patina is a thin, green colored layer that forms naturally on an article made of copper or bronze. Chemically, patina is basic copper carbonate, $2CuCO_3 \cdot Cu(OH)_2$. It is formed by the reaction of copper with atmospheric carbon dioxide and water vapors. This reaction is quite slow and, depending on the size of the artifact, takes months or even years to complete. However, in coastal areas or in humid climate patina develops rapidly.

Biocorrosion

The term biocorrosion refers to degradation of metals by anaerobic bacteria. This type of corrosion is observed mainly on iron pipes buried under soil, where moisture and nutrients are, no doubt, present, but oxygen content is negligible. Colonies of anaerobic bacteria—the microorganisms which thrive in absence of oxygen—generally inhabit such environments.

It is not that anaerobic bacteria do not need oxygen for sustenance. They do, but instead of procuring oxygen from atmosphere, they extract it from sulfate, nitrate and phosphate-based minerals in the soil. These minerals then decompose into monohydrogen sulfide ions (HS^-), ammonia and phosphine, respectively. Biocorrosion mostly takes place in presence of sulfate ions.

Like conventional corrosion, biocorrosion is an electrochemical phenomena. One portion of the iron pipeline acts as an anode, while another portion, which is embedded in sulfate mineral habitat of anaerobic bacteria, acts as a cathode. The following reactions occur on the surface of the iron accessory.

$$\text{At cathode:}\quad SO_4^{2-} + 5H_2O + 8e^- \xrightarrow{\text{Anaerobic bacteria}} HS^- + 9OH^-$$

$$\text{At anode:}\quad 4Fe + HS^- + 7OH^- \longrightarrow 3Fe(OH)_2 + FeS + H_2O + 8e^-$$

$$\text{Overall reaction:}\quad 4Fe + SO_4^{2-} + 4H_2O \xrightarrow{\text{Anaerobic bacteria}} 3Fe(OH)_2 + FeS + 2OH^-$$

The soft iron(II) sulfide flakes off, exposing fresh surface of iron to biocorrosion.

9. BIOINORGANIC CHEMISTRY

Introduction

Bioinorganic chemistry, at times also referred to as *biological inorganic chemistry*, is a discipline that examines the role of inorganic elements in life processes. The term, 'inorganic elements' includes mainly the metals and metal ions. The field of bioinorganic chemistry may be divided into the following two broad categories:

1. Natural phenomena, which include—
   - Behavior of metalloproteins
   - Metal-containing electron-transfer proteins
   - Metal ion transport and storage
   - Oxygen transport and activation proteins
2. Artificial processes, which include—
   - Role of metal ions in medicine
   - Role of chemotherapeutic reagents in combating metal-induced toxicity
   - Investigation of inorganic models or mimics that imitate the behavior of metalloproteins

A mix of biochemistry and inorganic chemistry, bioinorganic chemistry sheds light on how different organisms exploit the physical and chemical characteristics of the metals and metal ions in ways that are best suited to their biological systems.

Geochemical Effects on the Distribution of Elements

The naturally occurring elements are distributed over vast expanse of solid earth. The distribution, however, is not even and depends on several geographical and geological factors. Before we analyze these factors, it is pertinent to have a look at the internal structure of earth.

Zonal structure of the earth

The internal structure of earth is heterogenous and is segregated by two boundaries: The *Mohorovičić discontinuity* and the *Wiechert-Gutenberg discontinuity*. Also called *first order discontinuities*, these divide the earth into the following three zones:

1. *Crust*, from the surface of the earth to Mohorovičić discontinuity (from zero to 17 km deep). The predominant elements in the crust are oxygen silicon, aluminium and iron.

2. *Mantle*, from the Mohorovičić discontinuity to Wiechert-Gutenberg discontinuity (from 17 to 2900 km deep). The predominant elements in mantle are oxygen, followed by silicon, magnesium and iron.
3. *Core*, from Wiechert-Gutenberg discontinuity to the center of the earth (from 2900 to 6371 km deep). The predominant elements are iron and nickel.

The zonal structure of earth is shown in Fig. 9.1. The mantle is further divided into upper mantle and lower mantle by a transition zone. The transition zone also represents a discontinuity. However, since the variation of density between upper and lower mantles is not much, therefore, such a discontinuity is termed *second order discontinuity*. Likewise, the core too is sub-divided into outer core, which is in liquid state and inner core, which is in solid state, by a second order discontinuity.

The crust and the upper mantle together constitute the lithosphere. Some physical characters of different zones of earth are presented in Table 9.1.

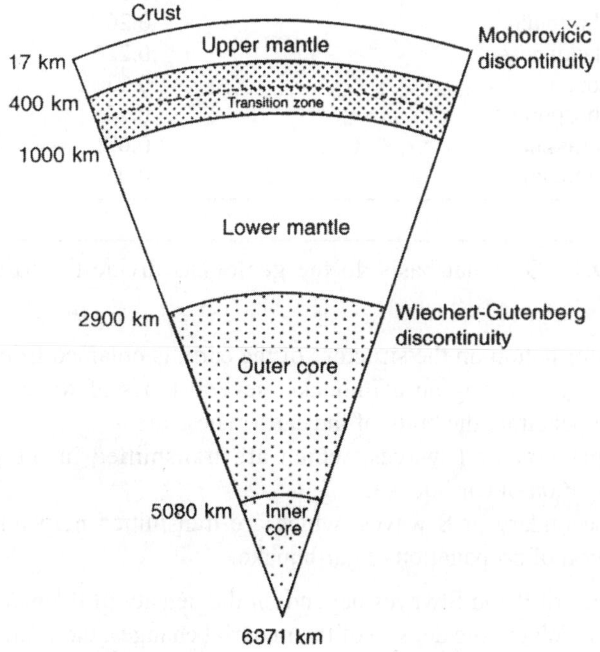

Fig. 9.1 The zonal structure of earth

Table 9.1 Physical characteristics of earth's zones

Zone	Thickness (Km)	Volume ($X10^{27}$ cm^3)	Mean density (g cm^{-3})	Mass ($X10^{24}$ Kg)	Mass (%)
Crust	17	0.008	2.8	0.024	0.4
Mantle	2900	0.899	4.5	4.075	68.1
Core	3471	0.175	10.7	1.876	31.5
Whole earth	6371	1.082	5.5	5.975	100.0

The composition of the earth is depicted in Table 9.2. The data reveals that 90% of the earth is made up of four elements: iron, oxygen, silicon and magnesium. The only other elements that are present in amounts greater than 1% are nickel, calcium, aluminium and sulfur.

Table 9.2 Composition of the earth

Element	Percent by weight
Iron	34.63
Oxygen	29.53
Silicon	15.20
Magnesium	12.70
Nickel	2.39
Sulfur	1.93
Calcium	1.13
Aluminium	1.09
Sodium	0.57
Chromium	0.26
Manganese	0.22
Cobalt	0.13
Phosphorus	0.10
Potassium	0.07
Titanium	0.05

Case Study 9.1 On what basis do the geologists divide the earth into three zones?

Most of the information on the structure of the earth is obtained from seismic (or earthquake) waves. During an earthquake, several types of waves originate, of which two can penetrate the body of the earth. These are—

1. The primary or P waves, which are transmitted in the direction of propagation of earthquake.
2. The secondary or S waves, which are transmitted perpendicular to the direction of propagation of earthquake.

The velocities of P and S waves depend on the density of the material through which they pass. Where the density of the material changes, the P and S waves are subjected to reflection and refraction. This causes a change in their velocity.

As the earthquake waves pass from crust to mantle, these encounter an increase in density from 2.8 to 4.5 g cm^{-3} (Table 9.1). The line of demarcation between crust and mantle is called Mohorovičić discontinuity. When the earthquake waves pass from mantle to core, through the Wiechert-Gutenberg discontinuity, these encounter an increase in density from 4.5 to 10.7 g cm^{-3}. The two boundaries, Mohorovičić discontinuity and Wiechert-Gutenberg discontinuity divide the earth into three Zones (Fig. 9.1).

Differentiation of elements

The terrestrial elements can be broadly classified into the following four categories:

1. *Siderophiles*– which tend to be associated with metallic iron.
2. *Chalcophiles*– which tend to be associated with sulfide minerals.
3. *Lithophiles*– which tend to be associated with oxide ions.
4. *Atmophiles*– which tend to occur as gaseous components of the atmosphere.

This four-fold geochemical classification is called differentiation of elements (Table 9.3). The relationship between geochemical classification of the elements and the periodic table is shown in Fig. 9.2.

Table 9.3 Differentiation of elements

Siderophiles	Chalcophiles	Lithophiles	Atmophiles
Fe, Co, Ni	(Cu), Ag	Li, Na, K, Rb, Cs	(H), N, (O)
Ru, Rh, Pd	Zn, Cd, Hg	Be, Mg, Ca, Sr, Ba	He, Ne, Ar, Kr, Xe
Os, Ir, Pt	Ga, In, Tl	B, Al, Sc, Y, La-Lu	
Au, Re, Mo	(Ge), (Sn), Pb	Si, Ti, Zr, Hf, Th	
Ge, Sn, W	(As), (Sb), Bi	P, V, Nb, Ta	
C, Cu, Ga	S, Se, Te	O, Cr, U	
Ge, As, Sb	(Fe), Mo, (Os)	H, F, Cl, Br, I	
	(Ru), (Rh), (Pd)	(Fe), Mn, (Zn), (Ga)	

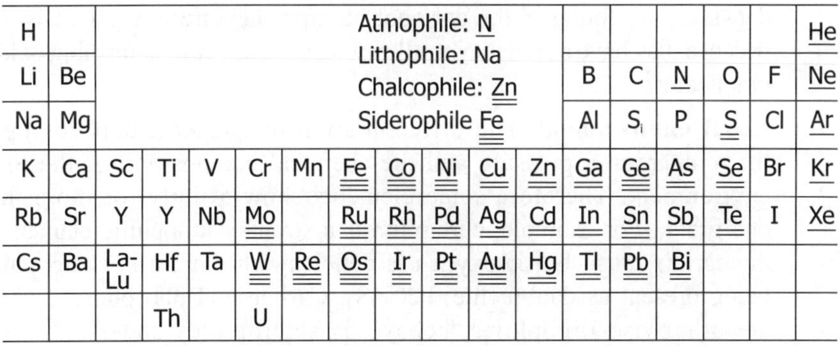

Fig. 9.2 Relationship between geochemical classification of elements and periodic table

The differentiation of elements depends on the following factors:

(a) **Chemical affinity** The siderophiles, such an nickel, gold and platinum have a tendency to exist in metallic forms. Moreover, these alloy readily with iron, hence, they are associated with iron in the earth's core. The chalcophiles, such an copper and silver are distinguished by their great affinity for sulfide ions. The lithophiles, such as alkali and alkaline earth metals, tend to form oxides or silicates. The atmophiles, which include noble gases and nitrogen, are inert gaseous substances, that do not have chemical affinity for iron, sulfide or oxide phases. Hence, they tend to concentrate in the atmosphere.

(b) Electronic configuration The lithophiles have eight electrons in the outermost shell of their ions. In case of chalcophiles, the outermost shell accommodates eighteen electrons. Siderophiles are mostly transition elements and have incomplete valence shells. The atmophiles have eight electrons in the outermost shell, but in the elemental rather than ionic state.

(c) Reduction potentials Lithophiles have highly negative reduction potentials. These elements are, therefore, easily converted into ionic states. Siderophiles have highly positive reduction potentials. These, therefore, exist in elemental states. The chalcophiles are intermediate in behavior. They do not exist in elemental states, but are converted into ions with difficulty.

(d) Density Liquid iron migrated to core of the earth because of its very high density. The less dense oxides, sulfides and silicates floated to the surface. However, density is not an important factor as far as the differentiation of elements is concerned. Although uranium and thorium have high densities, yet they are concentrated in the earth's crust. This has been attributed to their tendency to enter into combination with oxide or silicate ions. On the other hand, gold and platinum, which too have high densities, are found in the core not because they are heavy metals, but because they have a tendency to alloy with iron, rather than forming oxides or silicates.

(e) Local environment The differentiation of elements depends upon the temperature, pressure and the chemical composition of the local environment. Therefore some elements show affinity for more than one group. For example, chromium is a strongly lithophilic element in the earth's crust, but if oxygen is deficient, chromium is chalcophile, being present as daubreelite, $FeCr_2S_4$. Carbon and phosphorus, which are otherwise lithophiles, behave as siderophiles under oxidizing conditions. Similarly, iron is siderophile in the core, but lithophile in the crust.

Chemical Elements Essential to Life Processes

In the previous section, we discussed the geochemical distribution of elements. The need of such a discussion arose because crustal abundance is related to biological activity. In fact, life evolved utilizing those elements that were abundant and easily available. Subsequently, life processes became dependent on these very elements.

On the other hand, those elements that are rare and not easily accessible were not used by living systems. Neither did the life processes evolve mechanisms to deal with them. As a consequence, the elements with extremely small crustal abundance, such as lead and mercury, have proved to be toxic to living systems and hazardous to the environment. These are discussed in a later section of this chapter. It is pertinent to state that all abundant elements are not essential to living organisms. For example, titanium and zirconium, though relatively plentiful, have no biochemical role as they form extremely insoluble oxides at pH values normally encountered in living systems. Of the 92 naturally occurring elements, only about 26 are required by humans. Six of these, *viz.* oxygen, carbon, hydrogen, nitrogen, calcium and phosphorus are present in the biological systems in appreciable amounts. The other 20 elements that are essential for life are present in relatively small amounts and are often referred to as *dietary minerals* or *mineral nutrients*. A specific biological function is associated with every dietary mineral. A physiological deficiency appears when a particular element is not present in the diet. In most cases, the deficiency is relieved with the addition of that element in the dietary routine.

Chemical elements essential to life processes are classified into the following four types.

1. Bulk elements

Almost 96% mass of the human body is made up of just four elements—oxygen, carbon, hydrogen and nitrogen. If calcium and phosphorus are added to this list, then these six elements—designated as bulk elements—constitute about 98.5% of body weight.

The bulk elements are listed in Table 9.4 in order of their levels in human body, along with their main functions.

Table 9.4 Bulk elements and their functions

Element	**Functions**
Oxygen	Structural constituent of water and most biomolecules; required for respiration.
Carbon	Constituent of all biomolecules, including proteins, enzymes, nucleosides, nucleotides, and sugars.
Hydrogen	Constituent of water and biomolecules.
Nitrogen	Synthesis of proteins and nucleic acids.
Calcium	Builds bones and teeth; assists in blood clotting; needed for healthy muscles and heart.
Phosphorus	Constituent of bones and teeth; constituent of adenosine triphosphate, adenosine diphosphate and phospholipids.

2. Macro elements

The elements that are required by the humans in relatively large amounts for sustaining life processes are designated as macro elements. They are also referred to as *macro minerals*. In order of abundance in the human body, the macro elements are: potassium, sulfur, sodium, chlorine and magnesium. In biological systems, these elements are present as electrolytes. Table 9.5 lists their functions.

Table 9.5 Macro elements and their functions

Element	Functions
Potassium	Major electrolyte in blood and intracellular fluid; essential for transmission of nerve impulse and maintenance of osmotic balance.
Sulfur	Constituent of essential amino acids and vitamins; assists in fat metabolism and detoxification processes.
Sodium	Major electrolyte in blood and intracellular fluid; important in nerve functioning and maintenance of osmotic balance.
Chlorine	Major electrolyte in blood and extracellular and intracellular fluid; required for maintenance of osmotic balance; needed for production of hydrochloric acid in stomach to assist digestion.
Magnesium	A constituent of several enzymes that regulate sugar metabolism, energy production, cell membrane permeability and strength of muscles; important in building of bone structure.

3. Trace elements

Also called *micro elements*, trace elements are those which are required by living organisms in small amounts, usually as a part of a vital enzyme. In fact, the level of trace elements in humans is measured in terms of milligrams per kilogram of body weight. It must be emphasized that although trace elements are needed in small amounts, yet these are absolutely essential for normal growth and development. Their absence in the diet leads to deficiency diseases. On the other hand, when consumed even slightly above their optimum levels, these elicit toxicity. This concept is called *Bertrand's rule*. The important trace elements are iron, zinc, copper and manganese. Their functions are depicted in Table 9.6, along with the hazardous consequences associated with their insufficient and excessive uptake.

Table 9.6 Functions, deficiency diseases and toxicity associated with trace elements

Element	Functions	Insufficiency	Excess
Iron	Constituent of hemoglobin and myoglobin for transporting oxygen in blood.	Anemia	Damages liver; pervokes diabetes and cardiac failure
Zinc	Component of enzymes, such as carbonic anhydrase and carboxypeptidase; assists in blood sugar regulation, wound healing and regeneration of body tissue.	Vision impairment; reproductive system impairment	Vomiting; diarrhea; anemia; weak bones
Copper	Structural element in enzymes, such as ascorbic acid oxidase and tyrosinase; required for healthy arteries and blood formation.	Anemia	Depression; fatigue; migraine headache; Wilson's disease
Manganese	Component of the enzyme manganese superoxide dismutase; needed for fat and sugar metabolism.	Improper bone formation; reproductive disorders	Poor iron absorption

4. Ultratrace elements

The elements which are required for normal and healthy growth in extremely small quantities are called ultratrace elements. The estimated dietary requirement of such elements is usually less than one milligram per kilogram of body weight. A dietary deprivation and consumption beyond the optimum levels of these elements are harmful.

Amongst the ultratrace elements, the most important ones are iodine, selenium, molybdenum, chromium and cobalt. Each performs one or more specific functions, which have been reviewed in Table 9.7. The consequences associated with less-than-normal or more-than-normal uptake of these elements are also presented in Table 9.7.

Table 9.7 Functions, deficiency diseases and toxicity associated with ultratrace elements

Element	Functions	Insufficiency	Excess
Iodine	Required for production of thyroxine which plays an important role in metabolism.	Goiter—the enlargement of thyroid gland	Iodism—resulting in eruption of acne
Selenium	Constituent of antioxidant enzyme, glutathione peroxidase; vital for detoxification; assists in protein synthesis.	Keyshan disease—a form of heart ailment	Selenoisis—causing gastro-intestinal problems, fatigue and hair loss
Molybdenum	Constituent of the enzyme, xanthine oxidase.	No adverse effects are observed in healthy individuals; however, in patients on parenteral nutrition, deficiency of molybdenum increases heart and respiratory rates	Diarrhea
Chromium	Essential for insulin metabolism; lowers cholesterol.	Hyperglycemia—elevated blood sugar; glucosuria—glucose in urine	Carcinogenesis
Cobalt	Constituent of vitamin B12 which, in turn, assists in blood formation and in proper functioning of nervous system.	Anemia; dementia	Cardiac failure

Structure of the Cell

In the previous section, we reviewed the significance of those elements that are essential to life processes. The importance of these elements stems from the fact that their biochemical role is meticulously monitored and streamlined by the cells.

The cell is the basic unit of life. Though extremely small, it can carry out life processes by itself. For this reason, the molecular-level interactions that take place within a cell determine how the body functions as a whole.

The cell has three components: the cell membrane, the cytoplasm and the nucleus. These are depicted in Fig. 9.3 and briefly discussed below.

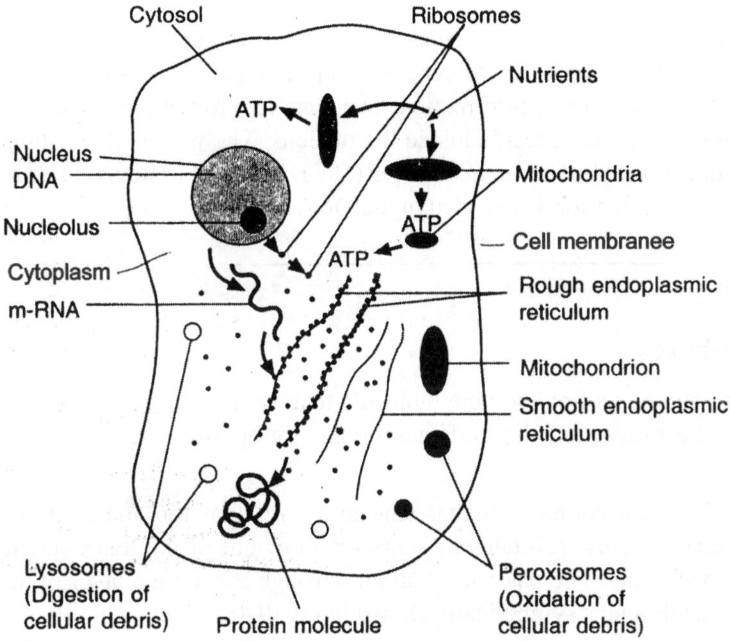

Fig. 9.3 The cell structure

The cell membrane

Each cell is enclosed by a *plasma membrane*, called the cell membrane. Though it separates the contents within the cell from the external environment, it is not a rigid and a passive barrier. It allows the entry and exit of chemical entities, so as to maintain a smooth pathway for the life processes.

The cytoplasm

Dispersed within the periphery of cell membrane, cytoplasm is composed of two constituents—cytosol and organelles. Cytosol is the liquid part of cytoplasm and contains water and water-soluble constituents including proteins, salts and minerals. The *organelles* are granular structures immersed in cytosol. The following are the important organelles.

- *Mitochondria*– its function is to produce adenosine triphosphate which, as we shall see in the next section, serves to store the energy of the cell.
- *Endoplasmic reticulum*– its function is to carry out biosynthesis of proteins and hormones.
- *Lysosome*– it degrades the cellular debris by hydrolysis.
- *Peroxisome*– it degrades the cellular debris by oxidation.
- *Golgi complex* or *Golgi apparatus*– it acts as the site for combination of proteins and polysaccharides to produce glycoproteins.

The nucleus

The nucleus, which houses the genetic material of the cell, is also an organelle. However, it merits a separate reference since it is the chief site for storage, replication and expression of genetic information. The genetic information, in turn, is stored in *chromosomes*, which reside inside the nucleus. The principal constituent of a chromosome is the *deoxyribonucleic acid* (*DNA*) bound to proteins called *histones*. The genetic information is encoded in the DNA.

Box 9.1

Ionophores

An ionophore is a lipid-soluble molecule that assists a selective ion to move across the bilayer of the cell membrane. Ionophores are classified into two types.

1. Chemical compounds that bind to a particular ion and neutralize its charge. This facilitates its transport in or out of the hydrophobic cell membrane. The passage of an ion through the double layer of the cell membrane is symbolically shown in Fig. 9.4.

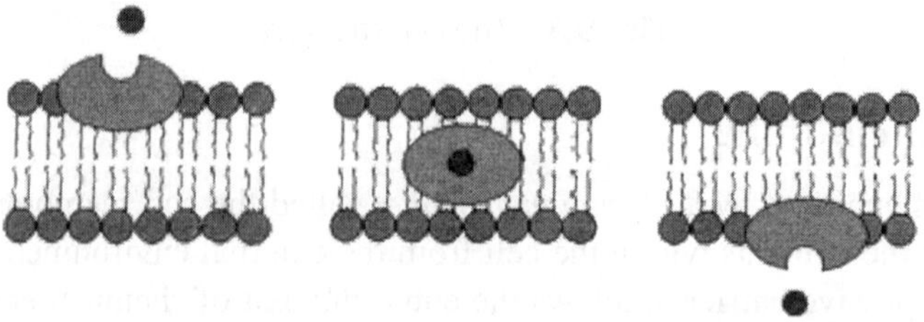

Fig. 9.4 Passage of an ion through the bilayer of cell membrane

2. Channel formers that introduce a hydrophilic pore into the cell membrane. The ions can pass through such pores without coming into contact with the inner, hydrophobic layer of the cell membrane.

Ionophores may be of natural origin or synthetic. Naturally occurring ionophores, also called *biological ionophores*, are mostly selective for a particular ion. For example, monensin allows the passage of sodium ions across the cell membrane, while enniatin ensures the passage of ammonium ions.

Synthetic ionophores are macrocyclic molecules, based on the crown ether or cryptand skeletons. These disrupt the concentration of ions within and without the cells of the microorganisms that infect the humans. As a result, the cells fail to perform their assigned functions and the microorganisms die. Thus the ionophores act as antibiotics.

Bioenergetics

We have seen that a cell is a highly organized system that contains many chemicals and structures. Yet, the general tendency of the universe is to move from an organized to a disorganized state. In fact, the second law of thermodynamics states that the disorderness of the universe increases steadily. The degree of disorderness is measured in terms of entropy. Greater the entropy, greater is the disorderness or randomness of the system. The cell must resist entropy and preserve its organized structure. For this it requires a continuous supply of energy.

The energy required by the cell is produced as a result of *catabolism*. The latter is a process of breaking down larger dietary molecules into smaller ones. For example, the proteins consumed by way of diet are cleaved into amino acids which, in turn, are used for biosynthesis of proteins that are essential to life processes. The catabolic reactions are exothermic in nature, and the energy so released is taken over by the cell.

The cell should also have a reservoir to store energy generated by the exothermic catabolic processes, so that it may use it as and when desired. The biomolecule, adenosine triphosphate, abbreviated as ATP, serves as a storage pool for energy. It is synthesized from adenosine diphosphate (ADP) and monohydrogen phosphate ions in an endothermic reaction (9.1).

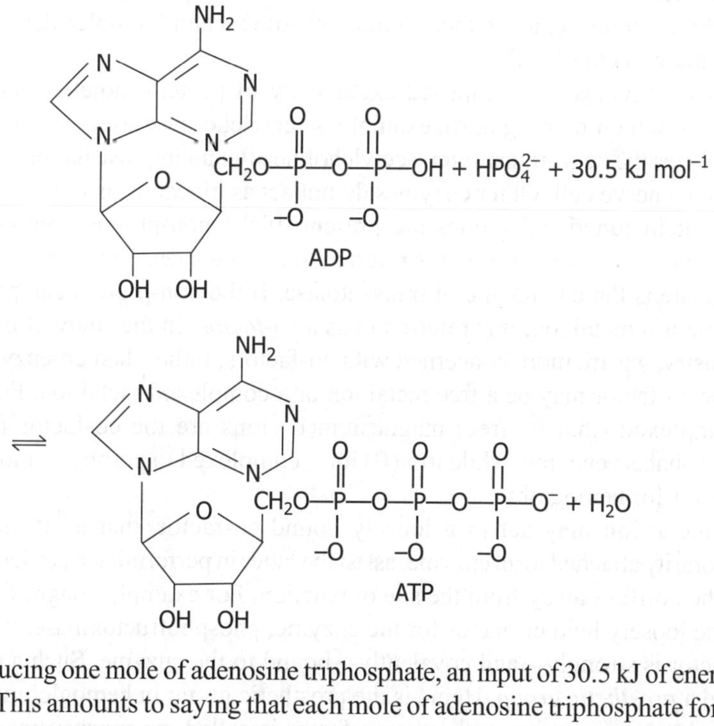

(9.1)

For producing one mole of adenosine triphosphate, an input of 30.5 kJ of energy is necessary. This amounts to saying that each mole of adenosine triphosphate formed in the body stores 30.5 kJ of useful energy. When cells need energy, adenosine triphosphate is hydrolyzed to adenosine diphosphate, that is, reaction (9.1) proceeds in a reverse manner, generating 30.5 kJ mol^{-1} of energy.

A part of this energy is used to preserve the organized structure of the cell. Rest is utilized for the following three functions:

1. Transport of chemical entities in and out of the cell membrane, as in the case of sodium-potassium pump.
2. Mechanical work, such as contraction of muscles.
3. Biosynthesis of molecules, such as DNA.

The reactions leading to the synthesis of biomolecules must proceed at an optimum rate. For this, energy input is an essential, but not the only requirement. A chemical transformation will take place at an appreciable rate only if the reactants are present in large amounts and the reaction temperature is high. Both these conditions are absent in the living systems. Hence, a suitable catalyst must be present to carry out a biochemical reaction. Such catalysts, which enhance the rate of biochemical reactions are called enzymes, and are discussed in the next section.

Enzymes

A chemical entity that acts as a catalyst in biochemical reactions and the structure of which is based on a protein skeleton is referred to as an enzyme. The enzymes are known to speed-up the reactions relevant to life processes 10^8 to 10^{20} times, at body temperature, using meager amounts of reactant biomolecules dissolved in the water content of cells.

Some enzymes are composed exclusively of protein moieties. These act as catalysts in their own right. An example is acetylcholinestrase—an enzyme which hydrolyzes the neurotransmitter, acetylcholine after an impulse has been conducted through a nerve cell. Other enzymes do not act as biocatalysts unless one or more non-protein functional groups are present. If the non-protein component of an enzyme is an organic group, it is referred to as *co-enzyme*. For example, thiamine phosphate is the co-enzyme of transketolase. If the non-protein component of an enzyme is a metal ion, it is referred to as a *co-factor*. In the study of bioinorganic chemistry, we are more concerned with co-factors, rather than co-enzymes.

The co-factor may be a free metal ion or a complexed metal ion. For example, uncomplexed (that is, free) magnesium(II) ions are the co-factor for glucose 6-phosphatase enzyme, while iron(II) ions, complexed to porphyrin moiety are the co-factor for hemoglobin.

A metal ion may act as a loosely bound co-factor, that is, it may become temporarily attached to an enzyme, assist the latter in performing a particular activity, and then diffuse away from the site of reaction. For example, magnesium(II) ions are the loosely held co-factor for the enzyme, phosphofructokinase. At times, the co-factor is strongly—and covalently—bound to the enzyme. Such a co-factor is called a *prosthetic group*. Heme is the prosthetic group in hemoglobin.

An inactive enzyme, without a co-factor is called an *apoenzyme*. An enzyme attached to a co-factor is called *holoenzyme*.

Apoenzyme + co-factor → Holoenzyme

Holoenzymes involving complexed or uncomplexed metal ions as prosthetic groups are called *metalloenzymes*.

Case Study 9.2 What is antabuse?

Antabuse is a drug that is used for de-addicting the alcoholics. It is generally administered without informing the addict.

The common name of antabuse is *disulfiram* and it has the following structure:

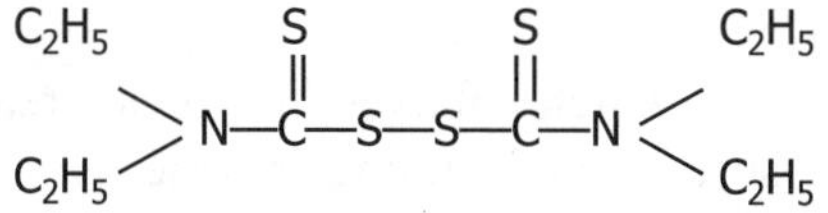

When alcohol is consumed, it initially metabolizes to acetaldehyde which is a mild poison and causes irritation. However, the enzyme aldehyde oxidase rapidly converts acetaldehyde to harmless acetic acid. Molybdenum(VI) ions act as co-factor for this enzyme and assist in instantaneous oxidation of acetaldehyde to acetic acid.

Antabuse drug coordinates to molybdenum(VI) ions through sulfur atoms and hinders the pathway by which acetaldehyde is rapidly oxidized. This causes severe unpleasant effects and the alcoholic is discouraged to consume more liquor.

Classification of Metallobiomolecules

Biomolecules associated with metal ions are called metallobiomolecules. Generally a biomolecule possesses donor atoms like oxygen, nitrogen and sulfur and these become coordinated to metal ions so as to perform their assigned roles in life processes. Metallobiomolecules are classified into three broad types.

1. Transport and storage proteins

The metal-containing proteins which keep in reserve the elements essential to life processes and transport these to the organs that feel their need come under this category. The transport and storage proteins are categorized into the following three sub-classifications:

- Oxygen binding. Their function is to coordinate with oxygen and then transfer it to a requisite site. Examples include hemoglobin and myoglobin. Both these are iron-containing proteins.
- Electron carriers. Their function is to release or capture electrons so as to facilitate electron transfer reactions in biological systems. Ceruloplasmin, a copper containing protein, is an example of an electron carrier protein.
- Metal storage and carriers. Their function is to store a particular metal and then carry it to the site that faces its deficiency. The two functions

are performed by different metalloproteins. For example, ferritin is concerned with storage of iron, while transferrin is responsible for its transportation.

2. Enzymes

We have stated in the previous section that biological catalysts containing complexed or uncomplexed metal ions as prosthetic groups are called metalloenzymes. Further, the non-metallic part of the enzyme is based on a protein framework. Depending on the type of biochemical reaction these catalyze, the metalloenzymes are sub-classified into three types.

- Hydrolases. Those which catalyze hydrolysis reactions. For example, carboxypeptidase, a zinc-containing enzyme.
- Oxidoreductases. Those which catalyze redox reactions. The cytochrome oxidase, for example, contains both iron and copper in its structure.
- Isomerases and synthestases. Those which speed-up isomerization reactions or synthetic pathways of biomolecules. Cobalamin, the cobalt-containing co-factor, is an example of this sub-classification.

3. Non-protein biomolecules

A metallobiomolecule in which a non-protein, organic functional group is coordinated to a metal ion comes under this category. The plant pigment chlorophyll, based on magnesium-porphyrin skeleton, is an example of a non-protein metallobiomolecule.

Zinc-Based Enzymes

The metalloenzymes based on zinc(II) ion are well suited for catalyzing acid-base reactions in biological systems. The presence of zinc(II) ion in the structure of these enzymes offers the following advantages:

- Zinc(II) ion forms strong bonds with the constituent atoms of proteins.
- Aqua ligands (H_2O), bound to zinc(II) ion are readily exchangeable with other coordinating groups.
- Zinc(II) ion does not possess a complicating redox chart. In fact, as an ion zinc exists exclusively in oxidation state (II), and therefore it is considered as redox inactive. If a metal ion exists in several oxidation states (as most of the transition metals do), the mechanistic pathway of the enzyme not only becomes complicated, but is dictated by the external conditions which might favor one or the other ionic state of the metal.
- The complexes of zinc with biological ligands, such as proteins and amino acids possess a flexible geometry. Hence their structure can easily adapt to the lock and key catalytic pathway which the enzymatic reactions follow.

The zinc(II) ion in metalloenzymes is coordinated to three amino acid ligands and one exchangeable water molecule (Fig. 9.5). Examples of zinc-based enzymes include carbonic anhydrase, carboxypeptidase, β-lactamase and alcohol dehydrogenase. The first two are discussed in sequal.

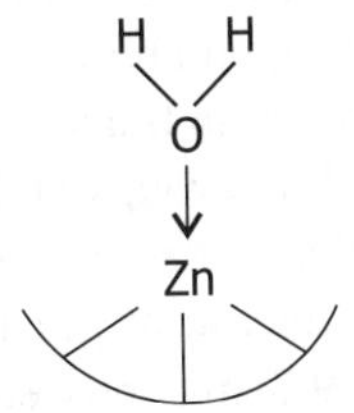

Fig. 9.5 Skeletal framework of zinc-based enzymes

1. Carbonic anhydrase

The carbonic anhydrase enzyme catalyzes the reaction of carbon dioxide and water to produce bicarbonate anion and proton.

$$CO_2 + H_2O \xrightarrow{\text{Carbonic anhydrase}} HCO_3^- + H^+$$

In the absence of carbonic anhydrase, and at pH 7, the above reaction is very slow, the rate constant being 10^{-3} s^{-1}. The turnover of carbon dioxide by biological systems is very high and a reaction with such a small rate constant cannot serve the purpose. In presence of carbonic anhydrase, however, the hydration of carbon dioxide proceeds with a rate constant of 10^6 s^{-1}, making it one of the fastest enzymatic reactions.

The reverse reaction,

$$HCO_3^- + H^+ \rightarrow H_2CO_3 \rightarrow CO_2 + H_2O$$

which involves dehydration of carbonic acid, is also catalyzed by carbonic anhydrase. For this reason, the latter is often referred to as *carbonate dehydratase*.

Structure The crystal structure of carbonic anhydrase reveals that the zinc(II) ion is placed in a conical cavity about 16 pm deep, which is lined up with several histidine residues. The zinc(II) ion is coordinated to three histidine ligands through nitrogen atoms and to one water molecule through oxygen atom. One such unit has a molecular weight of approximately 30,000. The skeletal structure, which has a near tetrahedral geometry, is shown in Fig. 9.6.

Mechanism The mechanism of the reaction catalyzed by carbonic anhydrase is outlined in Fig. 9.7. The attachement of three histidine nitrogens to zinc(II) ion induces a positive charge on the oxygen atom of the aqua ligand (a). This results in a loss of proton, so that instead of a water molecule, a hydroxide group now occupies the fourth coordination site of zinc (b). For this reason, this mechanism is called *zinc-hydroxide pathway*. Another water molecule becomes attached to the hydroxyl group through hydrogen bonding (c). A carbon dioxide molecule coordinates with intermediate (c). In this union, the carbon of carbon dioxide becomes attached to the oxygen of hydroxyl group, while one of the oxygen atoms of carbon dioxide becomes attached to the proton of the hydrogen-bonded water molecule (d). Next, the bond between oxygen of hydroxyl group and carbon of carbon dioxide strengthens at the cost of the bond between oxygen and hydrogen of the hydroxyl group (e). The intermediate (e) rearrange, so that in transition state (f), zinc is 5-coordinated, being firmly bound to three histidine groups and loosely to a bicarbonate group and an aqua ligand. Finally, the bicarbonate anion dissociates and the cycle starts once again. The net reaction is a release of bicarbonate anion and a proton. The mechanism, as highlighted in Fig. 9.7, may proceed in clockwise or anti-clockwise direction depending on the requirements of the biological system.

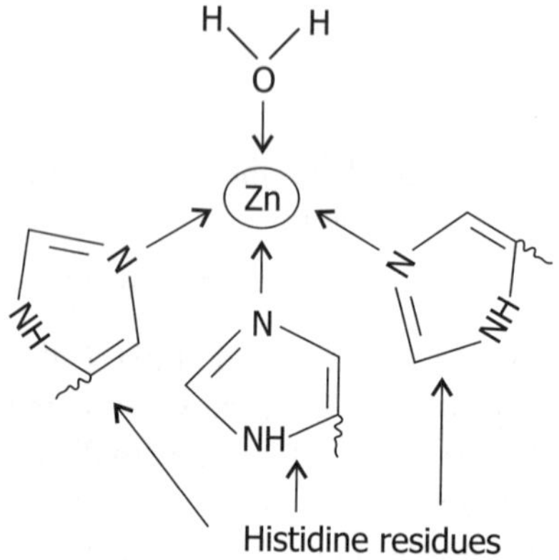

Fig. 9.6 Skeletal framework of carbonic anhydrase

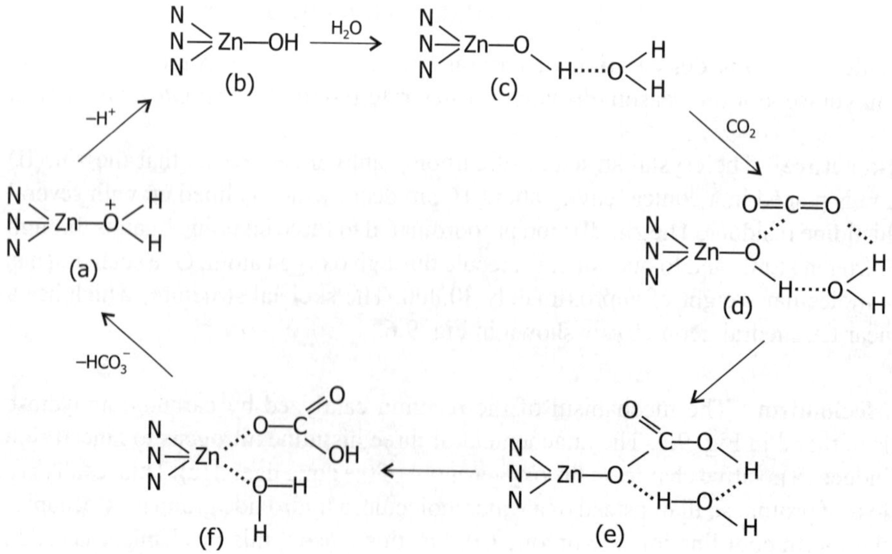

Fig. 9.7 Mechanistic pathway of carbonic anhydrase-catalyzed reaction

Functions The reaction catalyzed by carbonic anhydrase produces protons (acid) and bicarbonate ions (alkali). The reverse reaction, in which protons and bicarbonate ions are consumed, is also catalyzed by carbonic anhydrase. Thus the primary function of this enzyme is to maintain acid-base balance in blood and other tissues. Yet another function is to transport carbon dioxide out of the tissues.

The following are some of the examples of carbonic anhydrase-driven reactions.

- The parietal cells in the stomach secrete protons into the lumen for efficient digestion; simultaneously, a corresponding amount of bicarbonate ions are passed into blood.
- The pancreatic duct cells use the secretion of bicarbonate ions to impart alkaline character to the pancreatic juices. The protons are passed on to the blood stream.
- The acid-base balance in renal tubules and eyes helps maintain the water content in these organs at an optimal level.
- Carbon dioxide generated by metabolism in all cells is removed from the body by red blood cells that convert most of it into bicarbonate ions. Once transported to the lungs, it is reconverted to carbon dioxide and then exhaled.
- The carbon dioxide/bicarbonate ion equilibrium serves to regulate the pH of tissues.

Case Study 9.3 Why does one feel the fizzy sensation while drinking a carbonated beverage even if there is a long gap between opening of the bottle and consumption of the drink?

The fizziness one experiences while drinking a carbonated beverage arises due to decomposition of carbonic acid into carbon dioxide.

$$H_2CO_3 \rightarrow H_2O + CO_2$$

However, this reaction occurs at a slow rate. For this reason, a carbonated drink does not instantly lose carbon dioxide when the container is opened. When consumed, the carbonic acid comes in contact with carbonic anhydrase that is contained in saliva. The carbonic anhydrase catalyzes the decomposition reaction. The rapid evolution of carbon dioxide creates the fizzy sensation in mouth.

---

2. Carboxypeptidase

Carboxypeptidase, an enzyme synthesized in the pancreas and thereafter secreted into the small intestine, catalyzes the hydrolysis of proteins that are consumed as components of food into amino acids. During the hydrolysis reaction, a water molecule acts on the amide linkage of the protein unit and cleaves the carboxyl terminal of amino acid of the peptide chain. A representative example is shown in Fig. 9.8.

The enzyme preferentially cleaves those amino acids that have an aromatic group or a bulky aliphatic hydrocarbon chain.

Structure The central metal ion, zinc(II) in carboxypeptidase is coordinated in a tetrahedral geometry to two histidine units, one glutamate unit and one active site. The histidine units are bound to zinc(II) ion through nitrogen atoms, the glutamate unit through oxygen atom, while the active site is usually occupied by an exchangeable water molecule (Fig. 9.9).

The molecule of carboxypeptidase is nearly egg-shaped, with a maximum dimension of about 5000 pm, a minimum dimension of about 3800 pm and a molecular weight of approximately 34,600.

Mechanism During the hydrolytic cleavage of the peptide link, apart from the reactant protein and the catalyst carboxypeptidase, three other biomolecules are involved in the reaction. These are: arginine, tyrosine and glutamate. The hydrolysis reaction proceeds in the following steps (Fig. 9.10).

(i) The peptide (a), which is to be hydrolyzed, comes in the vicinity of enzyme, carboxypeptidase (b). Simultaneously, arginine (c) and tyrosine (d) come close to the peptide and become linked to it. The terminal carboxylate anion of the peptide interacts electrostatically with positively charged side chain of arginine; the NH group of peptide becomes linked to phenolic group of tyrosine through hydrogen bonding. The purpose of these two interactions is to impart a fixed conformation to the peptide, so that its carbonyl group becomes oriented towards the active center of enzyme (b).
The structural rearrangement of the participating molecules in the correct orientation is called *induced fit*. The induced fit state of this particular reaction is depicted as (e). The role of arginine ends here, although tyrosine is further involved in the reaction (step v).

(ii) The exchangeable water molecule at the active site of the enzyme is displaced and the carbonyl group of the peptide becomes coordinated to the zinc(II) ion, resulting in complex (f). For this reason this mechanism is often referred to as *zinc-carbonyl pathway*.

(iii) The carbonyl group is polarized due to electromeric effect, inducing a positive charge on the carbon atom. The resulting complex is designated (g) in Fig. 9.10.

(iv) The negatively charged glutamate ion removes a proton from a water molecule to produce a hydroxide ion.

(v) The nucleophilic hydroxide ion attacks the positively charged carbon atom of the intermediate (g) to yield (h). The tyrosine unit in the latter donates its proton to the NH bond of the peptide. The amino acid (i) dissociates from the peptide (a). Simultaneously, the carbonyl group dissociates from carboxypeptidase and a water molecule takens its place as the active site.

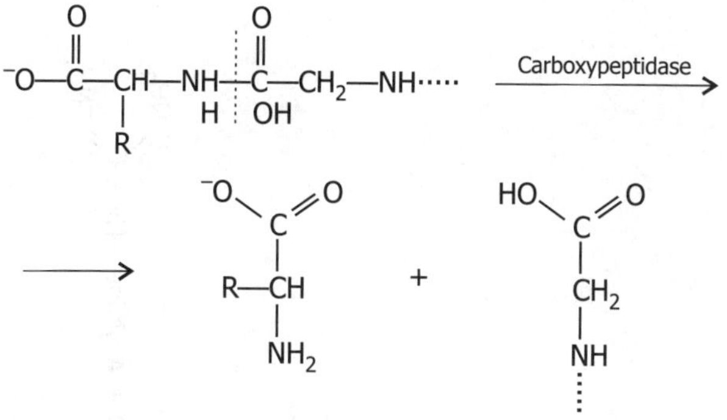

Fig. 9.8 Hydrolysis of protein in presence of carboxypeptidase

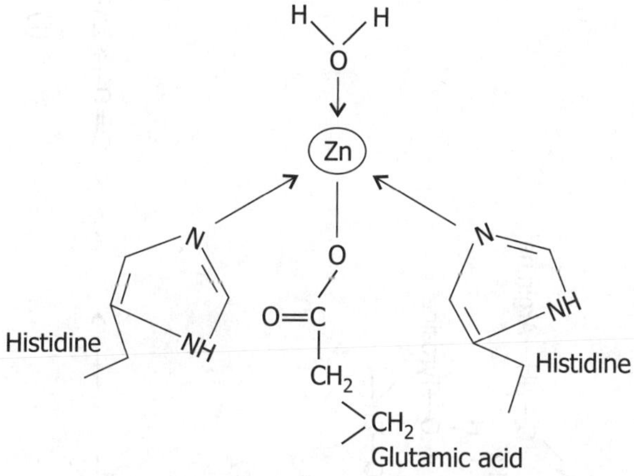

Fig. 9.9 Skeletal framework of carboxypeptidase

Functions The proteins perform a host of functions which are essential for the life processes. The biological system of humans, as well as of other organisms, has the capacity to biosynthesize the requisite proteins from the constituent amino acids. The source of amino acids are the proteins that are consumed as part of the diet. The main function of carboxypeptidase is to hydrolyze such proteins, so as to cleave their amino acids. The latter then can be used to synthesize the proteins which the body actually requires.

Carboxypeptidase also assists in blood clotting and wound healing. Moreover, it participates in the biosynthesis of insulin.

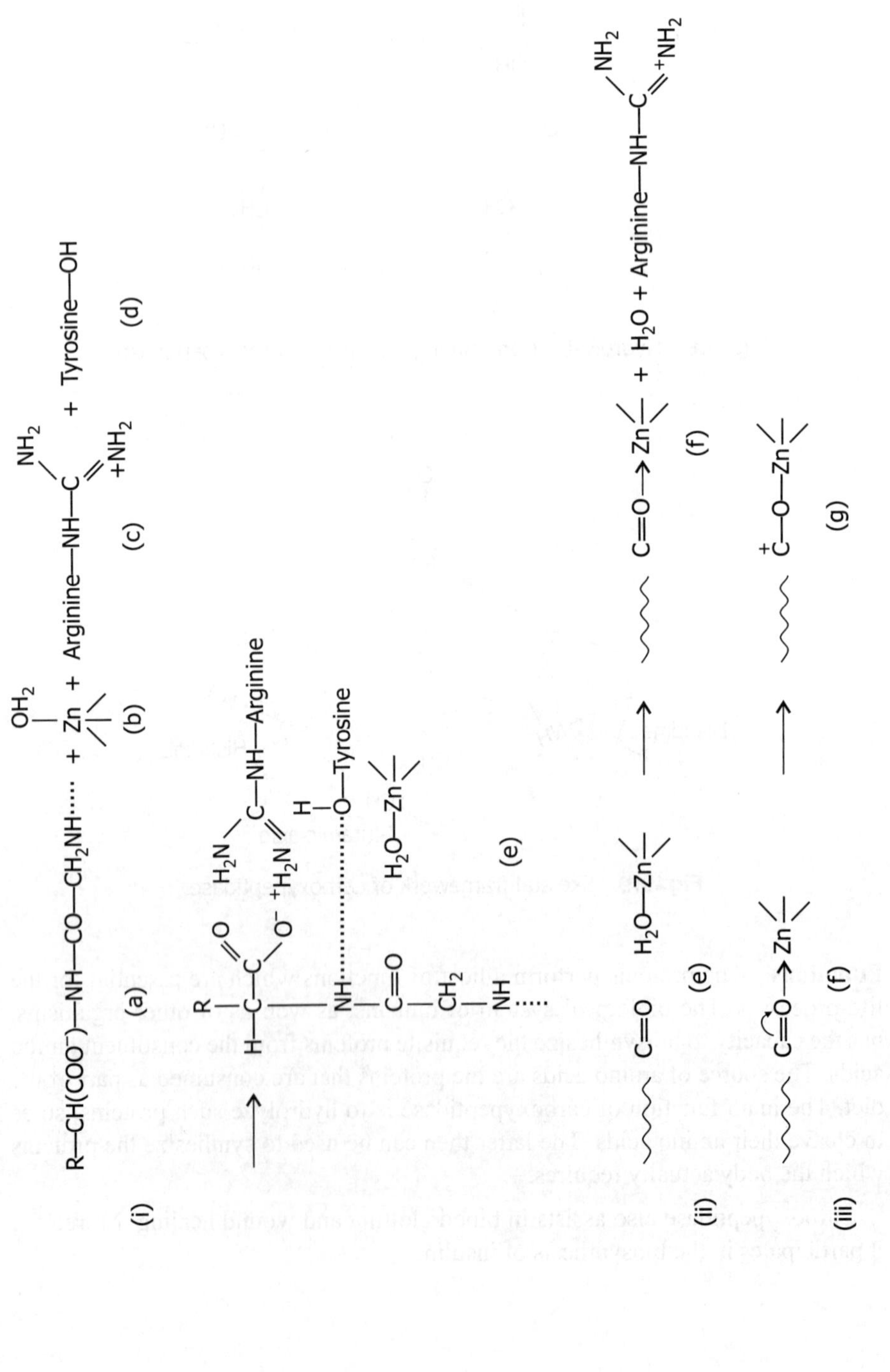
(i)
R—CH(COO⁻)—NH—CO—CH2NH···· + Zn + Arginine—NH—C + Tyrosine—OH
OH2
NH2
+NH2
(a)
(b)
(c)
(d)
R
H—C—C
O
O⁻
H2N
+H2N
C—NH—Arginine
H
O—Tyrosine
NH
C=O
CH2
NH····
H2O—Zn
(e)
(ii)
C=O
H2O—Zn
C=O→Zn + H2O + Arginine—NH—C
(e)
(f)
(iii)
C=O→Zn
C—O—Zn
(f)
(g)

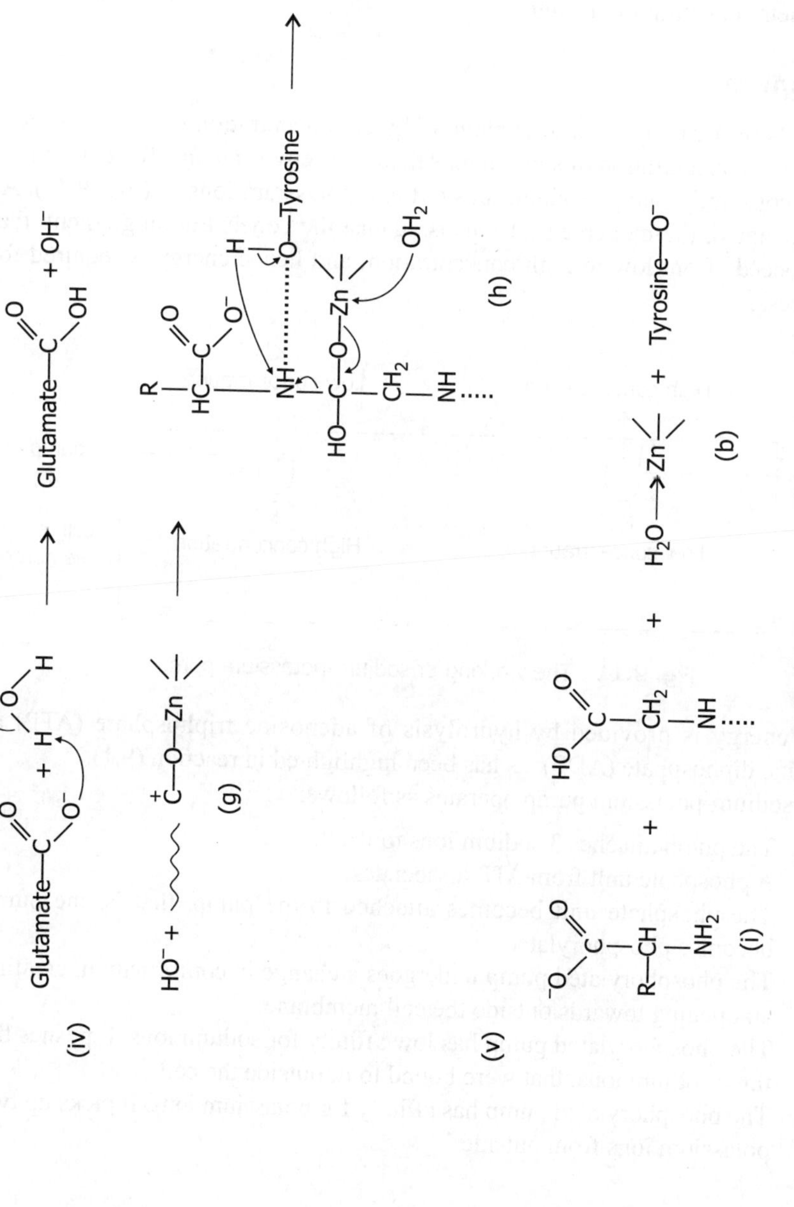

Fig. 9.10 Mechanistic pathway of carboxypeptidase-catalyzed reaction

Sodium-Potassium Pump

The sodium-potassium pump, more precisely referred to as sodium-potassium adenosine triphosphatase, $Na^+|K^+$–ATPase is an enzyme located in the cell membrane. It transports sodium and potassium ions across the cell membrane against their concentration gradient. Sodium ions are moved from the inside of the cell, where their concentration is low, to the extracellular fluid, where their concentration is much higher. Potassium ions are moved in the opposite direction, that is, from extracellular fluid, where their concentration is low to inside the cell, where their concentration is higher.

Mechanism

Most cells in the body need to contain a higher concentration of potassium ions and a lower concentration of sodium ions than their environment. To achieve this, the cell constantly pumps sodium ions out and potassium ions in (Fig. 9.11). As has been stated, the movement of ions is against the concentration gradient, that is, it proceeds from low to high concentration, and hence energy is required for the process.

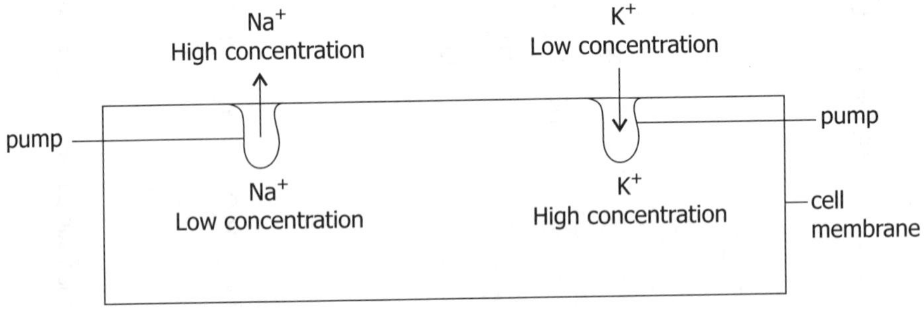

Fig. 9.11 The working of sodium-potassium pump

The energy is provided by hydrolysis of adenosine triphosphate (ATP) to adenosine diphosphate (ADP), as has been highlighted in reaction (9.1).

The sodium-potassium pump operates as follows:

1. The pump attaches 3 sodium ions to itself.
2. A phosphate unit from ATP dissociates.
3. The phosphate unit becomes attached to the pump, that is, the pump becomes phosphorylated.
4. The phosphorylated pump undergoes a change in conformation, creating an opening towards outside the cell membrane.
5. The phosphorylated pump has low affinity for sodium ions. It pushes the three sodium ions, that were bound to it, outside the cell.
6. The phosphorylated pump has affinity for potassium ions. It picks up two potassium ions from outside.

7. After the two potassium ions become attached to the pump, the latter sheds off the phosphate unit and becomes dephosphorylated.
8. The dephosphorylated pump re-orients, creating another opening which pushes two potassium ions inside the cell.
9. The pump binds three sodium ions from within the cell and the cycle restarts.

Thus in any one cycle, the sodium-potassium pump pushes three sodium ions outside the cell and brings two potassium ions inside the cell.

Functions

The sodium-potassium pump performs the following two functions:

1. To maintain cell potential A potential arises when there is an unequal distribution of charges in two adjacent compartments. In cells, the gradient is provided by sodium-potassium pump. After the completion of one cycle, three sodium ions go out of the cell and two potassium ions come inside the cell. This means that there is one less positive charge inside the cell than outside. Moreover, there are several anionic proteins inside the cell. As a result, the interior of the cell acquires an overall negative charge, while the exterior is positively charged.

In nerve cells (also called neurons), this potential (that is, negative charge inside and positive outside) is called *resting potential*. The resting potential of the neuron is –70 mV.

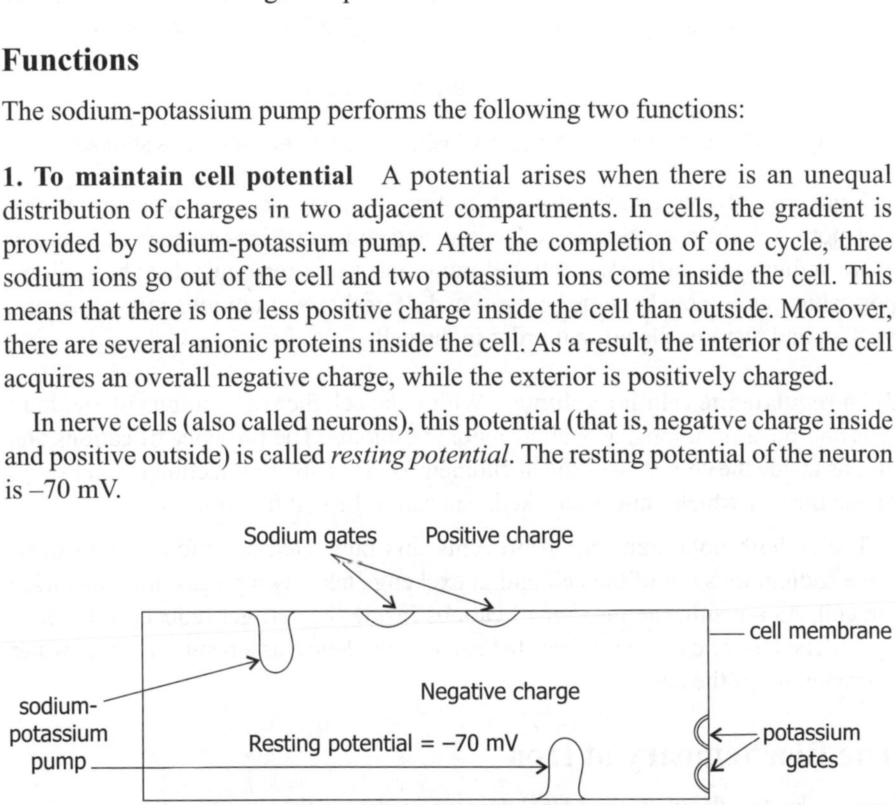

Fig. 9.12 A nerve cell is endowed with sodium and potassium gates in addition to sodium-potassium pumps

So long a nerve cell does not receive any stimules, it is said to be in a state of resting potential. The nerve cells have *sodium gates*, which allow sodium ions to pass inside and *potassium gates*, which allow potassium ions to pass outside (Fig. 9.12). *These gates are in addition to sodium-potassium pumps.*

When a nerve cell receives a stimules, the sodium gates open and sodium ions rush in. The cell potential starts rising from the value of –70 mV towards positive value. When it reaches the peak value of +30 mV, sodium gates close. Now the potassium gates open, the potassium ions start moving out, the potential decreases from +30 to –70 mV (Fig. 9.13).

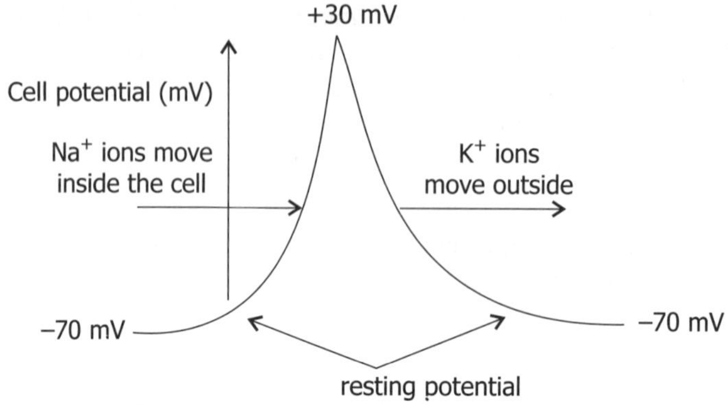

Fig. 9.13 Variation of cell potential when a nerve cell receives a stimulus

Since the cell once again acquires its resting potential, we do not feel the stimulus any more. Yet, the concentration of sodium and potassium ions are in reverse order. More sodium ions inside the cell and more potassium ions outside. The sodium-potassium pump now becomes operational. It pushes sodium ions out, potassium ions in, and restores the normal order in the cell.

2. To regulate the cellular volume Within the cell there are cations like sodium ions and potassium ions, as well as anionic proteins. The presence of cations and anions inside the cell induces the phenomena of osmosis. Extracellular water seeps inside the cell which, unless checked, can cause the cell to swell up.

The sodium-potassium pump prevents this unwanted situation. It transports three sodium ions out of the cell and in exchange takes two potassium ions inside the cell. As a result, the number of cations inside the cell get reduced after each cycle. The loss of cations reverses the osmotic tendency, as a result of which water is driven out of the cell.

The Biochemistry of Iron

Iron is the fourth most abundant element in the earth's crust and is, therefore, ubiquitous in living systems. Cells make use of iron in many ways and for many tasks. The biochemical importance of iron stems from two of its characteristics. Firstly, iron can easily shift between ferrous and ferric states and hence participate in the redox reactions within the biological systems. Secondly, nitrogen and oxygen atoms of several biomolecules, as well as molecular oxygen have ability to coordinate with iron. For example, hemoglobin is able to perform its function of transporting oxygen to the tissues since the iron(II) ions in the heme moieties easily complex with oxygen molecules and nitrogens of porphyrin rings.

However, iron also poses a problem. It can combine with water content of cells and oxygen content of air and get converted into insoluble iron(III) hydroxide. In an insoluble form, iron would fail to perform its assigned biochemical role. The living systems have evolved mechanisms by which iron is transported in necessary quantities in soluble forms. This job is carried out by a biomolecule called transferrin. The transferrin picks up iron from a reservoir created by another biomolecule called ferritin, converts it from a semi-crystalline state to a soluble state and then delivers it to the site that requires it.

Ferritin

Ferritin is an intracellular protein that stores iron and releases it in a controlled fashion either within the same cell or in other cells of the organism. It is distributed in virtually all the organs, although high concentrations are found in liver, spleen and bone marrow.

Structure Ferritin is composed of two units: a core and a protein sheath. The core is essentially spherical in geometry, with a diameter in the vicinity of 7000 pm. It is semi-crystalline in nature and its approximate composition is $(FeOOH)_8 \cdot FeO \cdot H_2PO_4$. Each core contains about 4500 iron(III) ions in a high spin state. The phosphate group serves to bind the (FeOOH) units with each other on one hand and the core with the surrounding protein sheath on the other.

The protein covering of the core is called *apoferritin*. It is made up of 24 subunits each of which is a peptide chain. The peptide chains are coiled around the core in such a manner that they form a hollow shell (Fig. 9.14).

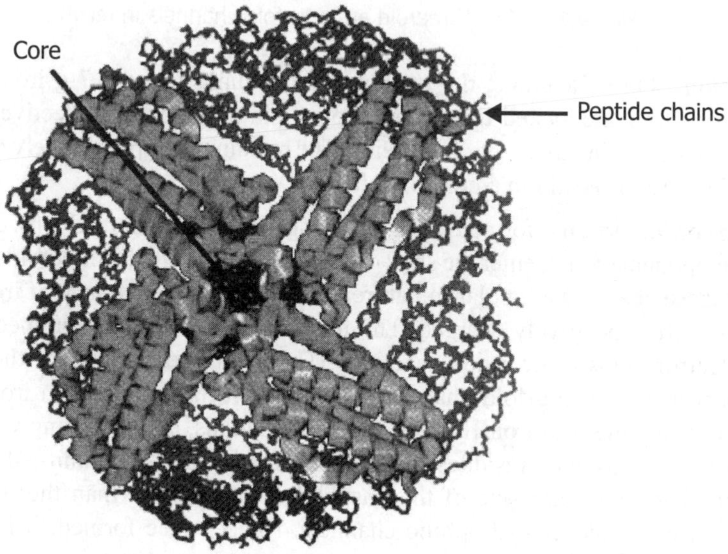

Fig. 9.14 The cavity in ferritin where iron is stored

The shell is endowed with channels which provide passage for iron in ionic form to enter in or to exit from ferritin. The channels are of two types: threefold channels and fourfold channels. The threefold channels are formed at the intersection of three peptide subunits and are polar or hydrophilic in nature. The fourfold channels are formed at the intersection of four peptide units and are non-polar or hydrophobic in nature. These channels are symbolically shown in Fig. 9.15. The molecular weight of ferritin is about 4,74,000.

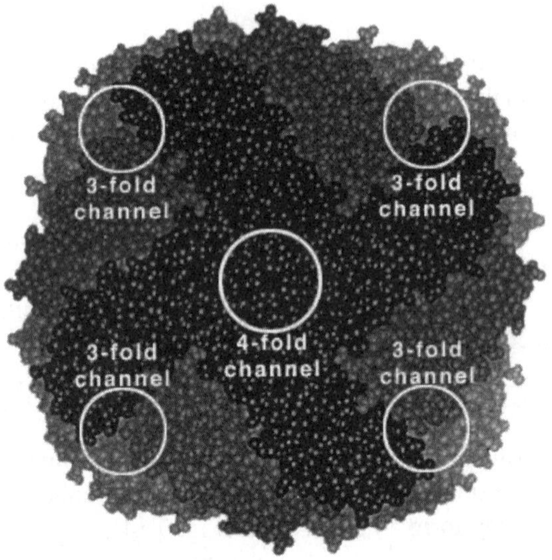

Fig. 9.15 The threefold and fourfold channels in ferritin

Function On an average, the human body contains about 3.7 g iron. Of this, 2.5 g and 0.2 g are locked-up as hemoglobin and myoglobin, respectively. About 0.02 g is present in other metalloproteins. The remaining, approximately one gram, is in the core of ferritin in a non-toxic form.

The core of ferritin is formed in the following manner. The iron(II) ions, consumed as a component of diet, enter the shell of apoferritin through the threefold channels. Since these channels are polar in nature, they facilitate the passage of iron(II) ion which itself is positively charged. The fourfold channel is programmed to carry out electron-transfer or redox reactions. The constituent atoms of the peptide subunits that form fourfold channels accept an electron from each iron(II) ion, thereby oxidizing it to iron(III) ion. The iron(III) ions, on interacting with water, oxygen and phosphate residues, form $(FeOOH)_8 \cdot FeO \cdot H_2PO_4$ unit—the core of ferritin. The molecular size of this insoluble unit is larger than the passage of either hydrophobic or hydrophilic channels. Hence, once formed, it is retained within the apoferritin.

When any organ faces a shortage of iron, the reserve in the core makes up the deficiency. However, as has been stated, the core is insoluble and cannot ooze out of the surrounding sheath. The fourfold, hydrophobic channel now returns the electron to iron(III) ion and reduces it to iron(II) ion. The iron(II) ion detaches itself form the core material, $(FeOOH)_8 \cdot FeO.H_2PO_4$ and, in aqueous state, passes out through the threefold, hydrophilic channel. On coming out of the ferritin molecule, the aqueous iron(II) ion interacts with more water molecules and is converted into hexaquairon(II) ion, $[Fe(H_2O)_6]^{2+}$. These ions enter the blood stream, where they encounter mildly alkaline conditions (pH 7.4). Under these conditions, the copper metalloenzyme, *ceruloplasmin*, oxidizes the hydrated iron(II) ions to iron(III) ions. The iron(III) ions are then taken over by transferrin and carried to the relevant organ.

Transferrin

Transferrin is a plasma glycoprotein that is widely distributed in the physiological fluids and cells. It transports iron through the blood to all tissues. Since it is a glycoprotein, it contains carbohydrate residues. The carbohydrate moiety of human serum transferrin represents about 6% of the weight of protein.

Overall, transferrin is a medium-sized protein, having a molecular weight of about 80,000. When not bound to iron, it is referred to as *apotransferrin*. The apotransferrin is synthesized by the liver and then secreted into the plasma. It then picks up iron(III) ions from the blood stream and delivers it to the site that experiences its shortfall.

Structure The apotransferrin molecule is made up of 678 amino acids and two glycan (carbohydrate) moieties. The amino acids are divided into two sets of polypeptides. One polypeptide set is constituted by amino acids 1 to 336 and is called *N-terminal domain*. The second polypeptide set is constituted by amino acids 337 to 678 and is called *C-terminal domain*. The two sets are arranged back-to-back, with each domain having a cleft (Fig. 9.16). The apex of each cleft coordinates to an iron(III) ion. Thus each transferrin molecule transports two iron(III) ions.

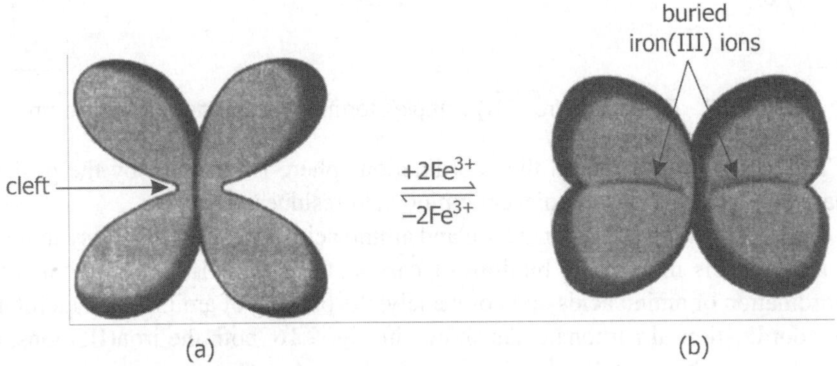

Fig. 9.16 Schematic representation of transferrin

The sequence of amino acids in N-terminal domain and C-terminal domain is about 40% identical. Yet, the amino acids which bind to iron(III) ions are identical in both the domains and are arranged at the clefts: two tyrosine, one aspartic acid and one histidine. These occupy four bonding sites around iron(III) ion. The fifth and the sixth bonding sites are filled up by a bidentate carbonate anion, which becomes, attached to the iron(III) ion through two oxygens. Thus there is octahedral coordination around each iron(III) ion of transferrin (Fig. 9.17).

Before becoming coordinated to iron(III) ion, the tyrosine is deprotonated at the phenolic group and is converted to the phenolate anion; it is designated as tyrosyl in Fig. 9.17. Likewise, the aspartic acid is deprotonated to carboxylate anion, and is referred to as aspartyl in Fig. 9.17. The histidine remains neutral. The combination of these four amino acid residues, along with carbonate anion, to the iron(III) ion imparts a dinegative charge to the coordination sphere.

$$Fe^{3+} + 2(\text{tyrosyl})^- + (\text{aspartyl})^- + \text{histidine} + CO_3^{2-} \rightarrow [Fe\ (\text{tyrosyl})_2\ (\text{aspartyl})\ (\text{histidine})\ CO_3]^{2-}$$

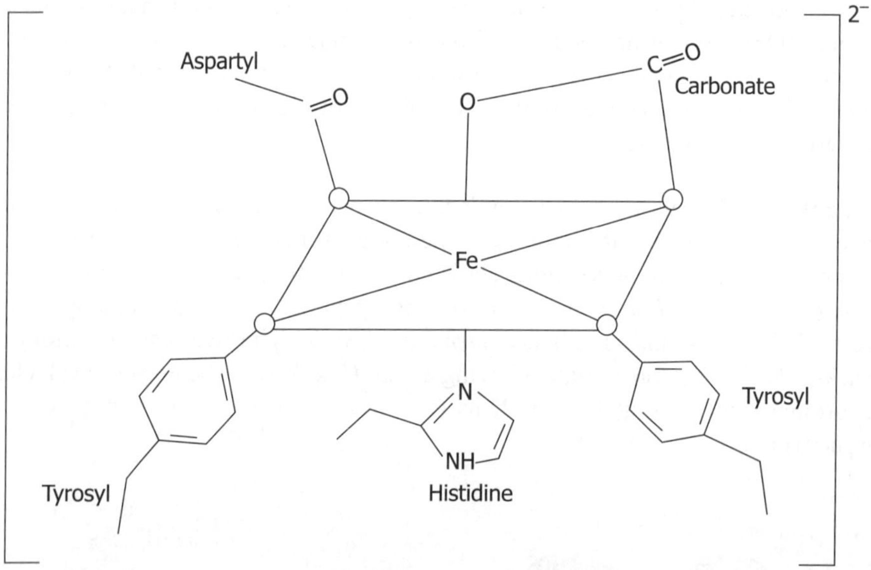

Fig. 9.17 Octahedral iron(III) complex formed at each cleft of transferrin

The dinegative charge on the coordination sphere is balanced by the positive charge on the nitrogens of adjacent amino acid residues.

The coordination of carbonate ion and amino acids to iron(III) ion is synergistic in nature. This means, the binding of carbonate to iron(III) ion facilitates the coordination of amino acids and conversely, the binding of amino acids facilitates the coordination of carbonate. As shown in Fig. 9.16, both the iron(III) ions, on complexation, become deeply buried inside the protein structure.

The transformation of iron(III) ions into anionic complex serves three purposes. Firstly, it renders iron(III) ions soluble under physiological conditions. Free (or uncomplexed) iron(III) ions, under these conditions, are precipitated as ion(III)

hydroxide. Secondly, iron is easily transported into the cells as a soluble complex. Thirdly, whereas the free iron(III) ions are toxic in nature, the complexed iron(III) ions are safe.

Mechanism The transferrin molecule, having two chelated iron(III) ions, flows through the blood until it finds a transferrin receptor on the surface of the cell. The regions on the surface of cells which house a large number of receptors are called *coated pits*. Transferrin binds to the receptor and is drawn into the cell. Once inside the cell, transferrin sheds its iron content. The mechanism by which iron is transported to the cell is called *receptor-mediated endocytosis of transferrin*. To understand this mechanism, one must first review the structure of transferrin receptor.

Abbreviated as TfR, the transferrin receptor is a dimer of two identical monomers. Each monomer is made up of 760 amino acids. The structure of this biomolecule may be divided into three parts: intracellular, middle and extracellular (Fig. 9.18). The intracellular part, which represents the tail or the lower portion of the receptor, is constituted by two strands, each built from amino acid residues 1 to 67. Both the strands have a terminal $-NH_2$ group which dips in the cytoplasmic region of the cell. The tail of the strands also has an acyl substituent which anchors the receptor to the plasma membrane.

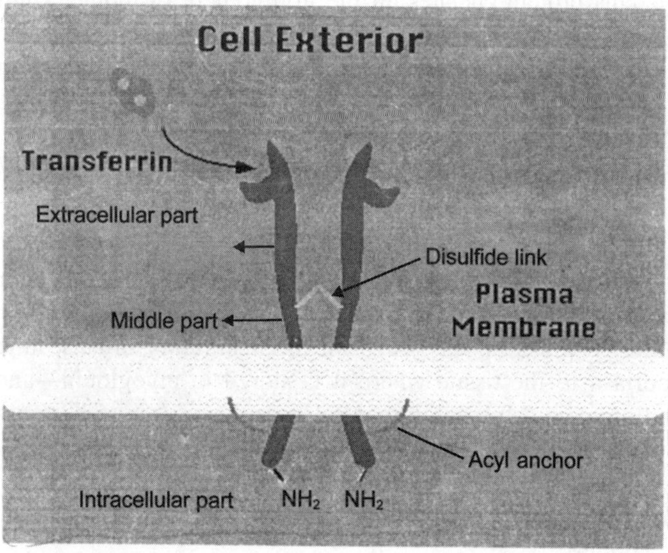

Fig. 9.18 Schematic representation of transferrin receptor

The middle portion of the receptor resembles a passage enclosed within two walls. Each wall represents a monomer strand and is built up by amino acid residues 67 to 88. The middle portion imparts transmembrane character to the receptor, meaning that the passage allows the flow of chemicals from outside to inside the cell and *vice-versa*.

The upper portion of the receptor is constituted by amino acid residues 89 to 760. In fact, the amino acid residue 89 in each strand is joined through a disulfide link, which unites the two monomers into a dimer. The upper portion of the receptor is extracellular, that is, extends outside the cell. It binds the transferrin molecule, pushes it through the passage-like middle portion to the intracellular tail.

Under physiological conditions, iron(III) ion is tightly bound to transferrin. In fact, the equilibrium constant for the reaction between an apotransferrin molecule and an iron(III) ion is 10^{24} mol^{-1}. However, the receptor delivers transferrin into an acidic compartment of the cell, where the pH is around 5.4. Under acidic conditions, iron(III) ion gets reduced to iron(II) ion. The latter does not have affinity for apotransferrin and, therefore, gets detached. The residual, apotransferrin, as well as the receptor are released for the next cycle.

Case Study 9.4 Why is mother's milk considered a healthier option for the infant than powdered milk?

It is a common knowledge that bottle-fed babies are more likely to suffer gastric infections than breast-fed toddlers. The mother's milk and the powdered milk both contain iron ions. The siderophore bacteria, which cause gastric infection in kids, thrive on iron ions. The mother's milk contains a transferrin-like protein called *lactoferrin*. This protein chelates all the iron content of mother's milk, leaving nothing for siderophores. The bacteria are starved of a vital resource and the chances of infection get reduced. The formula milk, on the other hand, does not contain human lactoferrin. The iron content of the feed is consumed by siderophores, which then tend to multiply and cause infection in the infant.

Hemoglobin

Hemoglobin, sometimes spelled as haemoglobin, and commonly abbreviated as Hb is an iron-containing oxygen-transport metalloprotein in the red blood cells of vertebrates. It picks up oxygen from the respiratory organs, lungs or gills, and transports it to the tissue where it is stored in myoglobin—another iron-based metalloprotein. Myoglobin then releases the oxygen to burn nutrients and provide energy necessary to sustain the organisms. In this process, carbon dioxide is released. It is collected by hemoglobin and carried back to the respiratory organs from where it is dispensed off. One liter of human blood contains about 150 g hemoglobin

Structure The basic unit in the structure of hemoglobin is the *heme* group. The heme group consists of an iron(II) ion, held in a chelating heterocyclic ring, called *porphyrin*. The porphyrin ring, in turn, consists of four pyrrole molecules cyclically linked together with conjugated double bonds (Fig. 9.19). Various organic substituents are attached to each pyrrole ring. A heme group is shown in Fig. 9.20.

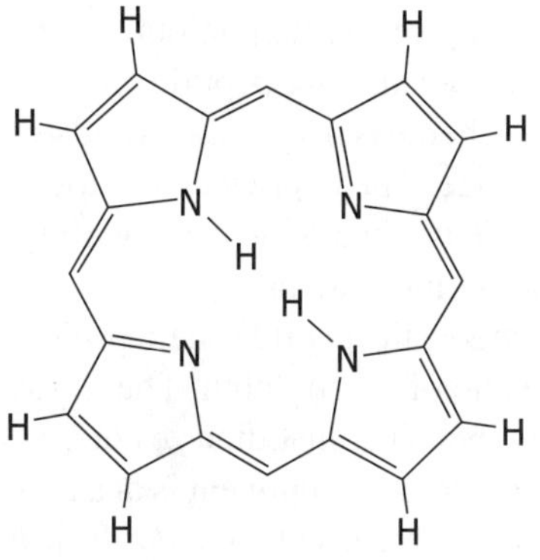

Fig. 9.19 A porphyrin skeleton

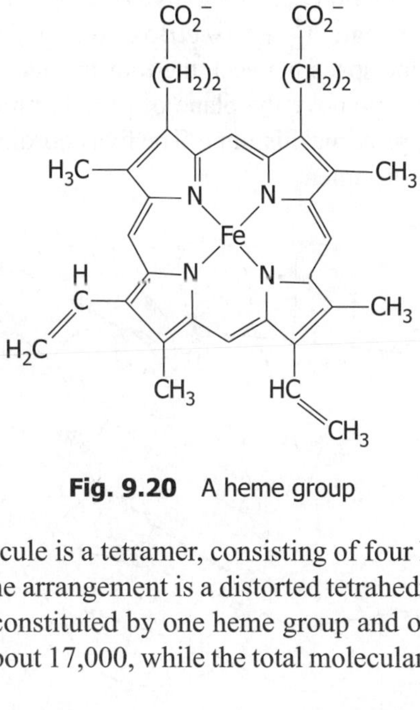

Fig. 9.20 A heme group

A hemoglobin molecule is a tetramer, consisting of four heme groups, bound to four protein chains. The arrangement is a distorted tetrahedron with a cavity in the center. Each subunit, constituted by one heme group and one protein chain, has a molecular weight of about 17,000, while the total molecular weight of hemoglobin is about 64,000.

Two subunits have a protein chain of 141 amino acids and are designated as α subunits; the remaining two have a protein chain of 146 amino acids and are designated as β subunits. For this reason, hemoglobin is often referred to as $\alpha_2\beta_2$ tetramer.

Hemoglobin transfers oxygen to tissue by a reversible mechanism. However, when a heme group, *under laboratory conditions*, interacts with oxygen, in presence of water, it is irreversibly oxidized into an oxo-bridged dimer, called *hematin*, $-Fe^{III}-O-Fe^{III}-$. *This does not happen in the living system.* Otherwise, all the hemoglobin would have been precipitated as hematin and rendered useless for oxygen transport.

In a cell, each heme group of the hemoglobin is engulfed in a peptide chain. Due to the presence of the surrounding peptide chain the heme group is held in a minimum possible space. If heme is irreversibly oxidized to dimeric hematin, it has to grow almost double its size. The peptide chain does not leave out space for the expansion of heme group. This type of steric hinderance prevents the conversion of heme to hematin in the living systems.

Two histidine components of the peptide chain project towards the iron(II) ion of each heme unit in a molecule of hemoglobin. The nitrogen of one histidine comes quite close to the iron(II) ion. This histidine residue is called *proximal histidine.* The nitrogen of the other histidine component is at a relatively larger distance from iron(II) ion. This histidine residue is called *distal histidine.*

As shown in Fig. 9.20, the iron(II) ion in a heme group is bound to four nitrogen atom. The nitrogen of proximal histidine too becomes attached to the iron(II) ion, raising the latter's coordination number to five (Fig. 9.21). The iron(II) ion, a $3d^6$ system, is in a high spin state, $t_{2g}^4\ e_g^2$ (*see* Box 26.2, p. 716) and its ionic radius (92 pm) is larger than the space between the four nitrogens of the heme group. It therefore lies about 42 pm above the plane of pyrrole nitrogens, being projected towards the nitrogen of proximal histidine. The five coordinated iron(II) ion is thus in a square pyramid environment.

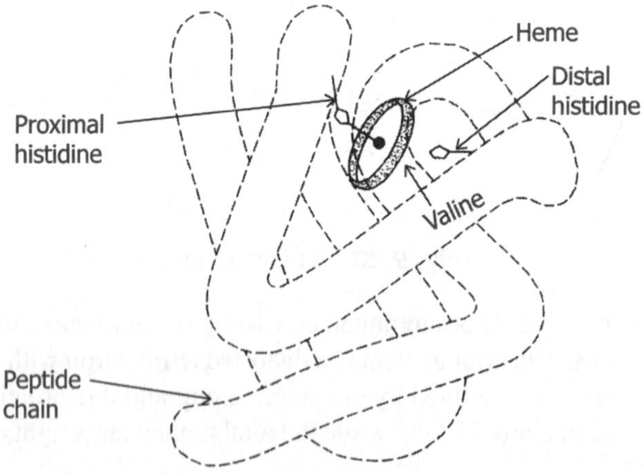

Fig. 9.21 Coordination of proximal histidine to iron(II) ion in heme

Mechanism The sixth coordination position can reversibly bind oxygen and transform the square pyramid geometry to an octahedral one. However, as shown in Fig. 9.22, a valine component of the peptide chain which engulfs the heme group tends to block the passage of oxygen molecule towards the sixth coordination site of iron(II) ion. For this reason, the deoxygenated hemoglobin is said to be in a *tense state* or *taut state*, often abbreviated as *T-state.*

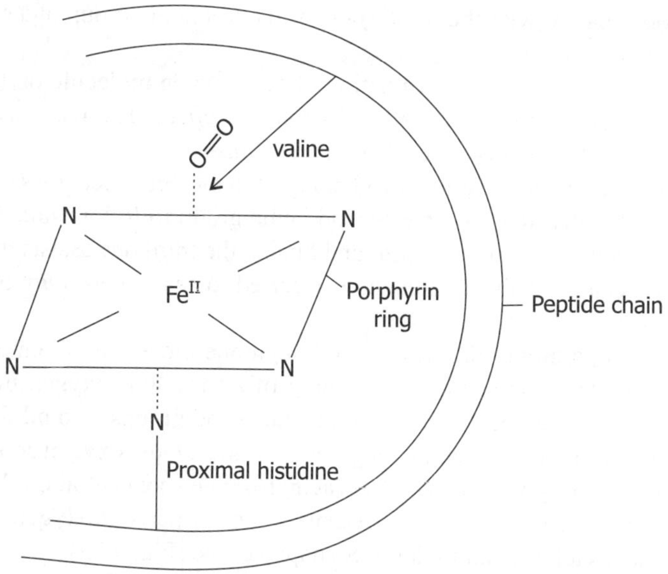

Fig. 9.22 Six coordinated iron(II) ion in heme

During respiration, the iron(II) ion in each of the four heme groups of a hemoglobin molecule becomes coordinated to an oxygen molecule. However, all four heme groups do not become oxygenated simultaneously. It is a stepwise phenomena. The coordination of the first oxygen molecule to a heme moiety is most difficult due to the blockade of the sixth coordination site by valine residue of the peptide chain (Fig. 9.22). Nevertheless, the coordination does take place. A fallout of the coordination is that iron(II) ion is transformed from a high spin state ($t_{2g}^4\ e_g^2$) to a low spin state ($t_2{}^6{}_g$). Its hydrated radius decreases (*see* Chapter 26, p. 724). As a result, the iron(II) ion, which had occupied a position above the plane of porphyrin ring, now tends to partially drop towards the space in between the four nitrogens. The proximal histidine nitrogen, moves closer to the iron(II) ion. The conformation of the entire peptide chain is disturbed.

Since the peptide chains of all heme groups are linked, the conformational change in the first heme moiety induces perturbations in the second heme moiety. As a result of these perturbations, the valine residue of the peptide chain engulfing the second heme group blocks the pathway of the second oxygen molecule to a lesser extent. The second oxygen molecule therefore becomes coordinated to the hemoglobin molecule more easily than the first one. The conformation of the peptide chain surrounding the third heme group alters and the valine residue moves further away from the iron(II) ion. The third oxygen molecule becomes attached to the hemoglobin molecule still more easily. Simultaneously, the iron(II) ions in each of the four heme groups drop to the plane of nitrogens in the porphyrin rings. The hemoglobin is now said to be in a *relaxed state* or *R-state*.

In the relaxed state, the valine residue of the peptide chain engulfing the fourth heme group moves far away from the iron(II) ion. For the fourth molecule of oxygen, coordination with the iron(II) ion of the last heme group in a molecule of hemoglobin is a cakewalk.

The influence exerted by coordination of one oxygen molecule on the binding of another oxygen molecule is called *homotropic effect*. The pathway by which homotropic effect occurs is called *Perutz mechanism*.

Thus once the first heme group in a hemoglobin molecule becomes oxygenated, albeit with difficulty, it assists the second heme group to follow suit. The second heme group, in turn, assists the third; and finally, the third one assists the fourth to become oxygenated. This phenomena is referred to as *cooperativity* between the heme groups.

The term cooperativity implies that although one molecule of hemoglobin can add on four oxygen molecules, the binding of all the four oxygen molecules is not independent. It is not the presence of four heme groups to bind four oxygen molecules that is important. What is significant is that at low oxygen concentrations hemoglobin is sparingly oxygenated and at high oxygen concentrations hemoglobin is excessively oxygenated. For this reason, a plot of percent oxygenation versus partial pressure yields a sigmoidal or S-shaped curve (Fig. 9.23).

Moreover, for the biochemical reaction that occurs in lungs (or gills),

$$Hb + 4O_2 \rightarrow Hb(O_2)_4$$

the equilibrium constant is expressed as,

$$KHb = \frac{[Hb(O_2)_4]}{[Hb]\, p^n_{O_2}}$$

The exponent n is called *Hill constant*. Its value would have been 4 if the four heme groups in a molecule of hemoglobin would have acted independently while absorbing oxygen. However, due to their cooperative action, the value of n is 2.8.

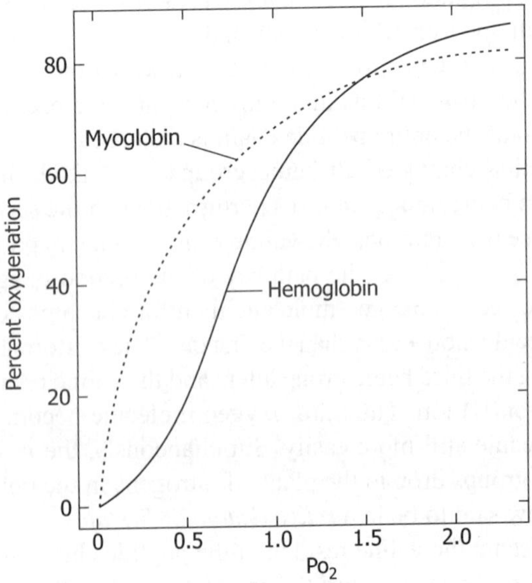

Fig. 9.23 Oxygen saturation curve for hemoglobin and myoglobin

The oxygen transported by hemoglobin to a specific tissue is utilized in a metabolic process. Carbon dioxide is released as a metabolite product. Its reaction with water, catalyzed by the enzyme carbonic anhydrase, yields carbonic acid. The latter dissociates into protons and bicarbonate ions.

$$CO_2 + H_2O \xrightarrow{\text{Carbonic anhydrase}} H_2CO_3$$

$$H_2CO_3 \rightarrow H^+ + HCO_3^-$$

Hence blood with high carbon dioxide levels has a low pH. When a fresh batch of oxygenated hemoglobin reaches the tissue, it encounters acidic conditions. Under these conditions, the protons and the bicarbonate ions become attached to the protein and amino acid components of hemoglobin respectively. The resultant, *carbaminohemoglobin* undergoes a conformational change which facilitates the release of oxygen. The decrease in hemoglobin's affinity for oxygen at low pH, due to bonding of protons and bicarbonate ions is called *Bohr effect.*

When the blood reaches lungs it comes across a carbon dioxide-deficient and oxygen-rich environment. Here hemoglobin sheds carbon dioxide, acquires affinity for oxygen and the cycle starts once again.

Myoglobin

An iron(II)-based metalloprotein, found in the muscles of vertebrates, myoglobin resembles hemoglobin both in structure and function. Whereas hemoglobin collects oxygen from lungs or gills and transports it to the blood capillaries of the tissue, myoglobin receives oxygen from the red blood cells of the tissue and transports it further to the mitochondria of muscle cells. Here oxygen is used in cellular respiration to produce energy.

Myoglobin not only supplies oxygen to the muscles, it acts as its reservoir as well. The organisms can withdraw oxygen from this reserve during increased metabolic activity. Thus under harsh conditions, myoglobin promptly releases oxygen and protects the body tissue from getting damaged.

If the concentration of myoglobin is high within the muscle cells, the organisms are able to hold their breath longer. Diving mammals like whales and seals have exceptionally high levels of myoglobin in their muscles.

Structure A molecule of myoglobin consists of a single heme group with an engulfing peptide chain containing 153 amino acid residues. It has a molecular weight of about 17,000.

As in the case of hemoglobin, the heme group is constituted by a porphyrin ring with an iron(II) ion in the center. The four nitrogens of porphyrin coordinate to the metal ion. However, the ionic radius of iron(II), which is in a high spin state ($t_{2g}^4\ e_g^2$) is greater than the cavity created by four pyrrole nitrogens of the porphyrin ring. For this reason, the pyrrole nitrogens and the iron(II) ion are not at the same level, the latter being about 42 pm above the plane of coordinating nitrogen atoms.

The heme group is embedded in a tight crevice of a hydrophobic protein chain having eight α helices. The engulfing protein chain prevents the irreversible oxidation of myoglobin to hematin. Two histidine residues protrude out of the protein chain and orient towards the iron(II) ion. One of these comes very close to the iron(II) ion and becomes coordinated to it. This is the *proximal histidine* residue. The other, called *distal histidine*, approaches the iron(II) ion from the end opposite to that of proximal histidine. As its name implies, it stations itself at an appreciable distance from the iron(II) ion and hence remains uncoordinated to it.

Thus, the iron(II) ion in myoglobin is 5-coordinated—being bound to four nitrogens of the porphyrin ring and the fifth nitrogen of proximal histidine residue (Fig. 9.24). In this state, myoglobin is designated as *deoxymyoglobin*. When an oxygen molecule becomes bound to the sixth coordination site of iron(II) ion, myoglobin is designated as *oxymyoglobin*.

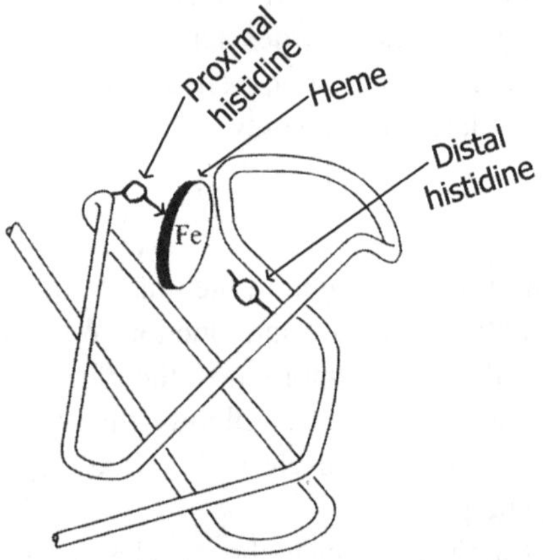

Fig. 9.24 Coordination of proximal histidine to iron(II) ion in myoglobin

As shown in Fig. 9.25, the oxygen molecule in oxymyoglobin is coordinated to the iron(II) ion in an angular or bent fashion. In hemoglobin too, there is bent coordination of oxygen to iron(II) ion.

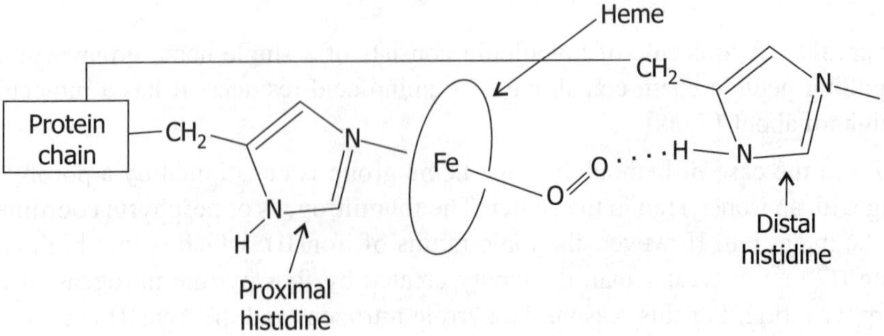

Fig. 9.25 Hydrogen bonding between distal histidine and angularly coordinated oxygen molecule in myoglobin

Even though the distal histidine is not coordinated to iron(II) ion of the heme group it serves three important functions.

1. It forms hydrogen bonds with the coordinated oxygen molecule (Fig. 9.25). As a result, the binding constant of oxygen to the heme group increases (both in hemoglobin and in myoglobin).
2. It prevents the coordination of carbon monoxide to the heme group. Carbon monoxide binds to iron(II) ion of both hemoglobin and myoglobin 200 times better than oxygen. Once it becomes coordinated to the iron(II) ion, the heme group loses its ability to transport oxygen to the tissue, resulting in episodes of carbon monoxide poisoning. The coordination of carbon monoxide to iron(II) ion is end-on, that is, results in a linear geometry. The distal histidine sterically hinders the linear attachment of carbon monoxide to iron(II) ion.
3. It prevents two oxymyoglobin molecules to come into close proximity of each other and thereby preclude their dimerization.

The archetype underlining the three functions of distal histidine—the first two applicable both to hemoglobin and myoglobin and the third one to myoglobin alone– is called *picket fence porphyrin model.*

Mechanism As has already been stated, hemoglobin gathers oxygen from lungs or gills and transfers it to the tissue from where it is picked up by myoglobin for respiration and/or metabolism. This means that myoglobin binds to oxygen more tightly than does hemoglobin. It is the difference in binding energy that is responsible for the movement of oxygen from hemoglobin to myoglobin or, in effect, from blood stream to the tissue.

The binding of oxygen to myoglobin follows the simple equilibrium reaction,

$$\mathrm{Mb} + \mathrm{O_2} \rightleftharpoons \mathrm{MbO_2} \tag{9.2}$$

The equilibrium constant for this reaction may be expressed as follows.

$$K_{Mb} = \frac{[\mathrm{MbO_2}]}{[\mathrm{Mb}]p_{O_2}} \tag{9.3}$$

Where p_{O_2} is the partial pressure of oxygen. The fraction of oxygen-bearing myoglobin may be mathematically written as,

$$f = \frac{[\mathrm{MbO_2}]}{\{[\mathrm{Mb}] + [\mathrm{MbO_2}]\}} \tag{9.4}$$

Expanding equation (9.4),

$$f[\mathrm{Mb}] + f[\mathrm{MbO_2}] = [\mathrm{MbO_2}]$$

or

$$f[\mathrm{Mb}] = (1 - f)\,[\mathrm{MbO_2}]$$

$$\frac{[MbO_2]}{[Mb]} = \frac{f}{1-f} \tag{9.5}$$

Substituting the value of $\frac{[MbO_2]}{[Mb]}$ from eq (9.5) in equation (9.3),

$$K_{Mb} = \frac{f}{(1-f)p_{O_2}}$$

or

$$f = K_{Mb}p_{O2} - fK_{Mb}p_{O2}$$

$$f(1 + K_{Mb}p_{O2}) = K_{Mb}p_{O2}$$

$$f = \frac{K_{Mb}\,p_{O_2}}{1 + K_{Mb}p_{O_2}}$$

$$\frac{1}{f} = 1 + \left(\frac{1}{K_{Mb}}\right)\frac{1}{p_{O_2}} \tag{9.6}$$

The myoglobin curve, shown in Fig. 9.23, corresponds to equation 9.6.

The shape of the myoglobin curve is different from that of the hemoglobin curve. The reason being that unlike hemoglobin, myoglobin does not show cooperative binding to oxygen. As has been explained in the previous section, hemoglobin is a tetramer. The four heme groups in a molecule of hemoglobin bind to oxygen by way of cooperativity, that is, one heme group assists the other in taking up oxygen. On the other hand, myoglobin is a monomer. One molecule of myoglobin has one heme group which can bind to only one molecule of oxygen. Hence the absence of cooperativity effect.

When the oxygen molecule becomes coordinated to the heme group of myoglobin, the iron(II) ion transitions from the high spin state ($t_{2g}^4\ e_g^2$) to a low spin state t_{2g}^6). Its size shrinks and it drops down to the plane of porphyrin nitrogens. It is in this state that myoglobin enters the tissue, where it encounters low partial pressure of oxygen. It now sheds the coordinated oxygen molecule. This may be explained by either of the following two interpretations.

(a) As shown in Fig. 9.23, myoglobin is less oxygenated at low partial pressures of oxygen. This means that when the partial pressure of oxygen is low, as it is within the tissue, oxymyoglobin releases oxygen.

(b) If Le Chatelier principle is applied to reaction 9.2, then at low oxygen concentration (read partial pressure), the equilibrium reaction shifts to left.

Whether one follows interpretation (a) or (b), the final outcome is same: that within the cell, oxymyoglobin (MbO_2) dissociates into myoglobin (Mb) and oxygen (O_2). This oxygen is used by the cell for its sustenance.

Box 9.2

An Alternative Approach to Oxygenation of Hemoglobin

We have stated that iron exists in a high spin oxidation state(II) in hemoglobin and that on oxygenation it is transformed to a low spin state, although its valency remains the same. The ionic radius of low spin iron(II) ion is smaller than that of high spin iron(II) ion. Therefore, when hemoglobin becomes oxygenated, the iron(II) ion tends to move down in the plane of pyrrole nitrogens of the heme group. This results in the cooperativity of heme groups to get a molecule of hemoglobin oxygenated.

There is an inherent drawback in the aforesaid postulate. Under laboratory conditions, oxyhemoglobin is found to be diamagnetic. No doubt, iron(II) ion, in a low spin state, is diamagnetic (t_{2g}^6 system), yet the coordinated oxygen molecule is paramagnetic with two unpaired electrons (*see* Table 3.4, p. 123), and, therefore, overall magnetic behavior of hemoglobin ought to be paramagnetic.

To overcome this discrepancy, an alternate model to the oxygenation of hemoglobin has been proposed. According to this model, the iron(II) ion in hemoglobin, upon oxygenation, is transformed to low spin iron(III) ion. The latter is a $3d^5$ system (t_{2g}^5 or $d_{xy}^2\ d_{xz}^2\ d_{yz}^1$; *see* Box 26.2, p. 716), having one unpaired electron. The oxygen molecule, on coordination to iron(III) ion is converted to a superoxide ion (O_2^-).

The superoxide ion too is paramagnetic, with one unpaired electron (*see* Table 3.4, p. 123). The unpaired electron on iron(III) ion couples with the unpaired electron on superoxide ion and thus oxyhemoglobin acquires diamagnetic character. Moreover, since iron(III) ion has a smaller ionic radius as compared to iron(II) ion (*see* Chapter 2, p. 60), it moves closer to the porphyrin plane of the heme group. This, in turn, results in the cooperativity of heme groups in getting the hemoglobin oxygenated.

The Biochemistry of Calcium

Calcium makes up about 2% of an adult body weight. Its distribution in the body is monitored by two hormones, parathormone and calcitonin as well as by vitamin D. The level of these substances in the blood controls—

- The amount of calcium absorbed in the intestine.
- The concentration of calcium in blood.
- The amount of calcium deposited in the bones and teeth.
- The amount of calcium excreted by the kidneys.

Calcium is important for the formation of strong bones and teeth. In fact, 99% of calcium in the body is found in bones and teeth as insoluble, inorganic salts, held in protein frameworks. The remaining 1% of calcium assists in—

- Transmission of nerve impulses.
- Contraction of muscles.
- Clotting of blood.

The biochemical functions of calcium are now reviewed.

1. Role of calcium in development of bones and teeth

To a layman, bones are static entities, where a very little change occurs. Yet, scientifically speaking, a bone is a dynamic part of the body and calcium is a key player in maintaining its health. There are two broad components of a bone: tissue and minerals.

The bone tissue are made up of special type of cells called *Osteoblasts*. The cells synthesize a fibrous protein called *collagen*. The fibers of collagen create a framework in which the second component of bones, that is, the minerals accumulate. This process is called *bone mineralization*. The main mineral present in the bone structure is hydroxyapatite, $Ca(OH)_2 \cdot 3[Ca_3(PO_4)_2]$.

The collagen framework and the bone minerals do not exist in isolation of each other. Rather, the two interact to create what is termed as the *bone matrix*. So strong is the interaction of collagen and minerals within the matrix that the bones acquire great tensile strength (Fig. 9.23a) and compressional strength (Fig. 9.23b). However, bones do not have a high level of torsional strength (Fig. 9.23c), that is, ability to endure twisting. In fact, bone fractures, often occur due to exertion of torsional forces on the limbs.

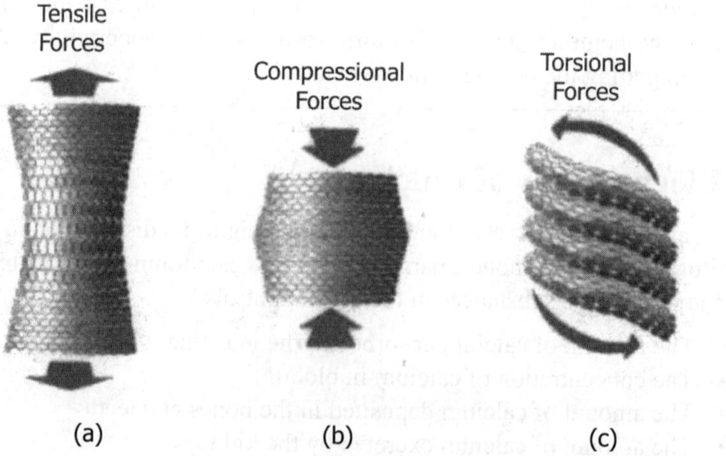

Fig. 9.26 Bones can withstand (a) tensile forces and (b) compressional forces, but not (c) torsional forces

Accumulation of calcium in the bones continues until it peaks in early adulthood, near the age of 35 years. Thereafter, the calcium content of the bones is extracted into blood to perform other essential functions of the body. The decline of calcium in the bones occurs at a rate of about 1% per year. The progressive weakening of bones, resulting in spontaneous fractures and compression of spine, is termed *osteoporosis*.

Like bones, teeth too have the protein framework incorporating calcium minerals. As shown in Fig. 9.27, enamel, dentine and cementum are the important components of a tooth. These are composed of the following calcium-based minerals.

Predominant: Hydroxyapatite, $Ca(OH)_2 \cdot 3[Ca_3(PO_4)_2]$

Trace: (i) Fluorapatite, $CaF_2 \cdot 3[Ca_3(PO_4)_2]$

(ii) Chlorapatite, $CaCl_2 \cdot 3[Ca_3(PO_4)_2]$

(iii) Dahllite, $CaCO_3 \cdot 3[Ca_3(PO_4)_2]$

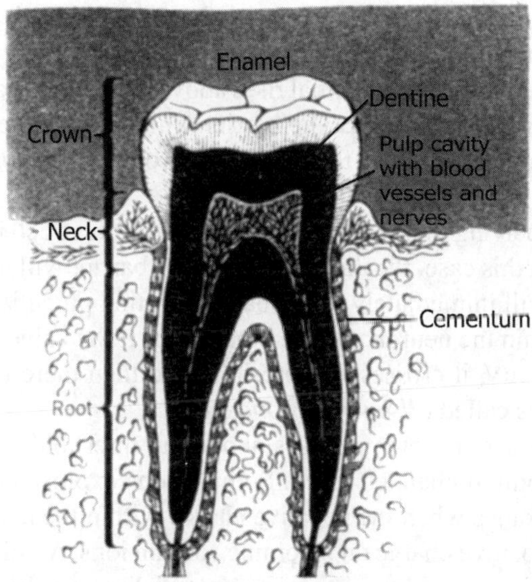

Fig. 9.27 The structure of a tooth

Just as loss of calcium from bones results in osteoporosis, its losts from teeth results in tooth decay.

2. Role of calcium in transmission of nerve impulse

The plasma membrane of nerve cells, also called *neurons*, is endowed with voltage-gated ion channels. These channels are permeable to a specific ion at a specific voltage. The voltage of the cell arises because there is a potential difference across the plasma membrane. The potential difference, in turn, originates because the nerve cell membrane is polarized, that is, its interior and exterior have different electrical charges.

The outside of a nerve cell contains excess of sodium ions and hence maintains a positive charge. The inside of the cell contains potassium ions which are outnumbered by negatively charged nucleic acid and protein residues. Thus inside the cell, the overall charge is negative.

When the nerve cell does not receive a stimulus, the potential difference between its exterior and interior is called *resting potential.* The value of resting potential is around –70 mV. This may seem to be a small value; however, since the size of the cell is very small, the voltage of this magnitude represents a reasonably strong electrical force.

When the nerve cell receives a stimulus, the voltage-gated channel, permeable to sodium ions, open. These channels are called *Hodgkin-Huxley* or *Nav*[1] channels, v denoting voltage. The sodium ions crossover from outside to inside the neuron, partly neutralizing the negative charge in the interior of nerve cell. As a result, the voltage of the cell begins to rise, that is, tends to move from negative to zero and then from zero to positive scale. Under these circumstances, the nerve cell is said to possess *action potential*, which may be defined as a short-lived event when the voltage of the neuron is higher than that in resting state.

However, only if the action potential rises above a threshold value, will the nerve impulse travel across the neuron. The barrier which must be crossed so that the impulse passes from the head to the tail of a neuron is called *threshold potential.* Its value is about –55 mV.

Suppose, on receiving a stimulus, the voltage of the nerve cell increases from –70 mV to –60 mV. In this case, the threshold potential barrier will not be crossed. The action potential will immediately revert back to resting potential. No impulse will be conducted within the neuron. On the other hand, if the value of action potential is more than –55 mV, it crosses the threshold and then there is no turning back. This phenomena is called *all-or-none concept.*

As soon as the action potential rises above the threshold value of –55 mV, a lot many more sodium channels open. Sodium ions explosively flood inside the neuron. A stage comes when the negative charge within the neuron is completely balanced by the positive charge on incoming sodium ions. At this stage, the neuron is said to be *depolarized* and its membrane potential has a valve of zero mV. Even after depolarization, sodium ions keep drifting inside the neuron and the membrane potential peaks to about +50 mV.

When the action potential reaches its peak, the sodium channels close and the potassium channels open. The sodium ions remain trapped inside the neuron while the potassium ions ooze out. The efflux of potassium ions decreases the action potential. Thus after rising to a maxima the action potential abruptly shoots down. Its value decreases from +50 mV to zero mV and then slides to negative scale.

When the voltage-gated potassium channels finally close, the neuron has slightly more potassium ions on the outside than it has sodium ions on the inside. This causes the membrane potential to drop below the value of resting potential. In this state, the plasma membrane is said to be *hyperpolarized.* It is so named because the value of its voltage temporarily becomes more negative than when it was in the resting state. Hyperpolarization ensures that the action potential does not travel back the way it came.

As the phase representing hyperpolarization touches the voltage minima, the sodium/potassium pumps become activated. The sodium ions and potassium ions are restored to their original sites, sodium ions outside and potassium ions inside the neuron. The action potential now becomes equal to the resting potential. The time lag between the beginning of hyperpolarization and restoration of resting potential is called *refractory period*. During this period, the neuron does not respond to a fresh stimulus. It is only when the action potential falls back to the resting potential and the refractory period ends that the neuron picks up a new stimulus.

An action potential is often referred to as *spike* and its passage through a neuron is called *spike train*. A spike train is depicted in Fig. 9.28. It accomplishes its passage through the neuron in 7 milliseconds—faster than the lightning strikes!

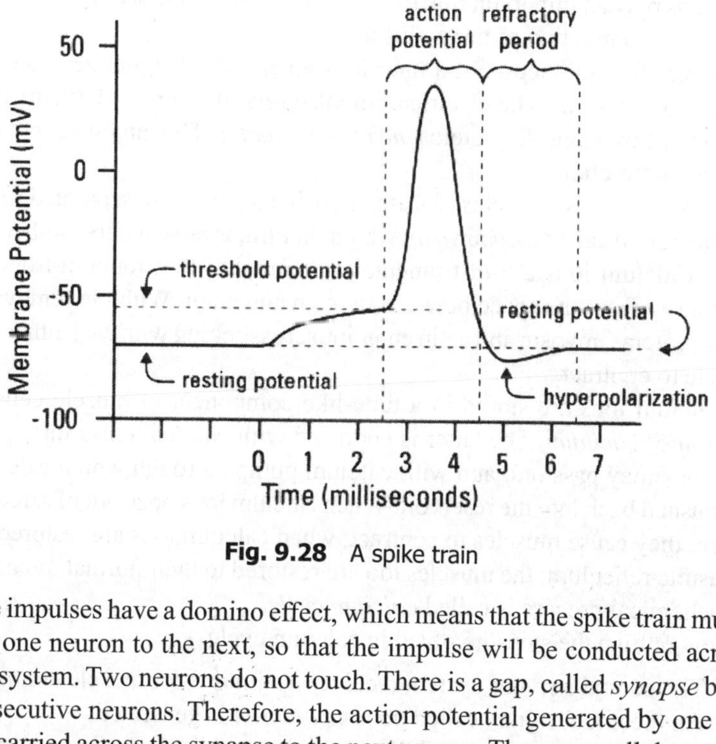

Fig. 9.28 A spike train

Nerve impulses have a domino effect, which means that the spike train must pass on from one neuron to the next, so that the impulse will be conducted across the nervous system. Two neurons do not touch. There is a gap, called *synapse* between two consecutive neurons. Therefore, the action potential generated by one neuron must be carried across the synapse to the next neuron. The nerve cell that transmits the impulse across the synapse is called *presynaptic neuron*; the one which receives the impulse is called *postsynaptic neuron*.

Calcium ions play a key role in transmitting the action potential from presynaptic to postsynaptic neuron. When the spike train reaches the tail end of the presynaptic neuron, the calcium channels in the plasma membrane open. Calcium ions rush inside the presynaptic neuron and release neurotransmitters into the synapse. The most important neurotransmitter found in the nervous system is acetylcholine (Fig. 9.29).

The neurotransmitter picks up the action potential from the presynaptic neuron and simultaneously binds to the neuroreceptor proteins housed in the postsynaptic neuron. Thus a passage is generated for the smooth transference of action potential from presynaptic to postsynaptic neuron. Thereafter, the postsynaptic neuron acts as a presynaptic cell for the next neuron.

After the neurotransmitter has performed its function, its constitution changes. It then breaks away from the neuroreceptor and re-enters the synapse, where its original structure is restored. Thereafter, it passes into the neuron and braces for the next spike train.

In conclusion, if calcium ions are not present in appropriate amount, neurotransmitter will not be released and there will be no pathway to get the impulse across the synapse—and hence across the nervous system.

3. Role of calcium in muscle contraction

Calcium ions play an important role in muscle contraction, specially in maintaining the rhythmeic contraction of heart muscles.

The trigger for contraction of a muscle is an electrical signal generated by the central nervous system. The electrical impulse sets of a series of events that lead to the union of two muscle proteins, *myosin* and *actin*. This union generates force that causes contraction.

When the muscle is in a relaxed state, myosin and actin are separated form each other by a protein called *tropomyosin* which, in turn, is attached to another protein, *troponin*. Calcium ions are instrumental in removing the troponin-tropomyosin combine away from the space between myosin and actin. With the removal of the buffering proteins, myosin and actin enter into crosslinking with each other, causing the muscle to contract.

The calcium ions are stored in a tube-like component of muscle cells, called *sarcoplasmic reticulum*. The latter is endowed with *calcium gates* through which calcium ions may pass out, and with calcium pumps through which calcium ions may be pushed back into the reservoir. When calcium ions ooze out of sarcoplasmic reticulum, they cause muscles to contract; when calcium ions are restored back to sarcoplasmic reticulum, the muscles too are restored to their normal, relaxed state. This biochemical process is called *calcium cycle*.

We now outline the steps involved in calcium cycle.

(i) The central nervous system receives a stimulus for muscle contraction.

(ii) An electrical signal, called *action potential* is generated.

(iii) The action potential travels down the nerve cell, also called *neuron*. There is a gap between the neuron and the muscle cell. This gap, called synapse, houses a chemical, acetylcholine, the structure of which is shown in Fig. 9.29.

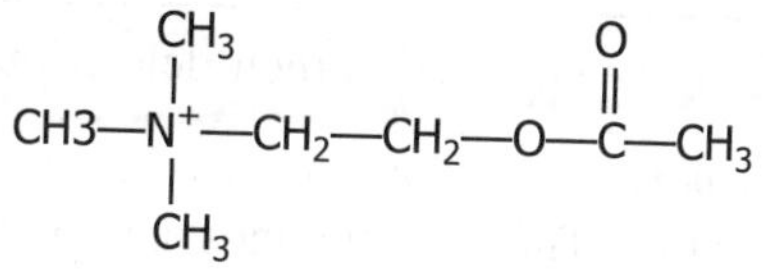

Fig. 9.29 Structure of acetylcholine

(iv) When the action potential reaches the tail of neuron, it releases a chemical message to acetylcholine.

(v) On receiving the message, acetylcholine transmits the action potential to the muscle cell (Fig. 9.30a). For this reason, acetylcholine is referred to as *neurotransmitter*.

(vi) On reaching the muscle cell, the action potential opens the calcium gates of sarcoplasmic reticulum releasing calcium ions (Fig. 9.30b).

(vii) The released calcium ions bind to troponin. The shape and the dimensions of troponin change due to its coordination to a metal ion. As a result of the alteration in the geometry of troponin, the attached tropomyosin slides away from the grove between myosin and actin. In the absence of any barrier, myosin and actin form crossbridges. Force is generated. The muscle cell shortens—and contracts (Fig. 9.30c).

(viii) After the action potential has passed, the calcium gates on sarcoplasmic reticulum close and calcium pumps become operational. The pumps put back calcium ions into the sarcoplasmic reticulum. On detaching from calcium ions, the troponin acquires its original configuration. Tropomyosin slide back to block the space between myosin and actin. The muscle relaxes (Fig. 9.30d).

(ix) Simultaneously, acetylcholine is degraded by the enzyme acetylcholinesterase. The link between the neuron and the muscle cell temporarily breaks down. The neuron replenishes synapse with acetylcholine (Fig. 9.30e).

(x) The cycle restarts.

A unique feature of muscle cells which distinguishes these from other types of cells is the presence of fiber strands called *myofilaments*. The myofilaments are classified into two types: fast-twitch and slow-twitch. *Fast-twitch* fibers are capable of developing greater forces. These contract faster and have greater anaerobic capacity. *Slow-twitch* fibers develop force slowly, can maintain contractions longer and have higher aerobic capacity. The sarcoplasmic reticulum in fast-twitch fibers is more developed than in slow-twitch fibers, enabling faster calcium release—and hence faster contraction. Muscles contain a mixture of fast-and slow-twitch fibers, although in any given muscle one type predominates.

Lower than optimum levels of calcium in blood cause *hypocalcemia*. The symptoms of hypocalcemia are muscle cramps and twitching of facial muscles. Higher than optimum levels of calcium in blood cause *hypercalcemia*. The symptoms of hypercalcemia are sluggishness and weakness of muscles.

Case Study 9.5 How do *nerve gases* elicit their deleterious effects?

Nerve gases, also called *nerve agents* are a set of compounds belonging to the organophosphate classification. These compounds disrupt the mechanism by which a nerve impulse is conducted from a neuron to an organ.

The gap between two nerve cells or between a nerve cell and the cell of a particular organ, called synapse, is filled with a chemical that acts as a neurotransmitter, that is, it passes the message from one neuron to the next or from a neuron to the cell of an organ. The most common neurotransmitter is acetylcholine. After providing a smooth passage to the nerve impulse, acetylcholine is degraded by an enzyme, acetylcholinesterase. Only after the neurotransmitter has been degraded, the organ which receives the impulse relaxes.

Nerve gases inhibit the function of acetylcholinesterase. They block those sites on the enzyme which are meant to bind the neurotransmitter and eventually degrade it. When the neurotransmitter does not degrade, the organ to which the nerve signal reaches fails to relax. The person suffers from contraction of pupils, profuse salivation, convulsions and eventual death by asphysxiation as control is lost over respiratory muscles.

Nerve gases were extensively used during World War I. Now these are classified as *weapons of mass destruction* by *UN resolution 687*. Their production and stockpiling is banned by the *Chemical Weapons Convention*, which came into force in April 1997.

4. Role of calcium in clotting of blood

The clotting of blood ensues whenever the blood vessels are cut or get damaged. If clotting, also called *coagulation*, does not occur, incessant bleeding will continue unabated, leading to shock, and possible death. A blood clot typically consists of platelets, trapped in a network of insoluble fibrin molecules. Together, these components form a mesh that stops the flow of blood. The process of blood coagulation involves a complex pathway which requires calcium ions at critical steps. If calcium is removed from the system, clotting does not occur. Therefore, blood banks add chelating agents like ethylenediamine tetraacetic acid (EDTA) to the blood received from donors. The chelating ligands, in turn, remove calcium ions from blood by way of complexation. The blood does not clot.

The following are the main steps involved in the clotting of blood.

(i) An injury to the tissue induces the production of a protein called *factor 10*. Calcium ions are necessary for the activity of factor 10.

(ii) Factor 10 binds to another protein, *Factor 5* which is already present in the blood. Together they form an enzyme called *prothrombinase*.

(iii) Prothrombinase catalyzes the conversion of prothrombin (which is already present in blood) to a protein called *thrombin*. This reaction is accelerated by the binding of calcium ions to both prothrombinase and prothrombin. If calcium ions are not present, the conversion will be too slow to be useful.

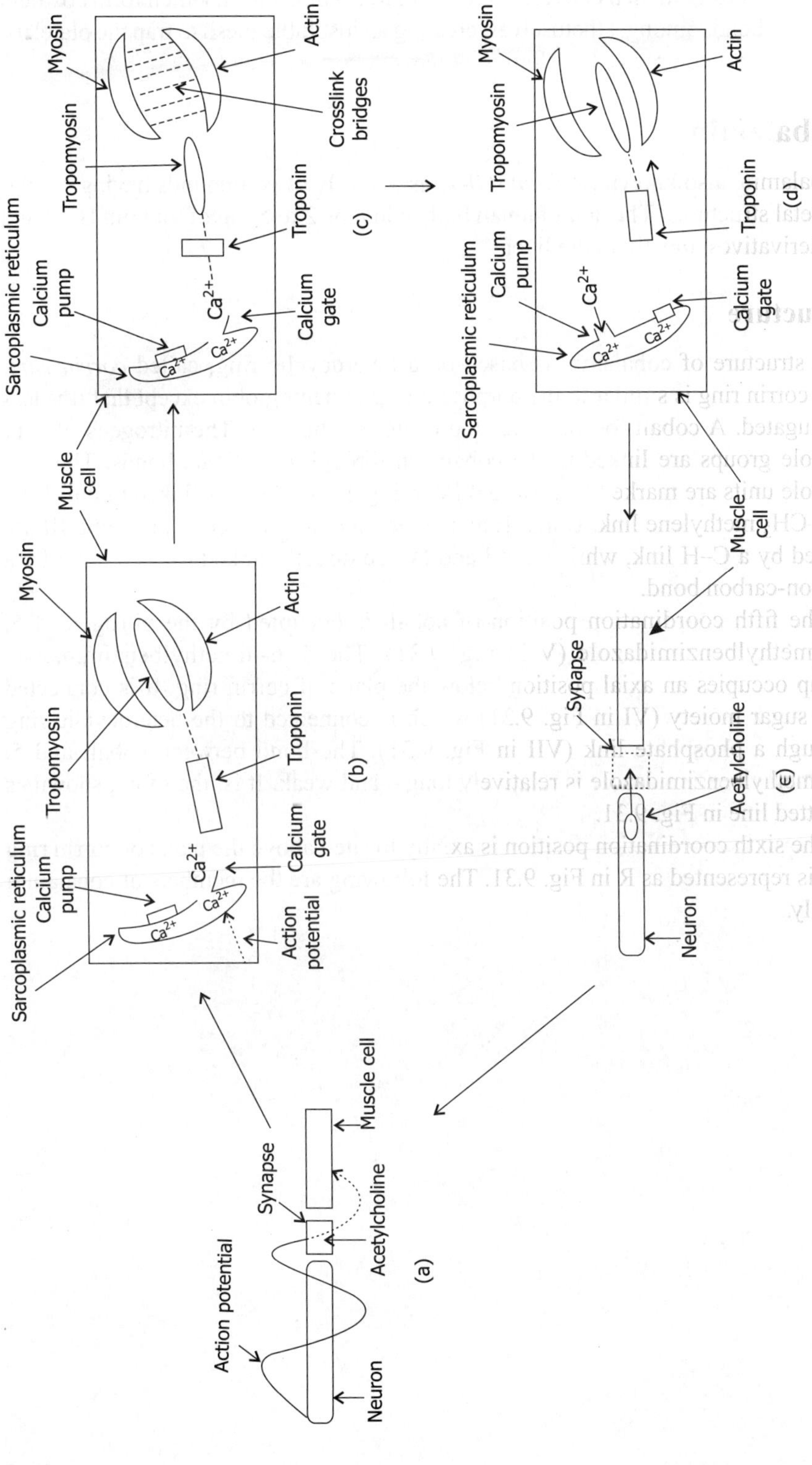

Fig. 9.30 The calcium cycle

(iv) Thrombin then converts fibrogen to fibrin molecules, which form covalent bonds amongst themselves, creating an insoluble mesh to trap the platelets. This completes the clotting process.

Cobalamin

Cobalamin, also known as *vitamin B-12* is a family of compounds having similar skeletal structures. The adult human body contains 2 to 5 mg of vitamin B-12 and its derivatives, mostly in the liver.

Structure

The structure of cobalamin is based on a heterocyclic ring, called *corrin ring.* The corrin ring is similar to the porphyrin ring in hemoglobin except that it is less conjugated. A cobalt ion occupies the center of the ring. The nitrogens of four pyrrole groups are linked to the cobalt ion through coordinate bonds. The four pyrrole units are marked I, II, III and IV in Fig. 9.31. Unit I is linked to unit II by a C–CH_3 methylene link. Units III and IV are similarly linked. Unit II and III are joined by a C–H link, while units I and IV are directly linked to each other by a carbon-carbon bond.

The fifth coordination position of cobalt is occupied by the nitrogen of 5, 6-dimethylbenzimidazole (V in Fig. 9.31). The 5, 6-dimethylbenzimidazole group occupies an axial position below the plane of corrin ring. It is connected to a sugar moiety (VI in Fig. 9.31) which is connected to the heterocyclic ring through a phosphate link (VII in Fig. 9.31). The bond between cobalt and 5, 6-dimethylbenzimidazole is relatively long—and weak. It is, therefore, shown as a dotted line in Fig. 9.31.

The sixth coordination position is axially located above the plane of corrin ring and is represented as R in Fig. 9.31. The following are the members of cobalamin family.

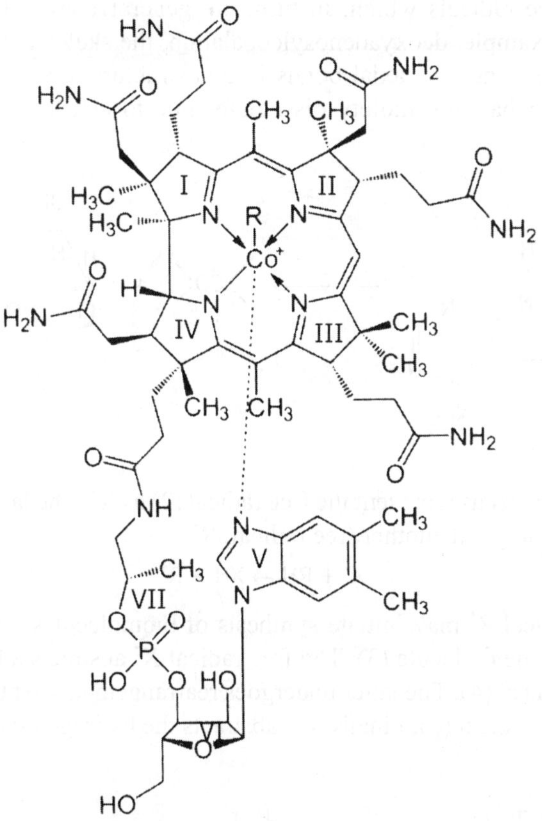

Fig. 9.31 Structure of cobalamin

- R = H_2O aquacobalamin
- R = OH^- hydroxocobalamin
- R = CN^- cyanocobalamin
- R = CH_3 methylcobalamin

In aquacobalamin, cobalt is present in oxidation state (II), while in rest of the cases, it is present in oxidation state (III). In another important derivative of cobalamin, the substituent R is 5′-deoxyadenosyl group. It is bound to cobalt through $-CH_2-$ group and hence deoxyadenosylcobalamin is classified as an organometallic compound. In fact, it is a rare example of a naturally occurring organometallic compound. The oxidation state of cobalt in this compound is (I).

Mode of action

Vitamin B-12 is a co-factor for a number of enzymes which catalyze the synthesis of biomolecules required for life processes. Such reactions are invariably initiated by free radicals which, in turn, are generated by cobalamin analogs. Consider, for example, deoxyadenosylcobalamin, the skeletal structure of which is shown as (1). Since the axial bonds in this structure are relatively weak, the deoxyadenosylcobalamin moiety dissociates due to cleavage of cobalt-carbon bond.

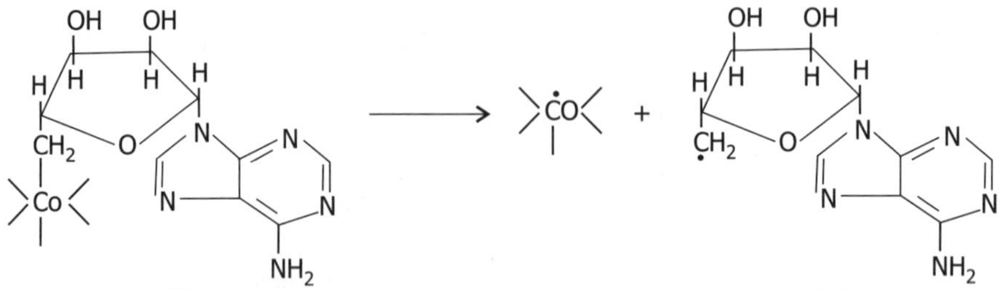

(1) (2)

For simplicity, let us represent the free radical (2) as $X^{\bullet}$. The latter can react with a substrate, RH to yield another free radical, $R^{\bullet}$.

$$X^{\bullet} + RH \rightarrow XH + R^{\bullet}$$

The free radical $R^{\bullet}$ may initiate synthesis of biomolecules which sustain life. Suppose RH is the molecule (3). The free radical $X^{\bullet}$ abstracts a hydrogen radical to yield free radical (4). The latter undergoes rearrangement first to free radical (5) and then to free radical (6). Finally, (6) abstracts the hydrogen radical from XH to form (7).

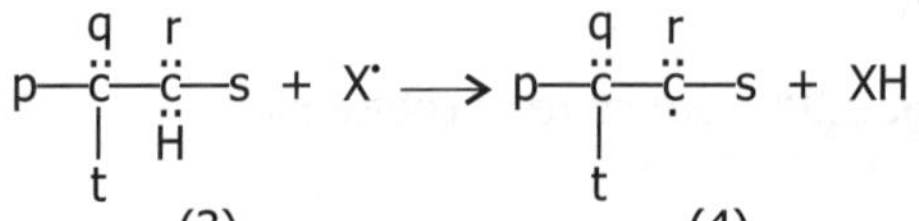

(3) (4)

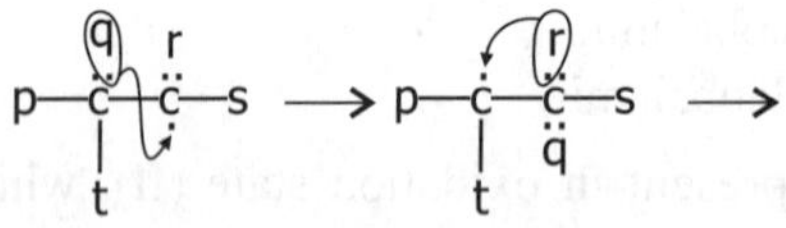

(5)

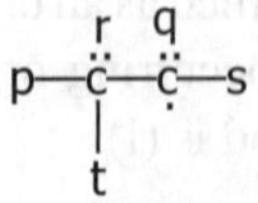

(6)

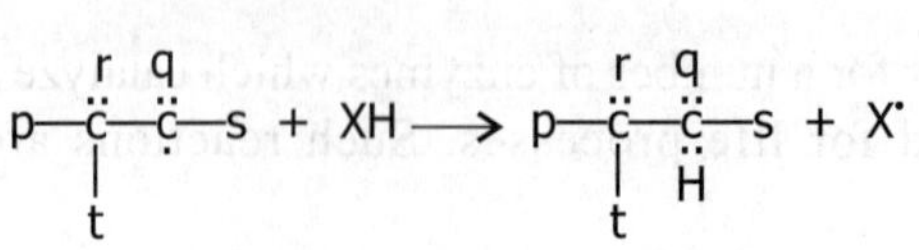

(7)

This rearrangement reaction has transformed (3) into its isomer (7). The enzymes which catalyze such isomerizations are classified as *mutases*. Suppose, in molecule (3) q = s = OH. In such cases, isomerization will be followed by dehydration.

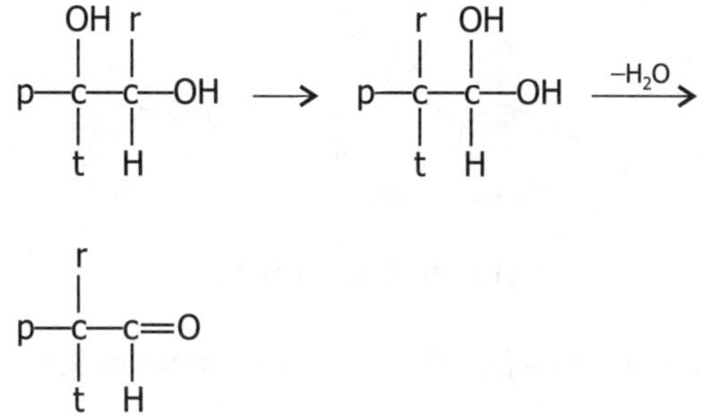

When q = OH and s = NH_2, isomerization will be followed by deamination.

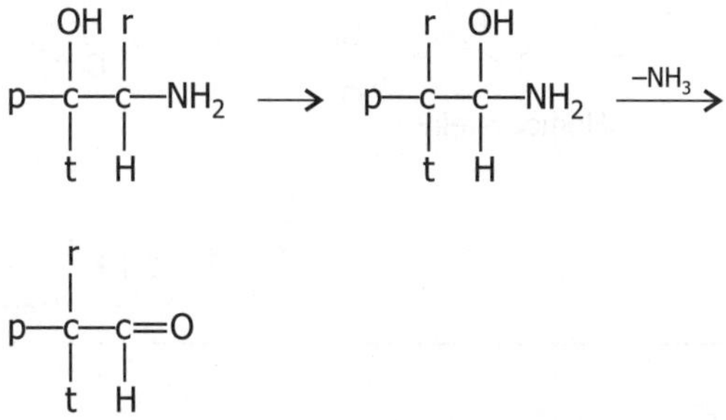

The enzymes which catalyze dehydration or deamination are classified as *lysases*. As a co-factor of mutases and lysases, cobalamin, through isomerization, dehydration and/or deamination, synthesizes biologically important molecules.

Functions

Cobalamin has many beneficial effects in regard to biological systems. It performs a key role in the metabolism of fatty acids which, in turn, is essential for the maintainence of myelin. Damage to myelin leads to paralysis and death.

Cobalamin aids in the synthesis of DNA. The production of blood cells in bone marrow requires an extremely high rate of DNA synthesis. If cobalamin is not present in optimum quantities, the DNA synthesis slows down—and so does the formation of red blood cells. The patient suffers from an ailment called *megaloblastic anemia*.

Lack of cobalamin also causes degenerative changes in spinal cord and nervous system. It increases the risk of neural tube defect in children.

Case Study 9.6 How does cobalamin carry out synthesis of essential amino acid, methionine?

As shown in Fig. 9.31, cobalamin has an octahedral geometry. However, the two axial bonds around the central cobalt ion are relatively long and weak. During synthesis of methionine, the ligands attached at axial positions dissociate, leaving a cobalt(I) ion in a square planar environment. This system has high nucleophilicity. The cobalt(I) ion abstracts the electrophilic methyl group from N-methyl hydrofolate.

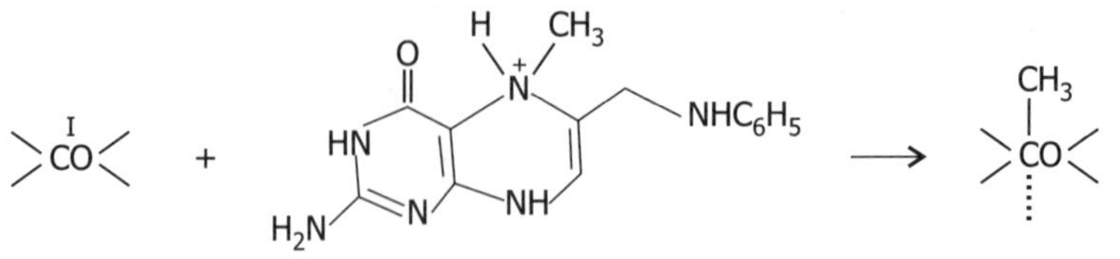

The resulting complex transfers the methyl carbonium ion to homocysteine to yield methionine.

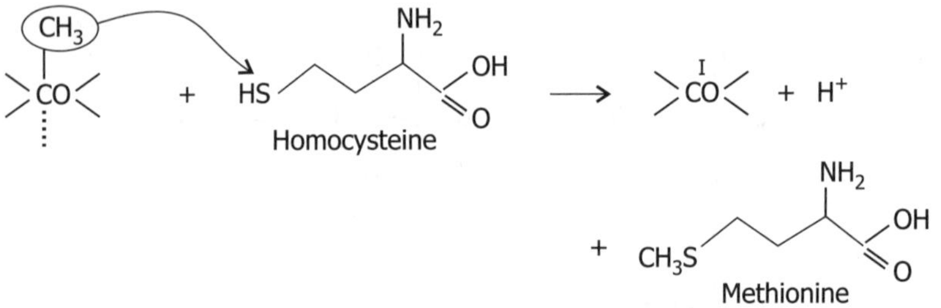

Photosynthesis

In the section on bioenergetics, we had defined catabolism as a phenomena that involves fission of large molecules into smaller chemical entities. The reverse of catabolism is *anabolism*, that is, synthesis of large molecules from relatively small and simple chemical species. Whereas catabolism is exothermic in nature, anabolism is endothermic.

Photosynthesis is an example of anabolism. In this phenomena, autotrophic or self-feeders, for example, plants, algae and a few bacteria species, take up simple inorganic molecules (water and carbon dioxide) and convert these into carbohydrate food material.

The overall reaction may be summarized by the following simplified equation.

$$6CO_2 + 6H_2O \xrightarrow{hv} C_6H_{12}O_6 + 6O_2 \quad (9.7)$$

Equation (9.7) represents a redox reaction in which carbon dioxide is reduced to glucose, while water is oxidized to oxygen.

Photosynthesis can proceed if the following conditions are met:

1. Presence of simple molecules. As shown in reaction (9.7), carbon dioxide and water vapors are the raw materials used by autotrophic organisms to prepare their foodstuff. These are captured by the autotrophs from the atmosphere.

2. Presence of a site for reaction. Plant cells contain special type of organelles called chloroplasts. These are not present in animal cells, but a typical plant cell contains 10–100 such organelles. Chloroplasts serve as a site for photosynthesis.
3. Source of electrons. As has been stated, reaction (9.7) involves reduction of carbon dioxide and oxidation of water. In biological systems, a redox reaction of this type can proceed only in the presence of a species that can donate electrons (causing reduction), as well as accept electrons (causing oxidation). This function is discharged by the co-enzyme, nicotinamide adenine dinucleotide phosphate (NADP), the structure of which is presented in Fig. 9.32.

$$\text{NADP} \rightleftharpoons \text{NADP}^+ + e^- \tag{9.8}$$

(The co-enzyme nicotinamide adenine dinucleotide phosphate is often abbreviated as NADPH. The last letter H stands for hydrogen, signifying that the co-enzyme is in a reduced state. However, since there are no replaceable hydrogens in the molecule, we feel that the abbreviations NADP and $NADP^+$ are better suited to represent respectively the reduced and oxidized states of the co-enzyme).

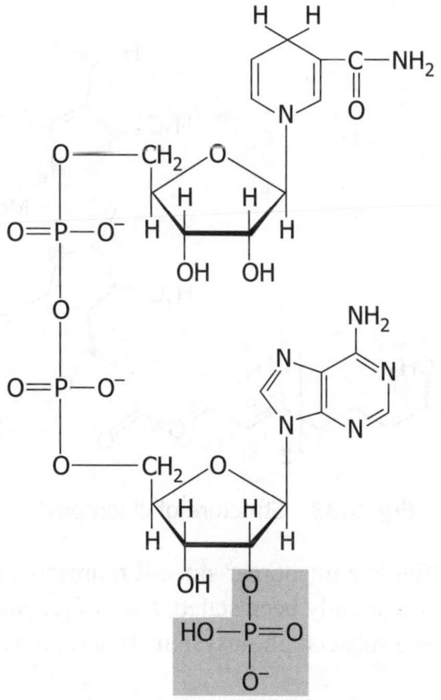

Fig. 9.32 Structure of nicotinamide adenine dinucleotide phosphate

4. Source of energy. As has been stated, photosynthesis is an endothermic reaction and, therefore, requires an input of energy. Equation (9.7) represents only a summary of the photosynthesis reaction. The actual process may be broadly divided into two sets: the light reaction and the dark reaction. The light reaction is so named because light—sunlight, to be more precise—serves as a source of energy for this set. In dark reaction adenosine triphosphate provides for the energy requirements. Paradoxically, adenosine triphosphate itself is synthesized in the light reaction.

The light reaction

Reduction of carbon dioxide requires electrons which water ought to donate. However, water is a poor electron donor, and hence a trifling reducing agent. It is the plant pigment *chlorophyll* that helps strip two electrons from one molecule of water to accomplish reduction of carbon dioxide.

The structure of chlorophyll is presented in Fig. 9.33. It has a porphyrin skeleton, attached to a phytyl group. The tetrapyrrole units of the porphyrin moiety are coordinated to a magnesium(II) ion. However, the coordinating nitrogen atoms of the pyrroles and magnesium(II) ion are not in the same plane, the latter being about 30 pm above the plane of the porphyrin function.

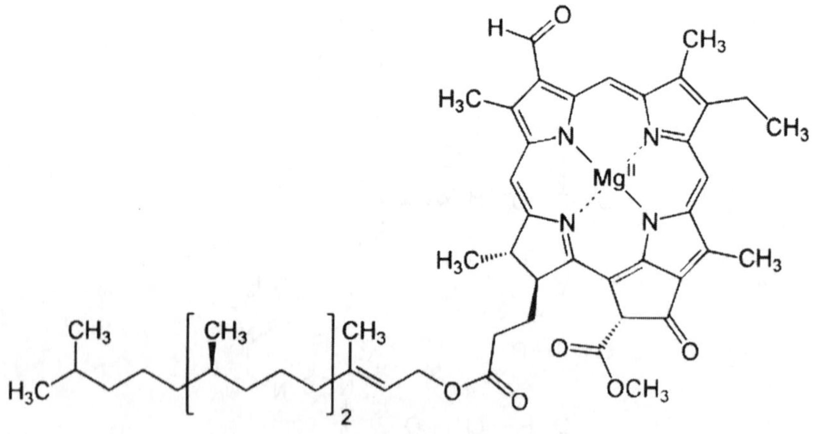

Fig. 9.33 Structure of chlorophyll

The term light reaction is a misnomer since it represents not a single chemical transformation but, as has already been stated, a set of reactions. The reactions are further classified into two subsets: photosystem-II and photosystem-I.

Photosystem-II The reactions under photosystem-II, abbreviated as PS-II begin with the absorption of visible radiation by chlorophyll. However, the concentration of chlorophyll in plant cells is quite small. Nevertheless, the plant cells contain highly conjugated carotene molecules which can efficiently absorb sunlight and then pass on its energy to chlorophyll.

The chlorophyll-carotene combine absorbs two photons of red component of visible light, having wavelength of 680 nm, and transmits green light—hence the green color of plants. The two photons first excite the chlorophyll molecule to a high energy state,

$$\text{Chla}_2 + h\upsilon \rightarrow \text{Chla}_2^* \tag{9.9}$$

and then causes its ionization

$$\text{Chla}_2^* \rightarrow \text{Chla}_2{}^+ + e^- \tag{9.10}$$

Here $Chla_2$ represents that site in the Chlorophyll molecule that absorbs radiation with wavelength 680 nm. The $Chla_2^+$ ion has the potential to strip two electrons from a molecule of water.

$$H_2O \rightarrow 2H^+ + \frac{1}{2}O_2 + 2e^- \tag{9.11}$$

One of the electron reduces chlorophyll cation

$$\text{Chla}_2^+ + e^- \rightarrow \text{Chla}_2 \tag{9.12}$$

The other electron released in equation (9.11), along with the one released in equation (9.10) travel through several electron carriers and ultimately reach photosystem-I.

The two photons absorbed in PS-II step have more than enough energy to oxidize chlorophyll molecule (reaction 9.10). The rest is used to synthesize adenosine triphosphate (ATP) from adenosine diphosphate (ADP).

$$\text{ADP} + \text{HPO}_4^- + \text{energy} \rightarrow \text{ATP} \tag{9.13}$$

[The precise chemical reaction for the synthesis of adenosine triphosphate from adenosine diphosphate is presented in equation (9.1, p. 319)].

Photosystem-I In photosystem-I, PS-I for short, another molecule of chlorophyll absorbs two photons, having wavelength 700 nm, gets excited,

$$\text{Chla}_1 + h\nu \rightarrow \text{Chla}_1^* \tag{9.14}$$

and finally ionized.

$$\text{Chla}_1^* \rightarrow \text{Chla}_1{}^+ + e^- \tag{9.15}$$

The $Chla_1$ is the site in chlorophyll molecule that absorbs radiation with 700 nm wavelength.

One of the two electrons brought forward from photosystem-II reduces chlorophyll cation,

$$\text{Chla}_1^+ + e^- \rightarrow \text{Chla}_1 \tag{9.16}$$

The other electron, along with the one released in equation (9.15) travel through a number of electron carriers and finally reduce the cationic co-enzyme, nicotinamide adenine dinucleotide phosphate ($NADP^+$, equation 9.8).

$$2\text{NADP}^+ + 2e^- \rightarrow 2\text{NADP} \tag{9.17}$$

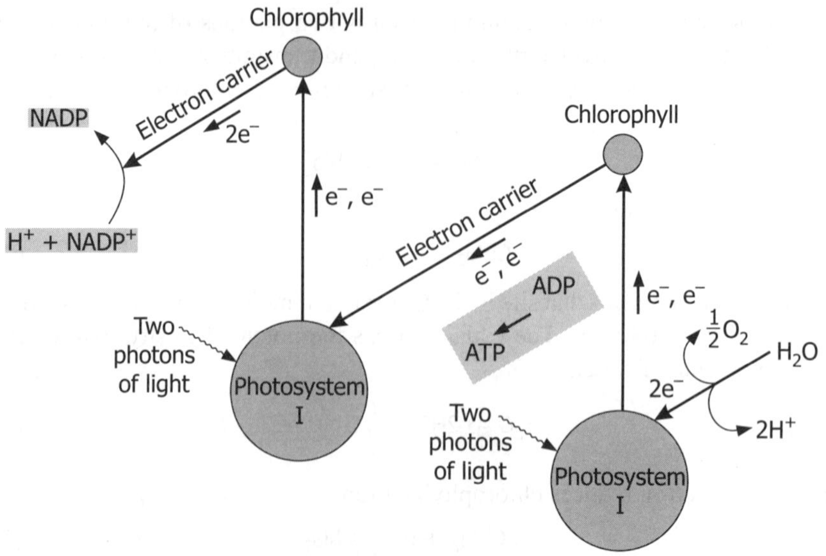

Fig. 9.34 The z-scheme

The flow diagram that emanates by combining photosystem-II and I has the shape of letter 'z'. Hence the overall light reaction is called *z-scheme* (Fig. 9.34).

Case Study 9.7 Why does the color of leaves changes from green to yellow or red during autumn?

Besides chlorophyll, leaves contain another class of pigments, celled *xanthophylls*. We have seen that chlorophyll absorbs radiations having wavelength 680 nm and 700 nm during photosystem-II and photosystem-I stages, respectively. When we refer to Table 26.2 (p. 728), we note that absorbed radiations are of red color, while the reflected (or complementary) color is green. Hence, leaves appear green in color. As autumn approaches, photosynthesis slows down and synthesis of chlorophyll abates. As a result, xanthophylls, which are present in minor amount and are masked by chlorophyll, become predominant during autumn. Xanthophylls absorb green and blue components of visible light and reflect yellow and red shades. For this reason the leaves appear yellow or red in autumn season.

The dark reaction

The dark reaction, also called *Calvin cycle*, involves several steps which culminate in the synthesis of glucose. Adenosine triphosphate, which was synthesized in the light reaction (equation 9.13) gets exothermally hydrolyzed to provide for the energy requirements of the endothermic steps of the dark reaction. The co-enzyme, nicotinamide adenine dinucleotide phosphate (NADP), which too was produced in the light reaction (9.17), is used as a raw material in one of the steps of dark reaction.

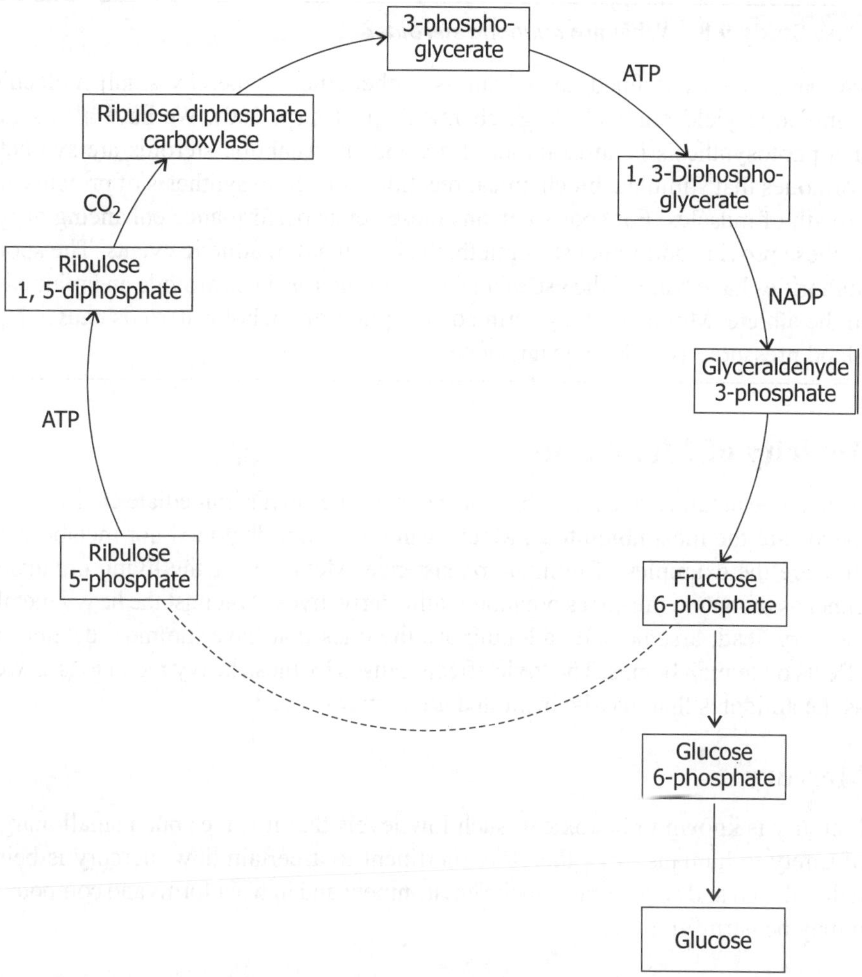

Fig. 9.35 The Calvin cycle

A flow diagram of the Calvin cycle is presented in Fig. 9.35. The cycle begins with the interaction of atmospheric carbon dioxide with ribulose 1, 5-diphosphate content of plant cells. The product, ribulose diphosphate carboxylase is unstable and decomposes into two molecules of 3-phosphoglycerate. The latter is transformed to 1,3-diphosphoglycerate. This is an endothermic step and hydrolysis of adenosinc triphosphate (ATP) provides for the requisite amount of energy to carry it through. The 1, 3-diphosphoglycerate reacts with nicotinamide adenine dinucleotide phosphate (produced in reaction 9.17), to yield glyceraldehyde 3-phosphate. A part of the latter is transformed successively into fructose 6-phosphate, glucose 6-phosphate and finally into glucose. The remaining part of glyceraldehyde 3-phosphate is converted into ribulose 5-phosphate which, using the energy provided by adenosine triphosphate (ATP), is transformed into ribulose 1, 5-diphosphate. With that the cycle begins again.

Case Study 9.8 What are *anabolic steroids?*

We have already defined anabolism as a phenomena whereby small molecules combine to yield relatively large chemical species (p. 360). We have also stated that photosynthesis is an example of anabolism. Anabolic steroids are synthetic hormones that stimulate biochemical reactions related to synthesis of proteins and growth of muscles. For sports persons these act as performance enhancing drugs, as these provide additional strength that is beneficial in athletic events. The sports authorities have banned the use of anabolic steroids as these provide an unfair edge to the athlete. Moreover, long-term consumption of anabolic steroids causes high blood pressure, liver damage and acne.

Toxicity of Metal Ions

Among the natural substances that man concentrates in his immediate environment, metals are the most ubiquitous. Metals can be neither degraded nor metabolized; they are the examples of ultimate persistence. Metals enter the living organisms either as inorganic salts or as organometallic derivatives. Amongst the heavy metals, mercury, lead, arsenic and cadmium are the ones that have the most deleterious effects on human beings. The toxic effects caused by these heavy metal ions as well as the antidotes that may be instituted are now reviewed.

Mercury

Mercury is known to be toxic at such low levels that it leaves but a small margin of safety to humans. It is, therefore, pertinent to ascertain how mercury is being redistributed and concentrated in the environment and in what forms and compounds it may be harmful to life.

Sources of mercury Mercury content of the environment has increased due to large scale urbanization and industrialization. The combination of physico-chemical properties, such as its liquid state at room temperature, uniform volume expansion over a wide temperature range, high surface tension and non-wettability to glass makes mercury uniquely useful for such devices as thermometers, barometers and manometers. The low electrical resistivity and high thermal conductivity make mercury an excellent electrical conductor and coolant respectively. Due to its capacity to absorb neutrons, mercury is employed as a shield against atomic radiations.

Biomethylation of mercury When inorganic or elemental mercury is discharged into an aquatic system, it is converted into highly toxic methylmercury(II) ion, CH_3Hg^+. Since chloride ions are present in every aquatic system, the methylmercury(II) ion is transformed into chloromethylmercury(II), CH_3HgCl. This transformation is called biomethylation of mercury and is brought about by a set of microorganisms called methanogenic bacteria.

Bioamplification of mercury Certain substances, of which chloromethylmercury(II) is an example, tend to concentrate among the food chains. Being more soluble in the tissue of simple organisms, than in the surrounding water, these substances penetrate into the bodies of organisms. Subsequently, they become a concentrated component of the more complex species which feed on the simpler ones. This phenomena is called bioamplification. As a result of bioamplification, chloromethylmercury(II) concentrates in the tissue of aquatic organisms, mainly fish.

The uptake of chloromethylmercury(II) is favored at low pH. Thus, fish in basic or neutral waters contain smaller mercury content than those in acidic waters. The extent of bioamplification increases with temperature and salinity of the water body. It also increases with the age, size and weight of the fish. In addition, the degree of bioamplification depends on the type of fish. Species like wall-eye, bass or trout which occupy a higher trophic level have greater amplification factors than blue-gill, carp or catfish, which occupy a lower level on the trophic scale. Thus, at each trophic level on the food chain, the mercury content progressively increases. Since humans are usually at the top of the food chain, high concentration of mercury may occur in human diets, with tragic results, such as Minamata disease (*see* box 9.3).

Box 9.3

Minamata Disease

One of the most visible tragedies caused by contamination of an aquatic system by mercury is the Minamata disease. Originating in 1960s in the Japanese fishing port of Minamata, the disease claimed about 50 lives and permanently paralyzed over 700 people. Many newborns sustained congenital defects.

The cause of the disease was traced to the discharge of mercury-rich effluents into Minamata Bay by Chisso chemical company. Its outbreak exposed two important facts: Firstly, the Chisso plant used mercury(II) chloride catalyst in acetaldehyde production and, therefore, only the comparatively non-toxic inorganic mercury was released in the industrial effluents. However, the sediments from the Minamata Bay were found to be rich in chloromethylmercury(II). It thus became evident that inorganic mercury was transformed to organomercurial through biomethylation. Secondly, the human deaths, in and around Minamata, coincided with large scale fish mortality and, therefore, it was inferred that chloromethylmercury(II) became concentrated in the body of the fish through the process of bioamplification.

Toxic effects of mercury Both inorganic mercury and chloromethylmercury(II) are poisonous in nature. Yet, the latter elicits more severe effects. The reasons for the greater toxic effects of chloromethylmercury(II) are manifold. Firstly, this compound is amphilitic in nature and, therefore, interacts favorably with the hypophilic cell structure, especially the cell membrane. Secondly, the methylmercury(II) ion can penetrate the membrane separating the blood stream from the brain and those in placenta through which the nutrients must pass to an unborn fetus. The latter was apparently the cause of congenital birth defects in Minamata episode. Thirdly, in chloromethylmercury(II), the metal ion is firmly bound to the carbon so that the molecule is not easily broken down and maintains its destructive action for weeks or months. In contrast, the injury caused by inorganic mercury is almost invariably reversible.

The toxicity of mercury manifests itself in the following forms:

1. *Enzyme inhibition.* Mercury reacts with a large number of enzymes, inhibiting catalysis of general metabolic reactions. Thus, mercury is often considered as an enzyme inhibitor. The chemical basis of this effect results from the mercury's strong affinity for the sulfhydryl (-SH) group in the enzymes.

$$CH_3Hg^+ + \text{enzyme} - SH \rightarrow CH_3Hg\text{–}S\text{–enzyme} + H^+$$

2. *Cellular dysfunction.* Mercury also reacts with sulfhydryl function in proteins. Bound to the proteins in a cell membrane, the mercury may alter the distribution of ions, change electrode potentials and thus interfere with the movement of fluids across the membrane. The binding of mercury to proteins also disturbs the normal operation of cell constituents, such as mitochondria and lysosomes.
3. *Neurological disorders.* The brain is the target organ in mercury poisoning, where essentially irreversible damage is done to cells of the nervous system. The first symptoms are loss of sensation at the extremities of hand and feet and areas around mouth (paresthesia). As the body burden of mercury increases, these symptoms are followed by loss of coordination in gait (ataxia), slurred speech (dysphasia) and loss of hearing. Severe poisoning can cause blindness, coma and death.
4. *Erethism.* When a person is exposed to relatively mild doses of mercury, the neurological disorders manifest themselves as psychic disorders. Such behavioural problems are called erethism and are characterized by self-consciousness, timidity, embarrassment, anxiety, indecision, lack of concentration, depression, despondency, irritability and excitability. Very often these symptoms cause a complete change in personality.
5. *Teratogenic effects.* Chloromethylmercury(II) readily penetrates the fetus through the placenta. The concentration of mercury in the blood and the brain of the fetus is about 20% higher than in the mother. Infants whose mothers are exposed to large amounts of chloromethylmercury(II) are likely to be afflicted with mental retardation, cerebral palsy and convulsions.

Antidotes to mercury poisoning As a general safeguard against mercury poisoning, it has been recommended that the weekly intake of mercury per person should be limited to 0.3 ppm, of which no more than 0.2 ppm should be in form of methylmercury(II) ion. The permissible upper limit of mercury in foods is 0.5 ppm. The level of mercury in blood should not be more than 1.0 ppm. Drinking water should have a maximum of 0.005 ppm mercury content.

When these limits are exceeded, the symptoms of mercury poisoning set in. As an antidote to poisoning effects, certain mercury-binding chemotherapeutic agents are administered to the patients. These agents tie-up mercury in form of complexes which are rapidly excreted from the body.

Some common chemotherapeutic agents are depicted in Fig. 9.36.

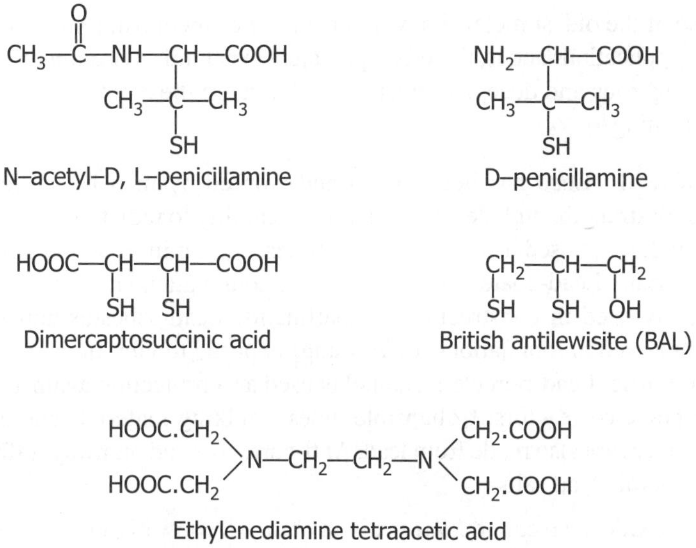

Fig. 9.36 Common therapeutic agents

Of these, N-acetyl-D,L-penicillamine and dimercaptosuccinic acid are most promising against methylmercury poisoning. These penetrate through the cell membrane and convert the intracellular methylmercury into a mobile complex that readily diffuses out of the cell. It is then passed out of the body as urinary or faecal excretions.

British antilewisite or BAL is very effective in treatment of inorganic mercury poisoning. However, it is not used in cases of methylemercury poisoning. The reason being that the complex which it forms with methylmercury crosses the blood-brain barrier very rapidly and becomes concentrated in the brain, rather than being excreted out of the body.

Case Study 9.9 What is the origin of the phrase, *as mad as a hatter?*

The phrase, as mad as a hatter, sometimes also stated as *mad hatter* or *mad as hatter*, referred to workers employed in felt hat manufacturing units. The curing of felt in these units was carried out by treatment with mercuric nitrate solution. The workers suffered occupational exposure to mercuric ions which, in turn, caused neurological damage. The symptoms included confused speech and distorted vision.

Mad hatter is also a fictional character in Lewis Carroll's, *Alice Adventures in the Wonderland.* Carroll grew up in Stockport, UK, where the main trade was hat making. Very often he used to see workers engaged in hat manufacturing in a disturbed and confused state of mind. Many met untimely deaths due to mercury poisoning. This prompted him to incorporate the character Mad Hatter in his world famous work.

Lead

Lead is one of the oldest metals known to man. Since medieval times it has found wide range applications and is, therefore, prevalent in natural environment. However, lead is a dangerous and deceptive poison. Children are the most common victims of lead poisoning.

Sources of lead Lead has been mined and worked by man for millennia. Its ductility, high strength, high density and impermeability to radiation make it a very useful metal. Lead is used in acid storage batteries, ammunition, solder devices and casting materials. Lead-coated steel sheets have sound attenuation properties and are commonly used in construction of apartments. Lead-asbestos anti-vibration pads are employed in foundations for buildings exposed to vibrations from nearby rail or road traffic. Lead-porcelain enamel is used as a protection against radiation in nuclear powered reactors. Collapsible tubes, caulking materials and corrosive liquid containers are also made from lead. Anti-corrosive and highway traffic safety paints incorporate lead salts.

Lead monoxide, also called *litharge*, is an orange-yellow pigment that is used in glazing pottery. When acidic food items, like fruit juices and pickles, are stored in improperly glazed earthenwares, lead slowly leaches from the glaze and becomes a component of the diet. Lead enters drinking water from old conduit water pipes.

Toxic effects of lead Lead poisoning, also called *plumbism* or *saturnism*, manifests itself in the following forms:

1. *Enzyme inhibition.* Like mercury, lead too coordinates with the sulfhydryl function in enzymes. This causes an inhibition of enzyme activity. The clearest manifestation of the inhibitory effect of lead is the disturbance in the biosynthesis of heme. The decrease in heme biosynthesis leads at first to a decrease in the life span of red cells and later to a decrease in the number of red cells. To compensate the shortage, the blood forming tissue steps up its production of red cells. However, the rate of production is less than the rate of disappearance. Therefore, in a person suffering from lead poisoning, immature red cells, called *reticulocytes* and *basophilic stippled cells*, appear in circulation, causing anemia.
2. *Kidney impairment.* Acute lead poisoning causes impairment of kidney functions. Lead tends to concentrate in the proximal renal tubules lining. There is a substantial loss of amino acids, glucose and phosphate in the urine because the damaged tubular cells fail to reabsorb these substances as completely as the normal tubular cells do. The energy storage system of the body suffers. Exposure to milder doses of lead causes *chronic nephritis*, a disease characterized by scarring and shrinking of kidney tissue.

3. *Neurological disorders.* Central nervous system effects are the most serious manifestation of lead poisoning. Continuous exposure to lead over prolonged periods causes *encephalopathy* or brain damage by two mechanisms. Firstly, the blood capillaries reaching the brain start leaking and cause edema, that is, swelling of the brain. There is not enough space to accommodate a swelled brain within the periphery of the rigid skull. Therefore, the brain cells gets damaged. Secondly, the cells constituting the motor nerve extremities are damaged and the conduction of nerve impulse is impaired.

   Symptoms of encephalopathy initiate with the onset of vomiting, ataxia, periods of alternative stupor and hyperirritability. Subsequently, these culminate in delirium, convulsions, coma and death.
4. *Behavioural problems.* When a person is exposed to milder doses of lead, the neurological damage manifests itself in form of behavioural problems. In such cases the symptoms include excitement, restlessness, insomnia, nightmares, impairment of memory and loss of concentration.
5. *Reproductive problems.* Women employed in pottery works, where exposure to lead is high, are more liable to be sterile than those in general population. If they become pregnant, the pregnancy is likely to result in miscarriage or stillbirth. If the child is born living, there is a higher infant mortality.
6. *Teratogenic effects.* Children born to women who were exposed to lead during pregnancy suffer from convulsions and a form of macrocephaly characterized by a square shaped head. It has also been observed that children born to women who suffered from plumbism during their childhood, may also show birth defects. This means that a girl growing up in a lead-polluted environment might, years later, pass that lead to her offspring. During adolescence, the lead becomes concentrated in the girl's bones. During pregnancy, calcium is removed from a woman's bones and passed to the blood stream. It then enters the placenta to help build the embryo's skeleton. In this process, lead also comes out of the bones, crosses the placenta and enters the fetus, where it elicits teratogenic effects.

Antidotes to lead poisoning The permissible limit of lead in food is 2 ppm. Lead content of blood should not exceed 0.25 ppm. Drinking water should not have more than 0.05 ppm lead content. Beyond these limits the symptoms of lead poisoning begin to manifest themselves.

As in the case of mercury, the treatment of lead poisoning is carried out by administering chemotherapeutic agents. These agents tend to sequester lead from the body tissues and bind it in form of soluble complexes. In soluble form, lead is excreted through the kidney. Initially, two agents are administered by injection: British antilewisite (BAL, Fig. 9.36) and ethylenediamine tetraacetic acid (Fig. 9.36). With these complexing reagents, very high tissue levels of lead are rapidly reduced. Thereafter, another chemotherapeutic agent, D-penicillamine (Fig. 9.36) may be administered orally as a follow-up therapy.

Case Study 9.10 Why is lead poisoning more dangerous to the children than the adults?

Plumbism has proved to be more dangerous to the children than the adults due to the following reasons. Firstly, once lead enters the human body it tends to concentrate in the bones. It remains in the bones in a relatively inert form, causing no ill effects. However, when the body feels shortage of an essential element like calcium or phosphorus, the blood leaches out these minerals from the bones and supplies it to the relevant organ. In this process, lead too becomes labile and enters the blood stream. It then concentrates in the tissue where it elicits toxic effects. Since the metabolic rates are higher in children than in adults, the chances of lead being leached out of the bones are more in case of infants.

Secondly, due to the higher metabolic rates in children, the ratio between the quantity of lead consumed through food to the body weight is more in case of infants. This means that the body burden of lead is more in children than in adults.

Thirdly, the opportunity for the oral intake of significant amounts of lead from non-food sources, for example, from lead-bearing dust, soil, painted walls, toys, and plasters, is considerably higher in small children. In fact, most of the lead poisoning incidents in children below the age of five are caused by eating flecks of lead-containing paints from the walls of deteriorating buildings. This behavior is called *pica*. Even otherwise, the habit of oral exploration, sucking fingers, mouthing and chewing objects is a natural activity amongst the toddlers. These habits contribute to the body's lead burden.

Arsenic

Arsenic is a metalloid, which exists in two oxidation states, (III) and (V). The compounds of arsenic, particularly in oxidation state (III) are deadly poisons. Just 0.1 g of the element is enough to kill a person. It is, therefore, not surprising that despite a virtual ban on its use in commercial products, arsenic continues to command a cautious respect.

Sources of arsenic A good number of arsenicals were used as pesticides in the past. Thus calcium arsenate, $Ca_3(AsO_4)_2$ controls boll-weevil in cotton fields, while lead arsenate, $Pb_3(AsO_4)_2$ is effective against cabbage worm and potato bug. Sodium arsinite, Na_3AsO_3 is an efficient herbicide that was extensively used to clear the weeds for laying railroads and telephone posts. It is also a fungicide and a wood preservative.

Arsenic is almost always embedded in gold and lead ores, and during the extraction of these metals is released to the atmosphere. Subsequently, it is rained down into the aquatic systems. The leachate (extract containing waste in dissolved form) from gold mines of yesteryears is still a source of arsenic pollution in water resources.

Arsenic compounds have been used in textile dyeing, calico printing and as paint pigments. These are efficient depilatory and bronzing agents. Other applications include preparation of opal glass and lead alloys for bullets.

At present, arsenic is considered to be *too hazardous to man and the environment.* Therefore, during the past few decades, environmental restrictions and regulations have limited the amount of arsenicals produced and used.

Box 9.4

The Healing Deities of China

During the third century BC, *Taoism* (also called *Daoism*) became the predominant religious sect of China. Several Taoist temples, adorned with a pantheon of deities, sprang up across the country (Fig. 9.37).

Fig. 9.37 A Taoist deity

During that era, a large section of Chinese population suffered from parasite–related ailments. When an ill person, while paying obeisance at the Taoist temple, touched a deity, he would get well. This was not a matter of belief, but a reality; a belief is based on faith, but the healing, in this case, was based on the principles of bioinorganic chemistry, although the scientific logic came to be understood many centuries later.

The Chinese deities were made of the mineral *realgar*, the main constituent of which is diarsenic disulfide, As_2S_2. When a devotee touched a statue, arsenic ions would get absorbed through the skin. The amount of arsenic that diffused inside the body was so meager that it did not elicit any toxic effects, yet it was enough to kill the parasites.

Biomethylation of arsenic In aqueous environment, arsenic gets biomethylated to methylarsenic compounds. This transformation is brought about by microorganisms, both under aerobic (that is, in presence of air) and anaerobic (that is, in absence of air) conditions. Thus *E-coli* bacterium and *S. brevicaulis* fungus biomethylate arsenic contaminants in water under aerobic conditions, while methanobacterial strains do so under anaerobic settings.

Arsinate and arsinite derivatives are transformed by microorganisms first to methylarsonic acid and then to cacodylic acid. Under aerobic conditions, cacodylic acid is converted to trimethylarsine oxide and finally to trimethylarsine. However, under anaerobic conditions, it is reduced to dimethylarsine (Fig. 9.38).

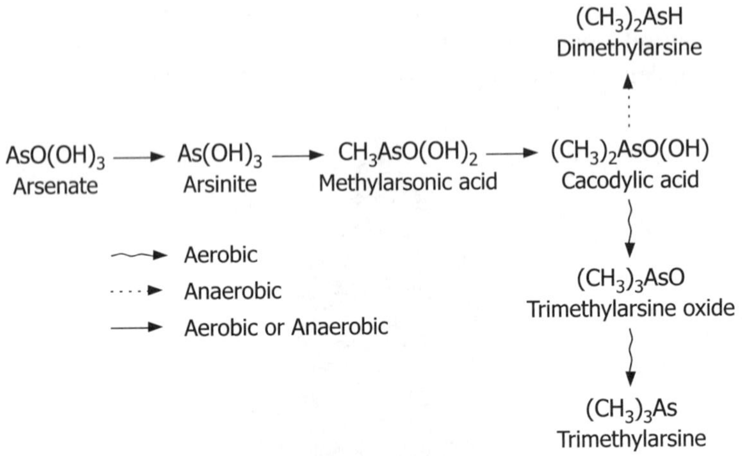

Fig. 9.38 Biomethylation of arsenic

Methylated forms of arsenic are absorbed by or incorporated into benthic organisms, algae, zooplankton and phytoplankton. These are then concentrated into benthic aquatic food chain and have been detected in egg shells, sea shells and fish. Marine organisms accumulate more arsenic than fresh water species. When humans consume contaminated sea food, they are likely to suffer from the symptoms of arsenic poisoning.

Toxic effect of arsenic Arsenic compounds are protoplasmic poisons that attack enzymes. Chronic exposure to arsenic leads to—

1. Cardiac damage, causing general weakness and respiratory distress.
2. Nervous disorders, causing decreased motor coordination and neuritis.
3. Gastrointestinal problems, causing abdominal pain, vomiting and diarrhea.
4. Skin infection, causing dermatitis and hair loss.

Arsenic is also known to cause lung cancer in humans. In fact, there is evidence that smoking and simultaneous exposure to environmental arsenic act synergistically in causing lung cancer, that is, their hazard taken together is greater than the sum of their individual effects, if each would have acted independently. Ingestion of arsenic through food and/or water is also suspected to cause liver and skin cancers.

Antidotes to arsenic poisoning The maximum allowable concentration of arsenic in drinking water is 0.05 ppm. Beyond this limit, the ill effects associated with arsenic poisoning become apparent.

As has been stated, arsenic(III) compounds (arsinites) are more toxic than arsenic(V) analogs (arsinates). This is due to the ability of arsenic(III) ion to be retained in the body longer, since it becomes bound to the sulfhydryl groups of enzymes. The reason for the spontaneous arsenic-enzyme interaction has been attributed to the soft acid nature of arsenic(III) ion and the soft base nature of the sulfhydryl function. Arsenic(V) ion, on the other hand, is a relatively hard acid and therefore, possesses only a meager affinity for thiol group. Hence it does not inhibit the enzyme functions and is readily excreted out of the body.

As soon as the symptoms of arsenic poisoning set in, chemotherapeutic, chelating agents, possessing soft ligating atoms, are administered to the patient. The arsenic(III) ions become bound to the chelating reagent in preference to the enzymes. The resulting soluble complex is excreted from the body. The most promising chemotherapeutic agent for treating arsenic poisoning is British antilewisite, BAL (Fig. 9.36).

Case Study 9.11 What is meant by the phrase, *death by wallpaper*?

After his defeat at the Battle of Waterloo in 1815, French emperor Napoleon Bonaparte was exiled to the British-ruled island of St. Helena in Atlantic ocean. During the next 5 years his health declined steadily and ultimately he died in 1821. Post-mortem revealed stomach cancer as the cause of death. The French, however, did not believe the official post-mortem and a rumour spread that Napoleon had been poisoned with arsenic.

The French quoted a medical report compiled 2 years before Napoleon's death which stated that the patient's gums were full of holes, and that he suffered from insomnia, swollen legs, attacks of migraine and hot flushes—all symptoms of arsenic poisoning. Moreover, when the body of the dead ruler was returned to France in 1840, the corpse inside the coffin had not decomposed. French doctors believed that Napoleon's body had been preserved by the arsenic that was used to kill him!

The English refuted this theory. They advocated that the damp climatic conditions on St. Helena would have retarded the decomposition of the corpse. Further, the body was preserved not in a single coffin, but in 3 progressively larger coffins, which formed an airtight multiple barrier, preventing the decay.

As late as 1994, analysis of Napoleon's hair samples revealed the presence of trace amounts of arsenic. However, by that time, much was known about the environmental behavior of this metalloid. It is believed that the wallpaper in Napoleon's bedroom was colored with arsenic-based paints. Fungus strains, which grow very fast in humid conditions, would have transformed the arsenic of wallpaper into trimethylarsine (Fig. 9.38)—an extremely poisonous gas. Chronic exposure to trimethylarsine would have, in all possibility, hastened the emperor's death. This phenomena has now come to be known as 'death by wallpaper'.

Cadmium

The compounds of cadmium, once ingested, have a tendency to bioaccumulate in liver and kidney. Hence the toxic nature of this heavy metal.

Sources of cadmium Cadmium resembles zinc in physical and chemical properties and for this reason, the ores of the two metals occur together in nature. During extraction of zinc, cadmium is released to the environment as a by-product. As a result, soil and water in the vicinity of zinc smelters have an abnormally high concentration of cadmium.

Cadmium is extremely resistant to corrosion. Before its toxic properties were documented, this heavy metal was extensively used for electroplating other metals—especially those metallic devices that were to remain in contact with sea water. Subsequently, cadmium leached into the sea and entered the marine food chain.

Plastic and ceramic items are colored yellow by cadmium sulfide and cadmium selenide pigments. Cadmium selenide is also used in photoelectric cells as well as in TV screens, fluorescent lamps and cathode ray tubes. Cadmium stearate is used as a stabilizer in polyvinyl chloride plastics.

Rechargeable nickel-cadmium batteries—commonly called *Nicad batteries*—are used in calculators, satellites, and missiles. Cadmium alloys find applications in solders, automatic sprinkler systems, fire detection apparatus, valve seals and high-pressure gas containers. When cadmium-based utilities have served their term, these are disposed of by burning in waste incinerators. As a result, cadmium is released to the atmosphere, but subsequently rained down into aquatic systems or soil.

Soil receives cadmium from two other sources. Firstly, phosphate fertilizers invariably contain cadmium as a natural contaminant. When land is sprayed with phosphates, cadmium becomes incorporated in the soil. Secondly, polluted irrigation water increases the cadmium load of soil. If the soil is acidic in nature, its cadmium content becomes solubilized and, therefore, is easily taken up by plants. From here it enters the food chain. Thus potatoes, wheat and rice are often contaminated with cadmium.

Toxic effects of cadmium The most notable example of cadmium poisoning occurred in Japan in the 1950s. About 200 persons living in Jintsu River Valley suffered from a disease which came to be called *itai itai* (which means, it hurts! it hurts!). The river received cadmium-containing effluents from a zinc refinery located along its coast. Downstream, the river water was used for irrigating rice fields. Cadmium thus became a component of diet. Once in the human body, cadmium replaces calcium in the bones. Symptoms similar to those of rheumatism set in. Subsequently, the bones softened and became susceptible to fractures. During *itai itai* episode, cadmium also damaged the kidneys of the victims, causing renal dysfunctions.

Consumption of cadmium-contaminated sea food causes enzyme poisoning. Cadmium replaces zinc in many vital enzymatic reactions, resulting in disruption or cessation of activity. This normally leads to acute gastroenteritis.

Epidemiological studies indicate that workers engaged in cadmium-related work are more likely to suffer from prostate and nasopharynx cancers than their counterparts engaged in other activities. Cancer risk increases with the length of employment.

Antidotes to cadmium poisoning The limit of cadmium in drinking water has been fixed at 0.01 ppm. The sea food containing more than 0.5 ppm cadmium is deemed unfit for human consumption. Beyond these limits, cadmium elicits toxic effects.

Nature has provided for a mechanism to protect humans against chronic exposure to low levels of cadmium. The human biological system contains a chelating agent, called *metallothionein*, which removes cadmium from the body. Cadmium coordinates to sulfhydryl groups of metallothionein and is then eliminated through urine. The rapid recovery from cadmium poisoning by natural means is called *emetic effect.*

However, if the amount of ingested cadmium is too high, as it was during *itai itai* outbreak, one cannot rely solely on natural means of recovery. Normally, British antilewisite (Fig. 9.36), a chelating, therapeutic reagent is used for removing heavy metal ions like mercury, lead and arsenic. However, this reagent fails to combat cadmium. It does coordinate to cadmium present in different tissues and transfers the entire complexed metal to the kidneys. However, rather than being eliminated out of the body, the complex dissociates in the kidneys and damages these organs. As a result, the condition of the person becomes worse as compared to what it was before medication.

The most effective antidote to cadmium toxicity is ethylenediamine tetraacetic acid (EDTA, Fig. 9.36). This therapeutic reagent chelates with cadmium, forming a stable and soluble complex that is rapidly eliminated from the body.

Case Study 9.12 What are *smelling salts*?

Smelling salts are a mixture of ammonium compounds, the main ingredient being ammonium carbonate. Also referred to as *spirit of hartsham* or *sal volatile*, these salts, when dissolved in water, yield vapors of ammonia. When a person who has been rendered unconscious due to an injury inhales ammonia emanating from smelling salts, he regains his consciousness. The vapors of ammonia cause the muscles that control breathing to work faster by irritating the mucous membranes of nose and lungs. This phenomena is called *inhalation reflux*. The inhalation reflux also elevates parameters like heart rate, blood pressure and brain activity, which normally decline in a fainted person.

Critics hold that unless absolutely necessary, smelling salts should not be administered since these inhibit or delay a thorough neurological diagnosis of the patient.

Metal-Based Pharmaceuticals

A good number of drugs contain metals—surprisingly, metals like platinum, gold and ruthenium, that are otherwise not present in biological systems. These metals are administered by way of their coordination compounds, but before attacking the target sites, their composition invariably alters. Metal-based compounds are also used in the diagnosis of disease. We now review the characteristics and modes of action of a few important metal-based pharmaceuticals. The structures of these drugs are depicted in Fig. 9.39.

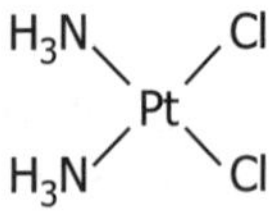

Cisplatin

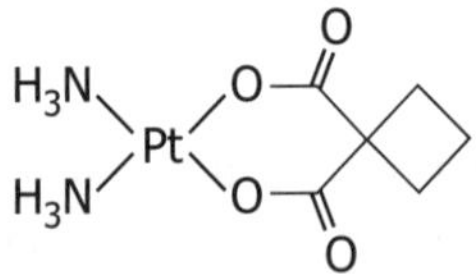

Carboplatin

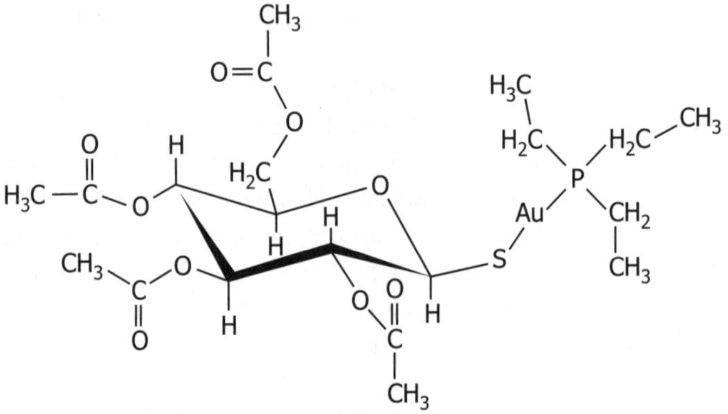

Auranofin

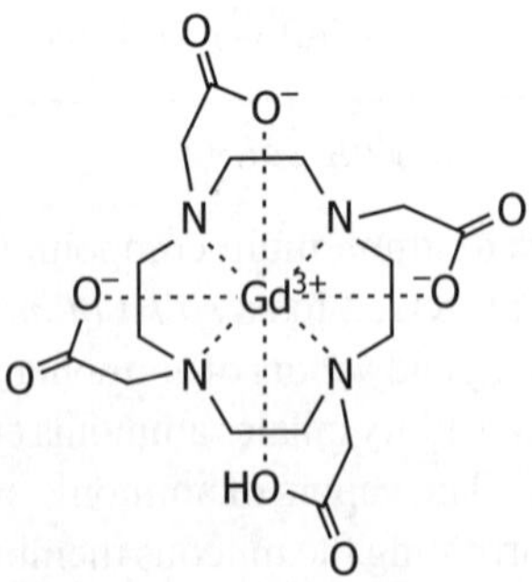

Gadoteric acid

Fig. 9.39 Metal-based pharmaceuticals

1. Cisplatin

Cisplatin, also called cisplatinum, but better named as *cis*-diamminedichloroplatinum(II), $[Pt(NH_3)_2Cl_2]$ (Fig. 9.39), is a chemotherapy drug. It is used to combat different types of cancers, including sarcomas, lymphomas, germ cell tumors and a few carcinomas, such as small cell lung cancer and ovarian cancer.

Mode of action The chemotherapeutic action of cisplatin is attributed to the formation of a stable complex between platinum(II) ion and the DNA. The complexation reaction results in crosslinking of DNA, which ultimately triggers the death of cancer cells.

Cisplatin is administered into the bloodstream of the patient. The plasma content of the bloodstream contains a high concentration of chloride ions. For this reason, cisplatin tends to remain as a neutral, molecular species, $[Pt(NH_3)_2Cl_2]$. Due to its electrical neutrality, the compound easily seeps through, first the cell membrane and then the nuclear membrane. These membranes provide an easy passage to cisplatin, since the lipid molecules, of which these are composed of, are electrically neutral.

On entering the cell, cisplatin encounters a substantially low concentration of chloride ions. As a result, one of the chloro ligand is substituted by water molecule, in a process called *aquation*.

$$[Pt(NH_3)_2Cl_2] + H_2O \rightarrow [Pt(NH_3)_2Cl(H_2O)]^+ + Cl^-$$

The product so formed, diammineaquachloroplatinum(II) ion is cationic in nature and is electrostatically attracted towards DNA. The guanine component of DNA (Fig. 9.40) becomes coordinated to the platinum(II) ion, resulting in the formation of $[Pt(NH_3)_2Cl\ (DNA\text{-}guanine)]^+$ complex ion. Subsequently, the second chloro ligand too is displaced by guanine moiety.

The mechanism of coordination of cisplatin with guanine components [G(N-7)] of DNA is shown in Fig. 9.41.

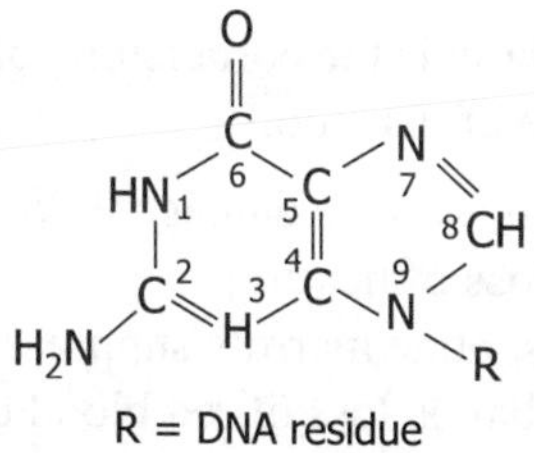

Fig. 9.40 Guanine component of DNA

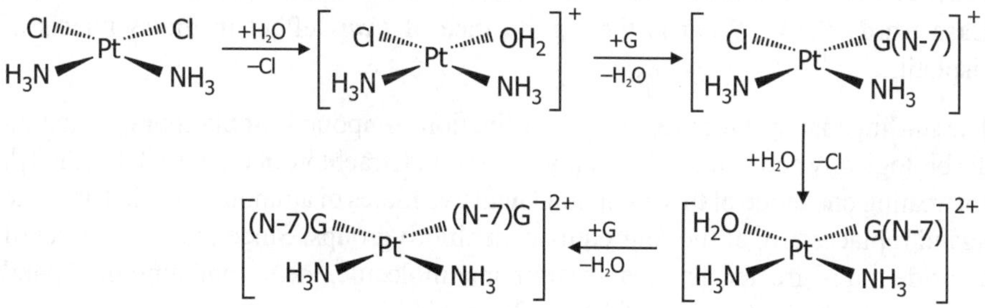

Fig. 9.41 Mechanism of coordination of cisplatin with DAN

It is to be emphasized that cisplatin binds to DNA by forming cross-links in which two chloro ligands are substituted by N-7 of two guanine residues on the *same* strand of DNA. The structure of the complex formed by interaction of one mole of cisplatin with two moles of guanine residue is shown in Fig. 9.42.

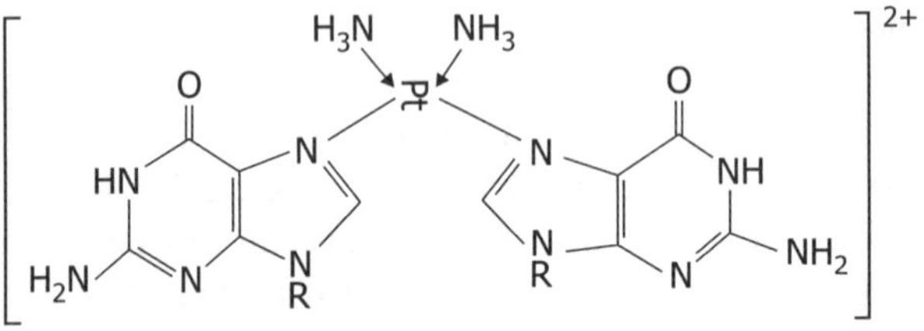

Fig. 9.42 Structure of cisplatin-guanine complex

As a result of this coordination, the DNA strand becomes distorted and is unable to participate in cell division. The cell is thus targeted to die. This process is called *apotosis*.

The mechanism of reaction between cisplatin and DNA requires that the two chloro ligands should be on the same side of the drug molecule. For this reason, transplatin, in which the two chloro groups are on the opposite side of the central atom, fails to inhibit growth of cancerous cells. Moreover, transplatin degrades before it has a chance to combine with the DNA of cancerous cells. Further, it is appreciably toxic in nature. In fact, before cisplatin is packaged as an anti-cancer drug, it is thoroughly tested for the absence of transplatin.

Side effects Although cisplatin is the cornerstone of treatment of many cancers, it produces the following side effects too.

- *Nephrotoxicity*, that is, serious damage to kidneys.
- *Ototoxicity*, that is, loss of hearing.
- *Myelotoxicity*, that is, bone marrow suppression.
- *Hemolytic anemia*, that is, loss of red blood cells.

Another platinum-based cancer drug, *carboplatin* (Fig. 9.39), has been found to possess high efficacy, but fewer side effects.

Case Study 9.13 What is the significance of *trans* effect in the synthesis of cisplatin?

The starting material for preparing coordination compounds of platinum, including the biologically active ones, is usually potassium tetrachloroplatinate(II), $K_2[PtCl_4]$. On treating one mode of this compound with two moles of ammonia, the first ammine ligand replaces any of the four equivalent chloro groups. Since the *trans* effect of chloride ion is greater than that of ammonia molecule, the second ammine ligand enters *cis* to the first one (*see* Chapter 28, p. 808).

However, the resulting *cis* isomer is contaminated with a small amount of *trans* analog. In order to avoid this problem and to ensure high yield and purity of the *cis* isomer, the potassium tetrachloroplatinate(II) is first converted into the corresponding tetraiodo complex by treatment with potassium iodide.

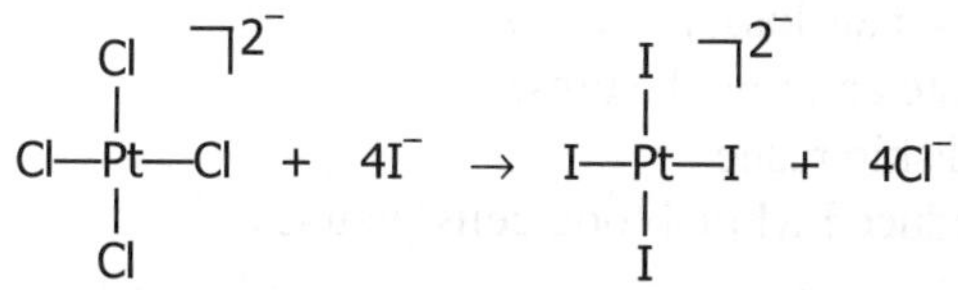

Since the *trans* effect follows the order, $I^- > Cl^-$ (*see* Chapter 28, p. 808), the reaction of potassium tetraiodoplatinate(II) with ammonia produces exclusively the *cis* product.

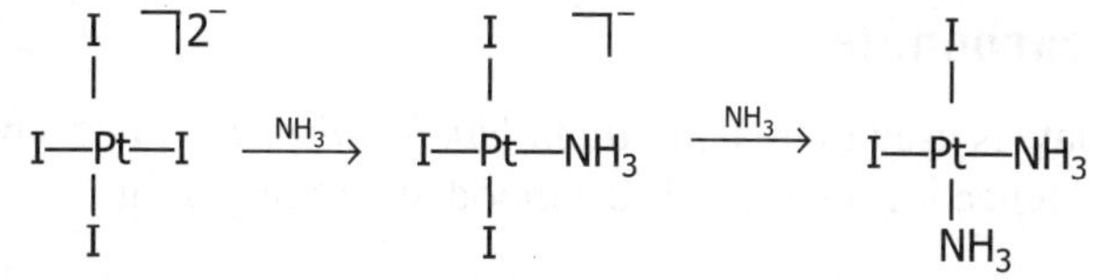

The *cis*-diamminediiodoplatinum(II) is hydrolyzed to *cis*-diamminediaqua-platinum(II) ion by aqueous silver nitrate.

$$[Pt(NH_3)_2I_2] + 2AgNO_3 + 2H_2O \rightarrow [Pt(NH_3)_2(H_2O)_2]^{2+} + 2AgI + 2NO_3^-$$

The precipitated silver iodide is isolated and the hydrolyzed product is treated with potassium chloride solution.

$$[Pt(NH_3)_2(H_2O)_2] + 2NO_3^- + 2KCl \rightarrow [Pt(NH_3)_2Cl_2] + 2KNO_3 + 2H_2O$$

2. Auranofin

Auranofin is a gold-based drug that is classified by the World Health Organization as an anti-rheumatic agent. It is used to treat rheumatoid arthritis and sold under the brand name *Ridaura*. Its structure is shown in Fig. 9.39.

Mode of action Auranofin is used in the treatment of active, progressive or destructive forms of rheumatoid arthritis. The latter is an inflammatory disease that affects the tissue around the joints. The inflammation of the tissue is caused by the action of hydrolytic enzymes, which are housed in lysosome organelles of the cells. These enzymes are meant to degrade the waste material in the cell and get rid of it. However, if these became over-active, they cause inflammation of the joint lining.

The drug auranofin contains gold(I) ion, a soft acid, which prefers to coordinate with the sulfur of the cysteine component of hydrolytic enzymes (*see* hard and soft acid base theory, Chapter 5, p. 245). As a result of this coordination, the activity of the hydrolytic enzymes is suppressed. The action of auranofin is threefold: Firstly, the inflammation of the lining of joints is reduced; secondly, it prevents the inflammation from destroying the bone and cartilage surrounding the joints; and thirdly, it decreases pain, swelling and morning stiffness in joints.

Side effects Auranofin causes the following side effects:

- Nausea, loss of appetite, heart burn and headache.
- Eye redness, hair loss, numbness.
- Metallic taste and mouth sores.
- Itching and skin rashes.
- Anemia, reduced white blood cells/platelets.

Gold compounds are also known to cause serious kidney and bone marrow problems. It is therefore pertinent that the patients being treated with auranofin be regularly monitored by screening of blood and urine.

3. Lithium carbonate

Lithium carbonate is used to treat *mania* in *bipolar disorders*. It is a potential anti-suicidal agent. Hence it is classified as a mood-stabilizing drug.

Mode of action Bipolar disorder is a mental illness characterized by cycles of extreme mood swings between mania and depression. Lithium carbonate combats mania and hence is considered as an anti-mania drug. Symptoms of mania include, hyperactivity, an unrealistically inflated self-image, excessive talkativeness and at times violent behavior. Maniac episodes cause the person to behave irrationally and with poor judgement.

Lithium carbonate acts as an anti-mania drug by three mechanisms. It is possible that all three mechanisms work together.

a. *Action on electrolytes.* Sodium ions assist the nerve cells to carry signals to brain. Excessive sodium ions pass on excessive (read unwanted) signals to the brain, resulting in mania. Lithium ions, replace sodium ions and, in consequence, disrupt their exchange and transport in nerves. This decreases the passage of excessive signals to the brain.
b. *Action on neurotransmitters.* Neurotransmitters are the chemicals which help in conveying the signals from nerve cells to target organs. During episodes of mania, the levels of three neurotransmitters are affected. While the amount of serotonin decreases, that of norepinephrine and dopamine increases. Lithium carbonate reverses this trend by increasing the level of serotonin and inhibiting norepinephrine and dopamine.
c. *Action on enzymes.* In cases of mania, the level of inositol phosphate in central nervous system increases. This chemical acts as a secondary messenger system and passes unwanted signals to the brain. Lithium carbonate blocks the action of GSK3β enzyme that is responsible for the synthesis of inositol phosphate, and thereby helps in stabilizing the mood.

Side effects Lithium carbonate is known to cause the following side effects:

- An overall dazed feeling and tremor of hands.
- Nausea and headache.

- Dehydration. Lithium carbonate reduces the action of vasopressin, the hormone that regulates the water levels in the body.

Most of the side effects which arise due to intake of lithium carbonate are reversed by consuming large amount of water. Yet, lithium ions are toxic in nature and, therefore, the patients undergoing treatment for mania should have regular blood tests to ensure that the concentration of lithium is within safe limits.

4. Gadoteric acid

Known largely by commercial names, *Artirem* or *Dotarem*, gadoteric acid is a macrocycle-structured gadolinium(III)-based magnetic resonance imaging (MRI) contrast agent. It is a complex of gadolinium(III) with the chelating ligand 1, 4, 7, 10-tetrazacyclododecane - N, N^{I}, N^{II}, N^{III}, N^{IIII} - tetraacetic acid, commonly abbreviated as DOTA. The structure of gadoteric acid is presented in Fig. 9.39.

Mode of action The gadolinium(III) ion has seven unpaired electrons in the valence 4f shell (*see* Table 30.1, p. 876). As a result, gadoteric, acid possesses a high magnetic moment and a strong paramagnetic effect (*see* Chapter 21, p. 611). The paramagnetism, in turn, shortens the longitudinal relaxation time of the protons in the water molecules present in tissues. In consequence, gadoteric acid is able to enhance the contrast between different tissues and highlight details such as the abnormalities of the blood-brain barrier.

Side effects Gadoteric acid is a relatively safe reagent for medical diagnosis. It is biologically inert and does not bind with plasma albumin. Moreover, it does not cross the normal blood-brain barrier. However, it is not suitable for patients with renal impairment. Such patients excrete the contrast reagent very slowly as compared to persons with normal renal function. As a result, patients suffering from renal ailments are likely to develop a serious connective tissue disease, *nephrogenic systemic fibrosis* (NSF) following the administration of gadoteric acid.

5. Technetium-99m exametazine

The complex, technetium-99m exametazine is a radiopharmaceutical that is used for medical imaging. The central metal in this coordination compound is technetium-99m, ${}^{99m}_{43}Te$ where m indicates that the nucleus of the isotope is in a metastable state. Its preparation involves two steps. In the first step, molybdenum-98 is bombarded with neutrons to produce molybdenum-99.

$${}^{98}_{42}Mo + {}^{1}_{0}n \rightarrow {}^{99}_{42}Mo$$

In the second step, molybdenum-99 decays with β-emission.

$${}^{99}_{42}Mo \rightarrow {}^{99m}_{43}Tc + {}^{0}_{-1}e$$

Mode of action The half-life of technetium-99m is only 6 hours, as a result of which it cannot be stored in stock. Moreover, by the time it is transported from a radioactivity processing unit to a hospital, it is highly probable that the isotope would have undergone disintegration. On the other hand, molybdenum-99 has a half-life of 66 hours, within which time it can easily reach a medical facility without undergoing appreciable disintegration. Thus whenever and wherever a patient requires technetium-99m, its percursor, molybdenum-99 in supplied.

The container in which molybdenum-99 is shipped to a hospital is called *technetium cow* or *moly cow* (this colloquial term signifies that while 'milk' cannot be carried to a patient, the 'cow' can be!). A variety of technetium cow, which is shown in Fig. 9.43, is made of thick lead so as to preclude leakage of radiation. Its inside walls are lined with alumina. It is also endowed with an extracting unit.

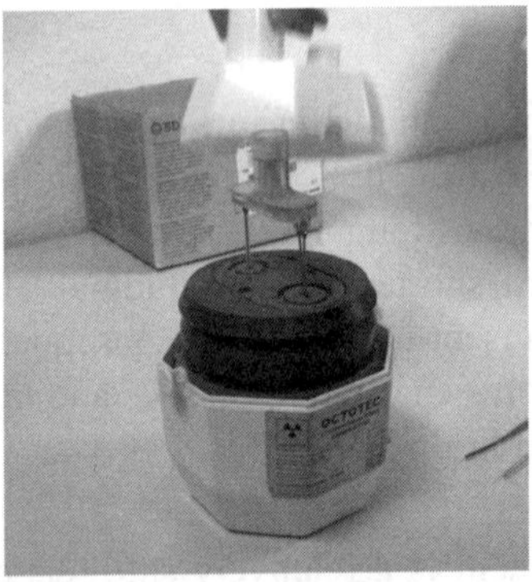

Fig. 9.43 A technetium cow

Within the technetium cow, a mixture of molybdate ions, MoO_4^{2-} and pertechnatate ions, TcO_4^- is generated. Since the molybdate ions have a greater negative charge, these get preferentially adsorbed on alumina living, as compared to pertechnatate ions, which carry a smaller negative charge. The soluble pertechnatate ions are extracted after every 6 hours and mixed with the ligand exametazine (Fig. 9.44a), in presence of tin(II) chloride and sodium chloride. The resulting, technetium-99m exametazine complex, which is sold under the commercial name *ceretec* (Fig. 9.44b), is administered to the patient as intravenous injection.

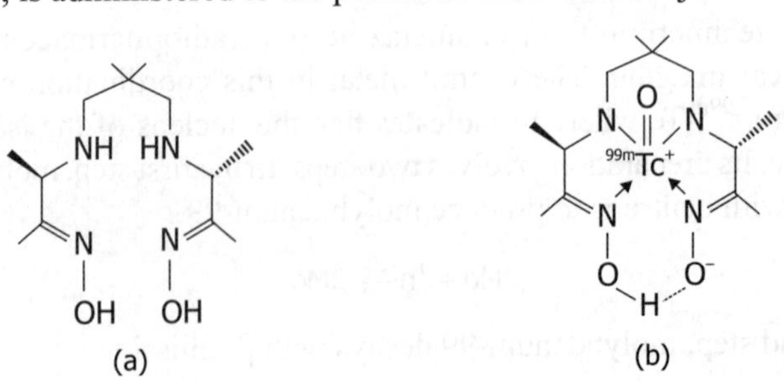

Fig. 9.44 Structure of (a) exametazine and (b) ceretec

The technetium-99m in ceretec decays by emitting γ-rays to technetium-99 in which the nucleus is in the normal ground state. The γ-rays, so emitted, form a picture of the organ in which ceretec is absorbed. This type of imaging evaluates the cerebral blood flow in patients who suffer stroke. It also labels, leukocytes in the investigation of infections of abdomen.

Side effects In a few cases, ceretec raises the blood pressure of patients. In extremely rare cases, it also causes fever, rashes and swelling of face.

10. INORGANIC POLYMERS

Introduction

Inorganic polymers are the macromolecules in which the skeletal backbone is made up of atoms of an element or elements other than carbon. Like their organic counterparts, the inorganic polymers too are industrially important compounds. In fact, where the organic polymers do not prove to be commercially viable, the inorganic polymers fill up the lacunae. Their unique physical characteristics, coupled with atypical structural features, has opened a vistas of applications—a trend that is anticipated to continue unabated in the years to come.

Types of Inorganic Polymers

Inorganic polymers may be classified either on the basis of the composition of backbone or with respect to their spatial structure. We describe both the systems of classification.

1. Classification on basis of composition of backbone

With reference to their key chain, inorganic polymers may be broadly divided into two types: homopolymers and heteropolymers. Homopolymers are those in which the entire backbone is made up of atoms of the same element. In heteropolymers, the repeating units have atoms of two or more elements. The former may be represented as $[-(M)_n-]$, the latter as $[-(M-M')_n-]$, where M and M′ are atoms of elements other than carbon. As displayed in Table 10.1, each of these broad divisions has several sub-divisions.

Table 10.1 Classification of inorganic polymers on the basis of sketetal backbone

Broad classification	Sub-classification	Representative example
Homopolymers	Pure elements	Black phosphorus
	Anionic polymers	Nonaiodide anion
	Cationic polymers	Octasulfide cation
	Backbone having side groups	Silanes and their derivatives
Heteropolymers	Backbone containing halogen	Antimony(V) fluoride
	Backbone containing sulfur	Silicon disulfide

Backbone containing nitrogen	Boron nitride
Backbone containing oxygen	Siloxanes
Backbone containing oxygen and phosphorus	Phosphorus(V) oxide
Backbone containing nitrogen and phosphorus	Phosphazenes

Homopolymers

There are four sub-divisions of this type of inorganic polymers.

a. A pure element exists in a polymeric state. Examples include black phosphorus (Fig. 17.3, p. 505) and plastic sulfur.
b. An anion exists in a polymeric state. Pentasulfide anion (Fig. 18.1c, p. 532) and nonaiodide anion (Fig. 19.2, p. 576) are examples.
c. A cation exists as a polymer, octasulfide cation (Fig. 18.2b, p. 533) being a representative example.
d. The atoms in the key chain are bound to side groups. Silanes, Si_nH_{2n+2} (Table 16.1, p. 486) are examples of this type of homopolymers. The methylsilanes too are considered as inorganic polymers, even though the side groups, in this case, are organic in nature.

Heteropolymers

As has been stated, heteropolymers are those in which the repeating units involve two or more atoms of different elements. The one that belongs to Group 15, 16 or 17 of the periodic table is taken as reference to define the sub-classification (Table 10.1). Antimony(V) fluoride and mercury(II) iodide are examples of inorganic polymers in which halogens are the heteroatoms (Fig. 10.1).

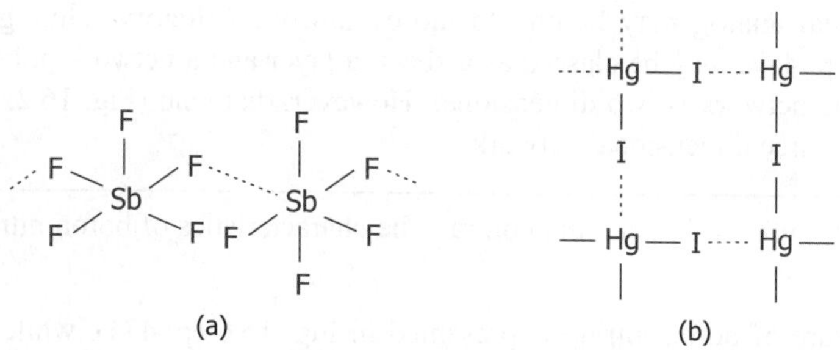

Fig. 10.1 Structure of (a) antimony(V) fluoride and (b) mercury(II) iodide

Silicates and silicones, discussed further in this chapter, belong to the sub-classification in which the heteroatom is oxygen. Likewise, in boron nitride (Fig. 15.6, p. 474), the heteroatom is nitrogen, while in silicon disulfide (Fig. 10.2) it is sulfur.

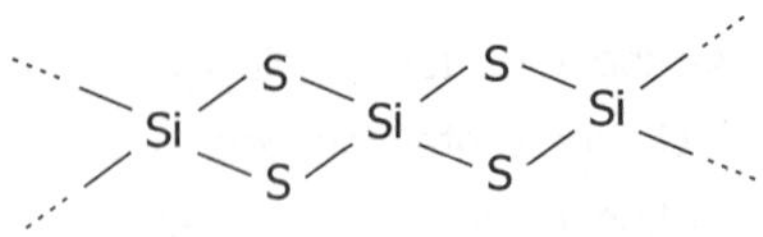

Fig. 10.2 Structure of silicon disulfide

Phosphorus(V) oxide (Fig. 17.4, p. 508) is an example of hetropolymers in which the main chain contains both phosphorus and oxygen. In phosphazenes, described later in this chapter, the heteroatoms are phosphorus and nitrogen.

2. Classification on basis of spatial structure

Inorganic polymers may also be classified on the basis of their skeletal framework. The major frameworks are chain, sheet, network, layers and cyclic. Representative examples of each type are cited in Table 10.2.

Table 10.2 Classification of inorganic polymers on the basis of spatial structure

Spatial arrangement	Representative example
Chain	Silicon disulfide
Sheet	Graphite
Network	Diamond
Layer	Boron nitride
Cyclic	Phosphorus(V) oxide

It is possible that one particular class of inorganic polymers may have different spatial arrangements. For examples, silicates, with which we will deal later in this chapter, exist as chain, sheet, layer and cyclic varieties. Phosphazenes may possess either a chain or a cyclic structure. At times, an individual analog may belong to more than one category. Thus graphite (Fig. 16.1, p. 487) may be classed as a sheet, a layer and a network polymer. In this case the network is two dimensional. However, diamond (Fig. 16.2, p. 488) possesses a three dimensional network.

Case Study 10.1 Compare and contrast the characteristics of boron nitride and graphite.

The structure of boron nitride is presented in Fig. 15.6 (p. 474), while that of graphite is shown in Fig. 16.1 (p. 487). The structures of both the polymers consist of sheets of hexagonal rings, stacked over each other in layered pattern. The bond distances and the densities of the two materials are also comparable. However, here the similarities end. A comparison of boron nitride and graphite is set out in Table 10.3.

Table 10.3 Comparison between boron nitride and graphite

Characteristic	Boron nitride	Graphite
Polymer type	Heteropolymer	Homopolymer
Structure	Layered sheets	Layered sheets
Stacking of layers	Boron atom in one sheet is stacked over a nitrogen atom in the sheet below	A carbon atom in one sheet lies over the center of the ring in the sheet below
Interlayer distance	333 pm	335 pm
Intra-layer distance	144 pm	142 pm
Density	2.29 g cm^{-3}	2.25 g cm^{-3}
Color	White	Black
Electrical property	Insulator	Conductor
Reactivity	Inert	Relatively reactive
Uses	Mainly as refractory material and abrasive	Mainly as a lubricant

Properties of inorganic Polymers

As has been stated, inorganic polymers are those macromolecules in which the repeating units are not based on carbon atoms. Initially, such polymers were studied merely from an academic point of view. It was widely believed that only the organic polymers have the potential to serve societal needs. It was only when the aircraft and aerospace industry demanded materials possessing low density and high temperature resistance that the inorganic polymers transitioned from the curiosity stage to commercial stage. As of now, these derivatives have permeated the arena of electronics, medicines, fibers, elastomers, insulators, and many more.

As compared to their organic counterparts, the inorganic macromolecules have a higher resistance to oxidative degradation. This characteristic prevails at higher temperatures as well. This means that inorganic polymers are thermally very stable and therefore find applications in those manufacturing processes which operate at elevated temperatures. Most analogs are not only thermally stable, but chemically inert too. These are not easily attacked by environmental chemicals and hence maintain their compositions for prolonged periods of time. Nevertheless, since inorganic polymers are more polar than their organic counterparts, some of the derivatives do undergo slow hydrolysis. However, by replacing the inorganic side groups by organic functions, the hydrolysis may be mitigated. The chain variety of inorganic polymers possess better plastic or elastomeric properties than the organic derivatives of same spatial conformation. The carbon atoms in the key chain of organic polymers can hold a fixed number of side groups. The elements that constitute the backbone of inorganic polymers have different atomic radii and oxidation states than carbon. As a result, depending on the desired application, the inorganic polymers can be tailor-made to possess a wide range of solubility and molecular flexibility.

Comparison of Inorganic Polymers with Organic Polymers

Inorganic polymers elicit physical and chemical characteristics which are quite different from those of organic polymers. A comparison between the two types of polymers is set out in Table 10.4.

Table 10.4 Comparison between inorganic polymers and organic polymers.

Organic polymers	Inorganic polymers
Not suitable for high temperature applications	Suitable for high temperature applications
Flexibility and torsional mobility of bonds more or less constant in all derivatives	Flexibility and torsional mobility of bonds may be varied, depending upon the nature of application
Prone to oxidative degradation	Not prone to oxidative degradation
Prone to attack by environmental chemicals like ozone	Inert to most environmental chemicals
Degraded by ultraviolet and gamma radiation	Not affected by ultraviolet and gamma radiation
Degraded by free radicals	Not degraded by free radicals
Mostly flammable	Mostly non-flammable
Dissolve, swell or soften when suspended in organic solvents	Inert to most organic solvents
Inert to hydrolysis	May be degraded by hydrolysis. However, hydrolysis may be mitigated by substituting organic side groups in the skeleton
Mostly derived from petroleum—a resource which is fast dwindling. Substitutes may be required in future	Synthesis does not involve scarce resources

Typical Examples

We now describe a few important examples of inorganic polymers, each based on a specific repeating unit.

1. Borazines

The preparation of borazine by the reaction of diborane and ammonia has been described in Chapter 15, p. 474. On heating strongly, borazine undergoes dehydrogenation and polymerization. Depending on the nature of substituents on boron and nitrogen, and also on temperature, the resulting polyborazines may acquire a structure analogous to naphthalene (Fig. 10.3a) or to biphenyl (Fig. 10.3b).

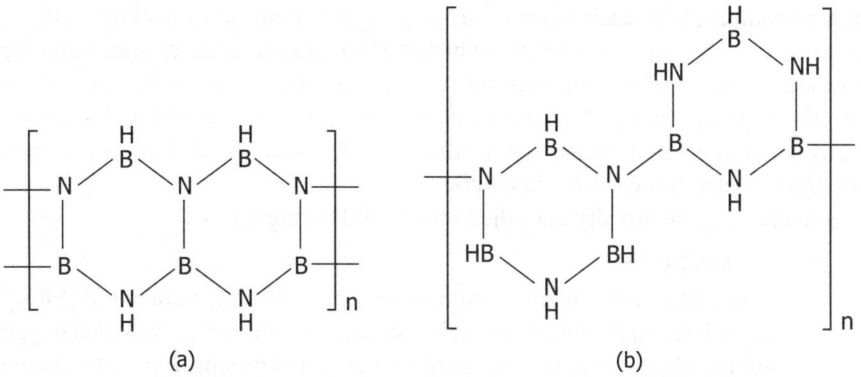

Fig. 10.3 Thermolysis of borazine yields (a) naphthalene-like and (b) biphenyl-like polymers

The thermolysis of unsubstituted borazine is quite difficult. The reason is that borazine is a volatile liquid, having a boiling point of 328 K. On heating, it tends to evaporate quickly. Hence pyrolysis has to be conducted in a pressure vessel having a system for the removal of hydrogen gas as the polymerization reaction proceeds.

A better method is to first prepare N-triphenylborazine by the reaction of diborane and triphenylamine,

$$3B_2H_6 + 6N(C_6H_5)_3 \rightarrow 2B_3(N\text{–}C_6H_5)_3H_3 + 12C_6H_6$$

and thereafter heat it to about 500 K, whereby a boron-boron linked polyborazine is formed.

All the polyborazines on fusion yield boron nitride. The latter exists in several crystalline forms. We have described the synthesis of this polymer by the reaction of diborane with ammonia in Chapter 15, p. 474. However, the boron nitride obtained by fusion of polyborazines is largely of h-type. This is the most important commercial variety of boron nitride. It is used as a refractory material, as an abrasive and in the manufacture of ceramics.

2. Silicates

Compounds in which silicon is coordinated tetrahedrally by four oxygen atoms to give [SiO_4] units are called silicates. The [SiO_4] units may exist as discrete tetrahedra or may combine with each other, by sharing oxygen atoms to form chains, rings or sheets. Depending on the manner in which the $[SiO_4]$ tetrahedra are linked together, a broad spectrum of silicate materials exist. Clay, stones and soils have silicate functional groups, and so have glass, talcum powder, asbestos and zeolites. While many silicates occur in nature, some are artificially synthesized. A few of them are soluble in water, but most are insoluble.

Silicates may be broadly classified into the following types:

a. Orthosilicates

Also called *nesosilicates,* orthosilicates have discrete tetrahedral $[SiO_4]^{4-}$ units. Each such unit retains an independent identity (Fig. 10.4). No oxygen atom is shared between two or more functional groups. Examples include Zircon, $ZrSiO_4$ and olivine, $(Mg, Fe)_2SiO_4$.

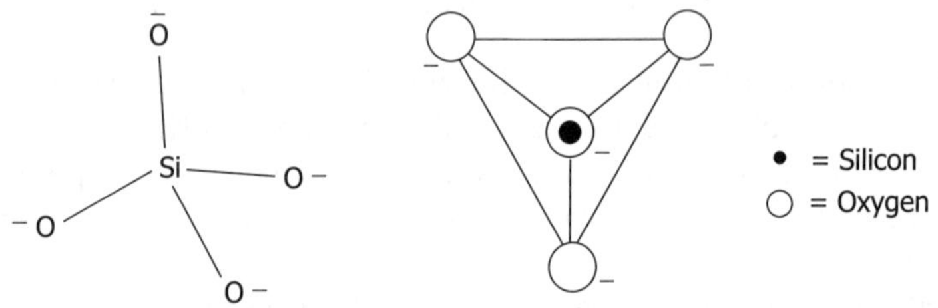

Fig. 10.4 Structure of an orthosilicate

b. Pyrosilicates

The pyrosilicates are also called *disilicates* or *sorosilicates.* They contain discrete $[Si_2O_7]^{6-}$ units (Fig. 10.5). Such units are formed by sharing of one oxygen atom between two tetrahedral $[SiO_4]$ units. Thortveitite, $(Sc, Y)_2Si_2O_7$ is an example of pyrosilicate.

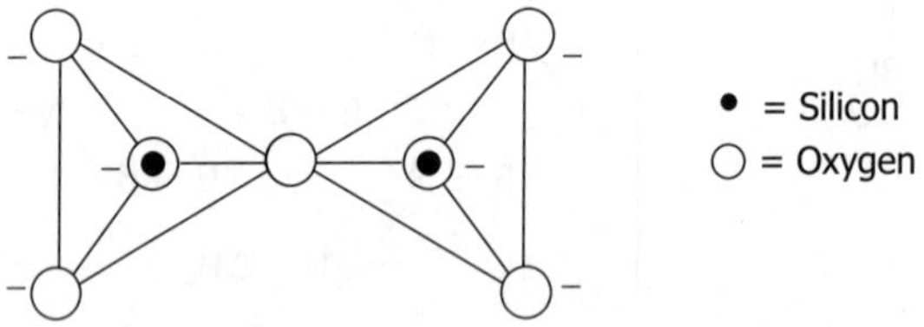

Fig. 10.5 Structure of a pyrosilicates

c. Cyclic silicates

If two oxygen atoms per $[SiO_4]$ tetrahedron are shared to give a ring structure, the resulting silicates are called cyclic silicates or *cyclosilicates* (Fig. 10.6). Their empirical formula is $(SiO_3)^{2-}$ and molecular formula is $(SiO_3)_n^{2n-}$. Generally n is either 3, for example, wollastonite, $CaSiO_3$ (Fig. 10.6) or 6, for example, beryl, $Be_3Al_2(Si_6O_{18})$.

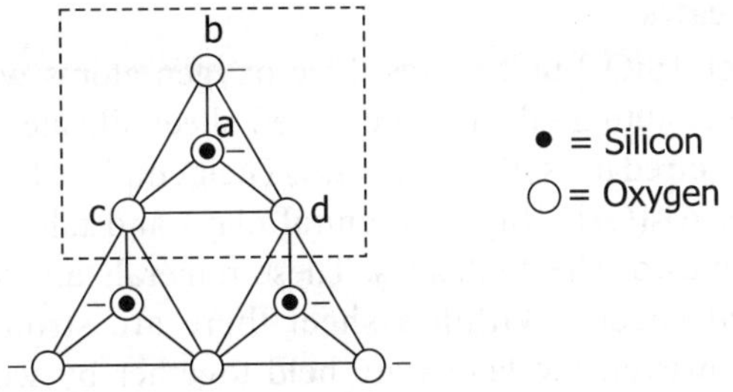

Fig. 10.6 Structure of a cyclic silicate

Oxygen atom	To be counted as
a	1
b	1
c	½
d	½

Number of charges on one tetrahedron = 2–
Empirical formula for one tetrahedron = $(SiO_3)^{2-}$
Molecular formula for n tetrahedra = $(SiO_3)_n^{2n-}$

d. Chain silicates

If two oxygen atoms per $[SiO_4]$ tetrahedron are shared to give a chain structure, the resulting silicates are called chain silicates or *inosilicates.* These too are based on $(SiO_3)^{2-}$ repeating units (Fig. 10.7). Enstatite, $MgSiO_3$, is an example of chain silicate.

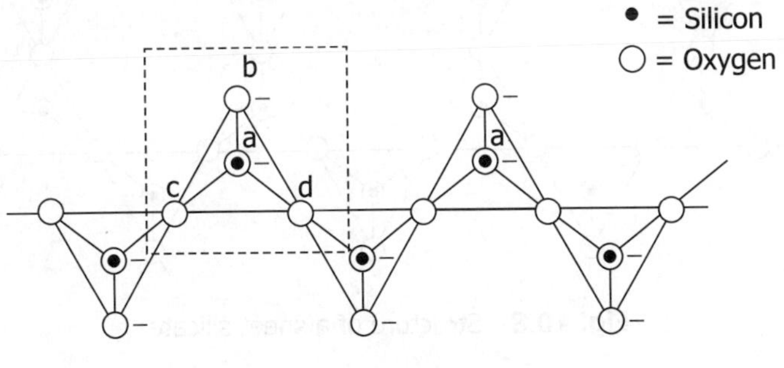

Fig. 10.7 Structure of a chain silicate

Oxygen atom	To be counted as
a	1
b	1
c	½
d	½

Number of charges on one tetrahedron = 2–
Empirical formula for one tetrahedron = $(SiO_3)^{2-}$

e. Sheet silicates

When each [SiO_4] unit shares three oxygen atoms with its neighboring units, the resulting silicates are called sheet silicates (Fig. 10.8). These are also referred to as *phyllosilicates*. Their empirical formula is $(Si_2O_5)^{2-}$. Kaolinite, $Al_2(OH)_4[Si_2O_5]$ (natural clay) and talc, $Mg_3(OH)_2[Si_4O_{10}]$ are examples of sheet silicates. These minerals are composed of layers of several sheets. Within a sheet there are strong silicon–oxygen bonds. However, the layers are held together by weak van der Waals forces of attraction. For this reason, the sheets slide past one another relatively easily and produce the slippery feel that is characteristic of talcum powder.

Clay minerals have enormous surface area in between the sheet layers. For this reason they can absorb large amount of water and nutrients in between the silicate sheets. That is why clays are useful growth media for plants.

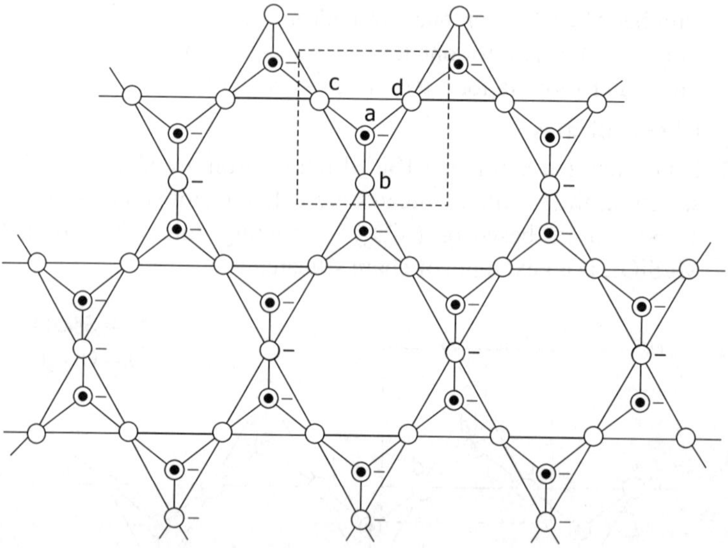

Fig. 10.8 Structure of a sheet silicate

Oxygen atom	To be counted as
a	1
b	½
c	½
d	½

Number of charges on one tetrahedron = 1–

Empirical formula for one tetrahedron = $(SiO_{2.5})^-$ or $(Si_2O_5)^{2-}$

Molecular formula for n tetrahedra = $(Si_2O_5)_n^{2n-}$

f. Three-dimensional silicates

Sharing of all four oxygen atoms of $[SiO_4]$ units gives rise to three-dimensional silicates or *tectosilicates*. Their empirical formula is SiO_2, but some of the silicon(IV) ions are replaced by aluminium(III) ions. For example, if one-quarter of silicon(IV) ions in $(SiO_2)_4$ units is replaced by aluminium(III) ions, a three-dimensional framework of $[AlSi_3O_8]^-$ units results. Feldspar, which belongs to the category of three-dimensional silicates has the composition, $K[AlSi_3O_8]$. Another variety of three-dimensional silicates, zeolites are formed by replacing two silicon(IV) ions by aluminium(III) ions in $(SiO_2)_5$ units.

Zeolites are a class of compounds which possess open channels in their skeletal structures. Two important analogs of this classification are natrolite, $Na_{16}(Al_2Si_3O_{10})_8{\cdot}16H_2O$ and mordenite, $Na_8(AlSi_5O_{12})_8{\cdot}24H_2O$. The sodium ions are exchangeable with other cations. When replaced by protons, the zeolites acquire acidic character, although they retain their original open-channelled structures. Such zeolites are called *solid acids*. These acids are used as clean and environment-friendly catalysts for such organic reactions as polymerization of lower hydrocarbons to isooctane.

Inference The silicates illustrate all types of bonding in solids. Covalent bond exists between silicon and oxygen in the tetrahedra, while van der Waals forces are prevalent between silicate sheets. When some of the silicon atoms are replaced by aluminium ions, there exists ionic bonding between the charged layers or chains. There is also hydrogen bonding between water molecules and silicate oxygen atoms in clay minerals.

Case Study 10.2 Comment on the structure of asbestos.

Asbestos is a form of silicate mineral. Its usefulness stems from its inertness to acids and alkalis, its fire-resistant character and its durability. There are two structural varieties of asbestos: amphibole and serpentine. Amphiboles are chain silicates. However, instead of a single chain, they possess a double chain structure. Most of the $[SiO_4]$ units share two oxygen atoms with their neighbors, but some of them also share three oxygen atoms (Fig. 10.9). The empirical formula of amphiboles is calculated by considering one tetrahedron that shares two oxygen atoms and another tetrahedron that shares three oxygen atoms. It works out to be $(Si_4O_{11})^{6-}$. An example of amphibole asbestos is crocidolite, $Na_2Fe_3^{II}Fe_2^{III}[(Si_4O_{11})_2](OH)_2$. It is also called *blue asbestos*.

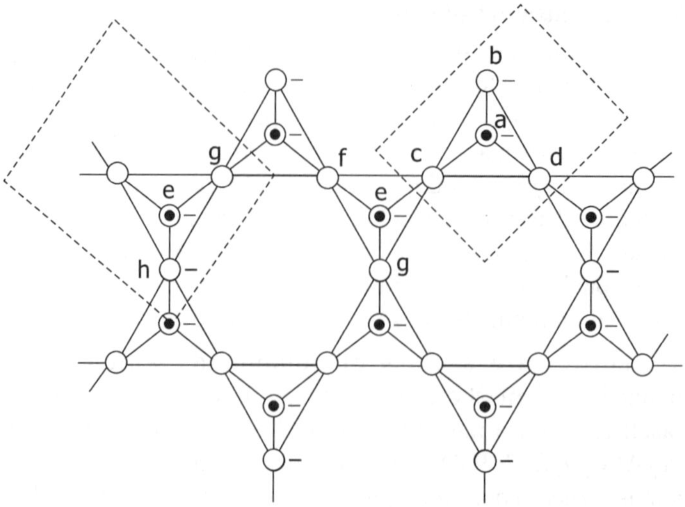

Fig. 10.9 Structure of amphibole asbestos

Oxygen atom	To be counted as
a	1
b	1
c	½
d	½
e	1
f	½
g	½
h	½

Total charges = 3– (on a, b and e)
Empirical formula = $(Si_2O_{5.5})^{3-}$ or $(Si_4O_{11})^{6-}$

The serpentine variety of asbestos has sheet-like silicate structure, with empirical formula, $(Si_2O_5)^{2-}$. Example is chrysotile, $Mg_3(OH)_4[Si_2O_5]$, also called *white asbestos*.

3. Silicones

The silicones – also called *siloxanes* – are a set of organosilicon polymers based on repeating siloxane $[R_2SiO]_n/[RSiO_2]_n$ units (R = aliphatic or aromatic group). These compounds have a wide range of commercial applications. The following properties of silicones are responsible for their vast industrial uses:

a. High thermal stability
b. Chemical inertness
c. Water repellency
d. Dielectric nature

e. Resistance to ultraviolet rays
f. Anti-foam characteristics
g. Non-toxicity

The high thermal and chemical stabilities of silicones are attributed to high strength of both silicon–carbon and silicon–oxygen bonds.

The first step in the manufacture of silicones is the preparation of chlorosilanes. These, in turn, may be prepared by one of the following methods:

(i) Reaction of Grignard reagent with silicon tetrachloride.

$$SiCl_4 + RMgCl \rightarrow RSiCl_3 + MgCl_2$$
$$RSiCl_3 + RMgCl \rightarrow R_2SiCl_2 + MgCl_2$$
$$R_2SiCl_2 + RMgCl \rightarrow \underset{\text{Chlorosilanes}}{R_3SiCl} + MgCl_2$$

(ii) Reaction of alkyl halides with silicon in the presence of copper catalyst at 573 K.

$$Si + 2RCl \xrightarrow[573\,K]{Cu} R_2SiCl_2$$

This method is called *Rochow direct process*. By changing the molar concentration of alkyl halide, other chlorosilanes may also be prepared. The reaction is exothermic and heat must be removed from the system to get better yield. Copper has a definite role in the mechanism of the reaction.

$$RCl + 2Cu \rightarrow [CuR] + [CuCl]$$
$$[CuR] \rightarrow Cu + [R^{\cdot}]$$
$$[CuCl] + Si \rightarrow Cu + [SiCl]$$
$$[R^{\cdot}] + [SiCl] \rightarrow RSiCl$$
$$RCl + 2Cu \rightarrow [CuR] + [CuCl]$$
$$[CuR] \rightarrow Cu + [R^{\cdot}]$$
$$[CuCl] \rightarrow Cu + Cl^{\cdot}$$
$$[R^{\cdot}] + Cl^{\cdot} + RSiCl \rightarrow R_2SiCl_2 + \text{Heat}$$

Once the chlorosilanes are prepared, these may be hydrolyzed to silanols.

$$RSiCl_3 + 3H_2O \rightarrow RSi(OH)_3 + 3HCl$$
$$R_2SiCl_2 + 2H_2O \rightarrow R_2Si(OH)_2 + 2HCl$$
$$R_3SiCl + H_2O \rightarrow \underset{\text{Silanols}}{R_3Si(OH)} + HCl$$

Silanols undergo spontaneous condensation polymerization. The disilanols, $R_2Si(OH)_2$ produce long-chain polymers,

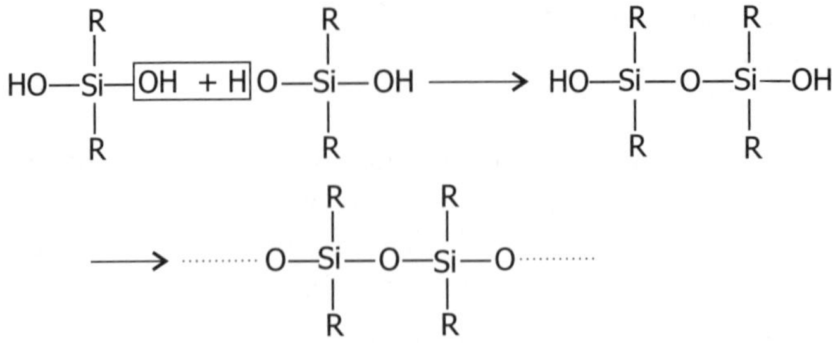

Long chain polymer

while the trisilanols, $RSi(OH)_3$ produce cross-linked polymers.

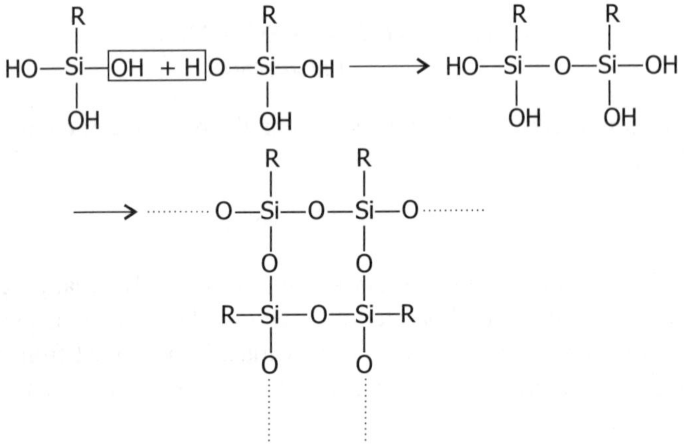

Cross-linked polymers

The monosilanol derivative, R_3SiOH, can undergo only dimerization.

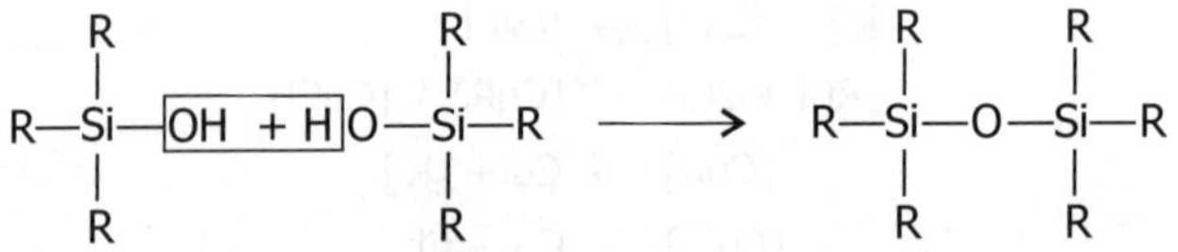

Dimer

Since the dimer has no silanol groups, (Si–OH), it cannot be polymerized further and hence has no industrial importance.

As the degree of polymerization increases, silicones transition from liquid state to solid state. By carefully controlling the extent of polymerization, silicones may be prepared as—

a. Liquids (oils)
b. Semi-solids (greases)
c. Elastomers (rubbers)
d. Solids (resins)

Uses of silicones

- Silicone oils are used as dielectric insulating media, hydraulic fluids and heat transfer media. Due to their low surface tension, these are used as anti-foam agents in textile dyeing. As they are non-toxic, they are added to cooking oil to prevent frothing.
- Silicone greases are used in car polishes, lubricants, suntan lotions and lipsticks.
- Silicone rubbers are used in cable insulations, electrical tapes, static and rotary seals, gaskets, rollers, medical tubings and space suits. They are also used for making molds for dentures.
- Silicone resins are used as laminates for printing circuit boards and for encapsulation of resistors. They are also used in high-temperature paints.

Case Study 10.3 What is *silly putty*?

Chemically, silly putty is a silicone polymer having a composition intermediate between silicone oils and silicone rubbers. On the one hand, it can be made to flow like a liquid, while on the other, it can be deformed like an elastic solid. It is used in children's toys.

Box 10.1
The Hair Conditioning Debate

Shampoos contain detergents to cleanse the hair of dirt and greasy matter. However, these very detergents also wash out the natural oils that strengthen and protect the hair. In order to overcome this problem, many brands incorporate silicone-based conditioners in their formulation. The role of the conditioner is to form a coating over the strands of hair so as to prevent the useful oils from being extracted out.

The routinely used conditioner by the cosmetic industry is polydimethylsiloxane (Fig. 10.10), though it is better known by its common name, *dimethicone* (Fig. 10.11).

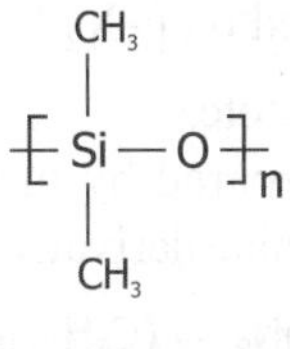

Fig. 10.10 Structure of polydimethylsiloxane

Contd...

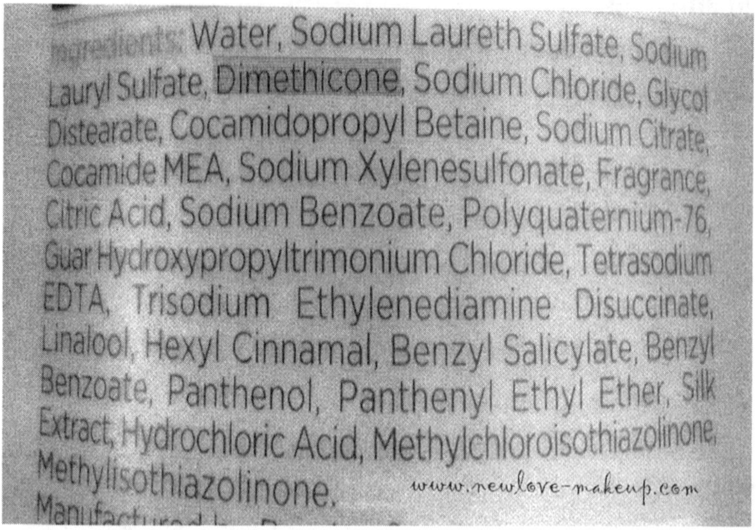

Fig. 10.11 Dimethicone is one of the shampoo additives

However, the utility of adding dimethicone in shampoos has been a matter of debate. How can one be sure that the silicone material will first form a protective layer on hair and then the detergent will take over the cleansing action? It can well be the other way round, that is, the cleansing action of detergent, and the unavoidable removal of hair-strengthening minerals may precede the coating action of dimethicone. For this reason many cosmetologists consider the addition of dimethicone to shampoos as a gimmick to boost the sale of the product.

4. Phosphazenes

Phosphazenes are characterized by repeating $-[-N-PX_2-]$ units in which nitrogen is trivalent and phosphorus pentavalent. These are classified into four types—

- Monophosphazenes, based on $X_3P=NR$ units
- Diphosphazenes, based on $[X_3P=N-PX_2-]$ units
- Cyclic phosphazenes, based on $[-X_2P=N-]$ units
- Chain phosphazenes, based on $[-X_2P=N-]$ units

Where $X = F, Cl, Br, CH_3, C_6H_5$ etc.

Monophosphazenes may be prepared by the reaction of an organic azide, for example, phenylazide with a phosphine derivative, for example, triphenylphosphine.

$$(C_6H_5)_3P + (C_6H_5)N_3 \rightarrow (C_6H_5)_3P = N(C_6H_5) + N_2$$

The same compound may also be prepared by the reaction of dichlorotriphenylphosphorus(V) with aniline.

$$(C_6H_5)_3PCl_2 + C_6H_5NH_2 \rightarrow (C_6H_5)_3P = N(C_6H_5) + 2HCl$$

This method of preparing monophosphazenes is called *Kirsanov reaction.*

Diphosphazene may be synthesized by the reaction of phosphorus(V) chloride with ammonium chloride in 1, 1, 2, 2-tetrachloroethane solvent.

$$3PCl_5 + NH_4Cl \rightarrow [Cl_3P = N - PCl_3]^+ [PCl_6]^- + 4HCl$$

The phosphorus-nitrogen single and double bonds are equivalent and hence written as,

$$[X_3P \overset{\cdots}{-} N \overset{\cdots}{-} PX_3]^+$$

Amongst the cyclic phosphazenes, the chloro derivatives are most common, and most easy to prepare. These are synthesized by reacting stoichiometric amounts of phosphorus(V) chloride and ammonium chloride in tetrachloroethane.

$$nPCl_5 + nNH_4Cl \rightarrow (PNCl_2)_n + 4nHCl$$

At 400 K, the trimer, $(PNCl_2)_3$ is the main product and the tetramer, $(PNCl_2)_4$ is obtained in minor amounts. By varying the temperature and molar concentrations of the reactants, phosphazene analogs with n = 5-8 can also be obtained.

The bromine-based phosphazenes may be prepared by the reaction between phosphorus(V) bromide and ammonium bromide. However, in this case, bromine has to be added to the reaction mixture so as to suppress the dissociation of phosphorus(V) bromide to phosphorus(III) bromide.

$$PBr_5 \rightleftharpoons PBr_3 + Br_2$$

The fluorine-based derivatives are prepared by fluorination of chlorophosphazenes.

$$(PNCl_2)_n + 2nNaF \rightarrow (PNF_2)_n + 2nNaCl$$

The chain polyphosphazenes may be prepared by heating the cyclic analogs. For example, when cyclic, trimeric chlorophosphazene is heated to 400 - 600 K, a number of chain polymeric products are obtained.

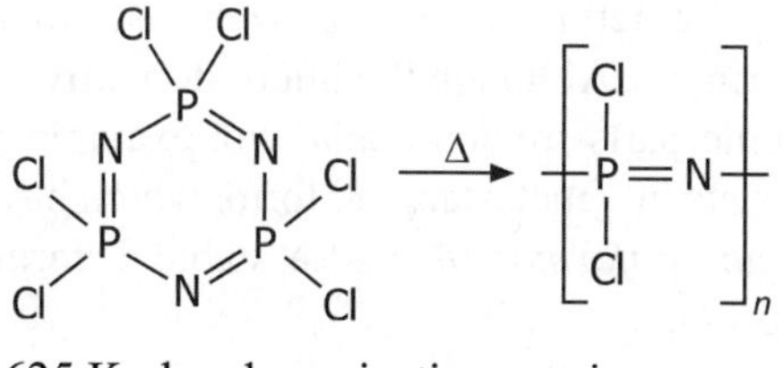

However, beyond 625 K, depolymerization sets in.

The halide trimer has a planar, six-membered ring (Fig. 10.12).

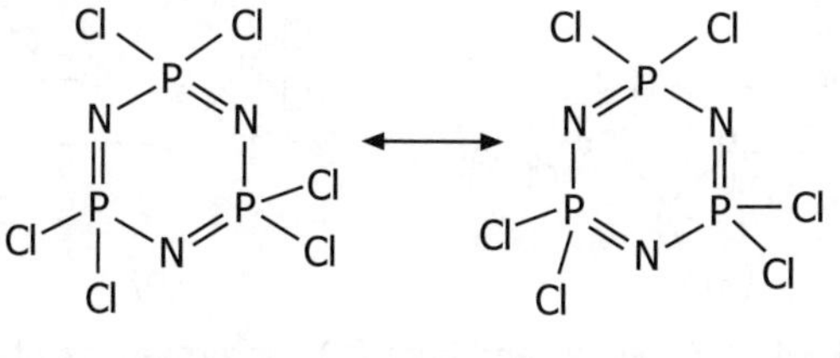

Fig. 10.12 Structrure of cyclic, trimeric chlorophosphazene

The nitrogen is sp^2 hybridized.

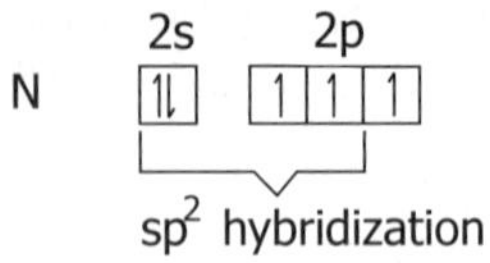

Two of the sp^2 hybrid orbitals contain one electron each and form bonds with phosphorus. The third contains a lone pair of electrons. This leaves out one electron in the unhybridized $2p_z$ orbital (Fig. 10.13).

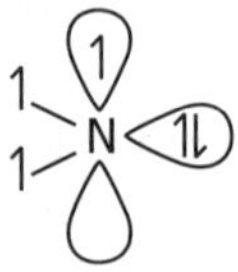

Fig. 10.13 Electron distribution around nitrogen

The phosphorus, after being promoted to excited state, undergoes sp^3 hybridization.

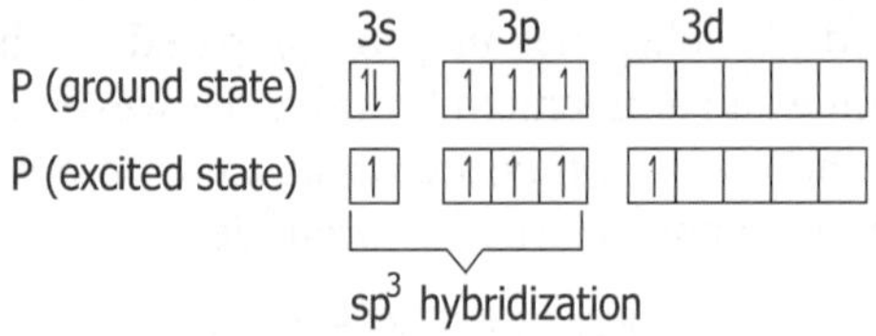

With the aid of the single electron in each of the four sp^3 hybrid orbitals, phosphorus forms four σ bonds—two with nitrogen and two with chlorine. The singly occupied unhybridized $3d_{xz}$ orbital of phosphorus overlaps with the singly occupied unhybridized $2p_z$ orbital of nitrogen, resulting in the formation of a π bond. Due to the pπ–dπ bonding, all phosphorus-nitrogen bond lengths in the cyclic, trimeric phosphazenes are equal. Moreover, as shown in Fig. 10.12, the compound is a resonance hybrid.

The structures of cyclic, tetrameric phosphazenes are more flexible than those of the corresponding trimers. Although the fluoro derivative, $(NPF_2)_4$ is planar, its chloro and bromo counterparts are non-planar. For example, the cyclic, tetrameric chlorophosphazene exists in a metastable K form, which has a *boat conformation* (also referred to as *tub conformation*) and a stable T form, which has a *chair conformation* (Fig. 10.14).

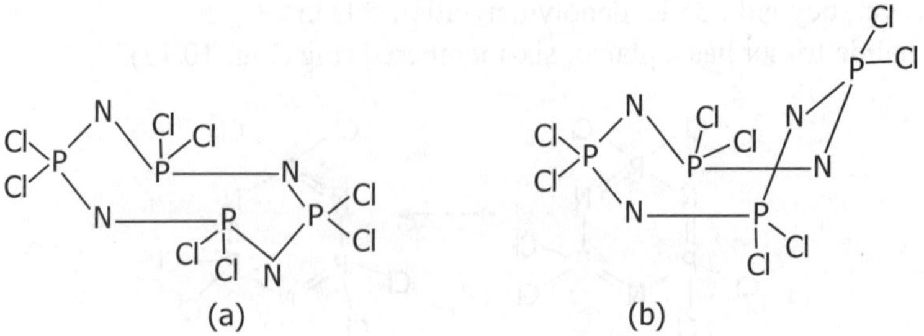

Fig. 10.14 Chair (a) and boat (b) conformations of cyclic, tetrameric chlorophosphazene

The chlorine atoms in cyclic chlorophosphazenes are reactive and may be replaced by other functional groups like alkyl, aryl, alkoxy or hydroxy. For example,

$$[NPCl_2]_3 + 6CH_3MgI \rightarrow [NP(CH_3)_2]_3 + 3MgCl_2 + 3MgI_2$$

$$[NPCl_2]_3 + 6C_6H_5Li \rightarrow [NP(C_6H_5)_2]_3 + 6LiCl$$

$$[NPCl_2]_3 + 6NaSCN \rightarrow [NP(SCN)_2]_3 + 6NaCl$$

Analogous reactions are also possible for chain polyphosphazenes. For example,

$$[PNCl_2]_n + 2nNaOCH_3 \rightarrow [PN(OCH_3)_2]_n + 2nNaCl$$

$$[PNCl_2]_n + 2n(CH_3)_2NH \rightarrow [PN(N(CH_3)_2)]_n + 2nHCl$$

All phosphazenes, whether cyclic or chain, have the following common features.

1. These contain the unsaturated group, $-N = PX_2-$ in which nitrogen is 2-coordinated and phosphorus is 4-coordinated.
2. The rings and the chains are very stable.
3. The phosphorus-nitrogen bond lengths are shorter than those expected for a covalent single bond between these two atoms.

Case Study 10.4 Comment on the applications of phosphazenes.

A good number of polyphosphazenes are biodegradable in nature. Their end products are harmless and their physical properties can be modified by changing the substituents on phosphorus atoms. For this reason, these are used as biodegradable supports for *in vivo* bone regeneration. The fibers made from alkoxy-substituted phosphazenes are best suited for such purposes as these can become bound to the calcium content of bones. The polymer fibers become integrated with the person's bone-making cells. As there cells multiply and fill the space between the fibers, the polymer begins to degrade.

Polyphosphazenes are also used as structural material for construction of heart valves and blood vessels. In drug delivery systems, the polyphosphazenes carry the bioactive molecules to the target cells and then degrade. The rate of degradation can be tuned by changing the substituents of phosphazenes.

Thin films of poly(aminophosphazenes) are used to bandage burn wounds and injures, since these prevent the loss of bodily fluids and keep germs out.

Phosphazenes may be molded into plastics and foams which are fire-proof, water-proof, and are unaffected by petrol, oil and solvents. They also form flexible plastics which are useful for making fuel hoses and gaskets since these retain their elasticity over a broad temperature range.

---

5. Polysulfates

Sulfate ion has a tetrahedral structure in which all sulfur-oxygen bonds are equivalent (Fig. 10.15a). The bond distance, 149 pm corresponds to a value in between that of a typical sulfur - oxygen single bond and a sulfur-oxygen double bond. The sulfate ion undergoes polymerization to yield polysulfate ions having the composition, $[S_nO_{3n+1}]^{2-}$.

For example, when two sulfate units combine by sharing an oxygen atom a disulfate ion, $[S_2O_7]^{2-}$ results. These are prepared by thermal dehydration of metal bisulfates, $MHSO_4$

$$2MHSO_4 \xrightarrow{\Delta} M_2S_2O_7 + H_2O$$

The trisulfate ion, $[S_3O_{10}]^{2-}$ and tetrasulfate ion, $[S_4O_{13}]^{2-}$ are also known. The structure of pentasulfate ion, $[S_5O_{16}]^{2-}$ is shown in Fig. 10.15b. Its skeleton has alternate long and short sulfur-oxygen bond lengths. The terminal SO_3 groups are nearly planar.

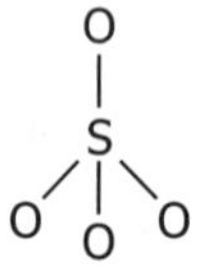

(a)

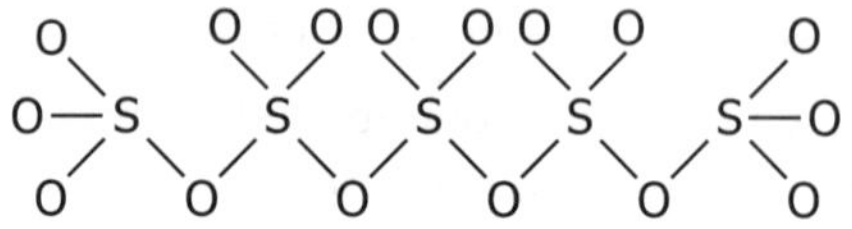

(b)

Fig. 10.15 Structure of (a) sulfate ion and (b) pentasulfate ion

11. HYDROGEN AND HYDRIDES

Introduction

Hydrogen is the most abundant element in the universe and the third most abundant element (after oxygen and silicon) on earth. It is also the simplest element, having electronic configuration $1s^1$. Yet it can exist in several forms: molecular hydrogen, H_2; atomic hydrogen, H; proton, H^+; hydride ion, H^-; molecular hydrogen ion, H_2^+; and triatomic-2 electron species, H_3^+. It also exists in three isotopic forms: protium, ${}^1_1H$; deuterium ${}^2_1H$ or D; and tritium ${}^3_1H$ or T. Further, each of the isotope, in diatomic molecular state, has two nuclear spin isomers, *ortho* and *para*.

Position in the Periodic Table

Hydrogen has just one electron in its outermost shell and, therefore, resembles alkali metals. It can lose the valence electron to form a univalent cation, like alkali metals. Moreover, like Group 1 elements, hydrogen is a good reducing agent. These factors favor the placement of hydrogen in Group 1 of the periodic table.

Hydrogen also resembles the halogens. It has a tendency to gain an electron to form hydride ion, just as halogens gain an electron to form halide ions. The hydrogen molecule, like halogen molecules, is diatomic. These factors favor the inclusion of hydrogen in Group 17 of the periodic table.

However, hydrogen differs from both alkali metals and halogens in several respects. The ionization energy of hydrogen is much higher than that of Group 1 elements. Therefore, the tendency to form univalent cation is much less in comparison to alkali metals. Further, the size of proton is smaller than that of lithium ion. Similarly, the electron affinity of hydrogen is much lower than that of halogens. Its tendency to form hydride ion is also less than that of halogens to form halide ions. This implies that hydrogen is less electropositive than alkali metals as well as less electronegative than halogens. It is, therefore, pertinent to treat hydrogen separately, without clubbing it with alkali metals or halogens.

Preparation of Hydrogen

The laboratory preparation of hydrogen involves the reaction of a metal with a dilute acid,

$$Zn + H_2SO_4 \rightarrow ZnSO_4 + H_2$$

or with an alkali,

$$2Al + 2NaOH + 6H_2O \rightarrow 2Na[Al(OH)_4] + 3H_2$$

or hydrolysis of an ionic hydride.

$$LiH + H_2O \rightarrow LiOH + H_2$$

On an industrial scale, hydrogen may be prepared by passing steam over red-hot iron.

$$\underset{\text{Steam}}{3Fe + 4H_2O} \rightarrow Fe_3O_4 + 4H_2$$

A still better method is the nickel-catalyzed reaction of coke with steam at 1800 K. Initially, water gas—a mixture of carbon monoxide and hydrogen—is produced.

$$\underset{\text{Steam}}{C + H_2O} \xrightarrow[1800\text{ K}]{\text{Ni}} CO + H_2$$

Carbon monoxide and hydrogen are difficult to separate because both are sparingly soluble in water. Therefore, the products are mixed with more steam and the mixture is passed over iron oxide catalyst at 700 K, whereby carbon monoxide is converted to carbon dioxide. The yield of hydrogen also improves.

$$CO + H_2 + \underset{\text{Steam}}{H_2O} \xrightarrow[700\text{ K}]{\text{Iron oxide}} CO_2 + 2H_2$$

This reaction is called *water gas shift reaction.* The mixture of carbon dioxide and hydrogen is easily separable because the former is highly soluble in water. Thus by spraying water on the products, carbon dioxide dissolves in aqueous solution, while hydrogen remains in the gaseous state.

Hydrogen may be produced commercially by the reaction of methane with steam at 1100 K, in the presence of nickel catalyst.

$$CH_4 + \underset{\text{Steam}}{H_2O} \xrightarrow[1100\text{ K}]{\text{Ni}} CO + 3H_2$$

Hydrogen is then separated from carbon monoxide by water gas shift reaction.

Another important industrial method for preparation of hydrogen is electrolysis of water containing a small amount of sodium hydroxide or potassium hydroxide. The alkali is added since pure water does not conduct electricity. The electrolytic cell uses nickel anode and iron cathode. The two electrodes are separated by a partition, otherwise hydrogen liberated at the cathode will explosively combine with oxygen liberated at the anode to yield water.

At cathode	: $2H_2O + 2e^- \rightarrow 2OH^- + H_2] \times 2$
At anode	: $4OH^- \rightarrow 2H_2O + O_2 + 4e^-$
Overall reaction	: $2H_2O \rightarrow 2H_2 + O_2$

Hydrogen may also be prepared by the reaction of phosphorus with steam—the *Liljenroth process*.

$$2P + \underset{\text{Steam}}{8H_2O} \rightleftharpoons 2H_3PO_4 + 5H_2$$

At higher temperatures, the equilibrium shifts from right to left. Therefore, reaction products are cooled as soon as they are formed.

Case Study 11.1 Hydrogen prepared by the reaction of zinc and dilute acids has an unpleasant odor. Why?

Hydrogen is an odorless gas. Yet an unpleasant smell is associated with hydrogen produced by the reaction of zinc and dilute sulfuric acid or dilute hydrochloric acid. This is because zinc used in the laboratory preparation of hydrogen is usually contaminated with phosphorus and arsenic. Thus, phosphine and arsine also evolve with hydrogen and the unpleasant odor is due to these side products. To remove these impurities, the reaction products are passed successively through potassium permanganate and potassium hydroxide. Finally, hydrogen is dried over calcium hydride.

Properties of Hydrogen

Hydrogen is the lightest gas. It is colorless, odorless and sparingly soluble in water. It is not very reactive since the hydrogen–hydrogen bond energy is appreciably high, and this bond has to be broken before the gas enters into a chemical reaction with other elements. Therefore, many important reactions involving hydrogen occur in the presence of catalysts, which lower the activation energy by weakening or breaking the H–H bond.

For example, in Haber's process, hydrogen combines with nitrogen at 700 K and 200 atmospheric pressure in the presence of an iron catalyst to yield ammonia.

$$N_2 + 3H_2 \xrightarrow[700\text{ K}]{\text{Fe}} 2NH_3$$

Its reaction with carbon monoxide occurs over copper catalyst at 600 K to produce methanol.

$$CO + 2H_2 \xrightarrow[600\text{ K}]{\text{Cu}} CH_3OH$$

The *hydrogenation reactions*, which convert unsaturated fatty acids (oils) into saturated analogs (fats) occur in the presence of finely divided nickel, palladium or platinum catalyst.

$$\underset{\text{Oil}}{CH_3.(CH_2)_nCH{=}CH.COOH} + H_2 \xrightarrow{Ni} \underset{\text{Fat}}{CH_3(CH_2)_n.(CH_2)_2.COOH}$$

The reaction between hydrogen and chlorine is catalyzed by sunlight.

$$H_2 + Cl_2 \xrightarrow{h\upsilon} 2HCl$$

Case Study 11.2 Is it possible to dry hydrogen gas by passing it over concentrated sulfuric acid?

No. When concentrated sulfuric acid absorbs moisture a very large amount of heat is generated. As a result, hydrogen is likely to catch fire and/or explode. The most promising reagent for drying hydrogen gas is calcium hydride.

Isotopes of Hydrogen

Hydrogen has three isotopes: (i) protium, ^1_1H or H; (ii) deuterium, ^2_1H or D; and (iii) tritium, ^3_1H or T. Each isotope contains one proton and 0, 1 or 2 neutrons, respectively. Protium has the highest natural abundance, 98.98%, followed by deuterium, 0.02% and tritium, 10^{-17}%. These isotopes show much greater differences in physical properties than isotopes of other elements. This is because the percentage difference in mass between protium; deuterium and tritium is much greater than that between the isotopes of other elements.

The physical properties of hydrogen and its isotopes are compared in Table 11.1. The values are higher for deuterium and tritium than for hydrogen. This may be attributed to the increase in van der Waals forces of attraction with increasing molecular size.

Table 11.1 Physical properties of hydrogen and its isotopes

Property	H_2	D_2	T_2	HD
Melting point (K)	13.95	18.73	20.62	16.60
Boiling point (K)	20.39	23.67	25.04	22.13
Heat of fusion (kJ mol^{-1})	0.117	0.197	0.250	0.159
Heat of vaporization (kJ mol^{-1})	0.904	1.226	1.393	1.075
Bond energy (kJ mol^{-1})	436	443	447	–

Forms such as hydrogen–deuterium (HD) and hydrogen–tritium (HT) possess properties intermediate between those of their components.

Since each of the three isotopes has one electron and one proton, it is expected that the best possible overlap would occur in molecular tritium, as its constituent atoms are of largest size. This is borne out by the high bond energy (447 kJ mol^{-1}) of tritium molecule (Table 11.1). In fact, tritium has the highest bond energy, not only as compared to deuterium or molecular hydrogen, but amongst all the homonuclear diatonic molecules having single bonds.

Deuterium is separated from protium by electrolysis of water. When water is electrolyzed, hydrogen is liberated at a faster rate than deuterium. Progressively, the remaining water becomes rich in heavy water, D_2O. About 29,000 liters of water must be electrolyzed to yield one liter of heavy water that is 99% pure. Further electrolysis of heavy water yields deuterium.

Deuterium undergoes exchange reactions, replacing protium in many compounds.

$$2NH_3 + 3D_2 \rightarrow 2ND_3 + 3H_2$$

$$CH_4 + 2D_2 \rightarrow CD_4 + 2H_2$$

However, heavy water is more often used than deuterium to prepare deuterated compounds.

$$Mg_3N_2 + 3D_2O \rightarrow 2ND_3 + 3MgO$$

$$P_4O_{10} + 6D_2O \rightarrow 4D_3PO_4$$

$$SO_3 + D_2O \rightarrow D_2SO_4$$

Tritium is prepared by irradiation of lithium with slow neutrons in a nuclear reactor.

$$^{6}_{3}Li + ^{1}_{0}n \longrightarrow ^{3}_{1}H + ^{4}_{2}He$$

It is radioactive and decays by β-emission.

$$^{3}_{1}H \longrightarrow ^{3}_{2}Hc + ^{0}_{-1}e$$

Its half-life is 12.26 years.

It may be stored by reacting it with finely divided uranium. The resulting hydride, UT_3, decomposes into tritium on heating above 673 K.

Tritiated compounds are generally prepared from tritium oxide.

$$Mg_3N_2 + 3T_2O \rightarrow 2NT_3 + 3MgO$$

Tritium is non-toxic, except when inhaled in large quantities.

Case Study 11.3 What are the advantages of tritium vis-à-vis other radioisotopes?

Tritium has been used in hydrological studies to follow the movement of groundwater and to determine the age of different aquatic systems. It is non-toxic and cost-effective. It emits low-energy β-rays and not harmful γ-rays. The β-rays emitted by tritium are stopped by a 6 mm layer of air or a 6 μm layer of water. Therefore, no shielding is required while working with tritium.

---

ortho- and *para*-Hydrogen

Just as an unpaired electron possesses a net spin, so do the nuclei of certain atoms. In a hydrogen molecule, the nuclei of the two constituent atoms may be spinning in the same or in the opposite directions. This gives rise to two forms of hydrogen molecule—*ortho* form, when the nuclear spins are parallel and *para* form, when the nuclear spins are anti-parallel (Fig. 11.1).

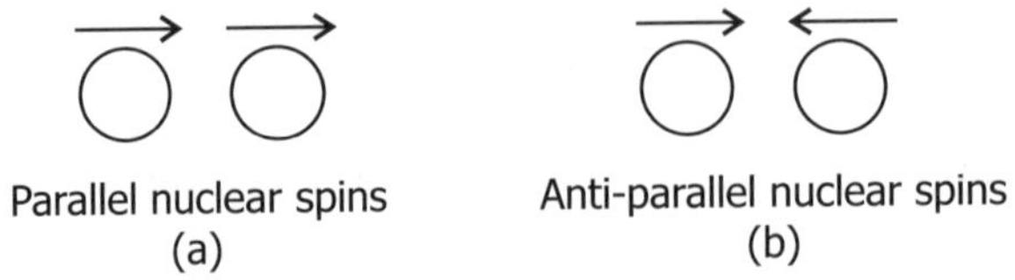

Fig. 11.1 (a) ortho and (b) para isomers of hydrogen

As shown in Table 11.2, the *ortho* form of hydrogen is in triplet state, while the *para* form is in singlet state. The triplet–singlet transition is forbidden and, therefore, occurs very slowly. Nevertheless, it may be catalyzed by metals such as iron, nickel and platinum, as well as by paramagnetic species such as oxygen and nitric oxide.

Table 11.2 Spin parameters of ortho- and para-hydrogen

Form	ortho-H_2	para-H_2
Resultant spin (S)	$\frac{1}{2} + \frac{1}{2} = 1$	$\frac{1}{2} + (-\frac{1}{2}) = 0$
Spin multiplicity (2S+1)	3 (Triplet)	1 (Singlet)

The *ortho*- and *para*-hydrogens are examples of *spin isomers*. Such isomers also exist for other symmetrical diatomic molecules, such as deuterium and nitrogen, whose nuclei have a net spin. These isomers differ in physical properties. Thus, *ortho*- and *para*-hydrogens possess different boiling points, specific heats and thermal conductivities.

The *para* form has a lower internal energy than the *ortho* form. Therefore, it is more stable at lower temperatures. In fact, at absolute zero, molecular hydrogen is made up of 100% *para* form. When the temperature increases, *para* form slowly converts into *ortho* form. At high temperatures, the gas may contain up to 75% *ortho* form. The *ortho* content cannot be increased beyond this level because any further increase in temperature would dissociate molecular hydrogen into atomic hydrogen. It is thus possible to prepare 100% *para*-hydrogen, but not 100% *ortho*-hydrogen.

The variation in the concentration of *ortho*-hydrogen with respect to temperature is shown in Fig. 11.2.

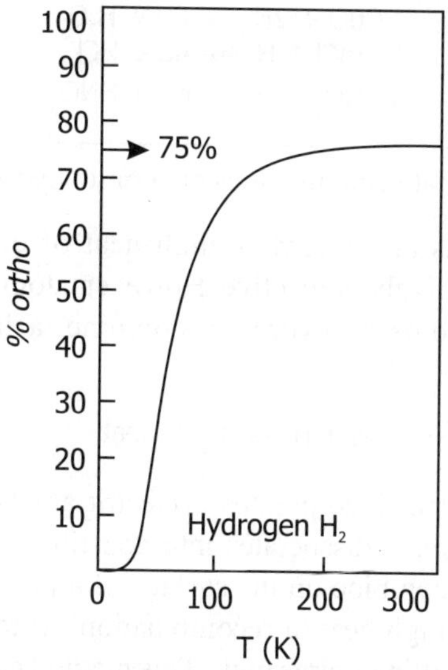

Fig. 11.2 Equilibrium between *ortho* and *para* forms of hydrogen

para-Hydrogen is prepared by passing molecular hydrogen through a tube containing charcoal, cooled to liquid air temperature. The *para* form may be preserved for several weeks, since its conversion to *ortho* form is slow.

Atomic Hydrogen

Due to its extremely high bond dissociation energy (435.88 kJ mol^{-1}), molecular hydrogen is relatively unreactive at room temperature. Nevertheless, when hydrogen gas at a low pressure is subjected to an electric discharge, it dissociates into constituent atoms, giving atomic hydrogen.

$$H_2 \xrightarrow[\text{discharge}]{\text{Electric}} 2H$$

Atomic hydrogen is very reactive. It combines directly with non-metals, to give their hydrides. For example,

$$2H + S \rightarrow H_2S$$
$$3H + P \rightarrow PH_3$$

It also combines with metals to yield ionic hydrides.

$$Li + H \rightarrow LiH$$
$$Ca + 2H \rightarrow CaH_2$$

Atomic hydrogen is a strong reducing agent. It readily converts metal salts into metals.

$$CuO + 2H \rightarrow Cu + H_2O$$
$$AgCl + H \rightarrow Ag + HCl$$
$$NaNO_3 + H \rightarrow Na + HNO_3$$

Case Study 11.4 On what principle does an atomic hydrogen torch work?

Hydrogen molecule has an extremely high heat of dissociation. This makes molecular hydrogen relatively unreactive. However, atomic hydrogen is extremely reactive. When two atoms of hydrogen combine, a large amount of heat is generated.

$$H + H \rightarrow H_2 + \text{Heat}$$

An atomic hydrogen torch incorporates an electric arc. When molecular hydrogen is passed through the arc, it dissociates into constituent hydrogen atoms. These atoms are allowed to recombine on the surface of a metal. The temperature rises to about 4000 K due to high heat of recombination. At this temperature, even the metals with very high melting points melt. This method can be used to weld metals such as tungsten and tantalum.

Hydrogen Economy

The dwindling reserves of fossil fuels and rising prices of oil demand alternative routes to overcome energy crisis. One possible route is a fuel economy based on hydrogen. Unlike fossil fuels, geothermal energy or solar energy, hydrogen is not a primary source of energy. It is an energy carrier. On combustion it produces only water. Hence, the advantage of hydrogen is its virtually pollution-free combustion. It is, therefore, destined to become the fuel of the future.

If hydrogen is to take over as a fuel of the future, it must be produced and stored in large quantities. The most practical and feasible method of producing hydrogen is electrolysis of water. No doubt, this method consumes electric power, yet during off-peak hours, when power is not being utilized to full extent, a part of the electrical energy may be directed to produce hydrogen. The hydrogen, in turn, can be used for producing electricity through fuel cells when the demand exceeds the supply or in remote places where power-generating resources are scarce.

In hydrogen-based fuel cells, an aqueous solution of sodium hydroxide or potassium hydroxide is used as an electrolyte. Hydrogen and oxygen are bubbled into the electrolyte. The following reactions take place:

$$\text{At cathode} \quad : O_2 + 2H_2O + 4e^- \rightarrow 4OH^-$$
$$\text{At anode} \quad : H_2 + 2OH^- \rightarrow 2H_2O + 2e^-] \times 2$$
$$\text{Overall reaction} : 2H_2 + O_2 \rightarrow 2H_2O$$

Since water is the only reaction product, fuel cells do not pollute the environment.

Thus, electricity and hydrogen-based economy complement each other. Electricity can be produced efficiently and without damage to the environment from hydrogen through fuel cells; hydrogen can be derived from electricity. The complementarity of hydrogen and electricity is further enhanced by the fact that hydrogen, unlike electricity, can be stored and transported over long distances.

In gaseous state, hydrogen has a relatively low density even at high pressures. Hence, very large containers are required to store it. Therefore, normally hydrogen is stored and transported in solid form as transition metal hydrides.

$$Ti + H_2 \rightleftharpoons TiH_2$$

$$4Pd + H_2 \rightleftharpoons 2Pd_2H$$

Hydrogen occupies interstitial sites in the metallic lattice of titanium or palladium. Therefore, such hydrides are called *interstitial hydrides*. The reactions leading to the formation of interstitial hydrides are exothermic in nature. Therefore, when the temperature is raised the hydrides reversibly decompose, releasing hydrogen.

Case Study 11.5 What is meant by metallization of hydrogen?

Hydrogen is unquestionably a non-metal in its natural gaseous state as well as in liquid state. However, when a thin (0.5 mm) layer of liquid hydrogen, contained between two single-crystal alumina anvils is subjected to shock compression by applying a pressure of 2 million atmosphere, it starts behaving like a metal. Its conductivity becomes comparable to rubedium and cesium. The transition from non-metallic to metallic properties, observed during the shock compression of fluid hydrogen is called metallization of hydrogen. Such phenomenon occurs on Jupiter. This planet, itself a fluid, is composed of an outer layer of molecular hydrogen that is continuously converted into metallic hydrogen in the core.

Hydrides

The binary compounds of hydrogen are called hydrides. These may be classified into four types: (i) ionic or salt hydrides; (ii) covalent or molecular hydrides; (iii) metallic or interstitial hydrides; and (iv) intermediate hydrides.

1. Ionic or salt hydrides

Ionic or salt hydrides contain H^- ion and are formed by highly electropositive elements of Groups 1 and 2 of the periodic table. In addition, europium and ytterbium of the lanthanoid series also form ionic hybrides (Fig. 11.3).

Ionic hydrides may be synthesized by heating the alkali or alkaline earth metals directly with hydrogen.

$$2Na + H_2 \rightarrow 2NaH$$

$$Ca + H_2 \rightarrow CaH_2$$

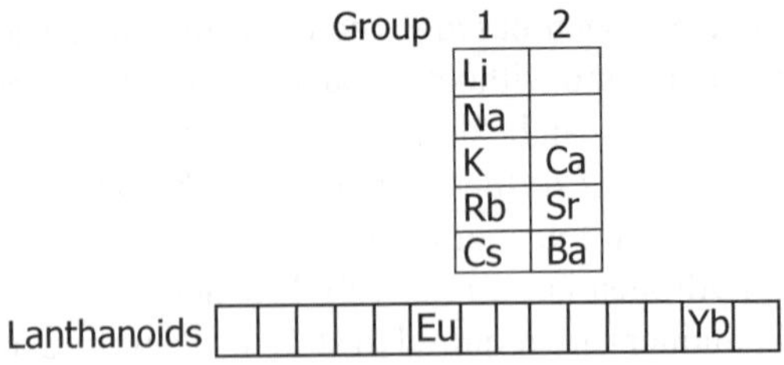

Fig. 11.3 Groups of elements forming ionic hydrides

Higher temperatures are required to prepare alkaline earth metal hydrides than alkali metal hydrides. With alkali metals, however, the reaction is slow and reversible and can be brought to completion either by increasing the presssure or better by volatilization of the hydride so formed. The ease of formation of the hydrides decreases down the group.

All ionic hydrides are either colorless or white compounds and are solids at room temperature, with high melting points and boiling points. These are true ionic compounds as is evident from the electrolysis of fused lithium hydride at a temperature slightly above its melting point, yielding lithium at the cathode and hydrogen at the anode.

Group 1 hydrides are more reactive than the corresponding Group 2 hydrides, and the reactivity increases from top to bottom in a group. Alkaline earth hydrides are more heat-stable and decompose at a higher temperature as compared to alkali metal hydrides. Within a group, the ease of thermal decomposition into constituent elements increases on moving from top to bottom. Lithium hydride does not decompose on heating.

Two of the lanthanoid series members, europium and ytterbium, also form ionic hydrides, with compositions EuH_2 and YbH_2, respectively. Unlike other lanthanoid hydrides which are non-stoichiometric and fall under the category of metallic hydrides, europium and ytterbium analogs are stoichiometric compounds. They are isomorphous with calcium hydride, CaH_2. Lanthanoids are typically trivalent, however, europium and ytterbium form divalent compounds because of the exceptional stability associated with half-filled [Eu(II): $4f^7$] and completely filled [Yb(II): $4f^{14}$] f orbitals.

Ionic hydrides are decomposed by water, giving the corresponding hydroxide and hydrogen.

$$NaH + H_2O \rightarrow NaOH + H_2$$

$$CaH_2 + 2H_2O \rightarrow Ca(OH)_2 + 2H_2$$

In fact, calcium hydride is used for preparing small quantities of highly pure hydrogen. Earlier, it was used as a safe and convenient means to inflate weather balloons. During World War II, it was sold under the commercial name, *hydrolith*, as a portable source of hydrogen for filling airships.

Ionic hydrides are strong reducing agents, especially at higher temperatures. For example, carbon monoxide, which itself is a reducing agent, is reduced to carbon by ionic hydrides.

$$2CO + NaH \rightarrow \underset{\text{Sodium formate}}{HCOONa} + C$$

The reaction between sodium hydride and carbon dioxide also yields sodium formate.

$$CO_2 + NaH \rightarrow HCOONa$$

The above reaction has an important consequence. The reaction of ionic hydrides with water is so vigorous that it often results in arson. The fire cannot be put off by conventional fire extinguishers as these are based on carbon dioxide. The ionic hydrides reduce carbon dioxide to metal formate and hence preclude its ability to extinguish fire. The best way to overcome this problem is to spray sand on the arson site.

Sulfur dioxide gets reduced to sodium thionite by sodium hydride.

$$2SO_2 + 2NaH \rightarrow \underset{\text{Sodium thionite}}{Na_2S_2O_4} + H_2$$

Similarly, silicon(IV) chloride is reduced to silane, while titanium(IV) chloride is reduced to metallic titanium.

$$SiCl_4 + 4NaH \rightarrow \underset{\text{Silane}}{SiH_4} + 4NaCl$$

$$TiCl_4 + 4NaH \xrightarrow{673\text{ K}} Ti + 4NaCl + 2H_2$$

Ionic hydrides are used to prepare important reagents such as lithium aluminium hydride, $LiAlH_4$,

$$4LiH + AlCl_3 \rightarrow LiAlH_4 + 3LiCl$$

or sodium borohydride, $NaBH_4$.

$$4NaH + B(OCH_3)_3 \rightarrow NaBH_4 + 3NaOCH_3$$

These complex hydrides also act as reducing agents.

Case Study 11.6 The size of hydride ion varies with the metal ion to which it is bound. Why?

The hydride ion has the same electronic configuration as helium. Yet, it is far less stable since the single proton in its nucleus has to control two electrons. This makes the hydride ion readily deformable. As a result, its radius varies considerably with the nature of metal ion to which it is bound. Thus in lithium hydride, sodium hydride and potassium hydride, the radius of hydride ion is 137, 146 and 152 pm, respectively.

---

2. Covalent or molecular hydrides

The p-block elements form covalent hydrides (Fig. 11.4).

Group	13	14	15	16	17
	B	C	N	O	F
	Al	Si	P	S	Cl
	Ga	Ge	As	Se	Br
	In	Sn	Sb	Te	I
		Pb	Bi	Po	

Fig. 11.4 Groups of elements forming covalent hydrides

Except boron, the general composition of Group 13 hydrides is XH_3. Boron forms a series of hydrides called boranes. The first member of the series is diborane, B_2H_6. The hydrides of other p-block elements have the formula XH_{18-n}, where n is the group of the periodic table to which the element X belongs.

Covalent hydrides may be prepared by the following methods:

(i) Direct reaction between the element and hydrogen.

$$H_2 + Cl_2 \rightarrow 2HCl$$

(ii) Treating binary compounds with dilute mineral acids.

$$2Mg_3B_2 + 4H_3PO_4 \rightarrow B_4H_{10} + 2Mg_3(PO_4)_2 + H_2$$

Magnesium boride

$$Zn_3As_2 + 6HCl \rightarrow 2AsH_3 + 3ZnCl_2$$

Zinc arsenide → Arsine

$$Al_4C_3 + 12HCl \rightarrow 3CH_4 + 4AlCl_3$$

Aluminium carbide → Methane

(iii) Reduction of a halide with lithium aluminium hydride in dry ether solvent.

$$SiCl_4 + LiAlH_4 \rightarrow SiH_4 + AlCl_3 + LiCl$$

(iv) Reaction of a binary compound of magnesium with ammonium chloride, in liquid ammonia.

$$Mg_3Bi_2 + 6NH_4Cl \rightarrow 2BiH_3 + 3MgCl_2 + 6NH_3$$

Magnesium bismuthite → Bismuth hydride

The covalent hydrides are volatile compounds and are gases at room temperature, except water, which is a liquid (due to hydrogen bonding). They are non-conducting in nature. In general, their stability decreases down a particular group. The melting points and boiling points increase in the same direction. There are only weak van der Waals forces acting between the molecules, and this is true for all the molecular hydrides, except for the most electronegative elements such as nitrogen, oxygen and fluorine (where relatively strong dipole–dipole attractions, in the form of hydrogen bonds, also exist).

In general, covalent hydrides are reducing agents. Some of these reduce metal salts.

$$4AgNO_3 + SiH_4 \rightarrow 4Ag + Si + 4HNO_3$$
$$3CuSO_4 + 2PH_3 \rightarrow Cu_3P_2 + 3H_2SO_4$$

Halogens convert covalent hydrides to their respective halides.

$$SiH_4 + 4Cl_2 \rightarrow SiCl_4 + 4HCl$$

In some cases, only partial substitution occurs.

$$GeH_4 + I_2 \rightarrow GeH_3I + HI$$

Some covalent hydrides have a tendency to form polynuclear compounds. They are generally formed as a result of direct bonding between atoms of the central element. Carbon, for example, forms a wide variety of hydrocarbons by virtue of its ability to enter into catenation. Silicon and germanium also show catenation, but to a much lesser extent.

C_nH_{2n+2}	Si_nH_{2n+2}	Ge_nH_{2n+2}
Alkanes	Silanes	Germanes

Nitrogen and oxygen also form polynuclear hydrides.

$NH_2–NH_2$	HO–OH
Hydrazine	Hydrogen peroxide

Polynuclear hydrides may also be formed without a direct bond between atoms of the central element. Boron can form a number of electron-deficient, polymeric hydrides in which the boron atoms are not linked directly, but through hydrogen bridges. Aluminium hydride is also polymeric, possessing hydrogen bridges (Fig. 11.5).

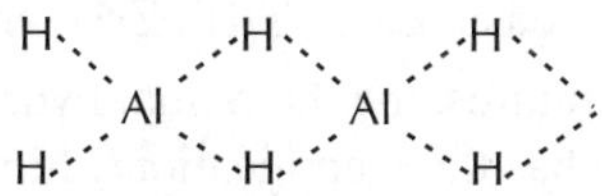

Fig. 11.5 Polymeric aluminium hydride showing hydrogen bridges

Covalent hydrides exist in the same molecular form in all states of aggregation. When solid methane is transformed successively into liquid and gaseous states, there is no rearrangement of carbon and hydrogen atoms relative to each other. The structure remains tetrahedral throughout.

3. Metallic or interstitial hydrides

Metallic or interstital hydrides are formed by the transition metals, lanthanoids and actinoids (Fig. 11.6).

Group	3	4	5	6	7	8	9	10
	Sc	Ti	V	Cr				Ni
	Y	Zr	Nb					Pd
	La	Hf	Ta					
	Ac							

Hydride gap

Lanthanoids	Ce	Pr	Nd		Sm		Gd	Tb	Dy	Ho	Er	Tm		Lu
Actinoids	Th	Pa	U	Np	Pu	Am								

Fig. 11.6 Groups of elements forming metallic hydrides

Elements in the middle of the d-block do not form hydrides. The absence of hydrides in this part of the periodic table is termed *hydride gap*. The heats of formation of these hydrides would be so small as to preclude their existence.

Interstitial hydrides are prepared by heating the metal with hydrogen under pressure.

$$Ti + H_2 \rightarrow TiH_2$$

$$2U + 3H_2 \rightarrow 2UH_3$$

The properties of interstitial hydrides are quite similar to their parent metals. They possess metallic luster and conduct electricity. In these compounds, hydrogen gets trapped in the interstices of the metal. The crystal lattice expands and, therefore, the hydrides are less dense than the corresponding metal. The incorporation of hydrogen in the metallic crystal distorts the lattice, as a result of which the hydrides become brittle. Thus, hydrides exist in a fine powder form.

Varying amounts of hydrogen are incorporated in the metallic lattice during the formation of interstitial hydrides. As a result, such compounds are non-stoichiometric in composition. The precise composition of lanthanum hydride is $LaH_{2.87}$; that of titanium hydride is $TiH_{1.8}$; of zirconium hydride is $ZrH_{1.9}$; and of niobium hydride is $NbH_{0.7}$. These are represented as $\overline{LaH_3}$, $\overline{TiH_2}$, $\overline{ZrH_2}$ and $\overline{NbH}$, respectively.

On heating at high temperatures, the interstitial hydrides decompose into metal and hydrogen. This reaction has three applications. First, it provides a convenient source of pure hydrogen. When the metal and hydrogen react to yield the hydride, most of the impurities do not enter the hydride. On heating the hydride, pure hydrogen evolves. Second, the metal obtained after decomposition of the hydride is in a finely divided and reactive form. It may be used as a catalyst for various industrial processes. Third, the interstitial hydrides constitute a safe form for

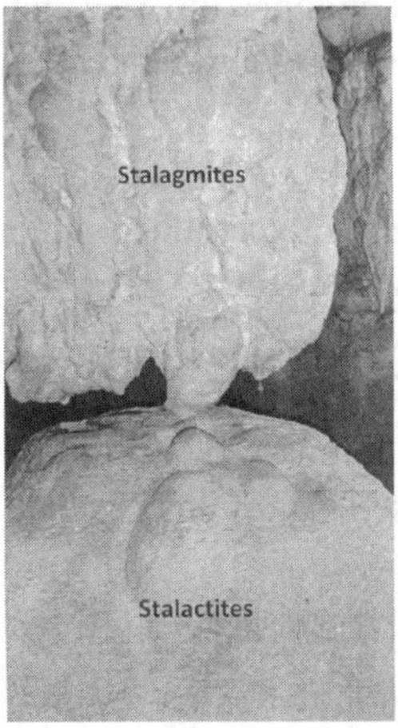

Fig. 13.4 A section of a limestone cave at Baratang Forest Division, Andaman and Nicobar Islands

In India, limestone caves are found at several places. Important amongst these are, Mawsmai caves, Meghalaya; Borra caves, Andhra Pradesh; Kutumsar caves, Chhattisgarh; and Baratang caves, Andaman and Nicobar Islands (Fig. 13.4).

Case Study 13.8 What is the origin of the phrase, *to be in the limelight*?

When lime is heated to about 2600 K, it emits an intense glow by the phenomena of incandescence. This form of glow is called limelight or *calcium light*. Prior to the invention of electrical bulbs, limelight was used as a means of stage lighting in theatres and music halls. The intense beam of light, emanating from heated lime, was cast on the stage performers, in the same manner as the modern spotlights. It was because of this illumination that the actors and actresses of that era became household names and celebrities. Thus originated the saying: to be in the limelight.

7. Complexes

Beryllium dialkyls, R_2Be may be synthesized by the reaction of a suitable Grignard reagent or alkyllithium with beryllium(II) chloride. The preparation, properties and structure of dimethylberyllium, $(CH_3)_2Be$, a polymeric solid, are discussed in Chapter 29, p. 846.

Diisopropylberyllium(II) may be prepared by a similar reaction.

$$BeCl_2 + 2(i\text{-}C_3H_7)MgBr \rightarrow (i\text{-}C_3H_7)_2Be + MgCl_2 + MgBr_2$$

When heated to 473 K, it yields a polymeric, non-volatile, colorless oil, hydridoisopropylberyllium(II).

$$(i\text{-}C_3H_7)_2Be \xrightarrow{323\text{ K}} (i\text{-}C_3H_7)BeH + C_3H_6$$

The dialkyls or diaryls of magnesium may be prepared by the reaction of organolithium compounds with Grignard reagent.

$$CH_3Li + CH_3MgBr \rightarrow (CH_3)_2Mg + LiBr$$

The organomercury compounds too produce similar complexes by direct reaction with magnesium metal in ether solvent.

$$(CH_3)_2Hg + Mg \rightarrow (CH_3)_2Mg + Hg$$

On an industrial scale, dialkylmagnesium complexes may be prepared by the reaction of an alkene with magnesium, in presence of hydrogen, at about 373 K and under high pressure.

$$2C_2H_4 + H_2 + Mg \rightarrow (C_2H_5)_2Mg$$

The corresponding compounds of calcium too are known, but these are very unstable.

Calcium(II) and magnesium(II) ions combine with ethylenediamine tetraacetic acid (EDTA) to yield extremely stable complexes. EDTA is commercially available as water-soluble disodium salt, Na_2H_2EDTA. However, during complexation, the reagent gets completely de-protonated to $(EDTA)^{4-}$ ion. The complexes of calcium(II) and magnesium(II) ions have the composition $[Ca(EDTA)]^{2-}$ and $[Mg(EDTA)]^{2-}$ respectively.

The structure of $(EDTA)^{4-}$ ion is displayed in Table 21.5, p. 623 wherein it is classified as a hexadentate ligand. This means that one $(EDTA)^{4-}$ ion becomes coordinated to a metal ion through six of its atoms. To be more precise, it complexes with metal ions through four oxygen and two nitrogen atoms. The structure of calcium(II)-EDTA complex is presented in Fig. 13.5. The magnesium complex has an analogous structure.

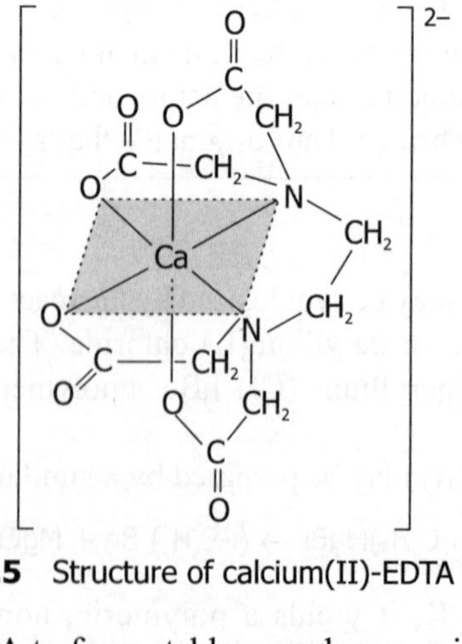

Fig. 13.5 Structure of calcium(II)-EDTA complex

The tendency of EDTA to form stable complexes assists not only in removing the calcium(II) and magnesium(II) ions from hard water, but also to carry out quantitative estimation of these alkaline earth metal ions in water samples.

EDTA also complexes with toxic metal ions like lead(II) and mercury(II) and removes these from living systems as soluble complexes. Moreover, it binds and removes those metal ions from eatables that catalyze the spoilage of food products.

Case Study 13.9 Account for the drying action of magnesium perchlorate.

Magnesium perchlorate, $Mg(ClO_4)_2$ is an example of an ionic compound in which the size of cation is incompatible with the size of anion—magnesium(II) ion being very small and the perchlorate ion being very large. For this reason, the lattice energy of this compound, in anhydrous state is quite less—and so is its stability. To overcome this problem, the magnesium(II) ion tends to absorb water molecules from the surrounding environment, whereupon the relatively stable hydrated magnesium perchlorate, having the composition, $[Mg(H_2O)_6](ClO_4)_2$, is formed. The size of the hydrated cation, $[Mg(H_2O)_6]^{2+}$ becomes compatible with that of perchlorate anion, and the lattice energy increases substantially. The tendency of magnesium(II) ion, in magnesium perchlorate, to absorb water is responsible for the drying action of this compound.

14. GROUP 12 ELEMENTS

Introduction

In the periodic table, the members of Group 12—zinc, cadmium and mercury—are placed at the interface of main group elements and transition elements. Not surprisingly, their characteristics show similarities to, as well as differences with the elements which flank these. For example, zinc resembles the main group elements in that it has a fixed oxidation state, its compounds are white or off-white in color and it forms a few complexes. On the other hand, the compounds of zinc are largely covatent—a characteristic which pushes it closer to the transition metals. The zinc(II) ion behaves essentially as a hard acid, mercury(II) ion as a typical soft acid, while cadmium(II) ion is a borderline case. While zinc and cadmium are quite reactive, mercury is relatively noble.

Physical Properties

The elements of Group 12 possess the valence shell configuration, $nd^{10}(n + 1)s^2$. Due to the completely filled d shell, only the s electrons are involved in metallic bonding. As a consequence, the metal-metal bonds are quite weak as compared to those in Group 11. A fallout is that Group 12 metals have weak tensile strength and low melting and boiling temperatures. At the extreme, mercury is a liquid under natural conditions. Another unique feature of mercury is that, apart from the noble gases, it is the only element whose vapors are monoatomic.

1. Metallic character

Normally, the metallic character increases from top to bottom is a group of periodic table. However, in Group 12, the trend is not so straightforward. Cadmium, no doubt has a greater metallic character than zinc, but that of mercury is least in the group. This is borne out by the values of ionization energies of these elements, as presented in Table 14.1.

Table 14.1. Some physical parameters of Group 12 elements

Property	Zinc	Cadmium	Mercury
1st Ionization energy (kJ mol^{-1})	906	876	1007
2nt Ionization energy (kJ mol^{-1})	1733	1631	1809
Standard reduction potential $M^{2+} + 2e^- \rightarrow M$ (V)	–0.76	–0.40	+0.85

As expected, the 1st and 2nd ionization energies of cadmium are lower than the corresponding valves for zinc. However, the analogous values for mercury are higher than those for zinc. This is attributed to the poor shielding of the outermost s electrons by the penultimate d and f shells of mercury. It is for this reason that mercury has a lesser metallic character than zinc and cadmium. For this very reason, mercury is relatively inert as compared to zinc and cadmium.

The standard reduction potential data, listed in Table 14.1, also supports this assertion. The negative values for zinc and cadmium point to the tendency of these metals to get relatively easily oxidized to divalent cationic states. The positive value of mercury indicates reluctance on the part of this heavy metal to undergo oxidation.

2. Oxidation states

Zinc exhibits oxidation state(II) in all its complexes and salts. In fact, many compounds of zinc are isomorphous to the corresponding analogs of magnesium. The predominant oxidation state for cadmium too is (II). However, when elemental cadmium is dissolved in molten cadmium chloride, several phases, having a range of colors, from dark red to black, are obtained. These melts are, in fact, mixed valence state complexes in which cadmium(I) and cadmium(II) ions are linked together by halogen bridges. The exact composition of each melt depends upon the relative concentrations of cadmium and cadmium chloride.

When aluminium(III) chloride is added to the cadmium–cadmium chloride melt, only one yellow colored compound, with empirical composition cadmium(I) tetrachloroaluminate(III), $Cd[AlCl_4]$ is formed. Its molecular formula is $[Cd_2]^{2+}[AlCl_4^-]_2$. It is diamagnetic in nature ($Cd[AlCl_4]$ would have been paramagnetic) and its cationic unit is formulated as $[Cd–Cd]^{2+}$, although the metal–metal bond is extremely weak.

The corresponding cationic unit of mercurous ion, Hg_2^{2+} has a relatively strong mercury(I)–mercury(I) bond. This is borne out by the fact that the metal–metal bond length in mercurous compounds lies in the range 250–270 pm, while in elemental mercury it is 300 pm. That the mercury(I) ion exists as a binuclear species is supported by the following evidence.

a. The mercury(I) compounds are diamagnetic. The mononuclear mercury(I) ion, with the configuration, $5d^{10}\,6s^1$ could have shown paramagnetic character.
b. X-ray picture of mercurous salts show the existence of Hg_2^{2+} ions, rather than Hg^+ ions.
c. Conductance measurements of the solutions of different mercurous salts correspond to the presence of binuclear cation and not of mononuclear cation.
d. It is possible to theoretically calculate the equilibrium constants for the following two reactions.

$$Hg + Hg^{2+} \rightleftharpoons Hg_2^{2+} \qquad (14.1)$$

$$Hg + Hg^{2+} \rightleftharpoons 2Hg^+ \qquad (14.2)$$

The experimental value tallies with the one calculated for reaction (14.1) and not for reaction (14.2).

e. The value of EMF determined for mercury(I) salts come out correctly only if the reduction potential is elucidated on the basis of equation (14.3) and not on the basis of equation (14.4).

$$Hg_2^{2+} + 2e^- \rightarrow 2Hg \qquad (14.3)$$

$$Hg^+ + e^- \rightarrow Hg \qquad (14.4)$$

The formation of binuclear mercury(I) cation may be explained by invoking molecular orbital theory. Mercury(I) ion has the configuration $5d^{10}\,6s^1$. The 6s orbitals of two mercury(I) ions interact to yield a σ_{6s} bonding molecular orbital and σ^*_{6s} anti-bonding molecular orbital (Fig. 14.1).

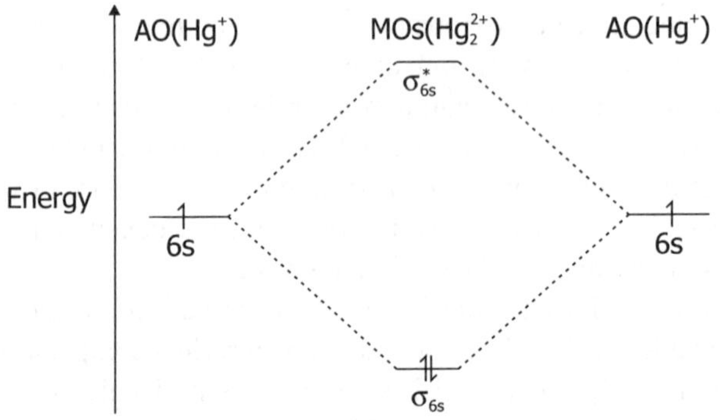

Fig. 14.1. Molecular orbital diagram for binuclear mercury(I) ion

The two electrons occupy the lower energy σ_{6s} bonding molecular orbital. The bond order is one. In other d-block compounds involving metal-metal bonds, the molecular orbitals are a hybrid of s and d orbitals. However, in case of mercury(I) ion, since the d orbitals are completely filled, the molecular orbitals are created from pure 6s orbitals.

Case Study 14.1 Cadmium(I) tetrachloroaluminate(III) is stable in solid state but not in aqueous solution. Why?

As has already been stated, cadmium(I) tetrachloroaluminate(III) ion is ionic in nature and exists as a dimer, $[Cd_2^{2+}][AlCl_4^-]_2$. An ionic compound possesses a high lattice energy and, in turn, appreciable stability only if the size of the cation and the anion are compatible. The sizes of tetrachloroaluminate(III) ion and dicadmium(I) ion are comparable and hence these unite to form a stable lattice. However, in aqueous solution, the complex ionizes into cationic and anionic components. As the stabilizing effect of the lattice wanes away, the cadmium(I) ions undergo disproportionation.

$$2Cd(I) \rightarrow Cd + Cd(II)$$

In fact, cadmium(I) ion does not exist in aqueous medium.

Chemical Properties

All three elements form a variety of alloys with other metals, brass, an alloy of zinc and copper, being the most well known example. The alloys of mercury are called *amalgams*. Sodium and zinc amalgams serve as important reducing agents. A number of amalgams have a fixed composition and a high heat of formation. This suggests that such entities are true chemical compounds and not merely solid solutions. Examples include pentasodium octamercury, Na_5Hg_8 and trisodium mercury Na_3Hg. These possess close packed lattices of mercury atoms, with the interstices being occupied by sodium atoms. Amalgams are largely formed by transition metals of second (4d) and third (5d) series. With the exception of manganese and copper, no metal of first (3d) transition series enters into amalgamation. For this reason, mercury is usually stored in iron containers.

1. Reaction with air

Zinc and cadmium rapidly tarnish in air to form a protective oxide layer. Pure mercury does not react with dry air at room temperature. However, impurities catalyze the reaction and an oxide deposition, in shape of a thread or a tail, is formed if impure mercury, placed in a glass container, is exposed to air.

Zinc and cadmium burn in air to yield the oxides, MO.

$$2Zn + O_2 \rightarrow 2ZnO$$

$$2Cd + O_2 \rightarrow 2CdO$$

Mercury forms the red mercury(II) oxide when heated in air to about 620 K.

$$2Hg + O_2 \xrightarrow{620\ K} \underset{\text{Red}}{2HgO}$$

However, when the temperature is raised to about 670 K, the oxide decomposes into the elements.

$$2HgO \xrightarrow{670\ K} 2Hg + O_2$$

A yellow mercury(II) oxide too is known and may be prepared by adding an alkali to mercury(II) chloride.

$$HgCl_2 + 2NaOH \rightarrow \underset{\text{Yellow}}{HgO} + 2NaCl + H_2O$$

The yellow and red oxides have the same lattice structures, but differ in size of constituent particles.

2. Reaction with acids

Zinc and cadmium react with non-oxidizing acids to yield hydrogen.

$$Cd + 2HCl \rightarrow CdCl_2 + H_2$$

Mercury, because of its positive standard reduction potential (Table 14.1), is inert to non-oxidizing acids.

All three elements react with oxidizing acids. For example, on heating with concentrated nitric acid, these yield nitrogen dioxide,

$$M + 4HNO_3 \rightarrow M(NO_3)_2 + 2H_2O + 2NO_2 \quad (M = Zn, Cd, Hg)$$

while on heating with concentrated sulfuric acid, sulfur dioxide gas is produced.

$$M + 2H_2SO_4 \rightarrow MSO_4 + 2H_2O + SO_2 \quad (M = Zn, Cd, Hg)$$

3. Reaction with alkalis

Zinc is the only element of Group 12 which reacts with alkalis to yield tetrahydroxozinate(II) ions (commonly called zincate ions) and hydrogen.

$$Zn + 2NaOH + 2H_2O \rightarrow \underset{\text{Sodium tetrahydroxozincate(II)}}{Na_2[Zn(OH)_4]} + H_2$$

4. Reaction with halogens

All three members of Group 12 react with halogens at elevated temperatures and in molar ratio 1:2 to yield halides of the composition MX_2.

$$M + X_2 \rightarrow MX_2 \qquad (M = Zn, Cd, Hg; X = F, Cl, Br, I)$$

The mercury(I) halides too may be obtained by directly heating mercury with halogens in 2:1 molar

$$2Hg + X_2 \rightarrow Hg_2X_2 \qquad (X = F, Cl, Br, I)$$

The iodides of mercury may also be obtained by vigorously agitating the metal and iodine in an alcohol suspension at room temperature. Depending on the proportion of the reactants, either olive green mercury(I) iodide, Hg_2I_2 or scarlet mercury(II) iodide, HgI_2 is isolated.

Case Study 14.2 Describe the application of zinc oxide in *xerography.*

The term xerography originated from the Greek *xeros* which mears dry and *graphein* which means writing. It is a dry process of photocopy in which zinc oxide plays a lead role as a photoconductor. A photoconductor is a substance which conducts electric current when illuminated by light and behaves as an insulator when the light is turned off.

The following steps, shown diagrammatically in Fig. 14.2, are involved in xerography.

a. A surface impregnated with zinc oxide is exposed to light. Zinc oxide becomes conducting due to ionization of zinc atoms.

$$Zn \rightarrow Zn^{2+} + Ze^-$$

The electrons move out, leaving the surface positively charged.

b. An image is placed on the positively charged surface.

c. When illuminated by light, the positive charges beneath the image are masked, but those on the remaining area are neutralized.

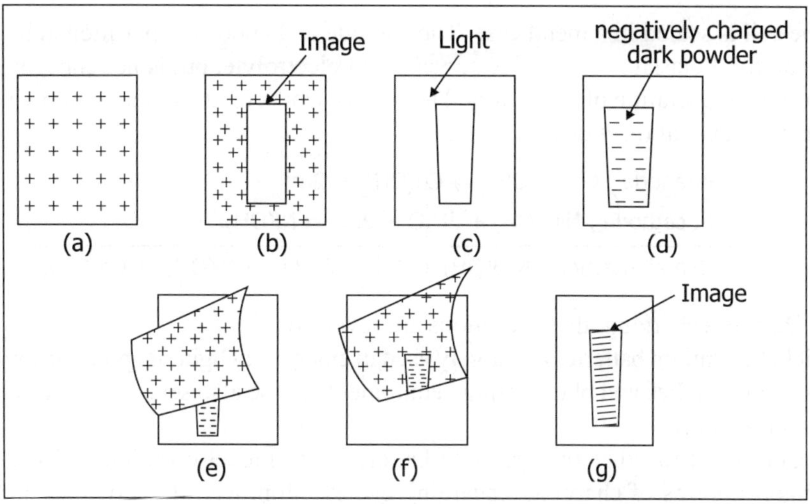

Fig. 14.2 Flow diagram of xerography

d. The image is removed and a negatively charged, dark powder is sprayed on the surface. The powder is attracted only to the positively charged portion of the surface that was masked by the image. This area becomes dark and now bears a negative charge.
e. A blank sheet of paper is pressed on the surface and given a positive charge.
f. The negatively charged powder over the image area is electrostatically lifted to the positively charged blank sheet. The temperature is raised.
g. The image gets imprinted on the blank sheet.

Box 14.1

Nickel-Cadmium Batteries

A nickel-cadmium battery is a rechargeable cell that uses nickel oxide hydroxide as cathode and metallic cadmium as anode. It is often abbreviated as NiCad battery. The device, which is commercially available in a range of sizes, has a cell potential of 1.2V. A sample is shown in Fig. 14.3.

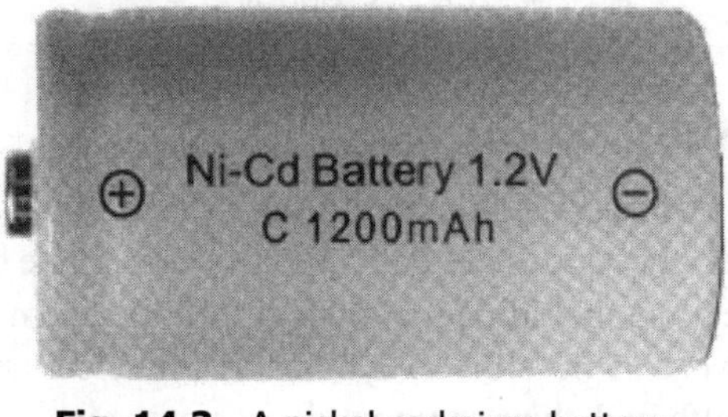

Fig. 14.3 A nickel-cadmium battery

Contd...

The cell is sealed in a metal case. The cathode and anode are partitioned by a separator. Potassium hydroxide serves as an electrolyte, but is not consumed during the operation of the battery. The battery works on the basis of following electrochemical reactions.

At anode : $Cd + 2OH^- \rightarrow Cd(OH)_2 + 2e^-$

At cathode : $NiO(OH) + 2H_2O + 2e^- \rightarrow Ni(OH)_2 + 2OH^-$

Overall reaction : $NiO(OH) + Cd + 2H_2O \rightarrow Ni(OH)_2 + Cd(OH)_2$

During recharging, these reactions are reversed.

Nickel-cadmium batteries are largely used in emergency lighting, portable toys, handheld torches and photographic equipment. These also serve as sources of standby power.

Compared to other rechargeable batteries, the nickel-cadmium cells last longer in terms of charging/recharging cycles. Moreover, these deliver their full rated capacity till it nears the discharge. These are more durable than other batteries and may be used under hostile conditions.

Nickel-cadmium batteries are costlier than other rechargeable cells. Once their life cycles are over, these devices pose a waste disposal problem as well.

Compounds

The mercury(I)–mercury(II) ions equilibrium may be represented as a disproportionation reaction.

$$Hg_2^{2+} \rightleftharpoons Hg + Hg^{2+}$$

Normally, mercury(I) ion does not undergo disproportionation. However, a reagent which precipitates mercury(II) ions out of the solution, tends to shift the equilibrium to the right, in turn, causing mercury(I) ions to disproportionate. Thus on adding sulfide ions to a solution of mercury(I) salt, a mixture of elemental mercury and extremely insoluble mercury(II) sulfide are formed.

$$Hg_2^{2+} + S^{2-} \rightarrow HgS + Hg$$

Similarly, addition of cyanide ions to a mercury(I) salt, results in the formation of mercury(II) cyanide, rather than mercury(I) cyanide.

$$Hg_2^{2+} + 2CN^- \rightarrow Hg(CN)_2 + Hg$$

1. Polycations of mercury

We have seen that mercury(I) ion actually exists as a binuclear divalent cation, Hg_2^{2+} in which the two metal ions are held together by a single bond. Several stable compounds of this cation, including mercury(I) chloride and mercury(I) nitrate are known. However, it is also possible to synthesize trinuclear and even tetranuclear cations of mercury, although their compounds are not as stable as those of binuclear mercury(I) cation.

When a mixture of elemental mercury, mercury(II) chloride and aluminium(III) chloride is heated to fusion, a compound containing trinuclear mercury cation, having the composition, $[Hg_3][AlCl_4]_2$ is obtained.

$$HgCl_2 + 2AlCl_3 + 2Hg \rightarrow [Hg_3][AlCl_4]_2$$

The trinuclear mercury cation has a linear geometry in which the two bond lengths are identical and the metal has an oxidation state of 2/3 (Fig. 14.4a).

$$[Hg \overset{255\ pm}{—} Hg \overset{255\ pm}{—} Hg]^{2+} \qquad [Hg \overset{255\ pm}{—} Hg \overset{262\ pm}{—} Hg \overset{259\ pm}{—} Hg]^{2+}$$

(a) (b)

Fig. 14.4. Structure of (a) trinuclear mercury cation and (b) tetranuclear mercury cation

When metallic mercury is treated with arsenic(V) fluoride in liquid sulfur dioxide solvent, at extremely low temperatures, a dark red compound containing tetranuclear mercury cation, $[Hg_4][AsF_6]_2$ is obtained.

$$4Hg + 3AsF_5 \rightarrow [Hg_4][AsF_6]_2 + AsF_3$$

In this reaction, arsenic(V) fluoride behaves both as a fluoride ion acceptor and an oxidizing agent, it itself being reduced to arsenic(III) fluoride. The cationic species in the complex is linear, with unequal bond lengths (Fig. 14.4b); mercury has an oxidation state of 1/2.

The tetranuclear mercury cation complex decomposes into an orange colored $[Hg_3][AsF_6]_2$ species.

$$[Hg_4][AsF_6]_2 \rightarrow [Hg_3][AsF_6]_2 + Hg$$

Finally, a colorless complex, $[Hg_2][AsF_6]_2$ is formed.

$$[Hg_3][AsF_6]_2 \rightarrow [Hg_2][AsF_6]_2 + Hg$$

It has a discrete binuclear mercurous cation, similar to the one present in mercury(I) chloride.

2. Organometallic compounds

Organozinc compounds of the type RZnX and R_2Zn, where R is an alkyl or aryl group and X is a halogen atom, are known. These may be prepared by the reaction of heated zinc or zinc-copper alloy with a boiling alkyl halide under nitrogen atmosphere. The most stable compounds are those in which the halogen atom is iodine.

$$C_2H_5I + Zn \rightarrow C_2H_5ZnI$$

On further heating, the derivative distils off.

$$2C_2H_5ZnI \rightarrow (C_2H_5)_2Zn + ZnI_2$$

The diaryls may be prepared by the reaction of dimethylzinc(II) with a suitable triarylboron(III). For example,

$$3(CH_3)_2Zn + 2(C_6H_5)_3B \rightarrow 3(C_6H_5)_2Zn + 2(CH_3)_3B$$

Organozinc compounds are either covalent liquids or solids with low melting and boiling temperatures. These possess a linear structure and are pyrophoric in nature.

The corresponding compounds of cadmium are prepared by the reaction of an anhydrous cadmium halide with Grignard reagent or with alkyllithium(I). For example,

$$CdCl_2 + 2CH_3MgCl \rightarrow (CH_3)_2Cd + 2MgCl_2$$

$$CdCl_2 + 2CH_3Li \rightarrow (CH_3)_2Cd + 2LiCl$$

The RCdX type of compounds may be prepared by the reaction of an anhydrous cadmium halide with dialkylcadmium(II) complexes.

$$(CH_3)_2Cd + CdCl_2 \rightarrow 2CH_3CdCl$$

The cadmium organometallics are thermally less stable than the corresponding zinc compounds, but are less reactive as well. One of their applications is the synthesis of ketones from acetylchloride. For example,

$$2CH_3COCl + (CH_3)_2Cd \rightarrow 2CH_3COCH_3 + CdCl_2$$

A Grignard reagent cannot be used for this preparation since it further reacts with the ketone.

The organomercury(II) compounds, RHgX and R_2Hg are more stable than their zinc or cadmium counterparts. These are prepared by the reaction of Grignard reagent with mercury(II) chloride in tetrahydrofuran solvent.

$$CH_3MgCl + HgCl_2 \rightarrow CH_3HgCl + MgCl_2$$

$$CH_3MgCl + CH_3HgCl \rightarrow (CH_3)_2Hg + MgCl_2$$

Compounds of the type RHgX are crystalline solids, while R_2Hg are covalent, toxic and pyrophoric liquids.

Case Study 14.3 Why are CFLs becoming more popular than conventional bulbs?

The CFL bulbs—or, to be more precise, the compact fluorescent light bulbs—are slowly replacing the incandescent lights, both in households and offices. This is mainly because the CFLs, one variety of which is shown in Fig. 14.5, consume less electricity. In fact, a 75 Watt conventional bulb gives as much light as a 20 Watt CFL. Moreover, the average lifetime of a CFL is 10–12 times longer than the conventional sources of light.

Fig. 14.5 A CFL

An incandescent lamp has a thin tungsten wire sealed under vacuum inside the glass bulb. When the lamp is switched on, current flows through the tungsten wire and electrical energy is converted into heat. The temperature of the wire rises to about 2600 K and white incandescence is emitted.

The CFLs work by a different mechanism. These light sources are endowed with a cathode and an anode, sealed in a glass bulb, along with mercury and argon. A fluorescent material is coated on the inside wall of the bulb. When light is switched on, electrons are emitted from the cathode to the anode. The temperature rises and mercury atoms vaporize. The electrons traversing from cathode to anode collide with mercury vapors. As a result of this impact, the valence electrons of mercury are raised to excited states. When the valence electrons fall back to the ground state, they emit their extra energy as ultraviolet light. The emitted ultraviolet rays strike the fluorescent coating on the inner surface of the bulb. The resulting fluorescence occurs in the visible range of the electromagnetic spectrum, causing illumination.

The mechanism by which CFLs produce light is more efficient and cost-effective than that of conventional bulbs. Hence the increasing market value of the CFLs.

15. GROUP 13 ELEMENTS

Introduction

Boron (B), aluminium (Al), gallium (Ga), indium (In) and thallium (Tl) constitute Group 13 of the periodic table. The first element, boron, is a non-metal, the others are all metals. The properties of these elements vary less regularly down the group than those of the s-block elements.

Physical Properties

Group 13 elements have valence shell electronic configuration of ns^2np^1. Their atomic radii do not increase regularly upon descending the group. The transition elements are located between calcium and gallium, as well as between strontium and indium. Between barium and thallium, both the transition elements and lanthanoids intervene. The increasing nuclear charge and poor shielding by d/f orbitals of the transition elements/lanthanoids result in the contraction of succeeding elements. This causes the radii of Ga, In and Tl to be smaller than expected.

The shrinkage in size also results in the valence electrons being held more strongly. As a result, ionization energy does not decrease substantially on moving down the group. This reduces the reactivities and metallic character of these elements vis-à-vis the s-block members.

1. Melting points and boiling points

Even though B is a non-metal, its melting point and boiling point are very high. This is attributed to the unusual crystal structure of B. Its structure is based on icosahedral units. Such units have a regular shape with 12 corners and 20 faces. Each corner is occupied by a B atom and each B atom is bound to 5 equivalent neighbors. This pattern results in strong bonding and hence B has a higher melting point and boiling point than Al.

If the melting point of B is unusually high, that of Ga is unusually low—just 303 K. Compared to Al, In and Tl which have close-packed structures, each atom of Ga is bound to just one neighbor at a smaller bond distance and six neighbors at larger bond distances. Lack of a compact crystal structure results in weak metallic bonds, which is reflected in the incredibly low melting point of Ga. However, the boiling point of Ga is not anomalous. It is of the same order as the boiling points of Al and In. This shows that the unusual pattern of Ga atoms in the solid crystal breaks down in the liquid state.

2. Inert pair effect

In the s-block, all Group 1 elements are univalent, while all Group 2 elements are divalent. Group 13 elements, with general configuration, ns^2np^1, are expected to be trivalent. In fact, all the elements of this group show oxidation state III by losing or sharing three valence electrons—two from the ns orbital and one from the np orbital. However, there is an increasing tendency to form univalent compounds on descending the group. Thus, compounds with gallium(I), indium(I) and thallium(I) are known. The stability of the univalent state increases down the group. While gallium(I) and indium(I) are less stable than their respective trivalent ions, thallium(I) is more stable than thallium(III).

The univalent state, i.e., the oxidation state two units less than the group valency, arises when only the np electron participates in bonding and the ns electron pair does not. The concept that the two paired-up electrons in the outermost s orbital do not participate in bonding is called *inert pair effect*.

Inert pair effect originates because of the interplay between the energy required during ionization and the energy released during compound formation. Let us compare the formation of thallium(III) chloride, $TlCl_3$ and thallium(I) chloride, TlCl. Although ionization energy decreases down the group but, as stated earlier, in the case of Group 13 the intervention of d- and/or f-block elements makes the decrease less significant. Therefore, creation of a thallium(III) ion requires substantially high amount of energy. According to Fajans' rules, thallium(III) chloride would be covalent in nature. The energy released during the formation of three thallium–chlorine bonds must compensate for the energy required to create a thallium(III) ion. However, since Tl is relatively large in size, its bonds with Cl are concomitantly long and hence weak. The bond energy released is not sufficient to compensate for the energy required for the formation of trivalent thallium.

According to Fajans' rules, thallium(I) chloride would be ionic in nature. During its formation, the energy released would be in the form of lattice energy. Further, lesser amount of energy is required for creating a thallium(I) ion than for creating a thallium(III) ion—first ionization energy is much less than third ionization energy. Thus, the lattice energy released when thallium(I) chloride is formed, easily compensates for the first ionization potential of Tl. This makes thallium(I) compounds more stable than thallium(III) counterparts.

Case Study 15.1 Comment on the conducting properties of aluminium.

Aluminium is quite soft and can be readily extruded into wires. It conducts only about 60% as much electrical current per unit volume as copper. However, its density is one-third that of copper. This means that aluminium can conduct twice as much current as the same mass of copper. It is, therefore, used in electrical transmission lines. Earlier, it was also used in wiring in homes. However, during high current flow, too much heat is generated at junctions where aluminium is connected to other metals. This causes fire hazards. For this reason, the use of aluminium cables for home wiring is now discouraged.

Chemical Properties

Pure, crystalline B is inert, but in a finely divided state, it is very reactive. Al is the most reactive of the post-transition metals. However, a thin, transparent film of aluminium oxide forms on the surface when Al comes in contact with air. It not only retards the reactivity but also prevents the oxidation and hence, corrosion of metallic layers below the outer surface. This property, coupled with the low density of Al, is responsible for structural applications of the metal. Many buildings are sheathed in Al since it is light weight and resists corrosion. No doubt, Al has low tensile strength but that can be taken care of by alloying it with Mg.

Aluminium items are often *anodized* for aesthetic reasons. This is done by electrolyzing dilute sulfuric acid in a cell endowed with aluminium anode. A relatively thick and rough layer of oxide deposits on the anode. This layer can adhere pigments of various hues, thus coloring, and decorating, aluminium.

1. Reaction with air

B burns in air to yield boron(III) oxide.

$$4B + 3O_2 \rightarrow 2B_2O_3$$

Al burns readily in air to give aluminium(III) oxide. The reaction is highly exothermic.

$$4Al + 3O_2 \rightarrow 2Al_2O_3 + \text{Heat}$$

Ga, In and Tl too are converted into their respective oxides but the reactions are very slow.

2. Reaction with halogens

B reacts with halogens at high temperatures to give boron(III) halides.

$$2B + 3X_2 \rightarrow 2BX_3 \text{ (X = F, Cl, Br or I)}$$

The other elements too react directly with halogens to yield trihalides. However, thallium(III) iodide is not known. On heating with I, Tl forms thallium(I) triiodide, $Tl^+(I_3)^-$. This is attributed to the reducing nature of iodide ion. When heated in a stream of F, Cl or Br, Tl forms a mixture of thallium(I) halide and thallium(III) halide.

3. Reaction with acids

Except B, all other elements of the group react with dilute mineral acids to yield hydrogen gas.

$$2Al + 6HCl \rightarrow 2AlCl_3 + 3H_2$$

When treated with concentrated nitric acid, Al is rendered passive due to the formation of an oxide layer on its surface.

4. Reaction with alkalis

On fusion with an alkali, B forms borates and liberates hydrogen gas.

$$2B + 6NaOH \rightarrow \underset{\text{Sodium borate}}{2Na_3BO_3} + 3H_2$$

Al dissolves in an aqueous alkali to form sodium tetrahydroxoaluminate(III), $Na[Al(OH)_4]$.

$$2Al + 2NaOH + 6H_2O \rightarrow 2Na[Al(OH)_4] + 3H_2$$

Case Study 15.2 Aluminium(III) fluoride is insoluble in anhydrous hydrogen fluoride, but dissolves if potassium fluoride is also present. Addition of boron(III) fluoride re-precipitates aluminium(III) fluoride. Why?

Anhydrous hydrogen fluoride does not dissociate into fluoride ions. However, if potassium fluoride is added, free fluoride ions are generated. Their combination with aluminium(III) fluoride leads to the formation of potassium hexafluoroaluminate(III), a complex which is soluble in anhydrous hydrogen fluoride.

$$AlF_3 + 3KF \rightarrow K_3[AlF_6]$$

On adding boron(III) fluoride, potassium hexafluoroaluminate(III) decomposes into aluminium(III) fluoride.

$$K_3[AlF_6] + BF_3 \rightarrow K[BF_4] + AlF_3 + 2KF$$

The above reaction is governed by hard and soft acid base concept (*see* Chapter 5, p. 245). The small fluoride ion, being a hard base prefers to combine with boron(III) ion, a hard acid, than aluminium(III) ion, a relatively soft acid.

Compounds

The oxides of Group 13 elements show a regular gradation in their properties. While boron(III) oxide is acidic in nature, aluminium(III) oxide is amphoteric and oxides of the remaining elements are basic in nature. Thallium(I) oxide is more basic than thallium(III) oxide. These observations are in line with the generalization that oxides of elements having small size and/or high charge possess acidic characteristics and those of elements having large size and/or low charge are basic in nature.

The trihalides of Group 13 elements are electron-deficient compounds. For example, in boron(III) fluoride, there are only six electrons—two less than the octet—in the outermost shell of the central atom. The compound has a plane triangular structure since B undergoes sp^2 hybridization.

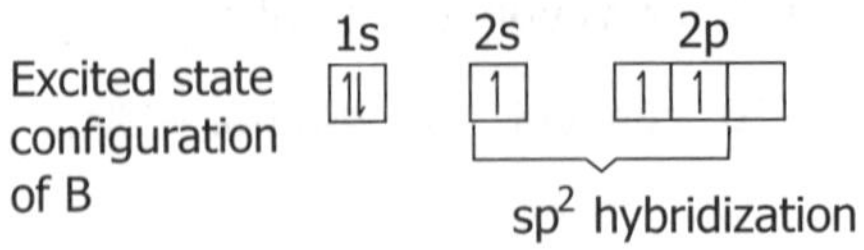

To achieve an eight-electron configuration and stability, the empty 2p orbital of B, the one that is not involved in hybridization, accepts an electron pair from any one of the three fluorine atoms (Fig. 15.1). This type of bonding is called *dative bonding* or *back bonding*. It is also referred to as *pπ–pπ bonding* because it results from the overlap of a completely filled p-orbital of fluorine with a vacant p-orbital of boron. The evidence for this type of bonding is provided by the fact that the boron–fluorine bond length is shorter than expected.

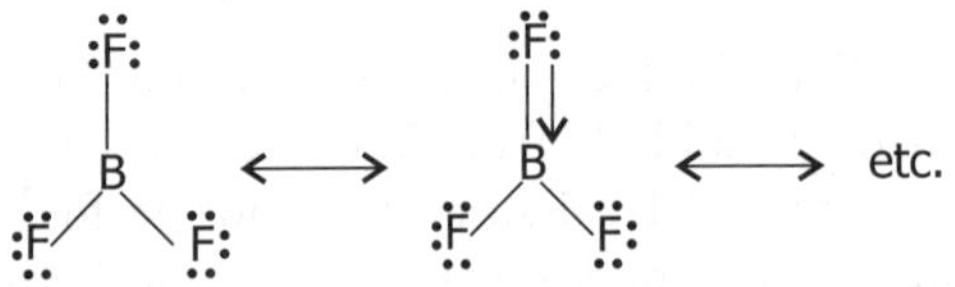

Fig. 15.1 Back bonding in boron(III) fluoride

Aluminium(III) fluoride is a typically ionic compound containing Al^{3+} and F^- ions. According to Fajans' rules, the remaining halides are covalent solids, and in all the cases the central atom, Al, is electron-deficient. To have eight-electron configuration around Al, aluminium(III) halides undergo dimerization. The dimeric structure of aluminium(III) chloride is depicted in Fig. 15.2.

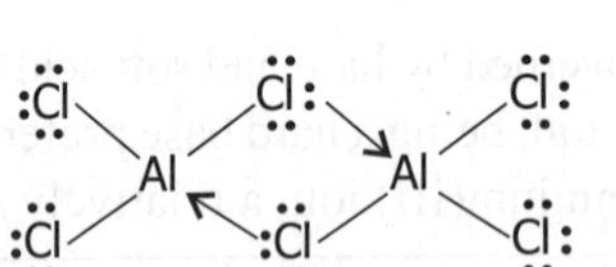

Fig. 15.2 Dimeric structure of solid aluminium(III) chloride

To attain the stable eight-electron configuration, Al does not invoke dative bonding. This is because the size of Al is larger than that of B. As the size increases, the bond distance between the aluminium and halogen also increases. Lateral overlap, leading to π-bonding, is possible only when the bond length between two connected atoms is short, as is the case with boron(III) halides. However, the corresponding aluminium compounds have relatively long bond lengths and this obviates the scope for pπ–pπ bonding.

On the other hand, boron(III) halides cannot form dimers analogous to those formed by aluminium(III) halides. During dimerization, the coordination number of Al increases from 3 to 4. Being larger in size, aluminium atom can easily accommodate 4 halogen atoms around it. However, with a small atom like B, coordination number 4 would lead to steric repulsion. When pπ–pπ bonding takes place (Fig. 15.1), there is an eight-electron configuration around B, but its coordination number is still 3.

As we have stated, aluminium(III) chloride is a covalent compound. The nature of bonding depends on the difference between ionization energy of the metal and electron affinity of the non-metal. If this difference is small, it can be compensated by the energy released in the form of lattice energy, and the compound is ionic in nature. If the difference is so large that it cannot be covered up by lattice energy, then the compound acquires covalent character. In solid state, aluminium(III) chloride is covalent because the difference between the third ionization potential of aluminium and the electron affinity of chlorine is too large to be compensated by lattice energy. However, when the compound is dissolved in water, a large amount of heat of hydration is released. This happens because aluminium ion is small in size and carries a high charge. The hydration energy compensates the difference of ionization potential and electron affinity. The compound becomes ionic. Its composition in aqueous state is $[Al(H_2O)_6]Cl_3$ or $[Al(H_2O)_6]^{3+}$ $3Cl^-$ and it has an octahedral structure (Fig. 15.3).

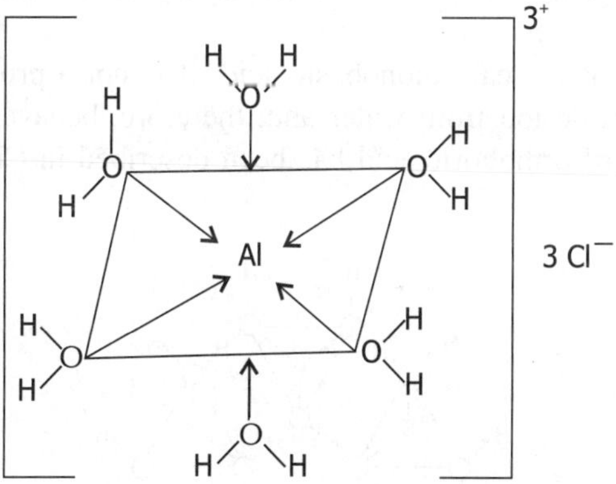

Fig. 15.3 Octahedral structure of aqueous aluminium(III) chloride

Thus, aluminium(III) chloride is a covalent compound in the solid state, with a dimeric structure (Fig. 15.2) and an ionic compound in the aqueous state, with an octahedral structure (Fig. 15.3).

Case Study 15.3 What is *fibrous alumina*?

If the surface of aluminium is rubbed with mercury, it is expected that a coherent oxide film, which protects the surface from attack by air and water, would be formed. It is not. Rather, white threads consisting of aluminium hydroxide evolve from the amalgamated aluminium. On ignition, these yield a variety of alumina which is extraordinarily voluminous and highly surface active. It is called fibrous alumina.

1. Orthoboric acid

Orthoboric acid, H_3BO_3 or $B(OH)_3$, is prepared either by the hydrolysis of boron(III) chloride,

$$BCl_3 + 3H_2O \rightarrow B(OH)_3 + 3HCl$$

or by treatment of borax with a mineral acid.

$$\underset{\text{Borax}}{Na_2B_4O_7.10H_2O} + 2HCl \rightarrow 2NaCl + 4H_3BO_3 + 5H_2O$$

The structure of orthoboric acid is based on plane triangular borate units, BO_3^{3-}, which are linked together by hydrogen bonds, giving a two-dimensional, sheet-like pattern (Fig. 15.4).

Orthoboric acid is a weak, monobasic acid. It is not a proton donor. Rather, it accepts a hydroxide ion from water and, therefore, behaves as a Lewis acid. The acidic nature of orthoboric acid has been described in Chapter 5 (*see* Case Study 5.7, p. 241).

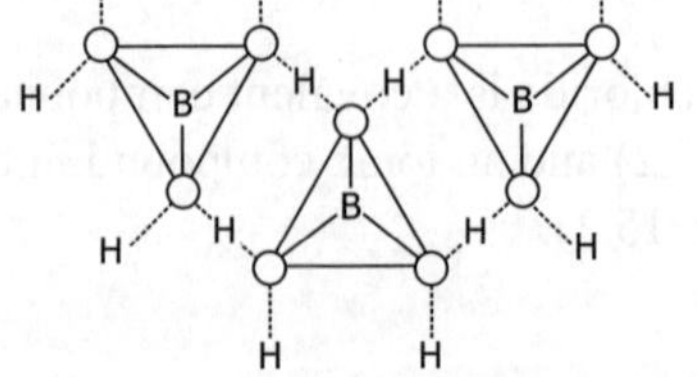

Fig. 15.4 Hydrogen-bonded structure of orthoboric acid

When boric acid is heated to 373 K, it becomes dehydrated and gets converted to metaboric acid, HBO_2.

$$H_3BO_3 \xrightarrow{373\text{ K}} HBO_2 + H_2O$$

On raising the temperature to red heat, two molecules of metaboric acid lose another molecule of water to yield boron(III) oxide.

$$2HBO_2 \xrightarrow{\text{Red heat}} B_2O_3 + H_2O$$

Boron(III) oxide may thus be considered the anhydride of boric acid. It is an acidic oxide.

2. Diborane

The most common hydride of boron, diborane (B_2H_6) is prepared by the reduction of boron(III) chloride with lithium aluminium hydride.

$$4BCl_3 + 3LiAlH_4 \rightarrow 2B_2H_6 + 3AlCl_3 + 3LiCl$$

Alternately, it may be prepared by treating boron(III) fluoride with lithium hydride,

$$8BF_3 + 6LiH \rightarrow B_2H_6 + \underset{\text{Lithium borofluoride}}{6Li[BF_4]}$$

or with sodium borohydride.

$$4BF_3 + 3Na[BH_4] \rightarrow 2B_2H_6 + \underset{\text{Sodium borofluoride}}{3Na[BF_4]}$$

Diborane is an electron-deficient compound. Stoichiometrically, it is analogous to ethane, C_2H_6. However, the total number of electrons involved in bonding is less in diborane.

B_2H_6	C_2H_6
$2B = 2 \times 3 = 6e^-$	$2C = 2 \times 4 = 8e^-$
$6H = 6 \times 1 = 6e^-$	$6H = 6 \times 1 = 6e^-$
Total $12e^-$	Total $14e^-$

Thus, unlike ethane, there are not enough valence electrons to form conventional two-electron bonds between all the adjacent pairs of atoms. To arrive at the structure of diborane, we begin with the valence electrons of the combining atoms.

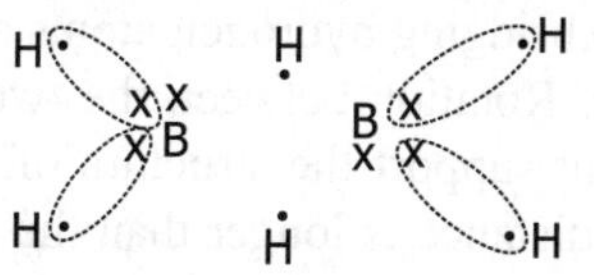

Each boron atom forms two normal bonds, i.e., two-electron–two-center bonds, with two hydrogen atoms. This leaves out one electron on each boron atom and one electron on each of the two remaining hydrogen atoms.

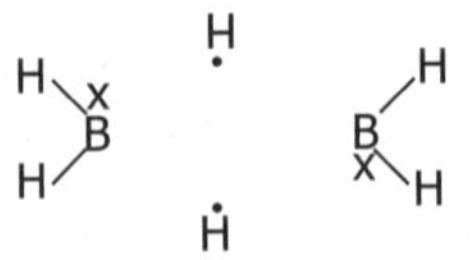

The remaining electron on the first boron atom couples with the valence electron of one of the free hydrogen atoms. However, this coupling leads to the formation of an abnormal bond—a bond which spans three atoms, boron–hydrogen–boron. It is called a two-electron–three-center bond. The second boron atom and the left out hydrogen atom combine in an analogous manner to form another two-electron–three-center bond.

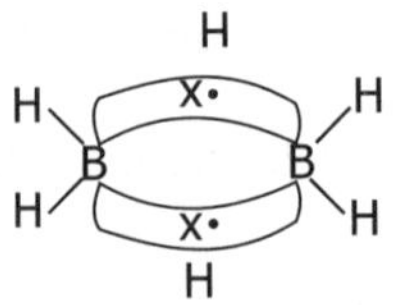

The hydrogen atoms linked to boron atoms by normal two-electron–two-center bonds are called *terminal hydrogen atoms*. Those linked to boron atoms by unusual two electron–three-center bonds are called *bridging hydrogen atoms*. The latter are so named because these act as links between the two boron atoms. The final structure of diborane is depicted in Fig. 15.5.

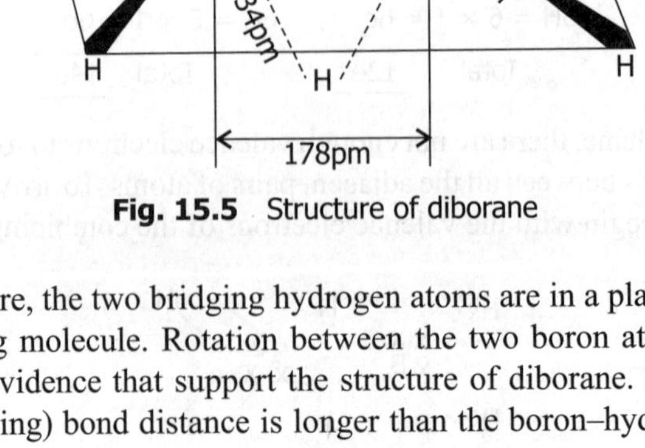

Fig. 15.5 Structure of diborane

In this structure, the two bridging hydrogen atoms are in a plane perpendicular to the remaining molecule. Rotation between the two boron atoms is hindered. There are two evidence that support the structure of diborane. First, the boron–hydrogen (bridging) bond distance is longer than the boron–hydrogen (terminal) bond distance. The bridging hydrogen atoms are linked to boron atoms by the two-electron–three-center bonds. These bonds are weaker, and hence longer, than the conventional two-electron–two center bonds. Second, when diborane is treated with a Grignard reagent, for example, methylmagnesium bromide, the hydrogen atoms are successively replaced by methyl groups.

$$B_2H_6 + CH_3MgBr \rightarrow B_2H_5(CH_3) + MgHBr$$

However, at the most, four hydrogen atoms are replaced by methyl groups. These are the four terminal hydrogen atoms. The bridging hydrogen atoms are trapped between the two boron atoms and hence cannot be exchanged by another functional group. This shows that four hydrogen atoms in diborane are in a different environment from the other two—a fact very well enunciated in Fig. 15.5.

Diborane burns in air to yield boron(III) oxide.

$$B_2H_6 + 3O_2 \rightarrow B_2O_3 + 3H_2O + \text{Heat}$$

The reaction is spontaneous and exothermic. It often results in an explosion. For this reason, diborane is preserved under vacuum.

The hydrolysis of diborane is also spontaneous.

$$B_2H_6 + 6H_2O \rightarrow 2H_3BO_3 + 6H_2$$

With lithium hydride, diborane yields lithium borohydride.

$$B_2H_6 + 2LiH \rightarrow 2Li[BH_4]$$

With hydrogen halides, substitution reactions occur.

$$B_2H_6 + HCl \rightarrow B_2H_5Cl + H_2$$

$$B_2H_6 + HI \rightarrow B_2H_5I + H_2$$

Similar products are obtained with boron(III) halides. For example,

$$B_2H_6 + BCl_3 \rightarrow B_2H_5Cl + BHCl_2$$

At low temperatures, chlorine reacts with diborane to yield boron(III) chloride.

$$B_2H_6 + 6Cl_2 \rightarrow 2BCl_3 + 6HCl$$

The reaction between sodium and diborane gives a mixture of sodium borohydride and sodium polyborohydride.

$$2B_2H_6 + 2Na \rightarrow Na[BH_4] + Na[B_3H_8]$$

At low temperatures, the reaction between diborane and ammonia produces an adduct.

$$B_2H_6 + 2NH_3 \rightarrow B_2H_6.2NH_3$$

In this reaction, diborane undergoes asymmetrical cleavage, and the adduct has the following ionic structure:

$$[H_3N \rightarrow BH_2 \leftarrow NH_3]^+ [BH_4]^-$$

Ammonia acts as a Lewis base, while boron acts as a Lewis acid.

At high temperatures, the reaction between diborane and ammonia produces boron nitride, a polymeric compound with composition $(BN)_x$.

$$B_2H_6 + NH_3 \rightarrow (BN)_x$$

The total number of valence electrons contributed by boron (= 3) and nitrogen (= 5) is the same as contributed by two carbon atoms (= 4 × 2). Therefore, boron nitride has a structure analogous to that of graphite (*see* Chapter 16, p. 487). The structure consists of sheets of hexagonal rings with alternate boron and nitrogen atoms (Fig. 15.6). A layer pattern results, with a boron atom in one sheet stacked over a nitrogen atom in the sheet below it.

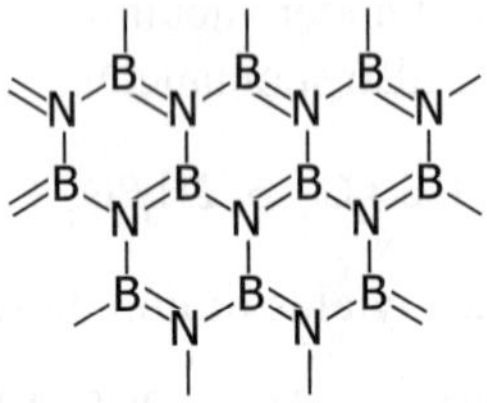

Fig. 15.6 Structure of boron nitride

Boron nitride is an inert material. It is neither attacked by oxygen, nor by steam. It is a good thermal insulator, has low thermal expansion and is resistant to thermal shock. For these reasons it is used to make high temperature crucibles. Powdered boron nitride is soft and non-toxic, as a result of which it finds applications in cosmetics. It induces a pearlscent sheen to personal care items like nail polishes and lipstics. It possesses potential to scatter light making wrinkles less noticeable. Therefore, it is incorporated in facial creams.

Diborane and ammonia, in molar ratio of 1:2, react at 473 K to yield borazine, $B_3N_3H_6$.

$$3B_2H_6 + 6NH_3 \xrightarrow{473\,K} 2B_3N_3H_6 + 12H_2$$

Borazine has a cyclic, hexagonal structure (Fig. 15.7) similar to benzene and for this reason, it is often called *inorganic benzene*.

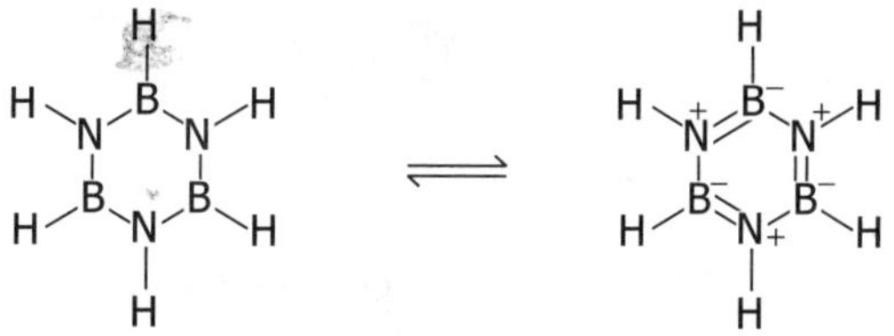

Fig. 15.7 Structure of borazine

Case Study 15.4 How far is it correct to refer borazine as inorganic benzene?

Borazine is isoelectronic with benzene. The two compounds have similar physical properties. However, one school of thought advocates that this similarity in physical properties has been over-emphasized and is responsible for the origin of the name "inorganic benzene" for borazine. The chemical properties of borazine and benzene are quite different. No doubt, both have an aromatic π-cloud, but the delocalization of π-electrons in borazine is not as symmetrical as in benzene. Due to difference in electronegativity, electron density is more localized on nitrogen than on boron in borazine. As a result, the π-bonding in borazine is much weaker than in benzene. Further, nitrogen has some degree of basic character, while boron has acidic character. Polar molecules, such as hydrogen chloride can, therefore, add on to the double bond. The chloride ion gets bonded to boron and the proton to nitrogen. No such addition reactions are possible for benzene.

3. Borax

A naturally occurring mineral, *tincal,* contains about 50% borax ($Na_2B_4O_7.10H_2O$). However, the compound is generally prepared from another mineral, *colemanite*, $Ca_2B_6O_{11}.5H_2O$. Colemanite is first pulverized and then boiled with a concentrated solution of sodium carbonate.

$$Ca_2B_6O_{11}.5H_2O + 2Na_2CO_3 \rightarrow Na_2B_4O_7 + 2NaBO_2 + 2CaCO_3 + 5H_2O$$

The precipitate of calcium carbonate is filtered out and carbon dioxide is passed through the remaining solution. This converts sodium metaborate ($NaBO_2$) into borax.

$$4NaBO_2 + CO_2 \rightarrow Na_2B_4O_7 + Na_2CO_3$$

Sodium carbonate, obtained in the above reaction, is recycled to produce more borax.

The composition of borax may be better written as $Na_2[B_4O_5(OH)_4].8H_2O$. In the anion, $[B_4O_5(OH)_4]^{2-}$, two boron atoms have a coordination number of 3, while the remaining two have a coordination number of 4 (Fig. 15.8).

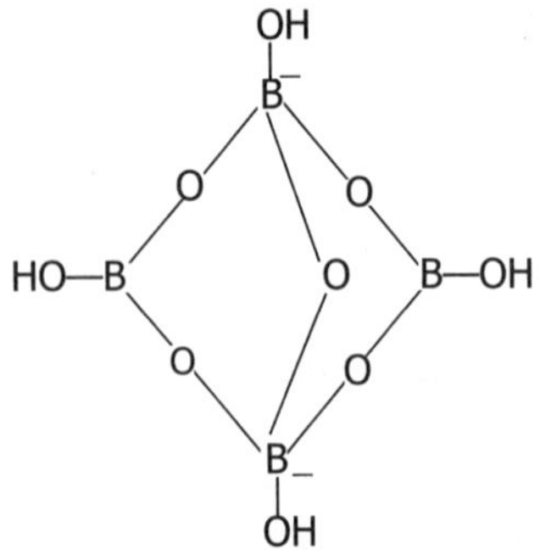

Fig. 15.8 Structure of borax

Thus, the structure consists of two triangular and two tetrahedral units.
An aqueous solution of borax is slightly alkaline due to hydrolysis.

$$Na_2B_4O_7 + 7H_2O \rightarrow 2NaOH + 4H_3BO_3$$

Borax reacts with alkalis to give metaborates.

$$Na_2B_4O_7 + 2NaOH \rightarrow 4NaBO_2 + H_2O$$

It reacts with acids to produce boric acid.

$$Na_2B_4O_7 + 2HCl + 5H_2O \rightarrow 2NaCl + 4H_3BO_3$$

$$Na_2B_4O_7 + H_2SO_4 + 5H_2O \rightarrow Na_2SO_4 + 4H_3BO_3$$

When heated with sulfuric acid, in the presence of ethyl alcohol, triethylborate is produced. This compound burns with a green-edged flame.

$$H_3BO_3 + 3C_2H_5OH \xrightarrow{H_2SO_4} \underset{\text{Triethylborate}}{B(OC_2H_5)_3} + 3H_2O$$

On heating in solid state, borax first loses water of crystallization,

$$Na_2B_4O_7.10H_2O \xrightarrow{\text{Heat}} Na_2B_4O_7 + 10H_2O$$

and then decomposes into a mixture of sodium metaborate and boron(III) oxide.

$$Na_2B_4O_7 \rightarrow 2NaBO_2 + B_2O_3$$

The mixture of sodium metaborate and boron(III) oxide swells in the form of a bead, called *borax bead.* When heated with a metal salt, the oxide component of borax bead gives colored metal metaborates. For example, cobalt salts impart a blue color to the borax bead.

$$CoSO_4 + B_2O_3 \rightarrow \underset{\substack{\text{Cobalt}\\\text{metaborate}\\\text{(Blue)}}}{Co(BO_2)_2} + SO_3$$

The borax bead test is thus used for qualitative detection of several metal ions.

4. Borides

Borides are a set of binary compounds whose compositions range from metal-rich M_5B to boron-rich MB_{66}. These compounds are extremely hard, inert, non-volatile and refractory materials with melting points and electrical conductivities exceeding those of their parent metal. Thus titanium diboride, TiB_2, has a melting point of 3253 K (for titanium, it is 1840 K) and conductivity five times that of titanium.

The diborides of titanium, zirconium and chromium, being hard, are used for making turbine blades, combustion chamber linings, rocket nozzles and ablation shields. Due to their ability to withstand high temperatures, they are used for making crucibles and thermocouples. As they are chemically inert and possess electrical conductivity, they are used as electrodes in industrial operations.

Borides may be prepared by direct combination of stoichiometric amounts of concerned elements,

$$Cr + 2B \xrightarrow{1400\ K} CrB_2$$

or by the reduction of a metal oxide with boron.

$$Sc_2O_3 + 6B \xrightarrow{2000\ K} 2ScB_2 + B_2O_3$$

Reduction of boron(III) halides with a suitable metal, in the presence of hydrogen, also yields the corresponding boride.

$$2BCl_3 + 2W + 2H_2 \xrightarrow{1500\ K} 2WB + 4HCl + Cl_2$$

Reduction of a metal oxide with boron(III) oxide, in the presence of carbon, is yet another method of preparing borides.

$$V_2O_5 + B_2O_3 + 8C \rightarrow 2VB + 8CO$$

A metal oxide may also be reduced by a combination of boron carbide and carbon to produce a boride.

$$2TiO_2 + B_4C + 3C \rightarrow 2TiB_2 + 4CO$$

The structures of borides show wide variations. The M_3B types of borides (for example, Co_3B) have structures based on discrete trigonal prism units (Fig. 15.9a). Each unit has a boron atom in the center. There is weak bonding between boron atoms of different trigonal prism units (Fig. 15.9b), as a result of which the boron–boron bond length is quite large, of the order of 210–330 pm.

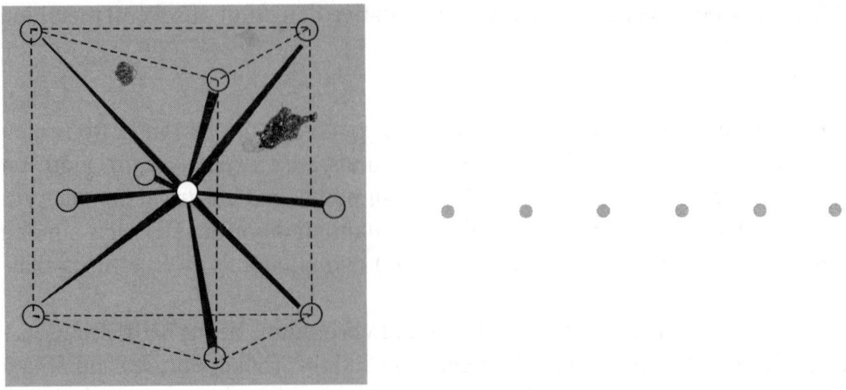

(a) (b)

Fig. 15.9 Structure of (a) discrete M_3B units and (b) position of boron atoms in several such units

As the concentration of boron in borides increases, the tendency of boron atoms to catenate also increases. The boron–boron interactions become relatively strong and the bond distance between two boron atoms gets reduced to the order of 175–180 pm.

Consider MB types of borides, taking iron boride (FeB) as a representative example. The iron atoms occupy the corners of trigonal prisms with boron atoms at the center (Fig. 15.10a). When several trigonal prisms are placed together, the boron atoms appear to be bonded together in zig-zag chains (Fig. 15.10b).

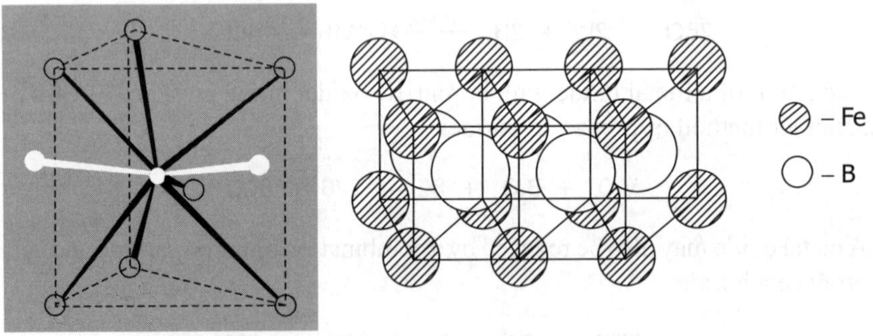

(a) (b)

Fig. 15.10 (a) Structure of iron boride based on trigonal prism units and (b) zig-zag bonding between boron atoms of different iron boride units

The MB_2 type of borides, of which aluminium diboride (AlB_2) is an example, have alternate layers of hexagons and triangles. The boron atoms are joined to form hexagons, while the aluminium atoms form triangles (Fig. 15.11). Each boron atom is in contact with 6 aluminium atoms, 3 each in the triangular layers above and below the hexagonal layer. In turn, each aluminium atom is in contact with 12 boron atoms, 6 each in the hexagonal layers above and below the triangular layer.

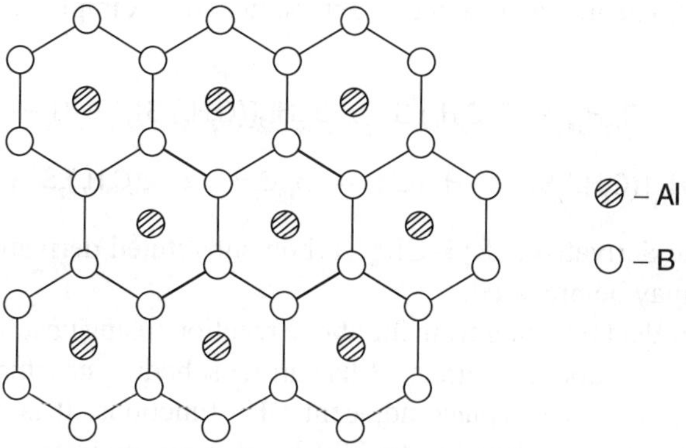

Fig. 15.11 Structure of aluminium diboride

In MB_6 borides, typified by calcium hexaboride (CaB_6), 6 boron atoms are arranged at the corners of an octahedron. The octahedron, in turn, is placed at the center of a cube, the corners of which are occupied by calcium atoms (Fig. 15.12).

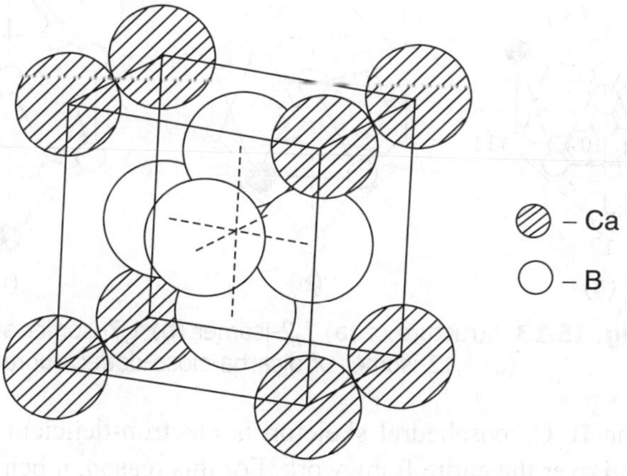

Fig. 15.12 Structure of calcium hexaboride

5. Carboranes

Carboranes, also called *carbaboranes*, are polyhedral compounds having the general composition, $B_{n-2}C_2H_n$. These are considered to be derived from *closo*-borane anions, $B_nH_{2n}^{2-}$ by replacing a BH^- unit with an isoelectronic CH unit. The most important carborane is dicarba-*closo*-dodecarborane, $B_{10}C_2H_{12}$. Its parent compound is *closo*-dodecaborane anion, $B_{12}H_{12}^{2-}$ (for this reason, the concerned carborane is referred to as dodecarborane, although there are 10 carbon atoms in its structure).

Dicarba-*closo*-dodecarborane ($B_{10}C_2H_{12}$) is synthesized by heating decaborane ($B_{10}H_{14}$) with ethyne in the presence of a thioether (for example, diethyl thioether) as catalyst.

$$B_{10}H_{14} + 2(C_2H_5)_2S \rightarrow B_{10}H_{12}[(C_2H_5)_2S]_2 + H_2$$

$$B_{10}H_{12}[(C_2H_5)_2S]_2 + H \equiv CH \rightarrow B_{10}C_2H_{12} + 2(C_2H_5)_2S + H_2$$

Using alkyne derivatives, $R^1 \equiv CR^2$, carbon substituted derivatives of the type $B_{10}C_2H_{10}R^1R^2$ may be prepared.

The initial product resulting from the above reaction (using unsubstituted ethyne) is 1,2-dicarba-*closo*-dodecarborane. It has an icosahedral structure (Fig. 15.13a) in which two CH groups replace adjacent BH^- functions. It is also referred to as the *ortho*-isomer. On heating to 700 K, it rearranges to 1,7-isomer, also called *meta*-isomer (Fig. 15.13b). If the temperature is further raised to 900 K, 1,12-isomer is formed. As shown in Fig. 15.13c, the two carbon atoms in 1,12-isomer occupy diametrically opposite corners. For this reason, it is also called *para*-isomer.

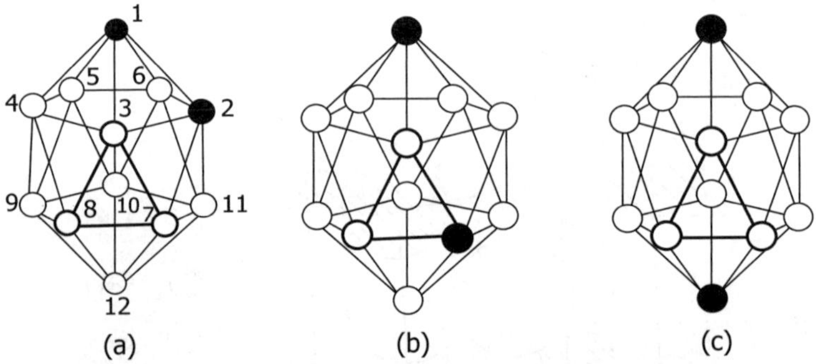

Fig. 15.13 Structure of (a) 1,2-isomer (b) 1,7-isomer and (c) 1,12-isomer of dicarba-closo-dodecarborane

Although the $B_{10}C_2$ icosahedral skeleton is electron-deficient, the electronic charge is spread over the entire framework. For this reason, it behaves as a three-dimensional aromatic unit. The $B_{10}C_2$ moiety is electron withdrawing in nature. As a result, the hydrogen atoms attached to carbon atoms are acidic and easily replaceable. For example, lithium ions substitute for hydrogen atoms when 1,2-dicarba-*closo*-dodecarborane is treated with alkyllithium(I).

$$B_{10}C_2H_{12} + 2C_2H_5Li \rightarrow B_{10}C_2H_{10}Li_2 + 2C_2H_6$$

The lithium derivative may be used to synthesize several other compounds.

$$B_{10}C_2H_{10}Li_2 + 2I_2 \rightarrow B_{10}C_2H_{10}I_2 + 2LiI$$

$$B_{10}C_2H_{10}Li_2 + 2NOCl \rightarrow B_{10}C_2H_{10}(NO)_2 + 2LiCl$$

$$B_{10}C_2H_{10}Li_2 + 2CO_2 + 2H^+ \rightarrow B_{10}C_2H_{10}(COOH)_2 + 2Li^+$$

$$B_{10}C_2H_{10}Li_2 + 2HCHO + 2H^+ \rightarrow B_{10}C_2H_{10}(CH_2OH)_2 + 2Li^+$$

$$B_{10}C_2H_{10}Li_2 + 2(C_6H_5)_2PCl \rightarrow B_{10}C_2H_{10}[(C_6H_5)_2P]_2 + 2LiCl$$

When 1,2-dicarba-*closo*-dodecarborane is treated with ethoxide ion in ethanol at 350 K, a *nido*-carborane anion, $B_9C_2H_{12}^-$ is formed.

$$B_{10}C_2H_{12} + C_2H_5O^- + 2C_2H_5OH \xrightarrow{350\ K} B_9C_2H_{12}^- + B(OC_2H_5)_3 + H_2$$

The *nido*-carborane anion may be deprotonated using a strong base.

$$B_9C_2H_{12}^- + OH^- \rightarrow B_9C_2H_{11}^{2-} + H_2O$$

The resulting dianion reacts with transition metal ions to yield *metallocarboranes*. For example,

$$4B_9C_2H^{2-}{}_{11} + 3Co^{2+} \rightarrow Co + 2Co(B_9C_2H_{11})_2^-$$

The structure of cobaltocarborane is depicted in Fig. 15.14.

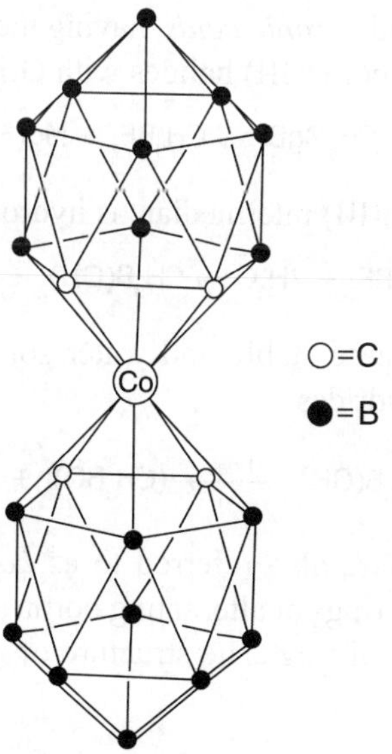

Fig. 15.14 Structure of cobaltocarborane

6. Complexes

Boron forms an extensive series of alkyl and arylorthoborates, having the composition, $B(OR)_3$. These are synthesized by the direct reaction of a boron(III) halide with alcohol. For example,

$$BCl_3 + 3C_2H_5OH \rightarrow B(OC_2H_5)_3 + 3HCl$$

The trialkyls and triaryls of the type, R_3B are monomeric compounds which may be prepared by the reaction of boron(III) fluoride-dimethylether complex with Grignard reagent.

$$2BF_3{\cdot}O(CH_3)_2 + 6CH_3MgBr \rightarrow 2(CH_3)_3B + 3MgF_2 + 3MgBr_2 + 2O(CH_3)_2$$

The lower trialkyls are very reactive and inflame in air, although their aryl counterparts are relatively stable. When sodium phenyl is used, instead of Grignard reagent, initially triphenylboron is formed.

$$BF_3{\cdot}O(CH_3)_2 + 3C_6H_5Na \rightarrow (C_6H_5)_3B + 3NaF + O(CH_3)_2$$

It reacts further with sodium phenyl to yield sodium tetraphenylborate(III), $Na[B(C_6H_5)_4]$.

$$(C_6H_5)_3B + C_6H_5Na \rightarrow Na[B(C_6H_5)_4]$$

The sodium analog is water soluble but the corresponding compounds with larger alkali metal ions are not. Hence sodium tetraphenylborate(III) is often used in the gravimetric determination of potassium, rubidium or cesium.

$$Na[B(C_6H_5)_4] + CsCl \rightarrow \underset{\text{Insoluble}}{Cs[B(C_6H_5)_4]} + NaCl$$

A set of compounds called *boronic acids*, having the composition, $RB(OH)_2$ are obtained by the reaction of boron(III) halides with Grignard reagent in dry ether.

$$BF_3{\cdot}O(C_2H_5)_2 + CH_3MgBr \rightarrow CH_3BF_2 + Mg(F)Br + O(C_2H_5)_2$$

The difluoromethylboron(III) intermediate is hydrolyzed.

$$CH_3BF_2 + 2H_2O \rightarrow CH_3B(OH)_2 + 2HF$$

The boronic acids are quite stable and water soluble. On heating, these are dehydrated to boronic anhydrides.

$$3CH_3B(OH)_2 \xrightarrow{\text{Heat}} (CH_3BO)_3 + 3H_2O$$

The anhydrides, which are also referred to as *boroxines*, possess a trimeric cyclic structure with planar rings of alternating boron and oxygen atoms. The alkyl groups too lie in the plane of ring. The structure of methylboroxine is shown in Fig. 15.15.

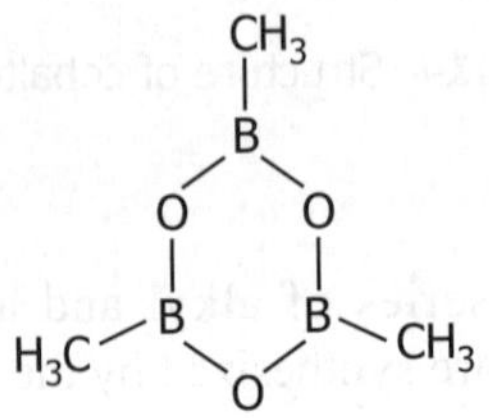

Fig. 15.15 Structure of methylboroxine

Boronous acids are a series of organoboron compounds having the composition, $R_2B(OH)$, where R is an alkyl or aryl group. These are prepared by the reaction of boron(III) halides with Grignard reagent in 1:2 molar proportion in dry ether,

$$BF_3{\cdot}O(C_2H_5)_2 + 2CH_3MgBr \rightarrow (CH_3)_2BF + MgF_2 + MgBr_2 + (C_2H_5)_2O$$

followed by the hydrolysis of the intermediate product.

$$(CH_3)_2BF + H_2O \rightarrow (CH_3)_2BOH + HF$$

One of the most important organometallic compounds of aluminium is trimethylaluminium(III). It exists as a dimer. The preparation, properties and structure of this compound are discussed in Chapter 29, p. 850.

Case Study 15.5 What is meant by the term, *knight move relationship*?

Knight move relationship means closeness in physical properties of one element to another element that is housed one period down and two groups to the right in the periodic table (Fig. 15.16).

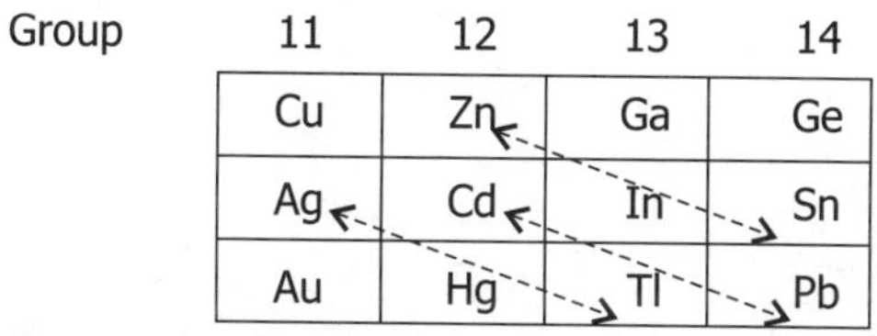

Fig. 15.16 Knight move relationship

The name originates from the movement of knight (Fig. 15.17) on the chessboard (horse, to the Indian chess player).

Fig. 15.17 The knight in the game of chess

Thus the compounds of zinc(II) and tin(II), as well as of cadmium(II) and lead(II) have similar physical properties. However, the similarities owing to knight move relationship is most apparent in compounds of thallium(I) and silver(I). For example—

1. Both the ions have almost equal charge densities.
2. The melting points of chlorides of thallium(I) and silver(I) are quite close to each other. So are the melting points of the nitrates of these elements.
3. Both the ions form normal oxides, Tl_2O and Ag_2O.
4. The fluorides of both ions are soluble in water.
5. The other halides of both ions are insoluble in water.
6. The chromates of both ions are brick-red in color and insoluble in water.

16. GROUP 14 ELEMENTS

Introduction

The elements, carbon (C), silicon (Si), germanium (Ge), tin (Sn) and lead (Pb), which constitute Group 14 of the periodic table, show less regular gradation in properties as compared to s-block elements. Nevertheless, in one aspect a clear line of demarcation may be drawn within the group. C is a non-metal; Si and Ge are metalloids—they show characteristics of both metals and non-metals; and Sn and Pb are metals. The elements are relatively unreactive, but reactivity increases down the group.

Physical Properties

The general configuration of Group 14 elements is ns^2np^2. The group oxidation state is IV. However, down the group, oxidation state II becomes more stable. This is attributed to inert pair effect—the electron pair in ns orbital remains inert and does not participate in bonding. For Sn, oxidation states II and IV are almost equally stable. However, for Pb, tetravalent state is strongly oxidizing, but divalent state is very stable. The energy required to remove all four valence electrons is extremely high. Hence, simple ionic compounds in tetravalent state are rare.

Although Sn and Pb are metals, their melting points and boiling points are lower than the non-metallic elements occupying higher positions in the group. This shows that both Sn and Pb do not use all the four valence electrons for metallic bonding.

1. Catenation

Carbon possesses a unique ability to bond to itself. This self-linkage is called catenation. Carbon atoms catenate to form long chains, branched chains and rings. This property of catenation is responsible for the occurrence of innumerable organic compounds.

Catenation is also shown by other elements of Group 14, but to a limited extent, and to a level that is far less than that of carbon. As shown in Table 16.1, the tendency for catenation decreases down the group. The number of alkanes is higher than the corresponding compounds of other elements.

Table 16.1 Tendency for catenation among Group 14 elements

Classification	General formula	Maximum value of n	Bond energy ($kJ\ mol^{-1}$)
Alkanes	C_nH_{2n+2}	>20	C–C: 348
Silanes	Si_nH_{2n+2}	8	Si–Si: 297
Germanes	Ge_nH_{2n+2}	6	Ge–Ge: 260
Stannanes	Sn_nH_{2n+2}	2	Sn–Sn: 240
Plumbane	Pb_nH_{2n+2}	1	–

The tendency of C to catenate originates because of the high carbon–carbon bond strength. As shown in Table 16.1, the variation in element–element bond energy follows the order:

$$C–C > Si–Si > Ge–Ge > Sn–Sn > Pb–Pb$$

For this reason, C undergoes catenation to the maximum extent, Pb to the minimum extent. Two reasons may be ascribed to high carbon–carbon bond energy. First, the size of C atom is smallest in the group. Therefore, the carbon–carbon bond length is also small. Conversely the tin–tin or lead–lead (hypothetical) bond distance is relatively large. Smaller the bond length, stronger is the bond. Second, the nuclear charge increases on moving down the group. This means that proton–proton repulsions in a C–C bond are relatively small as compared to the corresponding repulsions in Pb–Pb bond.

2. Allotropic forms of carbon

There are six crystalline allotropic forms of carbon: α-graphite, β-graphite, diamond, lonsdaleite, chaoite and carbon-VI.

Graphite is composed of planar, two-dimensional sheets of sp^2 hybridized carbon atoms. Each sheet is a network of fused, hexagonal rings of carbon atoms (Fig. 16.1). The layers are held together by relatively weak van der Waals forces of attraction. In α-graphite, the layers are arranged in the sequence ABAB..., i.e., the carbon atoms in alternate layers are vertically above each other. In β-graphite, the layers are stacked in the sequence ABCABC..., i.e., fourth layer is the repetition of the first one. The enthalpy difference between α- and β-graphites is very small and hence, the two forms are interconvertible; α-graphite may be changed into β-form by grinding. On heating to about 1300 K, β-form converts into α-form.

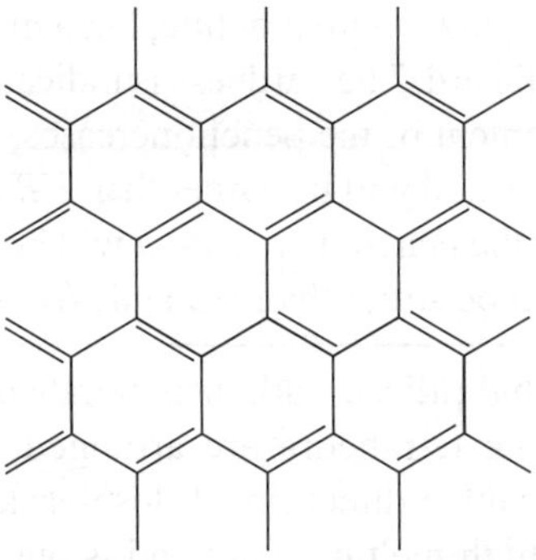

Fig. 16.1 Structure of graphite

Within each layer of α- or β-graphite, the C–C bond distance is 141 pm, while between the two layers, the distance is 355 pm. This relatively large inter-layer distance accounts for most of the remarkable properties of graphite. Due to weak bonding between the layers, graphite is soft and cleaves easily. Moreover, the layers easily slide over each other and thus graphite is used as a lubricant as well as an additive for motor oil. The wide inter-layer spacing increases the volume per unit mass. Therefore, graphite has lower density than diamond.

Graphite is thermodynamically the most stable allotropic form of carbon. It is, therefore, not possible to convert it into diamond, unless extreme conditions of high temperature (2000 K) and high pressure (50,000 atmosphere) are available. Conversely, diamond is thermodynamically unstable with respect to graphite.

$$C_{(diamond)} \rightarrow C_{(graphite)} \quad \Delta G = -3 \text{ kJ mol}^{-1}$$

The fact that thc above reaction proceeds with a negative value of change in free energy reveals that diamond may be easily transformed into graphite. Yet the diamond jewellery does not crumble into graphite or, for that matter, in any other allotropic form of carbon. The reason being that although the conversion of diamond into graphite is thermodynamically feasible, kinetically it is not. An extremely large amount of activation energy is required to re-orient the three dimensional network of diamond into the two-dimensional arrangement of graphite. Under ordinary conditions of temperature and pressure, it is not possible to cross such a huge activation energy barrier.

Although graphite is thermodynamically more stable than diamond, yet it is more reactive due to the vulnerable layered structure. In graphite, only three of the carbon's four valence electrons are involved in σ-bonding. The fourth one is involved in π-bonding.The π-electrons are delocalized over the whole structure. These are mobile and hence graphite can conduct electricity.

Case Study 16.1 What does the notation *HB* on lead pencils indicate?

The term lead pencil itself is a misnomer. The writing medium in pencils is graphite and not lead. The name originated because lead produces similar type of shading on paper (or any other surface) as does graphite.

The pencils do not incorporate pure graphite, but a mixture of graphite and clay. The latter makes the pencil hard. The hardness is indicated by letter *H*, the softness by letter *B*. As the clay content of the pencil increases, it will be marked as *2HB*, *3HB*, and so on. The *2HB* pencil will be harder than *HB* and *3HB* harder than *2HB*. As the graphite content of the pencil increases, it will be marked as *HB2*, *HB3*, and so on. The *HB2* pencil will be softer than *HB* and *HB3* softer than *HB2*.

In diamond, each sp^3 hybridized carbon is tetrahedrally coordinated to four equidistant neighbors. The tetrahedra are arranged to give a cubic unit cell (Fig. 16.2). This results in a three-dimensional close-packed arrangement, as a result of which diamond has a high melting point and is one of the hardest substances known. Due to compact packing, the density of diamond is also very high. Since there are no mobile electrons in the structure, diamond is an insulator. Yet, diamond is a good conductor of heat. The thermal conductivity results due to the compact packing of tetrahedral units in the structure of diamond. When a substance in which the constituent atoms are relatively loosely held is heated, the energy is absorbed. The atoms start oscillating. However, in the close-packed arrangement of carbon atoms in diamond, the absorption of heat energy is not possible as sufficient space for oscillation is not available to the constituents. The only possibility is that the heat is transferred from one tetrahedral unit to another, across the whole framework. Hence the thermal conductivity of diamond.

All naturally occurring diamonds are colored and contain trace amounts of nitrogen. Others contain a small amount of aluminium. Such diamonds are blue in color and are highly valued.

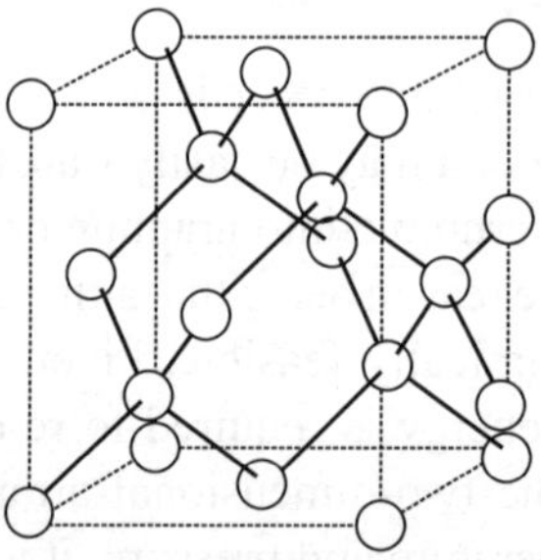

Fig. 16.2 Structure of diamond

Lonsdaleite form of carbon is closely related to diamond. Here also each carbon atom is sp^3 hybridized and tetrahedrally coordinated. However, the tetrahedra are arranged in the form of a hexagonal lattice rather than the cubic one of diamond. Although lonsdaleite occurs naturally at scanty locations, it may also be prepared artificially by subjecting α-graphite to high temperature and pressure.

Chaoite form of carbon is white in color. It may be synthesized as crystals of hexagonal symmetry by subliming graphite at about 2500 K under a low pressure of 10^{-4} mm.

Carbon-VI allotrope also possesses hexagonal symmetry. It may be obtained by heating graphite in argon atmosphere to about 2600 K, under a pressure of

10^{-4} mm range. The heating is carried out by radiations or by laser. The crystal structures of both chaoite and carbon-VI are based on carbyne-type skeletons, $-C \equiv C-C \equiv C-$. Both chaoite and carbon-VI are resistant to oxidation and reduction. In fact, these allotropic forms are more similar in characteristics to diamond than to graphite.

Case Study 16.2 What is *glanzkohlenstoff*?

If the flame of burning methane is impinged on glazed porcelain, heated to about 900 K, a highly lustrous form of carbon, called glanzkohlenstoff, is deposited. It consists of extremely minute crystals of graphite. These crystals have a greater absorption power and more hardness than graphite. However, these lack the characteristic cleavage property of graphite.

3. Allotropic forms of tin

Tin is a soft, malleable metal with a relatively low melting point (405 K). It has two allotropic forms: α-tin and β-tin. At normal temperatures and above, it exists as β-tin or *white tin*. This variety is metallic in nature and is used in alloys. The close packed arrangement of atoms results in a significantly high density of 7.31 g cm^{-3} (at 293 K) for white tin.

Below 286.2 K, β-tin is transformed to α-tin, also called *grey tin*. In this variety, the atoms do not retain their close packed arrangement, the density drops to 5.75 g cm^{-3}, the metallic properties are lost and the state changes from crystalline to powdered form. This causes the disintegration of items made from white tin if these are exposed to low temperatures for long periods.

The disintegrated, powdered form of tin is referred to as *tin pest*, *tin disease*, *tin blight* or *tin leprosy*. Although, tin pest is likely to develop when white tin is cooled to 286.2 K, the disintegration process is extremely slow due to high activation energy required for the transformation of β-tin to α-tin. Nevertheless, in presence of gallium or at temperatures in the vicinity of 240 K it speeds up.

The allotropic transformation of β-tin to α-tin is autocatalytic in nature, that is, it catalyzes itself, so that the reaction accelerates once it starts. This means that the origin of tin pest leads to more tin pest.

Case Study 16.3 Is it true that the Napoleon's Russian campaign failed due to the phenomena of tin pest?

The French invasion of Russia in 1812 dramatically reduced Napoleon's hegemony in Europe. During this campaign, the Russians used the Scorched-earth tactic, under which the army, along with the civilian population, retreated to the interior of their country, simultaneously poisoning the food items. As the French forces advanced further, the tin buttons of the coats, jackets and armours of the soldiers' uniform crumbled to powder. The extremely cold weather accelerated the transformation of β-tin to α-tin and the tin pest ate the buttons. The French troops found themselves in a pernicious state—meager food resources due to scorched-earth policy and biting cold, compounded by falling apart of the buttons of warm clothing—and that too many miles inside the enemy territory. The Russian troops, which till then were in withdrawal mode, now started to encircle the foe. For French army, it was doomsday.

Not everyone agrees with the theory of falling buttons. Critics hold that the tin used for making the buttons would not have been very pure. Impurities resist the allotropic transformation of β-tin to α-tin, thus retarding the development of tin pest. Moreover, the time required to develop tin pest under the weather conditions prevailing in Russia is believed to be much longer than the duration of Napoleon's campaign.

Chemical Properties

We have already mentioned that C, in the form of diamond, is unreactive, but as graphite is quite reactive. Si too is unreactive in crystalline state, but more reactive in molten form. The reactivity of Ge is more than that of Si. Similarly, Sn is more reactive than Ge, but it is also appreciably amphoteric in nature. Pb in a finely divided, powdered state is extremely reactive. In fact, it is pyrophoric, i.e., it catches fire spontaneously in air. However, subsequently the reactivity decreases due to the formation of a protective oxide layer on the surface.

1. Reaction with air

C burns in a limited quantity of air to yield carbon monoxide, and in excess air to yield cabon dioxide.

$$\underset{\text{Limited}}{2C + O_2} \rightarrow 2CO$$

$$\underset{\text{Excess}}{C + O_2} \rightarrow CO_2$$

Si combines with oxygen only above 1200 K and Ge reacts with it only at red heat. Sn shows analogous reaction at 1800 K.

$$Si + O_2 \xrightarrow{> 1200\ K} SiO_2$$

$$Ge + O_2 \xrightarrow{\text{Red heat}} GeO_2$$

$$Sn + O_2 \xrightarrow{1800\ K} SnO_2$$

Pb forms lead monoxide when heated in air at about 1000 K.

$$2Pb + O_2 \xrightarrow{1000\ K} 2PbO$$

2. Reaction with halogens

Carbon, as graphite, combines with fluorine at 700 K to yield graphite monofluoride, CF.

$$\underset{\text{Graphite}}{2C} + F_2 \xrightarrow{700\text{ K}} 2CF$$

Si combines with fluorine at room temperature, with chlorine at 500 K and with bromine or iodine at 800 K to yield the corresponding tetrahalide.

$$Si + 2X_2 \rightarrow SiX_4$$
$$(X = F, Cl, Br \text{ or } I)$$

Ge reacts with chlorine and bromine on moderate heating to yield the corresponding tetrahalide.

$$Ge + 2X_2 \rightarrow GeX_4$$
$$(X = Cl \text{ or } Br)$$

Sn reacts with chlorine or bromine at room temperature, and with fluorine or iodine on heating, to form tin(IV) halides.

$$Sn + 2X_2 \rightarrow SnX_4$$
$$(X = F, Cl, Br \text{ or } I)$$

Pb forms dihalides.

$$Pb + X_2 \rightarrow PbX_2$$
$$(X = F, Cl, Br \text{ or } I)$$

Case Study 16.4 How is a frosted design etched on glass?

Glass is a solid solution of mainly alkali and alkaline earth metal oxides in silica. The silica component reacts with hydrogen fluoride to produce hydrogen hexafluorosilicate(IV) which, being a white gel, imparts a frosted appearance to glass.

$$SiO_2 + 6HF \rightarrow H_2[SiF_6] + 2H_2O$$

A glass object, like a bulb, a vase or any other decoration piece is coated by wax on the inside surface. By piercing through the layer of wax a design is inscribed. Wax is shredded off the patterned area, rendering the glass surface bare. Thereafter, the object is treated with hydrogen fluoride. The latter is inert to wax, but reacts with the exposed, patterned lining of glass, yielding white hydrogen hexafluorosilicate(IV). In this manner a frosted design is etched on glass.

---

3. Reaction with acids

Carbon is oxidized to carbon dioxide by concentrated sulfuric acid.

$$\underset{\text{Conc.}}{C + 2H_2SO_4} \rightarrow CO_2 + 2SO_2 + 2H_2O$$

Si reacts with hydrochloric acid at 600 K.

$$Si + 3HCl \xrightarrow{600\ K} \underset{\text{Trichlorohydridosilicon(IV)}}{SiHCl_3} + H_2$$

Ge reacts in an analogous manner.

$$Ge + 3HCl \rightarrow \underset{\text{Trichlorohydridogermanium(IV)}}{GeHCl_3} + H_2$$

Sn liberates hydrogen when treated with dilute sulfuric acid; sulfur dioxide, when treated with concentrated sulfuric acid; and nitrogen dioxide, when treated with concentrated nitric acid.

$$Sn + \underset{\text{Dilute}}{H_2SO_4} \rightarrow SnSO_4 + H_2$$

$$Sn + \underset{\text{Conc.}}{2H_2SO_4} \rightarrow SnSO_4 + 2H_2O + SO_2$$

$$Sn + \underset{\text{Conc.}}{4HNO_3} \rightarrow \underset{\text{Metastanic acid}}{H_2SnO_3} + H_2O + 4NO_2$$

Pb does not react with dilute acids. It yields sulfur dioxide with concentrated sulfuric acid, and nitrogen dioxide with concentrated nitric acid.

$$Pb + \underset{\text{Conc.}}{2H_2SO_4} \rightarrow PbSO_4 + 2H_2O + SO_2$$

$$Pb + \underset{\text{Conc.}}{4HNO_3} \rightarrow Pb(NO_3)_2 + 2H_2O + 2NO_2$$

4. Reaction with alkalis

Si dissolves in hot, aqueous alkalis to liberate hydrogen.

$$Si + 4NaOH \rightarrow \underset{\text{Sodium silicate}}{Na_4SiO_4} + 2H_2$$

Ge reacts with fused alkalis to give germanates.

$$Ge + 4NaOH \rightarrow \underset{\text{Sodium germanate}}{Na_4GeO_4} + 2H_2$$

Sn and Pb react with hot, aqueous alkalis to yield stannates and plumbates, respectively.

$$Sn + 2KOH + 4H_2O \rightarrow \underset{\text{Potassium stannate}}{K_2[Sn(OH)_6]} + 2H_2$$

$$Pb + 2NaOH + 4H_2O \rightarrow \underset{\text{Sodium plumbate}}{Na_2[Pb(OH)_6]} + 2H_2$$

Case Study 16.5 On what principle does a lead-acid battery operate?

The lead-acid battery is the most successful rechargeable device. In its fully charged state, the active material on the cathode is lead(IV) oxide, PbO_2. The anode is made up of lead metal. The electrolyte is dilute sulfuric acid.

When the cell is producing current, the reaction at cathode involves reduction of lead(IV) to lead(II) ions. The latter is precipitated as lead(II) sulfate.

$$\text{At cathode: } PbO_2(s) + HSO_4^-(aq) + 3H^+(aq) + 2e^- \rightarrow PbSO_4(s) + 2H_2O(l)$$

The reaction at anode involves oxidation of elemental lead to lead(II) ion, which is also precipitated as lead(II) sulfate.

$$\text{At anode: } Pb(s) + SO_4^{2-}(aq) \rightarrow PbSO_4(s) + 2e^-$$

The net reaction is

$$PbO_2(s) + 2HSO_4^-(aq) + 2H^+(aq) + Pb(s) \rightarrow 2PbSO_4(s) + 2H_2O(l)$$

The potential difference of 2V is more than that required for oxidation of water to oxygen.

Compounds

Besides carbon monoxide and carbon dioxide, carbon forms another oxide, *carbon suboxide*, C_3O_2. It is prepared by dehydrating malonic acid in the presence of phosphorus(V) oxide at about 400 K.

$$\underset{\text{Malonic acid}}{HOOC.CH_2.COOH} \xrightarrow[400\ K]{P_4O_{10}} \underset{\text{Carbon suboxide}}{O{=}C{=}C{=}C{=}O} + 2H_2O$$

Case Study 16.6 Carbon dioxide is a gas but silicon dioxide is an infusible solid. Why?

Carbon dioxide has a linear, π-bonded structure:

$$\ddot{\underset{\cdot\cdot}{O}}{=}C{=}\ddot{\underset{\cdot\cdot}{O}}$$

Both C and O have small sizes and thus their orbitals undergo effective lateral overlap, leading to the formation of π-bonds. The mode of attraction between two or more molecules of carbon dioxide are weak van der Waals forces. Since strong intermolecular forces are lacking, carbon dioxide in its natural state is a gas. Si is larger in size than C. Its orbitals cannot have an effective sideways overlap with the orbitals of oxygen. This means that a single unit of SiO_2 would have the following structure:

$$^{-}O—Si^{2+}—O^{-}$$

Such a species would be very reactive and would combine with other such units, leading to a three-dimensional network of alternating silicon and oxygen atoms.

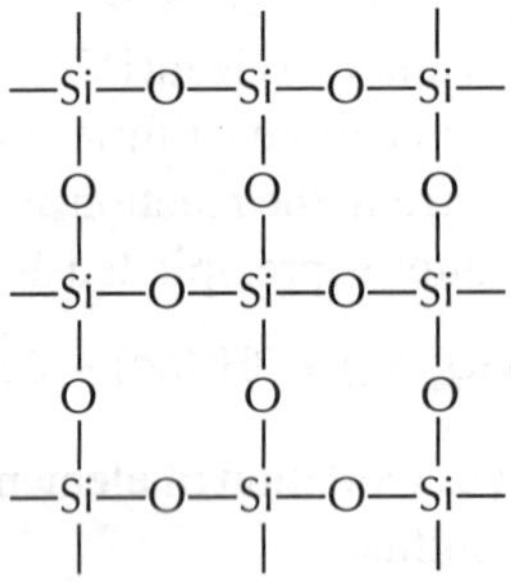

Strong forces of attraction between σ-bonded silicon and oxygen atoms push the compound into solid state. To melt silicon dioxide, a large number of strong silicon–oxygen bonds would have to be broken. This would require tremendously large amount of energy. For this reason, silicon dioxide is infusible in nature.

Silanes (Si_nH_{2n+2}) and alkanes (C_nH_{2n+2}) have same stoichiometries, but widely divergent reactivities. Silanes are highly reactive, whereas alkanes are relatively inert. The difference primarily arises because H is more electronegative than Si, but less electronegative than C. As a result, the signs of partial charges in a silicon–hydrogen bond are reverse of that in a carbon–hydrogen bond.

$$\overset{\delta+}{Si}—\overset{\delta-}{H} \qquad \overset{\delta-}{C}—\overset{\delta+}{H}$$

Si is thus more vulnerable to attack by nucleophilic reagents, including water. Further, the size of Si is larger than C. It can expand its coordination number and, therefore, add on molecules or ions. Moreover, such reactions take place by donation of electrons by the adding species to Si. Si can accept additional electrons because it has vacant d orbitals. C does not have vacant d orbitals.

Case Study 16.7 Silicon tetrachloride is rapidly hydrolyzed, but carbon tetrachloride is not. Why?

Carbon tetrachloride is an inert liquid. Even if it is kept in an open bottle, it would retain its composition for a prolonged period of time. On the other hand, silicon tetrachloride is so reactive that it starts giving out dense white fumes when exposed to air. The moisture content of air reacts spontaneously with silicon tetrachloride to give a set of mixed chloro-hydroxo compounds, which ultimately are transformed into silicon dioxide. The reaction may be depicted as follows:

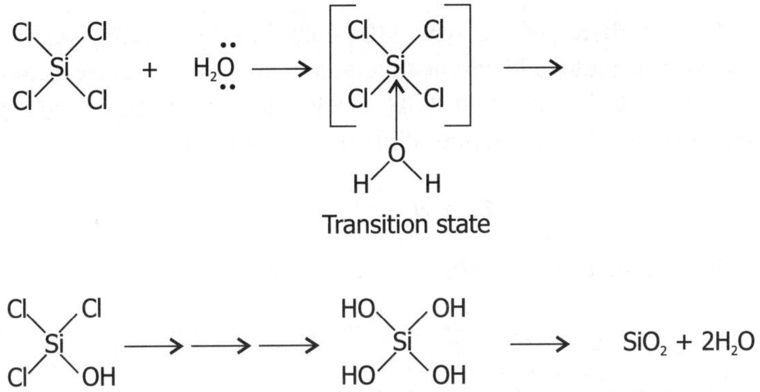

The key to the reactivity of silicon tetrachloride is to be found in the intermediate addition compound that forms in the transition state. The intermediate is shown in parenthesis because it has a transitionary existence—it decomposes almost as soon as it is formed. Two factors have contributed to the formation of this intermediate addition compound. First, silicon has a relatively large size and can increase its coordination number from 4 to 5. Second, silicon has vacant d orbitals which can accommodate the lone pair of electrons donated by the oxygen of water molecule through a coordinate bond. Carbon is smaller in size than silicon. It cannot have a coordination number 5. Moreover, it does not possess vacant d orbitals to accommodate electrons of the adding entities. Therefore, carbon tetrachloride is inert in nature.

1. Carbides

The binary compounds of carbon are called carbides. They may be classified into three types.

a. Ionic carbides

Also called *salt-like carbides*, the ionic carbides contain anionic carbon. The most common examples of ionic carbides are beryllium carbide, Be_2C and aluminium carbide, Al_4C_3. Beryllium carbide is prepared by heating beryllium oxide with C at about 2200 K,

$$2BeO + 2C \xrightarrow{2200\ K} Be_2C + CO_2$$

while aluminium carbide is prepared by heating the elements in an electric furnace.

$$4Al + 3C \rightarrow Al_4C_3$$

Beryllium carbide and aluminium carbide are together referred to as *methanides,* as they yield methane on hydrolysis.

$$Be_2C + 4H_2O \rightarrow 2Be(OH)_2 + CH_4$$

$$Al_4C_3 + 12H_2O \rightarrow 4Al(OH)_3 + 3CH_4$$

Ionic carbides of the type MC_2 contain ethyne group $(C \equiv C)^{2-}$. On hydrolysis, they yield ethyne (acetylene) and are, therefore, referred to as *acetylides*. Calcium carbide (CaC_2) belongs to this category and is prepared either by the direct reaction of elements at 1400 K,

$$Ca + 2C \xrightarrow{1400\ K} CaC_2$$

or by heating calcium oxide with carbon at 500 K.

$$CaO + 3C \xrightarrow{500\ K} CaC_2 + CO$$

On hydrolysis, calcium carbide yields ethyne.

$$CaC_2 + 2H_2O \rightarrow Ca(OH)_2 + C_2H_2$$

Magnesium forms two carbides, MgC_2 and Mg_2C_3. The latter on hydrolysis liberates propyne.

$$Mg_2C_3 + 4H_2O \rightarrow 2Mg(OH)_2 + CH_3.C \equiv CH$$

b. Covalent carbides

The covalent carbides, also called *molecular carbides,* are formed by the combination of carbon with non-metals or metalloids. The most common example of a covalent carbide is silicon carbide, SiC. It is a hard, infusible and inert material, which is used as an abrasive. It is prepared by heating sand with coke in an electric furnace, at about 2500 K. The reaction completes in two steps:

$$SiO_2 + 2C \xrightarrow{2500\ K} Si + 2CO$$

$$Si + C \rightarrow SiC$$

Pure silicon carbide is light yellow in color. However, commercial samples are purple or black due to the presence of iron impurities. It has a three-dimensional structure in which each Si atom is tetrahedrally surrounded by four carbon atoms and vice versa.

Boron carbide (B_4C) is another example of a covalent carbide. It is prepared by heating boron(III) oxide with carbon at 1800 K.

$$2B_2O_3 + 7C \xrightarrow{1800\ K} B_4C + 6CO$$

It is used as abrasive for polishing, in brake linings and for making bullet-proof jackets. For the last application, boron carbide threads are made by reacting boron(III) chloride with hydrogen and carbon yarn at 2000 K.

$$4BCl_3 + 6H_2 + C(\text{fiber}) \xrightarrow{2000\text{ K}} B_4C(\text{fiber}) + 12HCl$$

c. Interstitial carbides

Interstitial carbides are mostly formed by d- and f-block elements. These are infusible, extremely hard and refractory materials that retain the characteristic properties of metals, such as luster and conductivity. In these compounds, carbon atoms occupy the interstices in the metallic lattice. The atomic radius of the metal should be large enough to generate interstices of appropriate size which can accommodate carbon.

Interstitial carbides possess a wide range of stoichiometries. For example, vanadium can form carbides with compositions, V_2C, V_4C_3, V_6C_5 and V_8C_7.

As a result of their chemical inertness and extreme hardness, the interstitial carbides of tungsten, WC and tantalum, TaC find extensive use as high-speed cutting tools.

2. Graphite intercalation compounds

We have seen that graphite is composed of two-dimensional, parallel sheets of carbon atoms. The distance between two parallel sheets is relatively large and the inter-layer bonding is relatively weak. This enables a wide range of substances to intrude between the parallel planes in graphite to give compounds of variable compositions. These compounds are called *intercalation compounds*. These are sometimes also referred to as *lamellar compounds*.

When atoms, ions or molecules of a foreign substance are accommodated between the parallel sheets in graphite, the inter-layer distance increases. However, most of the characteristic properties of graphite are retained. The π-electron cloud continues to be dispersed over the entire structure and for this reason, intercalated graphite is generally conducting in nature. In case the added substance contributes its own electrons, the electrical conductivity increases. Intercalation is a reversible reaction—the foreign substance can be ejected out of the host lattice quite easily.

One of the most important intercalation compounds is prepared by heating graphite with vapors of potassium at about 600 K. Its composition is C_8K. Incorporation of potassium in graphite increases the inter-layer distance from 335 pm to about 1000 pm. The planes continue to be parallel but are now stacked vertically above each other, instead of the ABAB arrangement found in graphite. The lamellar compound conducts electricity better than graphite. This suggests that bonding involves the transfer of valence electron of potassium to the conduction band of graphite. The fact that graphite is diamagnetic but C_8K is paramagnetic supports this assertion.

When the potassium intercalation compound is heated to 700 K, under reduced pressure, the metal is driven out in stages to give a series of lamellar compounds with progressively decreasing metal content.

Composition	C_8K	$\rightarrow$	$C_{24}K$	$\rightarrow$ $C_{36}K$	$\rightarrow$	$C_{48}K$	$\rightarrow$	$C_{60}K$
Color	Bronze		Steel-blue	Dark blue		Black		Black

Apart from potassium, other alkali metals can also be intercalated into graphite lattice. Such compounds are highly reactive. They explode in water, giving the corresponding hydroxide, hydrogen and graphite.

A number of halides (for example, cadmium chloride), sulfides (for example, lead sulfide), metal oxides (for example, chromium(VI) oxide) and non-metal oxides (for example, sulfur trioxide) too form intercalation compounds. The composition of iron(III) chloride intercalate is roughly C_6FeCl_3. The iron(III) chloride molecules form a layer within the sheets of graphite host. As in the case of potassium intercalates, several intermediate lamellar compounds are formed at various stages. At each stage, charge is transferred from graphite to the guest molecule. Thus, the iron(III) chloride molecule acts as a Lewis acid. For example, intermediate compounds of the type $C^{+}_{70}Cl^{-}.FeCl_2.5FeCl_3$ have been identified when iron(III) chloride intrudes into graphite layers. With aluminium(III) chloride, intercalates with the composition C^{+}_{27} $AlCl_4^{-}.2AlCl_3$ are formed. These compounds contain positive holes in the valence band of graphite and are, therefore, conducting in nature.

The reaction of graphite with oxygen or fluorine gives non-conducting intercalation compounds. Graphite oxide is formed when graphite is treated with strong oxidizing agents such as concentrated nitric acid or potassium permanganate. It is a non-stoichiometric compound with approximate composition, C_2O. The oxygen forms bridging C—O—C linkages between layers of carbon atoms. All four carbon electrons are now involved in bonding and for this reason, the compound becomes non-conducting. Further, the graphite sheets do not remain planar but tend to be buckled. With fluorine, compounds of the type, CF, C_2F and C_4F are formed. These too have buckled layers and are non-conducting in nature.

3. Fullerenes

Fullerenes or, to be more precise, *buckminsterfullerenes*, are crystalline, allotropic forms of carbon. However, rather than clubbing them in the section on carbon allotropes, these merit a separate discussion. There are many reasons for this assertion. First, fullerenes are the latest additions to the allotropic forms of carbon. These were first synthesized in 1985. Second, analogous to graphite lamellar compounds, several foreign moieties may be intercalated in the cavities of fullerenes. Each intercalated species is important from a physicochemical and/or biochemical point of view. Third, fullerenes have a wide spectrum of applications. It is predicted that in times to come, their applications will increase manifold. Last, since most of them are non-toxic in nature, they will be easily acceptable and accessible to society.

The first fullerene to be synthesized had 60 carbon atoms, joined in a way to give rise to a symmetrical, round object, shaped like a soccer ball, hence the nickname *buckyball*. Its composition is given as C_{60}. The structure of this molecule is shown in Fig. 16.3.

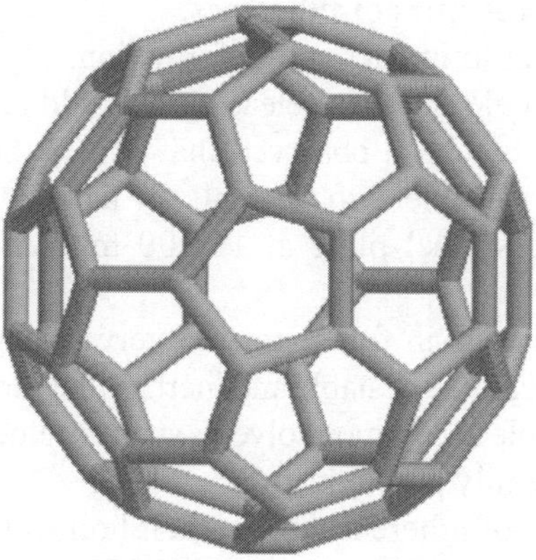

Fig. 16.3 A C_{60} fullerene

In this structure, hexagons and pentagons of carbon are linked together in a coordinated fashion to form a hollow, geodesic dome with bonding strains equally distributed among 60 carbon atoms. Some of the electrons are delocalized over the entire molecule. As a result, the molecule is conducting in nature. The C_{60} buckyball is extremely stable and can survive high temperatures and shock treatments.

Over a period of time, several other fullerenes came to be known. Those consisting of 70, 84 and 540 carbon atoms linked together were designated as C_{70}, C_{84} and C_{540}, respectively. Nevertheless, the C_{60} derivative remains the most important, and the most investigated, molecule of the fullerene family. The structure of C_{540} is shown in Fig. 16.4.

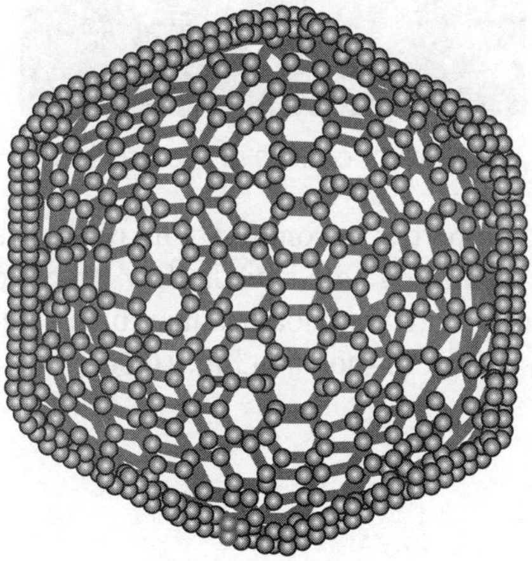

Fig. 16.4 A C_{540} fullerene

A common method used to produce fullerenes is to send a large current between two adjacent graphite electrodes in an inert atmosphere. On cooling, a sooty residue results from which many fullerenes may be isolated.

The geometrical shape of fullerenes is referred to as *geodesics*, or polyhedra that approximate a sphere. This unique shape is responsible for several interesting physical and chemical properties of fullerenes.

Fullerenes are resilient to impact and deformation. This means that squeezing a geodesic-shaped molecule and then releasing it would result in its popping back to its original shape. It has been observed that if it is thrown against an object it would bounce back, just like a soccer ball. In fact, a fullerene can withstand slamming on a stainless steel plate at 15,000 mph, merely bouncing back, unharmed.

The carbon–carbon bonds in fullerenes are very strong and this makes these carbon allotropes chemically very stable and inert. Fullerenes are the only allotropes of carbon which are soluble in common solvents such as toluene and carbon disulfide. Their solutions are generally purple in color.

An interesting feature of fullerenes is that their hollow structure allows them to hold other chemical species, such as drugs or metal ions inside them. The trapped moieties may be carried to the sites where these are desired. Figure 16.5 shows the presence of cavities in the structure of fullerene.

Fig. 16.5 Presence of cavities in the structure of fullerene

Fullerenes have two properties in common with carbon nanotubes: both are an assemblage of carbon atoms and both have hollow structures. In fact, nanotubes are cylindrical fullerenes. Carbon nanotubes may be single-walled, as shown in Fig. 16.6a or multiple-walled, as shown in Fig. 16.6b.

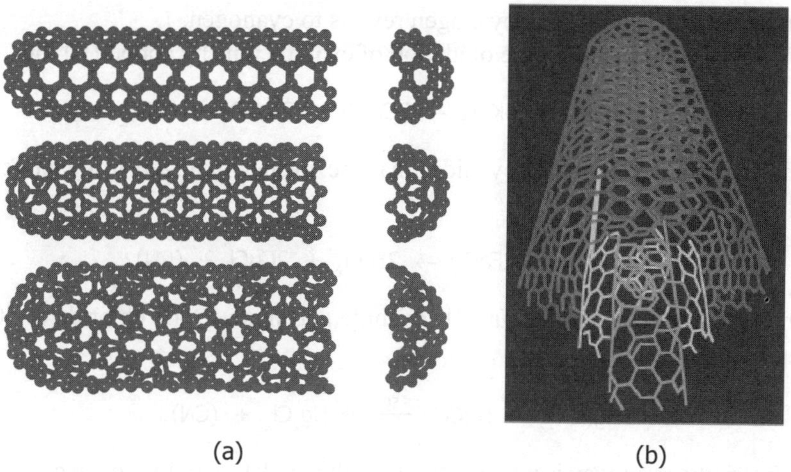

Fig. 16.6 Carbon nanotubes (a) Single-walled and (b) multiple-walled

Carbon nanotubes possess extraordinary properties, such as high tensile strength, high electrical conductivity, high ductility, high resistance to heat, and relative chemical inactivity.

Fullerenes are used as diagnostic tools for conducting MRI scans. They also act as drug carriers inside the body and help cure the patient faster. In the field of engineering, fullerenes are used in construction of scanning microscopes and electrical circuits. Fullerenes have the ability to adhere metals on to themselves and carry them close to other molecules during a chemical reaction. This property is responsible for fullerenes acting as catalysts in reactions.

On interacting with alkali metal ions, fullerenes start showing their metallic properties. Thus a C_{60} fullerene, containing potassium ions in its cavity behaves as a superconductor at 18 K. The conducting properties of fullerenes are used to make nanoscale wires. These wires exhibit electrical conductivity as high as copper, thermal conductivity as high as diamond and strength 100 times greater than steel.

4. Cyanogen

Cyanogen $(CN)_2$ is a colorless, poisonous gas which, when pure, is thermally very stable. However, in the presence of trace impurities, and at 500–700 K, it polymerizes to paracyanogen. Paracyanogen is a dark-colored solid having a condensed polycyclic structure (Fig. 16.7).

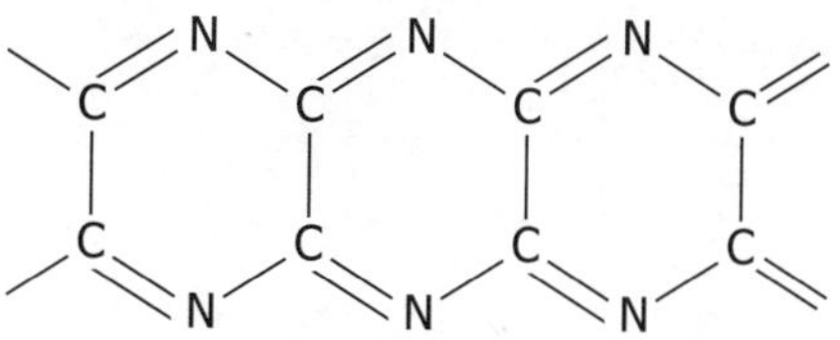

Fig. 16.7 Structure of paracyanogen

On heating to 1000 K, paracyanogen reverts to cyanogen.

Cyanogen is prepared by the oxidation of cyanide ions by copper(II) ions.

$$2CuSO_4 + 4KCN \rightarrow 2CuCN + 2K_2SO_4 + (CN)_2$$

The byproduct, copper(I) cyanide, may be further oxidized to cyanogen by iron(III) chloride.

$$2CuCN + 2FeCl_3 \rightarrow 2FeCl_2 + 2CuCl + (CN)_2$$

The reaction between mercury(II) cyanide and mercury(II) chloride at 500 K also produces cyanogen.

$$Hg(CN)_2 + HgCl_2 \xrightarrow{500\ K} Hg_2Cl_2 + (CN)_2$$

Industrially, cyanogen is manufactured by the oxidation of hydrogen cyanide by air over silver catalyst,

$$4HCN + O_2 \xrightarrow{\text{Ag catalyst}} 2H_2O + 2(CN)_2$$

or by the reaction of chlorine over activated charcoal.

$$2HCN + Cl_2 \xrightarrow{\text{Activated charcoal}} 2HCl + (CN)_2$$

Cyanogen has a linear structure.

$$N \equiv C–C \equiv N$$

The carbon–carbon bond length is shorter than expected, indicating delocalization of π-electrons over the entire framework of the molecule.

Cyanogen reacts with aqueous alkalis to yield a mixture of cyanide and cyanate.

$$2NaOH + (CN)_2 \rightarrow NaCN + NaCNO + H_2O$$

17. GROUP 15 ELEMENTS

Introduction

Group 15 elements—nitrogen (N), phosphorus (P), arsenic (As), antimony (Sb) and bismuth (Bi)— have the general configuration, ns^2np^3, with an unpaired electron in each of the p orbitals. The elements exhibit a maximum oxidation state V. However, on moving down the group, oxidation state III becomes more stable due to inert pair effect.

Physical Properties

The metallic character of Group 15 elements increases down the group. N and P are non-metals. As forms brittle steel-gray crystals, which have metallic appearance. However, it is non-ductile and non-conducting. Therefore, it is considered as a metalloid. Likewise, the crystals of Sb, no doubt, are lustrous, but it too is considered a metalloid. Bi is a true metal. Its crystals show metallic luster as well as appreciably high conductivity. It forms bismuth(III) ions and its oxide is basic in nature.

1. Bond strength

In its natural state, N is a gas, while all other elements of this group are solids. Gaseous nitrogen is inert since extremely high energy is required to break the triple bond holding the constituent atoms, $N\equiv N$. The symmetrical electron distribution also contributes to the stability and hence, inertness of nitrogen. Another factor responsible for the lack of reactivity of nitrogen is the absence of bond polarity. Carbon monoxide too has a triple bond, but is relatively more reactive because of its polar nature. Due to its non-reactive nature, nitrogen has high natural abundance—it is present to the extent of 78.1% in the atmosphere. In fact, it is the most abundant uncombined element available to man.

Phosphorus exists as P_4 units. Diatomic phosphorus (P_2) does exist, but unlike nitrogen, is unstable. This may be explained on the basis of relative strength of single and double bonds. For nitrogen, the triple bond is preferred because it has more than thrice the energy of the single bond. For phosphorus, the triple bond energy is less than three single bond energies. Therefore, a P_4 molecule, where each phosphorus is bound to three atoms of its own type through single bonds is more stable than a P_2 molecule, where there is a triple bond between two phosphorus atoms. Arsenic also exists as As_4 units.

2. Allotropic forms of phosphorus

There are many allotropic forms of phosphorus. The following three are most important:

a. **White phosphorus**

The most common allotropic form of phosphorus, having a tetrahedral structure (Fig. 17.1), is the white phosphorus. It is also the most volatile, most reactive and thermodynamically the least stable form of the element. It inflames in air and for this reason is generally stored under water. It is the slow phosphorescent oxidation of the vapors above this waxy substance that gave phosphorus its name. White phosphorus is insoluble in water, but soluble in carbon disulfide, benzene and liquid ammonia. It is highly toxic in nature.

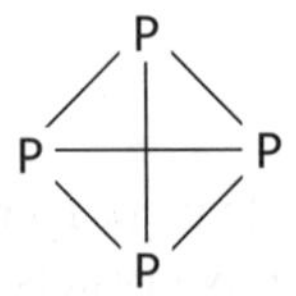

Fig. 17.1 Structure of white phosphorus

b. **Red phosphorus**

When white phosphorus is heated in the absence of air to about 500 K for several hours, it is converted into red phosphorus. When compared to white phosphorus, the red form is more dense and has a higher melting point. It is also much less reactive and, therefore, easy to handle. It has a polymeric structure formed by the union of several tetrahedral P_4 units (Fig. 17.2). During polymerization, one P–P bond in each unit breaks down. Red phosphorus is insoluble in organic solvents. It is non-toxic in nature.

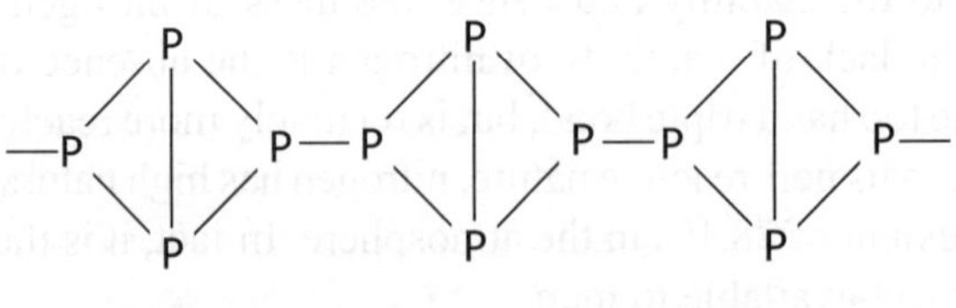

Fig. 17.2 Structure of red phosphorus

c. **Black phosphorus**

When white phosphorus is heated to 500 K, under a pressure of 12,000 atmosphere, it is transformed into black phosphorus. Black phosphorus has a layered structure, each layer being made up of a hexagonal network of phosphorus atoms (Fig. 17.3). It is thermodynamically the most stable form of phosphorus. It possesses luster, but has very low conductivity. The degree of polymerization is higher than that of red phosphorus. Accordingly, its density is also higher than that of the red allotropic form.

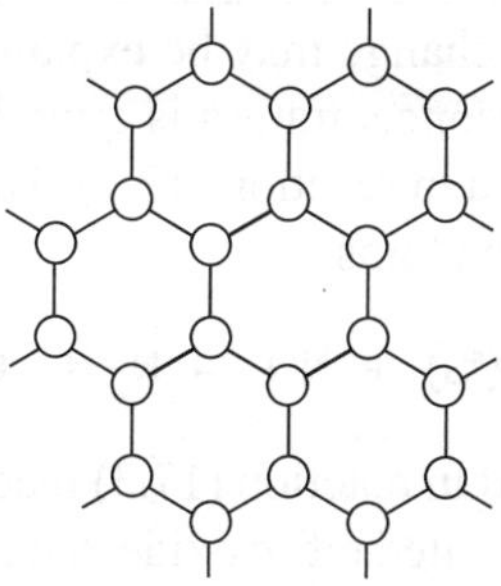

Fig. 17.3 Structure of black phosphorus

Box 17.1

The Nitrogen Afterglow

When an electric discharge is passed through nitrogen at a pressure of 1–2 mm Hg, atomic nitrogen, also called *active nitrogen*, is formed. The process is accompanied by a peach-yellow emission. This emission is called afterglow, as it persists for several minutes after the discharge has been stopped.

$$N_2 \xrightarrow{\text{Electric discharge}} 2N\ (^4S) \qquad (17.1)$$

The term S signifies that nitrogen atoms are in ground state. The superscript 4 refers to the spin multiplicity of nitrogen atom. Mathematically,

$$\text{Spin multiplicity} = 2ns + 1$$

where n is the number of unpaired electrons and s is the spin quantum number, ½. For a nitrogen atom, having the configuration,

$$1s^2 2s^2 2p_x^1 2p_y^1 2p_z^1$$

n = 3 and spin multiplicity = 4.

The afterglow is not unique to nitrogen alone. Hydrogen also shows the afterglow, but it ends with the discharge. With oxygen, it lasts for an infinitesimally brief period after the discharge has been cut-off. However, nitrogen atoms, in ground state, have a relatively long lifetime. Their recombination to yield first the excited nitrogen molecule and subsequently ground state molecular nitrogen is quite slow.

$$N\ (^4S) + N\ (^4S) \xrightarrow{M} N_2^* \rightarrow N_2 + h\upsilon\ \text{(peach-yellow)} \qquad (17.2)$$

The recombination takes place in the presence of a third body, M, which is either the surface of the reaction vessel or a molecule of nitrogen.

The concentration of the active nitrogen atom, N (^{4}S) may be determined by induction of nitric oxide gas, NO, in the discharge tube. The color of the afterglow changes successively from peach-yellow to (i) purple; (ii) blue; (iii) darkness; and (iv) yellowish-green.

Contd...

The mechanism of color change may be explained as follows: when nitric oxide is initially added, its concentration is quite low as compared to that of active nitrogen atoms formed in reaction (17.1). The reaction between the two species yields active oxygen atoms.

$$N\ (^4S) + NO \rightarrow N_2 + O \qquad (17.3)$$

The oxygen atoms formed in reaction (17.3) react with the remaining active nitrogen atoms to produce excited nitric oxide that passes on to the ground state with the emission of blue light.

$$N\ (^4S) + O \rightarrow NO^* \rightarrow NO + h\upsilon\ (\text{blue}) \qquad (17.4)$$

The mixture of peach-yellow and blue afterglow appears purple in color.

When the concentration of nitric oxide increases further, the concentration of active nitrogen atoms concomitantly reduces. Hence, the peach-yellow emission fades while the blue emission persists. A stage comes when the concentration of nitric oxide becomes equivalent to that of active nitrogen atoms. At this stage, all the active nitrogen atoms are eliminated. Neither reaction (17.2) nor reaction (17.4) is now possible. There is no afterglow. The darkness marks the end point of the reaction.

If the concentration of nitric oxide is increased still further, it combines with the active nitrogen atoms that are being produced in the discharge tube. This reaction is the same as that given in equation (17.3), the only difference is that the concentration of nitric oxide is now very high as compared to that of active nitrogen atoms. The active oxygen atoms do not get an opportunity to combine with active nitrogen atoms. Rather, they react with excess nitric oxide to yield excited nitrogen dioxide, which emits a yellowish-green light during its transformation to the ground state.

$$NO + O \rightarrow NO_2^* \rightarrow NO_2 + h\upsilon\ (\text{yellowish-green}) \qquad (17.5)$$

Chemical Properties

N and P essentially form covalent compounds. As, Sb and Bi show increasing tendency to cationic behavior. Due to its small size, N tends to form pπ–pπ bonds. No such bonds exist in the corresponding P compounds. P is appreciably larger than N and, therefore, lateral overlap involving p orbitals is not possible. On the other hand, there are weak-to-moderate dπ–pπ bonds in P compounds. These arise due to the lateral overlap of d orbitals of P with p orbitals of other atoms such as O or N. The d orbitals have more spatial extension than p orbitals and hence can enter into π-bond formation through lateral overlaps. N does not have d orbitals, as a result of which it does not show dπ–pπ bonding. Further, P possesses coordination number 5 or 6. The maximum coordination number possible for N is 4.

Case Study 17.1 Which is more reactive: trimethylamine oxide or trimethylphosphine oxide?

Trimethylamine oxide, $(CH_3)_3NO$, may be represented as:

$$(CH_3)_3N^+—O^-$$

However, in the corresponding phosphorus compound, two canonical forms are possible.

$$(CH_3)_3P^+—O^- \longleftrightarrow (CH_3)_3P{=}O$$

The multiple-bonded structure arises because phosphorus accepts charge density from oxygen and accommodates it in its vacant d orbitals. Nitrogen does not possess d orbitals and, therefore, cannot form the π-bonded structure. Thus when the nitrogen atom of a molecule donates its lone pair, there is no possibility of back donation. However, when phosphorus—or for that matter, even arsenic—donates an electron pair, there is a definite possibility of back bonding through dπ–pπ overlap.

Trimethylphosphine oxide is stable, first, because it has two resonating structures, and second, because the phosphorus–oxygen double bond requires very high energy to dissociate. Trimethylamine oxide, on the other hand, is reactive.

1. Reaction with air

Nitrogen forms nitric oxide when heated in air to about 3300 K.

$$N_2 + O_2 \xrightarrow{3300\ K} 2NO$$

When phosphorus is heated in a limited supply of air, it yields phosphorus(III) oxide, P_4O_6,

$$\underset{\text{Limited}}{P_4 + 3O_2} \rightarrow P_4O_6$$

and when heated in excess air, it yields phosphorus(V) oxide, P_4O_{10}.

$$\underset{\text{Excess}}{P_4 + 5O_2} \rightarrow P_4O_{10}$$

Both phosphorus(III) oxide and phosphorus(V) oxide have tetrahedral structures (Fig. 17.4). In phosphorus(III) oxide, an oxygen atom is trapped between each pair of phosphorus atoms. In phosphorus(V) oxide, an additional oxygen atom is bound to each phosphorus.

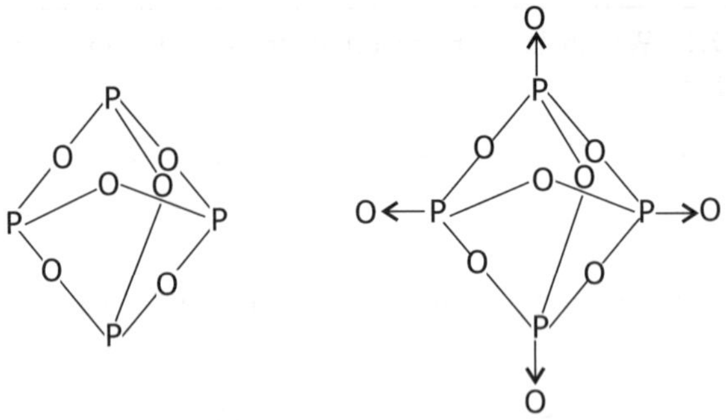

(a) (b)

Fig. 17.4 Structure of (a) phosphorus(III) oxide and (b) phosphorus(V) oxide

When heated in air, arsenic yields arsenic(III) oxide.

$$As_4 + 3O_2 \rightarrow 2As_2O_3$$

In the vapor state it exists as a dimer, As_4O_6.

Antimony(III) oxide is also obtained by direct reaction of the element with oxygen.

$$4Sb + 3O_2 \rightarrow 2Sb_2O_3$$

It also dimerizes in the vapor state to Sb_4O_6.

Bismuth(III) oxide may be prepared by directly heating bismuth in air. It is a basic oxide.

$$4Bi + 3O_2 \rightarrow 2Bi_2O_3$$

2. Reaction with halogens

Nitrogen does not react directly with halogens. The remaining elements form two series of halides—the trihalides and the pentahalides—by reacting directly with halogens. The trihalides may be prepared by the reaction of the elements with relevant halogen in stoichiometric ratios.

$$P_4 + 6X_2 \rightarrow 4PX_3$$
$$As_4 + 6X_2 \rightarrow 4AsX_3$$
$$2Sb + 3X_2 \rightarrow 2SbX_3$$
$$(X = F, Cl \text{ or } Br)$$

Bismuth trihalides cannot be prepared by this method. Phosphorus(III) fluoride too is preferably obtained by the reaction between phosphorus(III) chloride and arsenic(III) fluoride.

$$AsF_3 + PCl_3 \rightarrow AsCl_3 + PF_3$$

Direct reaction between phosphorus and fluorine is explosive in nature.

The pentahalides too may be prepared by direct union of elements.

$$P_4 + 10X_2 \rightarrow 4PX_5 \text{ (X = F, Cl or Br)}$$

The pentaiodide does not exist because iodide ion reduces phosphorus(V) to phosphorus(III), leading to the formation of phosphorus(III) iodide.

The only pentahalides that arsenic and bismuth form are the pentafluorides.

$$As_4 + 10F_2 \rightarrow 4AsF_5$$

$$2Bi + 5F_2 \rightarrow 2BiF_5$$

Antimony does not form pentahalides by direct reactions with halogens.

All trihalides are pyramidal molecules. All pentahalides are trigonal bipyramids. However, X-ray evidence reveals that phosphorus pentachloride is ionic and contains tetrahedral $[PCl_4]^+$ and octahedral $[PCl_6]^-$ units. Solid phosphorus pentabromide is made up of $[PBr_4]^+$ $[Br]^-$ ions.

Case Study 17.2 Comment on the non-existence of nitrogen(V) chloride and bismuth(V) chloride and on the instability of arsenic(V) chloride.

Since phosphorus(V) chloride has a trigonal bipyramid structure, the nitrogen(V) chloride, had it existed, would have had the same structure. However, unlike phosphorus, nitrogen does not possess d orbitals and thus, it can neither expand its octet nor can it enter into sp^3d hybridization. Moreover, coordination of five chlorine atoms around the small nitrogen atom would lead to steric over-crowdedness.

Bismuth(V) chloride does not exist because of inert pair effect. Bismuth(III) is more stable than bismuth(V) since the s electron pair tends not to participate in bonding. No doubt, bismuth(V) fluoride exists, and is stable. This is due to the high electronegativity of fluorine which can bring out the highest oxidation state of the metal.

Arsenic(V) chloride is extremely unstable. At one time its very existence was doubted. The highest oxidation state of p-block elements following the completion of the first transition series (3d series) is generally unstable. The 3d series traverses between phosphorus and arsenic. Due to the imperfect shielding of the nucleus by the penultimate d orbital, the valence electrons of arsenic are held very tightly. It is, therefore, not easy to involve $4s^2$ electrons in bonding.

---

3. Reaction with sulfur

Phosphorus and sulfur combine with each other when heated above 473 K to give a number of products, the important ones being, tetraphosphorus trisulfide (P_4S_3), tetraphosphorus pentasulfide (P_4S_5) and tetraphosphorus decasulfide (P_4S_{10}). Any of these compounds may be obtained in significant yield by heating red phosphorus with stoichiometric amounts of sulfur at a specific temperature. All the sulfides have tetrahedral skeletons (Fig. 17.5). Tetraphosphorus trisulfide is the most stable analog.

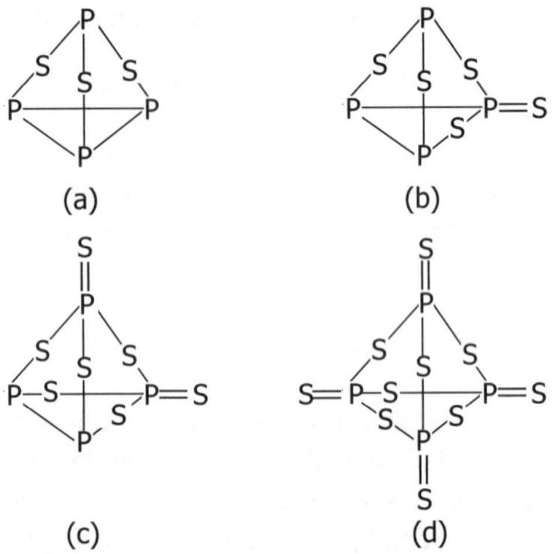

Fig. 17.5 Structure of (a) tetraphosphorus trisulfide (b) tetraphosphorus pentasulfide (c) tetraphosphorus heptasulfide and (d) tetraphosphorus decasulfide

Arsenic forms the following sulfides when heated with sulfur: tetraarsenic trisulfide (As_4S_3), tetraarsenic tetrasulfide (As_4S_4), diarsenic trisulfide (As_2S_3) and diarsenic pentasulfide (As_2S_5).

By direct combination with sulfur, antimony forms diantimony trisulfide (Sb_2S_3) and diantimony pentasulfide (Sb_2S_5).

Bismuth(III) sulfide, Bi_2S_3, cannot be prepared by direct combination of elements, but by reaction of hydrogen sulfide with aqueous bismuth(III) ions. For example,

$$2BiCl_3 + 3H_2S \rightarrow Bi_2S_3 + 6HCl$$

However, when bismuth and sulfur are heated directly to about 1500 K, under a pressure of 50 atmosphere, a disulfide, BiS_2, is obtained. Its precise composition is $Bi^{3+}[BiS_4]^{3-}$.

Case Study 17.3 Which chemical constituent of safety matches is replaced by another species in strike-anywhere matches?.

Safety matches can be struck only against the side of the box that contains the match sticks. The side surface of the box is coated with a mixture of red phosphorus, calcium carbonate, starch and powdered glass. The head of the match stick is composed of a mixture of potassium chlorate, sulfur, calcium carbonate, starch and powdered glass. When the match is struck, the heat generated during friction converts the red phosphorus coated on the box to white phosphorus. The latter ignites spontaneously in air.

$$P_4 + 5O_2 \rightarrow P_4O_{10}$$

The heat released due to the above reaction causes potassium chlorate to decompose.

$$2KClO_3 \rightarrow 2KCl + 3O_2$$

The sulfur then catches fire and ignites the wood.

$$S + O_2 \rightarrow SO_2$$

Strike-anywhere matches can be ignited by rubbing the head of the match stick on any frictional surface. For this reason, the chemicals that are coated on the box of safety matches are irrelevant for strike-anywhere matches. Nevertheless, phosphorus, in one form or the other, is required for lighting the match. Instead of the red phosphorus that is coated on the box of the safety matches, tetraphosphorus trisulfide (Fig. 17.5a), also called phosphorus sesquisulfide, is incorporated into the head of match stick. The remaining ingredients on the match stick are same as those in safety matches.

Phosphorus sesquisulfide easily ignites on friction and burns in air to yield phosphorus(V) oxide and sulfur dioxide.

$$P_4S_3 + 8O_2 \rightarrow P_4O_{10} + 3SO_2$$

The heat released in the above reaction causes potassium chlorate to decompose and sulfur to catch fire, much the same way as in safety matches

4. Reaction with metals

Nitrogen combines directly with metals to give nitrides. With highly electropositive metals, the compounds are ionic in nature and contain nitride anion, N^{3-}.

$$6Li + N_2 \rightarrow 2Li_3N$$

$$3Mg + N_2 \rightarrow Mg_3N_2$$

Similarly, phosphorus combines directly with alkali, alkaline earth and some of the lanthanoid metals to give ionic phosphides.

$$12Na + P_4 \rightarrow 4Na_3P$$

$$6Ca + P_4 \rightarrow 2Ca_3P_2$$

Arsenic, antimony and bismuth combine directly with alkali metals to yield arsenides, antimonides and bismuthites with the composition, M_3E (M = alkali metal, E = As, Sb or Bi). For example,

$$3Na + As \rightarrow \underset{\text{Sodium arsenide}}{Na_3As}$$

With alkaline earth metals, compounds of the composition, M_3E_2 are obtained. For example,

$$3Ca + 2Bi \rightarrow \underset{\text{Calcium bismuthite}}{Ca_3Bi_2}$$

Case Study 17.4 Which chemical species containing only nitrogen atoms are known?

There are three chemical species containing nitrogen and no other element. These are: molecular nitrogen, N_2; azide ion, N_3^-; and pentanitrogen cation, N_5^+. The last one is most unstable. It is prepared by the following reaction.

$$[N_2F]^+ [AsF_6]^- + HN_3 \rightarrow [N_5]^+[AsF_6]^- + HF$$

The cation is not stable at room temperature but can be preserved for long durations at 195 K.

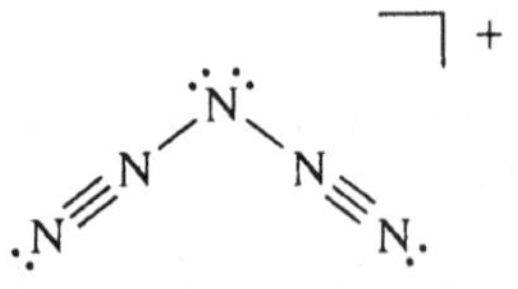

It has an angular structure, bent at the central nitrogen and linear at the neighboring atoms.

Compounds

The trihalides of Group 15 elements give different products on hydrolysis.

$$NCl_3 + 4H_2O \rightarrow NH_4OH + 3HOCl$$

Hypochlorous acid

$$PCl_3 + 3H_2O \rightarrow H_3PO_3 + 3HCl$$

Phosphorus acid

$$AsCl_3 + 3H_2O \rightarrow H_3AsO_3 + 3HCl$$

Arsenious acid

$$SbCl_3 + H_2O \rightarrow SbOCl + 2HCl$$

Antimony oxychloride

$$BiCl_3 + H_2O \rightarrow BiOCl + 2HCl$$

Bismuth oxychloride

All the elements form gaseous hydrides of the composition, MH_3. Their stability decreases down the group. Ammonia is prepared by *Haber–Bosch process,* by reacting nitrogen and hydrogen in the molar ratio of 1:3, at 700 K, under a pressure of 200 atmosphere, in the presence of iron catalyst.

$$N_2 + 3H_2 \xrightarrow[700\ K/200\ atm]{Fe\ catalyst} 2NH_3$$

Phosphine, PH_3, is prepared by the hydrolysis of calcium phosphide, Ca_3P_2.

$$Ca_3P_2 + 6H_2O \rightarrow 3Ca(OH)_2 + 2PH_3$$

The reduction of arsenious acid with zinc and dilute hydrochloric acid yields arsine, AsH_3.

$$3Zn + H_3AsO_3 + 6HCl \rightarrow 3ZnCl_2 + 3H_2O + AsH_3$$

Similarly, stibine, SbH_3 is prepared using antimonous acid, H_3SbO_3.

$$3Zn + H_3SbO_3 + 6HCl \rightarrow 3ZnCl_2 + 3H_2O + SbH_3$$

Bismuthine, BiH_3 may be prepared by the reduction of bismuth(III) chloride with sodium borohydride.

$$4BiCl_3 + 3NaBH_4 \rightarrow 4BiH_3 + 3NaCl + 3BCl_3$$

Case Study 17.5 What is 880 ammonia?

Ammonia is highly soluble in water. At ordinary temperatures and pressures, about 50 g ammonia dissolves in 100 g water. This reagent, marked as *aqueous ammonia* or *50% ammonia* is normally used for conventional chemistry experiments. The density of 50% ammonia solution is 0.880 g cm^{-3}. Hence, sometimes it is also referred to as 880 ammonia.

---

1. Oxides of nitrogen

Nitrogen exhibits a wide range of oxidation states, from I to V, in its oxides. Table 17.1 lists nitrogen oxides along with their structures and methods of preparation. There is a pπ–pπ multiple bonding between nitrogen and oxygen in all the oxides.

The synthesis of nitrous oxide, by thermal decomposition of ammonium nitrate, is an example of *conproportionation reaction*. The conproportionation concept has already been discussed in Case Study 8.10, p. 303, wherein it was stated that in such reactions a high and a low oxidation state of an element combine to give an intermediate oxidation state. In the present case, the reactant, ammonium nitrate has two nitrogen atoms. The oxidation state of ammonium nitrogen is (–III), while that of nitrate nitrogen is (V). The nitrogen in the product, nitrous oxide has an intermediate oxidation state of (I).

Nitrous oxide is used as an anesthetic by dentists. It is inert at 300 K, but at 773 K it decomposes into oxygen.

$$4N_2O \xrightarrow{773\,K} 3N_2 + O_2 + 2NO$$

It, therefore, supports combustion better than air. It is a neutral oxide.

Table 17.1 Oxides of nitrogen

Name	Composition	Oxidation state of nitrogen	Method of preparation	Structure
Nitrous oxide	N_2O	I	$NH_4NO_3 \xrightarrow{500\ K} 2H_2O + N_2O$	$\overset{-}{N}=\overset{+}{N}=O \longleftrightarrow N\equiv\overset{+}{N}-O^-$
Nitric oxide	NO	II	$3Cu + 8HNO_3 \rightarrow 3Cu(NO_3)_2 + 4H_2O + 2NO$	$N\overset{\cdot}{=\!=}O$
Dinitrogen trioxide	N_2O_3	III	$NO + NO_2 \xrightarrow{240\ K} N_2O_3$	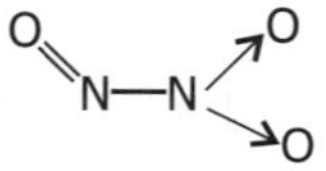
Nitrogen dioxide	NO_2	IV	$2Pb(NO_3)_2 \xrightarrow{673\ K} 2PbO + 4NO_2 + O_2$	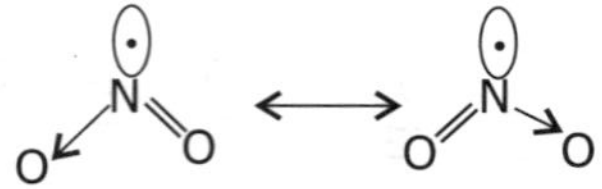
Dinitrogen tetraoxide	N_2O_4	IV	$NO_2 + NO_2 \xrightarrow{250\ K} N_2O_4$	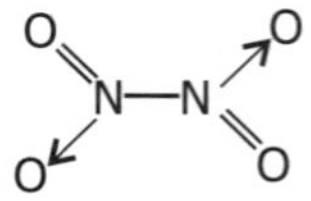
Dinitrogen pentaoxide	N_2O_5	V	$4HNO_3 + P_4O_{10} \xrightarrow{260\ K} 2N_2O_5 + 4HPO_3$	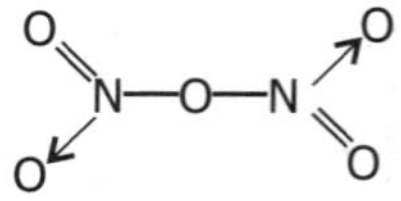

Nitric oxide is also a neutral oxide. It is a colorless gas, slightly heavier than air and sparingly soluble in water. It decomposes above 1273 K to nitrogen and oxygen and, therefore, supports combustion of only fiercely burning substances such as magnesium and sulfur. It is paramagnetic in nature, having one unpaired electron.

Dinitrogen trioxide is acidic in nature; it is the anhydride of nitrous acid, HNO_2.

$$N_2O_3 + H_2O \rightarrow 2HNO_2$$

Nitrogen dioxide is brown and paramagnetic. At low temperatures, it dimerizes to dinitrogen tetraoxide, which is colorless and diamagnetic. Dinitrogen tetraoxide is a mixed anhydride, as on hydrolysis it yields both nitrous acid and nitric acid.

$$N_2O_4 + H_2O \rightarrow HNO_3 + HNO_2$$

Dinitrogen pentaoxide is the anhydride of nitric acid.

$$N_2O_5 + H_2O \rightarrow 2HNO_3$$

With hydrogen peroxide, it gives pernitric acid, $HOONO_2$.

$$N_2O_5 + H_2O_2 \rightarrow HNO_3 + HOONO_2$$

It is an excellent oxidizing agent. For example, it oxidizes iodine to iodine pentaoxide.

$$5N_2O_5 + I_2 \rightarrow I_2O_5 + 10NO_2$$

Case Study 17.6 Why is it advisable to use gasoline mixed with nitrous oxide for racing cars?

In racing cars, the accelerator is required to be pressed to the maximum repeatedly. Under these circumstances, the engine acquires more power if liquid nitrous oxide is mixed with gasoline. This is attributed to the following reasons.

1. In the combustion chamber, liquid nitrous oxide is first converted into vapors. The latent heat of vaporization is acquired from the combustion chamber, as a result of which the engine cools down.
2. The vapors of nitrous oxide finally decompose into a mixture of nitrogen, oxygen and nitric oxide. In this decomposition reaction, which has been cited above, four moles of nitrous oxide decompose into six moles of gaseous products. As a result, the pressure inside the combustion chamber increases and, in turn, provides extra boost to the pistons.
3. Increased concentration of oxygen, as a result of decomposition of nitrous oxide, ensures more efficient fuel combustion.

A combination of these three factors increases the power output of the engine. However, both nitrous oxide and nitric oxide reduce the life of the catalytic convertor. Hence nitrous oxide is used as a gasoline additive only during short-term sports events.

---

2. Oxyacids of nitrogen

Following are some of the important oxyacids of nitrogen:

a. Hyponitrous acid

The oxidation state of nitrogen in hyponitrous acid, $H_2N_2O_2$ is I. The acid is prepared by the reaction of hydroxylamine and nitroethane in the presence of sodium ethoxide.

$$NH_2OH + C_2H_5NO_2 + 2C_2H_5ONa \rightarrow Na_2N_2O_2 + 3C_2H_5OH$$

$$Na_2N_2O_2 + \underset{\text{Dilute}}{2HCl} \rightarrow H_2N_2O_2 + 2NaCl$$

The acid is soluble in organic solvents. It is unstable and explosive, but its sodium salt is stable. The hyponitrite ion has a *trans* structure.

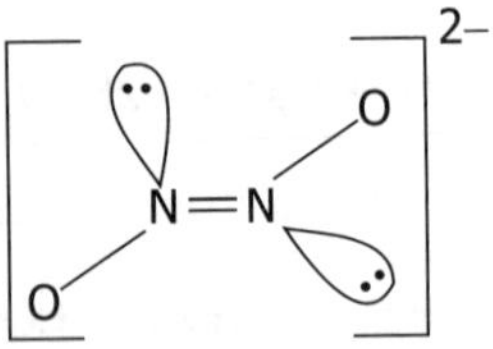

Hyponitrous acid is a reducing agent. It transforms iodine into hydrogen iodide, while oxidizing itself to nitric acid.

$$H_2N_2O_2 + 3I_2 + 3H_2O \rightarrow HNO_3 + HNO_2 + 6HI$$

b. Nitrous acid

In nitrous acid, the oxidation state of nitrogen is III. It is prepared by the reaction of barium nitrite with dilute sulfuric acid.

$$Ba(NO_2)_2 + H_2SO_4 \rightarrow 2HNO_2 + BaSO_4$$

Barium sulfate, being insoluble, is filtered off.

The structure of nitrous acid may be depicted as,

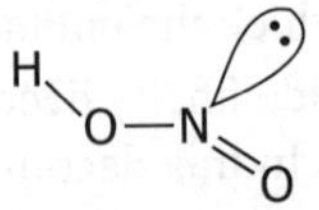

while that of nitrite ion may be indicated as,

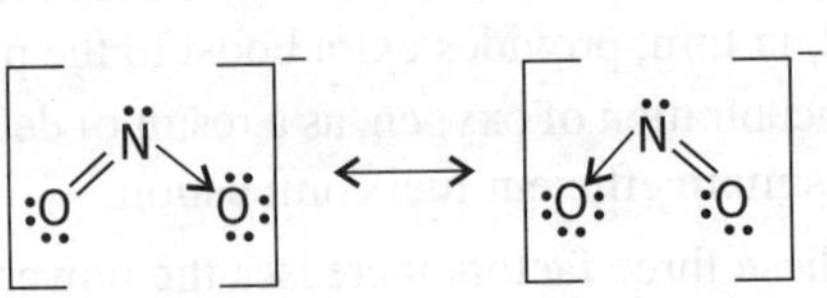

Nitrous acid is not very stable. It slowly converts into nitric acid.

$$3HNO_2 \rightarrow HNO_3 + H_2O + 2NO$$

It is a reducing agent. For example, it reduces silver bromate to silver bromide.

$$AgBrO_3 + 3HNO_2 \rightarrow AgBr + 3HNO_3$$

It is also a mild oxidizing agent. For example, it oxidizes hydrogen sulfide to sulfur,

$$2HNO_2 + H_2S \rightarrow 2H_2O + 2NO + S$$

and sulfurous acid to sulfuric acid.

$$2HNO_2 + H_2SO_3 \rightarrow H_2SO_4 + H_2O + 2NO$$

c. Nitric acid

The most important oxyacid of nitrogen is nitric acid. In this acid, nitrogen exhibits oxidation state V. Nitric acid is prepared by *Ostwald process.* The process initiates with the oxidation of ammonia to nitric oxide, using platinum, heated to 1000 K, as a catalyst.

$$4NH_3 + 5O_2 \xrightarrow{Pt/1000\ K} 6H_2O + 4NO$$

The nitric oxide is further oxidized to nitrogen dioxide,

$$2NO + O_2 \rightleftharpoons 2NO_2$$

which is then absorbed by water in the presence of excess oxygen.

$$4NO_2 + 2H_2O + O_2 \rightarrow 4HNO_3$$

The nitric acid so obtained has a concentration of about 60%. It is dehydrated with concentrated sulfuric acid and then distilled. Its concentration then increases to 98%. The structure of nitric acid may be depicted as,

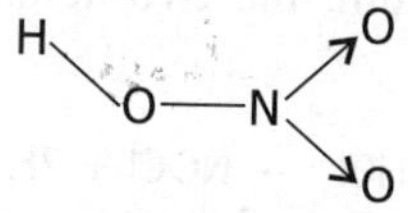

while that of nitrate ion may be represented in canonical forms.

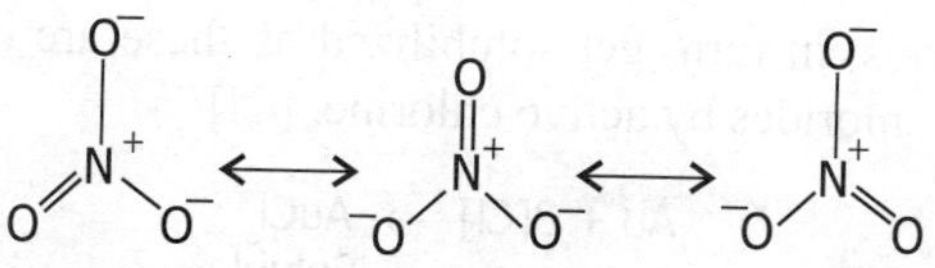

Nitric acid is a colorless liquid, but on standing it acquires a yellow tinge due to photochemical decomposition to nitrogen dioxide.

$$4HNO_3 \xrightarrow{h\nu} 4NO_2 + 2H_2O + O_2$$

Nitric acid is a strong oxidizing agent. It oxidizes phosphorus to phosphoric acid,

$$\underset{\text{Conc.}}{P_4 + 20HNO_3} \rightarrow 4H_3PO_4 + 4H_2O + 20NO_2$$

and iodine to iodic acid.

$$\underset{\text{Conc.}}{I_2 + 10HNO_3} \rightarrow 2HIO_3 + 4H_2O + 10NO_2$$

Concentrated nitric acid reacts with metals to yield nitrogen dioxide.

$$Cu + 4HNO_3 \rightarrow Cu(NO_3)_2 + 2H_2O + 2NO_2$$

$$Zn + 4HNO_3 \rightarrow Zn(NO_3)_2 + 2H_2O + 2NO_2$$

$$Pb + 4HNO_3 \rightarrow Pb(NO_3)_2 + 2H_2O + 2NO_2$$

The reaction of dilute nitric acid and metals generally yields nitric oxide.

$$3Cu + 8HNO_3 \rightarrow 3Cu(NO_3)_2 + 4H_2O + 2NO$$

$$3Pb + 8HNO_3 \rightarrow 3Pb(NO_3)_2 + 4H_2O + 2NO$$

However, zinc gives nitrous oxide on treatment with dilute nitric acid.

$$4Zn + 10HNO_3 \rightarrow 4Zn(NO_3)_2 + 5H_2O + N_2O$$

Magnesium reacts with cold and dilute nitric acid to yield hydrogen.

$$\underset{\text{Dilute}}{Mg + 2HNO_3} \rightarrow Mg(NO_3)_2 + H_2$$

A mixture of three parts concentrated hydrochloric acid and one part concentrated nitric acid is called *aqua regia.* In presence of noble metals, such as gold or platinum, the two acids react to yield active chlorine atoms.

$$HNO_3 + 3HCl \rightarrow \underset{\text{Nitrosyl chloride}}{NOCl} + 2H_2O + 2[Cl]$$

The metals, in turn, get solubilized as these are converted into their respective chlorides by active chlorine, [Cl].

$$Au + 3[Cl] \rightarrow \underset{\text{Soluble}}{AuCl_3}$$

$$Pt + 4[Cl] \rightarrow \underset{\text{Soluble}}{PtCl_4}$$

Case Study 17.7 Which products are expected to be formed when ammonia is burnt in air?

Theoretically, two reactions are possible when ammonia is burnt in air. In the first reaction, nitrogen and water vapors are produced.

$$4NH_3 + 3O_2 \rightarrow 2N_2 + 6H_2O \qquad \Delta G = -1306 \text{ kJ mol}^{-1} \qquad (17.6)$$

In the second reaction, nitric oxide and water vapors are produced.

$$4NH_3 + 5O_2 \rightarrow 4NO + 6H_2O \qquad \Delta G = -958 \text{ kJ mol}^{-1} \qquad (17.7)$$

Going by thermodynamic considerations, reaction (17.6) is more feasible as it involves a greater negative value of change in free energy. Yet, in the Ostwald process, cited above, we have proposed reaction (17.7) as the first step in the synthesis of nitric acid. It must be appreciated that apart from thermodynamic considerations, kinetic factors too are important in gauging the feasibility of chemical reactions. In the Ostwald process, the combustion of ammonia is carried out in presence of platinum, which serves as a catalyst. The catalyst lowers the activation energy of reaction (17.7) to a much greater extent than of reaction (17.6). Thus reaction (17.7), though thermodynamically less feasible than reaction (17.6), nevertheless prevails because in the presence of a catalyst, the kinetic factors overrule thermodynamic factors.

3. Hydrazine

Hydrazine, N_2H_4, is prepared by *Raschig process*. This process involves the oxidation of ammonia with aqueous sodium hypochlorite. The reaction takes place in two steps. In the first step, chloramine is produced.

$$NH_3 + NaOCl \xrightarrow{\text{Fast}} \underset{\text{Chloramine}}{NH_2Cl} + NaOH$$

In the second step, chloramine reacts with excess ammonia to yield hydrazine.

$$2NH_3 + NH_2Cl \rightarrow NH_2NH_2 + NH_4Cl$$

Glue or gelatin is added to the reaction mixture. It suppresses the side reactions, such as,

$$N_2H_4 + 2NH_2Cl \rightarrow 2NH_4Cl + N_2$$

Hydrazine is concentrated by distillation over sodium hydroxide.

The structure of hydrazine resembles that of hydrogen peroxide. The molecule has a relatively large dipole moment of 6.16×10^{-30} C m; that rules out the *trans* structure. Each nitrogen atom is tetrahedrally surrounded by one nitrogen atom, two hydrogen atoms and one lone pair of electrons. The molecule adopts a *gauche* conformation with an angle of rotation of 95° from a purely eclipsed position (Fig. 17.6).

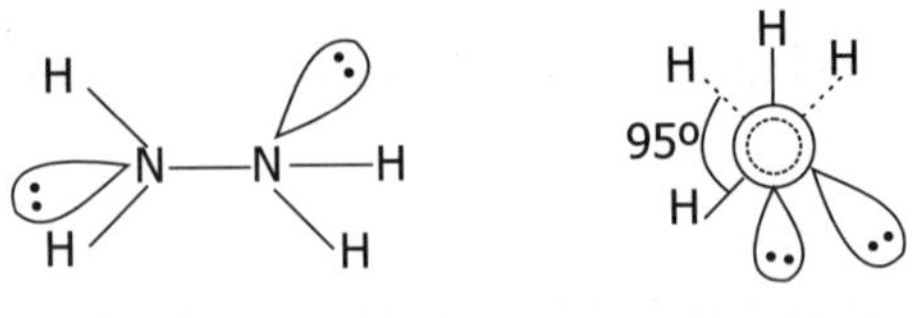

Fig. 17.6 Structure and conformation of hydrazine

As has already been discussed in Chapter 5 (*see* p. 242), hydrazine is a weaker base than ammonia. Each nitrogen in its structure possesses a lone pair and hence hydrazine is a diacidic base. It, therefore, forms two series of salts.

$$N_2H_4 + HCl \rightarrow N_2H_5^+ + Cl^-$$

$$N_2H_4 + 2HCl \rightarrow N_2H_6^{2+} + 2Cl^-$$

Hydrazine is an endothermic compound. It burns rapidly and completely in air with considerable evolution of heat.

$$N_2H_4 + \underset{\text{Air}}{O_2} \rightarrow N_2 + 2H_2O \; ; \; \Delta H = -620 \text{ kJ mol}^{-1}$$

For this reason, hydrazine is used as a rocket fuel in guided missiles, space shuttles and lunar missions.

It is a strong reducing agent, as is evident from the following reactions:

$$N_2H_4 + 2I_2 \rightarrow 4HI + N_2$$

$$N_2H_4 + 2S \rightarrow 2H_2S + N_2$$

$$N_2H_4 + 4CuO \rightarrow 2Cu_2O + 2H_2O + N_2$$

$$N_2H_4 + 4AgNO_3 \rightarrow 4Ag + 4HNO_3 + N_2$$

$$N_2H_4 + PtCl_4 \rightarrow Pt + 4HCl + N_2$$

$$N_2H_4 + 4FeCl_3 \rightarrow 4FeCl_2 + 4HCl + N_2$$

In the presence of a strong reducing agent, however, it acts as an oxidant.

$$N_2H_4 + Zn + 2HCl \rightarrow ZnCl_2 + 2NH_3$$

Case Study 17.8 Trimethylamine is pyramidal, while trisilylamine is plane triangular. Why?

Trimethylamine molecule has a pyramidal geometry, similar to that of ammonia, with a bond angle of about 108°. In case of trisilylamine, however, the lone pair of electrons on nitrogen enters the pure $2p_z$ orbital, while the nitrogen–silicon bonds are formed using sp_xp_y trigonal hybrid orbitals of the central atom. The lone pair on nitrogen prefers to occupy a pure p orbital because the empty 3d orbitals of silicon have appropriate symmetry and energy to interact with it and delocalize it over the entire system through dπ–pπ bonding.

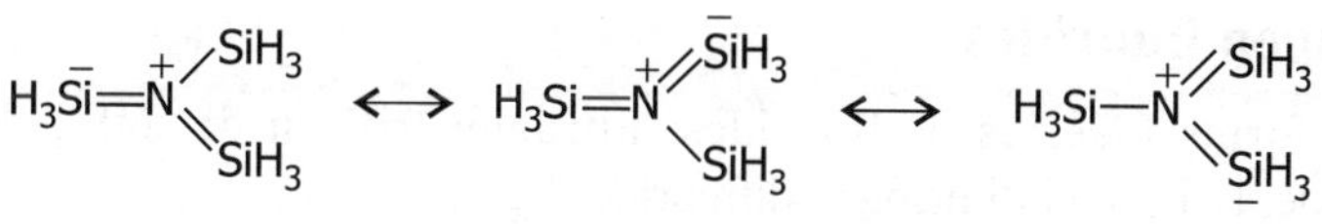

Additional bond strength, gained by dπ–pπ bonding, makes the molecule more stable and causes it to adopt a planar configuration. Carbon does not have d orbitals and for this reason, it is not possible for trimethylamine to attain a planar structure.

The delocalization of lone pair of electrons makes trisilylamine a weaker Lewis base than trimethylamine.

4. Hydroxylamine

Hydroxylamine, NH_2OH, is a weaker base than ammonia. It is more stable in the form of an acid salt, for example, either as hydroxylammonium chloride (commonly called hydroxylamine hydrochloride), $[NH_3OH]^+ Cl^-$,

$$NH_2OH + HCl \rightarrow [NH_3OH]^+ Cl^-$$

or as hydroxylammonium hydrogensulfate, $[NH_3OH]^+ HSO_4^-$,

$$NH_2OH + H_2SO_4 \rightarrow [NH_3OH]^+ HSO_4^-$$

It is prepared either by the reaction between ammonium nitrite and ammonium hydrogensulfite, in the presence of sulfur dioxide,

$$NH_4NO_2 + NH_4HSO_3 + SO_2 + 2H_2O \rightarrow [NH_3OH]^+HSO_4^- + (NH_4)_2SO_4$$

or simply by the reaction of sulfuric acid with nitromethane.

$$CH_3NO_2 + H_2SO_4 \rightarrow [NH_3OH]^+HSO_4^- + CO$$

Hydroxylamine can act as a reducing agent. For example, it reduces iron(III) chloride to iron(II) chloride under acidic conditions.

$$2NH_2OH + 4FeCl_3 \rightarrow 4FeCl_2 + 4HCl + H_2O + N_2O$$

It also reduces copper(II) oxide to copper(I) oxide.

$$2NH_2OH + 4CuO \rightarrow 2Cu_2O + 3H_2O + N_2O$$

It is an oxidizing agent too. It converts a suspension of iron(II) hydroxide, in sodium hydroxide, to iron(III) hydroxide.

$$2Fe(OH)_2 + NH_2OH + H_2O \xrightarrow{NaOH} 2Fe(OH)_3 + NH_3$$

It also oxidizes sodium arsinite to sodium arsenate.

$$NH_2OH + Na_3AsO_3 \rightarrow Na_3AsO_3 + NH_3$$

5. Nitrogen fluorides

Nitrogen forms a series of fluorides: nitrogen trifluoride, NF_3; dinitrogen tetrafluoride, N_2F_4; and dinitrogen difluoride, N_2F_2.

Nitrogen trifluoride is prepared by controlled fluorination of ammonia over copper catalyst.

$$4NH_3 + 3F_2 \xrightarrow{\text{Cu catalyst}} NF_3 + 3NH_4F$$

It is a colorless, odorless, thermodynamically stable and relatively unreactive gas with a pyramidal structure. The lack of reactivity is attributed to its low dipole moment (*see* Chapter 3, p. 132). Nevertheless, it reacts violently with hydrogen on sparking.

$$2NF_3 + 3H_2 \rightarrow N_2 + 6HF$$

Dinitrogen tetrafluoride, also called tetrafluorohydrazine, is prepared by passing nitrogen trifluoride over red hot copper.

$$2NF_3 + 2Cu \rightarrow N_2F_4 + 2CuF$$

Dinitrogen tetrafluoride is a colorless and reactive gas that exists in both *trans* and *gauche* conformations (Fig. 17.7).

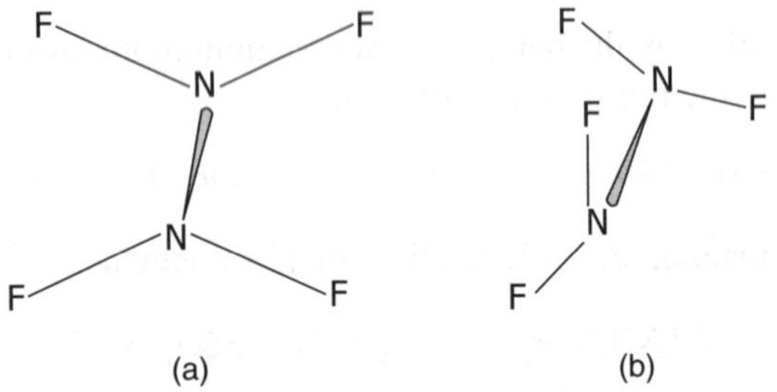

Fig. 17.7 Structure of (a) *trans* and (b) *gauche* forms of dinitrogen tetrafluoride

Dinitrogen tetrafluoride is a strong fluorinating agent.

$$SiH_4 + N_2F_4 \rightarrow SiF_4 + N_2 + 2H_2$$

$$10Li + N_2F_4 \rightarrow 4LiF + 2Li_3N$$

It forms adducts with fluoride ion acceptors.

$$AsF_5 + N_2F_4 \rightarrow [N_2F_3]^+[AsF_6]^-$$

At room temperature and above, it reversibly dissociates into blue-colored nitrogen difluoride radicals, NF_2. The latter are quite reactive.

$$2NF_2 + Cl_2 \rightarrow 2NF_2Cl$$

$$NF_2 + NO \rightarrow F_2NNO$$

Dinitrogen difluoride or difluorodiazine exists in *cis* and *trans* forms (Fig. 17.8).

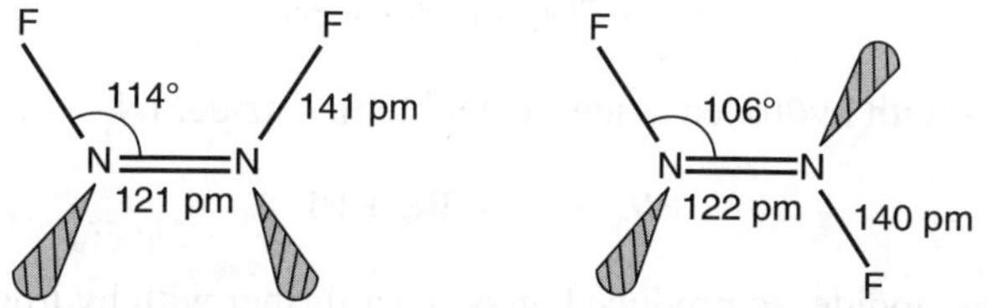

Fig. 17.8 Structure of (a) *cis* and (b) *trans* forms of dinitrogen difluoride

The *trans* form is prepared by the reaction of dinitrogen tetrafluoride with aluminium(III) chloride at 200 K.

$$2N_2F_4 + 2AlCl_3 \xrightarrow{200\ K} trans\text{-}N_2F_2 + 2AlF_3 + 3Cl_2 + N_2$$

When heated to 373 K, the *trans* form is converted into *cis* form. Dinitrogen difluoride is a fluoride ion donor.

$$N_2F_2 + SbF_5 \rightarrow [N_2F]^+ [SbF_6]^-$$

6. Hydrogen azide

Earlier known as hydrazoic acid, hydrogen azide, HN_3, is a colorless, poisonous liquid with an offensive odor. It is prepared by the reaction of nitrogen trichloride with hydrazine.

$$N_2H_4 + NCl_3 \rightarrow HN_3 + 3HCl$$

A better method of preparation involves the reaction of nitrous oxide with fused sodamide, under anhydrous conditions. Initially, sodium azide (NaN_3) and water vapors are produced. The latter hydrolyzes sodamide. The reactions may be represented as follows:

$$NaNH_2 + N_2O \rightarrow NaN_3 + H_2O$$
$$NaNH_2 + H_2O \rightarrow NaOH + NH_3$$

Overall reaction: $2NaNH_2 + N_2O \rightarrow NaN_3 + NaOH + NH_3$

The sodium azide is converted into hydrogen azide by treatment with dilute sulfuric acid followed by distillation.

$$2NaN_3 + H_2SO_4 \rightarrow 2HN_3 + Na_2SO_4$$

Hydrogen azide reacts with alkali metals to produce nitrogen. For example,

$$6HN_3 + 4Li \rightarrow \underset{\text{Lithium azide}}{4LiN_3} + 2NH_3 + 2N_2$$

With hypochlorous acid, chlorine azide, ClN_3, is formed,

$$HN_3 + HClO \rightarrow ClN_3 + H_2O$$

which reacts further.

$$HN_3 + ClN_3 \rightarrow HCl + 3N_2$$

Iodine reacts with hydrogen azide to yield iodine azide, IN_3.

$$HN_3 + I_2 \rightarrow IN_3 + HI$$

The hydrogen iodide, so produced, may react further with hydrogen azide.

$$HN_3 + 2HI \rightarrow I_2 + NH_3 + N_2$$

With nitrous acid, hydrogen azide yields a mixture of nitrous oxide and nitrogen.

$$HN_3 + HNO_2 \rightarrow H_2O + N_2O + N_2$$

Hydrogen azide reduces cerium(IV) salts to cerium(III) salts.

$$2CeCl_4 + 2HN_3 \rightarrow 2CeCl_3 + 2HCl + 3N_2$$

Case Study 17.9 Hydrogen azide is explosive, while sodium azide is not. Why?

Hydrogen azide, on slight heating, decomposes explosively into nitrogen and hydrogen.

$$2HN_3 \rightarrow 3N_2 + H_2$$

In fact, covalent azides are used as detonators and explosives.

The ionic azides, on the other hand, are quite stable. Sodium azide, for example, decomposes smoothly, and not explosively, when heated to 600 K.

$$2NaN_3 \xrightarrow{600\ K} 2Na + 3N_2$$

This is because azide anion has a linear structure to which three resonance forms contribute. The two nitrogen–nitrogen bond lengths are same.

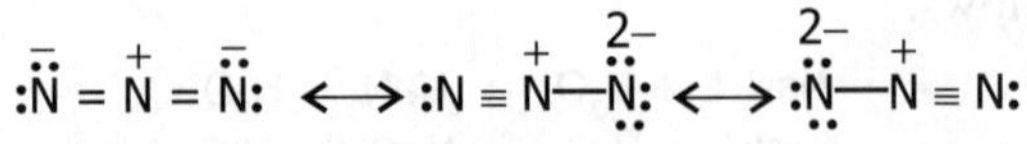

Covalent azides, including hydrogen azide, have a non-linear structure to which only two resonance forms contribute. The two nitrogen–nitrogen bond lengths are different.

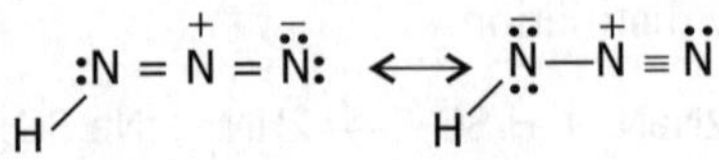

Therefore, in sodium azide, the structure is more symmetrical, and has three canonical forms, while in hydrogen azide, the structure is less symmetrical, and has two canonical forms. For this reason, hydrogen azide explodes violently, but sodium azide explodes smoothly.

Yet, the fact remains that sodium azide does detonate on impact, releasing nitrogen. This reaction is used to inflate the air bags in cars. During an accident, sodium azide decomposes and the nitrogen, so released, inflates the air bag, providing protection to the driver. Besides sodium azide, air bags also contain small amounts of potassium nitrate and silicon dioxide. During the decomposition of sodium azide, highly reactive sodium too is formed. It reacts with potassium nitrate to yield a mixture of sodium oxide and potassium oxide.

$$10Na + 2KNO_3 \rightarrow 5Na_2O + K_2O + N_2$$

The two oxides then react with silicon dioxide.

$$2Na_2O + SiO_2 \rightarrow Na_4SiO_4$$
$$2K_2O + SiO_2 \rightarrow K_4SiO_4$$

The mixture of sodium and potassium silicates, so formed, is unreactive—and safe.

7. Nitrides

The binary compounds of nitrogen are called nitrides. Nitrogen combines with virtually all the elements and hence a broad spectrum of nitrides exist. Such compounds may be classified into the following three categories.

a. Ionic nitrides

The ionic nitrides, also called *salt-like nitrides*, are formed with electropositive elements like lithium, magnesium, calcium, barium and zinc. The ionic formulations of these compounds are based on N^{3-} anion. Hence these may be represented as Li_3N and Ca_3N_2.

The ionic nitrides may be prepared by direct union of elements. For example,

$$3Ca + N_2 \rightarrow Ca_3N_2$$

Alternately, the metal may be heated with ammonia at about 1000 K to produce the nitride. For example,

$$3Mg + 2NH_3 \rightarrow Mg_3N_2 + 3H_2$$

Yet another method of preparing ionic nitrides is the thermal decomposition of metal amides. For example,

$$3Ba(NH_2)_2 \rightarrow Ba_3N_2 + 4NH_3$$

The ionic nitrides are readily hydrolyzed to metal hydroxide and ammonia.

$$Li_3N + 3H_2O \rightarrow 3LiOH + NH_3$$

b. Covalent nitrides

The covalent nitrides are formed by moderately reactive metals of Group 13 and by non-metals. Aluminium nitride may be prepared by the reaction of alumina with nitrogen in presence of coke.

$$Al_2O_3 + 3C + N_2 \rightarrow 2AlN + 3CO$$

Cyanogen $(CN)_2$ and tetrasulfur tetranitride (S_4N_4) are examples of non-metal nitrides. Boron nitride, $(BN)_x$ which has been discussed in Chapter 15, p. 474 has a polymeric structure, similar to graphite. Silicon nitride, Si_3N_4 has a diamond-like structure.

c. Interstitial nitrides

Also called *metallic nitrides*, the interstitial nitrides constitute the most extensive group of binary nitrogen compounds. These are formed by the transition and the inner transition elements. As the name suggests, nitrogen occupies some or all of the interstices in the metallic lattice. As a representative example, zirconium nitride may be prepared by the reaction of zirconium(IV) chloride with a mixture of nitrogen and hydrogen.

$$2ZrCl_4 + N_2 + 4H_2 \rightarrow 2ZrN + 8HCl$$

The interstitial nitrides are metallic in appearance, hardness and electrical conductivity. These have very high melting points and are chemically inert. For example, vanadium nitride reacts with sulfuric acid at an extremely slow pace.

$$2VN + 3H_2SO_4 \rightarrow V_2(SO_4)_3 + N_2 + 3H_2$$

Interstitial nitrides are used for making crucibles, high-temperature reaction vessels and thermocouple sheaths.

Case Study 17.10 What are *sialons*?

Compounds formed by combination of silicon, aluminium, oxygen and nitrogen are designated as sialons or Si-Al-O-N. The need for these compounds arose after it was observed that silicon nitride, Si_3N_4 can be used as an excellent ceramic material. Silicon nitride possesses high strength, a high decomposition temperature, resistance to oxidation and hostile environmental conditions and a low coefficient of friction. However, it has one drawback: it cannot be molded into desired shapes so as to serve as a starting material for manufacture of ceramics. The problem may be overcome if instead of silicon nitride, its solid solutions with alumina are used as ceramic materials. These solid solutions have a range of compositions, which may be expressed by the general formula, $Si_{6-0.75x}\,Al_{0.67x}\,O_x\,N_{8-x}$ $(0 < x < 6)$. Si-Al-O-N is an abbreviation for such solid solutions.

---

8. Phosphine

Phosphine is an extremely poisonous and a highly reactive gas. As has already been stated, it may be prepared by hydrolysis of calcium phosphide (p. 513).

Thermal decomposition of phosphorus acid at 473 K also yields phosphine.

$$4H_3PO_3 \rightarrow 3H_3PO_4 + PH_3$$

On an industrial scale, phosphine is synthesized by alkaline hydrolysis of white phosphorus.

$$P_4 + 3KOH + 3H_2O \rightarrow 3KH_2PO_2 + PH_3$$

As discussed in Case Study 2.3 (p. 70), the oxidation state of phosphorus in phosphine is (–III); in potassium dihydrogen phosphate, it is (I). In the above reaction, elemental phosphorus (oxidation state zero) is both reduced and oxidized. Therefore, this reaction is categorized as a disproportionation one.

Phosphine has a pyramidal structure, similar to ammonia. The HPH bond angle is 93.8°. It is a colorless gas with a faint garlic odor. The pure gas ignities at about 420 K, but when contaminated with diphosphorus tetrahydride, P_2H_4 it is spontaneously flammable.

$$4PH_3 + 8O_2 \rightarrow P_4O_{10} + 6H_2O$$

Phosphine also inflames in chlorine to produce phosphorus(III) chloride.

$$PH_3 + 3Cl_2 \rightarrow PCl_3 + 3HCl$$

It reacts explosively with nitrous oxide to yield phosphoric acid and nitrogen.

$$PH_3 + 4N_2O \rightarrow H_3PO_4 + 4N_2$$

Phosphine causes hydrophosphorylation of formic acid in acidic medium.

$$PH_3 + 4HCHO + HCl \rightarrow [P(CH_2OH)_4]Cl$$

The product, tetrakis(hydroxymethane)phosphonium chloride is mixed with urea-formaldehyde or melamine-formaldehyde resins for flameproofing of cotton cloth.

Phosphine is a strong reducing agent. It reduces the solutions of copper, silver and gold salts to metals. For example,

$$6Ag^+ + PH_3 + 3H_2O \rightarrow 6Ag + H_3PO_3 + 6H^+$$

When subjected to an electric arc, in presence of methane, phosphine gives the phosphorus analog of hydrogen cyanide.

$$PH_3 + CH_4 \xrightarrow{\text{Electric arc}} HC{\equiv}P + 3H_2$$

Unlike hydrogen cyanide, which is stable, the phosphorus counterpart is highly pyrophoric and can be preserved only below 140 K.

9. Oxyacids of phosphorus

Phosphorus forms two series of oxyacids: phosphorus acid series and phosphoric acid series. In each series, phosphorus is sp^3 hybridized. All oxyacids contain at least one phosphorus–oxygen double bond, P=O, implying the existence of $d\pi$–$p\pi$ bonding between the two atoms. The proton directly attached to phosphorus is non-ionizable. The proton attached to phosphorus through oxygen, P–O–H is, however, ionizable. The structures of some important oxyacids of phosphorus are depicted in Table 17.2.

Table 17.2 Some important oxyacids of phosphorus

Name	Composition	Oxidation state of phosphorus	Basicity	Structure
Phosphorus acid series				
(i) Hypophosphorus acid	H_3PO_2	I	Monobasic	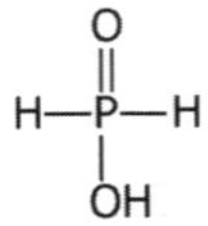
(ii) Phosphorus acid	H_3PO_3	III	Diabasic	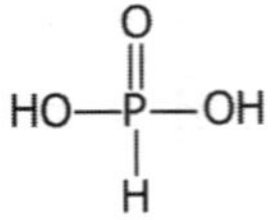
Phosphoric acid series				
(i) Hypophsphoric acid	$H_4P_2O_6$	IV	Tetrabasic	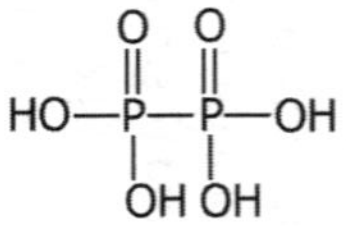
(ii) Orthophosphoric acid	H_3PO_4	V	Tribasic	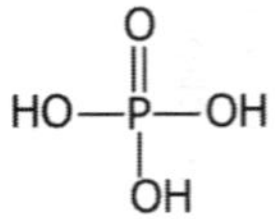
(iii) Pyrophophoric acid	$H_4P_2O_7$	V	Tetrabasic	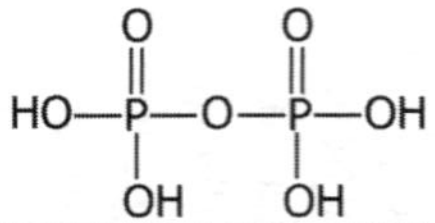

a. Phosphorus acid series

The following are the important members of this series:

(i) *Hypophosphorus acid,* H_3PO_2: It is prepared by heating white phosphorus in aqueous sodium hydroxide.

$$P_4 + 4NaOH + 4H_2O \rightarrow \underset{\text{Sodium hypophosphite}}{4NaH_2PO_2} + 2H_2$$

The free acid is obtained by treating sodium hypophosphite with a mineral acid.

$$NaH_2PO_2 + HCl \rightarrow H_3PO_2 + NaCl$$

The pure acid cannot be obtained by evaporating the aqueous solution because then it is oxidized to phosphoric acid. It is, therefore, extracted in ether and subsequently obtained in solid state by evaporation of solvent. On heating, hypophosphorus acid yields phosphine.

$$2H_3PO_2 \rightarrow H_3PO_4 + PH_3$$

It is a reducing agent. For example, it reduces silver nitrate to silver,

$$4AgNO_3 + H_3PO_2 + 2H_2O \rightarrow H_3PO_4 + 4HNO_3 + 4Ag$$

and mercury(II) chloride to mercury.

$$2HgCl_2 + H_3PO_2 + 2H_2O \rightarrow H_3PO_4 + 4HCl + 2Hg$$

(ii) *Phosphorus acid,* H_3PO_3: It is prepared by the hydrolysis of either phosphorus(III) chloride,

$$PCl_3 + 3H_2O \rightarrow H_3PO_3 + 3HCl$$

or phosphorus(III) oxide,

$$P_4O_6 + 6H_2O \rightarrow 4H_3PO_3$$

On heating, it decomposes into phosphine.

$$4H_3PO_3 \rightarrow 3H_3PO_4 + PH_3$$

It reduces copper(II) sulfate to elemental copper.

$$CuSO_4 + H_3PO_3 + H_2O \rightarrow H_3PO_4 + H_2SO_4 + Cu$$

b. Phosphoric acid series

The following are the important members of this series:

(i) *Hypophosphoric acid,* $H_4P_2O_6$: It is prepared by controlled oxidation of red phosphorus with sodium chlorite solution at room temperature. Disodium salt crystallizes out at pH 5.2.

$$2P + 2NaClO_2 + 8H_2O \rightarrow Na_2H_2P_2O_6.6H_2O + 2HCl$$

On acidification, the free acid is obtained.

$$Na_2H_2P_2O_6 + 2HCl \rightarrow H_4P_2O_6 + 2NaCl$$

Hypophosphorus acid undergoes slow hydrolysis to phosphorus and phosphoric acids.

$$H_4P_2O_6 + H_2O \rightarrow H_3PO_3 + H_3PO_4$$

(ii) *Orthophosphoric acid, H_3PO_4*: It is prepared by the hydrolysis of either phosphorus(V) chloride,

$$PCl_5 + 4H_2O \rightarrow H_3PO_4 + 5HCl$$

or phosphorus(V) oxide.

$$P_4O_{10} + 6H_2O \rightarrow 4H_3PO_4$$

On gentle heating to 500 K, phosphoric acid is transformed to pyrophosphoric acid.

$$2H_3PO_4 \xrightarrow{500\ K} H_4P_2O_7 + H_2O$$

On strong heating to 600 K, it dehydrates to metaphosphoric acid, $(HPO_3)_n$.

$$nH_3PO_4 \xrightarrow{600\ K} (HPO_3)_n + nH_2O$$

Metaphosphoric acids have cyclic structures. The most important and stable metaphosphoric acids are *cyclo*-trimetaphosphoric acid (n=3) and *cyclo*-tetrametaphosphoric acid (n = 4), the structures of which are depicted in Fig. 17.9.

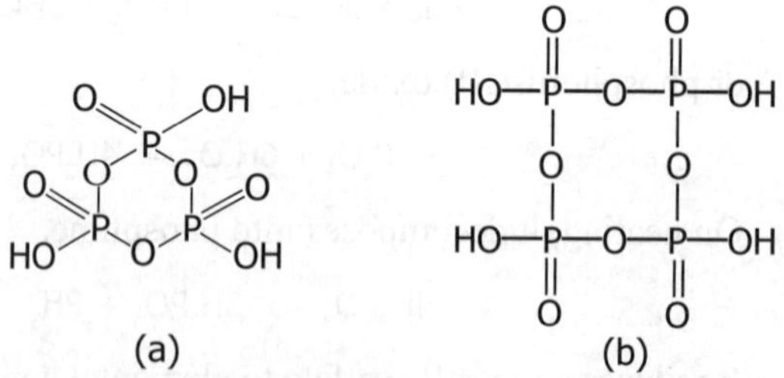

Fig. 17.9 Structure of (a) *cyclo*-trimetaphosphoric acid and (b) *cyclo*-tetrametaphosphoric acid

(iii) *Pyrophosphoric acid, $H_4P_2O_7$*: As has been outlined above, pyrophosphoric acid is prepared by heating orthophosphoric acid to 500 K. It is also called diphosphoric acid. On hydrolysis, it yields orthophosphoric acid.

$$H_4P_2O_7 + H_2O \rightarrow 2H_3PO_4$$

18. GROUP 16 ELEMENTS

Introduction

Group 16 elements—oxygen (O), sulfur (S), selenium (Se), tellurium (Te) and polonium (Po)—are collectively called *chalcogens* or the ore-forming elements, because a large number of metal ores are in the form of oxides or sulfides. Oxygen is the most abundant element on the surface of the earth. It reacts with almost all the other elements.

Physical Properties

The electronic configuration of Group 16 elements is ns^2np^4—two electrons short of the octet. Accordingly, these elements attain noble gas configuration either by gaining two electrons and forming a divalent anion, or by sharing two electrons and forming two covalent bonds. Most of the compounds of these elements are, however, covalent in nature.

The metallic character increases down the group. O and S are non-metals, Se and Te are metalloids, while Po is a metal. Oxygen can attain a maximum valency of 2 and a maximum coordination number 4. However, for other elements, oxidation states IV and VI are quite common and coordination number can be as high as 6 or 8. Sulfur invokes dπ orbitals to form multiple bonds.

Case Study 18.1 Outline the mechanism involved in the catalytic decomposition of potassium chlorate into oxygen.

Thermal decomposition of potassium chlorate at 550 K is a common method for the synthesis of oxygen on a laboratory scale. The reaction is catalyzed by manganese(IV) oxide, MnO_2, which participates in the mechanistic pathway. Potassium permanganate, $KMnO_4$ and potassium manganate, K_2MnO_4 are formed as intermediates. The reaction proceeds in the following steps.

$$2KClO_3 + 2MnO_2 \rightarrow 2KMnO_4 + Cl_2 + O_2$$

$$2KMnO_4 \rightarrow K_2MnO_4 + MnO_2 + O_2$$

$$K_2MnO_4 + Cl_2 \rightarrow 2KCl + MnO_2 + O_2$$

Overall reaction: $$2KClO_3 \rightarrow 2KCl + 3O_2$$

The evidence for this mechanism is provided by the fact that a faint odor of chlorine emanates during the course of the reaction. Moreover a purplish tinge of potassium permanganate appears in the reaction mixture.

---

1. Catenation

Of all the elements of the group, S has the strongest tendency to catenate. It forms several classes of catenated compounds, which have no corresponding analogs for the other congeners. This is attributed to very high sulfur–sulfur bond energy which, among the homonuclear single bonds, is exceeded only by hydrogen–hydrogen and carbon–carbon bonds. For example, sulfur forms a set of hydrides, called *polysulfanes,* having the composition, H_2S_n, where n = 2–8. The first member, hydrogen disulfane (H_2S_2) is colorless like hydrogen peroxide, H_2O_2. The remaining compounds are yellow in color. Polysulfanes are reactive liquids, the densities, viscosities and boiling points of which increase with molecular mass. These compounds have unbranched, linear chains of sulfur atoms. For example,

H—S—S—H

H—S—S—S—H

Sulfur also manifests catenation in polysulfide anions, $[S_n]^{2-}$. The most stable polysulfides are formed by large and electropositive alkali and alkaline earth metal ions. Some typical compositions are:

Na_2S_n n = 2–5

K_2S_n n = 2–6

BaS_n n = 2–4

The polysulfide unit forms non-linear chains (Fig. 18.1).

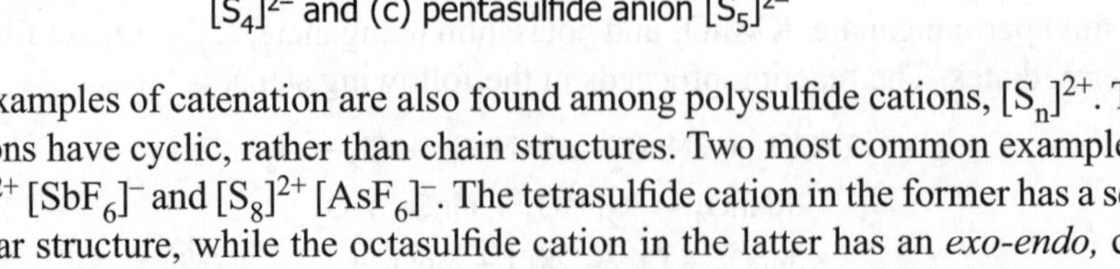

Fig. 18.1 Structure of (a) trisulfide anion, $[S_3]^{2-}$ (b) tetrasulfide anion, $[S_4]^{2-}$ and (c) pentasulfide anion $[S_5]^{2-}$

Examples of catenation are also found among polysulfide cations, $[S_n]^{2+}$. These cations have cyclic, rather than chain structures. Two most common examples are $[S_4]^{2+}[SbF_6]^-$ and $[S_8]^{2+}[AsF_6]^-$. The tetrasulfide cation in the former has a square planar structure, while the octasulfide cation in the latter has an *exo-endo,* cyclic structure (Fig. 18.2).

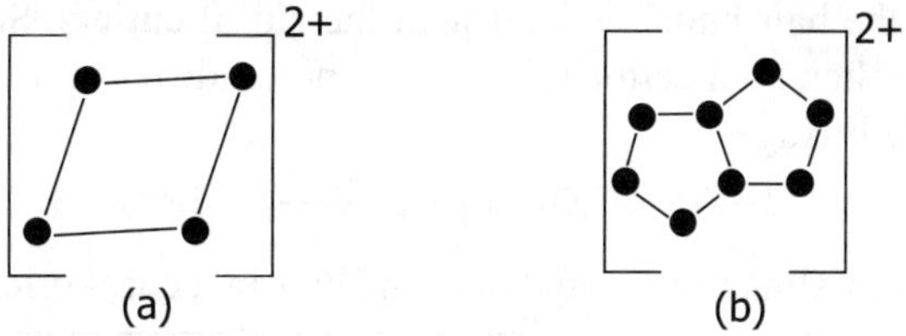

Fig. 18.2 Structure of (a) tetrasulfide cation, $[S_4]^{2+}$ and (b) octasulfide cation, $[S_8]^{2+}$

Selenium forms similar cations, for example, $[Se_4]^{2+}$ $[SO_3F]^-$ and $[Se_8]^{2+}$ $[AlCl_4]^-$. The structure of tetraselenide and octaselenide cations are cyclic and exactly analogous to those of the corresponding polysulfides (Fig. 18.2).

Polythionic acids, $H_2S_nO_6$, constitute yet another class of compounds in which sulfur exhibits catenation. The polythionate anion, $[S_nO_6]^{2-}$ has a non-cyclic, non-linear chain structure. The structures of trithionate ion, $[S_3O_6]^{2-}$; tetrathionate ion, $[S_4O_6]^{2-}$; and pentathionate ion, $[S_5O_6]^{2-}$ are presented in Fig. 18.3. The pentathionate ion exists in *cis* configuration.

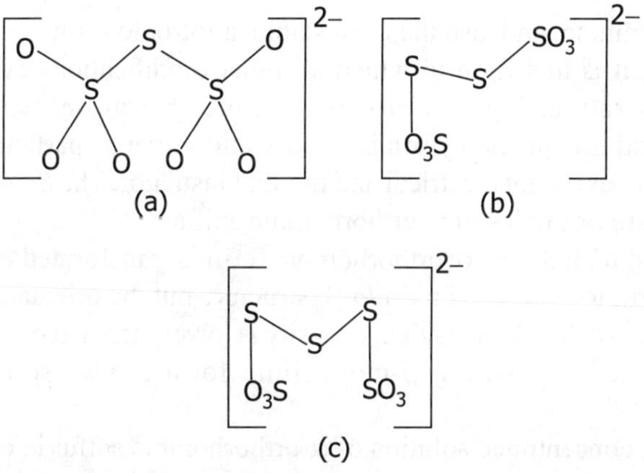

Fig. 18.3 Structure of (a) trithionate ion, $[S_3O_6]^{2-}$ (b) tetrathionate ion, $[S_4O_6]^{2-}$ and (c) pentathionate ion, $[S_5O_6]^{2-}$

Case Study 18.2 How is it possible to fashon straight hair into curly ones?

We have noted that sulfur shows the property of catenation and that the disulfide link is quite stable. This link is present in hair proteins. When thioglycollate ion, $HSCH_2COO^-$ is sprayed on straightened hair, the disulfide links are reduced to sulfhydryl functions, –SH.

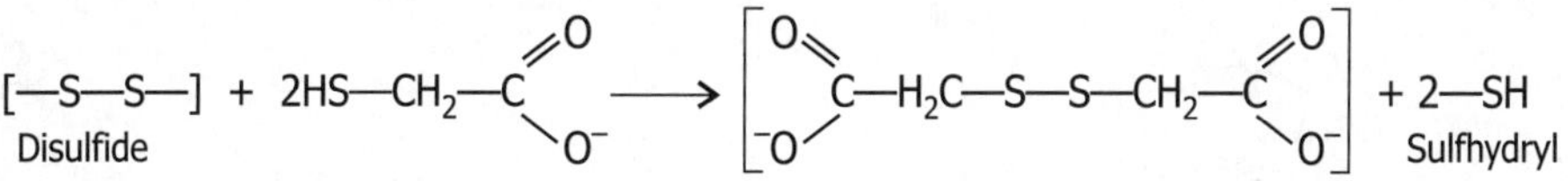

At this stage, the hair bunch is held with the aid of curlers. Soon it acquires a curly style. Thereafter, application of hydrogen peroxide re-oxidizes the sulfhydryl group to disulfide linkage.

$$2\text{—SH} + H_2O_2 \rightarrow [\text{—S—S—}] + 2H_2O$$

The disulfide crosslinks so formed hold the hair in a curly orientation.

2. Allotropic forms of sulfur

No other element has as many allotropic forms as sulfur. This is because a large variety of molecular forms can be achieved through catenation of sulfur atoms. Moreover, there are numerous ways in which the molecular forms can arrange within the crystal. Further, the sulfur–sulfur bond distances vary over an enormous range. Therefore, such bonds can be formed in legion of geometries. The bond angles between successive sulfur atoms too can vary a great deal. The sulfur–sulfur bonds are thus very flexible. The allotropic forms of sulfur may be broadly classified into two types: cyclic allotropes, designated as *cyclo*-S_n and chain allotropes, designated as *catena*-S_n. The former are more important.

The most common, and also the most stable, allotropic form is *α-orthorhombic sulfur*. In fact, it is this form to which all other modifications eventually revert at room temperature. It has a crown structure, based on *cyclo*-S_8 skeleton (Fig. 18.4a), and is appreciably soluble in organic solvents, particularly in carbon disulfide. It is an excellent electrical and thermal insulator. The *flower of sulfur* and *milk of sulfur* are examples of α-orthorhombic sulfur.

When heated to 368.5 K, α-orthorhombic form is transformed to *β-monoclinic sulfur*. This form too is based on *cyclo*-S_8 structure but the orientation is relatively disordered (Fig. 18.4b). As a result, its density is lower than that of α-orthorhombic form. However, like α-form, β-monoclinic form is also soluble in carbon disulfide.

When a hot, concentrated solution of α-orthorhombic sulfur in carbon disulfide is rapidly cooled, another allotropic form, called *γ-monoclinic sulfur,* crystallizes out. In this case too the *cyclo*-S_8 structure persists (Fig. 18.4c), but the packing is very compact, as a result of which its density is higher than that of α- or β-forms. It is sometimes referred to as *mother of pearl sulfur.*

Engel sulfur, also called *ε-sulfur*, is prepared by the reaction of tetrasulfane with disulfur dichloride in ether.

$$H_2S_4 + S_2Cl_2 \rightarrow S_6 + 2HCl$$

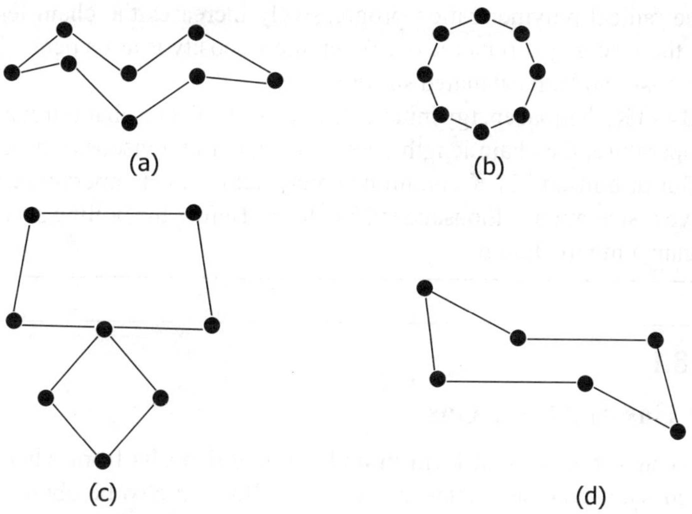

Fig. 18.4 Structure of (a) α-orthorhombic (b) β-monoclinic (c) γ-monoclinic and (d) engel form of sulfur

It has a *cyclo*-S_6 structure, in which the ring adopts the chair form (Fig. 18.4d).

The chain allotropes of sulfur, based on *catena*-S_n structures are prepared by quenching molten sulfur under different conditions. For example, *plastic* or *χ-sulfur* is obtained by pouring molten sulfur at 430 K into cold water. It has a helical structure. In *fibrous* or *Ψ-sulfur*, the helices are largely parallel, while in *lamina sulfur* these are partly cross-linked. These varieties are insoluble in organic solvents and possess high tensile strengths. However, with passage of time, all these modifications revert to α-orthorhombic form.

Case Study 18.3 Trace the variation in viscosity of liquid sulfur with temperature.

When α-orthorhombic sulfur is heated, it is converted to β-monoclinic form at 368.5 K. The melting point of β-monoclinic form is 392 K, but it melts a few degrees below due to disintegration of a few *cyclo*-S_8 structures. At its melting point, and a few degrees above it, liquid sulfur is yellow and transparent. It is mobile, i.e., its viscosity is very low. At 432 K, a remarkable transition in the properties of liquid sulfur occurs. Its color becomes brown. Its specific heat increases—and so does its viscosity. In fact, the viscosity steeply increases till about 473 K, when it becomes almost 10,000 times more than what it was at the melting point of sulfur. Within the 432–473 K temperature range, the *cyclo*-S_8 structure completely breaks down due to homolytic fission of sulfur–sulfur bonds. The ring structure is transformed into chain structure.

$$cyclo\text{-}S_8 \rightarrow catena\text{-}S_8$$

The chains are diradicals—each of the terminal sulfur atom is a free radical. The $\dot{S}\text{-}(S)_6\text{-}\dot{S}$ free radicals attack other chains and rings and grow in size.

$$catena\text{-}S_8 + cyclo\text{-}S_8 \rightarrow catena\text{-}S_{16}, \text{ etc.}$$

The free radical polymerization progressively increases the chain length and, therefore, the viscosity too increases. When the viscosity is at its peak, the chains involve almost two lakh catenated sulfur atoms.

Beyond 473 K, thermal energy initiates the cleavage of long chain structures. With rising temperature, the chain length becomes smaller and viscosity progressively reduces. Sulfur boils at 717 K and immediately below this temperature, catenated chains have just about one thousand sulfur atoms. Before the boiling sets in, sulfur is once again a mobile liquid.

Box 18.1

Sweet Gas and Sour Gas

Sulfur occurs in elemental form in underground deposits from where it can be mined. However, such sites are scarce. Sulfur can also be obtained from natural gas reservoirs which are scattered across the globe. Natural gas invariably contains hydrogen sulfide which, in turn, serves as a source of sulfur.

Natural gas mixed with low concentration of hydrogen sulfide is called sweet gas, while that containing a relatively large amount of hydrogen sulfide is called sour gas. For extracting sulfur, sour gas is the obvious choice.

Hydrogen sulfide is isolated from natural gas by adding ethanolamine.

$$HOCH_2CH_2NH_2 + H_2S \rightleftharpoons [HOCH_2CH_2NH_3]^+ HS^-$$

On warming, the equilibrium shifts towards the left.

A part of hydrogen sulfide is oxidized by air.

$$2H_2S + 3O_2 \rightarrow 2H_2O + 2SO_2$$

The remaining part is treated with sulfur dioxide formed in the above reaction,

$$4H_2S + 2SO_2 \rightarrow 4H_2O + 6S$$

This method of obtaining sulfur from natural gas is called *Claus process*.

Chemical Properties

Oxygen is a very reactive gas which vigorously oxidizes a large number of elements. Due to the presence of a double bond between two constituent atoms, the bond dissociation energy of oxygen molecule is appreciably high. Nevertheless, its reactions with other elements are largely exothermic and spontaneous.

Sulfur too is very reactive and combines directly with many elements. Its reactivity increases with temperature as a result of the cleavage of sulfur–sulfur bonds in the *cyclo*-S_8 structure. Selenium, tellurium and polonium are only moderately reactive. Yet, on the one hand, they combine as divalent anions with electropositive elements to yield selenides, tellurides and polonides, while on the other, they react with electronegative elements, such as oxygen and fluorine, exhibiting oxidation states II, IV and VI.

1. Reaction with air

On heating in air to about 500 K, the elements form the respective dioxides.

$$M + O_2 \xrightarrow{500\text{ K}} MO_2 \qquad (M = S, Se, Te \text{ or } Po)$$

2. Reaction with hydrogen

The reaction between oxygen and hydrogen spontaneously yields water. Sulfur forms hydrogen sulfide when heated to about 473 K. With selenium, the reaction occurs at about 600 K. Tellurium and polonium do not combine directly with hydrogen.

$$O_2 + 2H_2 \rightarrow 2H_2O$$

$$S + H_2 \xrightarrow{473\text{ K}} H_2S$$

$$Se + H_2 \xrightarrow{600\text{ K}} H_2Se$$

3. Reaction with halogens

Sulfur, selenium and tellurium all combine directly with fluorine to yield the respective hexafluorides.

$$M + 3F_2 \rightarrow MF_6 \quad (M = S, Se \text{ or } Te)$$

No other hexahalide of these elements is known. Direct reaction of sulfur with chlorine or bromine produces disulfur dichloride or disulfur dibromide.

$$2S + Cl_2 \rightarrow S_2Cl_2$$

$$2S + Br_2 \rightarrow S_2Br_2$$

Similarly, selenium reacts to give diselenium dichloride or diselenium dibromide.

$$2Se + Cl_2 \rightarrow Se_2Cl_2$$

$$2Se + Br_2 \rightarrow Se_2Br_2$$

The reaction of tellurium and chlorine under carefully controlled conditions in a sealed tube yields tritellurium dichloride.

$$3Te + Cl_2 \rightarrow Te_3Cl_2$$

Fluorination of selenium or tellurium at 273 K gives the corresponding tetrafluoride.

$$Se + 2F_2 \xrightarrow{273\text{ K}} SeF_4$$

$$Te + 2F_2 \xrightarrow{273\text{ K}} TeF_4$$

The tetrachlorides and tetrabromides may also be prepared by direct reactions with stoichiometric amounts of halogens. For example,

$$Se + 2Cl_2 \rightarrow SeCl_4$$

$$Te + 2Br_2 \rightarrow TeBr_4$$

4. Reaction with metals

Oxygen reacts directly with metals to form oxides,

$$2Mg + O_2 \rightarrow 2MgO$$

while sulfur forms sulfides.

$$Hg + S \rightarrow HgS$$

Selenium and tellurium react directly with alkali or alkaline earth metals to yield the corresponding selenides and tellurides. For example,

$$2Na + Se \rightarrow Na_2Se$$

$$Mg + Te \rightarrow MgTe$$

Transition metals too combine directly with selenium or tellurium, but the resulting compounds are non-stoichiometric. For example, titanium reacts with selenium, in the absence of air, at about 700 K, to yield a series of selenides, including Ti_2Se, Ti_3Se, $TiSe_2$ and $TiSe_3$.

5. Reaction with nitric acid

The chalcogens react with nitric acid to yield a mixture of dioxide and nitric oxide.

$$3M + 4HNO_3 \rightarrow 3MO_2 + 2H_2O + 4NO \quad (M = S, Se \text{ or } Te)$$

Box 18.2

Chevrel Phases

Polynuclear molybdenum chalcogenides, having the composition, $M_xMo_6X_8$ are called Chevrel phases (M is an electropositive metal and X is a chalcogen, generally oxygen or sulfur). A typical example is $PbMo_6S_8$. It is a superconductor below 13.3 K.

The structure of $PbMo_6S_8$ is shown in Fig. 18.5. It consists of an octahedral cluster of molybdenum atoms, surrounded by a cubic cluster of sulfur atoms which, in turn, is surrounded by a cubic cluster of lead atoms.

Contd...

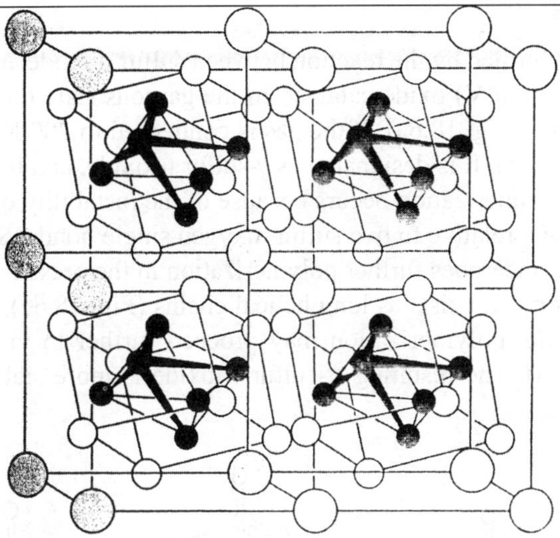

Fig. 18.5 Structure of $PbMO_6S_8$ (bold circles represent molybdenum atoms; unshaded circles represent sulfur atoms; and shaded circles represent lead atoms

The Mo_6S_8 unit cell is rotated with respect to the lead lattice. This is attributed to the repulsion between the negatively charged sulfur atoms in one cube with those in the other cube. If lead is replaced by a more electropositive metal like europium, the negative charge on sulfur atoms increases, and so does the degree of rotation.

Case Study 18.4 What is *mustard gas*?

Mustard gas is bis(2-chloroethyl)sulfide. It was used as a chemical weapon during World War I. It is prepared by the reaction of ethene with disulfur dichloride.

$$2C_2H_4 + S_2Cl_2 \rightarrow (ClCH_2CH_2)_2S + S$$

Although named mustard gas, this sulfur-based compound is actually a dense liquid, having a melting point and boiling point of 286 K and 488 K, respectively. When sprayed into the atmosphere, it stays close to the ground and is dispersed horizontally by wind action. When inhaled, it undergoes oxidation to its dioxo derivative in the living cells. The oxidized product eliminates hydrogen chloride to yield divinylsulfide.

$$(ClCH_2CH_2)_2S + 2(O) \rightarrow (ClCH_2CH_2)_2SO_2$$

$$(ClCH_2CH_2)_2SO_2 \rightarrow (CH_2CH)_2S + 2HCl + O_2$$

Divinylsulfide elicits toxic effects by entering into disruptive reactions with the cell proteins.

Compounds

Sulfur trioxide is formed by the reaction between sulfur dioxide and oxygen in the presence of vanadium(V) oxide catalyst. In the gaseous state the molecule has a triangular structure (Fig. 18.6a). If the gas is condensed to 200 K, a cyclic trimer is formed (Fig. 18.6b). It is designated as γ-sulfur trioxide, and has colorless, ice-like crystals. The trimerization occurs because of the instability of sulfur–oxygen double bonds, S=O, relative to two sulfur–oxygen single bonds, S–O. The γ-form of sulfur trioxide undergoes further polymerization in the presence of moisture to β-form. In this form, it exists as long helical chains (Fig. 18.6c). In the presence of excess moisture, polymerization may proceed further to yield cross-linked chains (Fig. 18.6d). The resulting α-sulfur trioxide is more stable as compared to γ- and β-forms.

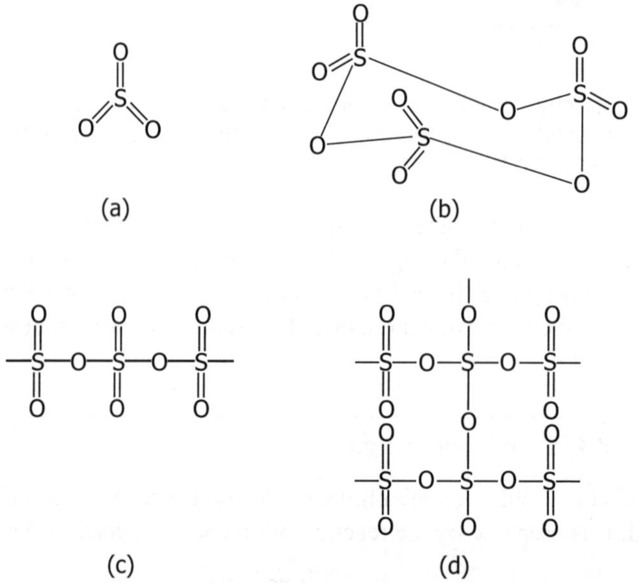

Fig. 18.6 Structure of (a) gaseous sulfur trioxide (b) γ-sulfur trioxide (c) β-sulfur trioxide and (d) α-sulfur trioxide

Ethanethiol, C_2H_5SH has earned the dubious distinction of being the most obnoxious smelling compound. About 50 parts per billion of this compound are added to natural gas so as to detect its leakage from the pipeline. The organosulfur molecule, propanethial S-oxide, the structure of which is depicted in Fig. 18.7, is the tear-inducing compound present in onions.

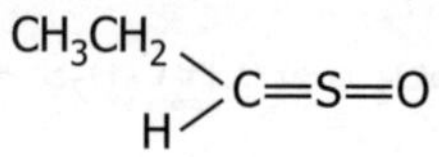

Fig. 18.7 Structure of propanethial S-oxide

At the other extreme, the luscious fragrance of marigold flowers arises due to a sulfur-based compound, 2,5-dithienylthiophene, the structure of which is shown in Fig. 18.8.

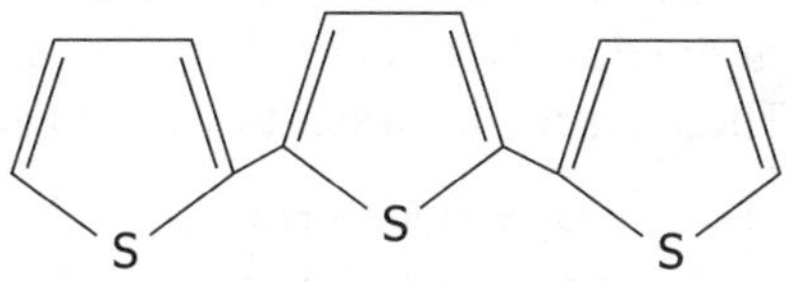

Fig. 18.8 Structure of 2,5-dithienylthiophene

Case Study 18.5 Why is sulfur hexafluoride chemically inert?

Sulfur hexafluoride is an extremely non-reactive compound. It remains unaffected when heated with hydrogen, oxygen, ammonia or halogens. Sodium metal can be melted in an atmosphere of sulfur hexafluoride and still there is no chemical reaction. Phosphorus can be distilled in it unchanged. It is inert towards fused potassium hydroxide or heated magnesium.

Thermodynamic calculations show that sulfur hexafluoride should be completely hydrolyzed at 300 K. Practically it does not. This means that the stability of sulfur hexafluoride is not due to thermodynamic parameters, but is a result of very high activation energy required for chemical transformation. This, in turn, may be attributed to the following factors. First, sulfur hexafluoride is a symmetrical molecule with octahedral geometry. Second, it does not possess dipole moment. Third, the sulfur atom is effectively shielded by six fluorine atoms, so that there is no position vulnerable to chemical attack. Fourth, no low-energy orbital is available with sulfur for bonding with other species, for example, water molecules.

Due to its chemical inertness, sulfur hexafluoride is used as an insulating gas in electrical equipment.

---

1. Oxides

Oxides are formed by virtually all the elements save the lighter noble gases. In terms of their chemical behavior, it is best to classify the oxides according to their acid-base characteristics. This criteria gives a fourfold classification of oxides.

a. **Basic oxides** The binary compounds possessing discrete oxide ions (O^{2-}) are grouped under this category. Basic oxides are formed by highly electropositive elements. Examples include sodium oxide, Na_2O and calcium oxide, CaO. These oxides react with water to yield the corresponding hydroxide.

$$Na_2O + H_2O \rightarrow 2NaOH$$

$$CaO + H_2O \rightarrow Ca(OH)_2$$

These also react with acids,

$$MgO + H_2SO_4 \rightarrow MgSO_4 + H_2O$$

b. Acidic oxides The covalent oxides of non-metals are largely acidic in nature. Nitrogen dioxide, NO_2 and sulfur dioxide, SO_2 are examples of acidic oxides. These react with water to yield an acidic solution.

$$CO_2 + H_2O \rightarrow H_2CO_3$$

$$SO_2 + H_2O \rightarrow H_2SO_3$$

For this reason, such oxides are often termed as acid anhydrides. Acidic oxides also react with alkalis.

$$Ca(OH)_2 + CO_2 \rightarrow CaCO_3 + H_2O$$

or with basic oxides;

$$CaO + SO_2 \rightarrow CaSO_3$$

c. Amphoteric oxides The oxides of less electropositive elements, which behave as acids towards strong bases and bases towards strong acids are called amphoteric oxides. Examples include beryllium oxide, BeO and aluminium oxide, Al_2O_3.

$$\underset{\text{Base}}{BeO} + \underset{\text{Acid}}{H_2SO_4} \rightarrow BeSO_4 + H_2O$$

$$\underset{\text{Acid}}{BeO} + \underset{\text{Base}}{2NaOH} \rightarrow Na_2BeO_2 + H_2O$$

d. Neutral oxides The oxides which react with neither water, nor acids, nor bases are termed neutral oxides. Carbon monoxide, CO and nitrous oxide, N_2O are neutral oxides.

A metal ion having a large size and/or a low oxidation state tends to form basic oxide. Such oxides are highly ionic in nature. In aqueous solution, the oxide ions undergo the following reaction.

$$O^{2-} + H_2O \rightarrow 2OH^-$$

The presence of hydroxide ions impart basic character to the solution.

Therefore, as the size of the metal ion increases, the chances of discrete oxide ions being furnished in the solution increase and so does the basic character of the oxide. Thus potassium oxide is more basic than sodium oxide which, in turn, is more basic than lithium oxide.

In Group 13 of the periodic table, one notes a transition from acidic oxide to amphoteric oxide to basic oxide on traversing from top to bottom.

B_2O_3	Al_2O_3	Ga_2O_3
Acidic	Amphoteric	Basic

On moving from left to right in a period, the size of the element decreases, the oxidation state increases and the character of oxide changes from basic, through amphoteric to acidic

Na_2O	MgO	Al_2O_3	SiO_2	P_4O_{10}
Basic	Less basic	Amphoteric	Less acidic	Acidic

If a metal ion exists in more than one oxidation state, the higher oxide is either less basic,

Tl_2O	Tl_2O_3
More basic	Less basic

or it is acidic.

CrO	Cr_2O_3	CrO_3
Basic	Amphoteric	Acidic

A metal ion which is small in size and/or is in a high oxidation state, tends to cleave the water molecule and become bound to the hydroxide ion

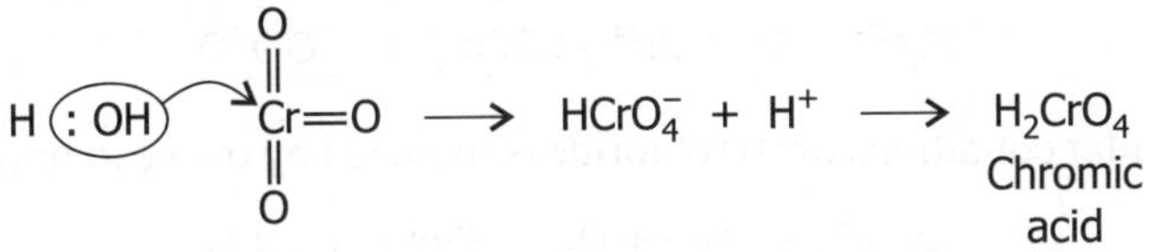

The presence of protons impart acidic character to the solution.

Case Study 18.6 How are rocks classified on the basis of acid-base character?

Rocks are made up of a mix of basic oxides like calcium oxide and acidic oxides like silicon dioxide. The classification of rocks on the basis their acid-base character depends on amount of silicon dioxide (or silica) content.

Silica content	Classification	Example
>50%	Acidic rock	Granite
45-50%	Basic rock	Basalt
<45%	Ultrabasic rock	Olivine

For example, olivine has the chemical composition $MgFeSiO_4$, which may be written as, $MgO{\cdot}FeO{\cdot}SiO_2$. The silica content of olivine is about 35%. Hence it is an ultrabasic rock.

2. Ozone

Ozone is the triatomic allotrope of oxygen. It is prepared by passing oxygen at 1 atmosphere pressure and 300 K through a system of concentric metalized glass tubes. A low-frequency power at 50-500 Hz and 10-20 kV is applied to the tubes so as to maintain the conditions of a silent electric discharge.

$$3O_2 \xrightarrow[\text{discharge}]{\text{Silent electric}} 2O_3$$

Ozone is a resonance hybrid of two canonical forms that have been depicted in Box 3.4, p. 137. It is an unstable, blue, diamagnetic gas with a pungent odor. It condenses to a deep-blue liquid and finally to a violet-black solid.

Ozone is a highly reactive gas and a strong oxidizing agent. For example, it oxidizes cyanide ions to cyanate ions,

$$CN^- + O_3 \rightarrow OCN^- + O_2$$

and nitrogen dioxide to dinitrogen pentoxide.

$$2NO_2 + O_3 \rightarrow N_2O_5 + O_2$$

It also oxidizes lead sulfide to lead sulfate.

$$PbS + 4O_3 \rightarrow PbSO_4 + 4O_2$$

In acidic medium, ozone transforms iodide ions to triiodide ions.

$$3I^- + O_3 + 2H^+ \rightarrow I_3^- + H_2O + O_2$$

Acidic solutions of cobalt(II) salts are oxidized to cobalt(III) analogs.

$$2Co^{2+} + O_3 + 2H^+ \rightarrow 2Co^{3+} + H_2O + O_2$$

Under similar conditions tin(II) chloride is oxidized by ozone to tin(IV) chloride.

$$3SnCl_2 + 6HCl + O_3 \rightarrow 3SnCl_4 + 3H_2O$$

The reaction of ozone with powdered hydroxides below 263 K, followed by extraction with liquid ammonia, yields the metallic ozonides, MO_3. For example,

$$2KOH + 5O_3 \rightarrow 2KO_3 + H_2O + 5O_2$$

The stability of alkali metal ozonides decreases in the order, Cs > Rb > K > Na. The alkaline earth metals too form ozonides, $M(O_3)_2$ under similar conditions. Their stability follows the order, Ba > Sr > Ca.

Case Study 18.7 What is *ozonised water*?

Water through which ozone has been bubbled in called ozonised water. It is also called *ozonated* water. Like mineral water, it is a purified and benevolent version of drinking water.

Ozone is not a very stable compound, and in aqueous medium readily dissociates into molecular oxygen and reactive oxygen atoms.

$$O_3 \rightarrow O_2 + (O)$$

The reactive oxygen atoms attack the cell walls of water-borne pathogens. Thus water is rid of the bacteria which cause cholera, dysentery and typhoid. The reactive oxygen atoms also remove harmful free radicals from the body. The molecular oxygen, resulting from the dissociation of ozone, enters the blood stream. The more oxygenated the blood, the better is its ability to perform its functions. Ozone-containing water is also known to clean up arteries and veins, break up red blood cell clumps, stimulate the immune system and reduce inflammation.

3. Water

Water is the only naturally occurring inorganic liquid. It is also the only chemical on planet earth that occurs naturally in all three states of matter—solid, liquid and gaseous. The large surface tension of water (73 dynes cm^{-1} at 293 K) and its ability to wet surfaces are the basis for the capillary action, which carries this liquid to the leaves of plants and trees. Water has a high heat capacity per unit weight. This means that it can absorb large quantities of heat without large changes in temperature. The structure of water (p. 110) and of ice (p. 167) has been described in Chapter 3.

Water reacts vigorously with alkali metals.

$$2Na + 2H_2O \rightarrow 2NaOH + H_2$$

The alkaline earth metals as well as zinc react with hot water to produce hydrogen.

$$Zn + 2H_2O \rightarrow Zn(OH)_2 + H_2$$

The reaction of steam with red-hot iron to yield hydrogen on an industrial scale has already been cited in Chapter 11, p. 406.

Oxides of active metals react with water to form the corresponding hydroxide,

$$BaO + H_2O \rightarrow Ba(OH)_2$$

while the oxides of non-metals form acids.

$$SO_3 + H_2O \rightarrow H_2SO_4$$

The decomposition brought about by water is called *hydrolysis*. The ionic carbides, belonging to methanide sub-classification, on hydrolysis, produce alkanes while those of acetylide sub-classification produce alkynes. These reactions have been stated in Chapter 16, p. 495. The products resulting from the hydrolysis of trihalides of Group 15 elements are depicted in Chapter 17, p. 512.

Lead(IV) sulfate and lead(IV) acetate are instantly decomposed by water to yield lead(IV) oxide.

$$Pb(SO_4)_2 + 2H_2O \rightarrow PbO_2 + 2H_2SO_4$$

$$(CH_3COO)_4Pb + 2H_2O \rightarrow PbO_2 + 4CH_3COOH$$

Hard water The high dielectric constant of water makes it an excellent solvent; it is often referred to as the *universal solvent*. As a result, natural water is never pure, but a solution of substances which come in contact with it.

The solvent action of water on rocks, soil as well as on atmospheric gases adds dissolved salts to the components of hydrosphere, such as streams, lakes and rivers. When water becomes contaminated with bicarbonate, chloride and sulfate of magnesium and/or calcium, it is said to have become hard.

Hard water does not readily form froth with soap. The latter are sodium or potassium salts of long-chain fatty acids. A representative example of soap is sodium stearate, $C_{17}H_{35}COONa$. The cleansing action of soap is attributed to its ability to form stable bubbles of froth by lowering the surface tension of water. However, the calcium and magnesium ions, present in hard water, interfere with the cleansing mechanism of soap. The stearate ions get precipitated as calcium or magnesium salts, so that the froth bubbles are either not formed at all or are formed to a meager extent.

$$Ca^{2+} + 2C_{17}H_{35}COO^- \rightarrow (C_{17}H_{35}COO)_2Ca(s)$$

The precipitated calcium or magnesium stearate gets deposited as scum on the material that was desired to be cleaned by soap.

Hardness of water is of two types: temporary hardness and permanent hardness.

Temporary hardness arises due to the presence of magnesium and calcium bicarbonates in the water supply. It can be removed by simply boiling water, whereby the soluble bicarbonates are decomposed into insoluble carbonates.

$$Mg(HCO_3)_2 \rightarrow MgCO_3 + H_2O + CO_2$$

$$Ca(HCO_3)_2 \rightarrow CaCO_3 + H_2O + CO_2$$

Since magnesium and calcium carbonates are insoluble in water, these do not interfere with the cleansing mechanism of soap.

Yet another method of eradicating temporary hardness is to treat the water with slaked lime, $Ca(OH)_2$. The hydroxide ions furnished by slaked lime convert the bicarbonate anions to carbonate anions.

$$HCO_3^- + OH^- \rightarrow CO_3^{2-} + H_2O$$

The carbonate anions precipitate the magnesium and calcium ions,

$$Mg^{2+} + CO_3^{2-} \rightarrow MgCO_3$$

$$Ca^{2+} + CO_3^{2-} \rightarrow CaCO_3$$

and remove these from solution.

The quantity of slaked lime must be carefully monitored. Excessive addition will introduce calcium ions into water, thereby negating the softening process.

Sodium carbonate too may be used for making hard water soft. The carbonate ions tie-up the magnesium and calcium ions present in hard water and remove these from solution as insoluble salts.

Permanent hardness arises due to the presence of chlorides and sulfates of magnesium and calcium in water sources. These salts enter into solution when slightly acidic water interacts with minerals like limestone and dolomite.

Sodium carbonate, which is effective in removing temporary hardness, can tackle permanent hardness too.

$$MgSO_4 + Na_2CO_3 \rightarrow MgCO_3 + Na_2SO_4$$

$$CaCl_2 + Na_2CO_3 \rightarrow CaCO_3 + 2NaCl$$

However, generally temporary and permanent hardness occur together. Hence hard water is treated with a mixture of slaked lime and sodium carbonate. The former nullifies temporary hardness, the latter permanent hardness.

The precipitation methods of nullifying hardness, irrespective whether it is of temporary or permanent type, suffer from an inherent drawback: on an industrial scale, the precipitated calcium and magnesium carbonates are retained within the system. These cause clogging of pipelines and boilers. This problem may be overcome by softening water by *ion exchange method.* The process involves replacement of objectionable cations by sodium ions, which do not give a precipitate with soap or form scales.

The ion exchanging materials are polymers having two types of ions: fixed and mobile. The mobile ions—also called free ions or counter ions—exchange with the cations which make water hard. An efficient variety of ion exchange media are naturally occurring sodium aluminium silicates, commonly referred to as zeolites and represented as Na_2Z. These possess a three-dimensional open structure, with negatively charged aluminosilicate framework, engulfed with positively charged sodium ions. The aluminosilicate particles constitute the fixed ions, while the sodium ions serve as the mobile species. When hard water is passed through a column of zeolite, the calcium and magnesium ions exchange with sodium ions and are thus removed from the liquid phase.

$$Na_2Z + Ca^{2+} \rightarrow CaZ + 2Na^+$$

$$Na_2Z + Mg^{2+} \rightarrow MgZ + 2Na^+$$

The exchange process is depicted in Fig. 18.9. Here R represents the fixed aluminosilicate ion. When its exchange capacity is exhausted, the zeolite can be regenerated by passing a concentrated solution of sodium chloride through the column.

$$CaZ + 2NaCl \rightarrow Na_2Z + CaCl_2$$

$$MgZ + 2NaCl \rightarrow Na_2Z + MgCl_2$$

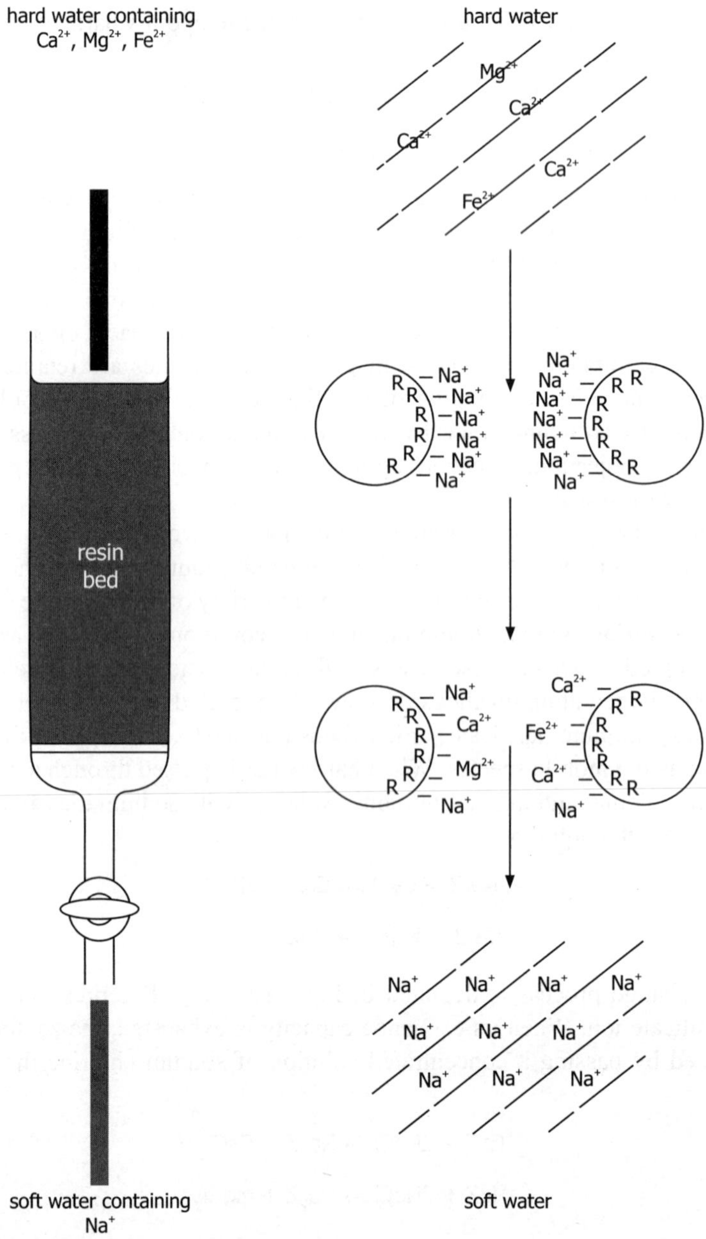

Fig. 18.9 Ion exchange method for softening hard water

4. Hydrogen peroxide

Hydrogen peroxide may be prepared by electrolysis of 50% sulfuric acid at a high current density. Peroxodisulfuric acid ($H_2S_2O_8$) is initially formed, which on hydrolysis yields hydrogen peroxide.

$$2HSO_4^{2-} \rightarrow H_2S_2O_8 + 4e^-$$

$$H_2S_2O_8 + 2H_2O \rightarrow 2H_2SO_4 + H_2O_2$$

A better method to manufacture hydrogen peroxide is the oxidation of 2-ethylanthraquinol in air. Hydrogen peroxide and 2-ethylanthraquinone are produced. The latter is dissolved in octanol-methylnaphthalene solvent mixture and reduced by Raney nickel hydrogenation to 2-ethylanthraquinol (Fig. 18.10). Being a cyclic process, this method of preparing hydrogen peroxide is commercially more viable.

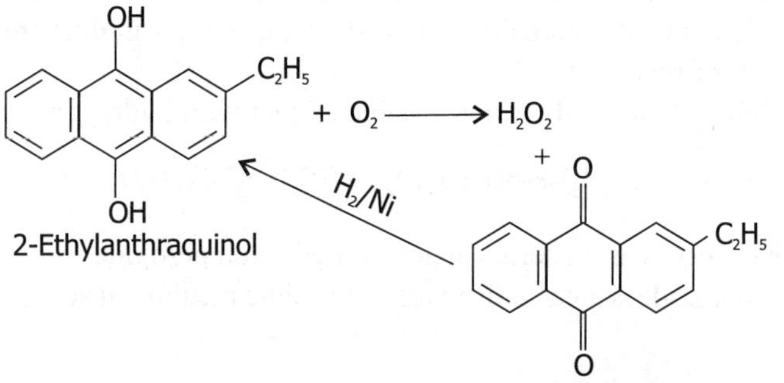

Fig. 18.10 Method for preparation of hydrogen peroxide

Hydrogen peroxide is extracted in water and concentrated to approximately 30% by distillation under reduced pressure.

In gaseous state, hydrogen peroxide has a skew configuration, with two hydrogen atoms adopting *trans* positions (Fig. 18.11). The same structure persists in liquid and solid states too, but the bond angles change from 94.8° to 102° due to hydrogen bonding.

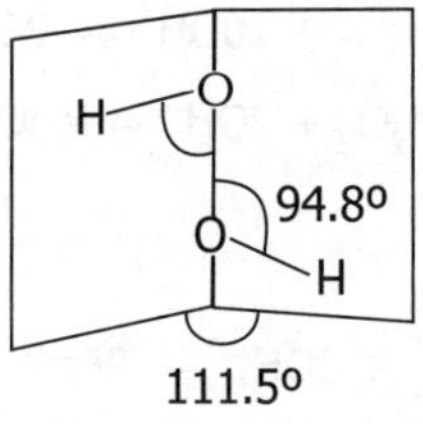

Fig. 18.11 Structure of hydrogen peroxide

The concentration of hydrogen peroxide is generally expressed in terms of *volume strength*—the volume of oxygen liberated at STP by the decomposition of 1 ml of hydrogen peroxide.

$$2H_2O_2 \rightarrow 2H_2O + O_2$$

Thus, an aqueous solution of hydrogen peroxide marked as "10 volume" means that 1 ml of the sample would liberate 10 ml of oxygen at STP.

Hydrogen peroxide is a colorless liquid. It has more extensive hydrogen bonding than water and, therefore, has a higher boiling point. It is also more dense than water. Its dielectric constant is also higher than water, but it has little solvating properties because it is easily decomposed into water and oxygen.

The decomposition reaction is catalyzed by metals like platinum and silver, by manganese dioxide or by alkalis. When hydrogen peroxide is placed in a glass bottle, trace amounts of alkalis leach out of the container material and rapidly decompose the compound. For this reason, hydrogen peroxide is stored in wax-coated or plastic containers, with stabilizers such as urea. Even a speck of dust can initiate explosive decomposition. Therefore, handling of anhydrous hydrogen peroxide or its concentrated solutions must be carried out in dust-free conditions and in absence of metal ions.

Triphenylphosphine oxide forms a stable adduct with hydrogen peroxide.

$$2(C_6H_5)_3P{=}O + H_2O_2 \rightarrow [(C_6H_5)_3PO]_2{\cdot}H_2O_2$$

This adduct serves as a useful carrier for hydrogen peroxide.

Hydrogen peroxide is a versatile reagent. In acidic medium, it acts as an oxidizing agent,

$$2I^- + H_2O_2 + 2H^+ \rightarrow I_2 + 2H_2O$$
$$2Fe^{2+} + H_2O_2 + 2H^+ \rightarrow 2Fe^{3+} + 2H_2O$$

as well as a reducing agent.

$$MnO_2 + H_2SO_4 + H_2O_2 \rightarrow MnSO_4 + 2H_2O + O_2$$
$$2Ce^{4+} + H_2O_2 \rightarrow 2Ce^{3+} + 2H^+ + O_2$$

Similarly, under alkaline conditions, it acts as an oxidizing agent,

$$2Cr^{3+} + 3H_2O_2 + 10OH^- \rightarrow 2CrO_4^{2-} + 8H_2O$$
$$Mn^{2+} + H_2O_2 + 2OH^- \rightarrow MnO_2 + 2H_2O$$

as well as a reducing agent.

$$2Fe^{3+} + H_2O_2 + 2OH^- \rightarrow 2Fe^{2+} + 2H_2O + O_2$$
$$2Ag^+ + H_2O_2 + 2OH^- \rightarrow 2Ag + 2H_2O + O_2$$
$$KIO_4 + H_2O_2 \rightarrow KIO_3 + H_2O + O_2$$

Hydrogen peroxide oxidizes even under neutral conditions. For example, lead sulfide may be oxidized to lead sulfate.

$$PbS + 4H_2O_2 \rightarrow PbSO_4 + 4H_2O$$

This reaction has an important application. Paintings in which lead(II) salts are used as white pigments tend to become dark on aging because the reaction with atmosphere hydrogen sulfide yields black lead sulfide. These may be restored to whiteness by treatment with hydrogen peroxide.

Hydrogen peroxide is a slightly stronger acid than water. It forms two series of salts.

$$NaOH + H_2O_2 \rightarrow \underset{\text{Sodium hydroperoxide}}{NaHO_2} + H_2O$$

$$2NaOH + H_2O_2 \rightarrow \underset{\text{Sodium peroxide}}{Na_2O_2} + 2H_2O$$

Case Study 18.8 Calculate the percentage strength, normality and strength in grams per liter of a 20 volume hydrogen peroxide solution.

One mL of 20 volume hydrogen peroxide solution, on decomposition, would liberate 20 mL oxygen at STP.

$$\underset{2(34)\,=\,68\text{ g}}{2H_2O_2} \rightarrow 2H_2O + \underset{22400\text{ mL}}{O_2}$$

Two moles (or 68 g) of hydrogen peroxide liberate one mole of oxygen. At STP, one mole of oxygen would occupy 22400 mL.

Or, 22400 mL of oxygen at STP is obtained from 68 g hydrogen peroxide.

20 mL of oxygen at STP would be obtained from $\frac{68}{22400} \times 20 = 0.06$ g hydrogen peroxide.

1 mL of 20 volume solution contains 0.06 g hydrogen peroxide.

100 mL of 20 volume solution contains $0.06 \times 100 = 6$ g hydrogen peroxide.

Strength = 6%

1000 mL of 20 volume solution contains $0.06 \times 1000 = 60$ g hydrogen peroxide

Strength in grams per liter = 60

$$\text{Normality} = \frac{\text{Strength (g L}^{-1})}{\text{Equivalent weight}}$$

Equivalent weight of hydrogen peroxide = 17

$$\text{Normality} = \frac{60}{17} = 3.52\text{ N}$$

A 20 volume solution of hydrogen peroxide has a strength of 6% or 60 g L^{-1} and normality, 3.52 N.

Box 18.3

Hydrogen Peroxide in Dental Care

Hydrogen peroxide finds application in cosmetic dentistry for restoring the whiteness of stained teeth. However, it is applied to teeth not as pure compound, but as a 1:1 adduct of urea. The adduct, called *carbamide peroxide* (Fig. 18.12), in prepared by dissolving urea in concentrated hydrogen peroxide. It crystallizes as a white solid.

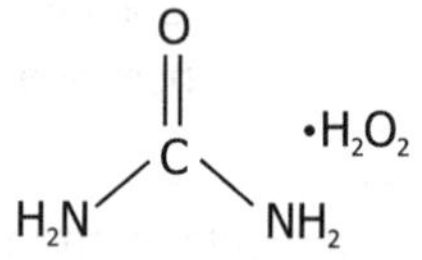

Fig. 18.12 Structure of carbamide peroxide

Carbamide peroxide acts as a carrier of hydrogen peroxide. When it comes in contact with teeth, its hydrogen peroxide constituent dissociates into reactive oxygen atoms.

$$H_2O_2 \rightarrow H_2O + (O)$$

The reactive oxygen atoms, in turn, oxidize the stains which become coated on the teeth of the persons who are habitual smokers or regular consumers of tobacco. The oxidation reaction bleaches the teeth and refurbishes their whiteness.

As a component of carbamide peroxide, hydrogen peroxide is also a pharmaceutical in dental care. It acts as a disinfectant and relieves minor inflammation of gums, as well as dental irritation.

5. Oxygen fluorides

The two important binary fluorides of oxygen are oxygen difluoride (OF_2) and dioxygen difluoride (O_2F_2), the former being more stable.

Oxygen difluoride is prepared by the reaction of gaseous fluorine with a 2% aqueous solution of sodium hydroxide.

$$2F_2 + 2NaOH \rightarrow 2NaF + OF_2 + H_2O$$

The alkali must be added in stoichiometric amounts; when present in excess, it may destroy a part of the product.

$$OF_2 + 2NaOH \rightarrow 2NaF + H_2O + O_2$$

Oxygen difluoride is a colorless, poisonous gas having an angular structure (Fig. 18.13a). It condenses to a pale yellow liquid. When preserved in a glass vessel, it is thermally stable up to 473 K; above this temperature, it decomposes into its constituent elements.

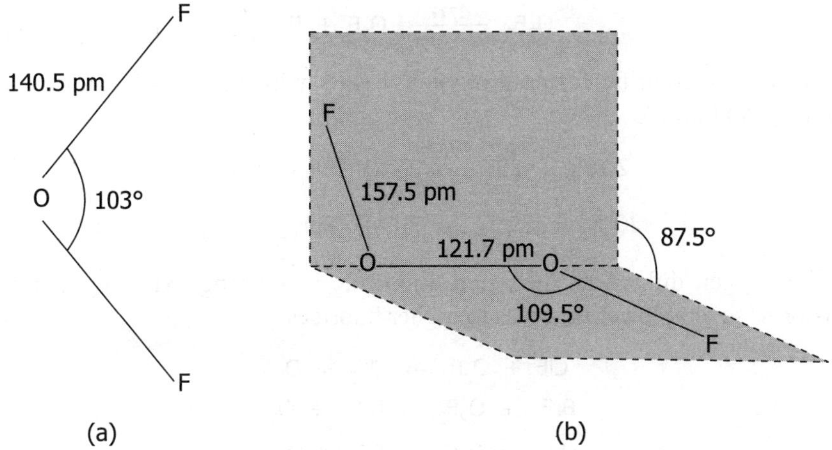

Fig. 18.13 Structure of (a) oxygen difluoride and (b) dioxygen difluoride

Oxygen difluoride is less reactive than fluorine, yet it is a strong oxidant as well as an excellent fluorinating agent.

$$2S + 2OF_2 \rightarrow SF_4 + SO_2$$
$$4P + 8OF_2 \rightarrow 2PF_5 + 2POF_3 + 3O_2$$
$$4Xe + 8OF_2 \rightarrow 2XeF_4 + 2XeOF_4 + 3O_2$$

When impinged with ultraviolet rays in an argon matrix at 4 K, oxygen difluoride dissociates into free radicals.

$$OF_2 \xrightarrow{h\upsilon} OF + F$$

Dioxygen difluoride, a yellow-colored compound, is prepared by passing a silent electric discharge through a mixture of oxygen and fluorine, maintained at 80 K and 10–20 mm Hg pressure.

$$O_2 + F_2 \rightarrow O_2F_2$$

The structure of dioxygen difluoride is similar to that of hydrogen peroxide and is presented in Fig. 18.13b. The oxygen–oxygen bond length is shorter than expected, while the oxygen–fluorine bond length is longer than expected. This may be attributed to the existence of canonical forms shown in Fig. 18.14.

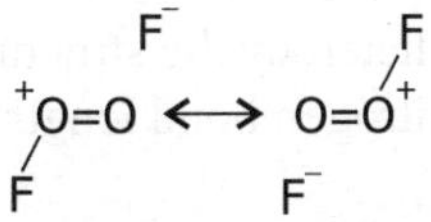

Fig. 18.14 Canonical forms of dioxygen difluoride

Due to longer, and hence, weaker bond between oxygen and fluorine, the dioxygen difluoride molecule dissociates into free radicals at 173 K.

$$O_2F_2 \xrightarrow{173\,K} O_2F + F$$

Dioxygen difluoride forms dioxygenyl salts with boron(III) fluoride and antimony(V) fluoride.

$$2O_2F_2 + 2BF_3 \rightarrow 2[O_2]^+ [BF_4]^- + F_2$$

$$2O_2F_2 + 2SbF_5 \rightarrow 2[O_2]^+ [SbF_6]^- + F_2$$

Like oxygen difluoride, dioxygen difluoride is a strong oxidizing agent. It transforms several lower fluorides to higher fluorides.

$$ClF + O_2F_2 \rightarrow ClF_3 + O_2$$

$$BrF_3 + O_2F_2 \rightarrow BrF_5 + O_2$$

$$SF_4 + O_2F_2 \rightarrow SF_6 + O_2$$

Sulfur hexafluoride may also be obtained by the reaction between dioxygen difluoride and hydrogen sulfide.

$$4O_2F_2 + H_2S \rightarrow SF_6 + 2HF + 4O_2$$

The reaction between dioxygen difluoride and xenon yields xenon tetrafluoride.

$$Xe + 2O_2F_2 \rightarrow XeF_4 + 2O_2$$

6. Sulfur nitrides

Sulfur and nitrogen are diagonally related in the periodic table. Their electronic charge densities and electronegativities are comparable. Therefore, sulfur and nitrogen combine covalently with each other, forming a class of compounds having chain, cyclic or polycyclic structures.

The most important analog of the series is tetrasulfur tetranitride, S_4N_4, as it is the starting material for the preparation of other compounds of this class. It is prepared by the reaction of disulfur dichloride either with ammonia,

$$6S_2Cl_2 + 16NH_3 \rightarrow S_4N_4 + 8S + 12NH_4Cl$$

or with ammonium chloride.

$$6S_2Cl_2 + 4NH_4Cl \rightarrow S_4N_4 + 8S + 16HCl$$

Tetrasulfur tetranitride has a heterocyclic structure in which the π-electrons are delocalized, so that all sulfur–nitrogen bond lengths are the same (Fig. 18.15).

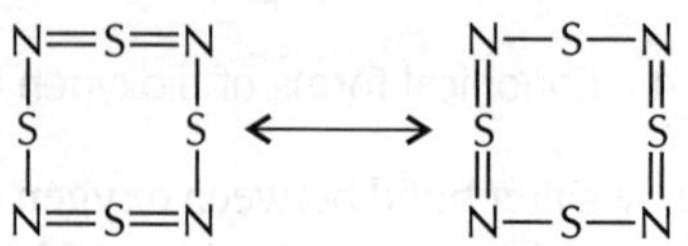

Fig. 18.15 Structure of tetrasulfur tetranitride

Tetrasulfur tetranitride is *thermochromic,* i.e., it changes color with temperature. For example, it is colorless at liquid nitrogen temperature, orange-yellow at room temperature and red at 373 K. It is kinetically stable, but endothermic with respect to the constituent atoms and, therefore, explodes on shock or heating. It is insoluble in water, but is hydrolyzed by aqueous sodium hydroxide, giving a mixture of thiosulfate and trithionate.

$$2S_4N_4 + 6NaOH + 9H_2O \rightarrow Na_2S_2O_3 + 2Na_2S_3O_6 + 8NH_3$$

When the vapors of tetrasulfur tetranitride are passed over silver, heated to about 550 K, under a pressure of 1 mm Hg, disulfur dinitride is formed.

$$S_4N_4 + 4Ag \xrightarrow{550\ K} S_2N_2 + 2Ag_2S + N_2$$

Disulfur dinitride is collected on a cold finger at 80 K. If it is sublimed at 273 K, it slowly polymerizes to a lustrous golden material, referred to as *polythiazyl* and designated as $(SN)_x$. This polymeric substance not only has metallic luster but, like metals, also possesses electrical and thermal conductivities. It is a super conductor below 0.26 K.

Polythiazyl is made up of coplanar chains of alternating sulfur and nitrogen atoms (Fig. 18.16).

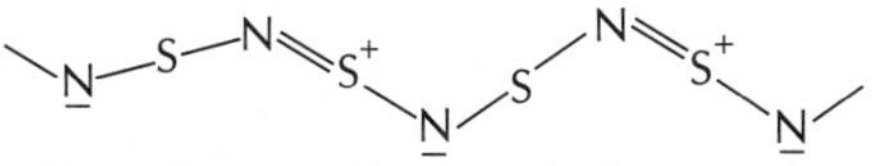

Fig. 18.16 Structure of polythiazyl

Although all sulfur and nitrogen atoms are stereochemically equivalent, there are two markedly different sulfur–nitrogen bond distances.

Case Study 18.9 Why is molybdenum(IV) sulfide an effective lubricant?

Molybdenum(IV) sulfide, MoS_2 has a layered structure. *Within* each layer, the covalent molybdenum-sulfur bonds are very strong. However, *in between* the adjacent layers, there are only weak S·····S interactions. Since the adjacent layers are held by weak forces of attraction, one layer can easily glide over another. As a result, molybdenum(IV) sulfide acquires slipperiness—and hence lubricating characteristics.

7. Oxyacids of sulfur

The important oxyacids of sulfur, along with their methods of preparation and structures, are listed in Table 18.1. Sulfuric acid is the most important member among the oxyacids of sulfur. On a small scale, it may be prepared by bubbling sulfur trioxide gas in water. On an industrial scale, however, it is manufactured by *contact process.* In this process, sulfur dioxide is first obtained, either by burning pure sulfur in air,

Table 18.1 Oxyacids of sulfur

Acid	Composition	Oxidation state of sulfur	Method of preparation	Structure
Sulfurous acid	H_2SO_3	IV	$SO_2 + H_2O \rightarrow H_2SO_3$	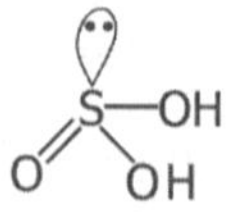
Disulfurous acid	$H_2S_2O_5$	III, V	$H_2SO_3 + SO_2 \rightarrow H_2S_2O_5$	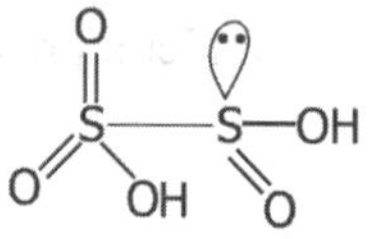
Sulfuric acid	H_2SO_4	VI	$SO_3 + H_2O \rightarrow H_2SO_4$	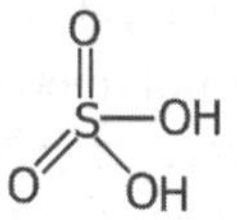
Disulfuric acid	$H_2S_2O_7$	VI	$H_2SO_4 + SO_3 \rightarrow H_2S_2O_7$	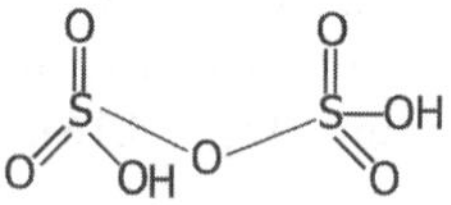
Thiosulfuric acid	$H_2S_2O_3$	– II, VI	$SO_3 + H_2S \rightarrow H_2S_2O_3$	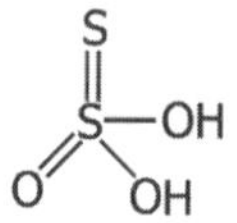
Dithionic acid	$H_2S_2O_6$	V	$H_2SO_3 + SO_3 \rightarrow H_2S_2O_6$	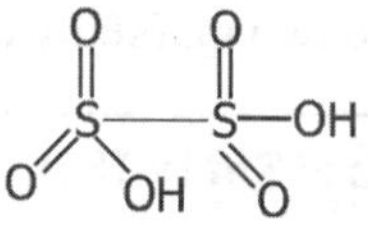

$$S + \underset{\text{Air}}{O_2} \rightarrow SO_2$$

or by roasting a sulfide ore, for example,

$$2ZnS + \underset{\text{Air}}{3O_2} \rightarrow 2ZnO + 2SO_2$$

Sulfur dioxide is then oxidized to sulfur trioxide.

$$2SO_2 + O_2 \rightleftharpoons 2SO_3 + \text{Heat}$$

According to Le Chatelier's principle, the yield of sulfur trioxide would improve at low temperatures. However, at low temperatures the reaction would become too slow to have practical utility. Thus, a catalyst is required to increase the yield of sulfur trioxide. Vanadium(V) oxide, on a silica support—sometimes also called *Kieselguhr support*—serves as an efficient catalyst. It works best in the temperature range 673–873 K. The reaction is generally carried out at 725 K.

On an industrial scale, sulfur trioxide cannot be absorbed directly in water. This is because it would first interact with water vapors above the liquid water and form a stable mist of sulfuric acid droplets, which become dispersed in the atmosphere. Therefore, sulfur trioxide is absorbed in 98% sulfuric acid to yield disulfuric acid, also called *pyrosulfuric acid* or *oleum.*

$$H_2SO_4 + SO_3 \rightarrow H_2S_2O_7$$

On hydrolysis, disulfuric acid yields sulfuric acid.

$$H_2S_2O_7 + H_2O \rightarrow 2H_2SO_4$$

Case Study 18.10 Can platinum be used instead of vanadium(V) oxide as a catalyst for the oxidation of sulfur dioxide to sulfur trioxide?

Air oxidation of sulfur dioxide to sulfur trioxide can be carried out more efficiently with platinum as a catalyst. However, platinum is readily poisoned by traces of sulfur which are invariably present in the reaction system. Vanadium(V) oxide is less efficient, but is not easily poisoned. Further, as compared to platinum, it is cost-effective. Therefore, on a long-term basis, vanadium(V) oxide proves to be a better catalyst.

Anhydrous sulfuric acid is a dense liquid and is readily miscible with water. However, the reaction between sulfuric acid and water is extremely exothermic. It can result in an explosion if water is added to the acid. Therefore, it is always safer to add acid to water, slowly, and with constant stirring. The anhydrous acid has a high dielectric constant. However, it is rarely used as a solvent for a reaction system since its boiling point is above 573 K and is, therefore, difficult to evaporate. Its high viscosity is attributed to the presence of strong hydrogen bonds. It has high electrical conductivity due to *autoprotolysis,* i.e., transfer of proton from one molecule to another.

$$H_2SO_4 + H_2SO_4 \rightarrow H_3SO_4^+ + HSO_4^-$$

The reaction between copper and sulphuric acid is normally considered to be a method for small scale production of sulfur dioxide gas.

$$Cu + 2H_2SO_4 \rightarrow CuSO_4 + 2H_2O + SO_2$$

However, careful analysis of reaction products reveals that some amount of copper(II) sulfide is also formed. The following reaction explains its formation:

$$SO_4^{2-} + 8H^+ + 8e^- \rightarrow S^{2-} + 4H_2O$$

Two factors assist in the culmination of the above reaction. First, a very high concentration of protons (H^+) pushes the reaction in the forward direction. Second, copper(II) sulfide is insoluble in water and, therefore, precipitates out of the solution.

Hence, the removal of sulfide ions in the above reaction, by precipitation, also takes the reaction in the forward direction.

Another oxyacid of sulfur is dithionous acid, $H_2S_2O_4$. We have not listed this analog in Table 18.1 since the free acid is non-existent. However, the dithionite ion, $S_2O_4^{2-}$ is relatively stable, and a strong reducing agent. It owes its reducing nature to its characteristic structure that is shown in Fig. 18.17.

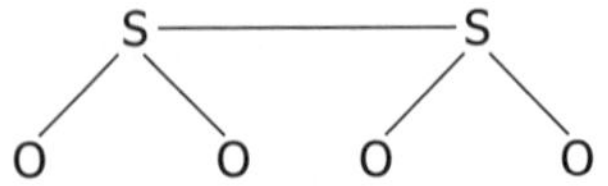

Fig. 18.17 Structure of dithionite ion

The oxygen atoms of the dithionite ion bear a considerable negative charge and are placed in close proximity due to the eclipsed configuration. Moreover, the repulsion due to the presence of lone pairs of electrons on sulfur atoms causes the sulfur-sulfur bond to lengthen—and weaken. The weak bonding in dithionite ion is responsible for its reducing characteristic.

Case Study 18.11 Sulfurous acid, on mild heating, decomposes into sulfur dioxide. However, sulfuric acid does not decompose into sulfur trioxide that readily. Why?

Sulfurous acid is formed by the combination of water and sulfur dioxide, just as sulfuric acid is formed by the addition of water to sulfur trioxide. However, the fact that sulfurous acid readily dissociates into sulfur dioxide, but sulfuric acid does not dissociate into sulfur trioxide so readily, shows that the mechanism of addition reactions are different. For the formation of sulfurous acid, the addition of sulfur dioxide and water may be pictured as follows:

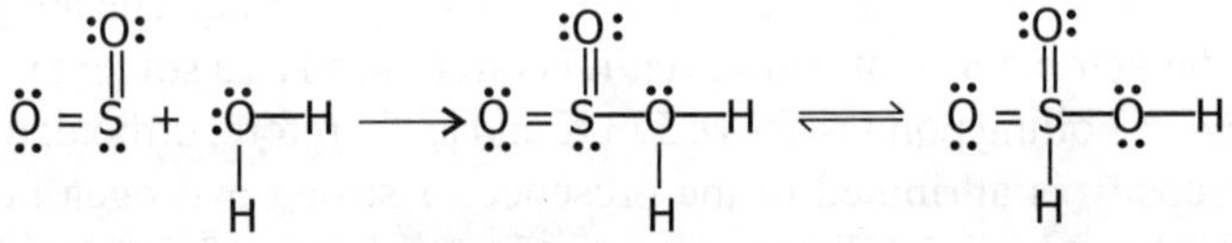

The bonds in sulfur dioxide are not substantially altered during the addition of water. Therefore, the sulfur dioxide component of sulfurous acid is not held very tightly. It is lost by mild heating.

On the other hand, the bonds in sulfur trioxide are substantially altered when water is added.

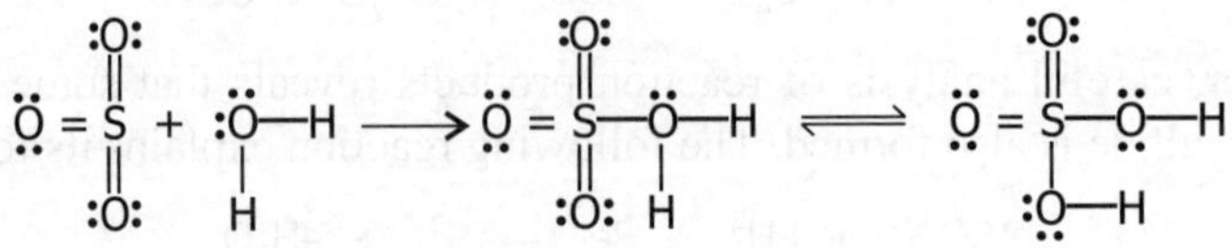

The transfer of a proton from water to an oxygen of sulfur trioxide disrupts one sulfur–oxygen double bond. This change is irreversible and, therefore, sulfur trioxide is not easily lost from sulfuric acid.

8. Peroxoacids of sulfur

There are two peroxoacids of sulfur: peroxomonosulfuric acid and peroxodisulfuric acid.

Peroxomonosulfuric acid, also called *Caro's acid*, has the composition, H_2SO_5. It is prepared by the reaction of chlorosulfuric acid and anhydrous hydrogen peroxide in 1:1 molar ratio.

$$HOOH + ClSO_2(OH) \rightarrow HOOSO_2(OH) + HCl$$

It may also be prepared by the reaction of anhydrous peroxodisulfuric acid, $H_2S_2O_8$ with anhydrous hydrogen peroxide.

$$H_2S_2O_8 + H_2O_2 \rightarrow 2H_2\ SO_5$$

The structure of peroxomonosulfuric acid is depicted in Fig. 18.18a

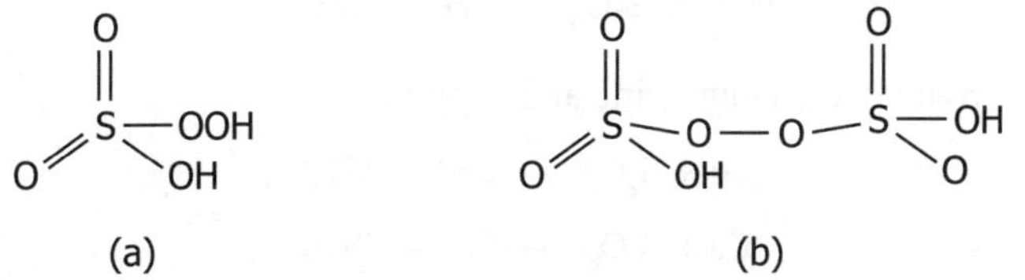

Fig. 18.18 Structure of (a) peroxomonosulfuric acid and (b) peroxodisulfuric acid

Peroxomonosulfuric acid is a colorless, crystalline solid, with a melting point of 318 K. It is explosive in nature. It is hydrolyzed by water.

$$H_2SO_5 + H_2O \rightarrow H_2SO_4 + H_2O_2$$

It oxidizes potassium iodide to iodine.

$$H_2SO_5 + 2KI \rightarrow K_2SO_4 + H_2O + I_2$$

Peroxodisulfuric acid, $H_2S_2O_8$ may be prepared by the electrolysis of ice-cold, 50% sulfuric acid with a high current density.

$$H_2SO_4 \rightarrow H^+ + HSO_4^-$$

At anode: $HSO_4^- \rightarrow HSO_4^{\cdot} + e^-$

$HSO_4^{\cdot} + HSO_4^{\cdot} \rightarrow H_2S_2O_8$

At cathode: $H^+ + e^- \rightarrow H$

$H + H \rightarrow H_2$

Hydrogen evolves at cathode which is separated from the anode to prevent the products from reacting with each other.

The reaction between chlorosulfuric acid and anhydrous hydrogen peroxide in 2:1 molar ratio is another synthetic route to peroxodisulfuric acid.

$$(HO)SO_2Cl + HOOH + ClSO_2(OH) \rightarrow (HO)SO_2\text{—O—O—}SO_2(OH) + 2HCl$$

Peroxodisulfuric acid, the structure of which is presented in Fig. 18.18b, is a colorless, hygroscopic, crystalline solid. It melts with decomposition at 338 K. It is hydrolyzed easily.

$$H_2S_2O_8 + H_2O \rightarrow H_2SO_5 + H_2SO_4$$

Like peroxomonosulfuric acid, peroxodisulfuric acid too oxidizes potassium iodide to iodine. However, this reaction proceeds at a comparatively slow speed since the first step is the hydrolysis of peroxodisulfuric acid to peroxomonosufuric acid.

$$H_2S_2O_8 + H_2O \rightarrow H_2SO_5 + H_2SO_4$$
$$H_2SO_5 + 2KI \rightarrow K_2SO_4 + H_2O + I_2$$
$$\overline{H_2S_2O_8 + 2KI \rightarrow K_2SO_4 + H_2SO_4 + I_2}$$

Peroxodisulfate ions, $S_2O_8^{2-}$ oxidize iron(II) ions to iron(III) ions.

$$2Fe^{2+} + S_2O_8^{2-} \rightarrow 2Fe^{3+} + 2SO_4^{2-}$$

These also oxidize elemental zinc and copper.

$$Zn + S_2O_8^{2-} \rightarrow Zn^{2+} + 2SO_4^{2-}$$
$$Cu + S_2O_8^{2-} \rightarrow Cu^{2+} + 2SO_4^{2-}$$

Likewise manganese(II) ions are oxidized to manganese(IV) oxide.

$$Mn^{2+} + S_2O_8^{2-} + 2H_2O \rightarrow MnO_2 + 4H^+ + 2SO_4^{2-}$$

Case Study 18.12 What is *lithopone*?

Zinc sulfide, when pure, is white in color and is used as an ingredient of paints, along with barium sulfate. The mixture of zinc sulfide and barium sulfate is called lithopone. It is obtained by the reaction of zinc sulfate and barium sulfide.

$$ZnSO_4(aq) + BaS(aq) \rightarrow ZnS(s) + BaSO_4(s)$$

Lithopone imparts luster to white paint.

---

9. Sulfur oxyhalides

Sulfur forms two series of oxyhalides: thionyl halides, SOX_2 and sulfuryl halides, SO_2X_2 (X = F, Cl or Br). The thionyl halides have a pyramidal structure, while the sulfuryl halides have a distorted tetrahedral structure (Fig. 18.19).

Fig. 18.19 Structure of (a) thionyl halides and (b) sulfuryl halides

Thionyl chloride is the most important compound in the thionyl halide series. It is prepared by the reaction between sulfur dioxide and phosphorus(V) chloride.

$$SO_2 + PCl_5 \rightarrow POCl_3 + SOCl_2$$

It may also be prepared by the interaction of sulfur trioxide with sulfur dichloride.

$$SO_3 + SCl_2 \rightarrow SOCl_2 + SO_2$$

Thionyl chloride reacts vigorously with water,

$$SOCl_2 + H_2O \rightarrow SO_2 + 2HCl$$

and is particularly suitable for dehydrating readily hydrolyzable salts. For example,

$$AlCl_3.6H_2O + 6SOCl_2 \rightarrow AlCl_3 + 6SO_2 + 12HCl$$

The most important sulfuryl halide, sulfuryl chloride (SO_2Cl_2) is prepared by the direct chlorination of sulfur dioxide in the presence of activated charcoal catalyst.

$$SO_2 + Cl_2 \rightarrow SO_2Cl_2$$

On hydrolysis, it yields sulfuric acid.

$$SO_2Cl_2 + 2H_2O \rightarrow H_2SO_4 + 2HCl$$

Increasing molecular mass among both thionyl halides and surfuryl halides is accompanied by decreasing thermal stability. Thus thionyl fluoride is very stable, but thionyl chloride begins to decompose above 425 K, and thionyl bromide decomposes even below its boiling point. Similarly, surfuryl fluoride is extremely stable, while surfuryl chloride decomposes easily into sulfur dioxide and chlorine. The tendency towards hydrolysis also increases with molecular mass. Thionyl fluoride and surfuryl fluoride react slowly with water, while the corresponding chlorides react vigorously.

Case Study 18.13 When a solution of silver nitrate is added to aqueous sodium thiosulfate, a white precipitate, **A**, is obtained. On adding excess sodium thiosulfate, **A** is converted into a soluble compound, **B**. When heated with water, **A** gives a black precipitate, **C**. After filtering out **C** and on addition of barium chloride to the filtrate, a white precipitate, **D**, is obtained. Identify **A, B, C** and **D**.

The reaction between silver nitrate and sodium thiosulfate yields a white precipitate of silver thiosulfate (**A**).

$$2AgNO_3 + Na_2S_2O_3 \rightarrow \underset{\substack{\text{Insoluble} \\ \textbf{A}}}{Ag_2S_2O_3} + 2NaNO_3$$

The white precipitate dissolves in excess sodium thiosulfate due to the formation of soluble sodium thiosulfatoargentate(I), (**B**).

$$Ag_2S_2O_3 + 3Na_2S_2O_3 \rightarrow \underset{\substack{\text{Soluble} \\ \textbf{B}}}{2Na_3[Ag(S_2O_3)_2]}$$

When silver thiosulfate (**A**) is heated with water, black silver sulfide (**C**) precipitates out.

$$\underset{\textbf{C}}{Ag_2S_2O_3 + H_2O \rightarrow Ag_2S \downarrow + H_2SO_4}$$

The filtrate, aqueous sulfuric acid, gives a white precipitate of barium sulfate (**D**) with barium chloride.

$$H_2SO_4 + BaCl_2 \rightarrow \underset{\textbf{D}}{BaSO_4} + 2HCl$$

Thus, **A** = $Na_2S_2O_3$, **B** = $Na_3[Ag(S_2O_3)_2]$, **C** = Ag_2S and **D** = $BaSO_4$

19. GROUP 17 ELEMENTS

Introduction

Group 17 elements, fluorine (F), chlorine (Cl), bromine (Br), iodine (I) and astatine (At), are called *halogens*, which means salt-forming. They are so named because of their tendency to react rapidly with metals to form salts. They form a wide spectrum of compounds with non-metals as well. F is the most reactive element known. The last halogen, At, is radioactive and is generally excluded from the comparative study of Group 17 elements.

Physical Properties

The general electronic configuration of halogens is ns^2np^5. They have one p electron less than the succeeding noble gases. Accordingly, these elements are highly electronegative—F being the most electronegative element in the periodic table. The electron affinity is maximum for Cl in this group; the value for F is intermediate between that of Cl and Br. This is attributed to the extremely small size of F and immense interelectronic repulsions among the valence shell electrons (*see* Chapter 2, p. 66).

All the halogens are volatile, diatomic molecules. The two constituent atoms are held together by covalent bonds. However, intermolecular attractions, in the form of van der Waals forces, increase with size. For this reason, F and Cl, in their natural states, are gases, Br is a liquid and I is a solid. The melting points and boiling points too increase with molecular mass. I_2 is sublimable. These elements are poor conductors of electricity. However, when crystals of I_2 are compressed, these acquire metallic characteristics and become conducting.

Halogens have high standard oxidation potentials, which decrease down the group. This indicates that F is a strong oxidizing agent, followed by Cl and Br, while I is a mild oxidizing agent. In fact, F brings out the highest oxidation states in the elements with which it reacts, for example, bismuth(V) fluoride.

Although all halide ions have a uninegative charge, yet the charge density on fluoride ion is much higher than that on the iodide ion. This implies that due to its large size, the negative charge on iodine is spread over a large surface area. On the other hand, since fluorine is small, the uninegative charge of fluoride ion is concentrated over a small surface area. As a result, the larger iodide ion is more easily polarizable than the smaller fluoride ion. Consequently, the metal iodides have greater covalent character, while the fluorides are essentially ionic.

1. Color

All halogens are colored. F is pale yellow, Cl is yellowish-green, Br is reddish-brown and I is violet. Quite obviously, the color intensity increases down the group. Color arises when visible light excites a valence electron to a higher energy level. When the electron descends to its original energy state, it emits its extra energy as colored light. On moving down the group, the promotional energy required to transfer the electron to a higher energy state decreases. Two reasons may be ascribed to this phenomenon. First, with increasing size, the valence electrons are held less tightly by the nucleus and, therefore, require lesser energy for excitation. Second, as the molecular mass increases, the difference between the energy levels reduces. Therefore, lesser energy is required to excite the electron. Consequently, lesser amount of energy is required to excite an electron in I than in F.

We have studied in Chapter 1 that energy and wavelength are inversely proportional to each other. Therefore, wavelength of the band in the visible spectrum of I is longer and the color is intense. Conversely, the wavelength of the band in the visible spectrum of F is shorter and the color is pale.

As the temperature is decreased, the intensity of color too tends to decrease. The various energy states are not affected by temperature. However, at lower temperature, the electron requires greater kinetic energy to get promoted to higher energy levels. This means that electronic transitions would be possible only if the molecule absorbs those photons of visible light that have appreciably high energy. Since such photons are relatively scarce, the color intensity registers a decline. Thus at 78 K, Cl is almost colorless, while Br has a pale yellow color.

2. Bond energy

Since halogens exist as diatomic molecules, it is expected that the bond energy would decrease as the size of the constituent atoms increases. This is because larger atoms undergo less effective overlap. In fact, the bond dissociation energy of Cl is higher than that of Br which, in turn, is higher than that of I. However, F is anomalous. Its bond energy is expected to be the highest, but it is the lowest among halogens.

The main reason for the low bond energy of F molecule is the mutual repulsion among the lone pairs of electrons on the two F atoms. Since the F–F bond distance is small, the repulsion between the non-bonding electrons on each atom is very strong. The non-bonding electrons, no doubt, are present on other halogen molecules as well. However, the longer bond lengths in other halogens separate the lone pairs by a greater distance, causing relatively weak repulsion.

Another reason for the anomalous character of F is that in other halogens some degree of pd hybridization occurs during the overlap of orbitals of constituent atoms. This type of hybridization allows, at least in part, multiple bonding between the halogen atoms, thus increasing the bond strength. In the case of a fluorine molecule, pd hybridization is not possible since F atoms do not possess d orbitals. This leads to weakening of F–F bond.

Case Study 19.1 Why are halogen bulbs more durable than conventional ones?

The conventional bulbs incorporate a tungsten wire which gets heated up when current is passed through it. As a result, an incandescent, white glow is produced. Even though tungsten has a very high melting point, 3695 K, yet, a part of tungsten undergoes sublimation. Over a period of time, the wire becomes thiner and finally breaks. Greater the wattage of the bulb, greater will be the rate of sublimation of tungsten, and earlier the bulb will fuse. Therefore, wherever high intensity light is required, it is pertinent to use halogen lamps in place of conventional lighting systems.

The halogen bulbs too use a tungsten filament, yet these are also filled with iodine. The iodine vapors react with the tungsten content that has left the wire through sublimation, forming gaseous tungsten(II) iodide.

$$W + I_2 \rightarrow WI_2$$

The vapors of tungsten(II) iodide, on coming in contact with the filament, undergo thermal decomposition.

$$WI_2 \rightarrow W + I_2$$

The tungsten gets re-deposited on the wire, restoring its thickness, and increasing the lifespan of the bulb.

Chemical Properties

The abnormally low bond dissociation energy of F makes it an extremely reactive element. It reacts with virtually all other elements, including some noble gases. Its reactions with some elements are vigorous, while with the others, explosive.

Even water vapors catch fire on interacting with fluorine, leading to the formation of hydrogen fluoride (*see* below). As a result, it is problematic to select the material for making a storage vessel for fluorine. The best option is to go for either nickel or copper vessel. It is not that these metals are unreactive towards fluorine. These do react, but are readily passivated by fluorine. A thin film of nickel fluoride or copper fluoride is formed on the inside surface of the container and it obviates further reaction between the metal and fluorine.

Even though the F–F bond is very weak, the bond formed between fluorine and other elements is very strong. This is because fluorine can bring out a high coordination number and a high oxidation state of the element with which it combines. Both the factors, in turn, ensure that the lattice energy, and hence the stability of the resulting compound is appreciably high.

1. Reaction with water

Fluorine, being a strong oxidant, oxidizes water to oxygen.

$$2F_2 + 2H_2O \rightarrow 4HF + O_2$$

However, the other halogens are not as strong oxidizing agents as fluorine. They form hypohalous acids (HOX) with water. For example,

$$Cl_2 + H_2O \rightarrow \underset{\text{Hypochlorous acid}}{HOCl} + HCl$$

Case Study 19.2 Which raw material serves as a substitute of chlorine for disinfecting swimming pools?

Swimming pools need to be disinfected at regular intervals since these are potential breeding sites of disease-causing bacteria. Earlier, chlorine was used as a bactericide. As has been cited above, in aqueous solution, it yields hypochlorous acid, which has anti-bacterial characteristics.

However chlorine has to be transported in cylinders, a mode which is both tedious and hazardous. Trichlorocyanuric acid serves as a substitute of chlorine. It reacts with water to produce hypochlorous acid.

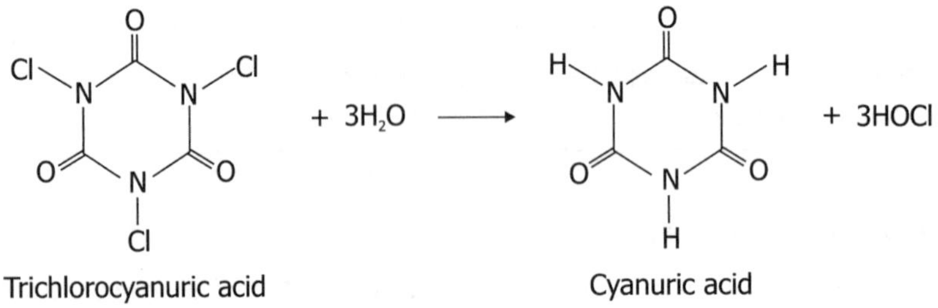

Trichlorocyanuric acid is commercially available in powdered form. It is safer to use and easier to handle than cylinder-packed chlorine.

2. Reaction with sodium hydroxide

All halogens react with aqueous sodium hydroxide or, for that matter, with any other alkali. Chlorine, bromine and iodine give the respective halide and hypohalite, when treated with cold, dilute sodium hydroxide solution. For example,

$$Cl_2 + \underset{\text{Cold and dilute}}{2NaOH} \rightarrow NaCl + \underset{\text{Sodium hypochlorite}}{NaOCl} + H_2O$$

With hot and concentrated alkali solution, halide and halate are formed. For example,

$$3Cl_2 + \underset{\text{Hot and conc.}}{6NaOH} \rightarrow 5NaCl + \underset{\text{Sodium chlorate}}{NaClO_3} + 3H_2O$$

The reaction of fluorine with sodium hydroxide yields oxygen difluoride.

$$2F_2 + 2NaOH \rightarrow 2NaF + H_2O + OF_2$$

Irrespective of whether cold/dilute or hot/concentrated sodium hydroxide is used, the reaction products, in this case, remain the same.

3. Reaction with hydrogen sulfide

All halogens oxidize hydrogen sulfide to sulfur.

$$X_2 + H_2S \rightarrow 2HX + S \quad (X = F, Cl, Br \text{ or } I)$$

4. Reaction with ammonia

Fluorine, chlorine and bromine react with ammonia to liberate nitrogen gas.

$$3X_2 + 8NH_3 \rightarrow 6NH_4X + N_2 \quad (X = F, Cl \text{ or } Br)$$

5. Reaction with carbon monoxide

Chlorine and bromine form the respective carbonyl halide with carbon monoxide.

$$X_2 + CO \rightarrow COX_2 \quad (X = Cl \text{ or } Br)$$

Case Study 19.3 What is the role of fluoride in dental care?

It has been observed that children who consume water containing approximately one part per million (ppm) of fluoride ions, during the period when their permanent teeth are growing, have fewer cavities than children who consume water devoid of fluoride ions. For this reason, community water supplies are generally fluoridated by adding a soluble fluoride salt, such as sodium fluoride. Another alternative is to use fluoride-based toothpastes. In these formulations, stannous fluoride is added to affect fluoridation.

The main constituent of tooth enamel is hydroxyapatite, $Ca_5(OH)(PO_4)_3$ which, in the presence of fluoride ions, is converted to harder, acid-resistant fluorapatite, $Ca_5F(PO_4)_3$, thus, preventing tooth decay.

$$Ca_5(OH)(PO_4)_3 + F^- \rightarrow Ca_5F(PO_4)_3 + OH^-$$

Paradoxically, consuming 2 ppm or more of fluoride is harmful, causing mottling of teeth.

Compounds

Iodine has an appreciable degree of covalent character and is, therefore, readily soluble in several organic solvents. The violet color of iodine, however, changes on dissolution. In aromatic hydrocarbons, such as benzene, the color changes to red, while in ethers or amines, it changes to brown. The change in color suggests that the halogen does not merely dissolve in the solvent in a physical sense, but enters into chemical interaction with it. In this interaction, the solvent acts as an electron donor and iodine as an electron acceptor.

The aromatic compounds donate electrons to the halogens through π-bonds, while ethers and amines do so through σ-bonds (Fig. 19.1). In either case, the transfer of electrons from the solvent to iodine perturbs the molecular energy levels of the latter. Now when an electron is excited by visible light from the highest occupied molecular orbital to the lowest unoccupied molecular orbital, the transition involves a different energy than that required for a similar transition in free halogens. As a result, the color changes. The compounds formed by the interaction of iodine with solvents are called *charge-transfer compounds* or *charge-transfer complexes.* The evidence that these compounds are formed by donor–acceptor interactions comes from the fact that when iodine is dissolved in a solvent such as carbon tetrachloride, which has no donor characteristics, no color change is involved.

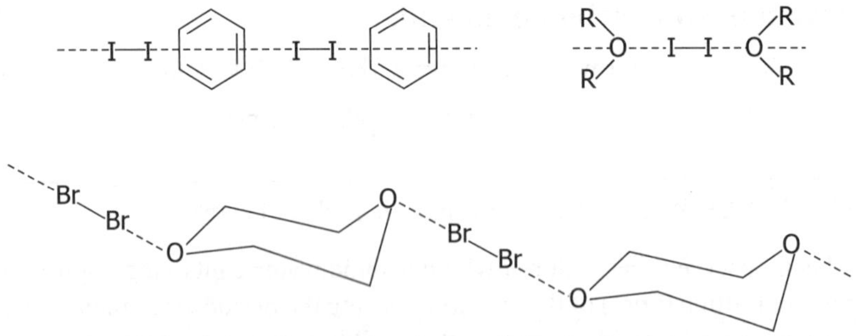

Fig. 19.1 Structure of some charge-transfer compounds

Charge-transfer complexes are not limited to iodine. Bromine, and to some extent chlorine, too form such compounds. For example, bromine forms a long-chain, charge-transfer compound when dissolved in dioxane. The structure of this compound is shown in Fig. 19.1.

The bond distance of the halogen in a charge-transfer complex is longer than that in a free compound. On mild heating, the complex dissociates into free halogen and the solvent. On cooling, the complex is formed once again.

As has been stated, the halogen molecule accepts electrons from organic molecules to form charge transfer complexes. It means that in these complexes, the halogens behave as Lewis acids. Such behavior is not restricted to charge transfer complexes alone. Molecular chlorine demonstrates Lewis acid character in its reaction with tetraphenylarsenic(V) chloride, $[(C_6H_5)_4As]Cl$ at 200 K

$$[(C_6H_5)_4As]Cl + Cl_2 \rightarrow [(C_6H_5)_4As]\,[Cl_3]$$

The trichloro anion in the product has a linear geometry.

Molecular chlorine also shows an unusual property when it acts as a Lewis base while reacting with a mixture of chlorine fluoride and antimony(V) fluoride.

$$ClF + Cl_2 + SbF_5 \rightarrow [Cl_3]\,[SbF_6]$$

In this reaction, a chlorine molecule combines with a chlorine cation, Cl^+.

Case Study 19.4 Which gaseous compound has the highest dielectric constant?

Perchloryl fluoride, $FClO_3$, has the highest dielectric constant of any known gas. It is prepared by the reaction of fluorosulfuric acid with sodium perchlorate.

$$\underset{\text{Sodium perchlorate}}{NaClO_4} + \underset{\text{Fluoro-sulfuric acid}}{2FSO_3H} \rightarrow \underset{\text{Disulfuric acid}}{H_2S_2O_7} + NaF + FClO_3$$

It is an inert and non-corrosive gas with a tetrahedral structure.

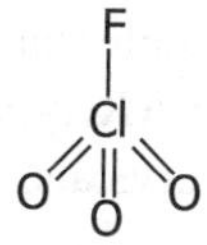

Perchloryl fluoride is used as a rocket propellent as well as an insulator in high-voltage systems.

1. Interhalogen compounds

The halogens combine with each other exothermically to give a set of compounds called interhalogen compounds. These exist in four stoichiometries: AX, AX_3, AX_5 and AX_7, where A is the larger halogen and X the smaller one. Since all interhalogen compounds have an even number of halogen atoms, these compounds are diamagnetic in nature. The A–X bonds are polar, whereas those in the halogen molecules are not. The A–X bond energies, melting points and boiling points increase as the difference in electronegativities of A and X increases. In general, the properties of interhalogen compounds are intermediate between those of constituent halogens, A_2 and X_2. Some interhalogen compounds are listed in Table 19.1.

Table 19.1 Interhalogen compounds

AX	**AX_3**	**AX_5**	**AX_7**
ClF	ClF_3	ClF_5	IF_7
BrF	BrF_3	BrF_5	
BrCl	IF_3	IF_5	
IF	ICl_3		
ICl			
IBr			

AX Most of the members of AX series are prepared by direct combination of halogens. For example,

$$Cl_2 + F_2 \xrightarrow{500\ K} 2ClF$$

$$I_2 + Cl_2 \xrightarrow{\text{Room temperature}} 2ICl$$

A few may be prepared by the reaction of halogens with halides,

$$I_2 + AgF \xrightarrow{273\ K} IF + AgI$$

or other interhalogen compounds.

$$Br_2 + BrF_3 \xrightarrow{600\ K} 3BrF$$

Being diatomic species, these molecules have linear geometry. Chlorine fluoride, a colorless gas, is the most volatile AX type of interhalogen compound, while iodine bromide, which forms black crystals, is the least volatile. The former is a very efficient fluorinating agent.

$$W + 6ClF \rightarrow WF_6 + 3Cl_2$$
$$Se + 4ClF \rightarrow SeF_4 + 2Cl_2$$

It can also act as a chlorofluorinating agent.

$$ClF + SF_4 \rightarrow SF_5Cl$$
$$ClF + CO \rightarrow COFCl$$

Further, it shows Lewis acid behavior.

$$ClF + NOF \rightarrow [NO]^+ [ClF_2]^-$$

Iodine chloride too shows Lewis acid behavior, the acid site being iodine.

$$CH_3CN + I\text{—}Cl \rightarrow CH_3CN\text{—}I\text{—}Cl$$
$$Cl^- + I\text{—}Cl \rightarrow [Cl\text{—}I\text{—}Cl]^-$$

Iodine fluoride is unstable and disproportionates rapidly into iodine pentafluoride and iodine.

$$5IF \rightarrow IF_5 + 2I_2$$

AX_3 The most important compound of AX_3 series of interhalogens is chlorine trifluoride. It is prepared by reacting stoichiometric quantities of constituent halogens at 500 K on copper surface.

$$Cl_2 + 3F_2 \xrightarrow[Cu]{500\ K} 2ClF_3$$

Bromine trifluoride is prepared similarly.

Iodine trifluoride, which is stable only below 243 K, may be synthesized by the reaction of iodine with xenon difluoride.

$$I_2 + 3XeF_2 \rightarrow 2IF_3 + 3Xe$$

The central atom, A in AX_3 type of interhalogen compounds, is sp^3d hybridized. Consider, for example, bromine trifluoride.

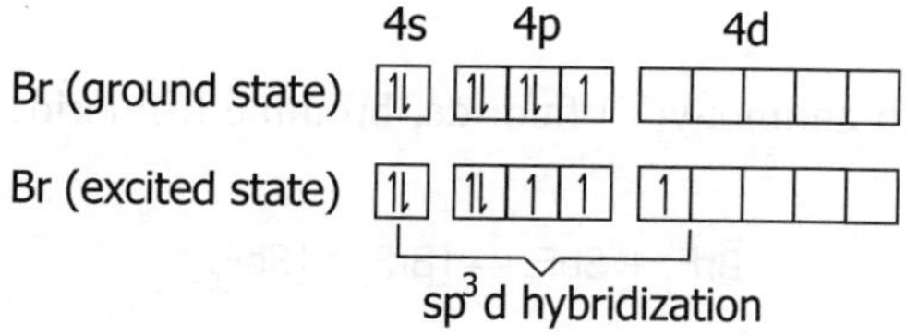

The two lone pairs occupy the equatorial positions of the trigonal bipyramid skeleton,

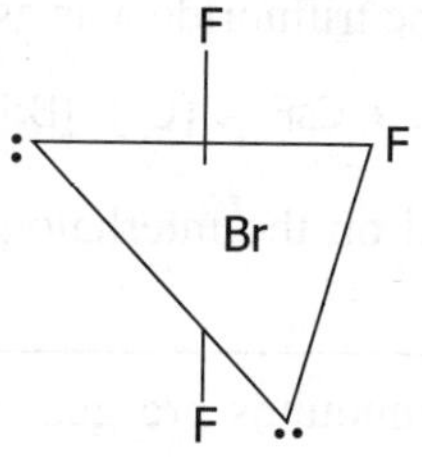

giving a 'T' shaped structure.

Chlorine trifluoride is an extremely reactive compound. It spontaneously ignites substances such as asbestos or wood.

It converts metal salts into fluorides.

$$4ClF_3 + 6MgO \rightarrow 6MgF_2 + 2Cl_2 + 3O_2$$
$$2ClF_3 + 2AgCl \rightarrow 2AgF_2 + Cl_2 + 2ClF$$

It converts metals to metal fluorides. This property is used for separating uranium and plutonium in a nuclear reactor.

$$3ClF_3 + U \rightarrow UF_6 + 3ClF$$
$$4ClF_3 + 3Pu \rightarrow 3PuF_4 + 2Cl_2$$

The uranium(VI) fluoride is volatile, while plutonium(IV) fluoride is not and hence the two are easily separated by diffusion technique.

Chlorine trifluoride can act as a fluoride ion donor (Lewis base),

$$ClF_3 + AsF_5 \rightarrow [ClF_2]^+ [AsF_6]^-$$

or a fluoride ion acceptor (Lewis acid).

$$ClF_3 + NOF \rightarrow [NO]^+ [ClF_4]^-$$

Case Study 19.5 Predict which of the following reagents will make bromine trifluoride a stronger Lewis acid or a stronger Lewis base: (a) antimony(V) fluoride; (b) cesium fluoride.

While reacting with antimony(V) fluoride, bromine trifluoride acts as a fluoride donor.

$$BrF_3 + SbF_5 \rightarrow [BrF_2]^+ [SbF_6]^-$$

The positive charge induced on the interhalogen moiety makes it a stronger Lewis acid.

With cesium fluoride, bromine trifluoride acts as a fluoride acceptor.

$$BrF_3 + CsF \rightarrow [Cs]^+ [BrF_4]^-$$

The negative charge induced on the interhalogen moiety makes it a stronger Lewis base.

AX_5 The AX_5 type of compounds are generally prepared either by direct fluorination of halogens,

$$Cl_2 + 5F_2 \xrightarrow{650\text{ K, }250\text{ atm}} 2ClF_5$$

$$Br_2 + 5F_2 \xrightarrow{450\text{ K}} 2BrF_5$$

$$I_2 + 5F_2 \xrightarrow{\text{Room temperature}} 2IF_5$$

or by fluorination of lower interhalogen compounds. For example,

$$ClF_3 + F_2 \xrightarrow{\text{Room temperature}} ClF_5$$

These may also be prepared by the reaction of fluorine with metal halides. For example,

$$KBr + 3F_2 \xrightarrow{300\text{ K}} KF + BrF_5$$

The central halogen in these compounds is sp^3d^2 hybridized. Consider the representative case of chlorine pentafluoride, ClF_5.

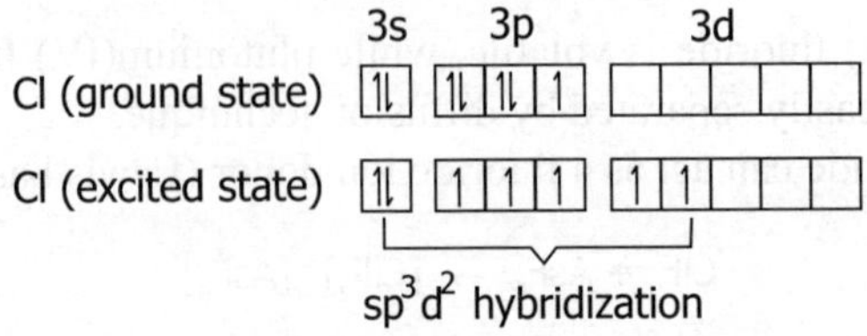

The structure is octahedral, with one position occupied by a lone pair or square pyramid. The central atom is slightly below the basal plane of four fluorine atoms.

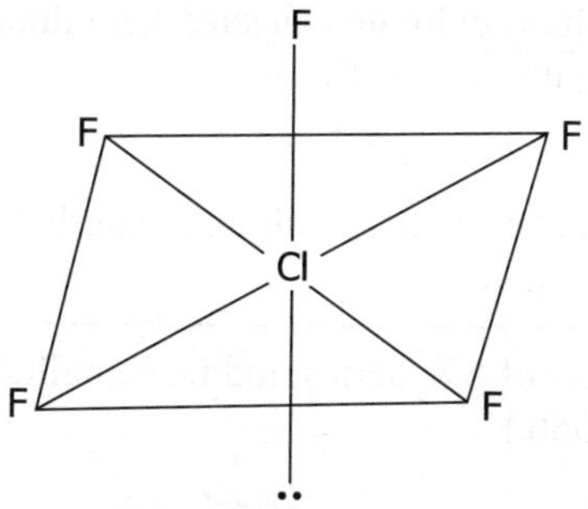

Chlorine and bromine pentafluorides are extremely reactive compounds and vigorous fluorinating agents, while iodine pentafluoride is relatively mild. Chlorine pentafluoride is vigorously hydrolyzed to yield chloryl fluoride, ClO_2F.

$$ClF_5 + 2H_2O \rightarrow ClO_2F + 4HF$$

It acts both as a fluoride donor,

$$ClF_5 + AsF_5 \rightarrow [ClF_4]^+ [AsF_6]^-$$

and as a fluoride acceptor.

$$ClF_5 + CsF \rightarrow [Cs]^+ [ClF_6]^-$$

Case Study 19.6 Outline the method for preparing very pure chlorine monofluoride, chlorine trifluoride and chlorine pentafluoride.

Chlorine and fluorine react with each other in the temperature range 473-573 K to yield a mixture of chlorine monofluoride, chlorine trifluoride and chlorine pentafluoride. All the three interhalogen compounds are gaseous and are extremely difficult to segregate. When the reaction is conducted at 523 K and chlorine and fluorine are mixed in 1:3 molar ratio, the predominant product is chlorine trifluoride, although small amounts of chlorine monofluoride and chlorine pentafluoride are also formed. To the mixture of these gases, a solution of potassium fluoride is added. Only chlorine trifluoride combines with potassium fluoride to yield potassium tetrafluorochlorate, $K[ClF_4]$. The latter is a solid and hence can easily be isolated from the gaseous mixture of chlorine monofluoride and chlorine pentafluoride.

$$ClF_3 + KF \rightarrow K[ClF_4]$$

On heating potassium tetrafluorochlorate to 410 K, the above reaction reverses,

$$K[ClF_4] \rightarrow ClF_3 + KF$$

yielding extremely pure chlorine trifluoride.

The chlorine trifluoride formed by this route is reduced by chlorine at 523 K to chlorine monofluoride.

$$ClF_3 + Cl_2 \rightarrow 3ClF$$

When potassium tetrafluorochlorate is heated with fluorine in a closed vessel at 473 K, pure chlorine pentafluoride is formed.

$$K[ClF_4] + F_2 \rightarrow ClF_5 + KF$$

By following this synthetic sequence, all three interhalogen compounds may be isolated in extremely pure states.

AX_7 The only member of AX_7 series, iodine heptafluoride (IF_7) is prepared by reacting the halogens at 550 K.

$$I_2 + 7F_2 \xrightarrow{550\ K} 2IF_7$$

It may also be prepared by the reaction of fluorine either with potassium iodide,

$$KI + 4F_2 \rightarrow KF + IF_7$$

or with lead iodide.

$$PbI_2 + 8F_2 \rightarrow PbF_2 + 2IF_7$$

As already discussed in Chapter 3 (p. 113), iodine heptafluoride has a pentagonal bipyramid structure, the iodine being sp^3d^3 hybridized.

Iodine heptafluoride is readily hydrolyzed to yield iodine oxopentafluoride, IOF_5.

$$IF_7 + H_2O \rightarrow IOF_5 + 2HF$$

The same compound is obtained with silicon dioxide.

$$2IF_7 + SiO_2 \rightarrow 2IOF_5 + SiF_4$$

Iodine heptafluoride can act as a fluoride donor,

$$IF_7 + AsF_5 \rightarrow [IF_6]^+ [AsF_6]^-$$

as well as a fluoride acceptor.

$$IF_7 + CsF \rightarrow [Cs]^+ [IF_8]^-$$

Case Study 19.7 Arrange the following in decreasing order of their reactivity: fluorine molecule, chlorine monofluoride and chlorine trifluoride.

The reactivity of a molecule is largely dependent on its bond energy. If the bond energy is low, the molecule readily dissociates and its reactivity towards other chemical species is high. If, on the other hand, its bond strength is high, it is non-reactive.

Fluorine is considered to be a very reactive halogen, but chlorine trifluoride is much more reactive. It reacts explosively with substances such as cotton, paper, wax and grease. This is because the energy required to break the $F_2Cl–F$ bond in chlorine trifluoride is lower than that required to break the F–F bond in fluorine molecule.

Between chlorine monofluoride and fluorine molecule, the former has higher bond energy than the latter. Hence, it is expected that chlorine monofluoride would be less reactive than fluorine molecule. However, the reverse is observed. Although chlorine monofluoride is less reactive than chlorine trifluoride, it is, nevertheless, more reactive than fluorine molecule.

The higher reactivity of chlorine monofluoride is attributed to three factors. First, the chlorine monofluoride molecule is less symmetrical than the fluorine molecule. Second, chlorine monofluoride is polar, while fluorine molecule is non-polar. Third, the chlorine in chlorine monofluoride bears a positive charge. A positively charged chlorine has a high oxidizing power and hence high reactivity.

The order of reactivity is,

$$ClF_3 > ClF > F_2$$

2. Polyhalides

Halide ions react with excess of halogens or with interhalogen compounds to yield polyhalide anions. These anions exist in three stoichiometries: $[A_n]^-$, $[AX_n]^-$ and $[AXY_n]^-$. Representative examples are presented in Table 19.2.

Table 19.2 Polyhalide anions

$[A_n]^-$	$[AX_n]^-$	$[AXY_n]^-$
Br_3^-	ClF_2^-	$IBrF^-$
I_3^-	IBr_2^-	$IClBr^-$
I_6^-	ICl_4^-	$IFCl_3^-$
I_7^-	BrF_6^-	
I_9^-		

The following chemical equations outline common preparation methods for polyhalides:

$$I_2 + I^- \rightarrow I_3^-$$

$$ICl + KCl \rightarrow K^+[ICl_2]^-$$

$$ICl + KBr \rightarrow K^+[IClBr]^-$$

$$ICl_3 + KCl \rightarrow K^+[ICl_4]^-$$

$$IF_5 + CsF \rightarrow Cs^+[IF_6]^-$$

In triatomic polyhalides, the central atom is sp^3d hybridized. For example, I_3^-, the structure of which is given in Fig. 19.2.

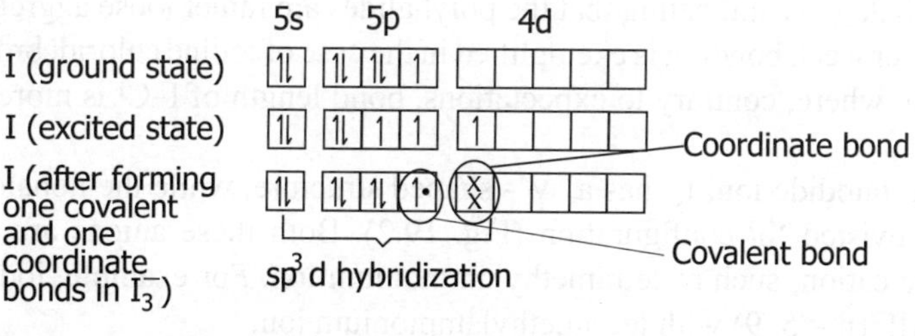

Three positions in the trigonal bipyramid structure are occupied by lone pairs, giving a linear or near linear geometry. Iodine dibromide ion has a similar structure (Fig. 19.2).

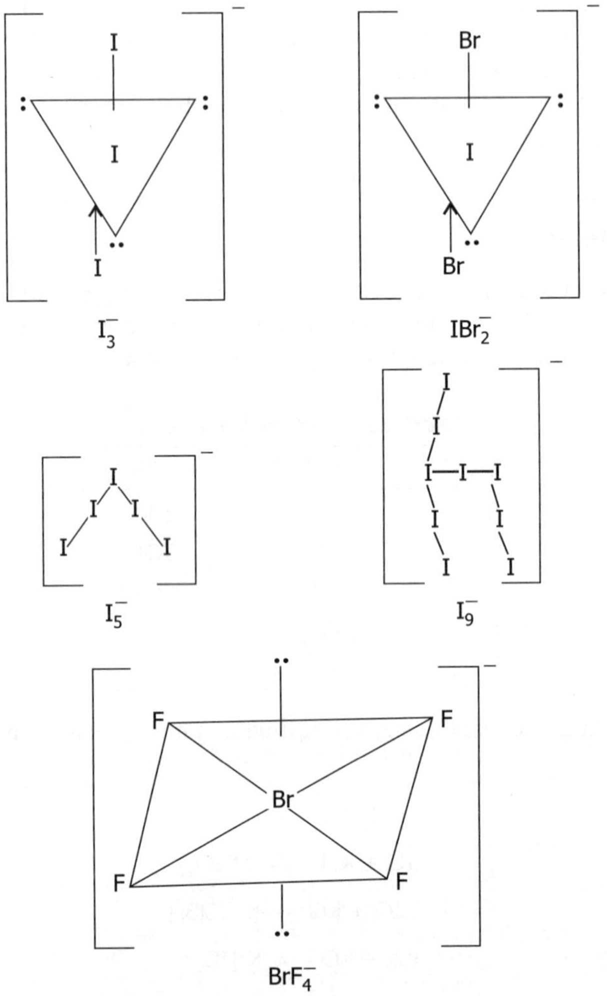

Fig. 19.2 Structure of some polyhalides

The heaviest atom in triatomic polyhalides is always in the middle. The interatomic distances are appreciably larger than those in the corresponding diatomic halogens, indicating that the polyhalides are rather loose aggregates. The evidence for weak bonding is exemplified in the case of iodine chloridebromide ion, $[Cl–I–Br]^-$ where, contrary to expectations, bond length of I–Cl is more than that of I–Br.

The pentaiodide ion, I_5^- has a 'V'- shaped structure, while the nonaiodide ion, I_9^- has a twisted 'h' configuration (Fig. 19.2). Both these anions are stabilized by a large cation, such as tetramethylammonium ion. For example, iodine forms $[(CH_3)_4N]I_n$ (n = 5, 9) with tetramethylammonium ion.

The pentatomic polyhalides of the AX_4^- types are square planar. The central atom is sp^3d^2 hybridized and two positions are occupied by lone pairs. Consider, for example, BrF_4^-, the structure of which is depicted in Fig. 19.2.

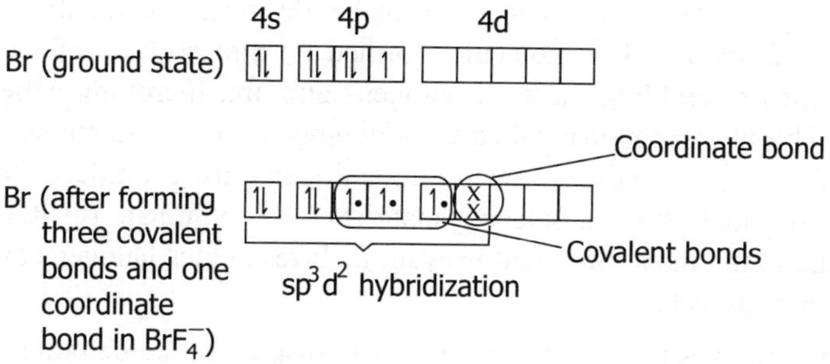

Some polyhalides, for example, tribromide ion, Br_3^- are stable only in solution. However, most of them, when bound to large cations, are stable in crystalline state as well. For a given cation, the more symmetrical the structure of a polyhalide anion, more will be its stability. For example, triiodide ion, I_3^-, forms more stable salts than iodine dibromide ion, IBr_2^-. Similarly, the polyhalides with larger central atoms are more stable. For example, the order of stability is,

$$I_3^- > Br_3^-$$

$$ICl_2^- > BrCl_2^-$$

The metal polyhalides are typical ionic compounds. They are soluble in water and conducting in nature. On heating, they decompose with the smaller halogen remaining bound to the metal. This is because compounds with smaller halide ions have higher lattice energy. For example,

$$Rb[ICl_2] \rightarrow RbCl + ICl$$

$$K[IClBr] \rightarrow KCl + IBr$$

Case Study 19.8 Are the two iodine–iodine bond lengths in triiodide ion equal?

In solid state, the triiodide ion exists in two forms: symmetrical, in which both bond lengths are same, and asymmetrical, in which the two bond lengths are different. A smaller cation tends to distort the structure of triiodide, making it asymmetrical. A cation whose size is compatible with that of the polyhalide anion, on the other hand, will not distort its structure, so that the two bond lengths would be the same. Thus in cesium triiodide, CsI_3, and in ammonium triiodide, NH_4I_3, the polyhalide anion is distorted and the bond distances are unequal. In tetraphenylarsenium triiodide, $[(C_6H_5)_4As]I_3$, the polyhalide anion is symmetrical, with both the iodine–iodine bond lengths being the same.

In solution state, the triiodide ion is always symmetrical, with equal bond distances.

3. Pseudohalogens and pseudohalides

A number of strongly bound radicals can form uninegative ions, similar to halide ions or may dimerize to form molecules, similar to diatomic halogens. The univalent anions have properties similar to those of halide ions, and for this reason, are termed *pseudohalides*. The molecules formed by dimerization of radicals have characteristics resembling those of halogens and are, therefore, called *pseudo-halogens*. This phenomenon is called *pseudohalogen concept*. Examples of radicals that can form pseudohalides and pseudohalogens are cyanide (CN), cyanate (CNO), thiocyanate (SCN), selenocyanate (SeCN), fulminate (ONC) and azide (N_3). Of these, the most important is cyanide. It resembles halogens or halides in the following respects:

a. It dimerizes to form cyanogen $(CN)_2$, just as the halide radical, X, forms the diatomic halogen, X_2. Cyanogen, like fluorine or chlorine, is a gas.
b. It forms hydrocyanic acid, HCN similar to the hydrogen halides, HX.
c. It forms soluble salts with alkali metals. The alkali metal halides are also soluble in water.
d. It forms insoluble salts with silver(I), mercury(I) and lead(II); the corresponding halides of these cations are insoluble too.
e. Silver cyanide is soluble in ammonia, and so is silver chloride.
f. Analogous to interhalogen compounds, pseudointerhalogen compounds such as iodine cyanide, ICN, can be synthesized.
g. It forms complexes with same stoichiometries as halide ions, for example, tetracyanozincate(II) ion, $[Zn(CN)_4]^{2-}$ and tetrachlorozincate(II) ion, $[ZnCl_4]^{2-}$.
h. Like the halides, cyanide can also act as a bridging group and cause polymerization. Of course, the degree of polymerization is different. For example, dialkylgold(III) chloride is dimeric, while dialkylgold(III) cyanide is tetrameric (Fig. 19.3).

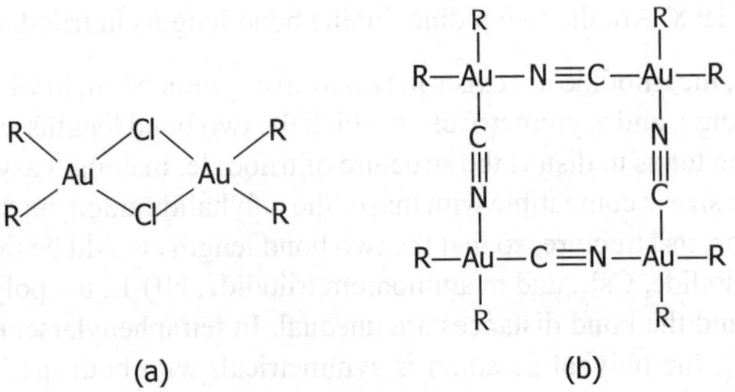

Fig. 19.3 Structure of (a) dialkylgold(III) chloride and (b) dialkylgold(III) cyanide

i. The halogens and the pseudohalogens show similar chemical behavior, as the following reactions exemplify.

(i) Reaction with hydroxide ions.

$$Cl_2 + 2OH^- \rightarrow Cl^- + OCl^- + H_2O$$

$$(CN)_2 + 2OH^- \rightarrow CN^- + OCN^- + H_2O$$

(ii) Reaction with manganese(IV) oxide under acidic conditions.

$$MnO_2 + 4H^+ + 2Cl^- \rightarrow Mn^{2+} + 2H_2O + Cl_2$$

$$MnO_2 + 4H^+ + 2SCN^- \rightarrow Mn^{2+} + 2H_2O + (SCN)_2$$

(iii) Thermal decomposition of lead(IV) compounds.

$$PbCl_4 \rightarrow PbCl_2 + Cl_2$$

$$Pb(SCN)_4 \rightarrow Pb(SCN)_2 + (SCN)_2$$

(iv) Addition to unsaturated hydrocarbons.

$$H_2C = CH_2 + Cl_2 \rightarrow H_2CCl \cdot CClH_2$$

$$H_2C = CH_2 + (SCN)_2 \rightarrow H_2CSCN \cdot CSCNH_2$$

Case Study 19.9 What are *halogenoids* and *interhalogenoids*?

Halogenoids is an alternative name for pseudohalogens. Molecules like cyanogen, $(CN)_2$ are called free halogenoids, while anions like cyanide, CN^- are called halogenoid ions.

Interhalogenoids are compounds formed by combination of different halogenoids. Examples include cyanogen thiocyanate, CNSCN and cyanogen azide, CNN_3. The former may be prepared by the reaction of thiocyanogen with mercury(II) cyanide, while the latter may be synthesized by the reaction of cyanogen bromide with sodium azide.

$$Hg(CN)_2 + 2(SCN)_2 \rightarrow Hg(SCN)_2 + 2CNSCN$$

$$NaN_3 + CNBr \rightarrow NaBr + CNN_3$$

Although both the compounds may be isolated as crystalline solids, these have low melting points.

---

4. Halogen cations

Similar to the trend in any other group, the ionization energy of the halogens decreases from top to bottom. Accordingly, the basic characteristics increase in this direction. As a result, iodine is known to exist in several cationic forms, such as iodine cations, I^+ or I^{3+}, diiodine cation, I_2^+ and triiodine cation, I_3^+.

For example, dipyridineiodine(I) nitrate, $[I(py)_2]^+ NO_3^-$ may be prepared by the reaction between stoichiometric amounts of silver nitrate and iodine in excess of pyridine (py).

$$AgNO_3 + I_2 + 2py \rightarrow [I(py)_2]^+NO_3^- + AgI$$

The same cation is obtained on electrolyzing iodine chloride in pyridine.

$$2ICl + 2py \rightarrow [I(py_2)]^+ + [ICl_2]^-$$

Addition of iodine to silver perchlorate, at 200 K, yields iodine(III) perchlorate.

$$3AgClO_4 + 2I_2 \xrightarrow{200\ K} I(ClO_4)_3 + 3AgI$$

The reaction between iodine and antimony(V) fluoride, in liquid sulfur dioxide medium, at 300 K, yields the diiodine cationic species.

$$2I_2 + 5SbF_5 \xrightarrow[SO_2(l)]{300\ K} 2[I_2]^+[Sb_2F_{11}]^- + SbF_3$$

The reaction of iodine with arsenic(V) fluoride in liquid sulfur dioxide solvent gives triiodine cation.

$$3I_2 + 3AsF_5 \rightarrow 2[I_3]^+[AsF_6]^- + AsF_3$$

The corresponding cations of bromine and chlorine are less stable. In fact, neither bromine(I) and chlorine(I) species nor dichlorine(I) species exist.

Dibromine cationic species may be prepared by the reaction of bromine with antimony(V) fluoride, in bromine pentafluoride solvent,

$$2Br_2 + 7SbF_5 \xrightarrow{BrF_5(l)} 2[Br_2]^+ [Sb_3F_{16}]^- + SbF_3$$

while tribromine cation is synthesized by the reaction between bromine, arsenic(V) fluoride and bromine trifluoride.

$$Br_2 + BrF_3 + AsF_5 \rightarrow [Br_3]^+ [AsF_6]^- + F_2$$

The analogous trichlorine cationic moiety is obtained at 200 K by reacting chlorine with arsenic(V) fluoride and chlorine fluoride.

$$Cl_2 + ClF + AsF_5 \xrightarrow{200\ K} [Cl_3]^+ [AsF_6]^-$$

5. Oxyacids of chlorine

Chlorine forms four oxyacids, the compositions, methods of preparation and structures of which are presented in Table 19.3. The variation in their acid strength has already been discussed in Chapter 5 (p. 230).

Table 19.3 Oxyacids of chlorine

Name	Composition	Oxidation state of chlorine	Method of preparation	Structure
Hypochlorous acid	HOCl	I	$2Cl_2 + 2HgO + H_2O \rightarrow HgO.HgCl_2 + 2HOCl$	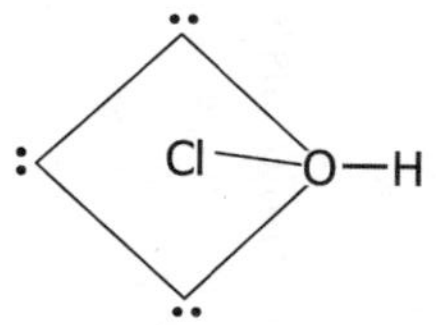
Chlorous acid	$HClO_2$	III	$Ba(OH)_2 + H_2O_2 + 2ClO_2 \rightarrow Ba(ClO_2)_2 + 2H_2O + O_2$ $Ba(ClO_2)_2 + H_2SO_4 \rightarrow BaSO_4 + 2HClO_2$ Barium chlorite	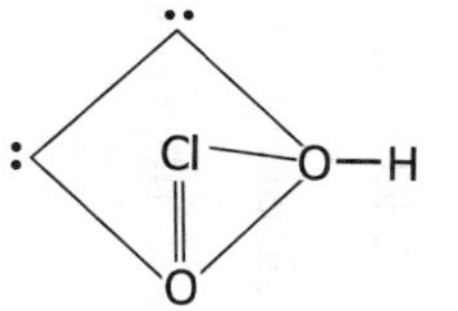
Chloric acid	$HClO_3$	V	$Ba(ClO_3)_2 + H_2SO_4 \rightarrow BaSO_4 + 2HClO_3$ Barium chlorate	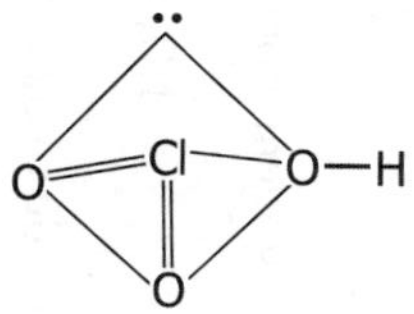
Perchloric acid	$HClO_4$	VII	$NH_4ClO_4 + HNO_3 \rightarrow NH_4NO_3 + HClO_4$ Dilute	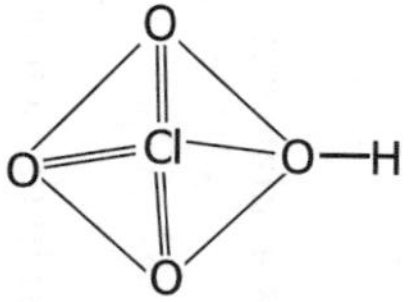

Hypochlorous acid readily decomposes into oxygen.

$$2HOCl \rightarrow 2HCl + O_2$$

It is a strong oxidizing agent. For example, it readily oxidizes nitrite ions to nitrate ions,

$$NO_2^- + HOCl \rightarrow NO_3^- + HCl$$

and hydrogen peroxide to oxygen.

$$H_2O_2 + HOCl \rightarrow HCl + H_2O + O_2$$

Chlorous acid is unstable and decomposes easily into chlorine dioxide, ClO_2.

$$5HClO_2 \rightarrow HCl + 4ClO_2 + 2H_2O$$

Similarly, chloric acid also tends to decompose in concentrated solutions.

$$4HClO_3 \rightarrow 4ClO_2 + 2H_2O + O_2$$

Perchloric acid is a stable species and a strong oxidizing agent. It gives hydrogen on reaction with electropositive metals.

$$Mg + 2HClO_4 + 6H_2O \rightarrow [Mg(H_2O)_6](ClO_4)_2 + H_2$$

It reacts with oxides of less basic metals to yield the corresponding perchlorates. For example,

$$Ag_2O + 2HClO_4 \rightarrow 2AgClO_4 + H_2O$$

While the perchlorate and periodate ions are very stable, the perbromate ion is not. It may be prepared by the oxidation of bromate ion with either xenon difluoride,

$$BrO_3^- + XeF_2 + H_2O \rightarrow BrO_4^- + 2HF + Xe$$

or fluorine

$$BrO_3^- + F_2 + 2OH^- \rightarrow BrO_4^- + 2F^- + H_2O$$

The instability of perbromate ion is attributed to its higher reduction potential as compared to perchlorate and periodate ions. Thus, only very strong oxidizing agents like xenon difluoride or fluorine have to be employed in the synthesis of perbromate ion. Once formed, it very easily gets reduced to bromate ion.

$$BrO_4^- + 2H^+ + 2e^- \rightarrow BrO_3^- + H_2O$$

Case Study 19.10 Which is more stable: chloric acid or perchloric acid?

Chloric acid and its salts are less stable than perchloric acid and its salts, since they easily undergo decomposition or disproportionation reactions. For such reactions, it is necessary that the molecules or ions collide with each other. The chlorine of chloric acid or chlorates is not well protected from every side and is, therefore, likely to be attacked by chemical species, including other chlorate ions.

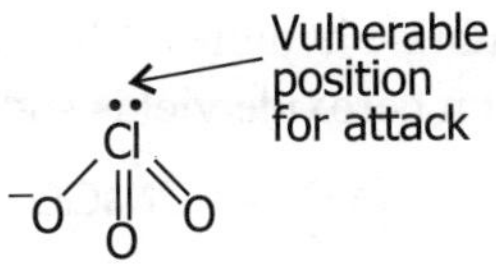

On the other hand, chlorine in perchloric acid or perchlorate ion is protected by oxygen atoms from all the sides.

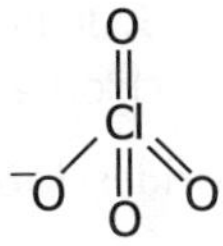

It is prevented from coming into contact with other ions or molecules. The greater symmetry of perchlorate ion further adds to its stability.

6. Chlorine dioxide

Chlorine dioxide is a yellow-colored, paramagnetic gas which explodes at pressures greater than 50 mm Hg. Despite this risk, it is used extensively for bleaching wood and pulp. The liquid explodes at 230 K and hence, does not have any commercial importance.

Chlorine dioxide is generally prepared by the reaction between sodium chlorate and sodium chloride under acidic conditions.

$$2ClO_3^- + 2Cl^- + 4H^+ \rightarrow 2ClO_2 + Cl_2 + 2H_2O$$

The chlorine byproduct is usually not removed since, as has been stated, chlorine dioxide is mainly used for bleaching purposes and chlorine too is a bleaching agent.

Extremely pure chlorine dioxide may be obtained by reducing silver chlorate with chlorine at 363 K.

$$2AgClO_3 + Cl_2 \xrightarrow{363\text{ K}} 2AgCl + 2ClO_2 + O_2$$

Chlorine dioxide is an angular molecule with the ClOCl bond angle of 117.6°. On photolysis or thermolysis it yields chlorine monoxide and oxygen radical.

$$ClO_2 \xrightarrow{h\upsilon \text{ or } \Delta} ClO + O$$

Subsequently,

$$ClO + ClO \rightarrow Cl_2 + O_2$$
$$ClO_2 + O \rightarrow ClO_3$$
$$ClO_3 + ClO_3 \rightarrow Cl_2O_6$$

Dichlorine hexaoxide, Cl_2O_6, may also be obtained by direct reaction between chlorine dioxide and ozone.

$$4ClO_2 + 2O_3 \rightarrow 2Cl_2O_6 + O_2$$

Chlorine dioxide forms clathrate hydrates, $ClO_2.nH_2O$ (n = 8–10) with water at 273 K. Its reaction with sodium peroxide yields sodium chlorite.

$$2ClO_2 + Na_2O_2 \rightarrow 2NaClO_2 + O_2$$

With sodium hydroxide, a mixture of sodium chlorite and sodium chlorate is formed.

$$2NaOH + 2ClO_2 \rightarrow \underset{\text{Sodium chlorite}}{NaClO_2} + \underset{\text{Sodium chlorate}}{NaClO_3} + H_2O$$

Arsenic(V) and antimony(V) fluorides form adducts with chlorine dioxide at 200 K.

$$ClO_2 + MF_5 \xrightarrow{200\text{ K}} ClO_2.MF_5 \quad (M = As \text{ or } Sb)$$

With dinitrogen tetraoxide, it forms chlorine nitrate.

$$2ClO_2 + N_2O_4 \rightarrow 2ClONO_2 + O_2$$

Case Study 19.11 Although both nitrogen dioxide and chlorine dioxide are odd electron molecules, yet the former dimerizes, the latter does not. Why?

The nitrogen in nitrogen dioxide is sp^2 hybridized. Two of these hybrid orbitals overlap with atomic orbitals of oxygen, the third houses the odd electron, all being in one plane (*see* Table 17.1, p. 514).

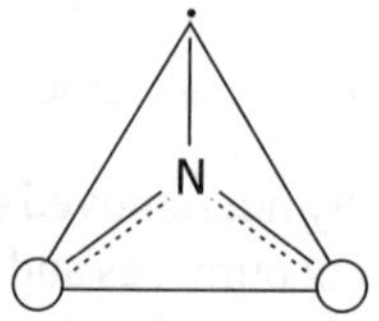

Since the odd electron is in the same plane as are the other atoms, it can couple with the odd electron of another nitrogen dioxide molecule, producing dimeric dinitrogen tetraoxide at 250 K.

The chlorine in chlorine dioxide is also sp^2 liybridized. Two of these hybrid orbitals overlap with atomic orbitals of oxygen, the third houses the lone pair. The odd electron, in this case is in a p orbital of chlorine. This orbital is positioned 90° with respect to the plane of the molecule.

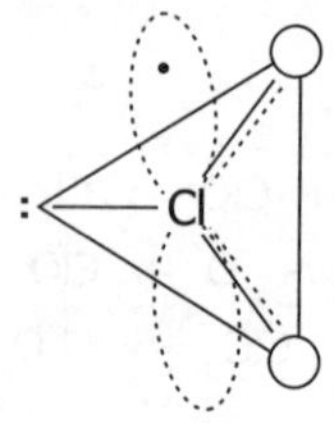

Due to its out-of-plane orientation, the odd electron on one chlorine dioxide molecule cannot couple with its counterpart on another molecule. Hence chlorine dioxide does not dimerize.

Box 19.1

Bleaching Action of Chlorine and Chlorine Dioxide: A Comparison

Both chlorine and chlorine dioxide serve as bleaches due to their oxidizing characteristics. Chlorine dioxide has two distinct advantages over chlorine. Firstly, chlorine dioxide is an excellent discoloring reagent for the wood pulp industry where it produces paper of appreciable strength. Chlorine, on the other hand, attacks the cellulose content of paper so that the mechanical strength of the paper is not up to the mark. Secondly, no toxic residues are produced during chlorine dioxide usage. With chlorine, toxic organic byproducts like dioxins are released to the environment.

7. Hypofluorous acid

The only known oxyacid of fluorine, hypofluorous acid (HOF) is prepared by the fluorination of ice at 230 K.

$$F_2 + H_2O(s) \xrightarrow{230\ K} HOF + HF \quad (19.1)$$

Hypoflourous acid is a white solid that melts at 156 K to a pale yellow liquid. It has an angular structure, with HOF bond angle of 97°. Of all the angular molecules in which oxygen is the central atom, hypofluorous acid has the smallest bond angle (in water, the HOH bond angle is 104.5°, while in oxygen difluoride, the FOF bond angle is 103°). This is because the fluorine atom in hypofluorous acid bears a partial negative charge and the hydrogen atom bears a partial positive charge. The mutual attraction between the two terminal atoms leads to a smaller bond angle.

Hypofluorous acid is an unstable compound. It reacts rapidly with water to produce hydrogen peroxide.

$$HOF + H_2O \rightarrow H_2O_2 + HF \quad (19.2)$$

Its reaction with hydrogen fluoride too is quite vigorous.

$$HOF + HF \rightarrow H_2O + F_2 \quad (19.3)$$

In the presence of moisture, it reacts with fluorine to yield oxygen difluoride.

$$HOF + F_2 \xrightarrow{Moisture} OF_2 + HF \quad (19.4)$$

Reactions 19.2–19.4 highlight the difficulty in preparing hypofluorous acid. As is clear from reaction 19.1, hypofluorous acid is prepared from two reactants, fluorine and ice (water). Apart from hypofluorous acid, reaction 19.1 also yields hydrogen fluoride. As shown in reactions 19.2–19.4, hypofluorous acid reacts with all the reactants and products involved in reaction 19.1. Therefore, it is pertinent to remove all the hypofluorous acid from the reaction as soon as it is formed. This is done by passing the contents successively through three U tubes made of Teflon and maintained at 230 K, 200 K and 90 K. The first removes water, the second hydrogen fluoride, while the third traps hypofluorous acid. The residual fluorine is passed through the first U tube, initiating reaction 19.1 and the cycle continues.

Case Study 19.12 Which of the following perchlorates are expected to be explosive in nature: NH_4ClO_4, $NaClO_4$, $Mg(ClO_4)_2$ and $[Fe(H_2O)_6]$ $[ClO_4]_2$?

Perchlorate ion is an extremely strong oxidizing agent and, when present in concert with a reducing species (that is, one which tends to get oxidized) does present an explosion hazard. Sodium(I) in sodium perchlorate and magnesium(II) in magnesium perchlorate are not prone to oxidation. Hence, $NaClO_4$ are $Mg(ClO_4)_2$ are safe. However, nitrogen(III) in ammonium perchlorate and iron(II) in hydrated ferrous perchlorate (IUPAC: hexaquairon(II) perchlorate) have a tendency to get oxidized to higher oxidation states. Hence NH_4ClO_4 and $[Fe(H_2O)_6]$ $(ClO_4)_2$ are both explosive in nature.

20. GROUP 18 ELEMENTS

Introduction

Group 18 of the periodic table comprises helium (He), neon (Ne), argon (Ar), krypton (Kr), xenon (Xe) and radon (Rn). Earlier, these elements were collectively referred to as *inert gases.* However, this is a misnomer since Xe is now known to form a good number of compounds with highly electronegative elements such as fluorine and oxygen.The name *rare gases* too is not appropriate since Ar makes up 0.9% of the atmosphere. A better common name is *noble gases* as, just like noble metals, such as gold and platinum, these elements are reluctant to enter into chemical combination, yet are not totally unreactive.

Physical Properties

The electronic configuration of He is $1s^2$, while that of remaining noble gases is ns^2np^6. In any case, these elements have a complete valence shell, and this is responsible for their non-reactive nature. The electron affinity of Group 18 members is zero and their ionization energies are very high. As a result, atoms of these elements have little tendency to either gain or lose electrons.

In their natural states, these elements are colorless, odorless and tasteless monoatomic gases. As already discussed in Chapter 3 (p. 172), the only source of attraction between the atoms of a particular noble gas are weak van der Waals forces. These increase with size, and hence the melting points and boiling points of Group 18 elements increase from He to Rn. In fact, the boiling point of He, 4 K, is the lowest of any known substance.

The noble gases are able to diffuse through materials such as rubber, polyvinyl chloride and even metals. Helium even diffuses out of glass, therefore, ordinary glassware is not suitable for cryogenic studies involving the gas in liquid state.

Case Study 20.1 What is the application of helium in diving operations?

In deep sea, a diver requires the same concentration of oxygen for breathing as is required on land. Therefore, divers carry with them a breathing apparatus containing nitrogen and oxygen in the same ratio as is present in normal air. When they get to the sea depths, the pressure increases steeply. The solubility of gases increases with increasing pressure. Therefore, as the divers go deeper into the sea, oxygen gets utilized in the respiratory system, but nitrogen accumulates, more and more in the body fluids. This does not cause much problem. However, when the divers reach the sea surface, the pressure suddenly reduces and the excess nitrogen, dissolved in their body, rapidly bubbles out. The bubble formation in the body leads to joint pain.

To avoid this problem, divers prefer to breathe a mixture of oxygen and helium. The latter is less soluble in body fluids than nitrogen. At greater depths, less amount of helium enters the body. Therefore, while surfacing, a less amount of helium comes out of the body, causing lesser irritation.

1. Liquid helium

When gaseous helium is cooled to 4.2 K, it liquifies. The liquid formed at this temperature has normal properties and is referred to as helium I. If it is cooled further, at one atmospheric pressure, it is expected to turn into a solid. However, it remains a liquid.

At 2.2 K, discontinuous and marked changes occur in the characteristics of liquid helium, and a new variety, called helium II, is formed. The temperature at which helium I transitions to helium II, 2.2 K, is called λ *point temperature.*

Helium II is not a normal liquid. It is a superfluid. Its viscosity is nearly zero. This means that if the liquid begins to flow, it shall continue flowing, with all the He atoms moving regularly, and in-phase with each other.

When it comes in contact with a surface, it can form an exceedingly thin film, only a few atoms thick, which flows without friction. Thus if helium II is placed in a vessel, it starts traveling upwards, along the walls, and against the gravity. Eventually, it will spill over the edge of the container until the levels inside and outside are the same.

The specific heat of helium II is ten times higher than that of helium I. Its thermal conductivity is 10^6 times more than that of helium I. In fact, helium II has 800 times greater thermal conductivity than copper. Helium II is also a superconductor, i.e., it shows nearly no electrical resistance causing, in turn, large amount of current to flow. The superconductivity arises because at temperatures approaching absolute zero, the system is so well ordered that the valence electrons of He move in pairs, not repelling each other, as these usually would.

Case Study 20.2 Which noble gas can best serve as lowest-temperature refrigerant?

The noble gas having the least boiling point in its group would be the best bargain for lowest-temperature refrigerant. Helium, with a boiling point of 4.2 K is the refrigerant of choice. In fact, it is already been used for extremely low temperature applications, such as cooling of superconducting magnets in nuclear magnetic resonance (NMR) spectrometers.

2. Clathrates

Substances formed by trapping of atoms or molecules in the cavities of a crystalline compound are called clathrates. There are no chemical bonds between the guest and the host species. Only physical attractions, manifesting in the form of van der Waals interactions, exist in clathrates.

The primary condition for the formation of clathrates is the existence of a host compound that can crystallize in an open structure containing cages or cavities. Such cages or cavities are generally formed due to hydrogen bonding between the molecules of host species. The secondary condition for clathrate formation dictates that the atomic or molecular size of the guest species should just match the size of the cage or cavity in the host entity. If the particle size of species to be trapped is very small as compared to that of the host lattice, it will simply pass out of it. If it is very large, it will not be accommodated in the first place.

One of the most stable clathrate systems involves the β-quinol host lattice. It results when quinol, 1,4-$C_6H_4(OH)_2$ is crystallized from an aqueous solution, in the presence of Ar, Kr or Xe, at a pressure of 10–40 atmosphere. In the β form, three molecules of quinol are held together by hydrogen bonds to produce one cavity having a diameter of approximately 400 pm (Fig. 20.1).

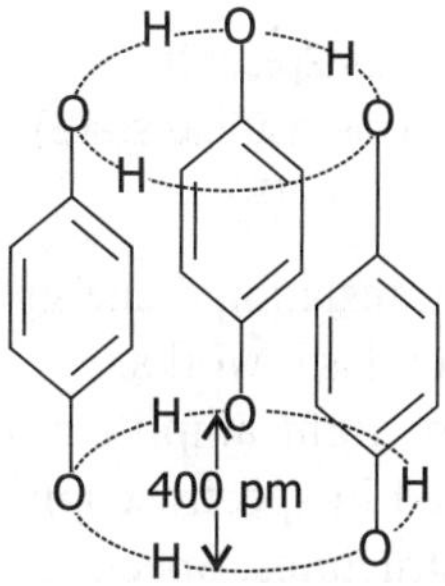

Fig. 20.1 Host lattice of β-quinol

The atoms of noble gases become trapped in the cavities during the process of crystallization, giving a composition, $[X\{C_6H_4(OH)_2\}_3]$, where X = Ar, Kr or Xe. However, the composition is non-stoichiometric since all the cavities of the host lattice are not occupied by noble gas atoms.

Another important type of clathrates are the noble gas hydrates. Thus when water is allowed to solidify under high pressure, in the presence of Ar, Kr or Xe, it forms open, hydrogen-bonded structures with cages in which the noble gas atoms become trapped. The volume of the crystal lattice increases due to the presence of guest atoms. Thus, the density of noble gas hydrates is less than that of ice. The average composition of these clathrates is $[X_8\{C_6H_4(OH)_2\}_{46}]$, where X = Ar, Kr or Xe.

The clathrates, once formed, are very stable. The trapped gaseous atoms cannot escape. Thus clathrates constitute a persistent, solid source of noble gases. He and Ne do not form clathrates since the atoms are too small and ooze out of the cavities as soon as they enter them. This differentiation may be used to affect the separation of noble gases. When quinol or water crytallizes in the presence of a mixture of Ne and Ar, only the latter forms the clathrate. This implies that a degree of selectivity is exhibited during clathrate formation.

When clathrates are dissolved in a solvent or are melted, the hydrogen-bonded structure breaks down and the trapped noble gas atoms are released.

Case Study 20.3 Explain the anesthetic action of xenon.

Xenon can cause absence of sensation and thus serve as an anesthetic agent. This property of xenon is somewhat surprising since the conditions prevailing in biological systems are not so aggressive as to effect chemical combination of the noble gas with one or more biomolecules. It is inferred that the structure of water in the biological fluids is altered due to the formation of xenon hydrates. This clathrate-type interaction is responsible for anesthetic action of xenon.

3. Helium–neon laser

When a He atom in an excited state ($1s^22s^12p^1$) strikes a Ne atom, energy transfer occurs. The Ne atom is transformed from its ground state (A) to a metastable state (B).

$$\underset{\text{(A)}}{\underset{\text{Ne (ground state)}}{1s^22s^22p^6}} \xrightarrow{\text{Energy gain}} \underset{\text{(B)}}{\underset{\text{Ne (metastable state)}}{1s^22s^22p^53s^03p^04s^1}} \xrightarrow{\text{Energy loss}} \underset{\text{(C)}}{\underset{\text{Ne (excited state)}}{1s^22s^22p^53s^03p^1}}$$

In the metastable state, the translational energy of Ne atom increases. It is on this principle that a helium–neon laser works.

Laser is an abbreviation for light amplification by stimulated emission of radiation. Such devices are used in spectroscopic techniques which, in turn, are used to deduce the molecular structures. Spectroscopic techniques require precise impinging of a radiation beam on a target molecule.

A helium–neon laser consists of a discharge tube containing He and Ne in the ratio of 10:1 at a pressure of 200 atmosphere. The discharge tube is closed at the ends with mirrors; at one end, the mirror allows a fraction of light to pass through it.

He atoms are excited by radiofrequency radiations. The energy is transferred to Ne atoms. Since the Ne atoms are fewer in number, almost all of them are transformed to the metastable state (B). Subsequently, the Ne atoms lose energy, emit light and pass on to an intermediate, excited state (C). The light emitted by this phenomenon is coherent in nature i.e., its waves have same wavelength, same direction of propagation and same phase angle. On being reflected from the mirror at one end of the tube, these waves multiply. The beam of light becomes progressively more intense. When it passes through the mirror at the opposite end, it can be directed at a target with minimum divergence and great accuracy. As a result, the structure of the target molecule can be elucidated.

Case Study 20.4 Which noble gas will provide least expensive inert atmosphere?

This depends on geographical location. Normally—and at a larger number of sites—argon is more plentiful relative to other noble gases. Hence it is cheaper too. However, at locations where large amounts of natural gas are produced, helium is a byproduct and is more abundant than argon. At such places, helium proves to be a less expensive alternative for provides inert atmosphere.

Chemical Properties

The chemical reactivity of noble gases is largely restricted to Kr, Xe and Rn. This concurs with the decreasing ionization energy and decreasing energy for unpairing the valence electrons on moving down the group. It is Xe which forms the largest number of compounds known for noble gases. It is conjectured that Rn, with lower ionization and promotional energies as compared to Xe, would form a still larger number of compounds. However, Rn is radioactive and has no stable isotope. This makes it difficult to isolate its compounds.

The compounds of Xe invariably involve bonds with fluorine or oxygen or both. The removal of electron from an atom of Xe, by way of ionization, or the promotion of electron to higher energy levels, by way of excitation, is not a one-way process. It has to be complemented by the atom(s) with which Xe reacts. Fluorine and oxygen have the highest and second highest electronegativities, respectively among all other elements. They assist in the removal of valence electrons of Xe from their ground states. Therefore, Xe prefers to combine with these atoms.

1. Reaction with platinum(VI) fluoride

The chemistry of Group 18 elements began with a reaction that had nothing to do with noble gases. This was the reaction between oxygen and platinum(VI) fluoride.

$$O_2 + PtF_6 \rightarrow [O_2]^+ [PtF_6]^-$$

In this reaction, molecular oxygen loses an electron and is transformed to molecular oxygen ion. The energy required for this transformation is comparable to the ionization energy of Xe.

$$O_2 \rightarrow O_2^+ + e^- ; 1175 \text{ kJ mol}^{-1}$$

$$Xe \rightarrow Xe^+ + e^- ; 1170 \text{ kJ mol}^{-1}$$

It was then speculated that Xe too should show a similar reaction. In fact, when Xe is mixed with red vapors of platinum(VI) fluoride, a yellow-colored solid is spontaneously obtained. Initially, it was suspected that the newly formed compound had the same stoichiometry as the oxygen derivative, i.e.,

$$Xe + PtF_6 \rightarrow \underset{\text{Xenon hexafluoroplatinate(V)}}{[Xe]^+ [PtF_6]^-}$$

This product is indeed obtained in the initial stages, but the reaction does not stop here. The xenon hexafluoroplatinate(V) undergoes the following stepwise transformations:

$$[Xe]^+ [PtF_6]^- + PtF_6 \xrightarrow{300\text{ K}} [XeF]^+ [PtF_6]^- + PtF_5$$

$$[XeF]^+ [PtF_6]^- + PtF_5 \xrightarrow{330\text{ K}} [XeF]^+ [Pt_2F_{11}]^-$$

Analogous reactions of Xe occur with other thermodynamically unstable metal hexafluorides, such as ruthenium(VI) fluoride, RuF_6, and rhodium(VI) fluoride, RhF_6. However, stable hexafluorides such as uranium(VI) fluoride, UF_6, and iridium(VI) fluoride, IrF_6, do not react with Xe.

2. Reaction with fluorine

Xe forms three fluorides, difluoride, tetrafluoride and hexafluoride, by direct reaction with fluorine in carefully controlled stoichiometries. The importance of the fluorides stems not only from the fact that they offer interesting structural and chemical properties in their own right, but also because they serve as starting materials for preparing other compounds of Xe.

Xenon difluoride is obtained by heating xenon and fluorine in 2:1 volume ratio at 673 K in a sealed nickel crucible.

$$\underset{2\ :\ 1}{Xe + F_2} \xrightarrow{673\ K} XeF_2$$

Xenon is taken in excess to avoid the simultaneous synthesis of xenon tetrafluoride which, if formed, would be extremely difficult to separate.

Xenon tetrafluoride is prepared by heating xenon and fluorine in a 1:5 volume ratio at 673 K in a nickel crucible. (A silica crucible cannot be used as fluorine reacts with this material.)

$$\underset{1\ :\ 5}{Xe + 2F_2} \xrightarrow{673\ K} XeF_4$$

In this case, fluorine is taken in excess to obviate the simultaneous formation of xenon difluoride.

Xenon hexafluoride can be prepared only under high pressure and for this reason, it is not formed along with the tetrafluoride, even though fluorine is present in excess.

In fact, xenon hexafluoride is prepared by heating xenon and fluorine in 1:20 volume mixture at 550 K, under 50–60 atmosphere pressure in a nickel vessel.

$$\underset{1\ :\ 20}{Xe + 3F_2} \xrightarrow{550\ K} XeF_6$$

The volume of fluorine in the reaction mixture is so much in excess that hexafluoride is not contaminated by the difluoride or tetrafluoride.

Compounds

Krypton difluoride (KrF_2) may be produced by cooling a mixture of krypton and fluorine at 80 K, and then irradiating it with X-rays. It is a thermally unstable compound which decomposes into its constituent elements at room temperature. When subjected to γ-radiation, it converts to krypton monofluoride, KrF.

$$2KrF_2 \rightarrow 2KrF + F_2$$

Krypton difluoride can form complexes by donating fluoride ion.

$$KrF_2 + AsF_5 \rightarrow [KrF]^+ [AsF_6]^-$$

Krypton difluoride is a better fluorinating agent than even fluorine. It is used to prepare compounds of metals in unusual oxidation states. For example, gold reacts with krypton difluoride to yield gold(V) fluoride.

$$8KrF_2 + 2Au \rightarrow 2[KrF]^+ [AuF_6]^- + 6Kr + F_2$$

$$[KrF]^+ [AuF_6]^- \xrightarrow{340\ K} AuF_5 + Kr + F_2$$

The usual oxidation states for gold are (I) and (III), while oxidation state (V) is an unusual one.

By far, the most important compounds of noble gases are those formed by Xe. In these compounds, Xe exhibits a range of oxidation states, from II to VIII. Some of these compounds are very stable and show an array of stereochemistries.

1. Xenon fluorides

We have already discussed the method for preparation of the difluoride, tetrafluoride and hexafluoride of Xe. We now turn to their structures, which we shall explain on the basis of valence bond theory.

In xenon difluoride, the central atom, Xe is sp^3d hybridized.

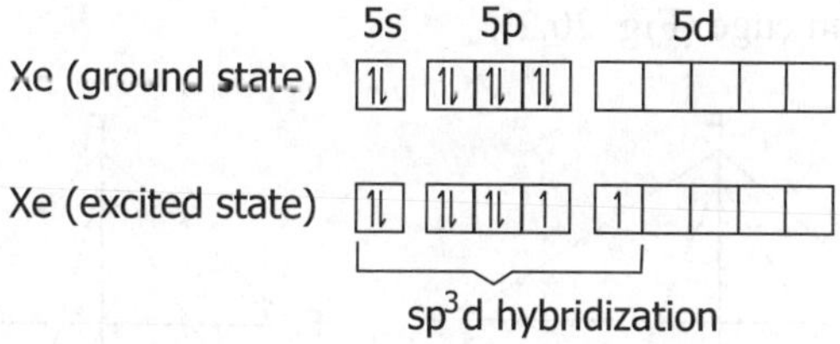

The structure may be described as either trigonal bipyramid, with three equatorial positions occupied by lone pairs of electrons or, more precisely, linear.

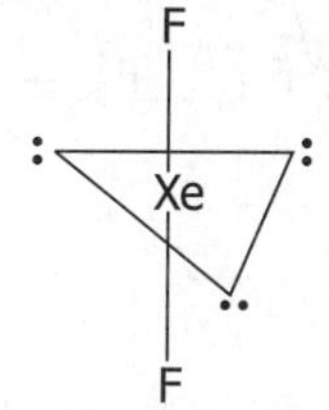

In xenon tetrafluoride, two electrons from the valence shell of Xe are promoted to the 5d orbital, before invoking sp^3d^2 hybridization.

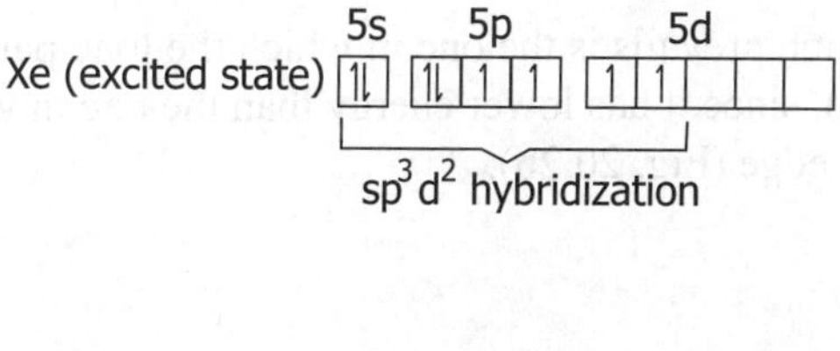

The structure is octahedral, with two axial positions occupied by lone pairs of electrons. It may also be described as square planar.

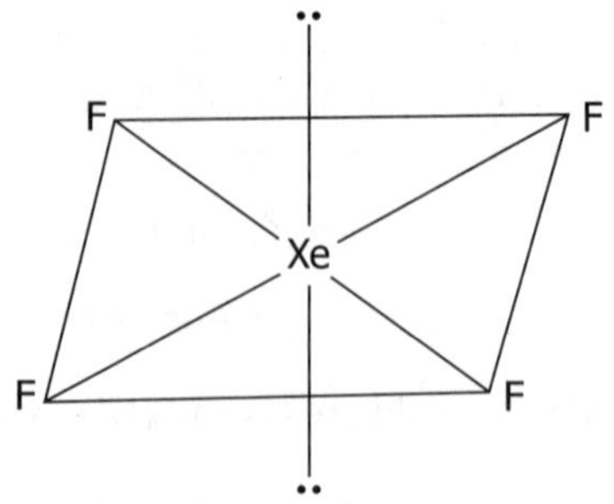

The bonding in xenon hexafluoride has been a subject of controversy. The xenon atom, in its excited state, promotes three electrons to the 5d orbital.

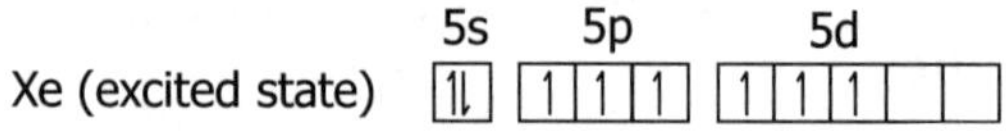

The six unpaired electrons form bonds with fluoride atoms. As a result, the final structure has six bond pairs and one lone pair of electrons. This type of geometry is described as distorted octahedron. The lone pair, which causes the distortion of the octahedral geometry, can have two configurations: it can extend either through a face or through an edge (Fig. 20.2).

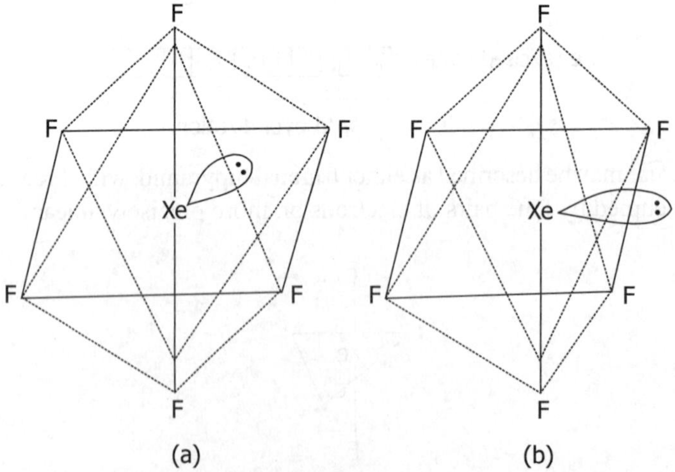

Fig. 20.2 In xenon hexafluoride, the lone pair may extend (a) through a face or (b) through an edge

The structure which prevails is the one in which the lone pair emerges through the face (Fig. 20.2a), since it has lower energy than the one in which the lone pair projects through an edge (Fig. 20.2b).

It is difficult to determine the accurate structure of xenon hexafluoride since the molecule is highly dynamic and passes through several conformations at a high rate. Moreover, unlike conventional molecules, the lone pair in xenon hexafluoride occupies lesser space than the bond pairs. For this reason, although the xenon–fluorine bonds near the lone pair are somewhat lengthened, with the F–Xe–F bond angles slightly less than the ideal value of 90°, yet the distortion is smaller than expected.

All the xenon fluorides are white, crystalline and sublimable solids. Xenon hexafluoride, however, is yellow in liquid and vapor states. The chemical properties of these compounds may be discussed under the following headings:

a. Fluorinating agents All the xenon fluorides are excellent fluorinating agents. Their ability to fluorinate other reagents varies in the order,

$$XeF_6 > XeF_4 > XeF_2$$

The following are some of the representative fluorinating reactions:

$$XeF_2 + 2NO \rightarrow \underset{\text{Nitrosyl fluoride}}{2NOF} + Xe$$

$XeF_2 + C_2H_4 \rightarrow$

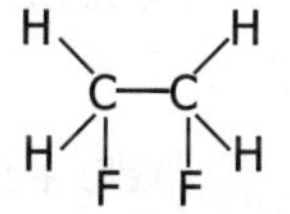

$+ Xe$

1,2-Difluoroethane

$$XeF_2 + (C_6H_5)_2S \rightarrow (C_6H_5)_2SF_2 + Xe$$

Difluorodiphenylsulfur(IV)

$$XeF_4 + 2Hg \rightarrow 2HgF_2 + Xe$$

$$XeF_4 + Pt \rightarrow PtF_4 + Xe$$

$$XeF_4 + 2SF_4 \rightarrow 2SF_6 + Xe$$

$$XeF_6 + 8NH_3 \rightarrow 6NH_4F + N_2 + Xe$$

b. Oxidizing agents The ability of xenon fluorides to act as oxidizing agents varies in the order,

$$XeF_6 > XeF_4 > XeF_2$$

Xenon difluoride, for example, oxidizes bromate ions to perbromate ions.

$$XeF_2 + BrO_3^- + H_2O \rightarrow BrO_4^- + 2HF + Xe$$

Similiarly, xenon tetrafluoride readily oxidizes iodide ions to iodine,

$$XeF_4 + 4KI \rightarrow 4KF + 2I_2 + Xe$$

while xenon hexafluoride oxidizes chloride ions to chlorine.

$$XeF_6 + 6HCl \rightarrow 6HF + 3Cl_2 + Xe$$

c. **Reaction with hydrogen** All the xenon fluorides react quantitatively with hydrogen.

$$XeF_2 + H_2 \rightarrow 2HF + Xe$$

$$XeF_4 + 2H_2 \rightarrow 4HF + Xe$$

$$XeF_6 + 3H_2 \rightarrow 6HF + Xe$$

d. **Reaction with water** All the xenon fluorides readily hydrolyze to give different products. Thus, xenon difluoride yields a mixture of hydrogen fluoride and oxygen on reacting with water,

$$2XeF_2 + 2H_2O \rightarrow 4HF + O_2 + 2Xe$$

while xenon tetrafluoride gives xenon trioxide as well.

$$6XeF_4 + 12H_2O \rightarrow 2XeO_3 + 24HF + 3O_2 + 4Xe$$

Xenon hexafluoride undergoes stepwise hydrolysis, yielding first xenon oxotetrafluoride,

$$XeF_6 + H_2O \rightarrow XeOF_4 + 2HF$$

and then xenon trioxide.

$$XeOF_4 + 2H_2O \rightarrow XeO_3 + 4HF$$

e. **Reaction with silica** Xenon hexafluoride reacts with silicon dioxide to yield successively, xenon oxotetrafluoride, xenon dioxodifluoride and xenon trioxide.

$$2XeF_6 + SiO_2 \rightarrow 2XeOF_4 + SiF_4$$

$$2XeOF_4 + SiO_2 \rightarrow 2XeO_2F_2 + SiF_4$$

$$2XeO_2F_2 + SiO_2 \rightarrow 2XeO_3 + SiF_4$$

For this reason, xenon hexafluoride cannot be stored in glass or quartz apparatus, as these materials have high silica content.

f. **Reaction with fluoride acceptors** Xenon difluoride reacts with pentafluorides of phosphorus, arsenic or antimony by donating a fluoride ion, and forming ionic complexes. For example,

$$XeF_2 + AsF_5 \rightarrow [XeF]^+ [AsF_6]^-$$

Xenon tetrafluoride forms analogous complexes.

$$XeF_4 + SbF_5 \rightarrow [XeF_3]^+ [SbF_6]^-$$

Similarly, xenon hexafluoride reacts not only with fluorides of Group 15 elements, but also with metal fluorides.

$$XeF_6 + FeF_3 \rightarrow [XeF_5]^+ [FeF_4]^-$$

g. **Reaction with fluoride donors** Xenon hexafluoride accepts fluoride ions from alkali metal fluorides to yield octafluoro complexes.

$$XeF_6 + 2MF \rightarrow M_2[XeF_8] \quad (M = Na, K, Rb \text{ or } Cs)$$

On moving from Na to Cs, the size of the cation becomes compatible with the size of octafluoroxenate(VI) anion, as a result of which the lattice energy and thermal stability increase. Thus the cesium complex decomposes at 673 K,

$$\underset{\text{Cesium octafluoroxenate(VI)}}{Cs_2[XeF_8]} \xrightarrow{673\text{ K}} 2CsF + XeF_6$$

while the sodium complex decomposes at 373 K.

The difluoride and tetrafluoride of xenon do not act as fluoride acceptors. This characteristic is utilized in purifying xenon hexafluoride in case it is contaminated with the other two fluorides. The mixture is sublimed and treated with sodium fluoride. Only xenon hexafluoride reacts to yield the solid, sodium octafluoroxenate(VI) complex.

$$XeF_6 + 2NaF \rightarrow Na_2[XeF_8]$$

The complex is separated from the gaseous difluoride and tetrafluoride. It is then heated to 373 K,

$$Na_2[XeF_8] \xrightarrow{373\text{ K}} 2NaF + XeF_6$$

and pure xenon hexafluoride is recovered.

h. **Substitution reactions** Xenon difluoride reacts with anhydrous perchloric acid to yield perchloratexenon fluoride,

$$XeF_2 + \underset{\text{Anhydrous}}{HClO_4} \rightarrow ClO_4XeF + HF$$

and with anhydrous fluorosulfuric acid to yield fluorosulfatexenon fluoride.

$$XeF_2 + \underset{\text{Anhydrous}}{HSO_3F} \rightarrow FSO_3XeF + HF$$

The structures of the products are depicted in Fig. 20.3.

(a) (b)

Fig. 20.3 Structure of (a) perchloratexenon fluoride and (b) fluorosulfatexenon fluoride

---

Case Study 20.5 Outline the molecular orbital picture of xenon difluoride.

The valence bond theory, based on the concept of hybridization, explains the structure of xenon difluoride quite satisfactorily, yet there is a problem related to the participation of 5d orbital in bonding. This orbital is much larger in size and much higher in energy as compared to the 5s and 5p orbitals. As a result, the effective mixing of valence shells, a prerequisite for hybridization, becomes questionable.

The molecular orbital theory, on the other hand, eschews the 5d orbital and involves only the 5p orbital of xenon in the bonding process. The valence shell configuration of xenon and fluorine are as follows:

The $5p_z$ orbital of xenon and the $2p_z$ orbitals of two fluorine atoms have comparable symmetries and, therefore, can overlap with each other. Thus, three orbitals are involved in the formation of xenon difluoride molecule—the $5p_z$ orbital of xenon and the $2p_z$ orbitals of each fluorine atom. On overlapping, three molecular orbitals are formed—one bonding, one non-bonding and one anti-bonding—their energies increasing in this order.

Four electrons are involved in bonding, two in the $5p_z$ orbital of xenon and one in the $2p_z$ orbital of each fluorine. In a molecule of xenon difluoride, two of them occupy the bonding molecular orbital, while the remaining two are accommodated in the non-bonding molecular orbital. Those occupying the bonding molecular orbital span three atoms, fluorine–xenon–fluorine. The energy of non-bonding molecular orbital is very close to that of the $2p_z$ atomic orbitals on fluorine atoms. For this reason, the electron pair in the non-bonding molecular orbital is largely concentrated on fluorine atoms, conferring ionic character to xenon–fluorine bonds. Given the large electronegativity difference between xenon and fluorine, the ionic character is not unexpected.

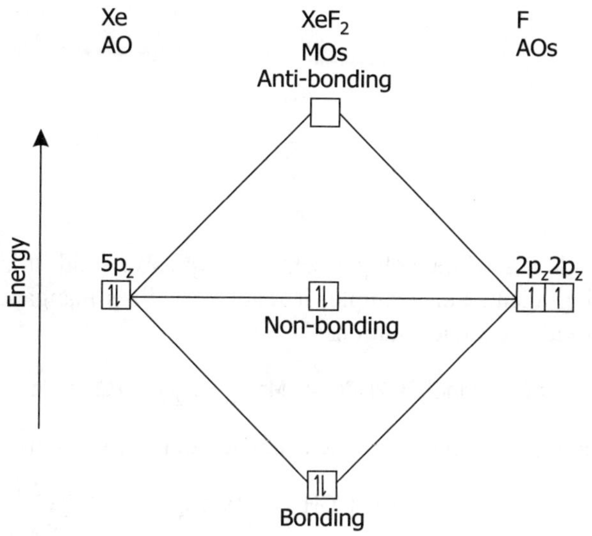

This type of combination is called *four-electron–three-center bond.* A linear arrangement of atoms gives the best possible overlap.

2. Xenon oxides

Xenon forms two oxides—xenon trioxide (XeO_3) and xenon tetraoxide (XeO_4) the latter being highly unstable.

Xenon trioxide is prepared by complete hydrolysis of xenon hexafluoride.

$$XeF_6 + 3H_2O \rightarrow XeO_3 + 6HF$$

This is a violent reaction and to control it, xenon hexafluoride is first mixed in a current of nitrogen and then slowly passed into water. The hydrogen fluoride is removed as insoluble magnesium fluoride, by adding magnesium oxide.

$$MgO + 2HF \rightarrow MgF_2 + H_2O$$

The xenon, in xenon trioxide, is sp^3 hybridized.

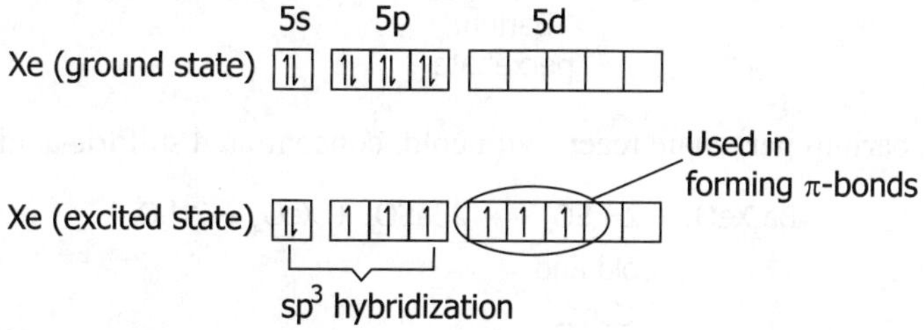

The structure may be described as tetrahedral, with one position occupied by a lone pair of electrons, or pyramidal.

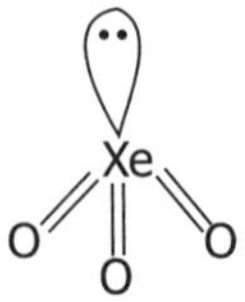

Xenon trioxide is a white, deliquescent and explosive solid. It is an extremely strong oxidizing agent. For example, it readily oxidizes manganese(II) ions to manganese dioxide in acidic solutions.

$$XeO_3 + MnCl_2 + 2HCl \rightarrow MnO_2 + H_2O + 2Cl_2 + Xe$$

In alkaline solutions, xenon trioxide is converted to xenate ions, $HXeO_4^-$.

$$XeO_3 + OH^- \rightarrow HXeO_4^-$$

Reaction of xenon trioxide with an alkali, in the presence of ozone, yields perxenate ions, XeO_6^{4-}.

$$XeO_3 + 4NaOH + O_3 \rightarrow \underset{\text{Sodium perxenate}}{Na_4XeO_6} + 2H_2O + O_2$$

In its reaction with potassium fluoride, xenon trioxide acts as a Lewis acid (a fluoride ion acceptor), forming fluorotrioxoxenate(VI) ion.

$$XeO_3 + KF \rightarrow K[XeO_3F]$$

The reaction of xenon oxotetrafluoride and xenon trioxide produces xenon dioxodifluoride.

$$XeO_3 + XeOF_4 \rightarrow 2XeO_2F_2$$

Xenon tetraoxide is prepared in two steps. First, barium perxenate is obtained by the reaction of xenon hexafluoride with barium hydroxide.

$$2XeF_6 + 8Ba(OH)_2 \rightarrow \underset{\text{Barium perxenate}}{Ba_2XeO_6} + 6BaF_2 + 8H_2O + O_2 + Xe$$

Second, barium perxenate reacts with cold, concentrated sulfuric acid.

$$Ba_2XeO_6 + \underset{\text{Cold and Conc.}}{2H_2SO_4} \rightarrow 2BaSO_4 + XeO_4 + 2H_2O$$

Xenon in xenon tetraoxide undergoes sp^3 hybridization.

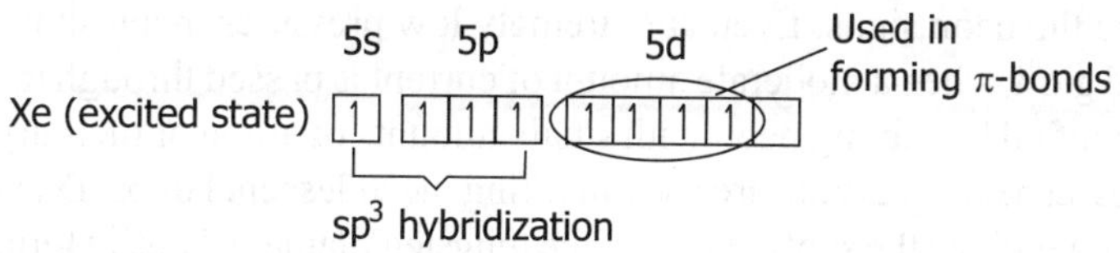

The structure is tetrahedral.

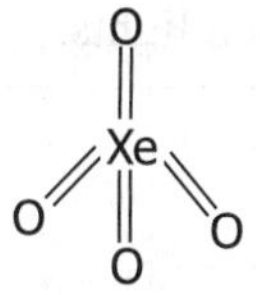

Xenon tetraoxide is an explosive and unstable gas. It reacts with xenon hexafluoride to yield xenon trioxodifluoride.

$$XeO_4 + XeF_6 \rightarrow XeO_3F_2 + XeOF_4$$

3. Xenon oxofluorides

As stated earlier, xenon oxotetrafluoride, $XeOF_4$ may be prepared by controlled hydrolysis of xenon hexafluoride. It may also be prepared by the reaction of xenon hexafluoride with either sodium nitrate or phosphorus oxotrifluoride.

$$XeF_6 + NaNO_3 \rightarrow XeOF_4 + NaF + FNO_2$$

$$XeF_6 + OPF_3 \rightarrow XeOF_4 + PF_5$$

Xenon oxatetrafluoride has a square pyramidal structure, with a non bonding electron pair protruding from the base of the pyramid.

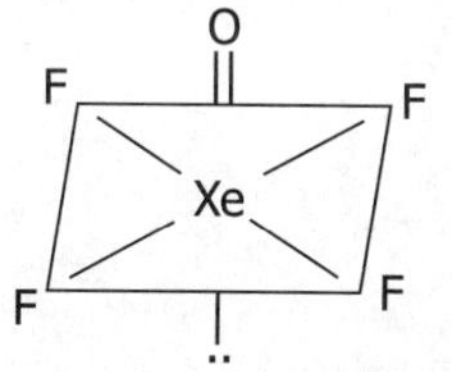

We have already cited the reaction of xenon oxotetrafluoride with xenon trioxide to yield xenon dioxodifluoride, XeO_2F_2. The same product may also be produced by the reaction of xenon oxotetrafluoride with cesium nitrate.

$$XeOF_4 + CsNO_3 \rightarrow XeO_2F_2 + CsF + FNO_2$$

Xenon dioxodifluoride has a tetrahedral structure.

Case Study 20.6 What are the applications of noble gases in lighting systems?

The applications of noble gases in lighting systems may be broadly classified into two types. First, they are used for decorative and advertisement purposes. A case in point are the neon signs. Even at extremely low pressures, neon shows a bright orange-red glow when a moderate amount of current is passed through it. The color may be modified by mixing neon with small amounts of argon or mercury. Second, noble gases, especially argon, are used in filling incandescent lamps. Due to its inert nature, argon inhibits the sublimation of the tungsten filament.This, in turn, prolongs the life of lamp and retards its blackening. Krypton serves the purpose even better but is not used in conventional lamps because it is expensive. Nevertheless, it is used in airport runways and approach lights.

21. THE TRANSITION ELEMENTS

Introduction

The d-block elements are those which have partially filled d orbitals in their elemental state or in an important valence state. In the periodic table, these elements are placed between s- and p-block elements. Therefore, they incorporate some properties of s-block elements and some of p-block elements. This implies that their properties are transitional between those of the s-block and p-block elements. For this reason, d-block elements are also called transition elements.

There are three series of d-block members in the periodic table. These are referred to as the 1st or 3d series, 2nd or 4d series and 3rd or 5d series (Table 21.1). The elements of these series have their valence electrons in 3d, 4d and 5d orbitals, respectively.

Table 21.1 The three series of d-block elements

Group	3	4	5	6	7	8	9	10	11	12
1st/3d Series	Sc	Ti	V	Cr	Mn	Fe	Co	Ni	Cu	Zn
2nd/4d Series	Y	Zr	Nb	Mo	Tc	Ru	Rh	Pd	Ag	Cd
3rd/5d Series	La	Hf	Ta	W	Re	Os	Ir	Pt	Au	Hg

We shall now discuss the general properties of d-block elements, with emphasis on 3d series members.

Physical Characteristics

The transition elements—

1. are all metals.
2. are strong, hard and brittle.
3. have high melting points and boiling points.
4. are moderately reactive.
5. form alloys with other transition elements (for example, brass is an alloy of copper and zinc) or with non-transition elements (for example, bronze is an alloy of copper and tin).

Case Study 21.1 Justify the inclusion of copper in the 3d transition series.

Strictly speaking, transition elements are those which have at least one unpaired electron in the valence d orbitals. Copper, with the electronic configuration, $[Ar]3d^{10}4s^1$, does not satisfy this criterion. However, copper shows more than one oxidation states (I, II and III)—a property characteristic of transition metals. The most important oxidation state for copper is II, as in this valence state it forms most numerous and most stable compounds. Copper(II) ion, like most other transition metal ions, is colored. It forms a large number of complexes like other members of the 3d series. In fact, the properties of copper are so similar to the other transition metals that it would be improper to exclude it from 3d series. Therefore, it is pertinent to modify the definition of transition metals to retain the membership of copper. In a broader concept, transition metals are those which have at least one unpaired electron either in their elemental state or in an important oxidation state. As has been stated, the most important oxidation state of copper is II, with electronic configuration, $[Ar]3d^9$. There is one unpaired electron in the valence shell of copper(II) ion and, therefore, the criterion of the broader definition of transition metals stands justified.

Electronic Configuration

Argon, with a complete octet, has the configuration $[Ne]3s^23p^6$. According to Aufbau principle, the valence electron of the next element should occupy the 3d orbital. However, the 4s orbital has a better penetration power than the 3d orbital. In fact, the penetration power varies in the order s > p > d > f. The driving force for the penetrating ability is provided by the fact that an orbital which reaches closer to the nucleus acquires more stability and lower energy. Since the 4s orbital can penetrate the Ar core to a better extent than the 3d orbital, the energy of the former falls below that of the latter.

As a result, when two more electrons are added to the Ar core, to give potassium and calcium, they enter the 4s and not the 3d orbital. The next electron enters the 3d orbital. Therefore, Sc, which marks the beginning of transition elements, possesses the configuration $[Ar]4s^23d^1$. This process continues until the entire 3d orbital is filled at zinc, with the configuration, $[Ar]4s^23d^{10}$ (Table 21.2). The atoms with unfilled, half-filled or completely filled orbitals have exceptionally high stability due to the large amount of exchange energy associated with their configurations (*see* Case Study 1.13, p. 47). Therefore, Cr and Cu show anomalous configurations.

Element	Expected configuration	Observed configuration
Cr	$[Ar]4s^23d^4$	$[Ar]4s^13d^5$
Cu	$[Ar]4s^23d^9$	$[Ar]4s^13d^{10}$

In these cases, an electron is borrowed out of turn from 4s orbital to 3d orbital so as to achieve a half-filled and a completely filled configuration, respectively.

We have seen that while electrons are being added, the 4s orbital becomes filled first and the 3d orbital later. However, when it comes to ionization, electrons are lost first from 4s orbital and then from 3d orbital. This happens because the filled 4s orbital is repelled by inner orbitals and hence loses its ability to get closer to the nucleus. It becomes lower in stability and higher in energy as compared to the 3d orbital. Therefore, with filled 4s orbital, the order of energy is,

$$E_{4s} > E_{3d}$$

Hence, during ionization, 4s electrons are lost before 3d electrons. This may be exemplified by considering the various oxidation states of titanium.

Ti	$[Ar]3d^2 4s^2$
Ti(II)	$[Ar]3d^2 4s^0$
Ti(III)	$[Ar]3d^1 4s^0$
Ti(IV)	$[Ar]3d^0 4s^0$

Since 4s electrons are lost first, it is advisable to write the configuration of transition metal atoms as $[Ar]3d^n 4s^2$ and not as $[Ar]4s^2 3d^n$.

Table 21.2 Electronic configuration of 3d series transition elements

	Sc	Ti	V	Cr	Mn	Fe	Co	Ni	Cu	Zn
4s	2	2	2	1	2	2	2	2	1	2
3d	1	2	3	5	5	6	7	8	10	10

Variable Oxidation States

The transition elements demonstrate the unique property of variable oxidation states, i.e., each element shows more than one valence state. The various oxidation states of the first series of transition elements are depicted in Fig. 21.1. Here, numerals within parentheses represent unusual oxidation states.

Sc	Ti	V	Cr	Mn	Fe	Co	Ni	Cu	Zn
III	II	II	II	II	II	II	II	I	II
	III	III	III	III	III	III	(III)	II	
	IV	IV	IV	IV	(IV)	(IV)	(IV)	(III)	
		V	V	V	(VI)				
			VI	VI					
				VII					

Fig. 21.1 Variable oxidation states of 3d series transition elements

The variable oxidation states arise because the energies of 3d and 4s orbitals are comparable, $E_{3d} \approx E_{4s}$ (although, to be precise, E_{4s} is slightly higher than E_{3d}; yet the energy difference is very small as compared to, say, that between 1s and 2s orbitals). Therefore, if the electrons can be lost from 4s orbital, these may also be lost, with nearly equal case, from 3d orbitals as well. Titanium shows an oxidation state of II by the ionization of the two 4s electrons. Removal of the two 4s electrons and one 3d electron gives rise to oxidation state III, while removal of all the electrons from 4s and 3d orbitals results in oxidation state IV. It is the participation of the 4s and 3d electrons in bonding that is responsible for variable oxidation states of transition elements.

After Mn, the number of oxidation states of the elements becomes less. This may be attributed to increasing nuclear charge which, in turn, increases the ionization energy. Due to high ionization energies of the later half elements, the number of oxidation states registers a decline.

Cu readily forms compounds in oxidation states I and II.

$$\text{Cu(I)} \quad [\text{Ar}]3d^{10}4s^0$$

$$\text{Cu(II)} \quad [\text{Ar}]3d^{9}4s^0$$

Going by the electronic configuration, it is expected that copper(I) compounds should be more stable than copper(II) compounds, as the former has a completely filled $3d^{10}$ configuration. However, copper(II) compounds are more stable. This is because copper(I) ion is larger in size than copper(II) ion and, therefore, its salts possess meager lattice energy. Whatever stabilization is afforded by favorable oxidation state is negated by unfavorable lattice energy.

Color

The transition metals mostly furnish colored ions. The color is attributed to the presence of one or more unpaired electrons in the valence shell of the concerned ion. The unpaired electrons absorb photons from visible light and are excited to higher energy levels. On reverting to their original states, they emit their extra energy as colored light. Hence, the ions appear colored.

Natural color is shown only by those substances which can absorb visible radiation and promote their electrons to higher energy states. Visible light cannot excite paired-up electrons. It does not possess as much energy to first unpair the two electrons and then push up one of them to a higher energy state. Therefore, color is shown only by those atoms, ions or molecules which possess unpaired electrons. Most transition metal compounds fall in this category and hence are colored. However, if a particular transition metal ion does not possess unpaired electrons, it is colorless or white.

For example, titanium(III) chloride is violet while titanium(IV) oxide is white. The titanium(III) ion, with a $3d^1$ configuration, possesses one unpaired electron. It absorbs visible radiation, is promoted to higher energy states and on falling back to its original position, emits its extra energy as violet-colored light. Therefore, it appears violet. Titanium(IV) ion, on the other hand, has a $3d^0$ configuration. There are no electrons in the valence shell and hence there is no scope for excitation.

Similarly, copper(I) chloride is white, whereas copper(II) chloride is green. The copper(I) ion has a $3d^{10}$ configuration, in which all the electrons are paired-up. They cannot be excited to higher energy levels by visible light and, therefore, the compound is white. The copper(II) ion has $3d^9$ configuration, with one unpaired electron. It can readily absorb visible light and hence is colored.

Potassium dichromate is orange in color, while potassium permanganate is purple. Both compounds are intensely colored, yet the metal ions do not possess any unpaired electron. The chromium(VI) ion, in potassium dichromate, has a $3d^0$ configuration, and so does the manganese(VII) ion in potassium permanganate. The color in these two cases is attributed to a phenomenon called *charge-transfer spectrum*.

Potassium dichromate and potassium permanganate contain seven and four oxygen atoms per molecule, respectively. Each oxygen atom, in turn, holds two pairs of non-bonding electrons. The oxygen atom is small and presence of two lone pairs of electrons causes electron–electron repulsions. When large number of oxygen atoms are present in a compound—as in these two cases—the total electron–electron repulsion becomes significantly high. To overcome this destabilizing phenomenon, the oxygen atoms tend to transfer some of their electronic charge to the metal ions. Both chromium(VI) and manganese(VII) have $3d^0$ configurations and, therefore, can accommodate the electrons donated by oxygen atoms. The transferred electrons are excited by visible light and are responsible for imparting color to these compounds.

The zinc(II) ion has a $3d^{10}$ configuration and the paired-up electrons cannot be excited by visible light. Since no other oxidation state is known for zinc, virtually all its compounds are white in color.

Case Study 21.2 Copper(I) chloride is white, whereas copper(I) oxide is red. Explain.

The copper(I) ion possesses $3d^{10}$ configuration and is not expected to show coloration. Accordingly, copper(I) chloride is white. However, the copper(I) oxide is red because this compound has a defect lattice. In its crystal structure, there are excess copper(I) ions occupying interstitial positions. Electron neutrality is maintained by the inclusion of interstitial electrons. These electrons absorb visible light and, therefore, the compound is colored.

Box 21.1

Charge-Transfer Phenomena

In most of the transition metal compounds or complexes, the color arises because visible light is able to promote an unpaired electron in a d orbital of the central ion to another d orbital. Such an excitation is called d–d transition. However, in a few cases, the color is attributed to charge-transfer phenomena, wherein the electron or electrons transition from one species to another. These transitions are of the following three types.

Contd...

1. Ligand-to-metal charge transfer

As has been cited above, the color of potassium permanganate and potassium dichromate arises due to transference of electrons from the filled orbitals of the ligands—oxygen atoms in this case—to the vacant d orbitals manganese(VII) ion or chromium(VI) ion. Such transitions usually occur when the metal ion is in a high oxidation state, that is, has very few or no electrons in its d orbitals.

2. Metal-to-ligand charge transfer

The metal-to-ligand charge transfer occurs in cases where the metal ion is in a low oxidation state. The transfer of electrons takes place from the filled metal d orbitals to the vacant π^* molecular orbitals of the ligands. The orange color of tris(dipyridyl)ruthenium(II) ion (Fig. 21.2) is attributed to metal-to-ligand charge transfer transitions.

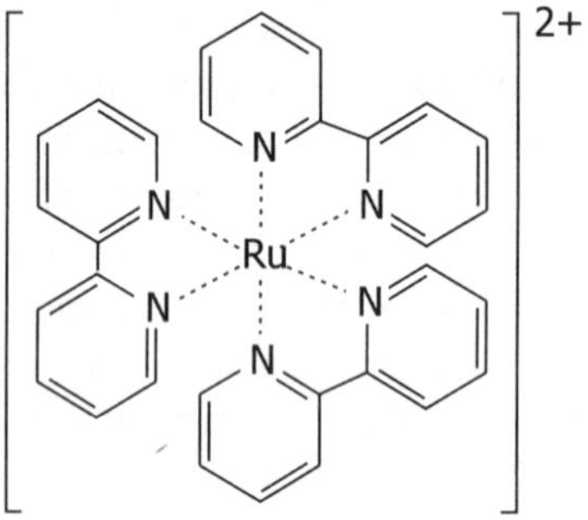

Fig. 21.2 Structure of tris(dipyridyl)ruthenium(II) ion

3. Metal-to-metal charge transfer

Prussian blue, $Fe_4^{III}[Fe^{II}(CN)_6]_3$ owes its color to metal-to-metal charge transfer spectra. As shown in Fig. 21.3, the cyano groups act as bridges between the iron(II) and iron(III) ions in the structure of Prussian blue.

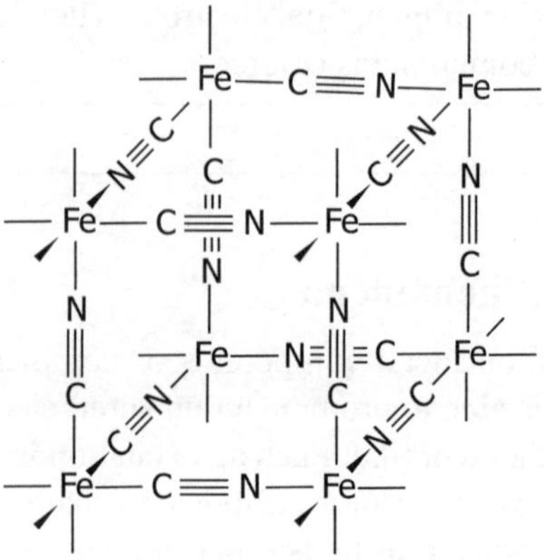

Fig. 21.3 Structure of Prussian blue

Contd...

An electron transfer occurs between the iron(II) and iron(III) moieties of Prussian blue, setting forth the following redox reaction.

$$Fe(II) + Fe^*(III) \rightarrow Fe(III) + Fe^*(II)$$

This reaction is induced by red, orange and yellow components of visible radiation, causing an electron to be transferred from iron(II) ion to iron(III) ion through a cyanide bridge. As a result of this electronic transition from one metal ion to another, the complex appears blue in color.

Catalytic Properties

Many transition elements and their compounds show catalytic properties. Some important examples are listed in Table 21.3.

Table 21.3 Processes catalyzed by transition elements/compounds

Element/Compound	Process
Iron	Haber's process for manufacture of ammonia
Nickel	Hydrogenation of unsaturated hydrocarbons
Copper	Manufacture of silicones
Platinum	Ostwald process for manufacture of nitric acid
Titanium(III) chloride	Production of polyethene
Vanadium(V) oxide	Contact process for manufacture of sulfuric acid
Manganese(IV) oxide	Laboratory preparation of oxygen

It is, in fact, very rare for a non-transition element or any of its compounds to act as a catalyst. This means that transition elements have some unique characteristic that is responsible for their catalytic properties. Before we examine this characteristic, we would review the two theories of catalysis. These are: activated complex theory and adsorption theory.

According to *activated complex theory,* a substance C can act as a catalyst only if it can react with the reactant A to form an intermediate [AC] having a lower activation energy, E_a', as compared to an uncatalyzed reaction.

Uncatalyzed reaction: $A \rightarrow [A^*] \rightarrow B$

Catalyzed reaction: $A + C \rightarrow [AC] \rightarrow B + C$

The intermediate [AC] is called activated complex and has only a transitionary existence. It decomposes into product B, simultaneously releasing the catalyst. In the case of an uncatalyzed reaction, the activated complex, [A*] has much higher activation energy, E_a (Fig. 21.4).

The reason why transition metals or their compounds act as catalysts is that these elements have variable oxidation states. There is, therefore, a finite probability that at least in one of the oxidation states, an activated complex with a lower activation energy will be formed.

According to *adsorption theory,* a substance acts as a catalyst only if it has the ability to adsorb the reactant molecules. This enables the reactant molecules to become localized on the surface of the catalyst and, in turn, facilitate their interaction.

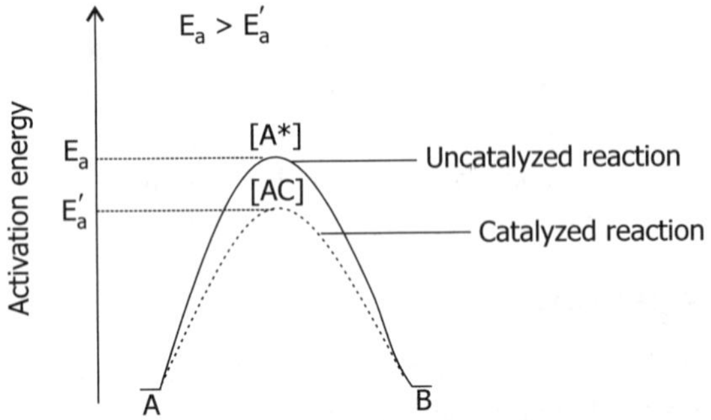

Fig. 21.4 Comparison of the activation energies of catalyzed and uncatalyzed reactions

For example, iron provides the surface for catalysis in the Haber's process.

$$N_2(g) + 3H_2(g) \overset{Fe}{\rightleftharpoons} 2NH_3(g)$$

In the absence of the catalyst, it is highly unlikely that one molecule of gaseous nitrogen and three molecules of gaseous hydrogen would collide and lead to the formation of ammonia. However, once the nitrogen and hydrogen molecules get adsorbed on the surface of iron, their random motion ceases and they get localized close to each other. The chances of their reaction increase manifold. Further, the ammonia so formed gets desorbed and evaporates, leaving the surface bare for more reactant molecules to settle.

Transition elements are better suited as adsorption catalysts since they possess rough surfaces. There are a large number of sites on their surface where the reactant molecules can get adsorbed. These sites are called *active centers.* The number of active centers on non-transition elements is comparatively less.

Case Study 21.3 Transition metals are not good reducing agents. Why?

As has been explained in Chapter 12 (p. 423), the electrode potential of a metal (E) depends on three factors: sublimation energy of metal (s); ionization energy of metal (I); and heat of hydration of metal ion (H). It is related to these parameters by the expression,

$$E = +S + I - H$$

Greater the negative value of E, stronger is the reducing character of metal. Since transition elements are small in size, their degree of hydrolysis is extensive. Therefore, heat of hydration has a large negative value. However, the sublimation energy and ionization energy of transition elements have a large positive value. As a result, the electrode potential has a smaller negative value as compared to that of alkali or alkaline earth metals. In effect, the transition metals are not as good reducing agents as are the s-block elements.

Magnetic Properties

The magnetic properties of a substance are directly dependent on the pattern of distribution of electronic charge in the atoms, ions or molecules of that substance. Any particle which is extremely small, carries a charge and is moving, behaves as a bar magnet. The electron fulfils all the three criteria: it is small, it holds a negative charge and it is mobile. There are two types of motions associated with the electron. First, the electron spins around its own axis. The magnetism resulting from this type of motion is called *spin moment.* Second, the electron revolves around the nucleus. The magnetism resulting from this type of motion is called *orbital moment.* The net magnetic properties of a substance are determined by the sum of spin and orbital moments of the electronic charge.

If two electrons are paired-up in an orbital, the spin moment of one gets canceled by that of the other. So does the orbital moment. Hence, magnetic properties are significant only if there are one or more unpaired electrons in the valence shell. Since transition metals, by definition, have unpaired electrons in their valence d shell, magnetic properties assume more importance for these elements than for the s- or p-block elements, in which the electrons are generally paired-up.

Thus, as far as magnetic behavior is concerned, substances may be classified into two broad types: those whose atoms, ions or molecules have one or more unpaired electrons and those in which there are no unpaired electrons. The former are called *paramagnetic* substances; the latter *diamagnetic.* A paramagnetic substance is weakly attracted by a magnetic field, while a diamagnetic substance is weakly repelled. Therefore, if a paramagnetic species is suspended from a spring balance in a magnetic field, it would be pulled downwards. It would weigh more in a magnetic field than in its absence. A diamagnetic substance, on the other hand, would be pushed up in a magnetic field and, therefore, weigh less (Fig. 21.5).

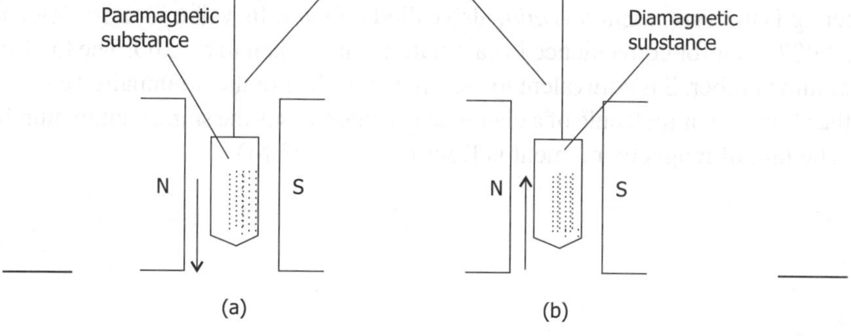

Fig. 21.5 (a) A paramagnetic substance weighs more in a magnetic field, while (b) a diamagnetic substance weighs less

Table 21.4 Difference between paramagnetic and diamagnetic substances

Paramagnetic substances	Diamagnetic substances
1. Those atoms, ions or molecules which have at least one unpaired electron	1. Those atoms, ions or molecules in which all the electrons are paired-up
2. Weakly attracted by magnetic field	2. Weakly repelled by magnetic field
3. Weigh more in a magnetic field	3. Weigh less in a magnetic field
4. Magnetic permeability is greater than unity	4. Magnetic permeability is less than unity
5. Magnetic susceptibility has a positive value	5. Magnetic susceptibility has a negative value

The differences in the properties of paramagnetic and diamagnetic substances are highlighted in Table 21.4. There are two parameters that define the magnetic properties of a chemical entity. These are: magnetic moment and magnetic susceptibility.

1. Magnetic moment

The magnetic moment is the sum of spin moment and orbital moment. However, the latter is very small as compared to the former. For this reason, magnetic moment is taken to be synonymous with spin moment. It describes the extent of paramagnetic character of a substance. An atom, ion or molecule with two unpaired electrons will obviously possess more paramagnetic character than its counterpart with only one unpaired electron. The degree by which the paramagnetic character of a species, which has a higher number of unpaired electrons, exceeds that of a species with a smaller number of unpaired electrons is defined by magnetic moment. The latter, however, is not a linear property. This means that magnetic moment of a compound with two unpaired electrons will not be double than that of a compound with only one unpaired electron.

Ignoring the orbital moment, the magnetic moment (spin only), μ_s is expressed by the relation,

$$\mu_s = g\sqrt{S\,(S + 1)} \tag{21.1}$$

where g is the *gyromagnetic ratio*, also called *g factor*. Its value for a free electron is 2.00023, but for convenience in calculations, it is taken to be 2.00. The total spin quantum number, S is equivalent to ns, where n is the number of unpaired electrons in the atom, ion or molecule of a chemical species and s is the spin quantum number, ½. The unit of magnetic moment is Bohr magneton (BM).

Case Study 21.4 Calculate the magnetic moment of copper(II) sulfate and manganese(II) sulfate.

$$\mu_s = g\sqrt{s\,(s+1)}$$

Copper(II) sulfate

	3d	4s
Cu	⇅ ⇅ ⇅ ⇅ ⇅	↑
Cu(II)	⇅ ⇅ ⇅ ⇅ ↑	☐

$n = 1$

$S = ns = \frac{1}{2}$

$g = 2.00$

$\mu_S = 2\sqrt{\frac{1}{2}\left(\frac{1}{2}+1\right)}$

$= 1.73$ BM

Manganese(II) sulfate

	3d	4s
Mn	↑ ↑ ↑ ↑ ↑	⇅
Mn(II)	↑ ↑ ↑ ↑ ↑	☐

$n = 5$

$S = ns = \frac{5}{2}$

$g = 2.00$

$\mu_S = 2\sqrt{\frac{5}{2}\left(\frac{5}{2}+1\right)}$

$= 5.91$ BM

Case Study 21.4 reveals that the magnetic moment of manganese(II) ion, which has five unpaired electrons, is not five times that of copper(II) ion, which has only one unpaired electron. The values of magnetic moment, so obtained, have been theoretically calculated.

As we shall see later, the magnetic moment values can also be experimentally determined. The calculated and experimental values are very close to each other, yet, in general, the latter are slightly higher. This is because while calculating magnetic moments of various species, we ignore the orbital moment contribution, as it is very small. However, the experimental values elucidate the sum of spin and orbital moments. Therefore, experimental μ is slightly higher than calculated μ.

A more accurate calculated value for magnetic moment is given by the expression,

$$\mu_{S+L} = \sqrt{4\,S\,(S+1) + L\,(L+1)} \tag{21.2}$$

where μ_{S+L} is the total magnetic moment, incorporating both spin moment and orbital moment. L represents the *orbital angular momentum quantum number*, and is a measure of the orbital moment. This expression, however, is rarely used for the calculation of magnetic moment since, unlike spin quantum number, the value of orbital angular momentum quantum number is not fixed. It varies from one chemical species to another and, therefore, is not very convenient to use. Moreover, as has been stated earlier, the orbital moment's contribution to net magnetism is very small.

If the value of L is assumed to be zero, equation (21.2) is reduced to,

$$\mu_S = \sqrt{4S(S+1)} = 2\sqrt{S(S+1)} \quad (21.3)$$

which is same as equation (21.1).

2. Magnetic susceptibility

The strength of a magnetic field, H, is related to the flux, B, i.e., the density of magnetic lines of forces and intensity of magnetization, I, i.e., the number of magnetic lines of forces per unit area, by the expression,

$$B = H + 4\pi I \quad (21.4)$$

Dividing throughout by H,

$$\frac{B}{H} = 1 + 4\pi\left(\frac{I}{H}\right) \quad (21.5)$$

The ratio between magnetic flux and strength of magnetic field, i.e., B/H, is called *magnetic permeability* (q). Magnetic permeability refers to the ability of a substance to allow magnetic lines of forces to pass through it.

Magnetic susceptibility (K) is defined as the ratio between intensity of magnetization and the strength of magnetic field, i.e., I/H. Equation (21.5) may now be written as,

$$q = 1 + 4\pi K \quad (21.6)$$

$$\text{or} \quad K = \frac{q-1}{4\pi} \quad (21.7)$$

For a paramagnetic entity, there is a greater flux of the lines of forces within the substance than without it. Therefore, the magnetic permeability (q) is greater than 1, and hence, according to equation (21.7), the susceptibility of a paramagnetic substance has a positive value. For a diamagnetic entity, there is smaller flux of lines of forces within the substance than without it. Therefore, the magnetic permeability is less than 1 and the susceptibility of a diamagnetic substance has a negative value (*see* Table 21.4).

The magnetic susceptibility, K, is expressed in terms of the volume of the substance. Therefore, it is often referred to as *volume susceptibility.* However, since transition metals and their compounds are all solids, it is more convenient to express the magnetic susceptibility in terms of weight. The *gram susceptibility,* χ is given by,

$$\chi = \frac{K}{d} \quad (21.8)$$

where d is the density of the substance.

A still better and convenient way to express magnetic susceptibility is in terms of the number of moles of the substance. This is called *molar susceptibility*, and is defined by the expression,

$$\chi_M = \chi . M \tag{21.9}$$

where M is the molecular weight of the substance.

According to *Curie law,* the molar susceptibility varies inversely with absolute temperature.

$$\chi_M \propto \frac{1}{T} \tag{21.10}$$

or $$\chi_M = \frac{C}{T} \tag{21.11}$$

Here T is the absolute temperature and C is a constant called *Curie constant.* Mathematically,

$$C = \frac{N\mu^2}{3K'} \tag{21.12}$$

where μ is the magnetic moment, N is the Avogadro number and K' is the Boltzmann constant.

From equations (21.11) and (21.12),

$$\chi_M = \frac{N\mu^2}{3K'\,T} \tag{21.13}$$

or $$\mu = \sqrt{\frac{3\,K'}{N}}.\sqrt{\chi_M T} \tag{21.14}$$

Equation (21.13) is called *Langevin expression.*

Substituting the values of Avogadro number and Boltzmann constant,

$$\mu = 2.84\sqrt{\chi_M T} \tag{21.15}$$

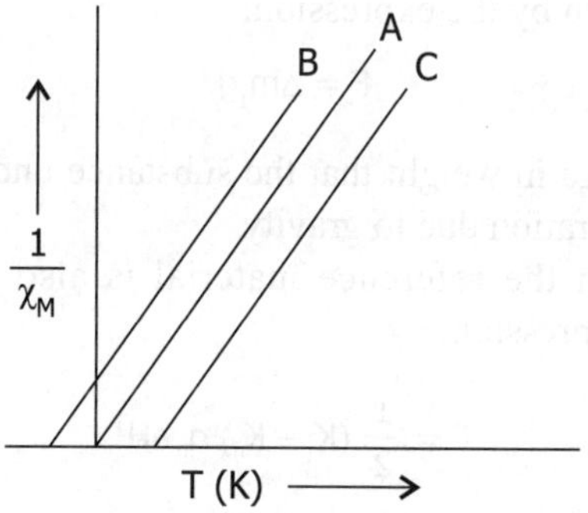

Fig. 21.6 Deviations from Curie law

It is possible to determine the values of molar susceptibility at various temperatures. If the inverse of molar susceptibility is plotted against absolute temperature, a straight line which intersects the origin results (line A in Fig. 21.6). The slope of this line equals C, the Curie constant (equation 21.11). However, there are many substances for which the straight line cuts the x-axis at a temperature above or below zero degree Kelvin (lines B and C in Fig. 21.6). Such lines cannot be represented by Curie law and, therefore, equation (21.11) is modified to the expression,

$$\chi_M = \frac{C}{T-\theta} \tag{21.16}$$

Equation (21.16) is the mathematical expression of *Curie–Weiss law* and θ is known as *Weiss constant.* The Curie–Weiss law emphasizes that the magnetic moment of a specific atom, ion or molecule is not independent of its immediate neighbors. Rather, the magnetic moment depends on the alignment of one atom, ion or molecule with respect to the orientations adopted by its surrounding particles.

Experimental determination of magnetic moment

Magnetic moment is a very important parameter. It tells us whether a particular compound is diamagnetic or paramagnetic, and if paramagnetic, how many unpaired electrons are present in the valence shell of its atom, ion or molecule. The magnetic moment may be calculated from equation (21.15), provided molar susceptibility, χ_M is known. The molar susceptibility, in turn, may be determined by a technique called *Gouy's method.*

In Gouy's method, a reference compound whose magnetic susceptibility is known, is taken. The most common reference is mercury tetrathiocyanato-Ncobaltate(II), $Hg[Co(NCS)_4]$. Finely divided form of this compound is filled in a narrow, flat-bottom sample tube, up to a calibration mark. The weight of the compound is recorded. The tube is then suspended in a magnetic field and the weight of the compound is recorded again. Since the reference compound is paramagnetic in nature, it undergoes an increase in weight in the magnetic field. The force acting on the substance is given by the expression,

$$F = \Delta m_1 g \tag{21.17}$$

where Δm_1 is the change in weight that the substance undergoes in the magnetic field and g is the acceleration due to gravity.

The force acting on the reference material is also related to the volume susceptibility by the expression,

$$F = \frac{1}{2}(K_1 - K_0)\, q_0 a H^2 \tag{21.18}$$

where K_1 = Volume susceptibility of mercury tetrathiocyanato-Ncobaltate(II) = 206.6×10^{-12} m^3 mol^{-1} at 293 K.

K_0 = Volume susceptibility of air = 0.364×10^{-12} m^3 mol^{-1} at 293 K.
q_0 = Permeability of free space = $4\pi \times 10^{-7}$ H m^{-1}.
a = Area of cross-section of sample tube.

H = Strength of magnetic field.

From equations (21.17) and (21.18),

$$\Delta m_1 g = \frac{1}{2}(K_1 - K_0)\, q_0\, aH^2 \tag{21.19}$$

or

$$K_1 = K_0 + \frac{2\Delta m_1 g}{q_0 aH^2} \tag{21.20}$$

From equation (21.20), the value of aH^2 is calculated. Next, the same sample tube is filled up to the same mark with the compound whose magnetic susceptibility is to be measured. It is then suspended in the same magnetic field, and the change in weight, Δm, is recorded. The volume susceptibility of unknown compound, K, is given as,

$$K = K_0 + \frac{2\Delta mg}{q_0 aH^2} \tag{21.21}$$

The value of aH^2, calculated from equation (21.20), is substituted in equation (21.21), and the value of K is determined. The volume susceptibility is converted into gram susceptibility, χ by dividing it by density of the substance (equation 21.8). The molar susceptibility, χ_M is obtained by multiplying gram susceptibility by molecular weight of the substance (equation 21.9). By substituting the value of molar susceptibility in equation (21.15), the magnetic moment may be elucidated.

Other forms of magnctism

While discussing paramagnetism, it is generally assumed that each metal ion is isolated from other metal ions, there being no interaction between the neighbors. This is relevant in cases where the paramagnetic metal ions are widely separated from one another by several diamagnetic entities. For example, in copper(II) sulfate.5-water, $CuSO_4.5H_2O$, the copper(II) ions are truly paramagnetic because of the intervention of diamagnetic water molecules of crystallization. In such cases, the magnetic dipoles of neighboring copper(II) ions are more or less randomized, as shown in Fig. 21.7a.

Yet, there are cases where the paramagnetic metal ions are either too close to each other, as in the case of a metallic lattice, or are separated by anions such as oxide/fluoride which can transmit magnetic interactions. In such cases, three other forms of magnetisms arise. These are: ferromagnetism, anti-ferromagnetism and ferrimagnetism. As depicted in Fig. 21.7, these differ from paramagnetism in alignment of magnetic moments.

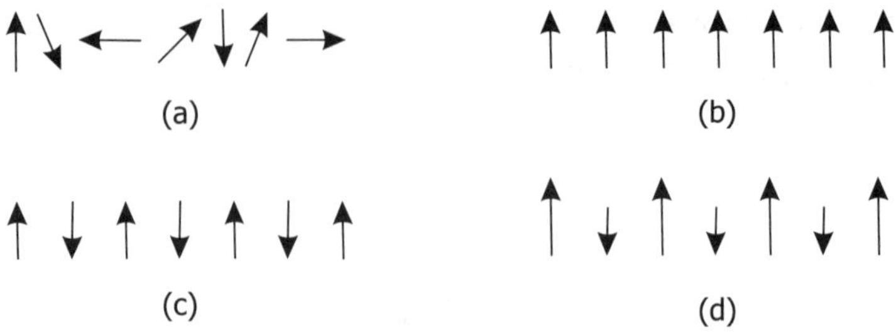

Fig. 21.7 Alignment of magnetic moments in (a) paramagnetic (b) ferromagnetic (c) anti-ferromagnetic and (d) ferrimagnetic substances

Ferromagnetic substances are those in which the magnetic moments of constitutent atoms, ions or molecules are aligned in the same direction. (Fig. 21.7b). The mutually aligned spins are called *magnetic domains*. Application of a magnetic field causes these domains to align with the magnetic field. As a result, the magnetic susceptibility becomes higher than expected. While the plot of magnetic susceptibility against temperature for a paramagnetic substance is a smooth curve (Fig. 21.8a), the corresponding plot for a ferromagnetic substance shows a discontinuity. The discontinuity is in the form of a transition temperature, called *Curie temperature,* T_C (Fig. 21.8b). Above the Curie temperature, a ferromagnetic substance shows typical paramagnetic character. However, below the Curie temperature, the magnetic moments of the individual constituents tend to align themselves parallel, reinforcing each other. As the temperature is lowered, the parallel alignments become more and more perfect. Therefore, the magnetic susceptibility increases steeply with decreasing temperature. At temperatures higher than Curie transition, thermal agitations randomize the magnetic moments so that their orientations are similar to that of a paramagnetic substance. Iron, cobalt and nickel are examples of ferromagnetic substances. The Curie temperature for iron is 1041 K.

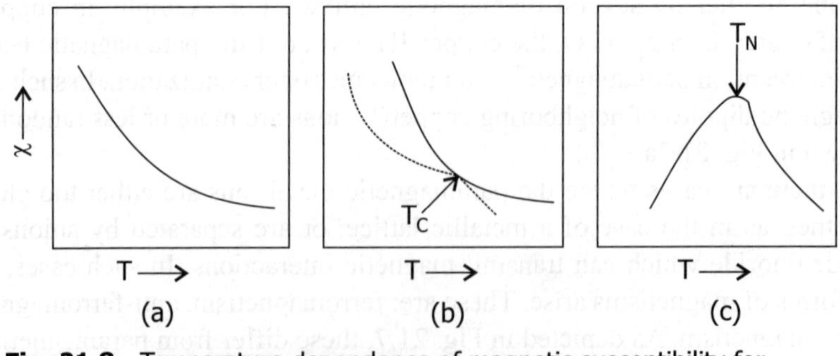

Fig. 21.8 Temperature dependence of magnetic susceptibility for (a) simple paramagnetism (b) ferromagnetism and (c) anti-ferromagnetism

Anti-ferromagnetic substances are those in which the magnetic moments of constituent atoms, ions or molecules are so aligned as to cancel each other out (Fig. 21.7c). As a result, the magnetic susceptibility becomes less than expected. The plot of magnetic susceptibility against temperature for an anti-ferromagnetic substance shows a discontinuity in the form of a transition temperature, called *Ne'el temperature,* T_N (Fig. 21.8c). Above the Néel temperature, an anti-ferromagnetic substance shows typical paramagnetic character. However, below the Néel temperature, the magnetic moments of the individual constituents tend to align anti-parallel to each other. As the temperature is lowered, the anti-parallel alignments become more and more perfect. Therefore, the magnetic susceptibility decreases as the temperature falls. At temperatures higher than Néel transition, thermal agitations randomize the magnetic moments at par with those of a paramagnetic substance. Nickel(II) oxide and manganese(II) oxide show anti-ferromagnetic behavior; the Néel temperature for the latter is 391 K.

Case Study 21.5 Manganese with five unpaired electrons is less ferromagnetic than iron with four unpaired electrons. Why?

The degree of ferromagnetic character depends on how effectively the magnetic moments of the constituent atoms, ions or molecules of a substance are aligned parallel to each other. The metallic radius of iron is smaller than that of manganese. Therefore, in a lattice of iron, the atoms are close to each other, as a result of which their individual magnetic moments can attain parallel orientations with respect to each other. Despite there being only four unpaired electrons in the valence shell, the magnitude of ferromagnetic character is quite large. The manganese atoms, because of their larger size, are physically separated to a larger extent in a metallic lattice. Therefore, their degree of ferromagnetic character is less, despite there being five unpaired electrons in the valence shell.

It must be emphasized that ferromagnetism and anti-ferromagnetism are properties of solids. In solution or gaseous phase, the magnetic species become physically separated from each other, the orientation of magnetic moments is lost and so is the ferromagnetic or anti-ferromagnetic characteristic.

Ferrimagnetism arises if some magnetic moments are systematically aligned so as to oppose others, but, in terms of numbers, a net magnetic moment results (Fig. 21.7d). An example of a ferrimagnetic substance is the mixed oxide of iron, triferric tetraoxide, Fe_3O_4, which has an inverse spinel structure (Chapter 26, p. 731). Both iron(II) and iron(III) ions are present in the structure of this compound. Half the iron(III) ions occupy tetrahedral sites, the other half occupy octahedral sites, while all the iron(II) ions occupy octahedral sites. The moments of iron(III) ions in tetrahedral and octahedral sites cancel each other. The observed magnetic moment, which imparts ferrimagnetic character to this oxide, results from iron(II) ions alone.

Case Study 21.6 The magnetic moment of potassium hexachloroosmate(IV) complex, $K_2[OsCl_6]$, increases when its solid solution is made in the diamagnetic potassium hexachloroplatinate(IV) complex, $K_2[PtCl_6]$. Why?

The osmium(IV) complex is anti-ferromagnetic in nature. The metal ions are so aligned that their individual magnetic moments cancel out. Therefore, the net magnetic moment is less than expected. When its solid solution is made in the diamagnetic platinum(IV) complex, the chlorine atoms of the latter intervene between the osmium(IV) ions. The metal ions become physically separated, and the anti-parallel orientations of the magnetic moments largely disappear. As a result, the anti-ferromagnetic character is lost and osmium(IV) ions behave as typical paramagnetic species. Therefore, the magnetic moment of the complex increases.

Complexation

Transition metals form a large number of stable complexes. The composition of a typical complex may be represented as $[Co(NH_3)_6]Cl_3$. In this compound, six ammonia molecules are linked to the cobalt(III) ion by coordinate bonds. The cobalt(III) ion and the ammonia molecules are enclosed in square brackets since both these entities lose their physical and chemical properties. The chloride ions, which are outside the square brackets, however, retain their physical and chemical properties.

The ammonia molecules are called ligands. A *ligand* is defined as a chemical species which combines with a metal ion to form a complex ion. In the example cited above, $[Co(NH_3)_6]^{3+}$ is the complex ion. A ligand may be a neutral molecule (for example, ammonia, NH_3; water, H_2O) or a negative ion (for example, fluoride, F^-; nitrite, NO_2^-). In rare cases it may be a positive ion (for example, nitrosonium, NO^+). If the ligand is a multiatomic species, then the atom that is directly linked to the metal is called *ligating atom* or *bonding site*. In the example below, nitrogen and oxygen are the ligating atoms of ammonia and water ligands, respectively.

The number of ligands attached to the metal in a complex ion is called the *coordination number*. The coordination number of metal in $[Co(NH_3)_6]^{3+}$ is six, while that in $[Cu(NH_3)_4]^{2+}$ is four. The complex species may be positively charged, e.g., $[Cr(H_2O)_6]^{3+}$, negatively charged, e.g., $[Cu(CN)_4]^{2-}$ or neutral, e.g., $[Pt(NH_3)_2Cl_2]$.

Thus, a complex is defined as a compound formed by the combination of a metal ion with a specific number of ligands in such a manner that both the metal ion and the ligands lose their characteristic properties. It is different from a double salt. When a double salt is dissolved in water, each cation and anion can be identified qualitatively. For example, Mohr salt, $FeSO_4.(NH_4)_2SO_4.6H_2O$, in aqueous solution ionizes into ferrous, ammonium and sulfate ions. Each of these ions retains its identity. However, in a complex, both the metal and the ligands lose their individual identities.

Transition metals have the unique ability to form complexes. A simple salt of transition metal gives a complex in aqueous solution, with water as the ligand. For example,

$$NiCl_2 + 6H_2O \rightarrow [Ni(H_2O)_6]Cl_2$$

Non-transition metals do not form as many complexes. The tendency of transition metals to form complexes so readily is attributed to three factors.

1. Small size of the metal ion

The driving force for complex formation is the attraction between the negative charge or the lone pair of electrons on the ligand with the nucleus of the metal ion. If the size of the metal ion is small, the internuclear distance between the metal and the ligand is also small. Conversely, the metal–ligand internuclear distance is large for a larger metal ion (Fig. 21.9).

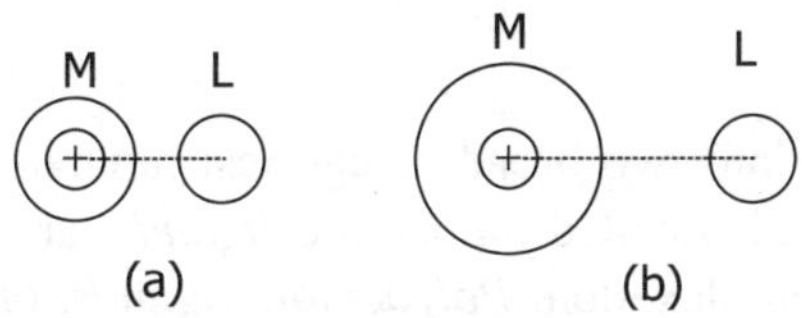

Fig. 21.9 Metal–ligand internuclear distance is (a) small for a smaller metal ion and (b) large for a larger metal ion
M= Metal ion; L = Ligand

According to Coulomb's law, the force of attraction is inversely proportional to the square of distance of separation. Thus, there is a greater degree of attraction between the metal and the ligands if the former has a small size. Hence, cobalt(II) ion forms a larger number of complexes than magnesium(II) ion.

2. High charge on the metal ion

Coulomb's law states that the force of attraction is directly proportional to the product of charges of the two interacting species. For metal–ligand interactions, higher the charge on metal ion, more will be the attraction for the ligand. Therefore, cobalt(III) ion forms complexes more readily than cobalt(II) ion.

3. Presence of vacant d orbitals on the metal ion

A complex is formed by the donation of electrons from the ligands to the metal ion. The metal should possess vacant orbitals of suitable energy to accommodate the electrons being donated by the ligands. Transition elements, by definition, possess partially filled d orbitals. These vacant d orbitals on metal are able to accept the electrons from the ligands, thereby assisting complex formation.

Case Study 21.7 On adding mercury(II) ions to aqueous tetrachlorocobaltate(II) ion, the color of the solution changes from blue to pink. On adding magnesium(II) ions to the pink solution, the color reverts back to blue. Explain these changes.

The color changes may be summarized in the following equilibrium reaction:

$$\underset{\text{Blue}}{[CoCl_4]^{2-}} + 6H_2O \underset{\text{Mg(II)}}{\overset{\text{Hg(II)}}{\rightleftharpoons}} \underset{\text{Pink}}{[Co(H_2O)_6]^{2+}} + 4Cl^-$$

These changes highlight the preference of a ligand to combine with a particular metal ion on the basis of hard and soft acid–base concept (*see* Chapter 5, p. 245). The mercury(II) ion is a soft acid. It prefers to combine with the chloride ions, which are soft bases. As a result, the reaction equilibrium shifts to the right, yielding pink-colored hexaquacobalt(II) ion. On the other hand, magnesium(II) ion is a relatively hard acid. It prefers to combine with water since the ligating atom of the latter (oxygen) is a hard base. This combination pushes the equilibrium to the left, resulting in the formation of blue-colored tetrachlorocobaltate(II) ion.

Types of Ligands

Ligands are classified into two broad categories: monodentate and polydentate. *Monodentate ligands,* also called *unidentate ligands,* are those which use only one donor atom for complexation. *Polydentate ligands,* on the other hand, have two or more atoms which can simultaneously serve as donors. The polydentate ligands are subdivided into bidentate, tridentate, tetradentate, pentadentate or hexadentate, depending on whether there are 2, 3, 4, 5 or 6 donor atoms, respectively. Representative examples are depicted in Table 21.5.

Another category of ligands is called *ambidentate,* or simply, *ambient* ligands. These ligands have two donor atoms, but use only one at a time. Therefore, these are monodentate ligands. Thiocyanate group is an example of ambidentate ligand. It can coordinate to the metal either through sulfur atom, in which case it is called thiocyanato-S or through nitrogen atom, in which case it is called thiocyanato-N.

$$\underset{\text{Thiocyanato-S}}{M \leftarrow SCN} \qquad \underset{\text{Thiocyanato-N}}{M \leftarrow NCS}$$

We shall learn more about this class of ligands in the chapters on nomenclature and isomerism.

Table 21.5 Different types of ligands

Monodentate	
Fluoride	$F^{-} \longrightarrow$
Water	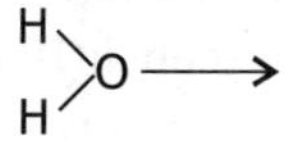
Ammonia	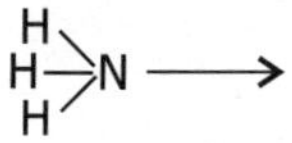
Bidentate	
Ethylenediamine (en)	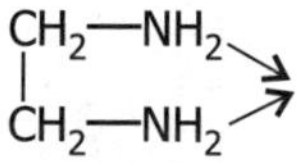
Oxalate	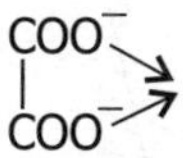
Tridentate	
Diethylenetriamine (dien)	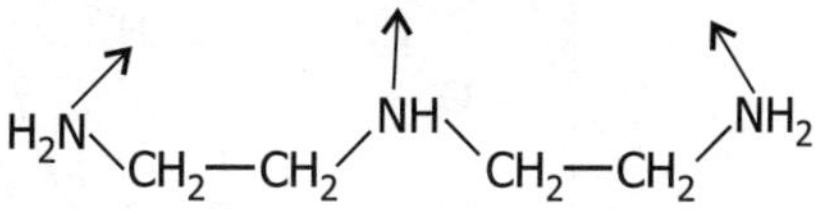
Tetradentate	
Nitrilotriacetic acid	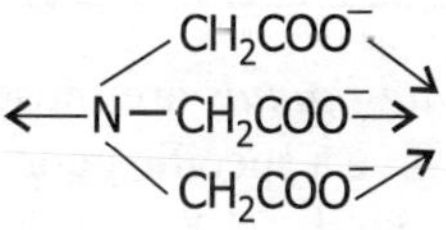
Pentadentate	
Ethylenediamine triacetic acid	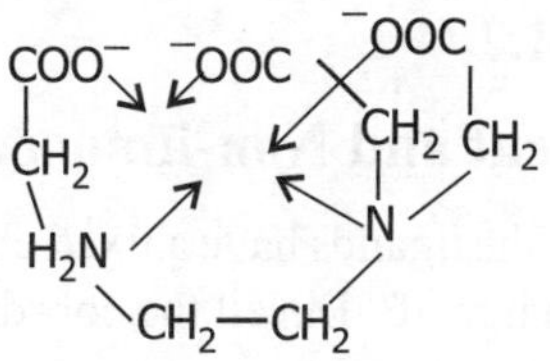
Hexadentate	
Ethylenediamine tetraacetic acid (EDTA)	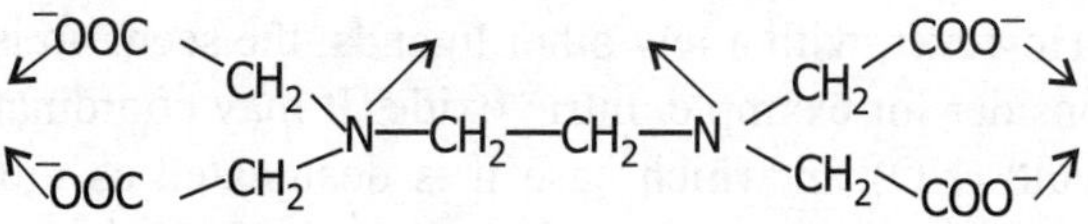

Case Study 21.8 Draw the structure of $[Co(EDTA)]^-$ complex.

EDTA stands for ethylenediamine tetraacetic acid, a hexadentate ligand (Table 21.5). It has six donor atoms—four oxygen atoms and two nitrogen atoms. Complexes with coordination number 6 possess an octahedral structure. The octahedral coordination around cobalt(III) ion in $[Co(EDTA)]^-$ complex may be represented as follows:

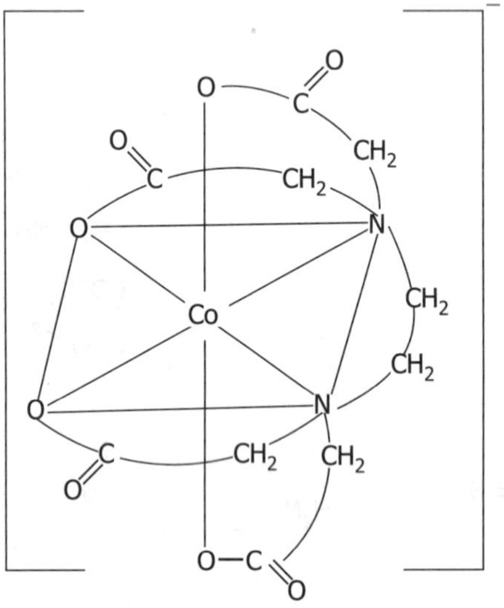

Flexidentate ligands are those which can flexibly wrap themselves around the metal ion to give a specific geometry. The ethylenediamine tetraacetic acid ligand, cited in Case Study 21.8, is a flexidentate ligand.

Box 21.2

Innocent and Non-innocent Ligands

Most of the ligands have a fixed charge over them. For example the fluoro ligand has a charge of −1 in all the coordination compounds, irrespective of the nature of metal ion or the composition of the complex ion. Likewise, the carbonyl ligand, CO is always neutral. Such ligands are called innocent ligands. It is not at all problematic to assign the oxidation state of the metal in complexes involving such ligands.

However, with a few other ligands, the scenario is not so straight forward. Consider for example, nitric oxide. It may coordinate to a metal as a neutral ligand (NO), in which case it is designated as nitrosyl group. Alternately, it may coordinate as a positive ligand (NO^+), in which case it is designated as nitrosonium group. It is also possible that both nitrosyl and nitrosonium

Contd...

groups may be present in a single complex (*see* Fig. 29.22b, p. 862). Unless it is precisely known whether nitric oxide coordinates as a neutral ligand or as a positively charged ligand, it is not possible to calculate the oxidation state of the metal. The ligands whose presence in the complex makes the assignment of oxidation state to the metal ambiguous are called non-innocent ligands.

Case Study 21.9 Cite an example where atoms of a noble gas act as ligands.

Xenon is known to act as a ligand, forming a complex of gold(II), having the composition, $[AuXe_4]^{2+}$. The four xenon atoms are arranged around gold(II) ion in a square planer configuration. The complex is prepared by the reduction of gold(III) fluoride by a mixture of hydrogen fluoride and antimony(V) fluoride, in presence of xenon. As stated in Chapter 5 (Box 5.1, p. 244), the hydrogen fluoride-antimony(V) fluoride combination acts as a superacid.

$$2HF + SbF_5 \rightarrow [H_2F]^+ [SbF_6]^-$$

$$2AuF_3 + 4[H_2F]^+ \rightarrow 2Au^{2+} + 8HF + F_2$$

$$Au^{2+} + 4Xe \rightarrow [AuX_4]^{2+}$$

The complex is isolated as dark red crystals and is stable up to 195 K.

Comparison of 3d Transition Series with 4d/5d Series

The 4d and 5d transition series elements of any specific group have similar characteristics, but both show typical differences from their lighter congener of the 3d series. Some of the important differences between the 3d series members vis-à-vis their 4d/5d counterparts are discussed below.

1. Size

On moving down a group in a periodic table, the ionic or atomic radii of the elements increases. Accordingly, the 4d elements are larger in size as compared to the 3d elements. However, the size of 4d and 5d elements of the same group are comparable. This is because between the 4d and 5d transition series there are 14 lanthanoid elements. The filling of the 4f valence shell of these elements causes a consistent decrease in the atomic and ionic radii. This decrease is called *lanthanoid contraction*. Due to lanthanoid contraction, the expected increase on moving from 4d to 5d elements is negated. Therefore, there is hardly any difference in atomic and ionic sizes between the two heavy elements of the same group, although the corresponding elements of the first transition series is significantly smaller. In fact, the resemblance between 4d and 5d series elements emanates from similarities in their sizes (for a more precise explanation, refer to the relativistic effect, cited in the next section).

2. Oxidation states

The 4d and 5d elements tend to exist in higher oxidation states than the 3d elements. This is because heavier elements have comparatively low ionization energies. For example, chromate ion, $[CrO_4]^{2-}$ is a strong oxidizing agent, while molybdate, $[MoO_4]^{2-}$ and tungstate, $[WO_4]^{2-}$ are stable ions. Similarly, permanganate, $[MnO_4]^-$ is a strong oxidizing agent, but pertechnetate $[TcO_4]^-$ and perrhenate, $[ReO_4]^-$ are stable. While chromium(VI) oxide, CrO_3 is unstable, molybdenum(VI) oxide, MoO_3 and tungsten(VI) oxide, WO_3 are stable. Complexes of 3d transition series in low oxidation states, such as $[CrCl_6]^{3-}$ and $[Co(NH_3)_6]^{3+}$, have no counterparts of equivalent oxidation states among heavier elements. Similarly, OsO_4 exists, FeO_4 does not; PtF_6 exists, NiF_6 does not (for a better explanation, refer to the relativistic effect, described in the next section).

Case Study 21.10 Comment on the oxidation states of the metal ions in MoS_2 and FeS_2.

There are two possible oxidation states for the metal ions.

(i) The metal is in oxidation state (IV), in concert with sulfide ions, S^{2-}.
(ii) The metal ion is in oxidation state (II), in concert disulfane ion, S_2^{2-}.

The sulfide ion is reducing in nature. It can combine with a metal ion in an oxidation state that is not easily reduced. For molybdenum, a 4d transition metal, oxidation state (IV) is stable. It can coexist with sulfide ions. For iron, a 3d transition metal, oxidation state (IV) is susceptible to reduction. It cannot coexist with sulfide ions. Therefore, MoS_2 is molybdenum(IV) sulfide, while FeS_2 is iron(II) disulfane.

---

3. Magnetic properties

According to crystal field theory (Chapter 26, p. 709), d orbitals in complexes split into t_{2g} and e_g sub-levels. The magnitude of splitting, Δ, for any given ligand is greater for a 4d or 5d member as compared to the 3d element of the same group. This is because 4d and 5d orbitals are larger in size than 3d orbitals, as a result of which ligands can approach more closer to the valence shell of heavier metals.

Since the magnitude of crystal field splitting is large for 4d and 5d elements, these metals form predominantly spin-paired complexes. Even with weak ligands, the complexes are diamagnetic in nature. The 3d elements, on the other hand, form spin-free, paramagnetic complexes with weak field ligands.

4. Metal–metal bonding

The 4d and 5d members have a greater tendency to form metal–metal bonds as compared to the 3d elements. As a result, a greater number of polynuclear and cluster compounds are known for the heavier elements than for the lighter ones. Examples of derivatives containing metal–metal bonds include $[Re_2Cl_8]^{2-}$, $[Mo_2Cl_9]^{3-}$, $Rh_4(CO)_{12}$ and $Ir_4(CO)_{12}$. Such compounds are rare for the first series transition elements.

5. Coordination number of complexes

For the 3d series metals, complexes with coordination number 6 are most common, followed by coordination number 4. Coordination numbers greater than 6 are uncommon for the complexes of these lighter elements. The heavier analogs, however, are larger in size and can accommodate a relatively large number of ligands around them. Therefore, the 4d and 5d elements often form complexes with coordination numbers greater than 6. Examples include $Na_3[ZrF_7]$ and $Cu_2[ZrF_8]$.

6. Abundance

The first series transition metals are more abundant in the earth's crust as compared to the heavier elements. The ten elements of the first series together make up about 6.8% of the earth's crust. The 20 elements of the second and third series, on the other hand, constitute only 0.025% of the earth's crust. The element technetium does not exist in nature.

Case Study 21.11 Compare the properties of chromium(VI) oxide with molybdenum(VI) oxide and tungsten(VI) oxide.

Chromium(VI) oxide (CrO_3) is unstable, volatile and soluble in water. The corresponding oxides of molybdenum and tungsten are stable, non-volatile and polymeric in nature.

The oxide of chromium is molecular in nature. There are multiple bonds between the metal and oxygen atoms. Further, since chromium is a member of 3d series, the high oxidation state of VI is unstable for it. However, for molybdenum and tungsten, the ionization energies are relatively low and, therefore, oxidation state VI is quite stable. Being larger than chromium, they cannot form multiple bonds. As a result, they acquire a polymeric structure. The polymeric structure would obviously be more stable, less volatile and less soluble.

Relativistic Effect and Anomalous Characteristics of Heavier Transition Elements

We have already discussed the significance of relativistic effect in Chapter 2, p. 74. Before applying this effect to explain the anomalous characteristics of heavier transition elements, let us review its fundamental principles.

1. The velocity with which an electron revolves around the nucleus increase with nuclear charge.
2. According to Einstein theory of relativity, the mass of the electron increases as its velocity approaches the speed of light.
3. The radius of the Bohr orbital decreases with the increasing mass of electron.
4. The s and p orbitals, which have lobes nearer to the nucleus, contract due to relativistic effect, their energy is lowered and they are stabilized. This is called relativistic contraction. The d and f orbitals, which have lobes further

away from the nucleus, do not experience relativistic effect. Rather, these orbitals are more effectively screened from the nucleus by contracted s and p orbitals and hence undergo expansion. Due to this relativistic expansion, these orbitals are raised in energy and are destabilized.

The relativistic effect is directly proportional to the square of nuclear charge, that is, Z^2. Hence it begins to become more important in the vicinity of $Z = 60 - 70$ and explains the following anamolous characteristics of heavier transition elements.

a. Comparable size of 4d and 5d elements

We have stated that the 4d and 5d elements of any one Group of the periodic table have almost same atomic radii. This is attributed to the intervention of the 4f elements in between the 4d and 5d transition series. Due to lanthanoid contraction, whatever increase in size is expected on moving from 4d to 5d elements is nullified. Lanthanoid contraction is the major, yet not the only reason for comparable sizes of second and third series transition elements.

As already described, the s and p orbitals contract, while the d and f orbitals expand in size due to relativistic phenomena. Since the outermost shells are ns and np rather than (n – 1)d or (n – 2)f, each heavier atom as a whole contracts. This factor cancels the expected increase on going from the 4d to 5d elements of each group.

Relativistic effect contributes to almost 10% extent in negating the increase in size that is normally expected on moving down a group.

b. Higher oxidation states of 5d elements

Due to relativistic effect, the 6s orbital tends to contract and get stabilized while the 5d orbitals tend to expand, and get destabilized.

The energy required to ionize 6s electrons is thus greater than required to ionize a 4s electron of a first series transition element. However, the energy expended to remove a 5d electron is less than that required to remove a 3d electron. The data presented in Table 21.6 supports this assertion.

Table 21.6 Comparison of ionization energies of titanium and hafnium

Element	Ionization energy ($kJ\ mol^{-1}$)		
	1st	2nd	3rd
Titanium	658	1310	2650
Hafnium	654	1440	2250

When the first two electrons residing in the valence s orbital have been ionized, it becomes more easy to remove electrons from penultimate 5d orbitals as compared to the 3d orbitals. For this reason, the 5d series transition elements exhibit higher oxidation states in their compounds. Relativistic effect is significant in second series transition elements too, although to a lesser extent as compared to the third series elements. Hence the 4d elements also form a good number of stable compounds in high valence states.

c. Unique properties of gold

With the progressive filling up of 5d shell and concomitant increase in atomic number, the relativistic effect manifests itself to a maximum extent at gold ($5d^{10}6s^1$) and mercury ($5d^{10}6s^2$). The contraction of 6s orbital is largest for these two elements. Their ionization energies, with respect to s electron(s) are also greater than expected. For gold, the combination of small size, high ionization energy and a vacancy in 6s shell leads to an increase in electron affinity. In fact, gold has the highest electron affinity of any atom other than halogens. The increased ionization energy and increased electron affinity of gold has two important implications. Firstly, due to greater than expected ionization energy, gold is quite inert (*see* Box 21.3). Unlike its group members, copper and silver, it does not corrode. Hence its ornamental value. Secondly, due to greater than expected electron affinity, gold readily forms the anion, Au^- which is comparable in size to bromide ion, Br^-. The compounds rubedium auride, Rb^+Au^- and cesium auride Cs^+Au^- are true ionic compounds and not alloys.

Box 21.3

The Golden Death Mask of Tut

The most stunning artifact that illustrates the inert nature of gold is the death mask of the Egyptian King Tut (Fig. 21.10). This king ascended the throne in 1332 BC when he was 10 years old and died in 1323 BC at the age of 19. In keeping with the Egyptian tradition, the gold mask was placed on the face of the mummy of Tut so as to ensure that whenever the spirit of the deceased returned it would recognize the body.

Fig. 21.10 The golden death mask of Tut

Gold was chosen to mould the mask, not because the Pharoh dynasty to which King Tut belonged was rolling in wealth, but because the Egyptians were aware that this precious metal neither tarnishes, nor deteriorates. Moreover, they credited the shining luster of gold with the power of sun god.

The mask which is 54 cm high, 39.3 cm wide, 49 cm deep and weights 10.32 kg was excavated in 1922. At present it adorns the *Cairo Museum*. After more than 33 centuries, it still shines like the sun!

Case Study 21.12 What is *aurophilicity*?

The tendency of elements to exhibit greater than expected combining capacity towards gold is called aurophilicity. For example, in the cation, $[C\{AuP(C_6H_5)_3\}_6]^{2+}$, carbon has the unusual covalency of six. This, so called, *hedgehog cation*, has Au(I)–Au(I) bonds in its structure. Other examples of aurophilic cations include $[N\{AuP(C_6H_5)_3\}_5]^{2+}$ and $[O\{AuP(C_6H_5)_3\}_4]^{2+}$.

Like silver(I), the gold(I) ion is expected to form linear complexes, with coordination number two. However, in aurophilic cations, gold(I) not only interacts with other gold(I) ions, but also with the non-metal (carbon, nitrogen or oxygen). Such behavior is the outcome of the relativistic effect. As has already been discussed, the radius of 6s orbital decreases and that of 5d orbitals increases due to relativistic phenomena. These two set of orbitals have almost the same radii, as a result of which the 5d electrons participate in bonding, unlike the 4d electrons of silver(I) ion. No doubt, the interactions of gold(I) with the non-metal atom are quite weak, yet the Au(I)–Au(I) interactions are relatively strong. The net result is a stable aurophilic cation.

22. WERNER COORDINATION THEORY

Introduction

Werner theory attempts to explain the nature of bonding in coordination compounds. It is based on the premise that metals possess two types of valencies: primary and secondary. In a chemical combination that leads to the formation of a coordination compound, first the primary valency is satisfied. For example, the cobalt(III) ion may combine with three chloride ions to satisfy its primary valency.

$$Co^{3+} + 3Cl^- \rightarrow CoCl_3 \qquad \text{[Combination through primary valency]}$$

However, even though the combining capacity of cobalt(III) ion has been exhausted, further combination may still take place by making use of secondary valency.

$$CoCl_3 + 6NH_3 \rightarrow CoCl_3.6NH_3 \qquad \text{[Combination through secondary valency]}$$

The primary and secondary valencies may be demarcated by depicting the addition compound, $CoCl_3.6NH_3$ as $[Co(NH_3)_6]Cl_3$. The chemical entities within the square brackets, i.e., ammonia molecules, satisfy the secondary valency of cobalt, while those without the square brackets, i.e., chloride ions, satisfy the primary valency of cobalt.

We now outline the postulates of the Werner theory.

Postulate 1

Metals possess two types of valencies—primary or ionizable valency and secondary or non-ionizable valency.

The combining power of a metal atom is divided into two spheres of attraction: the inner or the coordination sphere and the outer or the ionization sphere.

The ionization sphere constitutes the entities which satisfy the primary valency of the metal. Hence, it is also called *primary sphere*. The coordination sphere constitutes the entities which satisfy the secondary valency of the metal. Hence, it is also called *secondary sphere*. In the example of $[Co(NH_3)_6]Cl_3$ complex cited above, the two spheres of attraction may be indicated as follows:

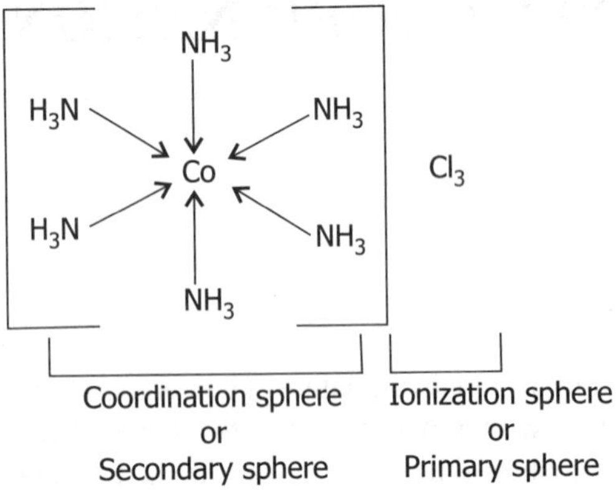

As has been stated, the addition of six molecules of ammonia takes place by secondary valency. In this combination, the physical and chemical characteristics of cobalt(III) ion and ammonia molecules are lost. However, the chloride ions retain their typical physical and chemical properties. This means that an aqueous solution of $[Co(NH_3)_6]Cl_3$ complex would give the tests for chloride ions, but not for cobalt(III) or ammonia species. Treating the complex with excess of silver nitrate solution would result in precipitation of three moles of silver chloride.

$$[Co(NH_3)_6]Cl_3 + 3AgNO_3 \rightarrow [Co(NH_3)_6](NO_3)_3 + 3AgCl$$

The reaction of $[Co(NH_3)_5Cl]Cl_2$ complex with silver nitrate will, however, yield only two moles of silver chloride.

$$[Co(NH_3)_5Cl]Cl_2 + 2AgNO_3 \rightarrow [Co(NH_3)_5Cl](NO_3)_2 + 2AgCl$$

The five ammonia, one chlorine and cobalt(III) ion, which together form the coordination sphere, cannot be detected by conventional tests, as all these chemical entities have lost their individual physical and chemical properties.

Case Study 22.1 Using following information:

Compound	Number of chloride ions precipitated
$PtCl_2.4NH_3$	2
$PtCl_2.3NH_3$	1
$PtCl_2.2NH_3$	0

write balanced equations for the reaction of these compounds with silver nitrate.

The compound $PtCl_2.4NH_3$ has two ionizable chlorine atoms. It may be represented as $[Pt(NH_3)_4]Cl_2$ and its reaction with silver nitrate may be written as:

$$[Pt(NH_3)_4]Cl_2 + 2AgNO_3 \rightarrow [Pt(NH_3)_4](NO_3)_2 + 2AgCl$$

There is only one ionizable chlorine in $PtCl_2.3NH_3$ compound. Its composition, $[Pt(NH_3)_3Cl]Cl$ indicates the presence of one chloride ion in the ionization sphere.

$$[Pt(NH_3)_3Cl]Cl + AgNO_3 \rightarrow [Pt(NH_3)_3Cl](NO_3) + AgCl$$

The fact that there are no ionizable chlorine atoms in $PtCl_2.2NH_3$ implies that there is no primary sphere in this complex. Accordingly, its composition is $[Pt(NH_3)_2Cl_2]$.

$$[Pt(NH_3)_2Cl_2] + AgNO_3 \rightarrow \text{No reaction}$$

Postulate 2

Every metal has a fixed number of secondary valencies. This number equals the number of coordinating groups directly linked to the metal. It is also called coordination number.

Neutral molecules as well as ions may become coordinated to the metal in the secondary sphere. The number of these coordinating groups may vary from 2 to 10. However, coordination numbers 4 and 6 are most common and most stable.

In the $[Co(NH_3)_6]Cl_3$ complex, for example, the coordination number of cobalt is 6. Loss of one ammonia would result in the transfer of a chloride ion from primary to secondary sphere, giving the complex $[Co(NH_3)_5Cl]Cl_2$. There are now five ammonia molecules and one chlorine in the coordination sphere, but the net coordination number is still 6. This complex may be represented as:

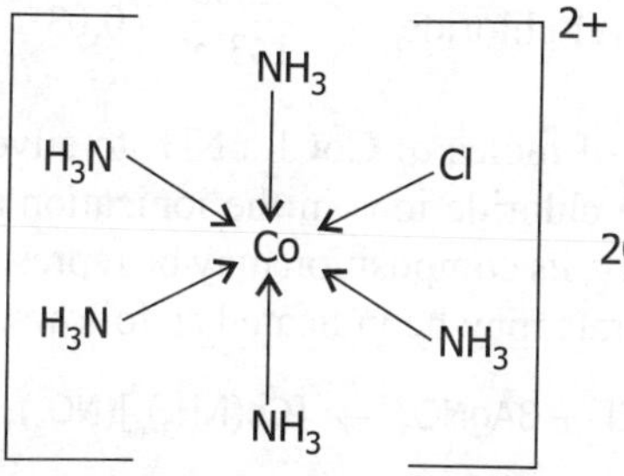

Removal of two ammonia molecules retains the same coordination number, and the resulting complex, $[Co(NH_3)_4Cl_2]Cl$ may be represented as:

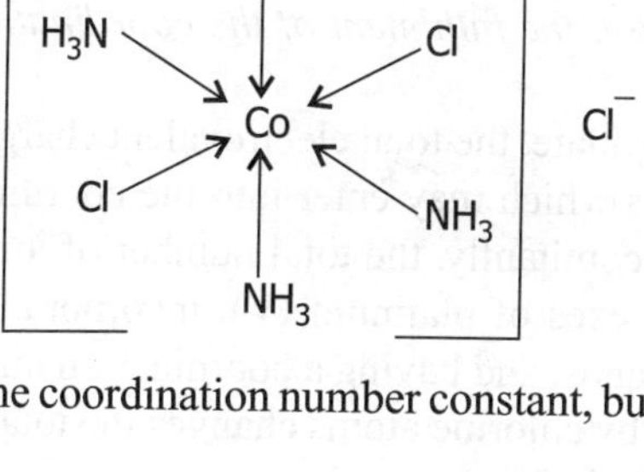

These changes keep the coordination number constant, but the number of primary valencies decreases.

Case Study 22.2 A solution containing 2.674 g of $CoCl_3.6NH_3$ was passed through a cation exchanger. The eluted solution gave 4.305 g precipitate of silver chloride with silver nitrate solution. Determine the composition of the complex in accordance with Werner theory (Mass number of Ag = 108, Cl = 35.5; Molecular mass of $CoCl_3.6NH_3$ = 267.4).

The passage of the solution of $CoCl_3.6NH_3$ through the cation exchanger yields those chloride ions free which constitute the ionization sphere. These chloride ions react with silver nitrate, giving a precipitate of silver chloride. The amount of silver chloride is indicative of the number of chloride ions in primary sphere. If three moles of silver chloride are obtained from one mole of $CoCl_3.6NH_3$, then there are three chloride ions in ionization sphere. If two moles of silver chloride are obtained, there are two ionizable chloride ions and if one mole of silver chloride is precipitated there is only one ionizable chloride ion.

Mass of $CoCl_3.6NH_3$ = 2.674 g
Molecular mass of $CoCl_3.6NH_3$ = 267.4

Number of moles of $CoCl_3.6NH_3$ = $\frac{2.674}{267.4} = 0.01$

Mass of silver chloride = 4.305 g
Molecular mass of silver chloride = 143.5

Number of moles of silver chloride = $\frac{4.305}{143.5} = 0.03$

Ratio between number of moles of $CoCl_3.6NH_3$ to silver chloride is 1:3. This means that there are three chloride ions in the ionization sphere of $CoCl_3.6NH_3$. According to Werner theory, its composition may be represented as $[Co(NH_3)_6]Cl_3$. Its reaction with silver nitrate may be indicated as follows:

$$\underset{\text{1 mole}}{[Co(NH_3)_6]Cl_3} + 3AgNO_3 \rightarrow [Co(NH_3)_6](NO_3)_3 + \underset{\text{3 moles}}{3AgCl}$$

Postulate 3

Primary valencies are satisfied by negative ions, whereas secondary valencies may be satisfied either by neutral molecules or by negative ions or, in rare instances, by positive ions. In any case, the fulfilment of the coordination number of the metal is essential.

According to this postulate, the total electrovalent charge on the complex varies with the number of ions which may enter into the coordination sphere to replace neutral molecules. Concomitantly, the total number of ions also varies. Consider, for example, the complexes of platinum(IV), incorporating ammonia molecules and chloride ions as ligands, and having a coordination number of 6. Replacement of ammonia molecules by chlorine atoms changes the total number of charges and total number of ions (Table 22.1).

Table 22.1 Number of charges and number of ions furnished by complexes

Compound	Number of charges			Number of ions		
	Cation	Anion	Total	Cation	Anion	Total
$[Pt(NH_3)_6]Cl_4$	4	4	8	1	4	5
$[Pt(NH_3)_5Cl]Cl_3$	3	3	6	1	3	4
$[Pt(NH_3)_4Cl_2]Cl_2$	2	2	4	1	2	3
$[Pt(NH_3)_3Cl_3]Cl$	1	1	2	1	1	2
$[Pt(NH_3)_2Cl_4]$	0	0	0	One neutral particle		

Herein lies the scope for the verification of Werner theory. As the number of charges increases, the molar conductivity of the complex also increases. Conversely, measurement of molar conductivity sheds light on the number of charges and hence on the possible structure of the complex.

The number of ions too provide a clue to structure determination. Cryoscopic measurements tell us that the freezing point depression is a colligative property, depending on the number of ions that a given complex produces. Larger number of ions causes a greater amount of depression in freezing point. Conversely, by determining the extent of freezing point depression, one may gauge the possible number of ions into which a complex would dissociate upon dissolution. This, in turn, indicates the plausible structure of the complex.

Molar conductivities and cryscopic measurements for chromium(III) complexes involving ammonia molecules and chloride ions, for example, help in demarcating the coordination sphere from ionization sphere (Table 22.2).

Table 22.2 Elucidation of structure on the basis of molar conductivity and cryoscopic measurements

Compound	Molar conductivity	Cryoscopic measurement	Structure
$CrCl_3.6NH_3$	6 charges	4 particles	$[Cr(NH_3)_6]^{3+} + 3Cl^-$
$CrCl_3.5NH_3$	4 charges	3 particles	$[Cr(NH_3)_5Cl]^{2+} + 2Cl^-$
$CrCl_3.4NH_3$	2 charges	2 particles	$[Cr(NH_3)_4Cl_2]^+ + Cl^-$
$CrCl_3.3NH_3$	0 charge	1 particle	$[Cr(NH_3)_3Cl_3]$

The molar conductivity and freezing point depression follow the order:

$$[Cr(NH_3)_6]Cl_3 > [Cr(NH_3)_5Cl]Cl_2 > [Cr(NH_3)_4Cl_2]Cl > [Cr(NH_3)_3Cl_3]$$

Case Study 22.3 Using the following information, establish the correct structure of the complexes.

Compound	Molar conductivity	Cryoscopic measurement
$Co(NO_2)_3.KNO_2.\ 2NH_3$	2 charges	2 particles
$Co(NO_2)_3.2\ KNO_2.\ NH_3$	4 charges	3 particles
$Co(NO_2)_3.3\ KNO_2$	6 charges	4 particles

On the basis of molar conductivity and cryoscopic measurement data, the structure of—

$Co(NO_2)_3.KNO_2.2NH_3$ is $K[Co(NH_3)_2(NO_2)_4]$

$Co(NO_2)_3.2\ KNO_2.NH_3$ is $K_2[Co(NH_3)(NO_2)_5]$

$Co(NO_2)_3.3\ KNO_2$ is $K_3[Co(NO_2)_6]$

For example, it has been observed that the freezing point of a solution of potassium iodide rises when mercury(II) iodide, which is almost insoluble in water, is added to the contents. The reaction between potassium iodide and mercury(II) iodide produces a coordination compound, potassium tetraiodomercurate(II).

$$2KI + HgI_2 \rightarrow K_2[HgI_4]$$

As is evident from the above chemical equation, two potassium ions and two iodide ions are consumed in the reaction. At the cost of these four ions, two potassium ions and one tetraiodomercurate(II) ion—a total of three ions—are produced. Since there is a net decrease in the number of ions in the solution, the freezing point rises.

Similarly, the freezing point of an aqueous solution of potassium cyanide rises on adding copper(I) cyanide to it. Copper(I) cyanide is insoluble in water, yet it forms a coordination compound, potassium tetracyanocuprate(I), on reacting with potassium cyanide.

$$3KCN + CuCN \rightarrow K_3[Cu(CN)_4]$$

In aqueous solution, potassium tetracyanocuprate(I) dissociates into four ions—three potassium ions and one tetracyanocuprate(I) ion. Against these, six ions—three potassium ions and three cyanide ions—are required for its formation. Since the total number of ions in the solution comes down from six to four, the freezing point of the mixture increases.

Postulate 4

The secondary valencies have a fixed orientation in space around the metal.

The secondary valencies are directional in nature. As a result, the coordinating groups in the secondary sphere are assigned definite positions in space. If a complex has coordination number 6, then irrespective of the nature of metal ion and the nature of coordinating groups, the secondary valencies would invariably be directed at the corners of an octahedron. Similarly, when four coordinating groups are attached to a metal ion, the geometry is either square planar or tetrahedral (Fig. 22.1).

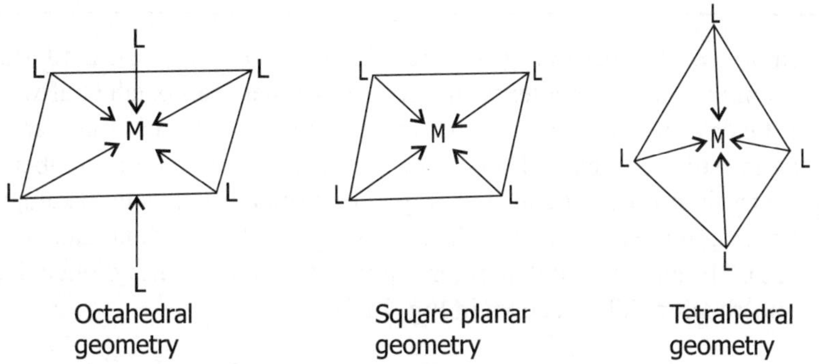

Fig. 22.1 The ligands, L, have a fixed orientation around the metal, M

The directional nature of secondary valencies makes possible the existence of stereoisomers among inorganic complexes. Thus the complex, $PtCl_2.2NH_3$ exists in two geometrical isomeric forms, *cis* and *trans*. Their structures may be represented as follows:

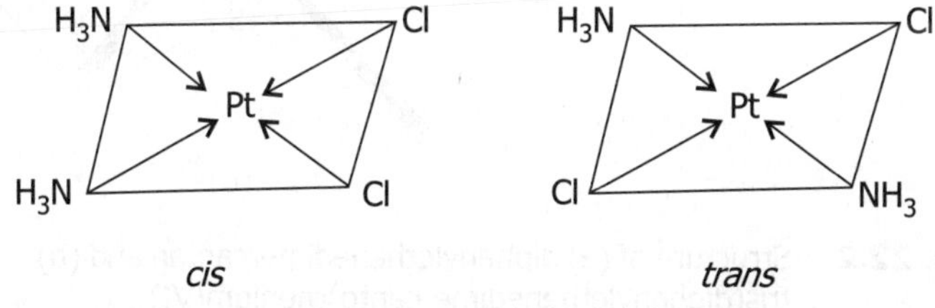

The isomeric forms arise because the four coordinating groups (two ammonia molecules and two chlorine atoms), rather than having random orientations around platinum, are directed at the corners of a square.

Case Study 22.4 One mole of a pink solid (**X**), having empirical formula, $CoCl_3.5NH_3.H_2O$ gives three moles of silver chloride on reaction with excess of silver nitrate solution. On heating, **X** loses a molecule of water to give a purple solid (**Y**) which has the same NH_3:Cl:Co ratio. Deduce the structures of **X** and **Y**.

Since three moles of silver chloride are obtained from one mole of **X**, it implies that all the three chlorine atoms are a part of the ionization sphere. The structure of **X** may, therefore, be represented as $[Co(NH_3)_5(H_2O)]Cl_3$.

$$[Co(NH_3)_5(H_2O)]Cl_3 + 3AgNO_3 \rightarrow [Co(NH_3)_5(H_2O)](NO_3)_3 + 3AgCl$$
1 mole 3 moles

On heating, **X** loses a water molecule from the coordination sphere. To retain the coordination number, one of the chlorine atoms from the primary sphere enters the coordination sphere to give purple-colored **Y**.

$$\underset{\textbf{X}\text{ (Pink)}}{[Co(NH_3)_5(H_2O)]Cl_3} \xrightarrow{\text{Heat}} \underset{\textbf{Y}\text{ (Purple)}}{[Co(NH_3)_5Cl]Cl_2} + H_2O$$

X and Y have the same NH_3:Cl:Co ratio.

For many years following Werner's theory, it was believed that for coordination number 6 complexes, no geometry other than octahedral would be possible. Over a period of time, however, it came to light that there are a few examples in which the central metal ion possesses coordination number 6, yet the geometry is not octahedral, but trigonal prismatic. One such example is a 6-coordinate complex of rhenium(VI) with diphenylethenedimercaptan ligand (Fig. 22.2a). The structure of this trigonal prismatic complex, tris(diphenylethene-dimercapto)rhenium(VI) is shown in Fig. 22.2b.

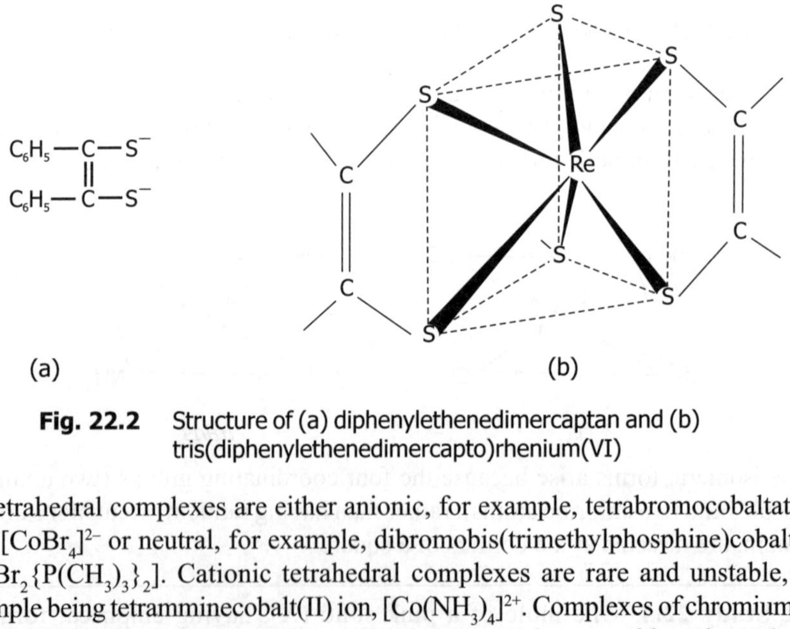

Fig. 22.2 Structure of (a) diphenylethenedimercaptan and (b) tris(diphenylethenedimercapto)rhenium(VI)

Tetrahedral complexes are either anionic, for example, tetrabromocobaltate(II) ion, $[CoBr_4]^{2-}$ or neutral, for example, dibromobis(trimethylphosphine)cobalt(II), $[CoBr_2\{P(CH_3)_3\}_2]$. Cationic tetrahedral complexes are rare and unstable, one example being tetramminecobalt(II) ion, $[Co(NH_3)_4]^{2+}$. Complexes of chromium(III) or cobalt(III) ions in tetrahedral coordination are not known, although octahedral coordination is very common for these two metal ions.

A large number of square planar complexes are known for palladium(II) and platinum(II) ions. These may be anionic, for example, tetrachloroplatinate(II) ion, $[PtCl_4]^{2-}$, cationic, for example, tetrammineplatinum(II) ion, $[Pt(NH_3)_4]^{2+}$ or neutral, for example, diamminedichloroplatinum(II), $[Pt(NH_3)_2Cl_2]$.

23. NOMENCLATURE OF COMPLEXES

Introduction

The International Union of Pure and Applied Chemistry (IUPAC) has adopted a set of norms for naming coordination compounds. These norms, commonly called IUPAC rules of nomenclature, are summarized below.

IUPAC Rules

1. The positive ion is named first followed by negative ion irrespective of whether coordination sphere is cationic or anionic.

Coordination sphere cationic	Coordination sphere anionic
$[Co(NH_3)_6]Cl_3$	$Na[Ag(CN)_2]$
To be named first	To be named first

2. Within the coordination sphere—

(a) Negative ligands are to be named by adding suffix "o" to their stem names. Names of some anionic ligands are listed in Table 23.1.

Table 23.1 IUPAC names of some anionic ligands

Anionic ligand	Name
CH_3COO^-	acetato
NH_2^-	amido
N_3^-	azido
Br^-	bromo
Cl^-	chloro
CN^-	cyano
F^-	fluoro
H^-	hydrido
OH^-	hydroxo
NH^{2-}	imido
I^-	iodo
HS^-	mercapto
N^{3-}	nitrido
$C_2O_4^{2-}$	oxalato

Contd...

Anionic ligand	Name
O^{2-}	oxo
O_2^{2-}	peroxo
SO_4^{2-}	sulfato
S^{2-}	sulfido
SO_3^{2-}	sulfito
O_2^-	superoxo
$S_2O_3^{2-}$	thiosulfato

The organic carbanions or radicals, however, are not to be named by adding suffix "o" to their stem names. Their original names are to be retained. For example,

$C_2H_5^{\cdot}$ or $C_2H_5^-$ ethyl

$CH_3^{\cdot}$ or CH_3^- methyl

(b) Neutral ligands have no special endings. Some neutral ligands with their names are listed in Table 23.2.

Table 23.2 Some neutral ligands and their IUPAC names

Neutral ligand	Name
NH_3	ammine
H_2O	aqua
CO	carbonyl
CH_3NH_2	methylamine
NO	nitrosyl
C_5H_5N (or py)	pyridine

(c) Positive ligands are named by adding suffix "ium" to their stem names. (Table 23.3)

Table 23.3 Cationic ligands and their IUPAC names

Cationic ligand	Name
$[NH_2{-}NH_3]^+$	hydrazinium
NO^+	nitrosonium

3. If there are more than one ligands of the same type, the prefixes listed in Table 23.4 are used to indicate their number.

Table 23.4 Prefixes for naming more than one ligands of the same type

Number of ligands	Prefix
2	bi or di
3	tri
4	tetra
5	penta
6	hexa
7	hepta
8	octa
9	nona
10	deca

4. If there are more than one ligands of different types, these are quoted in alphabetical order. For example, if a complex involves NH_3, CH_3COO and Cl^- ligands, these will be listed in the order, acetato–ammine–chloro. However, the prefixes di, tri, etc. are not to be included in alphabetizing.
5. After naming the ligands in alphabetical order and indicating their respective numbers, name the metal and mention its oxidation state in Roman number in simple parenthesis. [If the oxidation state of the metal is zero, then it is to be depicted in Arabic number.]

   Before proceeding further, let us review the rules we have discussed so far by taking a few examples.

   $[Co(NH_3)_6]Cl_3$: Hexamminecobalt(III) chloride

   Note carefully —

(a) Hexamminecobalt(III) is the name of the cation and chloride is the name of the anion. There has to be a gap between the two entities, just as we leave spacing between sodium and chloride when we write sodium chloride. However, there should be no gap within the name of the cation.

   Hexammine cobalt(III) is wrong.

(b) Do not leave space between the metal and its oxidation state.

   Cobalt (III) is wrong.

(c) There are three chloride ions in the ionization sphere. However, these are not to be named as trichloride, but simply as chloride. When the oxidation state of the cation is mentioned, the number of anions is self-evident. For example, when we write titanium(III) chloride, it is obvious that the composition is $TiCl_3$. The name *titanium(III) trichloride is wrong.*

$[Co(NH_3)_4Cl_2]Cl$: Tetramminedichlorocobalt(III) chloride

$[Co(H_2O)_2(NH_3)_2ClBr]I$: Diamminediaquabromochlorocobalt(III) iodide

The composition of the complex may be displayed only as a cationic coordination sphere, without specifying the anionic component. In such cases, name the cation and after leaving a space, write "ion". For example,

$[Co(NH_3)_6]^{3+}$: Hexamminecobalt(III) ion

(d) In some complexes, the coordination sphere is neutral. There are no simple ions. The complex, in such cases, is to be named as a single unit. For example,

$[Pt(NH_3)_2Cl_2]$: Diamminedichloroplatinum(II)

$[Hg(NH_2)Cl]$: Amidochloromercury(II)

6. If the coordination sphere is anionic, suffix "ate" is added to the stem name of the metal. For example,

$K_2[Ni(CN)_4]$: Potassium tetracyanonickelate(II)

$[CuCl_4]^{2-}$: Tetrachlorocuprate(II) ion

Generally, the suffix, "ate" is added to the English name of the metal. In some cases, this seems awkward and the Latin stem of the metal is used. However, when these metals are a part of cationic coordination sphere, their conventional, English names are quoted. The following comparisons shall clarify:

Cationic coordination sphere	**Anionic coordination sphere**
$[Ag(NH_3)_2]Cl$ Diamminesilver(I) chloride	$Na[Ag(CN)_2]$ Sodium dicyanoargentate(I)
$[Au(NH_3)_4](NO_3)_3$ Tetramminegold(III) nitrate	$[AuCl_4]^-$ Tetrachloroaurate(III) ion
$[Fe(H_2O)_6]SO_4$ Hexaquairon(II) sulfate	$K_4[Fe(CN)_6]$ Potassium hexacyanoferrate(II)

7. Some of the complexes involve *polysyllabic ligands*, i.e., the ligands whose names include a number. Examples of polysyllabic ligands are,

HSO_3^-	**bi**sulfito
C_5H_5 or $C_5H_5^-$	Cyclo**penta**dienyl
N_2	**di**nitrogen
O_2	**di**oxygen
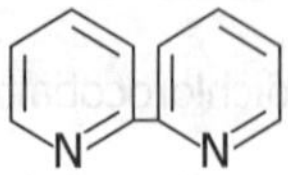 or dipy	**di**pyridyl

$CH_2—NH_2$ / $CH_2—NH_2$ or en	ethylene**di**amine
PF_3	**tri**fluorophosphine
$P(CH_3)_3$	**tri**methylphosphine
$P(C_6H_5)_3$ or PPh_3	**tri**phenylphosphine

(a) If more than one polysyllabic ligands of the same type are present in the coordination sphere, then the number is indicated by prefixes bis, tris, etc. instead of bi, tri, etc. (Table 23.5).

Table 23.5 Prefixes to represent number of polysyllabic ligands

Number of polysyllabic ligands	Prefix
2	bis
3	tris
4	tetrakis
5	pentakis
6	hexakis

(b) The name of the polysyllabic ligand is stated in single brackets. For example,

$[Co(en)_3]Cl_3$: Tris(ethylenediamine)cobalt(III) chloride

Even if one mole of a polysyllabic ligand is present in the coordination sphere, its name is still enclosed in parenthesis. For example,

$[Cu(en)(NH_3)_4]SO_4$: Tetrammine(ethylenediamine)copper(II) sulfate

The name of a polysyllabic ligand often begins with a number. For example, trifluorophosphine. The number is very much a part of ligand. Therefore, trifluorophosphine is considered to be beginning with alphabet 't'. Similarly, dinitrogen begins with alphabet 'd'.

$[Ru(NH_3)_5(N_2)]Cl_2$: Pentammine(dinitrogen)ruthenium(II) chloride

In exceptional cases, prefixes bis, tris, etc. may have to be used to indicate the number of ligands which are not polysyllabic. For example, the complex $[CuCl_2(CH_3NH_2)_2]$ may be named as dichlorodimethylaminecopper(II). Methylamine is a non-polysyllabic ligand and hence the prefix "di" ought to indicate its number in the present complex. However, the component dimethylamine inevitably has two meanings: it may imply two moles of methylamine, CH_3NH_2 or it may signify one mole of dimethylamine, $(CH_3)_2NH$. If "bis" is used instead of di, this confusion is averted. Therefore, the correct name of the complex is dichlorobis(methylamine) copper(II).

8. The ambidentate ligands are so named as to indicate the donor atom. For example,

$M\leftarrow NO_2^-$	nitro	$M\leftarrow ONO$	nitrito
$M\leftarrow SCN^-$	thiocyanato-S	$M\leftarrow NCS^-$	thiocyanato-N

That atom of the ambient ligand which is placed closer to the metal, in the composition of the complex, is to be considered the donor atom. For example,

$[Co(NH_3)_3(NO_2)_3]$: Triamminetrinitrocobalt(III)

$[Co(NH_3)_3(ONO)_3]$: Triamminetrinitritocobalt(III)

$[Co(NH_3)_5SCN]^{2+}$: Pentamminethiocyanato-Scobalt(III) ion

$[Co(NH_3)_5NCS]^{2+}$: Pentamminethiocyanato-Ncobalt(III) ion

Case Study 23.1 Write the IUPAC nomenclature for the following complex ion:

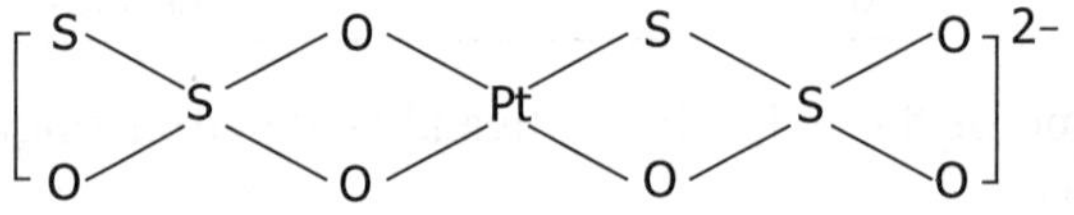

The coordinating group in this complex ion is the thiosulfate ligand. The latter is a negatively charged, chelating and ambient function which may coordinate to a metal ion either through two oxygen atoms, in which case it is referred to as thiosulfato-O,O- or through one oxygen and one sulfur atom, in which case it is referred to as thiosulfato-O,S-. In the given example, it coordinates to platinum(II) ion through both modes. Its nomenclature is –

Thiosulfato-O,O-thiosulfato-O,S-platinate(II) ion

9. If the composition of the complex involves solvents of crystallization, the name of the complex is to be successively followed by (i) number of molecules of the solvent of crystallization and (ii) name of the solvent. The latter two entities are to be spaced by a hyphen. For example,

$[Cr(H_2O)_4Cl_2]Cl.2H_2O$: Tetraquadichlorochromium(III) chloride.2-water

$[Co(NH_3)_5(SCN)_2]Cl.3CH_3OH$: Pentamminedithiocyanato-S cobalt(III) chloride.3-methanol

10. If the cation and anion are both complex ions, the metal in the negative coordination sphere is to be named as metal-ate. However, in such cases, it is difficult to evaluate the oxidation state of metals in the two coordination

spheres. Normally, it is assumed that in the anionic coordination sphere, cobalt, chromium and iron have oxidation states (III), while copper and platinum have oxidation states (II). For example, in the compound $[Co(NH_3)_6]$ $[Cr(CN)_6]$, the oxidation state of chromium may be presumed to be (III). The overall charge on the anion would, therefore, be –3, $[Cr(CN)_6]^{3-}$. Since the molecule is overall neutral, the charge on the cationic coordination sphere would be +3, $[Co(NH_3)_6]^{3+}$. As ammine is a neutral ligand, the oxidation state of cobalt is (III).

$[Co(NH_3)_6]$ $[Cr(CN)_6]$	:	Hexamminecobalt(III) hexacyanochromate(III)
$[Pt(NH_3)_4]$ $[PtCl_4]$	:	Tetrammineplatinum(II) tetrachloroplatinate(II)

In the examples cited above, the cationic and anionic coordination spheres are in 1:1 molar ratio. This may not always be the case. Consider, for example,

$$[Co(NH_3)_4Br_2]_2\ [ZnCl_4]$$

Here, the cation and anion are in 2:1 molar ratio. In such cases, split the two coordination spheres,

$$[Co(NH_3)_4Br_2]^{+} \quad [ZnCl_4]^{2-}$$

and name them individually, i.e.,
tetramminedibromocobalt(III) tetrachlorozincate(II).

11. In complexes with bridging ligands, the coordination sphere is divided into three hypothetical parts for the purpose of nomenclature.

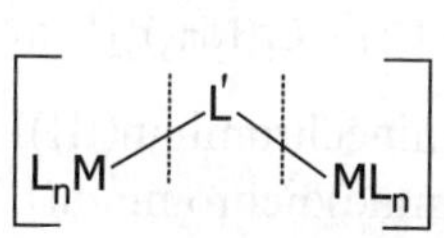

The entities to the left of bridging ligand are named first, the bridging ligand, prefixed with μ- is named next, and the moieties to the right of the bridging ligand are named last. The bridging ligand is separated from the units to its left and the units to its right by hyphens. For example,

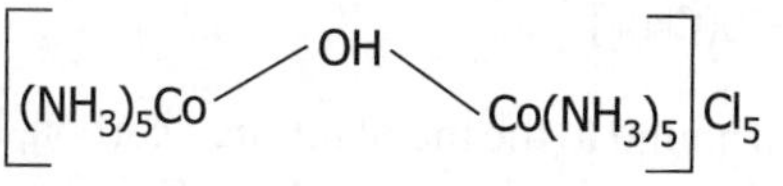

Pentamminecobalt(III)-μ-hydroxo-pentamminecobalt(III) chloride

If there are more than one bridging ligands, suffix di, tri, etc. are used to indicate their number. For example,

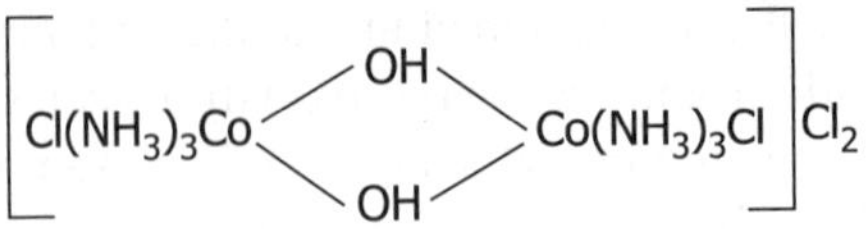

Triamminechlorocobalt(III)-μ-dihydroxo-triamminechlorocobalt(III) chloride

If there are more than one bridging ligands of different types, then they are named in the alphabetical order, each being prefixed with μ-. For example,

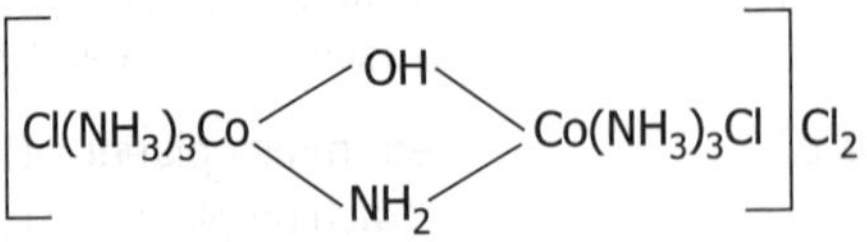

Triamminechlorocobalt(III)-μ-amido-μ-hydroxo-triamminechlorocobalt(III) chloride

If the bridging ligands are in anionic coordination sphere, the metals are named after adding suffix "ate" to their stem names.

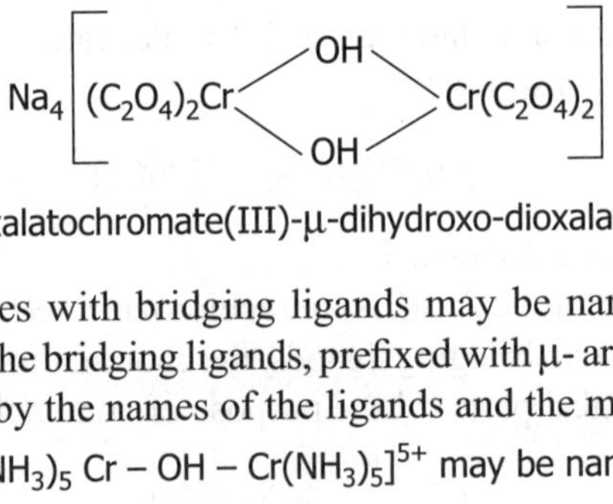

Sodium dioxalatochromate(III)-μ-dihydroxo-dioxalatochromate(III)

The complexes with bridging ligands may be named by an alternative method too. The bridging ligands, prefixed with μ- are named first, followed successively by the names of the ligands and the metal. For example,

$[(NH_3)_5\ Cr - OH - Cr(NH_3)_5]^{5+}$ may be named as,

μ-hydroxobis(pentamminechromium(III)) ion or as,
μ-hydroxobis(pentammine)dichromium(III) ion.

Case Study 23.2 Which of the following is the likely structural formula of pentacyanocobalt(III)-μ-cyano-pentamminecobalt(III)?

(a) $[(NH_3)_5Co\text{–}CN\text{–}Co(CN)_5]$

(b) $[(NH_3)_5Co\text{–}NC\text{–}Co(CN)_5]$

When a cyanide ligand is bound to the metal through carbon, M←CN, it is designated as cyano; when attached through nitrogen, M←NC, it is referred to as isocyano.

In the given case, the bridging group is coordinated as cyano to pentacyanocobalt(III) moiety. Therefore, the carbon of the bridging group should be attached to that cobalt(III) ion which holds the five cyano ligands. For this reason, structure (b) represents the correct structural formula.

---

12. In some cases a bridging complex contains a metal–metal bond too. In these examples, the nomenclature ends with a reference to the metal–metal bond.

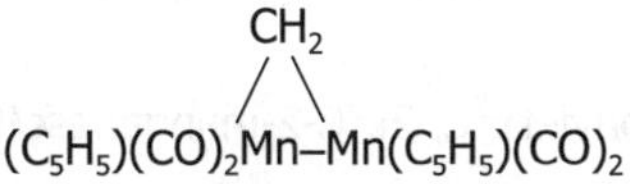

Dicarbonyl(cyclopentadienyl)manganese(I)-μ-carbene-dicarbonyl(cyclopentadienyl)manganese(I) (Mn–Mn)

13. A symmetrical, dinuclear complex (that is, one in which both the metal atoms are same and so are the number/type of ligands attached to each of them) without a bridging ligand may be named by either of the following two methods.
    (a) By dividing the skeletal structure of the complex into two equal halves,

    $$L_nM \vdots ML_n$$

    The half fragment is named, prefixed with bis and postfixed with (M–M). For example,

    $$[(CO)_5\,Mn - Mn(CO)_5]$$

    Bis(pentacarbonylmanganese(0)) (Mn–Mn)
    (b) By adding up the ligands attached to each metal atom, followed by the notation dimetal (M–M).
    The alternative name for the above complex is,
    Decacarbonyldimanganese(0) (Mn–Mn)
14. It is possible that the dinuclear complex may be unsymmetrical and have different metal atoms. In such cases, the metal atom having a higher electropositive character is accorded number 1, while the one having a lower electropositive character is accorded number 2. For example, in the dinuclear complex, $[(CO)_5Mn - Re(CO)_5]$, rhenium, being more electropositive will get priority over manganese. Before naming, the complex may be depicted as –

    $$[(CO)_5\underset{2}{Mn} - \underset{1}{Re}(CO)_5]$$

    The nomenclature of such complexes is derived by invoking *kappa convention*. Under this system, the ligating atom of the ligand is preceded by Greek letter kappa κ. In the above example, carbon is the ligating atom of the carbonyl ligand. Hence, according to kappa convention, the carbonyl group is referred to as κC. The number of carbonyl groups coordinated to a particular metal atom is indicated by right superscript to κ. The numerical locant (1 or 2) of the central atom is placed before κ. Thus $1\kappa^5C$ means that five carbonyl groups are attached to metal atom number 1, in this case rhenium. Likewise, $2\kappa^5C$ means that five carbonyl groups are attached to metal atom number 2, in this case manganese. In case of single-atom ligands, there is no need to indicate the ligating atom.

The names of the ligands are indicated in alphabetical order followed by the name of metal atoms, prefixed with their attached number, in alphabetical order. Finally the metal-metal bond is postfixed, once again in alphabetical order. The name of the mixed manganese-rhenium carbonyl complex, cited above, is,

Decacarbonyl-1κ^{5}C, 2κ^{5}C-2-manganese(0)-1-rhenium(0) (Mn–Re)

Likewise, the dinuclear complex, $[(CO)_4Co - Re(PF_3)_5]$ is named as follows.

$$[(CO)_4\underset{2}{Co} - \underset{1}{Re}(PF_3)_5]$$

Tetracarbonyl-2κ^{4}C-pentakis(triflurophosphine)-2κ^{5}P-2-cobalt(0)-1-rehnium(0) (Co – Re).

15. An unsymmetrical dinuclear complex may have identical metal atoms, but the type and/or number of ligands attached to each of these may be different. It is also possible that the two metal atoms may be in different oxidation states. Under these circumstances, metal atom 1 is designated in the following order of priority.
    (a) The one having a higher oxidation state.
    (b) The one having a higher coordination number.

    For example,

$$[Cl(CO)_4\underset{1}{Mn} - \underset{2}{Mn}(CO)_5]$$

Nonacarbonyl-1κ^{4}C, 2κ^{5}C-chloro-1κ-1-manganese(I)-2-manganese(0) (Mn – Mn)

$$[(CO)_5\,\underset{1}{Co} - \underset{2}{Co}(CO)_4]$$

Nonacarbonyl-1κ^{5}C, 2κ^{4}C-dicobalt(0) (Co – Co)

Representative Examples

We shall now enunciate nomenclature by taking examples of various classes of complexes. The classification, however, is arbitrary.

Complexes with neutral coordination sphere

$[Co(NH_3)_3ClCNNO_2]$	:	Triamminechlorocyanonitrocobalt(III)
$[Ni(CO)_4]$	:	Tetracarbonylnickel(0)
$[Fe(C_5H_5)_2]$	:	Bis(cyclopentadienyl)iron(II)
$[Ni(PF_3)_4]$	:	Tetrakis(trifluorophosphine)nickel(0)
$[CuCl_2\{O{=}C(NH_2)_2\}_2]$	:	Dichlorodiureacopper(II)
$[Rh(NCS)_3(NH_3)_3]$	:	Triamminetrithiocyanato-Nrhodium(III)
$[PdI_2(ONO)_2(H_2O)_2]$	:	Diaquadiiododinitritopalladium(IV)

$[Ru(HSO_3)_2(NH_3)_4]$	:	Tetramminebis(bisulfito)ruthenium(II)
$[Pt(en)_2Cl_2]$	:	Dichlorobis(ethylenediamine)platinum(II)
$[Hg(CH_3)_2]$	:	Dimethylmercury(II)
$[Hg(CH_3)Cl]$	:	Chloromethylmercury(II)
$[Co(NH_3)_3(NO_2)_3]$	:	Triamminetrinitrocobalt(III)
$[Rh(CH_3)\{(CH_3)_3P\}_4]$	:	Methyltetrakis(trimethylphosphine) rhodium(I)
$[Ir(CH_3)(CO)\{(C_6H_5)_3P\}_2Br_2]$	:	Dibromocarbonylmethylbis (triphenylphospine)iridium(III)
$[CuCl_2(CH_3NH_2)_2]$	:	Dichlorobis(methylamine)copper(II)
$[CoCl_3(NH_3)_2\{(CH_3)_2NH\}]$	:	Diamminetrichloro(dimethylamine)cobalt(III)
$[ReH_3\{P(C_6H_5)_3\}_5]$	:	Trihydriodopentakis(triphenylphosphine) rhenium(III)
$[PtCl_2\{P(C_6H_5)_3\}_2]$	:	Dichlorobis(triphenylphosphine)platinum(II)

Complexes with cationic coordination sphere

$[Cr(H_2O)_6](NO_3)_3$	:	Hexaquachromium(III) nitrate
$[Co(en)_2Br_2]Cl$	:	Dibromobis(ethylenediamine)cobalt(III) chloride
$[Cu(NH_3)_2(en)]Br_2$	:	Diammine(ethylenediamine)copper(II) bromide
$[Ni(NH_3)_2(en)](CH_3COO)_2$	:	Diammine(ethylenediamine)nickel(II) acetate
$Co(NH_3)_4(H_2O)Cl]Cl_2$	:	Tetrammineaquachlorocoball(III) chloride
$[Co(NH_3)_4SO_4]NO_3$	:	Tetramminesulfatocobalt(III) nitrate
$[Co(NH_2)_2(NH_3)_4]Br$	:	Diamidotetramminecobalt(III) bromide
$[Fe(CH_3NC)_6]Br_2$	:	Hexakis(methylisocyanide)iron(II) bromide
$[PtCl(CH_3NH_2)(NH_3)_2]Cl$	:	Diamminechloro(methylamine)platinum(II) chloride
$[Cr(NH_3)_2(H_2O)_3(OH)](NO_3)_2$	:	Diamminetriaquahydroxochromium(III) nitrate
$[Co(NH_3)_2(H_2O)_2(CN)_2]Cl$	:	Diamminediaquadicyanocobalt(III) chloride
$[Co(NH_2)_2(NH_3)_4]OC_2H_5$	:	Diamidotetramminecobalt(III) ethoxide
$[Co(NH_3)_5ONO]SO_4$	:	Pentamminenitritocobalt(III) sulfate
$[Co(en)_2(ONO)Cl]Cl$	:	Chlorobis(ethylenediamine)nitritocobalt(III) chloride
$[Ir(en)_2Cl_2]^+$	:	Dichlorobis(ethylenediamine)iridium(III) ion
$[Cr(NH_3)_3(H_2O)_3]Cl_3$	:	Triamminetriaquachromium(III) chloride
$[Cr(NH_2)_2(H_2O)_2(NH_3)_2]Cl$	:	Diamidodiamminediaquachromium(III) chloride
$[Cr(H_2O)_4Cl_2]NO_3$	:	Tetraquadichlorochromium(III) nitrate

$[Mn(CO)_6]ClO_4$	:	Hexacarbonylmanganese(I) perchlorate

Complexes with anionic coordination sphere

$K_2[Cu(CN)_4]$	:	Potassium tetracyanocuprate(II)
$K_4[Ni(CN)_4]$	:	Potassium tetracyanonikelate(0)
$K_4[Ni(CN)_2(C_2O_4)_2]$	:	Potassium dicyanodioxalatonickelate(II)
$K[Pt(NH_3)Cl_5]$	:	Potassium amminepentachloroplatinate(IV)
$[Cr(C_2O_4)_3]^{3-}$	:	Trioxalatochromate(III) ion
$Na_2[ZnCl_4]$	:	Sodium tetrachlorozincate(II)
$K_3[Fe(CN)_6]$	:	Potassium hexacyanoferrate(III)
$K_3[Fe(CN)_5NO]$	:	Potassium pentacyanonitrosylferrate(II)
$Na_2[OsCl_5N]$	:	Sodium pentachloronitridoosmate(VI)
$K[PtCl_3(C_2H_4)]$	:	Potassium trichloroetheneplatinate(II)
$Na[PtBrCl(ONO)(NH_3)]$	:	Sodium amminebromochloronitritoplatinate(II)
$K_2[Co(N_3)_4]$	:	Potassium tetrazidocobaltate(II)
$K_4[Mn(NCS)_6].3H_2O$	:	Potassium hexathiocyanato-N manganate(II).3-water
$Cs[TeF_5]$	:	Cesium pentafluorotellurate(IV)
$[(CH_3)_4N]_3\ [Cr(C_2O_4)_3]$	:	Tetramethylammonium trioxalatochromate(III)
$Na[Au(CN)_2]$	:	Sodium dicyanoaurate(I)
$Ca_2[Fe(CN)_6]$	:	Calcium hexacyanoferrate(II)
$[(CH_3)_4N]\ [Cr(SCN)_4(NO)_2]$	:	Tetramethylammonium dinitrosyltetrathiocyanato-Schromate(III)
$K_4[Mo(CN)_8]$	:	Potassium octacyanomolybdate(IV)
$(NH_4)_3[Co(C_2O_4)_3]$	:	Ammonium trioxalatocobaltate(III)
$NH_4[Cr(NCS)_4(NH_3)_2]$	:	Ammonium diamminetetrathiocyanato-N chromate(III)
$Cs[CrFCl_3]$	:	Cesium trichlorofluorochromate(III)
$K[Co(NH_3)_2Br_4]$	:	Potassium diamminetetrabromocobaltate(III)
$K[CrF_4O]$	:	Potassium tetrafluorooxochromate(V)
$Na[BH(OCH_3)_3]$	:	Sodium hydridotrimethoxyborate(III)
$K[SbCl_5C_6H_5]$	:	Potassium pentachlorophenylantimonate(V)
$K_2[SiF_6]$	:	Potassium hexaflurosilicate(IV)
$Ca_3[VCl_6]_2$	:	Calcium hexachlorovanadate(III)
$[(C_2H_5)_4N]_3\ [ZrI_7]$	:	Tetraethylammonium heptaiodozirconate(IV)

Complexes with cationic and anionic coordination spheres

$[Co(en)_3]\ [Cr(CN)_6]$	:	Tris(ethylenediamine)cobalt(III) hexacyanochromate(III)
$[Pt(py)_4]\ [PtCl_4]$	:	Tetrapyridineplatinum(II) tetrachloroplatinate(II)
$[Pt(NH_3)_4Cl_2]\ [PtCl_4]$	:	Tetramminedichloroplatinum(IV) tetrachloroplatinate(II)
$[Co(NH_3)_6]\ [Cr(NH_3)_2Cl_4]_3$	:	Hexamminecobalt(III) diamminetetrachlorochromate(III)
$[Fe(NH_3)_6]\ [Fe(CN)_6]$	:	Hexammineiron(III) hexacyanoferrate(III)
$[Cr(NH_3)_6]\ [CoF_6]$	:	Hexamminechromium(III) hexafluorocobaltate(III)
$[Ru(NH_3)_5(N_2)]\ [PtCl_4]$	:	Pentammine(dinitrogen)ruthenium(II) tetrachloroplatinate(II)
$[Cr(OH)_2(H_2O)_4]\ [AuCl_4]$	:	Tetraquadihydroxochromium(III) tetrachloroaurate(III)
$[Rh(NH_3)_4Cl_2]_2\ [Zn(CN)_4]$	:	Tetramminedichlororhodium(III) tetracyanozincate(II)

Complexes with bridging ligands

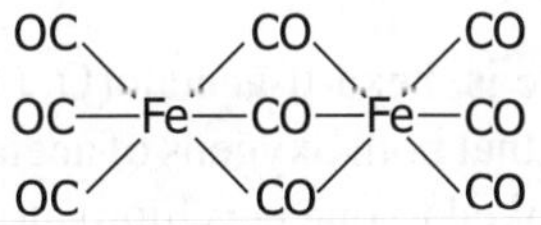

: Tricarbonyliron(0)-μ-tricarbonyl-tricarbonyliron(0)

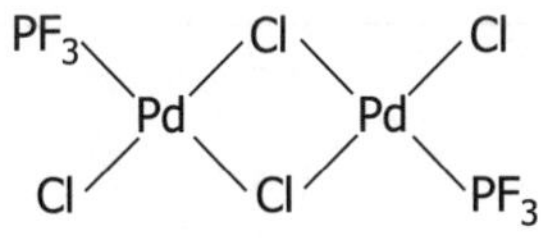

: Chloro(trifluorophosphine) palladium(II)-μ-dichloro-chloro(trifluorophosphine) palladium(II)

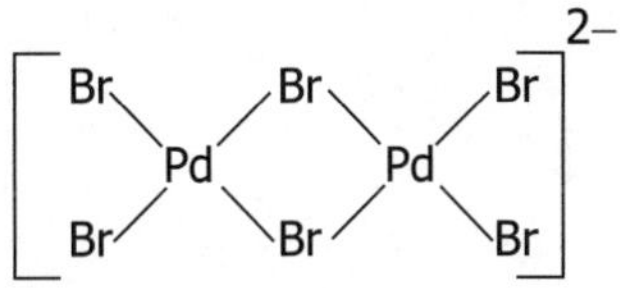

: Dibromopalladate(II)-μ-dibromo-dibromopalladate(II) ion

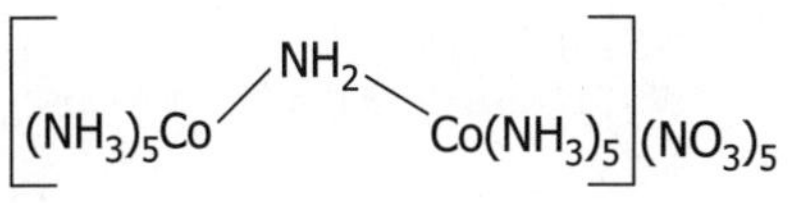

: Pentamminecobalt(III)-μ-amido-pentammine cobalt(III) nitrate

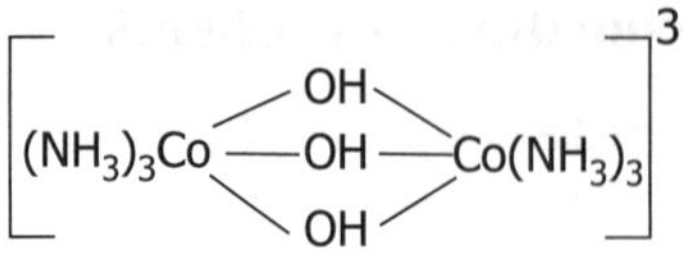

: Triamminecobalt(III)-μ-trihydroxo-triamminecobalt (III) ion

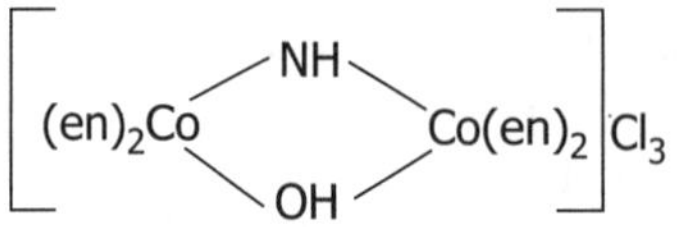

: Bis(ethylenediamine)cobalt(III)-μ-hydroxo-μ-imidobis(ethylenediamine) cobalt(III)chloride

Dinuclear complexes

$[Br_4Re - ReBr_4]^{2-}$	:	Bis(tetrabromorhenate(III)) (Re–Re) ion
$[\{(CH_3)_3P\}Au - Mn(CO)_5]$	:	Pentacarbonyl-2κ^5C-(trimethylphosphine) – 1κ^1P-1-gold(0)-2-manganese(0) (Au – Mn)

Case Study 23.3 Write the IUPAC nomenclature of basic beryllium acetate.

The structure of basic beryllium acetate, $[Be_4O(CH_3COO)_6]$, shown in Fig. 13.2 p. 448, has two bridging ligands: acetato and oxo. Each acetato ligand bridges two of the four beryllium atoms. Therefore, suffix μ is prefixed before the name of the bridging group. The oxo ligand, however, bridges all four beryllium atoms which are placed at the corners of a tetrahedron. For this reason, the oxo ligand is prefixed with suffix μ_4.

The IUPAC nomenclature of basic beryllium acetate is, hexa-μ-acetato(O, O')-μ_4-oxotetraberyllium(II). The notation (O, O') signifies that both oxygens of acetato moiety act as ligating atoms. The oxygen atom, O is linked to one beryllium atom, while the other oxygen atom, O' is bound to the adjacent beryllium atom.

24. ISOMERISM IN COMPLEXES

Introduction

Isomers are compounds having the same number and kinds of atoms arranged differently. The phenomenon of existence of isomers is called *isomerism*. Since their structural arrangements are different, the properties of isomers are also different. Isomers can be broadly classified into two types, each type having further sub-classifications.

Isomers

Structural Isomers

1. Ionization isomers
2. Hydrate isomers
3. Linkage isomers
4. Coordination isomers
5. Coordination position isomers
6. Ligand isomers
7. Polymerization isomers

Stereoisomers

1. Geometrical isomers
2. Optical isomers

Structural Isomers

Structural isomers are those which contain different atom-to-atom bonding sequences. They involve either more than one coordination sphere or different donor atoms of the same ligand in the same coordination sphere.

1. Ionization isomers

The isomers which involve exchange of ligands between coordination sphere and ionization sphere are called ionization isomers. For example,

$[Co(NH_3)_5Cl]SO_4$ and
Pentamminechlorocobalt(III) sulfate

$[Co(NH_3)_5SO_4]Cl$
Pentamminesulfatocobalt(III) chloride

In pentamminechlorocobalt(III) sulfate complex, the chloro ligand is in the coordination sphere, while the sulfate group is in the ionization sphere. An aqueous solution of this complex, on treatment with barium chloride, yields a white precipitate of barium sulfate.

$$[Co(NH_3)_5Cl]SO_4 + BaCl_2 \rightarrow [Co(NH_3)_5Cl]Cl_2 + BaSO_4\downarrow$$
White

In pentamminesulfatocobalt(III) chloride complex, the sulfato ligand is in the coordination sphere, while the chloride group is in the ionization sphere. With silver nitrate, an aqueous solution of this complex yields a white precipitate of silver chloride.

$$[Co(NH_3)_5SO_4]Cl + AgNO_3 \rightarrow [Co(NH_3)_5SO_4](NO_3) + AgCl\downarrow$$
White

A distinction between ionization isomers can thus be made by simple qualitative tests.

Other examples of ionization isomers include,

(a) $[Pt(NH_3)_4Cl_2]Br_2$ and
Tetramminedichloroplatinum(IV) bromide

$[Pt(NH_3)_4Br_2]Cl_2$
Tetramminedibromoplatinum(IV) chloride

(b) $[Co(en)_2NO_2Cl]SCN$
Chlorobis(ethylenediamine)nitrocobalt(III) thiocyanate

$[Co(en)_2NO_2SCN]Cl$ and
Bis(ethylenediamine)nitrothiocyanato-Scobalt(III) chloride

$[Co(en)_2SCNCl]NO_2$
Chlorobis(ethylenediamine)thiocyanato-Scobalt(III) nitrite

2. Hydrate isomers

When water molecules are exchanged between coordination sphere and ionization sphere, the resulting isomers are called hydrate isomers. For example,

$[Cr(H_2O)_6]Cl_3$
Hexaquachromium(III) chloride

$[Cr(H_2O)_5Cl]Cl_2.H_2O$ and
Pentaquachlorochromium(III) chloride.1-water

$[Cr(H_2O)_4Cl_2]Cl.2H_2O$
Tetraquadichlorochromium(III) chloride.2-water

An equimolar solution of these complexes yields three, two and one mole of silver chloride, when treated with excess silver nitrate solution.

$$[Cr(H_2O)_6]Cl_3 + 3AgNO_3 \rightarrow [Cr(H_2O)_6](NO_3)_3 + 3AgCl\downarrow$$

$$[Cr(H_2O)_5Cl]Cl_2.H_2O + 2AgNO_3 \rightarrow [Cr(H_2O)_5Cl](NO_3)_2 + 2AgCl\downarrow + H_2O$$

$$[Cr(H_2O)_4Cl_2]Cl.2H_2O + AgNO_3 \rightarrow [Cr(H_2O)_4Cl_2](NO_3) + AgCl\downarrow + 2H_2O$$

Thus from the number of moles of silver chloride formed, a distinction may be made among the hydrate isomers.

Other examples of hydrate isomers include,

(a) $[Cr(en)_2(H_2O)_2]Br_3$ and
Diaquabis(ethylenediamine)chromium(III) bromide

$[Cr(en)_2(H_2O)Br]Br_2.H_2O$
Aquabromobis(ethylenediamine)chromium(III) bromide.1-water

(b) $[Co(NH_3)_4(H_2O)Cl]Cl_2$ and
Tetrammineaquachlorocobalt(III) chloride

$[Co(NH_3)_4Cl_2]Cl.H_2O$
Tetramminedichlorocobalt(III) chloride.1-water

3. Linkage isomers

When there are two possible bonding sites in a ligand for coordination with a central metal atom, the resulting isomers are called linkage isomers. The coordinating group which can link to the metal through either of the two bonding sites is called an *ambidentate* or *ambient ligand.* For example,

$[Co(NH_3)_5NO_2]Cl$ and
Pentamminenitrocobalt(III) chloride

$[Co(NH_3)_5ONO]Cl$
Pentamminenitritocobalt(III) chloride

In the first case, ambidcntate NO_2 ligand is linked to cobalt through nitrogen, while in the second, it is linked through oxygen.

The nitro isomer is yellow in color. On exposure to ultraviolet radiation, it is converted to red-colored nitrito isomer. Conversely, the nitrito analog, when treated with warm hydrochloric acid, is transformed to nitro derivative.

The conversion of nitro complex into nitrito analog follows an intramolecular transition state.

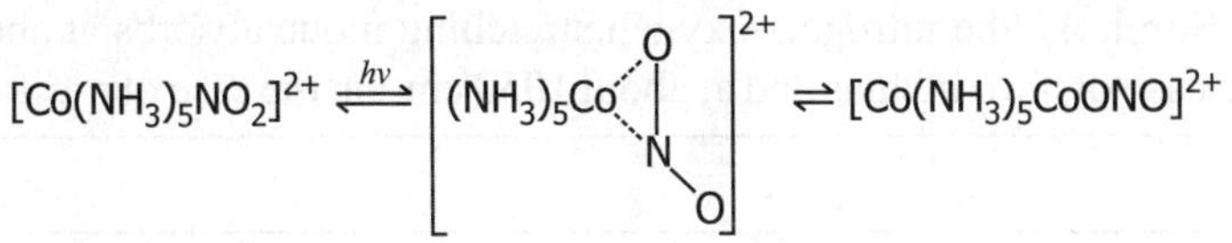

Linkage isomers are also called *salt isomers*. Another example of a linkage isomer is,

$[Co(NH_3)_5SCN]^{2+}$ and
Pentamminethiocyanato-Scobalt(III) ion

$[Co(NH_3)_5NCS]^{2+}$
Pentamminethiocyanato-Ncobalt(III) ion

In the first case, the thiocyanato ligand is linked to cobalt through sulfur, whereas in the second case, it is linked through nitrogen.

The following complex of palladium(II) offers yet another example of linkage isomerism.

$[Pd(SCN)_2\{As(C_6H_5)_3\}_2]$ and

Dithiocyanato-Sbis(triphenylarsine)palladium(II)

$[Pd(NCS)_2\{As(C_6H_5)_3\}_2]$

Dithiocyanato-Nbis(triphenylarsine)palladium(II)

In solvents of high dielectric constant, the S-coordination prevails, while in solvents of low dielectric constant, the N-coordination is stabilized.

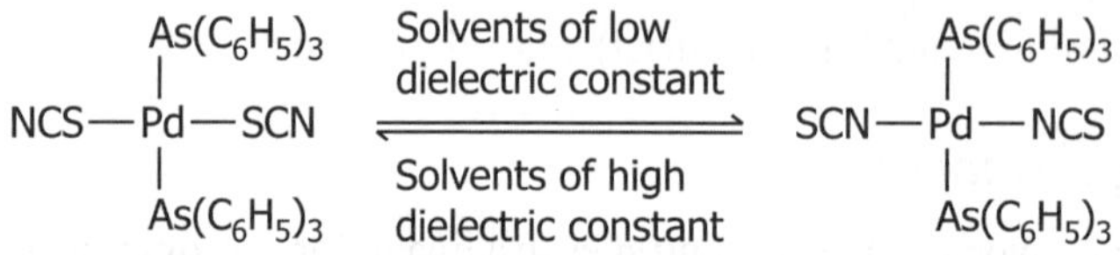

We now cite a case where linkage isomerism is observed in a coordination compound involving a bidentate ligand.

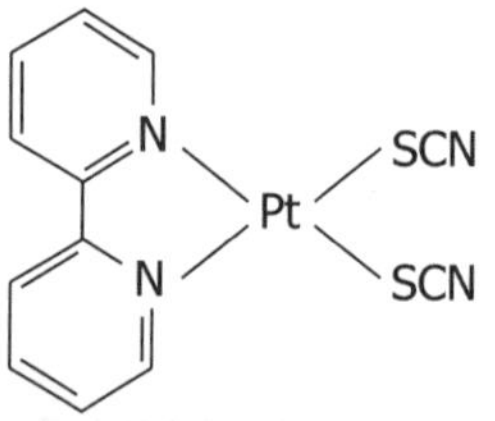

(Dipyridyl)dithiocyanato-Splatinum(II)

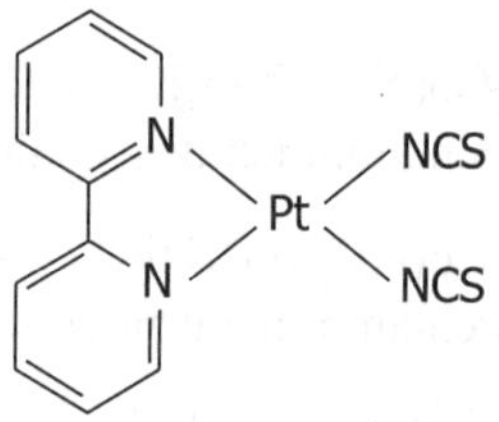

(Dipyridyl)dithiocyanato-Nplatinum(II)

Case Study 24.1 How are linkage isomers distinguished?

Infrared spectroscopy provides a simple method for making a distinction between linkage isomers. For example, in thiocyanato-S ligands, the carbon–sulfur stretching frequency occurs at about 700 cm^{-1}, while in thiocyanato-N ligands, it occurs at about 820 cm^{-1}. Similarly, the nitrogen–oxygen stretching mode absorbs at about 1300 cm^{-1} in the case of nitro ligands and at about 1050 cm^{-1} in the case of nitrito ligands.

Box 24.1

Symbiosis and Linkage Isomers

In Chapter 5, we had discussed the concept of symbiosis (*see* Case Study 5.12, p. 247). When applied to coordination compounds, symbiosis means that a metal ion bound to hard ligands will prefer to add on another hard group, while the

Contd...

one bound to soft ligands will add on another soft group. This criteria may be used to predict the bonding mode of the thiocyanate ligand (X) in complexes having the compositions, $[Co(NH_3)_5X]^{2+}$ and $[Co(CN)_5X]^{3-}$. In the first case, the metal ion has been rendered hard by coordination of five hard-base ammonia ligands. The thiocyanate moiety will coordinate though its hard nitrogen atom, giving pentamminethiocyanato-Ncobalt(III) ion, $[Co(NH_3)_5(NCS)]^{2+}$. In the second complex, the five soft cyano groups have imparted sufficient softness to the cobalt(III) ion. Hence the thiocyanate group will become coordinated through soft sulfur atom giving pentacyanothiocyanato-Scobaltate(III) ion, $[Co(CN)_5(SCN)]^{3-}$.

4. Coordination isomers

Isomers in which both the cation and anion are complex ions and involve an exchange of central metals between the two coordination spheres are called coordination isomers. Examples of coordination isomers include,

(a) $[Co(NH_3)_6]$ $[Cr(CN)_6]$ and
Hexamminecobalt(III) hexacyanochromate(III)

$[Cr(NH_3)_6]$ $[Co(CN)_6]$
Hexamminechromium(III) hexacyanocobaltate(III)

(b) $[Co(en)_3]$ $[Cr(C_2O_4)_3]$ and
Tris(ethylenediamine)cobalt(III) trioxalatochromate(III)

$[Cr(en)_3]$ $[Co(C_2O_4)_3]$
Tris(ethylenediamine)chromium(III) trioxalatocobaltate(III)

5. Coordination position isomers

This type of isomerism occurs in complexes possessing bridging ligands. Isomers involve an exchange of ligands between the two metal atoms that are linked through one or more bridging groups. For example,

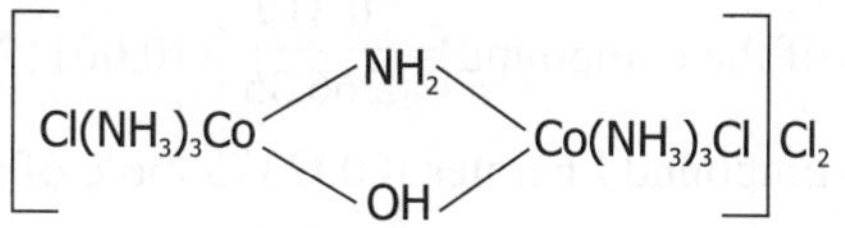

Triamminechlorocobalt(III)-μ-amido-μ-hydroxo-triamminechlorocobalt(III) chloride

and

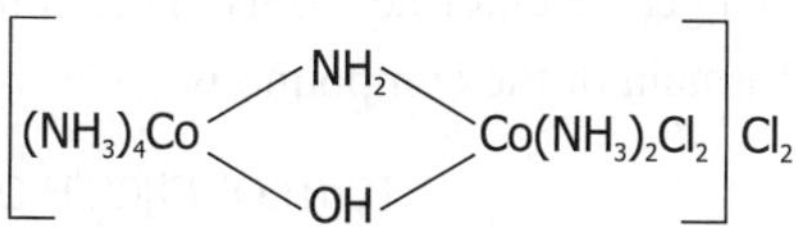

Tetramminecobalt(III)-μ-amido-μ-hydroxo-diamminedichlorocobalt(III) chloride

6. Ligand isomers

Some ligands exist in different isomeric forms. When they get coordinated to the same metal ion, the resulting complexes are also isomeric. This phenomenon is called *ligand isomerism*. For example, 1,2-diaminopropane (pn) and 1,3-diaminopropane, also called trimethylenediamine (tn)—are isomeric ligands.

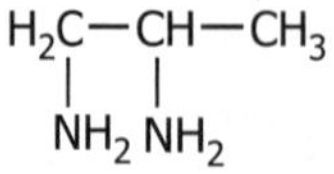

1,2-Diaminopropane (pn)

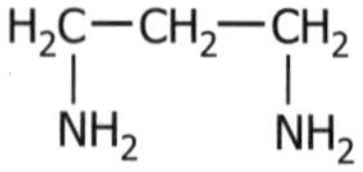

1,3-Diaminopropane
or
Trimethylenediamine (tn)

The following complexes involving these groups are an example of ligand isomers:

Dichlorobis(1,2-diaminopropane)cobalt(III) chloride, $[Co(pn)_2Cl_2]Cl$ and

Dichlorobis(1,3-diaminopropane)cobalt(III) chloride, $[Co(tn)_2Cl_2]Cl$

Case Study 24.2 A solution containing 0.319 g of a hydrate isomer with molecular formula $CrCl_3.6H_2O$ was passed through a cation exchange resin in acidic form, and the acid so liberated was neutralized with 19 ml of 0.125 N sodium hydroxide. Identify the isomer (Molecular weight of $CrCl_3.6H_2O$ = 266.35).

Number of millimoles of acid liberated = Number of millimoles of sodium hydroxide required for neutralization

$= N_1V_1$

$= 19 \times 0.125 = 2.375$

Number of moles of acid liberated $= \dfrac{2.375}{1000} = 0.002375$

Amount of compound passed through cation exchanger = 0.319 g

Molecular mass of the compound = 266.35

Number of moles of the compound $= \dfrac{0.319}{266.35} = 0.00119$

0.00119 mole of compound liberates 0.002375 mole of acid

1 mole of compound will liberate, $\dfrac{0.002375}{0.00119} = 2$ moles of acid

Number of exchangeable chlorine atoms in the compound = 2

The structural formula of the compound is

$$[Cr(H_2O)_5Cl]Cl_2.H_2O$$

Its IUPAC nomenclature is pentaquachlorochromium(III) chloride.1-water.

---

7. Polymerization isomers

Coordination compounds having the same empirical composition, but different molecular composition are referred to as polymerization isomers. It is pertinent to emphasize that these are not truly isomers, but by convention are considered so. The most common example of polymerization isomers is the square planar platinum complex,

$[Pt(NH_3)_2Cl_2]$
Diamminedichloroplatinum(II)
and its dimer, $[Pt(NH_3)_2Cl_2]_2$, better written as, $[Pt(NH_3)_4]$ $[PtCl_4]$
Tetrammineplatinum(II) tetrachloroplatinate(II)

The two may be distinguished by conductivity studies. Whereas, the empirical monomer is non-conducting, its isomeric dimer is conducting in nature, the conductivity value corresponding to four charges—two on cation, $[Pt(NH_3)_4]^{2+}$ and two on anion, $[PtCl_4]^{2-}$. Yet another example of polymerization isomers is the octahedral cobalt complex,

$[Co(NH_3)_3(NO_2)_3]$
Triamminetrinitrocobalt(III)
and its dimer, $[Co(NH_3)_3(NO_2)_3]_2$ or
$[Co(NH_3)_6]$ $(Co(NO_2)_6]$
Hexamminecobalt(III) hexanitrocobaltate(III)

The former is non-conducting, while the conductivity of the latter corresponds to the presence of three cationic changes, $[Co(NH_3)_6]^{3+}$ and three anionic charges, $[Co(NO_2)_6]^{3-}$.

Stereoisomers

Stereoisomers are those which have different orientations of ligands around the central metal atom. The atom-to-atom bonding sequence is, however, identical in these compounds. Among complexes, coordination number 4 and 6 are most frequently encountered and, therefore, stereoisomers shall be illustrated for these cases. Both types of stereoisomers, geometrical and optical, shall be discussed for each coordination geometry. A summary is presented in Table 24.1.

Table 24.1 Correlation between stereoisomers and geometry

Coordination number	Geometry	Geometrical isomers	Optical isomers
4	Tetrahedral	Not possible	Possible, but rare
	Square planar	Possible	Not possible
6	Octahedral	Possible	Possible

1. Geometrical isomers

The geometrical isomers are also called *cis–trans isomers*. The term *cis* means "adjacent to", while *trans* means "opposite to". Thus when two similar ligands are adjacent, the isomer is named *cis* and if they are opposite, the isomer is named *trans*. Geometrical isomers are distinguished by dipole moment studies. The *trans* isomer has zero or nearly zero dipole moment, while the *cis* analog possesses a positive dipole moment.

a. Coordination number 4: square planar complexes

For square planar complexes, *cis–trans* isomers exist for the following categories. Here, M refers to the metal atom; a, b, c and d are the ligands.

(i) *[Ma_2b_2]* Two isomers are possible. In *cis* complex, the two a (or b) ligands are adjacent, while in *trans*, the two a (or b) ligands are opposite.

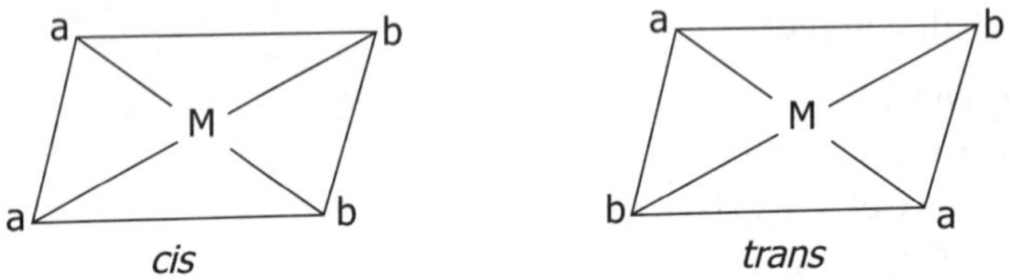

For example, diamminedichloroplatinum(II), [$Pt(NH_3)_2Cl_2$].

(ii) *[Ma_2bc]* Two isomers are possible. In *cis*, the two a ligands are adjacent, while in *trans*, they are opposite.

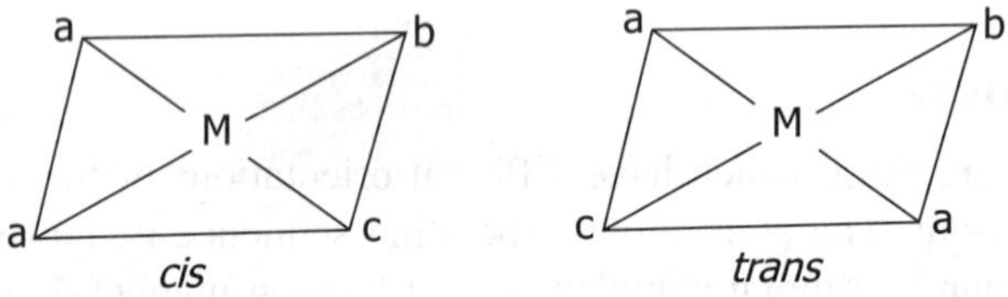

For example, diamminebromochloroplatinum(II), [$Pt(NH_3)_2ClBr$].

(iii) *[Mabcd]* Three isomers are possible. The position of any one ligand, say a, is fixed in the square planar skeleton, while b, c and d ligands are successively placed at the opposite end. The remaining two positions in each skeleton are randomly occupied by the remaining two ligands.

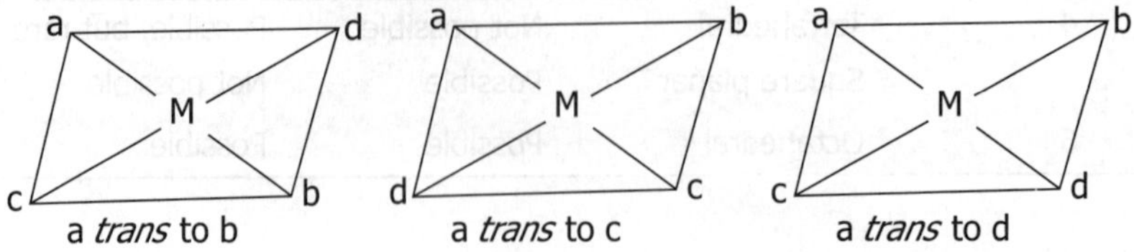

For example, amminebromochloropyridineplatinum(II), [$Pt(NH_3)(py)ClBr$].

Case Study 24.3 Write the IUPAC nomenclature of the following square planar complex I:

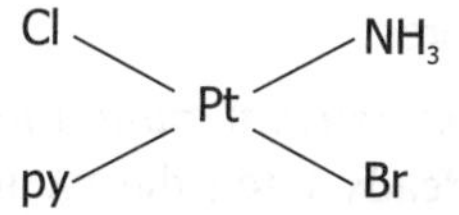

I

There are two plausible nomenclatures for complex I.

1. *trans*-Amminepyridine-*trans*-bromochloroplatinum(II)
2. *cis*-Amminebromo-*cis*-chloropyridineplatinum(II)

The first is correct. The second fails to distinguish between complex I and complex II shown below.

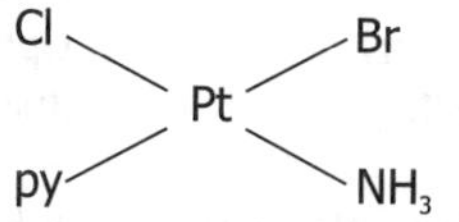

II

(iv) *Complexes with bridging ligands* Two isomers are possible. Similar ligands placed on the same side of two metal atoms of the bridging complex constitute the *cis* isomer. When placed on the opposite sides, *trans* isomer results. For example,

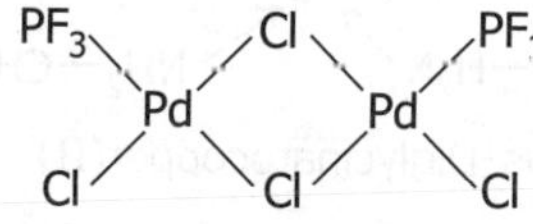

cis-Chloro(trifluorophosphine)palladium(II)-μ-dichloro-chloro(trifluorophosphine)palladium(II)

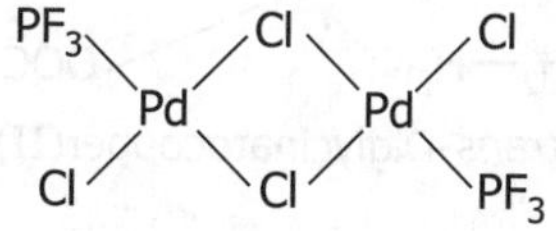

trans-Chloro(trifluorophosphine)palladium(II)-μ-dichloro-chloro(trifluorophosphine)palladium(II)

Strictly speaking, a third isomer is also possible.

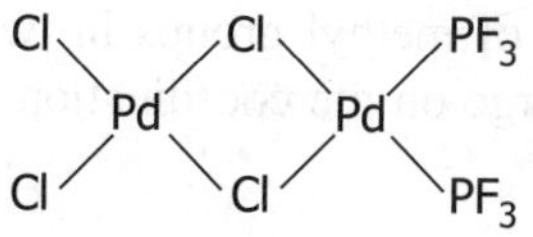

Dichloropalladium(II)-μ-dichloro-bis(trifluorophosphine)palladium(II)

This is also a *cis* isomer. However, due to steric repulsion, resulting from close proximity of two bulky trifluorophosphine groups, this complex (as well as similar other complexes) do not exist. This example is, therefore, only of theoretical interest.

(v) *Complexes with asymmetrical bidentate ligands* Two isomers are possible. We have already cited the example of ethylenediamine as a bidentate ligand. This ligand is a symmetrical one. If we divide it into two halves by an imaginary line, the upper half is a mirror image of the lower half. Such ligands do not show *cis–trans* isomerism in square planar geometry.

Ethylenediamine
Upper half is a mirror image of lower half.

Glycinate ion
Upper half is not a mirror image of lower half.

However, ligands such as glycinate ion, which are asymmetrical, i.e., their two halves are not mirror images of each other, form *cis* and *trans* isomers. In *cis* isomer, similar groups appear on the same side of the chelate ring, while in *trans* isomer, similar groups are positioned on the opposite sides of the chelate ring. For example,

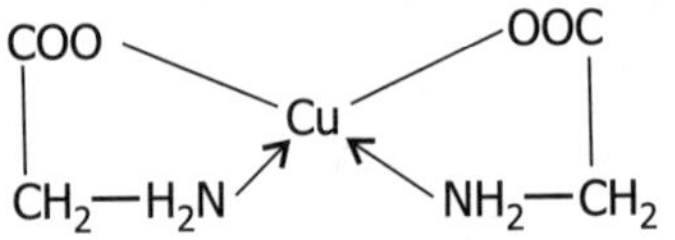

cis-Diglycinatocopper(II)

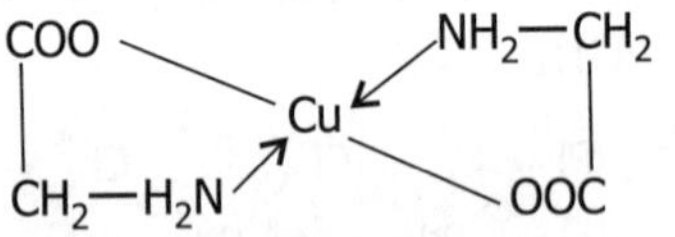

trans-Diglycinatocopper(II)

Case Study 24.4 How many geometrical isomers are expected to exist for the coordination compound, bis(methylethylenediamine)platinum(II) ion?

Four geometrical isomers are possible, as depicted in Fig. 24.1. They arise due to variations in the position of methyl groups in two chelate rings surrounding the platinum(II) ion (the charge on the coordination sphere has been ignored for convenience).

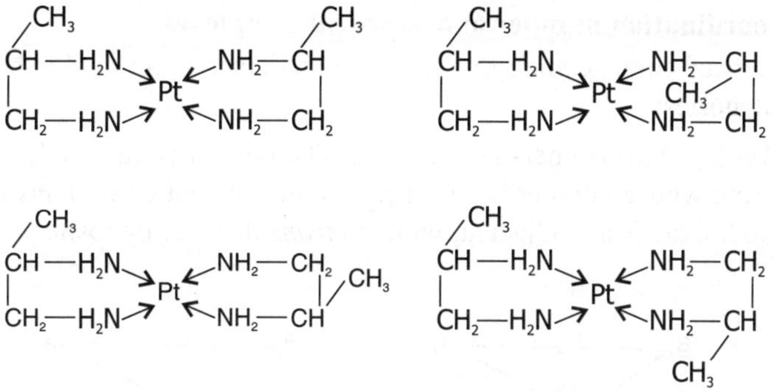

Fig. 24.1 Geometrical isomers of bis(methylethylenediamine)platinum(II) ion

Box 24.2

Distinguishing Geometrical Isomers by Chemical Method

As has been stated, geometrical isomers may be distinguished by the physical technique of dipole moment measurements. The *cis* isomer has a substantial value of dipole moment, which for the *trans* analog is either zero or nearly so.

The *cis* and *trans* isomers may also be distinguished by chemical evidence. For example, one mole of a *cis*, four coordinated complex, for example, *cis* diamminedichloroplatinum(II) reacts with one mole of oxalic acid.

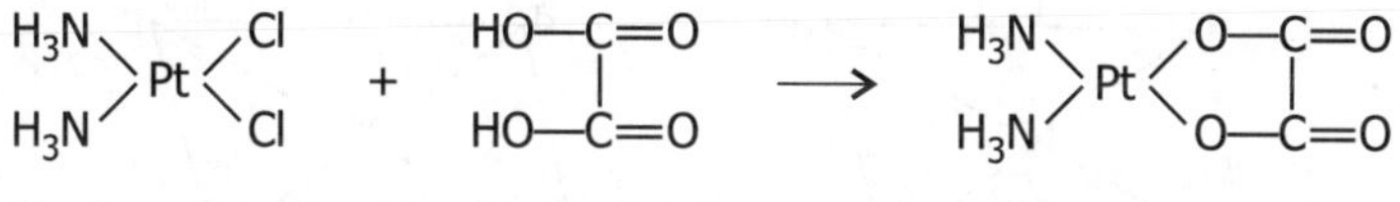

The *trans* isomer, however cannot lose both the chlorines with one mole of oxalic acid. Hence for reaction to reach completion, the molar ratio between the complex and oxalic acid must be 1 : 2.

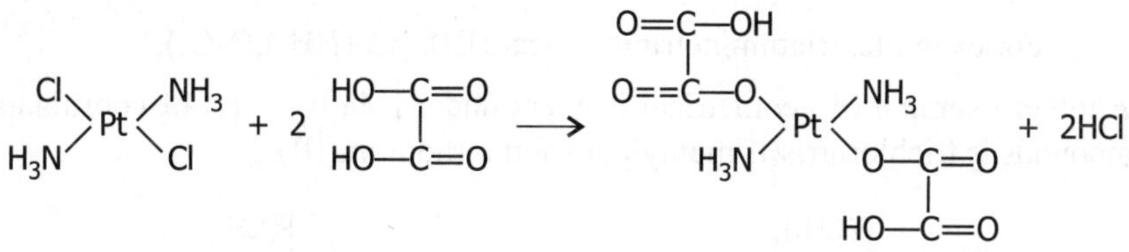

This method of making a distinction between *cis* and *trans* isomers is called *Grinberg test.*

The geometrical isomers of platinum may also be distinguished by *Kurnakov test*, which has been cited under *trans* effect in Chapter 28, p. 810.

b. Coordination number 6: octahedral complexes

For octahedral geometry, *cis–trans* isomerism is important for the following categories:

(i) *[Ma_4b_2]* Two isomers are possible. The type of isomer is decided by the ligand whose number is less, i.e., b in the present case. In *cis* isomer, the two b ligands are adjacent, while in *trans* they are opposite.

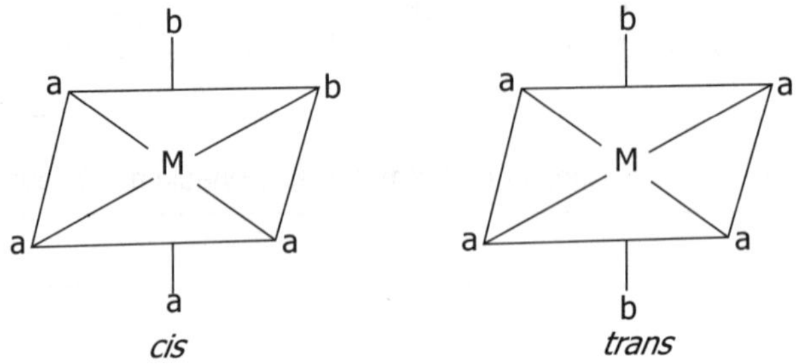

For example, tetramminedichlorocobalt(III) ion, $[Co(NH_3)_4Cl_2]^+$.

(ii) *[Ma_3b_3]* Two isomers are possible. In this case the two isomers are designated *facial* (or *fac*- in abbreviated form) and *meridional* (or *mer*-). These correspond to *cis* and *trans* geometries, respectively. In *facial*, all the three similar ligands (a or b) are placed at adjacent positions of the octahedral skeleton. In *meridional*, two a ligands are placed opposite to each other, two b ligands are placed opposite to each other and the remaining two positions are randomly occupied by a and b.

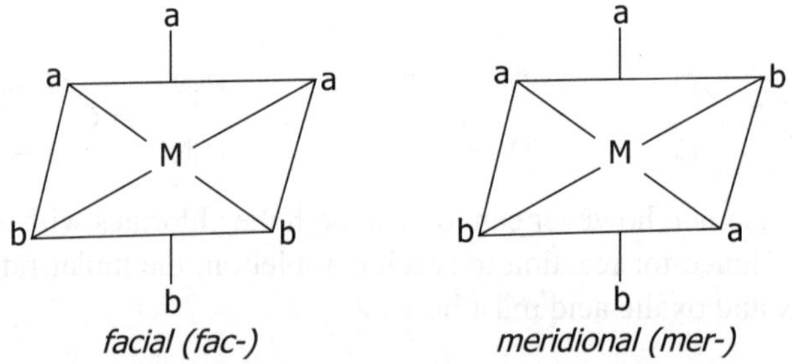

For example, triamminetrinitrocobalt(III), $[Co(NH_3)_3(NO_2)_3]$.

Another example of geometrical isomers under [Ma_3b_3] class of coordination compounds is trichlorotris(trimethylphosphine)rhodium(III).

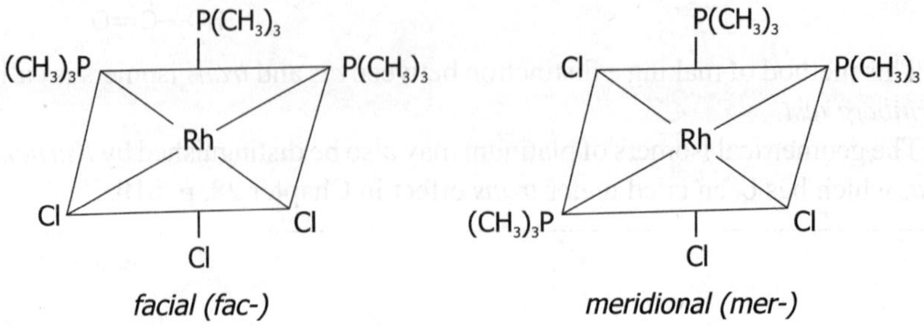

Case Study 24.5 The color of *trans*-bis(ethylenediamine)difluorocobalt(III) ion is pale, while that of the *cis* isomer as well as of the corresponding *trans*-dichloro complex is intense. Why?

The structures of *trans*-bis(ethylenediamine)difluorocobalt(III) ion (Fig. 24.2a), *cis*-bis(ethylenediamine)difluorocobalt(III) ion (Fig. 24.2b) and *trans*-dichlorobis(ethylenediamine)cobalt(III) ion (Fig. 24.2c) are shown below.

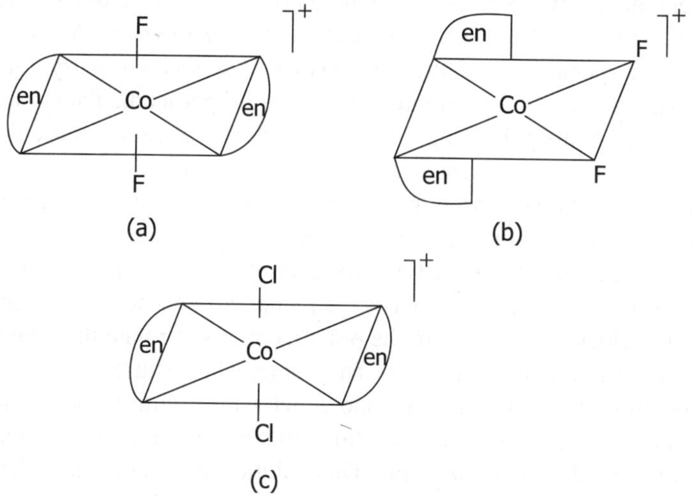

Fig. 24.2 Structure of (a) *trans*-bis(ethylenediamine)difluorocobalt(III) ion (b) *cis*-bis(ethylenediamine)difluorocobalt(III) ion and (c) *trans*-dichlorobis(ethylenediamine)cobalt(III) ion

The *trans*-$[Co(en)_2F_2]^+$ complex has a center of symmetry (a line drawn from the metal and extended along (x, x), (y, y) or (z, z) direction meets at similar atoms). As discussed in Case Study 26.9, p. 729, color in transition metal compounds arises due to d–d transitions and such transitions are forbidden in complexes having a center of symmetry—a phenomenon usually referred to as *Laporte forbidden transitions*. Hence, the complex has a pale color. The *cis*-$[Co(en)_2F_2]^+$ complex does not have a center of symmetry and, therefore, the Laporte rule does not apply in this case. The d–d transitions are allowed and the color is intense.

The *trans*-$[Co(en)_2Cl_2]^+$ complex possesses a center of symmetry. Since the d–d transitions are likely to be forbidden, its color, like that of *trans*-$[Co(en)_2F_2]^+$, is expected to be pale. Yet it is intense. In this case, the color originates due to the *charge transfer spectrum*. As stated in Chapter 21, (p. 607) with reference to the color of potassium dichromate and potassium permanganate, charge transfer phenomenon is attributed to transfer of electronic charge from the ligand(s) to the vacant d orbitals of the metal ion. The chloro ligand in the *trans*-$[Co(en)_2Cl_2]^+$ analog donates its excessive negative charge to the cobalt(III) ion, paving the way for charge transfer coloration. However, the fluoro groups are highly electronegative and do not part away with negative charge. Hence, the charge transfer phenomenon does not occur in *trans*-$[Co(en)_2F_2]^+$ complex.

2. Optical isomers

Those stereoisomers which fulfil the following criteria are designated as optical isomers or *enantiomers*:

* They are mirror images of each other.
* They are non-superimposable.
* They can rotate the plane of polarized light in opposite directions.

The phenomenon by which the plane of polarized light is rotated is called *optical activity*. It is measured by an instrument called *polarimeter*. A polarimeter is provided with a special type of filter called *polarizing filter*. Ordinary light consists of electromagnetic waves that vibrate in all directions perpendicular to the direction of travel. However, the polarizing filter absorbs all waves except those that vibrate in a single plane. The light consisting of such waves is called *plane-polarized light*. When this light is passed through an optically active compound, placed in the sample cell, its plane of polarization is rotated through an angle θ (Fig. 24.3). This phenomenon is called *optical rotation*. Optical isomers interact with polarized light in different ways. If one isomer rotates the plane of plane-polarized light through angle θ in the clockwise direction, the other isomer will rotate the plane by same degree, but in anti-clockwise direction. The isomer which causes clockwise rotation is called dextrorotatory (D-) and the one which causes anti-clockwise rotation is called levorotatory (L-). A single solution containing equimolar amounts of the two forms is called *racemic mixture*. This solution does not rotate the plane of plane-polarized light.

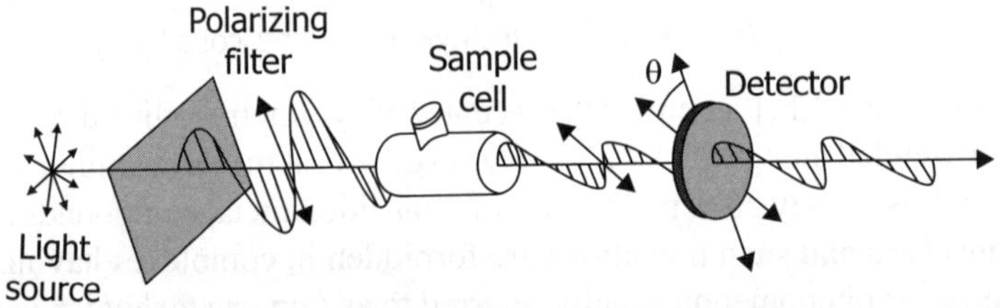

Fig. 24.3 Plane of polarization of plane-polarized light rotated through angle θ as it interacts with an optically active compound

Whether an optical isomer is dextrorotatory or levorotatory cannot be judged from the structure of the complex. Just as the order of a reaction cannot be predicted from the rate equation—it can be determined only experimentally—in the same way, the direction of rotation can be elucidated only by placing the isomer in the path of plane-polarized light and noting the angle and direction of optical rotation.

a. Coordination number 4: tetrahedral complexes

Optical isomers are observed among tetrahedral complexes if there is asymmetry around the metal atom, i.e., four different groups are tetrahedrally coordinated to the central metal. For example, ethylmethylpropylsulfidoarsenic(V).

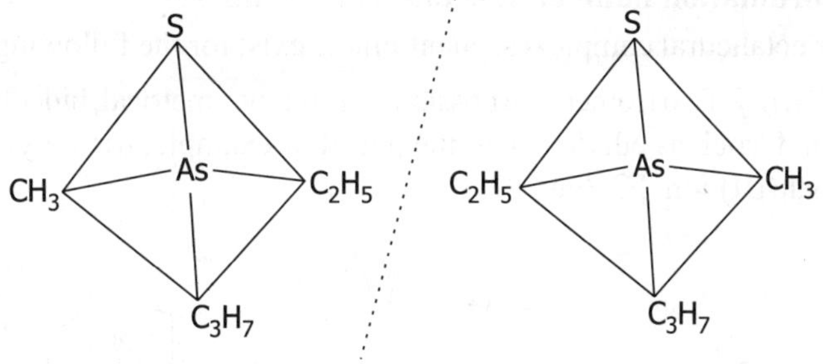

Optical isomers are also possible for aquabromochloroiodocobalt(III) complex, $[Co(H_2O)ClBrI]$

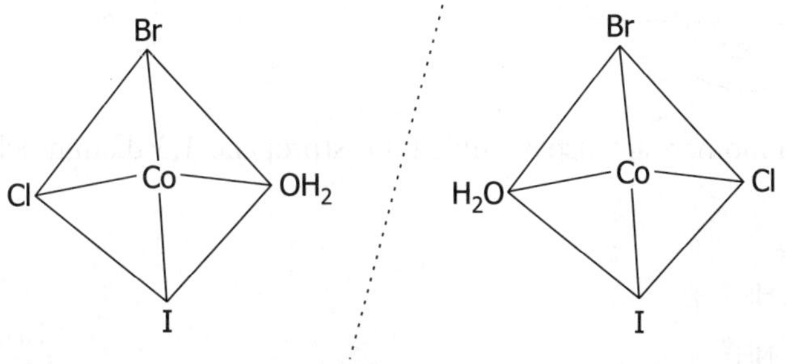

Another mixed halide complex of cobalt, bromochlorofluoroiodocobaltate(II) ion, $[CoFClBrI]^{2-}$ too shows optical isomerism.

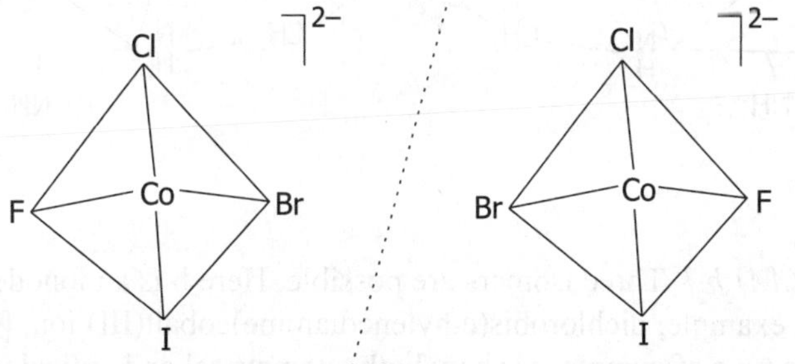

Tetrahedral complexes involving asymmetrical, bidentate ligands are resolvable into enantiomers. An example is bis(benzoacetonato)beryllium(II) complex, the structure of which is shown in Fig. 24.4.

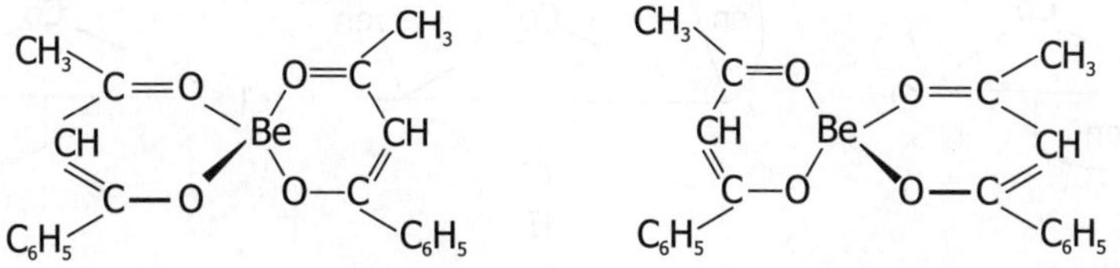

Fig. 24.4 Optical isomers of bis(benzoacetonato)beryllium(II) complex

b. Coordination number 6: octahedral complexes

For octahedral complexes, enantiomers exist for the following categories:

(i) *[M(AA)$_3$]* Two isomers are possible. AA is a symmetrical, bidentate chelating ligand, such as ethylenediamine (en). For example, tris(ethylenediamine) cobalt(III) ion, $[Co(en)_3]^{3+}$.

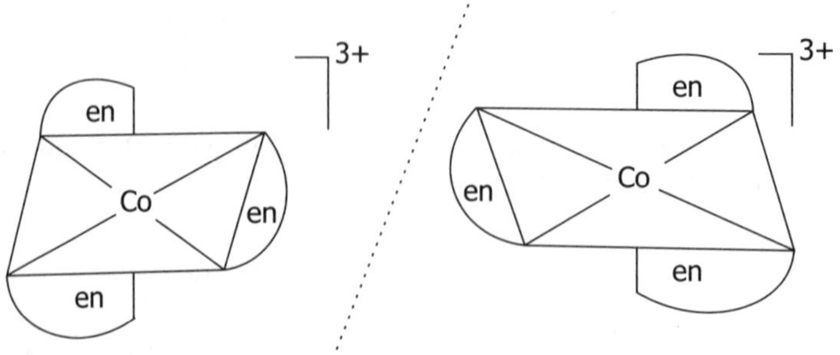

Enantiomers are also possible for tris(propane-1,3-diamine)chromium(III) ion

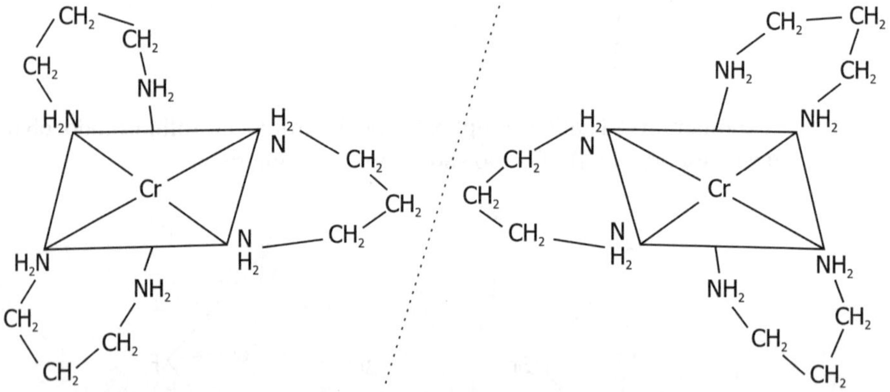

(ii) *[M(AA)$_2$b$_2$]* Three isomers are possible. Here b is a monodentate ligand. For example, dichlorobis(ethylenediamine)cobalt(III) ion, $[Co(en)_2Cl_2]^+$. This type of complexes show both geometrical and optical isomers.

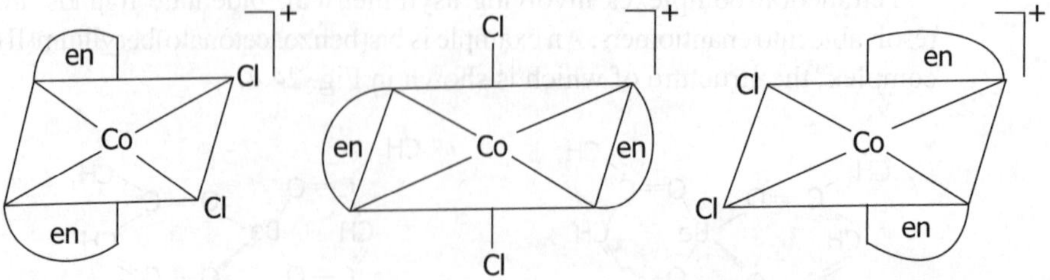

I II III

Complexes I and II are geometrical isomers—I is *cis* and II is *trans*. The *cis* isomer has a non-superimposable mirror image, III. Therefore, I and

III are optical isomers. A mirror image may be drawn for *trans* isomer, II as well. However, it shall be superimposable on II and, therefore, cannot be regarded as an enantiomer.

The isomers I–III may be distinguished by a combination of dipole moment and optical rotation studies. Isomer II will have a zero dipole moment, while isomers I and III will have a positive dipole moment. The latter two compounds may be segregated by measuring the direction of their optical rotation. One of them would be dextrorotatory, the other levorotatory.

Box 24.3

Cases Where Both Geometrical and Optical Isomers are Possible

We have already cited the example of $[M(AA)_2b_2]$ type of complexes for which both geometrical and optical isomers exist. There are many examples of such cases, although not as important. Octahedral complexes with asymmetrical bidentate ligands, such as triglycinatocobalt(III) exist in *cis* and *trans* forms (geometrical isomers), as shown in Fig. 24.5. Each of them has a resolvable mirror image (optical isomers).

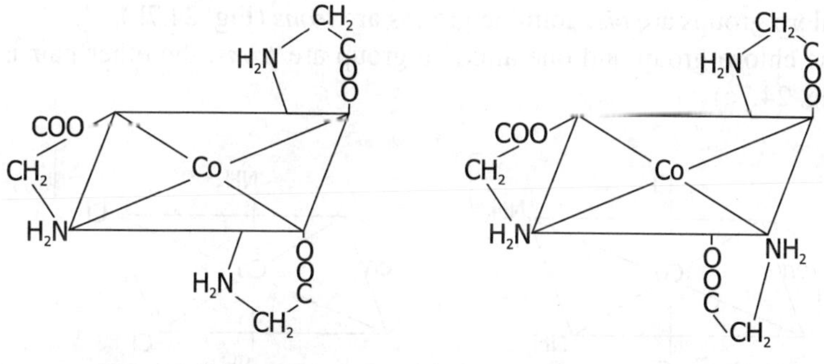

Fig. 24.5 Geometrical isomers of triglycinatocobalt(III)

Similarly, octahedral complexes, in which all the six ligands are different and unidentate, (Mabcdef) have, theoretically speaking, 15 geometrical isomers, each having a mirror image. For amminebromochloroiodonitropyridineplatinum(IV), $[PtCl(I)Br(NO_2)py(NH_3)]$ less than half of the possible geometrical isomers have been synthesized and none has been resolved.

The amminechlorobis(ethylenediamine)cobalt(III) ion exists in two geometrical forms—one in which the ammine and chloro groups are *trans* with respect to each other (Fig. 24.6a), and the other in which these two ligands are in *cis* position (Fig. 24.6b). The optical isomer of the latter too exists.

Contd...

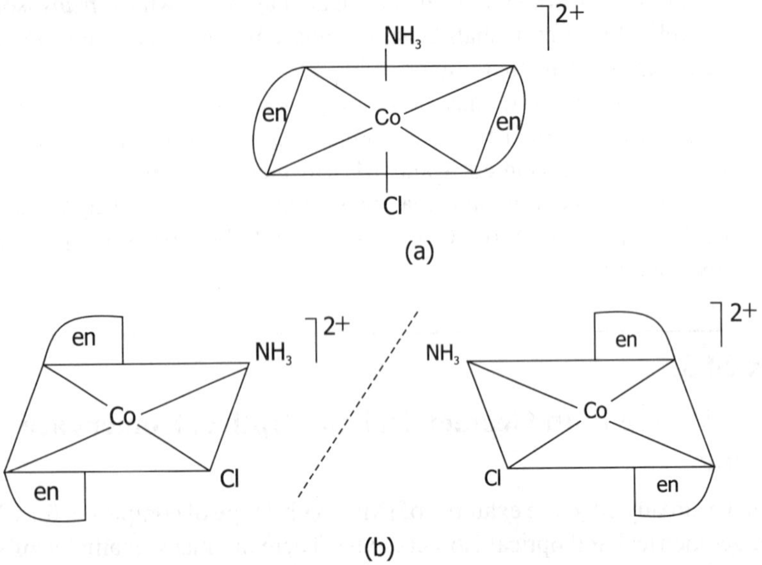

Fig. 24.6 Stereoisomers of amminechlorobis(ethylenediamine)cobalt(III) ion

In the case of diamminedichloro(ethylenediamine)cobalt(III) ion, the following three geometrical isomers exist:

1. Chloro groups are *trans*, ammine groups are *cis* (Fig. 24.7a).
2. Chloro groups are *cis*, ammine groups are *trans* (Fig. 24.7b).
3. One chloro group and one ammine group are *trans*, the other pair is *cis* (Fig. 24.7c)

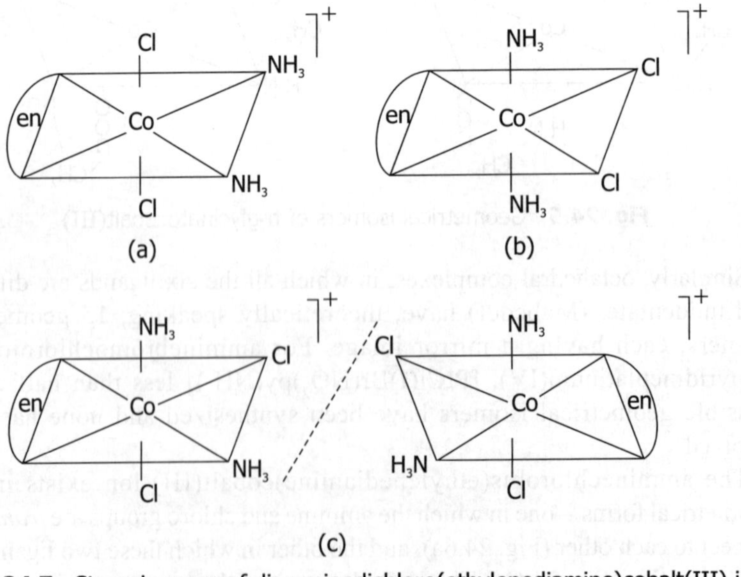

Fig. 24.7 Stereoisomers of diamminedichloro(ethylenediamine)cobalt(III) ion

The third isomer possesses a resolvable mirror image.

Case Study 24.6 A compound $Co(en)_2(NO_2)_2Cl$ has been prepared in three isomeric forms, **A**, **B** and **C**. **A** undergoes no reaction with $AgNO_3$ or en and is optically inactive. **B** reacts with $AgNO_3$, but not with en and is optically inactive. **C** is optically active and reacts with both $AgNO_3$ and en. Identify **A**, **B** and **C**.

Silver nitrate would react with the complex only if the chloride ion is in the ionization sphere. Ethylenediamine would react with the complex only if there are two monodentate ligands at adjacent positions in the coordination sphere. Ethylenediamine, which is a bidentate ligand, would replace the two monodentate ligands simultaneously.

For coordination number 6 complexes, having two bidentate ligands, the *cis* form is optically active, while the *trans* form is not.

Complex **A**

Property	**Inference**
It is optically inactive.	Structure is *trans*.
It does not react with $AgNO_3$.	Chlorine is in coordination sphere.
It does not react with en.	Chloro and nitro groups are not adjacent.

The structure of **A** is

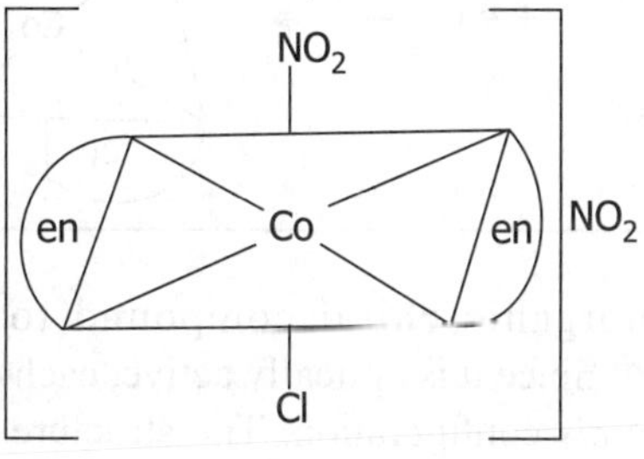

Complex **B**

Property	**Inference**
It is optically inactive.	Structure is *trans*.
It reacts with $AgNO_3$.	Chlorine is in ionization sphere.
It does not react with en.	The two nitro groups are not adjacent.

The structure of **B** is

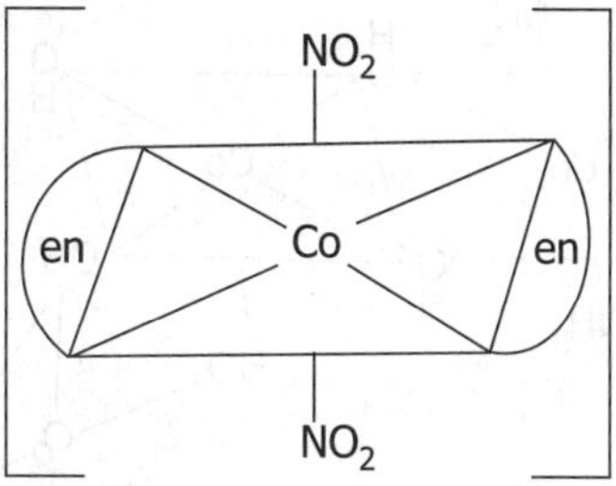

The reaction with silver nitrate may be indicated as,

$$[Co(en)_2(NO_2)_2]Cl + AgNO_3 \rightarrow [Co(en)_2(NO_2)_2](NO_3) + \underset{\text{White}}{AgCl\downarrow}$$

Complex **C**

Property	**Inference**
It is optically active.	Structure is *cis*.
It reacts with $AgNO_3$.	Chlorine is in ionization sphere.
It reacts with en.	The two nitro groups are adjacent.

The structure of **C** is,

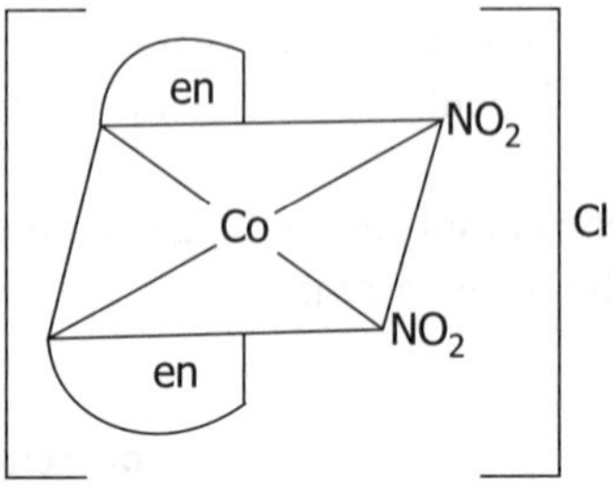

The reaction with ethylenediamine may be indicated as,

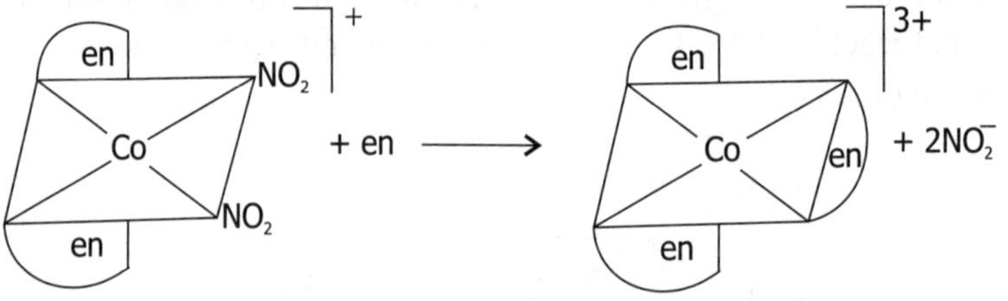

The first purely inorganic chiral compound to be synthesized was $[Co\{(OH)_2Co(NH_3)_4\}_3]^{6+}$. Since it is optically active, each of the $[Co(NH_3)_4(OH)_2]$ unit is expected to have *cis* configuration. The structure of the compound may, therefore, be depicted as follows:

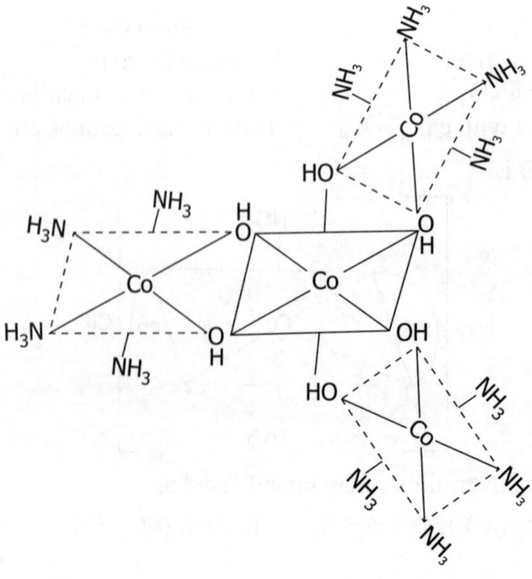

Case Study 24.7 When ethyldiphenylphosphine ($PEt\phi_2$) is added to a solution of nickel(II) bromide in carbon disulfide at 200 K, a red-colored diamagnetic complex is formed. When the solution is allowed to attain room temperature, the red complex isomerizes to a paramagnetic green complex. Identify the isomers.

On adding ethyldiphenylphosphine to nickel(II) bromide solution, a complex with composition, $[Ni(PEt\phi_2)_4]^{2+}$ is formed. When the ligand is bulky—as in this particular case—the four-coordinated complexes exist in two forms: low temperature, diamagnetic, square planar geometry and high temperature, paramagnetic, tetrahedral geometry.

Accordingly, the red complex, formed at 200 K, is square planar with a bond angle of approximately 90°. As the temperature increases, the bonds within the ligands start vibrating at a greater pace. The ligands, therefore, span greater volumes. To overcome steric over-crowdedness, the complex acquires a tetrahedral structure. In this geometry, the bond angle is approximately 109° and hence steric repulsions are comparatively less.

The two isomers may be represented as follows:

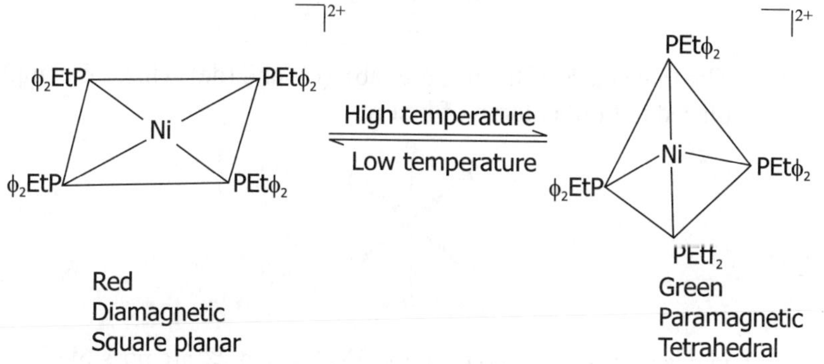

Such isomers are called *conformational isomers*.

Bailar Method of Isomeric Enumeration

For an octahedral coordination compound having six different monodentate ligands, [Mabcdef] a total of fifteen geometrical isomers are possible; each analog has an enantiomer. An example of such a complex is amminebromochloroiodonitropyridine platinum(IV), $[Pt(py)(NH_3)(NO_2)(Cl)(Br)(I)]$. The number of geometrical isomers for this complex are listed in Table 24.2, and may be worked out as follows.

1. The ligands are arbitrarily designated by different notations. Let us assume, (py) = a; (NH_3) = b; (NO_2) = c; (Cl) = d; (Br) = e; (I) = f.
2. The ligands which are trans to each other are grouped together. For example, (ab) means that b is trans to a.
3. The position of any one ligand, say a, is fixed.
4. Place b *trans* to a. With the position (ab) fixed, the following three geometrical isomers are possible.

(a) One pair of ligands is (ab). The other two pairs are (cd) and (ef). The three pairs are listed one under the other in alphabetical order. The complex represented by the notation (ab/cd/ef) is placed in the first column of first row in Table 24.2. It is designated as 1X.

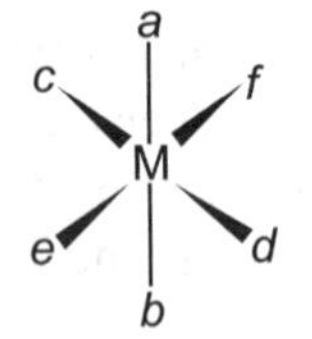

(b) Apart from the pair (ab), the other two pairs of ligands are (ce) and (df). This complex, designated as 1Y occupies the second column and first row in Table 24.2.

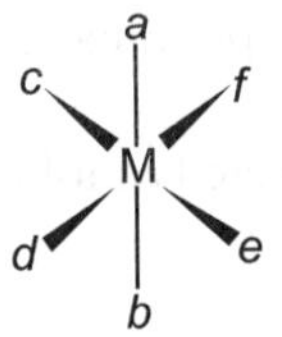

(c) The three pairs of ligands are (ab), (cf) and (de). This, 1Z complex is placed in third column of first row.

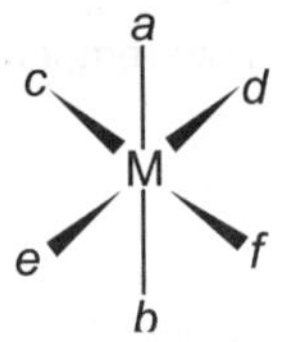

5. When c is trans to a, three isomers, 2X, 2Y and 2Z are possible.
Place d, e, and f successively *trans* to a. For each set three isomers are possible.
For every geometrical isomer, an enantiomer in form of its mirror image is possible, as has been showcased in Fig. 24.8.

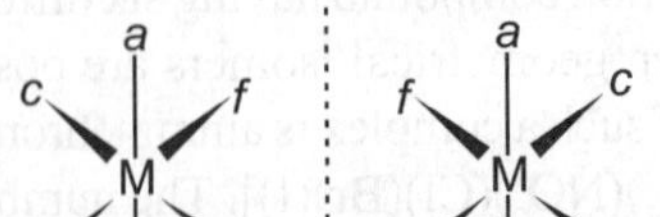

Fig. 24.8 Mirror image of a geometrical isomer

This system of preparing a schedule of stereoisomers is referred to as Bailar method of isomeric enumeration.

Table 24.2 Enumeration of geometrical isomers possible for [Mabcdef] complex

	X	Y	Z
1.	ab	ab	ab
	cd	ce	cf
	ef	df	de
2.	ac	ac	ac
	bd	be	bf
	ef	df	de
3.	ad	ad	ad
	bc	be	bf
	ef	cf	ce
4.	ae	ae	ae
	bc	bd	bf
	df	cf	cd
5.	af	af	af
	bc	bd	be
	de	ce	cd

25. STABILITY OF COMPLEXES

Introduction

The terms, stable and unstable, indicate thermodynamic characteristics. A stable complex is one which has a large negative value of free energy, ΔG for a reaction. The one which has a low negative or positive free energy is unstable.

The terms, labile and inert, indicate kinetic characteristics. An inert complex is one which has high activation energy, E_a for a reaction. The one with low activation energy is labile. In general, a complex which reacts completely within one minute, at 298 K is labile. Others are inert.

Tetracyanonickelate(II) ion, $[Ni(CN)_4]^{2-}$, is an example of a complex which is thermodynamically stable, but kinetically labile.

Hexamminecobalt(III) ion, $[Co(NH_3)_6]^{3+}$, is an example of a complex which is thermodynamically unstable, but kinetically inert.

Tris(ethylenediamine)cobalt(III) ion, $[Co(en)_3]^{3+}$, is an example of a complex which is thermodynamically stable and kinetically inert.

Tetraquacobalt(II) ion, $[Co(H_2O)_4]^{2+}$, is an example of a complex which is thermodynamically unstable and kinetically labile.

Criteria to Gauge Lability

The important factors which govern the lability of complexes are as follows.

1. Lability and inertness are related to the strength of metal–ligand bonds. A weak bond makes the complex more reactive, and hence more labile. The outer sphere complexes have relatively long bond lengths than inner sphere complexes (*see* Chapter 26, p. 702). This means that the metal–ligand bonds of outer sphere complexes are relatively weak, and such compounds are labile. For this reason, hexafluorocobaltate(III) ion, $[CoF_6]^{3-}$ is labile, while hexamminecobalt(III) ion, $[Co(NH_3)_6]^{3+}$ is inert.
2. The complexes of 4d and 5d series transition metals have longer, and hence weaker, bond lengths, as compared to the corresponding complexes of 3d series transition metals. Therefore, complexes of 4d and 5d series metals are relatively labile than their 3d series counterparts. Hexacyanomolybdate(III) ion, $[Mo(CN)_6]^{3-}$ is labile, while hexacyanochromate(III) ion, $[Cr(CN)_6]^{3-}$ is inert.
3. If there are vacant orbitals in the valence shell of central metal ion in the complex, it is more likely to be acted upon by the constituents of atmosphere—mainly water vapors and carbon dioxide—or with any other reagent and hence would be labile. If, on the other hand, all the valence shell

orbitals of the central metal ion are filled up, it is unlikely to be attacked by a chemical reagent and hence would be inert.

The hexamminevanadium(III) ion, $[V(NH_3)_6]^{3+}$ has vacant orbitals in the valence shell of vanadium and is, therefore labile. The hexamminecobalt(III) ion, $[Co(NH_3)_6]^{3+}$ does not have vacant orbitals in the valence shell of cobalt and is, therefore, inert. A more precise explanation has been provided in Chapter 26 (*see* Case Study 28.3, p. 803).

Case Study 25.1 Which is more labile: $[Ni(bipy)_3]^{2+}$ or $[Ni(bipy)_2(H_2O)_2]^{2+}$?

In tris(bipyridyl)nickel(II) ion, $[Ni(bipy)_3]^{2+}$, the metal ion is so well shielded by three chelate rings that further reaction with any other chemical species is extremely difficult.

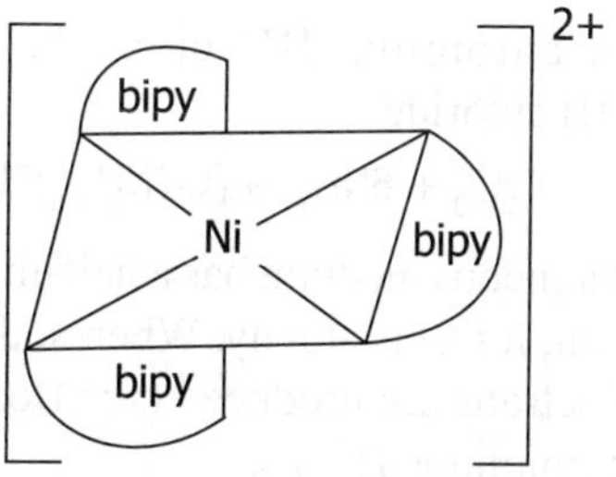

Therefore, the complex is inert.

In diaquabis(bipyridyl)nickel(II) ion, the water molecules may be easily replaced by another ligand.

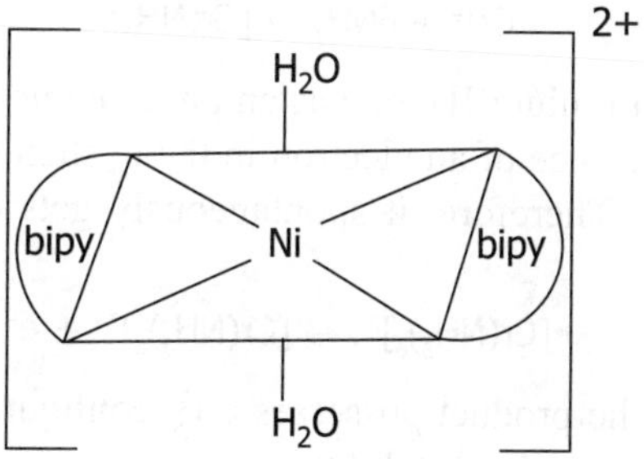

Therefore, the complex is labile.

The aqua ligand is quite versatile and coordinates with a large number of metal ions to form complexes of varied compositions, but largely six-coordinated, $[M(H_2O)_6]^{n+}$. As noted in Chapter 5, p. 234, such hydrated cations give acidic solution.

$$[M(H_2O)_6]^{n+} \rightleftharpoons [M(H_2O)_5(OH)]^{(n-1)+} + H^+$$

This reaction shows that the acidity of a compound (read ligand) increases on coordination. Water is a weak acid and hence the above reaction proceeds with a moderate rate. The aqua complexes, therefore, are not labile. On the other hand, hydrogen sulfide is more acidic than water. If it gets coordinated to a metal ion, the reaction leading to the release of proton will occur at a relatively very fast rate. Such complexes will be so labile that it would not be possible to isolate them. Thus, complexes involving hydrogen sulfide ligand are not known.

For the same reason, the complexes of phosphine are non-existent, although ammonia is a versatile coordinating group. Phosphine is less basic—and hence more acidic—than ammonia. Its complexes are extremely labile and defy synthesis.

Case Study 25.2 The reaction between chromium(III) chloride and ammonia gets catalyzed by adding sodium metal to the contents. Why?

The reaction between chromium(III) chloride and ammonia yields the hexamminechromium(III) chloride.

$$CrCl_3 + 6NH_3 \rightarrow [Cr(NH_3)_6]Cl_3$$

Chromium(III) ion in aqueous medium has a half-filled t_{2g} shell (t^3_{2g}) and hence is quite inert. For this reason, it reacts slowly. When sodium is added to the reaction mixture, ammoniated electrons are produced (*see* Box 6.2, p. 253). These reduce chromium(III) ions to chromium(II) ions.

$$Cr^{3+} + e^- \rightarrow Cr^{2+}$$

The chromium(II) ion, with the configuration $t^3_{2g}\,e^1_g$ is labile and therefore reacts with ammonia at a fast pace.

$$Cr^{2+} + 6NH_3 \rightarrow [Cr(NH_3)_6]^{2+}$$

The hexamminechromium(II) ion has an unfavorable crystal field stabilization energy due to the presence of an electron in the e_g shell of chromium(II) ion (*see* Chapter 26, p. 715). Therefore, it spontaneously gets oxidized to hexammine-chromium(III) ion.

$$[Cr(NH_3)_6]^{2+} \rightarrow [Cr(NH_3)_6]^{3+} + e^-$$

The metal ion in the product possesses a t^3_{2g} configuration, a favorable crystal field stabilization energy—and stability.

Criteria to Gauge Stability

Enunciated below are the important factors which govern the stability of complexes.

1. Chelate effect

The term chelate effect states that complexes having chelate rings in their structure are more stable than complexes having no rings. Chelate rings are formed only in those complexes which contain one or more polydentate ligands. Therefore, it may be advocated that complexes with polydentate ligands are more stable than complexes with monodentate ligands. Further, higher the number of chelate rings, greater the stability of the complex.

A comparison of the stability constants (β) of chelated and non-chelated complexes, as given in Table 25.1, justifies the significance of chelate effect (greater the value of stability constant, higher the stability of the complex). The stability constant of a chelated complex is higher than that of a non-chelated complex. As the number of rings increases, the stability of the complex also increases. Thus, $[Ni(en)_3]^{2+}$ with three chelate rings is more stable than $[Ni(H_2O)_2(en)_2]^{2+}$ with two chelate rings which, in turn, is more stable than $[Ni(H_2O)_4(en)]^{2+}$ with only one chelate ring (Fig. 25.1).

Table 25.1 Comparison of stability constants of chelated and non-chelated complexes

Chelated systems		Non-chelated systems	
Complex	**Log β**	**Complex**	**Log β**
$[Ni(H_2O)_4(en)]^{2+}$	7.51	$[Ni(H_2O)_4(NH_3)_2]^{2+}$	5.00
$[Ni(H_2O)_2(en)_2]^{2+}$	13.86	$[Ni(H_2O)_2(NH_3)_4]^{2+}$	7.87
$[Ni(en)_3]^{2+}$	18.28	$[Ni(NH_3)_6]^{2+}$	8.61

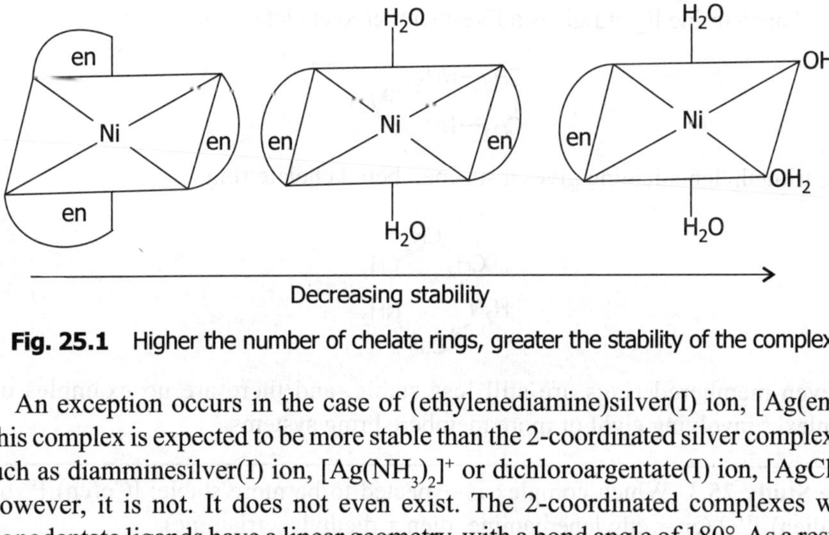

Fig. 25.1 Higher the number of chelate rings, greater the stability of the complex

An exception occurs in the case of (ethylenediamine)silver(I) ion, $[Ag(en)]^+$. This complex is expected to be more stable than the 2-coordinated silver complexes, such as diamminesilver(I) ion, $[Ag(NH_3)_2]^+$ or dichloroargentate(I) ion, $[AgCl_2]^-$. However, it is not. It does not even exist. The 2-coordinated complexes with monodentate ligands have a linear geometry, with a bond angle of 180°. As a result, the bond pair–bond pair repulsions are minimized.

$$\overset{180^\circ}{[NH_3-Ag-NH_3]^+} \qquad \overset{180^\circ}{[Cl-Ag-Cl]^-}$$

The chelated complex is expected to have a much smaller bond angle and a non-linear geometry.

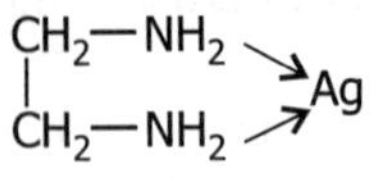

The bond pair–bond pair repulsions in such a complex would be so strong that these negate the stability which would have been afforded by the chelate effect.

Five-membered chelate rings are most stable as their formation leads to a minimum strain. A comparison of the stability constants (β) of five-membered and six-membered rings is given in Table 25.2.

Table 25.2 Relative stabilities of five-membered and six-membered chelate rings

Five-membered chelate ring		Six-membered chelate ring	
Complex	**Log β**	**Complex**	**Log β**
$[Cu(H_2O)_4(en)]^{2+}$	10.72	$[Cu(H_2O)_4(tn)]^{2+}$	9.98
$[Cu(H_2O)_2(en)_2]^{2+}$	20.03	$[Cu(H_2O)_2(tn)_2]^{2+}$	17.17

en = ethylenediamine, $NH_2(CH_2)_2NH_2$
tn = trimethylenediamine, $NH_2(CH_2)_3NH_2$

Ethylenediamine ligand gives a five-membered chelate ring,

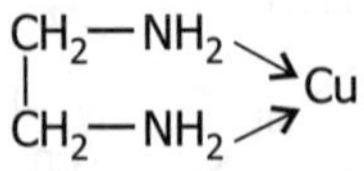

while trimethylenediamine gives a six-membered chelate ring.

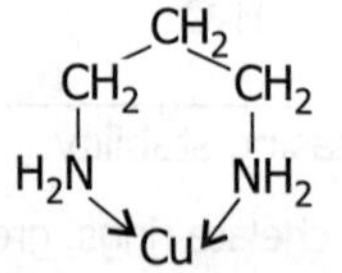

Seven-membered rings are still less stable, and there are no examples of complexes involving eight or more membered ring systems.

Case Study 25.3 Which complex is expected to be more stable: $[Co(en)_3]^{3+}$ or $[Co(dien)_2]^{3+}$? (en = ethylenediamine; dien = diethylenetriamine).

Ethylenediamine is a bidentate ligand,

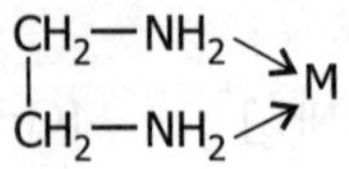

while diethylenetriamine is a tridentate ligand.

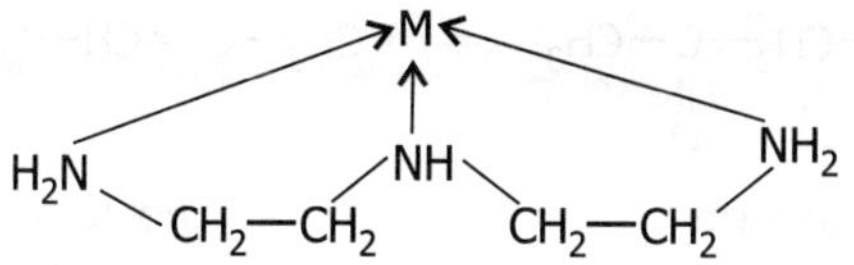

Both form five-membered chelate rings. The structures of the complexes in question may be depicted as follows:

$[Co(en)_3]^{3+}$

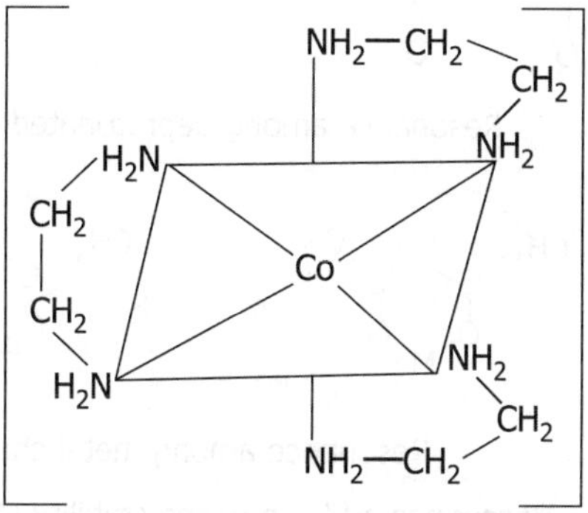

$[Co(dien)_2]^{3+}$

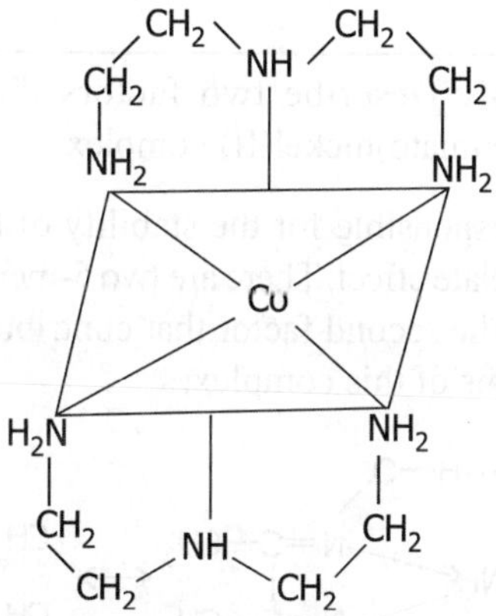

We find that there are three 5-membered rings in $[Co(en)_3]^{3+}$ complex and four 5-membered rings in $[Co(dien)_2]^{3+}$ complex. Therefore, $[Co(dien)_2]^{3+}$ is more stable.

If two or more resonating structures are possible for a chelate complex, the stability is further enhanced. For example, acetylacetonate ligand, in its enol form, gives a six-membered chelate ring with metals. Even though the six-membered chelates are less stable than their five-membered counterparts, acetylacetonato complexes are extremely stable. These are known for all the transition metals and for most of the lanthanoids and actinoids. The six-membered chelate structure is stabilized by resonance (Fig. 25.2).

Keto form Enol form

Tautomeric forms of acetylacetonate

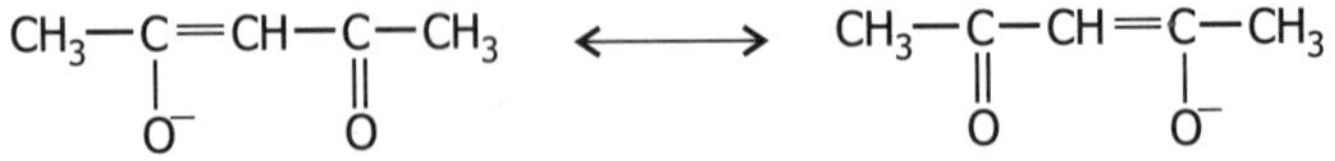

Resonance among deprotonated enol form

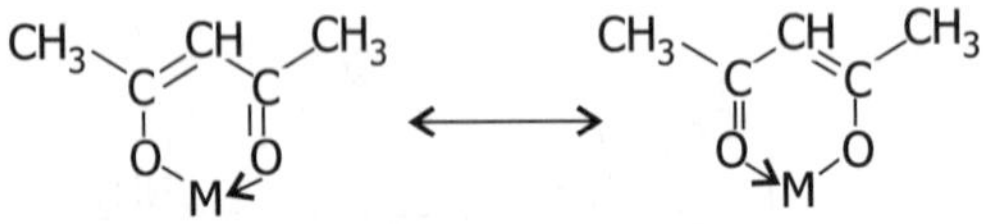

Resonance among metal chelates

Fig. 25.2 Resonance adds on to the stability afforded by chelate effect

---

Case Study 25.4 Describe two factors that contribute to the stability of bis(dimethylglyoximato)nickel(II) complex.

The first factor responsible for the stability of bis(dimethylglyoximato)nickel(II) complex is the chelate effect. There are two 5-membered chelate rings in the structure of the complex. The second factor that contributes to the stability is the existence of resonating forms of this complex.

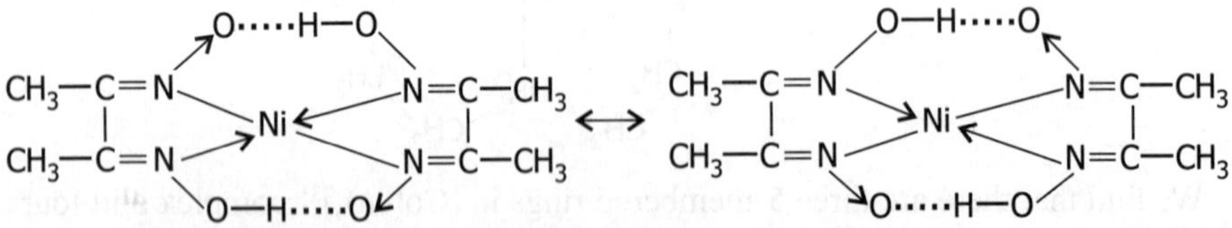

If resonance increases the stability of chelates, steric factors reduce it. For example, N,N′-dimethylethylenediamine forms less stable complexes than ethylenediamine.

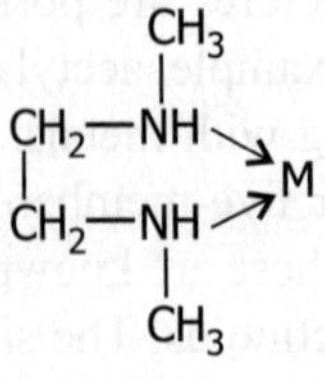

N,N′-dimethylethylenediamine

The two methyl substituents cause steric over-crowdedness near the metal, leading to loss of stability. For the same reason, complexes of 8-hydroxyquinoline are more stable than those of 8-hydroxy-2-methylquinoline.

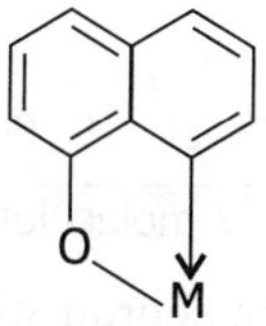

8-hydroxyquinoline chelate

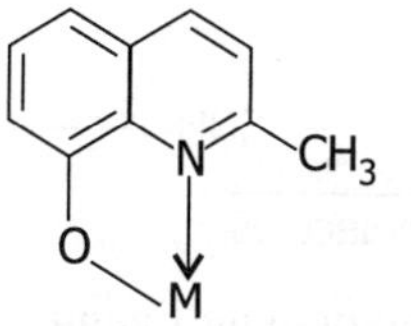

8-hydroxy-2-methylquinoline chelate

Case Study 25.5 Consider the following two isomeric ligands:

$$H_2\ddot{N}CH_2CH_2\ddot{N}HCH_2CH_2\ddot{N}HCH_2CH_2\ddot{N}H_2$$

A

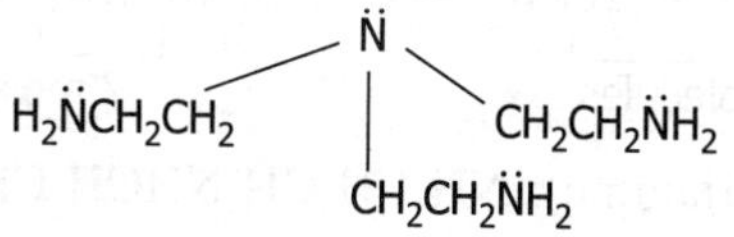

B

Which of these will form a more stable square planar complex with copper(II) ion?

The straight chain ligand A can coordinate its four nitrogen atoms at the corners of a square, but the branched chain ligand B cannot. Thus, A would form the more stable complex with copper(II) ion.

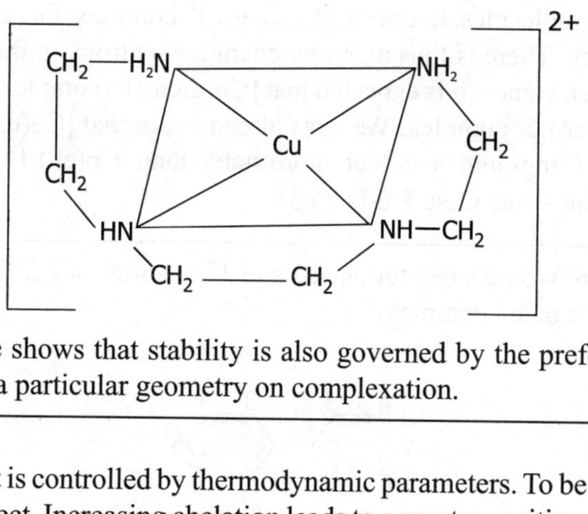

This example shows that stability is also governed by the preference of metal ions to achieve a particular geometry on complexation.

Chelate effect is controlled by thermodynamic parameters. To be more precise, it is an entropy effect. Increasing chelation leads to a greater positive value of change in entropy, ΔS. This may be explained qualitatively by assuming that when ligands enter the coordination sphere, water molecules are displaced. When monodentate ligands coordinate to the metal, an equal number of water molecules are released. This means that the total number of reactant molecules is equal to the total number of product molecules.

$$\underbrace{[Co(H_2O)_6]^{3+} + 6NH_3}_{\text{7 molecules}} \longrightarrow \underbrace{[Co(NH_3)_6]^{3+} + 6H_2O}_{\text{7 molecules}}$$

However, for each polydentate ligand entering the coordination sphere, two or more molecules of water are released. The number of product molecules is now more than the number of reactant molecules.

$$\underbrace{[Co(H_2O)_6]^{3+} + 3en}_{\text{4 molecules}} \longrightarrow \underbrace{[Co(en)_3]^{3+} + 6H_2O}_{\text{7 molecules}}$$

where en (ethylenediamine), $NH_2CH_2CH_2NH_2$, is a bidentate ligand.

$$\underbrace{[Co(H_2O)_6]^{3+} + 2dien}_{\text{3 molecules}} \longrightarrow \underbrace{[Co(dien)_2]^{3+} + 6H_2O}_{\text{7 molecules}}$$

where dien (diethylenetriamine), $NH_2CH_2CH_2NHCH_2CH_2NH_2$, is a tridentate ligand.

An increase in the number of product molecules over the reactant molecules represents an increase in disorderness which, in turn, implies a greater positive value of ΔS. Therefore, chelate complexes are thermodynamically more stable than non-chelate ones.

It must also be noted that during the formation of $[Co(en)_3]^{3+}$ complex, seven product molecules are released for every four molecules of reactants. There is a net increase of three molecules. In case of $[Co(dien)_2]^{3+}$ complex, there is a net increase of four molecules. There is thus a greater change in entropy in the latter example than in the former. Hence, it is expected that $[Co(dien)_2]^{3+}$ complex should be more stable than $[Co(en)_3]^{3+}$ complex. We have already seen that $[Co(dien)_2]^{3+}$ complex has four chelated rings and is, in fact, more stable than $[Co(en)_3]^{3+}$complex, which has only three rings (*see* Case Study 25.3).

Case Study 25.6 Would substitution at 3 and 3′ positions of 2,2′-bipyridine affect the stability of its metal complex?

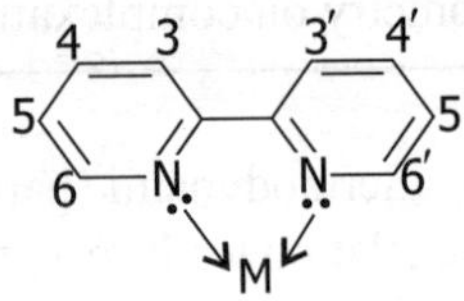

Substitution at 6 and 6′ positions lowers the stability of the complex because of steric over-crowdedness. However, substitution at 3 and 3′ positions, which are far away from the coordination sites, also lowers the stability of the complex. This is because the two substituents at 3 and 3′ positions make the rings puckered. As a result, the rings do not lie in the same plane and chelation becomes difficult.

2. Nature of metal ions and the ligands

The stability of the complexes also depends on the degree of hardness or softness of both the metal ions and the ligating atom(s) of the coordinating groups. We have already discussed hard and soft acid base (HASB) theory in Chapter 5, p. 245. Suffice it to say here that during the formation of a coordination compound, the metal ion acts as an acid, while the ligands act as bases.

A metal ion having a small size and/or high charge, such as titanium(IV) or cobalt(III), behaves as a hard acid. Conversely, a metal ion having a relatively large size and/or small charge, such as gold(I) or mercury(II) behaves as a soft acid. Likewise, a small, highly electronegative ligating atom, such as nitrogen, oxygen or fluorine, acts as a hard base, while a large, comparatively less electronegative atom, such as carbon, phosphorus or sulfur, acts as a soft base.

The hard-hard or soft-soft combination is preferred over a hard-soft/soft-hard interaction. For example, the ambidentate thiocyanato ligand, SCN^- would prefer to coordinate to hard cobalt(III) ion through nitrogen and to soft mercury(II) ion through sulfur.

A metal ion coordinated to hard ligands acquires hard character and prefers to combine further with hard coordinating groups. A metal ion coordinated to soft ligands acquires soft character and prefers to combine further with soft ligands. This phenomena is called *symbiosis* and is also discussed in Chapter 5 (*see* Case Study 5.12, p. 247).

For example, in the pentamminefluorocobalt(III) ion, $[Co(NH_3)_5F]^{2+}$ the coordination of five nitrogen atoms of ammine ligands imparts hardness to the cobalt(III) ion, which now readily combines with the hard fluoro group. The complex owes its stability to the presence of all-hard ligands around cobalt(III) ion. However, the pentacyanofluorocobaltate(III) ion, $[Co(CN)_5F]^{3-}$ is unstable. The coordination of five soft carbon atoms of cyano groups imparts an appreciable degree of softness to the cobalt(III) ion. It now does not accommodate the hard fluoro ligand at its sixth coordination site. Nevertheless, the pentacyanoiodocobaltate(III) ion, $[Co(CN)_5I]^{3-}$ is stable since all six ligands are soft bases.

Case Study 25.7 What is template effect?

Template effect refers to a situation whereby a coordinated metal ion reacts with a suitable organic reagent to form a macrocyclic complex. An example is the reaction between a nickel dithiolate coordination compound and 1,2-di(bromomethyl) benzene,

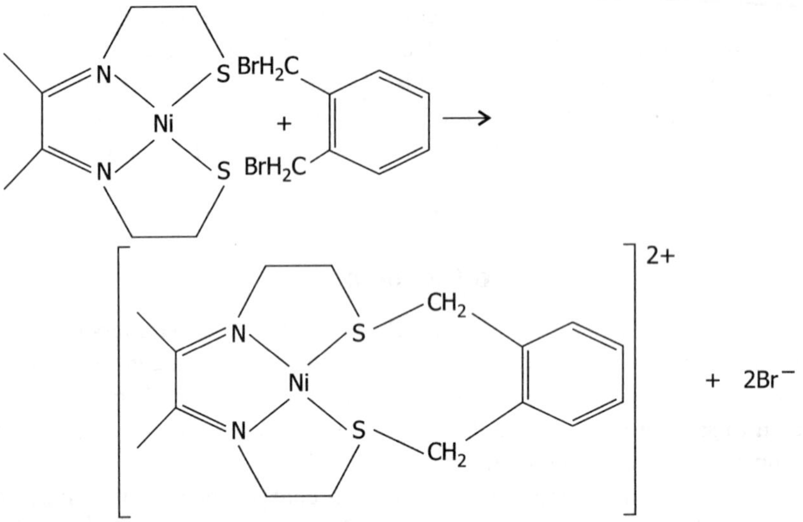

The reaction leading to the macrocyclic complex basically occurs between the groups already coordinated to the metal ion and the organic reagent. However, in the absence of the metal ion, the species would have produced an entirely different product.

3. Formation constants of complexes

Another index of the stability of complexes in solution is the formation constant. Suppose a metal M (charge ignored for convenience) undergoes stepwise combination with a ligand L in aqueous medium. The equilibrium expressions for the reactions may be written as follows.

$$M + L \rightleftharpoons ML \qquad K_1 = \frac{[ML]}{[M][L]}$$

$$ML + L \rightleftharpoons ML_2 \qquad K_2 = \frac{[ML_2]}{[ML][L]}$$

$$ML_2 + L \rightleftharpoons ML_3 \qquad K_3 = \frac{[ML_3]}{[ML_2][L]}$$

and so on. The K_1, K_2, K_3, ... K_f are called *stepwise formation constants* or *stepwise stability constants*.

These equilibrium expressions may be expressed in the following manner too.

$$M + L \rightleftharpoons ML \qquad \beta_1 = \frac{[ML]}{[M][L]}$$

$$M + 2L \rightleftharpoons ML_2 \qquad \beta_2 = \frac{[ML_2]}{[M][L^2]}$$

$$M + 3L \rightleftharpoons ML_3 \qquad \beta_3 = \frac{[ML_3]}{[M][L^3]}$$

and so on. The β_1, β_2, β_3, ... β_f are called *overall formation constants* or *overall stability constants*.

The stepwise and overall stability constants are related to each other. Consider the expression for β_3. On multiplying the numerator and denominator by [ML] [ML_2], and then rearranging, we get,

$$\beta_3 = \frac{[ML_3]}{[M][L_3]} \cdot \frac{[ML][ML_2]}{[ML][ML_2]}$$

$$\beta_3 = \frac{[ML]}{[M][L]} \cdot \frac{[ML_2]}{[ML][L]} \cdot \frac{[ML_3]}{[ML_2][L]}$$

$$\beta_3 = K_1 \,.\, K_2 \,.\, K_3$$

In general

$$\beta_f = K_1 K_2 K_3 ... K_f$$

A higher value of overall stability constant, β_f is achieved if the concentration of the complex species, ML_n is appreciably greater than the concentration of the constituents, [M] and [L]. This, in turn, implies that a complex with a higher value of β_f will be more stable than the one which has a lower value of the corresponding parameter.

However, in general, the values of stepwise stability constants, K_1, K_2, K_3, etc. register a progressive decline. This is borne out by the following example.

$$Cd^{2+} + CN^- \rightleftharpoons [Cd(CN)]^+ \qquad K_1 = 10^{5.48}$$

$$[Cd(CN)]^+ + CN^- \rightleftharpoons [Cd(CN)_2] \qquad K_2 = 10^{5.12}$$

$$[Cd(CN)_2] + CN^- \rightleftharpoons [Cd(CN)_3]^- \qquad K_3 = 10^{4.63}$$

$$[Cd(CN)_3]^- + CN^- \rightleftharpoons [Cd(CN)_4]^{2-} \qquad K_4 = 10^{3.55}$$

$$\beta_4 = K_1 \cdot K_2 \cdot K_3 \cdot K_4 = 10^{18.78}$$

The decrease in the values of stepwise stability constants is attributed to one or a combination of the following three factors.

a. Statistical factors It need be emphasized that the equilibrium expressions cited for stepwise stability constants are over simplified. As has been stated, these reactions occur in aqueous medium. Before the metal ion, M can react with the ligand, L, it becomes solvated to yield $[M(H_2O)_n]$, where n is generally 4 or 6. The reaction between metal and ligand actually involves replacement of aqua moiety, H_2O with the coordinating group, L.

When the first ligand molecule tends to enter the coordination sphere, it has the choice to replace any one of the n aqua ligands. When the second ligand molecule coordinates, the choice reduces to (n – 1) aqua ligands. Thus the probability of substitution statistically declines—and so does the stepwise stability constant.

b. Steric factors When the coordinating groups are bulkier than aqua ligand, steric over-crowdedness will increase with each enterant. As a result of steric hinderance, the values of stepwise stability constants tend to fall.

c. Coulombic factors If the ligands are charged, as are the cyano groups in the example cited above, the inter-ligand repulsions will become important. With each entering group, the inter-ligand repulsion will increase and, in turn, the stepwise stability constant will decrease.

Case Study 25.8 For the silver(I) – ammine system, the second stepwise stability constant has a higher value than the first one. Why?

Silver(I) ion, in dissolved phase, forms a four-coordinated, tetrahedral aqua complex.

$$Ag^+ + 4H_2O \rightleftharpoons [Ag(H_2O)_4]^+$$

The ammine groups then replace aqua ligands in a stepwise manner.

$$[Ag(H_2O)_4]^+ + NH_3 \rightleftharpoons [Ag(H_2O)_3(NH_3)]^+ + H_2O \qquad K_1 = 2.3 \times 10^3$$

$$[Ag(H_2O)_3(NH_3)]^+ + NH_3 \rightleftharpoons [Ag(NH_3)_2]^+ + 3H_2O \qquad K_2 = 6.9 \times 10^3$$

Stability constant K_1 represents the formation of tetrahedral amminetri-aquasilver(I) ion, $[Ag(H_2O)_3(NH_3)]^+$, while K_2 represents the formation of a linear, diamminesilver(I) ion $[Ag(NH_3)_2]^+$. Normally, K_2 is less than K_1. However, in this case K_2 is greater than K_1. The reason being that as the tetrahedral complex formed in the first step, converts to a linear complex in the second step, the bond angles increase from approximately 109° to 180°, the steric repulsion decreases and the stability increases.

---

4. Dissociation constants of complexes

Dissociation constant, K_d is inverse of the formation constant K_f. That is,

$$K_d = \frac{1}{K_f}$$

It represents the dissociation of a complex into its components.

$$[ML] \rightleftharpoons [M] + [L]$$

We had stated in the previous section that a complex having a higher value of formation constant (also called stability constant) is more stable than the one which has a lower value. The reverse is true when stability is to be gauged on the basis of dissociation constant: a lower value of K_d implies greater stability.

For the equilibrium reaction cited above, the dissociation constant may be expressed as,

$$K_d = \frac{[M][L]}{[ML]}$$

A low value of K_d implies that the concentration of the complex, [ML] is greater than the concentration of its constituents, [M] and [L]. This means that the complex dissociates into the concerned metal and the ligands to a very small extent which, in turn, reflects its stability.

As a sample case, let us calculate the stepwise and overall dissociation constants of diamminesilver(I) ion, $\{Ag(NH_3)_2\}^+$. The values are as follows.

Step 1 dissociation: $\{Ag(NH_3)_2\}^+ \rightleftharpoons \{Ag(NH_3)\}^+ + NH_3$

$$K_{d1} = \frac{\left[\{Ag(NH_3)\}^+\right]\left[NH_3\right]}{\left[\{Ag(NH_3)_2\}^+\right]} \tag{25.1}$$

Step 2 dissociation: $\{Ag(NH_3)\}^+ \rightleftharpoons Ag^+ + NH_3$

$$K_{d2} = \frac{\left[Ag^+\right]\left[NH_3\right]}{\left[\{Ag(NH_3)\}^+\right]} \tag{25.2}$$

Overall dissociation: $\{Ag(NH_3)_2\}^+ \rightleftharpoons Ag^+ + 2NH_3$

$$K_d = \frac{\left[Ag^+\right]\left[NH_3\right]^2}{\left[\{Ag(NH_3)_2\}^+\right]} \tag{25.3}$$

On multiplying equation (25.1) by equation (25.2), we get,

$$K_{d1} \times K_{d2} = \frac{\left[\{Ag(NH_3)\}^+\right]\left[NH_3\right]}{\left[\{Ag(NH_3)_2\}^+\right]} \times \frac{\left[Ag^+\right]\left[NH_3\right]}{\left[\{Ag(NH_3)\}^+\right]}$$

$$K_{u1} \times K_{u2} = \frac{\left[Ag^+\right]\left[NH_3\right]^2}{\left[\{Ag(NH_3)_2\}^{'}\right]} \tag{25.4}$$

From equation (25.3) and (25.4),

$$K_d = K_{d1} \times K_{d2}$$

It has been determined experimentally that, $K_{d1} = 1.4 \times 10^{-4}$; $K_{d2} = 4.3 \times 10^{-4}$

$$\text{Therefore, } K_d = (1.4 \times 10^{-4})\,(4.3 \times 10^{-4}) = 6.0 \times 10^{-8}$$

The formation constant of diamminesilver(I) ion, $\{Ag(NH_3)_2\}^+$,

$$K_f = \frac{1}{K_d} = 1.6 \times 10^9$$

Case Study 25.9 Elucidate the concentration of Ag^+, $\{Ag(NH_3)\}^+$ and $\{Ag(NH_3)_2\}^+$ ions in an aqueous solution containing 0.50 M ammonia and 0.04 M silver nitrate. The first step and overall dissociation constants of diamminesilver(I) ion are 1.4×10^{-4} and 6.0×10^{-8}, respectively.

Since the concentration of aqueous ammonia is far more than that of silver nitrate, it is pertinent to assume that silver(I) ions in the solution would be completely transformed into diamminesilver(I) ion, $\{Ag(NH_3)_2\}^+$.

Therefore, $[\{Ag(NH_3)_2\}^+] = 0.04$ M

In this transformation, two moles of ammonia are consumed for one mole of silver(I) ion.

Concentration of ammonia consumed = (0.04 × 2) = 0.08 M

Concentration of unreacted ammonia = (0.50 – 0.08) = 0.42 M

The overall dissociation constant may be expressed as,

$$K_d = \frac{[Ag^+][NH_3]^2}{[\{Ag(NH_3)_2\}^+]} = 6.0 \times 10^{-8}$$

$$\frac{[Ag^+](0.42)^2}{(0.04)} = 6.0 \times 10^{-8}$$

$$[Ag^+] = 1.36 \times 10^{-8} \text{ M}$$

The first step dissociation constant may be expressed as,

$$K_{d1} = \frac{[\{Ag(NH_3)\}^+][NH_3]}{[\{Ag(NH_3)_2\}^+]} = 1.4 \times 10^{-4}$$

$$\frac{[\{Ag(NH_3)\}^+](0.42)}{(0.04)} = 1.4 \times 10^{-4}$$

$$[\{Ag(NH_3)\}^+] = 1.47 \times 10^{-3} \text{ M}$$

The concentration of the ions present in silver(I) ion-ammonia system are as follows: $[Ag^+] = 1.36 \times 10^{-8}$ M; $[\{Ag(NH_3)\}^+] = 1.47 \times 10^{-3}$ M; and $[\{Ag(NH_3)_2\}^+] = 0.04$ M.

When the relative stabilities of two or more complexes are to be compared, dissociation constant is a better parameter than the formation constant. The dissociation constant depends on the following factors.

a. Oxidation state of the metal ion A metal ion in a higher valent state holds the ligands more tightly than the one in a lower valent state. As a result, a complex in which the metal is in a higher oxidation state will have a smaller dissociation constant, and greater stability, as compared to an analog in which the metal is in a lower oxidation state. For example,

Complex	$[Fe(C_2O_4)_3]^{4-}$	$[Fe(C_2O_4)_3]^{3-}$
K_d	6×10^{-6}	3×10^{-21}

The trioxalatoiron(III) ion is more stable than the trioxalatoiron(II) ion.

Case Study 25.10 The dissociation constants, K_d of $[CdCl_4]^{2-}$ and $[CdI_4]^{2-}$ are 9.3×10^{-3} and 8×10^{-7} respectively. Explain the variation.

As has already been stated, a complex in which a soft ligand is coordinated to a soft metal ion will be more stable than the one in which there is hard-soft or soft-hard interaction between the metal ion and the ligands. The cadmium(II) ion is quite soft. Its complex with the soft iodide ion has a smaller dissociation constant and greater stability as compared to the corresponding complex with relatively hard chloride ion.

b. Chelation Chelated complexes have a greater stability than non-chelated ones. The dissociation constants for the former are smaller as compared to the latter. For example,

Complex	$[Cd(en)_2]^{2+}$	$[Cd(CH_3NH_2)_4]^{2+}$
K_d	2.60×10^{-11}	2.82×10^{-7}

The bis(ethylenediamine)cadmium(II) ion is more stable than the tetrakis(methylamine)cadmium(II) ion.

Case Study 25.11 Which complex is expected to be more stable: diglycinatozinc (II) or ethylenediaminetetraacetatozincate(II) ion?

Both the complexes, diglycinatozinc(II), $[Zn(gly)_2]$ and ethylenediamine tetraacetatozincate(II) ion, $[Zn(EDTA)]^{2-}$ have five-memberd chelate rings in their structures. Yet, the latter complex is more stable than the former. This is attributed to two reasons.

Firstly, glycinato ligand is bidentate, while ethylenediaminetetraacetato ligand is hexadentate. Chelate effect states that greater the polydentate character of the ligand, greater will be the stability of its complex.

Secondly, while glycinato ligand forms non-fused rings, the ethylenediaminetetraacetato ligand forms fused rings. Complexes with non-fused rings are relatively more prone to dissociation than their counterparts with fused rings. A comparison of dissociation constants of the two complexes supports this assertion.

Complex	$[Zn(gly)_2]$	$[Zn(EDTA]^-$
K_d	1.1×10^{-10}	2.63×10^{-17}

The $[Zn(EDTA)]^-$ complex has smaller dissociation constant and greater stability than $[Zn(gly)_2]$ complex.

5. Effective atomic number rule

The effective atomic number rule or EAN rule states that those complexes which have as many electrons as noble gases are very stable. If the number of electrons donated by the ligands plus the number of electrons on the metal add on to 18 (argon), 36 (krypton), 54 (xenon) or 86 (radon), then the EAN rule is said to be obeyed.

Consider the application of EAN rule to carbonyl complexes of chromium and nickel.

$Cr(CO)_6$			$Ni(CO)_4$		
Cr	=	$24e^-$	Ni	=	$28e^-$
$6CO : 6 \times 2e^-$	=	$12e^-$	$4CO : 4 \times 2e^-$	=	$8e^-$
		$36e^-$			$36e^-$

Alternately, consider only the valence electrons of the metal rather than the total number of electrons.

$Cr(CO)_6$			$Ni(CO)_4$		
$Cr : 3d^54s^1$	=	$6e^-$	$Ni : 3d^84s^2$	=	$10e^-$
$6CO : 6 \times 2e^-$	=	$12e^-$	$4CO : 4 \times 2e^-$	=	$8e^-$
		$18e^-$			$18e^-$

When this criterion is followed, all the complexes which obey the EAN rule possess 18 electrons. For this reason, this rule is sometimes also referred to as the *18-electron rule.*

The EAN rule is obeyed frequently by carbonyl compounds. In fact, it may be used to predict the composition of the carbonyl complexes. For example, iron carbonyl, $Fe(CO)_n$ shall follow EAN rule and, therefore, possess stability if the value of n is five.

$Fe(CO)_5$

Fe	=	$26e^-$	or	$Fe : 3d^64s^2$	=	$8e^-$
$5CO : 5 \times 2e^-$	=	$10e^-$		$5CO : 5 \times 2e^-$	=	$10e^-$
		$36e^-$				$18e^-$

Examples of complexes which follow EAN rule are highlighted in Box 25.1.

Box 25.1

Complexes Obeying the EAN Rule

$[Co(NH_3)_6]^{3+}$

Co(III)	=	$24e^-$
$6NH_3 : 6 \times 2e^-$	=	$12e^-$
		$36e^-$

$[Fe(CN)_6]^{4-}$

Fe(II)	=	$24e^-$
$6CN : 6 \times 2e^-$	=	$12e^-$
		$36e^-$

$[Cr(C_6H_6)_2]$

Cr(0)	=	$24e^-$
$2C_6H_6 : 2 \times 6e^-$	=	$12e^-$
		$36e^-$

$[Fe(C_5H_5)_2]$

Fe(0)	=	$26e^-$
$2C_5H_5 : 2 \times 5e^-$	=	$10e^-$
		$36e^-$

$[CoF_6]^{3-}$

Co(III)	=	$24e^-$
$6F : 6 \times 2e^-$	=	$12e^-$
		$36e^-$

$[NiCl_4]^{2-}$

Ni(II)	=	$28e^-$
$4Cl : 4 \times 2e^-$	=	$8e^-$
		$36e^-$

$[Zn(CN)_4]^{2-}$

Zn(II)	=	$28e^-$
$4CN^- : 4 \times 2e^-$	=	$8e^-$
		$36e^-$

$[Pd(NH_3)_6]^{4+}$

Pd(IV)	=	$42e^-$
$6NH_3 : 6 \times 2e^-$	=	$12e^-$
		$54e^-$

$[Rh(H_2O)_6]^{3+}$

Rh(III)	=	$42e^-$
$6H_2O : 6 \times 2e^-$	=	$12e^-$
		$54e^-$

$[Ru(NH_3)_5(N_2)]^{2+}$

Ru(II)	=	$42e^-$
$5NH_3 : 5 \times 2e^-$	=	$10e^-$
N_2	=	$2e^-$
		$54e^-$

$[PtCl_6]^{2-}$

Pt(IV)	=	$74e^-$
$6Cl : 6 \times 2e^-$	=	$12e^-$
		$86e^-$

$[IrCl_6]^{3-}$

Ir(III)	=	$74e^-$
$6Cl : 6 \times 2e^-$	=	$12e^-$
		$86e^-$

Case Study 25.12 Which is expected to be more stable: $[Cu(NH_3)_4]^{2+}$ or $[Cu(CN)_4]^{3-}$?

The stability of the two complexes may be gauged on the basis of EAN rule.

$[Cu(NH_3)_4]^{2+}$

Cu(II)	=	$27e^-$
$4NH_3 : 4 \times 2e^-$	=	$8e^-$
		$35e^-$

$[Cu(CN)_4]^{3-}$

Cu(I)	=	$28e^-$
$4CN : 4 \times 2e^-$	=	$8e^-$
		$36e^-$

The $[Cu(CN)_4]^{3-}$ complex obeys the EAN rule and, therefore, is more stable.

Metals or metal ions with odd number of electrons cannot satisfy the EAN rule. This is because each ligand normally donates two electrons to the metal. If the metal has an odd atomic number, the total number of electrons in the complex would also be odd, irrespective of the number of ligands. The noble gas configuration requires an even number of electrons.

Such complexes tend to satisfy the EAN rule by varied options. Consider, for example, the complex species, $[Mn(CO)_5]$.

$[Mn(CO)_5]$

Mn	=	$25e^-$
5CO : $5 \times 2e^-$	=	$10e^-$
		$35e^-$

There are a total of 35 electrons in this complex—one less than that demanded by EAN rule. For this reason, this complex is so unstable that it does not exist. However, it may accept an electron from a reducing agent and form an anion, $[Mn(CO)_5]^-$ which obeys the EAN rule.

$[Mn(CO)_5]^-$

Mn	=	$25e^-$
5CO : $5 \times 2e^-$	=	$10e^-$
Negative charge	=	$1e^-$
		$36e^-$

Alternately, it may combine with an alkyl radical ($R^{\cdot}$) or a halide radical ($X^{\cdot}$).

$[RMn(CO)_5]$

Mn	=	$25e^-$
5CO : $5 \times 2e^-$	=	$10e^-$
$R^{\cdot}$	=	$1e^-$
		$36e^-$

$[Mn(CO)_5X]$

Mn	=	$25e^-$
5CO : $5 \times 2e^-$	=	$10e^-$
$X^{\cdot}$	=	$1e^-$
		$36e^-$

Another option is that two $[Mn(CO)_5]$ units, each possessing an odd electron, dimerize. One electron on each manganese atom is involved in the formation of metal–metal bond.

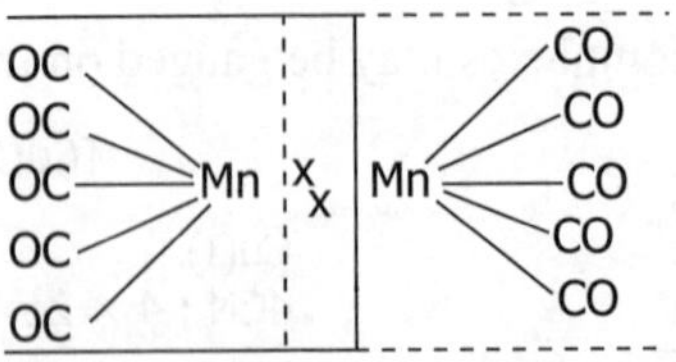

Mn	=	$24e^-$
5CO : $5 \times 2e^-$	=	$10e^-$
Mn–Mn bond	=	$2e^-$
		$36e^-$

Mn	=	$24e^-$
5CO : $5 \times 2e^-$	=	$10e^-$
Mn–Mn bond	=	$2e^-$
		$36e^-$

There is thus noble gas configuration around each manganese. This means that each $[Mn(CO)_5]$ unit obeys EAN rule.

Hexacarbonylvanadium(0), $V(CO)_6$ is not a very stable complex. It decomposes at 350 K. This is because it does not obey the EAN rule.

V	=	$23e^-$
6CO : $6 \times 2e^-$	=	$12e^-$
		$35e^-$

The complex does not dimerize, first because too much energy is required to rearrange the stable octahedral geometry and second coordination number around vanadium would increase to seven causing, in turn, steric over-crowdedness. Nevertheless, the complex readily reacts with sodium.

$$Na + V(CO)_6 \rightarrow Na^+[V(CO)_6]^-$$

The resulting anion follows the EAN rule.

$[V(CO)_6]^-$

V	=	$23e^-$
6CO : $6 \times 2e^-$	=	$12e^-$
Negative charge	=	$1e^-$
		$36e^-$

Similarly, $[Co(CO)_4]$ which does not follow EAN rule is non-existent, but $[HCo(CO)_4]$, $[Co(CO)_4]^-$ and $[Co_2(CO)_8]$ species, which obey the rule, are stable.

$$[HCo(CO)_4] \xleftarrow{H^{\bullet}} [Co(CO)_4] \xrightarrow{Na} Na^+[Co(CO)_4]^-$$

Co	=	$27e^-$	Co	=	$27e^-$	Co	=	$27e$
4CO : $4 \times 2e^-$	=	$8e^-$	4CO : $4 \times 2e^-$	=	$8e^-$	4CO : $4 \times 2e^-$	=	$8e^-$
H·	=	$1e^-$			$35e^-$	Negative charge	=	$1e^-$
		$36e^-$						$36e^-$

$[Co(CO)_4]$ ↓ Dimerization

$[Co_2(CO)_8]$

Each $[Co(CO)_4]$

Co	=	$26e^-$
4CO : $4 \times 2e^-$	=	$8e^-$
Co–Co bond	=	$2e^-$
		$36e^-$

Case Study 25.13 What are *piano stool compounds*?

A number of mixed hydrocarbon–carbonyl complexes possess a molecular skeleton that resembles a piano stool. These complexes are, therefore, often referred to as piano stool compounds. The most common example of such compounds is tricarbonyl(cyclopentadienyl)manganese(I).

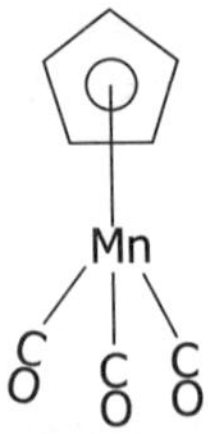

It follows EAN rule, and, therefore, is very stable.

Mn(I)	= 24e$^-$
$C_5H_5^-$	= 6e$^-$
3CO : 3 × 2e$^-$	= 6e$^-$
	36e$^-$

6. Sixteen electron rule

We have already stated that complexes having a total of 18 valence electrons about the metal or metal ion possess the effective atomic number of the next higher inert gas and hence are very stable. It has been observed that complexes having sixteen electrons about the metal are often as stable or, at times, more stable than 18-electron molecules. This is referred to as the 16-electron rule.

The 16-electron rule is predominantly followed by complexes with metal or metal ions belonging to the right end of 4d and 5d transition element series. These include rhodium, palladium, iridium and platinum. These metals prefer to form complexes with coordination number 4, rather than coordination number six.

Consider, for example, the complexes of iridium(I), palladium(II) and platinum(II) ions. These metal ions have a d^8 configuration.

During complexation, a ligand normally donates two electrons to the metal ion. This means that if a metal ion with d^8 configuration has to have 18 electrons in its valence shell, it must combine with five ligands. However, in general, complexes with coordination number 4 (square planar or tetrahedral) are more stable than their counterparts with coordination number 5 (trigonal bipyramid or square pyramid). Therefore, these metal ions prefer to form complexes with ML_4, rather than ML_5, compositions. The stability associated with coordination number 4 validates the 16-electron rule.

Here are enunciated, a few examples of stable complexes which follow 16-electron rule.

a. Carbonylchlorobis(triphenylphosphine)iridium(I), $[IrCOCl\{P(C_6H_5)_3\}_2]$

Ir(I)	= 8e$^-$
CO	= 2e$^-$
Cl$^-$	= 2e$^-$
$2P(C_6H_5)_3$: 2 × 2e$^-$	= 4e$^-$
	16e$^-$

b. Diamminedichloroplatinum(II), $[Pt(NH_3)_2Cl_2]$

Pt(II)	=	$8e^-$
$2NH_3 : 2 \times 2e^-$	=	$4e^-$
$2Cl^- : 2 \times 2e^-$	=	$4e^-$
		$16e^-$

c. Trichloro(ethene)platinate(II) ion (Zeise's salt), $[PtCl_3(C_2H_2)]^-$

Pt(II)	=	$8e^-$
$3Cl^- : 3 \times 2e^-$	=	$6e^-$
C_2H_2	=	$2e^-$
		$16e^-$

In some cases, the 16-electron configuration imparts more stability to the complex than 18-electron configuration. For example, the square planar, tetrakis(triethylphosphine)platinum(0) complex, which follows the 18-electron rule, readily decomposes with the loss of one triethylphosphine ligand. The resulting, tris(triethylphosphine) complex has a trigonal geometry and follows 16-electron rule.

$$Pt[\{P(C_6H_5)_3\}_4] \longrightarrow Pt[\{P(C_6H_5)_3\}_3] + P(C_6H_5)_3$$

Square planar			Triangular		
Pt(0)	=	$10e^-$	Pt(0)	=	$10e^-$
$4P(C_6H_5)_3 : 4 \times 2e^-$	=	$8e^-$	$3P(C_6H_5)_3 : 3 \times 2e^-$	=	$6e^-$
$18e^-$					$16e^-$

In this case, steric factor plays an overriding part. The ligand-ligand repulsions between bulky triphenylphosphine groups goes in favor of coordination number 3—and 16-electron rule. (*see* Case Study 25.14).

Case Study 25.14 What is *Tolman cone angle*?

Tolman cone angle is a measure of the bulkiness of a ligand. It refers to the angle subtended by the skeleton of the coordination group. As the size of the ligand increases, the Tolman cone angle becomes larger (Fig. 25.3).

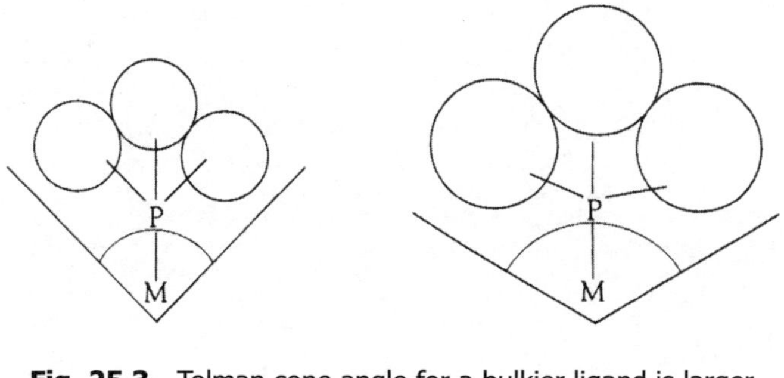

Fig. 25.3 Tolman cone angle for a bulkier ligand is larger than that of a smaller ligand

Table 25.3 lists the Tolman cone angles for a few representative phosphine-based ligands.

Table 25.3 Tolman cone angles for phosphine-based ligands

Ligand	Angle (°)
PF_3	104
$P(CH_3)_3$	118
$P(CH_3)_2 (C_6H_5)$	122
$P(CH_3) (C_6H_5)_2$	136
$P(C_6H_5)_3$	145

It is evident that the Tolman cone angle of triphenylphosphine is relatively large. It is for this reason that the tetrakis(triphenylphosphine)platinum(0) complex is under substantial steric stress and, therefore, despite being an 18-electron system, easily decomposes into tris(triphenylphosphine)platinum(0) (reaction cited above).

26. THEORIES OF METAL–LIGAND BONDING

Introduction

In this chapter, we outline three theories which explain the nature of metal–ligand bonding in complexes. These are, valence bond theory, crystal field theory and molecular orbital theory. In this order, each theory represents an improvement over its predecessor.

Valence Bond Theory

The valence bond theory of coordination compounds is based on the concept of hybridization. It propounds that a few vacant atomic orbitals on the central metal atom first hybridize, and then accept a suitable number of lone pairs of electrons from the ligands. The theory also explains the geometry and magnetic properties of complexes.

The types of hybridization that occur among complexes with coordination numbers 4 and 6 are listed in Table 26.1

Table 26.1 Correlation of hybridization with coordination number and geometry of complexes

Coordination number	Geometry	Hybridization
6	Octahedral	d^2sp^3 or sp^3d^2
4	Square planar	dsp^2
	Tetrahedral	sp^3

The total number of metal atomic orbitals which undergo hybridization is equivalent to the coordination number of the complex. Thus for coordination number 6 complexes, two d, one s and three p orbitals, i.e., a total of six orbitals of the metal are involved in hybridization. These orbitals should be vacant, since only then would each of them be able to accommodate a pair of electrons from each ligand. Similarly, for square planar geometry, one vacant d, one vacant s and two vacant p orbitals of the metal, i.e., a total of four orbitals, are involved in hybridization.

Coordination number 6

The valence bond theory proposes the formation of a complex in a number of steps. For example, consider hexamminecobalt(III) ion, $[Co(NH_3)_6]^{3+}$.

Step 1. Write the valence shell configuration of the metal.

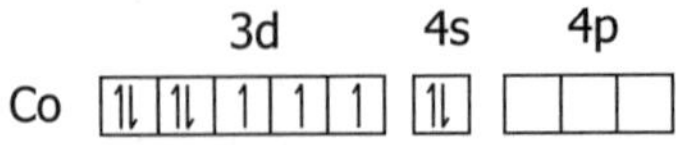

Step 2. Write the valence shell configuration of the metal ion.

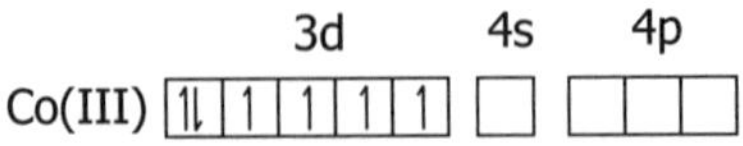

Step 3. Write the configuration of the complex. It is evident from the composition of the complex that the coordination number of metal atom is 6. There has to be either d^2sp^3 or sp^3d^2 hybridization of atomic orbitals on the metal. The requisite number of vacant s and p orbitals are available for hybridization. However, there are no vacant d orbitals, and two of them are required for octahedral coordination.

The valence bond theory assumes that the metal atom, in this case cobalt, is positioned at the center of an octahedron. The six ligands, in this case ammonia molecules, approach the metal from infinity along arbitrarily chosen x, y and z axes (Fig. 26.1).

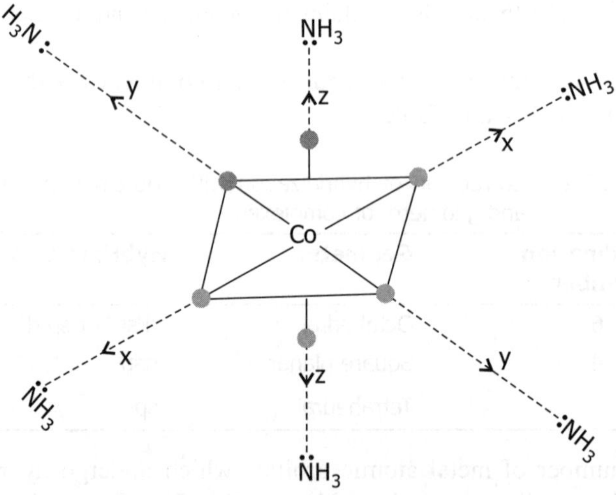

Fig. 26.1 Structure of hexamminecobalt(III) ion

Ultimately these ligands will settle down at the corners of the octahedron (shown as circles in Fig. 26.1). Each ammonia molecule has a lone pair of electrons on the nitrogen. The lone pair shall be donated to the vacant hybrid orbitals on the metal. There are six pairs of electrons and these can be accommodated only if there are six vacant orbitals on the metal.

As the ammonia molecules come closer to the corners of the octahedron, the six electrons in the 3d orbital of Co re-align themselves. This re-alignment is against

Hund's rule, but it renders two d orbitals vacant. The electronic configuration of the metal ion, immediately before hybridization, may be pictured as follows:

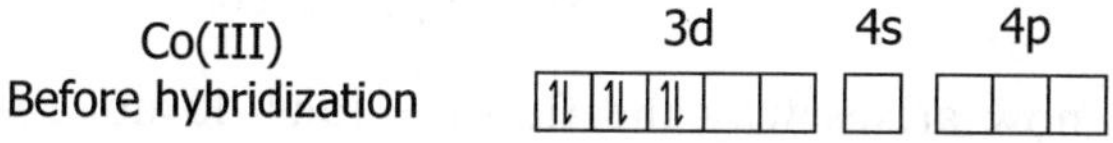

Two d, one s and three p orbitals then undergo d^2sp^3 hybridization. Six degenerate d^2sp^3 hybrid orbitals are formed. Each of them accepts the lone pair of electrons on nitrogen of each ammonia molecule. The final picture may be envisioned as follows:

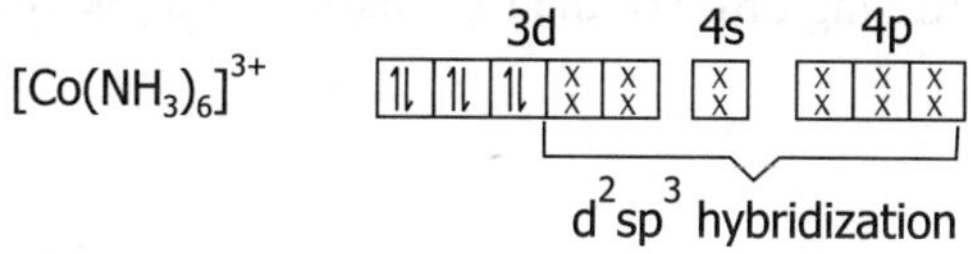

The three steps for arriving at the valence bond configuration of the complex are hypothetical. Hypotheses are an integral part of any theory. However, the hypotheses are validated only if theory can be experimentally verified. In the present case, X-ray crystallography proves that the complex is indeed octahedral and magnetic moment studies indicate that the complex is diamagnetic.

Let us now consider another complex, hexafluorocobaltate(III) ion, $[CoF_6]^{3-}$.

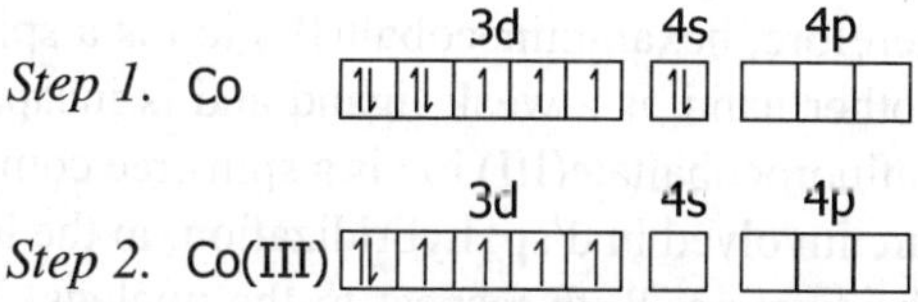

Step 3. Before we theorize the configuration of the complex, we would examine what the experimental findings tell us. X-ray results reveal that there is octahedral coordination around the metal. The magnetic moment measurements tally with the presence of four unpaired electrons. This means that we cannot re-align the four unpaired electrons in the 3d orbital of cobalt(III) ion. Yet we need two vacant d orbitals for octahedral coordination. It is, therefore, pertinent that we use two d orbitals of the 4d subshell.

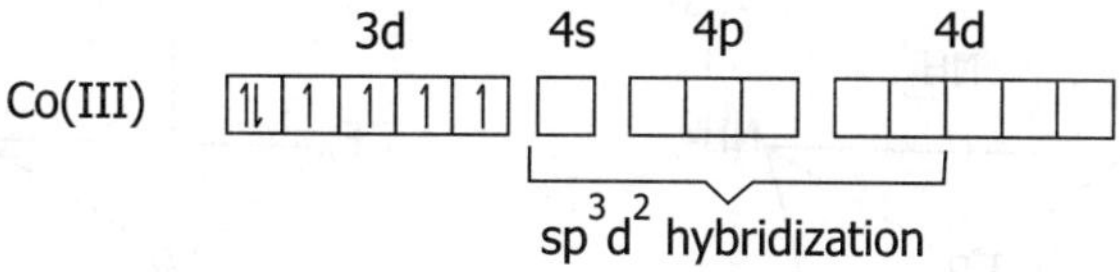

In this case, the metal atomic orbitals undergo sp^3d^2 hybridization. The fluoride ion has four pairs of valence electrons. One of them is donated by each fluoride to each of the sp^3d^2 hybrid orbital. The configuration of the complex may be written as,

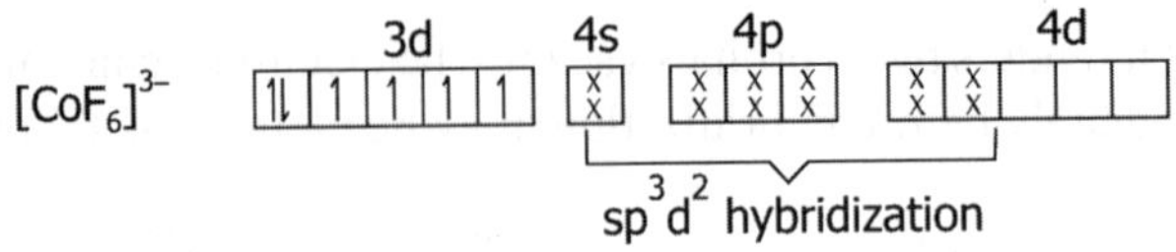

The question now arises why the fluoride ions do not re-align the 3d electrons against Hund's rule, much the same way as ammonia molecules do in hexamminecobalt(III) ion? It is because ammonia is a strong ligand and can pair-up the electrons, while fluoride being weak cannot do so. This begets another question: how to judge which ligands are strong and which are weak? The *spectrochemical series* assists us in resolving this issue. The series is an arrangement of ligands in the decreasing order of their strength. It may be roughly depicted as follows:

$$CO > CN^- > NO_2^- > en > py \approx NH_3 > H_2O > OH^- > F^- > Cl^- > Br^- > I^-$$

Strong ligands | Weak ligands

While extending the valence bond theory to the complexes, we are not interested in knowing the relative strength of one ligand with respect to the other. We need only to club the ligands into two broad categories: strong and weak. Ligands to the left of water are strong, while water and ligands to its right are weak.

Ammonia, being a strong ligand, can pair-up the electrons in the valence shell of cobalt(III) ion. Therefore, hexamminecobalt(III) ion is a spin-paired complex. Fluoride ion, on the other hand, is a weak ligand and is incapable of pairing the spins. Therefore, hexafluorocobaltate(III) ion is a spin-free complex. In the former case, the 3d orbitals are involved in d^2sp^3 hybridization; in the latter, 4d orbital are involved in sp^3d^2 hybridization. With respect to the nucleus, 3d is inside the 4d orbital. Therefore, hexamminecobalt(III) ion is an *inner sphere* (or *inner orbital*) complex, while hexafluorocobaltate(III) ion is an *outer sphere* (or *outer orbital*) complex.

We shall review the parameters of the two cases to which we have applied the valence bond theory.

$[Co(NH_3)_6]^{3+}$

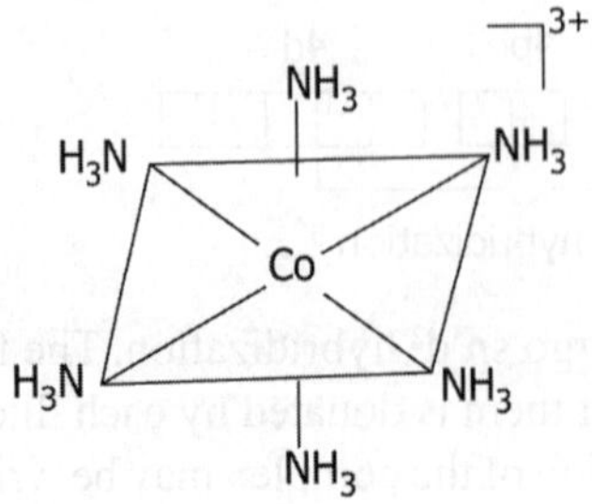

$[CoF_6]^{3-}$

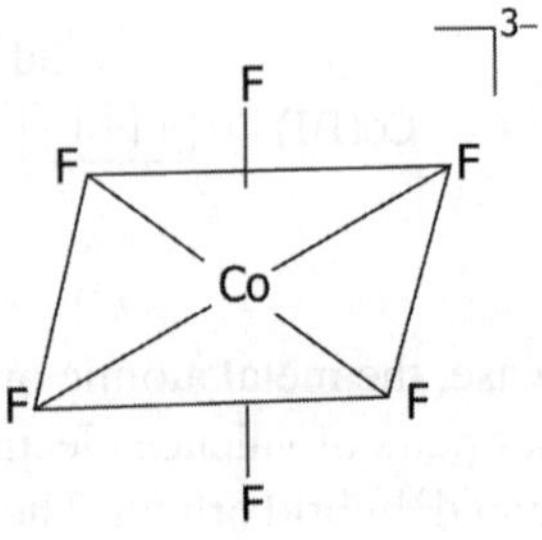

- Octahedral
- Diamagnetic
- Spin-paired complex
- Inner sphere complex

- Octahedral
- Paramagnetic, with four unpaired electrons
- Spin-free complex
- Outer sphere complex

We shall now study a few more cases in the light of valence bond theory. However, henceforth we shall not number the steps necessary for reaching the configuration of the complex.

1. Hexacyanoferrate(II) ion, $[Fe(CN)_6]^{4-}$

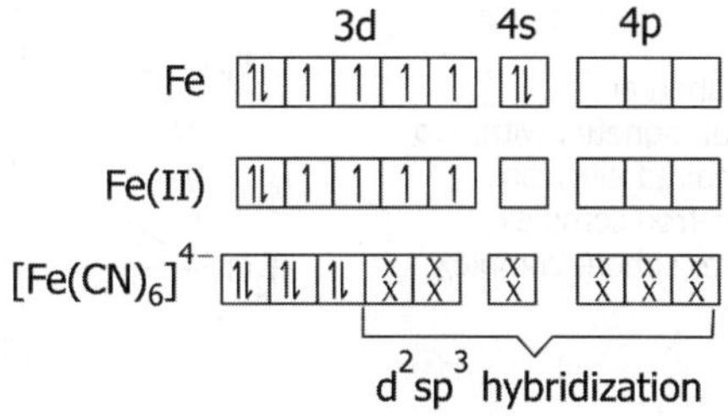

- Octahedral
- Diamagnetic
- Spin-paired complex
- Inner sphere complex

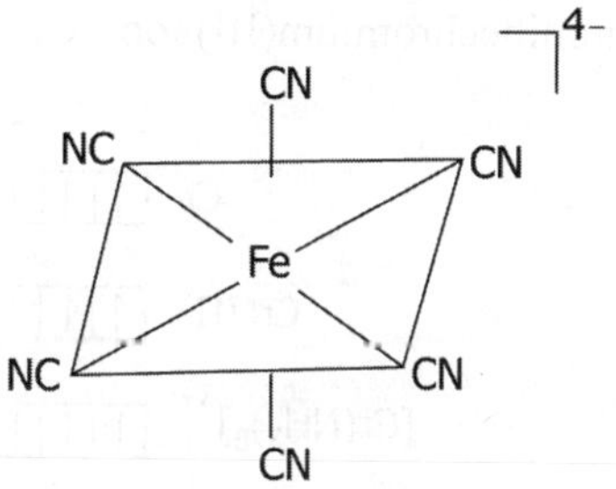

2. Hexacyanoferrate(III) ion, $[Fe(CN)_6]^{3-}$

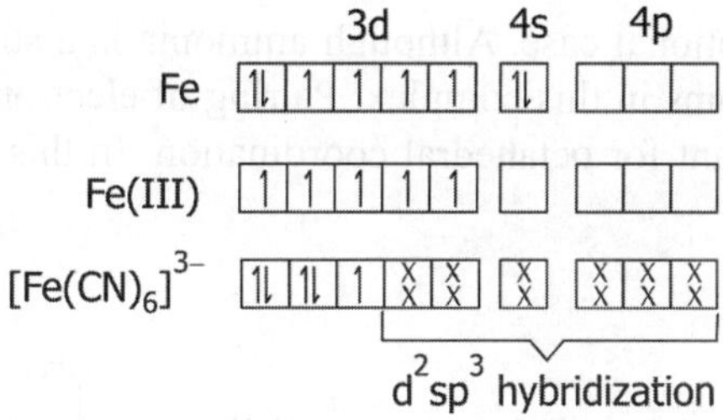

- Octahedral
- Paramagnetic, with one unpaired electron
- Spin-paired complex
- Inner sphere complex

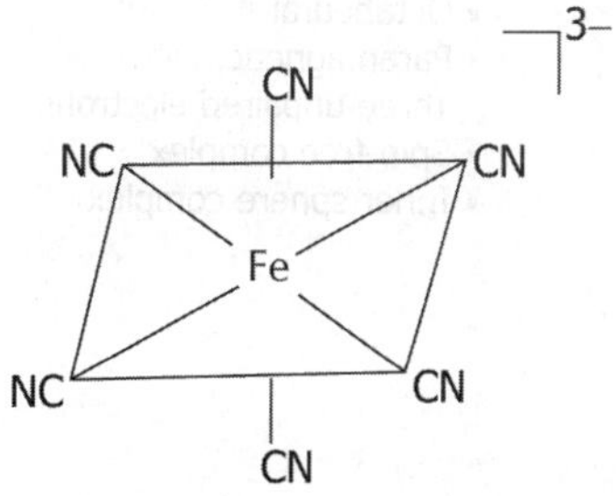

3. Hexaquanickel(II) ion, $[Ni(H_2O)_6]^{2+}$

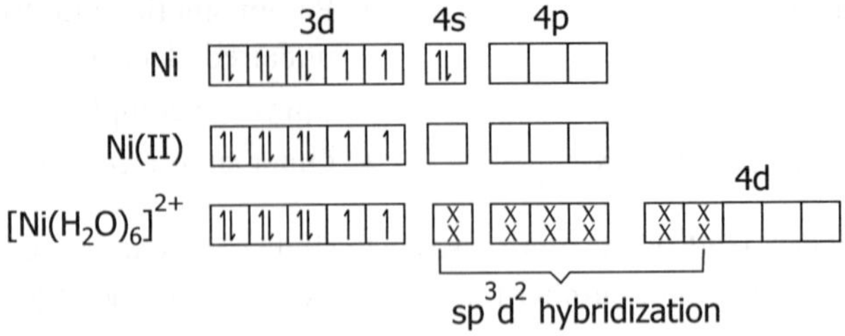

- Octahedral
- Paramagnetic, with two unpaired electrons
- Spin-free complex
- Outer sphere complex

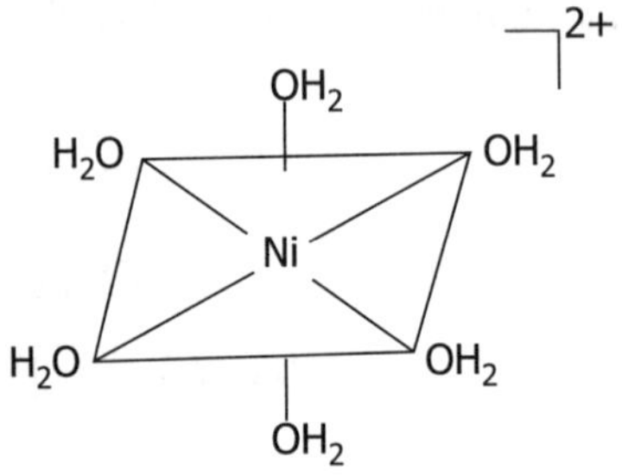

4. Hexamminechromium(III) ion, $[Cr(NH_3)_6]^{3+}$

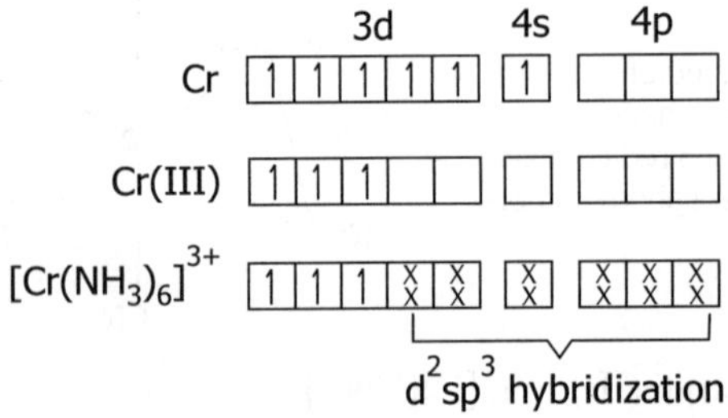

This is an exceptional case. Although ammonia is a strong ligand, it does not cause pairing of spins in this complex. Pairing of electrons takes place to render two d orbitals vacant for octahedral coordination. In this case, these are already vacant.

- Octahedral
- Paramagnetic, with three unpaired electrons
- Spin-free complex
- Inner sphere complex

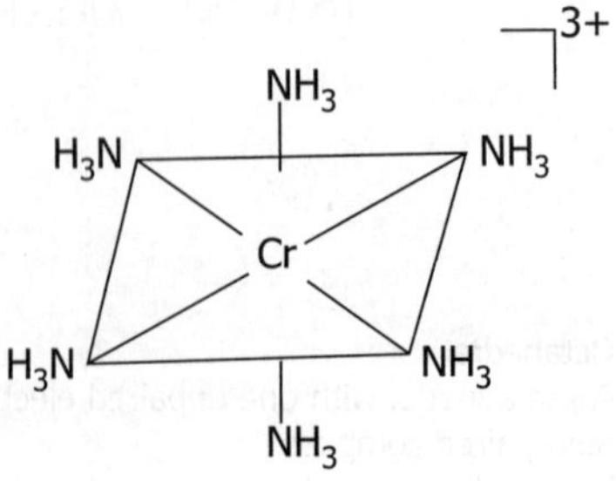

5. Hexamminenickel(II) ion, $[Ni(NH_3)_6]^{2+}$

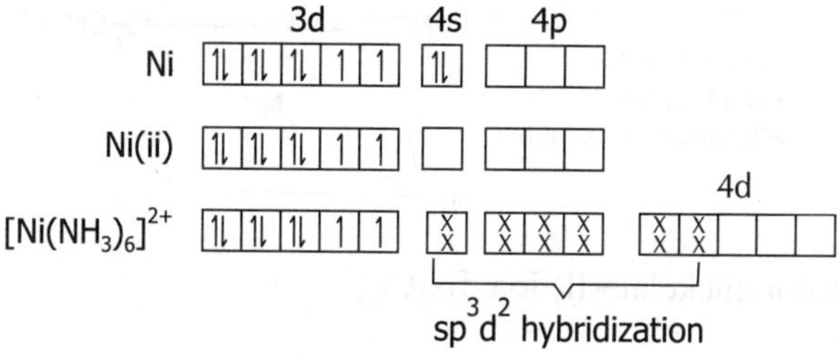

This too is an exceptional case. Even if ammonia ligands were to pair-up the spins, only one 3d orbital would have been rendered vacant. Since two vacant d orbitals are necessary for octahedral coordination, an outer orbital complex is formed, utilizing two d orbitals of the 4d subshell.

- Octahedral
- Paramagnetic, with two unpaired electrons
- Spin-free complex
- Outer sphere complex

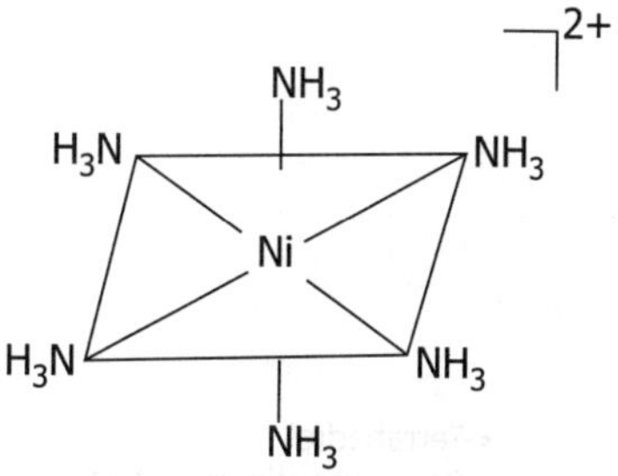

Coordination number 4

Four ligands around a central metal atom can give either a square planar or a tetrahedral arrangement. The geometry which will prevail depends on the nature of the ligand and configuration of the metal atom. In general, strong ligands tend to pair-up metal electrons and give square planar complexes, while weak ligands give high-spin, tetrahedral complexes. However, there are notable exceptions to this generalization. We will now study a few examples.

1. Tetracyanonickelate(II) ion, $[Ni(CN)_4]^{2-}$

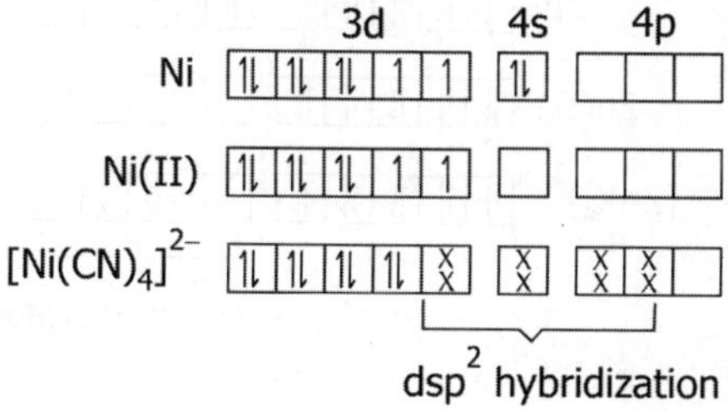

- Square planar
- Diamagnetic
- Spin-paired complex

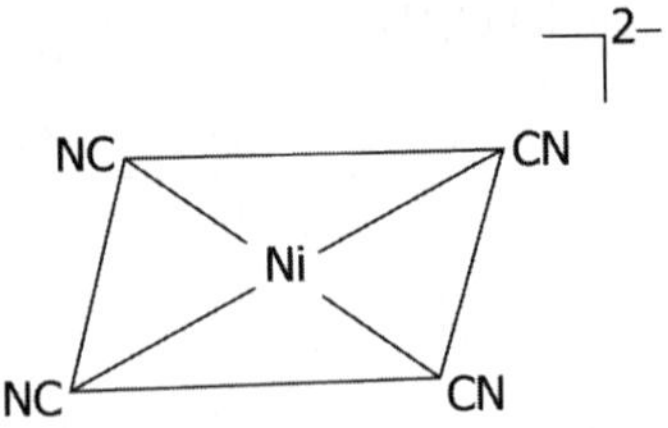

2. Tetrachloronickelate(II) ion, $[NiCl_4]^{2-}$

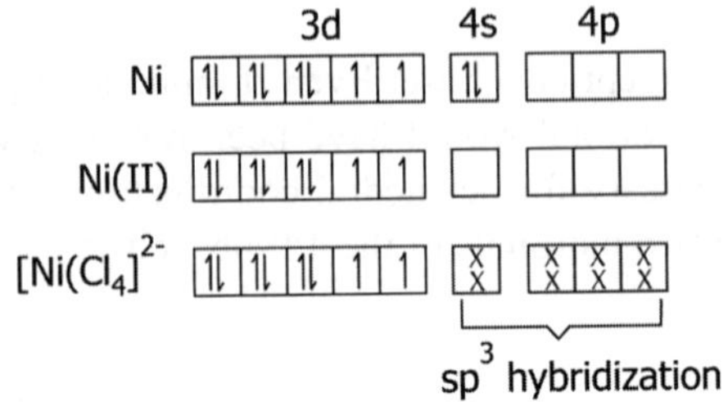

- Tetrahedral
- Paramagnetic, with two unpaired electrons
- Spin-free complex

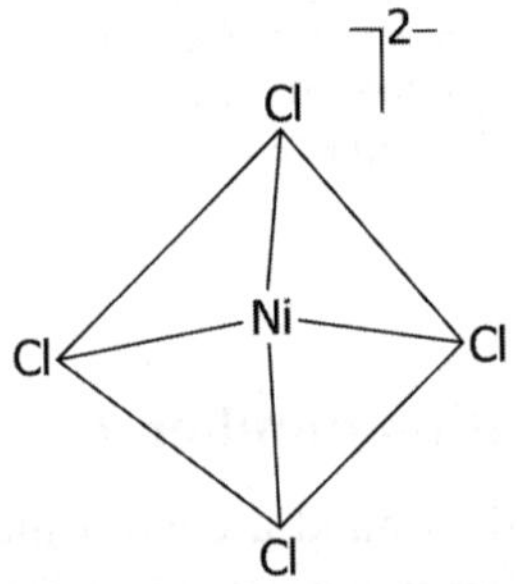

3. Tetracyanozincate(II) ion/tetrachlorozincate(II) ion, $[ZnX_4]^{2-}$ (X = CN or Cl) This case will demonstrate that the geometry of the complex also depends on the electronic configuration of the metal.

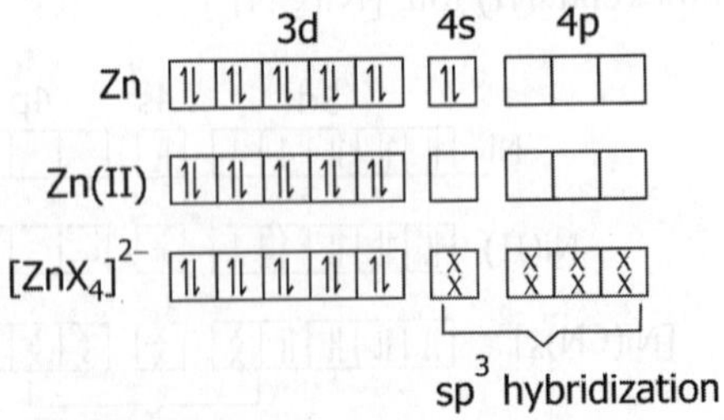

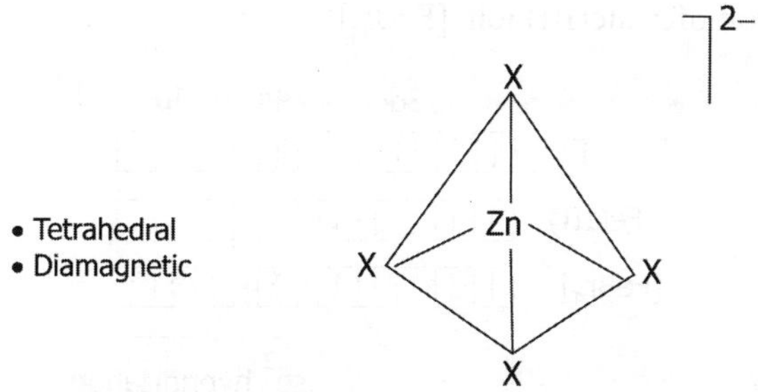

Since the 3d orbitals of zinc(II) ion are completely filled, dsp^2 hybridization is not possible. All the complexes of Zn, with coordination number 4, are tetrahedral in geometry, irrespective of whether the ligands are strong or weak.

4. Tetracarbonylnickel(0), $[Ni(CO)_4]$
   The carbonyl group is a very strong ligand and, therefore, it is expected that the complex would be square planar. However, the spin pairing caused by carbonyl units results in a completely filled 3d orbital, ruling out the possibility of dsp^2 hybridization. Thus, the complex has tetrahedral geometry.

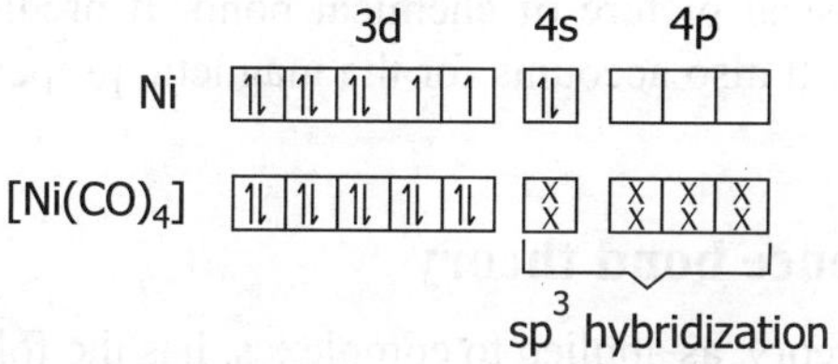

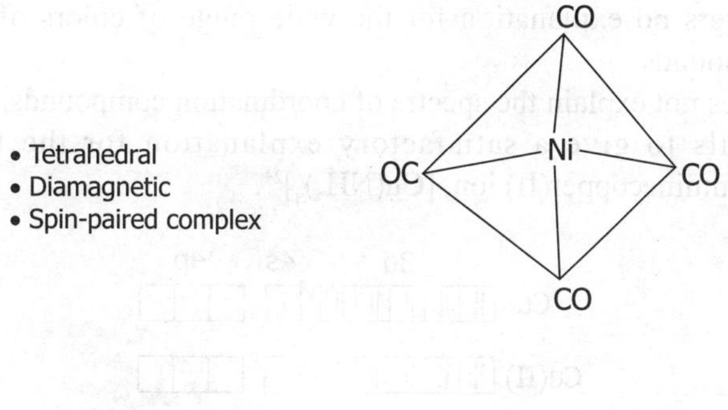

4. Tetrabromoferrate(III) ion, $[FeBr_4]^-$

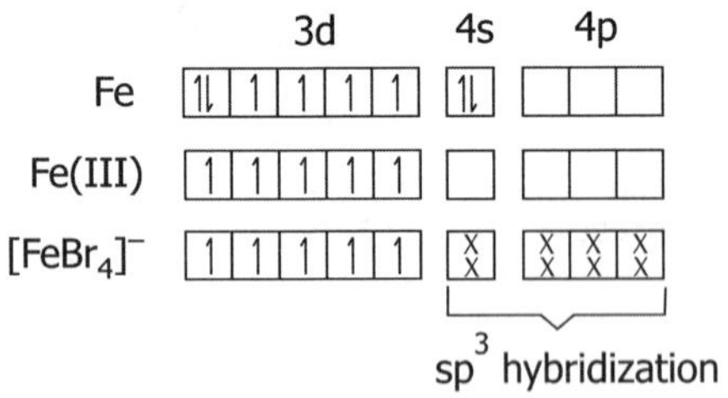

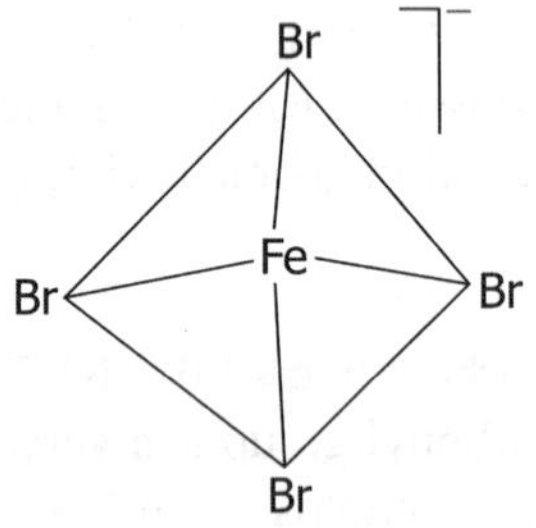

- Tetrahedral
- Paramagnetic, with five unpaired electrons
- Spin-free complex

Merits of valence bond theory

The valence bond theory presents a simple description of metal–ligand bonding, in terms of the classical picture of chemical bond. It predicts the geometry of complexes correctly. It also accounts for the magnetic properties of coordination compounds.

Demerits of valence bond theory

The valence bond theory, as applied to complexes, has the following drawbacks:

1. It offers no explanation for the wide range of colors of coordination compounds.
2. It does not explain the spectra of coordination compounds.
3. It fails to give a satisfactory explanation for the formation of tetramminecopper(II) ion, $[Cu(NH_3)_4]^{2+}$.

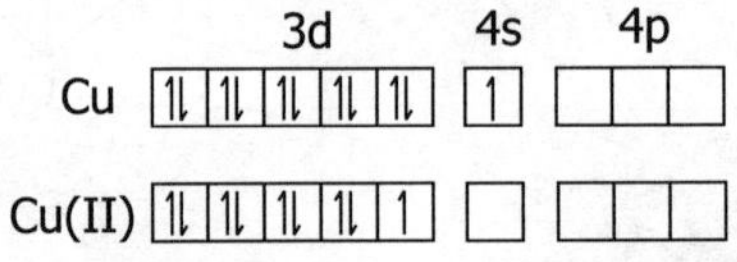

Expected configuration of $[Cu(NH_3)_4]^{2+}$ ion,

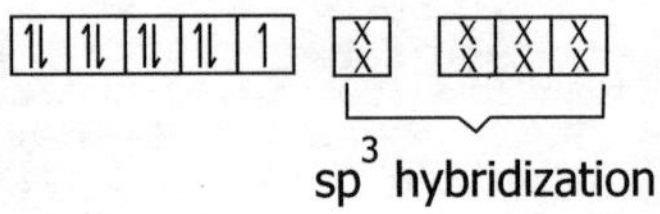

Since the five 3d orbitals are either completely or partially filled, it is expected that the metal orbitals would undergo sp^3 hybridization. The complex would be tetrahedral and paramagnetic, with one unpaired electron. It has been experimentally determined that the complex is indeed paramagnetic with one unpaired electron, but its geometry is square planar. This means that the metal's atomic orbitals are dsp^2 hybridized. To account for the experimental findings, the valence bond theory postulates that the single electron in the 3d orbital is excited to the 4p orbital. This vacates one d orbital and paves the way for dsp^2 hybridization.

Observed configuration of $[Cu(NH_3)_4]^{2+}$ ion,

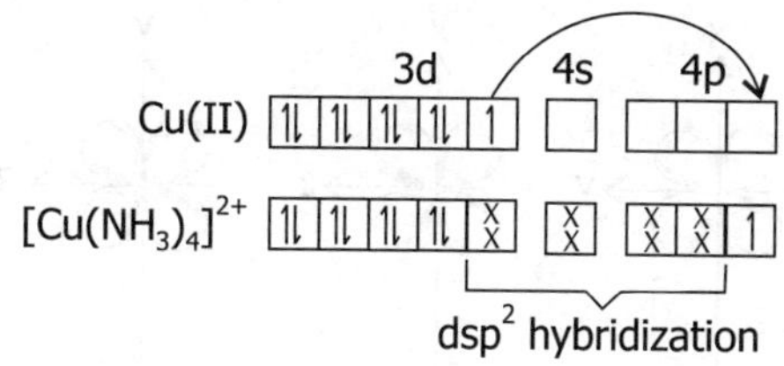

- Square planar
- Paramagnetic, with one unpaired electron

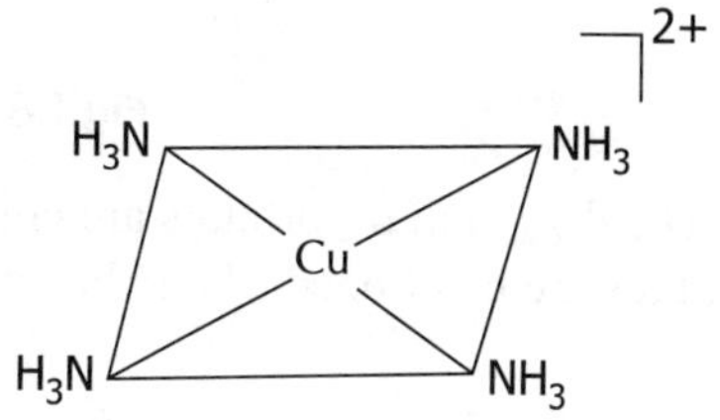

Exciting an electron from 3d to 4p orbital is not energetically favorable. Valence bond theory offers no explanation for this controversial step.

Crystal Field Theory

The crystal field theory is based on ionic model. It considers the attraction between the metal and the ligands in a complex to be purely electrostatic in nature. The metal has a positive charge equivalent to its oxidation state. If the coordinated groups are negatively charged, for example, chloride ions, there is ion–ion attraction between the metal and the ligands. If the coordinating groups are neutral, for example, ammonia, the negative end of their dipole is oriented towards the metal. In this case, there are ion–dipole attractions between the metal and the ligands. In addition, there are repulsive forces between the electrons of metal and those of ligands.

The crystal field theory thus treats ligands as point charges. There is no overlap between the orbitals of the metal and ligands, just as there is no overlap between the orbitals of sodium ion and chloride ion in a molecule of sodium chloride. The theory further states that the d orbitals on the *free metal ion* are degenerate, i.e., all the five d orbitals have the same energy. However, as the ligands approach the central metal to form the complex, the degeneracy of the d orbitals is resolved. The ligands create a crystal field around the metal and perturb the energy of the d orbitals. One set of d orbitals is repelled more strongly by the ligand electrons than the other set. The two sets, therefore, acquire different energies. The pattern of energy of d orbitals varies with the type of crystal field the ligands create around the central metal atom which, in turn, varies with the geometry of the complex.

The five d orbitals are designated as d_{xy}, d_{xz}, d_{yz}, $d_{x^2-y^2}$ and d_{z^2}. These may be broadly divided into two types. The d_{xy}, d_{xz} and d_{yz} orbitals are collectively called t_{2g} *orbitals*. The lobes of these orbitals are positioned *in between* the coordinates of x, y and z axes (Fig. 26.2).

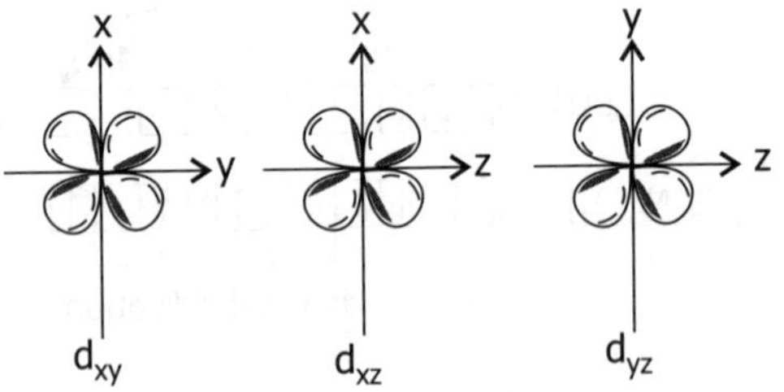

Fig 26.2 The t_{2g} orbitals

The $d_{x^2-y^2}$ and d_{z^2} orbitals are collectively called e_g *orbitals*. The lobes of these orbitals are positioned *along* the coordinates of x, y and z axes (Fig. 26.3).

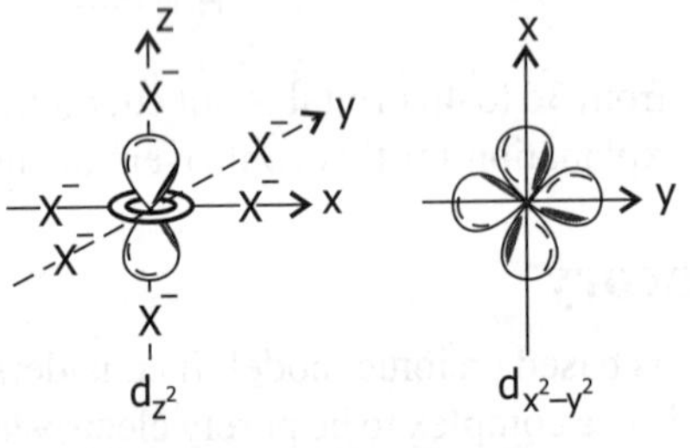

Fig. 26.3 The e_g orbitals

The d_{z^2} orbital is made up of a linear combination of two orbitals, $d_{z^2-x^2}$ and $d_{z^2-y^2}$ (Fig. 26.4).

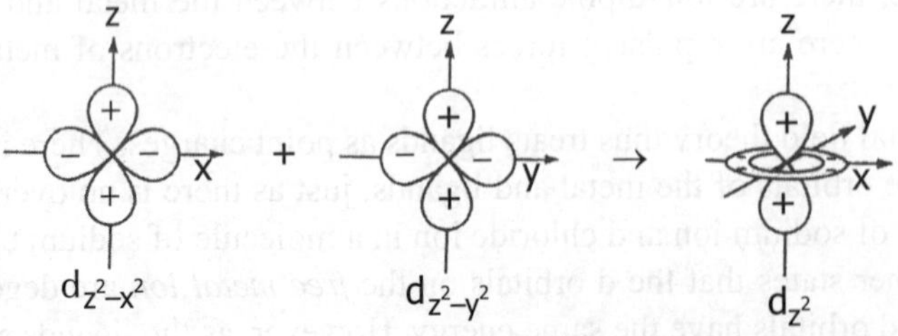

Fig. 26.4 Formation of d_{z^2} orbital by linear combination of $d_{z^2-x^2}$ and $d_{z^2-y^2}$ orbitals

The angular probability distribution of electrons within an orbital is directly proportional to the length of a line drawn from the center of the orbital to its circumference (*see* Chapter 1, p. 42–43). For d orbitals, the angular probability distribution is maximum at the tips of the lobes and goes on decreasing towards the meeting point of the two dumbbells. This aspect is important here because on its basis we can gauge which set of d orbitals is repelled more by the ligand electrons and which is repelled less.

We now extend the crystal field theory to explain the bonding between the central metal and the ligands in octahedral, tetrahedral and square planar complexes.

Octahedral complexes

In octahedral complexes, the metal atom is at the center of an octahedron, while the six ligands approach it along the arbitrarily chosen x, y and z axes. Ultimately, the ligands occupy the corners of the octahedron. There is, however, no overlap between the orbitals of the metal and the ligands. Their respective orbitals are held at a suitable distance from each other.

A ligand approaching a t_{2g} orbital of the metal may be visualized as depicted in Fig. 26.5.

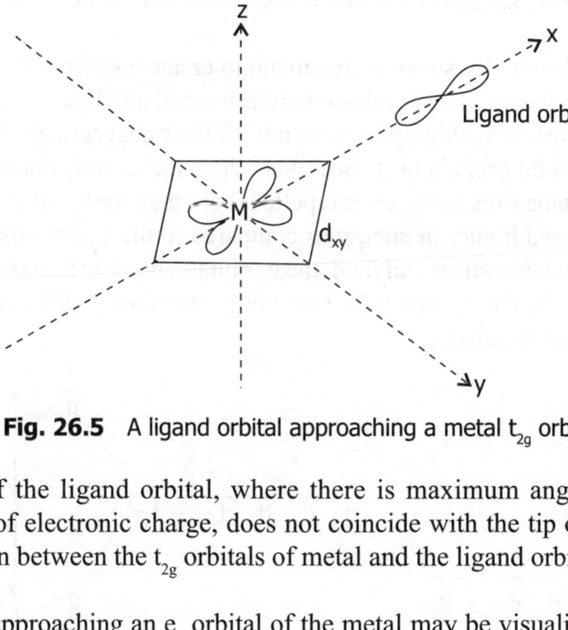

Fig. 26.5 A ligand orbital approaching a metal t_{2g} orbital

The tip of the ligand orbital, where there is maximum angular probability distribution of electronic charge, does not coincide with the tip of the t_{2g} orbital. The repulsion between the t_{2g} orbitals of metal and the ligand orbitals is relatively less.

A ligand approaching an e_g orbital of the metal may be visualized as shown in Fig. 26.6.

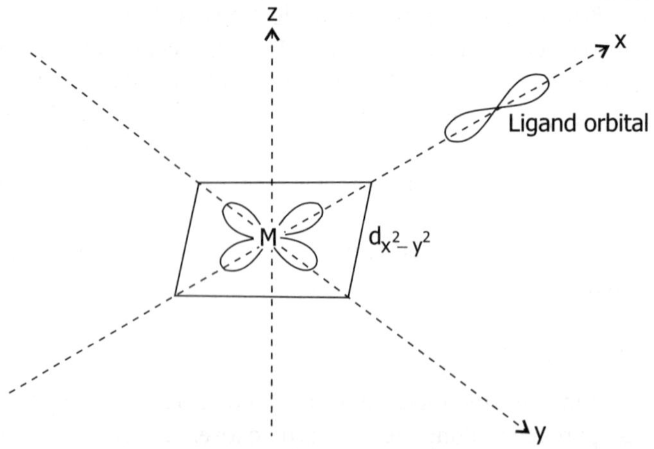

Fig. 26.6 A ligand orbital approaching a metal e_g orbital

The tips of the ligand and metal orbitals, in this case, are in one line. The extent of repulsion is greater in this case. If these two orbitals had overlapped end-to-end, the repulsion would have ceased due to neutralization of spins and formation of a σ-bond. However, as has been stated earlier, crystal field theory is not based on overlap model.

The six ligands initially surround the metal to create a spherically symmetrical field of negative charges. The repulsion between metal and ligand electrons raises the energy of 3d orbitals, although it does not lift their degeneracy (this concept is relevant to 4d and 5d orbitals too). However, when the ligands finally occupy the corners of the octahedron, the e_g set is repelled more than the t_{2g} set. As a result, the e_g orbitals are raised higher in energy as compared to the t_{2g} orbitals. Thus under the influence of octahedral crystal field, the d orbitals split into two sets of different energies (Fig. 26.7). The e_g set has higher energy than the t_{2g} set by an amount Δ_O, where O stands for octahedral.

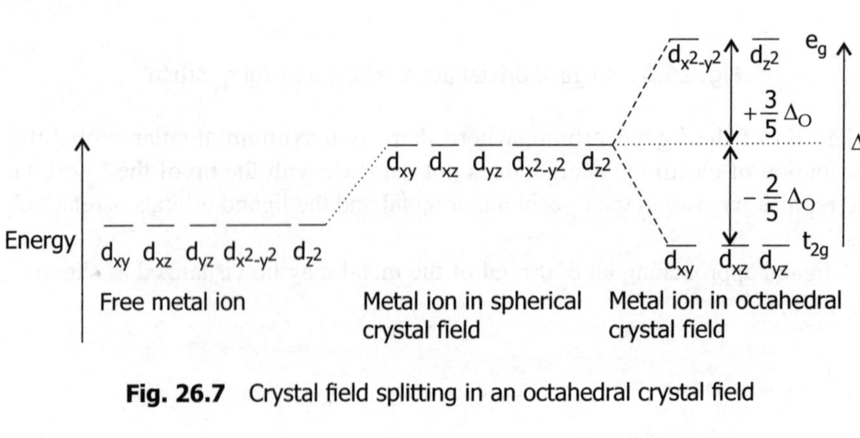

Fig. 26.7 Crystal field splitting in an octahedral crystal field

The t_{2g} orbitals are stabilized by the amount, –2/5 Δ_O, while the e_g orbitals are destabilized by +3/5 Δ_O, with respect to the average energy of degenerate 3d orbitals in a spherical crystal field.

The ligands which are high on the spectrochemical series create a stronger crystal field than those which are down the series (*see* Box 26.1). The crystal field splitting, Δ_o is accordingly higher for strong ligands, such as ammonia, than for weak ligands, such as fluoride ion.

Therefore, in the hexamminecobalt(III) ion, $[Co(NH_3)_6]^{3+}$, the six electrons of the cobalt(III) ion tend to occupy the t_{2g} orbitals, leading to diamagnetism.

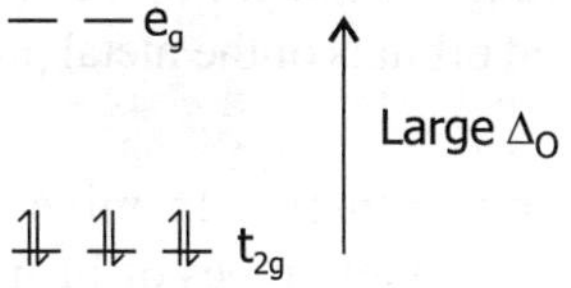

In hexafluorocobaltate(III) ion, $[CoF_6]^{3-}$, the crystal field splitting is small and it is energetically more favorable if the six electrons are spread out on both t_{2g} and e_g orbitals. The complex is paramagnetic, with four unpaired electrons.

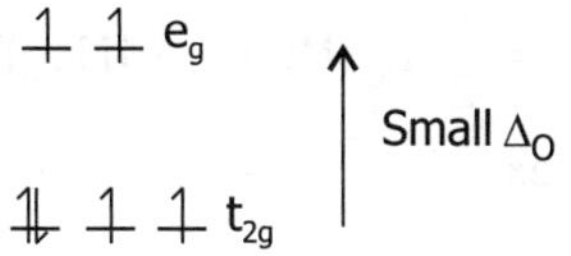

Box 26.1

Factors Affecting Crystal Field Splitting

The magnitude of crystal field splitting depends on the following factors:

1. **Nature of the ligands** The strong ligands on the spectrochemical series can approach the central metal more closely than their weak counterparts. Therefore, strong ligands produce greater crystal field splitting than weak ones.
2. **Geometry of the complexes** The complexes with higher coordination number cause a greater amount of crystal field splitting. Thus, a larger crystal field is generated by octahedral coordination than that by tetrahedral coordination. In fact, crystal field splitting in tetrahedral complexes (Δ_t) is nearly half that of octahedral complexes.

$$\Delta_t = \frac{4}{9}\Delta_O \approx \frac{1}{2}\Delta_O$$

3. **Oxidation state of the metal** A metal with a higher positive charge would be able to draw ligands more close to itself than if it has a lower positive charge. Therefore, higher the oxidation state of the metal, greater the crystal field splitting, i.e.,

$$M(IV) > M(III) > M(II)$$

Contd...

4. **Series to which the metal belongs** Crystal field splitting also depends on the position of the metal in the d-block of the periodic table. Provided the oxidation state of the metal remains the same, the crystal field splitting within a group varies in the order:

$$5d > 4d > 3d$$

The 5d orbitals are larger in size as compared to the 4d orbitals which, in turn, are larger than 3d orbitals. Therefore, in complexes of 5d series metals, the ligands are more close to the d orbitals than these are in the corresponding complexes of 4d or 3d metals. The closeness of the ligands to the 5d orbitals of the metal produces larger crystal field splitting.

5. **Rule of average environment** In mixed ligand complexes, i.e., complexes having two or more types of ligands, the extent of splitting depends on the weighted average of the strength of various ligands. For example, the crystal field splitting in octahedral $[ML_3\,L^1{}_3]$ complex will be the mean of Δ_o for $[ML_6]$ and $[ML^1{}_6]$, i.e.,

$$\Delta_o = \frac{1}{2}\{\Delta_o\,[ML_6] + \Delta_o[ML^1{}_6]\}$$

For $[ML_4L^1{}_2]$ complex,

$$\Delta_o = \frac{1}{3}\{2\Delta_o\,[ML_6] + \Delta_o\,[ML^1{}_6]\}$$

Case Study 26.1 Why do ammonia molecules produce a greater crystal field splitting than fluoride ions?

Crystal field splitting arises due to interaction of the ligand orbitals with the metal orbitals. To achieve a greater crystal field splitting, it is desirable that the ligand should focus its negative charge on the metal orbitals. A ligand with one electron pair (such as ammonia) can do this more readily than a species with four electron pairs (such as fluoride ion).

As far as magnetic properties are concerned, the same results were obtained when these two complexes are treated on the basis of valence bond theory. We may generalize that in crystal field terminology the low-spin and the high-spin complexes correspond, respectively to the inner sphere and outer sphere complexes in valence bond theory.

Unlike valence bond theory, where the ligand electrons are finally placed in the hybrid orbitals of the metal, the crystal field theory is concerned with distribution of metal electrons alone in the t_{2g} and e_g sets of orbitals. The ligand electrons do not enter the metal orbitals. The ligands are mere point charges which retain their electrons on to themselves.

Case Study 26.2 Both manganese(II) and iron(III) ions have $3d^5$ configurations. Most of the complexes of manganese(II) are high-spin, whereas those of iron(III) are low-spin. Why?

The magnitude of the crystal field splitting, Δ_O depends, apart from other factors, on the oxidation state of the metal (*see* Box 26.1). Higher the oxidation state, larger is the crystal field splitting. Iron(III) ion has a higher oxidation state than manganese(II) ion, despite both possessing $3d^5$ configuration. The difference in energy between the e_g orbitals and the t_{2g} orbitals of iron(III) octahedral complexes is, therefore, much higher than that of manganese(II) complexes of the same geometry. As a result, the valence electrons tend to pair-up in t_{2g} orbitals in case of iron(III) complexes.

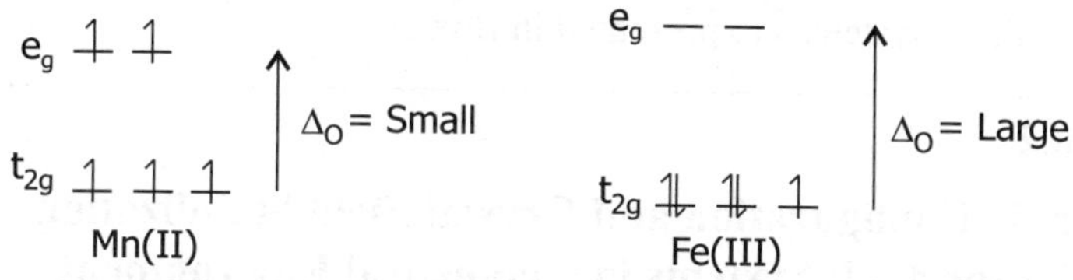

Crystal field stabilization energy

The stabilization gained by a complex as a result of splitting of d orbitals in a crystal field is called crystal field stabilization energy (CFSE). It primarily depends upon the extent of crystal field splitting, Δ_O and the number of electrons occupying t_{2g} and e_g orbitals. However, it also depends on whether pairing of electrons takes place or not.

As already mentioned the t_{2g} orbitals are stabilized to the extent of $-2/5\ \Delta_O$, while the e_g orbitals are destabilized to the extent of $+3/5\ \Delta_O$, as compared to the energy of five degenerate d orbitals in a spherical crystal field (Fig. 26.7). The crystal field stabilization energy is the sum of the energies of all the metal ion electrons. For a metal ion having the configuration $t^x_{2g}\ e_g^y$,

$$\text{CFSE} = x\left(-\frac{2}{5}\Delta_O\right) + y\left(+\frac{3}{5}\Delta_O\right) + zP$$

Here, P is the pairing energy and z is the number of electron pairs in the configuration.

Consider, for example, a metal ion with d^4 configuration in a crystal field. There can be two arrangements of electrons:

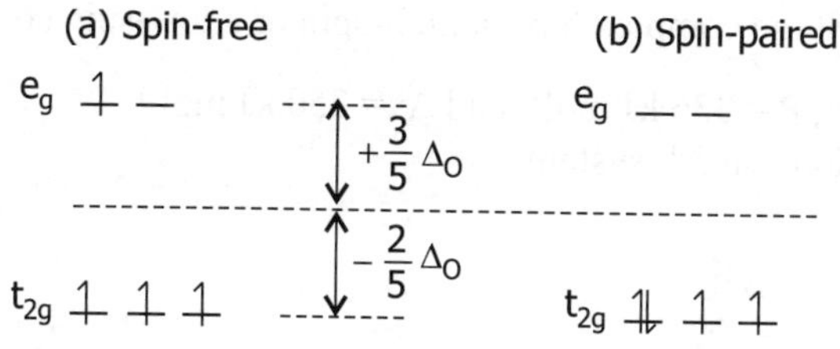

$$\text{CFSE} = 3\left(-\frac{2}{5}\Delta_O\right) + 1\left(+\frac{3}{5}\Delta_O\right) \qquad \text{CFSE} = 4\left(-\frac{2}{5}\Delta_O\right) + P$$

$$= -\frac{3}{5}\Delta_O \qquad\qquad = -\frac{8}{5}\Delta_O + P$$

The configuration which gives a greater negative value of CFSE will prevail. This, in turn, depends on the relative values of Δ_O and P.

When P > Δ_O, high-spin configuration results.
When Δ_O > P, low-spin configuration results.

The high-spin and low-spin states are possible not only for a d^4 system, but for d^5–d^7 systems as well. The electronic configurations and crystal field stabilization energies for these systems are presented in Box 26.2.

Box 26.2

Electronic Configuration and Crystal Field Stabilization Energies for d^4–d^7 Systems in Octahedral Environment

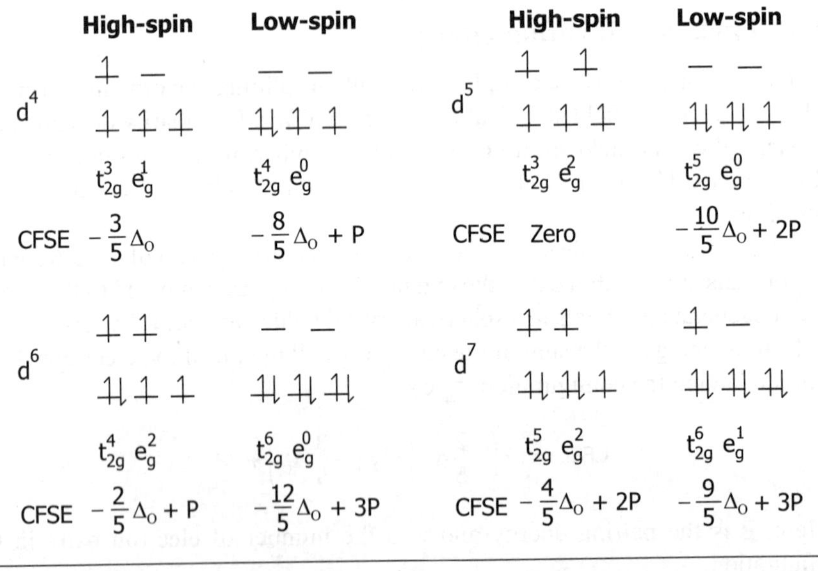

Case Study 26.3 For manganese(III) ion, the electron pairing energy is 336 kJ mol^{-1}. The crystal field splitting for hexaquamanganese(III) ion is 250 kJ mol^{-1}. Will the complex have a high-spin or a low-spin configuration?

For $[Mn(H_2O)_6]^{3+}$, P = 336 kJ mol^{-1} and Δ_O = 250 kJ mol^{-1}.
Manganese(III) ion is a $3d^4$ system.

(a) High-spin configuration (b) Low-spin configuration

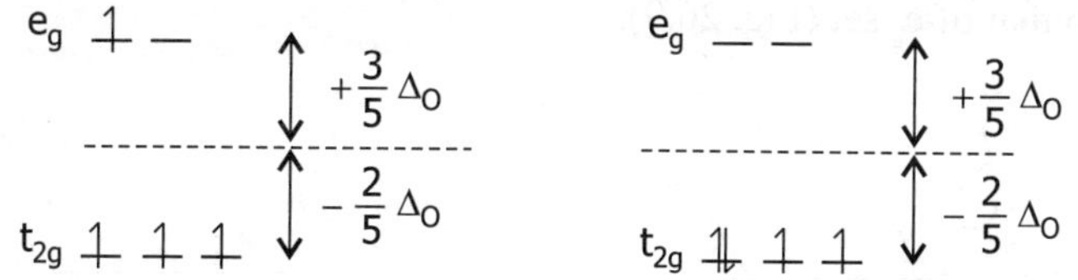

(a)

$$\text{CFSE} = 3\left(-\frac{2}{5}\Delta_o\right) + 1\left(+\frac{3}{5}\Delta_o\right)$$

$$= -\frac{3}{5}\Delta_o$$

$$= -\frac{3}{5}(250)$$

$$= -150 \text{ kJ mol}^{-1}$$

(b)

$$\text{CFSE} = 4\left(-\frac{2}{5}\Delta_o\right) + P$$

$$= -\frac{8}{5}\Delta_o + P$$

$$= -\frac{8}{5}(250) + 336$$

$$= -64 \text{ kJ mol}^{-1}$$

The complex will exist in high-spin configuration since in this state it has a greater negative value of CFSE.

Tetrahedral complexes

A tetrahedral complex may be visualized as a cube, with the metal occupying its center and the ligands occupying four out of eight corners. The directions of the x, y and z axes point to the centers of the faces of the cube. The four ligands approach the metal along the lines which *do not* coincide with the coordinates of x, y and z axes (Fig. 26.8).

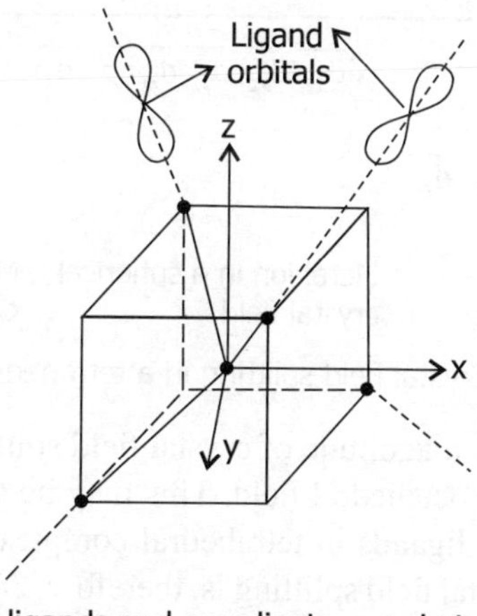

Fig. 26.8 Position of ligands and coordinate axes in tetrahedral environment

As the four ligands come near the metal, first a spherical field is created. The energy of the metal d orbitals is raised, although all five of them remain degenerate. When the ligands are finally positioned at the four corners of the cube, the t_{2g} orbitals of the metal are repelled more than the e_g orbitals. This is because the tips of ligand orbitals and those of t_{2g} set are nearly on one line. The region of maximum angular probability distribution of ligand orbitals coincides quite closely with that of t_{2g} set, but not with that of e_g set (Fig. 26.9).

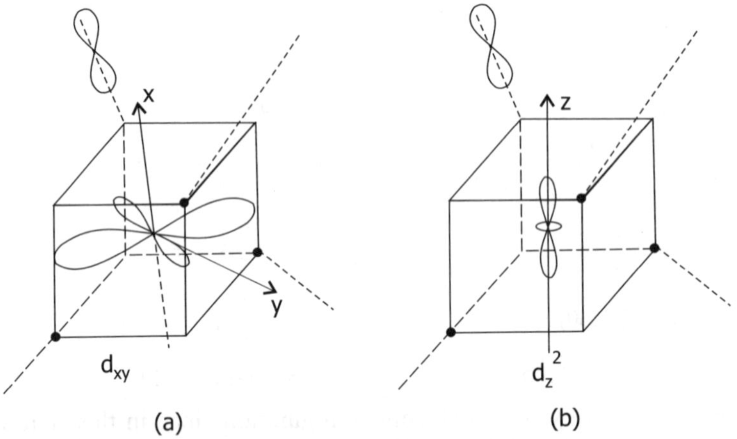

Fig. 26.9 Position of the tips of ligand orbitals with respect to (a) t_{2g} orbital and (b) e_g orbital

The difference in energy between t_{2g} and e_g orbitals is called *crystal field splitting for tetrahedral complexes* and is designated as Δ_t. The e_g orbitals are stabilized to the extent of $-3/5\Delta_t$, while the t_{2g} orbitals are destabilized to the extent of $+2/5\Delta_t$, as compared to the energy of degenerate d orbitals (Fig. 26.10).

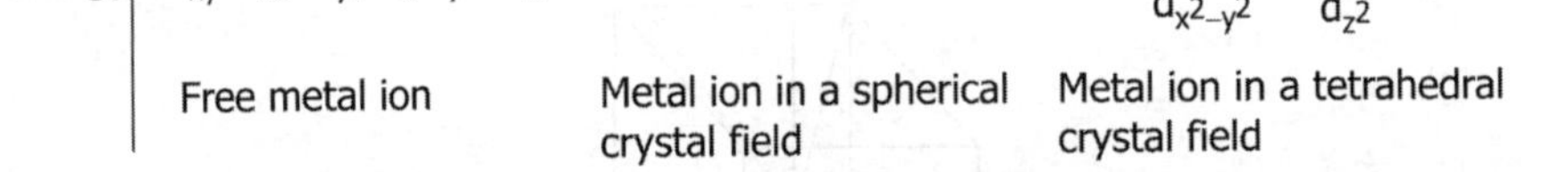

Fig. 26.10 Crystal field splitting in a tetrahedral crystal field

As has been stated, the magnitude of crystal field splitting in a tetrahedral field is smaller than that in an octahedral field. This may be attributed to two reasons. First, there are only four ligands in tetrahedral complexes as compared to six in octahedral ones. The crystal field splitting is, therefore, 2/3 times less in tetrahedral geometry. Second, unlike octahedral complexes, the line of approach of ligands towards the central atom does not coincide with the coordinates of x, y and z axes in tetrahedral geometry. This reduces the crystal field splitting further by 2/3 times. Thus, the tetrahedral crystal field splitting is 2/3 × 2/3 or 4/9 of octahedral crystal field splitting, i.e.,

$$\Delta_t = \frac{4}{9}\Delta_o$$

Since the difference in energy between e_g and t_{2g} orbitals is small in tetrahedral geometry, such complexes largely exist in high-spin state. Moreover, CFSE of tetrahedral complexes is less than that of octahedral complexes. Therefore, the octahedral complexes are more stable and more numerous than the tetrahedral analogs. Nevertheless, tetrahedral geometry is favored under the following conditions:

1. When the ligands are bulky, steric hindrance makes octahedral complexes unstable. Tetrahedral coordination is relatively more stable.
2. When, after being split in a tetrahedral field, t_{2g} and e_g orbitals are either unfilled or half-filled or completely filled, the tetrahedral geometry is favored. It is not necessary that both e_g and t_{2g} sets should be half-filled or completely filled. Even if one is half-filled and the other is completely filled, tetrahedral coordination is favored. The d^0, d^2, d^5, d^7 and d^{10} configurations satisfy this criterion and, therefore, metal ions with these configurations give stable tetrahedral complexes.

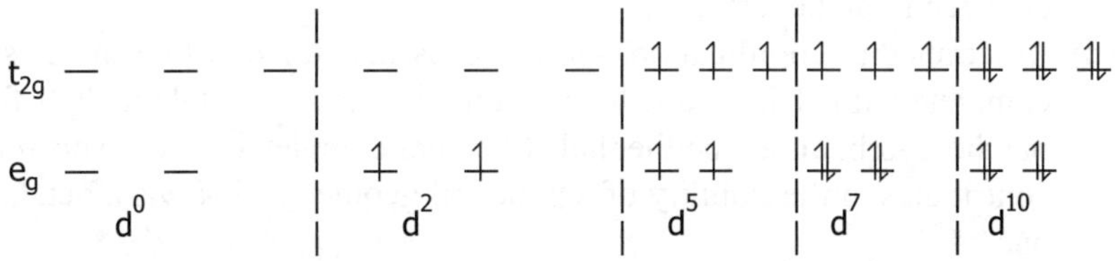

3. When the electronic configuration of the central metal ion is d^0, d^5 and d^{10}, there is an additional reason favoring tetrahedral coordination: there is no CFSE associated with these configurations. Hence there is no question of loss in CFSE if tetrahedral (rather than octahedral) coordination results.
4. When the ligands are weak field, the loss in CFSE is smaller, if tetrahedral complexes are formed instead of octahedral ones.
5. When the metal is in a low oxidation state, the magnitude of crystal field splitting is small and so is the CFSE, if tetrahedral complexes are formed instead of octahedral ones.
6. When the difference between the crystal field stabilization energies of octahedral and tetrahedral coordinations is quite small. This is true for d^1, d^2, d^6 and d^7 configurations.

Case Study 26.4 Why does cobalt(II) ion form predominantly tetrahedral complexes, especially with weak field ligands?

Let us calculate the crystal field stabilization energy for cobalt(II) ion, a $3d^7$ system, in octahedral and tetrahedral geometries.

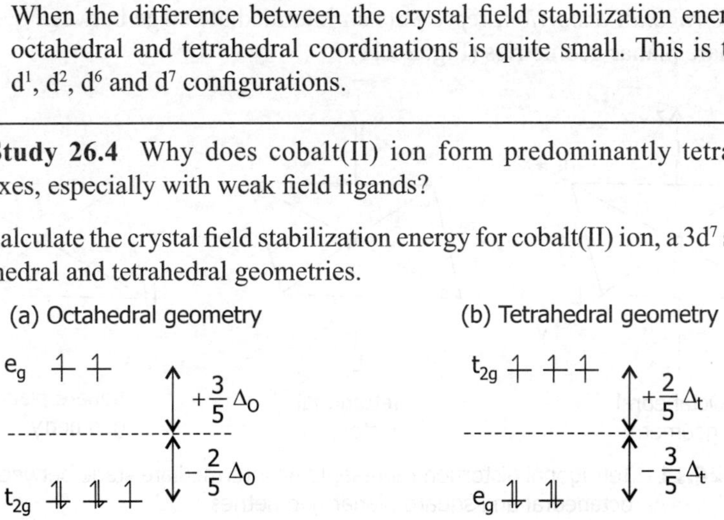

$$\text{CFSE} = 5\left(-\frac{2}{5}\Delta_o\right)+2\left(+\frac{3}{5}\Delta_o\right)+2P \qquad \text{CFSE} = 4\left(-\frac{3}{5}\Delta_t\right)+3\left(+\frac{2}{5}\Delta_t\right)+2P$$

$$= -0.8\Delta_o + 2P \qquad = -1.2\Delta_t + 2P$$

$$= -0.53\,\Delta_o + 2P\ [\text{Since } \Delta_t = \frac{4}{9}\Delta_o]$$

Difference in crystal field stabilization energies = $-0.27\Delta_o$

The loss of CFSE is small in the following cases:

(a) Tetrahedral, instead of octahedral, complex is formed. There is less steric repulsion in tetrahedral complexes as compared to octahedral complexes.

(b) The metal is in a low oxidation state. This factor reduces the energy difference between t_{2g} and e_g orbitals which, in turn, reduces the loss of CFSE.

(c) The ligand is a weak field one. This factor too decreases the energy gap between t_{2g} and e_g orbitals.

(d) In tetrahedral coordination, the t_{2g} set is half-filled, while the e_g set is completely filled. In octahedral coordination, the e_g orbitals are half-filled, but the t_{2g} orbitals are neither half-filled nor completely filled. This further contributes to the stability of tetrahedral geometry vis-à-vis a octahedral one.

All these factors favor tetrahedral geometry.

Square planar complexes

A square planar complex may be considered as an extreme case of octahedral geometry. Suppose two *trans* ligands on z-axis of an octahedral complex are slowly withdrawn towards infinity. As the distance of these two ligands from the central metal ion becomes appreciably larger than that of the other four ligands along x and y axes, the geometry of the complex becomes distorted. This irregular geometry, called *tetragonal distortion*, represents an intermediate stage between octahedral and square planar geometries (Fig. 26.11).

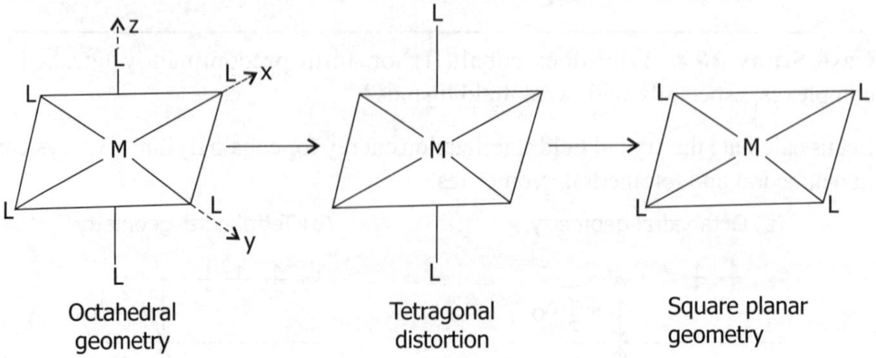

Fig. 26.11 Tetragonal distortion represents an intermediate stage between octahedral and square planar geometries

At the stage of tetragonal distortion, the degeneracy within the e_g and t_{2g} sets of orbitals in octahedral geometry (Fig. 26.12a) is further resolved. Those orbitals which have a z component (d_{z^2} in e_g set; d_{xz} and d_{yz} in t_{2g} set) become more stable, since the metal–ligand repulsions are progressively reduced along the z-axis. Simultaneously, the orbitals which have x and y components ($d_{x^2-y^2}$ in e_g set; d_{xy} in t_{2g} set) become de-stabilized since the repulsive influence of metal electrons on the ligands, which was earlier directed on six coordinating units, is now being felt by four groups. Therefore, in the e_g set, the $d_{x^2-y^2}$ orbital is raised in energy, while the d_{z^2} orbital is lowered and in the t_{2g} set, the d_{xy} orbital acquires a higher energy than the d_{xz} and d_{yz} orbitals (Fig. 26.12b).

When the two ligands along the z-axis are completely withdrawn, the tetragonal distortion geometry is transformed into a square planar one. The repulsions felt by the orbitals having z component are now negligible. Accordingly, the d_{z^2} orbital becomes lower in energy as compared to the d_{xy} orbital (Fig. 26.12c).

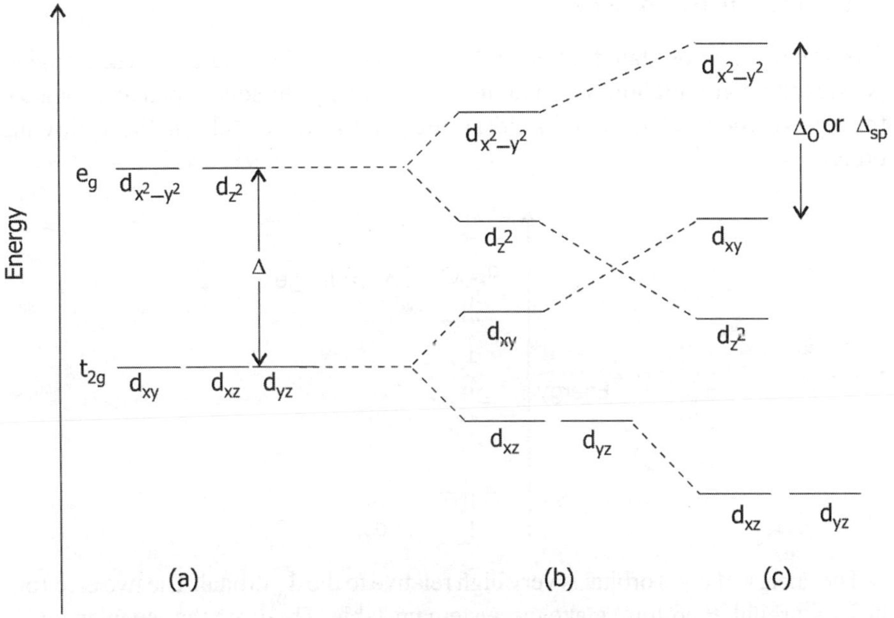

Fig. 26.12 Splitting of d orbitals in (a) octahedral (b) tetragonal distortion and (c) square planar environments

As usual, the crystal field splitting is taken to be the energy difference between e_g and t_{2g} orbitals. In this case, it corresponds to the difference in the energy of $d_{x^2-y^2}$ orbital and d_{xy} orbital. The value, Δ_{sp} is the same as for octahedral coordination, Δ_O.

Let us now frame the crystal field picture of tetracyanonickelate(II) ion, $[Ni(CN)_4]^{2-}$. We have seen that the valence bond theory predicted this complex to be square planar and diamagnetic. The nickel(II) ion has $3d^8$ configuration. The crystal field pattern of energy levels in square planar geometry may be depicted as follows:

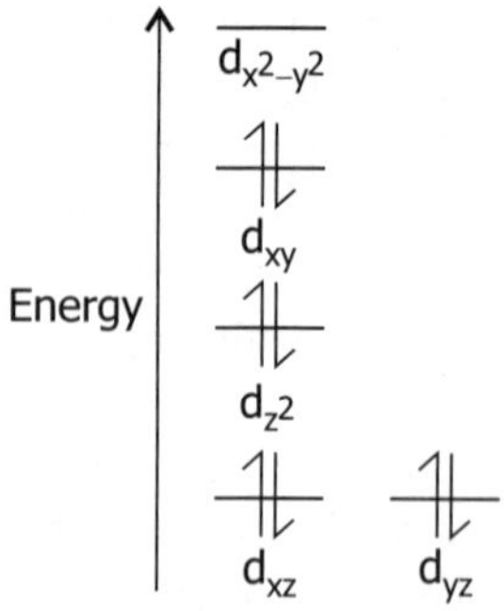

The eight valence electrons on nickel(II) ion are paired-up in the first four orbitals, giving a diamagnetic complex.

Case Study 26.5 Several square planar complexes are known for gold(III) ion, but none for silver(III) ion. Why?

The group oxidation state for both gold and silver is I. In this state, these elements possess the configuration $5d^{10}$ and $4d^{10}$, respectively. In square planar geometry, the ten valence electrons of gold(I) occupy the d orbitals in the following order:

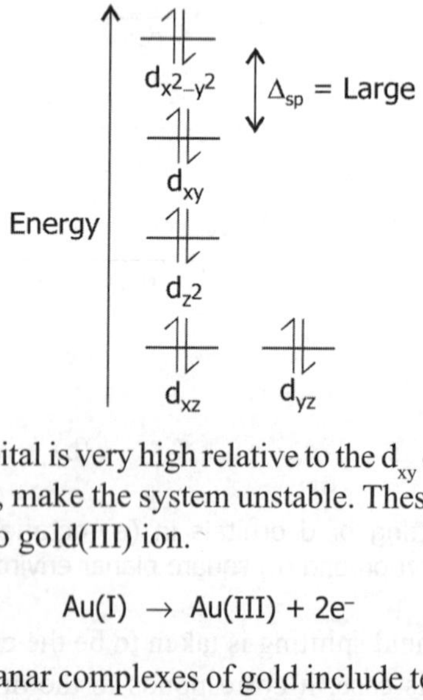

The energy of $d_{x^2-y^2}$ orbital is very high relative to the d_{xy} orbital. The two electrons in $d_{x^2-y^2}$ orbital, therefore, make the system unstable. These are thus spontaneously lost, converting gold(I) to gold(III) ion.

$$\text{Au(I)} \rightarrow \text{Au(III)} + 2e^-$$

Examples of square planar complexes of gold include tetrachloroaurate(III) ion, $[AuCl_4]^-$ and tetramminegold(III) ion, $[Au(NH_3)_4]^{3+}$.

Silver, which belongs to 4d series of transition metals, gives a smaller splitting when placed in a crystal field than gold, which belongs to the 5d series. The magnitude of Δ_{sp} is smaller for complexes of silver than that of gold. This means that the two electrons in $d_{x^2-y^2}$ orbital of silver(I) ion are not lost spontaneously. Hence, oxidation state (III) is uncommon and rare for silver.

Applications of crystal field theory

The crystal field theory is successful in explaining the following physical parameters:

1. **Magnetic properties**

   The magnetic behavior of transition metal complexes is aptly gauged by the crystal field theory. For the 3d series metals, if the ligands produce a weak crystal field, the complexes are largely paramagnetic and if the ligands create a strong crystal field, the complexes are invariably diamagnetic. This is because for strong ligands the crystal field splitting is large and the valence electrons of the metal ion tend to pair-up. For weak ligands, on the other hand, the crystal field splitting is small and the valence electrons of the metal ion tend to remain unpaired by occupying both t_{2g} and e_g sets of orbitals.

   For example, consider the following two complexes of cobalt(III) ion [electronic configuration, $3d^6$]:

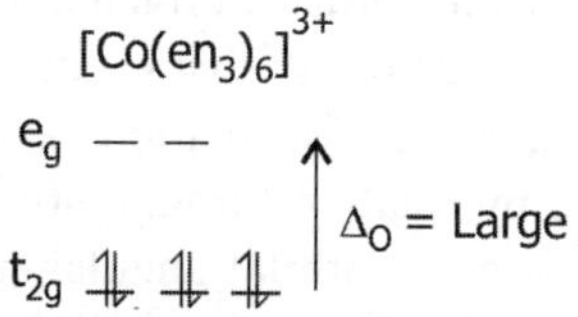

Strong field ligand
Diamagnetic

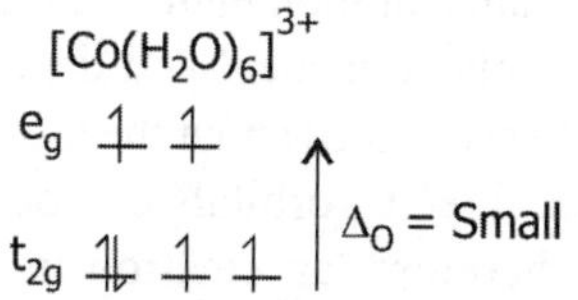

Weak field ligand
Paramagnetic, with
four unpaired electrons

   When the metal belongs to 4d or 5d transition series, the magnitude of crystal field splitting is larger as compared to a 3d series metal. Therefore, even weak field ligands form diamagnetic complexes. Take, for example, hexachloroiradate(III) ion, $[IrCl_6]^{3-}$. The iridium(III) ion, with $5d^6$ configuration, produces large crystal field splitting, even though the chloro groups are weak ligands. Accordingly, the valence electrons tend to remain paired-up in t_{2g} orbitals.

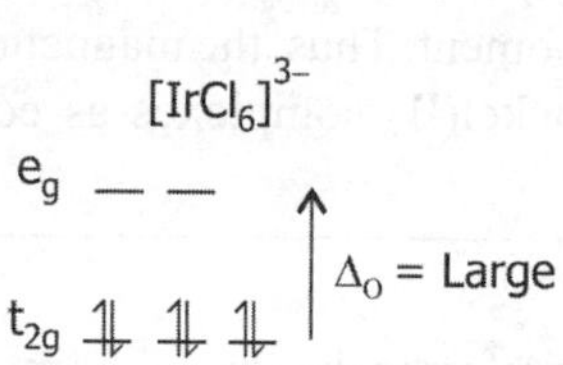

   Similarly, in hexachloroferrate(III) ion, $[FeCl_6]^{3-}$ the valence electrons of iron(III) [$3d^5$] are in a high-spin configuration, $t_{2g}^3 e_g^2$, while in hexachloroosmate(III) ion, $[OsCl_6]^{3-}$, the valence electrons of osmium(III) [$5d^5$] are in a low-spin configuration, $t_{2g}^5 e_g^0$.

Case Study 26.6 Octahedral, high-spin nickel(II) complexes have magnetic moments of the order of 3 BM, while for the corresponding tetrahedral complexes, the value is about 4 BM. Explain the discrepancy.

As discussed in Chapter 21 (p. 613), the precise magnetic moment of a chemical species is given by the expression,

$$\mu_{S+L} = \sqrt{4S\,(S+1) + L\,(L+1)}$$

The term μ_{S+L} is the total magnetic moment, including both the spin moment (S), as well as the orbital moment (L). Normally L, the orbital angular momentum quantum number, is taken as zero since the orbital contribution to magnetic moment is generally quenched. In such cases, the above expression reduces to,

$$\mu_S = \sqrt{4S\,(S+1)}$$

However, the orbital moment is not always quenched. For an electron to have orbital angular momentum, it must be possible to transform the orbital it occupies into an entirely equivalent and degenerate orbital by rotation. In that event, the electron effectively rotates about the axis, which, in turn, is used for rotation of the orbital.

The three t_{2g} orbitals can be interconverted by rotation through 90° (Fig. 26.2) and, therefore, an electron in a t_{2g} orbital possesses orbital angular momentum, and the term L has a positive value. The two e_g orbitals have different shapes (Fig. 26.3) and cannot be interconverted by rotation. An electron in e_g orbital does not have orbital angular momentum. In such cases, L = 0.

Nevertheless, if all t_{2g} orbitals are singly occupied, an electron in, say, d_{xy} orbital cannot be transferred to d_{xz} or d_{yz} orbital since they already contain an electron having the same spin. Similarly, if all the t_{2g} orbitals are doubly occupied, the mutual transfer of electron is not possible. Thus for t_{2g}^3 and t_{2g}^6 systems, there is no orbital angular momentum, i.e., L = 0.

For a nickel(II) ion in an octahedral environment, the configuration is $t_{2g}^6 e_g^2$. Only the spin moment contributes to the magnetic moment. In a tetrahedral environment, the configuration of nickel(II) ion is $e_{2g}^4 t_{2g}^4$. Both spin moment and orbital moment contribute to the magnetic moment. Thus, the magnetic moment has a greater value for high-spin tetrahedral nickel(II) complexes as compared to their octahedral counterparts.

2. **Hydrated radius**

   The hydrated radius refers to the size of the metal ion when it is coordinated to water molecules. It is calculated by subtracting the atomic radius of oxygen (the donor atom of aqua ligand) from the metal ion–oxygen internuclear distance.

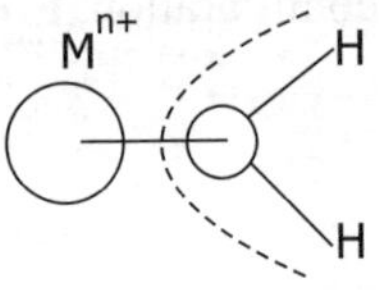

Hydrated radius is not as accurate as atomic or ionic radius since the oxygen of water molecule is not in contact with the metal ion. Its value is, therefore, on the higher side of ionic radius. The closer the ligand is to the metal ion, smaller is the value of hydrated radius.

Let us now investigate the variation in hydrated radii of divalent metal ions, M(II) of the first transition series. As with atomic or ionic radius, the hydrated radius is expected to decrease from left to right, with increasing atomic number. As shown in Fig. 26.13, a smooth decrease occurs only for divalent calcium, manganese and zinc. These ions have $3d^0$, $3d^5$ and $3d^{10}$ configurations, respectively. As a result, they have an almost spherical crystal field of six water molecules around them. The other ions of their series do not possess a spherical crystal field. The d electrons are not uniformly distributed in the t_{2g} and e_g orbitals. Therefore, the decrease in hydrated radii of the remaining metal ions is not regular.

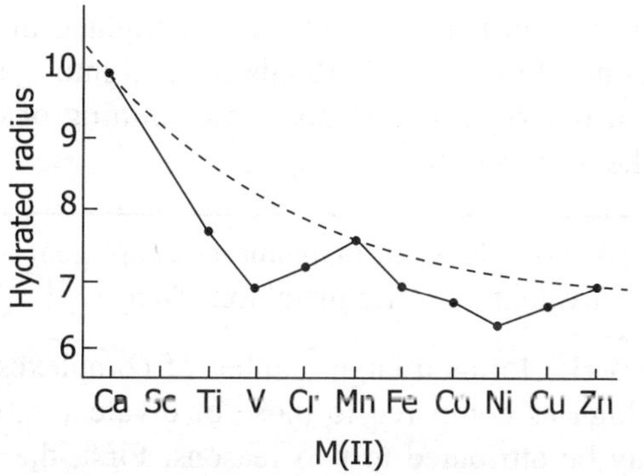

Fig. 26.13 Relative hydrated radii of divalent metal ions of 3d series

The scandium(II) ion has the configuration t_{2g}^1. Since the lobes of t_{2g} orbitals are projected in between the axes, it implies that the negative charge of the electron is concentrated in a region away from the line along which the ligands (H_2O) approach the scandium(II) ion. If the electronic charge had been distributed spherically around the metal nucleus, it would have repelled the incoming ligand. However, since it is concentrated away from the metal–ligand intermolecular axis, it provides a little shielding between scandium(II) ion and the aqua coordinating groups. As a result, the ligands are drawn closer to the metal, reducing the hydrated radius. In titanium(II) ion (t_{2g}^2), the same effect is observed to an even greater degree since not only both the valence electrons are placed in t_{2g} orbitals, but the nuclear charge has also increased by one unit. Higher the nuclear charge, smaller is the radius. The next metal ion, vanadium(II), with the configuration t_{2g}^3, has smaller radius than titanium(II) ion, since the nuclear charge has further increased.

For the chromium(II) ion, the configuration is $t_{2g}^3\ t_g^1$. The e_g orbitals have their lobes projected along the axis, i.e., their lobes lie directly on the line along which the ligand is approaching the metal ion. The e_g electron of chromium(II), therefore, repels the incoming ligand. The internuclear metal–ligand distance increases and so does the hydrated radius. In case of manganese(II) ion, there are two electrons in e_g orbitals ($t_{2g}^3\ t_g^2$). The metal–ligand repulsion is more than that in case of chromium(II) ion and, therefore, the hydrated radius of manganese(II) ion is concomitantly larger.

In the next three metal ions, the valence electrons once again enter the t_{2g} set: iron(II)–$t_{2g}^4\ t_g^2$; cobalt(II)–$t_{2g}^5\ t_g^2$; and nickel(II)–$t_{2g}^6\ t_g^2$. The ligands are progressively drawn closer to the metal ions and there is a gradual decrease in hydrated radii. In copper(II) and zinc(II) ions, the additional electrons enter the e_g orbitals, giving the configuration, $t_{2g}^6\ t_g^3$ and $t_{2g}^6\ t_g^4$, respectively. The ligands are repelled and the value of hydrated radii once again increases.

The variation in hydrated radii, as highlighted in Fig. 26.13, can be explained only if the metal d orbitals are split into t_{2g} and e_g sets. It is the crystal field theory which proposes the splitting of d orbitals and thus explains the observed trend.

Case Study 26.7 The complexes of manganese(II) are generally less stable than those of subsequent divalent 3d series metal ions. Why?

Although manganese(II) forms a large number of complexes, their equilibrium constants are not high as compared to those of divalent cations of succeeding elements. This may be attributed to two reasons. First, the hydrated radius of manganese(II) ion is larger than that of cations following it. A smaller cation forms more stable complexes. Second, no crystal field stabilization energy is associated with manganese(II) complexes.

---

3. **Enthalpies of hydration**

The enthalpy or heat of hydration refers to the energy released when one mole of a metal ion gets hydrated.

$$M^{2+} + nH_2O \rightarrow [M(H_2O)_n]^{2+}$$

The enthalpies of hydration of the divalent metal ions of first transition series are plotted in Fig 26.14. The values for calcium(II) ($3d^0$), manganese(II) ($3d^5$) and zinc(II) ($3d^{10}$) lie on a smooth curve. The remaining ions have higher enthalpies of hydration than these three cations.

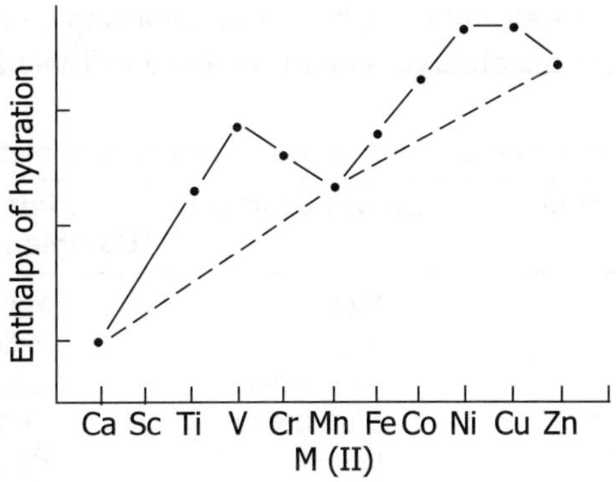

Fig. 26.14 Relative enthalpies of hydration of divalent 3d metal ions

The enthalpy of hydration is inversely proportional to the size of the metal ion. Smaller the cation, higher is the heat of hydration. As shown in Fig. 26.14, metal ions that have a large hydrated radii possess a small heat of hydration and vice versa. For this reason, the curves of hydrated radii (Fig. 26.13) and enthalpies of hydration (Fig. 26.14) are actually mirror image of each other. Once again, the observed trend of enthalpy of hydration can be explained only if it is assumed that the d orbitals split into t_{2g} and e_g sets—the fundamenatal premise of crystal field theory.

Case Study 26.8 Account for the anomalous trend of lattice energy values of the following transition metal oxides.

TiO	VO	MnO	
3878	3913	3810	kJ mol^{-1}

The lattice energy increases with decreasing size of metal ions. Hence on moving from left to right across the d block, this parameter should increase. As a result, the lattice energy of manganese(II) oxide should have been greater than that of vanadium(II) oxide. However, manganese(II) ion has a crystal field stabilization energy of zero. The crystal field stabilization energy values of titanium(II) oxide and vanadium(II) oxide are $-\frac{4}{5}\Delta_o$ and $-\frac{6}{5}\Delta_o$, respectively. It is pertinent that the greater lattice energies of titanium(II) oxide and vanadium(II) oxide arise from the contribution of crystal field stabilization energy.

---

4. **Color**

The color of the transition metal complexes arises due to absorption of visible light by unpaired electrons. The electrons are then excited from one set of d orbitals to another. All the wavelengths of visible light are not absorbed. Those which are not absorbed are transmitted or reflected. They tend to have a color that is complementary to the color of absorbed wavelength. The color of the metal complex corresponds to the wavelength of reflected or transmitted light. The relationship between wavelength of absorbed light and observed color is outlined in Table 26.2.

Table 26.2 Colors absorbed and observed in transition metal complexes

Wavelength range (nm)	Color absorbed	Color observed (Complementary color)
400–435	Violet	Yellow-green
435–480	Blue	Yellow
480–490	Green-blue	Orange
490–500	Blue-green	Red
500–560	Green	Purple
560–580	Yellow-green	Violet
580–595	Yellow	Blue
595–650	Orange	Green-blue
650–750	Red	Green

The wavelength of the light absorbed (λ) depends on the magnitude of crystal field splitting (Δ).

$$\Delta \propto \frac{1}{\lambda}$$

The magnitude of crystal field splitting for a specific metal ion and a specific geometry, in turn, depends on the strength of the ligand.

Consider, for example, hexamminechromium(III) ion, $[Cr(NH_3)_6]^{3+}$. Ammonia is a strong ligand which causes a large crystal field splitting. The configuration of the metal ion ($3d^3$) in ground state and excited state may be represented as follows:

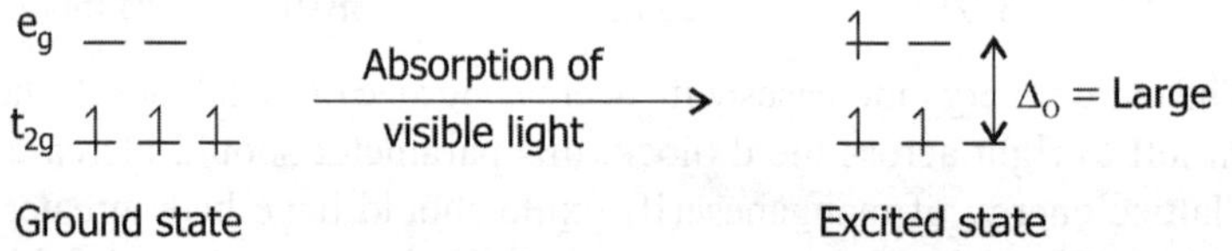

Since the value of crystal field splitting in octahedral geometry (Δ_o) in this case is large, excitation of electron from t_{2g} to e_g orbitals occurs with the absorption of a shorter wavelength of visible component. In fact, this complex absorbs visible light in blue region and reflects yellow light. The compound appears yellow in color.

Consider now hexaquachromium(III) ion, $[Cr(H_2O)_6]^{3+}$. Water is a weak ligand which causes a small crystal field splitting. Since the value of Δ_o is small, excitation of electron from t_{2g} to e_g orbitals requires absorption of a longer wavelength in visible component. This complex absorbs yellow light and appears blue.

The complex hexamminecoblat(III) ion $[Co(NH_3)_6]^{3+}$ is yellow in color. It thus absorbs wavelengths in blue region of the spectrum. If one ammine ligand is replaced by aqua group, the resulting pentammineaquacobalt(III) ion, $[Co(NH_3)_5(H_2O)]^{3+}$ will have a different color. Since water is a weaker ligand than ammonia, the energy gap, Δ_o between the e_g and the t_{2g} orbitals will become smaller. This means that a lower amount of energy need be absorbed for the d–d transition to take place. In fact, this complex absorbs wavelengths in the green-blue region of the spectrum and reflects orange light. Accordingly, this complex is orange in color.

Case Study 26.9 Although manganese(II) ion has five unpaired electrons, hexaquamanganese(II) ion, $[Mn(H_2O)_6]^{2+}$ is not intensely colored. Why?

Color arises due to transition of electron from one energy level to another. There are two selection rules which govern electronic transitions.

1. Transitions between states of different spin multiplicities are *spin forbidden.* Spin multiplicity is defined as,

$$\text{Spin multiplicity} = 2S + 1$$

where S is the net spin quantum number of the electrons involved in transition. For example, the transition $s^2 \rightarrow s^1p^1$ will be allowed only if the spins of electrons in s^1p^1 state are $+\frac{1}{2}$ and $-\frac{1}{2}$.

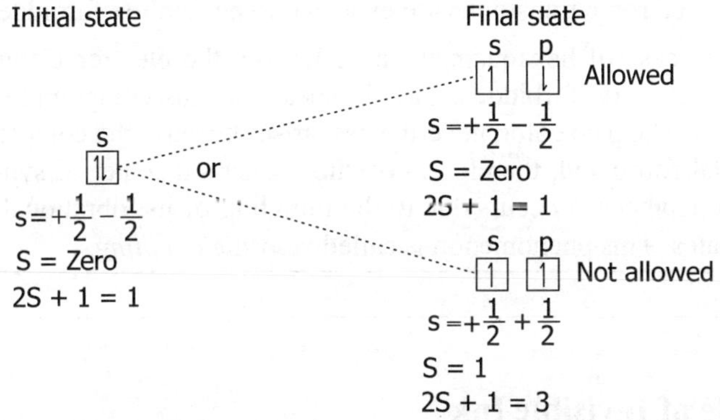

In case of hexaquamanganese(II) ion, the d orbitals are split into t_{2g} and e_g sets. The metal ion is a $3d^5$ system and since aqua ligand is a weak one, the ground state configuration may be written as $t_{2g}^3\, e_g^2$. When one of the t_{2g} electrons is excited to e_g level, the spin multiplicity changes, as shown in Fig. 26.15.

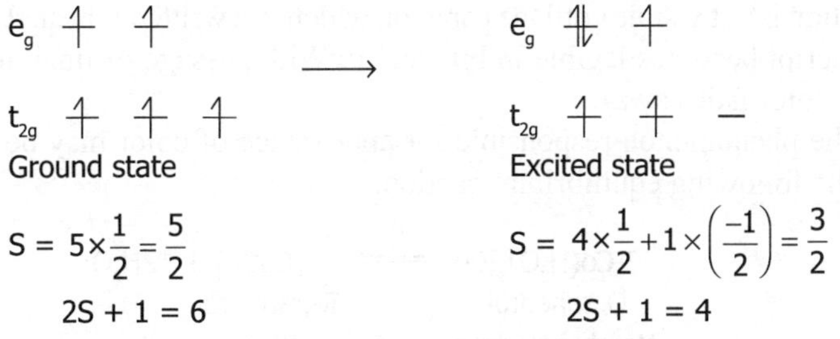

Fig. 26.15 Spin multiplicity forbidden transition

Since the spin multiplicity changes, this transition is forbidden. Nevertheless, the following transition is allowed (Fig. 26.16)

$$S = 4\times\frac{1}{2}+1\times\left(\frac{-1}{2}\right)=\frac{3}{2} \qquad S = 4\times\frac{1}{2}+1\left(\times\frac{-1}{2}\right)=\frac{3}{2}$$

$$2S + 1 = 4 \qquad 2S + 1 = 4$$

Fig. 26.16 Spin allowed transition

However, the probability that the metal ion will have the ground state configuration shown on the left side of Fig. 26.16 is quite low. Therefore, the hexaquamanganese(II) ion, no doubt, is colored, but the shade is quite pale.

2. If the molecule has center of symmetry, the transitions within the same energy state are *Laporte forbidden.*

   The molecular ion, $[Mn(H_2O)_6]^{2+}$ is supposed to possess center of symmetry (a line drawn from the center of the ion and extended equally to left and right meets at identical groups, water molecules in this case). The electronic transitions are expected to be Laporte forbidden.

Thus in the case of hexaquamanganese(II) ion, the electronic transitions are both spin and Laporte forbidden. Strictly speaking, this compound should have been colorless. The pale color, nevertheless, arises because the complex ion is not a perfect octahedron and, therefore, does not possess a center of symmetry. The distortion of octahedron occurs due to the mingling of its vibrational states with electronic states. This phenomenon is called *vibronic coupling.*

Box 26.3

Principle of Invisible Ink

The mechanism of invisible ink is based on the concept that conversion of one complex into another is often accompanied by a striking change in color. Invisible ink may be prepared by dissolving cobalt(II) chloride in water. Hexaquacobalt(II) chloride, an octahedral, pink colored complex is formed. In dilute solutions, the complex is almost colorless. What is written with this solution is not visible until the paper on which it is written is heated. On heating, the script becomes legible in blue color. With passage of time, however, the blue color fades away.

The phenomenon responsible for appearance of color may be represented by the following equilibrium reaction:

$$\underset{\text{Nearly colorless}}{\underset{\text{Octahedral}}{2[Co(H_2O)_6]Cl_2}} \underset{}{\overset{\Delta}{\rightleftharpoons}} \underset{\text{Blue}}{\underset{\text{Tetrahedral}}{Co[CoCl_4] + 12H_2O}}$$

Contd...

On heating, the octahedral complex changes into tetrahedral cobalt(II) tetrachlorocobaltate(II). Simultaneously, the magnitude of crystal field splitting gets reduced due to two reasons. First, the splitting is less for tetrahedral than for octahedral complex. Second, chloro is a weaker ligand than aqua. Excitation of electron is, therefore, easier in the case of tetrahedral complex. In fact, it absorbs yellow light and appears blue. Hence, the blue-colored script. On standing, the blue, tetrahedral complex absorbs water vapors from the atmosphere and reverts back to the colorless octahedral analog. Therefore, the writing again becomes invisible.

This concept has yet another application. Drying agents like silica gel and calcium chloride are colored blue by tetrachlorocobaltate(II) ion. So long the blue color is retained, the effectiveness of the drying reagent prevails. The appearance of pink coloration indicates that tetrachlorocobaltate(II) ion has taken up enough moisture and has been transformed into hexaquacobalt(II) chloride. This, in turn, implies that the drying agents have absorbed water to their maximum capacity. Therefore, these should be either replaced by a fresh batch, or heated to expel the absorbed water.

5. The glitter of gemstones

We had earlier explained the blue shade of sapphire on the basis of the redox reaction between titanium(IV) and iron(II) ions which are present as impurities in the lattice of *corundum*, Al_2O_3 (*see* Chapter 8 p. 287). The luster of ruby and emerald, on the other hand, is attributed to the crystal field splitting within the chromium(III) oxide impurity, present in the base materials of these gemstones.

Ruby contains about 1% chromium(III) oxide dispersed in *corundum*. The valence shell configuration of chromium(III) ion is $3d^3$. In terms of crystal field theory, the three electrons are spread out in t_{2g} orbitals. One or more of these electrons absorb blue-green light (λ = 490–500 nm) (*see* Table 26.2) gets promoted to e_g level and reflects red light—the characteristic color of ruby.

Emerald too contains chromium(III) oxide impurity, but the base material in this case is *beryl*, $Be_3Al_2Si_6O_{18}$. The presence of beryllium and silicon makes the bonding in *beryl* weaker than in *corundum*. This, in turn, creates a weaker crystal field around chromium(III) oxide. As a result, the crystal field splitting in chromium(III) ion in emerald is less than what is in ruby. Thus it requires light of lower energy (that is, of longer wavelength) to promote the t_{2g} electrons to e_g orbitals. In fact, chromium(III) ion in emerald absorbs orange and red components of visible light (λ = 595–750 nm) and reflects green light, with a tinge of blue shade. For this reason, the green-blue hue is often referred to as *emerald green*.

6. Structure of spinels

Mixed metal oxides having the composition $A^{II}B_2^{III}O_4$ are called spinels. A and B may be different metals or the same metal in different oxidation states.

Spinels are classified into two types: normal spinels, $A^{II}[B_2^{III}]O_4$ and inverse spinels, $B^{III}[A^{II}\,B^{III}]O_4$. In normal spinels, A occupies tetrahedral interstices, while B occupies octahedral interstices. The compound $NiAl_2O_4$ is an example of normal spinel. In inverse spinels, half B's occupy tetrahedral sites and half B's octahedral sites, while A is in octahedral interstices. An example of inverse spinel is Fe_3O_4, which may be represented as, $Fe^{III}[Fe^{II}Fe^{III}]O_4$.

Whether a particular spinel will acquire normal structure or inverse structure largely depends on CFSE. That structure prevails which has a greater negative value of CFSE. For example, consider the mixed manganese oxide, Mn_3O_4. If it is a normal spinel, its CFSE may be calculated as follows:

$Mn^{II}[Mn_2^{III}]O_4$

Mn^{II} $(3d^5)$
Tetrahedral site

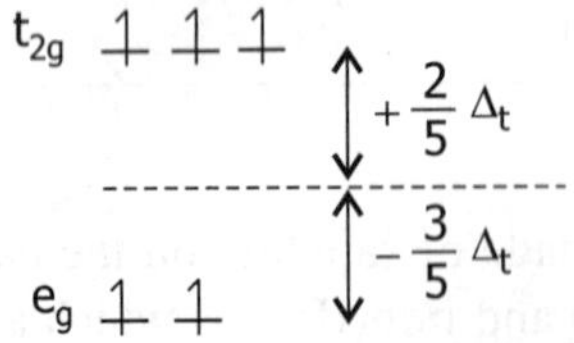

$$\text{CFSE} = 2\left(-\frac{3}{5}\Delta_t\right) + 3\left(+\frac{2}{5}\Delta_t\right)$$

$$= 0$$

Mn^{III} $(3d^4)$
Octahedral site

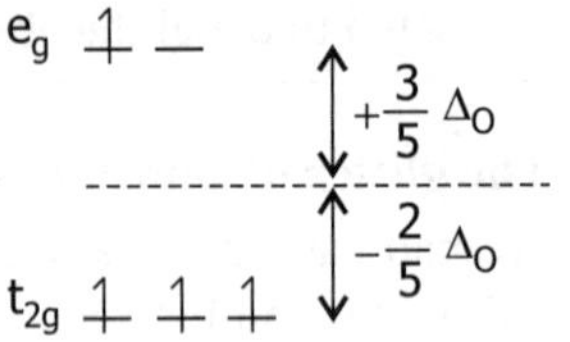

$$\text{CFSE} = 3\left(-\frac{2}{5}\Delta_o\right) + 1\left(+\frac{3}{5}\Delta_o\right)$$

$$= \left(-\frac{3}{5}\Delta_o\right)$$

$$= \left(-\frac{3}{5}\Delta_o\right)\times 2 = -\frac{6}{5}\Delta_o$$

(Since there are two Mn^{III} ions)

$$\text{Total CFSE} = -\frac{6}{5}\Delta_o$$

If it is an inverse spinel, the crystal field stabilization energy may be calculated as follows:

$Mn^{III}[Mn^{II}Mn^{III}]O_4$

Mn^{III} $(3d^4)$
Tetrahedral site

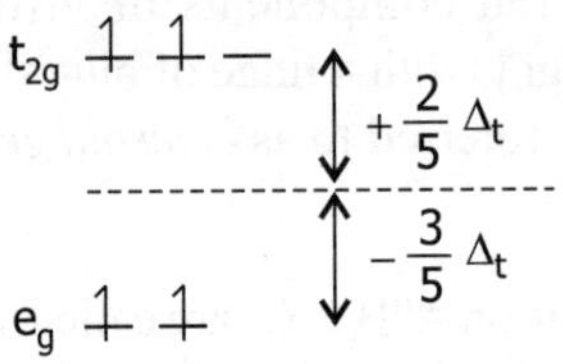

Mn^{II} $(3d^5)$
Octahedral site

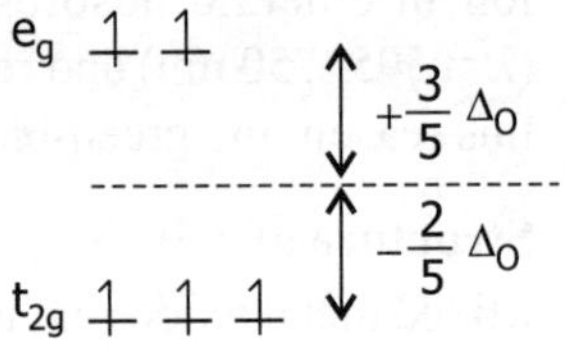

$$\text{CFSE} = 2\left(-\frac{3}{5}\Delta_t\right) + 2\left(+\frac{2}{5}\Delta_t\right) \qquad \text{CFSE} = 3\left(-\frac{2}{5}\Delta_o\right) + 2\left(+\frac{3}{5}\Delta_o\right)$$

$$= -\frac{2}{5}\Delta_t \qquad\qquad = 0$$

$$= -\frac{1}{5}\Delta_o \ (\text{Since } \Delta_t \approx \frac{1}{2}\Delta_o)$$

Mn^{III} ($3d^4$)
Octahedral site

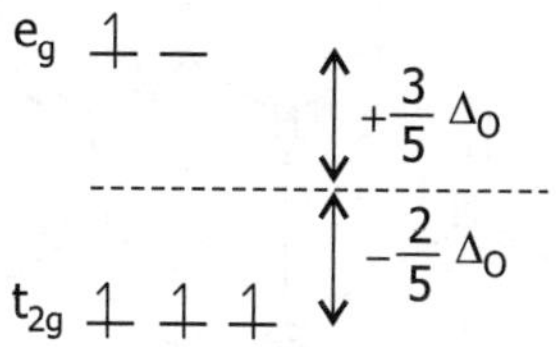

$$\text{CFSE} = 3\left(-\frac{2}{5}\Delta_o\right) + 1\left(+\frac{3}{5}\Delta_o\right)$$

$$= -\frac{3}{5}\Delta_o$$

$$\text{Total CFSE} = -\frac{1}{5}\Delta_o + 0 + \left(-\frac{3}{5}\Delta_o\right) = -\frac{4}{5}\Delta_o$$

The normal spinel has a greater negative CFSE and hence this structure prevails.

It must, however, be emphasized that CFSE is not the only factor that dictates the structure of spinels. Other parameters, such as ion–ion interactions and crystal packing, are also important. Therefore, the structure of every spinel cannot be predicted on the basis of CFSE calculations alone.

7. **Reaction mechanisms**

The complexes show a wide variation in the rate of substitution reactions. Some reactions are slow to occur, while others are appreciably rapid. When one ligand is replaced by another, the magnitude of crystal field splitting changes and so does the CFSE. This means that the stabilization afforded due to the splitting of d orbitals in the crystal field is directly related to the activation energy of the complex. For example, in the case of an iron(III) complex in high-spin state, the crystal field stabilization energy is zero. Replacement of one ligand by another will cause neither gain nor loss of crystal field stabilization energy. Therefore, the complexes of iron(III) undergo substitution at a fast rate. On the other hand, substitution of one ligand by another in chromium(III) [t_{2g}^3—half-filled] and cobalt(III) [t_{2g}^6—completely-filled] complexes is an extremely slow process. These complexes possess very high crystal field stabilization energies; substitution of ligand brings down the magnitude of crystal field stabilization energies and, therefore, the reactions do not occur smoothly.

Case Study 26.10 The tris(1, 10 phenanthroline)iron(II) complex, $[Fe(phen)_3]^{2+}$ is more stable than the tetraqua(1, 10-phenanthroline)iron(II) complex, $[Fe(phen)(H_2O)_4]^{2+}$. Why?

There are two reasons for the tris(1, 10-phenanthroline) complex being more stable than its mono counterpart. Firstly, 1, 10-phenanthroline is a bidentate chelating ligand.

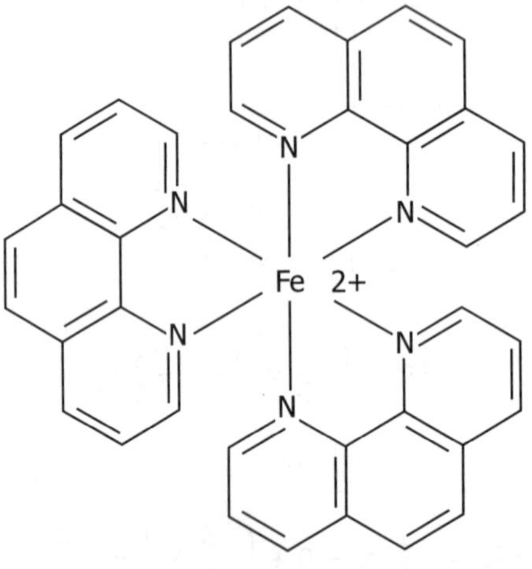

Tris(1, 10-phenanthroline)iron(II)

The $[Fe(phen)_3]^{2+}$ complex has three chelating rings, while the $[Fe(phen)(H_2O)_4]^{2+}$ has only one chelating ring. According to chelate effect, the coordination compound having a larger number of chelate rings is more stable as compared to the one that has a smaller number of such rings in its structure.

The second—and equally important—reason for the greater stability of $[Fe(phen)_3]^{2+}$ complex is attributed to its favorable crystal field stabilization energy. The ligand 1, 10-phenanthroline is quite strong and hence the energy difference between t_{2g} and e_g orbitals of iron(II) ion is appreciably large. The configuration of iron(II) ion (a $3d^6$ system) in the complex is t_{2g}^6.

The $[Fe(phen)(H_2O)_4]^{2+}$ complex, on the other hand, involves four weak aqua ligands. In this case the energy difference between t_{2g} and e_g orbitals is relatively small. The configuration of iron(II) ion in this complex is $t_{2g}^4\ e_g^2$. The presence of electrons in e_g orbitals makes the crystal field stabilization energy less favorable and, in turn, reduces the stability of the complex.

Drawbacks of crystal field theory

The crystal field theory suffers from the following drawbacks:

1. **Classification of ligands**

   As already mentioned, the crystal field theory is based on ionic model. The halide ions are known to form simple compounds which are largely ionic in nature. Yet, with reference to complexation, they are considered weak ligands. Since the halide ions are low on the spectrochemical series, they produce small crystal field splitting. On the other hand, the ligands which are high on spectrochemical series create strong crystal fields. These ligands—for example, ammonia, carbonyl, nitrosyl, pyridine—are neutral and are known to form covalent, rather than ionic compounds.

2. **Nephelauxetic effect**

Nephelauxetic literally means "atmospheric cloud expansion." When extended to chemical species, it implies electron cloud expansion. Nephelauxetic effect is particularly important for complexes. It has been noted that the repulsions between the d^n electrons in the free metal ion are more than that in a complex. Since the crystal field theory treats the ligands as point charges, the electron–electron repulsions in a free metal ion and a complexed one should be essentially constant. However, the fact that the repulsions get reduced on complexation indicates that the mean distance between the d electrons of metal has increased.This is possible only if there is an expansion of d electron cloud. The expansion of electron cloud, in turn, is possible only if the d orbitals on metal overlap with valence orbitals of ligands, providing a path for the electrons to spread out. This refutes the validity of crystal field theory, which states that there is only electrostatic force of attraction between the metal ions and the ligands.

Case Study 26.11 Which ligand is expected to have a larger nephelauxetic effect: ammine or cyano?

According to molecular orbital theory, the ammine ligand enters into complexation with metal ions without involving π bonding. It forms molecular orbitals only with metal e_g orbitals and not with t_{2g} orbitals (*see* p. 743). The cyano ligand, on the other hand, is a π-acid (much like carbon monoxide). During complexation, it forms molecular orbitals both with e_g and t_{2g} orbitals of the metal. Thus electron cloud expansion is more pronounced with cyano ligand than with ammine group. Hence the former has a larger nephelauxetic effect as compared to the latter.

---

3. **Anti-ferromagnetic coupling**

Anti-ferromagnetic compounds are those which have a less than expected magnetic susceptibility. In such compounds, roughly one-half of the ions have their magnetic moments lined-up in the opposite direction to those of the other half. An example of an anti-ferromagnetic compound is nickel(II) oxide, a crystal of which consists of an array of nickel(II) ions separated by oxide ions. This system may be visualized as,

$$----Ni^{2+}---O^{2-}---Ni^{2+}----$$

The metal ions possess unpaired electrons in d orbitals, while the oxide ions have paired-up electrons in p orbitals.

The anti-ferromagnetic behavior may be explained by assuming that there is an overlap between the metal ions' d orbitals and the oxide ions' p orbitals. The unpaired electron on one metal ion shall pair-up with one of the p electrons of oxygen, provided their spins are opposite. The other p electron can pair with the unpaired electron of the second metal ion. This is, however, possible only if the unpaired electrons on each of the two metal ions have opposite spins (circled).

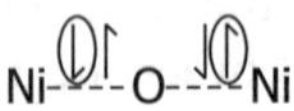

The net result is that by intervention of oxide ion, the metal ions acquire anti-parallel alignment of magnetic moment. If overlapping of metal and oxide is possible in a molecule like nickel oxide, which has a substantial degree of ionic character, it will be all the more probable in complexes. Yet the very concept of metal orbitals overlapping with ligand orbitals is antithesis to crystal field theory.

4. **Electron spin resonance spectrum**

The electron spin resonance is observed when atoms, ions or molecules having one or more unpaired electrons are placed in a magnetic field. The simplest species showing this effect is the hydrogen atom, having one unpaired electron. Under normal circumstances, the energy levels corresponding to two spin angular momentum quantum numbers ($m_s = +1/2$ and $m_s = -1/2$) are degenerate, i.e., there are equal chances for an electron to possess a spin of +1/2 or –1/2. However, in the magnetic field, the degeneracy of the two m_s states is resolved (Fig. 26.17). The energy of the anti-clockwise spin state is lowered and that of the clockwise spin state is raised. The electron occupies the low energy state.

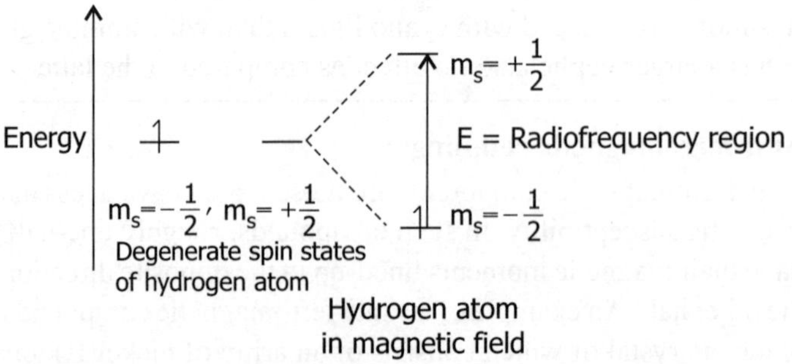

Fig. 26.17 Resolution of spin degeneracy in a magnetic field

If energy equivalent to that of a radiofrequency radiation is provided to the electron, it is excited to the upper energy level. Since only one electron is being excited, the spectrum for this change is observed as a single peak (Fig. 26.18).

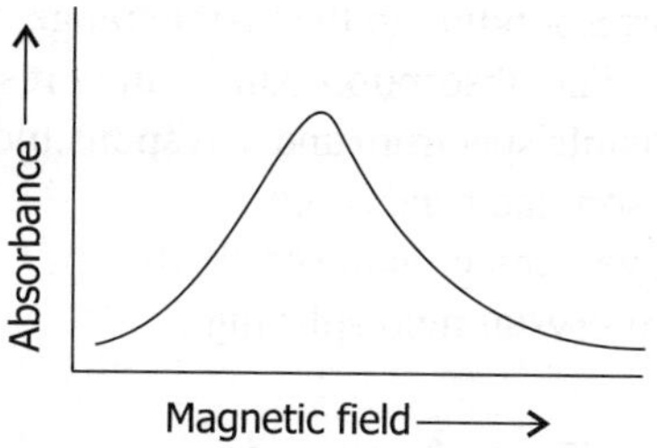

Fig. 26.18 Electron spin resonance spectrum of hydrogen atom

We shall now extend this concept to a complex, potassium hexachloroiradate(IV), $K_2[IrCl_6]$. The iridium(IV) ion has the configuration $5d^5$. Although chloro is a weak field ligand, the magnitude of crystal field splitting is substantially high, since iridium is a 5d series metal. Therefore, the complex is spin-paired.

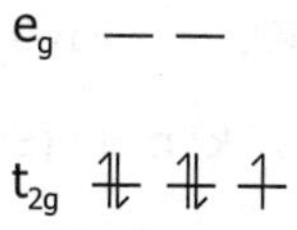

As in the case of hydrogen atom, there is just one unpaired electron, which may be excited from $m_s = -1/2$ state to $m_s = +1/2$ state when radiofrequency radiation is absorbed. We should expect a single peak in the electron spin resonance spectrum, much like that in the case of hydrogen. However, complex pattern of shoulder bands is observed instead (Fig. 26.19). The shoulder bands are called *hyperfine splittings.* These splittings may be explained by assuming that the iridium orbitals overlap with chlorine orbitals, so that the unpaired electron on the metal is delocalized on to the six ligands. If the crystal field theory had been completely valid, the electron spin resonance spectrum of iridium(IV) ion would have shown a single, unsplit absorption band.

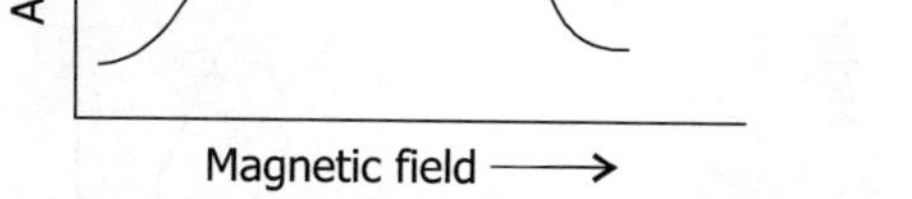

Fig. 26.19 Electron spin resonance spectrum of hexachloroiradate(IV) ion

5. Visible spectrum

Let us consider the hexaquatitanium(III) ion, $[Ti(H_2O)_6]^{3+}$. The titanium(III) ion is a $3d^1$ system with the configuration, $t_{2g}^1e_g^0$. Upon irradiation with light of frequency corresponding to the visible range, the t_{2g} electron is excited to the e_g orbital. The absorption band which results from this excitation appears in the visible spectrum and is responsible for the violet color of the complex. The absorption band is shown in Fig. 24.20. The peak of the band occurs at the wave number value of 20,400 cm^{-1}. This value is identical to the magnitude of crystal field splitting.

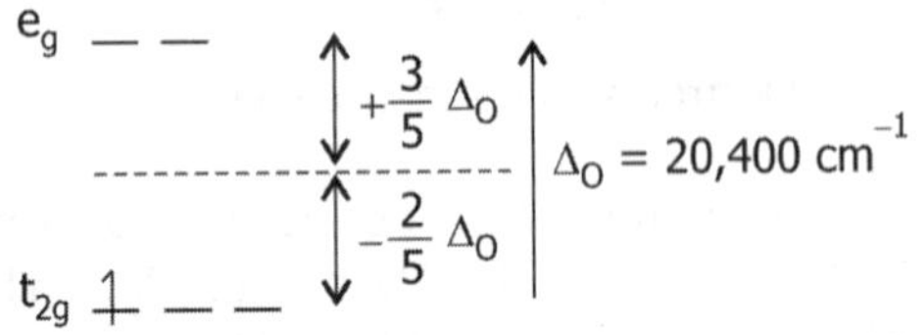

$$\text{CFSE} = 1\left(-\frac{2}{5}\Delta_o\right) = -\frac{2}{5}\Delta_o$$

$$= -\frac{2}{5}(20{,}400) = -\,8160\ \text{cm}^{-1}$$

$$\text{CFSE} \approx -97\ \text{kJ mol}^{-1}\ (\text{Since } 1\ \text{kJ} = 84\ \text{cm}^{-1})$$

The experimental value of crystal field stabilization energy is nearly the same.

As seen in Fig. 26.20, the absorption band is quite broad. The absorption band of uncomplexed titanium(III) ion, on the other hand, is very sharp. If crystal field theory would have projected a perfect picture of metal–ligand relationship, the absorption band of titanium(III) ion in the complex would have also been in the form of a sharp peak. The fact that the band is quite broad proves that the aqua ligand molecules are not point charges. Their orbitals are overlapping with the orbitals of metal ion. Rather than being purely ionic, the metal–ligand bonds have appreciable covalent character.

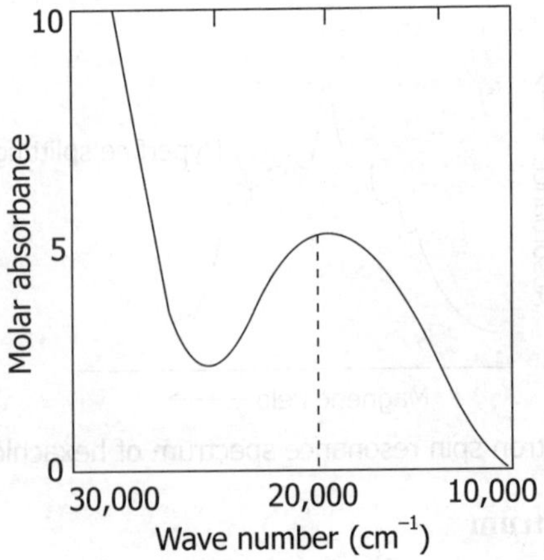

Fig. 26.20 Visible spectrum of hexaquatitanium(III) ion

Case Study 26.12 A complex ion having the composition, $[TiL_6]^{3+}$ (where L is a neutral ligand) shows an absorption maxima at 500 nm. Calculate the value of crystal field splitting, Δ_o for this complex ion.

We have seen in the last section that the value of absorption maxima (in terms of wavelength, λ) may be taken as equivalent to crystal field splitting, Δ_o. For the given complex,

$$\lambda = 500 \text{ nm} = 500 \times 10^{-7} \text{ cm}$$

The wavelength is related to the wave number ($\bar{\upsilon}$) by the following expression (*see* Equation 1.23, p. 11).

$$\bar{\upsilon} = \frac{1}{\lambda}$$

$$\bar{\upsilon} = \frac{1}{500 \times 10^{-7}} = 20{,}000 \text{ cm}^{-1}$$

The value of Δ_o for this complex is 20,000 cm^{-1}.

6. Intensities of d–d transitions

The visible spectrum of hexaquatitanium(III) ion (Fig. 26.20) depicts that the absorption band has only a moderate intensity. Although the intensity is not zero, from theoretical considerations it is expected to be zero. The selection rules for visible spectrum permit only those transitions for which the change in azimuthal quantum number is an integer, i.e., the transitions for which $|\Delta l| = 0$ are forbidden, where Δl is the difference in the azimuthal quantum numbers of orbitals which an electron occupies in ground state and excited state.

In an uncomplexed titanium(III) ion, the 3d orbitals are not split into t_{2g} and e_g sets. On absorbing visible radiation, the single electron in 3d orbital is excited to the next higher, 4s orbital. This transition, called d–s transition, is allowed since $|\Delta l| \neq 0$ (Fig. 26.21). As a result, a very intense absorption band is observed.

In a complexed titanium(III) ion, on the other hand, the 3d orbitals are split into t_{2g} and e_g sets. On absorbing visible radiation, the single electron in t_{2g} orbital is excited to the next higher, e_g orbital. This transition, called d–d transition, is not allowed since $|\Delta l| = 0$ (Fig. 24.21).

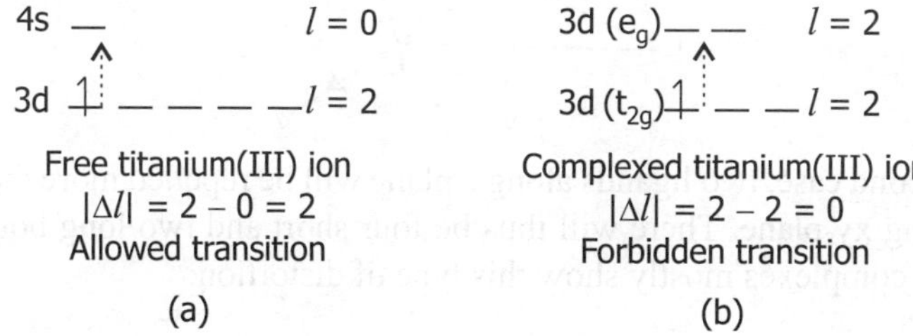

Fig. 26.21 Electronic excitation in (a) free and (b) complexed titanium(III) ion

The fact that, despite being forbidden, the d–d transition still occurs proves that either the t_{2g} orbitals or the e_g orbitals or both are not pure d orbitals. The imperfection in their symmetry can arise only if these orbitals mix up with the ligand orbitals. The mixing of metal and ligand orbitals is contrary to the electrostatic model on which the crystal field theory is based.

Jahn–Teller Theorem

Jahn–Teller theorem, also called *Jahn–Teller effect*, states that any non-linear molecular system in a degenerate electronic state shall be unstable. It will undergo some form of distortion that will lower its symmetry and resolve the degeneracy. The theorem is a fallout of the structural effects of crystal field splitting. It applies to those complexes in which the t_{2g} and e_g sets of d orbitals are asymmetrically filled. For example, consider copper(II) ion ($3d^9$ system) in an octahedral environment. The valence electrons shall possess the configuration, $t_{2g}^6 e_g^3$.

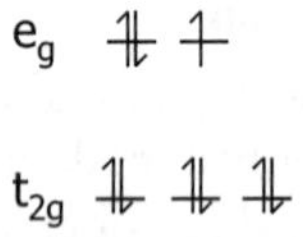

When we consider the individual orbitals in the e_g set, there are two possible configurations:

(i) $t_{2g}^6, d_{x^2-y^2}^2 d_{z^2}^1$

(ii) $t_{2g}^6, d_{x^2-y^2}^1 d_{z^2}^2$

In the first case, four ligands along xy plane will be repelled more than the two ligands along z plane. The perfectly octahedral structure will be distorted, giving four long and two short bond lengths.

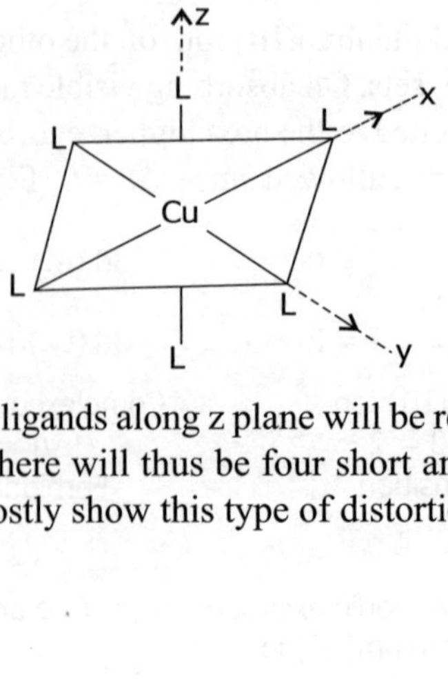

In the second case, two ligands along z plane will be repelled more than the four ligands along xy plane. There will thus be four short and two long bond lengths. The copper complexes mostly show this type of distortion.

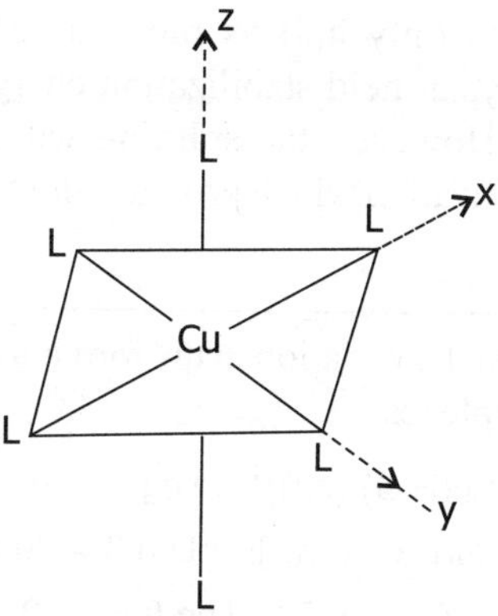

In either case, the unequal bond lengths lower the symmetry below that expected for a regular octahedral structure. Jahn–Teller theorem predicts that the distortion would occur, but does not tell us how large it would be. Further, as a result of distortion, the degeneracy within the t_{2g} and e_g sets of orbitals is lifted. The energy pattern, therefore, changes. When we consider the t_{2g}^6, $d_{x^2-y^2}^1 d_{z^2}^2$ case, we find that ligands along the z plane are at a greater distance from the metal ion than ligands along the xy plane. Therefore, the orbitals having a z component, both in the e_g and t_{2g} sets, are stabilized with respect to orbitals having x and y components. As a result, degeneracy is resolved within the e_g set as well as within the t_{2g} set (Fig. 26.22). Both the splittings are much smaller than Δ_O—the energy gap between e_g and t_{2g} orbitals.

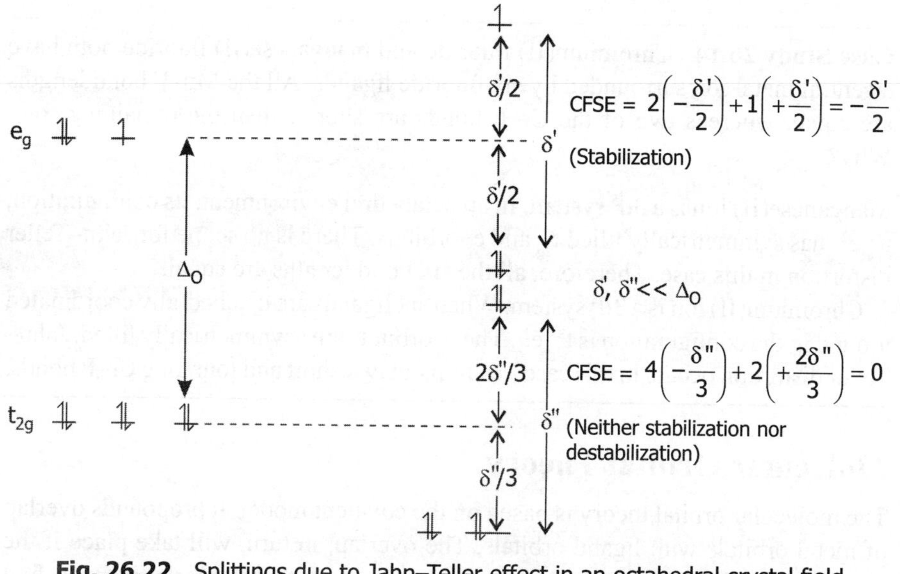

Fig. 26.22 Splittings due to Jahn–Teller effect in an octahedral crystal field

The two e_g orbitals are split in such a manner that one is destabilized as much as the other is stabilized. The t_{2g} orbitals are split in such a manner that two orbitals are stabilized only half as much as one orbital is destabilized. There is no net gain in crystal field stabilization energy as far as the splitting of t_{2g} orbitals is concerned. However, the splitting within the e_g set results in net stabilization. It is this stabilization which provides the driving force for Jahn–Teller distortion.

Case Study 26.13 Explain the variation in the formation constants of the following octahedral copper(II) complexes:

$$[Cu(H_2O)_4(en)]^{2+} : \text{Log } \beta = 10.72$$
$$[Cu(H_2O)_2(en)_2]^{2+} : \text{Log } \beta = 20.03$$
$$[Cu(en)_3]^{2+} : \text{Log } \beta = -1.0$$

The formation constant reflects the stability of a complex. Higher the value of formation constant, greater the stability of the complex.

The higher stability of $[Cu(H_2O)_2(en)_2]^{2+}$ complex as compared to $[Cu(H_2O)_4(en)]^{2+}$ analog is attributed to chelate effect. The former compound has two chelate rings, while the latter has only one. Going by this trend, $[Cu(en)_3]^{2+}$ is expected to be the most stable of the three complexes, as it has three chelate rings in its structure. However, the three chelate rings make the structure so rigid that there is no scope for distortion. Therefore, the stability associated with Jahn–Teller effect is lost.

Jahn–Teller effect is observed for $t_{2g}^6 e_g^1$, $t_{2g}^6 e_g^3$ and $t_{2g}^3 e_g^1$ cases. It is not observed for $t_{2g}^0 e_g^0$, $t_{2g}^6 e_g^2$ and $t_{2g}^6 e_g^4$ configurations.

Case Study 26.14 Chromium(II) fluoride and manganese(II) fluoride both have a central metal ion surrounded by six fluoride ligands. All the Mn–F bond lengths are equal, whereas two of the Cr–F bonds are shorter than the remaining four. Why?

Manganese(II) ion is a $3d^5$ system. In an octahedral environment, its configuration, $t_{2g}^3 e_g^2$, has symmetrically filled t_{2g} and e_g orbitals. There is no scope for Jahn–Teller distortion in this case. Therefore, all the six bond lengths are equal.

Chromium(II) ion is a $3d^4$ system. When six ligands are octahedrally coordinated around it, the configuration is $t_{2g}^3 e_g^1$. The e_g orbitals are asymmetrically filled. Jahn–Teller distortion occurs in this case, resulting in two short and four long Cr–F bonds.

Molecular Orbital Theory

The molecular orbital theory is based on the covalent model. It propounds overlap of metal orbitals with ligand orbitals. The overlap, in turn, will take place if the concerned orbitals have appropriate symmetry. It is, therefore, pertinent to find out which orbitals can overlap and which cannot. In this context, the application of molecular orbital theory to metal complexes may be considered under two categories:

1. Complexes in which there is no π-bonding between the metal and the ligands.
2. Complexes in which there is π-bonding between the metal and the ligands.

We shall consider both the cases for octahedral complexes, and present them in a qualitative manner. The symmetry requirements of the orbitals can be worked out by the application of group theory, an endeavor that is beyond the scope of the present text. Therefore, we shall not extend the molecular orbital theory to tetrahedral and square planar complexes.

1. Octahedral complexes with no π-bonding

In an octahedral complex, each of the six ligands possesses one σ orbital which overlaps with one of the metal atomic orbitals to give two molecular orbitals—one bonding and one anti-bonding. In a metal of first transition series, for example, there are nine valence shell orbitals: five 3d, one 4s and three 4p orbitals. Six of them—$3d_{x^2-y^2}$, $3d_{z^2}$ (collectively, e_g orbitals), 4s, $4p_x$, $4p_y$ and $4p_z$—have lobes along the metal–ligand bond directions. They can undergo end-to-end overlap with ligand orbitals and are, therefore, suitable for σ-bonding (Fig. 26.23).

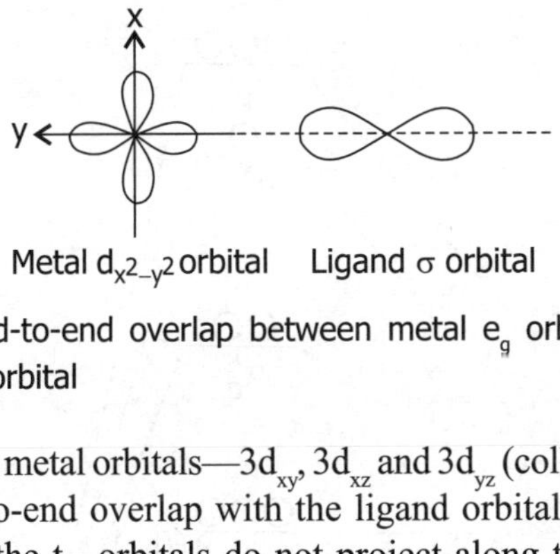

Fig. 26.23 End-to-end overlap between metal e_g orbitals and ligand σ orbital

The remaining three metal orbitals—$3d_{xy}$, $3d_{xz}$ and $3d_{yz}$ (collectively, t_{2g} orbitals)—cannot undergo end-to-end overlap with the ligand orbitals (Fig. 26.24). This is because the lobes of the t_{2g} orbitals do not project along the metal–ligand bond directions.

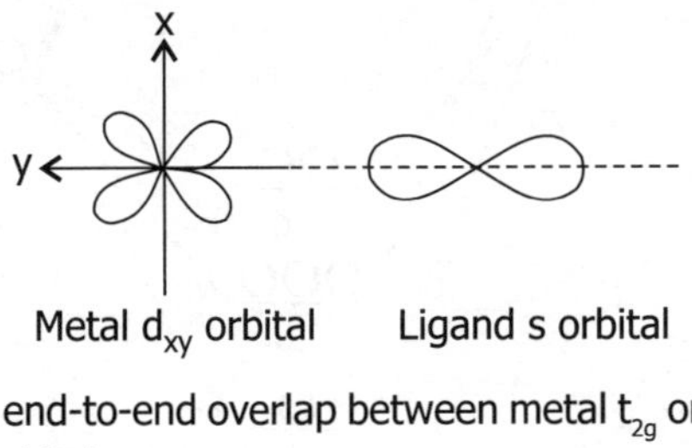

Fig. 26.24 No end-to-end overlap between metal t_{2g} orbitals and ligand σ orbital

These may, however, have lateral overlap with the ligands, provided the latter possess suitable π orbitals, in addition to σ orbitals, for bonding. Some ligands do possess π orbitals of appropriate symmetry and, therefore, can participate in π-bonding. We shall consider such cases in the next section.

Other ligands do not have suitable orbitals for π-bonding and, therefore, can enter into σ-bonding alone with the metal. In such complexes, the three t_{2g} orbitals of the metal are non-bonding in nature. Ammonia and aqua complexes come under this category.

For example, consider the case of hexaquacobalt(III) ion, $[Co(H_2O)_6]^{3+}$. The cobalt(III) ion has six valence orbitals which can participate in the construction of molecular orbitals. These are, $3d_{x^2-y^2}$, $3d_{z^2}$, 4s, $4p_x$, $4p_y$ and $4p_z$. There are six ligand orbitals—one contributed by each coordinating group—that can enter into σ-bonding with metal orbitals. The combination of six metal orbitals and six ligand orbitals gives twelve molecular orbitals. Six of the twelve molecular orbitals have lower energy than the parent orbitals; these are designated as bonding molecular orbitals. The remaining six molecular orbitals have higher energy than the parent orbitals; these are designated as anti-bonding molecular orbitals. The three t_{2g} orbitals on cobalt(III) ion—d_{xy}, d_{xz}, d_{yz}—remain non-bonding. Their energy is neither lowered nor raised.

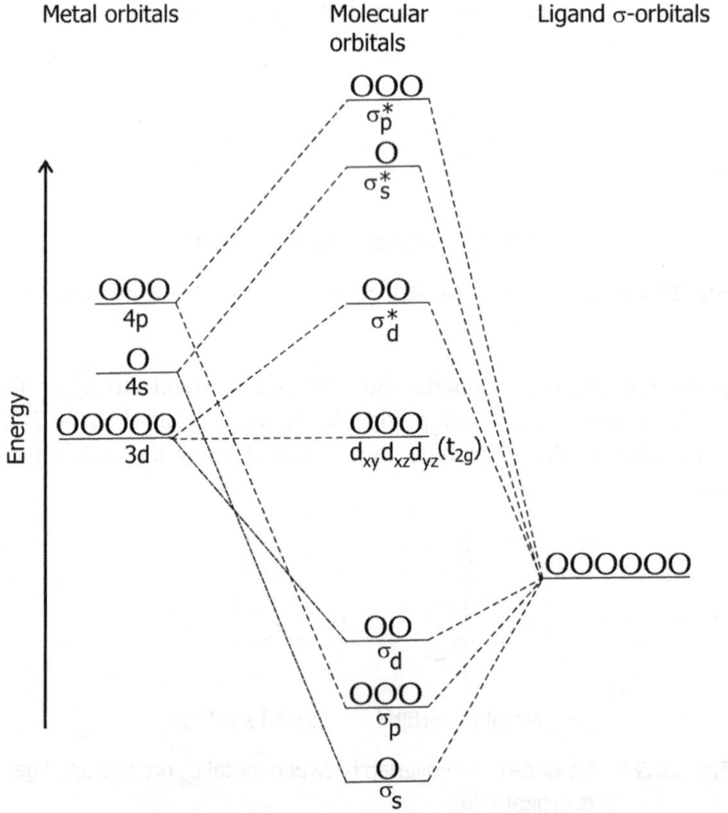

Fig. 26.25 Molecular orbitals for octahedral complexes

The pattern of the energies of various molecular orbitals is depicted in Fig. 26.25. We shall now calculate the number of valence electrons that are involved in metal–ligand bonding.

Co(III), $3d^6$:	Number of electrons	=	$6e^-$
$6 \times H_2O$:	Number of electrons contributed for bonding	=	$12e^-$
		Total	$18e^-$

The first 12 electrons are accommodated in the bonding molecular orbitals, σ_s, σ_p and σ_d. The remaining six electrons are distributed in the three degenerate, non-bonding orbitals of t_{2g} set and the two degenerate anti-bonding orbitals of σ^*_d set. The difference in the energy between these two levels corresponds to the crystal field splitting, Δ_O. In molecular orbital theory, water is considered a weak ligand since its orbitals overlap poorly with the metal orbitals. The magnitude of Δ_O is, therefore, small and for this reason, the hexaquacobalt(II) ion is a spin-free complex.

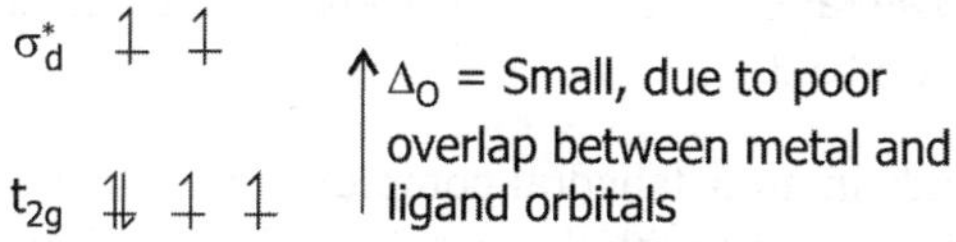

We now consider hexamminecobalt(III) ion, $[Co(NH_3)_6]^{2+}$. This complex too has 18 electrons and the first 12 are accommodated in the bonding molecular orbitals, in the same way as in the hexaquacobalt(III) ion. However, the remaining six are paired up in the non-bonding t_{2g} orbitals. As compared to water, the orbitals of ammonia have a more efficient overlap with the metal orbitals. For this reason, ammonia is considered a strong ligand in the molecular orbital framework. Since the value of Δ_O is high, the complex is spin-paired.

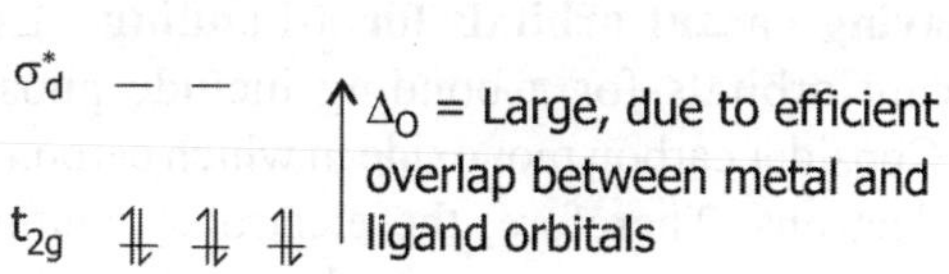

Thus the hexaquacobalt(III) ion is paramagnetic, with four unpaired electrons, while the hexamminecobalt(III) ion is diamagnetic. The crystal field theory, when applied to these two cases, gives the same results. However, in crystal field theory, Δ_O represents the difference in energy between e_g and t_{2g} sets of metal orbitals. This difference, in turn, arises because t_{2g} orbitals are repelled less than the e_g orbitals by any particular ligand. A stronger ligand will cause the two sets of orbitals to be separated by a greater energy gap than a weaker ligand. In molecular orbital theory, Δ_O represents the energy difference between anti-bonding, $e_g(\sigma_d^*)$ molecular orbitals and the non-bonding t_{2g} atomic orbitals. A stronger ligand is one whose orbitals overlap more efficiently with the metal orbitals, giving a larger value of Δ_O. The orbitals of a weaker ligand overlap poorly with the metal orbitals, resulting in a smaller value of Δ_O.

2. Octahedral complexes with π-bonding

If the ligands have π orbitals then lateral overlap of these with the t_{2g} orbitals of the metal is possible (Fig. 26.26). Such an overlap gives rise to π-bonding between the metal and the ligands, in addition to the existing σ-bonding.

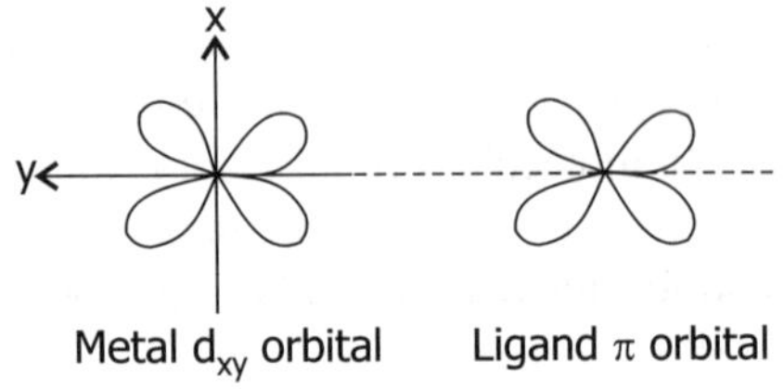

Fig. 26.26 Lateral overlap between metal t_{2g} orbitals and the ligand π orbital

If all the six ligands in an octahedral complex are of the same type, each shall possess a π orbital for overlap with the metal t_{2g} orbitals. However, at a particular moment, at the most three ligands can participate in π-bonding. This is because only three of the metal orbitals—d_{xy}, d_{xz} and d_{yz}—are available for π-bonding. This does not mean that only three of the six metal–ligand bonds have π character. The π electron cloud is not static, but covers all the metal–ligand bonds by way of resonance.

The molecular orbital picture of complexes with π-bonding may be considered under two sub-categories:

a. **Ligands having vacant orbitals for π-bonding** Examples of ligands having vacant orbitals for π-bonding include phosphine and carbon monoxide. Consider carbon monoxide in which carbon has six and oxygen has eight electrons. Therefore, the electronic configuration of carbon monoxide, with fourteen electrons in all, may be written as follows:

$$\text{CO} : \sigma_{1s}^{2}\ \sigma_{1s}^{*2}\ \sigma_{2s}^{2}\ \sigma_{2s}^{*2} \begin{bmatrix} \pi_{2p_y}^{2} \\ \pi_{2p_z}^{2} \end{bmatrix} \sigma_{2p_x}^{2} \begin{bmatrix} \pi_{2p_y}^{*0} \\ \pi_{2p_z}^{*0} \end{bmatrix}$$

Vacant orbitals for participation in π-bonding

The π^*_{2py} and π^*_{2pz} orbitals of carbon monoxide are vacant and these are suitable for π overlap with metal t_{2g} orbitals. This is called *back bonding* or *metal-to-ligand π-bonding*. The three t_{2g} orbitals of metal combine with three π orbitals of the three ligands to give six molecular orbitals—three bonding and three anti-bonding. The σ_d^* orbitals do not have appropriate symmetry to overlap with the ligand π orbitals. Hence, their energy is not perturbed.

The sequence of various molecular orbitals for π complexes remains the same as for σ only complexes. We, therefore, reproduce only the central part of Fig. 26.25—the portion which highlights the t_{2g} and σ_d^* orbitals—and see how the magnitude of Δ_o is affected (Fig. 26.27).

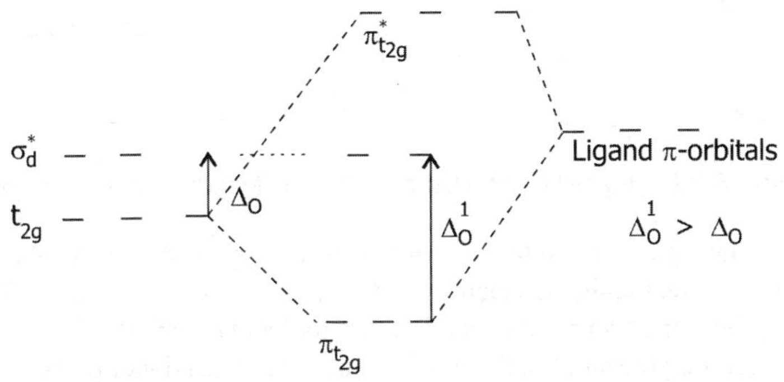

Fig. 26.27 Interaction of vacant π orbitals of ligands with metal t_{2g} orbitals

The ligand π orbitals have energies higher than the metal t_{2g} orbitals. When the two sets combine, the bonding $\pi_{t_{2g}}$ molecular orbitals are stabilized with respect to the σ_d^* molecular orbitals. The value of Δ_o changes to Δ_o^1, the latter having a larger magnitude. The extent of metal–ligand overlap increases. For this reason, ligands which possess vacant orbitals for π-bonding occupy a very high position in the spectrochemical series. The electrons in the non-bonding t_{2g} orbitals get transferred to $\pi_{t_{2g}}$ orbitals. No more electrons are added to the system as the ligands use only vacant orbitals for π-bonding.

b. Ligands having filled orbitals for π-bonding The halide ligands have filled orbitals for π-bonding. Consider, for example, the fluoride ion.

$$F^- : \quad 1s^2\, 2s^2\, (2p_x^2\, 2p_y^2)\, 2p_z^2$$

Filled orbitals
for participation
in π-bonding

The filled $2p_x$ and $2p_y$ orbitals of the ligands overlap with t_{2g} orbitals of the metal. As a result, the electrons are delocalized from ligands to the metal. This is called *ligand-to-metal π-bonding.* It occurs in complexes in which the metal is in a high oxidation state.

The filled ligand π orbitals are lower in energy than metal t_{2g} orbitals. On combination, these two sets produce bonding $\pi_{t_{2g}}$ molecular orbitals and anti-bonding $\pi^*_{t_{2g}}$ molecular orbitals. The latter are destabilized with respect to t_{2g} orbitals and, therefore, the magnitude of splitting is reduced from Δ_o to $\Delta_o{}^1$, (Fig. 26.28).

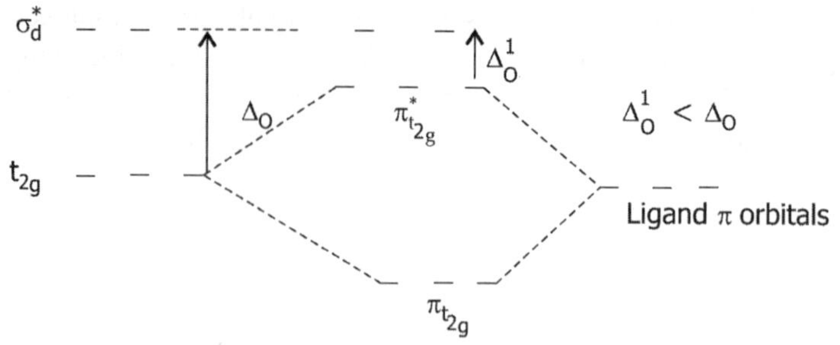

Fig. 26.28 Interaction of filled π orbitals of ligands with metal t_{2g} orbitals

The ligand π orbitals are nearer in energy to $\pi_{t_{2g}}$ molecular orbitals. Therefore, the ligand electrons enter the $\pi_{t_{2g}}$ molecular orbitals. The metal t_{2g} electrons occupy anti-bonding $\pi^*_{t_{2g}}$ molecular orbitals.

Among the halides, fluoride ion has the highest donor power and iodide the lowest. For this reason, π-bonding is more feasible with fluoride than with iodide ligands. Hence, on the spectrochemical scale, fluoride is a stronger ligand than iodide.

Comparison of Crystal Field and Molecular Orbital Theories

The crystal field theory assumes that there is ionic bonding between the metal and ligands. Molecular orbital theory advocates for covalent bonding. The two theories represent two extremes. The metal–ligand bonds are neither purely ionic nor completely covalent. This may be verified by determining the values of some of the bonding parameters.

Let us consider the hexamminecobalt(III) ion, $[Co(NH_3)_6]^{3+}$. If the bonding between the cobalt(III) ion and ammonia ligands is electrostatic, as the crystal field theory would propound, the charge on the metal ion would be +3. If, on the other hand, the metal–ligand bonding is covalent, as the molecular orbital theory would propagate, the charge on the metal would be zero. The experimental value of charge on cobalt, in this particular complex, has been elucidated to be +0.3, showing that the cobalt–ammonia bonds possess intermediate character between the two extreme types of bondings.

The intermediate bond character is also confirmed by determining the value of *spin–orbit coupling constant*. In a 100% covalent bond, complete coupling of spin and orbital moments takes place and the value of spin–orbit coupling constant is one. In a 100% ionic bond, neither the spin moments nor the orbital moments couple with each other and the value of spin–orbit coupling constant is zero. In hexamminecobalt(III) ion, it is 0.7.

The quantitative values of charge and spin–orbit coupling constant are more close to those predicted by molecular orbital theory. This indicates that the bonding between cobalt and ammonia is predominantly covalent. Yet, it is not 100% covalent. This is true for most other complexes too.

The orbitals of the metal and of the ligands for sure overlap with each other. However, the extent of overlap is small. Therefore, the crystal field theory cannot be completely written off. Some amount of ionic bonding does exist.

The theory which takes into account the partly ionic and partly covalent character of metal–ligand bonds is called *ligand field theory* or *adjusted crystal field theory.* It elucidates the percentage ionic (or covalent) character of the metal–ligand bonds in a particular complex. It then attempts to explain the values of various bonding parameters. The details of the ligand field theory are, however, beyond the scope of this textbook.

Case Study 26.15 Comment on the asymmetrical structure of manganese(III) fluoride.

In a crystal of manganese(III) fluoride, several $[MnF_6]$ units are linked together, creating an octahedral coordination of fluoride ions around manganese(III) ion. The interesting feature of this compound is that there are three different manganese–fluorine bond distances, as shown in Fig. 26.29.

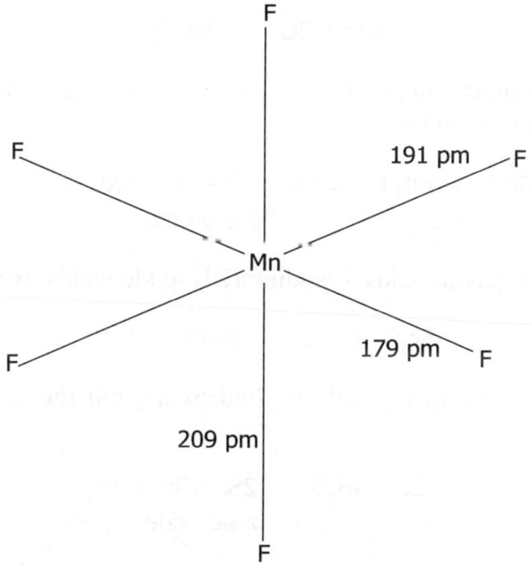

Fig. 24.29 Structure of manganese(III) fluoride

The manganese(III) ion is a $3d^4$ system and has the configuration, $t_{2g}^3 e_g^1$. Due to asymmetrical filling of e_g orbitals, it undergoes Jahn–Teller distortion. This is reflected in the relatively long bond length (209 pm) along the z-axis. The unequal bond lengths along the x- and y-axes, 191 pm and 179 pm, respectively may be explained as follows. If one fluorine ion, common to two $[MnF_6]$ units is 191 pm away from two manganese(III) ions, while the other is 179 pm away, it results in a close-packed arrangement of octahedral skeletons. It is the packing of anions in the crystal lattice, in addition to the Jahn–Teller distortion that makes the molecular structure of manganese(III) fluoride relatively asymmetrical.

---

27. CHEMISTRY OF 3d SERIES TRANSITION ELEMENTS

Introduction

The chemistry of 3d transition elements in various oxidation states is reviewed in this chapter.

Scandium

The only important oxidation state of scandium is III, in which it forms numerous salts. For example, *scandium(III) oxide* (Sc_2O_3) is formed by heating the metal in air.

$$4Sc + 3O_2 \rightarrow 2Sc_2O_3$$

Anhydrous scandium chloride may be prepared by heating scandium oxide with ammonium chloride at 600 K.

$$Sc_2O_3 + 6NH_4Cl \xrightarrow{600\text{ K}} \underset{\text{Anhydrous}}{2ScCl_3} + 6NH_3 + 3H_2O$$

However, with aqueous acids, scandium(III) oxide yields oxochloride.

$$Sc_2O_3 + 2HCl \rightarrow 2ScOCl + H_2O$$

Scandium does not form a definite hydroxide, but the basic oxide is well established.

$$2Sc + 4H_2O \rightarrow \underset{\text{Basic oxide}}{2ScO.OH} + 3H_2$$

Hexafluoroscandate(III) ion is formed by the combination of scandium(III) fluoride with fluoride ions.

$$ScF_3 + 3NH_4F \rightarrow (NH_4)_3[ScF_6]$$

In *pentanitratoscandate(III) ion,* $[Sc(NO_3)_5]^{2-}$, four nitrate ligands are bidentate, and one is monodentate, giving a total coordination number of 9.

Case Study 27.1 When scandium(III) oxide is fused with sodium oxide, a mixed oxide, **A** is formed. **A** gets readily hydrolyzed to yield a basic oxide, **B**. When **B** is treated with an alkali, an octahedral complex, **C** is formed, but when treated with hydrochloric acid, followed by ammonium thiocyanate, a yellow precipitate, **D** is formed. Identify **A**, **B**, **C** and **D**.

Scandium(III) oxide (Sc_2O_3) on fusion with sodium oxide gives a mixed oxide of sodium and scandium, called sodium dioxoscandate(III), **A**.

$$\underset{}{Sc_2O_3} + Na_2O \rightarrow \underset{\textbf{A}}{2Na[ScO_2]}$$

On hydrolysis, the mixed oxide yields basic scandium oxide, **B**.

$$\underset{\textbf{A}}{Na[ScO_2]} + H_2O \rightarrow NaOH + \underset{\textbf{B}}{ScO.OH}$$

When the basic oxide is treated with sodium hydroxide, an octahedral complex, sodium hexahydroxoscandate(III), **C** is formed.

$$\underset{\textbf{B}}{ScO.OH} + 3NaOH + H_2O \rightarrow \underset{\textbf{C}}{Na_3[Sc(OH)_6]}$$

With hydrochloric acid, the basic scandium oxide first gives scandium(III) chloride,

$$\underset{\textbf{B}}{ScO.OH} + 3HCl \rightarrow ScCl_3 + 2H_2O$$

which then gives a yellow precipitate of scandium(III) thiocyanate, **D**, with ammonium thiocyanate.

$$ScCl_3 + 3NH_4SCN \rightarrow \underset{\textbf{D}}{Sc(NCS)_3}\downarrow + 3NH_4Cl$$

Thus, **A** = $Na[ScO_2]$, **B** = ScO.OH, **C** = $Na_3[Sc(OH)_6]$ and **D** = $Sc(NCS)_3$.

Titanium

Titanium forms compounds in oxidation states II, III and IV. Oxidation state IV is the most stable and common. In fact, the lower valent states are easily oxidized to titanium(IV).

Oxidation state IV

Titanium(IV) chloride ($TiCl_4$) is prepared by passing chlorine gas over heated titanium(IV) oxide, in the presence of carbon.

$$TiO_2 + C + 2Cl_2 \rightarrow TiCl_4 + CO_2$$

It is a colorless liquid with a pungent odor. It fumes in moist air and gets converted into titanium(IV) oxide.

$$TiCl_4 + 2H_2O \rightarrow TiO_2 + 4HCl$$

With concentrated hydrochloric acid, it yields hexachlorotitanate(IV) ion.

$$\underset{\text{Conc.}}{TiCl_4 + 2HCl} \rightarrow \underset{\text{Soluble}}{H_2[TiCl_6]}$$

This ion exists in aqueous solution, but may be precipitated if hydrogen ions are replaced by a large cation. For example,

$$H_2[TiCl_6] + 2CsCl \rightarrow \underset{\text{Insoluble}}{Cs_2[TiCl_6]} + 2HCl$$

$$H_2[TiCl_6] + 2NH_4Cl \rightarrow \underset{\text{Insoluble}}{(NH_4)_2[TiCl_6]} + 2HCl$$

Titanium(IV) chloride is a Lewis acid. It readily combines with Lewis bases.

$$TiCl_4 + 2R_2O \rightarrow (R_2O)_2TiCl_4 \quad (R = CH_3 \text{ or } C_2H_5)$$

Titanium(IV) chloride may be readily transformed into the corresponding fluoride or bromide.

$$TiCl_4 + 4HF \rightarrow TiF_4 + 4HCl$$

$$TiCl_4 + 4HBr \rightarrow TiBr_4 + 4HCl$$

Titanium(IV) oxide, also called titanium dioxide, (TiO_2), is amphoteric in nature.

$$TiO_2 + H_2SO_4 \rightarrow \underset{\text{Titanyl sulfate}}{TiOSO_4} + H_2O$$

$$TiO_2 + 2NaOH \rightarrow \underset{\text{Sodium titanate}}{Na_2TiO_3} + H_2O$$

Titanium(IV) solutions develop intense yellow-orange color on addition of hydrogen peroxide. The color is attributed to peroxotitanyl ion, TiO_2^{2+}.

$$Ti^{4+} + O_2^{2-} \rightarrow TiO_2^{2+}$$

The reaction may be used for colorimetric estimation of either titanium(IV) ions or peroxide ions.

Titanium(IV) alkoxides are prepared by the reaction of titanium(IV) chloride with the corresponding alcohol, in alkaline medium.

$$TiCl_4 + 4ROH + 4NH_3 \rightarrow Ti(OR)_4 + 4NH_4Cl$$

Ammonia, being basic, assists the deprotonation of alcohol and thus increases the rate of reaction.

The alkoxides are readily hydrolyzed. The initial hydrolytic step involves the coordination of a water molecule to the titanium(IV) ion. This is followed by the interaction of a hydrogen atom of water with alkoxy oxygen through hydrogen bonding.

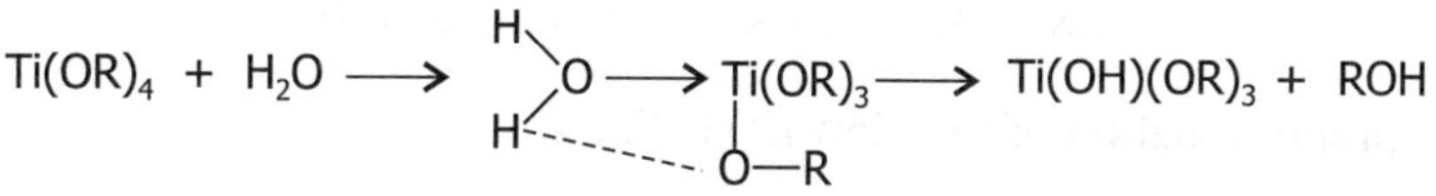

Titanium and barium form a mixed oxide called barium titanate ($Ba_2[TiO_4]$). It is prepared by fusing a mixture of barium carbonate and titanium(IV) oxide.

$$2BaCO_3 + TiO_2 \rightarrow Ba_2[TiO_4] + 2CO_2$$

It has a spinel structure with discrete $[TiO_4]$ units.

Barium titanate shows remarkable ferroelectric nature, i.e., it has an electric dipole moment even in the absence of an external electric field. The electric dipole moment arises when the center of positive charge in a crystal does not coincide with the center of negative charge. In barium titanate, there is an enormous size difference between large barium(II) ion and small titanium(IV) ion. The charges on the barium and titanate ions do not coincide with each other. As a result, titanium(IV) ion is pushed off the center of titanate ion. This phenomenon is often referred to as rattling around of titanium(IV) ion within the $[TiO_4]$ unit. Due to this phenomenon, barium titanate has an appreciably high dipole moment. When an electric field is applied, the titanium(IV) ions are further pushed to one side of the titanate ion. The compound acquires a high electrical polarization and is thus used in capacitors.

Oxidation state III

Titanium(III) chloride ($TiCl_3$) is obtained by passing vapors of titanium(IV) chloride and hydrogen through a red-hot tube at about 900 K.

$$2TiCl_4 + H_2 \xrightarrow{900\ K} 2TiCl_3 + 2HCl$$

In an aqueous solution, it forms hexaquatitanium(III) chloride.

$$TiCl_3 + 6H_2O \rightarrow [Ti(H_2O)_6]Cl_3$$

Titanium(III) oxide (Ti_2O_3) is obtained by heating a mixture of titanium(IV) oxide and titanium(IV) chloride in a stream of hydrogen at 1300 K.

$$3TiO_2 + TiCl_4 + 2H_2 \xrightarrow{1300\ K} 2Ti_2O_3 + 4HCl$$

Titanium(III) oxide is more basic than titanium(IV) oxide.

Oxidation state II

For titanium, oxidation state II is unstable and uncommon.

Titanium(II) halides are obtained by reduction of titanium(IV) halides with titanium,

$$TiX_4 + Ti \rightarrow 2TiX_2 \qquad (X = Cl, Br \text{ or } I)$$

or by disproportionation of titanium(III) halides.

$$TiX_3 + TiX_3 \rightarrow TiX_4 + TiX_2$$

Titanium(II) chloride may be prepared by reduction of titanium(IV) chloride with hexamethyldisilane. The latter acts both as a reducing agent and as a non-aqueous solvent.

$$(CH_3)_3Si - Si(CH_3)_3 + TiCl_4 \rightarrow TiCl_2 + 2(CH_3)_3SiCl$$

In fact, all the reactions involving titanium(II) ions must be carried out in non-aqueous medium. In aqueous solution titanium(II) ions are not stable as these are readily oxidized by water. Nevertheless, if aqueous titanium(II) ions are required, the only way to produce these is to bombard a solution of titanium(III) ions with electrons. This technique is called *pulse radiolysis*.

Titanium(IV) oxide is reduced to *titanium(II) oxide* (TiO) by titanium.

$$TiO_2 + Ti \rightarrow 2TiO$$

Titanium(II) alkoxides are useful reagents for nitrogen fixation. These are obtained by reduction of titanium(IV) alkoxides with sodium.

$$Ti(OR)_4 + 2Na \rightarrow Ti(OR)_2 + 2NaOR$$

$$Ti(OR)_2 + N_2 \rightarrow [Ti(OR)_2N_2]$$

$$[Ti(OR)_2N_2] + 4Na + 6ROH \rightarrow Ti(OR)_4 + 4RONa + 2NH_3$$

Atmospheric nitrogen can thus be converted into ammonia which, in turn, may be transformed into fertilizers.

Case Study 27.2 A colorless, pungent liquid, **A**, on heating with hydrogen, yields a violet solid, **B**, which acts as Ziegler–Natta catalyst for polymerization of ethene to polyethene. **B** on treatment with aqueous iron(III) chloride yields an oxo salt, **C**. Identify **A**, **B** and **C**.

The Ziegler–Natta polymerization catalyst is titanium(III) chloride, the violet solid, **B**. It is obtained by reduction of titanium(IV) chloride, a colorless, pungent liquid, with hydrogen at about 900 K. Therefore, **A** is titanium(IV) chloride.

$$\underset{\mathbf{A}}{2TiCl_4} + H_2 \xrightarrow{900\text{ K}} \underset{\mathbf{B}}{2TiCl_3} + 2HCl$$

Titanium(III) chloride **B** reacts with a solution of iron(III) chloride as follows:

$$FeCl_3 + \underset{\textbf{B}}{TiCl_3} + H_2O \rightarrow FeCl_2 + TiOCl_2 + 2HCl$$

The oxo salt, **C**, is titanyl chloride, $TiOCl_2$.
Thus, **A** = $TiCl_4$, **B** = $TiCl_3$ and **C** = $TiOCl_2$.

Vanadium

Vanadium exhibits oxidation states II to V. The widest range of compounds is known in oxidation state V.

Oxidation state V

Vanadium(V) oxide (V_2O_5) is prepared by heating ammonium metavanadate (NH_4VO_3).

$$2NH_4VO_3 \xrightarrow{\text{Heat}} V_2O_5 + 2NH_3 + H_2O$$

It is an amphoteric oxide.

$$V_2O_5 + 6HCl \rightarrow \underset{\text{Vanadyl chloride}}{2VOCl_2} + 3H_2O + Cl_2$$

$$V_2O_5 + 6NaOH \rightarrow \underset{\text{Sodium vanadate}}{2Na_3VO_4} + 3H_2O$$

Vanadium(V) oxide reacts with halogens to yield vanadium oxotrihalides. For example,

$$2V_2O_5 + 6Cl_2 \rightarrow 4VOCl_3 + 3O_2$$

$$2V_2O_5 + 6Br_2 \rightarrow 4VOBr_3 + 3O_2$$

The oxotrihalides are readily hydrolyzed to yield metavanadic acid (HVO_3).

$$VOCl_3 + 2H_2O \rightarrow HVO_3 + 3HCl$$

Vanadium(V) fluoride is the only halide formed by vanadium in oxidation state V. It is a colorless liquid which may be prepared by heating vanadium in a stream of fluorine at 600 K,

$$2V + 5F_2 \xrightarrow{600\text{ K}} 2VF_5$$

or by disproportionation of vanadium(IV) fluoride at 900 K, in an atmosphere of nitrogen.

$$VF_4 + VF_4 \xrightarrow{900\text{ K}} VF_5 + VF_3$$

Oxidation state IV

Vanadium(IV) oxide (VO_2) is prepared by reducing vanadium(V) oxide with oxalic acid.

$$V_2O_5 + (COOH)_2 \rightarrow 2VO_2 + H_2O + 2CO_2$$

On treatment with sulfuric acid, vanadium(IV) oxide yields vanadyl sulfate ($VOSO_4$).

$$VO_2 + H_2SO_4 \rightarrow VOSO_4 + H_2O$$

Vanadyl sulfate reacts with sodium hydroxide to give a yellow precipitate of vanadyl hydroxide, $VO(OH)_2$.

$$VOSO_4 + 2NaOH \rightarrow VO(OH)_2\downarrow + Na_2SO_4$$

Aqueous acids transform vanadyl hydroxide to deep blue pentaquaoxovanadium(IV) ion, $[VO(H_2O)_5]^{2+}$.

$$VO(OH)_2 + 2H^+ + 3H_2O \rightarrow [VO(H_2O)_5]^{2+}$$

Vanadium(IV) oxide is amphoteric in nature.

$$VO_2 + 2HCl \rightarrow \underset{\text{Vanadyl chloride}}{VOCl_2} + H_2O$$

$$VO_2 + 2NaOH \rightarrow \underset{\text{Sodium hypovanadate}}{Na_2VO_3} + H_2O$$

Vanadyl halides (VOX_2) may also be obtained by reduction of the corresponding oxotrihalides with hydrogen. For example,

$$2VOCl_3 + H_2 \rightarrow 2VOCl_2 + 2HCl$$

Among *vanadium(IV) halides*, vanadium(IV) fluoride is the most stable. It is prepared by the reaction of hydrofluoric acid with vanadium.

$$V + 4HF \rightarrow VF_4 + 2H_2$$

Vanadium(IV) chloride is prepared by the direct reaction of chlorine with vanadium at 450 K.

$$V + 2Cl_2 \xrightarrow{450\ K} VCl_4$$

Oxidation state III

Vanadium(III) oxide (V_2O_3) may be prepared by the reduction of vanadium(V) oxide with carbon monoxide.

$$V_2O_5 + 2CO \rightarrow V_2O_3 + 2CO_2$$

It is a non-stoichiometric oxide and, therefore, represented as $\overline{V_2O_3}$. It is a basic oxide and dissolves in aqueous acids to yield hexaquavanadium(III) ion, $[V(H_2O)_6]^{3+}$. It may be blue or green in color depending on the anion.

$$V_2O_3 + 9H_2O + 6H^+ \rightarrow 2[V(H_2O)_6]^{3+}$$

The hexaquavanadium(III) ion, on standing, slowly decomposes into oxovanadium(III) ion, VO^+ and hydroxovanadium(III) ion, $V(OH)^{2+}$.

$$2[V(H_2O)_6]^{3+} \rightarrow VO^+ + V(OH)^{2+} + 3H^+ + 10H_2O$$

All the *vanadium(III) halides*, except the fluoride, are synthesized by heating vanadium with the halogen. For example,

$$2V + 3Cl_2 \rightarrow 2VCl_3$$

$$2V + 3Br_2 \rightarrow 2VBr_3$$

Vanadium(III) fluoride is prepared by the reaction of hydrogen fluoride with vanadium(III) chloride.

$$VCl_3 + 3HF \rightarrow VF_3 + 3HCl$$

Vanadium(III) fluoride, on being crystallized from water, gives triaquatrifluorovanadium(III) complex, $[VF_3(H_2O)_3]$.

$$VF_3 + 3H_2O \rightarrow [VF_3(H_2O)_3]$$

The chloride and bromide, however, yield hexaquavanadium(III) ion. For example,

$$VCl_3 + 6H_2O \rightarrow [V(H_2O)_6]^{3+} + 3Cl^-$$

Vanadium(III) iodide disproportionates into vanadium(II) iodide and vanadium(IV) iodide.

$$VI_3 + VI_3 \rightarrow VI_2 + VI_4$$

Oxidation state II

The non-stoichiometric *vanadium(II) oxide* is prepared by reducing vanadium(V) oxide with hydrogen at 2000 K.

$$V_2O_5 + 3H_2 \xrightarrow{2000\text{ K}} 2VO + 3H_2O$$

It is a basic oxide and dissolves in aqueous acids to give violet-colored hexaquavanadium(II) ion, $[V(H_2O)_6]^{2+}$.

$$VO + 5H_2O + 2H^+ \rightarrow [V(H_2O)_6]^{2+}$$

Vanadium(II) halides may be obtained by the reduction of vanadium(IV) halides with hydrogen in a red-hot tube. For example,

$$VCl_4 + H_2 \rightarrow VCl_2 + 2HCl$$

On hydrolysis, vanadium(II) halides yield vanadium(III) oxohalides. For example,

$$2VCl_2 + 2H_2O \rightarrow 2VOCl + 2HCl + H_2$$

Case Study 27.3 When an acidified solution of ammonium metavanadate is treated with sulfur dioxide, a deep blue solution, **A** is obtained. On adding zinc amalgam to **A**, a violet solution, **B** results. When solutions **A** and **B**, containing equal amounts of vanadium are mixed, a green solution, **C** is obtained. Identify **A, B** and **C**.

Ammonium metavanadate, on acidification, yields vanadium(V) oxide, which on treatment with sulfur dioxide is reduced to vanadium(IV) oxide. The latter, in the presence of an acid, gets converted into deep blue pentaquaoxovanadium(IV) ion, $[VO(H_2O)_5]^{2+}$ (**A**).

$$2NH_4VO_3 + H_2SO_4 \rightarrow (NH_4)_2SO_4 + V_2O_5 + H_2O$$

$$V_2O_5 + SO_2 + H_2O \rightarrow 2VO_2 + H_2SO_4$$

$$VO_2 + 2H^+ + 4H_2O \rightarrow \underset{\mathbf{A}}{[VO(H_2O)_5]^{2+}}$$

Zinc amalgam, in the presence of an acid, liberates hydrogen which, in turn, reduces pentaquaoxovanadium(IV) ion to violet, hexaquavanadium(II) ion, $[V(H_2O)_6]^{2+}$ (**B**).

$$\underset{\mathbf{A}}{[VO(H_2O)_5]^{2+}} + H_2 \xrightarrow{Zn(Hg)/H^+} \underset{\mathbf{B}}{[V(H_2O)_6]^{2+}}$$

When equimolar quantities of **A** and **B** are mixed, green-colored hexaquavanadium(III) ion, $[V(H_2O)_6]^{3+}$ **C** is formed.

$$\underset{\mathbf{A}}{[VO(H_2O)_5]^{2+}} + \underset{\mathbf{B}}{[V(H_2O)_6]^{2+}} + 2H^+ \rightarrow \underset{\mathbf{C}}{2[V(H_2O)_6]^{3+}}$$

Thus, **A** = $[VO(H_2O)_5]^{2+}$, **B** = $[V(H_2O)_6]^{2+}$ and **C** = $[V(H_2O)_6]^{3+}$.

Chromium

Chromium shows an array of oxidation states, ranging from II to VI. The highest oxidation state, VI, however, exists only in oxo species.

Oxidation state VI

Chromium(VI) oxide (CrO_3) may be prepared by adding concentrated sulfuric acid to a saturated solution of potassium dichromate.

$$\underset{\text{Conc.}}{K_2Cr_2O_7 + H_2SO_4} \rightarrow K_2SO_4 + 2CrO_3 + H_2O$$

It dissolves in water to yield chromic acid (H_2CrO_4).

$$CrO_3 + H_2O \rightarrow H_2CrO_4$$

It is an acidic oxide and, therefore, reacts readily with alkalis.

$$CrO_3 + 2NaOH \rightarrow \underset{\text{Sodium chromate}}{Na_2CrO_4} + H_2O$$

With hydrochloric acid, it liberates *chromyl chloride gas* (CrO_2Cl_2).

$$CrO_3 + 2HCl \rightarrow CrO_2Cl_2 + H_2O$$

Chromium(VI) oxide is a powerful oxidizing agent. It can oxidize tin(II) chloride to tin(IV) chloride.

$$3SnCl_2 + 2CrO_3 + 12HCl \rightarrow 3SnCl_4 + 2CrCl_3 + 6H_2O$$

Potassium dichromate ($K_2Cr_2O_7$) is prepared by fusing chromite ore, FeO. Cr_2O_3 with sodium carbonate, in the presence of air. Sodium chromate (Na_2CrO_4) is initially formed.

$$4FeO.Cr_2O_3 + 8Na_2CO_3 + 7O_2 \rightarrow 2Fe_2O_3 + 8Na_2CrO_4 + 8CO_2$$

Sodium chromate is then treated with sulfuric acid to form sodium dichromate ($Na_2Cr_2O_7$).

$$2Na_2CrO_4 + H_2SO_4 \rightarrow Na_2SO_4 + Na_2Cr_2O_7 + H_2O$$

Finally, sodium dichromate is reacted with a solution of potassium chloride to yield potassium dichromate.

$$Na_2Cr_2O_7 + 2KCl \rightarrow K_2Cr_2O_7 + 2NaCl$$

Potassium dichromate is an orange solid and is soluble in water. It is an analytical reagent and is preferred over sodium dichromate since the latter is hygroscopic. Its structure consists of two tetrahedral chromate units, linked together by a common oxygen atom (Fig. 26.1).

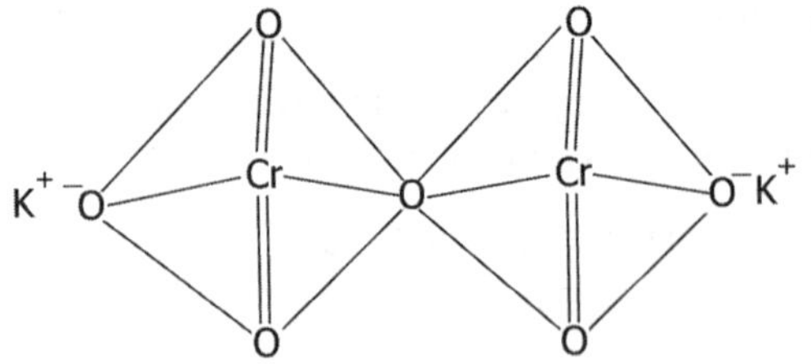

Fig. 27.1 Structure of potassium dichromate

Some of the important properties of potassium dichromate are listed below:

1. On heating, potassium dichromate decomposes into potassium chromate.

$$4K_2Cr_2O_7 \xrightarrow{\text{Heat}} 4K_2CrO_4 + 2Cr_2O_3 + 3O_2$$

2. The action of an alkali also transforms orange-colored potassium dichromate into yellow potassium chromate.

$$K_2Cr_2O_7 + 2KOH \rightarrow 2K_2CrO_4 + H_2O$$

The potassium chromate may be reconverted into potassium dichromate by treating it with a dilute acid.

$$2K_2CrO_4 + \underset{\text{Dilute}}{H_2SO_4} \rightarrow K_2Cr_2O_7 + K_2SO_4 + H_2O$$

Thus, there exists an equilibrium between dichromate and chromate ions.

$$\underset{\text{Yellow}}{2CrO_4^{2-}} + H^+ \rightleftharpoons \underset{\text{Orange}}{Cr_2O_7^{2-}} + OH^-$$

3. Potassium dichromate is a reagent for qualitative estimation of chloride ions in the chromyl chloride test. When solid potassium dichromate and a metal chloride, for example, sodium chloride, are strongly heated along with concentrated sulfuric acid, a red-colored gas, called chromyl chloride (CrO_2Cl_2) is liberated.

$$K_2Cr_2O_7 + 4NaCl + \underset{\text{Conc.}}{6H_2SO_4} \xrightarrow{\text{Heat}} 4NaHSO_4 + 2KHSO_4 + 2CrO_2Cl_2 + 3H_2O$$

On passing chromyl chloride gas through a solution of sodium hydroxide, sodium chromate is formed.

$$CrO_2Cl_2 + 4NaOH \rightarrow 2NaCl + Na_2CrO_4 + 2H_2O$$

Addition of acidified lead acetate yields a yellow precipitate of lead chromate ($PbCrO_4$).

$$Na_2CrO_4 + \underset{\text{Acidified}}{(CH_3COO)_2Pb} \rightarrow \underset{\text{Yellow}}{PbCrO_4\downarrow} + 2CH_3COONa$$

4. When potassium dichromate is boiled with concentrated hydrochloric acid, chlorine gas is liberated.

$$\underset{\text{Conc.}}{K_2Cr_2O_7 + 14HCl} \rightarrow 2KCl + 2CrCl_3 + 7H_2O + 3Cl_2$$

5. Potassium dichromate is a strong oxidizing agent in acidic medium. Its oxidizing action may be represented by the following half-equation:

$$Cr_2O_7^{2-} + 14H^+ + 6e^- \rightarrow 2Cr^{3+} + 7H_2O$$

It oxidizes iodide ions to iodine.

$$\begin{array}{l} Cr_2O_7^{2-} + 14H^+ + 6e^- \rightarrow 2Cr^{3+} + 7H_2O \\ \qquad 2I^- \rightarrow I_2 + 2e^- \quad] \times 3 \\ \hline Cr_2O_7^{2-} + 6I^- + 14H^+ \rightarrow 2Cr^{3+} + 3I_2 + 7H_2O \\ \hline \end{array}$$

Similarly, it oxidizes iron(II) ions to iron(III) ions.

$$\begin{array}{l} Cr_2O_7^{2-} + 14H^+ + 6e^- \rightarrow 2Cr^{3+} + 7H_2O \\ \qquad Fe^{2+} \rightarrow Fe^{3+} + e^- \quad] \times 6 \\ \hline Cr_2O_7^{2-} + 6Fe^{2+} + 14H^+ \rightarrow 2Cr^{3+} + 6Fe^{3+} + 7H_2O \\ \hline \end{array}$$

The oxidizing action of potassium dichromate may also be expressed in terms of molecular equation involving nascent oxygen.

$$K_2Cr_2O_7 + 4H_2SO_4 \rightarrow K_2SO_4 + Cr_2(SO_4)_3 + 4H_2O + 3(O)$$

For example, hydrogen sulfide is oxidized to sulfur by acidified potassium dichromate.

$$\begin{array}{l} K_2Cr_2O_7 + 4H_2SO_4 \rightarrow K_2SO_4 + Cr_2(SO_4)_3 + 4H_2O + 3(O) \\ \qquad H_2S + (O) \rightarrow H_2O + S \quad] \times 3 \\ \hline K_2Cr_2O_7 + 3H_2S + 4H_2SO_4 \rightarrow K_2SO_4 + Cr_2(SO_4)_3 + 3S + 7H_2O \\ \hline \end{array}$$

Lead monoxide may be oxidized to lead dioxide.

$$\begin{array}{l} K_2Cr_2O_7 + 4H_2SO_4 \rightarrow K_2SO_4 + Cr_2(SO_4)_3 + 4H_2O + 3(O) \\ \qquad PbO + (O) \rightarrow PbO_2 \quad] \times 3 \\ \hline K_2Cr_2O_7 + 3PbO + 4H_2SO_4 \rightarrow K_2SO_4 + Cr_2(SO_4)_3 + 3PbO_2 + 4H_2O \\ \hline \end{array}$$

Similarly, sodium sulfite is oxidized to sodium sulfate.

$$\begin{array}{l} K_2Cr_2O_7 + 4H_2SO_4 \rightarrow K_2SO_4 + Cr_2(SO_4)_3 + 4H_2O + 3(O) \\ \qquad Na_2SO_3 + (O) \rightarrow Na_2SO_4 \quad] \times 3 \\ \hline K_2Cr_2O_7 + 3Na_2SO_3 + 4H_2SO_4 \rightarrow K_2SO_4 + Cr_2(SO_4)_3 + 3Na_2SO_4 + 4H_2O \\ \hline \end{array}$$

Chromium peroxides are a set of compounds formed by the reaction of chromates or dichromates with hydrogen peroxide under acidic or alkaline conditions. These compounds are largely unstable, both in solid state and solution phase, and decompose with evolution of oxygen. Most of them are explosive and inflammable in air. They are generally deep blue, violet or red in color. The oxidation state of chromium in the peroxides ranges from IV to VI. However, we have considered the peroxides in this section because the parent compound, chromium peroxide (CrO_5), has the metal in oxidation state VI.

Chromium peroxide is better depicted as $CrO(O_2)_2$, implying that there are two peroxo and one oxo units in its structure. It is, therefore, named as oxodiperoxochromium(VI) (chromium peroxide is the common name). It is prepared by the reaction of hydrogen peroxide with acidified chromate ions.

$$CrO^{2-}_4 + 2H_2O_2 + 2H^+ \rightarrow CrO(O_2)_2 + 3H_2O$$

It is blue in color and not very stable. This is because chromium is 5-coordinated in this compound (Fig. 27.2).

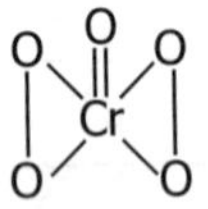

Fig. 27.2 Structure of chromium peroxide

On treatment with hydroxide ions or pyridine, the coordination number of the metal increases to 6 (Fig. 27.3) and, therefore, the adduct is more stable than the parent compound. Both the addition compounds are violet in color.

$$CrO(O_2)_2 + OH^- \rightarrow [CrO(O_2)_2(OH)]^-$$

Hydroxooxodiperoxochromate(VI) ion

$$CrO(O_2) + C_6H_5N \rightarrow [CrO(O_2)(C_6H_5N)]$$

Oxodiperoxopyridinechromium(VI)

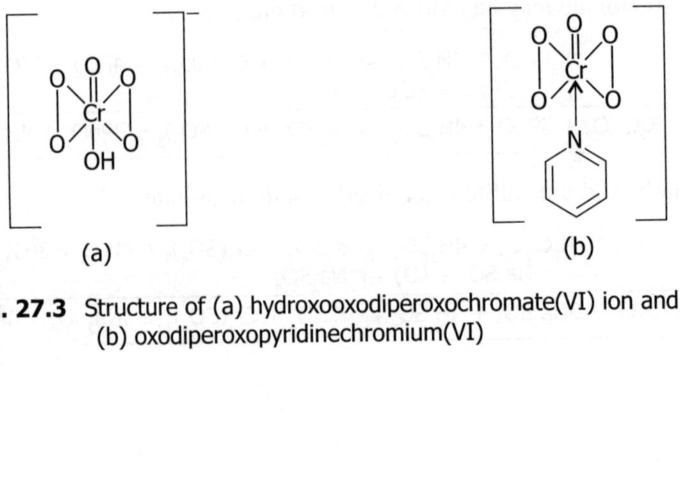

Fig. 27.3 Structure of (a) hydroxooxodiperoxochromate(VI) ion and (b) oxodiperoxopyridinechromium(VI)

When chromium peroxide is treated with hydrogen peroxide and potassium hydroxide, red-brown $K_3[CrO_8]$ or $K_3[Cr(O_2)_4]$ (Fig. 27.4a) is formed.

$$2CrO_5 + 3H_2O_2 + 6KOH \rightarrow \underset{\text{Potassium tetraperoxochromate(V)}}{2K_3[CrO_8]} + 6H_2O$$

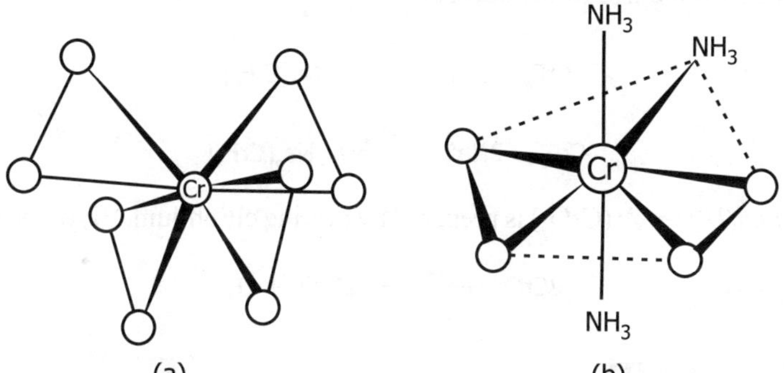

Fig. 27.4 Structure of (a) $K_3[CrO_8]$ and (b) $[(NH_3)_3CrO_4]$

The corresponding ammonium complex on heating decomposes into triamminediperoxochromium(IV), $[(NH_3)_3CrO_4]$ or $[(NH_3)_3Cr(O_2)_2]$ (Fig. 27.4b).

$$4(NH_4)_3[CrO_8] \xrightarrow{\text{Heat}} 4[(NH_3)_3CrO_4] + 6H_2O + 5O_2$$

Oxidation state V

A limited number of compounds of chromium are known in the oxidation state V. *Chromium(V) fluoride* (CrF_5) is the only binary compound known for pentavalent chromium. It is prepared by heating the elements at high temperature and under high pressure.

$$2Cr + 5F_2 \rightarrow 2CrF_5$$

Chromium oxotrifluoride may be prepared by the reaction of chromium(VI) oxide with chlorine trifluoride.

$$CrO_3 + ClF_3 \rightarrow CrOF_3 + ClO_2$$

Oxofluorochromates(V) ($MCrOF_4$) may be prepared from chromium(VI) oxide. For example,

$$3CrO_3 + 3KCl + 4BrF_3 \rightarrow 3KCrOF_4 + 2Br_2 + 3ClO_2$$

Oxochloro compounds ($M_2[CrOCl_5]$) too are synthesized from chromium(VI) oxide. For example,

$$2CrO_3 + 4NaCl + 8HCl \rightarrow 2Na_2[CrOCl_5] + 4H_2O + Cl_2$$

Oxidation state IV

Chromium(IV) fluoride (CrF_4) is formed by heating the elements at 600 K.

$$Cr + 2F_2 \xrightarrow{600\ K} CrF_4$$

It may be transformed into complex fluorides by reaction with alkali metal fluorides in bromine trifluoride solvent.

$$CrF_4 + NaF \xrightarrow{BrF_3} Na[CrF_5]$$

$$CrF_4 + 2NaF \xrightarrow{BrF_3} Na_2[CrF_6]$$

Chromium(IV) oxide (CrO_2) is prepared by heating chromium(VI) oxide at 500 K.

$$2CrO_3 \xrightarrow{500\ K} 2CrO_2 + O_2$$

Oxidation state III

Chromium(III) oxide (Cr_2O_3) is generally prepared by thermal decomposition of ammonium dichromate.

$$(NH_4)_2Cr_2O_7 \xrightarrow{Heat} Cr_2O_3 + 4H_2O + N_2$$

It is an amphoteric oxide.

$$Cr_2O_3 + 9H_2O + 6H^+ \rightarrow \underset{\text{Hexaquachromium(III) ion}}{2[Cr(H_2O)_6]^{3+}}$$

$$Cr_2O_3 + 3H_2O + 2OH^- \rightarrow \underset{\text{Tetrahydroxochromate(III) ion}}{2[Cr(OH)_4]^-}$$

Chromium(III) sulfide (Cr_2S_3) cannot be precipitated from aqueous solution since it is susceptible to hydrolysis. However, it may be prepared by passing hydrogen sulfide gas over chromium(III) chloride at red heat.

$$2CrCl_3 + 3H_2S \rightarrow Cr_2S_3 + 6HCl$$

Chromium(III) chloride ($CrCl_3$) in anhydrous state is prepared by the reaction of chlorine with chromium,

$$2Cr + 3Cl_2 \rightarrow \underset{\text{Anhydrous}}{2CrCl_3}$$

or by refluxing the hydrated chloride with thionyl chloride.

$$[Cr(H_2O)_6]Cl_3 + 6SOCl_2 \rightarrow CrCl_3 + 12HCl + 6SO_2$$

In an atmosphere of chlorine, chromium(III) chloride sublimes at 900 K. However, in the absence of chlorine, it decomposes into chromium(II) chloride.

$$2CrCl_3 \xrightarrow{900\text{ K}} 2CrCl_2 + Cl_2$$

Chromium(III) chloride has an appreciable covalent character and, therefore, is only sparingly soluble in water. Chromium(II) chloride, on the other hand, is ionic and is water-soluble. When chromium(II) ions are added to a suspension of chromium(III) chloride, the latter solubilizes. This happens because the soluble chromium(II) ion transfers an electron via a chlorine bridge to chromium(III) ion in the crystal. This reduces the chromium(III) ion to chromium(II) ion which, being soluble, leaves the crystal and enters the solution. It then reacts with more chromium(III) ions in the crystal and as a result, the chromium(III) chloride dissolves.

The compound, ammonium diamminetetrathiocyanato-Nchromate(III). 1-water, $NH_4[Cr(NH_3)_2(NCS)_4]\cdot H_2O$ is called *Reinecke's salt.* The large anionic group of this salt is used to precipitate large cations.

Oxidation state II

Chromium(II) compounds are strong reducing agents with a tendency to get oxidized to chromium(III). They are rapidly oxidized by atmospheric oxygen and, therefore, have to be preserved under nitrogen. Aqueous chromium(II) salts may be prepared by the reaction of dilute acids with extremely pure chromium metal. Impure metal is unsatisfactory since the contaminants tend to catalyze the oxidation of chromium(II) to chromium(III) ions.

$$\underset{\text{Dilute}}{Cr + H_2SO_4} + 5H_2O \rightarrow CrSO_4.5H_2O + H_2$$

$$\underset{\text{Dilute}}{Cr + 2HCl} + 4H_2O \rightarrow CrCl_2.4H_2O + H_2$$

Chromium(II) chloride may be prepared either by the reaction of hydrogen chloride gas with chromium heated to 1000 K,

$$Cr + 2HCl \xrightarrow{1000\text{ K}} CrCl_2 + H_2$$

or by reduction of anhydrous chromium(III) chloride with hydrogen at 800 K.

$$\underset{\text{Anhydrous}}{2CrCl_3 + H_2} \xrightarrow{800\text{ K}} 2CrCl_2 + 2HCl$$

Chromium(II) acetate, which exists as a dimeric dihydrate, $Cr_2(CH_3COO)_4.2H_2O$, is the most stable compound of chromium(II). It owes its stability to the chelate effect of bidentate acetate ligands (Fig. 27.5).

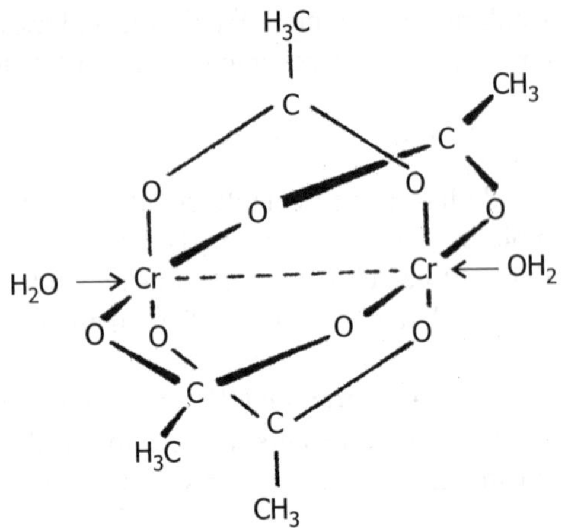

Fig. 27.5 Structure of chromium(II) acetate dimer

Chromium(II) acetate is precipitated as a red solid when a solution of chromium(II) chloride is added to a concentrated solution of sodium acetate.

$$\underset{\text{Conc.}}{2CrCl_2 + 4CH_3COONa} + 2H_2O \rightarrow Cr_2(CH_3COO)_4.2H_2O\downarrow + 4NaCl$$

The bond distance between the two chromium atoms is shorter than expected. This indicates very strong interactions between the two metal ions. As a result of the strong interactions, the four unpaired electrons on each chromium(II) ion [Cr(II): $3d^4$] get paired-up. Hence, chromium(II) acetate is a diamagnetic compound.

Case Study 27.4 A strong oxidizing agent, **A**, on heating with potassium chloride and concentrated sulfuric acid gives a red-colored gas, **B**, which on reaction with potassium hydroxide yields a yellow product, **C**. The compound **C** may also be obtained by heating **A** with an alkali. On adding hydrogen peroxide to an acidified solution of **C**, a blue, unstable product, **D** is obtained. Identify **A**, **B**, **C** and **D**.

When a metal chloride is heated with potassium dichromate and concentrated sulfuric acid, red-colored chromyl chloride gas is evolved. Therefore, the oxidizing agent **A** is potassium dichromate ($K_2Cr_2O_7$) while the red gas, **B**, is chromyl chloride (CrO_2Cl_2).

$$\underset{\mathbf{A}}{K_2Cr_2O_7} + 4KCl + \underset{\text{Conc.}}{6H_2SO_4} \rightarrow 6KHSO_4 + 3H_2O + \underset{\mathbf{B}}{2CrO_2Cl_2}\uparrow$$

When chromyl chloride gas (**B**) is passed through a solution of potassium hydroxide, yellow-colored potassium chromate (**C**) is formed.

$$4KOH + \underset{\mathbf{B}}{CrO_2Cl_2} \rightarrow \underset{\mathbf{C}}{K_2CrO_4} + 2KCl + 2H_2O$$

Potassium chromate may also be obtained by heating potassium dichromate with potassium hydroxide.

$$\underset{\mathbf{A}}{K_2Cr_2O_7} + 2KOH \rightarrow \underset{\mathbf{C}}{2K_2CrO_4} + H_2O$$

Acidified chromate ions, on reaction with hydrogen peroxide, yield blue, unstable chromium peroxide, CrO_5 (**D**).

$$CrO^{2-}_4 + 2H^+ + 2H_2O_2 \rightarrow \underset{\mathbf{D}}{CrO_5} + 3H_2O$$

Thus, **A** = $K_2Cr_2O_7$, **B** = CrO_2Cl_2, **C** = K_2CrO_4 and **D** = CrO_5.

Manganese

Of all the 3d elements, manganese shows the widest range of oxidation states. Oxidation state II is most stable, as the manganese(II) ion possesses half-filled, $3d^5$ configuration. The highest oxidation state, VII, occurs only in oxo compounds.

Oxidation state VII

Potassium permanganate ($KMnO_4$) is the most important compound of heptavalent manganese. It is prepared by fusing solid manganese dioxide with solid potassium hydroxide, in the presence of air. The reaction leads to the formation of an intermediate compound, potassium manganate (K_2MnO_4).

$$2MnO_2 + 4KOH + O_2 \rightarrow 2K_2MnO_4 + 2H_2O$$

Potassium manganate may then be oxidized to potassium permanganate by a mild oxidizing agent such as chlorine.

$$2K_2MnO_4 + Cl_2 \rightarrow 2KMnO_4 + 2KCl$$

The purple-colored permanganate ion has tetrahedral coordination of oxygen atoms around manganese (Fig. 27.6).

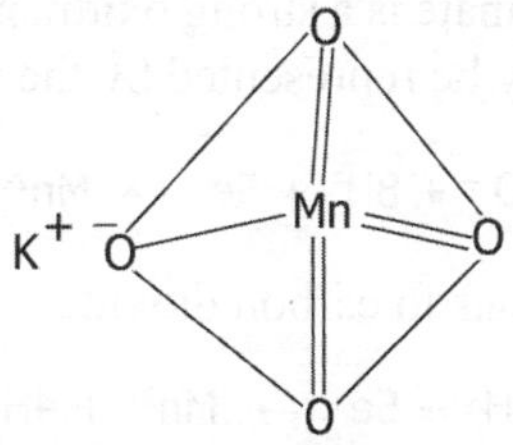

Fig. 27.6 Structure of potassium permanganate

The important properties of potassium permanganate are as follows:

1. On heating, it decomposes into manganese dioxide.

$$2KMnO_4 \xrightarrow{\text{Heat}} K_2MnO_4 + MnO_2 + O_2$$

2. On heating with potassium hydroxide, it is reduced to potassium manganate.

$$4KMnO_4 + 4KOH \rightarrow 4K_2MnO_4 + 2H_2O + O_2$$

3. Its reaction with concentrated sulfuric acid may be carried out under two conditions.
   a. When potassium permanganate is in excess, manganese heptoxide is initially formed, which then explosively decomposes into manganese dioxide.

$$2KMnO_4 + \underset{\text{Conc.}}{H_2SO_4} \rightarrow \underset{\substack{\text{Manganese}\\ \text{heptaoxide}}}{Mn_2O_7} + K_2SO_4 + H_2O$$

$$2Mn_2O_7 \xrightarrow{\text{Explosion}} \underset{\substack{\text{Manganese}\\ \text{dioxide}}}{4MnO_2} + 3O_2$$

   b. When sulfuric acid is in excess, green-colored trioxomanganese(VII) ion (MnO_3^+) is formed.

$$KMnO_4 + \underset{\text{Conc.}}{3H_2SO_4} \rightarrow K^+ + MnO_3^+ + H_3O^+ + 3HSO_4^-$$

4. Acidic solution of potassium permanganate slowly decomposes into manganese dioxide.

$$4MnO_4^- + 4H^+ \xrightarrow{\text{Sunlight}} 4MnO_2 + 2H_2O + 3O_2$$

   The reaction is catalyzed by sunlight and for this reason, potassium permanganate is stored in colored bottles.

5. Potassium permanganate is a strong oxidizing agent in acidic medium. Its oxidizing action may be represented by the following half-equation:

$$MnO_4^- + 8H^+ + 5e^- \rightarrow Mn^{2+} + 4H_2O$$

   It oxidizes oxalate ions to carbon dioxide.

$$\begin{array}{rcll} MnO_4^- + 8H^+ + 5e^- & \rightarrow & Mn^{2+} + 4H_2O &] \times 2 \\ C_2O_4^{2-} & \rightarrow & 2CO_2 + 2e^- &] \times 5 \\ \hline 2MnO_4^- + 5C_2O_4^{2-} + 16H^+ & \rightarrow & 2Mn^{2+} + 8H_2O + 10CO_2 & \end{array}$$

Similarly, it oxidizes iron(II) to iron(III) ions.

$$MnO_4^- + 8H^+ + 5e^- \rightarrow Mn^{2+} + 4H_2O$$
$$Fe^{2+} \rightarrow Fe^{3+} + e^- \quad] \times 5$$
$$\overline{MnO_4^- + 5Fe^{2+} + 8H^+ \rightarrow Mn^{2+} + 5Fe^{3+} + 4H_2O}$$

The oxidizing action of potassium permanganate may also be expressed in terms of molecular equation involving nascent oxygen.

$$2KMnO_4 + 3H_2SO_4 \rightarrow K_2SO_4 + 2MnSO_4 + 3H_2O + 5(O)$$

It oxidizes hydrogen sulfide to sulfur.

$$2KMnO_4 + 3H_2SO_4 \rightarrow K_2SO_4 + 2MnSO_4 + 3H_2O + 5(O)$$
$$H_2S + (O) \rightarrow H_2O + S \quad] \times 5$$
$$\overline{2KMnO_4 + 5H_2S + 3H_2SO_4 \rightarrow K_2SO_4 + 2MnSO_4 + 5S + 8H_2O}$$

Similarly, it oxidizes sulfite ions to sulfate ions.

$$2KMnO_4 + 3H_2SO_4 \rightarrow K_2SO_4 + 2MnSO_4 + 3H_2O + 5(O)$$
$$SO_3^{2-} + (O) \rightarrow SO_4^{2-} \quad] \times 5$$
$$\overline{2KMnO_4 + 5SO_3^{2-} + 3H_2SO_4 \rightarrow K_2SO_4 + 2MnSO_4 + 5SO_4^{2-} + 3H_2O}$$

Hydrogen peroxide, which is normally considered as an oxidizing agent, is also oxidized by acidified potassium permanganate.

$$2KMnO_4 + 3H_2SO_4 \rightarrow K_2SO_4 + 2MnSO_4 + 3H_2O + 5(O)$$
$$H_2O_2 + (O) \rightarrow H_2O + O_2 \quad] \times 5$$
$$\overline{2KMnO_4 + 5H_2O_2 + 3H_2SO_4 \rightarrow K_2SO_4 + 2MnSO_4 + 8H_2O + 5O_2}$$

6. Potassium permanganate also acts as an oxidizing agent in alkaline medium. Its oxidizing action under basic conditions may be outlined as follows:

$$2KMnO_4 + 2KOH \rightarrow 2K_2MnO_4 + H_2O + (O)$$
$$K_2MnO_4 + H_2O \rightarrow MnO_2 + 2KOH + (O) \;] \times 2$$
$$\overline{2KMnO_4 + H_2O \rightarrow 2MnO_2 + 2KOH + 3(O)}$$

It must be noted that although the reaction is to be conducted under alkaline conditions, the alkali, potassium hydroxide, appears on the product side of the above equation. This is because more potassium hydroxide is being formed in this reaction than was originally added.

Alkaline potassium permanganate oxidizes iodide ions to iodate ions, IO_3^-.

$$2KMnO_4 + H_2O \rightarrow 2MnO_2 + 2KOH + 3(O)$$
$$I^- + 3(O) \rightarrow IO_3^-$$
$$\overline{2KMnO_4 + I^- + H_2O \rightarrow 2MnO_2 + IO_3^- + 2KOH}$$

7. Potassium permanganate acts as an oxidizing agent even under neutral conditions. For example, it oxidizes hydrogen sulfide to sulfur.

$$2KMnO_4 + 4H_2S \rightarrow K_2SO_4 + 2MnS + S + 4H_2O$$

The products formed in this reaction are different from those resulting from the oxidation of hydrogen sulfide by acidic potassium permanganate.

Neutral potassium permanganate oxidizes manganese(II) sulfate to manganese(IV) oxide.

$$2KMnO_4 + 3MnSO_4 + 2H_2O \rightarrow 5MnO_2 + K_2SO_4 + 2H_2SO_4$$

It also oxidizes sodium thiosulfate to sodium sulfate.

$$8KMnO_4 + 3Na_2S_2O_3 + H_2O \rightarrow 8MnO_2 + 3K_2SO_4 + 3Na_2SO_4 + 2KOH$$

Oxidation state VI

The only prominent compound involving manganese(VI) is *potassium manganate* (K_2MnO_4) As has been described in the preparation of potassium permanganate, this compound may be prepared by fusing together potassium hydroxide and manganese dioxide, in the presence of air.

Potassium manganate is stable only under alkaline conditions. In acidic or neutral medium, it decomposes into manganese dioxide.

$$3MnO_4^{2-} + 4H^+ \rightarrow MnO_2 + 2MnO_4^- + 2H_2O$$

$$3MnO_4^{2-} + 2H_2O \rightarrow MnO_2 + 2MnO_4^- + 4OH^-$$

Carbon dioxide oxidizes it to potassium permanganate.

$$3K_2MnO_4 + 2CO_2 \rightarrow 2KMnO_4 + MnO_2 + 2K_2CO_3$$

Oxidation state V

Potassium hypomanganate (K_3MnO_4), a bright blue salt, is the only compound of manganese in oxidation state V. It is prepared by the reaction of potassium permanganate with excess sodium sulfite.

$$3KMnO_4 + 4Na_2SO_3 \rightarrow K_3MnO_4 + 2MnO_2 + 4Na_2SO_4$$

Oxidation state IV

Manganese(IV) fluoride (MnF_4) is formed by direct reaction between manganese and fluorine.

$$Mn + 2F_2 \rightarrow MnF_4$$

It is the highest halide formed by manganese. It is volatile, blue in color and extremely reactive.

Manganese(IV) oxide or manganese dioxide (MnO_2) occurs naturally as black-colored *pyrolusite*.

On heating to 1300 K, it forms trimanganese tetraoxide.

$$3MnO_2 \xrightarrow{1300\ K} Mn_3O_4 + O_2$$

Under acidic conditions, it oxidizes oxalate ions to carbon dioxide.

$$MnO_2 + C_2O_4^{2-} + 4H^+ \rightarrow Mn^{2+} + 2H_2O + 2CO_2$$

It reacts with concentrated hydrochloric acid to yield chlorine.

$$\begin{array}{rcl} MnO_2 + 4HCl & \rightarrow & MnCl_4 + 2H_2O \\ MnCl_4 & \rightarrow & MnCl_2 + Cl_2 \\ \hline MnO_2 + \underset{\text{Conc.}}{4HCl} & \rightarrow & MnCl_2 + 2H_2O + Cl_2 \\ \hline \end{array}$$

With sulfuric acid, it liberates oxygen.

$$2MnO_2 + 2H_2SO_4 \rightarrow 2MnSO_4 + 2H_2O + O_2$$

Oxidation state III

Manganese(III) ion is not very stable and disproportionates in aqueous solution.

$$2Mn^{3+} + 2H_2O \rightarrow Mn^{2+} + MnO_2 + 4H^+$$

Manganese(III) oxide (Mn_2O_3) occurs in nature as *braunite.* It may be prepared by heating manganese dioxide to 800–900 K.

$$4MnO_2 \xrightarrow{Heat} 2Mn_2O_3 + O_2$$

It reacts with hydrofluoric acid to yield *manganese(III) fluoride* (MnF_3).

$$Mn_2O_3 + 6HF \rightarrow 2MnF_3 + 3H_2O$$

Manganese(III) fluoride reacts with alkali metal fluorides to give tetrafluoro and pentafluoro complexes.

$$MnF_3 + NaF \rightarrow Na[MnF_4]$$

$$MnF_3 + 2NaF \rightarrow Na_2[MnF_5]$$

No simple chloride, bromide or iodide of manganese(III) are known. This is attributed to the tendency of halide ions to reduce manganese(III) to manganese(II).

Manganese(III) acetate, $Mn(CH_3COO)_3.2H_2O$ is the most stable salt of trivalent manganese. It owes its stability to the chelating ability of acetate ligands. It is prepared by oxidation of manganese(II) acetate by chlorine.

$$3Mn(CH_3COO)_2 + Cl_2 \rightarrow 2Mn(CH_3COO)_3 + MnCl_2$$

Oxidation state II

The manganese(II) ion exists in aqueous solution as the stable, pale pink, hexaqua complex, $[Mn(H_2O)_6]^{2+}$.

Manganese(II) oxide (MnO) is obtained as green powder by heating manganese(II) carbonate.

$$MnCO_3 \xrightarrow{\text{Heat}} MnO + CO_2$$

Manganese(II) hydroxide, $Mn(OH)_2$, is precipitated as a pink solid when aqueous manganese(II) salts are treated with sodium hydroxide.

$$MnCl_2 + 2NaOH \rightarrow Mn(OH)_2\downarrow + 2NaCl$$

It is amphoteric in nature.

$$Mn(OH)_2 + 2HCl \rightarrow MnCl_2 + 2H_2O$$

$$Mn(OH)_2 + NaOH \rightarrow \underset{\text{Sodium trihydroxomanganate(II)}}{Na[Mn(OH)_3]}$$

On standing, it turns black due to atmospheric oxidation to manganese dioxide.

$$2Mn(OH)_2 + O_2 \rightarrow 2MnO_2 + 2H_2O$$

Case Study 27.5 A black oxide, **A**, on treatment with concentrated hydrochloric acid, yields an aqueous salt solution, **B** and a yellowish-green gas, **C**. On passing **C** through a solution of potassium iodide, a violet solid, **D** is obtained. When **B** is heated with sodium bismuthate and dilute nitric acid, a purple-colored compound, **E** is formed. Identify **A**, **B**, **C**, **D** and **E**.

When chlorine gas is passed through aqueous potassium iodide, violet-colored iodine is formed. Therefore, the yellowish-green gas, **C**, is chlorine and the violet solid, **D**, is iodine.

$$2KI + \underset{\mathbf{C}}{Cl_2} \rightarrow 2KCl + \underset{\mathbf{D}}{I_2}$$

Chlorine, in turn, is released by the reaction of black manganese dioxide **(A)** and concentrated hydrochloric acid. Simultaneously, manganese(II) chloride **(B)** is formed as an aqueous salt solution.

$$\underset{\textbf{A}}{MnO_2} + \underset{\text{Conc.}}{4HCl} \rightarrow \underset{\textbf{B}}{MnCl_2} + 2H_2O + \underset{\textbf{C}}{Cl_2}$$

Manganese(II) chloride, when heated with sodium bismuthate and dilute nitric acid, yields purple-colored permanganic acid, $HMnO_4$ **(E)**.

$$\underset{\textbf{B}}{2MnCl_2} + 5NaBiO_3 + \underset{\text{Dilute}}{16HNO_3} \rightarrow \underset{\textbf{E}}{2HMnO_4} + 5Bi(NO_3)_3 + NaNO_3 + 4NaCl + 7H_2O$$

Thus, **A** = MnO_2, **B** = $MnCl_2$, **C** = Cl_2, **D** = I_2 and **E** = $HMnO_4$.

Iron

Most of the compounds of iron exist in oxidation states II and III, although a few are also known in unusual oxidation states, IV and VI.

Oxidation state VI

Tetraoxoferrate(VI) ion is obtained by suspending hydrous iron(III) oxide in concentrated alkali and passing chlorine.

$$Fe_2O_3.nH_2O + 10NaOH + 3Cl_2 \rightarrow 2Na_2[FeO_4] + 6NaCl + (n+5)H_2O$$

The potassium salt may be prepared by fusing iron filings with potassium nitrate and dissolving the melt in water.

$$2KNO_3 + Fe \rightarrow K_2[FeO_4] + 2NO$$

The ferrate(VI) ion is stable under alkaline conditions, but decomposes in acidic or neutral medium.

$$4FeO_4^{2-} + 20H^+ \rightarrow 4Fe^{3+} + 10H_2O + 3O_2$$

$$4FeO_4^{2-} + 10H_2O \rightarrow 4Fe^{3+} + 20OH^- + 3O_2$$

Ferrate(VI) ion is an extremely powerful oxidizing agent—even stronger than potassium permanganate. It oxidizes ammonia to nitrogen.

$$2FeO_4^{2-} + 2NH_3 + 2H_2O \rightarrow 2Fe^{3+} + 10OH^- + N_2$$

Oxidation state IV

Tetraoxoferrate(IV) ions exist as stable strontium or barium salts. The latter may be prepared by the oxidation of barium hexahydroxoferrate(III) with molecular oxygen, in the presence of barium hydroxide.

$$2Ba_3[Fe(OH)_6]_2 + 2Ba(OH)_2 + O_2 \rightarrow 4Ba_2[FeO_4] + 14H_2O$$

The strontium compound is prepared by an analogous method.

Oxidation state III

Iron(III) ions in aqueous solution are readily reduced by many reducing agents. Addition of iodide ions results in the formation of iron(II) iodide, rather than iron(III) iodide (*see* below). Addition of sulfide ions to iron(III) ions yields a mixture of iron(II) sulfide and colloidal sulfur. Iron(III) carbonate too cannot be precipitated. Addition of sodium carbonate or bicarbonate to an iron(III) solution precipitates the hydrated iron(II) oxide.

The *iron(III) halides*, FeX_3 (X = F, Cl or Br) may be prepared by heating the metal with the halogen. For example,

$$2Fe + 3Cl_2 \xrightarrow{\text{Heat}} 2FeCl_3$$

In vapor state, iron(III) chloride exists as a dimer (Fig. 27.7a), while in hydrated state, it forms a 6-coordinated, *trans* complex (Fig. 27.7b).

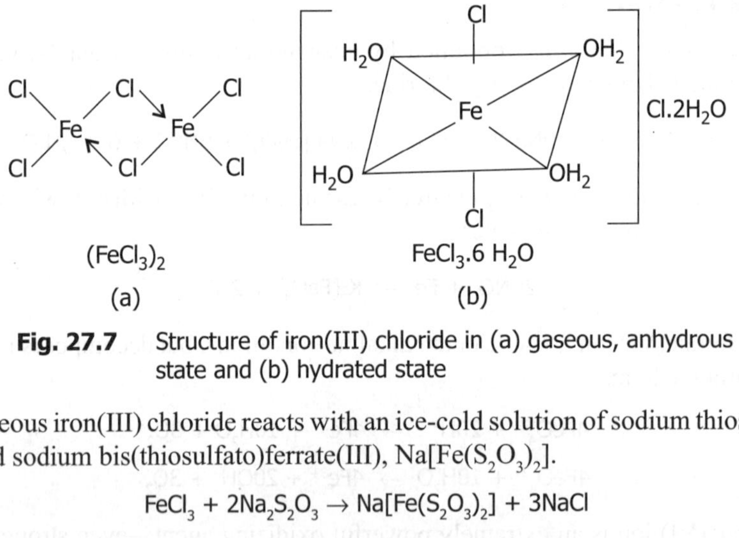

Fig. 27.7 Structure of iron(III) chloride in (a) gaseous, anhydrous state and (b) hydrated state

Aqueous iron(III) chloride reacts with an ice-cold solution of sodium thiosulfate to yield sodium bis(thiosulfato)ferrate(III), $Na[Fe(S_2O_3)_2]$.

$$FeCl_3 + 2Na_2S_2O_3 \rightarrow Na[Fe(S_2O_3)_2] + 3NaCl$$

When the contents are warmed, the products react with unused iron(III) chloride, producing sodium tetrathionate and iron(II) chloride.

$$Na[Fe(S_2O_3)_2] + NaCl + FeCl_3 \rightarrow 2FeCl_2 + Na_2S_4O_6$$

Iron(III) iodide is not known since a strong oxidizing agent like iron(III) ion cannot coexist with a strong reducing agent like iodide ion. In fact, iodide ions reduce iron(III) ions to iron(II) ions.

$$2Fe^{3+} + 2I^- \rightarrow 2Fe^{2+} + I_2$$

There is no evidence for the existence of iron(III) hydroxide. Addition of an alkali to an iron(III) salt yields *hydrous iron(III) oxide*, $Fe_2O_3.nH_2O$.

$$2FeCl_3 + 6NaOH + (n-3)H_2O \rightarrow Fe_2O_3.nH_2O + 6NaCl$$

Fusion of iron(III) oxide with sodium carbonate yields sodium dioxoferrate(III), $Na[FeO_2]$.

$$Fe_2O_3 + Na_2CO_3 \rightarrow 2Na[FeO_2] + CO_2$$

Potassium ferricyanide, $K_3[Fe(CN)_6]$, which is better known as potassium hexacyanoferrate(III), is prepared by the oxidation of potassium ferrocyanide.

$$2K_4[Fe(CN)_6] + Cl_2 \rightarrow 2K_3[Fe(CN)_6] + 2KCl$$

$$2K_4[Fe(CN)_6] + H_2O_2 \rightarrow 2K_3[Fe(CN)_6] + 2KOH$$

There is an octahedral coordination of cyano ligands around iron(III) ion. The structure has been discussed in the section dealing with valence bond theory of complexes (*see* Chapter 26, p. 703).

Potassium ferricyanide reacts with iron(II) salts to give a blue-colored complex, called *Turnbull's blue.* It is, therefore, used for qualitative detection of iron(II) ions.

$$K_3[Fe(CN)_6] + FeCl_2 \rightarrow \underset{\text{Turnbull's blue}}{K\{\overset{II}{Fe}[\overset{III}{Fe}(CN)_6]\}} + 2KCl$$

Potassium ferricyanide is an oxidizing agent in alkaline medium. Its oxidizing action may be represented as follows:

$$2K_3[Fe(CN)_6] + 2KOH \rightarrow 2K_4[Fe(CN)_6] + H_2O + (O)$$

It oxidizes hydrogen sulfide to sulfur.

$$\begin{aligned} 2K_3[Fe(CN)_6] + 2KOH &\rightarrow 2K_4[Fe(CN)_6] + H_2O + (O) \\ H_2S + (O) &\rightarrow H_2O + S \\ \hline 2K_3[Fe(CN)_6] + 2KOH + H_2S &\rightarrow 2K_4[Fe(CN)_6] + S + 2H_2O \end{aligned}$$

It oxidizes lead monoxide to lead dioxide.

$$\begin{aligned} 2K_3[Fe(CN)_6] + 2KOH &\rightarrow 2K_4[Fe(CN)_6] + H_2O + (O) \\ PbO + (O) &\rightarrow PbO_2 \\ \hline 2K_3[Fe(CN)_6] + PbO + 2KOH &\rightarrow 2K_4[Fe(CN)_6] + PbO_2 + H_2O \end{aligned}$$

It oxidizes hydrazine to nitrogen.

$$\begin{array}{rcl} 2K_3[Fe(CN)_6] + 2KOH & \rightarrow & 2K_4[Fe(CN)_6] + H_2O + (O)] \times 2 \\ N_2H_4 + 2(O) & \rightarrow & 2H_2O + N_2 \\ \hline 4K_3[Fe(CN)_6] + N_2H_4 + 4KOH & \rightarrow & 4K_4[Fe(CN)_6] + 4H_2O + N_2 \end{array}$$

Oxidation state II

Iron(II) halides, FeX_2 (X = F, Cl, Br or I) exist in both anhydrous and hydrated states. The bromide and iodide may be synthesized by direct reaction of the elements.

$$Fe + Br_2 \rightarrow FeBr_2$$

$$Fe + I_2 \rightarrow FeI_2$$

The fluoride and chloride, however, cannot be prepared by this method since the direct reaction of elements yields iron(III) halides. They may be prepared by the reaction of hydrogen fluoride or hydrogen chloride with iron filings.

$$Fe + 2HF \rightarrow FeF_2 + H_2$$

$$Fe + 2HCl \rightarrow FeCl_2 + H_2$$

Iron(II) chloride may also be prepared by the reduction of iron(III) chloride.

$$2FeCl_3 + H_2 \rightarrow 2FeCl_2 + 2HCl$$

Iron(II) oxide (FeO) is a non-stoichiometric compound. It is a basic oxide and is prepared by thermal decomposition of iron(II) oxalate.

$$Fe(C_2O_4) \xrightarrow{\text{Heat}} FeO + CO_2 + CO$$

Iron(II) hydroxide, $Fe(OH)_2$ may be prepared by the reaction of iron(II) salts with an alkali. For example,

$$FeSO_4 + 2NaOH \rightarrow Fe(OH)_2 + Na_2SO_4$$

It is instantly converted to hydrous iron(III) oxide, $Fe_2O_3.nH_2O$ by atmospheric oxidation.

Iron(II) sulfate heptahydrate may be prepared by the action of dilute sulfuric acid on iron filings.

$$Fe + H_2SO_4 + 7H_2O \rightarrow FeSO_4{\cdot}7H_2O + H_2$$

Iron(II) sulfate is not a very stable compound. It is easily oxidized to iron(III) sulfate by atmospheric oxidation. Moreover, it effloresces, that is loses water of crystallization, when exposed to air. Hence it cannot be used as an analytical reagent. Nevertheless, its double salt with ammonium sulfate, ammonium iron(II) sulfate hexahydrate ($FeSO_4{\cdot}(NH_4)_2SO_4{\cdot}6H_2O$), commonly called Mohr's salt, has a stable lattice. It neither gets oxidized, nor effloresces in presence of air, and is, therefore, used in redox titrations.

Potassium ferrocyanide, $K_4[Fe(CN)_6]$—more appropriately named as potassium hexacyanoferrate(II) by the IUPAC system of nomenclature—may be prepared, on a laboratory scale, by adding potassium cyanide solution to aqueous iron(II) sulfate. Initially, insoluble iron(II) cyanide is formed, which dissolves in excess of potassium cyanide and yields potassium ferrocyanide.

$$FeSO_4 + 2KCN \rightarrow \underset{\text{Insoluble}}{Fe(CN)_2} + K_2SO_4$$

$$Fe(CN)_2 + 4KCN \rightarrow \underset{\text{Soluble}}{K_4[Fe(CN)_6]}$$

On an industrial scale, it is manufactured by making use of hydrogen cyanide which, in turn, is a byproduct in the preparation of coal gas. Hydrogen cyanide is reacted with hydrated iron oxides—a mixture of iron(II) and iron(III) oxides.

$$FeO.nH_2O + 2HCN \rightarrow Fe(CN)_2 + (n+1)H_2O$$

$$Fe_2O_3.nH_2O + 6HCN \rightarrow 2Fe(CN)_3 + (n+3)H_2O$$

The iron(II) cyanide and iron(III) cyanide then react with each other to yield ferri-ferrocyanide, which, on shaking with lime, yields calcium ferrocyanide.

$$3Fe(CN)_2 + 4Fe(CN)_3 \rightarrow \underset{\text{Ferri-ferrocyanide}}{\overset{\text{III}}{Fe_4}[\overset{\text{II}}{Fe}(CN)_6]_3}$$

$$Fe_4[Fe(CN)_6]_3 + 6Ca(OH)_2 \rightarrow \underset{\text{Calcium ferrocyanide}}{3Ca_2[Fe(CN)_6]} + 4Fe(OH)_3$$

Calcium ferrocyanide, on treatment with a solution of potassium carbonate, gives potassium ferrocyanide.

$$Ca_2[Fe(CN)_6] + 2K_2CO_3 \rightarrow K_4[Fe(CN)_6] + 2CaCO_3$$

Ferrocyanide ion has an octahedral structure, which is described in the section on valence bond theory, as applied to the complexes (*see* Chapter 26, p. 703).

Ferrocyanide ion is non-toxic, whereas ferricyanide ion is poisonous. This is because ferrocyanide ion is kinetically inert and, therefore, reacts very slowly. In fact, sodium ferrocyanide is added to common salt, to the extent of 0.01%, so as to prevent lump formation in the latter. Thus even in humid climate, the common salt maintains its free-flow characteristics. The ferricyanide ion, however, is kinetically labile and dissociates easily into cyano compounds, which are hazardous.

Some of the important properties of potassium ferrocyanide are as follows:

1. On heating, it decomposes into iron carbide.

$$K_4[Fe(CN)_6] \xrightarrow{\text{Heat}} 4KCN + FeC_2 + N_2$$

2. On boiling with dilute sulfuric acid, it yields potassium ferro-ferrocyanide, $K_2Fe[Fe(CN)_6]$.

$$2K_4[Fe(CN)_6] + \underset{\text{Dilute}}{3H_2SO_4} \rightarrow K_2\overset{\text{II}}{Fe}[\overset{\text{II}}{Fe}(CN)_6] + 3K_2SO_4 + 6HCN$$

3. With concentrated sulfuric acid it gives a mixture of sulfates—ammonium sulfate, potassium sulfate, ferrous sulfate—and carbon monoxide.

$$K_4[Fe(CN)_6] + \underset{\text{Conc.}}{6H_2SO_4} + 6H_2O \rightarrow 3(NH_4)_2SO_4 + 2K_2SO_4 + FeSO_4 + 6CO$$

4. It gives a chocolate-brown precipitate of copper ferrocyanide, $Cu_2[Fe(CN)_6]$ on reaction with copper(II) ions.

$$K_4[Fe(CN)_6] + 2CuSO_4 \rightarrow Cu_2[Fe(CN)_6]\downarrow + 2K_2SO_4$$

5. When iron(III) ions are added to potassium ferrocyanide, a blue-colored complex, called *Prussian blue*, is obtained.

$$FeCl_3 + K_4[Fe(CN)_6] \rightarrow \underset{\text{Prussian blue}}{K\{\overset{\text{III}}{Fe}[\overset{\text{II}}{Fe}(CN)_6]\}} + 3KCl$$

This compound was used to dye the uniform of Prussian soldiers. Hence the name Prussian blue.

6. Zinc(II) ions yield an off-white precipitate of zinc ferrocyanide, $Zn_2[Fe(CN)_6]$ with potassium ferrocyanide.

$$K_4[Fe(CN)_6] + 2ZnSO_4 \rightarrow Zn_2[Fe(CN)_6]\downarrow + 2K_2SO_4$$

7. The following reactions exemplify the tendency of potassium ferrocyanide to act as a reducing agent. Potassium ferrocyanide itself gets oxidized to potassium ferricyanide:

$$2K_4[Fe(CN)_6] + Cl_2 \rightarrow 2K_3[Fe(CN)_6] + 2KCl$$

$$2K_4[Fe(CN)_6] + H_2O_2 \rightarrow 2K_3[Fe(CN)_6] + 2KOH$$

$$5K_4[Fe(CN)_6] + KMnO_4 + 8HCl \rightarrow 5K_3[Fe(CN)_6] + 6KCl + MnCl_2 + 4H_2O$$

$$3K_4[Fe(CN)_6] + 4HNO_3 \rightarrow 3K_3[Fe(CN)_6] + 3KNO_3 + 2H_2O + NO$$

Both ferrocyanide and ferricyanide ions are thermodynamically very stable due to the occurrence of π-bonding between iron and cyano ligands. The *hexammineiron(II) ion*, $[Fe(NH_3)_6]^{2+}$, on the other hand, has no π-bonding between the metal and the ligands. It is, therefore, not very stable. It may be prepared by the reaction of gaseous ammonia with iron(II) halides.

$$FeCl_2 + 6NH_3 \rightarrow [Fe(NH_3)_6]Cl_2$$

However, it can be preserved only in saturated ammonia solution. In water, it dissociates with the release of ammonia.

Sodium nitroprusside, $Na_2[Fe(CN)_5NO]$ has iron in oxidation state (II). This is because the nitric oxide unit bears a positive charge, NO^+. According to IUPAC nomenclature, it is named sodium pentacyanonitrosoniumferrate(II). This complex is synthesized by oxidation of potassium ferrocyanide by nitric acid, in the presence of sodium carbonate. It is a three-step reaction.

(i) $3K_4[Fe(CN)_6] + 4HNO_3 \rightarrow 3K_3[Fe(CN)_6] + 3KNO_3 + 2H_2O + NO$

(ii) $2K_3[Fe(CN)_6] + 3Na_2CO_3 \rightarrow 2Na_3[Fe(CN)_6] + 3K_2CO_3$

(iii) $Na_3[Fe(CN)_6] + NO \rightarrow Na_2[Fe(CN)_5NO] + NaCN$

Sodium nitroprusside may also be prepared by the reaction of sodium ferrocyanide with sodium nitrite, followed by hydrolysis.

$$Na_4[Fe(CN)_6] + NaNO_2 \rightarrow Na_4[Fe(CN)_5(NO_2)] + NaCN$$

$$Na_4[Fe(CN)_5(NO_2)] + H_2O \rightarrow Na_2[Fe(CN)_5NO] + 2NaOH$$

Sodium nitroprusside has an octahedral structure, as shown below:

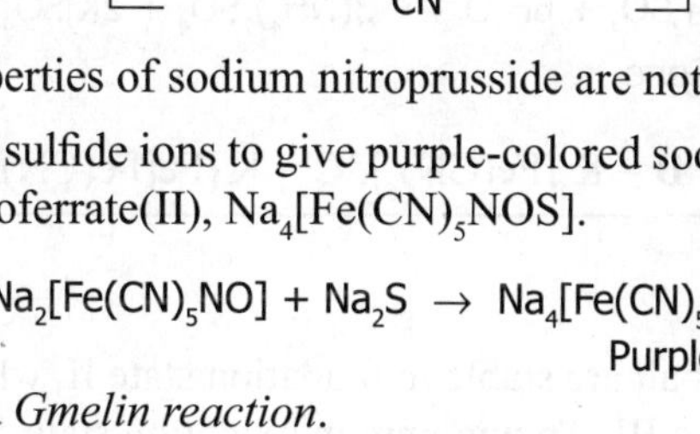

The following properties of sodium nitroprusside are noteworthy:

1. It reacts with sulfide ions to give purple-colored sodium pentacyanonitrosoniumsulfidoferrate(II), $Na_4[Fe(CN)_5NOS]$.

$$Na_2[Fe(CN)_5NO] + Na_2S \rightarrow \underset{\text{Purple}}{Na_4[Fe(CN)_5NOS]}$$

This is called *Gmelin reaction.*

2. It reacts with sulfite ions to give red-colored sodium pentacyanonitrosoniumsulfitoferrate(II), $Na_4[Fe(CN)_5NO(SO_3)]$.

$$Na_2[Fe(CN)_5NO] + Na_2SO_3 \rightarrow \underset{\text{Red}}{Na_4[Fe(CN)_5NO(SO_3)]}$$

This is called *Bödecker reaction.*

3. It reacts with silver ions to give a buff-colored (skin-colored) precipitate of silver pentacyanonitrosoniumferrate(II), $Ag_2[Fe(CN)_5NO]$.

$$\underset{}{Na_2[Fe(CN)_5NO]} + 2AgNO_3 \rightarrow \underset{\text{Buff}}{Ag_2[Fe(CN)_5NO]\downarrow} + 2NaNO_3$$

Case Study 27.6 On adding a solution of potassium cyanide to an aqueous metal salt, a green precipitate, **A** is initially obtained which subsequently dissolves in excess of cyanide to give a yellow complex, **B**. Addition of iron(III) chloride solution to **B** gives a blue compound, **C**. On heating **B** with concentrated sulfuric acid, a poisonous gas **D** is evolved. Identify **A**, **B**, **C** and **D**.

Iron(III) ions give a blue-colored complex, called Prussian blue, with potassium ferrocyanide. Therefore, the yellow complex **(B)** is potassium ferrocyanide, while the blue compound **(C)** is Prussian blue.

$$\underset{\textbf{B}}{K_4[Fe(CN)_6]} + FeCl_3 \rightarrow \underset{\textbf{C}}{K\{Fe[Fe(CN)_6]\}} + 3KCl$$

Potassium ferrocyanide, in turn, may be obtained by adding potassium cyanide to an iron(II) salt. Initially iron(II) cyanide is precipitated as a green solid, **A**.

$$FeSO_4 + 2KCN \rightarrow \underset{\textbf{A}}{Fe(CN)_2\downarrow} + K_2SO_4$$

On adding more of potassium cyanide, the precipitate of iron(II) cyanide dissolves to yield potassium ferrocyanide **B**.

$$\underset{\textbf{A}}{Fe(CN)_2} + 4KCN \rightarrow \underset{\textbf{B}}{K_4[Fe(CN)_6]}$$

On heating potassium ferrocyanide with concentrated sulfuric acid, carbon monoxide—a poisonous gas **(D)**—is released.

$$\underset{\textbf{B}}{K_4[Fe(CN)_6]} + \underset{\text{Conc.}}{6H_2SO_4} + 6H_2O \rightarrow 3(NH_4)_2SO_4 + 2K_2SO_4 + FeSO_4 + \underset{\textbf{D}}{6CO}$$

Thus, **A** = $Fe(CN)_2$, **B** = $K_4[Fe(CN)_6]$, **C** = $K\{Fe[Fe(CN)_6]\}$ and **D** = CO.

Cobalt

The simple salts of cobalt are stable in oxidation state II, while the complexes are stable in oxidation state III. An uncommon oxidation state, IV, also exists.

Oxidation state IV

Cesium hexafluorocobaltate(IV), $Cs_2[CoF_6]$, may be prepared by the fluorination of cesium tetrachlorocobaltate(II), $Cs_2[CoCl_4]$.

$$Cs_2[CoCl_4] + 3F_2 \rightarrow Cs_2[CoF_6] + 2Cl_2$$

Barium tetraoxocobaltate(IV), $Ba_2[CoO_4]$, may be obtained by the reaction of barium hydroxide and cobalt(II) hydroxide, in the presence of air, at about 1300 K.

$$8Ba(OH)_2 + 8Co(OH)_2 + \underset{\text{Air}}{3O_2} \xrightarrow{1300\text{ K}} 4Ba_2[CoO_4] + 4Co(OH)_3 + 10H_2O$$

Cobalt(IV) oxide (CoO_2) is prepared by the oxidation of alkaline cobalt(II) ions with chlorine.

$$CoCl_2 + 4NaOH + Cl_2 \rightarrow CoO_2 + 4NaCl + 2H_2O$$

A binuclear complex, *pentamminecobalt(III)-μ-peroxo-pentamminecobalt(III) ion*, is obtained when oxygen (air) is bubbled through a solution of cobalt(II) ions, in the presence of ammonia and ammonium chloride. (The same reaction if conducted in the presence of gelatin or activated charcoal, yields hexamminecobalt(III) chloride, $[Co(NH_3)_6]Cl_3$.)

$$2CoCl_2 + 8NH_3 + 2NH_4Cl + \underset{\text{Air}}{O_2} \rightarrow [(NH_3)_5Co{-}O{-}O{-}Co(NH_3)_5]Cl_4 + 2HCl$$

This brown-colored complex turns green when reacted with persulfate ions.

$$\underset{\text{Brown, diamagnetic}}{2[(NH_3)_5\overset{\text{III}}{Co}{-}O{-}O{-}\overset{\text{III}}{Co}(NH_3)_5]^{4+}} + S_2O_8^{2-} \rightarrow \underset{\text{Green, paramagnetic}}{2[(NH_3)_5\overset{\text{III}}{Co}{-}O{-}O{-}\overset{\text{IV}}{Co}(NH_3)_5]^{5+}} + 2SO_4^{2-}$$

The green complex, pentamminccobalt(III)-μ–peroxo-pentamminecobalt(IV) ion, has one cobalt in oxidation state III and the other in oxidation state IV. Since ammonia is a strong ligand, it creates a low-spin state in two cobalt ions. According to crystal field theory, low-spin cobalt(III) ion ($3d^6$) is diamagnetic, while low-spin cobalt (IV) ion ($3d^5$) is paramagnetic.

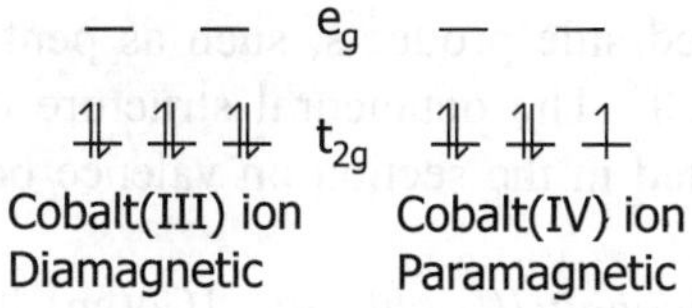

Therefore, the brown complex is diamagnetic, while the green one is paramagnetic. It must, however, be emphasized that both the cobalt ions in the green complex are absolutely identical. The unpaired electron spends equal time on both the cobalt ions.

Oxidation state III

The oxidation potential of cobalt(III) ion, in aqueous solution, is so high that it oxidizes water, and itself gets reduced to cobalt(II) ion. In the presence of ligands, the oxidation potential becomes very low and, therefore, cobalt(III) ion is extremely stable when complexed. Due to this reason, cobalt forms only a few simple salts, but a large number of stable complexes in oxidation state III.

Cobalt(III) fluoride (CoF_3), in anhydrous state, is a brown solid which may be prepared by the reaction of fluorine with cobalt(II) fluoride. It hydrolyzes in water.

$$2CoF_2 + F_2 \rightarrow 2CoF_3$$

Cobalt(III) sulfate, $Co_2(SO_4)_3$ is prepared by the reaction of an oxidizing agent, such as ozone, with a cobalt(II) salt, in the presence of sulfuric acid. It is also stable as long as it is dry.

$$2CoCl_2 + 3H_2SO_4 + O_3 \rightarrow Co_2(SO_4)_3 + 4HCl + H_2O + O_2$$

Anhydrous *cobalt(III) nitrate*, $Co(NO_3)_3$ may be prepared by the reaction of cobalt(III) fluoride with sodium nitrate in a non-aqueous solvent, such as dinitrogen tetraoxide, N_2O_4 at low temperatures.

$$CoF_3 + 3NaNO_3 \xrightarrow{N_2O_4} \underset{\text{Anhydrous}}{Co(NO_3)_3} + 3NaF$$

Each nitrate ion becomes attached to cobalt through two oxygen atoms, giving an octahedral structure.

Cobalt(III) ion complexes with a wide range of nitrogen donor ligands, including ammonia, ethylenediamine, nitro and thiocyanato-N. The initial step in the synthesis of these complexes generally involves oxidation of cobalt(II) ion with air.

For example, *hexamminecobalt(III) chloride* may be synthesized by oxidizing cobalt(II) chloride in the presence of ammonia and ammonium chloride.

$$4CoCl_2 + 4NH_4Cl + 20NH_3 + O_2 \rightarrow 4[Co(NH_3)_6]Cl_3 + 2H_2O$$

The reaction is conducted in the presence of gelatin or activated charcoal. These suppress unwanted side products, such as pentamminechlorocobalt(III) chloride, $[Co(NH_3)_5Cl]Cl_2$. The octahedral structure of hexamminecobalt(III) chloride has been depicted in the section on valence bond theory of complexes (*see* Chapter 26, p. 702).

Tris(ethylenediamine)cobalt(III) chloride, $[Co(en)_3]Cl_3$ is similarly prepared by air oxidation of cobalt(II) chloride, in the presence of ethylenediamine and its hydrochloride.

$$4CoCl_2 + 8en + 4en.HCl + O_2 \rightarrow 4[Co(en)_3]Cl_3 + 2H_2O$$

Sodium hexanitrocobaltate(III), $Na_3[Co(NO_2)_6]$ is prepared by mixing a solution of sodium nitrite in acetic acid and cobalt(II) nitrate with water and then blowing air for a few hours.

$$
\begin{array}{rcl}
Co^{2+} & \rightarrow & Co^{3+} + e^- \\
NO_2^- + 2H^+ + e^- & \rightarrow & H_2O + NO \\
Co^{3+} + 6NO_2^- & \rightarrow & [Co(NO_2)_6]^{3-} \\
\hline
Co^{2+} + 7NO_2^- + 2H^+ & \rightarrow & [Co(NO_2)_6]^{3-} + H_2O + NO
\end{array}
$$

It is an orange-colored complex. When treated with a potassium salt, it yields a yellow precipitate of potassium hexanitrocobaltate(III).

$$\underset{\text{Orange}}{Na_3[Co(NO_2)_6]} + 3KCl \rightarrow \underset{\text{Yellow}}{K_3[Co(NO_2)_6]\downarrow} + 3NaCl$$

It is, therefore, used for qualitative detection of potassium ions. It has an octahedral structure.

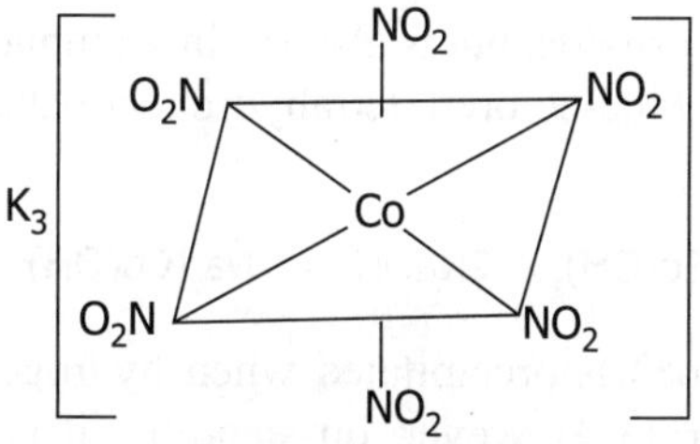

When air is passed through an aqueous solution of cobalt(II) nitrate, in presence of ammonia and ammonium nitrate at 310 K, a mixture of two bridging complexes is obtained.

$$8Co(NO_3)_2 + 2NH_4NO_3 + 34NH_3 + 2O_2 \rightarrow$$

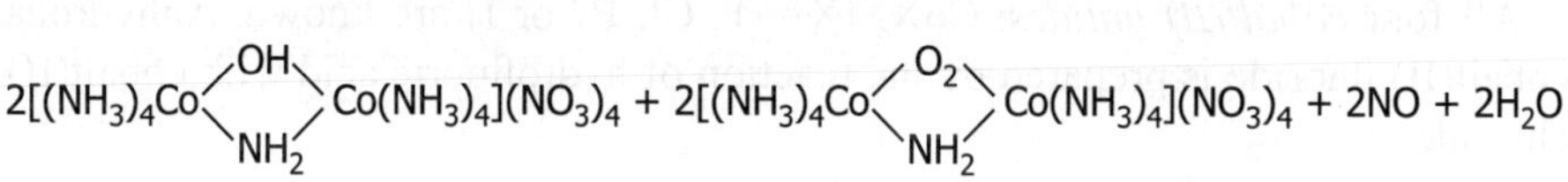

On acidifying the reaction mixture with dilute sulfuric acid, the two bridging complexes are isolated as sulfate salts.

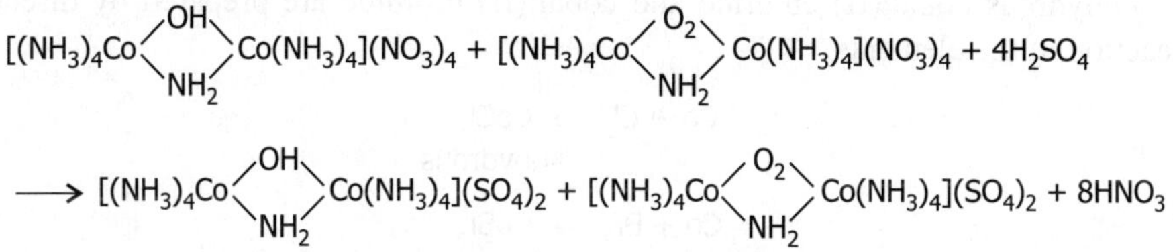

The mixture of tetramminecobalt(III)-μ-amido-μ-hydrxo-tetramminecobalt(III) sulfate and tetramminecobalt(III)-μ-amido-μ-superoxo-tetramminecobalt(III) sulfate is called *Vortmann's sulfate*.

Oxidation state II

Cobalt(II) oxide (CoO) may be prepared by thermal decomposition of cobalt carbonate.

$$CoCO_3 \xrightarrow{\text{Heat}} CoO + CO_2$$

On heating to 750 K in the presence of atmospheric oxygen, CoO gives the mixed oxide, Co_3O_4.

$$6CoO + O_2 \xrightarrow{750\text{ K}} 2Co_3O_4$$

Cobalt(II) hydroxide, $Co(OH)_2$ is obtained as a pink solid when cobalt(II) chloride is treated with an alkali.

$$CoCl_2 + 2NaOH \rightarrow Co(OH)_2 + 2NaCl$$

It is largely a basic hydroxide, but is also slightly amphoteric. Thus, it dissolves in excess alkali to yield blue-colored, tetrahydroxocobaltate(II) ion, $[Co(OH)_4]^{2-}$. For example,

$$Co(OH)_2 + 2NaOH \rightarrow Na_2[Co(OH)_4]$$

Cobalt(II) sulfide (CoS) is precipitated when hydrogen sulfide gas is passed through cobalt(II) chloride. However, on standing, it is oxidized to cobalt(III) hydroxosulfide, Co(OH)S.

$$CoCl_2 + H_2S \rightarrow CoS\downarrow + 2HCl$$

$$4CoS + 2H_2O + O_2 \rightarrow 4Co(OH)S$$

All four *cobalt(II) halides*, CoX_2 (X = F, Cl, Br or I) are known. Anhydrous cobalt(II) fluoride is prepared by the reaction of hydrofluoric acid with cobalt(II) chloride.

$$CoCl_2 + 2HF \rightarrow \underset{\text{Anhydrous}}{CoF_2} + 2HCl$$

Anhydrous cobalt(II) chloride and cobalt(II) bromide are prepared by direct reaction of the elements.

$$Co + Cl_2 \rightarrow \underset{\text{Anhydrous}}{CoCl_2}$$

$$Co + Br_2 \rightarrow \underset{\text{Anhydrous}}{CoBr_2}$$

Anhydrous cobalt(II) iodide is prepared by heating the metal with hydrogen iodide.

$$Co + 2HI \xrightarrow{\text{Heat}} \underset{\text{Anhydrous}}{CoI_2} + H_2$$

Cobalt(II) acetate, $Co(CH_3COO)_2.4H_2O$ is obtained by dissolving cobalt(II) carbonate in acetic acid.

$$CoCO_3 + 2CH_3COOH + 3H_2O \rightarrow Co(CH_3COO)_2.4H_2O + CO_2$$

Although cobalt(II) salts are very stable and difficult to oxidize to cobalt(III) compounds, the complexes of cobalt(II) are readily oxidized. For example,

$$[Co(NH_3)_6]^{2+} \xrightarrow{\text{Air}} [Co(NH_3)_6]^{3+} + e^-$$

The reason for this transformation is the greater crystal field stabilization energy (CFSE) (*see* Chapter 26, p. 715) of cobalt(III) ion ($3d^6$ system) as compared to that of cobalt(II) ion ($3d^7$ system). Addition of potassium cyanide to cobalt(II) solution yields potassium pentacyanocobaltate(II) complex, $K_3[Co(CN)_5]$.

$$CoCl_2 + 5KCN \rightarrow K_3[Co(CN)_5] + 2KCl$$

This green-colored complex is paramagnetic, with one unpaired electron. Being 5-coordinated, its geometry is square pyramidal.

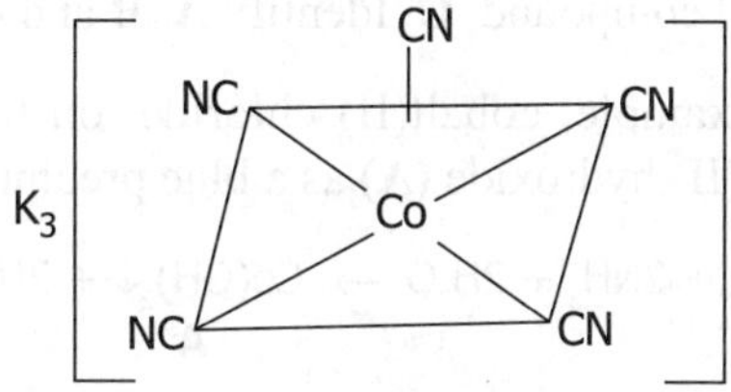

Unlike other cobalt(II) complexes, this one is not octahedral. If it had been octahedral, its configuration would have been $t^6_{2g}\,e_g^{\,1}$. However, since cyano ligands have the ability to enter into π-bonding with the metal ions, the crystal field splitting, Δ_O for this complex would have been very large. Accordingly, the e_g orbitals would have had very high energy. Therefore, the octahedral complex would have been unstable.

Addition of ethanol to potassium pentacyanocobaltate(II) precipitates a purple-colored, diamagnetic dimer, designated as potassium decacyanodicobaltate(II).

$$\underset{\text{Green solution}}{2K_3[Co(CN)_5]} \xrightarrow{\text{Ethanol}} \underset{\text{Purple precipitate}}{K_6[Co_2(CN)_{10}]\downarrow}$$

The structure of the purple dimer may be elaborated as follows:

$$K_6[(CN)_5Co–Co(CN)_5]$$

The green solution may be oxidized to yield brown-colored, potassium pentacyanocobaltate(III)-μ-peroxo-pentacyanocobaltate(III).

$$2K_3[Co(CN)_5] + O_2 \rightarrow K_6[(CN)_5\overset{\text{III}}{Co}–O–O–\overset{\text{III}}{Co}(CN)_5]$$

The peroxo complex may be further oxidized to red-colored potassium pentacyanocobaltate(III)-μ-superoxo-pentacyanocobaltate(III) complex by bromine.

$$\underset{\text{Peroxo}}{2K_6[(CN)_5Co\text{–}O\text{–}O\text{–}Co(CN)_5]} + Br_2 \rightarrow \underset{\text{Superoxo}}{2K_5[(CN)_5Co\text{–}O\text{–}O\text{–}Co(CN)_5]} + 2KBr$$

When green potassium pentacyanocobaltate(II) is boiled with hydrogen peroxide, it gets oxidized to yellow potassium hexacyanocobaltate(III), $K_3[Co(CN)_6]$.

$$3K_3[Co(CN)_5] + H_2O_2 \rightarrow 2K_3[Co(CN)_6] + Co(OH)_2 + 3KCN$$

Cobalt(III) ion, in a low-spin state, has $3d^6$ configuration. Unlike cobalt(II) ion, it does not have an e_g electron. Therefore, octahedral geometry is favored even with π-bonding cyano ligands.

Case Study 27.7 When a pink, aqueous solution of cobalt(II) salt is treated with aqueous ammonia, a blue precipitate, **A** is initially formed, which then dissolves in excess of the reagent, forming a yellow-brown solution, **B**. On standing in air, this solution yields a red compound, **C**. Identify **A**, **B** and **C**.

A cobalt(II) salt, for example, cobalt(II) chloride, on treatment with aqueous ammonia, yields cobalt(II) hydroxide **(A)** as a blue precipitate.

$$CoCl_2 + 2NH_3 + 2H_2O \rightarrow \underset{\mathbf{A}}{Co(OH)_2\downarrow} + 2NH_4Cl$$

In excess of ammonia, cobalt(II) hydroxide dissolves to give yellow-brown hexamminecobalt(II) ion, **B**.

$$Co(OH)_2 + 6NH_3 \rightarrow \underset{\mathbf{B}}{[Co(NH_3)_6]^{2+}} + 2OH^-$$

If the yellow-brown solution is allowed to stand in air, it gets oxidized to a red-colored, binuclear, bridging complex, tetramminecobalt(III)-μ-dihydroxo-tetramminecobalt(III) ion, **C**.

$$\underset{\mathbf{B}}{4[Co(NH_3)_6]^{2+}} + 10H_2O + O_2 \longrightarrow \underset{\mathbf{C}}{2[(NH_3)_4Co(\mu\text{-}OH)_2Co(NH_3)_4]^{4+}} + 8NH_4OH$$

Thus, $\mathbf{A} = Co(OH)_2$, $\mathbf{B} = [Co(NH_3)_6]^{2+}$ and $\mathbf{C} = [(NH_3)_4Co(\mu\text{-}OH)_2Co(NH_3)_4]^{4+}$.

Nickel

Oxidation state II dominates the chemistry of nickel. Oxidation states III and IV also exist, but are uncommon.

Oxidation state IV

The hydrated *nickel(IV) oxide,* $NiO_2.nH_2O$, may be prepared by fusing a nickel(II) salt with sodium hydroxide, in the presence of air.

$$2NiCl_2 + 4NaOH + O_2 \rightarrow 2NiO_2 + 4NaCl + 2H_2O$$

It oxidizes water.

$$2NiO_2 + 2H_2O \rightarrow 2Ni(OH)_2 + O_2$$

Potassium hexafluoronickelate(IV), $K_2[NiF_6]$, is prepared by fluorination of a mixture of nickel(II) chloride and potassium chloride.

$$NiCl_2 + 2KCl + 3F_2 \rightarrow K_2[NiF_6] + 2Cl_2$$

Oxidation state III

Nickel(III) oxohydroxide, NiO(OH), is obtained by oxidation of nickel(II) chloride with bromine in sodium hydroxide.

$$2NiCl_2 + 4NaOH + Br_2 \rightarrow 2NiO(OH) + 4NaCl + 2HBr$$

If nickel metal is fused with sodium hydroxide, in the presence of oxygen, *sodium dioxonickelate(III)*, $Na[NiO_2]$ is formed.

$$4NaOH + 4Ni + 3O_2 \rightarrow 4Na[NiO_2] + 2H_2O$$

Potassium hexafluoronickelate(III), $K_3[NiF_6]$, is prepared by fluorination of nickel(II) chloride and potassium chloride at high temperature and pressure.

$$2NiCl_2 + 6KCl + 6F_2 \rightarrow 2K_3[NiF_6] + 5Cl_2$$

Oxidation state II

Nickel(II) oxide, a green solid, is prepared by thermal decomposition of nickel(II) carbonate.

$$NiCO_3 \xrightarrow{\text{Heat}} NiO + CO_2$$

Nickel(II) hydroxide, $Ni(OH)_2$, is obtained as a green precipitate by the addition of sodium hydroxide to an aqueous solution of a nickel(II) salt.

$$NiCl_2 + 2NaOH \rightarrow \underset{\text{Green}}{Ni(OH)_2\downarrow} + 2NaCl$$

Nickel(II) sulfide, NiS, is precipitated as a black solid when hydrogen sulfide gas is passed through a solution of nickel(II) ions.

$$NiCl_2 + H_2S \rightarrow NiS\downarrow + 2HCl$$

It is soluble in aqueous acids, but is very quickly oxidized by air to insoluble nickel(III) hydroxosulfide, Ni(OH)S.

$$4NiS + 2H_2O + O_2 \rightarrow \underset{\text{Insoluble}}{4Ni(OH)S}$$

On adding potassium cyanide to a nickel(II) salt solution, *nickel(II) cyanide*, $Ni(CN)_2$ is precipitated. It is green in hydrated form and yellowish-brown in anhydrous state.

$$NiSO_4 + 2KCN \rightarrow Ni(CN)_2\downarrow + K_2SO_4$$

It dissolves in excess of potassium cyanide to yield square planar, yellow, potassium tetracyanonickelate(II) complex, $K_2[Ni(CN)_4]$.

$$Ni(CN)_2 + 2KCN \rightarrow K_2[Ni(CN)_4]$$

The red-colored nickel(II) dimethylglyoximato complex, formed by the reaction of nickel(II) ions with dimethylglyoxime in ammonical medium, is also square planar. Its structure is given in Chapter 25 (*see* Case Study 25.4, p. 682).

Case Study 27.8 Addition of potassium cyanide to aqueous nickel(II) chloride yields a green precipitate, **A**, which dissolves in excess of the reagent to give a yellow compound, **B**. When solid potassium cyanide is added to a solution of **B**, an unstable, red compound, **C** is obtained. On boiling **B** with sodium hydroxide and bromine water, a black solid, **D**, precipitates out. Identify **A**, **B**, **C** and **D**.

When potassium cyanide is added to nickel(II) chloride solution, a green precipitate of nickel(II) cyanide, **A**, is obtained.

$$NiCl_2 + 2KCN \rightarrow \underset{\mathbf{A}}{Ni(CN)_2\downarrow} + 2KCl$$

In excess of potassium cyanide, the precipitate of nickel(II) cyanide dissolves to yield yellow-colored potassium tetracyanonickelate(II) complex, **B**.

$$\underset{\mathbf{A}}{Ni(CN)_2} + 2KCN \rightarrow \underset{\mathbf{B}}{K_2[Ni(CN)_4]}$$

Addition of solid potassium cyanide (or a concentrated solution thereof) to potassium tetracyanonickelate(II) gives red colored potassium pentacyanonickelate(II) complex, **C**.

$$\underset{\mathbf{B}}{K_2[Ni(CN)_4]} + KCN \rightarrow \underset{\mathbf{C}}{K_3[Ni(CN)_5]}$$

The complex C is not very stable. It exists only in solution state and decomposes into **B** in solid state.

When a solution of **B** is boiled with sodium hydroxide and bromine water, black nickel(II) oxohydroxide, NiO(OH), **D** is precipitated.

$$2K_2[Ni(CN)_4] + 4NaOH + Br_2 \rightarrow \underset{\mathbf{D}}{2NiO(OH)\downarrow} + 4KCN + 4NaCN + 2HBr$$

Thus, **A** = $Ni(CN)_2$, **B** = $K_2[Ni(CN)_4]$, **C** = $K_3[Ni(CN)_5]$ and **D** = NiO(OH).

Copper

For copper, oxidation state II is most stable, I is largely unstable and III is unusual.

Oxidation state III

Potassium dioxocuprate(III), $K[CuO_2]$ may be obtained by fusing potassium hydroxide with copper(II) oxide, in the presence of oxygen.

$$4KOH + 4CuO + O_2 \rightarrow 4K[CuO_2] + 2H_2O$$

Potassium hexafluorocuprate(III), $K_3[CuF_6]$ is prepared by treating a mixture of potassium chloride and copper(I) chloride with fluorine.

$$3KCl + CuCl + 3F_2 \rightarrow K_3[CuF_6] + 2Cl_2$$

Potassium bis(periodato)cuprate(III), $K_7[Cu(IO_6)_2]$ may be synthesized by oxidizing an alkaline copper(II) salt with periodic acid, $HIO_4.2H_2O$ (or H_5IO_6).

$$4CuCl_2 + 28KOH + 12H_5IO_6 \rightarrow 4K_7[Cu(IO_6)_2] + 2I_2O_5 + 8HCl + 40H_2O + O_2$$

Oxidation state II

Copper(II) oxide (CuO) is obtained as black powder by heating copper(II) nitrate.

$$2Cu(NO_3)_2 \xrightarrow{\text{Heat}} 2CuO + 4NO_2 + O_2$$

It gets reduced to elemental copper when heated to 500 K in a stream of hydrogen.

$$CuO + H_2 \xrightarrow{500\text{ K}} Cu + H_2O$$

Copper(II) hydroxide, $Cu(OH)_2$ is prepared by adding an alkali to a copper(II) salt. For example,

$$CuCl_2 + 2NaOH \rightarrow Cu(OH)_2 + 2NaCl$$

On warming, it is dehydrated to the oxide.

$$Cu(OH)_2 \rightarrow CuO + H_2O$$

With a concentrated alkali, it gives deep blue polynuclear complexes.

$$nCu(OH)_2 + 2NaOH \rightarrow Na_2[Cu_n(OH)_{2n+2}]$$

Anhydrous *copper(II) nitrate*, $Cu(NO_3)_2$ may be obtained in a non-aqueous medium by the action of dinitrogen tetraoxide (N_2O_4) dissolved in ethylacetate, on copper.

$$Cu + 3N_2O_4 \rightarrow Cu(NO_3)_2.N_2O_4 + 2NO$$

The adduct, $Cu(NO_3)_2.N_2O_4$ has the composition, $[NO]^+[Cu(NO_3)_3]^-$. On heating to 360 K, it yields anhydrous copper(II) nitrate.

$$[NO]^+[Cu(NO_3)_3]^- \xrightarrow{\text{Heat}} \underset{\text{Anhydrous}}{Cu(NO_3)_2} + 2NO_2$$

Hydrated *copper(II) sulfate*, $CuSO_4.5H_2O$ has an octahedral structure, with four water molecules occupying equatorial, planar positions. The oxygen atoms of sulfate anions occupy axial positions. The remaining water molecule is hydrogen bonded between an oxygen atom of any one sulfate anion and the adjacent, planar water molecule (Fig. 27.8).

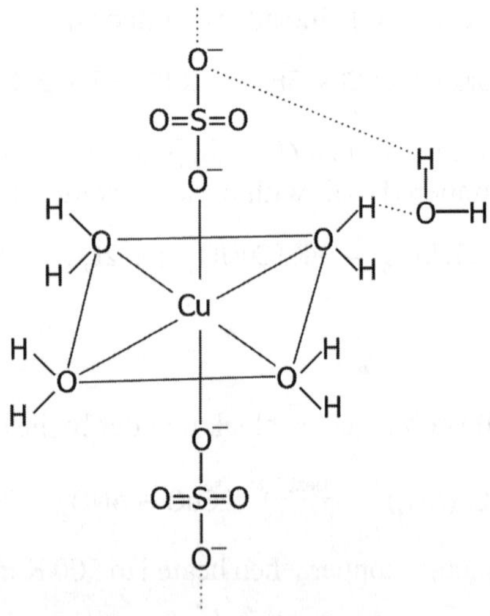

Fig. 27.8 Structure of solid copper(II) sulfate.5-water

When copper(II) sulfate, or for that matter any oxyacid salt of copper(II) ion, is dissolved in water, blue-colored hexaqua ion, $[Cu(H_2O)_6]^{2+}$ is formed. It has a distorted octahedral structure—with two long and four short copper–oxygen bonds. This is attributed to Jahn–Teller distortion (*see* Chapter 26, p. 740).

Addition of ligands to hexaquacopper(II) ion leads to the formation of complexes with stepwise replacement of water molecules. With ammonia, for example, four water molecules in the coordination sphere may be successively displaced to yield tetramminediaquacopper(II) ion, $[Cu(NH_3)_4(H_2O)_2]^{2+}$. The replacement of fifth and sixth water molecules is difficult. In fact, hexamminecopper(II) ion, $[Cu(NH_3)_6]^{2+}$ can be prepared only in liquid ammonia solvent. The reason for this unusual behavior is the Jahn–Teller effect. The fifth and sixth bonds are so long, and so weak (even for water ligand), that the formation constants for the addition of fifth and sixth ligands are extremely small.

The *copper(II) halides*, CuX_2 (X = F, Cl or Br) have polymeric structures with octahedral coordination. The halogen bridges in one chain form relatively long bonds to copper in other chains. The octahedral is thus distorted—another case of Jahn–Teller distortion (Fig. 27.9).

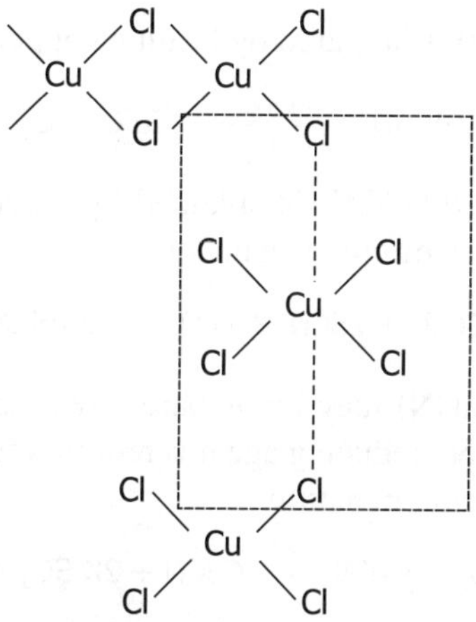

Fig. 27.9 Structure of copper(II) chloride

Copper(II) acetate is dimeric and hydrated, with the composition, $Cu_2(CH_3COO)_4.2H_2O$. The structure is similar to that of dimeric, dihydrated chromium(II) acetate. However, in the latter compound, the chromium–chromium bond length is much shorter than expected, suggesting strong interactions between the two metal atoms. This is also borne out by the diamagnetic nature of chromium(II) acetate. In copper(II) acetate, on the other hand, the distance between the two copper atoms is relatively long, suggesting the absence of metal–metal bond. The compound is paramagnetic in nature.

This does not mean that there is absolutely no interaction between the metal atoms. If it had been so, the magnetic moment of this compound would have been of the order of 1.73 BM per copper atom—the value expected for a $3d^9$ metal ion. However, the experimental magnetic moment of copper(II) acetate is 1.4 BM per copper atom. This is indicative of weak coupling between the two copper atoms. In fact, the $3d_{x^2-y^2}$ orbital of one copper undergoes a weak lateral overlap with the corresponding orbital of the other copper atom. This type of overlap is called *δ-bonding*.

Oxidation state I

Copper(I) compounds are diamagnetic and colorless, as expected for the $3d^{10}$ system. Since copper(I) ion has a relatively large size, the lattice energy values of its compounds are concomitantly small. Therefore, most of the compounds are unstable, despite the completely filled $3d^{10}$ configuration of the metal ion. In fact, copper(I) readily disproportionates in an aqueous solution.

$$Cu^+ + Cu^+ \rightleftharpoons Cu^{2+} + Cu$$

The only copper(I) compounds which are stable are those that are insoluble and, therefore, precipitate out of aqueous solution as soon as they are formed. These include copper(I) oxide, copper(I) thiocyanate, copper(I) cyanide and copper(I) halides.

Copper(I) oxide (Cu_2O) is prepared by heating copper(II) oxide to 1000 K.

$$4CuO \xrightarrow{1000\ K} 2Cu_2O + O_2$$

Copper(I) thiocyanate (CuSCN) is obtained by reducing copper(II) ions with sulfite ions, in the presence of thiocyanate ions.

$$2Cu^{2+} + SO_3^{2-} + 2SCN^- + H_2O \rightarrow 2CuSCN + H_2SO_4$$

Copper(I) cyanide (CuCN) may be prepared by adding potassium cyanide to aqueous copper(II) ions. No reducing agent is required for this reaction as cyanide ions themselves are reducing in nature.

$$2CuSO_4 + 4KCN \rightarrow 2CuCN + 2K_2SO_4 + (CN)_2$$

With excess of potassium cyanide, the insoluble copper(I) cyanide yields the soluble, tetrahedral, potassium tetracyanocuprate(I), $K_3[Cu(CN)_4]$.

$$CuCN + 3KCN \rightarrow K_3[Cu(CN)_4]$$

Copper(I) halides, CuX (X = Cl, Br or I) may be prepared by one of the following methods (copper(I) fluoride does not exist):

1. By boiling an acidic solution of copper(II) chloride with copper turnings. Initially, dichlorocuprate(I) ion is produced, which, on dilution, gives copper(I) chloride.

$$CuCl_2 + Cu + 2HCl \rightarrow 2H[CuCl_2]$$

$$H[CuCl_2] \xrightarrow{Dillution} CuCl + HCl$$

   Copper(I) bromide may also be prepared by a similar method.

2. By reducing a solution of copper(II) ions with sulfite ions, in the presence of sodium chloride or sodium bromide.

$$2Cu^{2+} + SO_3^{2-} + 2Cl^- + H_2O \rightarrow 2CuCl + H_2SO_4$$

3. Copper(I) iodide may be prepared by the reaction of potassium iodide with aqueous copper(II) sulfate.

$$2CuSO_4 + 4KI \rightarrow 2CuI + 2K_2SO_4 + I_2$$

Iodide ions reduce copper(II) to copper(I). Since chloride or bromide ions are not strong reducing agents, it is not possible to prepare copper(I) chloride or copper(I) bromide by this method.

The copper(I) halides have a cyclic, trimeric structure in vapor state. In solid state, they possess a zincblende structure in which the metal ion is tetrahedrally surrounded by halide ions. The structure of copper(I) chloride in vapor and solid states is given in Fig. 27.10.

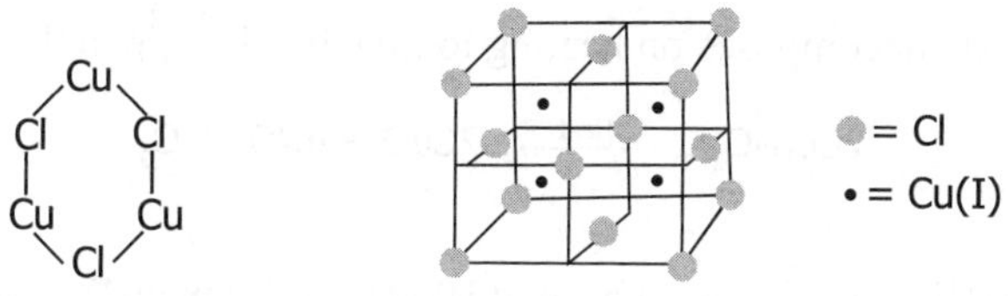

(a) (b)

Fig. 27.10 Structure of copper(I) chloride in (a) vapor state and (b) solid state

We now list the important properties of copper(I) chloride. The bromide and iodide have analogous characteristics.

1. It reacts with ammonia to form diamminecopper(I) chloride, $[Cu(NH_3)_2]Cl$.

$$CuCl + 2NH_3 \rightarrow [Cu(NH_3)_2]Cl$$

2. It dissolves in hydrochloric acid to yield successively, dichlorocuprate(I) ion, $[CuCl_2]^-$ and tetrachlorocuprate(I) ion, $[CuCl_4]^{3-}$.

$$CuCl + HCl \rightarrow H[CuCl_2]$$

$$CuCl + 3HCl \rightarrow H_3[CuCl_4]$$

3. Ammonical copper(I) chloride absorbs carbon monoxide and yields carbonylchlorocopper(I), Cu(CO)Cl.

$$CuCl + CO \rightarrow Cu(CO)Cl$$

4. It also absorbs ethyne to yield chloro(ethyne)copper(I), $Cu(C_2H_2)Cl$.

$$CuCl + C_2H_2 \rightarrow Cu(C_2H_2)Cl$$

In the presence of chloride ions, dichloro(ethyne)cuprate(I) ion, $[Cu(C_2H_2)Cl_2]^-$ is formed.

$$CuCl + C_2H_2 + Cl^- \rightarrow [Cu(C_2H_2)Cl_2]^-$$

5. It reacts with sodium hydroxide to yield copper(I) oxide.

$$2CuCl + 2NaOH \rightarrow Cu_2O + 2NaCl + H_2O$$

6. It yields copper(I) sulfide with hydrogen sulfide.

$$2CuCl + H_2S \rightarrow Cu_2S + 2HCl$$

Case Study 27.9 When copper(II) nitrate is heated, a black solid, **A**, is obtained. As the temperature is increased beyond 1000 K, **A** is transformed into a red solid, **B**. Action of dilute hydrochloric acid on **B** yields a white compound, **C**. The compound **C** may also be obtained by passing a toxic gas, **D**, with smell of burning sulfur, through the solution of copper(II) nitrate, in the presence of sodium chloride. When **C** is treated with excess of hydrochloric acid it first gives a linear complex, **E** and then a tetrahedral complex, **F**. Identify **A, B, C, D, E** and **F**.

Copper(II) nitrate decomposes on heating to give black copper(II) oxide, **A**.

$$2Cu(NO_3)_2 \xrightarrow{\text{Heat}} \underset{\textbf{A}}{2CuO} + 4NO_2 + O_2$$

When copper(II) oxide is heated beyond 1000 K, red copper(I) oxide, **B** is formed.

$$\underset{\textbf{A}}{4CuO} \xrightarrow{> 1000\text{ K}} \underset{\textbf{B}}{2Cu_2O} + O_2$$

With dilute hydrochloric acid, **B** yields copper(I) chloride, **C**.

$$\underset{\textbf{B}}{Cu_2O} + \underset{\text{Dilute}}{2HCl} \rightarrow \underset{\textbf{C}}{2CuCl} + H_2O$$

The toxic gas, with a smell of burning sulfur, is sulfur dioxide **D**. It reduces copper(II) ions to copper(I) ions.

$$SO_2 + 2H_2O \rightarrow SO_4^{2-} + 4H^+ + 2e^-$$

$$[Cu^{2+} + e^- \rightarrow Cu^+] \times 2$$

$$2Cu^{2+} + 2H_2O + \underset{\textbf{D}}{SO_2} \rightarrow 2Cu^+ + SO_4^{2-} + 4H^+$$

In the presence of sodium chloride, copper(I) chloride, **C** is precipitated.

$$Cu^+ + NaCl \rightarrow \underset{\textbf{C}}{CuCl\downarrow} + Na^+$$

When copper(I) chloride is treated with hydrochloric acid, it solubilizes and yields hydrogen dichlorocuprate(I) complex, **E**.

$$CuCl + HCl \rightarrow \underset{\textbf{E}}{H[CuCl_2]}$$

Complex **E** has a linear structure.

$$\underset{\textbf{E}}{[Cl{-}Cu{-}Cl]^-}$$

Further addition of hydrochloric acid gives hydrogen tetrachlorocuprate(I) complex, **F**.

$$\underset{\mathbf{E}}{H[CuCl_2]} + 2HCl \rightarrow \underset{\mathbf{F}}{H_3[CuCl_4]}$$

Complex F has a tetrahedral structure.

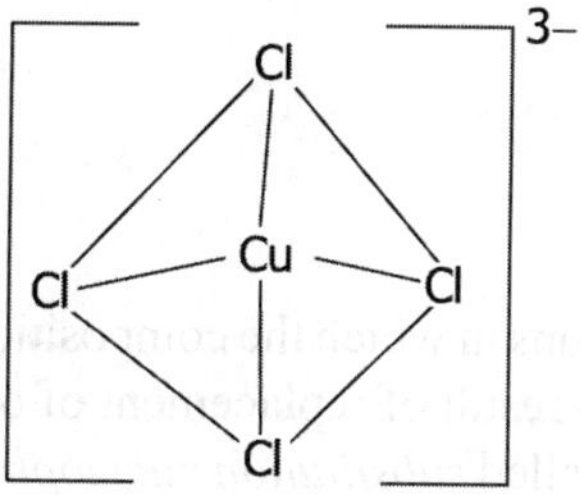

Thus, **A** = CuO, **B** = Cu_2O, **C** = CuCl, **D** = SO_2, **E** = $H[CuCl_2]$ and **F** = $H_3[CuCl_4]$.

28. KINETIC ASPECTS OF COMPLEXES

Introduction

There are a number of reactions in which the composition of the coordination sphere of the complexes alters as a result of replacement of one or more ligands by other groups. Such reactions are called *substitution nucleophilic reactions* and designated as S_N. The ability of a complex to undergo such reactions is called its *lability*. Those complexes in which these reactions occur rapidly are called *labile*, while those in which these proceed at a relatively slow rate are called *inert*. Labile and inert are kinetic terms and have already been discussed in Chapter 25.

Mechanisms of Substitution Reactions

The substitution reactions of complexes follow two extreme mechanistic pathways: *substitution nucleophilic unimolecular reactions* and *substitution nucleophilic bimolecular reactions*. These are now generalized with reference to an octahedral complex, $[L_5MX]^{n+}$, where the ligand X is deemed to be substituted by another ligand Y.

1. Substitution nucleophilic unimolecular reactions

In this form of substitution, the complex dissociates, losing the ligand X and the vacant site is subsequently taken up by the new ligand, Y. The mechanism may be represented as follows:

$$\text{(i)}\quad [L_5MX]^{n+} \xrightarrow{\text{Slow}} \underset{\substack{\text{5-Coordinated}\\ \text{intermediate}}}{[L_5M]^{(n+1)+}} + X^-$$

$$\text{(ii)}\quad [L_5M]^{(n+1)+} + Y^- \xrightarrow{\text{Fast}} [L_5MY]^{n+}$$

The loss of ligand X is the slow, rate-determining step. The 5-coordinated intermediate $[L_5M]^{(n+1)+}$ instantaneously reacts with the new ligand Y. The rate of reaction, v is directly proportional to the concentration of the original complex and is independent of the concentration of ligand Y, i.e.,

$$v = k\{[L_5MX]^{n+}\}$$

where k is the rate constant for the substitution reaction. Since the rate of reaction is dependent on the concentration of only one species, the reaction is unimolecular. Such reactions are designated as S_N^1 reactions. The substitution nucleophilic unimolecular reaction pathway is also referred to as *dissociative mechanism*.

2. Substitution nucleophilic bimolecular reactions

The second pathway of nucleophilic substitution also occurs in two steps.

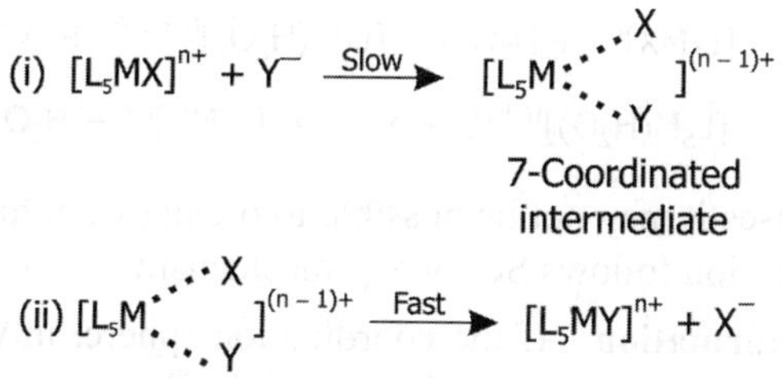

The intermediate is a 7-coordinated species in which the ligand X has not completely detached from the metal ion, while the ligand Y has not completely attached to the metal ion. The formation of this transition state is a slow step. Accordingly, the rate of reaction, v is given as,

$$v = k\{[L_5MX]^{n+}\}\{[Y^-]\}$$

The rate is dependent on two concentrations: that of the original complex and that of the entering ligand, Y. Thus, the reaction is bimolecular and is designated as S_N^2. The 7-coordinated intermediate sheds off ligand X rapidly, completing the substitution process. This type of reaction pathway is also called *associative mechanism.*

Case Study 28.1 Which substitution mechanism is more plausible in cases where the ligands cause steric over-crowdedness around the metal ion?

It has been observed that the rate of substitution reactions in derivatives of dichlorobis(ethylenediamine)cobalt(III) ion, $[Co(en)_2Cl_2]^+$, increases as the hydrogen atoms on carbon of ethylenediamine ligands are progressively replaced by methyl groups, i.e., the complexes containing substituted ethylenediamine ligands react faster than the one containing unsubstituted ligands. The replacement of hydrogen atoms by methyl groups makes the ligand bulky which, in turn, makes it more difficult for an entering group to approach the metal ion. Thus, steric over-crowdedness should inhibit S_N^2 mechanism. The increase in reaction rate with increasing size of the ligand implies that the mechanism is predominantly S_N^1. The removal of one ligand (chloro in this case) reduces the congestion around the metal, relieves the steric strain and increases the rate of reaction.

Factors Complicating Substitution Reactions

Reactions which follow purely S_N^1 or S_N^2 pathways are rare. In most cases, the transition state is neither a truly 5-coordinated species nor a 7-coordinated entity. This is because a number of side reactions preclude the substitution to occur by simple S_N^1 or S_N^2 mechanism. There are three factors which complicate the substitution reactions.

a. Solvent interaction The bulk of substitution reactions have been studied in water, which is a ligand in its own right. It is thus possible that solvolysis may occur before the actual substitutions.

$$[L_5MX]^{n+} + H_2O \rightarrow [L_5M(H_2O)]^{(n+1)+} + X^-$$

$$[L_5M(H_2O)]^{(n+1)+} + Y^- \rightarrow [L_5MY]^{n+} + H_2O$$

In such cases it may not be possible to predict from the rate law whether the substitution follows S_N^1 or S_N^2 mechanism.

b. Ion pair formation If the coordination sphere, in which replacement reaction is to occur, carries a high positive charge and the entering group carries a high negative charge, then ion pair combination may take place before actual substitution.

$$[L_5MX]^{n+} + Y^{m-} \rightarrow \{[L_5MX][Y]\}^{(n-m)+}$$

It is not possible to determine whether the ion pair transfers itself into the desired substituted product by S_N^1 or S_N^2 mechanism.

c. Conjugate base formation Consider the reaction between pentamminechlorocobalt(III) ion and hydroxide ion. The rate of the reaction is directly proportional to the product of the concentration of the complex and the hydroxide ion, i.e.,

$$v \propto [\text{complex}][OH^-]$$

This indicates that the reaction follows an S_N^2 mechanism.

$$[Co(NH_3)_5Cl]^{2+} + OH^- \xrightarrow{\text{Slow}} [Co(NH_3)_5Cl(OH)]^+$$

$$[Co(NH_3)_5Cl(OH)]^+ \xrightarrow{\text{Fast}} [Co(NH_3)_5(OH)]^{2+} + Cl^-$$

Yet, it is possible that the hydroxide ion, which is a strong base, rather than simply replacing the chloro ligand, may abstract a proton from one of the ammine ligands, forming a conjugate base. The reaction occurs in three steps.

(i) $\underset{\text{Complex}}{[Co(NH_3)_5Cl]^{2+}} + OH^- \overset{\text{Fast}}{\rightleftharpoons} \underset{\text{Conjugate base}}{[Co(NH_3)_4(NH_2)Cl]^+} + H_2O$

(ii) $[Co(NH_3)_4(NH_2)Cl]^+ \xrightarrow{\text{Slow}} [Co(NH_3)_4(NH_2)]^{2+} + Cl^-$

(iii) $[Co(NH_3)_4(NH)_2]^{2+} + H_2O \xrightarrow{\text{Fast}} [Co(NH_3)_5OH]^{2+}$

This pathway is called *substitution nucleophilic unimolecular conjugate base mechanism*, and designated as S_N^{1CB}. It is unimolecular since step (ii), being the slowest, is the rate-determining step, and it involves only one reactant species. The rate of reaction is thus proportional to the concentration of that one species, the conjugate base of the complex.

$$v \propto [\text{conjugate base}] \tag{28.1}$$

The equilibrium constant K, for step (i) may be written as,

$$K = \frac{[\text{Conjugate base}][H_2O]}{[\text{Complex}][OH^-]}$$

Since water is present in a large amount, its concentration is taken as constant. Therefore,

$$[\text{conjugate base}] = \text{constant } [\text{complex}][OH]^- \tag{28.2}$$

From equations (28.1) and (28.2),

$$v \propto [\text{complex}][OH^-]$$

We get the same rate law as for a simple S_N^2 mechanism. Thus, it is not possible to predict whether the reaction is actually proceeding by an authentic S_N^2 mechanism or by an S_N^{1CB} mechanism.

Case Study 28.2 The reaction,

$$[Cr(NH_3)_5Cl]^{2+} + NH_3 \rightarrow [Cr(NH_3)_6]^{3+} + Cl^-$$

when conducted in liquid ammonia, is catalyzed by potassium amide. Why?

Potassium amide, in liquid ammonia, ionizes into potassium and amide ions.

$$KNH_2 \rightleftharpoons K^+ + NH_2^-$$

Amide ion, in liquid ammonia medium, is as strong a base as hydroxide ion in aqueous medium. The substitution reaction is accomplished in the following three steps:

(i) $[Cr(NH_3)_5Cl]^{2+} + NH_2^- \xrightleftharpoons{\text{Fast}} [Cr(NH_3)_4(NH_2)Cl]^+ + NH_3$

(ii) $[Cr(NH_3)_4(NH_2)Cl]^+ \xrightarrow{\text{Slow}} [Cr(NH_3)_4(NH_2)]^{2+} + Cl^-$

(iii) $[Cr(NH_3)_4(NH_2)]^{2+} + 2NH_3 \xrightarrow{\text{Fast}} [Cr(NH_3)_6]^{3+} + NH_2^-$

The reaction proceeds by S_N^{1CB} mechanism. It is the amide ion which carries through step (i) at a fast pace. Hence, it catalyzes the substitution reaction.

Box 28.1

Eigen-Wilkins Mechanism and Fuoss-Eigen Equation

According to Eigen-Wilkins mechanism, an intermediate complex, in form of an ion pair, is formed in a pre-equilibrium step.

$$\{ML_5X\}^{n+} + Y^{m-} \rightleftharpoons \{L_5MXY\}^{(n-m)+}$$

Where X is the leaving ligand and Y is the entering ligand. The intermediate then decomposes into the products in the subsequent rate determining step.

$$\{L_5MXY\}^{(n-m)+} \xrightarrow{\text{Slow}} \{L_5MY)^{n+} + X^{m-}$$

The equilibrium constant, K_{eq} for the reaction leading to the formation of the intermediate complex may be expressed as,

$$K_{eq} = \frac{[\{L_5MXY\}^{(n-m)+}]}{[\{ML_5X\}^{n+}][Y^{m-}]}$$

or

$$[\{L_5MX\}^{(n-m)+}] = K_{eq}[\{ML_5X\}^{n+}][Y^{m-}] \quad (28.3)$$

The rate, ν of the reaction representing the decomposition of counter ion to the products may be expressed as,

$$\nu = \kappa[\{L_5MXY\}^{(n-m)+}] \quad (28.4)$$

From equations (28.3) and (28.4) we get,

$$\nu = \kappa K_{eq}[\{ML_5X\}^{n+}]\,[Y^{m-}] \quad (28.5)$$

However, the term $[\{ML_5X\}^{n+}]$ must take into account the fact that a part of it is present as the ion pair, $\{ML_5XY\}^{(n-m)+}$.

$$[M]_T = [\{ML_5XY\}^{(n-m)+}] + [\{ML_5X\}^{n+}]$$

or

$$[\{ML_5X\}^{n+}] = [M]_T - [\{ML_5XY\}^{(n-m)+}] \quad (28.6)$$

Where $[M]_T$ is the total concentration of the reactant complex and the intermediate complex. Substituting the value of $[\{ML_5X\}^{n+}]$ from equation (28.6) in equation (28.5).

$$\nu = \kappa K_{eq}([M]_T - [\{ML_5XY\}^{(n-m)+}])\,[Y^{m-}]$$

$$\nu = \kappa K_{eq}[M]_T\,[Y^{m-}] - \kappa K_{eq}[\{ML_5XY\}^{(n-m)+}]\,[Y^{m-}] \quad (28.7)$$

From equations (28.4) and (28.7),

$$\nu = \kappa K_{eq}[M]_T\,[Y^{m-}] - \nu K_{eq}[Y^{m-}]$$

or

$$\nu + \nu K_{eq}[Y^{m-}] = \kappa K_{eq}[M]_T\,[Y^{m-}]$$

$$\nu = \frac{\kappa K_{eq}[M]_T\,[Y^{m-}]}{1 + K_{eq}[Y^{m-}]} \quad (28.8)$$

Contd...

If the concentration of the entering group, Y is very small, then $K_{eq}[Y^{m-}] << 1$. Equation (28.8) reduces to,

$$\nu = \kappa K_{eq}[M]_T [Y^{m-}] \quad (28.9)$$

The rate ν, $[M]_T$, and $[Y^{m-}]$ can be determined experimentally. However, the rate constant κ can be elucidated only if the value of equilibrium constant, K_{eq} is known. The Fuoss-Eigen equation provides an estimate of this parameter, taking into account the strength of Coulombic attraction between the reactant and the entering group. The equation is expressed as,

$$K_{eq} = \frac{4}{3}\pi a^3 N\, e^{-V/RT} \quad (28.10)$$

Here, a is the distance of closest approach of the counter ions in the intermediate complex, N is the Avogadro number, R is the gas constant and T is the absolute temperature. The factor V represents the Coulombic potential energy, and is expressed mathematically as,

$$V = \frac{Z_1 Z_2 e^2}{4\pi\varepsilon a}$$

where Z_1 and Z_2 are the charges on the reactant complex and entering group, respectively, e is the electronic charge and ε is the permittivity of the solvent in which the reaction is carried out.

The Fuoss-Eigen equation predicts that if the two reacting ions are large in size, the value of a will be large and so will be the chances of forming the ion pairs. Likewise, if the values of Z_1 and Z_2 are large, the possibility of ion pair formation will be high. On the other hand, if the entering ligand is neutral (for example, ammine, NH_3), then the Coulombic potential energy, V will be zero. In such a case, equation (28.10) will reduce to,

$$K_{eq} = \frac{4}{3}\pi a^3 N \quad (28.11)$$

The value of equilibrium constant K_{eq} drastically reduces and so does the chances of counter complex formation.

Factors Affecting Rates of Substitution Reactions

Rates of substitution reactions among complexes vary over a wide range and depend on a number of parameters.

1. **Oxidation state of the metal** For a metal ion with a high oxidation state, the rates of substitution reactions are slower than that for an isoelectronic metal with a lower oxidation state. This is because metal–ligand bond strength increases with the charge on the metal ion. Therefore, it becomes difficult to replace one ligand with another. Consider the optically active

ligand, 1,2-propylenediaminetetraceticacid, the structure of which is shown in Fig. 28.1.

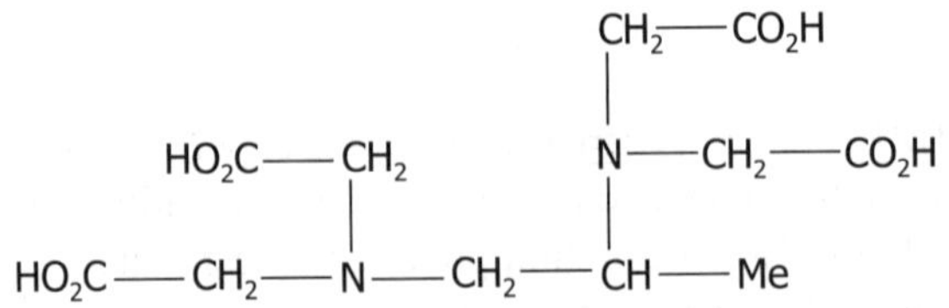

Fig. 28.1 Structure of 1,2-propylenediaminetetraceticacid

This ligand is commonly abbreviated as H_4pdta. The d-isomer of the ligand forms a complex with manganese(II) ion, the composition being Mn(d-H_2pdta). The latter reacts with the l-form of the free ligand in aqueous medium at pH 6.3.

$$\text{Mn(d-H}_2\text{pdta)} + \text{l-H}_4\text{pdta} \rightarrow \text{Mn(l-H}_2\text{pdta)} + \text{d-H}_4\text{pdta}$$

At 293 K, a racemic mixture of d- and l-forms of manganese(II) complexes results and the optical activity drops to zero within 5 minutes. The corresponding complex of iron(III), Fe(d-Hpdta), does not form a racemic mixture under similar conditions even after a lapse of 48 hours. Although both Mn(II) and Fe(III) are $3d^5$ systems, the higher oxidation state of the latter retards the rate of reaction.

2. **Electronic configuration of the metal ion** Reactions of octahedral complexes of chromium(III) and low-spin cobalt(III) ions are slow. Those of nickel(II) analogs too are not very fast. All other arrangements of d electrons, particularly low-spin d^4 and d^5 systems are associated with kinetic lability. Chromium(III) ion in an octahedral environment has a stable, half-filled t_{2g}^3 configuration. Low-spin cobalt(III) in a similar environment has a stable, completely filled t_{2g}^6 configuration. Nickel(II) ion has a $3d^8$ system with $t_{2g}^6e_{2g}^2$ configuration in octahedral complexes. In this case, the t_{2g} orbitals are completely filled while the e_g orbitals are half-filled. Due to the favorable electronic configurations, these metal ions have a high negative value of crystal field stabilization energy (CFSE). When their complexes undergo substitution reactions the pathway is either through a 5-coordinated transition state, if S_N^1 mechanism is followed, or through a 7-coordinated transition state if S_N^2 mechanism is followed. In either case, the favorable CFSE is lost in forming these unfavorable coordinated intermediates. For this reason, the complexes of these metal ions are rather inert.

No doubt, zinc(II) ion in octahedral geometry too has a very stable $t^6{}_{2g}e^4{}_{2g}$ configuration. However, there is no CFSE associated with zinc(II) ions. Hence, the question of losing CFSE does not arise. Zinc(II) complexes are rather labile. The octahedral complexes with metal ions having d^4 and d^5 configurations owe their exceedingly labile nature to their tendency to undergo redox reactions. The d^4 metal ions tend to undergo oxidation, so as to acquire t^3_{2g} configuration, while the d^5 metal ions tend to undergo reduction, so as to acquire t^6_{2g} configuration. We have discussed kinetic aspects of redox reactions in a later section of this chapter.

Case Study 28.3 The hexamminevanadium(III) ion is labile, but the hexamminecobalt(III) ion is inert. Why?

As has been stated in Chapter 25, (p. 676), the complexes which have vacant orbitals in their valence shell are prone to react fast with atmospheric constituents or other reagents and hence are labile in nature. Those which do not have vacant orbitals in their valence shell are inert. If we invoke the valence bond theory (*see* Chapter 26, p. 699), the valence shell configuration of hexamminevanadium(III) ion may be written as,

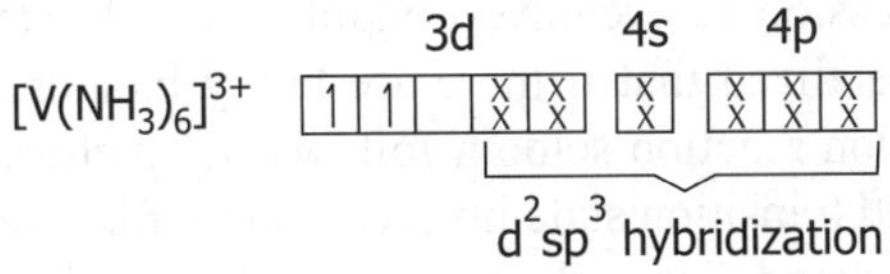

For hexamminecobalt(III) ion, the configuration is,

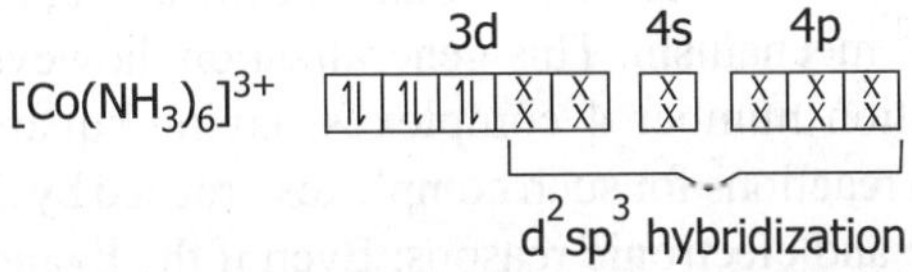

The former has a vacant orbital in its valence shell and hence is labile. The latter does not have any vacant orbital in its valence shell and hence is inert.

---

3. Nature of the ligands Consider a reaction of the type,

$$[Co(en)_2ClX]^+ + H_2O \rightarrow [Co(en)_2X(H_2O)]^{2+} + Cl^-$$

The rate of such a reaction decreases in the order,

$$X = NH_2^- > SCN^- > NO_2^-$$

The amide ion, NH_2^- has a tendency to form π-bonds with the metal ion by donating p electrons. This reduces the quantum of positive charge on the metal and thus, facilitates the cleavage of Co–Cl bond. Moreover, the π-bonding stabilizes the 5-coordinated transition state. The π-bonding ability of thiocyanate ion is negligible and hence, the reaction proceeds slowly. The nitro ligand on the other hand, tends to increase the positive charge on metal by way of electron-withdrawing inductive effect. The

Co–Cl bond gets strengthened and the rate of reaction gets drastically reduced.

4. **Stereochemical arrangement of ligands** The reaction,

$$[Co(en)_2Cl(OH)]^+ + H_2O \rightarrow [Co(en)_2(H_2O)(OH)]^{2+} + Cl^-$$

is faster if the chlorobis(ethylenediamine)hydroxocobalt(III) ion has *cis* rather than *trans* geometry. The hydroxo ligand is in a better position to donate electrons to empty orbitals of the metal if it has a *cis* position with respect to the leaving chloro group. As a result of this donation, the charge on the metal ion is reduced and the Co–Cl bond cleaves easily.

If the hydroxo group is replaced by an electron-withdrawing ligand, such as nitro (NO_2^-), the substitution reaction for the *trans* isomer is faster than that for the *cis* isomer.

5. **Geometry of the complex** The factors affecting kinetics of complexes that we have stated so far are applicable to virtually all the complexes, yet the discussion has revolved mainly around octahedral geometry. It is often generalized that if there are bulky ligands around a metal ion, the substitution reaction seldom follows S_N^2 pathway. This is because a 7-coordinated transition state involving large functional groups would be sterically hindered. For such complexes, the reactions invariably proceed via S_N^1 mechanism. If, on the other hand, the ligands are not so bulky as to cause steric over-crowdedness around the metal ion, the substitution takes place by S_N^2 mechanism. This generalization, however, is not applicable to coordination number 4 complexes having square planar geometry. Substitution reactions for such complexes proceed by S_N^2 mechanism both due to steric and electronic reasons. Even if the ligands are large-sized, a transition state involving expansion of coordination number to 5, is not sterically improbable.

As far as electronic factors are concerned, complexes with square planar geometry are formed largely by metal ions having a low-spin d^8 configuration, and such metal ions have a vacant p_z orbital of relatively low energy which can accommodate the pair of electrons donated by the entering ligand, even if the leaving ligand has not been completely shed off.

The support for S_N^2 mechanism is provided by the fact that substitution in square planar complexes occurs with high degree of stereospecificity, i.e., the reaction of *cis*-$[PtA_2LX]$ complex with entering group Y would yield *cis*-$[PtA_2LY]$ derivative. The *trans* isomer would give the corresponding *trans* product. Stereospecific displacement can be explained only if a 5-coordinated trigonal bipyramidal transition state—the one required by S_N^2 mechanism—is invoked.

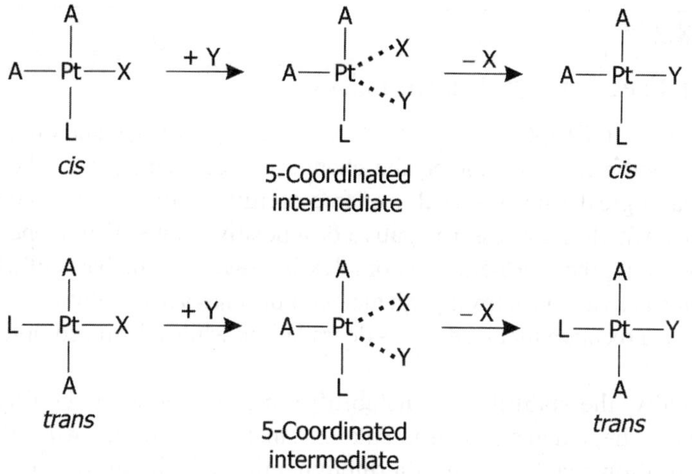

If the reaction had proceeded via S_N^1 mechanism, retention of stereospecific configuration would not have been possible. This is because a 3-coordinated species with a near planar geometry would have been formed in the transition state under these circumstances. Irrespective of the geometry (*cis* or *trans*) of original complex, the ligand Y could have entered the coordination sphere either adjacent to L or opposite to it with equal probability, giving *cis* or *trans* product, respectively.

Case Study 28.4 Arrange the following complexes in decreasing order of their rate of substitution by aqua ligand: $[Co(NH_3)_6]^{3+}$, $[Rh(NH_3)_6]^{3+}$, $[Ir(NH_3)_6]^{3+}$, $[Mn(H_2O)_6]^{2+}$, $[Ni(H_2O)_6]^{2+}$.

The three ammine complexes will undergo substitution with a lower rate as compared to the two aqua complexes. This is because the oxidation state of metal ions in the ammine complexes is III, while in the aqua complexes, it is II. Among the ammine complexes, the metals cobalt, rhodium and iridium belong to the 3d, 4d and 5d series of transition elements, respectively. Each of these metal ion has d^6configuration. The crystal field splitting in octahedral field (Δ_o) for any group of d-block elements follows the order, 5d>4d>3d. Thus, the CFSE will be highest for $[Ir(NH_3)_6]^{3+}$ complex and lowest for $[Co(NH_3)_6]^{3+}$ complex. Accordingly, the former will undergo substitution at a slower pace as compared to the latter.

Among the aqua complexes, the coordination sphere of manganese(II) ion will undergo substitution more readily than the one of nickel(II) ion. This is because manganese(II) ion, with the configuration $t_{2g}^3e_{2g}^2$ has zero CFSE, while nickel(II) ion with the configuration $t_{2g}^6e_{2g}^2$ is appreciably stabilized by crystal field effects.

The order of rate of substitution of the complexes in question would be,

$$[Mn(H_2O)_6]^{2+} > [Ni(H_2O)_6]^{2+} > [Co(NH_3)_6]^{3+} > [Rh(NH_3)_6]^{3+} > [Ir(NH_3)_6]^{3+}$$

Box 28.2

Linear Free Energy Relationship

As discussed in Chapter 25, the stability of a complex is gauged on the basis of a thermodynamic parameter, the change in free energy, ΔG. The analog which has a greater negative value of free energy is more stable as compared to the one which has a lesser negative or a positive value of free energy. On the other hand, the inertness of a complex is gauged on the basis of a kinetic parameter, activation energy, E_a. An inert complex has a high value of activation energy for a reaction that leads to its degradation, while a labile complex has a low value.

Normally, the stability (or instability) and inertness (or lability) of a complex are treated in a compartmentalized manner. However, when it comes to substitution reactions, the thermodynamic and kinetic considerations often go together. For example, the rate of substitution of a ligand—a kinetic factor—depends largely on the strength with which it is bound to the metal—a thermodynamic factor. The combination of kinetic and thermodynamic effects for a substitution reaction is referred to as linear free energy relationship. According to this concept, a plot of the logarithm of rate constant (kinetic parameter) for different leaving groups in a substitution reaction against the logarithm of the equilibrium constant (thermodynamic parameter) for the same ligands in similar complexes is linear.

The rate constant of a reaction is given by Arrhenius equation,

$$K = A\,e^{-\frac{E_a}{RT}}$$

or

$$\log K = \log A - \frac{E_a}{2.303\,RT} \qquad (28.12)$$

Here K is the rate constant, A is the frequency factor, R is the gas constant and T is the absolute temperature.

The equilibrium constant, K_{eq} is related to change in free energy by the expression,

$$\Delta G = -RT \ln K_{eq} \qquad (28.13)$$

The change in free energy is related to the change in entropy by the expression,

$$\Delta G = \Delta H - T\Delta S \qquad (28.14)$$

where ΔH is the change is enthalpy that occurs during the course of reaction. From equations (28.13) and (28.14),

$$\ln K_{eq} = -\frac{\Delta H}{RT} + \frac{\Delta S}{R}$$

Contd...

or $$\log K_{eq} = -\frac{\Delta H}{2.303\,RT} + \frac{\Delta S}{2.303\,R} \quad (28.15)$$

Equation (28.12) reflects the kinetic behavior of the complex, while equation (28.15) reflects the thermodynamic behavior. If the frequency factor, A and the change in entropy, ΔS are nearly constant, a plot of log K *versus* log K_{eq} will give a straight line.

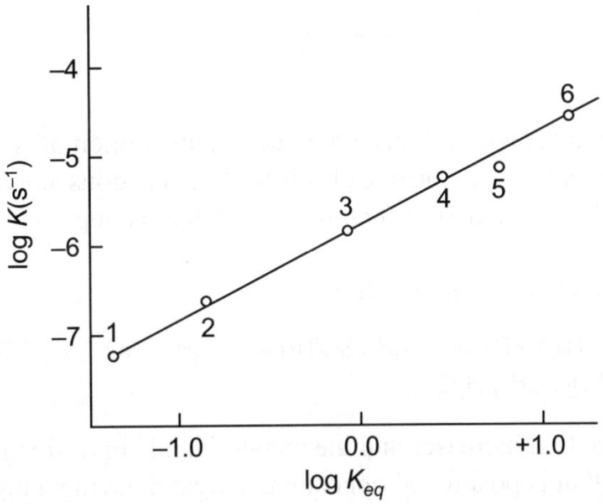

Fig. 28.2 Linear relationship between rate constant and equilibrium constant

Fig. 28.2 depicts such a plot for the substitution reaction,

$$[Co(NH_3)_5X]^{2+} + H_2O \rightarrow [Co(NH_3)_5(H_2O)]^{3+} + X^-$$

where the leaving group, X represents the following: 1 = F^-; 2 = $H_2PO_4^-$; 3 = Cl^-; 4 = Br^-; 5 = I^-; and 6 = NO_3^-.

The linear curve shows that there is a strong influence of thermodynamic parameter, ΔH on the kinetic parameter, E_a. It may thus be generalized that the stronger the bond between the metal and the leaving group, greater will be the activation energy for the substitution reaction.

The *trans* Effect

We have already stated that in square planar complexes the substitution reaction follows S_N^2 mechanism. The rate of reaction in such complexes is influenced not only by the mechanistic pathway but also by the nature of the ligand positioned *trans* to the leaving group. This concept is called *trans* effect and is defined as the influence of a coordinated group upon the rate of substitution reaction of a ligand opposite to it in a metal complex.

Thus, *trans* effect states that the kinetics of substitution of a ligand in a coordination compound is determined by the nature of substituent at the opposite end of the diagonal. This means that the stability of the bond between the central metal ion and a leaving group is little affected by the characteristics of the neighboring *cis* moieties, but is greatly influenced by the one at distant *trans* position. Thus for a complex of the type,

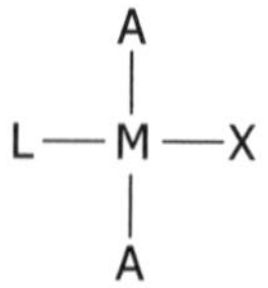

the ligand L and not A will influence the rate of substitution of X.

A ligand which influences the rate of substitution reactions, but does not undergo substitution itself is called *spectator ligand.* In the example cited above, L is a spectator ligand.

The order of *trans* effect is as follows:

$CO, CN^-, C_2H_4 > PR_3, H^- > CH_3, SC(NH_2)_2 > C_6H_5^-, NO_2^-, I^-, SCN^- > Br^- >$
$Cl^- > py, NH_3, OH^-, H_2O$

The rate of reaction increases and the metal–ligand bond strength decreases for a leaving group that is positioned opposite to a ligand having a high *trans* effect.

Applications of *trans* effect

The *trans* effect has the following applications to the physico-chemical characterization of complexes:

1. **Synthesis of geometrical isomers**

   The *trans* effect has proved useful for devising synthetic routes for synthesis of different geometrical isomers of complexes. Consider, for example, the synthesis of *cis* and *trans* isomers of diamminedichloroplatinum(II), $[Pt(NH_3)_2Cl_2]$. The *cis* analog is prepared by the reaction of ammonia with tetrachloroplatinate(II) ion, $[PtCl_4]^{2-}$.

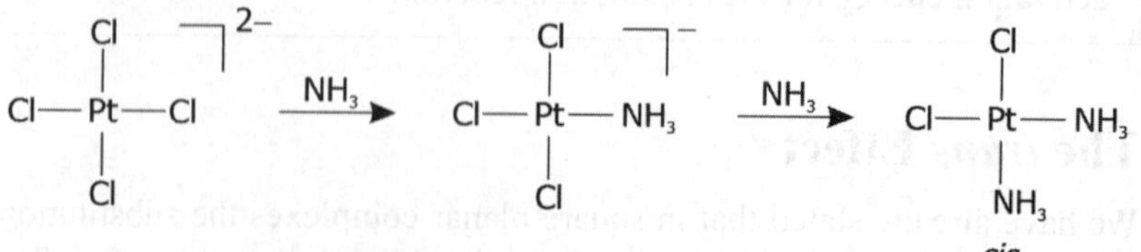

cis

   Since the *trans* directing influence of chloride ion is greater than that of ammonia, the least reactive group in the intermediate $[Pt(NH_3)Cl_3]^-$ is the one opposite to the ammonia. Therefore, the second ammonia molecule enters a *cis* position.

The *trans* isomer may be prepared by the reaction of tetrammineplatinum(II) ion with chloride ions.

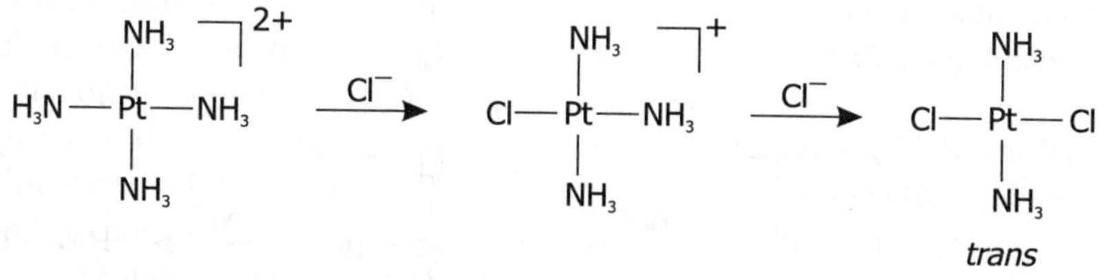

The most labile ammine group in the intermediate $[Pt(NH_3)_3Cl]^+$ is the one opposite to the chloro group. The superior *trans*-directing influence of chloro ligand pushes out the labile ammine ligand, producing the *trans* isomer.

Similarly, *cis*-amminedichloronitroplatinate(II) ion may be obtained by reacting tetrachloroplatinate(II) ion successively with ammonia and nitrite ion.

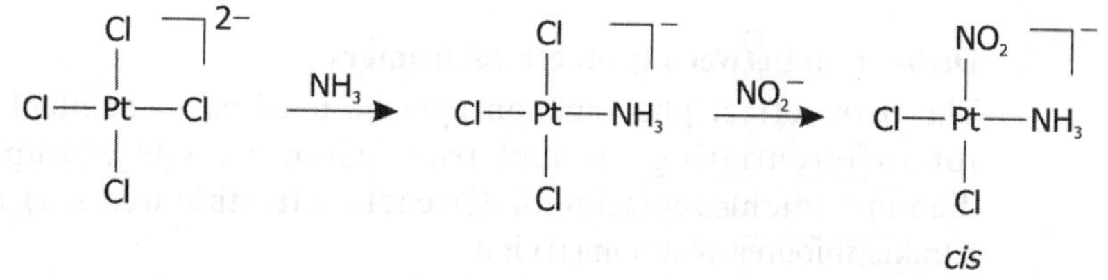

The *trans* isomer is obtained by reversing the order of introduction of the entering groups.

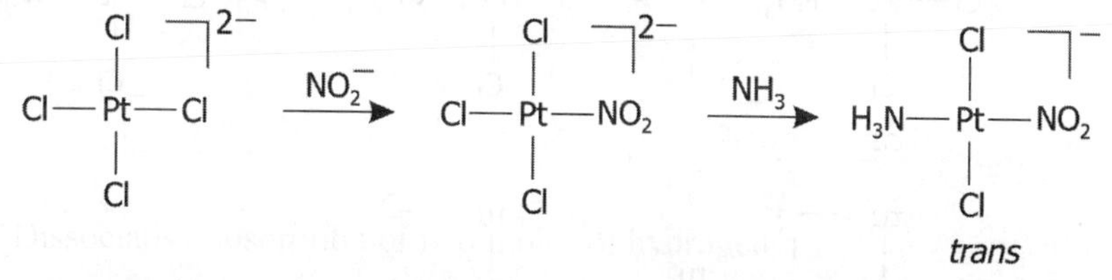

The directive force of these reactions is governed by the fact that the *trans*-directing influence of nitro group is greater than that of chloride ion which, in turn, is greater than that of ammonia molecule.

Similarly, the superior *trans*-directing ability of bromide ion as compared to chloride ion, and the weaker *trans* effect of ammonia or pyridine molecules guides the synthesis of three isomeric amminebromochloropyridineplatinum(II) complexes.

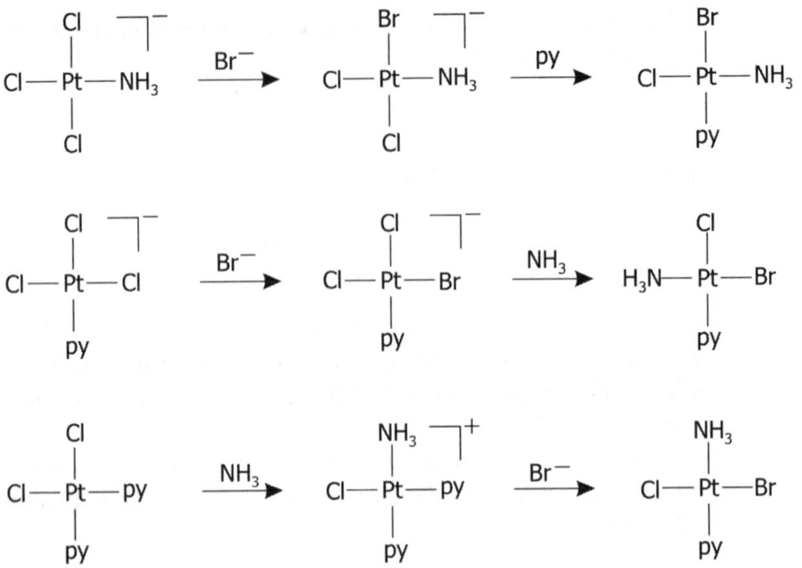

2. **Distinction between geometrical isomers**

The *trans* effect phenomenon may be used as a chemical method for differentiating *cis* and *trans* isomers. For example, *cis*-diamminedichloroplatinum(II) reacts with thiourea (tu) to yield tetrakis(thiourea)platinum(II) ion.

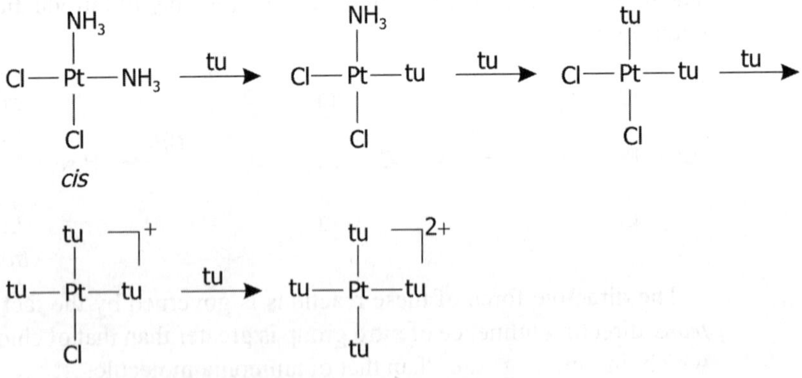

The *trans*-diamminedichloroplatinum(II) isomer, however, yields diamminebis(thiourea)platinum(II) ion.

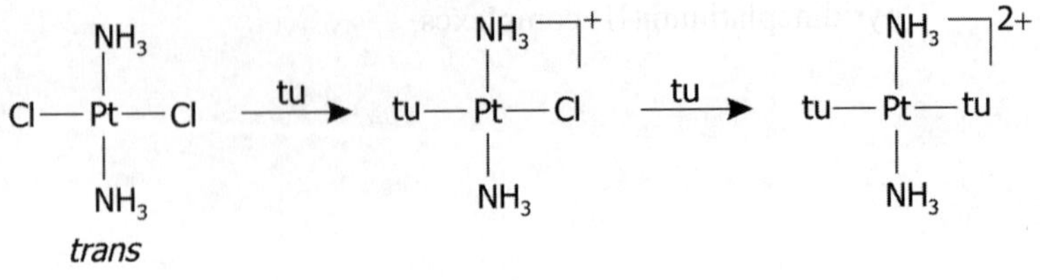

The *trans*-directing influence of thiourea exceeds that of ammine and chloro ligands. This method of distinguishing *cis*- and *trans*- platinum(II) complexes is called *Kurnakov test.*

Thiosulfate ion too has a superior *trans*-directing ability as compared to ammine and chloro groups. Therefore, the reaction of *cis*-diamminedichloroplatinum(II) with thiosulfate ion gives tetrakis(thiosulfato)platinate(II) ion, $[Pt(S_2O_3)_4]^{6-}$. The *trans* isomer yields diamminebis(thiosulfato)platinate(II) ion, $[Pt(NH_3)_2(S_2O_3)_2]^{2-}$.

Case Study 28.5 Design a route for the synthesis of *cis*- and *trans*-amminedichloro-(triphenylphosphine)platinum(II), starting from tetrachloroplatinate(II) ion.

Reaction of one mole of tetrachloroplatinate(II) ion with one mole of ammonia, followed by one mole of triphenylphosphine results in the *cis* isomer.

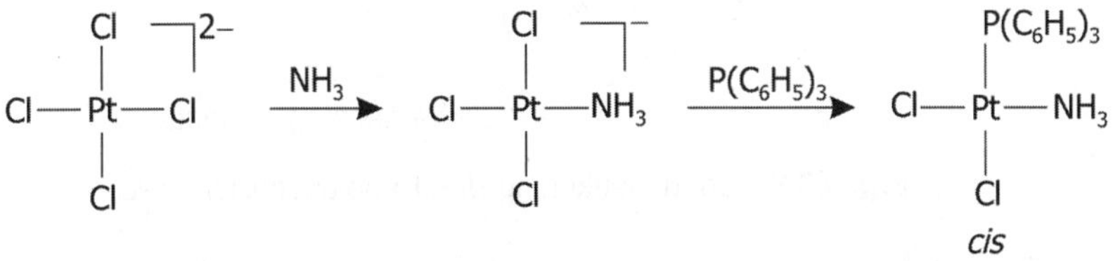

If the entering groups react in a reverse order, the *trans* isomer is formed.

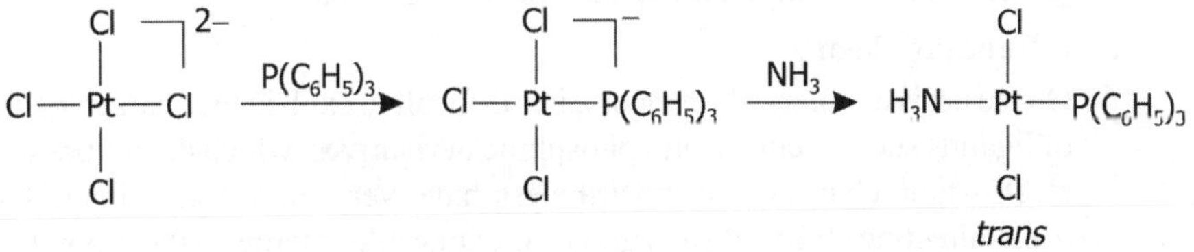

These synthetic routes are guided by the fact that the *trans* effect of the ligands follows the order:

$$P(C_6H_5)_3 > Cl^- > NH_3$$

Theories of *trans* effect

There are two theories which explain the *trans*-directing ability of ligands in the course of substitution reactions: polarization theory and π-bonding theory. The former stresses the importance of ground state contribution to the rates of reactions, while the latter is primarily concerned with the transition state.

1. **Polarization theory**

   The polarization theory is based on the thermodynamic consideration that the *trans*-directing ligand weakens the bond between the metal ion and the leaving group at the opposite end of the diagonal.

Consider a complex having the composition $[MA_2LX]$, where X is the leaving group and L is a ligand having appreciably high *trans*-directing influence. The primary charge on the metal would induce a dipole in L, which, in turn, would induce a dipole in the metal. The direction of this dipole, as shown in Fig. 28.3, would be such that the negative end of L would point towards the positive center and away from the negative center of metal ion. This is understandable since a ligand becomes coordinated to a metal by donation of electrons (negative charge). However, the leaving ligand, X, too is bound to the metal through electron donation. As shown in Fig. 28.3, the ligand X has no choice, but to orient its negative end towards the center of the negative charge on the metal ion.

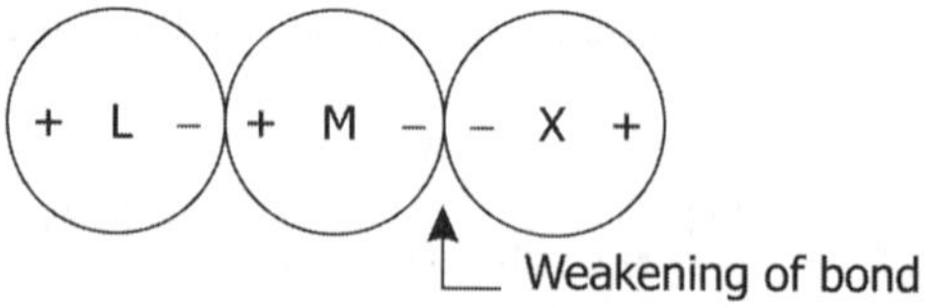

Fig. 28.3 Bond weakening due to polarizability effect

The net result is that ligand X experiences greater repulsion and the bond weakens. Thus, iodide ion has greater polarizability, and consequently a greater *trans*-directing ability than the chloride ion.

2. **π-Bonding theory**

On the basis of polarization theory, it is difficult to explain the characteristics of ligands such as ethene and phosphine derivatives, which do not possess electrostatic charges, but nevertheless have very high *trans* effect. The *trans*-directing ability of such groups is explicable in terms of the π-bonding theory.

This theory is based on the kinetic consideration that 4-coordinated complexes undergo substitution largely by S_N^2 mechanism. Suppose a ligand L in a square planar complex, $[MA_2LX]$, has the tendency to accept electrons, originally occupying the d_{xz} or d_{yz} orbitals of the central metal ion, M, into vacant π orbitals of its own. Under these circumstances, the electron density above and below the bond between the metal ion and the ligand X, positioned *trans* to L gets reduced. The entering ligand, Y is now induced to coordinate to the metal ion across the region of smaller electron density, i.e., along the direction of M–X bond (Fig. 28.4).

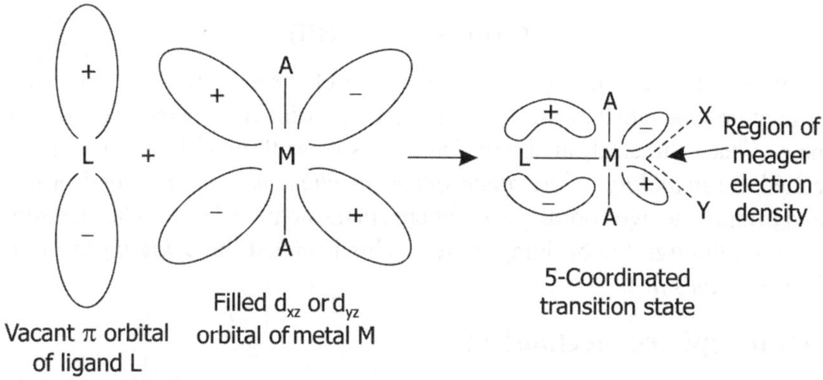

Fig. 28.4 5-coordinated transition state and S_N^2 mechanism

Momentarily, the entering group, Y and the leaving group, X are loosely coordinated at the same side of metal ion, supporting the S_N^2 mechanism and leading to the formation of a 5-coordinated transition state. The 5-coordinated, trigonal bipyramid species, in turn, is stabilized due to transfer of electronic charge from metal M to ligand L (dispersal of charge on metal leads to a more stable transition state). Finally, ligand X, which occupies *trans* position with respect to L is eliminated.

When X is lost, the configuration of the product is retained.

Case Study 28.6 For diamminedichloroplatinum(II) complex, the *trans* isomer is more stable than the *cis* isomer. Why?

For explaining the greater thermodynamic stability of *trans*-diamminedichloroplatinum(II) complex vis-à-vis its *cis* analog, two parameters have to be taken into consideration. First, the *trans* effect of chloro ligand is greater than that of ammine and second, Pt–N bond strength is greater than Pt–Cl bond strength.

In the *cis* isomer, each chloro ligand induces a bond-weakening effect on the Pt–N bond opposite to it. As a consequence, the Pt–Cl bond gets strengthened at the cost of the Pt–N bond. However, since Pt–Cl bond strength is less than that of Pt–N bond, the overall stability of the complex will be affected more by the changes in the latter than in the former. In the *trans* isomer, the strength of the Pt–N bond is not reduced. Hence, it is more stable than its *cis* counterpart.

Redox Reactions

In the context of complexes, a redox reaction involves interaction between two coordination spheres, whereby the metal ion of one is oxidized and that of the other is reduced. For example, a reaction between a cobalt(III) and a chromium(II) complex may yield a cobalt(II) and a chromium(III) complex. The changes in oxidation states of the concerned metal ions reflect the transfer of an electron.

$$Cr(II) \rightarrow Cr(III) + e^-$$
$$Co(III) + e^- \rightarrow Co(II)$$

There are two mechanisms by which the electron is transferred from one coordination sphere to the other. In the first, the coordination sphere of each complex remains intact; the electron simply tunnels its ways through both the shells. This is called the *tunneling* or the *outer sphere mechanism.* In the second process, a ligand bridges the two coordination spheres in the activated state. The electron gets transferred through the bridging moiety. This is called the *bridging* or the *inner sphere mechanism.*

1. Outer sphere mechanism

In outer sphere mechanism, there is direct electron transfer from one intact coordination sphere to another. To ensure that the reaction occurs through electron transfer alone, it is pertinent that the two coordination spheres should be inert with respect to ligand exchange. Consider, for example, hexacyanoferrate(II) ion, $[Fe(CN)_6]^{4-}$ and hexacyanoferrate(III) ion, $[Fe(CN)_6]^{3-}$. Both the complexes are inert. They exchange cyanide ligands extremely slowly with solutions containing cyanide ions. Yet the reaction,

$$[{}^*Fe(CN)_6]^{4-} + [Fe(CN)_6]^{3-} \rightarrow [{}^*Fe(CN)_6]^{3-} + [Fe(CN)_6]^{4-} \qquad (28.16)$$

in which *Fe represents radioactively labeled iron, does occur—and it occurs at a fast rate. Since the repulsion between iron(II) and iron(III) ions prevents a close approach of the two coordination spheres, the only possibility is that the electron is transferred from one species to another via a tunneling or outer sphere mechanism.

The outer sphere reactions are first order with respect to each coordination sphere and, therefore, second order overall. If cationic species are present in the reaction medium, these tend to reduce the repulsion between the reacting coordination spheres, bring them close together and hence catalyze their reaction. For example,

$$[Fe(CN)_6]^{4-} \ .... \ M^+ \ .... \ [Fe(CN)_6]^{3-}$$

If the cation is a metal ion, then larger its size, the better will be its catalytic activity. A smaller metal ion tends to hold the solvent molecules in preference to the coordination spheres. A larger cation is not heavily solvated and hence, can act as a better bridge between the two reacting complex ions. Thus, cesium ion is a better catalyst for the iron(II)–iron(III) outer sphere reaction than lithium ion. However, if the cationic species is non-metallic in nature, the smaller its size, the better will be its catalytic action. Thus, tetramethylammonium ion, $(CH_3)_4N^+$ is a more efficient catalyst for the iron(II)–iron(III) reaction than tetrapropylammonium ion, $(C_3H_7)_4N^+$, although both the cations have a unit positive charge. In the latter species, the positive charge is dispersed over a larger volume element and, therefore, the electrostatic field around it is not very effective in linking the two anionic coordination spheres.

Another example of the outer sphere mechanism is the reaction between hexamminecobalt(II) ion, A and hexamminecobalt(III) ion, B.

$$\underset{A}{[Co(NH_3)_6]^{2+}} + \underset{B}{[Co(NH_3)_6]^{3+}} \rightarrow [Co(NH_3)_6]^{3+} + [Co(NH_3)_6]^{2+} \quad (28.17)$$

The Co(II)–N bond distance in A is lengthier than the Co(III)–N bond distance in B. Thus, when A is converted to B, the metal–ligand bond distance is reduced. Conversely, when B is converted to A, the metal–ligand bond distance is increased. In the transition state, the bond distance should be same. However, these readjustments of bond lengths have to occur *before* the electron is transferred from one coordination sphere to the other. If electron transfer takes place after readjustment of bond lengths in the transition state, it would result in a cobalt(II) complex having bond lengths typical of a cobalt(III) complex and a cobalt(III) complex with bond lengths typical of a cobalt(II) complex. Both the products would be in a high energy state. They would then relax with release of energy. This violates the first law of thermodynamics. Thus, the readjustments of bond lengths have to occur before electron transfer.

Let us refer once again to reaction (28.16). In this reaction,

- The Fe(II)–C and Fe(III)–C bond lengths are nearly same.
- Both iron(II) and iron(III) are in low-spin states.
- The electronic configurations of iron(II) and iron(III) ions are t_{2g}^6 and t_{2g}^5, respectively. None of the metal ion has electrons in e_g orbitals.

With reference to reaction (28.17), the following are noteworthy:

- The Co(II)–N and Co(III)–N bond lengths differ appreciably.
- The cobalt(II) ion is in a high-spin state, while the cobalt(III) ion is in a low-spin state.
- The configuration of cobalt(II) is $t_{2g}^5e_g^2$ and that of cobalt(III) is t_{2g}^6.

It is pertinent that in the transition state a greater degree of readjustments is required for carrying out reaction (28.17) than for reaction (28.16). It implies that the activation energy of the former reaction is appreciably high; that of the latter is quite low. Thus, although both the reactions proceed by outer sphere mechanism, the iron(II)–iron(III) reaction is very fast, having a rate constant of 10^5 L mol^{-1} s^{-1} at 298 K, while the cobalt(II)–cobalt(III) reaction is extremely slow, having a rate constant of 10^{-4} L mol^{-1} s^{-1} at 298 K.

2. Inner sphere mechanism

In inner sphere mechanism, the liaison between the two coordination spheres is brought about through a bridging ligand. The electron is exchanged between the two complex ions through this common group. In most of the cases, the bridging ligand itself gets exchanged. Unlike the outer sphere mechanism, the inner sphere process involves bond making and bond breaking, in addition to electron transfer.

A classic example of an inner sphere process is the reaction between non-labile, pentamminechlorocobalt(III) ion and labile hexaquachromium(II) ion. In the transition state, the chlorine atom, while being bound to cobalt(III) ion, displaces a water molecule from the labile chromium(II) coordination sphere, and forms an intermediate.

$$[Co(NH_3)_5Cl]^{2+} + [Cr(H_2O)_6]^{2+} \rightarrow [(H_3N)_5Co\cdots Cl\cdots Cr(OH_2)_5]^{4+} + H_2O \quad (28.18)$$

The redox reaction takes place within the chlorine-bridged dinuclear complex, resulting in the formation of reduced cobalt(II) and oxidized chromium(III) ions.

$$[(H_3N)_5Co\cdots Cl\cdots Cr(OH_2)_5]^{4+} \rightarrow [(H_3N)_5Co]^{2+} + [ClCr(OH_2)_5]^{2+} \quad (28.19)$$

In this redox reaction, the electron is transferred from chromium(II) to cobalt(III), whereby the former is oxidized to chromium(III), while the latter is reduced to cobalt(II). As cobalt gets reduced, its bond with chlorine becomes labile. As chromium gets oxidized, its bond with chlorine becomes non-labile.

The pentamminecobalt(II) ion, formed in reaction (28.19), is rapidly hydrolyzed to hexaquacobalt(II) ion.

$$[Co(NH_3)_5]^{2+} + 5H^+ + 6H_2O \rightarrow [Co(H_2O)_6]^{2+} + 5NH_4^+ \quad (28.20)$$

The overall process may be written by combining equations pertaining to reactions (28.18) to (28.20).

$$\underset{A}{[Co(NH_3)_5Cl]^{2+}} + \underset{B}{[Cr(H_2O)_6]^{2+}} + 5H^+ + 5H_2O \rightarrow \underset{C}{[Co(H_2O)_6]^{2+}} + \underset{D}{[Cr(H_2O)_5Cl]^{2+}} + 5NH_4^+ \quad (28.21)$$

The overview of the inner sphere mechanism for reaction (26.21) may be stated as follows: a non-labile cobalt(III) complex, A reacts with a labile chromium(II) complex, B to yield a labile cobalt(II) complex, C and a non-labile chromium(III) complex, D. The transfer of chloride ion from A to D is quantitative. If radioactive chloride ions are added to the reaction medium, the chromium(III) complex remains devoid of radioactivity. Thus, the chloro ligand is transferred directly from coordination sphere A to D and not by way of solvent.

For a generalized system of the type $[Co(NH_3)_5X]^{2+}$, the rate of reaction following inner sphere mechanistic pathway decreases in the order,

$$X = N_3^- > I^- > Br^- > Cl^- > F^-$$

This is because the ease of transferring the electron from one coordination shell to another increases with polarizability of the bridging ligand. In the above series, azide ion (N_3^-) is the most polarizable while fluoride ion (F^-) is the least polarizable. Conversely, if a redox reaction involving a complex ion having azide ligand proceeds at a fast pace, it indicates that inner sphere course is being followed.

If the ligand X is an organic functional group such as oxalate (Fig. 28.5a) or maleate (Fig. 28.5b), the reaction rate is still faster. These groups provide an unhindered pathway of conjugated π-bonds between the two metal ions and thus, facilitate the electron exchange.

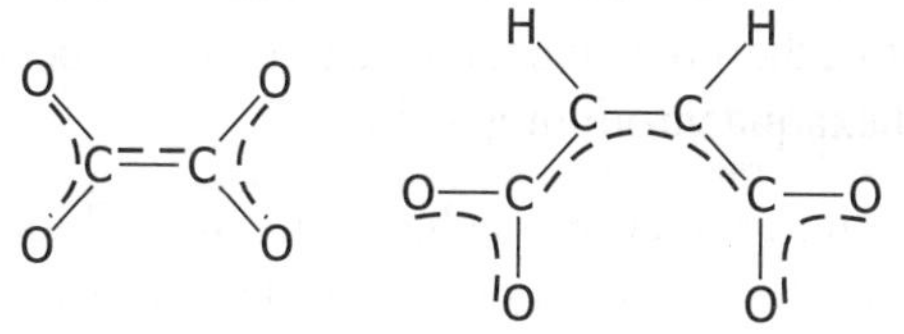

(a) (b)

Fig. 28.5 Structure of (a) oxalate and (b) maleate functional groups

In rare instances, the inner sphere mechanism proceeds without transfer of the bridging ligand from one coordination sphere to another. For example, there is a definite evidence that the reaction between hexacyanoferrate(III) ion and pentacyanocobaltate(II) ion involves a binuclear transition state.

$$[Fe(CN)_6]^{3-} + [Co(CN)_5]^{3-} \rightarrow [(NC)_5Fe... CN...Co(CN)_5]^{6-}$$

In fact, the anionic, dinuclear intermediate can be isolated as its barium salt. This indicates that the reaction is indeed accomplished by the inner sphere pathway. Yet, the intermediate decomposes without transfer of the bridging cyano group.

$$[(NC)_5Fe...CN...Co(CN)_5]^{6-} \rightarrow [Fe(CN)_6]^{4-} + [Co(CN)_5]^{2-} \quad (28.22)$$

If the reaction had proceeded through conventional pathway, the products, $[Fe(CN)_5]^{3-}$ and $[Co(CN)_5(NC)]^{3-}$ would have been formed. These are less stable than those actually formed in reaction (28.22).

The inner sphere mechanism explains the solubilizing effect of aqueous chromium(II) ions on chromium(III) chloride. The latter does not dissolve in water to a significant extent. However, in the presence of chromium(II) ions or a reducing agent such as tin(II) chloride which can generate some chromium(II) ions from chromium(III) chloride, it dissolves readily. This is due to transfer of electron from aqueous chromium(II) ion to crystalline chromium(III) ion via a chlorine bridge—as conceptualized by inner sphere mechanism.

$$[Cr(II)............ Cl .... Cr(III)]$$

The chromium(III) ions leave the crystal and move into the aqueous solution.

The following are the consequences of the inner sphere mechanism:

a. Transfer of ligand invariably occurs from one coordination sphere to another.
b. The rate of redox process is equal to or less than the rate of exchange of ligands.
c. The reaction is zero order with respect to one coordination sphere. If the rate-determining step is the dissociation of ligand from one complex ion, the reaction will be first order in that species and zero order in the second species. If, on the other hand, the rate-determining step is the attack of the second complex ion, the reaction will be first order in it and zero order in the first complex.
d. The rate constant depends on the nature of the bridging ligand.

Case Study 28.7 What are the products of the reaction between pentamminethiocyanato-Scobalt(III) ion and hexaquachromium(II) ion.

The reaction follows inner sphere mechanism. Since thiocyanato group is an ambidentate ligand, there are two possible transition states for this reaction. (i) The chromium(II) ion may become bound to the nitrogen atom of the bridging thiocyanato ligand; and (ii) the chromium(II) ion may become bound to the sulfur atom of the bridging ligand.

$$\text{(i)}\quad [(NH_3)_5Co(SCN)]^{2+} + [Cr(H_2O)_6]^{2+} \rightarrow [(NH_3)_5Co\text{–}SCN\text{–}Cr(H_2O)_5]^{4+} + H_2O$$

$$[(NH_3)_5Co\text{–}SCN\text{–}Cr(H_2O)_5]^{4+} \rightarrow [Co(NH_3)_5]^{2+} + [Cr(H_2O)_5NCS]^{2+}$$

$$\text{(ii)}\quad [(NH_3)_5Co(SCN)]^{2+} + [Cr(H_2O)_6]^{2+} \rightarrow$$ 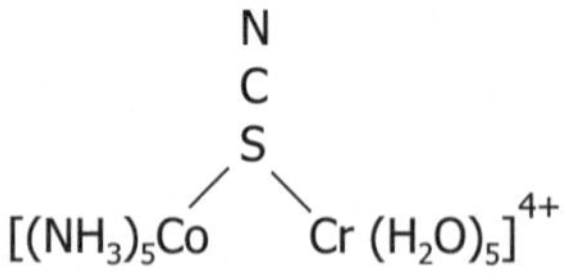 $$+ H_2O$$

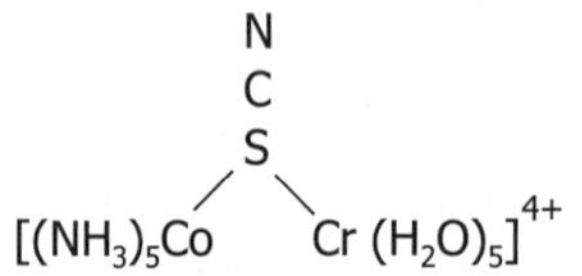 $$\rightarrow [Co(NH_3)_5]^{2+} + [Cr(H_2O)_5SCN]^{2+}$$

The products of the reaction are pentaquathiocyanato-Nchromium(III) ion [reaction (i)] and pentaquathiocyanato-Schromium(III) ion [reaction (ii)]. These are formed to the extent of 70% and 30%, respectively.

3. Complementary reactions

Redox reactions of complexes in which the oxidant gains two electrons and the reductant loses two electrons are called complementary reactions. These reactions involve those elements which have stable oxidation states differing by two units, such as platinum(II)–platinum(IV) and thallium(I)–thallium(III) systems. The following are examples of complementary reactions:

$$Sn(II) + Tl(III) \rightarrow Sn(IV) + Tl(I)$$

$$Sn(II) + Hg(II) \rightarrow Sn(IV) + Hg(0)$$

$$V(II) + Tl(III) \rightarrow V(IV) + Tl(I)$$

In complementary reactions, the molar ratio of the two reacting species is 1:1.

4. Non-complementary reactions

Redox reactions in which the oxidation states of the reactants change by a variable number of units are called non-complementary reactions. In such reactions, different number of molecules of oxidant and reductant are involved in the stoichiometric equations. For example, the reaction of tin(II) and iron(III) ions takes place as follows:

$$Sn(II) + Fe(III) \rightarrow Sn(III) + Fe(II)$$

$$Sn(III) + Fe(III) \rightarrow Sn(IV) + Fe(II)$$

The evidence for tin(III) species is indirect. If trioxalatocobaltate(III) ion, $[Co(C_2O_4)_3]^{3-}$ is added to the reaction medium of tin(II)–iron(III) system, it gets rapidly reduced. On the other hand, if this complex is added to the tin(II)–thallium(III) system, cited under the category of complementary reactions, it remains unaffected.

It has also been observed that potassium dichromate [containing Cr(VI)] oxidizes iodide ions to iodine in acidic medium very slowly. However, in the presence of iron(II) ions, the reaction is catalyzed. It is inferred that chromium(V) and chromium(IV) ions, formed as a result of non-complementary reactions, catalyze the oxidation process.

$$Cr(VI) + Fe(II) \rightarrow Cr(V) + Fe(III)$$

$$Cr(V) + Fe(II) \rightarrow Cr(IV) + Fe(III)$$

$$\text{Overall reaction: } Cr(VI) + 2Fe(II) \rightarrow Cr(IV) + 2Fe(III)$$

If iron(III) ions are added to potassium dichromate–iodide–acid mixture, no catalytic effect is observed. This provides further support for oxidizing action of chromium(V) and chromium(IV) ions and, in turn, for the non-complementary reaction between chromium(VI) and iron(II) ions. Similarly, the non-complementary reaction between thallium(III) and iron(II) ions, produces the unusual thallium(II) ions as an intermediate.

$$Tl(III) + Fe(II) \rightarrow Tl(II) + Fe(III)$$

$$Tl(II) + Fe(II) \rightarrow Tl(I) + Fe(III)$$

$$\text{Overall reaction: } Tl(III) + 2Fe(II) \rightarrow Tl(I) + 2Fe(III)$$

The oxidation of chromium(III) to chromium(VI) by peroxodisulfate ion ($S_2O_8^{2-}$) is catalyzed by silver(I) ion. Initially, the peroxodisulfate ion oxidizes silver(I) to silver(II). The latter then transforms chromium(III) to chromium(VI) ions by a non-complementary reaction.

$$Ag(I) + S_2O_8^{2-} \rightarrow Ag(III) + 2SO_4^{2-}$$

$$Ag(III) + Ag(I) \rightarrow 2Ag(II)$$

$$Cr(III) + 3Ag(II) \rightarrow Cr(VI) + 3Ag(I)$$

Case Study 28.8 Comment on the nature of the following reaction:

$$[Co(NH_3)_5(CO_3)]^+ + 2H_3O^+ \rightarrow [Co(NH_3)_5(H_2O)]^{3+} + 2H_2O + CO_2$$

At the first glance it appears to be a substitution reaction, where the carbonato ligand is replaced by aqua group. However, it is not. No metal–ligand bond is broken in this case. The reaction initiates with the protonation of one of the oxygen atoms of the carbonate group.

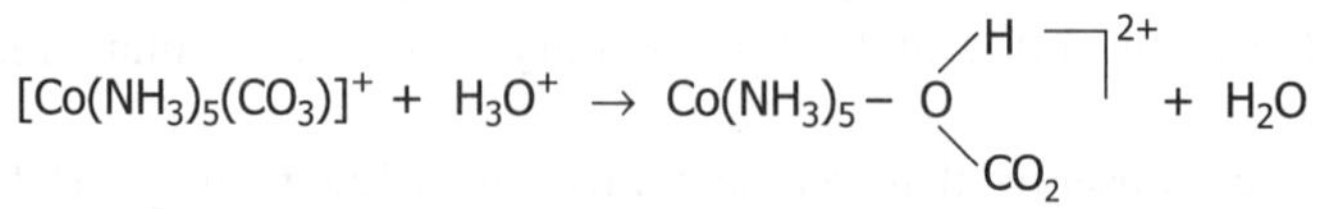

Next, the protonated species releases carbon dioxide.

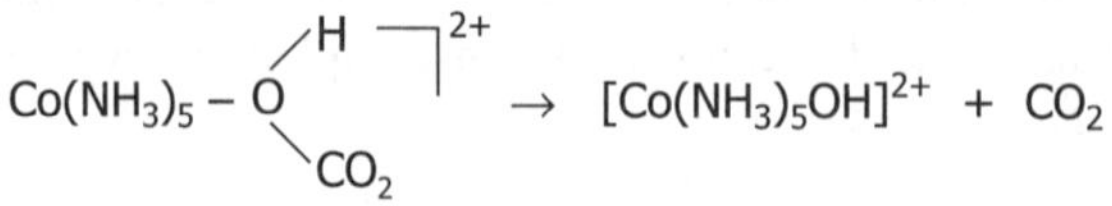

Further, protonation follows.

$$[Co(NH_3)_5OH]^{2+} + H_3O^+ \rightarrow [Co(NH_3)_5(H_2O)]^{3+} + H_2O$$

Since cleavage of the metal–ligand bond does not occur in this case, the reaction proceeds at a very high rate.

29. ORGANOMETALLIC COMPOUNDS

Introduction

Organometallic compounds are the chemical species in which one or more carbon atoms of an organic functional group are directly bound to metal. In a broader sense, the term organometallic compounds also includes—

1. Metal carbonyl compounds such as tetracarbonylnickel(0), $Ni(CO)_4$ although carbon monoxide is not an organic group.
2. Analogs with boron–carbon bonds, for example, trimethylboron(III), $(CH_3)_3B$, as well as the derivatives with silicon–carbon bonds, for example, tetraphenylsilicon(IV), $(C_6H_5)_4Si$, although both boron and silicon are not metals in the true sense of the term.

However, the term organometallic compounds does not include—

1. The traditional ionic compounds such as sodium cyanide, Na^+CN^-.
2. The compounds that are formed by the combination of an organic moiety and a metal but contain no metal–carbon bond, for example, diacetatocopper(II), $[Cu(OOCCH_3)_2]$ and tetraethoxytitanium(IV), $[Ti(OC_2H_5)_4]$. These compounds, no doubt, contain a metal ion and an organic moiety, but they have metal–oxygen bonds rather than metal–carbon bonds in their structures. Such derivatives may be designated as metal–organic compounds, but not as organometallic compounds.

Classification of Organometallic Compounds

The organometallic compounds are broadly classified into the following four categories:

1. Ionic organometallics

Highly electropositive metals, mainly of Groups 1 and 2, form ionic organometallic compounds. Such compounds are very reactive and quite unstable. Being polar, they are conducting in nature. Their electrolysis yields a mixture of hydrocarbon gases at the anode and metal at the cathode.

Ionic organometallic compounds are soluble in polar organic solvents such as tetrahydrofuran and dimethylformamide, but insoluble in non-polar ones such as benzene or toluene. In fact, many such compounds, for example, sodium cyclopentadienide (Fig. 29.1a), are stable only in solution state. Similarly, sodium naphthyl (Fig. 29.1b), a dark green, highly reactive, paramagnetic adduct is stable only in solution state. A few ionic organometallics, however, may be isolated as crystalline salts. Sodium triphenylmethyl (Fig. 29.1c) is one such compound.

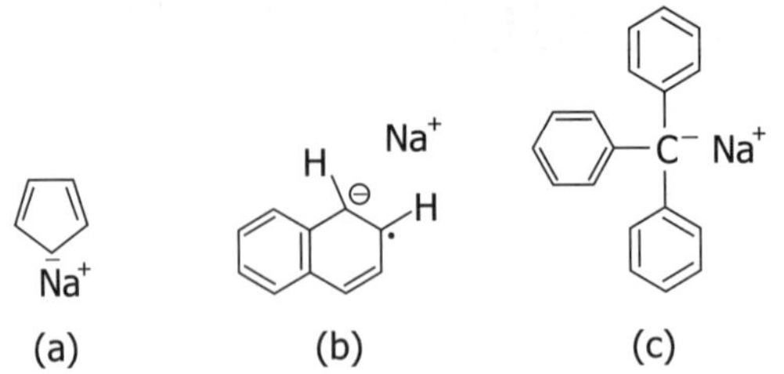

Fig. 29.1 Structure of (a) sodium cyclopentadienide, (b) sodium naphthyl and (c) sodium triphenylmethyl

2. σ-Bonded organometallics

Compounds in which the carbon of the organic moiety is bound to a metal of relatively low electropositive character by a normal two-electron–two-center bond are classified as σ-bonded organometallics. Analogous compounds with metalloid–carbon bonds too fall under this category. Some compounds contain exclusively the metal and hydrocarbon radicals. Trimethylgallium(III), $(CH_3)_3Ga$ and tetraphenylsilicon(IV), $(C_6H_5)_4Si$ are examples of such compounds. In other compounds, partial substitution of non-organic groups may occur, as in chlorotrimethyltin(IV), $(CH_3)_3SnCl$ or chloromethylmercury(II), CH_3HgCl.

Dialkyls of zinc, cadmium and mercury, which may also be listed as σ-bonded organometallics, are colorless, low viscosity, high density, intensely refractive liquids. These have a linear geometry, are highly reactive and spontaneously catch fire on exposure to air. The mercury compounds are very poisonous. Besides being accidentally inhaled or ingested, these may also be absorbed through the skin, and they elicit toxic effects on entering the body.

3. Multicentered-bonded organometallics

Beryllium forms a number of polymeric dialkyls (Fig. 29.2a) while aluminium forms a few dimeric trialkyls (Fig. 29.2b) involving electron-deficient bridging structures and multicenter bonding. The aluminium organometallics retain their dimeric structures both in the solution state as well as in the vapor phase. However, the alkaline earth alkyls are polymeric only in solid state. In vapor phase, they possess a linear, monomeric geometry.

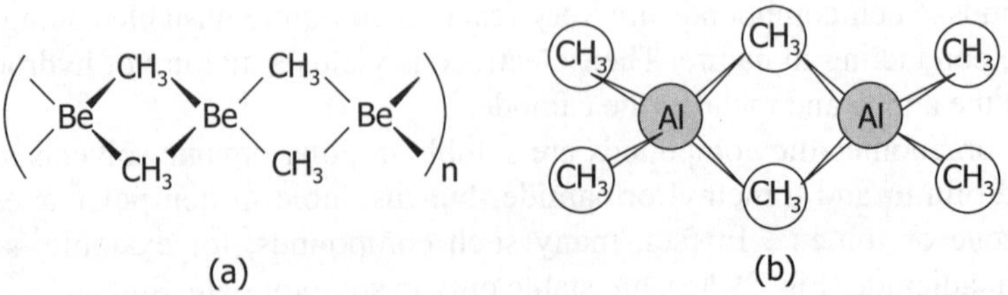

Fig. 29.2 Structure of (a) dimethylberyllium(II) and (b) trimethylaluminium(III)

4. π-Bonded organometallics

There are a good number of organometallic derivatives which entail a π-bond between the metal and the carbon atom of the ligand. Such compounds, which are largely formed by transition elements, may be subdivided into two types.

The first type of π-bonded organometallics are those in which a normal σ-bond is formed initially by donation of ligand electrons to the metal. This is followed by back donation of metal electrons on to the vacant π orbitals of the ligand, resulting in the formation of a π-bond. This phenomenon is called *back bonding* or *dative bonding*. Back bonding will occur if the valence shell of the metal is nearly complete; only then it will back donate its valence electrons to the ligand. Further, the ligand should have vacant π orbitals; only then it will accept the electrons from the metal.

The ability of a ligand to enter into back bonding by accepting electron density from the metal into its vacant π orbitals is called π *acidity*. Carbonyl (CO) and cyanide (CN^-) ligands have high π-acidities. They stabilize low oxidation states of the metal in the complexes. For example, the oxidation states of metals in $[Ni(CO)_4]$ and $[Mn(CO)_5]^-$ are zero and –1, respectively.

The second type of π-bonded organometallics are those in which donation of electrons from the ligand to the metal as well as back donation of electrons from the metal to the ligand are both accomplished by π orbitals of the ligand. Such complexes are designated as *π-complexes*. Ethene ($H_2C{=}CH_2$) and cyclopentadienide anion ($C_5H_5^-$) form π-complexes with metals, the representative examples being tris(π-ethene)nickel(0) (Fig. 29.3a) and ferrocene (Fig. 29.3b), respectively.

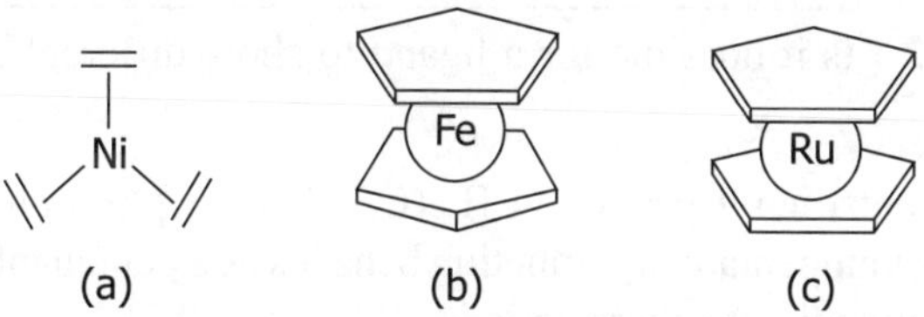

Fig. 29.3 Structure of (a) tris(π-ethene)nickel(0) (b) ferrocene and (c) ruthenocene

Ferrocene was the first dicyclopentadienyl–metal compound to be prepared. Since both donation and back acceptance of electrons take place through π orbitals of the cyclopentadienyl groups, the composition of ferrocene is often expressed as $(\pi\text{-}C_5H_5)_2Fe$. The two cyclopentadienyl rings have staggered configuration with respect to each other. The corresponding ruthenium compound, i.e., ruthenocene, $(\pi\text{-}C_5H_5)_2Ru$, has eclipsed configuration of cyclopentadienyl rings (Fig. 29.3c). Compounds such as ferrocene and ruthenocene, in which the metal atom lies in between the two rings, are called *sandwich compounds*.

Case Study 29.1 What are super Grignard reagents?

Lithium aryls or alkyls, RLi (R = CH_3, C_2H_5, C_6H_5, etc.) are often referred to as super Grignard reagents. These are so named because they are 100 times more reactive than the conventional Grignard reagents. Whenever a particular Grignard reagent is difficult to prepare or it reacts very slowly, a corresponding lithium compound serves the purpose. For example, phenyllithium(I) reacts one hundred times as fast as phenylmagnesium bromide, C_6H_5MgBr. A large number of alkyllithium derivatives are used not only in organic synthesis but also in polymerization processes.

Hapticity of Organometallic Compounds

The term hapticity refers to the number of carbon atoms of a ligand that are attached to the metal atom in an organometallic compound. Hapticity may vary from one to eight. It is designated as η^1, η^2......η^8. Consider, for example, the case of ferrocene. The π-electron cloud of each cyclopentadienyl group in ferrocene (Fig. 29.3b) is delocalized over all the five carbon atoms of the ring. This implies that all five carbon atoms of the cyclopentadienyl ligand are bound to iron. Therefore, the cyclopentadienyl moiety is a pentahapto ligand. For this reason, the composition of ferrocene is better expressed as $(\eta^5\text{-}C_5H_5)_2Fe$, rather than $(\pi\text{-}C_5H_5)_2Fe$. Similarly, in Zeise's salt [trichloroetheneplatinate(II) ion], the π-electron cloud of the ethene group, which is delocalized over both the carbon atoms, is involved in bonding. Therefore, ethene is a dihapto ligand and the composition of Zeise's salt is expressed as $[Pt(\eta^2\text{-}C_2H_4)Cl_3]^-$.

Case Study 29.2 Is it possible for a ligand to show different hapticities in one complex?

The two cyclopentadienyl groups in $[Fe(C_5H_5)_2(CO)_2]$ complex have different hapticities. One cyclopentadienyl function behaves as a pentahapto ligand whereas the other shows monohapto characteristic.

The pentahapto and monohapto cyclopentadienyl groups act as six- and two-electron donors, respectively. The resulting complex is very stable since it follows the 18-electron rule.

Fe(II)	=	$6e^-$
2CO : $2 \times 2e^-$	=	$4e^-$
$(\eta^5\text{-}C_5H_5)^-$	=	$6e^-$
$(\eta^1\text{-}C_5H_5)^-$	=	$2e^-$
		$18e^-$

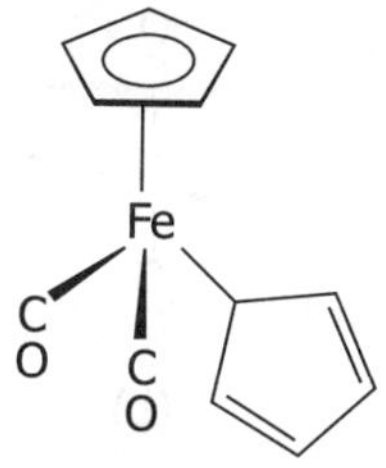

The correct composition of the complex may be written as [Fe(η^1-C_5H_5)(η^5-C_5H_5)$(CO)_2$]. Its IUPAC nomenclature is dicarbonyl(η^1-cyclopentadienyl)(η^5-cyclopentadienyl)iron(II).

Carbonyl Organometallics

Nearly all the transition metals react with carbon monoxide to yield carbonyl compounds. Such compounds exist in a wide array of compositions, ranging from simple monomeric derivatives to cluster complexes. In addition, there are a large number of complexes in which the carbonyl group is present in concert with other ligands.

Carbonyl complexes

Carbonyl compounds are hydrophobic in nature, are highly toxic and are soluble in non-polar solvents. These are either inflammable liquids or readily combustible solids. For example, tetracarbonylnickel(0), $Ni(CO)_4$ and pentacarbonyliron(0), $Fe(CO)_5$ are both liquids and their vapors form explosive mixtures with air.

Besides the mononuclear derivatives, for example, $Cr(CO)_6$; binuclear carbonyl compounds, for example, $Mn_2(CO)_{10}$; trinuclear carbonyl compounds, for example, $Os_3(CO)_{12}$; and tetranuclear carbonyl compounds, for example, $Co_4(CO)_{12}$, are also known.

Structures

The carbonyl ligand can enter into three types of bonding with metals—terminal, double-bridged and triple-bridged. These three bonding modes are shown in Fig. 29.4.

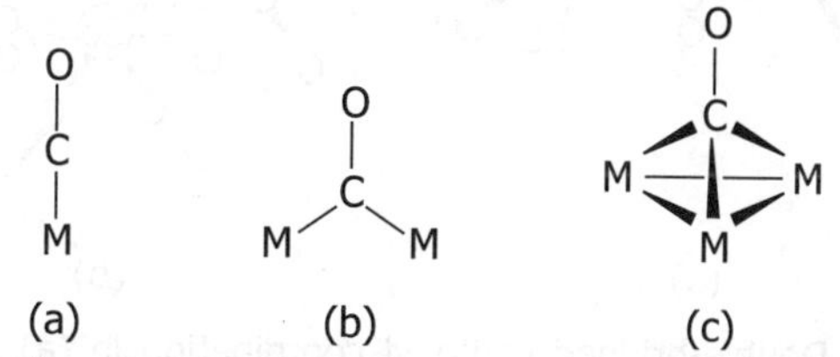

Fig. 29.4 Carbonyl group bound to a metal M as (a) terminal ligand (b) double-bridged ligand and (c) triple-bridged ligand

The terminal carbonyl coordination is the most common. Here, the ligating carbon atom of the ligand donates two electrons to the metal and the M–C–O skeleton aligns linearly. The structures of some organometallic compounds having terminal carbonyl ligands are shown in Fig. 29.5.

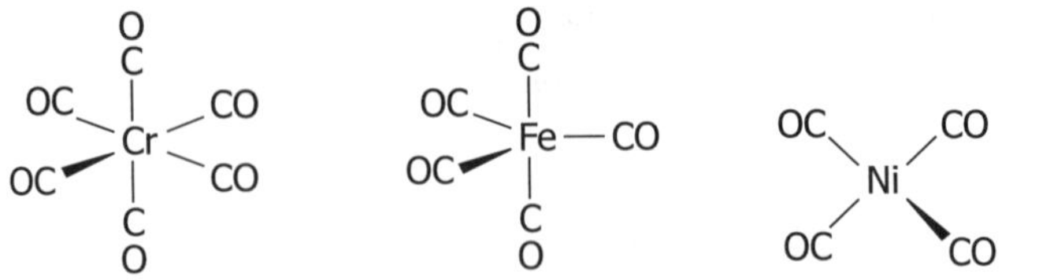

(a) (b) (c)

Fig. 29.5 Terminal carbonyl groups in (a) hexacarbonylchromium(0), (b) pentacarbonyliron(0) and (c) tetracarbonylnickel(0)

Terminal carbonyl bonding is also noted in polynuclear complexes. The examples include dodecacarbonyltriosmium(0) and dodecacarbonyltetrairidium(0) (Fig. 29.6).

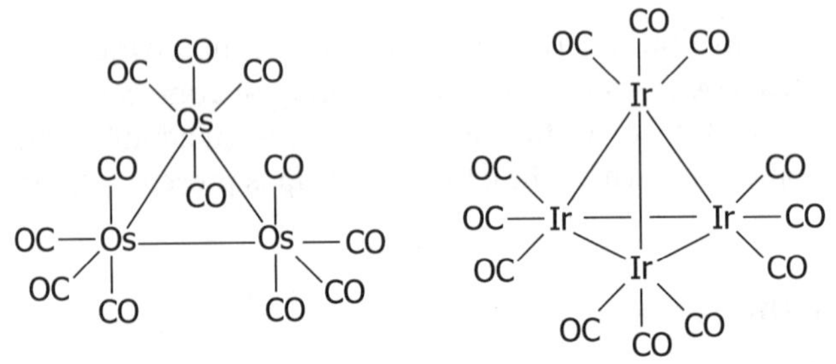

Fig. 29.6 Terminal carbonyl bonding in polynuclear complexes

In double-bridging bonding, also called *ketonic bonding*, each metal–carbon bond is formed by one electron donated by metal and one donated by carbon. This type of coordination is noted in $Fe_2(CO)_9$ (Fig. 29.7a) and $Co_4(CO)_{12}$ (Fig. 29.7b).

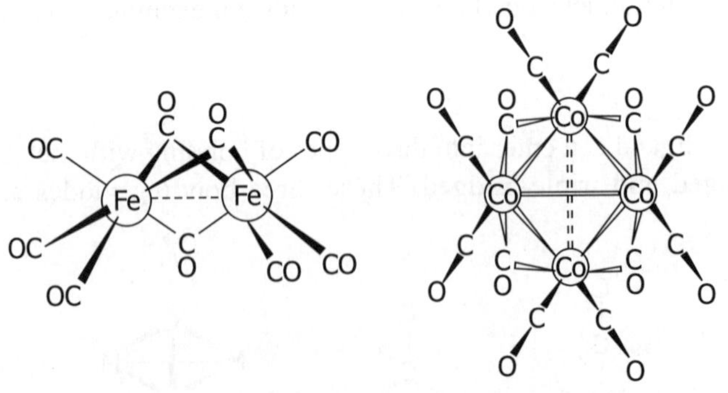

(a) (b)

Fig. 29.7 Double-bridged carbonyl coordination in (a) $Fe_2(CO)_9$ and (b) $Co_4(CO)_{12}$

The complexes involving triple-bridging carbonyl groups are rare. In this type of bonding, carbon of the bridging group shares one of its electrons with an electron contributed by one metal atom and another with an electron contributed by the second metal atom. The third metal–carbon bond is formed by donation of a π-electron pair of carbon monoxide to the metal atom. In $[Rh_6(CO)_{16}]$ complex, for example, four carbonyl groups are triple-bridging. These are marked with an asterisk (*) in Fig. 29.8. For this reason, the composition of the complex is better represented as $[Rh_6(\mu_3\text{-}CO)_4(CO)_{12}]$.

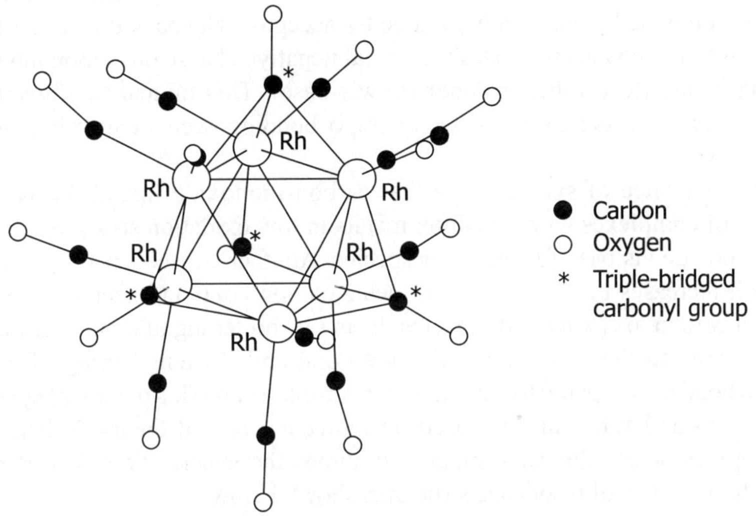

Fig. 29.8 Structure of $[Rh_6(CO)_{16}]$ complex

Bonding

The highest occupied molecular orbital of carbon monoxide, the σ_{2pz} orbital, holds two electrons. This electron pair is loosely held and hence is available for donation to metal. The bonding takes place when the filled σ_{2pz} molecular orbital of carbon monoxide overlaps with an empty orbital of the metal. As a result, a normal ligand-to-metal σ-bond is formed (Fig. 29.9a).

Carbon monoxide also has two degenerate, empty, anti-bonding π orbitals, $\pi^*_{2p_y}$ and $\pi^*_{2p_z}$ orbitals. These are the lowest unoccupied molecular orbitals of carbon monoxide, mutually perpendicular to each other. The overlap of one of these vacant anti-bonding molecular orbitals with a filled metal orbital of π-symmetry (generally d_{xz} or d_{yz}) results in the formation of a second bond called *dative bond* (Fig. 29.9b). The dative bond serves to pass the excessive negative charge (that accumulates on the metal due to the formation of σ-bond) to the ligand.

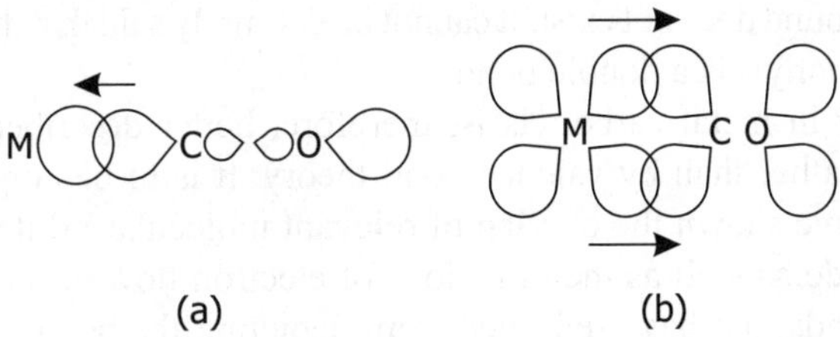

Fig. 29.9 Bonding in metal–carbonyl complexes (a) normal bonding and (b) dative bonding

Back donation of electrons from the metal to the ligands is possible only if the metal is in a low oxidation state. Hence, carbon monoxide ligand stabilizes low oxidation states of metals in its complexes.

The σ- and π-bondings mutually supplement each other. The donation of electrons to metal, during the formation of σ-bond, increases the partial positive charge on the carbon monoxide ligand, making it a better acceptor. The back donation of metal electrons to the ligands increases the partial negative charge on carbon monoxide, making it, in effect, a better donor (Lewis base). This mutual reinforcement is called *synergism*. As a result of synergism, σ-bonding strengthens π-bonding and vice versa.

A consequence of synergism is that carbon monoxide ligand forms a large number of complexes with transition metals in low oxidation states, even though it is a poor Lewis base for the other species. Another consequence of synergism is that it increases the bond order of metal–carbon bond and decreases the bond order of carbon–oxygen bond. This results in (a) shortening of metal–carbon bond as compared to the metal–carbon single bond and (b) lengthening of carbon–oxygen bond as compared to carbon–oxygen triple bond. This picture of synergism between σ- and π-bonding is based on molecular orbital theory. Valence bond theory gives nearly the same picture. It shows the synergistic effect of σ- and π-bonding by way of resonance structures shown below:

$$\underset{\text{I}}{M{-}C{\equiv}O} \longleftrightarrow \underset{\text{II}}{M{=}C{=}O}$$

The problem with valence bond concept is that it describes discrete metal–carbon and carbon–oxygen double bonds in resonance hybrid form II. There is a lack of experimental evidence for this type of formulism. The carbon–oxygen bond is not sensitive to bond order; it does not vary to a significant extent with bond order. Hence, it cannot be said conclusively that the carbon–oxygen bond in metal carbonyl compounds is a double bond. The metal–carbon bond length, however, varies significantly with bond order. Theoretical calculations show that the average bond length of a metal–carbon double bond, M=C is 30–40 pm shorter than that of a metal–carbon single bond.

Carbonyls are the only compounds in which the bond distance between zerovalent metal and carbon, M(0)–C, can be measured. That this bond has a bond order of two can be ascertained only if a comparison can be made with a reference compound containing a single bond between zerovalent metal and carbon. Since such a reference compound does not exist, it cannot be definitely said that the metal–carbon bond in the carbonyls is a double bond.

The bonding in metal carbonyls is, therefore, better described by molecular orbital theory rather than by valence bond theory. It is to be emphasized that in Fig. 27.9 we have shown the overlap of relevant molecular orbitals of metal and carbon monoxide as well as the directions of electron flow during the formation of σ- and π-bonds, but have refrained from depicting the bond orders of metal–

carbon or carbon–oxygen bonds. Nevertheless, infrared spectroscopy sheds light on this aspect and gives a better picture of bond orders in carbonyl complexes. It is discussed in the next section.

Infrared spectroscopy of carbonyl complexes

Infrared spectroscopy, also called *vibrational spectroscopy*, throws light on several structural aspects of metal carbonyls. For a M–CO system, the vibrational spectrum may be analyzed with respect to metal–carbon stretching frequency or carbon–oxygen stretching frequency. However, practically, better information is obtained from carbon–oxygen vibrational mode. This mode is depicted as a clear, sharp and high intensity band in the infrared spectrum. The metal–carbon stretching, on the other hand, often gives low intensity, diffused and broad bands in the spectral charts. Moreover, the wave number range, in which the metal–carbon stretching mode absorbs, contains bands belonging to vibrational modes of other functional groups as well. It is often not possible to demarcate the metal–carbon band from the peaks of other absorptions. The carbon–oxygen stretching mode, however, is easily identifiable since no other prominent band occurs in its vicinity.

From the point of view of metal carbonyls, the infrared spectroscopy gives two vital informations.

1. It predicts the position of attachment of carbonyl ligands to the metal in the complex. We have already stated that the carbonyl groups may be terminal, double-bridged or triple-bridged (*see* p. 825). In each of these bonding modes, the carbon–oxygen stretching absorbs at a characteristic frequency (Table 29.1).

Table 29.1 Stretching frequencies of carbonyl groups in different bonding modes

Carbonyl groups	**Stretching frequency**
Free carbon monoxide	2143 cm^{-1}
Terminal carbonyl	2150–1950 cm^{-1}
Double-bridged carbonyl	1850–1750 cm^{-1}
Triple-bridged carbonyl	1720–1620 cm^{-1}

Thus, the infrared spectrum of pentacarbonyliron(0) (Fig. 29.5b), which has only terminal carbonyl groups, shows a single (but split) band at about 2000 cm^{-1} (Fig. 29.10a). On the other hand, tricarbonyliron(0)-μ-tricarbonyl-tricarbonyliron(0) (Fig. 29.7a), which has both terminal and bridging carbonyl ligands, shows two bands in its infrared spectrum (Fig. 29.10b). The one at 2050 cm^{-1} corresponds to the terminal carbonyl groups while the other at 1850 cm^{-1} corresponds to the bridging carbonyl groups.

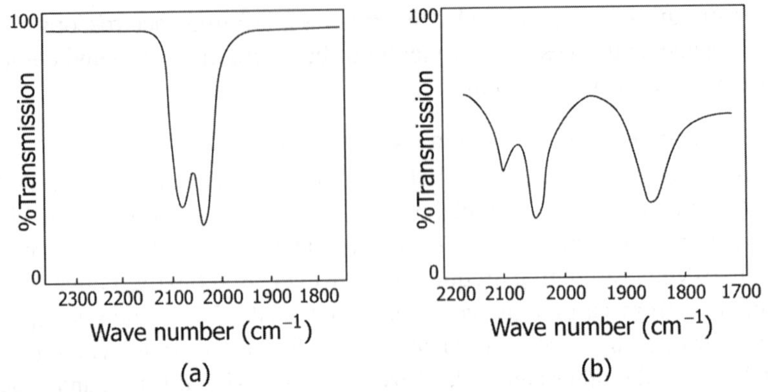

Fig. 29.10 Infrared spectrum of (a) pentacarbonyliron(0) and (b) tricarbonyliron(0)-μ-tricarbonyl-tricarbonyliron(0)

2. It predicts the existence and extent of back bonding in metal carbonyls. As has been discussed earlier, back bonding leads to an increase in the metal–carbon bond order and a decrease in the carbon–oxygen bond order. This is well reflected in the infrared spectra of carbonyl complexes. Greater the bond order, higher the wave number at which the absorption would occur (Box 29.1).

Box 29.1

Fundamentals of Infrared Spectroscopy

A molecule is not a static entity. The relative positions of atoms in a molecule are not exactly fixed nor is the bond that holds the two atoms together exactly rigid. The bond acts like a spring and the position of the atoms that it unites continuously fluctuates due to vibrations. There are two types of vibrations—stretching and bending. In *stretching vibrations*, the distance between the two atoms increases or decreases but the atoms remain positioned along the same bond axis. In *bending vibrations*, the bond angle increases or decreases and the atoms change position with respect to bond axis. The stretching and bending vibrations of a bond occur at quantized frequencies. When infrared light of that frequency falls on the molecule, energy is absorbed and the amplitude of vibration increases. On reverting from the excited state to the ground state, the absorbed energy is released as heat.

The stretching frequency, in terms of wave number ($\bar{\upsilon}$), of an x–y bond is designated as υ(x–y) and expressed in units of cm^{-1}. Mathematically, it is given as,

$$\bar{\upsilon} = \frac{1}{2\pi c}\sqrt{\frac{k}{\mu}}$$

Contd...

where c is the velocity of light, k is the force constant in Nm^{-1} and μ is the reduced mass. The latter is expressed as,

$$\mu = \frac{A_x A_y}{A_x + A_y}$$

where A_x and A_y are the mass numbers of atoms x and y, respectively. The reduced mass of metal–carbon bond in organometallic compounds of iron, cobalt and nickel will vary in the order,

$$\mu(\text{Ni–C}) > \mu(\text{Co–C}) > \mu(\text{Fe–C})$$

The metal–carbon stretching frequency would, therefore, follow the order,

$$\upsilon(\text{Fe–C}) > \upsilon(\text{Co–C}) > \upsilon(\text{Ni–C})$$

The force constant k is directly proportional to the bond strength which, in turn, is directly proportional to the bond order. Greater the bond order, greater is the value of force constant and higher is the value of absorption frequency.

Uncomplexed carbon monoxide has a bond order of three. Its stretching mode occurs at 2143 cm^{-1}. In double-bridged carbonyl groups, the carbon–oxygen bond order is two. The force constant, k is, therefore, less than that of free carbon monoxide and the stretching frequency comes down to 1850–1750 cm^{-1} range. The bond order of a terminal carbonyl group is less than three but more than two. Its stretching frequency has an intermediate value between 2150 cm^{-1} and 1950 cm^{-1}.

As the extent of back bonding increases, the metal–carbon bond order increases. The bond order of carbon–oxygen bond, however, decreases correspondingly. The value of υ(C–O) too is reduced.

As the negative charge on the coordination sphere of a carbonyl complex increases, the extent of back bonding also increases. This, in turn, leads to an increase in metal–carbon bond order and a concomitant decrease in carbon–oxygen bond order. The carbon–oxygen stretching frequency υ(C–O), therefore, shifts to a lower wave number. The examples in Table 29.2 demonstrate this variation.

Table 29.2 Variation of C–O stretching frequency with charge on coordination sphere

Complex	**υ(C–O)**
$Ni(CO)_4$	2060 cm^{-1}
$[Co(CO)_4]^-$	1890 cm^{-1}
$[Fe(CO)_4]^{2-}$	1790 cm^{-1}

Case Study 29.3 Explain the variation of carbon–oxygen stretching frequency in the following carbonyl compounds:

$[Mn(CO)_6]^+$, 2090 cm^{-1}; $[Cr(CO)_6]$, 2000 cm^{-1}; and $[V(CO)_6]^-$, 1860 cm^{-1}.

The carbon–oxygen stretching frequency, υ(C–O), varies with the bond order. Higher the bond order, greater the frequency at which the infrared absorption takes place. The carbon–oxygen bond order in complexes, in turn, depends on the quantum of metal–carbon π-bonding. A large amount of metal–carbon π-bonding means smaller bond order of C–O bond and a lower value of υ(C–O) in infrared spectrum.

Due to the presence of positive charge on metal in $[Mn(CO)_6]^+$, the extent of back bonding is curtailed to a significant extent. The bond order of C–O bond is quite high and so is the value of absorption frequency.

The metal in $[V(CO)_6]^-$ carries a negative charge. It can participate in back bonding very effectively. As a result, the C–O bond order is reduced, and the stretching frequency occurs at a low frequency.

Since the metal in $[Cr(CO)_6]$ is neutral, the extent of π-bonding is neither too large nor too small. Accordingly, the carbonyl frequency occurs at an intermediate value.

A parallel trend in the carbon-oxygen stretching frequency is observed in the following mixed ligand complexes.

$$[Ni(CO)_3\{P(CH_3)_3\}] : 2060\ cm^{-1}$$
$$[Ni(CO)_3(PF_3)] : 2110\ cm^{-1}$$

The trimethylphosphine group in tricarbonyl(trimethylphosphine)nickel(0) complex is electron-donating in nature. It induces a negative charge on the metal, increasing the chances of back bonding and thereby reducing the bond order of carbonyl group. The infrared absorption therefore occurs at a lower frequency.

The trifluorophosphine function in tricarbonyl(trifluorophosphine)nickel(0) is electron-withdrawing in nature. It induces a positive charge on the metal, decreasing the extent of back bonding and, in turn, increasing the bond order of the carbonyl group. The infrared absorption therefore occurs at a higher frequency.

Synthesis

There are four general methods for the preparation of carbonyl complexes.

1. Direct reaction

$$Ni + 4CO \xrightarrow{300\ K} Ni(CO)_4$$
$$Fe + 5CO \xrightarrow{475\ K} Fe(CO)_5$$

2. Reductive carbonylation

$$OsO_4 + 9CO \xrightarrow[350\ atmosphere]{525\ K} Os(CO)_5 + 4CO_2$$
$$Re_2O_7 + 17CO \xrightarrow[350\ atmosphere]{525\ K} Re_2(CO)_{10} + 7CO_2$$

$$WCl_6 + 3Fe(CO)_5 \xrightarrow{375\ K} W(CO)_6 + 3FeCl_2 + 9CO$$

$$CrCl_3 + Al + 6CO \xrightarrow{Benzene} Cr(CO)_6 + AlCl_3$$

3. Photolysis

$$2Fe(CO)_5 \xrightarrow{h\nu} Fe_2(CO)_9 + CO$$

4. Pyrolysis

$$2Co_2(CO)_8 \xrightarrow{250\ K} Co_4(CO)_{12} + 4CO$$

Reactions

A carbonyl ligand can become coordinated to a very large number of metals. Moreover, it can form organometallic compounds of varying compositions with any one metal. It is, therefore, obvious that the chemistry of metal carbonyls is quite extensive. We describe here some important reactions of carbonyl organometallic compounds.

1. Substitution

$$Cr(CO)_6 + C_6H_6 \rightarrow [Cr(C_6H_6)(CO)_3] + 3CO$$

$$Fe(CO)_5 + P(C_6H_5)_3 \rightarrow [Fe\{P(C_6H_5)_3\}(CO)_4] + CO$$

$$Ni(CO)_4 + 2PF_3 \rightarrow [Ni(PF_3)_2(CO)_2] + 2CO$$

2. Reductive decarbonylation

$$Fe(CO)_5 + 4OH^- \rightarrow [Fe(CO)_4]^{2-} + CO_3^{2-} + 2H_2O$$

$$Fe(CO)_5 + BH_4^- \rightarrow [HFe(CO)_4]^- + BH_3 + CO$$

3. Oxidative decarbonylation

$$Fe(CO)_5 + I_2 \rightarrow [Fe(CO)_4I_2] + CO$$

4. Disproportionation

$$3Mn_2(CO)_{10} + \underset{\text{Pyridine}}{12py} \rightarrow 2[Mn(py)_6]^{2+} + 4[Mn(CO)_5]^- + 10CO$$

Case Study 29.4 With reference to carbonyl complexes, what is meant by matrix isolation?

Matrix isolation is a method of synthesizing unstable metal carbonyls such as tetracarbonylpalladium(0), $[Pd(CO)_4]$, tetracarbonylplatinum(0), $[Pt(CO)_4]$ and hexacarbonyltitanium(0), $[Ti(CO)_6]$. It involves reaction of carbon monoxide with metal atoms dispersed in a noble gas matrix at very low temperatures. For example, hexacarbonyltitanium(0) is prepared by reacting titanium with carbon monoxide in argon atmosphere at 200 K.

Pentacarbonyliron(0)

As a representative example of metal carbonyls, we now discuss the characteristics of pentacarbonyliron(0). We have chosen this particular compound since it is one of the most important derivatives of the metal carbonyl series. What is being described in this unit may partly overlap with the contents of the previous section wherein the example of pentacarbonyliron(0) appears both in the general methods of preparation and common reactions of metal carbonyls.

Pentacarbonyliron(0) may be prepared by the following two methods:

1. Direct reaction between iron, heated to 473 K, and carbon monoxide at 200 atmosphere pressure.

$$Fe + 5CO \rightarrow Fe(CO)_5$$

2. By the reaction of carbon monoxide at 200 atmosphere pressure with a mixture of iron(II) iodide and copper at 473 K.

$$FeI_2 4CO \rightarrow [Fe(CO)_4I_2]$$

$$[Fe(CO)_5I_2] + 2Cu + CO \rightarrow Fe(CO)_5 + Cu_2I_2$$

Pentacarbonyliron(0) has a triangular bipyramid structure (Fig. 29.5b). It is a toxic liquid which forms an explosive mixture with air. It is soluble in common organic solvents like acetone, methanol, ether and benzene.

Pentacarbonyliron(0) reacts with a strong base to yield carbonylate anion.

$$Fe(CO)_5 + 4OH^- \rightarrow [Fe(CO)_4]^{2-} + CO_3^{2-} + 2H_2O$$

With a weak base, like ammonia, hydrogen tetracarbonyliron(-II) is obtained, along with carbamic acid.

$$Fe(CO)_5 + NH_3 + H_2O \rightarrow H_2[Fe(CO)_4] + NH_2COOH$$

Sulfuric acid decomposes pentacarbonyliron(0) to ferrous sulfate.

$$Fe(CO)_5 + H_2SO_4 \rightarrow FeSO_4 + 5CO + H_2$$

With nitric oxide, tetranitrosyliron(0) is formed.

$$Fe(CO)_5 + 4NO \rightarrow Fe(NO)_4 + 5CO$$

Partial substitution occurs with triphenylphosphine.

$$Fe(CO)_5 + P(C_6H_5)_3 \rightarrow [Fe\{P(C_6H_5)_3\}(CO)_4] + CO$$

With non-metal halides it is oxidized to iron(II) ion.

$$Fe(CO)_5 + 2CCl_4 \rightarrow FeCl_2 + C_2Cl_6 + 5CO$$

$$Fe(CO)_5 + SO_2Cl_2 \rightarrow FeCl_2 + SO_2 + 5CO$$

A similar oxidation reaction also occurs with tungsten(VI) chloride at 375 K.

$$3Fe(CO)_5 + WCl_6 \rightarrow W(CO)_6 + 3FeCl_2 + 9CO$$

(In the previous section, this reaction was classified as reductive carbonylation since tungsten(VI) chloride is reduced to hexacarbonyltungsten(0)).

Pentacarbonyliron(0) is also oxidized by tin(IV) chloride, although the product in this case is not iron(II) chloride, but tetracarbonyldichloroiron(II).

$$Fe(CO)_5 + SnCl_4 \rightarrow [Fe\{CO)_4Cl_2] + SnCl_2 + CO$$

The two products further combine to yield an adduct.

$$[Fe(CO)_4Cl_2] + SnCl_2 \rightarrow (CO)_4FeCl_2.\ SnCl_2$$

Analogous reactions also occur with antimony(V) chloride.

$$Fe(CO)_5 + SbCl_5 \rightarrow [Fe\{CO)_4Cl_2] + SbCl_3 + CO$$

$$[Fe(CO)_4Cl_2] + SbCl_3 \rightarrow (CO)_4FeCl_2.\ SbCl_3$$

Pentacarbonyliron(0) undergoes photolysis to yield the bridged dimer, tricarbonyliron(0)-μ-tricarbonyl-tricarbonyliron(0), the structure of which is depicted in Fig. 27.7a.

$$2Fe(CO)_5 \xrightarrow{h\nu} Fe_2(CO)_9 + CO$$

Carbonylate anion complexes

Anionic carbonyl complexes are called carbonylate organometallics. These are also referred to as *carbometallates*. Many carbonylate compounds are isoelectronic and isostructural with carbonyl complexes. For example, tetracarbonylferrate (–II) ion, $[Fe(CO)_4]^{2-}$ and tetracarbonylcobaltate(–I) ion, $[Co(CO)_4]^-$ are isoelectronic and isostructural with tetracarbonylnickel(0). The structures of a few representative carbonylate anion complexes are shown in Fig. 29.11.

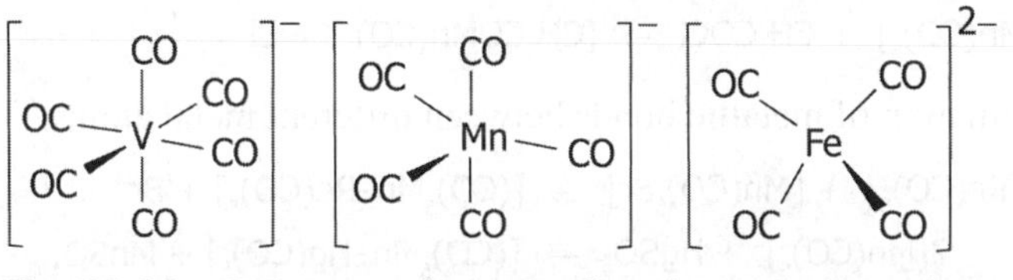

Fig. 29.11 Structures of some carbonylate anion complexes

The carbonylate complexes are easily oxidized by air. The alkali metal salts of these anions, for example, sodium tetracarbonylferrate(–II), $Na_2[Fe(CO)_4]$, are soluble in water. Most of these compounds are prepared *in situ* and used for further reactions without isolation. However, if needed, these may be precipitated as salts of large cations; for example, tetraphenylarsenic(V) tetracarbonylferrate (–II), $[(C_6H_5)_4As]_2\,[Fe(CO)_4]$. The sodium salt, $Na_2[Fe(CO)_4]$ is called *Collman's reagent*. Like Grignard reagent, the Collman's reagent can be used to synthesize a variety of organic compounds, including aldehydes, carboxylic acids and ethers.

Carbonylate anions serve as starting materials and intermediates in the preparation of several carbonyl compounds.

Synthesis

Carbonylate compounds are synthesized by the following methods:

1. Base-induced redox reactions of metal carbonyls

$$Fe(CO)_5 + 4OH^- \rightarrow [Fe(CO)_4]^{2-} + CO_3^{2-} + 2H_2O$$

$$3Mn_2(CO)_{10} + \underset{\text{pyridine}}{12py} \rightarrow 2[Mn(py)_6]^{2+} + 4[Mn(CO)_5]^- + 10CO$$

2. Reaction of metal carbonyls with alkali metals

$$Fe(CO)_5 + 2Na \xrightarrow{NH_3(l)} Na_2[Fe(CO)_4] + CO$$

$$Cr(CO)_6 + 2Na \xrightarrow[\text{Tetrahydrofuran}]{Hg} Na_2[Cr(CO)_5] + CO$$

3. Substitution reactions involving metal carbonyls

$$Mo(CO)_6 + Br^- \rightarrow [Mo(CO)_5Br]^- + CO$$

$$[Re(CO)_5Cl] + 2CN^- \rightarrow [Re(CO)_4(CN)_2]^- + Cl^- + CO$$

Reactions

Some of the important reactions of carbonylate compounds are as follows:

1. With alkyl halides

$$Na_2[Fe(CO)_4] + CH_3Cl \rightarrow Na[Fe(CH_3)(CO)_4] + NaCl$$

$$[Re(CO)_5]^- + CH_3I \rightarrow [Re(CH_3)(CO)_5] + I^-$$

2. With acyl halides

$$Na_2[Fe(CO)_4] + CH_3COCl \rightarrow Na[CH_3COFe(CO)_4] + NaCl$$

$$[Mn(CO)_5]^- + CH_3COCl \rightarrow [CH_3COMn(CO)_5] + Cl^-$$

3. Formation of metallic bonds between different metal atoms

$$[Re(CO)_5]^- + [Mn(CO)_5Br] \rightarrow [(CO)_5Mn\text{–}Re(CO)_5] + Br^-$$

$$2[Mn(CO)_5]^- + HgSO_4 \rightarrow [(CO)_5Mn\text{–}Hg(CO)_5] + MnSO_4$$

Case Study 29.5 What are *isolobal fragments*?

Chemical fragments whose frontier orbitals have similar symmetries, energies and shapes as well as same number of electrons, are called isolobal fragments. Such entities exhibit some common characteristics.

Consider, for example, $[Mn(CO)_5]^{\cdot}$, a fragment of decacarbonyldimanganese(0), $Mn_2(CO)_{10}$ and $Cl^{\cdot}$, a fragment of chlorine molecule, Cl_2. There are a number of similarities between the two fragments.

1. The carbonyl radical has 17 valence electrons, one less than the requirement of effective atomic number (EAN) rule; the chlorine radical has 7 valence electrons one short of the octet.
2. Both $[Mn(CO)_5]^{\cdot}$ and $Cl^{\cdot}$ tend to dimerize, forming $Mn_2(CO)_{10}$ and Cl_2, respectively.

3. Both may be reduced to anions, $[Mn(CO)_5]^-$ and Cl^-.
4. Both form hydrides, $[HMn(CO)_5]$ and HCl.
5. Both combine with species having unpaired electrons, such as iodine radical, to yield $[Mn(CO)_5I]$ and ICl.

In light of these similarities, $[Mn(CO)_5]^{\cdot}$ and $Cl^{\cdot}$ may be regarded as isolobal fragments.

Olefin Organometallics

The first transition metal olefin complex to be synthesized was Zeise's salt, potassium trichloro(η^2-ethene)platinate(II).1-water, $K[Pt(\eta^2\text{-}C_2H_4)Cl_3].H_2O$ (Fig. 29.12a). Thereafter, a wide array of olefin complexes were isolated and characterized. Some of them, especially those belonging to Groups 3–5 of periodic table, were not very stable. However, the complexes of later transition metals, including those of copper(I), silver(I) and gold(I) were stable and presented interesting structural features. The dihapto olefin moiety is coordinated to the metal through the double bond.

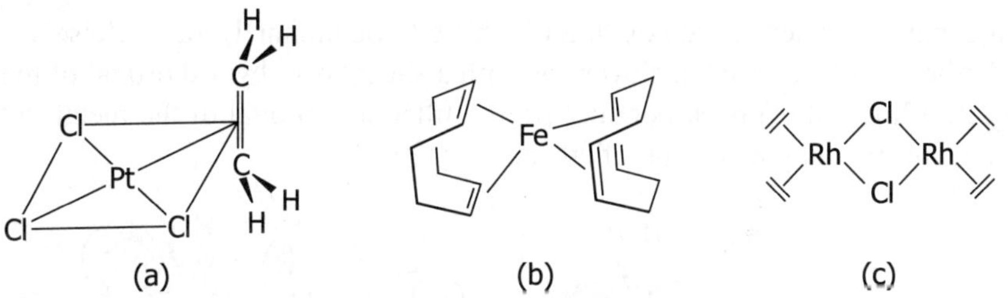

Fig. 29.12 Structures of some olefin complexes

Complexes in which the olefin ligand is chelating are also known. Bis(η^4-cycloocta-1,5-diene)iron(0) is an example of such a complex. Its structure is depicted in Fig. 29.12b. Dimeric bridging complexes of olefins too exist. In these, the alkene ligands occupy terminal positions, for example, bis(η^2-ethene)rhodium(I)-μ-dichloro-bis(η^2-ethene)rhodium(I). It is a chloro-bridged complex involving terminal ethene ligands (Fig. 29.12c).

Olefin complexes are largely diamagnetic in nature.

Structure

While discussing the structure of olefin complexes, we shall consider Zeise's salt as the representative example. Aspects arising out of the attachment of ethene ligand to the metal ion in this compound may be extrapolated to other olefin complexes as well.

In Zeise's salt, the central platinum(II) ion undergoes dsp^2 hybridization using 5d, 6s and two of its 6p orbitals. Accordingly, there is square planar coordination around platinum(II) ion. In this, three coordination sites are occupied by chlorine atoms and the fourth one by ethene group. The carbon–carbon double bond, (C=C) of the latter is perpendicular to the plane of $[PtCl_3]$ residue. As a result, the repulsion between hydrogen atoms of ethene and chlorine atoms at the remaining three coordination sites is minimized.

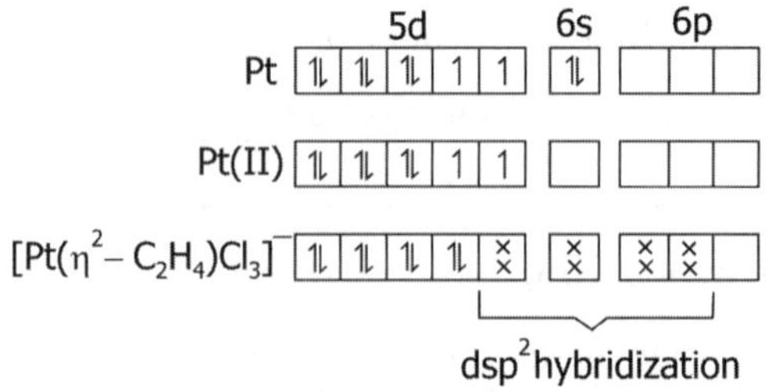

Both carbon atoms of ethene are equidistant from the central platinum(II) ion. The carbon–carbon bond distance is increased from 133.7 pm in free ethene to 137.5 pm in the complex. Moreover, the originally planar ethene molecule becomes slightly distorted on complexation; its hydrogen atoms are tilted backwards with respect to the square planar arrangement, as shown in Fig. 29.12a.

Bonding

Two components define the bonding of ethene to platinum(II) ion in Zeise's salt: (a) A filled π-orbital of ethene overlaps with a vacant dsp^2 hybrid orbital of metal (Fig. 29.13a); and (b) back bonding from a filled dxy orbital of the metal to the vacant, anti-bonding π* orbital of ethene (Fig. 29.13b).

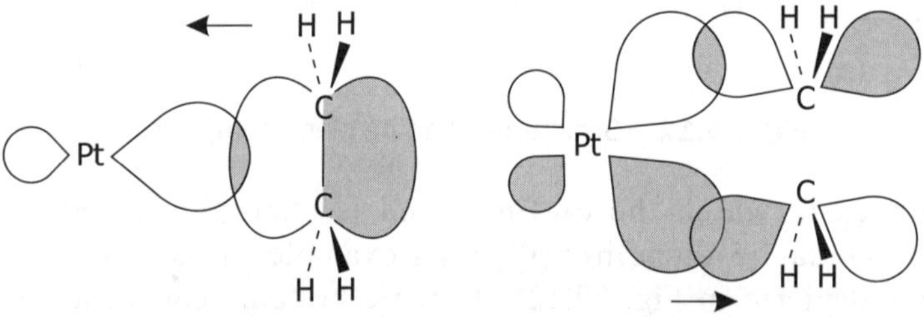

(a) (b)

Fig. 27.13 Bonding in Zeise's salt

The two components involved in metal–olefin bonding reinforce each other synergistically. The perpendicular orientation of the ethene group, with respect to the square planar skeleton, enables it to adjust its filled π and empty π* orbitals for effective overlap with the metal orbitals. Since the anti-bonding π* orbitals of ethene are used in π-bonding, the carbon–carbon bond in the complex weakens and, as stated earlier, lengthens.

The degree of back bonding depends, to quite some extent, on the nature of substituents on ethene ligands. Electron-withdrawing substituents increase the ability of the olefin to accept electrons from the metal, thereby strengthening the metal–ligand bond. Therefore, tetrafluoroethene forms more stable complexes as compared to ethene itself. The alkyl substituted olefins, on the other hand, form less stable complexes as compared to unsubstituted analogs. Two reasons may be ascribed to this observation. First, the alkyl groups, being electron-releasing, decrease the ability of olefin to enter into π-bonding; second, replacement of hydrogen atoms by relatively large alkyl groups results in steric over-crowdedness.

Synthesis

Some common and general methods for the preparation of olefin complexes are as follows:

1. Substitution reactions

$$K_2[PtCl_4] + C_2H_4 + H_2O \rightarrow K[Pt(\eta^2\text{-}C_2H_4)Cl_3].H_2O + KCl$$

(Fig. 29.12a)

$$[Re(CO)_5Cl] + C_2H_4 + AlCl_3 \rightarrow [Re(\eta^2\text{-}C_2H_4)(CO)_5]^+[AlCl_4]^-$$

2. Addition reactions

$$AgNO_3 + 2C_2H_4 \rightarrow [Ag(\eta^2\text{–}C_2H_4)_2]NO_3$$

3. Reduction reactions

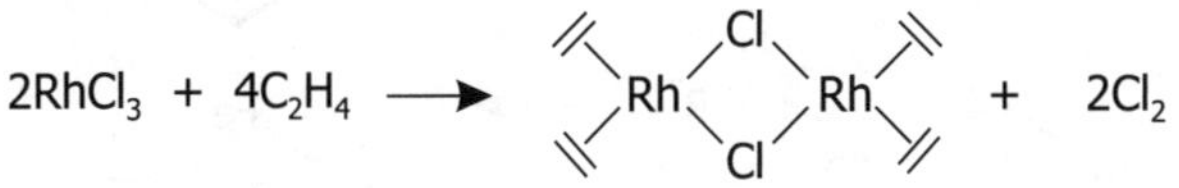

(Fig. 29.12c)

4. Vapor phase condensation reactions

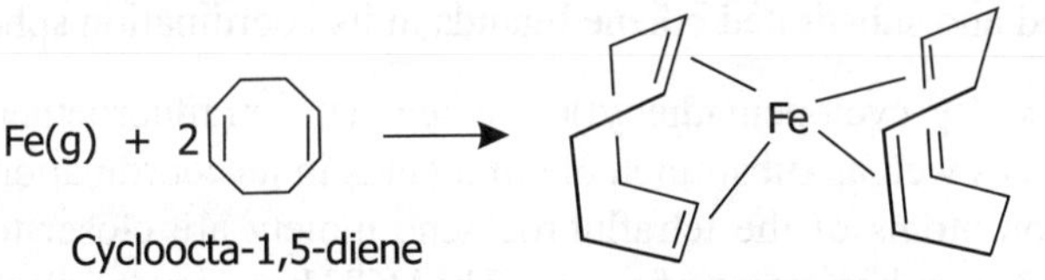

(Fig. 29.12b)

Reactions

Important reactions of olefin complexes are described below:

1. Redox reactions

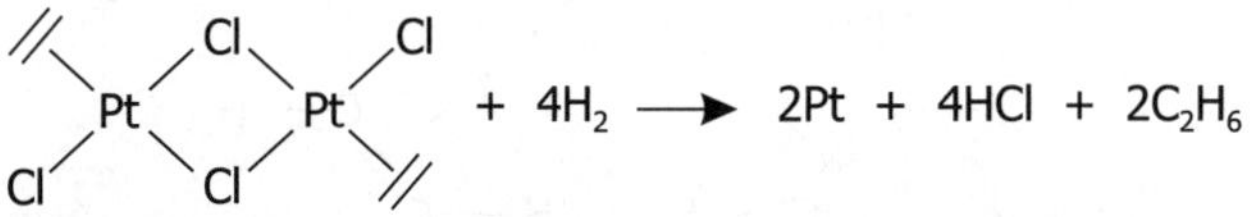

2. Substitution reactions

$$[Pt\{(C_6H_5)_3P\}_2(C_2H_4)] + C_{60} \rightarrow [Pt\{(C_6H_5)_3P\}_2(\eta^2\text{-}C_{60})] + C_2H_4$$

C_{60} is an allotropic form of carbon, called *buckminsterfullerene* (*see* Chapter 16, p. 498). Like olefins, it becomes coordinated to the metal through a double bond.

$$[Ni\{(C_6H_{11})_3P\}_2(\eta^2\text{-}C_2H_4)] + O_2 \rightarrow [Ni\{(C_6H_{11})_3P\}_2(O_2)] + C_2H_4$$

3. Nucleophilic attack
When olefins become coordinated to the metal, their characteristics change. Free olefins undergo electrophilic attack rather than nucleophilic one. However, on complexing with the metal, they become susceptible to nucleophilic attack rather than electrophilic attack. This is because the carbon atoms of olefins acquire positive charge as a result of donation of π-electrons to the metal.

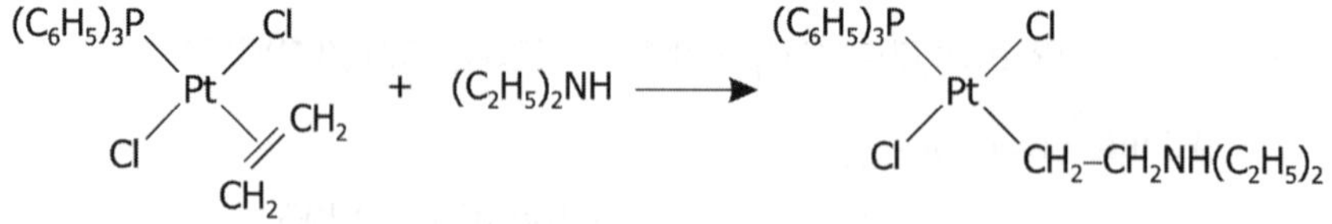

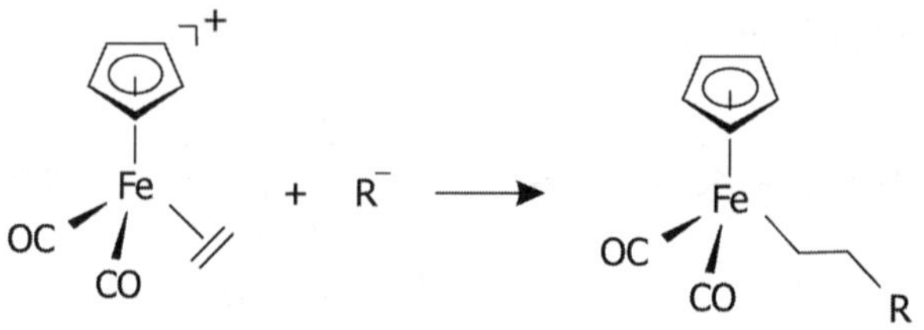

Case Study 29.6 Give an example of an organometallic compound which has both unsubstituted and substituted ethene ligands in its coordination sphere.

The complex, (η^5-cyclopentadienyl)(η^2-ethene)(η^2-tetrafluoroethene)rhodium(I), involves free as well as substituted olefin ligands in its coordination sphere.

The carbon atoms of the tetrafluoroethene moiety are closer to the metal as compared to the carbon atoms of ethene. The HCH bond angle is larger (138°) than the FCF bond angle (106°).

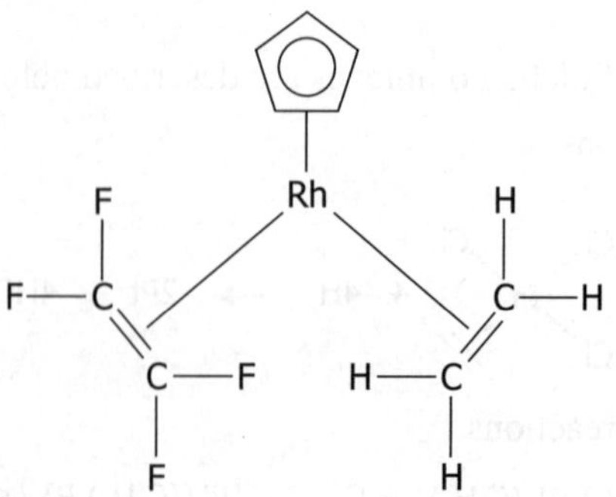

The rotational barrier along the Rh–C_2F_4 bond axis is larger than that along the Rh–C_2H_4 bond axis. This is because the electronegative fluoro substituents cause the extent of back bonding from the metal to the ligand to enhance, as compared to the back bonding in the corresponding ethene ligand. The double bond character in tetrafluoroethene ligand is higher than that in the unsubstituted ethene ligand. Therefore, the rotational barrier in the former is larger.

Alkyl Organometallics

Both transition and non-transition metals form alkyl complexes. However, the alkyl compounds of non-transition metals are more in number and more stable as compared to their counterparts of transition metals. Moreover, the non-transition metal–alkyls are not only commercially important compounds but are also intermediates in the synthesis of other organometallics. Their utility stems from the fact that the metal–carbon bond is quite weak as compared to metal–oxygen or metal–halogen bonds. This, in turn, allows transfer of alkyl group to other chemical entities.

Considering methyl group as a representative alkyl function and considering lithium, beryllium and aluminium as the representative metals of Group 1, 2 and 13, respectively, we describe here the chemistry of methyllithium(I), dimethylberyllium(II) and trimethylaluminium(III).

Methyllithium(I)

Methyllithium(I) is a colorless liquid soluble in hydrocarbon and non-polar solvents. It shows the properties of a typical covalent compound. It has better thermal stability as compared to other methyl derivatives of alkali metals. It ignites spontaneously in air, water and even in water vapors.

Structure

Methyllithium(I), in its natural liquid state as well as in the solid state, possesses a body-centered cubic (BCC) structure (*see* Chapter 3, p. 148). The building blocks of the structure are tetrameric $[CH_3Li]_4$ unit cells. In each unit cell, the lithium atoms are sp^3 hybridized.

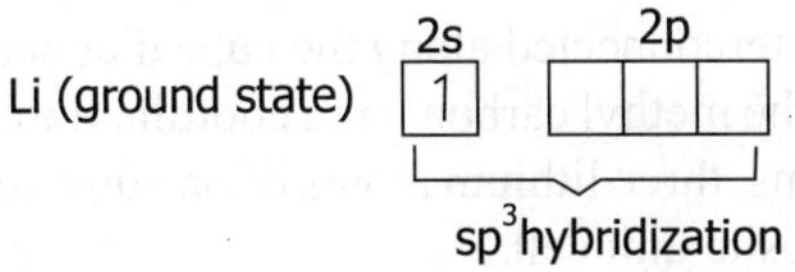

The four lithium atoms in each $[CH_3Li]_4$ unit cell are held at the edges of a tetrahedron. Four electrons (one contributed by each lithium atom) are delocalized throughout the tetrahedral skeleton. This skeleton, like the skeleton of any other tetrahedron, has four triangular faces. A methyl group is positioned above the center of each triangular face.

Figure 29.14 shows the position of one methyl carbon over one of the tetrahedral faces. The methyl carbon has one electron in a sp^3 hybrid orbital which it contributes for bonding. The three lithium atoms on the edges of the triangular face of tetrahedron contribute one electron (from the four delocalized electrons in the tetrahedron) for bonding.

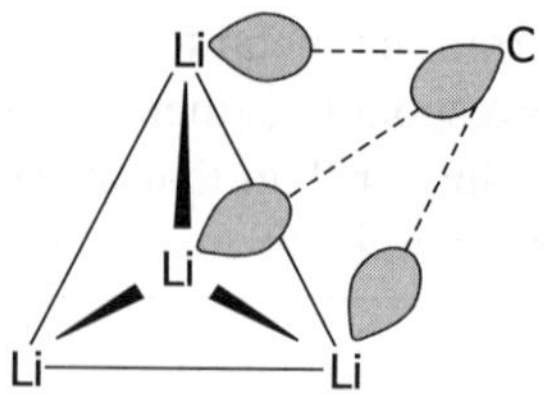

Fig. 29.14 Bonding between one methyl group and three lithium atoms in a $[CH_3Li]_4$ cluster

The two electrons span over four atoms, one carbon and three lithium atoms, of a triangular face of the tetrahedron. Accordingly, a two-electron–four-center bond is formed. A similar multicenter bond is formed on the remaining three faces of the tetrahedron as well.

As a result of the two-electron–four-center bonding, the lithium–carbon and lithium–lithium bonds are held together in a cluster. The unit cell emerging from the $[CH_3Li]_4$ cluster is shown in Fig. 29.15.

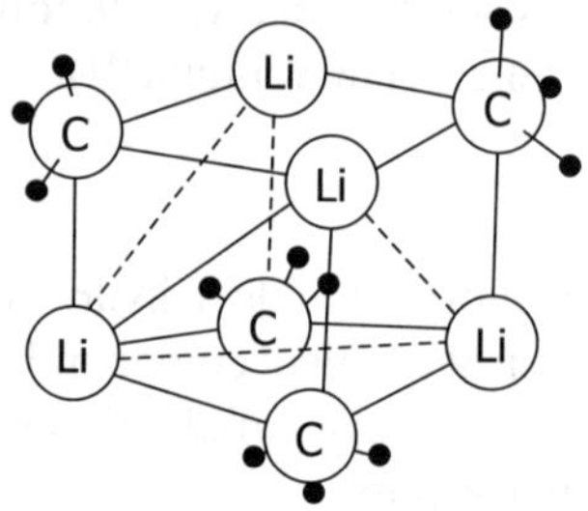

Fig. 29.15 Bonding in a $[CH_3Li]_4$ unit cell

In each unit cell, the methyl carbon has a coordination number of 6; it is bound to three hydrogen atoms and three lithium atoms of any one face of the tetrahedron.

Nine unit cells are packed into a BCC arrangement. Eight of them occupy the edges of the cube and the ninth one occupies the center of the cube. This arrangement is shown in Fig. 29.16.

The unit cells are interconnected along the cube diagonals through the bridging methyl groups. Now, the methyl carbon has a coordination number of 7; it is bound to three hydrogen atoms, three lithium atoms of one unit cell and one lithium atom of the diagonally opposite unit cell.

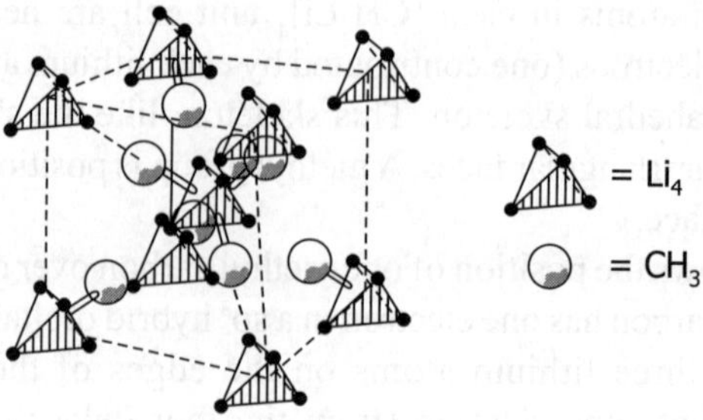

Fig. 29.16 Body-centered cubic structure of methyllithium(I)

Bonding

Figure 29.15 shows that one methyl group and three lithium atoms of a $[CH_3Li]_4$ unit cell are held together by a two-electron–four-center bond. There are four such multicentered bonds in each unit cell. While the methyl group contributes one electron for bonding, the other comes from the delocalized cloud spanning over the lithium tetrahedron.

The two-electron–four-center bonding may be explained on the basis of molecular orbital theory. An sp^3 hybrid orbital of methyl carbon combines with the sp^3 hybrid orbitals of three lithium atoms (of any one face of the tetrahedron) to give one bonding molecular orbital (a) two non-bonding molecular orbitals (b and c) and one anti-bonding molecular orbital (d). The energy pattern of these orbitals is shown in Fig. 29.17.

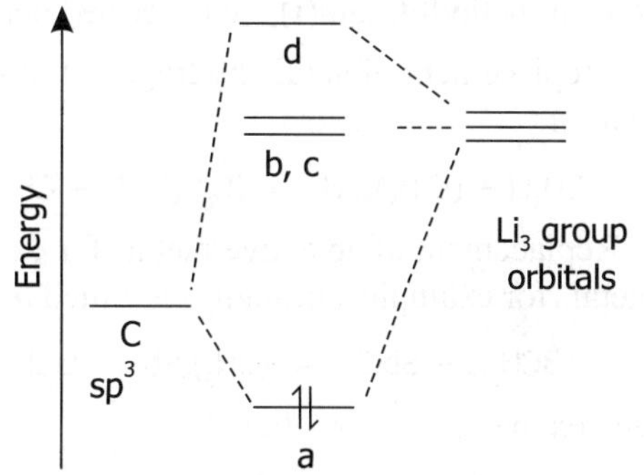

Fig. 29.17 Molecular orbital diagram for one of the four two-electron–four-center bonds in $[CH_3Li]_4$ unit cell

The two electrons that hold the four atoms together occupy the bonding molecular orbital, a. Although too few electrons are available for bonding yet the unit cells are stable. The stability is attributed to two factors. First, the bonding molecular orbital has very low energy as compared to the non-bonding and anti-bonding molecular orbitals. Second, only the lowest energy bonding molecular orbital is occupied; the relatively high-energy, non-bonding and anti-bonding molecular orbitals are vacant.

Synthesis

Important methods for synthesizing methyllithium(I) are as follows:

1. **Direct reaction** The reaction between bromomethane and lithium in dry ether at room temperature is a simple method for the preparation of methyllithium(I).

$$CH_3Br + 2Li \xrightarrow[\text{Dry ether}]{\text{Room temperature}} CH_3Li + LiBr$$

2. **Transmetallation reactions** These involve displacement of a less active metal (for example, mercury) from its organo-derivative by a more active metal (for example, lithium).

$$(CH_3)_2Hg + 2Li \rightarrow 2CH_3Li + Hg$$

Case Study 29.7 Can alkyl iodides be used to prepare alkyllithium(I) compounds?

Alkyllithium(I) compounds are synthesized by the reaction of lithium metal with an alkyl chloride or alkyl bromide in a non-polar solvent (benzene or cyclohexane) under dry nitrogen atmosphere. When alkyl iodides are used for preparing lithium alkyls, a coupling product is formed by the combination of unreacted alkyl iodide and alkyllithium(I).

$$CH_3I + 2Li \rightarrow CH_3Li + LiI$$
$$CH_3Li + CH_3I \rightarrow C_2H_6 + LiI$$

Therefore, alkyl iodides cannot be used for synthesizing alkyllithium(I) compounds.

Reactions

The important reactions of methyllithium(I) are described below:

1. Metallation Replacement of acidic hydrogen atoms of hydrocarbons is called *metallation*.

$$CH_3Li + (C_6H_5)_3CH \rightarrow (C_6H_5)_3CLi + CH_4$$

2. Metathesis Replacement of an active metal (for example, lithium) by a less active metal (for example, antimony) is called *metathesis*.

$$3CH_3Li + SbCl_3 \rightarrow (CH_3)_3Sb + 3LiCl$$

3. Metal–halogen exchange

$$CH_3Li + C_6H_5Br \rightarrow C_6H_5Li + CH_3Br$$

4. Addition to multiple bond

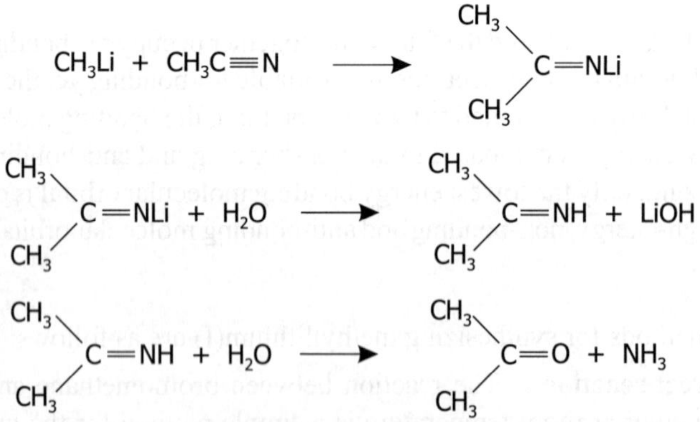

5. Deprotonation of organophosphonium compounds

$$[(C_6H_5)_3PCH_3]^+Cl^- + CH_3Li \rightarrow (C_6H_5)_3P{=}CH_2 + LiCl + CH_4$$

Triphenylphosphonium methylide

6. Hydrolysis

$$CH_3Li + H_2O \rightarrow LiOH + CH_4$$

Box 29.2
Grignard Reagent and Schlenk Equilibrium

The most common organometallic compound of magnesium—and also the most versatile in terms of its applications to synthetic organic chemistry—is the Grignard reagent, commonly formulated as RMgX, where R is an alkyl or aryl group and X is a halogen. It is prepared by adding an alkyl or aryl halide to a suspension of magnesium in dry etherial solvent.

$$RX + Mg \xrightarrow{\text{Ether}} RMgX$$

The above reaction occurs by a free radical mechanism.

$$R - X + \dot{M}g \rightarrow R^{\bullet} + X^{-} + \dot{M}g^{+}$$

$$R^{\bullet} + \dot{M}g^{+} \rightarrow R - Mg^{+}$$

$$R - Mg^{+} + X^{-} \rightarrow R - Mg - X$$

The first step is the slowest, rate-determining one.

The carbon of the R group is nucleophilic in nature. This characteristic is responsible for most of the applications of Grignard reagent to organic synthesis. Yet, due to this very characteristic, Grignard reagent cannot be prepared in aqueous medium. The reason being that the nucleophilic carbon of alkyl or aryl group becomes protonated in presence of water.

$$\overset{\delta^-}{H_3C} - \overset{\delta^+}{Mg} - \overset{\delta^-}{X} + H - OH \rightarrow CH_4 + Mg(OH)X$$

As has been stated, Grignard reagents are prepared in ether medium. Ether not only acts as a solvent, but also as a Lewis base. Two molecules of ether become coordinated to the electrophilic magnesium.

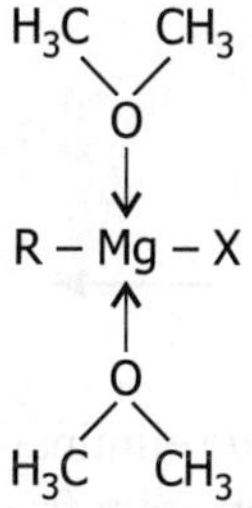

Hence it is pertinent to formulate Grignard reagent as $RMgX(L_2)$, where L is a molecule of ether.

In a solution of Grignard reagent, the following equilibrium exists.

$$2RMgX \rightleftharpoons MgX_2 + MgR_2$$

This state is called Schlenk equilibrium.

Contd...

Normally a simple ether like dimethyl ether or diethyl ether is used in the synthesis of Grignard reagent. The Schlenk equilibrium lies predominantly to the left. However, if a relatively strong Lewis base (B), like dioxane, pyridine or diglyme is added to a solution of Grignard reagent, the equilibrium tends to shift to the right. The reason being that these reagents coordinate to magnesium halide and precipitate it as MgX_2B species. The removal of one of the products pushes the Schlenk equilibrium to right, whereby the dialkylmagnesium or diarylmagnesium may be isolated.

If the concentration of Grignard reagent is high, it may exist in equilibrium with its dimer.

$$2RMgX \rightleftharpoons R-Mg\begin{smallmatrix}X\\ \\X\end{smallmatrix}Mg-R \rightleftharpoons RMg^+ + RMgX_2^-$$

The dimer, in turn, is in equilibrium with organomagnesium cation (RMg^+) and organomagnesium dihalide anion ($RMgX_2^-$).

Case Study 29.8 Titanium(IV) chloride, on reacting with methylmagnesium bromide gives a stable product, while with ethylmagnesium bromide gives an unstable product. Why?

The reaction between titanium(IV) chloride and methylmagnesium bromide produces tetramethyltitanium(IV).

$$TiCl_4 + 4CH_3MgBr \rightarrow Ti(CH_3)_4 + 4Mg(Cl)Br$$

The analogous reaction with ethylmagnesium bromide yields tetraethyltitanium(IV).

$$TiCl_4 + 4C_2H_5MgBr \rightarrow Ti(C_2H_5)_4 + 4Mg(Cl)Br$$

Ethyl group, with a hydrogen atom on a carbon adjacent to the one that is directly linked to the metal, is susceptible to undergo a low energy β-elimination decomposition reaction

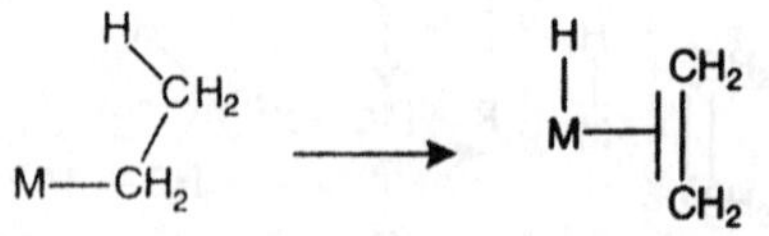

The reaction is known as β-hydride elimination because the hydrogen atom that is eliminated is on the second carbon atom from the metal ion.

The methyl group does not possess a β hydrogen atom. Hence tetramethyltitanium(IV) complex remains stable.

Dimethylberyllium(II)

Dimethylberyllium(II) is a white, reactive solid. It is spontaneously inflammable in air and is vigorously hydrolyzed by water.

Structure

In vapor state, dimethylberyllium(II) is monomeric and possesses a linear geometry. However, in solid state it has a long-chain polymeric structure. The metal atoms in the polymeric chain are tetrahedrally coordinated to bridging methyl groups. The bridged, polymeric structure is shown in Fig. 29.2a.

Bonding

Dimethylberyllium(II) is an electron-deficient molecule and, as has been stated, belongs to the group of multicenter-bonded organometallic compounds. The central atom is sp^3 hybridized.

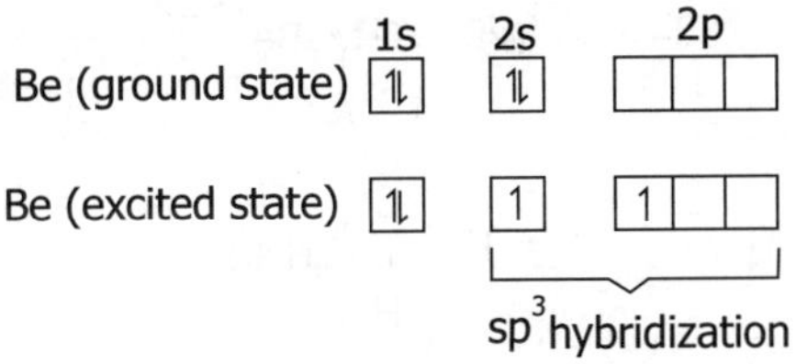

Beryllium has four sp^3 hybrid orbitals; two of them contain one electron each and two are vacant.

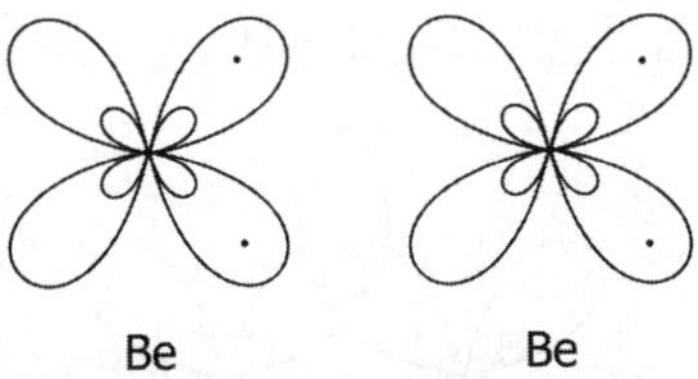

The carbon of the methyl group has an sp^3 hybrid orbital occupied by one electron. This hybrid orbital of carbon overlaps simultaneously with the filled sp^3 hybrid orbital of one beryllium and the vacant sp^3 hybrid orbital of adjacent beryllium.

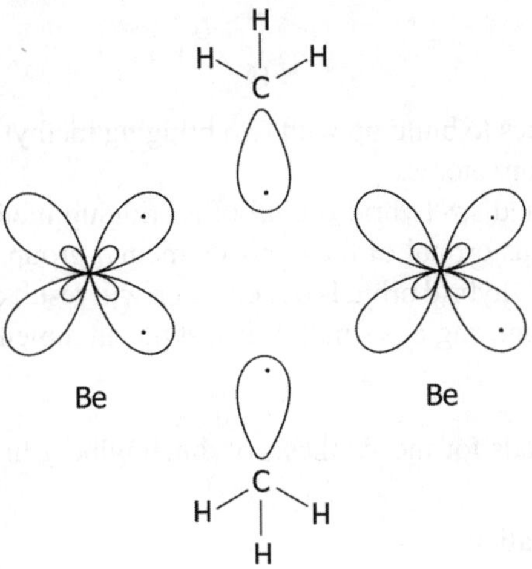

Thus, a two-electron–three-center bond, covering beryllium–carbon–beryllium atoms, results.

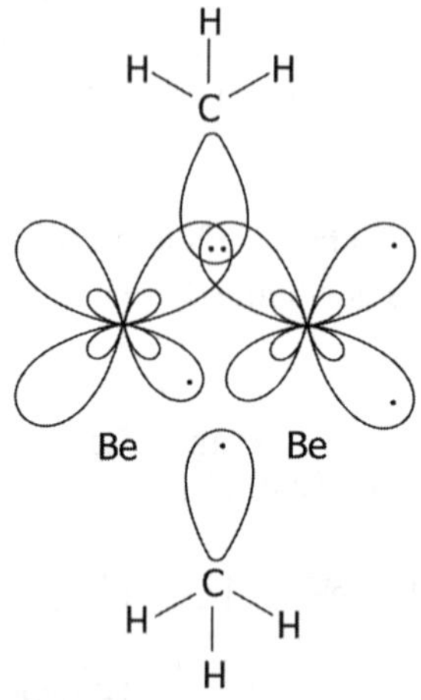

The second bridging methyl then enters into a similar overlap, forming another two-electron–three-center bond.

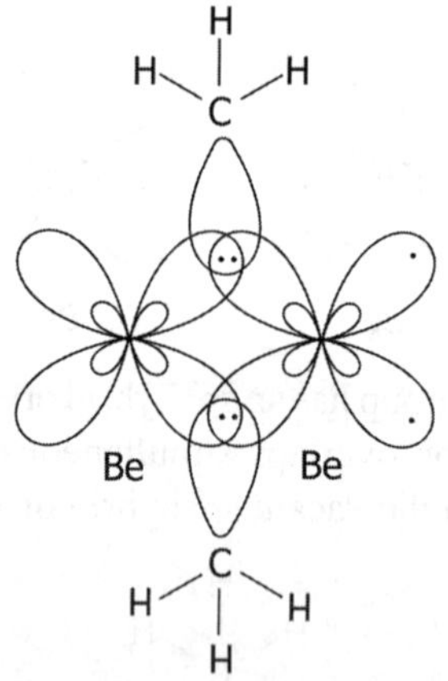

The chain continues to build up with two bridging methyl groups held between two adjacent beryllium atoms.

The singly occupied sp^3 hybrid orbital of a bridging methyl group is oriented towards the analogous orbital of the opposite methyl group. Effective overlap of these orbitals with the hybrid orbitals of beryllium will result only if the beryllium–carbon–beryllium bond angle is small. It is just 66° in dimethylberyllium(II).

Synthesis

The important methods for the synthesis of dimethylberyllium(II) are enunciated below:

1. Transmetallation

$$(CH_3)_2Hg + Be \rightarrow (CH_3)_2Be + Hg$$

2. Metathesis

$$2CH_3Li + BeCl_2 \rightarrow (CH_3)_2Be + 2LiCl$$

Reactions

Some important reactions of dimethylberyllium(II) are given below:

1. With alcohols

$$(CH_3)_2Be + \underset{\text{n-Butanol}}{C_4H_9OH} \rightarrow [Be(OC_4H_9)(CH_3)] + CH_4$$

The resulting alkoxide exits as a tetramer, $[Be(OC_4H_9)(CH_3)]_4$.

2. With epoxides

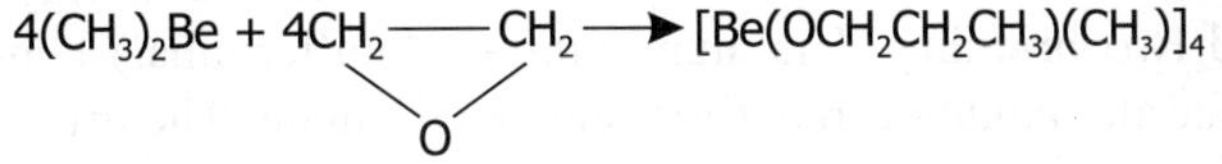

3. With beryllium halides in the presence of a reducing agent

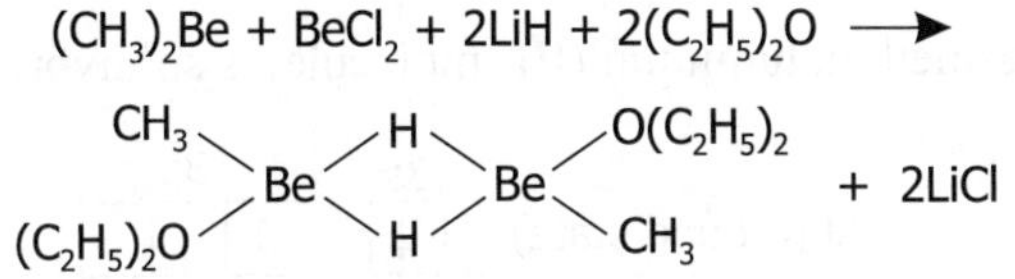

4. With beryllocene

$$(CH_3)_2Be + (\eta^2\text{-}C_5H_5)_2Be \rightarrow 2[Be(\eta^5\text{-}C_5H_5)(CH_3)]$$

5. With Lewis bases

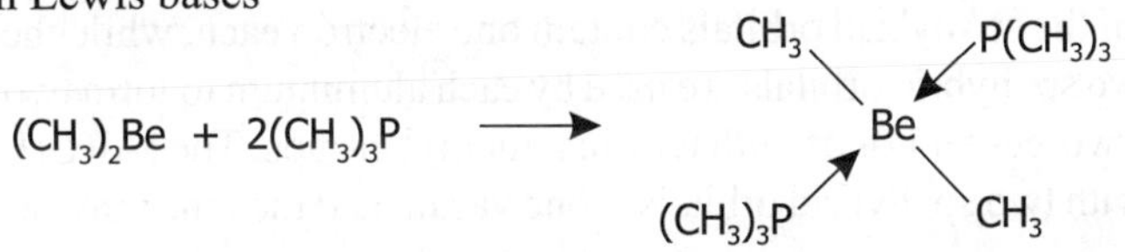

Case Study 29.9 Outline the importance of diethylzinc(II) in preservation of archeological scripts.

The additives used during the manufacture of paper release acidic chemicals over a period of several years. The acids, in turn, cause degradation of paper by attacking its cellulose content. The older the paper sample, greater are its chances of decaying. To overcome this problem, the archeological scripts are treated with vapors of diethylzinc(II). This organometallic compound neutralizes the acidic residues by reacting with hydrogen ions.

$$Zn(C_2H_5)_2 + 2H^+ \rightarrow Zn^{2+} + 2C_2H_6$$

It also reacts with the moisture that the manuscripts may have absorbed from air.

$$Zn(C_2H_5)_2 + H_2O \rightarrow ZnO + 2C_2H_6$$

The zinc oxide, so formed, is basic in nature and hence serves to neutralize the acidic compounds that may be released by the paper in the years to come, thereby ensuring long-term durability of the manuscripts.

Trimethylaluminium(III)

Trimethylaluminium(III) is a colorless, mobile liquid. It is an exceptionally reactive compound. It is pyrophoric in nature, i.e., it catches fire in air. It reacts explosively with water. For these reasons, it is preserved under the protective atmosphere of argon.

Structure

Trimethylaluminium(III) has a dimeric structure, the composition being $[Al_2(CH_3)_6]$. Each aluminium atom is tetrahedrally coordinated to four methyl groups, two terminal and two bridging. The aluminium–carbon(terminal) bond lengths are shorter than the aluminium–carbon(bridging) bond lengths. The structure is depicted in Fig 29.2b.

Bonding

Aluminium, in trimethylaluminium(III) molecule, is sp^3 hybridized.

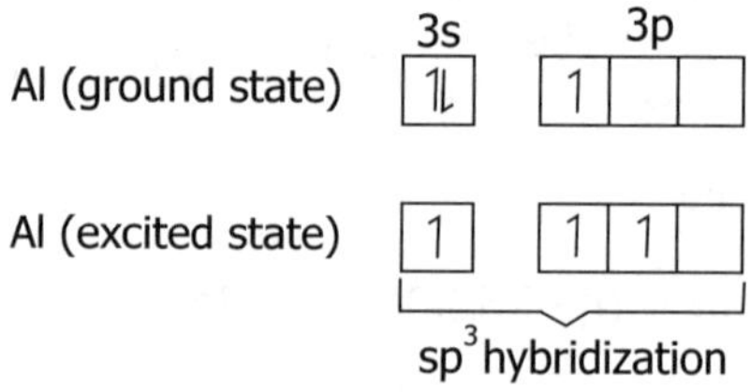

Three of the sp^3 hybrid orbitals contain one electron each, while the fourth one is vacant. Two sp^3 hybrid orbitals are used by each aluminium to form two normal, two-electron–two-center bonds with terminal methyl groups. The $[Al(CH_3)_2]$ fraction is now left with two sp^3 hybrid orbitals—one vacant and the other having one electron.

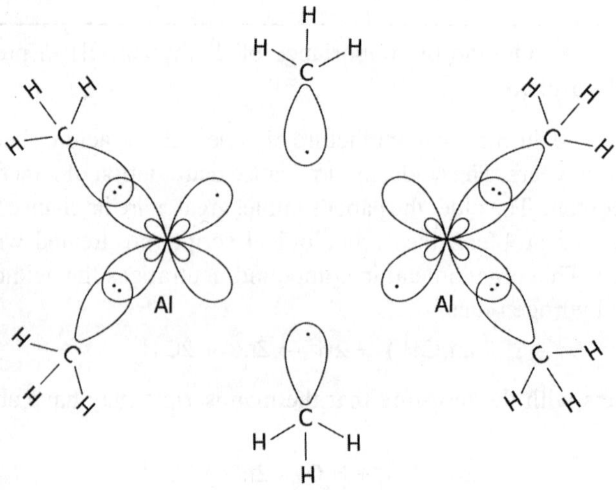

Each of the two bridging methyl groups has one unbound, singly occupied sp^3 hybrid orbital on its carbon atom (the remaining three being bound to hydrogen atoms). This methyl orbital overlaps with a vacant sp^3 hybrid orbital of one aluminium as well as with the singly occupied sp^3 hybrid orbital of the other aluminium. Thus, a two-electron–three-center bond, spanning over aluminium–carbon–aluminium, results.

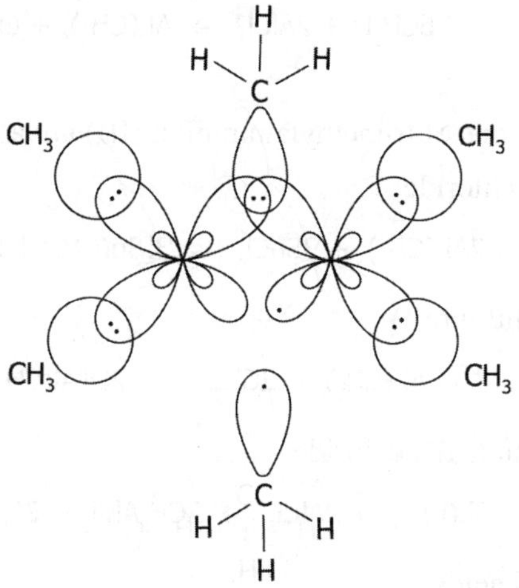

The second bridging methyl group then enters into a similar overlap, resulting in another two-electron–three-center bond.

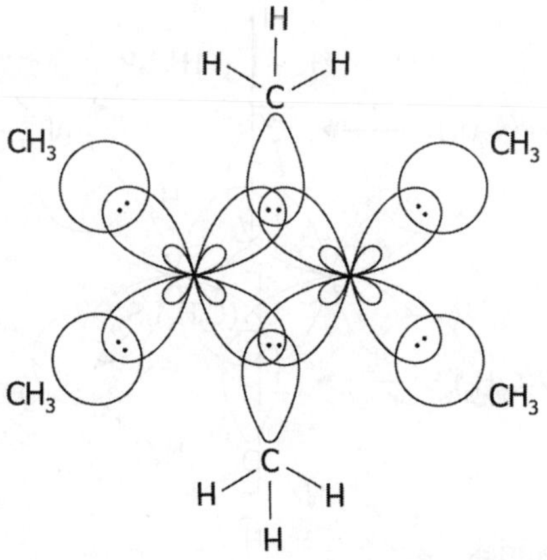

The two multicentered-bonded sp^3 orbitals of the bridging methyl groups are directed towards each other, i.e., these are oriented along the carbon–carbon axis. This means that the two-electron–three-center bond forming sp^3 hybrid orbital of one methyl group is directed along the line joining the carbon of other methyl group. Effective overlap of these orbitals with sp^3 hybrid orbitals of aluminium is possible only if the bond angle is not very large. Accordingly, the aluminium–carbon–aluminium bond angle in trimethylaluminium(III) is only 70°.

Synthesis

Important synthetic routes to trimethylaluminium(III) are now discussed.

1. Transmetallation

$$3(CH_3)_2Hg + 2Al \rightarrow Al_2(CH_3)_6 + 3Hg$$

2. Metathesis

$$6CH_3Li + 2AlCl_3 \rightarrow Al_2(CH_3)_6 + 6LiCl$$

Reactions

Some important reactions of trimethylaluminium(III) are as follows:

1. With tin(IV) chloride

$$2Al_2(CH_3)_6 + 3SnCl_4 \rightarrow 3[Sn(CH_3)_4] + 4AlCl_3$$

2. With methyllithium(I)

$$Al_2(CH_3)_6 + 2CH_3Li \rightarrow 2Li[Al(CH_3)_4]$$

3. With aluminium(III) chloride

$$Al_2(CH_3)_6 + 2AlCl_3 \rightarrow 2CH_3AlCl_2 + 2(CH_3)_2AlCl$$

4. With mineral acids

$$Al_2(CH_3)_6 + 6HCl \rightarrow 2AlCl_3 + 6CH_4$$

5. With Lewis bases

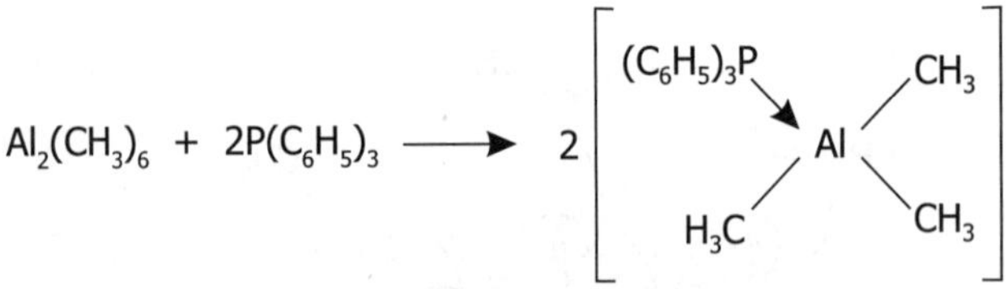

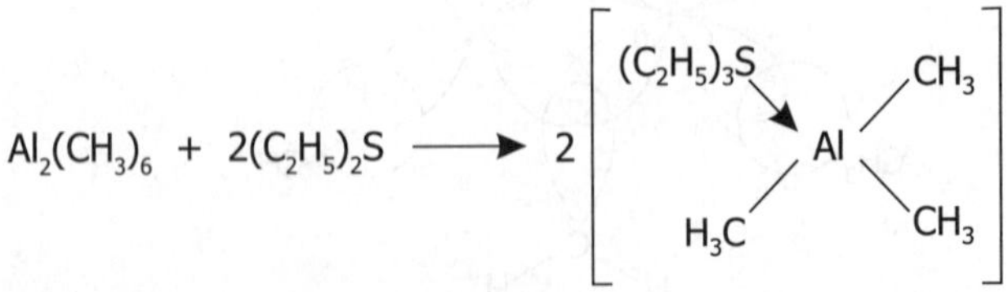

6. With alkyl halides

$$Al_2(CH_3)_6 + 2C_2H_5Cl \rightarrow 2(CH_3)_2AlCl + 2C_2H_5CH_3$$

7. Hydrolysis

$$Al_2(CH_3)_6 + 6H_2O \rightarrow 2Al(OH)_3 + 6CH_4$$

8. Oxidation

$$Al_2(CH_3)_6 + 3O_2 \rightarrow 2Al(OCH_3)_3$$

9. Reduction

$$Al_2(CH_3)_6 + 2H_2 \xrightarrow[150\text{ atmosphere}]{325\text{ K}} 2(CH_3)_2AlH + 2CH_4$$

Case Study 29.10 Which is a better solvent for alkyllithium(I) compounds: heptane or diethylether?

Alkyllithium(I) compounds are not isolable in solid state. Hence, these are preserved in a suitable solvent. Heptane is a better solvent than diethylether since the alkyllithium(I) compounds remain unchanged in heptane for prolonged periods. These compounds cannot be preserved for long in diethylether. This is because the alkyl carbanion, which is a very strong base, abstracts α-hydrogen of ether.

$$C_4H_9Li + CH_3CH_2OCH_2CH_3 \rightarrow CH_3CH(Li)OCH_2CH_3 + C_4H_{10}$$

Hence, if lithium alkyls are kept in diethylether, their concentration registers a decline with passage of time.

Cyclopentadienyl Organometallics

Cyclopentadiene (C_5H_6) is a weak acid. On reaction with a base, it deprotonates forming a symmetrical cyclopentadienyl anion, $C_5H_5^-$. The latter possesses a sextet of π-electrons and hence is aromatic in nature. All the five carbon–carbon bond lengths in cyclopentadienyl anion are equivalent which shows that the π-electron cloud is delocalized throughout the pentagonal ring.

Cyclopentadienyl is a versatile ligand. It forms complexes with a wide spectrum of metals, both transition and non-transition ones. Metal cyclopentadienyl compounds may be classified into three broad types:

1. **Ionic derivatives** These are formed from $C_5H_5^-$ and M^{n+} units. The two chemical entities are held together by electrostatic forces of attraction. Examples include sodium cyclopentadienide, $C_5H_5^-\ Na^+$ and calcium cyclopentadienide, $(C_5H_5)_2Ca$.
2. **Covalent derivatives** In these derivatives, the cyclopentadienyl moiety behaves as a monohapto (η^1) ligand and is σ-bonded to the metal. Cyclopentadienylthallium(I), $(C_5H_5)Tl$, belongs to this category.
3. **π-bonded derivatives** In these compounds, all the five carbon atoms of cyclopentadienyl ring are bound to the metal atom through π-electron cloud. The cyclopentadienyl group behaves as a pentahapto (η^5) ligand. Ferrocene or bis(η^5-cyclopentadienyl)iron(II) (Fig. 27.3b), belongs to this category. It is unquestionably the most important metallocene. Its chemistry is described in the next section.

Ferrocene

Ferrocene is an orange, crystalline solid that exists in monoclinic form at room temperature. Between 110 K and 164 K, it is triclinic, while below 110 K, it exists in orthorhombic form. It melts at 446 K, sublimes at 673 K and decomposes above

743 K. It is diamagnetic in nature, dissolves in most organic solvents but is insoluble in water. It remains stable in air.

Structure

As shown in Fig 29.3b, the iron atom in ferrocene is sandwiched between the two cyclopentadienyl rings. The π-electron cloud of each ring is delocalized over all five carbon atoms. The two cyclopentadienyl rings have staggered configuration relative to each other. However, in gaseous state, the two cyclopentadienyl groups acquire eclipsed configuration with respect to each other. The energy barrier to ring rotation is very small (of the order of 5 kJ mol^{-1}). The hydrogen atoms of the ring are bent towards the iron atom.

Bonding

The bonding in ferrocene is best explained by molecular orbital theory. The construction of molecular orbitals for this compound proceeds in the following three steps:

Step 1 *Combination of atomic orbitals of a cyclopentadienyl ring to form molecular orbitals*

The five p_z orbitals (one of each carbon) of the cyclopentadienyl ring combine with each other to give five molecular orbitals. These five molecular orbitals may be classified into three types on the basis of their symmetry. The lowest energy molecular orbital has a perfect cylindrical symmetry and enters into σ-bonding with metal orbitals. It has no nodes and is designated by symbol a. Next higher in energy are two degenerate molecular orbitals, each having one node. These enter into π-bonding with metal and are designated as e_1 orbitals. The remaining two molecular orbitals are the least symmetrical, each having two nodes. These are the highest energy molecular orbitals, which enter into δ-bonding with metal, and are designated as e_2 orbitals. The energy pattern of the cyclopentadienyl molecular orbitals along with their shapes is depicted in Fig. 29.18.

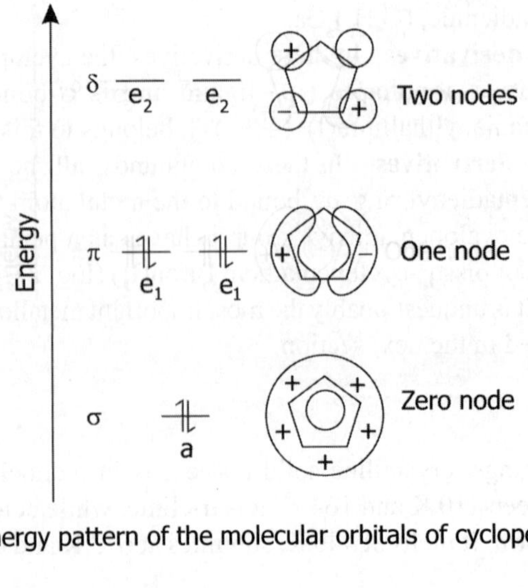

Fig. 29.18 Energy pattern of the molecular orbitals of cyclopentadienyl group

The cyclopentadienyl anion is a 6-electron donor. These electrons occupy the a and two e_1 molecular orbitals. Therefore, the electronic configuration of each cyclopentadienyl ring in ferrocene may be written as, $a^2\ e_1^4\ e_2^0$.

Step 2 *Combination of molecular orbitals of two cyclopentadienyl rings to form ligand group orbitals*

The five molecular orbitals of one cyclopentadienyl ring combine in pairs with a similar set of the second cyclopentadienyl ring to give ten *ligand group orbitals* (LGOs). In this combination, the wave functions of the molecular orbitals of the two cyclopentadienyl rings may be added to or subtracted from each other. Addition of wave functions leads to *gerade* (*g*) LGOs, while subtraction leads to *ungerade* (*u*) LGOs. For example, combination of lowest energy molecular orbitals, each designated as a, yields two ligand group orbitals, $LGOa_{1g}$, formed by addition of wave functions of two a orbitals and $LGOa_{2u}$, formed by subtraction of wave functions of two a orbitals. Similarly, the two degenerate e_1 molecular orbitals of one cyclopentadienyl ring combine with their counterparts of the other cyclopentadienyl ring to yield two degenerate $LGOe_{1g}$ orbitals and two degenerate $LGOe_{1u}$ orbitals. Similarly, two degenerate $LGOe_{2g}$ and two degenerate $LGOe_{2u}$ orbitals are also formed.

The combinations involved in the formation of LGOs are shown in Table 29.3.

Step 3 *Combination of ligand group orbitals with atomic orbitals of iron to form molecular orbitals of ferrocene*

The molecular orbital diagram of ferrocene is presented in Fig. 29.19. Each ligand group orbital merges with a suitable atomic orbital of iron, the one with which its symmetry matches, to yield the molecular orbitals of ferrocene. The $LGOa_{1g}$ orbital, for that matter, has the correct symmetry to combine with 4s and $3d_{z^2}$ orbitals of iron. These three orbitals overlap with each other and then split into three molecular orbitals, a_{1g} (bonding), a'_{1g} (non-bonding) and a^*_{1g} (anti-bonding).

The $LGOa_{2u}$ orbital can formally combine with $4p_z$ orbital of iron. There is a wide difference between the energies of these two orbitals. Nevertheless, the combination does take place and results in the formation of two molecular orbitals, a_{2u} (bonding) and a^*_{2u} (anti-bonding). However, the a_{2u} orbital has a very little component of $4p_z$ orbital and, for this reason, we have not drawn a line joining these two orbitals. Similarly, the a^*_{2u} molecular orbital has a very little contribution from the $LGOa_{2u}$ and, therefore, we have refrained from joining these orbitals in Fig. 29.19.

The two degenerate ligand group orbitals, $LGOe_{1g}$, have appropriate symmetry to enter into combination with $3d_{xz}$ and $3d_{yz}$ orbitals of iron. Two degenerate bonding molecular orbitals, designated as e_{1g} and two degenerate anti-bonding molecular orbitals, designated as e^*_{1g} are formed as a result of this combination. Similarly, the two degenerate ligand group orbitals, $LGOe_{1u}$, combine with $4p_x$ and $4p_y$ orbitals of iron, giving two degenerate bonding molecular orbitals, e_{1u} and two degenerate anti-bonding molecular orbitals, e^*_{1u}.

The two degenerate ligand group orbitals, $LGOe_{2g}$, have the same symmertry as the $3d_{xy}$ and $3d_{x^2-y^2}$ orbitals of iron. Their combination gives four molecular

Table 29.3 Formation of ligand group orbitals of cyclopentadienyl rings and molecular orbitals of ferrocene

Bond type	Ligand orbitals $C_5H_5^- + C_5H_5^-$	Ligand group orbitals	Metal orbitals	Molecular orbitals of ferrocene
σ	a + a	$LGOa_{1g}$	4s, $3d_{z^2}$	a_{1g} – Bonding a'_{1g} – Non-bonding a^*_{1g} – Anti-bonding
		$LGOa_{2u}$	$4p_z$	a_{2u} – Bonding a^*_{2u} – Anti-bonding
π	$e_1 + e_1$	$LGOe_{1g}$ $LGOe_{1g}$	$3d_{xz}$ $3d_{yz}$	e_{1g} e_{1g} – Bonding e^*_{1g} e^*_{1g} – Anti-bonding
	$e_1 + e_1$	$LGOe_{1u}$ $LGOe_{1u}$	$4p_x$ $4p_x$	e_{1u} e_{1u} – Bonding e^*_{1u} e^*_{1u} – Anti-bonding
δ	$e_2 + e_2$	$LGOe_{2g}$ $LGOe_{2g}$	$3d_{xy}$ $3d_{x^2-y^2}$	e_{2g} e_{2g} – Bonding e^*_{2g} e^*_{2g} – Anti-bonding
	$e_2 + e_2$	$LGOe_{2u}$ $LGOe_{2u}$	No suitable orbitals	e_{2u} e_{2u} – Non-bonding

orbitals, two bonding, represented as e_{2g} and two anti-bonding, represented as e^*_{2g}. The symmetry of ligand group orbitals, $LGOe_{2u}$, does not match with any of the metal orbitals. Accordingly, they remain in non-bonding mode and are designated as e_{2u} molecular orbitals.

The combinations involved in the formation of molecular orbitals of ferrocene from ligand group orbitals of cyclopentadienyl groups and atomic orbitals of iron are shown in Table 29.3.

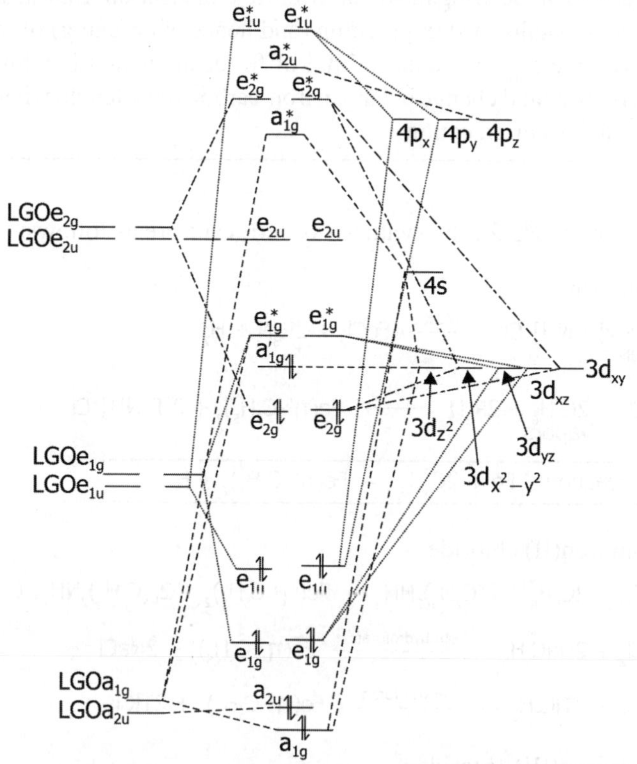

Fig. 29.19 Molecular orbital diagram of ferrocene

Ferrocene is an 18-electron system as illustrated below.

$$
\begin{aligned}
Fe(II) &= 6e^- \\
2C_2H_5 : 2 \times 6e^- &= 12e^- \\
\hline
&\quad 18e^-
\end{aligned}
$$

According to Fig. 29.19, the electronic configuration of ferrocene may be written as,

$$a_{1g}^2 a_{2u}^2 e_{1g}^4 e_{1u}^4 e_{2g}^4 a_{1g}'^2$$

This indicates the diamagnetic nature of ferrocene.

Case Study 29.11 Will the oxidation of ferrocene to ferrocenium cation cause a substantial change in iron-carbon bond length?

Ferrocene is an 18-electron system and, as shown in Fig. 29.19, the highest occupied molecular orbital is a'_{1g}. Oxidation of ferrocene results in removal of one of the two electrons in a'_{1g} orbital, leaving the 17-electron ferrocenium cation, $[Fe(\eta^5\text{-}C_5H_5)_2]^+$. Removal of an electron from a bonding molecular orbital leads to weakening (and hence lengthening) of metal-ligand bond, while loss of electron from an anti-bonding molecular orbital results in strengthening (and hence, shortening) of metal-ligand bond. However, the a'_{1g} molecular orbital of ferrocene is non-bonding in nature. Therefore, no substantial change occurs in iron-carbon bond length when ferrrocene is oxidized to ferrocenium cation.

Synthesis

The important methods for the synthesis of ferrocene are as follows:

1. From iron

$$\underset{\text{Powder}}{Fe} + 2[R_3NH]^+Cl^- \xrightarrow{773\text{ K}} FeCl_2 + 2R_3N + H_2$$

$$FeCl_2 + \underset{\text{Vapors}}{2C_5H_6} + 2R_3N \longrightarrow Fe(\eta^5\text{-}C_5H_5)_2 + 2[R_3NH]^+Cl^-$$

$$\text{Net reaction: } Fe + 2C_5H_6 \rightarrow Fe(\eta^5\text{-}C_5H_5)_2 + H_2$$

2. From iron(II) chloride

$$FeCl_2 + 2C_5H_6 + 2(C_2H_5)_2NH \rightarrow Fe(\eta^5\text{-}C_5H_5)_2 + 2[(C_2H_5)_2NH_2]^+Cl^-$$

$$FeCl_2 + 2NaC_5H_5 \xrightarrow{\text{Tetrahydrofuran}} Fe(\eta^5\text{-}C_5H_5)_2 + 2NaCl$$

$$FeCl_2 + 2TlC_5H_5 \xrightarrow{\text{Tetrahydrofuran}} Fe(\eta^5\text{-}C_5H_5)_2 + 2TlCl$$

3. From iron(III) chloride

$$2FeCl_3 + 2C_5H_5MgCl \rightarrow 2[Fe(\eta^5\text{-}C_5H_5)]^+Cl^- + 2MgCl_2 + Cl_2$$

$$[Fe(\eta^5\text{-}C_5H_5)]^+Cl^- + C_5H_5MgCl \rightarrow Fe(\eta^5\text{-}C_5H_5)_2 + MgCl_2$$

Reactions

The following are the important reactions of ferrocene:

1. **Friedel–Crafts acylation** In presence of aluminium(III) chloride catalyst, ferrocene undergoes Friedel–Crafts acylation to yield 1,1′-diacetylferrocene. Since acyl group is ring deactivating, 1,2-diacetylferrocene is not formed.

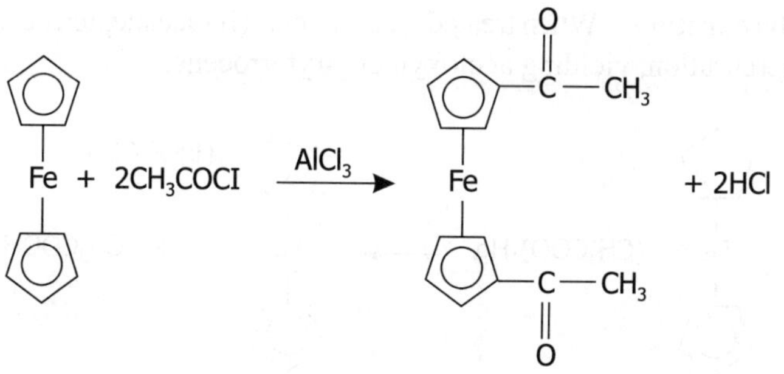

2. **Mannich reaction** Analogous to aromatic compounds, ferrocene undergoes Mannich reaction on treatment with formaldehyde and a primary or secondary amine. The reaction results in amino alkylation of ferrocene.

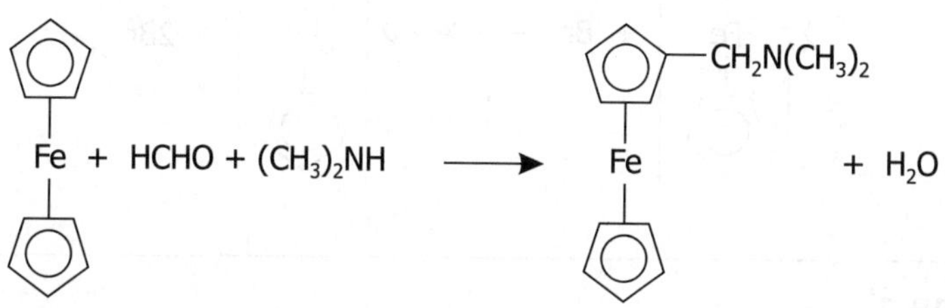

3. **Metallation** Ferrocene reacts with methyllithium(I) in 1:2 molar ratio to produce 1,1′-dilithioferrocene.

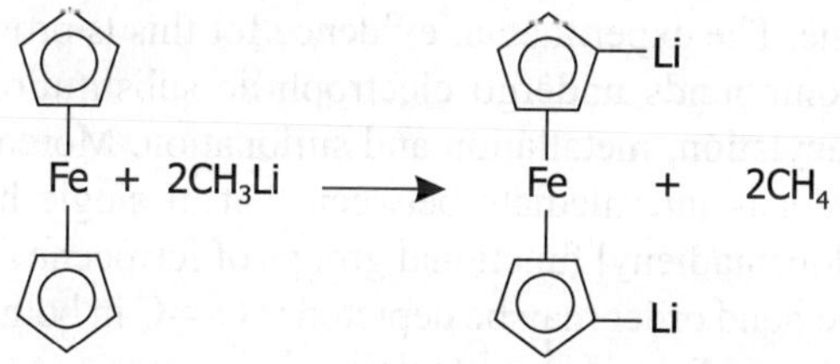

4. **Sulfonation** Ferrocene reacts with concentrated sulfuric acid to give ferrocenesulfonic acid.

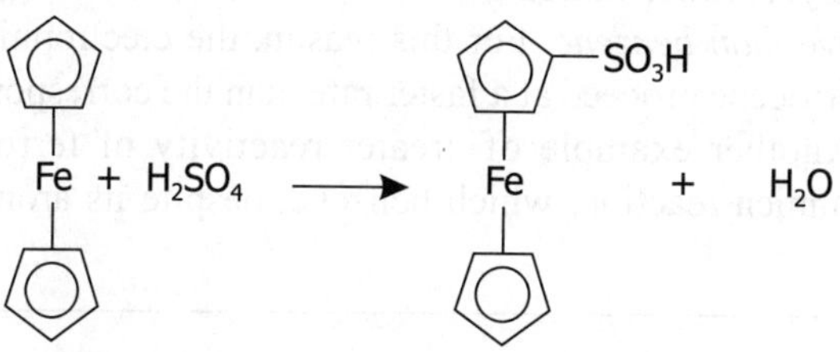

5. **Mercuration** When treated with mercury(II) acetate, ferrocene undergoes mercuration, yielding acetoxymercuryferrocene.

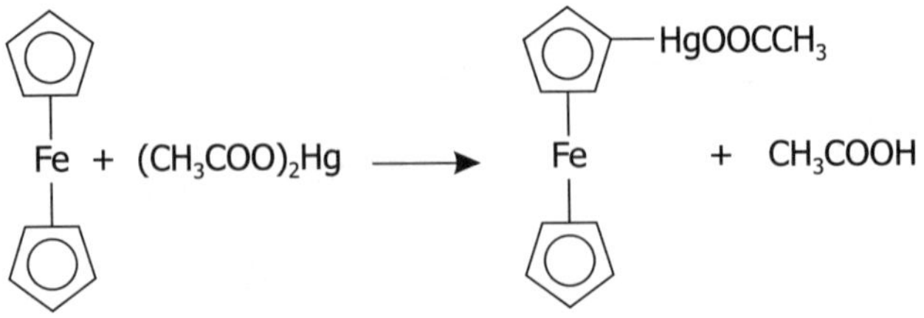

6. **Oxidation** Bromine oxidizes ferrocene to ferrocenium ion.

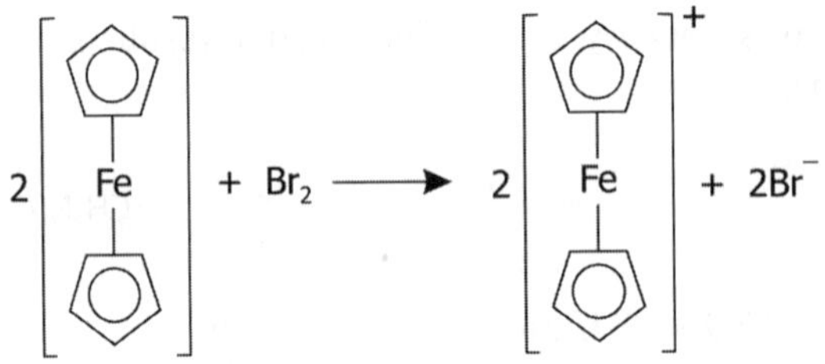

Box 29.3

A Comparison of Ferrocene and Benzene

The cyclopentadienyl groups in ferrocene possess aromatic character similar to that in benzene. The experimental evidence for this is provided by the fact that both the compounds undergo electrophilic substitution reactions like Friedel–Crafts acylation, metallation and sulfonation. Moreover, the carbon-carbon bond order is intermediate between that of single bond and double bond in the cyclopentadienyl functional groups of ferrocene and benzene. It is pertinent that the bond order may be depicted is C$\overset{\cdots}{=}$C in both the compounds. Further, like benzene, ferrocene is stable in air.

Both benzene and cyclopentadienyl rings possess 6π electrons. However, the π-cloud in benzene is spread over six carbon atoms, while that in cyclopentadienyl is concentrated over five carbon atoms. As a result, *ferrocene is more reactive than benzene*. For this reason, the electrophilic substitution reactions in ferrocene proceed at a faster rate than the corresponding reactions in benzene. Another example of greater reactivity of ferrocene is that it undergoes Mannich reaction, which benzene, despite its aromatic character, does not.

Case Study 29.12 Why are cobaltocene and nickelocene less stable than ferrocene?

The total number of electrons in the three metallocenes are counted as follows:

$[Fe(\eta^5\text{-}C_5H_5)_2]$		$[Co(\eta^5\text{-}C_5H_5)_2]$		$[Ni(\eta^5\text{-}C_5H_5)_2]$	
Fe(II) : $3d^6$	= $6e^-$	Co(II) : $3d^7$	= $7e^-$	Ni(II) : $3d^8$	= $8e^-$
$2C_5H_5^-$: 2 x $6e^-$	= $12e^-$	$2C_5H_5^-$: 2 x $6e^-$	= $12e^-$	$2C_5H_5^-$: 2 x $6e^-$	= $12e^-$
	18e		$19e^-$		$20e^-$

Obviously, ferrocene is stable since it follows the 18-electron rule. In contrast, cobaltocene and nickelocene are 19-and 20-electron systems, respectively. Figure 29.19 shows that the 19th and 20th electrons would be placed in high energy, anti-bonding e^*_{1g} orbitals. These electrons are easily oxidizable. Hence, cobaltocene and nickelocene are less stable than ferrocene.

Box 29.4

Fluxional Behavior of Organometallics

Several organometallic compounds are stereochemically nonrigid. Their molecules can rearrange to more than one equivalent configurations having comparable energies. When the energy barrier between two or more configurations can be overcome thermally, the system is said to be fluxional.

For example, in the complex, bis(η^1-cyclopentadienyl)bis(η^5-cyclopentadienyl)titanium(IV), $[(\eta^1\text{-}C_5H_5)_2(\eta^5\text{-}C_5H_5)_2Ti]$, there is a rapid interconversion between monohapto and pentahapto cyclopentadienyl ligands at room temperature. It is only at about 200 K, that the interconversions stop and two monohapto and two pentahapto ligands can be identified.

Likewise, when the complex, tris(cyclopentadienyl)scandium(III), $[Sc(C_5H_5)_3]$ is characterized at 240 K, it becomes evident that there are two pentahapto and one monohapto ligands in its structure. At room temperature, there is continuous switching between η^1-and η^5-cyclopentadienyl rings, so that all three ligands appear to be the same.

As depicted in Fig. 29.20, the metal ion in tetracarbonyl(tetramethylallene) iron(II), migrates over the allene π system, thus showing fluxional behaviour.

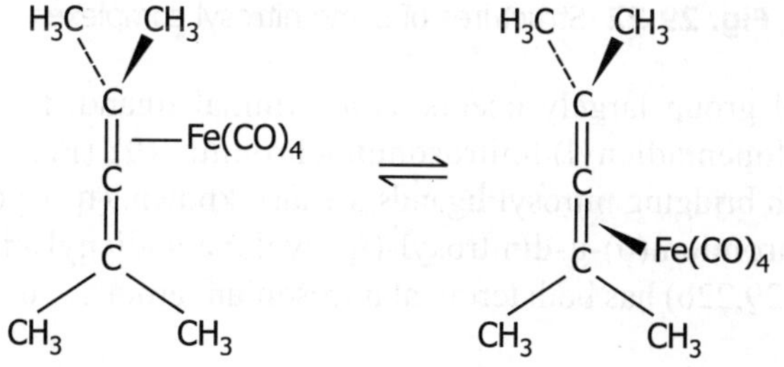

Fig. 29.20 Migration of iron(II) ion over π-allene system

Contd...

At about 210 K, the two structures retain their respective identities.

In (η^1-cyclopentadienyl)trimethylgermanium(IV), $[Ge(\eta^1-C_5H_5)(CH_3)_3]$, the metal ion continuously—and rapidly—shifts form one carbon atom to the adjacent carbon atom (Fig. 29.21).

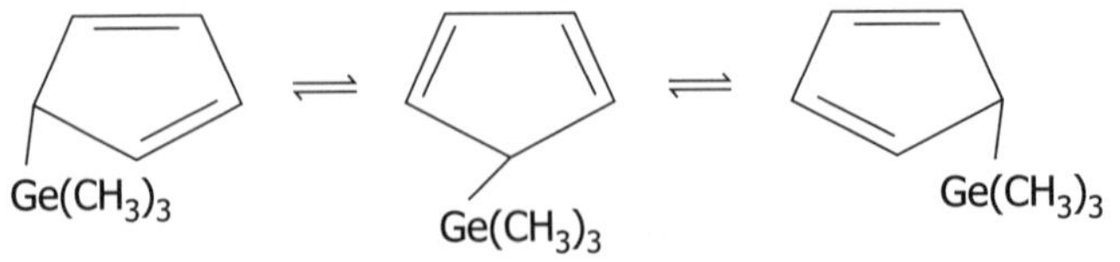

Fig. 29.21 Fluxional process in germanium(IV) complex

This, so called 1, 2-shift, gives rise to several configurations—and fluxionality.

Nitrosyl Organometallics

Ideally, nitrosyl complexes do not come under the category of organometallic compounds. The nitrosyl ligand coordinates to the metal through nitrogen, whereas in organometallic compounds, the presence of a metal–carbon bond is essential. Yet, for two reasons, the metal nitrosyls are included in the organometallic classification. First, the nitrosyl cation, NO^+, is isoelectronic with carbonyl group, CO. Therefore, the bonding modes of the two groups are quite alike. Second, complexes involving only nitrosyl groups are extremely rare; however, mixed carbonyl–nitrosyl complexes are quite common. Examples include pentacarbonylnitrosoniumvanadium(–I), $[V(NO)(CO)_5]$, tetracarbonylnitrosoniummanganese(–I), $[Mn(NO)(CO)_4]$ and tricarbonylnitrosoniumcobalt(–I), $[Co(NO)(CO)_3]$. Such compounds do come within the ambit of organometallics.

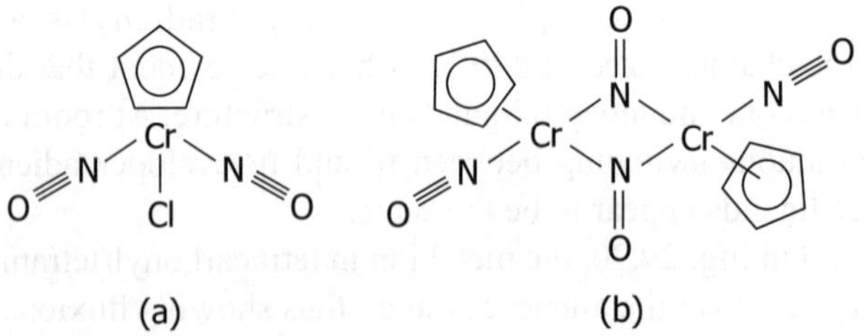

Fig. 29.22 Structures of some nitrosyl complexes

The nitrosyl group largely occurs as a terminal ligand, for example, as in chloro(η^5-cyclopentadienyl)dinitrosoniumchromium(0) (Fig. 29.22a). A few complexes with bridging nitrosyl ligands are also known. (η^5-Cyclopentadienyl)-nitrosoniumchromium(0)-μ-dinitrosyl-(η^5-cyclopentadienyl)nitrosoniumchromium(0) (Fig. 29.22b) has both terminal nitrosonium groups and bridging nitrosyl groups.

Structures

When a nitrosyl group becomes coordinated to a metal, the following skeletal structures may prevail:

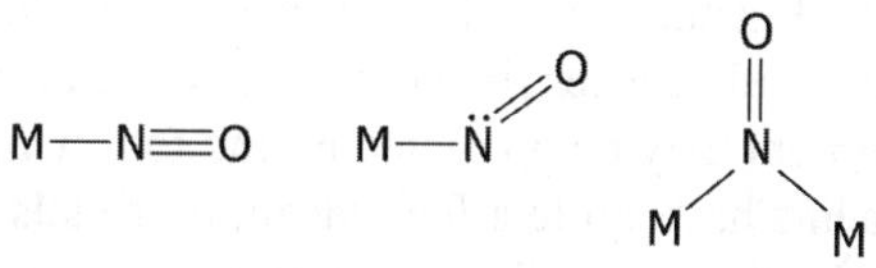

I II III

In structure I, the nitrogen of the nitrosyl group is sp hybridized. The ligand has a linear geometry. In structure II, the nitrogen is sp^2 hybridized and the nitrosyl group is bent. In structure III too, the nitrogen is sp^2 hybridized and the nitrosyl ligand is bridging. The linear and bent nitrosyl groups always have terminal positions. It is possible that one particular complex may have both linear and bent nitrosyl groups. Chlorodinitrosoniumbis(triphenylphosphine)ruthenium(0) ion, $[RuCl(NO)_2\{P(C_6H_5)_3\}_2]^+$ (Fig. 29.23) is an example of such a complex.

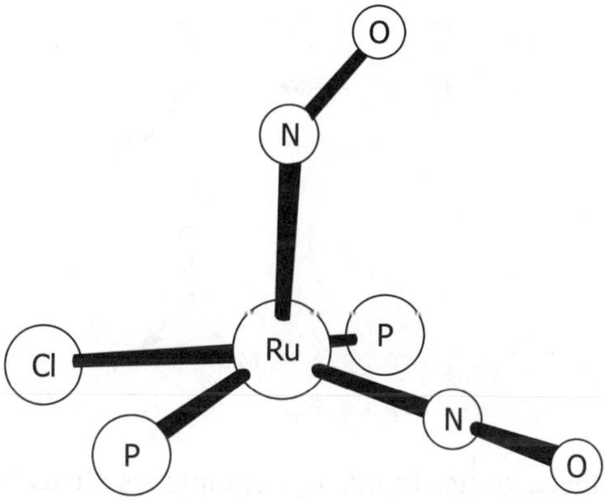

Fig. 29.23 A complex having both linear and bent nitrosyl groups

Bonding

In nitrosyl group, the constituent atoms, nitrogen and oxygen, have lone pairs and can be potential donors towards the metal. However, it is the nitrogen that acts as the ligating atom. This is because a formal positive charge would be placed on oxygen if it coordinates to the metal. Positive charge on a highly electronegative element such as oxygen would lead to instability.

In most complexes, the nitrosyl group has a linear geometry (structure I). In such cases, the ligand is a 3-electron donor. Initially, one electron is transferred to the metal. The latter is reduced from M^{n+} to $M^{(n-1)+}$ state. The nitrosyl ligand is converted into nitrosonium cation, NO^+. Next, the nitrogen atom of the nitrosonium cation donates two electrons to the metal, forming a σ-bond. This is followed by back donation of two electrons from the metal to the ligand, resulting in a π-bond. This phenomenon may be termed as *two-electron–σ-donor, two-electron–π-acceptor concept.*

In cases where nitrosyl group has a bent geometry (structure II), the nitrogen atom is a 1-electron donor. The metal too donates one electron and by mutual sharing a σ-bond is formed. The remaining two electrons remain localized on nitrogen. In fact, the presence of these two electrons on nitrogen is responsible for the bent structure of the ligand, just as the presence of two lone pairs of electrons on oxygen of water molecule causes the latter to be angular. Whether the nitrosyl group will be linear or bent may be predicted by the following generalization: the earlier transition elements have quite a few vacant d orbitals and, therefore, have the capacity to accept electrons. In such cases the nitrosyl ligand acts as a 3-electron donor and has a linear geometry. In the latter transition elements, the d orbitals are nearly filled-up. In such cases, the nitrosyl ligand acts as a 1-electron donor and has a bent geometry.

When nitrosyl acts a bridging ligand (structure III), its nitrogen atom is sp^2 hybridized. Here, the ligand is a 3-electron donor. It becomes bound to one of the metal atoms by sharing two electrons—one donated by that metal atom and the other by nitrogen. It then donates the remaining two electrons to the second metal atom and completes the bridge.

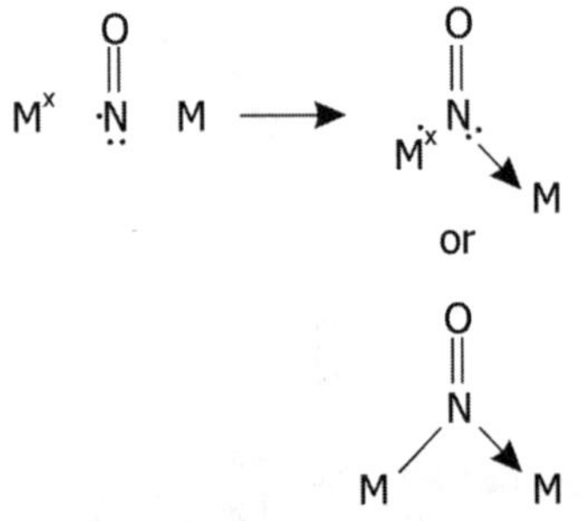

Synthesis

Some important synthetic routes to nitrosyl complexes are as follows:

1. Replacement reactions

$$Fe(CO)_5 + 2NO \rightarrow [Fe(CO)_2(NO)_2] + 3CO$$
$$Co_2(CO)_8 + 2NO \rightarrow 2[Co(CO)_3(NO)] + 2CO$$

2. Addition reactions

$$K_2[PtCl_4] + NOCl \rightarrow K_2[Pt(NO)Cl_5]$$
$$[Co\{P(C_6H_5)_3\}_2Cl_2] + NO \rightarrow [Co\{P(C_6H_5)_3\}_2Cl_2(NO)]$$

3. Oxide ion abstraction

$$[Fe(CN)_5(NO_2)]^{4-} + 2H^+ \rightarrow [Fe(CN)_5(NO)]^{2-} + H_2O$$

4. Oxygen atom abstraction

$$[Fe(CO)_5] + KNO_2 \rightarrow K[Fe(CO)_3(NO)] + CO + CO_2$$

5. Reductive nitrosylation

$$2CoCl_2 + 2CH_3OH + 2CH_3NH_2 + 6NO \rightarrow [CoCl(NO)_2]_2 + 2[CH_3NH_3]^+Cl^- + 2CH_3ONO$$

Case Study 29.13 What are Roussin's salts?

Roussin's salts refer to two series of iron complexes involving nitrosyl and sulfido ligands or nitrosyl and sulfido ester ligands. The composition of the first series or red salts is $[Fe(NO)_2S)]_2^{2-}$ or $[Fe(NO)_2(SR)]_2^{2-}$. The composition of the second series or black salts is $[Fe_4(NO)_7S_3]^-$ or $[Fe_4(NO)_7(SR)_3]^-$ (R is an alkyl or an aryl group). Roussin's salts are diamagnetic in nature.

Reactions

Nitrosyl complexes show the following types of reactions:

1. **Substitution reactions** As has been stated earlier, the nitrosyl ligand is a 3-electron donor, while the carbonyl group is a 2-electron donor. In electron-counting terms, two nitrosyl groups are equivalent to three carbonyl groups, i.e., 2NO = 3CO. This relationship holds good in reactions in which the carbonyl groups are replaced by nitrosyl ligands.

$$[Co(CO)_3(NO)] + 2NO \rightarrow [Co(NO)_3] + 3CO$$
$$[Mn(CO)_4(NO)] + 2NO \rightarrow [Mn(CO)(NO)_3] + 3CO$$

In mixed carbonyl–nitrosyl complexes, it is always the carbonyl ligand which is replaced during the substitution reactions.

$$[RhCl_2(NO)(CO)_2] + 2P(C_6H_5)_3 \rightarrow [RhCl_2(NO)\{P(C_6H_5)_3\}_2] + 2CO$$

This proves that the metal–NO bond has greater bond energy than the metal–CO bond.

2. **Depolymerization reactions**

$$[Fe(NO)_2Cl]_2 + 2P(C_6H_5)_3 \rightarrow 2[Fe(NO)_2Cl\{P(C_6H_5)_3\}]$$
$$[Ni(NO)Cl]_4 + 4P(C_6H_5)_3 \rightarrow 2[Ni(NO)Cl\{P(C_6H_5)_3\}]_2$$

Case Study 29.14 What is the composition of the brown-ring compound formed in the qualitative analysis of nitrate anion?

The nitrate anion is tested by adding an aqueous solution of an iron(II) salt followed by concentrated sulfuric acid. The following reactions take place:

$$2NaNO_3 + \underset{\text{Conc.}}{H_2SO_4} \rightarrow Na_2SO_4 + 2HNO_3$$

$$6FeSO_4 + \underset{\text{Conc.}}{3H_2SO_4} + 2HNO_3 \rightarrow 3Fe_2(SO_4)_3 + 4H_2O + 2NO$$

$$FeSO_4 + 5H_2O + NO \rightarrow \underset{\text{Brown ring}}{[Fe(H_2O)_5NO]SO_4}$$

The brown-ring compound involves nitrosonium ligand, NO^+. Here, iron is in oxidation state I and thus the brown-ring compound is named pentaquanitrosoniumiron(I) sulfate. Its color is attributed to charge-transfer phenomenon.

Organometallic Catalysts

Several organometallic derivatives serve as catalysts for reactions that have industrial and/or commercial importance. As with other examples of catalysis, the organometallic catalysts are grouped in two categories: homogenous and heterogenous. In homogenous systems the catalyst is in solution state and so are all the reactants and the products. In heterogenous systems, the reactants and the products are in dissolved/vapor states, while the catalysts are solid materials. Such catalysts have active sites on their surface which serve as a point of contact for the reactant molecules.

Important examples of organometallic catalysis may be classified as follows. Homogenous catalysis: Wacker process; Hydrogenation by Wilkinson's catalyst; Hydroformylation process; Production of synthesis gas using carbonyl compounds.

Heterogenous catalysis: Ziegler-Natta polymerization; Fischer-Tropsch process.

The mechanisms of these industrially important processes are now described.

Wacker process

Also called *Smidt process*, the Wacker process is used to synthesize acetaldehyde from ethene. The overall process may be depicted as follows.

$$H_2C = CH_2 + H_2O \xrightarrow{[PdCl_4]^{2-}} CH_3CHO$$

Tetrachloropalladate(II) ion, $[PdCl_4]^{2-}$ acts as a catalyst in this synthesis. It forms as organometallic compound, trichloro(η^2-ethene)palladate(II) ion, by reacting with ethene. The intermediate organometallic derivative plays a decisive role in the synthetic procedure. Copper(II) chloride is also one of the raw materials which assists in the regeneration of the catalyst.

The following steps are involved in the preparation of acetaldehyde from ethene by Wacker process.

a. Substitution of ethene

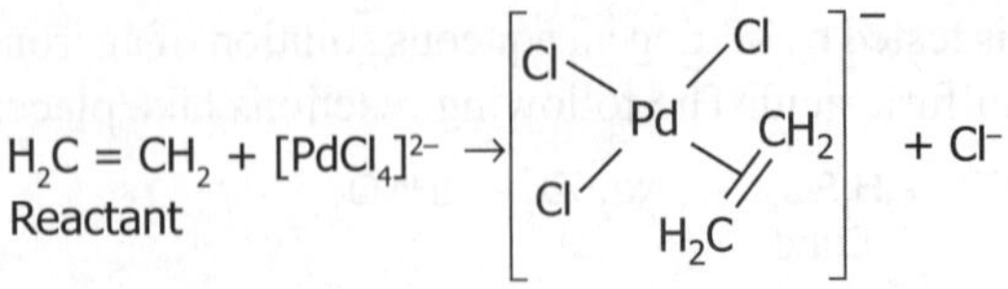

b. Substitution of water

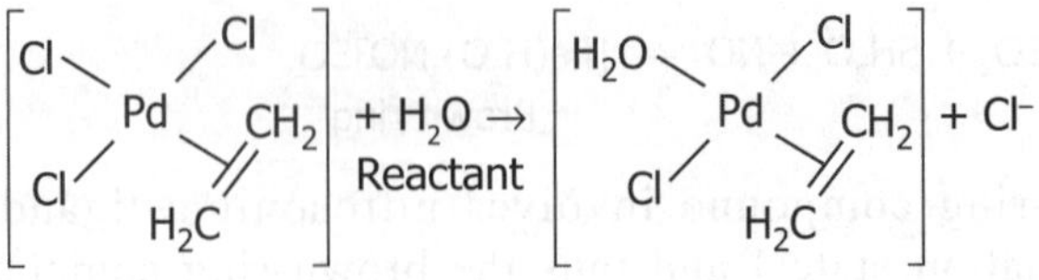

c. Addition to double bond

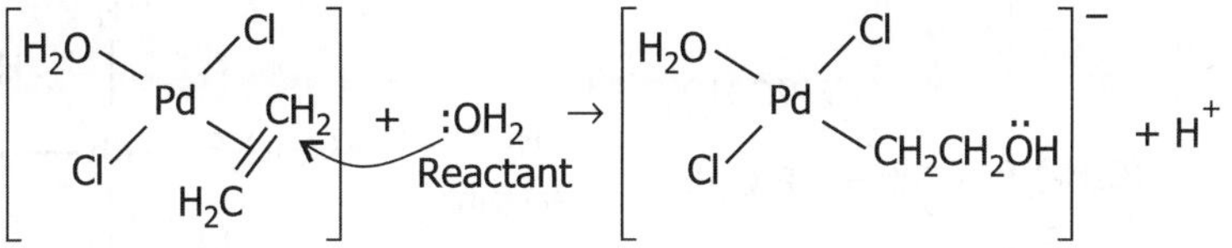

d. Elimination

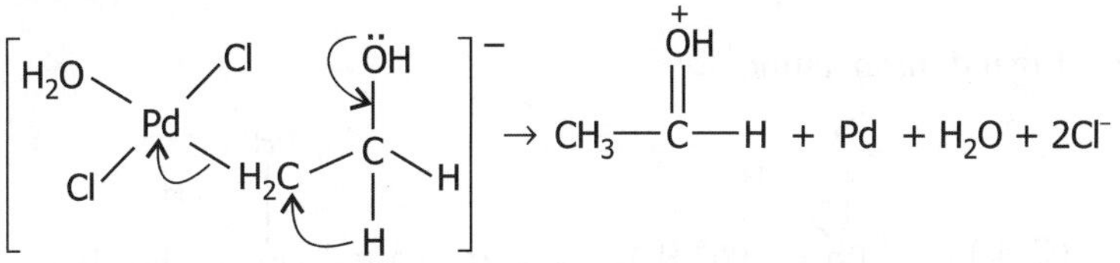

e. Rearrangement

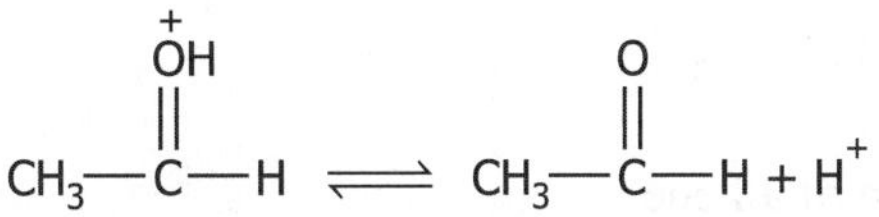

f. Catalyst regeneration

A mixture of copper(II) chloride and hydrochloric acid regenerates the catalyst by the following series of reactions.

$$CuCl_2 + 2Cl^- \rightarrow [CuCl_4]^{2-}$$

$$Pd + 2[CuCl_4]^{2-} \rightarrow [PdCl_4]^{2-} + [CuCl_2]^-$$

(From step d)

$$2[CuCl_2]^- + \frac{1}{2}O_2 + 2H^+ + 4Cl^- \rightarrow 2[CuCl_4]^{2-} + H_2O$$

Alkene hydrogenation by Wilkinson catalyst

Wilkinson catalyst, chlorotris(triphenylphosphine)rhodium(I), $ClRh\{P(C_6H_5)_3\}_3$ is not an organometallic compound. However, its characteristics are similar to many 4-coordinated organometallics. For this reason the reaction between alkenes and hydrogen, the rate of which it enhances, is classified under organometallic catalysis. The overall reaction may be displayed as,

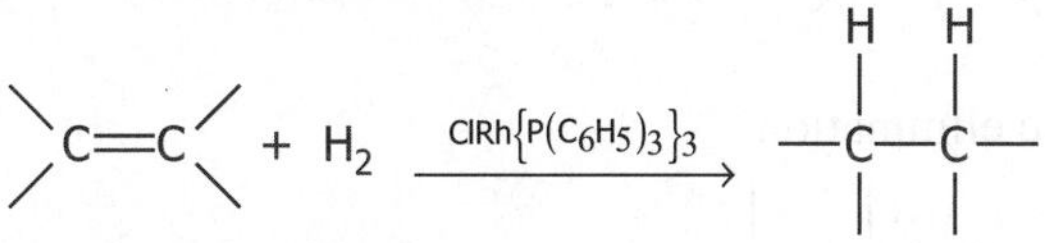

The reaction proceeds in the following sequence.

a. Oxidative addition

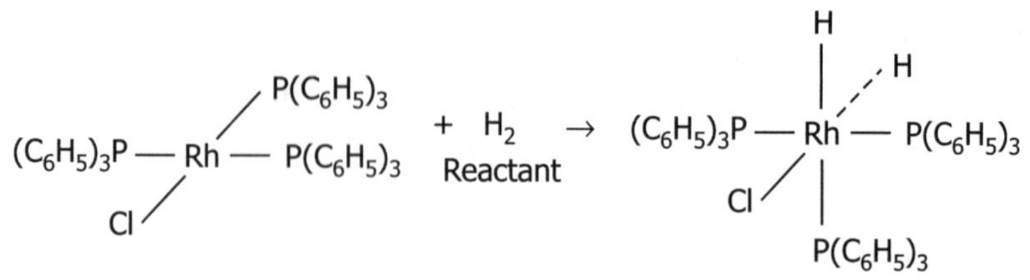

b. Ligand dissociation

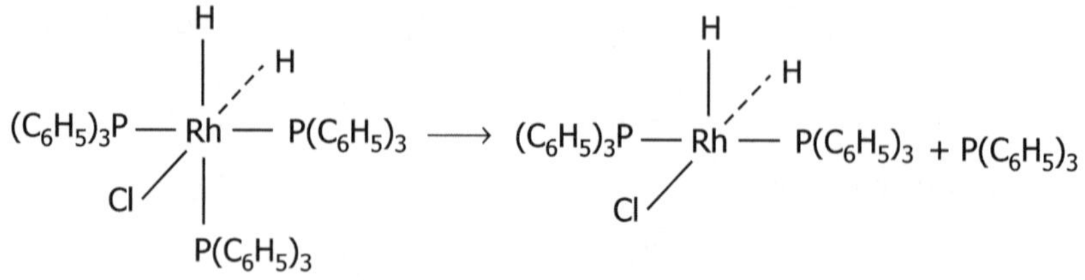

c. Coordination of alkene

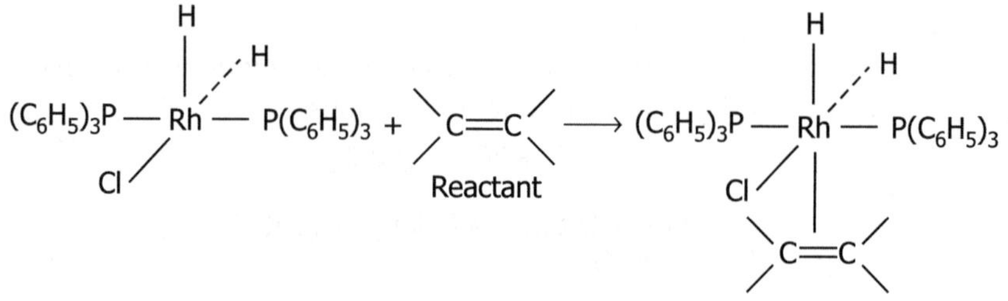

d. 1, 2-Insertion

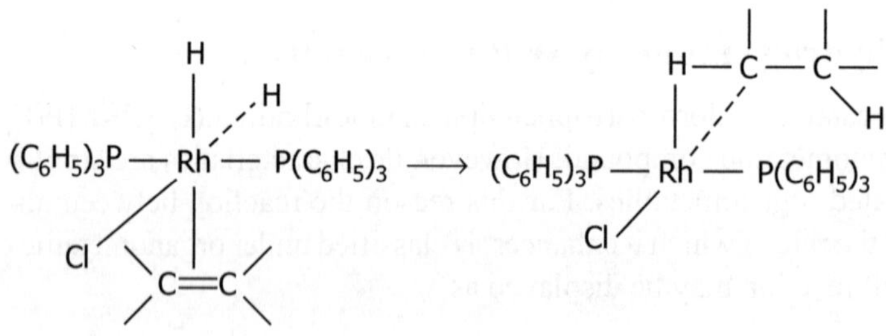

e. Reductive elimination

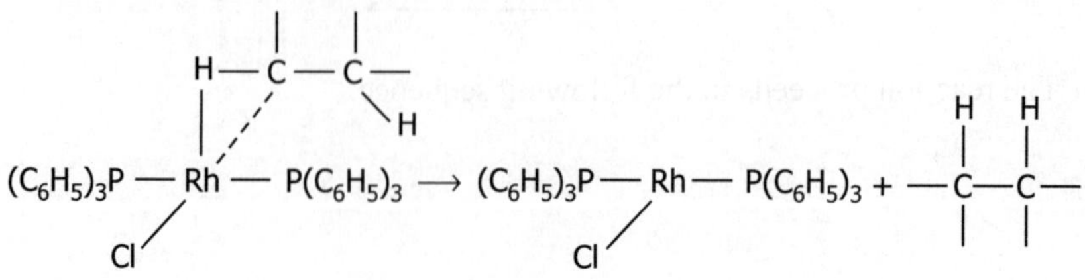

f. Catalyst regeneration

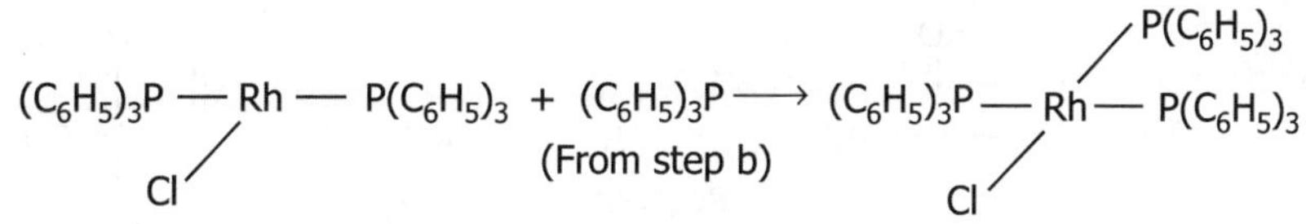

Step d, which involves 1, 2-insertion, is the slowest, rate determining step. In alkenes containing several double bonds, the one which is least hindered will be hydrogenated by wilkinson catalyst.

Hydroformylation process

The hydroformylation process, which is sometimes also referred to as the *oxo process* is a synthetic procedure for converting an alkene into an aldehyde. The aldehyde so formed has one more carbon atom than the alkene. The reaction between the alkene, carbon monoxide and hydrogen is carried out at high temperatures and pressure. It is catalyzed by tetracarbonnylhydridocobalt(I), $HCo(CO)_4$. The overall reaction may be written as,

$$R_2C = CH_2 + CO + H_2 \xrightarrow{HCo(CO)_4} R_2CH{-}CH_2CHO$$

It is carried out in the following steps.

a. Elimination of carbon monoxide

$$HCo(CO)_4 \rightarrow HCo(CO)_3 + CO$$

b. Coordination of alkene

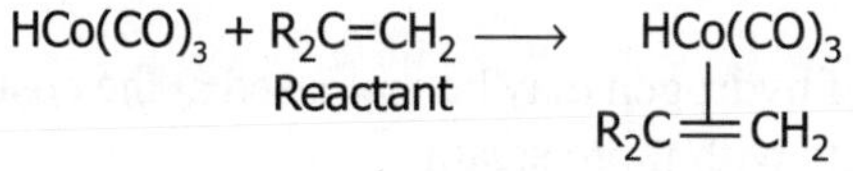

c. 1, 2-Insertion

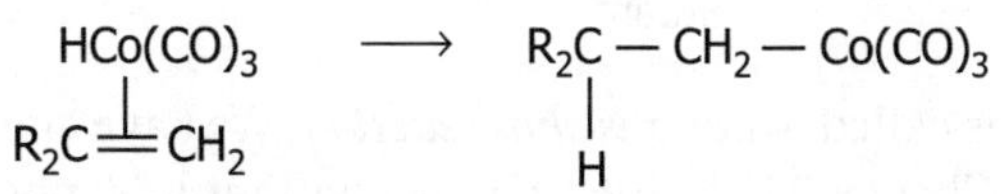

d. Addition of carbon monoxide

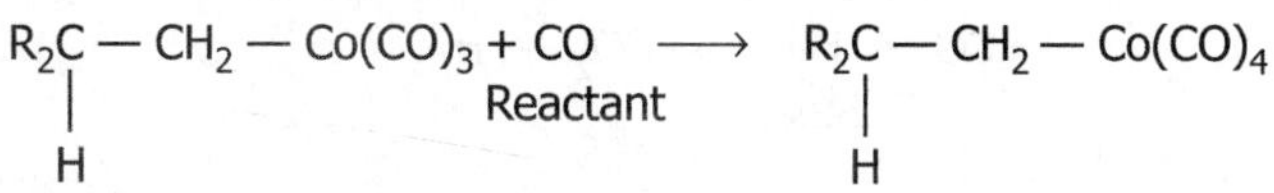

e. Rearrangement

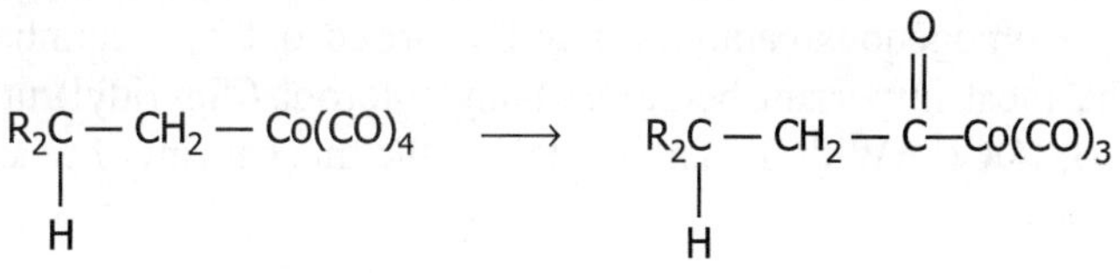

f. Oxidative addition of hydrogen

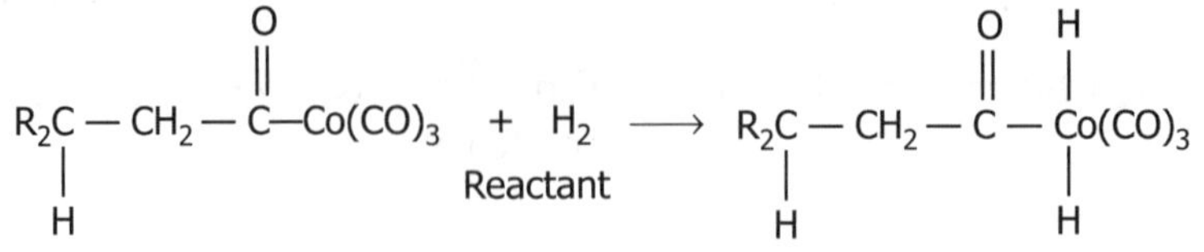

g. Reductive elimination

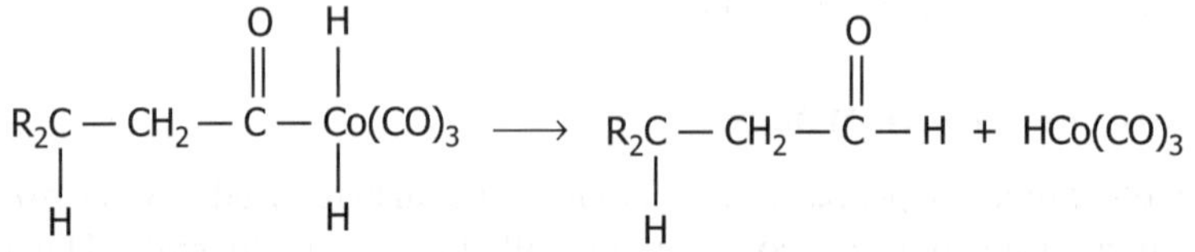

h. Catalyst regeneration

$$\underset{}{HCo(CO)_3} + \underset{\text{(From step a)}}{CO} \longrightarrow HCo(CO)_4$$

Step b in this sequence follows first order kinetics and is the slowest, rate-determining step.

Production of synthesis gas

Synthesis gas is a mixture of carbon monoxide and hydrogen. It is used both as a fuel and as a reducing agent in metallurgical operations. It is prepared by the reaction of steam on red-hot coke.

$$C + H_2O \rightarrow CO + H_2$$

The concentration of hydrogen may be increased at the cost of carbon monoxide by reacting the products with more steam.

$$CO + H_2 + \underset{\text{Steam}}{H_2O} \rightarrow CO_2 + 2H_2 \qquad (29.1)$$

Reaction (29.1) is called *water gas shift reaction*. We have cited the significance of this reaction in Chapter 11, p. 406. We may write it in a more simplified form as,

$$CO + \underset{\text{Steam}}{H_2O} \rightarrow CO_2 + H_2 \qquad (29.2)$$

Reaction (29.2) may be speeded up both by heterogenous catalysts and homogenous catalysts. An example of heterogenous catalyst is the mixed iron oxide, Fe_3O_4. Homogenous catalysis may be carried out by a number of metal carbonyls, the most important being, carbonylchlorobis(dipyridyl)ruthenium(II) ion, $[Ru(dipy)_2COCl]^+$. With the latter catalyst, the mechanism of reaction (29.2) is as follows.

a. Substitution of water

$$[Ru(dipy)_2COCl]^+ + H_2O \rightarrow [Ru(dipy)_2CO(H_2O)]^{2+} + Cl^-$$

b. Substitution of carbon monoxide

$$[Ru(dipy)_2CO(H_2O)]^{2+} + CO \rightarrow [Ru(dipy)_2(CO)_2]^{2+} + H_2O$$

c. Reaction with water

$$[(CO)(dipy)_2Ru - C \equiv O]^{2+} + H_2O \rightarrow \left[(CO)(dipy)_2\,Ru - \underset{\displaystyle OH}{\underset{|}{C}} = O\right]^+ + H^+$$

d. Elimination of carbon dioxide

$$\left[(CO)(dipy)_2\,Ru - \underset{\displaystyle OH}{\underset{|}{C}} = O\right]^+ \rightarrow [Ru(dipy)_2(CO)H]^+ + CO_2$$

e. Substitution of chloride ion

$$\underset{\text{(From step a)}}{[Ru(dipy)_2(CO)H]^+ + Cl^-} \rightarrow [Ru(dipy)_2(CO)Cl]^+ + H^-$$

f. Production of hydrogen

$$\underset{\text{(From step c)}}{H^+} + \underset{\text{(From step e)}}{H^-} \rightarrow H_2$$

Alkene polymerization by Ziegler-Natta catalyst

Polyalkenes are a class of synthetic polymers having broad spectrum applications. The alkene polymerization is largely carried out with the aid of Ziegler-Natta catalyst. The catalyst itself is formed by the combination of titanium(IV) chloride and triethylaluminium(III). The product is a fine powder which acts as a heterogenous catalyst. The vacant coordination sites on titanium(IV) ion, which are marked by a circle, play an important role in the mechanistic pathway of alkene polymerization.

a. Alkylation of titanium(IV) chloride

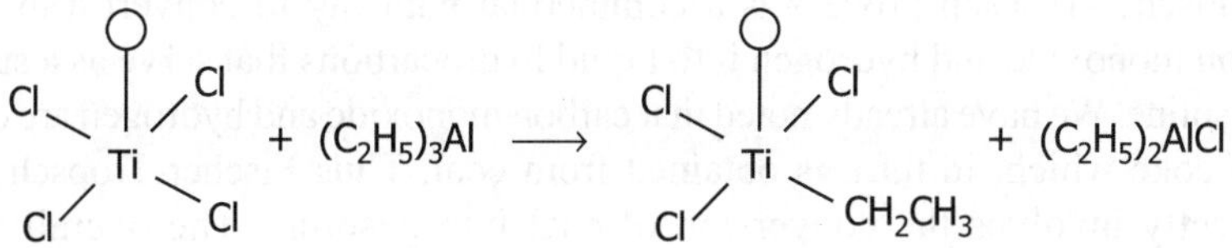

Catalyst

b. Coordination of alkene monomer at vacant site

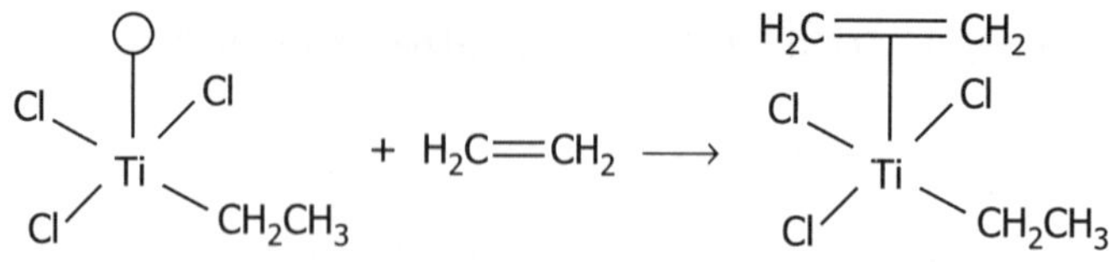

c. Migratory insertion

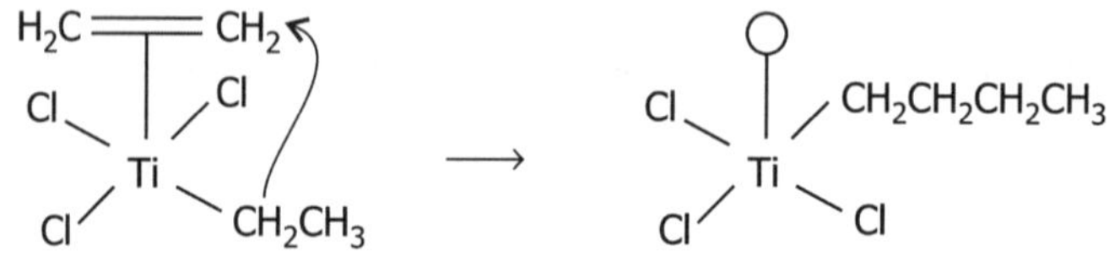

d. Chain propagation

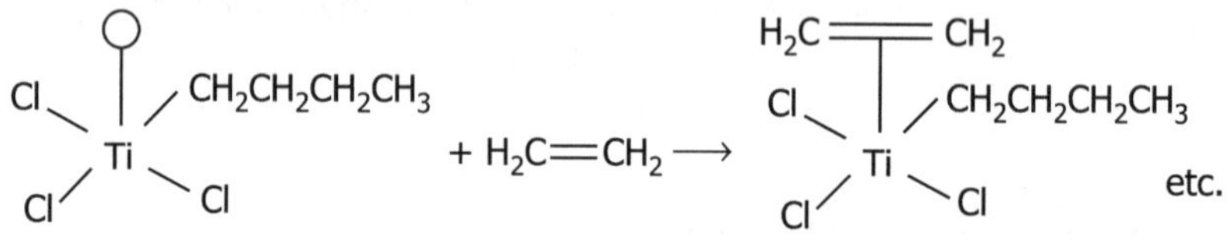

e. β-hydrogen elimination

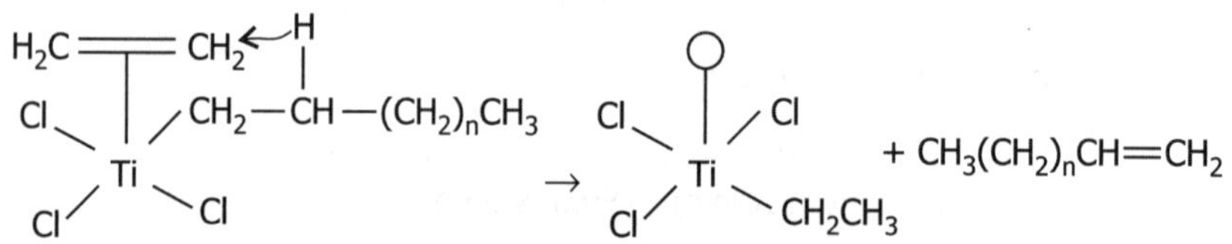

f. Catalyst regeneration

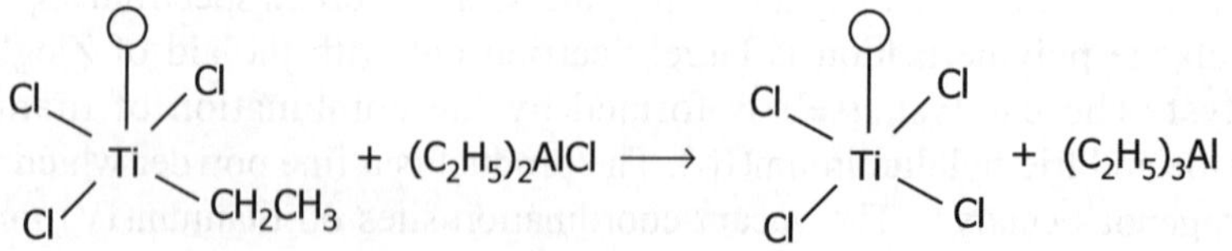

Fischer-Tropsch process

The Fischer-Tropsch process is a commercial pathway to convert a mixture of carbon monoxide and hydrogen into liquid hydrocarbons that serve as a substitute of gasoline. We have already noted that carbon monoxide and hydrogen are obtained from coke which, in turn, is obtained from coal. Thus Fischer-Tropsch process indirectly involves the conversion of coal into gasoline. The overall reaction underlining this methodology may be written as,

$$nCO + (2n + 1)H_2 \xrightarrow{\text{Catalyst}} C_nH_{2n+2} + nH_2O \qquad (29.3)$$

where n varies from 10 to 20. Most of the alkanes, so produced, are straight-chain hydrocarbons that are suitable as diesel fuel. Side reactions, producing alkenes, alcohols and other oxygenated organic compounds, also occur.

Reaction (29.3) is carried out in the temperature range 423–573 K and at a pressure of about 100 atmosphere. The reacting gases must be desulfurized since sulfur tends to poison the catalyst.

A variety of catalysts may be used in Fischer-Tropsch process, the most common ones being cobalt, iron and ruthenium. It need be emphasized that Fischer-Tropsch catalysts are not organometallic compounds, but transition metals. However, during the course of reaction (29.3), these metallic catalysts are converted into metal carbonyls—a reason why Fischer-Tropsch process is dealt with in the section of organometallic catalysis.

Reaction (29.3) may be dissected into the following steps.

a. Associative adsorption of carbon monoxide on catalyst.

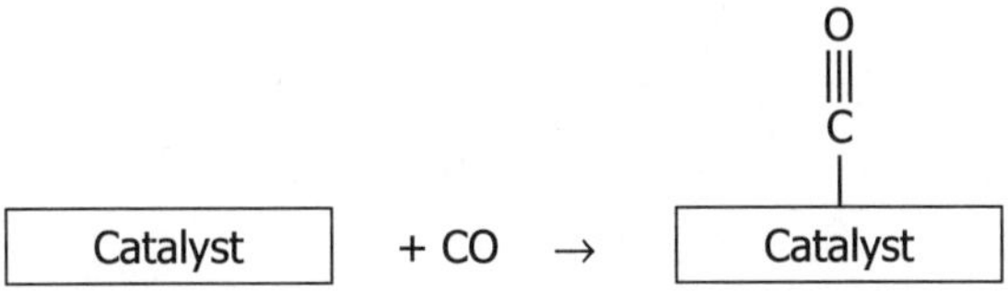

b. Splitting of carbon-oxygen bond to yield metal carbide and metal oxide.

c. Dissociative adsorption of two moles of hydrogen.

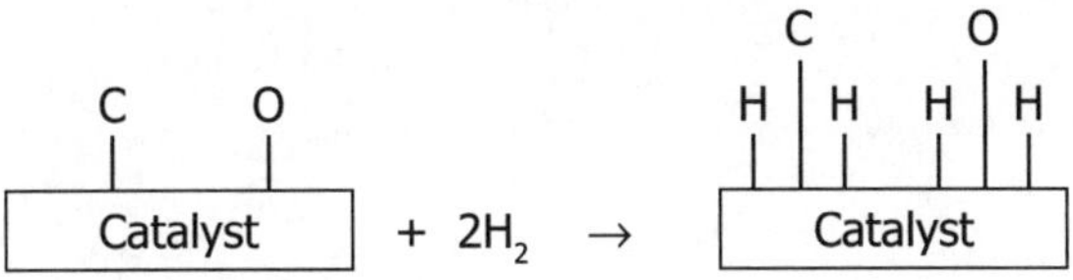

d. Transfer of two hydrogen atoms to oxygen to yield water and transfer of two hydrogen atoms to carbon to yield carbene. Simultaneous desorption of water.

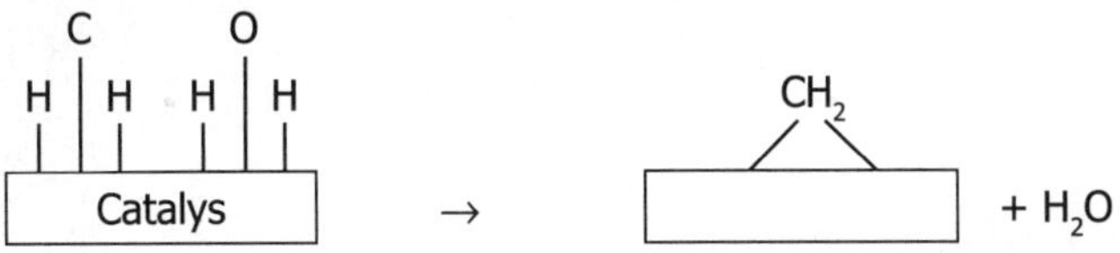

e. Propagation, involving migratory insertion of carbene units.

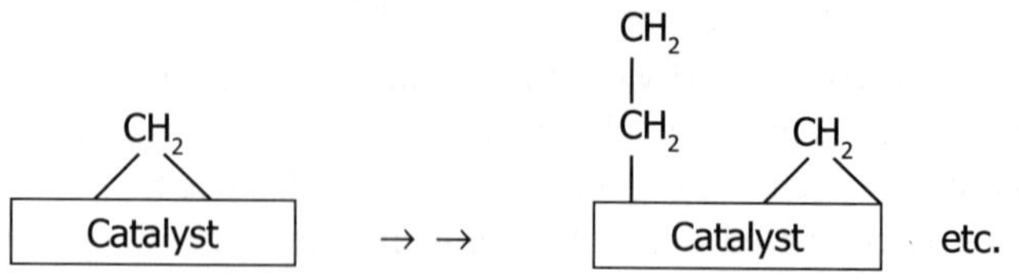

Termination involves the desorption of hydrocarbon mixture and regeneration of catalyst.

30. THE LANTHANOIDS

Introduction

The lanthanoids are those elements in which the 4f shell is progressively filled. Earlier, these elements were called *lanthanides*, but the nomenclature was changed since the *-ide* ending normally means an anion, for example, chloride or cyanide. The suggestion that these be named as *lanthanons* too had a few takers since the ending *-on* refers to metalloids, for example, boron or silicon.

There are a total of 14 lanthanoid elements, beginning with cerium and ending at lutetium. Being a part of the f-block elements, these are placed at the bottom of the main periodic table. However, the term "lanthanoid series" also includes lanthanum, which otherwise is housed in Group 3 of the periodic table.

The lanthanoids are a close knit family of elements—their physical and chemical properties being very similar to each other. The elements are highly electropositive and form predominantly ionic compounds. Sometimes these elements are referred to as *rare earths*, but the term is a misnomer. The lanthanoids are present in greater abundance than several other elements, such as bismuth, arsenic, or mercury, which are not considered rare. The lanthanoid elements occur together as a mixture of orthophosphates in a mineral called *monazite sand*. Lanthanum, cerium, praseodymium and neodymium constitute about 90% of monazite sand.

General Characteristics

Like most metals, the lanthanoids have a bright silvery appearance. Five of the elements—lanthanum, cerium, praseodymium, neodymium and europium—are extremely reactive. These combine with atmospheric oxygen to form an oxide coating that tarnishes the surface. For this reason, these metals are preserved under mineral oil. The remaining elements are relatively unreactive. For example, gadolinium and lutetium retain their silvery luster for a long time.

The lanthanoids react slowly with cold water and more rapidly with hot water to yield hydrogen gas. They burn readily in air to form the respective oxides.

Electronic Configuration

The electronic configurations of lanthanoids are depicted in Table 30.1. Lanthanum has the configuration, $[Xe]5d^1 6s^2$. The valence electron of cerium enters the 4f orbital, giving the configuration, $[Xe]5d^1 4f^1 6s^2$. Thereafter, it becomes energetically more favorable to move the 5d electron into the 4f orbital. The 5d electron reappears in case of gadolinium and lutetium. Gadolinium has a $5d^1$ arrangement because this results in a stable, half-filled $4f^7$ configuration. Lutetium has a $5d^1$ arrangement because the 4f shell is already completely filled.

Oxidation States

The characteristic group valency of lanthanoids is III. Oxidation states II and IV also occur in some cases, but these are always less stable than the group valency. Table 30.1 lists the possible oxidation states of lanthanoids.

Oxidation states II and IV are associated with the special stability of unfilled, half-filled and completely filled penultimate 4f shell. For example, cerium(IV) and terbium(IV) ions are quite stable because they possess $4f^0$ and $4f^7$ configurations, respectively. Similarly, both europium and ytterbium show oxidation state II because in this valence state they possess the stable $4f^7$ and $4f^{14}$ configurations, respectively. Further, II and IV valencies also occur for elements whose configurations in these states are close to $4f^0$, $4f^7$ or $4f^{14}$ arrangements. For example, samarium(II), with $4f^6$ configuration and thulium(II), with $4f^{13}$ configuration, are stable ions. So are praseodymium(IV) and neodymium(IV) with $4f^1$ and $4f^2$ arrangements, respectively.

Table 30.1 Electronic configurations and oxidation states of lanthanoids

Element	Symbol	Atomic number	Electronic configuration	Oxidation state(s)
Lanthanum	La	57	$[Xe]5d^16s^2$	III
Cerium	Ce	58	$[Xe]4f^15d^16s^2$	III, IV
Praseodymium	Pr	59	$[Xe]4f^36s^2$	III, IV
Neodymium	Nd	60	$[Xe]4f^46s^2$	III, IV
Promethium	Pm	61	$[Xe]4f^56s^2$	III
Samarium	Sm	62	$[Xe]4f^66s^2$	II, III
Europium	Eu	63	$[Xe]4f^76s^2$	II, III
Gadolinium	Gd	64	$[Xe]4f^75d^16s^2$	III
Terbium	Tb	65	$[Xe]4f^96s^2$	III, IV
Dysprosium	Dy	66	$[Xe]4f^{10}6s^2$	III
Holmium	Ho	67	$[Xe]4f^{11}6s^2$	III
Erbium	Er	68	$[Xe]4f^{12}6s^2$	III
Thulium	Tm	69	$[Xe]4f^{13}6s^2$	II, III
Ytterbium	Yb	70	$[Xe]4f^{14}6s^2$	II, III
Lutetium	Lu	71	$[Xe]4f^{14}5d^16s^2$	III

Case Study 30.1 What is *mischmetal*?

Mischmetal is an alloy of rare earth elements. It consists of approximately 50% cerium, 25% lanthanum, 15% neodymium and 10% of other lanthanoids. It is also called *cerium mischmetal* or *rare earth mischmetal*. It is used as a deoxidizer during the production of other alloys. It also serves to remove oxygen in vacuum tubes. When blended with iron oxide and magnesium oxide, its hardness increases. The harder material is called *ferrocerium*. The latter is used in the flint ignition device of lighters and torches.

Lanthanoid Contraction

The term lanthanoid contraction means a significant and steady decrease in the size of atoms and ions of lanthanoids, with increasing atomic number. Thus lanthanum has the largest atomic radius, while lutetium has the smallest. Lanthanoid contraction arises due to the imperfect shielding of one electron by the other in the same subshell. While traversing the lanthanoid series, the nuclear charge and the number of 4f electrons increase by one unit at each step. Since f orbitals have the lowest symmetry as compared to all other orbitals, they shield the nuclear charge imperfectly. Therefore, all the electrons are pulled closer to the nucleus, causing a shrinkage of size. The net decrease in size from lanthanum to lutetium is the lanthanoid contraction.

Case Study 30.2 Lanthanoid contraction is better reflected in trivalent ions than in neutral atoms of lanthanoids. Why?

The 4f orbitals are deeply buried beneath the 5s/5p and 6s orbitals and hence the screening power of 4f electrons is unity. Therefore, addition of an electron to the 4f orbital is exactly balanced by the addition of nuclear charge. No appreciable change in $Z_{effective}$ occurs. In case of trivalent ions, the 4f shell becomes a penultimate one and hence the screening power of a 4f electron is no longer unity. As a result, each time an electron is removed from the 4f shell, $Z_{effective}$ increases prominently, causing significant contraction.

Consequences of lanthanoid contraction

Following are the fallouts of lanthanoid contraction:

1. As the size decreases, the attraction between atoms in a metallic lattice increases. Consequently, the hardness, and the melting points and boiling points of lanthanoids increase from lanthanum to lutetium.
2. Lutetium(III) ion, being smallest in the series, is the most heavily hydrated, while lanthanum(III) ion, being the largest, is the least hydrated.
3. The ability to form complexes is the highest for lutetium(III) ion and the least for lanthanum(III) ion.

4. Lanthanum(III) hydroxide is the most basic, while lutetium(III) hydroxide is the least basic in the series.
5. The sizes of the 4d and 5d transition series elements of the same group are identical. Lanthanoids intervene the 4d and 5d series. Due to lanthanoid contraction, whatever increase is expected on moving from a 4d element to a 5d element of the same group, becomes negated. Thus, zirconium and hafnium, niobium and tantalum, molybdenum and tungsten have similar sizes. Their physical and chemical properties, consequently, are also similar.

Complexes

Lanthanoids form fewer complexes as compared to the transition elements. This may be attributed to two reasons. First, the lanthanoids are larger in size as compared to the transition elements. A smaller metal ion can interact better with a ligand and, therefore, forms complexes more readily. Second, the valence electrons of lanthanoid ions are present in 4f orbital and this orbital is buried beneath 5s and 5p orbitals. As a result, the metal–ligand interactions are relatively weak. In fact, lanthanoid complexes with cyano and carbonyl ligands are extremely rare. These ligands can enter into π-bonding with most transition elements. However, the 4f orbital in lanthanoids is so well shielded by overlying orbitals that it cannot participate in π-bonding with these ligands.

The amine complexes of lanthanoids cannot be synthesized in aqueous medium because water is a stronger base than amines. The most stable complexes are formed by chelating ligands such as ethylenediamine tetraacetic acid and acetylacetone. In contrast to transition element complexes, where coordination number 6 is most common, the lanthanoid complexes invariably possess coordination numbers 7, 8 or even 9. The relatively large size of lanthanoid ions makes possible the coordination of a greater number of ligands.

Color

Many lanthanoid(III) ions are colored in solid state as well as in aqueous solution. The color depends on the number of unpaired electrons. Ions with (n)f electrons and (14 – n)f electrons have similar colors. This means that the color sequence in the lanthanum to gadolinium series is repeated in the lutetium to gadolinium series (Table 30.2).

In lanthanoid ions, the 4f orbital is split into two sets of f orbitals, separated by a small energy difference, much the same way as transition metal ions show variation in energy between t_{2g} set and e_g set of d orbitals. Color arises when an unpaired electron absorbs visible light and is transferred from one 4f orbital to another, higher energy 4f orbital. This transfer is called f–f transition.

Table 30.2 Colors of trivalent lanthanoid ions

Ion	Number of 4f electrons	Color	Ion	Number of 4f electrons
La(III)	0	Colorless	Lu(III)	14
Ce(III)	1	Colorless	Yb(III)	13
Pr(III)	2	Green	Tm(III)	12
Nd(III)	3	Lilac	Er(III)	11
Pm(III)	4	Pink, Yellow	Ho(III)	10
Sm(III)	5	Yellow	Dy(III)	9
Eu(III)	6	Pale pink	Tb(III)	8
Gd(III)	7	Colorless	Gd(III)	7

The colors originating from f–f transitions are generally pale. This is because f–f transitions are forbidden—these are referred to as *Laporte forbidden transitions*. Such transitions do not involve a change in azimuthal quantum number, i.e., $|\Delta l| = 0$ (Fig. 30.1a). An allowed transition is one which involves a change in azimuthal quantum number, i.e., $|\Delta l| \neq 0$ (Fig. 30.1b). Such transitions give intense colors.

Transition type

(a)	(b)
4f → 4f	4f → 5d
l 3 3	l 3 2
$\lvert\Delta l\rvert = 0$	$\lvert\Delta l\rvert = 1$
Forbidden transition	Allowed transition

Fig. 30.1 (a) Transitions involving no change in azimuthal quantum number are forbidden, while (b) those involving a change in azimuthal quantum number are allowed

The 4f orbitals are seated deep inside the 5s and 5p orbitals. These are not affected by the nature and number of ligands. Lanthanum(III) ion has $4f^0$ configuration, while lutetium(III) ion has $4f^{14}$ configuration. Therefore, these ions are colorless.

Case Study 30.3 Cerium(III) ion with a $4f^1$ configuration is colorless, while cerium(IV) ion with a $4f^0$ configuration is red. Explain.

Color is generally associated with those chemical entities which possess unpaired electrons. On absorbing visible light, the unpaired electron is transitioned to a higher energy level and on reverting to its original state, emits its energy as colored light. The color of transition metal ions is attributed to d–d transitions, while that of lanthanoid ions to f–f transitions.

In case of cerium(III) ion, which has $4f^1$ configuration, the f–f transitions do not take place. Rather, the single 4f electron is excited to 5d orbital. There are two reasons for 4f to 5d transition. First, the f–f transitions are Laporte forbidden, while f–d transitions are allowed. Second, and more important, when the electron is transferred to 5d orbital, the 4f orbital becomes vacant—and special stability is associated with $4f^0$ configuration.

The 4f to 5d transition, however, requires more energy than f–f transition. The energy difference between 5d and 4f orbitals falls in the ultraviolet region, and not in the visible region of light spectrum. Since natural color is shown by those species whose unpaired electrons can be excited by visible light, cerium(III) ion is colorless, despite having an unpaired electron in its 4f valence shell.

The color of cerium(IV) ion, which has $4f^0$ configuration is assigned to *charge-transfer spectrum*. In this phenomenon, electrons are first transferred from the ligands to the metal ion and then excited by visible light to undergo f–f transition. Charge transfer spectrum is observed only if the metal ion is in a high oxidation state, so that it can attract the ligand electrons. Charge transfer is thus more probable for cerium(IV) ion than for cerium(III) ion.

Magnetic Properties

The precise magnetic moment of a paramagnetic atom, ion or molecule is given by the expression,

$$\mu_{S+L} = \sqrt{4\,S\,(S+1) + L\,(L+1)}$$

which incorporates the contribution of spin moment—represented by the resultant spin quantum number (S) as well as orbital moment—represented by resultant orbital angular momentum quantum number (L). While calculating the magnetic moment of 3d transition elements and their compounds, the orbital moment is generally ignored, as its contribution is negligible. However, in case of lanthanoid ions it is not negligible. This is because the 4f orbitals are shielded from external influence by 5s and 5p orbitals. The orbital moment is, therefore, not quenched by the external fields of the ligands or neighboring ions.

For $4f^0$, $4f^7$ and $4f^{14}$ configurations, however, there is no orbital angular momentum, i.e., L = 0. In such cases, the above expression reduces to,

$$\mu_S = \sqrt{4\,S\,(S+1)}$$

and magnetic moment may be calculated on the basis of spin-only value. Here, S = ns, where n is the number of unpaired electrons and s is the spin quantum number, ½. The $4f^0$ and $4f^{14}$ cases are diamagnetic (n = 0) and their magnetic moment, in any case, is zero. For $4f^7$ arrangement, of which gadolinium(III) is the most important example, the magnetic moment may be calculated as follows:

$$n = 7, \qquad s = \frac{1}{2}, \qquad S = \frac{7}{2}$$

$$\mu_S = \sqrt{4.\ \frac{7}{2}\left(\frac{7}{2} + 1\right)} = \sqrt{63} = 7.9 \text{ BM}$$

For other lanthanoid ions, the spin contribution and orbital contribution must be integrated with each other. This integration, called *Russell–Saunders coupling*, gives rise to another quantum number, the *total angular momentum quantum number*, designated as J. The value of J may be calculated as follows:

$J = L - S$ when the orbital is less than half-filled.

$J = L + S$ when the orbital is more than half-filled.

When the orbital is exactly half-filled, as in the case of gadolinium(III) ion, the magnetic moment may be calculated by ignoring the orbital moment.

In case of lanthanoid elements, the ground J state is separated from the higher ones by large energy gaps. Therefore, the electrons tend to occupy the ground J level, the higher levels being virtually unpopulated. As a result, the calculated and experimental values of magnetic moments are comparable. For samarium(III) and europium(III) ions, however, the energy gap between ground and higher J levels is relatively less. The higher levels too are, therefore, significantly populated. The calculated values of magnetic moment are based on ground J levels and these, in case of samarium(III) and europium(III) ions, do not match with the experimental values.

Case Study 30.4 What is *NIB*?

NIB is a very strong permanent magnet, so named because it is made of an alloy of Neodymium-Iron-Boron. Also referred to as *neo magnet* or *neodymium magnet*, its exact composition is $Nd_2Fe_{14}B$.

Neo magnets may be cast in form of small spheres, which may then be assembled in different shapes (Fig. 30.2).

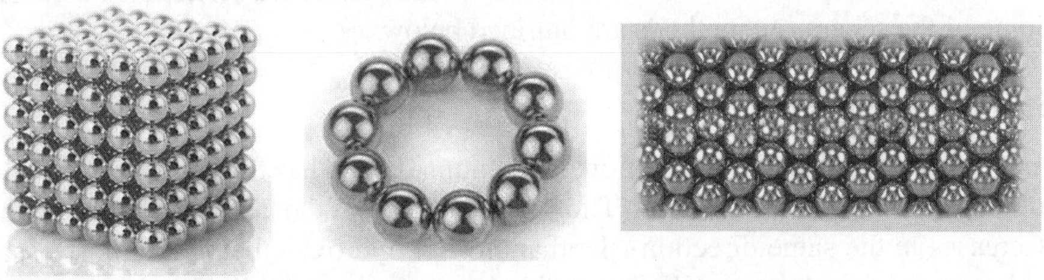

Fig. 30.2 An assembly of neo magnetic spheres

These type of magnets are a component of hard disks, electric motors, electric generators for wind turbines and magnetic resonance imaging devices. These are also used as educational toys, magnetic fastners, stress relief commodities and Jewelery items.

Spectral Properties

As has been stated earlier, the 4f orbitals are deeply buried beneath the 5s and 5p orbitals. Therefore, these are not much affected by the external influence of ligands or surrounding ions. The 4f orbitals do split up, just as the d orbitals segregate into

t_{2g} and e_g sets, under the influence of crystal field. However, the f orbital splitting is very small—of the order of 100 cm^{-1}—as compared to d orbital splitting. Therefore, when f–f transitions occur, the absorption bands are extremely sharp. These are, in fact, similar to those of free atoms, where there is no ligand field effect. In contrast, the d–d transitions result in broad bands, since the d orbital effectively interacts with ligands. Moreover, the position of absorption band varies from ligand to ligand in case of transition metal complexes. In case of lanthanoid complexes, the position of absorption band is independent of the nature of ligands, as the ligands do not have an effective interaction with 4f electrons.

Case Study 30.5 The absorption band of cerium(III) ion is broad, while that of other lanthanoid ions is sharp. Explain.

The absorption bands resulting from f–f transitions are sharp because the 4f orbital is shielded from the influence of other ions by overlying 5s and 5p orbitals. In the case of cerium(III) ion, f–d transition, rather than f–f transition is more favorable. Cerium(III) ion has a $4f^1$ configuration. When this single electron is transmitted to 5d orbital, cerium(III) ion acquires the stable $4f^0$ configuration. The 5d level is profoundly influenced by the ligand field; there is a more effective interaction between ligands and 5d electron, than that between ligands and 4f electron. As a result of this interaction, cerium(III) ion shows a broad band.

Separation of Lanthanoids

In nature, the lanthanoids occur together as monazite sand. It is an exceedingly tedious task to separate individual members of the lanthanoid series from this ore. Some of the separation methods are outlined below.

1. Precipitation

As a result of lanthanoid contraction, lanthanum(III) ion has the largest size, while lutetium(III) ion has the smallest. The basic character of the lanthanoid hydroxides decreases in the same direction—lanthanum(III) hydroxide being the most basic and most soluble, lutetium(III) hydroxide being the least basic and least soluble. Thus, when specific amounts of hydroxide ions are added to a mixture of lanthanoid nitrates, the heavier analogs are precipitated, while the lighter ones remain in solution. The precipitates of heavier lanthanoids are redissolved in nitric acid and small amounts of hydroxide ions are added. The resulting precipitates contain mainly lutetium hydroxide. After many such operations, the elements are separated from each other.

2. Fractional crystallization

The salts of lanthanum are most soluble, while those of lutetium are least soluble. Therefore, lutetium salts crystallize out first from an aqueous solution of lanthanoid ions. The lanthanum salts crystallize last. Efficient separation by fractional crystallization cannot be achieved in a single step. However, repeated operations can sort out the individual members of this series.

3. Thermal decomposition

If a mixture of solid lanthanum nitrates is heated, the lutetium nitrate decomposes into the corresponding oxide at the lowest temperature, while lanthanum nitrate decomposes at the highest temperature. The oxides of all lanthanoids are insoluble in water whereas all the nitrates are soluble. In thermal decomposition, a mixture of lanthanoid nitrates is heated to a specific temperature. The contents are shaken with water. The insoluble oxides are separated from soluble nitrates by filtration. The oxides are redissolved in nitric acid, converted to solid nitrates and the latter are re-heated. After each operation, the purity of any particular element increases.

4. Solvent extraction

The salts of heavier lanthanoids are soluble in organic solvents, while those of lighter ones are soluble in water and other polar solvents. Thus if a solution of lanthanoid nitrates in nitric acid is shaken with tri-*n*-butylphosphate, the heavier nitrates are extracted into the solvent. This step is repeated several times to separate the individual elements.

5. Ion exchange

The most efficient method of separating the lanthanoids is the ion exchange technique. In fact, this technique has rendered all other separation methods obsolete.

A mixture of hydrated salts of lanthanoid elements is prepared. The individual salts are hydrated to varying degrees. Lutetium, being the smallest, is the most heavily hydrated, while lanthanum, being the largest, is the least hydrated.

$LuX_3.nH_2O$ $\quad$ $LaX_3.nH_2O$

n = large $\quad$ n = small

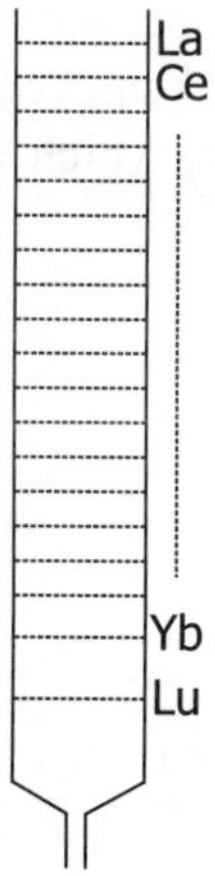

Fig. 30.3 Differentiation of lanthanoids in an ion exchange column

This means that the hydrated radius of lanthanum is the smallest, while that of lutetium is the largest. The hydrated mixture is passed down a chromatographic column containing synthetic ion exchange resin. The resin possesses sulfonic acid functional groups, SO_3H, the hydrogen atoms of which are exchanged with lanthanoid ions.

The binding of lanthanoid cations to the anionic sites of the exchange resin involves electrostatic attractions. Hence, smaller the hydrated radius of the ion, stronger the binding. The lanthanum hydrated ion, being the smallest, becomes attached to the resin immediately, i.e., it gets absorbed at the top of the column. The lutetium hydrated ion, being the largest, gets absorbed at the bottom of the column (Fig. 30.3).

The equilibrium between the lanthanoid ions in aqueous solution and the lanthanoid ions attached to the resin may be expressed as:

$$3H^+(R) + M^{3+}(aq) \rightleftharpoons M^{3+}(R) + 3H^+(aq)$$

According to law of mass action, the equilibrium constant, K, is given by:

$$K = \frac{[H^+(aq)]^3}{[H^+(R)]^3} \cdot \frac{[M^{3+}(R)]}{[M^{3+}(aq)]}$$

where $H^+(R)$ and $M^{3+}(R)$ represent hydrogen ions and lanthanoid ions bound to the resin. The equilibrium constant decreases with increasing atomic number because the factor $[M^{3+}(R)]/[M^{3+}(aq)]$ decreases.

Once the ions become attached to the resin, they get rid of the water of crystallization. Now lanthanum(III) ion is the largest, while lutetium(III) ion is the smallest. An eluting reagent is then run down the column. This reagent is a chelating ligand—generally ethylenediamine tetraacetic acid or citric acid–ammonium citrate buffer at pH 3—which complexes with the lanthanoid ions. The ease of complexation, however, varies with the size of the metal ion. The smallest lutetium ion forms the complex most readily, while the largest lanthanum ion complexes least readily. The lutetium complex elutes out first whereas the lanthanum complex elutes out last. The separation is not achieved in one step. Each eluting fraction is transformed into the hydrated salt, which is processed again through the ion exchange column.

31. THE ACTINOIDS

Introduction

The elements in which the valence electrons are added in 5f orbital are called actinoids. Although the 5f orbital is not filled up in actinium and thorium, these elements are considered a part of the actinoid series. In fact, the series derives its name from actinium—its first member. All actinoids exihibit radioactivity. Only the first four elements of the series—actinium, thorium, protactinium and uranium—occur in nature. The remaining are synthesized by artificial transmutation.

General Properties

All elements of this group are metals. They are silver colored. The melting points of the metals are moderately high, yet lower than those of transition elements. Actinoids exist in a variety of structures. All except californium have more than one crystalline form; plutonium has six. All the metallic lattices have close-packed arrangements.

Due to wide variations in metallic structures, the atomic radii of actinoids show irregular trends. Their ionic radii are comparable to those of lanthanoid ions. As a result, the chemical characteristics of the 4f and 5f elements are quite similar. Nevertheless, actinoids have higher densities than lanthanoids. They also possess a greater abılıty to form complexes as compared to lanthanoids.

Case Study 31.1 The reaction of actinoids with hot water is faster than expected. Why?

With boiling water or steam, an actinoid forms the corresponding oxide and hydrogen. For example,

$$Th + 2H_2O \rightarrow ThO_2 + 2H_2$$

The hydrogen so liberated attacks the unreacted metal to yield its hydride.

$$Th + H_2 \rightarrow ThH_2$$

The hydride itself reacts with water.

$$ThH_2 + 2H_2O \rightarrow ThO_2 + 3H_2$$

Thus, instead of a single reaction, two side reactions also take place simultaneously. The metal is consumed at a fast pace and thus the reaction completes rapidly.

Electronic Configuration

The valence shell electronic configurations of actinoids are depicted in Table 31.1. The two elements immediately preceding the actinoid series, francium and radium, possess the valence shell electronic configuration, $[Rn]7s^1$ and $[Rn]7s^2$, respectively. Actinium, the leading element of actinoid series, has the configuration $[Rn]6d^1 7s^2$. Its properties are similar to that of lanthanum, which has the configuration, $[Xe]5d^1 6s^2$. In the case of lanthanoid series, the 4f orbital is regularly filled up in the elements following lanthanum. By analogy, it might be expected that the 5f orbital would be filled up in elements succeeding actinium. However, for the next four elements—thorium, protactinium, uranium and neptunium—the energy difference between 5f and 6d orbitals is meager. The electrons may, therefore, occupy either of these orbitals. From plutonium onwards, the 5f orbitals become appreciably lower in energy as compared to the 6d orbitals and hence get successively filled up.

Table 31.1 Electronic configurations and oxidation states of actinoids

Element	Symbol	Atomic number	Electronic configuration	Oxidation state(s)
Actinium	Ac	89	$[Rn]6d^1 7s^2$	III
Thorium	Th	90	$[Rn]6d^2 7s^2$	III, IV
Protactinium	Pa	91	$[Rn]5f^2 6d^1 7s^2$	III, IV, V
Uranium	U	92	$[Rn]5f^3 6d^1 7s^2$	III, IV, V, VI
Neptunium	Np	93	$[Rn]5f^4 6d^1 7s^2$	III, IV, V, VI, VII
Plutonium	Pu	94	$[Rn]5f^6 7s^2$	III, IV, V, VI, VII
Americium	Am	95	$[Rn]5f^7 7s^2$	II, III, IV, V, VI
Curium	Cm	96	$[Rn_n]5f^7 6d^1 7s^2$	III, IV
Berkelium	Bk	97	$[Rn]5f^9 7s^2$	III, IV
Californium	Cf	98	$[Rn]5f^{10} 7s^2$	II, III
Einsteinium	Es	99	$[Rn]5f^{11} 7s^2$	II, III
Fermium	Fm	100	$[Rn]5f^{12} 7s^2$	II, III
Mendelevium	Md	101	$[Rn]5f^{13} 7s^2$	II, III
Nobelium	No	102	$[Rn]5f^{14} 7s^2$	II, III
Lawrencium	Lr	103	$[Rn]5f^{14} 6d^1 7s^2$	III

Oxidation States

Like lanthanoids, all actinoids show the oxidation state of III. However, unlike lanthanoids, oxidation state III is not the most stable one for actinoids. Oxidation state II is not very stable, except for americium, for which loss of two valence electrons leaves a stable, half-filled, $5f^7$ configuration. The most stable oxidation states for the first four elements, actinum, thorium, protactinium and uranium are III, IV, V and VI, respectively. These involve loss of all valence electrons.

The elements uranium, neptunium, plutonium and americium show a large number of oxidation states. For these elements, the 5f, 6d and 7s orbitals are of comparable energies. Bonding can, therefore, involve any or all of these orbitals. Moreover, during compound formation, an electron may shift from one shell to another. Therefore, the electronic configuration of the ions of these four members, in particular, and other actinoids, in general, varies from compound to compound. Sometimes, it is impossible to elucidate as to which orbitals are involved in bonding. The important oxidation states of actinoids are listed in Table 31.1.

Compounds

The compounds of actinoids with non-metals are prepared usually by heating the elements together. Hydrides of the type, AnH_2 (An = Th, Np, Pu, Am or Cm) and AnH_3 (An = Pa, U, Np, Pu or Am) exist, although they are not thermally very stable. Moreover, they are rapidly acted upon by air and moisture. In comparison, the borides, carbides and nitrides of actinoids are more stable.

The oxides are basic in nature and possess refractory properties. Thorium(IV) oxide has the highest melting point (3663 K) of any metal oxide. As compared to lanthanoids, a larger number of actinoid oxides have non-stoichiometric composition. This is because each lanthanoid member usually forms only one stable oxide. On the other hand, each actinoid metal possesses a range of oxidation states of comparable stabilities. Therefore, on heating with oxygen, actinoids form a number of oxides with variable compositions. This results in non-stoichiometry.

Mixed metal oxides are obtained by heating alkali metal oxides and actinoid oxides in the presence of oxygen.

$$10Li_2O + 4AnO_2 + 3O_2 \rightarrow 4Li_5[AnO_6] \text{ (An = Np or Pu)}$$

The neptunium(VII) and plutonium(VII) ions are stabilized as octahedral $[AnO_6]^{5-}$ units.

The halides of actinoids exist in compositions ranging from AnX_2 to AnX_6. The hexahalides are confined to hexafluorides of uranium, neptunium and plutonium and hexachloride of uranium. All these are strong oxidizing agents.

Case Study 31.2 The ions $[NpO_6]^{5-}$ and $[PuO_6]^{5-}$ can be synthesized only in strongly alkaline solutions. Why?

The higher oxides of actinoids are unstable with respect to decomposition into lower oxides. The decomposition takes place readily in acidic medium,

$$4[AnO_6]^{5-} + 28H^+ \rightleftharpoons 4[AnO_2]^{2+} + 14H_2O + O_2$$

as well as under neutral conditions,

$$4[AnO_6]^{5-} + 14H_2O \rightleftharpoons 4[AnO_2]^{2+} + 28OH^- + O_2 \text{ (An = Np or Pu)}$$

It is only under strongly alkaline conditions that $[NpO_6]^{5-}$ and $[PuO_6]^{5-}$ ions remain stable.

1. Thorium(IV) nitrate

The only stable oxidation state of thorium is (IV). Within this state, the most common compound and the usual starting material for preparing other salts of thorium is the nitrate, $Th(NO_3)_4 \cdot 5H_2O$. The latter is soluble in water as well as in alcohols and ketones.

Thorium(IV) nitrate reacts with sodium hydroxide to yield insoluble thorium(IV) hydroxide.

$$Th(NO_3)_4 + 4NaOH \rightarrow Th(OH)_4 + 4NaNO_3$$

With sodium fluoride, thorium(IV) fluoride is precipitated.

$$Th(NO_3)_4 + 4NaF \rightarrow ThF_4 + 4NaNO_3$$

On heating, thorium(IV) nitrate decomposes into the oxide.

$$Th(NO_3)_4 \xrightarrow{\text{Heat}} ThO_2 + 4NO_2 + O_2$$

With magnesium nitrate, thorium(IV) nitrate produces the complex, magnesium hexanitratothorate(IV). 8-water.

$$Th(NO_3)_4 + Mg(NO_3)_2 + 8H_2O \rightarrow Mg[Th(NO_3)_6] \cdot 8H_2O$$

This complex has a coordination number of 12, showing that the nitrate ligands act as bidentate, chelating groups.

2. Uranium(III) hydride

Uranium metal reacts readily with hydrogen at about 550 K to give uranium(III) hydride.

$$2U + 3H_2 \rightarrow 2UH_3$$

On raising the temperature, the reaction reverses to yield extremely reactive, finely divided uranium.

Uranium(III) hydride is a reactive compound, and a more suitable starting material for preparing other compounds of uranium than the metal itself. It inflames in air to produce uranium(IV) oxide.

$$4UH_3 + \underset{Air}{7O_2} \rightarrow 4UO_2 + 6H_2O$$

It reacts with hydrogen chloride at 550 K to yield uranium(III) chloride,

$$UH_3 + 3HCl \rightarrow UCl_3 + 3H_2$$

and with chlorine at 473 K to give uranium(IV) chloride.

$$2UH_3 + 7Cl_2 \rightarrow 2UCl_4 + 6HCl$$

With hydrogen sulfide, the reaction occurs at 700 K to produce uranium(IV) sulfide.

$$2UH_3 + 4H_2S \rightarrow 2US_2 + 7H_2$$

3. Uranium oxides

The uranium-oxygen system is complicated since a number of oxides, of nearly equal stability exist. These oxides are non-stoichiometric compounds. The important ones are, uranium(IV) oxide, UO_2; uranium(VI) oxide, UO_3; and a mixed uranium(IV)–uranium(VI) oxide, U_3O_8. These are brown-black, greenish-black and orange-yellow in colors, respectively.

Uranium(VI) oxide may be prepared by the thermal decomposition of uranyl nitrate at about 650 K,

$$2UO_2(NO_3)_2 \xrightarrow{650\ K} 2UO_3 + 4NO_2 + O_2$$

or by thermal decomposition of ammonium diuranate

$$(NH_4)_2U_2O_7 \xrightarrow{650\ K} 2UO_3 + H_2O + 2NH_3$$

When uranium(VI) oxide is heated to 1000 K, it yields the mixed oxide.

$$6UO_3 \xrightarrow{1000\ K} 2U_3O_8 + O_2$$

Uranium(VI) oxide may be reduced to uranium(IV) oxide by carbon monoxide at 650 K.

$$UO_3 + CO \rightarrow UO_2 + CO_2$$

All oxides dissolve in nitric acid to give uranyl nitrate, $UO_2(NO_3)_2 \cdot nH_2O$

$$UO_3 + 2HNO_3 \rightarrow UO_2(NO_3)_2 + H_2O$$

The uranyl unit has the composition, $[O = U = O]^{2+}$.The number of molecules of water of crystallization (n) in uranyl nitrate may be 2, 3, or 6, depending on whether it is prepared from fuming, concentrated or dilute nitric acid, respectively.

Case Study 31.3 Uranium(VI) fluoride gets rapidly hydrolyzed, while sulfur(VI) fluoride resists hydrolysis. Why?

Normally, the initiation step in a hydrolysis reaction is the coordination of water molecule to the central atom. During this step, the coordination number of the central atom increases by one unit. Uranium can expand its coordination number to eight, although a few ten-coordinated compounds are also known for this metal. Hence the transition state, $[H_2O \rightarrow UF_6]$ having a coordination number of seven is possible and this initiates the hydrolysis reaction.

For sulfur, the maximum coordination number is six. For this reason, the transition state $[H_2O \rightarrow SF_6]$ is not possible. As a result, sulfur(VI) fluoride is not susceptible to hydrolysis. In fact, as stated in Case Study 18.5 p. 541, this compound is chemically inert.

Actinoid Contraction

There is a steady and consistent decrease in the ionic radii of actinoids with increasing atomic number. The phenomenon is analogous to lanthanoid contraction and is, therefore, referred to as actinoid contraction. The decrease in size is attributed to the imperfect shielding of one 5f electron by another in the same shell. While traversing the actinoid series, the nuclear charge and valence electrons both increase by one unit at each step. Due to the lack of symmetry of 5f orbitals, the electrons do not shield the nuclear charge efficiently. The valence electrons are strongly pulled towards the nucleus and the atomic size decreases.

Magnetic Properties

The experimental and calculated values of magnetic moment of lanthanoid ions usually match with each other. However, the observed magnetic moments of actinoids are appreciably lower than the calculated ones. This is because, as compared to the 4f electrons, the 5f electrons are less effectively shielded from the crystal field as a result of which the orbital contribution to the magnetic moment gets quenched. Further, the values of magnetic moment of actinoid ions are more temperature dependent than those of the lanthanoid ions.

Electronic Spectra

The electronic spectra of actinoid complexes arise from the following three types of transitions.

1. **f–f transitions** These are strictly forbidden, yet they do occur because of deviations from perfect symmetry around the metal ion. The bands are narrow, intense and absorb in visible and ultraviolet regions, giving rise to color in aqueous solutions.

2. **5f–6d transitions** These are allowed transitions and hence are more intense and broader than the f–f transitions. Such transitions are confined to ultraviolet region of the spectrum and, therefore, do not affect the color.
3. **Charge transfer transitions** These occur due to transfer of charge from the ligand to the metal. Such transitions are allowed; their bands are intense and broad. They occur in ultraviolet region but generally "tail off" in the visible part of the spectrum. These produce intense coloration among the complexes.

Since the 5f orbitals interact with the ligands to an appreciable extent, the spectra of actinoids are sensitive to the crystal field and, unlike lanthanoids, change drastically from one compound to another.

Complexes

The 5f orbitals can spatially extend outside the 6s and 6p orbitals to a much greater degree than 4f orbitals can extend outside 5s and 5p orbitals. As a result, the valence orbitals of actinoids can interact more strongly with the ligands than the valence orbitals of lanthanoids. In consequence, the actinoids are more prone to form complexes than lanthanoids. Halide and oxoanionic ligands, such as oxalate, sulfate and phosphate form complexes with actinoids readily. Even π-bonding ligands, such as phosphines and thioethers can form complexes with actinoids. Since the energies of 5f, 6d and 7s shells are very close to each other, hybridization is possible between these orbitals. Thus, actinoid complexes are known in a range of stereochemistries. Examples include, octahedral, tetrachlorobis(triethylphosphine)uranium(IV), $[UCl_4\{P(C_2H_5)_3\}_2]$ and dodecahedral, tetraoxalatothorate(IV) ion.4-water, $[Th(C_2O_4)_4]^{4-}.4H_2O$.

Case Study 31.4 Higher oxidation states are more common for actinoids than for lanthanoids. Why?

Actinoids have larger atomic radii as compared to lanthanoids. Moreover, the energy levels of the valence electrons are more close to each other in actinoids than in lanthanoids. Therefore, the 5f electrons are more easily removed in comparison to 4f electrons and consequently, for actinoid elements, the lower oxidation states are less important and higher oxidation states are more common as compared to the lanthanoid elements. For example, for curium, oxidation state V is more stable, while for its lanthanoid counterpart, gadolinium, oxidation state III is more stable.

---

32. Qualitative Analysis

Introduction

Qualitative analysis is the branch of chemistry that deals with identification of unknown substances. The substance in question may belong to inorganic or organic classification. The inorganic qualitative analysis involves identification of ions in a salt or in a mixture of salts or in a mineral. In organic chemistry, qualitative analysis helps ascertaining the identity of a specific member of a carbon-based homologous series.

Qualitative analysis does not concern itself with the amount or concentration of a particular ion in a mineral or of a specific organic compound in a petroleum product. This is the purview of quantitative analysis, where the final result is in terms of a numeral—a mathematical quantity. Inorganic qualitative analysis directs us to a set of tests by which we may elucidate whether a given cation or anion is present in the unknown sample. These tests, in turn, are based on different types of chemical reactions—precipitation, acid-base, redox, complexation. Thus qualitative analysis gives us a chance to correlate what we learn in theory with what we practice in the laboratory.

Experimental Techniques

Inorganic qualitative analysis is mainly concerned with identification of anions and cations in ores, minerals, salts and mixtures of salts. The exercise is conducted usually at a semimicro level, whereby small amounts of reagents are used in detection procedures.

The main equipment for carrying out the analysis is a *semimicro kit*, a sample of which is shown in Fig. 32.1.

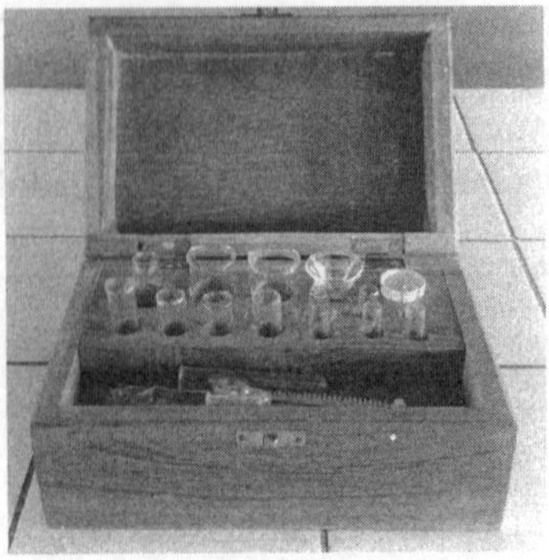

Fig. 32.1 A semimicro kit

The kit contains a few test tubes for carrying out a legion of tests. It also houses centrifuge tubes which are used for segregating the precipitates by centrifugation. The boiling tubes are used for preparing extracts. Alternately, a beaker or a conical flask of semimicro grade may be used for this purpose. Evaporation of solvent, whenever required, is carried out in the crucible.

Another component of the semimicro kit is the U tube, which is shown separately in Fig. 32.2. It is used for directing gases or vapors to reagents.

Fig. 32.2. A U tube

Whenever elevated temperatures are required, a heating block (Fig. 32.3) is used. It has depressions to accommodate test tubes, beakers and crucible in which the contents may be warmed or boiled.

Fig. 32.3. A heating block

The spot tests are generally carried out on a spot plate, one variety of which is shown in Fig. 32.4. The plate is made of glazed porcelain and has 9–12 depressions for carrying out spottings.

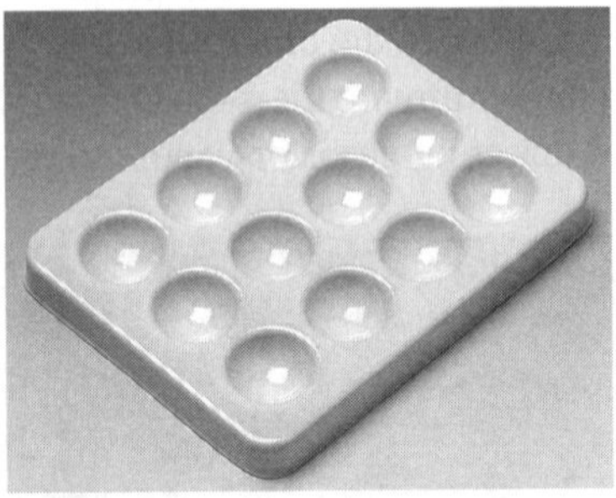

Fig. 32.4. A spot plate

Analysis of Anions

To ascertain the presence of a specific anion in an inorganic substance, two types of tests need be performed: preliminary tests and confirmatory tests. While the former are only indicative of the presence of an anion or anions, the latter provide the final validation. At times, if two or more specific anions are present together in an inorganic mixture, their confirmatory tests may have to be modified. Under these circumstances, one may have to invoke another analytical procedure, called combination tests.

Preliminary tests

The anions may be classified into the following three broad groups.

1. ***Dilute hydrochloric acid/dilute sulfuric acid group*** The anions which react with either of these acids at room temperature to yield a gaseous product fall in this group. These are, carbonate (CO_3^{2-}), sulfide (S^{2-}), sulfite (SO_3^{2-}), thiosulfate ($S_2O_3^{2-}$), nitrite (NO_2^-) and acetate ($CHCOO^-$).
2. ***Concentrated sulfuric acid group*** The anions which react with concentrated sulfuric acid, on warming, are members of this group. These are, chloride (Cl^-), bromide (Br^-), iodide (I^-), fluoride (F^-), nitrate (NO_3^-) and oxalate ($C_2O_4^{2-}$).
3. ***Inert to acids group*** The anions which do not react with dilute hydrochloric acid/dilute sulfuric acid or with concentrated sulfuric acid are clubbed in this group. These are, sulfate (SO_4^{2-}), phosphate (PO_4^{3-}) and borate (BO_3^{3-}).

The preliminary tests endeavor to identify the gas which evolves when a particular anion reacts with a dilute or concentrated acid. The identity of the gas, in turn, narrows down the possibility of concerned anion as being a constituent of the inorganic material. Since no gas evolves when sulfates, phosphates or borates are treated with an acid, there are no preliminary tests for these anions.

1. Carbonates, CO_3^{2-} Place about 100 mg of the salt or the mixture in a U tube and add about 2 mL dilute hydrochloric acid or dilute sulfuric acid. Carbon dioxide gas evolves. It is a colorless and odorless gas.

$$CO_3^{2-} + 2H^+ \rightarrow H_2O + CO_2$$

Pass the gas through lime water. It turns milky due to the formation of white, turbid, calcium carbonate. Prolonged passage of carbon dioxide causes the milkiness to disappear due to the formation of soluble calcium bicarbonate.

$$\underset{\text{Lime water}}{Ca(OH)_2} + CO_2 \rightarrow \underset{\text{White turbidity}}{CaCO_3} + H_2O$$

$$CaCO_3 + H_2O + \underset{\text{Excess}}{CO_2} \rightarrow \underset{\text{Soluble}}{Ca(HCO_3)_2}$$

2. Sulfides, S^{2-} On treatment with dilute acids, the sulfides produce hydrogen sulfide gas, which is colorless and has an odor of rotten eggs.

$$S^{2-} + 2H^+ \rightarrow H_2S$$

A moist lead acetate paper, when exposed to hydrogen sulfide gas, turns black due to the formation of lead sulfide.

$$(CH_3COO_2)Pb + H_2S \rightarrow \underset{\text{Black}}{PbS\downarrow} + 2CH_3COOH$$

Alternately, hydrogen sulfide gas, when passed though a solution of copper sulfate, produces a black precipitate of copper sulfide.

$$CuSO_4 + H_2S \rightarrow \underset{\text{Black}}{CuS\downarrow} + H_2SO_4$$

3. Sulfites, SO_3^{2-} and thiosulfates, $S_2O_3^{2-}$ Either of these two anions, when treated with a dilute acid yield sulfur dioxide gas.

$$SO_3^{2-} + 2H^+ \rightarrow H_2O + SO_2$$

$$S_2O_3^{2-} + 2H^+ \rightarrow S + H_2O + SO_2$$

Sulfur dioxide is a colorless gas, having a pungent odor of burning sulfur. With the aid of a U tube, the gas may be passed through a dilute solution of potassium permanganate, acidified with dilute sulfuric acid. The purple color of potassium permanganate discharges.

$$2KMnO_4 + 5SO_2 + 2H_2O \rightarrow K_2SO_4 + 2MnSO_4 + 2H_2SO_4$$

Likewise, when sulfur dioxide gas is passed through an acidified solution of potassium dichromate, the orange color turns green due to the formation of chromium(III) sulfate. However, the green color is so pale that it almost appears colorless.

$$K_2Cr_2O_7 + H_2SO_4 + 3SO_2 \rightarrow K_2SO_4 + \underset{\text{Pale green}}{Cr_2(SO_4)_3} + H_2O$$

4. Nitrites, NO_2^- The nitrites react with either dilute hydrochloric acid or dilute sulfuric acid, to produce nitrogen dioxide gas.

$$NO_2^- + H^+ \rightarrow HNO_2$$

$$3HNO_2 \rightarrow HNO_3 + H_2O + 2NO$$

$$2NO + O_2 \rightarrow 2NO_2$$

Nitrogen dioxide is a brown colored gas, having a pungent odor.

5. Acetates, CH_3COO^- The reaction of dilute acids with acetates produces vapors of acetic acid.

$$CH_3COO^- + H^+ \rightarrow CH_3COOH$$

These vapors are colorless, and smell of vinegar.

6. Chlorides, Cl^- Place about 100 mg of the given salt or mixture in a test tube and add about 2 mL concentrated sulfuric acid. On warming, hydrogen chloride gas evolves.

$$Cl^- + H_2SO_4 \xrightarrow{\text{Warm}} HSO_4^- + HCl$$

Hydrogen chloride gas is colorless and has a pungent odor. At a preliminary level, it may be identified by bringing a glass rod dipped in aqueous ammonia at the mouth of the test tube, whereupon dense white fumes of ammonium chloride are generated.

$$NH_3 + HCl \rightarrow \underset{\text{White}}{NH_4Cl\uparrow}$$

7. Bromides, Br^- The bromides, on warming with concentrated sulfuric acid, produce vapors of bromine.

$$Br^- + H_2SO_4 \xrightarrow{\text{Warm}} HSO_4^- + HBr$$

$$2HBr + H_2SO_4 \rightarrow 2H_2O + SO_2 + \underset{\text{Red-brown}}{Br_2\uparrow}$$

Bromine may be discerned by its red-brown vapors.

8. Iodides, I^- The reaction between iodides and concentrated sulfuric acid at about 330 K produces vapors of iodine.

$$I^- + H_2SO_4 \xrightarrow{\text{Warm}} HSO_4^- + HI$$

$$2HI + H_2SO_4 \rightarrow 2H_2O + SO_2 + \underset{\text{Violet}}{I_2\uparrow}$$

Iodine vapors are easily identifiable because of their violet color.

9. Fluorides, F^- All fluorides produce colorless vapors of hydrogen fluoride on warming with concentrated sulfuric acid.

$$F^- + H_2SO_4 \xrightarrow{\text{Warm}} HSO_4^- + HF$$

When a moist glass rod is held over the mouth of the test tube, hydrogen fluoride reacts with the silica content of glass to yield sulfur tetrafluoride which hydrolyzes to silicic acid, H_2SiO_3.

$$SiO_2 + 4HF \rightarrow SiF_4 + 2H_2O$$

$$3SiF_4 + 3H_2O \rightarrow 2H_2[SiF_6] + \underset{\text{White}}{H_2SiO_3\downarrow}$$

A white deposition of silicic acid on glass rod serves as a preliminary test of fluoride ions.

10. Nitrates, NO_3^- Nitrates react with concentrated sulfuric acid to produce nitrogen dioxide gas.

$$4NO_3^- + 2H_2SO_4 \xrightarrow{\text{Warm}} 2SO_4^{2-} + 2H_2O + 4NO_2\uparrow + O_2\uparrow$$

Nitrogen dioxide is a reddish-brown gas having a pungent odor.

11. Oxalates, $C_2O_4^{2-}$ On warming with concentrated sulfuric acid, oxalates produce carbon dioxide gas.

$$C_2O_4^{2-} + H_2SO_4 \xrightarrow{\text{Warm}} SO_4^{2-} + H_2O + CO_2\uparrow + CO\uparrow$$

Carbon dioxide gas is identified by lime water test which we have already described in context of carbonate ions.

Conformatory tests

We have noted that the preliminary tests of anions are conducted by treating the solid inorganic substance with a suitable acid. Most conformatory tests, on the other hand, are performed with a solution of anions. The anions may be solubilized as sodium carbonate extract or as water extract. We now describe the procedures for preparing these extracts.

Preparation of sodium carbonate extract The usefulness of sodium carbonate extract stems from the fact that all salts of sodium are soluble in water. It is, therefore, pertinent to convert the anions present in the given material into their respective sodium salts. This is done as follows.

Take about 100 mg of the inorganic substance in a beaker and add about 300 mg of sodium carbonate. Suspend the mixture in about 5 mL distilled water and boil for 5–7 minutes. Centrifuge and discard the residue. The centrfugate serves as the sodium carbonate extract.

Preparation of water extract Certain anions form water-soluble salts irrespective of the cations to which these are bound. Example include, thiosulfates

($S_2O_3^{2-}$), nitrites (NO_2^-), acetates (CH_3COO^-) and nitrates (NO_3^-). It is, therefore, worthwhile to prepare a water extract for analyzing these anions. The procedure is as follows.

Place about 100 mg of the substance in a beaker and add about 5 mL distilled water. Boil the contents for 5–7 minutes. Centrifuge and discard the residue. Treat the centrifugate as water extract.

We now describe the confirmatory tests for the anions.

1. Carbonates, CO_3^{2-} The detection of carbonates present a paradoxy. Since all carbonates are not soluble in water, a water extract cannot be used for performing confirmatory tests. A sodium carbonate extract too cannot be used since it contains unreacted sodium carbonate and would therefore test positive for carbonate ions. In view of these limitations, the outcome of the preliminary test may be taken as a proof of final confirmation. Thus if the substance reacts with dilute hydrochloric acid or dilute sulfuric acid to produce carbon dioxide which, in turn, produces milkiness with lime water, we may conclude that the ion in question is carbonate.

Nevertheless, if it is known for sure that the substance under examination contains a soluble carbonate, the following confirmatory tests may be performed using the water extract.

a. *Barium chloride solution* A white precipitate of barium carbonate results.

$$CO_3^{2-} + BaCl_2 \rightarrow \underset{\text{White}}{BaCO_3\downarrow} + 2Cl^-$$

This precipitate is soluble in dilute hydrochloric acid.

$$BaCO_3 + 2HCl \rightarrow BaCl_2 + H_2O + CO_2\uparrow$$

b. *Silver nitrate solution* A white precipitate of silver carbonate is formed.

$$CO_3^{2-} + 2AgNO_3 \rightarrow \underset{\text{White}}{Ag_2CO_3\downarrow} + 2NO_3^-$$

This precipitate dissolves in aqueous ammonia due to the formation of soluble diamminesilver(I) ion.

$$Ag_2CO_3 + 4NH_3 \rightarrow 2[Ag(NH_3)_2]^+ + CO_3^{2-}$$

2. Sulfides, S^{2-} Acidify the sodium carbonate extract with acetic acid.

a. *Copper sulfate solution* A black precipitate of cupric sulfide is formed.

$$Na_2S + CuSO_4 \rightarrow \underset{\text{Black}}{CuS\downarrow} + Na_2SO_4$$

b. *Sodium nitroprusside solution* A purple coloration develops.

$$Na_2S + Na_2[Fe(CN)_5NO] \rightarrow \underset{\text{Purple}}{Na_4[Fe(CN)_5NOS]}$$

3. Sulfites, SO_3^{2-} Acidify the sodium carbonate extract with dilute hydrochloric acid.

a. *Barium chloride solution and bromine water* A white precipitate of barium sulfate is obtained.

$$Na_2SO_3 + BaCl_2 \rightarrow BaSO_3 + 2NaCl$$

$$Br_2 + H_2O \rightarrow 2HBr + (O)$$

$$BaSO_3 + (O) \rightarrow \underset{\text{White}}{BaSO_4\downarrow}$$

b. *Lead acetate solution* On adding a solution of lead acetate to acidified sodium carbonate extract and heating strongly a white precipitate of lead sulfate is formed.

$$Na_2SO_3 + (CH_3COO)_2Pb \rightarrow PbSO_3 + 2CH_3COONa$$

$$2PbSO_3 + O_2 \rightarrow \underset{\text{White}}{2PbSO_4\downarrow}$$

4. Thiosulfates, $S_2O_3^{2-}$ To be performed with water extract.

a. *Silver nitrate solution* A white precipitate of silver thiosulfate results.

$$S_2O_3^{2-} + 2AgNO_3 \rightarrow \underset{\text{White}}{Ag_2S_2O_3\downarrow} + 2NO_3^-$$

On standing the precipitate turns yellow/orange and finally black.

$$Ag_2S_2O_3 + H_2O \rightarrow \underset{\text{Black}}{Ag_2S\downarrow} + H_2SO_4$$

b. *Lead acetate solution* A white precipitate of lead thiosulfate is obtained.

$$S_2O_3^{2-} + (CH_3COO)_2Pb \rightarrow \underset{\text{White}}{PbS_2O_3\downarrow} + 2CH_3COO^-$$

On boiling, black lead sulfide gets precipitated.

$$PbS_2O_3 + H_2O \rightarrow \underset{\text{Black}}{PbS\downarrow} + H_2SO_4$$

5. Nitrites, NO_2^- To be performed with water extract.

a. *Potassium iodide solution* Addition of potassium iodide solution, along with a few drops of dilute sulfuric acid, to the water extract, yields iodine.

$$2NO_2^- + 2KI + 4H^+ \rightarrow 2K^+ + I_2 + 2H_2O + 2NO\uparrow$$

Iodine produces a blue color with starch.

b. *Sulfanilic acid-1-naphthylamine reagent* Place a drop of water extract on a spot plate. Over it place one drop of sulfanilic acid, followed by 1-naphthylamine. A red colored azo dye forms.

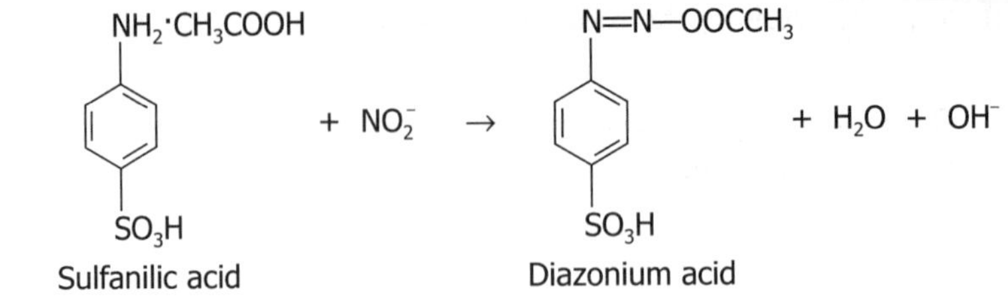

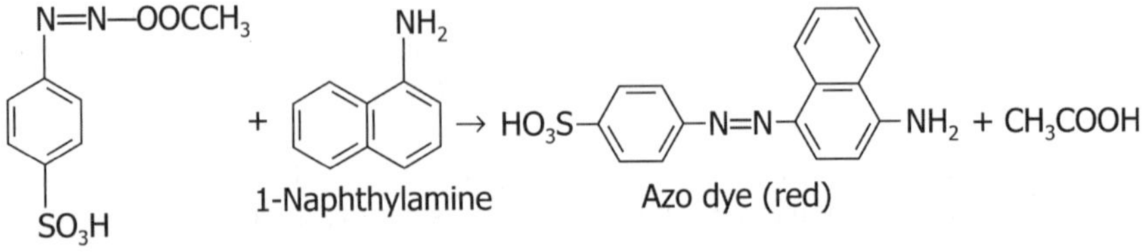

6. Acetates, CH_3COO^- To be performed with water extract.

a. *Neutral ferric chloride solution* The bench ferric chloride solution is acidic due to the presence of hydrochloric acid. Before performing the test for acetate anion, the reagent must be rendered neutral. This is done as follows. To about 2 mL of bench ferric chloride solution, ammonium hydroxide is added drop wise. When all of hydrochloric acid is neutralized, the next drop of ammonium hydroxide forms a red-brown turbidity of ferric hydroxide.

$$NH_4OH + HCl \rightarrow NH_4Cl + H_2O$$

$$3NH_4OH + FeCl_3 \rightarrow \underset{\text{Red-brown}}{Fe(OH)_3\downarrow} + 3NH_4Cl$$

The addition of ammonium hydroxide is stopped and the clear supernatant liquid is decanted. It is then added to a few drops of water extract containing acetate ions. A red coloration of hexacetatoferrate(III) ions develops.

$$6CH_3COO^- + FeCl_3 \rightarrow \underset{\text{Red}}{[Fe(CH_3COO)_6]^{3-}} + 3Cl^-$$

On adding a solution of sodium fluoride, the red color discharges due to the formation of hexafluoroferrate(III) ions.

$$[Fe(CH_3COO)_6]^{3-} + 6NaF \rightarrow \underset{\text{Colorless}}{[FeF_6]^{3-}} + 6CH_3COONa$$

b. *Ethanol-concentrated sulfuric acid* To about 2 mL of water extract, add equal amount of ethanol and approximately 0.5 mL concentrated sulfuric acid. Warm gently for about 2 minutes. A pleasant, fruity odor of ethyl acetate emanates.

$$CH_3COO^- + H_2SO_4 \rightarrow CH_3COOH + HSO_4^-$$

$$CH_3COOH + C_2H_5OH \rightarrow \underset{\text{Fruity odor}}{CH_3COOC_2H_5} + H_2O$$

7. Chlorides, Cl^- Test (a) is preformed with acidified sodium carbonate extract, test (b) with solid salt or a mixture of salts.

a. *Silver nitrate solution* Acidify the sodium carbonate extract with dilute nitric acid and add a few drops of silver nitrate solution. A white precipitate of silver chloride is obtained.

$$NaCl + AgNO_3 \rightarrow \underset{\text{White}}{AgCl\downarrow} + NaNO_3$$

The precipitate is soluble in aqueous ammonia due to the formation of diamminesilver(I) chloride.

$$AgCl + 2NH_3 \rightarrow \underset{\text{Soluble}}{[Ag(NH_3)_2]Cl}$$

b. *Chromyl chloride test* Place about 25 mg of the solid material in a U tube and add an equal amount of solid potassium dichromate. Pour about 2 mL concentrated sulfuric acid and heat strongly. Red vapors of chromyl chloride gas, CrO_2Cl_2 evolve. Pass the vapors through a solution of sodium hydroxide which had earlier been placed in the U section of the tube. Thereafter, add successively a few drops of acetic acid and lead acetate solution to the U section. A deep yellow precipitate of lead chromate, $PbCrO_4$ confirms the presence of chloride ions.

Taking sodium chloride as a representative case, the reaction sequence is as follows.

$$4NaCl + K_2Cr_2O_7 + 6H_2SO_4 \rightarrow 4NaHSO_4 + 2KHSO_4 + 3H_2O + \underset{\text{Red}}{2CrO_2Cl_2\uparrow}$$

$$4NaOH + CrO_2Cl_2 \rightarrow \underset{\text{Sodium chromate}}{Na_2CrO_4} + 2NaCl + 2H_2O$$

$$Na_2CrO_4 + (CH_3COO)_2Pb \rightarrow \underset{\text{Yellow}}{PbCrO_4\downarrow} + 2CH_3COONa$$

8. Bromides, Br^- While test (a) is performed with acidified sodium carbonate extract, test (b) is performed without acidification.

a. *Silver nitrate solution* Sodium carbonate extract, acidified with dilute nitric acid, yields a pale yellow precipitate of silver bromide.

$$NaBr + AgNO_3 \rightarrow \underset{\text{Pale yellow}}{AgBr\downarrow} + NaNO_3$$

This precipitate is partially soluble in aqueous ammonia.

b. *Layer test* To about 2 mL of sodium carbonate extract add an equal amount of carbon tetrachloride. The latter, being immiscible with water, forms a separate lower layer in the test tube. Add a few drops of chlorine water. Agitate the test tube and allow it to stand for about one minute.

$$2NaBr + Cl_2 \rightarrow 2NaCl + \underset{\text{Red-brown}}{Br_2}$$

Bromine, being soluble in carbon tetrachloride, diffuses into the lower, organic layer and imparts a red-brown color to it.

9. Iodides, I^- Test (a) is performed with acidified sodium carbonate extract, test (b) without acidification.

a. *Silver nitrate solution* Sodium carbonate extract, acidified with dilute nitric acid, yields a yellow precipitate of silver iodide with silver nitrate solution.

$$\underset{}{NaI} + AgNO_3 \rightarrow \underset{\text{Yellow}}{AgI\downarrow} + NaNO_3$$

The yellow precipitate is insoluble in aqueous ammonia.

b. *Layer test* Add about 2 mL carbon tetrachloride to an equal amount of sodium carbonate extract. Dropwise add chlorine water.

$$2NaI + Cl_2 \rightarrow 2NaCl + \underset{\text{Violet}}{I_2}$$

Shake the test tube vigorously and allow it to stand. Iodine solubilizes in the lower, carbon tetrachloride layer and imparts a violet color to it.

10. Fluorides, F^- Test (a) may be performed with solid material, test (b) with acidified sodium carbonate extract.

a. *Sand-concentrated sulfuric acid* To about 100 mg of solid substance add an equal amount of sand and approximately 3 mL concentrated sulfuric acid. Heat to boiling.

$$2F^- + H_2SO_4 \rightarrow SO_4^{2-} + 2HF$$

$$4HF + \underset{\text{Sand}}{SiO_2} \rightarrow SiF_4 + 2H_2O$$

$$3SiF_4 + 3H_2O \rightarrow \underset{\text{Oily droplets}}{H_2SiO_3} + 2H_2[SiF_6]$$

The silicic acid, H_2SiO_3 forms oily droplets which persist even when the heating is stopped.

b. *Calcium chloride solution* Acidify the sodium carbonate extract with acetic acid and add a few drops of calcium chloride solution. A white precipitate of calcium fluoride is obtained.

$$2NaF + CaCl_2 \rightarrow \underset{\text{White}}{CaF_2\downarrow} + 2NaCl$$

11. Nitrates, NO_3^- To be performed with water extract.

a. *Ring test* To the water extract add a saturated solution of ferrous sulfate. Incline the test tube at an angle and add a few drops of concentrated sulfuric acid slowly and along the wall of the test tube. A brown ring is formed at the junction of two liquids.

$$NO_3^- + H_2SO_4 \rightarrow HNO_3 + HSO_4^-$$

$$6FeSO_4 + 3H_2SO_4 + 2HNO_3 \rightarrow 3Fe_2(SO_4)_3 + 4H_2O + 2NO$$

$$\underset{\text{Brown ring}}{FeSO_4 + 5H_2O + NO \rightarrow [Fe(H_2O)_5NO]SO_4}$$

b. *Reduction to nitrites* To the water extract add a few specks of zinc dust, followed by a few drops of dilute sulfuric acid. The nitrate ions get reduced to nitrite ions.

$$Zn + H_2SO_4 \rightarrow ZnSO_4 + 2[H]$$

$$NO_3^- + 2[H] \rightarrow NO_2^- + H_2O$$

The test solution is placed in the depression of a spot plate and one drop each of sulfanilic acid and 1-naphthylamine is added. A red colored azo dye forms. The reaction sequence is same as that described for the confirmatory test of nitrite ions (p. 900).

12. Oxalates, $C_2O_4^{2-}$ To be performed with sodium carbonate extract, acidified with acetic acid.

a. *Calcium chloride solution* Acidify the sodium carbonate extract with acetic acid and add a few drops of calcium chloride solution.

$$Na_2C_2O_4 + CaCl_2 \rightarrow \underset{\text{White}}{CaC_2O_4\downarrow} + 2NaCl$$

A white precipitate of calcium oxalate results.

b. Take one drop of bench potassium permanganate and add about 10 drops of distilled water. Add one drop of this diluted solution to acidified sodium carbonate extract. The contents acquire a pink color. On warming, the pink color discharges.

$$5C_2O_4^{2-} + 2MnO_4^- + 16H^+ \rightarrow 2Mn^{2+} + 8H_2O + 10CO_2\uparrow$$

13. Sulfates, SO_4^{2-} To be performed with acidified sodium carbonate extract.

a. *Barium chloride solution* Acidify sodium carbonate extract with dilute hydrochloric acid. Add a few drops of barium chloride solution. A white precipitate of barium sulfate confirms the presence of sulfate ions.

$$Na_2SO_4 + BaCl_2 \rightarrow \underset{\text{White}}{BaSO_4\downarrow} + 2NaCl$$

b. *Lead acetate solution* Acidify sodium carbonate extract with acetic acid and add a few drops of lead acetate solution. A white precipitate of lead sulfate is obtained.

$$Na_2SO_4 + (CH_3COO)_2Pb \rightarrow \underset{\text{White}}{PbSO_4\downarrow} + 2CH_3COONa$$

14. Phosphates, PO_4^{3-} Test (a) is performed with solid material, test (b) with acidified sodium carbonate extract.

a. *Ammonium molybdate test* To about 50 mg of solid salt or mixture, add about 3 mL concentrated nitric acid. Heat to boiling. Add a few drops of ammonium molybdate solution. A yellow precipitate confirms the presence of phosphate ions.

$$PO_4^{3-} + 3HNO_3 \rightarrow H_3PO_4 + 3NO_3^-$$

$$H_3PO_4 + 12(NH_4)_2MoO_4 + 21HNO_3 \rightarrow \underset{\text{Yellow}}{(NH_4)_3PO_4{\cdot}12MoO_3\downarrow} + 21NH_4NO_3 + 12H_2O$$

b. *Ferric chloride solution* Acidify the sodium carbonate extract with acetic acid and add a few drops of neutral ferric chloride solution (the procedure for preparing neutral ferric chloride solution has been described in the section dealing with confirmatory tests for acetate ions, p. 900).

$$Na_3PO_4 + FeCl_3 \rightarrow \underset{\text{Yellowish-white}}{FePO_4\downarrow} + NaCl$$

A yellowish-white precipitate of ferric phosphate is obtained.

15. Borates, BO_3^{3-} Test (a) is performed with solid substance, test (b) with sodium carbonate extract.

a. *Green-edged flame test* Place about 200 mg of solid salt or a mixture of salts in a china dish. Add a few drops of concentrated sulfuric acid, followed by a few drops of ethanol. Warm the contents and then ignite.

$$2BO_3^{3-} + 3H_2SO_4 \rightarrow \underset{\text{Boric acid}}{2H_3BO_3} + 3SO_4^{2-}$$

$$H_3BO_3 + 3C_2H_5OH \rightarrow B(OC_2H_5)_3 + 3H_2O$$

Boron ethoxide, $B(OC_2H_5)_3$, so formed, burns with a green-edged flame inside the china dish. The flame may later be put off by placing a wire gauze over the china dish.

b. *Silver nitrate solution* Acidify the sodium carbonate extract with acetic acid and add a few drops of silver nitrate solution. A white precipitate of silver metaborate $AgBO_2$ is obtained.

$$Na_3BO_3 + 2CH_3COOH \rightarrow NaBO_2 + 2CH_3COONa + H_2O$$

$$NaBO_2 + AgNO_3 \rightarrow \underset{\text{White}}{AgBO_2\downarrow} + NaNO_3$$

The white precipitate may not persist on shaking or on adding excess acetic acid. Nevertheless, irrespective whether the precipitate remains or not, when the contents are boiled, a black precipitate of silver oxide is formed.

$$2AgBO_2 + 3H_2O \rightarrow \underset{\text{Black}}{Ag_2O\downarrow} + 2H_3BO_3$$

Box 32.1

The Mystery of Adolf Hitler's Diaries

During the early 1980s, a man from East Germany, who identified himself as Dr. Fischer, contacted the editorial office of *Der Stern Magazine* in West Germany. He claimed that he had in his possession 64 handwritten diaries of Adolf Hitler, which he was willing to sell at a price of USD 3.8 million. When asked where from did he procure those diaries, Dr. Fischer advocated that he recovered these from the site of a plane crash near Dresden in April 1945. The editorial team of *Der Stern* could confirm that a plane carrying personal belongings of Hitler did crash over Dresden in April 1945. In fact, after the crash Hitler had issued a statement expressing sorrow for the loss of his personal archives.

Before making the payment, *Der Stern* asked for the first volume of the manuscript, which Dr. Fischer promptly delivered. The editors got it screened by three renowned handwriting experts who certified that the calligraphic script indeed was of Hitler. Satisfied that the journals were genuine, the staff of *Der Stern* made payment to Dr. Fischer, for they felt sure that once they would bring out a series of stories on Hitler, many big publishing houses would purchase copyrights for reprinting the articles and, in effect, they would earn many times more than the initial investment.

The first story of the proposed series appeared in the April 25, 1983 issue of *Der Stern*. The cover page, shown in Fig. 32.5, carried the imprint, *Hitler's Tagebücher Entdeckt,* which means, Hitler's Diary Discovered. As anticipated, several international newspapers and magazines offered huge sums of money to *Der Stern* to reproduce that and subsequent write-ups. However, what was not anticipated was that after reading that article, many former aides of Hitler, including his ex-secretary, came forward to testify that the Nazi leader never wrote a diary.

Fig. 32.5 The April 25, 1983 issue of *Der Stern*

Contd..

Finally, West Germany's *Federal Archives* stepped in and decided to ascertain the authenticity of the diaries on the basis of a chemical test. This test is similar to the one used for qualitative detection of chloride ions. The fountain pen ink contains chloride ions. When ink is applied on a paper, the constituent chloride ions begin to migrate through the capillaries of the document. The maximum distance of migration is 3 mm from the point of ink application. It takes 2 years for chloride ions to cover this distance.

The scientists at *Federal Archives* sprayed a solution of silver nitrate on a scripted portion of the diary. The chloride content of the ink got precipitated as white silver chloride. When exposed to ultraviolet radiation (or even sunlight), silver chloride decomposed into finely divided, black, metallic silver and chlorine.

$$2AgCl \xrightarrow{h\upsilon} 2Ag + Cl_2$$

The black spots of metallic silver appeared less than 3 mm from the nearest letters. This proved that the diaries were written less than two years earlier—and hence were fraudulent.

Postscript: Dr. Fischer turned out to be a petty criminal, named Konrad Kujau, who specialized in forgery. He confessed to have forged several other documents, letters and even paintings of Hitler. He was sent to prison for 42 months. A journalist, Gerd Heidemann, who acted as a go-between Kujau and *Der Stern* received the same punishment. The court noted that Heidemann's assets had increased exponentially after he met Kujau and held it as a proof that all along he was aware that the diaries were forged. What about the three handwriting experts? When they had compared handwriting in the diaries to that in letters written by Hitler, they had pronounced it identical. Indeed it was. The letters which were given to them for comparison were not written by Hitler, but were earlier forgeries by Kujau!

Case Study 32.1 Why is it essential to acidify the sodium carbonate extract while performing the confirmatory tests for most anions?

The amount of sodium carbonate used for preparing the said extract is almost three times that of the inorganic mixture. A part of it remains unreacted and interferes with the confirmatory tests of many anions. For example chloride ions are detected by adding a solution of silver nitrate to the sodium carbonate extract. A white precipitate of silver chlorite is obtained. It is soluble in aqueous ammonia. The unreacted sodium carbonate in the extract would also react with silver nitrate solution to produce a while precipitate of silver carbonate. Even this precipitate is soluble in aqueous ammonia. Therefore it is pertinent that before adding the silver nitrate reagent, the unwanted sodium carbonate in the extract be neutralized with dilute nitric acid.

Likewise, sulfate ions are identified by the white precipitate of barium sulfate which results when a solution of barium chloride is added to the sodium carbonate extract. Suppose sulfate ions are absent and the extract is not acidified before adding barium chloride solution. A white precipitate of barium carbonate would be obtained and would misguide the analyst. Thus for virtually all the confirmatory tests, the excess sodium carbonate component of the extract must be removed by neutralizing it with a suitable acid.

Combination tests

A conformatory test may fail to validate an anion when the latter is present in combination with another specific anion. For such combinations, the analytical procedures have to follow a modified course.

1. Oxalate in presence of carbonate Both oxalates and carbonates yield white precipitates with calcium chloride solution.

$$C_2O_4^{2-} + CaCl_2 \rightarrow \underset{\text{White}}{CaC_2O_4\downarrow} + 2Cl^-$$

$$CO_3^{2-} + CaCl_2 \rightarrow \underset{\text{White}}{CaCO_3\downarrow} + 2Cl^-$$

No doubt, calcium oxalate is insoluble in acetic acid, while calcium carbonate is soluble, yet if the acidification is not done properly, the results may be misleading. It is therefore better to first add dilute sulfuric acid to the mixture and wait for the effervescence of carbon dioxide to cease. When a pinch of solid manganese dioxide is added, a fresh stream of carbon dioxide evolves and serves as a conformatory test for oxalate ions.

$$C_2O_4^{2-} + MnO_2 + 4H^+ \rightarrow Mn^{2+} + 2H_2O + 2CO_2\uparrow$$

2. Nitrate in presence of nitrite Nitrates are commonly confirmed by ring test. In this test, a saturated solution of ferrous sulfate is added to the water extract, containing nitrate ions. Thereafter, concentrated sulfuric acid is carefully added, whereupon a brown ring develops at interface of two liquids. If the water extract contains nitrite ions too, a similar ring may be formed with dilute sulfuric acid.

$$NO_2^- + H_2SO_4 \rightarrow HNO_2 + HSO_4^-$$

$$3HNO_2 \rightarrow HNO_3 + H_2O + 2NO$$

$$FeSO_4 + 5H_2O + NO \rightarrow \underset{\text{Brown ring}}{[Fe(H_2O)_5NO]SO_4}$$

In order to avoid this confusion, nitrite ions may first be confirmed by sulfanilic acid-1-naphthylamine test, and then decomposed to nitrogen by boiling with dilute sulfuric acid and urea.

$$NO_2^- + H_2SO_4 \rightarrow HNO_2 + HSO_4^-$$

$$2HNO_2 + NH_2CONH_2 \rightarrow 3H_2O + 2N_2 + CO_2$$

Nitrate may then be confirmed by ring test.

3. Bromide in presence of iodide Iodide may be identified by layer test. Thereafter, it may be removed as iodine vapors by boiling the sodium carbonate extract with sodium nitrite and dilute sulfuric acid.

$$2I^- + 2NO_2^- + 4H^+ \rightarrow 2H_2O + \underset{\text{Violet}}{I_2\uparrow} + 2NO\uparrow$$

Once the violet vapors of iodine cease to evolve, the layer test for bromide ions may be performed with the remaining extract.

4. Oxalate in presence of fluoride Oxalate ions do not interfere with detection of fluoride ions. The solid inorganic mixture, along with sand, is boiled in concentrated sulfuric acid. The appearance of oily droplets of silicic acid confirms fluoride ions.

Once the fluoride ions are confirmed, a few drops of calcium chloride solution are added to sodium carbonate extract, acidified with acetic acid. A white precipitate of calcium fluoride is obtained.

$$2F^- + CaCl_2 \rightarrow \underset{\text{White}}{CaF_2\downarrow} + 2Cl^-$$

If oxalate ions are also present, the precipitate, so obtained, is a mixture of calcium fluoride and calcium oxalate.

$$C_2O_4^{2-} + CaCl_2 \rightarrow \underset{\text{White}}{CaC_2O_4\downarrow} + 2Cl^-$$

In order to ascertain the presence of oxalate ions, the precipitate is collected and a few drops of acetic acid are added to it. Thereafter, two drops of dilute potassium permanganate solution are added. The contents acquire pink color. On warming, the pink color discharges.

Calcium fluoride does not participate in this reaction. However, the discoloration of potassium permanganate is brought about by calcium oxalate.

$$\begin{array}{rl} MnO_4^- + 8H^+ + 5e^- & \rightarrow Mn^{2+} + 4H_2O] \times 2 \\ C_2O_4^- & \rightarrow 2CO_2 + 2e^-] \times 5 \\ \hline 2MnO_4^- + 5C_2O_4^- + 16H^+ & \rightarrow 2Mn^{2+} + 8H_2O + 10CO_2\uparrow \end{array}$$

Thus if the pink color is discharged, the presence of oxalate ions is confirmed.

Analysis of Cations

For the purpose of systematic analysis, cations are divided into seven groups, numbered zero, I to VI. A sequence of tests is followed, starting from Group zero and ending at Group VI. These tests confirm the presence or absence of specific cations. However, before resorting to the stepwise analysis, it is better to perform a few preliminary procedures which, rather than conforming, just hint at the possibility of certain cations as being the constituents of the material under examination. In this section, we first describe these preliminary procedures and then present the scheme of conformatory tests. This is followed by a description of spot tests.

Borax bead test

The borax bead test is a means to examine the presence of certain cations in an inorganic material, albeit at a preliminary level; it should not be considered a proof of final validation. The test is conducted with the aid of a platinum wire, the free end of which is coiled into a loop (Fig. 32.6).

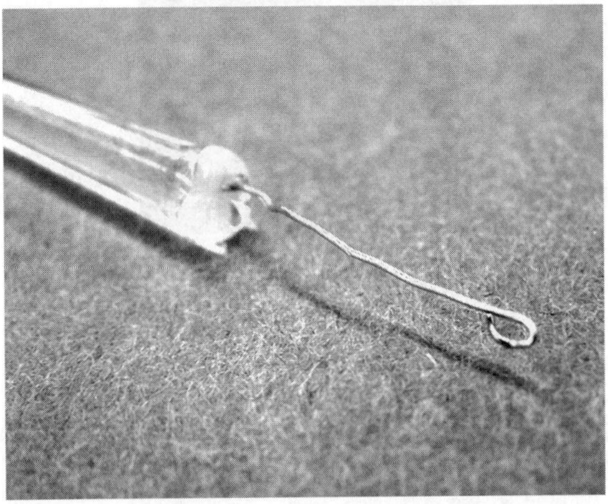

Fig. 32.6 A platinum wire

The loop is heated strongly till it becomes red-hot. Thereafter, it is dipped in borax, $Na_2B_4O_7{\cdot}10H_2O$ and held once again in the flame, whereupon it loses water of crystallization and swells.

$$Na_2B_4O_7{\cdot}10H_2O \xrightarrow{\text{Heat}} Na_2B_4O_7 + 10H_2O$$

On heating further, anhydrous borax decomposes into sodium metaborate, $NaBO_2$ and boron(III) oxide, B_2O_3.

$$Na_2B_4O_7 \xrightarrow{\text{Heat}} 2NaBO_2 + B_2O_3$$

The mixture of these two compounds forms a colorless, transparent, glass-like bead on the loop.

The bead is sprayed with 2–3 drops of water and then dipped into the test material. It must be ensured that only a meager amount of the salt(s) under question adhere to the bead. Otherwise, in the subsequent heating step, the bead darkens, as a result of which it becomes difficult to discern the presence of a particular cation.

The colorless, glassy bead, along with the adhering substance is once again injected into the flame, whereupon beads of characteristic colors are formed on the loop of the platinum wire. The color of the bead, in turn, depends on whether heating is carried out in the outer, oxidizing section of the flame (zone A in Fig. 32.7) or in the middle, reduction section (zone B in Fig. 32.7). In the oxidizing zone, metaborates of cations in higher oxidation states are formed; in reducing zone, either the metaborates of cations in lower oxidation state are formed or, at times, the free metal is deposited.

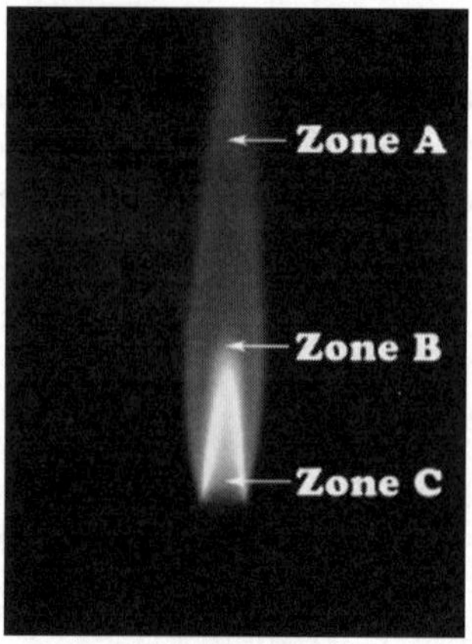

Zone A : Oxidizing
Zone B : Reducing
Zone C : Unburned gases and air

Fig. 32.7 Zones of the bunsen burner flame

For example, when a salt of copper is subjected to borax bead test, initially cupric oxide is formed.

$$CuCO_3 \xrightarrow{\text{Heat}} CuO + CO_2$$

In the oxidation zone of the flame, its reaction with boron(III) oxide yields a blue colored bead of copper(II) metaborate.

$$CuO + B_2O_3 \rightarrow \underset{\text{Blue bead}}{Cu(BO_2)_2}$$

In the reducing flame, copper(II) metaborate decomposes into metallic copper and the bead turns red and opaque.

$$Cu(BO_2)_2 + 2NaBO_2 + C \rightarrow Na_2B_4O_7 + \underset{\text{Red bead}}{Cu} + CO$$

The carbon which reduces copper(II) metaborate to copper is a component of unburned gases present in the inner, zone C of the flame (Fig. 32.7).

Likewise, the iron salts produce a yellow bead of iron(III) metaborate, $Fe(BO_2)_3$ in oxidizing flame and a green bead of iron(II) metaborate, $Fe(BO_2)_2$ in reducing flame.

$$4FeO + 6B_2O_3 + O_2 \rightarrow 4Fe(BO_2)_3$$

$$FeO + B_2O_3 \rightarrow Fe(BO_2)_2$$

Table 32.1 presents a schedule of the color of beads produced by a few cations in oxidizing and reducing zones, of the flame.

Table 32.1 Borax bead test

Metal ion	Color of the bead	
	Oxidizing flame	**Reducing flame**
Copper	Blue	Red
Iron	Yellow	Green
Chromium	Yellow	Green
Manganese	Pink	Colorless
Cobalt	Dark blue	Dark blue
Nickel	Yellow-brown	Grey

After each test, the platinum wire is cleansed by heating the bead to fusion and plunging it into water.

Microcosmic bead test

The microcosmic bead test is quite similar to the borax bead test, except that microcosmic salt, sodium ammonium hydrogen phosphate, $Na(NH_4)HPO_4{\cdot}4H_2O$ is used as the starting material. When this salt is heated on the loop of a platinum wire, it first loses the water of crystallization and then decomposes to produce a colorless, transparent bead of sodium metaphosphate, $NaPO_3$. For this reason, this investigation is also referred to as the *phosphate bead test*.

$$Na(NH_4)HPO_4{\cdot}4H_2O \xrightarrow{\text{Heat}} Na(NH_4)HPO_4 + 4H_2O$$

$$Na(NH_4)HPO_4 \xrightarrow{\text{Heat}} NaPO_3 + H_2O + NH_3$$

The phosphate bead is moistened with 2–3 drops of water and dipped in the material under examination. It is then held successively in the oxidizing and reducing zones of the flame (Fig. 32.7). As in the borox bead test, the salt first decomposes into the oxide, which reacts chemically with the sodium metaphosphate bead to yield colored bimetallic orthophosphates.

For example, if the microcosmic bead test is performed with a cobalt salt, a blue bead of sodium cobalt(II) orthophosphate is produced.

$$CoCO_3 \xrightarrow{\text{Heat}} CoO + CO_2$$

$$NaPO_3 + CoO \rightarrow \underset{\text{Blue bead}}{NaCoPO_4}$$

The sodium cobalt(II) orthophosphate is very stable and hence retains its composition in the reducing flame as well. For this reason, no change in color is observed when the bead is held in the reducing zone of the flame. However, the corresponding copper compound decomposes into metallic copper in the reducing flame to yield a red bead.

$$NaCuPO_4 + C \rightarrow \underset{\text{Red bead}}{Cu} + NaPO_3 + CO$$

Carbon is drawn from the inner zone of the flame (Fig. 32.7). Likewise, the brown bead of sodium nickel(II) orthophosphate, which results when a nickel salt is subjected to microcosmic test, turns grey in the reducing flame due to deposition of nickel on the loop of platinum wire.

$$NaNiPO_4 + C \rightarrow \underset{\text{Grey bead}}{Ni} + NaPO_3 + CO$$

With manganese salts, a similar reaction occurs in the reducing zone of the flame. A silver colored bead of metallic manganese is anticipated in this case. However, the silver colored particles of manganese get enmashed into sodium metaphosphate, as a result of which the bead appears colorless.

$$NaMnPO_4 + C \rightarrow \underbrace{\underset{\text{Silver}}{Mn} + NaPO_3}_{\text{Colorless}} + CO$$

The results of microcosmic bead test are depicted in Table 32.2.

Table 32.2 Microcosmic bead test

Metal ion	Color of the bead	
	Oxidizing flame	**Reducing flame**
Copper	Blue	Red
Iron	Yellow	Green
Chromium	Green	Green
Manganese	Violet	Colorless
Cobalt	Blue	Blue
Nickel	Brown	Grey

When the oxidation state of metal is (III), a mixture of metal(III) phosphate and sodium phosphate is produced in the oxidizing flame.

$$Cr_2O_3 + 3NaPO_3 \rightarrow \underset{\text{Green bead}}{2CrPO_4} + Na_3PO_4$$

Chrominum(III) phosphate produces a green bead, which remains unchanged in the reducing flame as chromium(III) ion is resistant to reduction. The iron salts produce a yellow bead of iron(III) phosphate in the oxidizing flame.

$$Fe_2O_3 + 3NaPO_3 \rightarrow \underset{\text{Yellow bead}}{2FePO_4} + Na_3PO_4$$

In reducing flame, the following reaction takes place.

$$6FePO_4 + 2Na_3PO_4 + 7C \rightarrow \underset{\text{Green bead}}{2Fe_3(PO_4)_2} + 3Na_2O + P_4O_6 + 7CO$$

The iron(II) phosphate produces a green bead on the platinum wire.

Sodium carbonate bead test

The sodium carbonate bead test endeavors to examine the presence of manganese and chromium ions in the given inorganic material. The base chemical in this test is sodium carbonate, which is fused to form a bead on the loop of a platinum wire.

To about 10 mg of sodium carbonate, a crystal of potassium nitrate is added. When the mixture is heated on the platinum wire loop, a white, opaque bead is formed. Potassium nitrate is added to lower the melting point of sodium carbonate, as a result of which the mass fuses into a bead more easily. The bead is moistened with 2–3 drops of water and then dipped into the material under examination. The contents are inserted in the oxidation zone of the flame and the color of the bead is recorded.

If the given substance happens to be a salt of manganese, it first decomposes into manganese(II) oxide and then combines with fused sodium carbonate, in presence of air, to yield a green bead of sodium manganate, Na_2MnO_4.

$$MnO + Na_2CO_3 + O_2 \rightarrow \underset{\text{Green bead}}{Na_2MnO_4} + CO_2$$

In case of a chromium salt, the chromium(III) oxide, formed as a result of initial decomposition, reacts with fused sodium carbonate, in an oxidizing environment, to produce a yellow bead of sodium chromate, Na_2CrO_4.

$$2Cr_2O_3 + 4Na_2CO_3 + 3O_2 \rightarrow \underset{\text{Yellow bead}}{4Na_2CrO_4} + 4CO_2$$

Charcoal cavity test

The charcoal cavity test is based on the premise that metallic carbonates, when strongly heated, decompose into the corresponding oxides and in some cases, to metal itself. The color or the physical state of the product assists in establishing the identity of the metal ion.

The test is performed on a charcoal block, in the center of which a cavity has been created. About 10 mg of the substance is placed in the cavity and mixed with about double the amount of sodium carbonate. The mixture is heated by directing the flame of the bunsen burner into the cavity with the aid of a blow pipe (Fig. 32.8). A crust of metal oxide or of metal becomes coated on the inside walls of the cavity. In some cases the deposition is a mixture of oxide crust and metal.

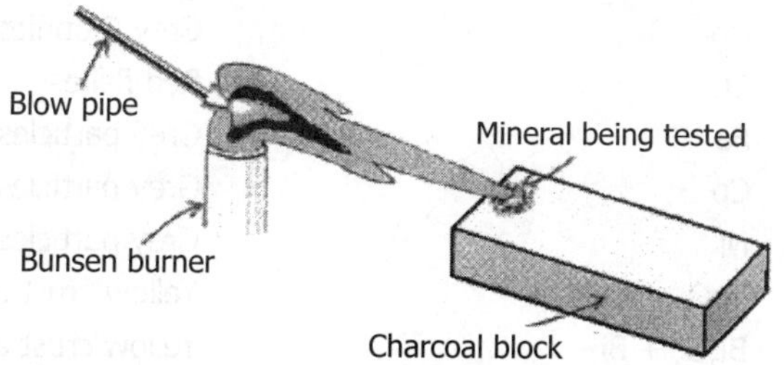

Fig. 32.8 Equipment for charcoal cavity test

Suppose a salt of zinc is subjected to charcoal cavity test. Initially zinc carbonate is formed, which subsequently decomposes into zinc oxide.

$$ZnSO_4 + Na_2CO_3 \xrightarrow{\text{Fuse}} ZnCO_3 + Na_2SO_4$$

$$ZnCO_3 \xrightarrow{\text{Heat}} ZnO + CO_2$$

The crust of zinc oxide is yellow when hot and white when cold (*see* Case Study 3.5, p. 96). A cadmium salt under similar circumstances yields a brown crust.

Copper salts yield the oxide, which is further reduced to metallic copper by the charcoal content of the cavity.

$$CuSO_4 + Na_2CO_3 \xrightarrow{\text{Fuse}} CuCO_3 + Na_2SO_4$$

$$CuCO_3 \xrightarrow{\text{Heat}} CuO + CO_2$$

$$CuO + C \rightarrow \underset{\text{Red flakes}}{Cu} + CO$$

Copper becomes deposited in the cavity as red flakes. Likewise, lead salts yield a grey bead of metallic lead, along with a yellow crust of lead oxide.

$$Pb(NO_3)_2 + Na_2CO_3 \xrightarrow{\text{Fuse}} PbCO_3 + 2NaNO_3$$

$$PbCO_3 \xrightarrow{\text{Heat}} \underset{\text{Yellow crust}}{PbO} + CO_2$$

$$PbO + C \rightarrow \underset{\text{Grey bead}}{Pb} + CO$$

The results of the charcoal cavity test for a few metal ions are set out in Table 32.3.

Table 32.3 Charcoal cavity test

Metal ion	Deposition in charcoal cavity	Color/Physical state
Zinc	ZnO	Crust; Yellow while hot, white while cold
Cadmium	CdO	Brown crust
Aluminium	Al_2O_3	White crust
Calcium	CaO	White crust
Strontium	SrO	White crust
Barium	BaO	White crust
Magnesium	MgO	White crust
Mercury	Hg	Grey Globules
Copper	Cu	Red flakes
Iron	Fe	Grey particles
Cobalt	Co	Grey particles
Nickel	Ni	Grey particles
Lead	PbO + Pb	Yellow crust and grey bead
Bismuth	Bi_2O_3 + Bi	Yellow crust and silver bead

The cavity is cleansed by scrapping the adhering material with a spatula.

Ash test

Also called *cobalt nitrate test,* ash test may be considered as an extension of charcoal cavity procedure. As depicted in Table 32.3, the salts of aluminium, zinc and magnesium, when subjected to charcoal cavity test leave a white crustal deposition of their respective oxides. In the ash test, the white deposition in the cavity is treated with two drops of a solution of cobalt(II) nitrate and heating is continued with the aid of the blow pipe. Cobalt(II) nitrate decomposes into cobalt(II) oxide,

$$2Co(NO_3)_2 \xrightarrow{\text{Heat}} 2CoO + 4NO_2 + O_2$$

which then combines with the crustal oxide deposition.

If aluminium oxide is present in the charcoal cavity, its reaction with cobalt(II) nitrate yields a solid solution having the composition, $Al_2O_3{\cdot}CoO$.

$$Al_2O_3 + CoO \rightarrow \underset{\text{Blue}}{Al_2O_3{\cdot}CoO}$$

This solid solution is blue in color and is called *Thenard's blue*. With zinc oxide, a green colored solid solution, called *Rinmann's green* results.

$$ZnO + CoO \rightarrow \underset{\text{Green}}{ZnO{\cdot}CoO}$$

Likewise, magnesium oxide gives a pink mass having analogous composition.

$$MgO + CoO \rightarrow \underset{\text{Pink}}{MgO{\cdot}CoO}$$

Flame test

The flame of the bunsen burners has appropriate energy to excite the electrons of some of the metal ions to higher stationary states. The energy difference between the stationary states is quantized, and corresponds to a specific wavelength. When the electron reverts to its original position, the energy that it gained during excitation is emitted as radiation. If the wavelength of the emitted radiation falls in the visible region of the electromagnetic spectrum (λ = 400–700 nm), a particular color is imparted to the flame. The flame coloration serves as a means of identification of these cations (*see* Chapter 12, p. 422). The cations which may be detected on the basis of this criteria are listed in Table 32.4, along with the color that these impart to the flame.

Table 32.4 Flame coloration of a few cations

Cation	Flame coloration
Sodium	Golden-yellow
Calcium	Brick-red
Strontium	Crimson
Barium	Green
Potassium	Lilac (pink-violet)

Flame tests are performed with a platinum wire. The latter is washed successively with concentrated hydrochloric acid and distilled water. Its free end is then bent to form a loop, as shown in Fig. 32.6. The substance in question is placed in a watch glass and a few drops of concentrated hydrochloric acid are added to form a paste. The loop of the platinum wire is first dipped into the paste and then held in the blue zone of the bunsen flame. The color imparted to the flame is recorded.

The flame coloration is a means to further authenticate what has already been confirmed on the basis of chemical tests. The chemical test of a particular ion usually culminates in the appearance of a precipitate, the color of which is specific to that ion. As indicated in Tables 32.13, later in this chapter, the flame tests should be performed with these precipitates and not with the mixture of salts. For example, a mixture of calcium and strontium salts, when subjected to flame test, will give a coloration bordering orange, as a result of which it will not be possible to validate either of the two cations.

Further, while performing the flame tests, the platinum wire should not be substituted by a glass rod. Glass is predominantly sodium silicate. When held inside the flame, it gives a very bright golden-yellow flame which is attributed to sodium. It masks the actual flame coloration by which a particular cation may be identified.

Schematic analysis of cations

The first step in the schematic detection of cations is to solubilize these in a suitable solvent. In the order of preference, the following solvents need be tried out for checking the solubility of the given inorganic material.

1. Distilled water, cold
2. Distilled water, hot
3. Dilute hydrochloric acid, cold
4. Dilute hydrochloric acid, hot
5. Concentrated hydrochloric acid, cold
6. Concentrated hydrochloric acid, hot
7. Dilute nitric acid, cold
8. Dilute nitric acid, hot
9. Concentrated nitric acid, cold
10. Concentrated nitric acid, hot
11. Aqua regia

The solubility may be tested by placing about 15 mg of the unknown substance in a semi-micro test tube and adding about 5 mL of the solvent. Once the appropriate solvent is established, a saturated solution is prepared as follows. About 8 mL of the solvent is added to about 1 g of the given salt or a mixture of salts. The contents are thoroughly agitated and centrifuged. The centrifugate serves as the original solution, so named because the entire procedure of cations' identification has to be carried out by using a single aliquot.

If the solvent of choice is nitric acid (dilute or concentrated) or aqua regia, an additional step need be followed. The solution is boiled gently to complete evaporation. The residue is dissolved in distilled water. If any residue still remains it is filtered and discarded. The centrifugate so obtained now serves as the original solution (*see* Case Study 32.2).

Case Study 32.2 When the analyte is soluble in nitric acid or aqua regia, why is the final test solution prepared in distilled water?

If the material in question is soluble in dilute or concentrated nitric acid or in aqua regia, the solution, so prepared, is boiled to dryness and the residue is dissolved in distilled water. As we shall see further, the precipitating reagent for the cations of both Group II and Group IV is hydrogen sulfide. In presence of nitric acid, hydrogen sulfide is oxidized to colloidal sulfur.

$$2HNO_3 + H_2S \rightarrow S + 2H_2O + 2NO_2$$

The colloidal sulfur remains suspended in the solution and is very difficult to get rid off by centrifugation. On the other hand, by boiling the inorganic salts in nitric acid, all the cations therein are transformed to their respective nitrates. We have already stated that all nitrates are soluble in water. Hence it is pertinent to prepare the final test solution in distilled water and proceed with the analysis.

Group zero cation

The lone member of Group zero, the ammonium ion NH_4^+ may be identified by heating the solid salt or a mixture of salts with sodium hydroxide solution. The ammonium ion, if present, decomposes into gaseous ammonia.

$$NH_4^+ + NaOH \rightarrow Na^+ + H_2O + NH_3$$

Ammonia may be tested by bringing a glass rod dipped in dilute hydrochloric acid at the mouth of the test tube, whereupon dense white fumes of ammonium chloride are generated.

$$NH_3 + HCl \rightarrow \underset{\text{White}}{NH_4Cl\uparrow}$$

The presence of ammonium ions may be further confirmed by Nessler's reagent test. Nessler's reagent is an alkaline solution of potassium tetraiodomercurate(II) ($K_2[HgI_4]$ + KOH). When this reagent is added to the water extract of an unknown salt(s), a brown precipitate of basic amidoiodomercury(II) is formed if ammonium ions are one of the constituents of the substance in question.

$$NH_4^+ + 2[HgI_4]^{2-} + 4OH^- \rightarrow \underset{\substack{\text{Basic amidoiodomercury(II)}\\ \text{Brown}}}{HgO \cdot Hg(NH_2)I\downarrow} + 7I^- + 3H_2O$$

Group I cations

The three cations of Group I are—lead (Pb^{2+}), silver (Ag^{+}) and mercurous (Hg_2^{2+}). On adding dilute hydrochloric acid, these are precipitated as respective chlorides, all of which are white in color. If the original solution has been prepared in concentrated hydrochloric acid, then rather than adding the precipitating reagent (dilute hydrochloric acid), the contents should be diluted with distilled water. Excess hydrochloric acid should be avoided (*see* Case Study 32.3). The contrifugate is retained for the analysis of Group II. The residue, comprising one or more chlorides of Group I cations is treated in the sequence described in Table 32.5.

Table 32.5 Analysis of Group I cations

<table>
<tr><td colspan="3">A. $PbCl_2$ (white), AgCl (white), Hg_2Cl_2 (white).
Suspend the precipitate in distilled water and boil for about 5 minutes.
Centrifuge while hot.</td></tr>
<tr><td rowspan="2">B. Centrifugate. $PbCl_2$, dissolved in hot water. Divide into 3 parts.
(i) Add acetic acid and potassium chromate solution.
Yellow precipitate of lead chromate.
(ii) Add potassium iodide solution.
Yellow precipitate of lead iodide.
(iii) Add dilute sulfuric acid.
White precipitate of lead sulfate.
Pb^{2+} confirmed</td><td colspan="2">C. Residue. AgCl (white), Hg_2Cl_2 (white). Add a few drops of 1:1 ammonia solution and centrifuge.</td></tr>
<tr><td>D. Centrifugate. $[Ag(NH_3)_2]Cl$. Divide into 2 parts.
(i) Add potassium chromate solution.
Red precipitate of silver chromate.
(ii) Add potassium iodide solution.
Yellow precipitate of silver iodide.
Ag^{+} confirmed</td><td>E. Residue. $Hg(NH_3)Cl$ + Hg (black).
Dissolve the residue in aqua regia. Discard the suspended material, if any. Evaporate to dryness. Dissolve the residue in distilled water. Add stannous chloride solution.
White precipitate (Hg_2Cl_2) turning grey (Hg).
Hg_2^{2+} confirmed</td></tr>
</table>

Reactions involved in Table 32.5

Section A

$$Pb^{2+} + 2HCl \rightarrow PbCl_2\downarrow + 2H^{+}$$

$$Ag^{+} + HCl \rightarrow AgCl\downarrow + H^{+}$$

$$Hg_2^{2+} + 2HCl \rightarrow Hg_2Cl_2\downarrow + 2H^{+}$$

Section B

(i) $PbCl_2 + K_2CrO_4 \rightarrow PbCrO_4\downarrow + 2KCl$

Lead chromate
Yellow

(ii) $PbCl_2 + 2KI \rightarrow PbI_2\downarrow + 2KCl$

Yellow

(iii) $PbCl_2 + H_2SO_4 \rightarrow PbSO_4\downarrow + 2HCl$

Section C

$AgCl + 2NH_3 \rightarrow [Ag(NH_3)_2]Cl$

Diamminesilver(I) chloride

$Hg_2Cl_2 + 2NH_3 \rightarrow Hg\downarrow + Hg(NH_2)Cl + NH_4Cl$

Amidochloromercury(II)

Black

Section D

(i) $2[Ag(NH_3)_2]Cl + K_2CrO_4 \rightarrow Ag_2CrO_4\downarrow + 2KCl + 4NH_3$

Silver chromate
Red

(ii) $[Ag(NH_3)_2]Cl + KI \rightarrow AgI\downarrow + KCl + 2NH_3$

Yellow

Section E

$Hg(NH_2)Cl + Hg + 2HNO_3 + 6HCl \rightarrow 2HgCl_2 + NH_4NO_3 + NCl_3 + 3H_2O$

$2HgCl_2 + SnCl_2 \rightarrow Hg_2Cl_2\downarrow + SnCl_4$

White

$Hg_2Cl_2 + SnCl_2 \rightarrow 2Hg\downarrow + SnCl_4$

Grey

Case Study 32.3 Why excess of dilute hydrochloric acid should not be added for precipitating Group I cations?

As has already been discussed, the cations of Group I—Pb^{2+}, Ag^+ and Hg_2^{2+}—are precipitated as their chlorides—$PbCl_2$, AgCl and Hg_2Cl_2—by adding dilute hydrochloric acid to the original solution. However, if excess hydrochloric acid is added, silver chloride and lead chloride tend to form complexes.

$AgCl + HCl \rightarrow [AgCl_2]^- + H^+$

Dichloroargentate(I) ion

$PbCl_2 + 2HCl \rightarrow [PbCl_4]^{2-} + 2H^+$

Tetrachloroplumbate(II) ion

The dichloroargentate(I) ions and tetrachloroplumbate(II) ions, so formed, are soluble in water. As a result, these are carried along with the original solution to Group II. Hence by adding excess hydrochloric acid, silver and lead ions may be missed out in Group I.

Excess hydrochloric acid poses yet another problem. As has been depicted in Section B of Table 32.5, one of the confirmatory tests of lead ions involve addition of a solution of potassium chromate, whereby a yellow precipitate of lead chromate is obtained. In strongly acidic solutions, the chromate ions are converted to dichromate ions.

$$\underset{\text{Chromate ion}}{2CrO_4^{2-}} + 2H^+ \rightarrow \underset{\text{Dichromate ion}}{Cr_2O_7^{2-}} + H_2O$$

As a result, lead ions will not be precipitated as lead chromate.

Group II cations

Group II has the distinction of having the largest number of cations in its ambit. These are—mercuric (Hg^{2+}), lead (Pb^{2+}), bismuth (Bi^{3+}), cupric (Cu^{2+}), cadmium (Cd^{2+}), arsenic (As^{3+}), antimony (Sb^{3+}) and tin (Sn^{2+}).

The precipitating reagent is hydrogen sulfide gas, in presence of dilute hydrochloric acid. If one has arrived at Group II after systematically analyzing the cations of Group I then hydrochloric acid need not be added. The reason being that dilute hydrochloric acid was the precipitating reagent for Group I cations and the original solution would already be acidic. In any case, excess of hydrochloric acid must be avoided (*see* Case Study 32.3).

Once the cations of Group II have been precipitated as respective sulfides, the centrifugate is retained for analysis of Group III. The sulfides of Group II cations are further split up into Group IIA and Group IIB

Group IIA: HgS (black) PbS (black) Bi_2S_3 (brown) Cus (black) CdS (yellow)

Group IIB: As_2S_3 (yellow) Sb_2S_3 (orange) SnS (yellow)

The segregation into Group IIA and Group IIB is brought about by adding yellow ammonium sulfide solution. The sulfides of Group IIB cations are soluble in yellow ammonium sulfide due to the formation of ammonium thio salts; those of Group IIA are not. The steps involved in the precipitation of Group II cations as sulfides and their separation into Group IIA and Group IIB are listed in Table 32.6.

Table 32.6 Analysis of Group II cations

A. HgS (black), PbS (black), Bi_2S_3 (brown), CuS (black), CdS (yellow). As_2S_3 (yellow), Sb_2S_3 (orange), SnS (yellow). Wash the precipitates with distilled water and add yellow ammonium sulfide solution. Agitate and centrifuge.	
B. Residue. HgS (black), PbS (black), Bi_2S_3 (brown), CuS (black), CdS (yellow).	C. Centrifugate. $(NH_4)AsS_3$, $(NH_4)SbS_3$, $(NH_4)_2SnS_3$.
Continued in Table 32.8, p. 923	*Continued in Table 32.9, p. 927*

Reactions involved in Table 32.6

Section A

$$Hg^{2+} + H_2S \rightarrow HgS\downarrow + 2H^+$$

$$Pb^{2+} + H_2S \rightarrow PbS\downarrow + 2H^+$$

$$2Bi^{3+} + 3H_2S \rightarrow Bi_2S_3\downarrow + 6H^+$$

$$Cu^{2+} + H_2S \rightarrow CuS\downarrow + 2H^+$$

$$Cd^{2+} + H_2S \rightarrow CdS\downarrow + 2H^+$$

$$2As^{3+} + 3H_2S \rightarrow As_2S_3\downarrow + 6H^+$$

$$2Sb^{3+} + 3H_2S \rightarrow Sb_2S_3\downarrow + 6H^+$$

$$Sn^{2+} + H_2S \rightarrow SnS\downarrow + 2H^+$$

Section B Nil

Section C

Yellow ammonium sulfide, also called ammonium polysulfide is a mixture of normal ammonium sulfide and sulfur. Its composition is written as $(NH_4)_2S_x$, but normally taken as $(NH_4)_2S_2$. It is oxidizing in nature and transforms the cations of Group IIB into higher oxidation states. Thus arsenic(III) sulfide is oxidized to soluble ammonium trithioarsenate(V), $(NH_4)AsS_3$; arsenic(III) sulfide to ammonium trithioantimonate, $(NH_4)SbS_3$; and tin(II) sulfide to ammonium trithiostannate(IV), $(NH_4)_2SnS_3$.

$$As_2S_3 + 2(NH_4)_2S_2 \rightarrow 2(NH_4)AsS_3 + (NH_4)_2S$$

$$Sb_2S_3 + 2(NH_4)_2S_2 \rightarrow 2(NH_4)SbS_3 + (NH_4)_2S$$

$$SnS + (NH_4)_4S_2 \rightarrow (NH_4)_2SnS_3$$

Case Study 32.4 For the precipitation of Group II cations, why hydrogen sulfide ought to be passed through the original solution in presence of dilute hydrochloric acid?

According to one of the most fundamental principles of ionic equilibria, precipitation takes place only when the ionic product of an electrostatic compound exceeds its solubility product. Ionic product is the product of the concentration of ions furnished by a dissolved salt at a specific temperature. Solubility product is the product of concentration of ions furnished by a saturated solution of a salt at a particular temperature.

If hydrogen sulfide gas alone is passed through the original solution during cation analysis, the solubility product of Group II metal ions, no doubt, will be exceeded by their ionic products and these will be precipitated as their respective sulfides. However, the solubility products of the sulfides of Group III, IV, etc. cations will also be exceeded by their respective ionic products, as a result of which these too will get precipitated along with Group II metal sulfides.

When hydrogen sulfide gas is passed through the original solution in presence of dilute hydrochloric acid, the following ions are generated in the reaction system.

$$H_2S \rightleftharpoons 2H^+ + S^{2-}$$

$$HCl \rightarrow H^+ + Cl^-$$

The protons are the common ions in these two reactions. As a result of common ion effect, the concentration of sulfide ions is suppressed, and in consequence, the value of the ionic product is also reduced. Nevertheless, the ionic product can still exceed the solubility product of Group II metal sulfides and precipitate these. However, the solubility products of the metal sulfides of subsequent groups are appreciably high and cannot be exceeded by their ionic products. Hence these are not precipitated along with Group II cations.

Box 32.2

The Dual Membership of Lead Ions

Lead ions find a place in both Group I and Group II. Ideally, these should be precipitated as insoluble lead chloride in Group I, along with the chlorides of the other two members, silver ions and mercurous ions. However, as the data in Table 32.7 reveals, there is enormous difference between the solubility of lead chloride on one hand and the solubility of silver chloride and mercurous chloride on the other.

Table 32.7 Solubilities of Group I chlorides at 298 K

Salt	Solubility product	Molar solubility	Solubility (g/L) in water
$PbCl_2$	1.7×10^{-5}	1.6×10^{-2} M	4.5
$AgCl$	1.8×10^{-10}	1.3×10^{-5} M	1.9×10^{-3}
Hg_2Cl_2	1.1×10^{-18}	6.5×10^{-7} M	3.1×10^{-4}

Since lead chloride is much more soluble in water than silver chloride and mercurous chloride, chances are that it might not get precipitated in Group I, and pass on to Group II. It is for this reason that lead ions are housed both in Group I and Group II.

Group IIA cations

Once the cations of Group II have been broadly separated into Group IIA and Group IIB, it is pertinent to proceed to the analysis of the former. As has been depicted in Section B of Table 32.6, the Group IIA cations are precipitated as their respective sulfides. Save mercuric sulfide, all other sulfides of this group are soluble in dilute nitric acid. The mercuric sulfide, on the other hand, may be solubilized in aqua regia. The sequence of operations to systematically analyze these cations is displayed in Table 32.8.

Table 32.8 Analysis of Group IIA cations

<table>
<tr><td colspan="4">A. Continued from Section B, Table 32.6
HgS (black), PbS (black), Bi_2S_2 (brown), CuS (black), CdS (yellow).
Add about 3 mL dilute nitric acid. Boil gently and centrifuge.</td></tr>
<tr><td rowspan="3">B. Residue. HgS (black). Dissolve in aqua regia. Discard the suspended matter, if any. Evaporate to dryness. Dissolve the residue in distilled water. Add stannous chloride solution.
White precipitate (Hg_2Cl_2) turning grey (Hg).

Hg^{2+} confirmed</td><td colspan="3">C. Centrifugate. $Pb(NO_3)_2$, $Bi(NO_3)_3$, $Cu(NO_3)_2$, $Cd(NO_3)_2$. Separate 5 drops of the solution and add 2 drops of dilute sulfuric acid to it. If a white precipitate is obtained, add dilute sulfuric acid to the bulk solution and proceed to Section D. Otherwise skip this step and go directly to Section E.</td></tr>
<tr><td rowspan="2">D. Residue. $PbSO_4$ (white). Add ammonium acetate solution. Lead sulfate partially dissolves as lead acetate. Centrifuge and divide the centrifugate into 2 parts.
(i) Add acetic acid and potassium chromate solution.
Yellow precipitate of lead chromate.
(ii) Add potassium iodide solution.
Yellow precipitate of lead iodide.

Pb^{2+} confirmed</td><td colspan="2">E. Centrifugate. $Bi(NO_3)_3/Bi_2(SO_4)_3$, $Cu(NO_3)_2/CuSO_4$, $Cd(NO_3)_2/CdSO_4$.
Add excess of 1:1 ammonia solution and centrifuge.</td></tr>
<tr><td>F. Residue. $Bi(OH)_3$ (white). Dissolve the residue in dilute hydrochloric acid and divide into 2 parts.
(I) Add excess water. White precipitate of bismuth oxo chloride.
(ii) Add sodium stannite reagent. Black precipitate of elemental bismuth.

Bi^{3+} confirmed</td><td>G. Centrifugate. $[Cu(NH_3)_4]^{2+}$ (blue), $[Cd(NH_3)_4]^{2+}$ (colorless). Divide into 2 parts.
(i) Add acetic acid and potassium ferrocyanide solution. Brown precipitate of copper ferrocyanide.

Cu^{2+} confirmed

(ii) Dilute the centrifugate almost 5 times and pass hydrogen sulfide gas.
Yellow precipitate of cadmium sulfide.

Cd^{2+} confirmed</td></tr>
</table>

Reactions involved in Table 32.8

Section A

$$3PbS + 8HNO_3 \rightarrow 3Pb(NO_3)_2 + 3S + 4H_2O + 2NO$$

$$Bi_2S_3 + 8HNO_3 \rightarrow 2Bi(NO_3)_3 + 3S + 4H_2O + 2NO$$

$$3CuS + 8HNO_3 \rightarrow 3Cu(NO_3)_2 + 3S + 4H_2O + 2NO$$

$$3CdS + 8HNO_3 \rightarrow 3Cd(NO_3)_2 + 3S + 4H_2O + 2NO$$

Section B

$$3HgS + 2HNO_3 + 6HCl \rightarrow 3HgCl_2 + 3S + 4H_2O + 2NO$$

$$HgCl_2 + SnCl_2 \rightarrow \underset{\text{White}}{Hg_2Cl_2\downarrow} + SnCl_4$$

$$Hg_2Cl_2 + SnCl_4 \rightarrow \underset{\text{Grey}}{2Hg\downarrow} + SnCl_4$$

Section C

$$Pb(NO_3)_2 + H_2SO_4 \rightarrow \underset{\text{White}}{PbSO_4\downarrow} + 2HNO_3$$

If lead ions are absent, dilute sulfuric acid need not be added to the bulk solution. The remaining cations continue to be present as nitrates. If lead ions are present, and sulfuric acid is added to the bulk solution the remaining cations will be transformed to their respective sulfates.

$$2Bi(NO_3)_3 + 3H_2SO_4 \rightarrow Bi_2(SO_4)_3 + 6HNO_3$$

$$Cu(NO_3)_2 + H_2SO_4 \rightarrow CuSO_4 + 2HNO_3$$

$$Cd(NO_3)_2 + H_2SO_4 \rightarrow CdSO_4 + 2HNO_3$$

Section D

$$PbSO_4 + 2CH_3COONH_4 \rightarrow (CH_3COO)_2Pb + (NH_4)_2SO_4$$

$$\text{(i) } (CH_3COO)_2Pb + K_2CrO_4 \rightarrow \underset{\text{Yellow}}{PbCrO_4\downarrow} + 2CH_3COOK$$

$$\text{(ii) } (CH_3COO)_2Pb + 2KI \rightarrow \underset{\text{Yellow}}{PbI_2\downarrow} + 2CH_3COOK$$

Section E

$$Bi^{3+} + 3NH_3 + 3H_2O \rightarrow \underset{\text{White}}{Bi(OH)_3\downarrow} + 3NH_4^+$$

$$Cu^{2+} + 4NH_3 \rightarrow \underset{\text{Blue}}{[Cu(NH_3)_4]^{2+}}$$

$$Cd^{2+} + 4NH_3 \rightarrow \underset{\text{Colorless}}{[Cd(NH_3)_4]^{2+}}$$

Section F

$$Bi(OH)_3 + 3HCl \rightarrow BiCl_3 + 3H_2O$$

(i) $BiCl_3 + H_2O \rightarrow BiOCl\downarrow + 2HCl$
White

(ii) Sodium stannite reagent may be prepared by dropwise adding sodium hydroxide solution to stannous chloride solution. Initially a white precipitate of stannous hydroxide is obtained, which dissolves in excess sodium hydroxide to give sodium stannite.

$$SnCl_2 + 2NaOH \rightarrow \underset{\text{White}}{Sn(OH)_2\downarrow} + 2NaCl$$

$$Sn(OH)_2 + 2NaOH \rightarrow \underset{\text{Sodium stannite}}{Na_2SnO_2} + 2H_2O$$

$$2BiCl_3 + 3Na_2SnO_2 + 3H_2O \rightarrow \underset{\text{Sodium stannate}}{3Na_2SnO_3} + 6HCl + \underset{\text{Black}}{2Bi\downarrow}$$

Section G

If the color of centrifugate is blue, cupric ions are certainly present, cadmium ions may or may not be present. If the centrifugate is colorless, cupric ions are absent, cadmium ions may or may not be present.

(i) $[Cu(NH_3)_4]^{2+} + 4CH_3COOH \rightarrow Cu^{2+} + 4CH_3COONH_4$

$$2Cu^{2+} + K_4[Fe(CN)_6] \rightarrow \underset{\text{Brown}}{Cu_2[Fe(CN)_6]\downarrow} + 4K^+$$

(ii) $[Cd(NH_3)_4]^{2+} + H_2S \rightarrow \underset{\text{Yellow}}{CdS\downarrow} + 2NH_4^+ + 2NH_3$

Box 32.3

When Cupric and Cadmium Ions are Present Together

As depicted in Section G of Table 32.8, if either cupric ions or cadmium ions are present in a mixture of salts, then the former may be identified by potassium ferrocyanide test, while the latter may be identified by diluting the test solution and passing hydrogen sulfide gas. If cupric ions are absent, and cadmium ions are present, addition of potassium ferrocyanide solution yields a white precipitate of cadmium ferrocyanide.

$$2Cd^+ + K_4[Fe(CN)_6] \rightarrow \underset{\text{White}}{Cd_2[Fe(CN)_6]\downarrow} + 4K^+$$

The problem arises when cupric and cadmium ions are present together in a mixture of salts. Under these circumstances, cupric ions may still be detected by adding potassium ferrocyanide, whereby a brown precipitate of cupric ferrocyanide in obtained.

Contd...

$$2Cu^{2+} + K_4[Fe(CN)_6] \rightarrow \underset{\text{Brown}}{Cu_2[Fe(CN)_6]\downarrow} + 4K^+$$

However, the brown precipitate masks the white precipitate of cadmium ferrocyanide, thereby precluding the identification of cadmium ions.

In presence of cupric ions, cadmium ions may be detected by the appearance of the yellow precipitate of cadmium sulfide when hydrogen sulfide gas is passed through the diluted test solution. However, before performing this test, cupric ions must be removed from the solution. Otherwise, when hydrogen sulfide gas is passed a black precipitate of copper sulfide will be obtained which will mask the yellow precipitate of cadmium sulfide.

Hence when cupric and cadmium ions are present in combination, one has to proceed as follows. As indicated in Section G of Table 32.8, divide the test solution into two parts. In the first part ascertain the presence of cupric ions by potassium ferrocyanide test. In the second part, remove the cupric ions by either of the following two methods.

1. Add sodium dithionite solution to the centrifugate containing a mixture of tetramminecopper(II) ions, $[Cu(NH_3)_4]^{2+}$ and tetramminecadmium(II) ions, $[Cd(NH_3)_4]^{2+}$. The former is reduced to metallic copper.

$$[Cu(NH_3)_4]^{2+} + S_2O_4^{2-} + 2H_2O \rightarrow 4NH_4^+ + 2SO_3^{2-} + Cu\downarrow$$

The sodium dithionite reagent, however, does not react with tetramminecadmium(II) ions. The precipitated copper is removed as residue and hydrogen sulfide gas is passed through the centrifugate, whereby a yellow precipitate of cadmium sulfide confirms the presence of cadmium ions.

$$[Cd(NH_3)_4]^{2+} + H_2S \rightarrow \underset{\text{Yellow}}{CdS\downarrow} + 2H^+ + 4NH_3$$

2. Add a solution of potassium cyanide to the test solution. The metal ion in tetramminecopper(II) ion is reduced from oxidation state (II) to (I), followed by the formation of an extremely stable dicyanocuprate(I) ion.

$$2[Cu(NH_3)_4]^{2+} + 6CN^- \rightarrow 2[Cu(CN)_2]^- + \underset{\text{Cyanogen}}{(CN)_2} + 8NH_3$$

The dicyanocuprate(I) ion does not react with hydrogen sulfide gas.

$$[Cu(CN)_2]^- + H_2S \rightarrow \text{No reaction}$$

The tetramminecadmium(II) ion reacts with potassium cyanide reagent to yield a relatively unstable tetracyanocadmate(II) ion, which reacts with hydrogen sulfide gas to yield a yellow precipitate of cadmium sulfide.

$$[Cd(NH_3)_4]^{2+} + 4CN^- \rightarrow [Cd(CN)_4]^{2-} + 4NH_3$$

$$[Cd(CN)_4]^{2-} + H_2S \rightarrow \underset{\text{Yellow}}{CdS\downarrow} + 2HCN + 2CN^-$$

Thus cadmium ions may be detected in presence of cupric ions.

Group IIB cations

We have seen that whereas the sulfides of Group IIB cations are soluble in yellow ammonium sulfide, those of Group IIA are not. This helps in isolating the cations of Group IIB from those of Group IIA. The solubility of Group IIB sulfides, AS_2S_3, Sb_2S_3 and SnS in yellow ammonium sulfide, in turn, is attributed to the formation of ammonium trithioarsenate(V), $(NH_4)AsS_3$, ammonium trithioantimonate(V) $(NH_4)SbS_3$ and ammonium trithiostannate(IV), $(NH_4)_2SnS_3$, respectively.

The first step in the analysis of Group IIB is to re-precipitate the cations as sulfides. Thereafter, the steps listed in Table 32.9 are invoked to identify the concerned cations.

Table 32.9 Analysis of Group IIB cations

<table>
<tr><td colspan="2">A. Continued from Section C, Table 32.6
$(NH_4)AsS_3$, $(NH_4)SbS_3$, $(NH_4)_2SnS_3$.
Treat with dilute hydrochloric acid and centrifuge.</td></tr>
<tr><td>B. Residue. As_2S_5 (yellow), Sb_2S_3 (orange), SnS_2 (yellow).
Add about 3 mL concentrated hydrochloric acid. Boil gently and centrifuge.</td><td>Centrifuge. Discard</td></tr>
<tr><td>C. Residue. As_2S_5 (yellow).
Add concentrated nitric acid. Boil and add ammonium molybdate solution.
Yellow precipitate of ammonium arsenate-molybdenum trioxide adduct appears after about one minute.
As^{3+} confirmed</td><td>D. Centrifugate. $SbCl_3$, $SnCl_4$.
Divide into 2 parts.
(i) Render mildly alkaline by adding 1:1 ammonia solution. Add a pinch of solid oxalic acid, boil and pass hydrogen sulfide gas.
Orange precipitate of antimony sulfide.
Sb^{3+} confirmed
(ii) Add a piece of zinc granule and warm. Add mercuric chloride solution.
White precipitate (Hg_2Cl_2) turning grey (Hg).
Sn^{2+} confirmed</td></tr>
</table>

Reactions involved in Table 32.9

Section A

$$2(NH_4)AsS_3 + 2HCl \rightarrow As_2S_5\downarrow + 2NH_4Cl + H_2S$$

$$2(NH_4)SbS_3 + 2HCl \rightarrow Sb_2S_3\downarrow + 2NH_4Cl + 2S + H_2S$$

$$(NH_4)_2SnS_3 + 2HCl \rightarrow SnS_2\downarrow + 2NH_4Cl + H_2S$$

Section B

$$Sb_2S_3 + 6HCl \rightarrow 2SbCl_3 + 3H_2S$$

$$SnS_2 + 4HCl \rightarrow SnCl_4 + 2H_2S$$

Section C

$$3As_2S_5 + 28HNO_3 + 4H_2O \rightarrow 6H_3AsO_4 + 9H_2SO_4 + 6S + 28NO$$

$$H_3AsO_4 + 12(NH_4)_2MoO_4 + 21HNO_3 \rightarrow \underset{\text{Yellow}}{(NH_4)_3AsO_4 \cdot 12MoO_3\downarrow} + 21NH_4NO_3 + 12H_2O$$

Section D

(i) $$2SbCl_3 + 3H_2S \xrightarrow{\text{Acidic medium}} \underset{\text{Orange}}{Sb_2S_3} + 6HCl$$

(ii) $$SnCl_4 + Zn \rightarrow SnCl_2 + ZnCl_2$$

$$SnCl_2 + 2HgCl_2 \rightarrow SnCl_4 + \underset{\text{White}}{Hg_2Cl_2\downarrow}$$

$$SnCl_2 + Hg_2Cl_2 \rightarrow SnCl_4 + \underset{\text{Grey}}{2Hg\downarrow}$$

Removal of interfering anions

Interfering anions are so named because they interfere with the analysis of cations listed in Group III and beyond. These anions readily combine with one or more cations of Group III–VI, either precipitating these as insoluble salts, or by forming stable complexes. In either case, the metal ion is removed from the solution and is rendered unavailable for further analysis.

The four most important interfering anions are fluoride, oxalate, borate and phosphate. The reasons why these interfere with cation analysis and the procedure of their removal from the original solution are described in Table 32.10.

Table 32.10 Interfering anions and their removal

Interfering anion	Reasons for interference	Procedure of removal
Fluoride	In Group V, calcium ions are rendered insoluble as calcium fluoride, $Ca^{2+} + 2F^- \rightarrow CaF_2\downarrow$ while in Group VI, magnesium ions are precipitated as magnesium fluoride. $Mg^{2+} + 2F^- \rightarrow MgF_2\downarrow$ Even though magnesium fluoride is not completely insoluble, yet in presence of fluoride ions, the concentration of magnesium ions is rendered so low that conventional tests fail to detect it.	Boil the centrifugate from Group II to near dryress. Add concentrated hydrochloric acid. Boil to near dryness. Repeat the procedure two times. Fluoride ions are converted into hydrogen fluoride which volatilizes with hydrochloric acid. $F^- + H^+ \rightarrow HF\uparrow$ Dissolve the residue in distilled water and proceed to Group III.

Oxalate	Oxalate ions form stable complexes with cations of Group III, for example, $Fe^{3+} + 3C_2O_4^{2-} \rightarrow [Fe(C_2O_4)_3]^{3-}$ Group V cations are precipitated as insoluble oxalates, for example, $Ca^{2+} + C_2O_4^{2-} \rightarrow CaC_2O_4\downarrow$ In Group VI, magnesium ions too are precipitated as magnesium oxalate. $Mg^{2+} + C_2O_4^{2-} \rightarrow MgC_2O_4\downarrow$	Boil the centrifugate from Group II to near dryness. Add concentrated nitric acid. Boil to near dryness. Repeat the procedure two times. Oxalate ions decompose into carbon dioxide. $2C_2O_4^{2-} + 4HNO_3 \rightarrow 4NO_2^- + 2H_2O + 4CO_2 + O_2$ Dissolve the residue in distilled water and proceed to Group III.
Borate	Borate ions form insoluble salts with cations of Group III, for example, $Fe^{3+} + BO_3^{3-} \rightarrow FeBO_3\downarrow$ cations of Group IV, for example, $3Zn^{2+} + 2BO_3^{3-} \rightarrow Zn_3(BO_3)_2\downarrow$ and cations of Group V, for example $3Ca^{2+} + 2BO_3^{3-} \rightarrow Ca_3(BO_3)_2\downarrow$	Same as for fluoride ions. On treatment with concentrated hydrochloric acid, borate ions are converted into boric acid which volatilizes with steam. $BO_3^{3-} + 3H^+ \rightarrow H_3BO_3\uparrow$ Dissolve the residue in distilled water and proceed to Group III.
Phosphate	Phosphate ions form insoluble salts with cations of Group III, for example $Fe^{3+} + PO_4^{3-} \rightarrow FePO_4\downarrow$ cations of Group IV, for example, $3Zn^{2+} + 2PO_4^{3-} \rightarrow Zn_3(PO_4)_2\downarrow$ and cations of Group V, for example, $3Ca^{2+} + 2PO_4^{3-} \rightarrow Ca_3(PO_4)_2\downarrow$	Boil the centrifugate from Group II till the volume is reduced to about one-half. Add a few drops of dilute hydrochloric acid and a pinch of ammonium chloride. Warm and dropwise add zirconyl nitrate reagent, $ZrO(NO_3)_2$. $2PO_4^{3-} + 4H^+ + ZrO(NO_3)_2 \rightarrow ZrO(H_2PO_4)_2\downarrow + 2NO_3^-$ Discard the white precipilate of zirconyl phosphate and proceed with the centrifugate for the analysis of Group III cations.

Group III cations

Before beginning the analysis of Group III cations, the unreacted hydrogen sulfide carried forward from Group II must be removed from the original solution. Otherwise there is a likelihood that it might precipitate the cations of Group III and of subsequent groups as sulfides. The solubility of gases is inversely proportional to temperature. Thus by simply boiling the centrifugate from Group II for about 2 minutes, the unwanted hydrogen sulfide is pushed out of the solution. It is also pertinent to state that if the salt or mixture contains an interfering anion, which has been removed subsequent to Group II analysis, the 'boiling off' step may be skipped. The reason is that while removing any one or more of the interfering anions, the protocol demands that the original solution be boiled for quite some time.

The cations of Group III are ferric (Fe^{3+}), aluminium (Al^{3+}) and chromium (Cr^{3+}). It is possible that by the time the analysis reaches Group III stage, the ferric ions may have got reduced to ferrous ions. Hence first 1-2 drops of concentrated nitric acid are added to the centrifugate from Group II, so as to re-oxidize ferrous to ferric ions. Thereafter, solid ammonium chloride is added and the solution is heated to boiling. Next, 1:1 ammonia is added till the solution smells of gaseous ammonia. The three cations are precipitated as their respective hydroxides: $Fe(OH)_3$ (red-brown), $Al(OH)_3$ (white) and $Cr(OH)_3$ (green). The centrifugate is retained for analysis of Group IV.

The steps listed in Table 32.11 are then followed to confirm the presence of one or more of these cations in the given inorganic material.

Case Study 32.5 For the precipitation of Group III cations, ammonium hydroxide is added to the original solution in concert with ammonium chloride. Why?

Even though ammonium hydroxide is a weak base, it produces sufficient hydroxide ions so as to cause the ionic product to exceed the solubility product of not only Group III metal hydroxides, but those of the subsequent groups too. In presence of ammonium chloride, the concentration of hydroxide ions is suppressed due to common ion effect. The value of ionic product deceases.

$$NH_4OH \rightleftharpoons NH_4^+ + OH^-$$

$$NH_4Cl \rightarrow NH_4^+ + Cl^-$$

As a result, the ionic products of only Group III metal hydroxides surpass their solubility products. The solubility products of Group IV, V, etc. metal hydroxides are relatively large and are not exceeded by their ionic products when the concentration of hydroxide ions is low. Thus only the Group III cations are precipitated as hydroxides, while those of the subsequent groups are not.

Table 32.11 Analysis of Group III cations

<table>
<tr><td colspan="2">A. $Fe(OH)_3$ (red-brown), $Al(OH)_3$ (white), $Cr(OH)_3$ (green).
Wash the precipitates with ammonium chloride solution. Add a few drops of hydrogen peroxide and sodium hydroxide solution. Boil and centrifuge.</td></tr>
<tr><td>B. Residue. $Fe(OH_3)$ (red-brown).
Dissolve in dilute hydrochloric acid. Discard the residue, if any. Divide into 2 parts.
(i) Add potassium ferrocyanide solution.
Prussian blue coloration.
(ii) Add potassium thiocyanate solution.
Blood red coloration, which discharges on adding sodium fluoride solution.
Fe^{3+} confirmed</td><td>C. Centrifugate. $NaAlO_2$, Na_2CrO_4.
Divide into 4 parts.
(i) Add dilute hydrochloric acid, followed by excess ammonium hydroxide. Boil.
Gelatinous white precipitate of aluminium hydroxide.
Al^{3+} confirmed
(ii) Acidify with acetic acid and add lead acetate solution.
Yellow precipitate of lead chromate.
(iii) Acidify with acetic acid and add barium chloride.
Yellow precipitate of barium chromate.
(iv) Acidify with dilute nitric acid. Add amyl alcohol, followed by hydrogen peroxide. Shake and allow the two layers to separate.
Blue coloration in upper amyl alcohol layer due to chromium peroxide.
Cr^{3+} confirmed</td></tr>
</table>

Reactions involved in Table 32.11

Section A

$$Fe^{3+} + 3NH_4OH \rightarrow Fe(OH)_3\downarrow + 3NH_4^+$$

$$Al^{3+} + 3NH_4OH \rightarrow Al(OH)_3\downarrow + 3NH_4^+$$

$$Cr^{3+} + 3NH_4OH \rightarrow Cr(OH)_3\downarrow + 3NH_4^+$$

Section B

$$Fe(OH)_3 + 3HCl \rightarrow FeCl_3 + 3H_2O$$

(i) $FeCl_3 + K_4[Fe(CN)_6] \rightarrow K\{Fe[Fe(CN)_6]\} + 3KCl$

Prussian blue

(ii) $FeCl_3 + 6KSCN \rightarrow K_3[Fe(SCN)_6] + 3KCl$

Potassium hexathiocyanato-Sferrate(III)

Blood-red

$K_3[Fe(SCN)_6] + 6NaF \rightarrow K_3[FeF_6] + 6NaSCN$

Potassium hexafluoroferrate(III)

Colorless

Section C

$$Al(OH)_3 + NaOH \rightarrow \underset{\text{Sodium aluminate}}{NaAlO_2} + 2H_2O$$

$$2Cr(OH)_3 + 4NaOH + 3H_2O_2 \rightarrow \underset{\text{Sodium chromate}}{2Na_2CrO_4} + 8H_2O$$

(i) $$NaAlO_2 + 4HCl \rightarrow AlCl_3 + NaCl + 2H_2O$$

$$AlCl_3 + 3NH_4OH \rightarrow \underset{\text{Gelatinous white}}{Al(OH)_3\downarrow} + 3NH_4Cl$$

(ii) $$Na_2CrO_4 + (CH_3COO)_2Pb \rightarrow \underset{\substack{\text{Lead chromate}\\ \text{Yellow}}}{PbCrO_4\downarrow} + 2CH_3COONa$$

(iii) $$Na_2CrO_4 + BaCl_2 \rightarrow \underset{\substack{\text{Barium chromate}\\ \text{Yellow}}}{BaCrO_4\downarrow} + 2NaCl$$

(iv) $$Na_2CrO_4 + 2HNO_3 + 2H_2O_2 \rightarrow \underset{\substack{\text{Chromium peroxide}\\ \text{Blue}}}{CrO_5} + 2NaNO_3 + 3H_2O$$

Case Study 32.6 Can a combination of sodium hydroxide and sodium chloride be used to precipitate Group III cations instead of ammonium hydroxide – ammonium chloride pair?

If instead of ammonium hydroxide – ammonium chloride combine, a mixture of sodium hydroxide and sodium chloride is used for precipitation of Group III metal hydroxides, the purpose of schematic analysis is defeated. Both the reagents are strong electrolytes and are completely dissociated.

$$NaOH \rightarrow Na^+ + OH^-$$

$$NaCl \rightarrow Na^+ + Cl^-$$

Despite the fact that sodium ions are the common product of both these reactions, the concentration of hydroxide ions is very large. The ionic products of metal hydroxides will be so high that these will exceed the solubility products of not only Group III hydroxides, but also the hydroxides of the subsequent groups. Hence all these will get precipitated in Group III itself.

Group IV cations

Cobalt (Co^{2+}), nickel (Ni^{2+}), zinc (Zn^{2+}) and manganese (Mn^{2+}) are the members of Group IV. The precipitating reagent for this group is hydrogen sulfide in presence of ammonium hydroxide.

Add about 0.5 g ammonium chloride to the contrifugate brought forward from Group III and boil. Add ammonium hydroxide solution and pass hydrogen sulfide gas. The cations of Group IV get precipitated as CoS (black), NiS (black), ZnS (off-white) and MnS (buff) and are further processed by following the procedure described in Table 32.12. The centrifugate is reserved for analysis of Group V.

Table 32.12 Analysis of Group IV

<table>
<tr><td colspan="3">A. CoS (black), NiS (black), ZnS (off-white), MnS (buff).
Wash the precipitates with a solution of ammonium chloride and add dilute hydrochloric acid.</td></tr>
<tr><td rowspan="2">B. Residue. CoS (black), NiS (black).
Dissolve in aqua regia. Evaporate the solution to dryness. Dissolve the residue in distilled water and divide into 2 parts.
(i) Add potassium thiocyanate solution, followed by amyl alcohol. Agitate. Blue color of potassium tetrathiocyanato-Scobaltatc(II) in upper layer.
Co^{2+} confirmed
(ii) Add ammonium hydroxide and dimethylglyoxime solution. Red precipitate of dimethylg-lyoximatonickel(II)
Ni^{2+} confirmed</td><td colspan="2">C. Centrifugate. Zn^{2+}, Mn^{2+}.
Add sodium hydroxide solution. Boil and centrifuge.</td></tr>
<tr><td>D. Centrifugate. Na_2ZnO_2.
Divide into 2 parts.
(i) Pass hydrogen sulfide gas. Off-white precipitate of zinc sulfide.
(ii) Add acetic acid and potassium ferrocyanide. White precipitate of zinc ferrocyanide.
Zn^{2+} confirmed</td><td>E. Residue. $Mn(OH)_2$ (black).
Dissolve in concentrated nitric acid and warm. Divide into 2 parts.
(i) Add solid red lead and boil. Allow the solid to settle. Purple color of supernatant liquid.
(ii) Add sodium bismuthate solution and boil. Purple color of supcrnatant liquid.
Mn^{2+} confirmed</td></tr>
</table>

Reactions involved in Table 32.12

Section A

$$Co^{2+} + H_2S \rightarrow CoS\downarrow + 2H^+$$

$$Ni^{2+} + H_2S \rightarrow NiS\downarrow + 2H^+$$

$$Zn^{2+} + H_2S \rightarrow ZnS\downarrow + 2H^+$$

$$Mn^{2+} + H_2S \rightarrow MnS\downarrow + 2H^+$$

Section B

$$3CoS + 2HNO_3 + 6HCl \rightarrow 3CoCl_2 + 3S + 4H_2O + 2NO$$

$$3NiS + 2HNO_3 + 6HCl \rightarrow 3NiCl_2 + 3S + 4H_2O + 2NO$$

(i) $CoCl_2 + 4KSCN \rightarrow K_2[Co(SCN)_4] + 2KCl$

Potassium thiocyanato-Scobaltate(II)
Blue

(ii) 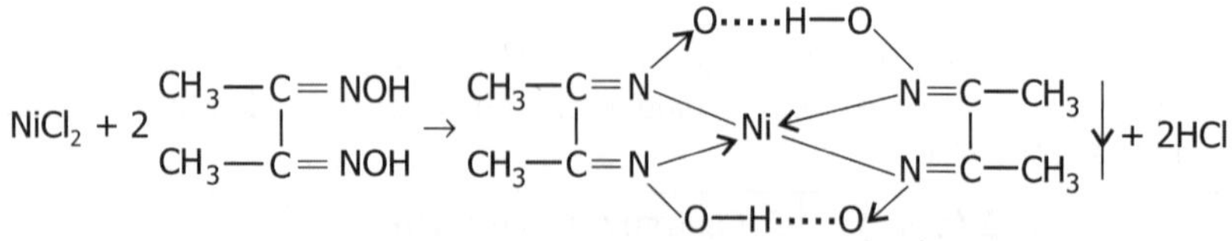

Dimethylglyoxime   Dimethylglyoximatonickel(II)
Red

Section C

$$ZnS + 2HCl \rightarrow ZnCl_2 + H_2S$$

$$MnS + 2HCl \rightarrow MnCl_2 + H_2S$$

Section D

$$ZnCl_2 + 2NaOH \rightarrow Zn(OH)_2 + 2NaCl$$

$$Zn(OH)_2 + 2NaOH \rightarrow Na_2ZnO_2 + 2H_2O$$

Sodium zincate

(i) $Na_2ZnO_2 + H_2S \rightarrow ZnS\downarrow + 2NaOH$

Off-white

(ii) $Na_2ZnO_2 + 4CH_3COOH \rightarrow 2CH_3COONa + (CH_3COO)_2Zn + 2H_2O$

$$2(CH_3COO)_2Zn + K_4[Fe(CN)_6] \rightarrow Zn_2[Fe(CN)_6]\downarrow + 4CH_3COOK$$

White

Section E

$$MnCl_2 + 2NaOH \rightarrow Mn(OH)_2\downarrow + 2NaCl$$

Black

$$Mn(OH)_2 + 2HNO_3 \rightarrow Mn(NO_3)_2 + 2H_2O$$

(i) $2Mn(NO_3)_2 + 5PbO_2 + 6HNO_3 \rightarrow 2HMnO_4 + 5Pb(NO_3)_2 + 2H_2O$

Permanganic acid
Purple

(ii) $2Mn(NO_3)_2 + 5NaBiO_3 + 16HNO_3 \rightarrow 2HMnO_4 + 5NaNO_3 + 5Bi(NO_3)_3 + 7H_2O$

Purple

Case Study 32.7 For the precipitation of Group IV metal sulfides, why is hydrogen sulfide gas passed through the original solution under alkaline conditions?

The solubility products of Group IV metal sulfides are appreciably high. However, hydrogen sulfide is weakly dissociated and produces only meager amount of sulfide ions.

$$H_2S \rightleftharpoons 2H^+ + S^{2-} \tag{32.1}$$

The value of ionic product is concomitantly low—so low that it cannot exceed the value of solubility product of any of the Group IV metal sulfide. Hence these do not get precipitated by hydrogen sulfide alone. However, when hydrogen sulfide is used as a precipitating agent in presence of a mild alkali, like ammonium hydroxide, the concentration of sulfide ions increases. Ammonium hydroxide dissociates into hydroxide ions.

$$NH_4OH \rightleftharpoons NH_4^+ + OH^- \tag{32.2}$$

The hydrogen cations generated in reaction (32.1), combine with the hydroxide ions furnished in reaction (32.2).

$$H^+ + OH^- \rightarrow H_2O$$

As a result, the hydrogen cations are removed from the equilibrium (32.1) and in accordance with Le Chatelier's principle, the reaction is pushed in the forward direction. This increases the concentration of sulfide ions, and, in turn, the value of ionic product. The solubility products of Group IV metal sulfides are now exceeded and these get precipitated.

Group V cations

The cations included in Groups V are—barium (Ba^{2+}), strontium (Sr^{2+}) and calcium (Ca^{2+}). The precipitating reagent for this group is ammonium carbonate in mildly alkaline medium.

The centrifugate carried forward from Group IV in first boiled for about 2 minutes to push out dissolved hydrogen sulfide gas. About 0.5 g ammonium chloride and a few drops of ammonium hydroxide are added to it. This is followed by a saturated solution of ammonium carbonate, whereupon the cations of Group V get precipitated as respective carbonates—$BaCO_3$ (white), $SrCO_3$ (white) and $CaCO_3$ (white). These are isolated and subjected to the sequence of operations described in Table 32.13. The centrifugate is retained for analysis of Group VI.

Table 32.13 Analysis of Group V

<table>
<tr><td colspan="3">A. $BaCO_3$ (white), $SrCO_3$ (white), $CaCO_3$(white).
Wash the precipitates with warm water, and thereafter dissolve these in acetic acid. Separate 5 drops of the solution and add 2 drops of potassium chromate solution to it. If a yellow precipitate is obtained, add potassium chromate to the bulk solution and proceed to Section B. Otherwise skip this step and go directly to Section C.</td></tr>
<tr><td rowspan="2">B. Residue. $BaCrO_4$ (yellow).
Make a paste of the residue in concentrated hydrochloric acid. Apply the paste on a platinum wire and perform the flame test.
Grassy-green flame.
Ba^{2+} confirmed</td><td colspan="2">C. Centrifugate. $(CH_3COO_2)Sr$, $(CH_3COO)_2Ca$.
Add a few drops of ammonium sulfate and centrifuge.</td></tr>
<tr><td>D. Residue. $SrSO_4$ (white).
Perform flame test with the residue.
Crimson flame.
Sr^{2+} confirmed</td><td>E. Centrifugate. $CaSO_4$(aq)
Add a few drops of ammonium oxalate solution.
White precipitate of calcium oxalate.
Perform flame test with above precipitate.
Red flame.
Ca^{2+} confirmed</td></tr>
</table>

Reactions involved in Table 32.13

Section A

$$Ba^{2+} + (NH_4)_2CO_3 \rightarrow BaCO_3\downarrow + 2NH_4^+$$

$$Sr^{2+} + (NH_4)_2CO_3 \rightarrow SrCO_3\downarrow + 2NH_4^+$$

$$Ca^{2+} + (NH_4)_2CO_3 \rightarrow CaCO_3\downarrow + 2NH_4^+$$

$$BaCO_3 + 2CH_3COOH \rightarrow (CH_3COO)_2Ba + H_2O + CO_2$$

$$SrCO_3 + 2CH_3COOH \rightarrow (CH_3COO)_2Sr + H_2O + CO_2$$

$$CaCO_3 + 2CH_3COOH \rightarrow (CH_3COO)_2Ca + H_2O + CO_2$$

Section B

$$(CH_3COO)_2Ba + K_2CrO_4 \rightarrow \underset{\text{Barium chromate}}{BaCrO_4\downarrow} + 2CH_3COOK$$

Section C Nil

Section D

$$(CH_3COO)_2Sr + (NH_4)_2SO_4 \rightarrow \underset{\text{White}}{SrSO_4\downarrow} + 2CH_3COONH_4$$

Section E

$$(CH_3COO)_2Ca + \underset{\text{Ammonium oxalate}}{(NH_4)_2C_2O_4} \rightarrow \underset{\text{Calcium oxalate, White}}{CaC_2O_4\downarrow} + 2CH_3COONH_4$$

Case Study 32.8 Explain the significance of ammonium carbonate-ammonium chloride-ammonium hydroxide mixture as the precipitating reagent for Group V cations.

The cations of Group V—barium, strontium, calcium—may be precipitated as carbonates by ammonium carbonate alone. However, the carbonate ions furnished by ionization of ammonium carbonate are in appreciably high concentration. The ionic product of Group V metal carbonates, no doubt, surpasses the solubility products of these salts and, in effect, precipitates them out, but simultaneously, magnesium carbonate too gets precipitated. As we shall see, magnesium ion belongs to Group VI of cation analysis. The solubility product of magnesium carbonate is greater than those of Group V metal carbonates. When a mixture of ammonium carbonate, ammonium hydroxide and ammonium chloride is used as the precipitating reagent, the concentration of carbonate ions is suppressed due to common ion effect.

$$(NH_4)_2CO_3 \rightarrow 2NH_4^+ + CO_3^{2-}$$

$$NH_4Cl \rightarrow NH_4^+ + Cl^-$$

$$NH_4OH \rightleftharpoons NH_4^+ + OH^-$$

The ionic product is now reduced to a value which exceeds the solubility products of barium, strontium and calcium carbonates, but not of magnesium carbonate. Hence the magnesium ions gets segregated from the cations of Group V.

Group VI cations

Group VI includes magnesium (Mg^{2+}) and potassium (K^+) ions. However, by the time we arrive at this last cationic group, the original solution becomes so dilute that the two ions may defy detection. It is therefore pertinent to concentrate the original solution before targeting the identification of its members. This is done by gently heating the solution till its volume is reduced to about one-half of the original. Thereafter it is divided into three parts.

(i) Add a few drops of ammonium hydroxide, a pinch of ammonium chloride and solution of disodium hydrogen phosphate. Scratch the sides of the test tube with a glass rod.

Appearance of a white precipitate of magnesium ammonium phosphate confirms the presence of magnesium ions.

$$Mg^{2+} + Na_2HPO_4 + NH_4OH \rightarrow \underset{\substack{\text{Magnesium ammonium phosphate} \\ \text{White}}}{Mg(NH_4)PO_4\downarrow} + 2Na^+ + H_2O$$

(ii) Add an ammonical solution of oxine and a few drops of ammonium chloride solution.

A yellow precipitate of dioxinatomagnesium(II) further confirms the presence of magnesium ions.

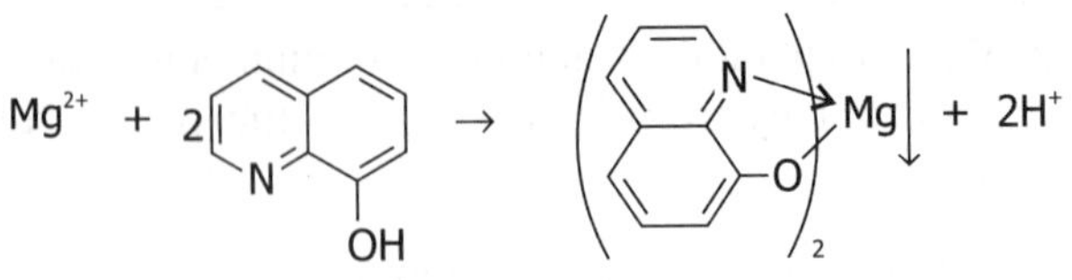

Dioxinatomagnesium(II)
Yellow

(iii) Acidify with acetic acid and add a few drops of sodium hexanitrocobaltate(III) solution.

A yellow precipitate of potassium hexanitrocobaltate(III) confirms the presence of potassium ions.

$$3K^+ + Na_3[Co(NO_2)_6] \rightarrow K_3[Co(NO_2)_6]\downarrow + 3Na^+$$

Potassium hexanitrocobaltate(III)
Yellow

Perform flame test with the precipitate. A lilac (pink-violet) flame further confirms the presence of potassium ions.

Spot tests

Two advantages are associated with spot tests. Firstly, these are so sensitive that only microgram quantities of the substance would suffice for analysis. Secondly, and this may be considered as a fallout of the first advantage, these conserve materials. A spot test may be carried out either on a filter paper or, better, in a spot plate (Fig. 32.4).

When one is endeavoring to analyze a mixture of cations, spot tests should ideally be performed in the respective group to which the metal ion belongs. However, when one is sure that no such ion which may despoil the coloration of the spot is present, the test may be performed directly with the original solution. For this reason, while describing the spot tests for individual cations, we have indicated which ions are likely to cause interference.

1. Silver ions, Ag^+ Place one drop of the test solution in a groove of the spot plate and add successively, one drop of dilute nitric acid and one drop of *rhodanine* reagent (I; about 0.3% by weight in acetone). A red-violet stain of silver–rhodanine complex (II) appears.

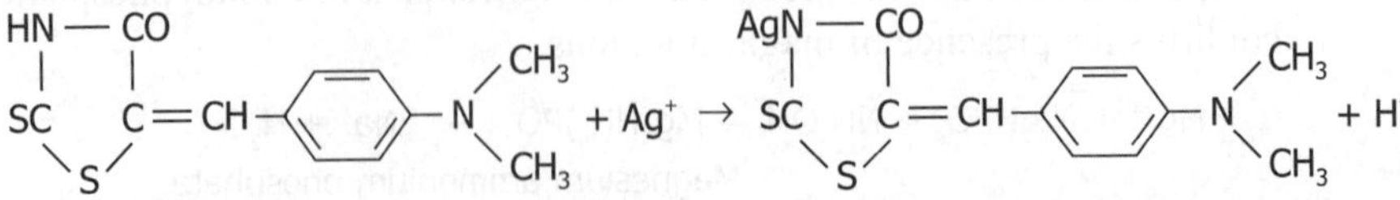

(I) (II)

Mercuric (Hg^{2+}) and cupric (Cu^{2+}) ions interfere with this test.

2. Cupric ions, Cu^{2+} Place one drop of the solution of a cupric salt on a filter paper. Spot one drop of dilute hydrochloric acid over it, followed by one drop of *benzoin-α-oxime* (I), also called *cupron* reagent (5% by weight in alcohol). Hold the filter paper over ammonia vapors. A green spot of copper–benzoinoxime complex(II) results.

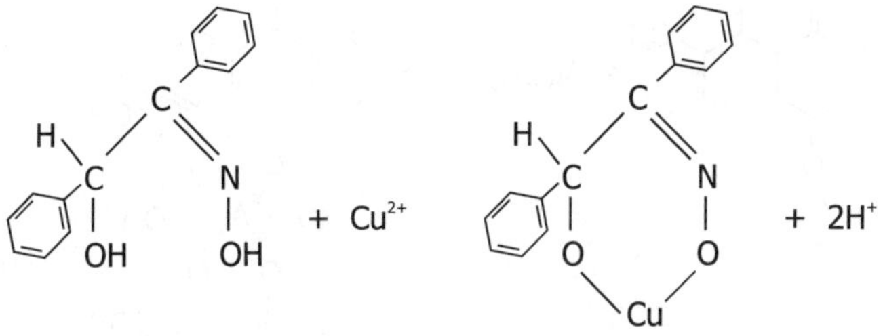

(I) (II)

Ferric ions (Fe^{3+}) interfere with this test.

3. Ferric ions, Fe^{3+} One drop of test solution is placed on a spot plate. A drop of dilute hydrochloric acid is added, followed by a drop of *ferron* reagent (I). A green or greenish-blue spot of ferric–ferron complex(II) appears.

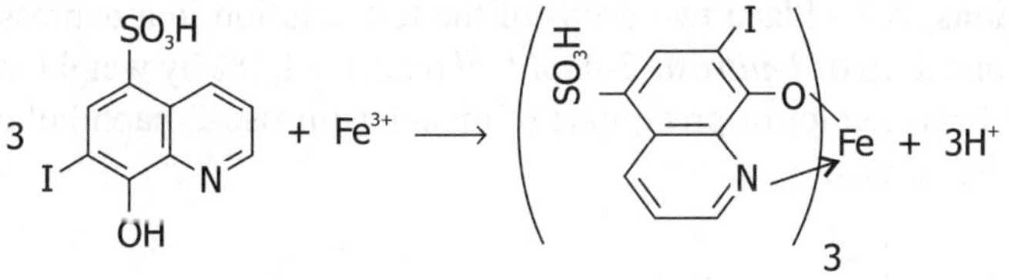

(I) (II)

Cupric ions (Cu^{2+}) interfere with this test.

4. Aluminium ions, Al^{3+} One drop of a saturated alcoholic solution of *alizarin* (I) is spotted on a filter paper. A drop of test solution is placed over it and the filter paper is exposed to ammonia vapors. Two reactions occur simultaneously. Firstly, aluminium ions are transformed to aluminium hydroxide (reaction 32.3) and secondly, alizarin is converted into violet colored ammonium alizarinate (II; reaction 32.4, forward).

$$Al^{3+} + 3NH_3 + 3H_2O \rightarrow Al(OH)_3 + 3NH_4^+ \quad (32.3)$$

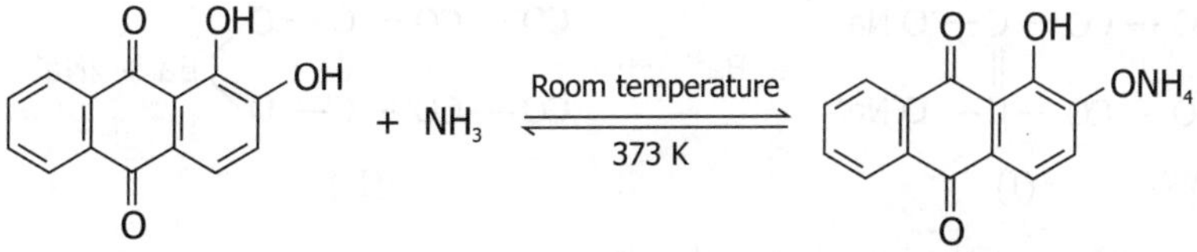

(32.4)

(I) (II)

The filter paper is dried at about 373 K, whereupon ammonium alizarinate reversibly decomposes into alizarin (reaction 32.4, reverse). Next, free alizarin reacts with aluminium hydroxide, formed in reaction (32.3) and yields the complex (III) which has the appearance of a red lake.

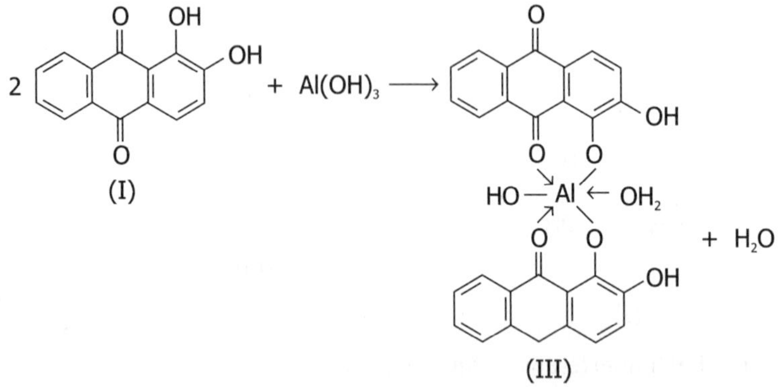

(32.5)

The red lake formed at the culmination of reaction (32.5) identifies aluminium ions.

Ferric (Fe^{3+}), chromium (Cr^{3+}) and manganese (Mn^{2+}) ions interfere with this test.

5. Nickel ions, Ni^{2+} Place two drops of the test solution in a depression of spot plate. Add one drop of *1-nitroso-2-naphthol* reagent (1.1% by weight in acetone or ethanol). A brown color or precipitate of nickel-1-nitroso-2-naphthol complex(II) identifies nickel ions.

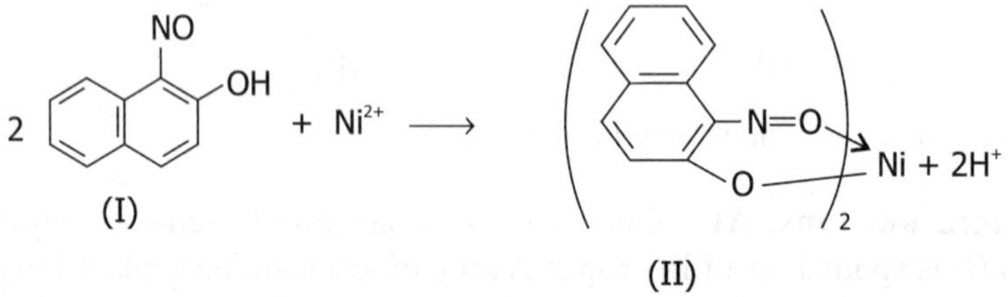

Cobalt ions (Co^{3+}) interfere with this test.

6. Barium ions, Ba^{2+} Place one drop of the test solution on a filter paper. Add a drop of *sodium rhodizonate* reagent (I; 0.5% by weight in water) over it. A reddish-brown spot of barium-rhodizonate (II) appears.

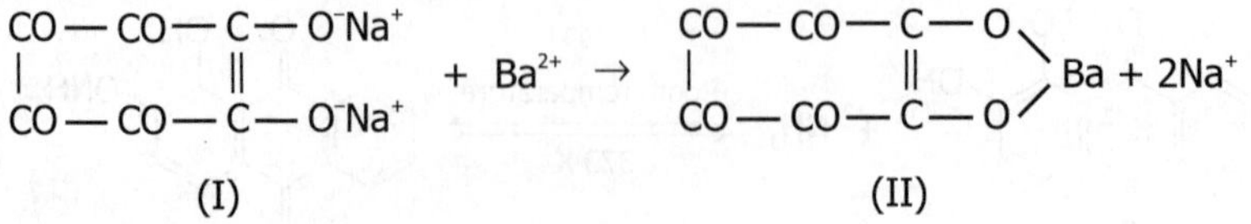

Strontium ions (Sr^{2+}) interfere with this test.

7. Strontium ions, Sr^{2+} Spot a drop of the test solution on a filter paper. Add a drop of *sodium rhodizonate* reagent (I; 0.5% by weight in water) over it. A reddish-brown spot of strontium-rhodizonate complex (II) appears.

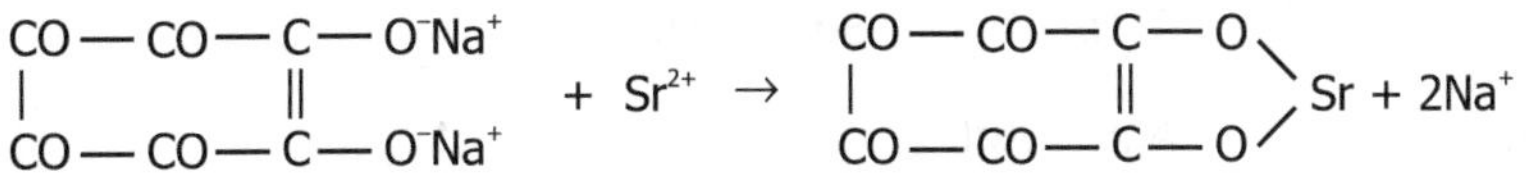

(I) (II)

Barium ions (Ba^{2+}) interfere with this test.

8. Magnesium ions, Mg^{2+} Place one drop of the test solution on the spot plate and add one drop of *magneson-I* reagent (I; 0.5% by weight in 10 mL sodium hydroxide solution (2 M) and 90 mL water). Further add 1 drop of sodium hydroxide solution. The magnesium ions get converted to magnesium hydroxide which, in turn, gets adsorbed on magneson-I reagent. A blue lake like appearance develops on the spot plate as a result of this physical change.

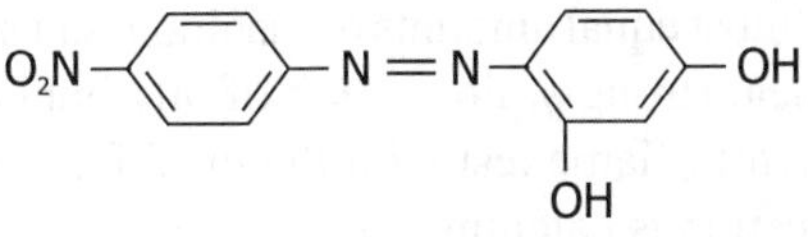

(I)

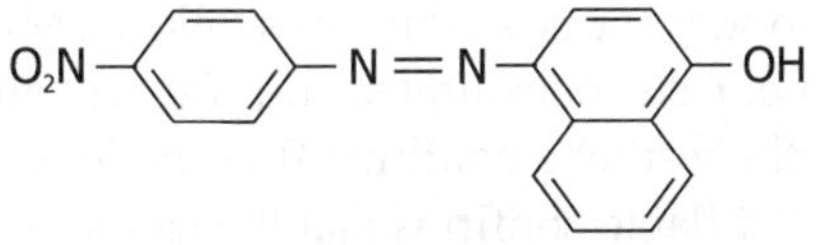

(II)

Magneson-II reagent (II; 0.5% by weight in 10 mL sodium hydroxide solution (2 M) and 90 mL water) when used in an analogous manner proves more sensitive than magneson-I reagent, although in this case too a blue lake like appearance develops.

Calcium (Ca^{2+}) and aluminium (Al^{3+}) ions interfere with this test.

Analysis of Insoluble Compounds

A good number of salts and metal oxides are insoluble in water, acids, and even aqua regia. After all other anions and cations have been detected in a mixture of salts or a mineral aggregate, a different procedure has to be called into action to identify the insoluble constituent(s). A schedule of these salts and oxides is displayed in Table 32.14. If an inorganic mixture or a mineral contains only one insoluble component, a preliminary idea about its composition may be gauged by observing its color. The colors of some common insoluble inorganic compounds too are presented in Table 32.14.

Table 32.14 Common insoluble inorganic compounds

Color	Insoluble compound
White	AgCl CaF_2 $PbSO_4$ $SrSO_4$ $BaSO_4$. Al_2O_3 As_2O_3 Sb_2O_3 SnO_2
Yellow	AgBr AgI
Green	Cr_2O_3
Red-brown	Fe_2O_3

The procedure for analyzing these compounds is now described.

Calcium fluoride We have already stated that the confirmatory test of fluoride may be performed with a solid salt (p. 902). The most common—and also the most important—insoluble fluoride is fluorspar, CaF_2. In fact, it is from this naturally occurring mineral that fluorine is extracted. Thus if the insoluble salt gives a positive test for fluoride ions, it may be concluded that the compound in question is calcium fluoride. The procedure is as follows. Mix almost equal amounts of insoluble salt and sand. Add concentrated sulfuric acid and heat strongly. Formation of oily bubbles in the test tube confirms fluoride ions. Perform flame test with the insoluble salt. A red flame confirms that the cation in question is calcium ion.

Lead sulfate It has already been discussed in context of Group II cation analysis that lead sulfate is partially soluble in ammonium acetate (p. 923). To the insoluble compound add a concentrated solution of ammonium acetate and warm. Centrifuge immediately and divide the centrifugate into two parts. Add a solution of potassium chromate to the first part and a solution of potassium iodide to the second part. Appearance of a yellow precipitate in both the tests—lead chromate in the first case, lead iodide in the second—confirms that the insoluble compound is lead sulfate.

Silver chloride Add 1:1 ammonia solution to the insoluble part. If the insoluble compound is silver chloride, it will completely dissolve. Acidify the clear solution with dilute nitric acid. The appearance of a white precipitate further confirms that the insoluble salt is silver chloride.

Silver bromide Add a few particles of zinc power and dilute sulfuric acid to the insoluble portion and heat very strongly. Silver bromide gets reduced to metallic silver by nascent hydrogen generated *in situ*.

$$Zn + H_2SO_4 \rightarrow ZnSO_4 + 2[H]$$

$$AgBr + [H] \rightarrow Ag + HBr$$

The hydrogen bromide so formed reacts with zinc.

$$Zn + 2HBr \rightarrow ZnBr_2 + H_2$$

When the contents are filtered, finely divided silver collects as the residue and zinc bromide passes into the centrifugate.

Perform layer test with the centrifugate. A red-brown coloration in the organic layer confirms bromide ions (p. 901). Add concentrated nitric acid to the residue. Silver gets converted into silver nitrate.

$$6Ag + 8HNO_3 \rightarrow 6AgNO_3 + 4H_2O + 2NO$$

Addition of dilute hydrochloric acid gives a white precipitate of silver chloride (p. 918).

The appearance of white precipitate, coupled with the results of layer test, confirm that the insoluble salt is silver bromide.

Silver iodide As an insoluble salt, silver iodide is identified by following the same series of steps as described for silver bromide. In this case too, metallic silver collects as the residue.

$$Zn + H_2SO_4 \rightarrow ZnSO_4 + 2[H]$$

$$AgI + [H] \rightarrow Ag + HI$$

However, the centrifugate contains zinc iodide instead of zinc bromide.

$$Zn + 2HI \rightarrow ZnI_2 + H_2$$

Perform layer test with the centrifugate. A violet, organic layer confirms the presence of iodide ions (p. 902). The residue is treated with concentrated nitric acid, followed by dilute hydrochloric acid, whereupon a white precipitate of silver chloride in obtained. The positive results of these twin tests confirm that the insoluble compound is silver iodide.

Iron(III) oxide Mix the insoluble material with solid sodium hydroxide and fuse in a nickel crucible till the mass turns liquid. Cool and add dilute hydrochloric acid. Insoluble iron(III) oxide gets converted into soluble, iron(III) chloride.

$$2Fe_2O_3 + 8NaOH + 3O_2 \xrightarrow{\text{Fuse}} \underset{\text{Sodium tetraoxoferrate(VI)}}{4Na_2[FeO_4]} + 4H_2O$$

$$2Na_2[FeO_4] + 16HCl \rightarrow 2FeCl_3 + 4NaCl + 8H_2O + 3Cl_2$$

The iron(III) chloride gives prussian blue coloration with potassium ferrocyanide and blood red coloration with potassium thiocyanate (p. 931). The positive results of both these tests confirm that the insoluble compound is iron(III) oxide.

Remaining compounds Once all the above insoluble compounds have been identified, or all these are found to be absent, the remaining analogs may be analyzed by following the sequence depicted in Table 32.15.

Table 32.15 Analysis of remaining insoluble compounds

<table>
<tr><td colspan="2">A. $SrSO_4$, $BaSO_4$, As_2O_3, Sb_2O_3, SnO_2, Al_2O_3, Cr_2O_3.
Mingle one part of the insoluble compound with 3 parts of sodium carbonate, 3 parts of potassium carbonate and one crystal of potassium nitrate (fusion mixture). Heat the blend strongly in a silica crucible till a liquid layer appears on the top of the mass. Shake with boiling distilled water and centrifuge.</td></tr>
<tr><td>B. Residue. $SrCO_3$, $BaCO_3$
Dissolve in dilute hydrochloric acid.

Sr^{2+}, Ba^{2+}

Analyze by the scheme described in Table 32.13 for Group V cations (p. 936).</td><td>C. Centrifugate. $NaAsO_2$, $NaSbO_2$, Na_2SnO_3 $NaAlO_2$, Na_2CrO_4.
Add concentrated hydrochloric acid.

As^{3+}, Sb^{3+}, Sn^{2+}, Al^{3+}, Na_2CrO_4

Analyze by the scheme described in Table 32.9 for Group IIB cations (p. 927), followed by the sequence discussed in section C of Table 32.11 for Group III cations (p. 931).</td></tr>
</table>

Reactions involved in Table 32.15

Section A

$$SrSO_4 + Na_2CO_3 \xrightarrow{\text{Fuse}} SrCO_3 + Na_2SO_4$$

$$BaSO_4 + Na_2CO_3 \xrightarrow{\text{Fuse}} BaCO_3 + Na_2SO_4$$

$$As_2O_3 + Na_2CO_3 \xrightarrow{\text{Fuse}} \underset{\text{Sodium arsinite}}{2NaAsO_2} + CO_2$$

$$Sb_2O_3 + Na_2CO_3 \xrightarrow{\text{Fuse}} \underset{\text{Sodium antimonite}}{2NaSbO_2} + CO_2$$

$$SnO_2 + Na_2CO_3 \xrightarrow{\text{Fuse}} \underset{\text{Sodium stannate}}{Na_2SnO_3} + CO_2$$

$$Al_2O_3 + Na_2CO_3 \xrightarrow{\text{Fuse}} \underset{\text{Sodium aluminate}}{2NaAlO_2} + CO_2$$

$$2Cr_2O_3 + 4Na_2CO_3 + 3O_2 \xrightarrow{\text{Fuse}} \underset{\text{Sodium chromate}}{4Na_2CrO_4} + 4CO_2$$

(The corresponding potassium salts too may be formed).

Section B

$$SrCO_3 + 2HCl \rightarrow Sr^{2+} + 2Cl^- + H_2O + CO_2$$

$$BaCO_3 + 2HCl \rightarrow Ba^{2+} + 2Cl^- + H_2O + CO_2$$

Section C

$$NaAsO_2 + 4HCl \rightarrow As^{3+} + Na^+ + 4Cl^- + 2H_2O$$

$$NaSbO_2 + 4HCl \rightarrow Sb^{3+} + Na^+ + 4Cl^- + 2H_2O$$

$$Na_2SnO_3 + 6HCl \rightarrow Sn^{2+} + 2Na^+ + 4Cl^- + 3H_2O + Cl_2$$

$$NaAlO_2 + 4HCl \rightarrow Al^{3+} + Na^+ + 4Cl^- + 2H_2O$$

APPENDIX

Table 1 Units and their symbols

Unit	Symbol
atomic mass unit	amu
Becquerel	Bq
Bohr magneton	BM
Coulomb	C
Coulomb meter	C m
Debye	D
eletron volts	ev
electrostatic unit	esu
fentometer	fm
Joule	J
kilo Joule	kJ
million electron volts	MeV
mole	mol
Newton	N

Table 2 Symbols and values of physical constants

Physical constant	Symbol	Value
Atomic mass unit	amu	1.661×10^{-27} kg
Avogadro number	N	6.023×10^{23}
Charge on an electron	e	1.60219×10^{-19} C
		4.8×10^{-10} esu
Gas constant	R	8.3143 J deg^{-1} mol^{-1}
Mass of electron	m_e	9.10952×10^{-31} kg
Mass of neutron	m_n	1.6749×10^{-27} kg
Mass of proton	m_p	1.672×10^{-27} kg
Planck's constant	h	6.63×10^{-34} J s
		6.63×10^{-27} erg s
Velocity of light	c	3×10^{8} m s^{-1}

SUGGESTED READING

Atkins, P., Overton, T., Rourke, J., Welle, M. and Armstrong, F. (2006), *Shriver & Atkins Inorganic Chemistry*, 4th Edition, Oxford Press, New Delhi.

Brown, D.A. (1967), *Comparative Inorganic Chemistry*, McGraw-Hill Publishing Co. Ltd., London.

Brown, G.I. (1973), *A New Guide to Modern Valence Theory*, Longman, London.

Companion, A.L. (1979), *Chemical Bonding*, Tata McGraw-Hill Publishing Co. Ltd., New Delhi.

Cotton, F.A., Wilkinson, G. and Gaus, P.L. (2003), *Basic Inorganic Chemistry*, 3rd Edition, John Wiley & Sons, Inc., New York.

Douglas, B., McDaniel, D. and Alexander, J. (1994), *Concepts and Models of Inorganic Chemistry*, 3rd Edition, Wiley, New Delhi.

Duffy, J.A. (1974), *General Inorganic Chemistry*, Longman, London.

Emeleus, H.J. and Sharpe, A.G. (1973), *Modern Aspects of Inorganic Chemistry*, Routledge & Kegan Paul, London.

Heslop, R.B. and Jones, K. (1976), *Inorganic Chemistry*, Elsevier Scientific Publishing Co., Amsterdam.

Huheey, J.E., Keiter, E.A., Keiter, R.L. and Medhi, O.K. (2006), *Inorganic Chemistry: Principles of Structure and Reactivity*, 4th Edition, Pearson, New Delhi.

Gilreath, E.S. (1963), *Fundamental Concepts of Inorganic Chemistry*, McGraw-Hill, Auckland.

Greenwood, N.N. and Earnshaw, A. (2005), *Chemistry of the Elements*, 2nd Edition, Butterworth-Heinemann, Oxford.

Jolly, W.L. (2007), *Modern Inorganic Chemistry*, 2nd Edition, Tata McGraw-Hill Publishing Co., New Delhi.

Lee, J.D. (2006), *Concise Inorganic Chemistry*, 5th Edition, Blackwell, Oxford.

Mackay, K.M. and Mackay, R.A. (1989), *Introduction to Modern Inorganic Chemistry*, 4th Edition, Prentice Hall, New Jersey.

Miessler, G.L. and Tarr, D.A. (2009), *Inorganic Chemistry*, 3rd Edition, Pearson, New Delhi.

Moody, B. (1996), *Comparative Inorganic Chemistry*, 3rd Edition, CBS Publishers & Distributors, New Delhi.

Rayner-Canham, G. and Overton, T. (2003), *Descriptive Inorganic Chemistry*, 3rd Edition, W.H. Freeman & Co., New York.

Sanderson, R.T. (1960), *Chemical Periodicity*, Van Nostrand Co., New York.

Svehla, G. and Sivasankar, B. (2013), *Vogel's Qualitative Inorganic Analysis,* Pearson, Chennai.

Weller, M., Overton, T., Rourke, J. and Armstrong, F. (2015), *Inorganic Chemistry*, Oxford University Press, new Delhi.

INDEX